World Regional Geography

NINTH EDITION

World Regional Geography

A Development Approach

Edited by

David L. Clawson
University of New Orleans

Douglas L. Johnson
Clark University

Viola Haarmann
Clark University

Merrill L. Johnson
University of New Orleans

Contributors

Christopher A. Airriess
Ball State University

Robert Argenbright
University of North Carolina Wilmington

Samuel Aryeety-Attoh
Loyola University

Bella Bychkova-Jordan
University of Texas at Austin

Terry G. Jordan-Bychkov
University of Texas at Austin

William C. Rowe
Louisiana State University

Jack F. Williams
Michigan State University

PEARSON

Prentice
Hall

Upper Saddle River, NJ 07458

Library of Congress Cataloging-in-Publication Data

World regional geography : a development approach / edited by David L. Clawson ... [et al.] ; contributors Christopher A. Airriess ... [et al.]. — 9th ed.

   Includes bibliographical references and index.
   ISBN: 0-13-149703-0
   1. Economic development.   2. Economic history.   3. Economic geography.
   4. Developing countries—Economic conditions.   I. Clawson, David L. (David Leslie).
   II. Airriess, Christopher A.
     HD 82.G39 2007
     33 0.9—dc22

2006009284

Acquisition Editor: Jeff Howard
Editor in Chief, Science: Dan Kaveney
Project Manager: Amanda Brown
Production Editor: Debra A. Wechsler
Executive Managing Editor: Kathleen Schiaparelli
Assistant Managing Editor: Beth Sweeten
Senior Managing Editor, Science and Math Media:
   Nicole M. Jackson
Editor in Chief, Development: Carol Trueheart
Development Editor: Barbara Muller
Manufacturing Manager: Alexis Heydt-Long
Manufacturing Buyer: Alan Fischer
Director of Creative Services: Paul Belfanti
Creative Director: Juan R. López
Art Director: Kenny Beck
Interior and Cover Design: John Christiana
Senior Managing Editor, Art Production and Management:
   Patricia Burns

Manager, Production Technologies: Matthew Haas
Managing Editor, Art Management: Abigail Bass
AV Production Editor: Eric A. Day; Jessica Einsig
Illustrations: Argosy Publishing/MapQuest
Director, Image Resource Center: Melinda Reo
Manager, Rights and Permissions: Zina Arabia
Manager, Visual Research: Beth Brenzel
Cover Image Specialist: Karen Sanatar
Photo Researcher: Yvonne Gerin
Image Permission Coordinator: Carolyn Gauntt
Copy Editor: Marcia Youngman
Editorial Assistant: Margaret Ziegler
Cover: (front) Yann Arthus-Bertrand, Jordan, Maan Province, Center-pivot irrigation near Wadi Rum (Région Maan—Champs circulaires près de Wadi Rum—Jordan) © Altitude; (back) Yann Arthus-Bertrand, Banlieue de Copenhague—Quartier de Brondby—Danemark © Altitude.

ISBN 0-13-149703-0

Pearson Education LTD., *London*
Pearson Education Australia PTY, Limited, *Sydney*
Pearson Education Singapore, Pte. Ltd
Pearson Education North Asia Ltd, *Hong Kong*
Pearson Education Canada, Ltd., *Toronto*
Pearson Educación de Mexico, S.A. de C.V.
Pearson Education — Japan, *Tokyo*
Pearson Education Malaysia, Pte. Ltd.

Brief Contents

Contents

PART FOUR
Europe 244
Bella Bychkova-Jordan and Terry G. Jordan-Bychkov

Chapter 10
The European Habitat 246

Chapter 11
Europe: Culture, Society, Economy 264

Chapter 12
Europe: Political Geography 286

PART FIVE
Russia and Central Eurasia 306
Robert Argenbright and William C. Rowe

Chapter 13
The In-Between Countries of Eurasia: Physical Geography and Historical Context 308

PART EIGHT
The Middle East and North Africa   518
Douglas L. Johnson and Viola Haarmann

Chapter 22
The Middle East and North Africa: Physical and Cultural Environments   520

We live in a most remarkable age when satellites can circle the globe in a matter of minutes and electronic messages can be sent halfway around the world in just a few seconds. For many, these are the best of times with rapid advances in health care extending human life expectancies to record high levels and material comforts that were unheard of even a generation ago now being commonplace. It is easy in such conditions to overlook the fact that most of the world's people live in less industrialized regions characterized by hunger, malnutrition, poor health, and limited educational and economic opportunities. Many also live under circumstances of restricted personal freedom owing to political, racial, religious, or gender-based prejudice. What is even more troubling is the fact that the income gap between the technologically more developed and less developed nations is increasing.

The purpose of this volume is to introduce students to the geographical foundations of development and underdevelopment, and to help them recognize the contributions that the study of geography can make to environmentally and culturally sustainable development. As we study the lives of others through this text, we will not only learn about them but also come to better understand ourselves. We will come to realize further that development or the improvement of the human condition, consists of far more than increased economic output and that each person and society contributes in a positive way to the cultural diversity and richness of the earth.

College students are in a unique position to increase their understanding of the world and to use that knowledge to benefit themselves and others. Through the study of world regional geography, we begin to comprehend the issues involved in the pursuit of world peace, preservation of the environment, improved health, and higher levels of living. In fact, the Association of American Geographers, the National Geographic Society, the American Geographical Society, and the National Council for Geographic Education have devoted significant resources and effort to improving geographic awareness. Even the U.S. Congress has cited geographic education as critical to understanding our increasingly interdependent world.

This ninth edition of *World Regional Geography* is dedicated to college students who are seeking a better understanding of our complex and challenging world. It is written for both majors and nonmajors and does not require an extensive background in geography. The basic regional structure of the eighth edition has been retained, including a multiple author approach that permits each region to be discussed by a scholar who is an expert in that area. Although our regional specializations vary, we are united in our dedication to expanding geographical awareness and to contributing to the development of the peoples of both the more and less industrialized nations.

World Regional Geography opens with three chapters that consider geography as a discipline and examine issues related to the nature of development, the impact of globalization, and the importance of sustainability. Eight additional parts—organized into twenty-four chapters—follow, structured within a regional framework. Within this regional framework, students are introduced to basic geographical and developmental concepts.

Although the use of this text as a discrete and unified entity is encouraged, its organization allows a variety of teaching strategies. As time requires, sections treating historical or environmental processes, or specific regions, can be selectively emphasized or deleted. In a two-term sequence, the book facilitates consideration along continental or regional divisions or along the lines of more industrialized–less industrialized countries. The suggested further readings for each part of the text are grouped at the end of the volume where they can be used to supplement the text; table and figure references can encourage students to pursue external data sources and analysis.

World Regional Geography contains numerous features that are designed to assist and stimulate students:

- More than 250 full-color maps and diagrams have been rendered by a professional cartographic studio, incorporating the latest boundary and name changes and physical, cultural, and economic data. The maps are carefully integrated into the text to convey spatial relationships and strengthen our understanding of geographic patterns and concepts.

- More than 300 specially chosen color photographs help communicate the physical, cultural, and economic nature of individual regions.

- Informative tables, graphs, and charts supplement textual material.

- Boxed features highlight subjects of special regional and topical significance, such as cultural diversity, gender and development, environmental issues, health and nutrition, migration, natural source utilization, or distinctive regional features.

- Key terms are presented in boldface type within the text and at the end of each chapter; a convenient glossary is provided at the end of the book.

- Review and Geographer's Craft questions are placed at the end of each part to help students identify key issues and to enable them to understand how geography can contribute to the solution of development challenges.

- The Further Readings listed at the end of the regions in the eighth edition have been moved and consolidated at the back of the book in this edition.

New to This Edition

The ninth edition of *World Regional Geography* represents a thorough revision. Every part of the book was examined carefully with the dual goals of keeping topics and data current, and improving the clarity of the text and the graphics. The ninth edition features a continuing updating of all of the data, maps, and illustrative material. Readers will note especially the comprehensive reworking of the introductory chapters and the Russian and Central Eurasian unit.

The ninth edition also reflects significant changes in the list of contributors and in the makeup of the editorial team. We are excited to announce that Douglas L. Johnson and Viola Haarmann of Clark University have joined Merrill L. Johnson and David L. Clawson as co-editors. They bring to the project considerable editorial experience and a wealth of geographical insight; you will notice that they also write the Middle Eastern and North African chapters. David Clawson retired from the University of New Orleans last May after almost three decades of service, but he remains active in his academic pursuits and, thankfully, will continue as the book's Latin American expert. The list of contributors has also changed. We are pleased to welcome Robert Argenbright and William C. Rowe as the new writers of the Russian and Central Eurasian chapters, replacing Roger L. Thiede; we want to thank Professor Thiede for his excellent contributions to multiple editions of this book. Finally, we sadly note the passing of Terry G. Jordan-Bychkov of the University of Texas at Austin who, with Bella Bychkova-Jordan, wrote the Europe chapters. Professor Jordan-Bychkov was one of American geography's most gifted scholars and teachers, and his passing has left a large hole in the discipline. We are grateful to Bella Bychkova-Jordan for agreeing to continue as the book's Europe expert.

Supplements

The book includes a complete supplements program for both students and teachers.

For the Student

Mapping Workbook (0-13-154775-5).   The Mapping Workbook features political and physical outline maps for each chapter, based on the maps in the text, ready for student use in identification exercises. If your instructor chooses, the Map Workbook can be packaged with the text.

World Regional Geography Website (*www.prenhall .com/clawson*).   The website tied chapter by chapter to the text provides students with three main resources for further study:

- *Online Study Guide.* The online study guide provides immediate feedback to study questions, access to current geographical issues, and links to interesting and relevant sites on the Web. Additionally, instructors can create a customized syllabus that links directly to the site.
- *Virtual Field Trips.* Integrated with the book's website, these field trips give students the on-the-ground feeling of visiting a place. The field trips include information about the history, places, people, and current events of places in each world region.
- *Country-by-Country Data.* Also integrated with the website, this material summarizes the vital statistics for each country within a region.

Science on the Internet: A Student's Guide (0-13-028253-7).   This unique resource helps students locate and explore the myriad educational resources on the Internet. It also provides an overview of the Internet, general navigation strategies, and brief student activities. It is available *free* when packaged with the text.

Rand McNally Atlas of World Geography.   This atlas includes 126 pages of up-to-date regional maps and 20 pages of illustrated world information tables. It is available *free* when packaged with *World Regional Geography*. Please contact your local Prentice Hall representative for details.

For the Instructor

Instructor's Resource Manual with Tests (0-13-101546-X).   The Instructor's Resource Manual, intended as a resource for both new and experienced teachers, includes a variety of lecture outlines, additional source materials, teaching tips, advice on how to integrate visual supplements, answers to the end-of-chapter exercises, and various other ideas for the classroom. Further, the manual contains an extensive array of test questions that accompany the book. These questions are available in hard copy and also on disks for Windows and Macintosh (0-13-101535-4).

Overhead Transparencies (0-13-154772-0).   The transparencies feature more than 150 illustrations from the text, all enlarged for excellent classroom visibility.

Instructor's Resources CD-ROM (0-13-154769-0).   Provides high-quality electronic versions of select photos and all of the illustrations from the text, as well as customizable PowerPoint presentations and MS Word files of the Instructor's Resource Manual with Tests. This is a powerful resource for anyone who is using electronic media in the classroom. Create your own PowerPoint™ presentation with the materials provided or customize PowerPoint™ with your own materials.

The ultimate purpose of this book is to help students develop an increased understanding of the geographic diversity of our world, both cultural and physical, and a knowledge of how each of us can contribute to the betterment of humankind.

Acknowledgments

We express gratitude to all who assisted with the preparation of this ninth edition of *World Regional Geography*. The suggestions and comments of the reviewers and previous readers have been indispensable, prompting, among other things, a greater emphasis on globalization, sustainability, gender, and environmental awareness. Although we owe a debt of gratitude to the many reviewers who have offered input over the course of many editions, we specifically thank those reviewers who took the time to critique the eighth edition of the book prior to the revision and writing that has produced the ninth edition. Here we thank Brian W. Blouet, the College of William and Mary; Stanley D. Brunn, University of Kentucky; Tarek A. Joseph, Henry Ford Community College; Robert G. Kremer, Metropolitan State College of Denver; Paul R. Larson, Southern Utah University; James Penn, Grand Valley State University; Stephen E. Podewell, Western Michigan University; Susan C. Slowey, Bline College; and Jacob R. Sowers, Kansas State University.

Neither original nor revised editions are accomplished by editors and authors working in isolation. We are especially grateful to the Prentice Hall publishing team for their patience, friendship, and professionalism. Specific members include Dan Kaveney, Jeff Howard, Debra Wechsler, Marcia Youngman, Barbara Muller, Margaret Ziegler, Yvonne Gerin, Amanda Brown, and Renee Hanenberg.

Finally, we are grateful to our families for their loving and patient support of our endeavors on behalf of this textbook, especially in those moments when we seemed to be hopelessly distracted by our thoughts and musings about world regional geography.

Merrill L. Johnson

Authoritative and Substantive Coverage

Our multiple-author approach allows us to present coverage of the regions of the world by experts on their lands and people. To ensure consistency, the academic editors carefully edit every chapter for language and approach.

Douglas L. Johnson

Editor
Part One, *Basic Concepts and Ideas,*
and **Part Eight,** *The Middle East and North Africa*

Douglas Johnson is Professor of Geography at Clark University. From the beginning of his geographical career, he was fascinated with the North African and Middle Eastern culture realm. Both his master's thesis and doctoral dissertation dealt with the history and spatial implications of nomadism. He received his Ph.D. from the University of Chicago. Studying the North African and Middle Eastern cultural and physical environments and teaching students about the complexities of this region have been a central focus of his work. In addition to lengthy field periods in Libya, Sudan, and Morocco, he has held visiting appointments in the Middle Eastern Center at the University of California, Berkeley and at Al-Akhawayn University in Morocco. His research is focused on issues of land degradation and desertification, arid land management, pastoral nomadism, and the cultural ecology of animal keeping. He is an editorial board member of *The Columbia Gazetteer of the World* (Columbia University Press, 1998) and co-author of *Land Degradation: Creation and Destruction,* 2nd edition (Rowman & Littlefield, 2006). For the last three years he has served as co-editor of *The Geographical Review.* In light of recent events in the Middle East, contributing to a world regional geography textbook that helps promote better understanding of one of the most conflicted areas in the world seems more important than ever.

Viola Haarmann

Editor
Part One, *Basic Concepts and Ideas,*
and **Part Eight,** *The Middle East and North Africa*

Viola Haarmann is a Research Fellow in the George Perkins Marsh Institute at Clark University. From an early age she was at home both in Europe and North America as she grew up in Germany and Canada. She earned a dual Master's degree in English and Geography, and also received her D.Sc. from Hamburg University, Germany, after carrying out fieldwork on land-use potential and change in the southern Sahel zone of Sudan's Darfur Province. For many years she held a Sahel project coordinator position at the Hamburg geography department. Since moving to the United States, she has operated as an independent editor and scholar. She is an editorial board member of *The Columbia Gazetteer of the World* (Columbia University Press, 1998), and for the last three years has served as co-editor of *The Geographical Review.* Her current research interests focus on the geography of food and agriculture, with an emphasis on the development of sustainable, small-scale alternatives to corporate farming.

Merrill L. Johnson

Editor
Part One, *Basic Concepts and Ideas,*
and **Part Two,** *United States and Canada*

Merrill Johnson is Associate Dean and Professor of Geography in the College of Liberal Arts, University of New Orleans. He has a B.A. in International Relations from West Texas State University, an M.A. in Geography from Arizona State University, and a Ph.D. in Geography from the University of Georgia. He joined the faculty of the University of New Orleans in 1981, was Chair of the department from 1989 to 2000, and was appointed an associate dean of the College in 2001. In addition, he is Director of Latin American Outreach for the University. His academic interests focus on the United States South, with emphasis on its industrial and political geography. Secondary interests include cartography and Latin America. While he has taught a variety of undergraduate and graduate courses in political, economic, and regional geography, and in cartography and geographic thought, he remains passionate about teaching introductory World Regional Geography. Indeed, he has taught several thousand students in more than 60 world-regional sections in his career at the University of New Orleans.

David L. Clawson

Editor
Part One, *Basic Concepts and Ideas,*
and **Part Three,** *Latin America and the Caribbean*

David Clawson is a Professor Emeritus of Geography in the College of Liberal Arts, University of New Orleans. He has a B.A. from the University of Utah, and an M.A. and Ph.D. from the University of Florida. He has served as Chair of the Department of Geography and as Director of the Latin American Studies Program at the University of New Orleans. His research interests center on third world agrarian development, the geography of religions and belief systems, and geographic education and thought. In addition to his work on *World Regional Geography,* he is the author of *Latin America and the Caribbean: Lands and Peoples,* 3rd ed. (McGraw-Hill, 2004), has served as editor of a number of scholarly volumes, and has published numerous journal articles. In addition to World Regional Geography, he taught courses in geographic education, Latin America, agricultural, cultural, and environmental geography. He has held Fulbright grants to Mexico and the Philippines and was also a recipient of the University of New Orleans Alumni Association's "Excellence in Teaching Award."

Bella Bychkova-Jordan

Part Four, *Europe*

Bella Bychkova-Jordan was born in the remote Siberian village of Djarkhan, near the Arctic Circle in the ethnic republic of Sakha/Yakutia, Russia. Her people, the Sakha, are sometimes called the "Turks of the Subarctic" and have the distinction of being the northernmost cattle and horse raisers in the world. She left that ethnic setting to acquire an education at Yakutsk and Moscow, as holder of a Lenin Stipend, the highest scholarship award that could be bestowed upon a student in the Soviet Union. In 1994 she emigrated to the United States and settled in Austin, Texas. There she earned both a Master's and Ph.D. in cultural geography from the University of Texas at Austin. A specialist in Eastern Europe and Russia, she is presently Lecturer in the Center for Russian, East European, and Eurasian Studies at the University of Texas. She is the principal author of *Siberian Village: Land and Life In the Sakha Republic* (University of Minnesota Press, 2001) and co-author of *The European Culture Area*, 4th edition (Rowman & Littlefield, 2002). Her hobbies include travel, watercolor painting, and classic literature. She became an American citizen in 2003.

Terry G. Jordan-Bychkov

Part Four, *Europe*

The late Terry Jordan-Bychkov was the Walter Prescott Webb Professor in the Department of Geography at the University of Texas at Austin. A native of Dallas and a sixth-generation Texan, he earned his doctorate in cultural and historical geography at the University of Wisconsin at Madison. He served as honorary president of the Association of American Geographers and received an Honors Award from that professional organization. Terry Jordan-Bychkov was listed in *Who's Who In America* since 1987. His main research interest was in European culture and its transfer to overseas lands. He authored and co-authored over 25 books and more than 100 scholarly articles. These include *The Human Mosaic: A Thematic Introduction to Human Geography*, 9th edition (W. H. Freeman, 2003); *The European Culture Area*, 4th edition (Rowman & Littlefield, 2002); *Anglo-Celtic Australia: Colonial Immigration and Cultural Regionalism* (Center for American Places, 2002); *The Upland South: The Making of an American Folk Region and Landscape* (Center for American Places, 2003). His hobbies were genealogy and travel, and he prided himself in having visited 64 different countries.

Robert Argenbright

Part Five, Chapters 13 and 14.
The In-Between Countries of Eurasia

Robert Argenbright is an Associate Professor of Geography in the Earth Sciences Department of the University of North Carolina Wilmington. He earned a B.A., M.A., and Ph.D., all in Geography, from the University of California, Berkeley. His research focuses on the historical geography of the Soviet Union and the current transformation of Moscow. A brief tour of Moscow and Leningrad in 1975 first piqued his interest in Russia. Since then, he has returned to the region 15 times to conduct research. He has published articles on the USSR and post-Soviet Russia in such journals as *Eurasian Geography and Economics*, *The Geographical Review*, *Political Geography*, *Revolutionary Russia*, and *The Russian Review*. Currently he is writing a book with the working title *Moscow Under Construction*. Included among the many courses he has taught at UNCW are Geography of Post-Soviet Eurasia and World Regional Geography.

William C. Rowe

Part Five, Chapter 15,
Central Asia and Afghanistan

William Rowe, a native of upper East Tennessee and sixth-generation Appalachian tobacco farmer, has traveled, worked, and studied in the Muslim world for over twenty years. He received his Bachelor of Science in Languages (BSLA) from Georgetown University, with concentrations in Arabic and French. He worked at the American University in Cairo before returning to the United States for his Master's on water and population in Southern and Eastern Morocco at the University of Texas at Austin. For his dissertation his focus shifted to the newly independent Muslim nations of Central Asia with an emphasis on Tajikistan and its language, Tajiki. He spent two years on research in Tajikistan as it was emerging from its devastating civil war. Since receiving his Ph.D., he has continued his work in Central Asia, most notably in the Tajik regions of Afghanistan and the Zerafshan River Valley of Uzbekistan and has been able to witness the changes in the region since American involvement in the wake of 9/11. His research concentrates on the economic and environmental transformations that have occurred since independence in Central Asia and in post 9/11 Afghanistan as well as Muslim identity in post-Soviet Central Asia. He is currently a visiting professor at Louisiana State University.

Jack F. Williams

Part Six, *Australia, New Zealand, and the Pacific Islands,* and **Part Seven, Chapter 18,** *Japan*

Jack Williams is Professor of Geography, and former Director of the Asian Studies Center, at Michigan State University in East Lansing. He received his university training in geography at the University of Washington and the University of Hawaii. An urban/economic geographer, he has published widely in that subfield, and is co-editor of a popular text in world regional urban development, now in its third edition. First and foremost, he considers himself a regional geographer, a specialist in East Asia, and has traveled widely over the years in the Pacific Rim region, including Japan and Australia, since his first experience in Asia in the early 1960s (Peace Corps, Malaysia). A native of Seattle, he has always been interested in the Pacific and Asia. He has been fortunate to observe firsthand the extraordinary growth and development of this key region over the past 40 years, and incorporates that experience into his chapters in this book on world regional geography.

Christopher A. Airriess

Part Seven, *Asia* (except Chapter 18, *Japan,* authored by Jack Williams)

Christopher Airriess is a Professor of Geography at Ball State University, Muncie, Indiana. He earned his B.A. and M.A. in Geography at Louisiana State University, and his Ph.D. in Geography at the University of Kentucky in 1989. Born on Long Island, New York, he spent his childhood in Singapore and Malaysia. While he has traveled throughout much of the western Pacific Rim, his favorite places are southern China, Indonesia, Malaysia, and Singapore. His research interests include development, the geography of ports and maritime transport, and the human dimensions of ethnic-Southeast Asians in North America. He is the recipient of two Fulbright Awards that allowed extended visits to Indonesia in 1987, and Hong Kong in 2000. In addition to regularly teaching World Regional Geography, he also teaches Human Geography, and Geography of Asia. He believes that a World

Regional Geography course provides an essential piece to the undergraduate educational experience by imparting an understanding of the connections between the environment, culture, society, and economic and political systems within the context of real places where the process of globalization directly and indirectly impacts people's lives.

Samuel Aryeetey-Attoh
Part Nine, *Africa South of the Sahara*

Samuel Aryeetey-Attoh is Dean of the Graduate School at Loyola University, Chicago. He received his Ph.D. from Boston University and his M.A. from Carleton University, Ottawa. He also earned a B.A. degree with Honors from the University of Ghana, Legon. His research and teaching interests are in Urban and Regional Planning, Housing and Community Development, and the Geography of Development in Africa. Besides a book on the *Geography of Sub-Saharan Africa,* 2nd edition (Prentice Hall, 2002), he has published many articles in geographical journals and has received numerous research grants. He is past chair of the *African Specialty Group of the AAG,* co-editor of the *African Geographical Review,* and a member of the *American Council on Education Fellows,* and the *American Institute of Certified Planners.* He believes a course in World Regional Geography presents instructors with an opportunity to showcase the integrative and holistic human and physical dimensions of geography and to demonstrate how they relate to real world situations—whether social, political, environmental or economic in nature.

Basic Approach and Graphics Program

The ninth edition of *World Regional Geography* retains the popular economic development theme that has characterized previous editions of the book. This approach helps professors focus their course on a more conceptual treatment of world geography rather than the more traditional descriptive treatment. In the ninth edition the introductory unit has been reduced to three chapters, and the development framework is now concentrated. Discussions of gender issues, belief systems, socioeconomic class structures, and political institutions have been integrated throughout.

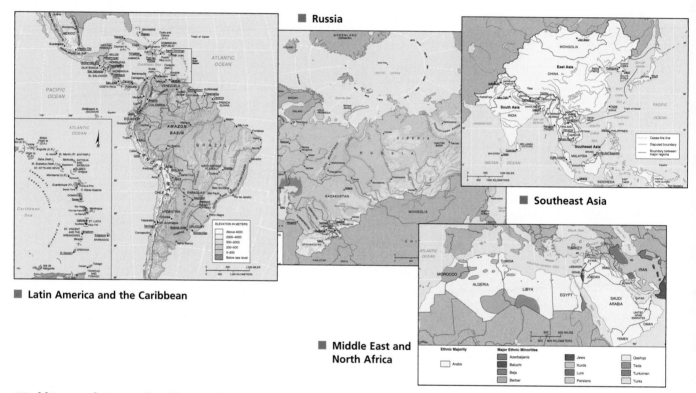

■ **Russia**

■ **Southeast Asia**

■ **Latin America and the Caribbean**

■ **Middle East and North Africa**

World Regional Geography offers a rich and diverse cartographic and photo program. Hundreds of featured maps and photographs help professors teach their students the important spatial elements inherent to the study of geography—and illustrate the diversity of the world's landscapes and populations.

■ **Great Mosque at Mecca**

■ **Mali, West Africa**

■ **Oil tanker, Qatar**

Instructor Resources

Instructor's Resource CD-ROM
Everything you need to prepare for class all
in one place.

- All of the maps and art from the book
- 100% of the photos from the book
- Customizable PowerPoint presentations
- Instructor's Manual formatted in Microsoft Word
- Test Item File formatted in Microsoft Word

Transparencies
The transparencies feature over 150 illustrations from the
text, all enlarged for excellent classroom visibility.

Instructor's Manual
The Instructor's Manual, intended as a resource
for both new and experienced instructors, includes a
variety of lecture outlines, additional source materials, teaching tips, advice on how to integrate visual
supplements, and various other ideas for the classroom.

Test Item File
The Test Item File contains over 1500 questions covering
multiple-choice, true/false, and short-answer questions and
is available in print or electronic formats via the Instructor's
Resource cd-rom.

Test Gen EQ
Test Gen is a computerized text generator that lets you view
and edit test bank questions, transfer questions to tests, and
print the text in a variety of customized formats. Included
in each package is the QuizMaster EQ program that lets
you administer tests on a computer network, record student
scores, and print diagnostic reports.

Student Resources

On-line Study Guide
Tied chapter-by-chapter to the text. The on-line study
guide provides students with immediate feedback to study
questions, access to current geographical issues, and
provides links to other interesting and relevant websites.
www.prenhall.com/clawson

Mapping Workbook
The Mapping Workbook features political and physical outline maps for each chapter, based on the maps in the text,
ready for student use in identification exercises. If your instructor chooses, the Map Workbook can be packaged with
the text. To order this discounted package use ISBN 0-13-
104249-1.

Rand McNally Atlas of World Geography
If your instructor chooses, the *Rand McNally Atlas of World
Geography* can be packaged with the text. To order this dis-
counted package use ISBN 0-13-104249-1.

Geography Coloring Book, 3/e by Wynn Kapit
This unique educational tool introduces the countries of the
world and the states of the United States to students. If your
instructor chooses, the *Geography Coloring Book* can be
packaged with the text. To order this discounted package
use ISBN 0-13-104250-5.

Building Geographic Literacy: An Interactive Approach, 5e by Charles A. Stansfield, Jr.
Innovative in its approach, this combined textbook and
interactive workbook enables students to learn place
geography—and, as they learn it, to reinforce and apply
their knowledge by constructing thematic maps which
convey physical, economic, cultural, or political characteristics of places. *Building Geographic Literacy* can be packaged with the text. To order the discounted package use
ISBN 0-13-104251-3.

This political map reveals a highly compartmentalized world. The numerous political entities range in size from the vast area of Russia to minute but significant countries such as Singapore, Malta, and Grenada. The names of those political entities evoke images of different environments, peoples, cultures, and levels of well-being. However, the political boundaries that segregate approximately 6 billion inhabitants do not fully reflect the geographic complexities of our world. Numerous other sets of boundaries could be imposed: boundaries that delineate multinational alliances, boundaries that classify economic and agricultural environments, boundaries that outline the world's myriad peoples, languages, religious faiths, and political ideologies. Unraveling the complexities of our world requires intellectual attention to many questions relating to geography. This map and the chapters that follow are intended to start us on a journey toward understanding our diverse and increasingly interdependent world.

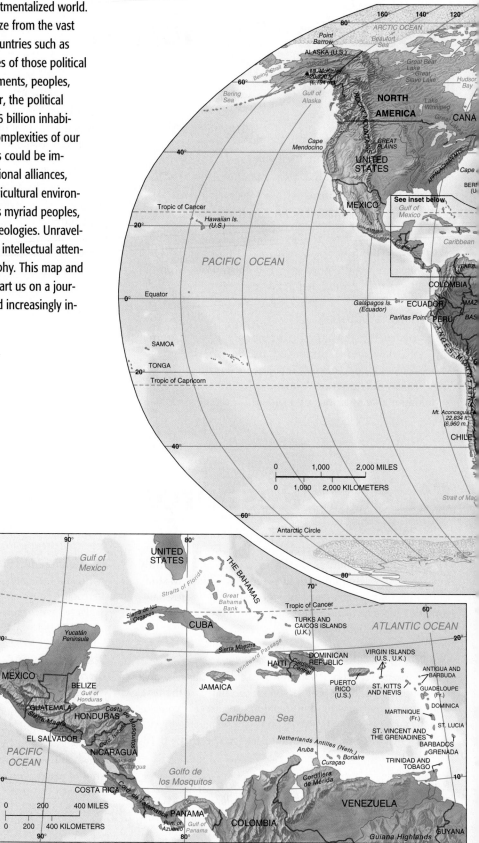

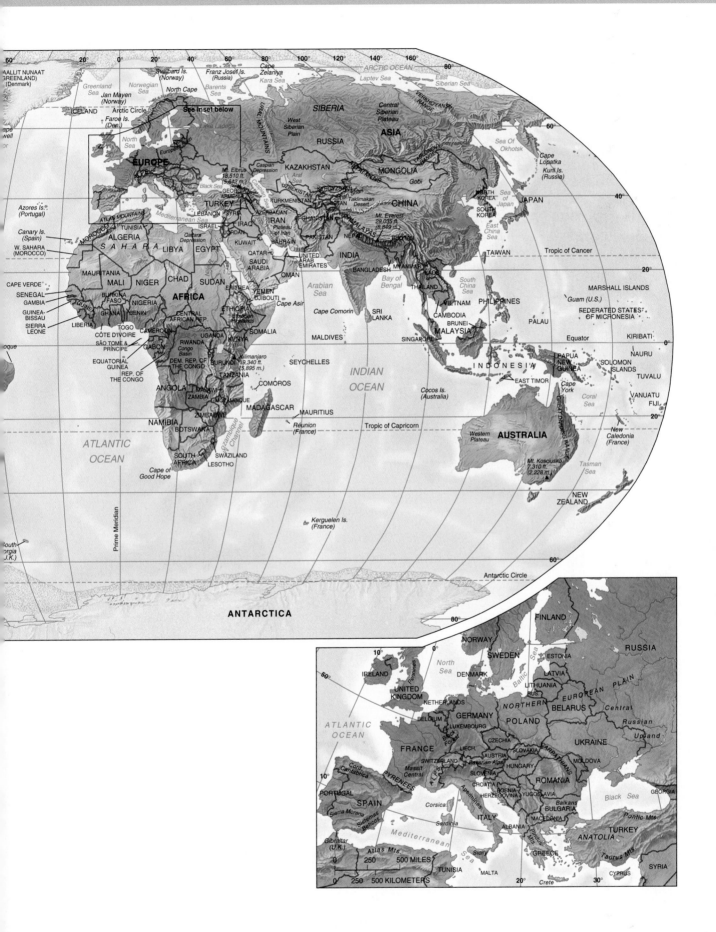

World Regional Geography

PART ONE

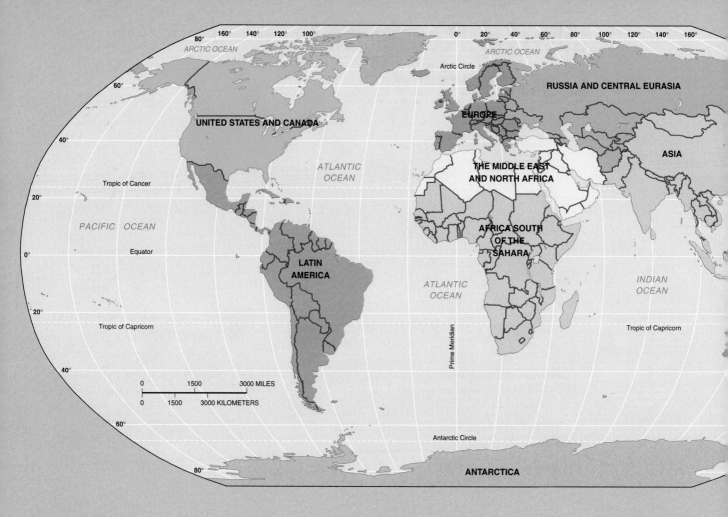

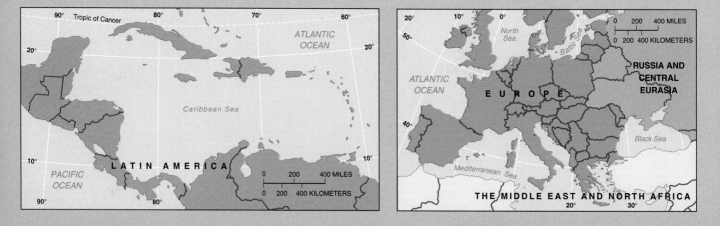

Basic Concepts and Ideas

Merrill L. Johnson, Douglas L. Johnson, Viola Haarmann, and David. L. Clawson

This book provides an introduction to world regional geography, focusing on where places and activities are located, and why. The hope is that this book will acquaint the reader with the world in which we live, and then organize that knowledge within a regional framework of development and underdevelopment, broadly defined.

The word **development** is used in a variety of ways. Basically, it denotes a progressive improvement of the human condition, in both material and nonmaterial ways. Economic development signifies a process of long-term advancement in the physical and material quality of life. Development and underdevelopment also have nonmaterial dimensions, many of which relate to the achievement of personal fulfillment through the exercise of individual freedoms. Because most nonmaterial aspects of development and underdevelopment cannot be easily measured in a numerical or statistical sense, this volume is organized primarily on the basis of economically developed and less developed regions. We will, nevertheless, also address other aspects of development, including social class structure, health, educational achievement, gender relations, and political and religious freedom. As we shall see, economic development can be measured in several ways—by income, energy use, employment, and various other indicators. Such measures must be kept in proper perspective, for each may tell us little about other aspects of an area, a nation, or a people. For example, differences in income may simply reflect more basic differences in cultural goals and values.

An extended discussion of what geographers do is included in Chapter 1. Chapter 2 provides an overview of the meaning of development in an age of globalization. Chapter 3 considers the natural environment, primarily from the standpoint of resources, and also attends to elements of culture, especially those that influence development. These chapters use measures of economic well-being to define the more developed and less developed regions of the world, discuss their characteristics, and present some theories of development. They set the stage for further exploration of the more developed and less developed regions of the world.

CONTENTS

San Francisco Bay, California Intensely urbanized areas surround most of the Bay's shoreline, while the more rugged topography of the region's hills is much less densely occupied. Moist and more richly vegetated near the coast, aridity increases dramatically as one moves inland. The trend line of the San Andreas fault is clearly observable in Marin County, north of the entrance to the Bay.

Chapter 1

What Is Geography and Should We Care?

- **Geography Studies the Why of Where**
- **Roots: Whence We Came**
- **The Modern Practice of Geography**
- **Geography and Its Disciplinary Neighbors**
- **Understanding Change in Economy and Society**

Geography Studies the Why of Where

The late James A. Michener, a noted American novelist, was a staunch supporter of the study of geography. He believed that his novels—including *Hawaii, Caravans, The Covenant, Centennial, Chesapeake, Poland, Texas,* and *Caribbean*—enjoyed great success because he fixed them firmly within a regional geographic context. Not surprisingly, he relied heavily on geographic works when he did research for his novels so that his readers would have a thorough and accurate knowledge of the locations in which his elaborate stories were set.

As Michener's novels suggest, **geography** is fundamentally the study of location—the location of physical features, economic activities, human settlement patterns, and anything else that a person finds on a map. In other words, geographers have a keen interest in understanding what defines terrestrial space (as opposed to "outer" space), and the interactions between people and their environments within that space. For that reason, geography is known as a "spatial" science.

Most geographers agree that, at the most basic level, students of geography are called on to address three questions:

- **What is located where**? This is the map or the "location list" question, a type of question that may take students back to their earliest school days when they were called upon to identify state capitals, locate rivers, and name exotic foreign locales—all in the name of geographic literacy. For example, we may ask where the mountains are found in the eastern United States (answer: the Appalachian system running from Maine to Alabama). Or we may ask where in Europe ethnic tensions have created a fractured political geography (answer: the Balkan Peninsula). The "where" question is important. Asking where people, places, and activities are located provides context and creates a basic knowledge of locations that can be used as a basis for more detailed examination of places later. It is equally important for the student to know what is geographically incorrect, as for example a student of African geography believing that Brazil is an African country, or a student of United States geography thinking that the Mississippi River flows through New York City. On the other hand, rote memorization of location lists to the exclusion of all else can lead to the incorrect conclusion that geography is nothing more than a memory exercise more appropriate to parlor games and long car trips than serious study. "What is the capital city of South Dakota?" asks the driver of his passenger in an attempt to make the trip go by faster (by the way, the answer is Pierre). Fortunately, geography is more than just memorizing place names.

- **Why are things located where they are**? Geographers primarily want to understand why things are located where they are. This is the explanation question, or the question that looks for the processes that produce a particular geographic pattern. To continue with the examples used above, we know that the Appalachian mountain system runs the length of the eastern United States. Why are these mountains where they are? Our search for an answer would begin with an investigation into the history of colliding continents in the world, a story of plate tectonics that has contributed to the formation of numerous mountain chains worldwide. Along the same lines, the Balkan Peninsula has a lengthy and unfortunate tradition of ethnic strife leading to political fracturing. Why is this? We would want to start our inquiry by looking at the highly conflicted territorial histories of the south Slavs and the influences of outside empires. But these are not the only types of *why* questions of interest to geographers. In other cases, geographical understanding may require examining environmental interactions. Ecologists and environmental geographers, for instance, may seek to understand the relationship between cropping patterns and the expansion of deserts or the role of city structures in modifying climate (for example, creating urban heat islands).

- **So what?** Now that we know where things are, and why they are where they are, we ask the age-old question, "Who cares?" This is the "significance" question, or the question that asks the geographer to find meaning in the geographical phenomena that he or she has identified. We may know where the Appalachians are located and how they came into being, but do we understand their significance to settlement patterns in the early United States (that is, colonial clustering along the eastern seaboard), to weather and climate patterns (orographic effects producing massive rainfall quantities in the southern Appalachians), or to cultural patterns (think of the very different cultures that developed in the remote valleys of the Appalachians). By the same token, we should be reminded that the fractured political geography of the Balkans contributed to the outbreak of World War I, a war in which the United States ultimately became involved; and that, as of this writing, political distress has led to the stationing of U.S. troops in parts of the Balkan Peninsula as peacekeepers to separate ethnic groups engaged in bloody ethnic conflict. For many Americans, the significance is quite personal.

In summary, the goal of this book is to help students know and understand where places are, why they are there, and their significance. This goal is addressed using an economic-development perspective. Above all, we want students to know that *geography matters.*

Roots: Whence We Came

Geography has a rich, ancient, and varied heritage. Its solid foundation rests on the works of ancient scholars, who recorded the physical and cultural characteristics of lands near and far. The study of geography has evolved in many civilizations, with the first surviving maps appearing on clay tablets in ancient Iraq six millennia before the present. Folk cultures have also developed their own pragmatic body of geographic knowledge—for example, producing maps of wave patterns around Pacific Ocean islands that, in combination with knowledge of the location of stars and constellations, made possible navigation across great stretches of otherwise featureless ocean. But it was the Greeks who made the most enduring contributions to geography's early formal development. In fact, the term *geography* comes to us from the Greek words *geo* ("the earth") and *graphos* ("to write about or describe").

Contributions of the Greeks

The early Greeks studied the same kinds of geographic problems that confront us today, but without the benefit of modern knowledge and technologies. **Herodotus** (fifth century B.C.), called by some the father of geography as well as the father of history, placed historic events in their geographic settings in his famous *Historia,* which he wrote in the mid-fifth century B.C. He both described and explained the physical and human geography of his day, with particular emphasis on the seasonal flows of the Nile and Ister (Danube) rivers. Herodotus was also one of the earliest Greek geographers to map and name the continents of Europe, Asia, and Africa (which he called Libya). **Aristotle** (384–322 B.C.) discussed the physical characteristics of the earth, including temperature, wind, alluvial or stream deposition, and vulcanism, in his *Meteorologica.* Aristotle was also the first Greek geographer to divide the world into three broad climatic zones, which he called the "torrid" (tropical), the "ekumene" (literally, "the home of man," which corresponded to the mid-latitudes), and the "frigid," or polar realms (Figure 1–1). Although Aristotle erroneously believed that neither the torrid nor frigid zones permitted the full development of human potential, owing to their perpetually hot and cold temperatures, his discussion of the influences of physical environments on their human occupants was reflective of one of the principal concerns of geographic inquiry in all ages.

Other Greek scholars were concerned about the size and shape of the earth and its relationship to the rest of the cosmos. Which methods, they wondered, could be used to show where places are in relation to one another and what people do in the various parts of the world? The Greeks did not answer all of their own questions, but they provided a perspective for seeking the answers.

Another Greek geographer, **Eratosthenes** (276?–196? B.C.), lived in Alexandria, Egypt, and, among other things,

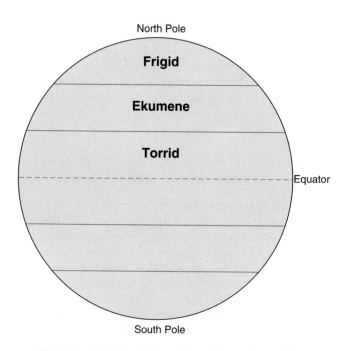

■ **FIGURE 1-1 Aristotle's zones of habitability.** Aristotle believed that human economic and cultural development was controlled by one's physical environment. Today we recognize that human behavior is not controlled or predestined by local physical geography.

measured the earth's circumference. He had learned that on only one day of the year (the Northern Hemisphere's summer solstice) did the noon sun shine directly down a well near what is now the city of Aswan. On that special day, Eratosthenes measured the noon sun's angle at Alexandria, some 500 miles (805 kilometers) north of Aswan, and found that there the sun's rays were not vertical but cast a shadow of 7.2° from a pole projecting straight up from the earth. Reasoning that the sun's rays are parallel to one another, he applied the geometric rule that when a straight line intersects two parallel lines, the alternate interior angles are equal. Therefore, he concluded, the distance between Aswan and Alexandria of 500 miles (805 kilometers) must be equal to a 7.2° arc of the earth's surface. With that information Eratosthenes then computed the value of a full circle's 360°, estimating the earth's circumference to be 25,200 miles (40,554 kilometers). Today we know that the circumference at the equator is quite close to that estimate: It is actually 24,901.5 miles (40,073.9 kilometers).

Eratosthenes and other Greeks recognized the need for maps to show the relationships between one place or region and another. They wanted a means of locating themselves on the earth and of describing their location to other people. To understand the challenge of their task, think of a mark on a smooth, uniform ball and then try to describe the position of the mark. Fortunately, the earth rotates around an axis that intersects the surface at two known points—the North Pole and the South Pole. With those points, the Greek geographers established two reference lines: the equator halfway between the poles and another line extending from pole to pole. They then drew a grid of latitude and longitude lines from those geographic reference points, thereby permitting any point on earth to be located by just two numbers. For example, the location of Washington, D.C., is 38°50′N, 77°00′W.

The Greeks' next step was to use the geographic grid to construct a map whereby all or part of the earth could be reduced to a two-dimensional plane and the relative positions of places and regions could be established. The Greek geographers—indeed, all geographers down through the ages—were particularly fond of maps, and maps became the tools used by geographers to depict spatial relationships.

Ptolemy (ca. A.D. 150), a Greek astronomer of Alexandria, was an early mapmaker. He designed a map of the world and then, using a coordinate system, compiled the location of 8,000 known places (Figure 1–2). This early work of the Greeks was not flawless. For instance, Christopher Columbus, accepting Ptolemy's view of the world, which estimated the earth's circumference to be much smaller than it actually is, thought he had arrived off the coast of Asia when he had really reached the Americas.

Geography thrived during Greek and Roman times. New lands were discovered, and inventories of their resources and characteristics had great practical importance. One compiler was **Strabo** (64 B.C.–ca. A.D. 23), whose *Geographia* was a description of the then known world and revealed a fascination with the distinctive physical and human characteristics of local places that make each site unique. Part gazetteer, part travel guide, part handbook for government officials, Strabo's work was an effort to view places in a holistic, multipurpose fashion. The accomplishments of Strabo, as well as Herodotus, Aristotle, Eratosthenes, Ptolemy, and others, established geographers as leading figures of their times. Collectively, their most enduring contribution was the development of a scholarly approach that emphasized the importance of describing the world, viewing the world from a spatial perspective, and employing a holistic recognition of the interdependence of the physical and cultural elements of the world. Geography's position was one of acclaim until the Roman Empire began to decline.

The Renaissance and the Age of Discovery

The Renaissance and the Age of Discovery marked a resurgence of geography and other sciences. New routes to the Orient and the Americas were opened, first under the sponsorship of **Prince Henry the Navigator** of Portugal and later by seamen sailing for Spain, Holland, England, and France. A new era of the geographer-explorer began, which was to include Christopher Columbus, Ferdinand Magellan, and James Cook, among others. Renaissance geographers produced increasingly accurate maps (Figure 1–3). The physical and cultural characteristics of foreign lands were described, and the processes that created differences and similarities between one place and another were analyzed in greater detail.

With increased knowledge, European scholars began to question age-old concepts in light of discoveries in other parts of the world. An age of scientific reasoning began, with experimentation and the testing of hypotheses. In the natural sciences, new explanations were presented that challenged old ideas about the origin of continents and oceans, the formation of landforms, and the evolution of plants and animals.

After the early explorers came scientific travelers—students of natural history—who sought evidence of and explanation for the varied world around them. Among those was the great German geographer **Alexander von Humboldt** (1769–1859). Humboldt traveled widely in Europe and Latin America. His curiosity, careful observation, and broad background of study in botany, physics, chemistry, Greek, archaeology, and geology enabled him to synthesize information from a variety of fields into a coherent geographic composite. In his most celebrated work, *Kosmos* (1845–1862), he attempted a comprehensive description of the earth.

Karl Ritter (1779–1859), a contemporary of Humboldt's, first studied geography as a basis for understanding history, but he eventually found that geography itself could provide an understanding of the human dimension of the world. His great work, *Die Erdkunde*, though never completed, included nineteen volumes on Africa and Asia. Ritter is generally recognized as having held the first chair of geography in Germany, at the University of Berlin in 1820.

By the middle of the nineteenth century, geography was a respected discipline in European universities, and

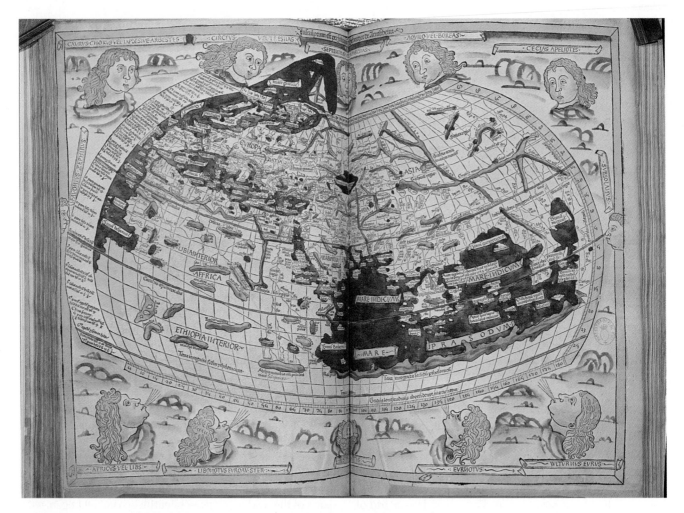

■ **FIGURE 1-2 Ptolemy's world map.** Ptolemy brought the classical period of Greek and Roman geography to a close. His world map, while containing numerous errors, was nevertheless a reflection of the remarkable geographical achievements of the Greek geographers. It stood as the standard reference map of the Western world for more than 1,300 years, until the Age of Exploration brought increased geographical knowledge of the world.

■ **FIGURE 1-3 A world map (1571) by Flemish cartographer (mapmaker) Abraham Ortelius (1527–1598).** By the late sixteenth century, increased exploration was leading to more accurate renderings of world maps. By comparing this view of the east and west coasts of the Americas or the west coast of Africa with the South Pacific, we can see the effect of exploration.

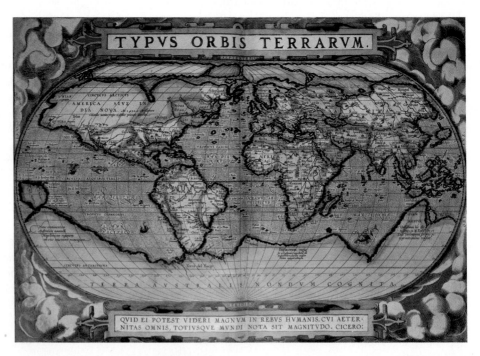

Table 1-1   Establishment of Geographical Societies

Year of Founding	Name of Society	Country
1821	Société de Géographie	France
1828	Gesellschaft für Erdkunde	Germany
1830	Royal Geographical Society	Great Britain
1838	Instituto Histórico e Geográfico	Brazil
1839	Sociedad Mexicana de Geografía	Mexico
1845	Russian Geographical Society	Russia
1851	American Geographical Society	United States
1888	National Geographic Society	United States
1904	Association of American Geographers	United States

Source: Compiled from Geoffrey J. Martin and Preston E. James, *All Possible Worlds: A History of Geographical Ideas,* 3rd ed. (New York: John Wiley and Sons, 1993), 161.

geographical societies served as important meeting places for scholars of all disciplines who were interested in the world around them. From Europe, a new age of geography spread around the world (Table 1–1).

In the United States, geography found a receptive audience. Because Americans were eager for knowledge about their country, especially the frontier regions, geographical literature was particularly popular. In 1851, the American Geographical Society was formed, followed by the National Geographic Society in 1888 and the Association of American Geographers in 1904. Geography as an academic field of study began to flower in the latter part of the nineteenth century, and in subsequent years it has spread from a few centers to almost every major college and university.

Mainstream Western geographical thought can correctly be traced to the ancient Greeks, but other centers of geographical thought existed as well. As already noted, Islamic explorers and scholars served as a bridge from ancient to modern thought. Ancient China, too, was a major center of geographical scholarship and exploration, with Chinese travel books dating back to A.D. 1000. Although Chinese geography did not eclipse the work of Greek scholars during the classical period, it thrived from the fifth to the fifteenth centuries. During that millennium, Chinese geographers traveled through southern Asia, the Mediterranean, and western Europe. They established human geography, completed regional studies inside and outside China, studied geomorphic processes, and wrote geographical encyclopedias.

The Modern Practice of Geography

Modern geography has grown beyond a simple description of the earth. Today's geographers not only describe through words, maps, and statistics, but also analyze interrelationships between physical and cultural phenomena to explain why things are distributed over the earth as they are.

Modern Geography

Modern geography is best understood as the study of how the physical and cultural attributes of the earth interact to form spatial or regional patterns. Modern geography has improved our ability to explain the world by utilizing four traditional areas of study:

1. the location of physical and cultural features and activities (spatial distributions)
2. the relationships between people and the lands that support them
3. the existence of distinctive areas or regions, including analysis and explanation of how they came to be formed
4. the physical characteristics of the earth, perhaps the oldest of all geographic traditions

Interestingly, the focus of each of these traditions as described by William Pattison[1] is evident in the work of the early Greek scholars. The **spatial tradition**, with its concern for distance, geometry, and movement, can be seen in the work of Ptolemy. The writings of Hippocrates were concerned with the relationship of human health to the surrounding environment, a theme common to the **man-land tradition**. Similarly, the **area studies tradition**, with its concern for the nature of places and for understanding the "where" of places, is evident in Strabo's *Geographia.* Lastly, the **earth science tradition**, as a study of the earth and its environments, is identifiable in the work of Aristotle and his students. To these classical traditions can readily be added emerging **perspectives**, such as the **behavioral** or **feminist** or **post-modern**, which enrich the traditional approaches to geographic scholarship by incorporating new knowledge and insights into the geographer's understanding of how the world works.

The Subdivisions of Geography

Geography has many subdivisions. The principal ones are physical geography, human geography, systematic geography, and regional geography.

Physical geography is the study of the environment from the viewpoint of distribution and process. For example, landform geographers, or geomorphologists, are concerned with the location of terrain features and with the ways in which those features have acquired their shapes and forms. Geomorphologists might study the impact of stream deposition in a floodplain, the effect of wind erosion in a dry land, or the formation of coral reefs around a tropical island. Biogeographers are interested in the distribution of plants and animals, the ways organisms live together, the processes (both natural and people-induced) that affect the biological earth, and the effect of changes on human life. Climatologists study the long-term characteristics of the atmosphere and any climatic

---

[1]William D. Pattison, "The Four Traditions of Geography," *Journal of Geography* 63 (May 1964): 211–216; reprinted in the *Journal of Geography* 89 (September–October 1990): 202–206.

differences created by temperature or energy and moisture conditions in various parts of the earth. Physical geography, which has blossomed in recent years, emphasizes the interdependence of people and the physical earth. Such contemporary problems as ozone depletion, acid precipitation, desertification, and rain-forest removal are of particular interest.

Human geography consists of the study of various aspects of our occupancy of the earth. Urban geographers, for example, examine the location and spatial structure of cities in an attempt to explain why urban areas are distributed as they are and to account for patterns of settlement and economic and cultural activity within cities. Urban geographers are interested in the process of urban growth and decline, in the types of activities carried on in cities, and in the movement of goods and people within urban settings. Cultural geographers examine the ways in which groups of people organize themselves; they study such cultural institutions as language and religion, as well as social and political structures. Economic geography involves the study of systems of livelihood, especially the distribution of related activities and explanations for such distribution. Economic geographers are concerned with analysis of natural and cultural resources, with their utilization, and with the structures of power and control over and patterns of access to resources at various scales that determine the equity with which resources are shared.

Systematic, or topical, **geography** consists of the study of specific subjects. Historical geographers, for example, study past landscapes and the changes that have taken place. How did the people of the Great Plains organize themselves in 1870, as compared with their organization in 1935 in the midst of the Great Depression? What past characteristics have persisted, and what effect do they have on present-day distributional patterns? Michener used a concept of historical geography called **sequential occupancy** as the organizing theme of some of his novels. He described the geography of an area during succeeding periods, gradually building an image of how it is organized today. Historical geography thus adds depth perception to time, facilitating an explanation of present patterns and their reasons for being. Systematic geographers normally study one aspect of the field—landforms, economic activities, or urban places, for example.

Regional geography involves the analysis of environmental and human patterns within a single area. A wide variety of facts are placed into a coherent form in order to explain how a region is organized and how it functions. A regional geographer, in essence, is an expert on a particular area of the world, applying systematic approaches to an understanding of that area (Figure 1–4). The regional approach provides the framework for this text.

All fields of geography, despite focusing on different sets of phenomena, share the geographic viewpoint; that is, all geographers analyze spatial arrangements (distributions) and search for explanations of the patterns and interrelationships among those and other phenomena. All geographers rely on maps as analytical tools, and many have added computers and remote-sensing techniques to aid in recording and

■ **FIGURE 1-4 Swiss Alpine Valley.** The study of man-land relationships is one of the four traditions in geographical scholarship. This scene is from Grindelwald, Switzerland. Note the tiny farmstead in the lower left portion of the image.

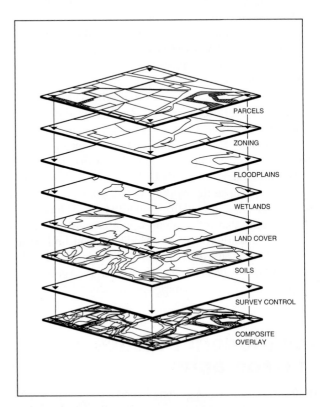

■ **FIGURE 1-5 A hypothetical GIS design.** Within a GIS, environmental data attached to a common terrestrial reference system, such as latitude/longitude, can be stacked in layers for spatial comparison and analysis.

analyzing data. The rapid growth in the use of computers has resulted in the burgeoning use of geographic information systems to analyze a wide range of geographical issues.

The Geographic Information Science Explosion

Twenty-first-century maps are more than just maps—they are analytical tools referred to as **geographic information systems (GIS)** and are part of a larger field of study called *geographic information science* (GIScience). Broadly defined, GIS is a digital representation of the earth's surface (a site, region, or country) that can be used to describe landscape features (roads, boundaries, mountains, rivers) and can support analysis of these features. In a sense, a GIS is like having a whole atlas in a single computer presentation with the ability to relate different pages of the atlas to each other.

Key to our understanding of GIS is the information layer—a map page showing a specific type of information, such as political boundaries, physical features, economic activities, cultural attributes, or any of a large number of other possibilities (Figure 1–5). A GIS project will have multiple layers of information that can be called on to answer a specific question or to address a specific issue. For example, layers may be combined to show electoral districts and income levels in a city; another set of layers may be used to show fire hazard potential in a national forest, given certain vegetation types and climatic conditions (Figure 1–6). Distances, areas, and volumes can be computed; searches can be conducted; optimal routes can be selected; and facilities can be located at the most suitable sites. In a fully integrated GIS implementation, some layers may contain satellite imagery or other remotely sensed information, and other layers line data (road

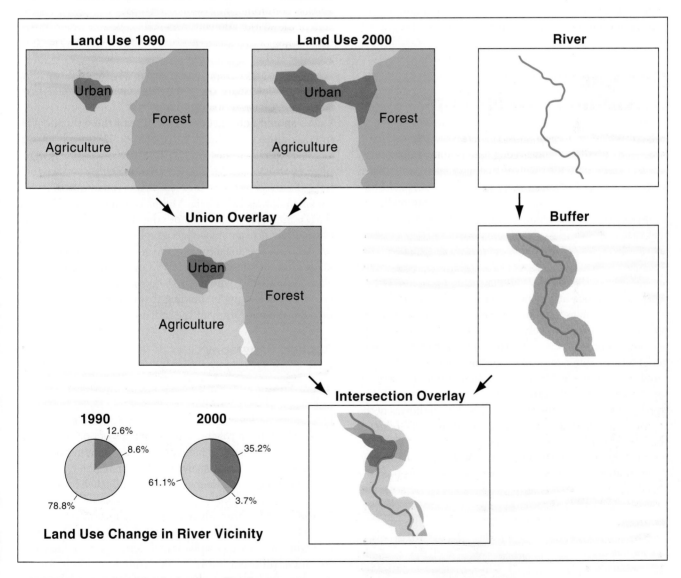

▪ **FIGURE 1-6 A hypothetical GIS application.** This illustration shows how various types of map information can be electronically combined into a single presentation. Note how the final overlay shows how much urban growth between 1990 and 2000 occurred in a buffer zone adjacent to the river. This buffer zone could describe a floodplain, indicating that recent urban development is exposed to a flood hazard. Computer technology can be used to modify this presentation in ways limited only by the data available and interest of the user.
Source: This illustration is courtesy of Dr. André Skupin, San Diego State University.

networks or county boundaries, for instance) that can be superimposed on the imagery. Within a GIS, projections provide an important mechanism for linking data from different sources, and scale provides the basis for size calculations of all sorts. Mastery of both projection and scale becomes the user's responsibility when the map user is also the mapmaker, as is often the case with GIS.

GIS use is exploding. Governments use GIS to track everything from power lines to demographic profiles. Businesses find geographic information helpful to locate facilities and to develop markets. Law enforcement agencies employ electronic maps to identify crime "hot spots" and to build "geographic profiles" of criminals. Militaries of the twenty-first century rely on GIS for terrain analysis and battlefield information. Even political campaigns have discovered the value of mapped profiles of potential voters and the issues that interest them. The list is almost endless. Geographers find that their GIS skills are in high demand by employers in both the public and private sectors.

Geography and Its Disciplinary Neighbors

As noted above, geography is a spatial science that focuses on the location of human and physical phenomena, and their interactions within terrestrial space, which of course explains geographers' fascination with maps. Geographers are more interested in understanding space than in examining the institutional or physical characteristics of objects that occupy that space. This spatial empasis has two implications for understanding geography's relationship with neighboring disciplines. First, geography is a bridge between the social and physical sciences, and possesses many characteristics of both areas. Take a look at a typical road map, for example, and note the references to mountains, rivers, cities, boundaries, and of course, roads—all on the same map. While most maps are not so encompassing, the road map demonstrates how geography requires an understanding of both human and physical phenomena. Or, to cite a different example, a geographer who studies the growing of wheat on the Great Plains needs information on the climate, soils, and landforms of the region (the physical sciences), as well as knowledge of the farmers' cultural characteristics, the transport network, the costs of wheat farming in relation to other economic opportunities, and a host of other socioeconomic factors (the social sciences). For that reason, geography is referred to as a **holistic discipline** that synthesizes knowledge from many fields.

Consequently, and this is the second implication, geography touches many related disciplines in both the social and physical sciences (Figure 1–7). For example, a political geographer may have interests shared with political scientists, a historical geographer with historians, a biogeographer with biologists, and a GIS person with computer scientists. The list goes on. Remember, however, that the geographer is distinguished from these neighboring disciplines by his or her spatial perspective; other disciplines do not have a spatial starting point, although they may deal with some of the same objects or phenomena.

Here is an example of how geographers interact with related disciplines, using the case of the economic geographer. Economists are interested in the production, distribution, and consumption of goods and services. They study how people use resources to earn a livelihood, investigating such topics as the costs and benefits of resource allocation, the causes of changes in the economy, the impact of monetary policies, the workings of different economic systems, the problem of supply and demand, and the dynamics of business cycles and forecasting. Economic geographers are also concerned with how people earn their livelihoods, but economic geographers look at where the economic activity takes place and what factors—such as the availability of labor, raw materials, and markets—along with certain physical attributes, influence that location. For instance, economic geographers are less interested in learning about the economics of the auto industry than in understanding why auto manufacturing is located where it is, keeping in mind that economic considerations can hardly be ignored in understanding location. They want to know about the historical, physical, social, as well as economic contexts of auto plant location.

We should note that geography is not the only integrating discipline; so is the study of history. History, however, uses a **chronological** (time) **framework**, whereas geography's perspective is **chorologic** (place). Neither can be studied effectively without a knowledge of the other. Isaiah Bowman, former president of Johns Hopkins University, said that "a man [or woman] is not educated who lacks a sense of time [history] and place [geography]."[2] By integrating information in a regional context, the geographer pulls together knowledge shared with a variety of disciplines into a single, all-encompassing, whole. In so doing, geography can provide insights and understanding that would not be available through separate study of the individual elements.

Applied Geography

Modern geographers differ from their late nineteenth and early twentieth century predecessors in emphasizing explanation rather than description. They ask not only "where" questions but "why" questions. This shift in emphasis has increased geography's utility in solving many problems of our contemporary world. The result is increased employment in roles such as market and location analysts, urban or regional planners, cartographers, environmental analysts, and teachers.

Education   Traditionally many geographers have been employed as teachers, although geography as a specific subject is taught with varying emphasis in different parts of the country and at different levels in the curriculum. Many Americans have little contact with geography as a formal subject after middle school, and the general public's geographic

---

[2]Quoted in Alfred H. Meyer and John H. Strietelmeyer, *Geography in World Society* (Philadelphia: Lippincott, 1963), 31.

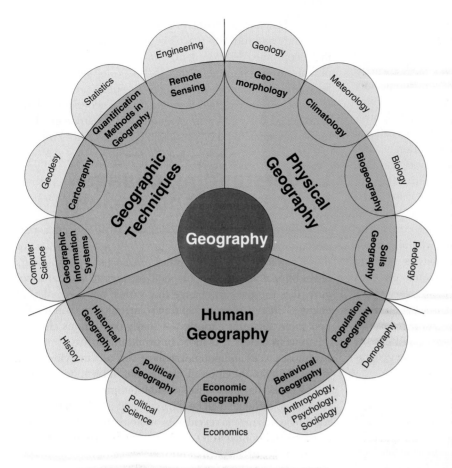

■ FIGURE 1-7 The scope of geography. Geography is a synthesizing and integrating discipline. This diagram shows that geography interrelates with many fields of study, including the physical sciences, engineering, social sciences, and humanities.

knowledge and skills are often inadequate. Reading a map is a basic skill, yet many people have difficulties performing this simple task. Far too many citizens find it difficult to locate other countries on a map and frequently know abysmally little about the cultures, values, and beliefs of other communities. The *Goals 2000: The Educate America Act,* passed by Congress in 1994, designated geography as one of several fundamental subjects that deserved more attention in school curricula. National Geography Awareness Week has been promoted since the late 1980s as a way to encourage a wider positive perception of geography's contemporary importance. Since 2001, The College Board has included geography in its advanced placement course system and has offered advance placement geography exams as part of its SAT (scholastic apptitude test) array. In 2006, nearly 20,000 prospective college students are expected to take this adjunct to the college admissions application process. This reflects a remarkable growth from the 3,272 students who took the exam in the first year it was offered! Where there are increasing numbers of students taking geography courses, there are also opportunities for larger numbers of geographically trained teachers! This development is good for both American education and for newly minted teachers, since the ways in which the United States is connected to the larger world—from the need for products and materials not readily available in the United

States, to migrants fleeing adverse economic conditions in their homelands, to foreign policy entanglements—will only grow greater as we move deeper into the twenty-first century.

Business Many of the skills that geographers possess, particularly those related to location, are very useful in the business world. Many geographers have found employment as locational analysts or environmental consultants. Banks would be foolish to make loans to new businesses or to established companies wishing to start new outlets or franchises without first determining whether the proposed location is an economically advantageous one. It is as important to the financial institution lending start-up money as it is to the business owner to know that the new restaurant or comic book store or gift shop will make money. How else can the bank be sure that the principle and interest on its loan will be repaid? Many large companies retain their own internal group of locational analysts who travel from site to site, study the relative benefits of central city as opposed to suburban locations, examine the purchasing power of the citizens of a particular community or neighborhood, assess traffic patterns to ensure a new outlet is accessible to a high volume of potential customers, evaluate the likely competition that rival firms might represent in the area, and predict the direction that patterns of growth and decline in the regional economy are likely to experience. Once all the data are assembled, it is possible to determine whether a proposed location is a good one or not.

The skills of physical geographers and students of environmental hazards also have application to real world problems. Concerns about the environmental impacts of economic development have required the preparation of environmental impact statements before a project can proceed. Federal law requires that changes with impacts on the environment, such as new housing developments, industrial plant expansion plans, road construction, wetland drainage, and many other activities, must be identified, their scope assessed, and potential damages mitigated before a project can begin. Geographers have often played a major role in enviromental impact assessment as well as in the emergency management of hazards, both natural and human-caused.[3] Geographers have also employed their general knowledge of foreign areas by helping companies understand the challenges and opportunities that businesses in those places are likely to encounter, as well as providing cartographic, climatological, geographic information science analysis, travel agent, and consumer

---

[3]David Alexander, *Principles of Emergency Planning and Management* (New York: Oxford University Press, 2002).

behavior services to firms and individuals wishing to conduct business outside the North American continent.

Government   Second to teaching, more geographers probably apply their skills to government agencies than to any other area. At a local level, many are municipal or regional planners, charged with facilitating orderly residential, business, and industrial growth and redevelopment. On a national level, knowledge of distant places and cultures, often in combination with remote sensing, map interpretation, and GIS skills, provide geographers with analytical abilities that are much in demand in government agencies. The Office of the Geographer in the Department of State is a focus of geographic activity, as is service in the diplomatic corps. The Agency for International Development also contains many geographers whose expertise in resource analysis, regional development, planning, and sustainability science provides valued services. Internationally, geographers often work closely with agencies of the United Nations such as FAO, UNEP, and the United Nations University, as well as with international financial institutions such as the World Bank and the Inter-American Development Bank. For those geogra-

■ FIGURE 1-8 Geography at work in disaster aid.
Disaster victims often return to their homes in need of food, shelter, and medical attention. Furthermore, they are frequently depressed and disoriented, and do not know where help is available. This is where Global MapAid (GMA) steps in. GMA is a not-for-profit organization that designs and reproduces specialist maps of disaster areas that are distributed free of charge to disaster victims. These maps are produced using on-site mapping teams and sophisticated computer technology; but are simple in design and clearly indicate the locations of nearby medical facilities, government offices, Red Cross centers, open stores, and anything else of practical value to returnees. These maps help to reorient people faced with the daunting task of rebuilding homes and reconstructing lives. GMA has a global reach, but most recently was involved in Hurricane Katrina assistance. The work of GMA demonstrates one of the many ways in which geography can be applied to real-world situations and can make a difference in human lives.
(*Source:* Logo used with permission. See *www.globalmapaid.rdvp.org* or *www.globalmapaid.org*.)

phers who view governmental agencies as excessively bureaucratic and too "top down" in their approach to promoting economic development, NGOs (non-governmental organizations) engaged in grassroots development activities have provided a productive outlet for energy and initiative (Figure 1–8). Increasingly large numbers of geographers have been engaged in employing geographic skills learned in school to the pragmatic world of improving people's daily lives.

Understanding Change in Economy and Society

All cultures undergo change. Regardless of how conservative and rigid a culture or society may appear, regardless of the viewpoints voiced and positions advocated by proponents of traditional values and perspectives, change invariably occurs. It is the pace of change that varies, not the reality of change. Change is more rapid in some cultures than in others; some societies experience nearly continuous change, whereas in others it is more intermittant or less apparent. For most places and peoples, collectively and individually, it is usually a combination of factors that determine whether in any particular period progress and improvement or stagnation and difficulty will be the dominant condition that they must face. Change is revealed by more than just the patterns of increased connectivity and interaction that we call globalization. At least three patterns of change with significant implications for development merit our attention.

Economic Growth and Decline   No civilization has successfully created an economy that has grown continuously. For all economies, periods of growth are offset by occasional episodes in which the economy stagnates or experiences even more profound retrenchment. Often these episodes of growth and decay are the product of other processes linked to and interacting with the economic activity. The quality of political leadership that a country enjoys often has a profound effect on the vitality of its economic institutions. Failure to provide an environment that is conducive to investment, for example, can encourage the flight of capital to other political jurisdictions and investment opportunities. Elite selfishness can extract too large a share of national productivity, channeling capital into the hands of an advantaged class while diminishing opportunities for the majority and undermining social solidarity and community spirit. Competition among rival factions in the ruling class for access to power and wealth can fragment the stability needed for orderly change and development. Insistence on a narrowly defined orthodoxy or lack of the vision needed to foresee future positive conditions and to work toward them can prevent orderly adaptation to changing conditions. In combination with the normal fluctuations and cycles present in the natural environment, as well as unforeseen extreme events, these negative tendencies can provide feedback that discourages economic growth and destabilizes societies.

Examples of these processes abound. In Pharaonic Egypt, political stability was based on the annual flood of the Nile, which brought water to the fields (Figure 1–9). The gradual development and improvement of lift devices to raise water during low-flow periods of the year to irrigate land between the floods permitted intensified agricultural production and population growth, but was insufficient to sustain the population if the flood failed. The Pharaoh derived substantial political power from predicting, accompanied by appropriate rituals, when the life-giving floodwaters would arrive. Occasionally a low-flood year would occur, but stored reserves usually were sufficient to tide people over the deficit period. But on several occasions, regional fluctuations in climate, when compared to the long-term record of river flow, produced decade- to century-long moisture shortfalls and consequently severe declines in the level of the Nile's flood. Less water resulted in fewer fields being cultivated, fewer fields produced lower overall yields and economic stagnation, more hungry people generated political unrest, and political conflict and frequent regime change resulted in conditions not conducive to economic growth.

The contemporary United States provides numerous examples of how public policies can influence long-term economic growth and decline. One such example is the national government's energy policy, which has continued to place primary reliance on fossil fuels at the expense of alternative energy sources. This decision favors existing energy suppliers and has many implications for long-term environmental quality. Maintaining the existing energy mix is favored by a political environment that makes it difficult to curb energy-consumptive motor vehicles such as SUVs (sports utility vehicles) and discourages energy conservation in general. The unfounded conviction that as yet unproven new sources of oil and natural gas have the potential to make significant contributions to meeting long-term domestic energy demands produces a refusal to invest in alternative sustainable energy technologies such as wind, hydrogen, and solar power. This reluctance to envisage an alternative energy future and to fund research to develop technology that would support a new, more sustainable energy environment may impact significantly the growth potential of the American economy, especially relative to other countries where energy policies have embraced a wider array of options.

A dramatic example of how economic change can be rapid and profound, and how public policies can be both helpful and harmful, occurred along the United States Gulf Coast in the aftermath of Hurricane Katrina in August of 2005. The storm devastated deltaic Louisiana and adjacent parts of Mississippi and Alabama, creating one of the most tragic and expensive natural disasters in the history of the country (see *Geography in Action* boxed feature, The Latest Battle for New Orleans, Chapter 4). It was apparent early in the tragedy that the response of the national government to the disaster was something less than quick-witted and nimble-footed, although state and local governments also came under fire. Government aid eventually flowed in, but was dominated at the outset by lucrative, no-bid contracts to favored private companies in an effort to get recovery activities started quickly. Long-term reconstruction plans were influenced by philosophical debates about the value of

■ **FIGURE 1-9 Agriculture along the Nile River, Egypt.**
Where the Nile's waters reach, rice, vegetables, dates, and fruit trees can be grown. Land not commanded by the river's life-giving water remains desolate. Fishing from small boats is also a traditional activity.

private- versus public-sector solutions, tax incentives, enterprise zones, and the wisdom of rebuilding in hazardous and flood-prone areas. At the same time, policymakers were haunted by the knowledge that a relatively small public investment in levee improvements and other mitigation strategies in the years before the storm could potentially have saved hundreds of billions of dollars in reconstruction costs after the storm. What Katrina really revealed, however, was how vulnerable cities can be in the face of nature's wrath, and how policymakers do not always get it right, requiring new policies to fix the damage. Meanwhile, hundreds of thousands of people found themselves suddenly without homes and jobs, and a devastated city of New Orleans discovered how true it is that economies and societies can change literally overnight.

Land Use/Land Cover Change   Human activity has resulted in massive alteration of the world's **land use** and **land cover**. Humankind benefits in many ways from disturbance in the landscape, since without directing energy away from other species and toward human sustenance our species could not long support current population levels. Agriculture requires removing trees and grass in order to grow the sun-loving plants we prefer to eat. Mining requires digging in the ground to find the ores that we can process and turn into energy and tools. Housing requires assembling construction materials from forests, rock, sand, and metallic ores to build the final shelter. The problem with the land use / land cover changes that our species introduces is less the initial disturbance, although these impacts can be severe and degrading, and is more the mess we leave behind when we are done. This "mess" includes not only the altered environmental conditions in which we often live, but also the problems that change in one area of life and livelihood cause for other people, other places, and future generations.

■ FIGURE 1-10 Deforestation in the Brazilian Amazon.
Roads open the rain forest to lumbering operations and agricultural settlement. Clear-cutting reduces the ability of the forest to grow back, and farmers often find that maintaining cultivation in the face of declining soil fertility is difficult. Invasion by grasses and the expansion of cattle herding is often the result.

Expanding agriculture has cleared large tracts of the earth's land surface of woody vegetation and substituted agroecosystems that people control. **Deforestation** has proceeded rapidly in many parts of the world, nowhere more rapidly in the last third of the twentieth century than in the Amazon basin (Figure 1–10). Efforts to promote settlement by landless farmers from densely populated areas has removed forest cover and converted these areas to agricultural uses. When low fertility soils are unable to sustain an intensive agricultural regime, land abandonment results. In many such areas, forest fails to regenerate and grasses invade the available space. This shift in land cover is encouraged by cattle ranchers, who see an opportunity to assemble large tracts of land into ranches that can sustain a domesticated animal population that requires very modest labor inputs and ultimately can generate a marketable product in demand in both local and global markets. Meat off-take may be low, but the profit margin can be large for the landowner with sufficient land. Cattle reach market weight in about three years, the capital represented by the herd multiplies much more rapidly than money invested in stocks and bonds, and a small number of people can do very well financially when grassland replaces forest in a tropical setting. Where land pressure is intense in Africa and Latin America, this change in land cover can proceed rapidly. Deforestation is linked to global geochemical cycles because it contributes to carbon emissions (burning the woody trash that remains after cutting timber; burning to promote growth of grasses) and removes one of the major sinks (forest growth) for carbon capture from the atmosphere. Estimates suggest that from 1960 to 1990, 20 percent of the world's tropical forests were removed and converted to other uses.[4] This has serious implications for the animals as well as the plants that are adapted to spending all or an important part of their life cycle in the habitat that is in the process of being transformed.

[4]Howard Youth, "Watching Birds Disappear," in *State of the World 2003: A Worldwatch Institute Report on Progress Toward a Sustainable Society,* edited by Linda Starke, (New York: W. W. Norton, 2003), 14–37.

But in many places forests do return. Poorer farm and pasture lands that are no longer economically productive in competition with more favored areas frequently show signs of forest regeneration. Many of Vermont's dairy farms have stopped operations because they cannot achieve the economies of scale (often units of 1,000 milking cows or more) that are increasingly typical of farms in Wisconsin, Minnesota, and California. Land no longer mowed rapidly reverts to forest. Migration from the Appalachian Mountains to seek jobs in big cities outside the region has withdrawn labor from farm operations and encouraged forest growth at the expense of former farmland.

Wetlands have also experienced significant change. Although the spatial scale of wetland modification is less extensive than that of woodland conversion, the impact is very important ecologically. Even wetland areas such as the Florida Everglades, which have received considerable protection, are threatened by human activity. This involves a process of nibbling away at the wetland around the edges as developers place pressure on local planning authorities to authorize residential construction on land traditionally deemed too wet for housing. Part of the transformation process also includes the diversion of water away from the wetland as areas on its margin seek protection from flooding. Loss of moisture means an increased frequency of fires, a progressive drying of the wetland habitat, and significant transformation of the land cover. Coastal marshes important to offshore fishing resources, because the juvenile stages of fish growth occur there, are frequently converted to marinas or prime housing sites with access to nearby beaches. These transformations solidify the human presence in a very dynamic and unstable natural habitat and expose both structures and people to a high level of risk whenever a high-energy storm tracks along the coast and produces storm surges and heightened erosion.

Thus changes in land use and land cover are risky business. Many such changes are necessary to sustain human populations. But many, particularly in richer, more heavily industrialized parts of the world, are not linked to enhanced economic growth in agriculture or industry; rather they are frequently recreational and self-indulgent in character (Figure 1–11). The changes often come with an expensive future price tag in terms of damage from natural hazards and diminished biodiversity.

Environmental Change Many of the transformations that people produce are beneficial when viewed from a human perspective, at least, and if they exhibit long-term sustainability features, they are changes that many plants and animals can adjust to with considerable success. Habitat changes that disadvantage open-field, edge-environment songbirds, for instance, are offset to some degree in suburban America by the propensity of homeowners to erect feeding stations that sustain the birds when naturally occurring food supplies dwindle. Coyotes, when attacked as a menace to domesticated animals and hunted vigorously, shifted from a more visible social, pack-oriented behavior pattern

■ FIGURE 1-11 Beach front homes undermined by erosion. Home location close to the ocean is a major amenity factor for which people pay a high price. Not only are house lots costly, but also the threat of erosion is an enormous future risk. Near Panama City Beach, Florida, accelerated erosion associated with hurricane Dennis in July, 2005 removed the buffer between houses and the ocean. Efforts to harden such coastal sites with walls and rip-rap are seldom effective in the long term.

to a largely solitary existence—and survived. But the record for most species is that rapid, human-induced changes in habitat involve a rapid drop in population numbers that threaten survival.

An increasingly powerful technology, employed without careful analysis of the potential side effects of its use, often has enhanced human impacts on the global habitat. If ancient peoples were able to substitute time for technological power and slowly transform their surroundings, the gradual pace of change meant that there was a greater ability for other species to adjust to these human driven changes. This grace period, this lag time, no longer exists. The litany of dramatic, rapid, and generally undesirable transformations is well known. Despite countervailing arguments about the reliability of the evidence, the overwhelmingly held scientific understanding of human impacts on global climate is that significant warming is taking place. If the concerted efforts on an international scale are not taken, serious consequences, ranging from rising sea levels, melting of glaciers and polar icecaps, and coastal erosion to changing patterns of drought and rainfall, are likely outcomes. Release of CFCs (chlorofluorocarbons) into the atmosphere is another problem. CFCs escape from human products such as refrigerators, air-conditioning units, aerosol sprays, solvents, and combustion by airplanes and terrestrial vehicles. They have had an unexpected impact on gases in the stratosphere. Here near the top of the atmosphere, CFCs attacked ozone, a naturally occurring gas that is important in atmospheric dynamics and regulation of solar radiation. The result was to produce "holes" in the ozone layer or **ozone holes** over the poles and a thinning elsewhere.

Efforts to stabilize ozone levels have been partially successful. That is, further decreases in atmospheric ozone appear to have stopped. But it will be a very long time before chemical reactions in the atmosphere will restore the losses that were sustained in a very short time in the second half of the twentieth century.

Economic development inevitably produces environmental change. Constant awareness of this simple fact, rigor in planning efforts to identify and avoid potentially harmful impacts, vigilance in monitoring and measuring critical parameters that warn about adverse changes, and resolute commitment to implementing countermeasures when changes threaten sustainability are essential if the benefits of economic development are to be disseminated to the many rather than being reserved for the few.

▶▶ Summary

This chapter explains what geography is all about. We learned that geography is a spatial science that focuses on the location of human and physical phenomena, and their interactions within space. These interactions include the relationships between people and their environments. We learned that central questions for geographers include the "where," "why," and "so what" questions. We discovered that geography is a holistic discipline that is integrative in its spatial approach and touches on a variety of sister disciplines. Also in this chapter we studied the nature of change. We examined how societies and economies never remain the same permanently, and how land-use and environmental modifications—some beneficial, some not—accompany social and economic change. This understanding of what geographers do and the nature of geographical change informs the contents of the chapters that follow. Above all in this chapter, we learned that geography matters!

▶▶ Key Terms

Alexander von Humboldt	holistic discipline
area studies tradition	human geography
Aristotle	land use / land cover
behavioral perspective	Karl Ritter
chorologic framework	man-land tradition
chronological framework	ozone holes
deforestation	physical geography
development	post-modern perspective
earth science tradition	Prince Henry the Navigator
Eratosthenes	Ptolemy
feminist perspective	regional geography
geographic information systems (GIS)	sequential occupancy
	spatial tradition
geography	Strabo
Herodotus	systematic geography

■ **Favela Morumbi in Sao Paulo.** Densely settled Favela (shantytown) Morumbi exists in close spatial proximity to the luxury high-rise apartments of the upscale Morumbi district.

Geography and Development in an Era of Globalization

- **Development—What's in a Name?**
- **Human Transitions and Development Processes**
- **Globalization as a Human Transition?**
- **Sustainability Issues**

I t is trite but true to remark that the world is getting smaller. Not only are we bombarded daily with news from such places as Iraq, Northern Ireland, Afghanistan, South Africa, China, Venezuela, Israel, Russia, the former Yugoslavia, and Japan, but events in those and other countries influence our daily lives. When the Organization of Petroleum Exporting Countries (OPEC) increased crude oil prices fivefold during the 1970s, the United States was abruptly awakened to the extent of its reliance on other countries to supply its energy needs. A similar realization followed the Iraqi invasion and attempted annexation of Kuwait in 1990. A coffee crop failure, the development of a new high-yielding variety of wheat, the discovery of a new chemical process, an outbreak of a new flu strain, and scores of other distant events may all materially affect the way we live. As world population grows and levels of living improve, the degree of international interdependence also increases. Consequently, we are forced to know more about the world in which we live in order to respond wisely to changing circumstances.

As the title indicates, this book takes a developmental approach to the understanding of world regional geography. One of the most serious ethical issues of our time is the great disparity in material and nonmaterial well-being that exists among the world's societies. With a more intimate world brought about by better communication and transportation, knowledge of how others live is at our fingertips. Peoples that are economically less fortunate often wish to emulate their materially richer neighbors but may be frustrated in their attempts by cultural, social, and political constraints. The disparity in economic achievement is widening, and the social and political ramifications affect us all. The causes and consequences of that disparity provide the focus and theme of this volume. By way of introduction, however, it is necessary to examine the various meanings associated with de-

velopment, how the term is used in this book, and the manner in which development should be examined in an era of globalization.

Development—What's in a Name?

The term development is complex in its meanings and implications. First, development embraces more than economic improvement, although the economic component figures prominently in this book and normally comes to mind when hearing the word development. Scholars also study issues related to social and political development, and even on occasion consider a psychological component. Second, the word "development" is only one of many terms applied to the processes of change, or lack of change, that have come to distinguish the economic and political circumstances of different countries. Indeed, there is a tyranny of labels, with each producing connotations that may or may not be helpful in understanding the change process. The expression "undeveloped," for example, suggests a total absence of those characteristics that we associate with modern economies and societies, which is an unfair characterization of most countries. The term "stagnant" is better only in the sense that there is no presumption that an economy or society is completely lacking—it simply is not going anywhere. The term "emerging" is more optimistic, perhaps excessively so, with its expectation that an economy or society is on the right path, and that it is only a matter of time before a country will escape its past and reap the benefits of its improved circumstances, just as a beautiful butterfly emerges from its cocoon. A more benign or clinical term preferred by scholars and policymakers is "**less-developed country (LDC),**" which avoids many of the negative connotations or the inflated expectations of the preceding terms. All of these terms apply primarily to the economic parameters of a society, usually measured in crude economic terms such as average annual per capita income or gross national product. Although a country's economic performance may be deficient when measured by these standards in comparison to post-industrial Western countries, the same country may be very highly developed in other dimensions of human activity, such as art, homeopathic medicine, environmental values, or care of the elderly. In this book, we will generally use expressions such as "developed," "developing," or "less-developed" in describing the material and nonmaterial aspects of growth and change in countries and regions.

Development—Toward a Definition

Development can mean different things to different people; but for purposes of this textbook, development refers to a progressive improvement of the human condition in both material and nonmaterial ways. As we contemplate this path to improvement, we should be aware of certain assumptions that guide our understanding of what this path looks like. More to the point:

- Development is a process of change that may or may not embrace the developmental ideals associated with North America and Europe; in other words, the paths to modern economies followed by Americans and Europeans may not be workable and/or desirable in other countries, given their specific cultural and environmental contexts. For example, American farmers use tractors and combines to achieve a high level of agricultural productivity per farm. Are tractors and combines really needed to increase productivity and improve well-being in many peasant-farm settings? If so, what will be the consequences for densely populated rural areas?

- Development is a process of change that is not necessarily synonymous with economic growth. In the market economies of North America, Europe, and Asia, a high priority is placed on increasing output—more stuff is good! But improvement of the human condition may or may not be immediately associated with growth in the production of goods and services. Improvements in welfare may come from such simple expedients as the provision of clean water to an isolated rural community, which may make village families happy and healthy, but is unlikely to cause a spike in the local stock market. In addition, as discussed below, some cultures may object to the materialistic impulses that create economic growth.

- Development is a process of change that is sustainable in the dual sense that improvements in human welfare today should not be achieved at the expense of future generations, and should be made in a manner sensitive to environmental impacts. For example, exploitation of natural resources to extract as much value in as short a time as possible without regard to long-term implica-

tions would not be regarded as a sustainable development strategy, even though short-term profits and economic growth might be impressive.

Consequently, and to build on the definition that we started above, we can define development as a process of change that leads to improved well-being in people's lives, which takes into account the needs of future generations and is compatible with local cultural and environmental contexts.

Consistent with this definition and the assumptions that support it, we can identify four components that figure prominently in the development process, particularly economic development:

- An important component of the economic development process is *people*—their numbers, distribution, consumption, production, and technology base. Improved sanitation and medical science have lowered death rates sharply over the past two centuries and caused unprecedented world population growth. If the resultant additional workers prove, through the adoption of improved technologies, to be more productive than their forefathers, material levels of living will increase despite population growth. If, however, production levels remain constant or decrease, living levels will decline.

- The *natural environment* of a country or region provides both the stage for development and the materials necessary to achieve developmental goals. Some environments have abundant supplies of raw materials that can be used for economic gain. For example, a well-watered alluvial plain coupled with a long growing season provides many opportunities for productive agriculture (Figure 2–1). Similarly, highly mineralized areas with easily extractable ores offer other means of livelihood. Conversely, areas with steeply sloping land, thin soils, moisture deficiency, or few minerals provide relatively few opportunities for growing crops or mining and thus require alternative development strategies.

- *Culture* plays an important role. The way in which a society organizes itself in terms of beliefs, values, customs, and lifestyles greatly influences both the direction and the degree of economic development. Many Western cultures embrace materialistic values, wherein the acquisition of material wealth is viewed as an index of individual worth and success. Although not Western in tradition or culture, the Japanese and many other East Asian peoples have similar value systems. Other cultures, however, do not place so high a priority on material achievement (Figure 2–2). The social and political structure of a society also has a direct influence on development. Some nations have achieved a relatively high level of social and political equality, thereby assuring the fullest possible development of their human resources. Others are characterized by rigid social stratification, gender inequalities, and political

■ **FIGURE 2-1 Alluvial plain in South China.** The application of large amounts of labor and water management technology makes the river floodplains of South China extremely productive agricultural environments.

control by elites that prevent large portions of the population from achieving their true potential. Similarly, the collective values of some cultures encourage the adoption of new ideas and technologies, while those of others discourage experimentation and change.

• Never forget *history.* That the past is a key to the present and a guidepost to the future is well demonstrated in the formation of the world's various cultures and their economic activities. Economic development is not a short-term process. In most nations that are now undergoing rapid change or that have attained a high level of economic well-being, the necessary founda-

tions or prerequisites for economic development were laid decades, even centuries, ago. For example, the cornerstones of Europe's Industrial Revolution, which began in the middle of the eighteenth century, were formed during the Renaissance, with beginnings in Roman and Greek times and even earlier. A more recent example is Taiwan, where many of the foundations for the island's recent economic growth were laid in the early part of the twentieth century. As we strive to promote economic and social development, it is important to recognize that lasting, substantive changes usually occur gradually over long periods of time. Such a perspective will encourage the laying of necessary foundations and the patience and perseverance needed to complete the transformation.

One final comment regarding terminology. Since World War II, countries have been unofficially divided into separate "worlds," reflecting both geopolitical and developmental considerations. The First World is a Cold War designation that initially included the countries of North America, Western Europe, Australia, New Zealand, and Japan—in other words, the "West," or those countries that stood as the front line of defense against expansion by the Soviet Union and China. While these countries were highly developed, it was the geopolitical role that they performed that defined their status as First World. The Second World was comprised of the old Cold War "enemies" of the West, particularly the Soviet Union and its client states of Eastern Europe, and the People's Republic of China. It is noteworthy that the former Soviet Union was a significant industrial power in its own right, at least in terms of heavy industry, and for that reason

■ **FIGURE 2-2 People watching solar-powered TV in Niger, Africa.** As economic development and technological change have spread, traditional cultures can be dramatically impacted over a short time.

could less easily be distinguished economically from the First World, unless of course we focused on differences in economic philosophy—capitalist versus communist. **The Third World**, however, was constructed almost solely in response to economic considerations. The Third World embraced those countries in Latin America, Africa, Asia, and the Middle East that we normally view as less developed or economically distressed. More recently, some scholars have added a Fourth World to include countries in which economic conditions are especially desperate, the so-called "basket cases" of the world. With the collapse of the Soviet Union and the end of the Cold War, such a division of the globe into "worlds" makes less sense. For one thing, many of the Second World countries of what used to be called Eastern Europe—those countries that were tied to the Soviet bloc—have entered into political and economic relationships with the West that makes them very much a part of the First World. Poland, for example, is now a member of both the North Atlantic Treaty Organization and the European Community. The small states of Estonia, Latvia, and Lithuania, which used to be a part of the Soviet Union, take great pains to downplay their Russian connections and to promote their ties to North America and Western Europe. Even though these terms may have outlived their usefulness, they are still part of our vocabulary; the term Third World, in particular, remains a popular synonym for underdevelopment.

In an attempt to avoid the Cold War connotations of the various "worlds" described above, scholars have found new ways to group countries based more strictly on developmental conditions. For example, terms such as "industrialized" (United States), "transitional" (Poland), "newly industrializing" (South Korea), and "less-industrialized" (Ethiopia) are used. In some cases, we see expressions such as "**old industrial countries (OICs)**" juxtaposed with "**newly industrialized countries (NICs)**," with the clear connotation that the former have grown old and stale and the latter young and dynamic. We can save further discussion of development terminology until later.

Human Transitions and Development Processes

The Significance of Economic Revolutions

As pointed out above, the development process involves a change or a transition of some sort—preferably a change that leads to economic, political, and/or social improvement. Typically such change is slow and plodding, and hardly perceptible over time. At times, however, change occurs rapidly and massively, leading to a fundamental transformation of society. Such a transformation is commonly referred to as a revolution. It may be a political, economic, or social revolution; but in every instance it involves a transition that is fundamental, a change that transforms society to its core. The American Revolution, for example, led to a fundamental change in the structure of U.S. politics and government. Some would argue that Canada and the United States experienced a social revolution in the 1960s that fundamen-

tally transformed the way in which society looked at itself. Both countries have passed through economic revolutions that transformed the ways in which goods were produced, services were rendered, and/or information was made available. Indeed, to understand the development process worldwide, it is important to have an appreciation for the power of revolutions to restructure societies and economies.

The Agricultural Revolutions   Perhaps 10,000 years ago, people learned how to domesticate plants and animals. This turned out to be one of the greatest innovations in human history, leading to one of the world's most transforming economic revolutions. Instead of wandering around eating whatever Mother Nature's bounty provided, people gradually settled down and started raising crops. Instead of chasing game and fish, they began herding livestock and breeding animals. In addition, new technologies were introduced, from the simple hoe to more complex irrigation systems. The ancient Egyptians, for example, knew how to divert water from the Nile River to irrigate their fields, an innovation that thoroughly transformed their capacity to produce crops and create surpluses.

A key to understanding the meaning of the **Agricultural Revolution** is to understand the significance of **agricultural surpluses**. Before the domestication of plants and animals, people lived literally hand-to-mouth, and everybody was involved in the process of finding food. With domestication, agricultural productivity soared to the point that some farmers produced more than they consumed, creating surpluses. Consequently, not everyone had to devote full-time to providing food and fiber, and not everyone had to live on the farm. Towns and cities appeared, supported by the farmer's surpluses.

With the appearance of towns and cities, living arrangements became more congested. The first real towns and cities appeared perhaps 8,000 years ago in those parts of the world where agricultural productivity was greatest. Life in settlements was somewhat easier and more secure than that in the countryside. Permanent homes, even substantial houses, replaced the crude huts or caves that may have served as temporary lodgings. Many tools and other large and small luxuries were acquired, such as chairs, tables, and beds, which previously may have been impractical because of the migratory way of life. Materialism may have had its true beginning with the development of agriculture. Possessions could be accumulated and passed on to new generations.

Thanks to surpluses and the growth of cities, economies and societies became more complex. Expanded population, production, and interpersonal contact created a need for increased group action and led to the growth of secular leadership organizations. A political organization was established to settle disputes, govern, and provide leadership for collective action in warfare and in such public works as irrigation, drainage, and road building. The formation of a priestly class helped formalize religion. Religious leaders were frequently the holders of both philosophical and practical knowledge, often serving as medical men and weather forecasters. In the Mayan civilization of southern Mexico and Guatemala, for example, priests developed an agricultural calendar based on the progression of the sun, the planets, and the stars. It predicted the beginning of wet and dry seasons and told farmers when to prepare the land for planting to take full advantage of the seasonal rains. As increased production per worker yielded more than a family unit needed, a portion of the labor force was freed not only for government and religious activities but also for activities such as pottery making, metallurgy, and weaving (Figure 2–3).

Of course, the connection between the production of surpluses and the appearance of complex, urban-oriented societies occurred only in the most environmentally favored locations—the Nile River Valley, Mesopotamia, and the Indus River Valley to name only a few locations. Elsewhere, the benefits of the Agricultural Revolution were limited to a more sedentary version of a subsistence existence, which was still an improvement over the migratory hunter-gatherer societies that existed previously.

In a sense, the agricultural revolutions have never completely ended. A second revolution began in Europe in the mid-seventeenth century in which new technologies and procedures were applied to farming, partly to increase farming's commercial potential. By the twentieth century, still another revolution occurred that replaced most farm labor with machinery, and eventually led to a corporate takeover of many family farms. A single farm could now support dozens of nonfarmers.

The Industrial Revolution Remember what a revolution is: it is a fundamental change in the way that something is done. By the mid-eighteenth century, it was time for a fundamental change in the way that goods were made—it was time for a manufacturing or **Industrial Revolution**.

Actually, manufacturing is a human activity that has existed since before the beginning of civilization. The prehis-

toric cave dweller who chipped a piece of flint into a spear point was involved in manufacturing. So were the ancient Greeks and Romans when they shaped pottery out of clay and transformed ore into metal ornaments. But this manufacturing process was primitive, relying on manual labor (hence the term "manu" facturing) and the most unsophisticated forms of technology. In the medieval period, for example, the shoemaker depended on his hammer, shaping tools, and a strong arm to make shoes; the blacksmith had his fire, bellows, anvil, hammer, and an even stronger arm to bend iron; and the tailor made use of his needle, thread, and the digital dexterity of his hands (no strong arm required here) to stitch cloth. This type of handicraft manufacturing was slow and inefficient, although quality could be, and often still is, very high.

Everything changed beginning in the mid-1700s. Europe and North America entered an age of mass production of manufactured goods, using much more sophisticated technology and large industrial workforces. Coal became the industrial fuel of choice, replacing animate sources of power such as animals and people. This fundamental change in manufacturing represented a response to at least two major forces: First, an age of innovation had dawned thanks to the thinking and discoveries of the Renaissance, which led to the introduction of new technologies to speed up the production of goods. In 1769, for example, James Watt developed a practical steam engine. His innovation paved the way for a wholesale mechanization of the manufacturing process and for the steam engine to become the main source of power for the Industrial Revolution in its first century and a half. In addition, new furnace-based technologies emerged to replace the time-tested, but highly inefficient, charcoal method of iron processing. New spinning and weaving technologies made the volume manufacturing of cloth much more efficient. In brief, machinery replaced muscle power and inanimate energy replaced animate energy. Second, the way in which production was organized changed drastically. Manufacturing no longer was a single activity done by a craftsman in his shop, but a volume enterprise carried out in a factory—a stand-alone building containing a full-time industrial workforces and an array of machines. Demand for processed goods exceeded what the cottage industries and craft shops of the day could produce, and the more efficient factory was the only way to supply this demand. Over time, craftsmen and small guilds gradually disappeared from the scene or remained in isolated pockets as quaint reminders of a pre-Revolutionary past. Indeed, the factory came to epitomize all that the Industrial Revolution was about, as well as to symbolize the social and environmental price paid in the name of production efficiency.

It is important to point out that not every country or region experienced the type of industrial revolution described above; the Industrial Revolution was initially a European and North American phenomenon, with Japan and selected other countries joining

■ **FIGURE 2-3 Mayan ruins at Palenque, Chiapas, Mexico.** Surpluses produced by agricultural productivity enabled Mayan communities to support large, urban settlements with spectacular public monuments.

in later. One of the markers of underdevelopment today is the relative absence of modern manufacturing capability.

The Industrial Revolution spawned companion revolutions. For example, development of the steam engine led to a transportation revolution (Figure 2–4). Steam power was first applied to maritime shipping in 1807; by 1829, steam railway locomotives were being built. Steam power significantly diminished the **friction of distance**, or the difficulty of moving from place to place. Before railroads, a traveler walking the 948 miles from New York to St. Louis took more than six weeks to make the journey; by 1870, with railroads, the trip became a matter of about three days; today a commercial airliner can cover the distance in under three hours flying time! Transportation efficiency contributed to production efficiency in manufacturing centers. To cite a second example, and to build on a point made above about agricultural revolutions, the Industrial Revolution created a new way of doing things down on the farm. Tractors, combines, and other equipment quickly replaced sweat and muscle in the cultivation of crops, which in turn created a large farm population looking for another line of work. Most of that work took the form of factory jobs in cities, which contributed to an Urban Revolution as people swarmed into the cities in search of factory jobs. The impact often was profound. The village of Essen, Germany, for example, had 4,000 people in 1800; by 1920, there were 439,000 people working in a highly industrialized city. Similar stories can be told for the United States. In 1800, the vast majority of people worldwide lived in the countryside. By 2000, an overwhelming majority of people in developed countries lived in towns and cities.

The Information Revolution   The world's great revolutions are not yet finished. Indeed, as you read this text you are taking part in an **Information Revolution** that is reshaping societies and economies, not to mention personal lives. Who in the twenty-first century can imagine living without a cell phone or laptop computer? We are part of what increasingly is called the Information Age and/or the New Economy.

Of course, information has been around since the beginning of human communication—even prehistoric cave dwellers conveyed information when they painted pictures of animals on cave walls. It is a matter of how the information is produced, stored, accessed, and applied that creates a revolution; and in the Information Age it has all boiled down to a single transforming technology: the microprocessor. Mechanical computers have existed for some time—even the ancient abacus can be considered a type of computer—but it took the miniaturization and integration of electronic circuits to create the fundamental changes that led to the current Information Revolution. The first true electronic computers appeared after World War II and required a whole room to do what a good PDA does today. More space-efficient mainframes followed, and by the 1990s desktop and portable computers became everyday appliances. In the twenty-first century, thanks to constant innovation, hardly anything is done without computer involvement (Figure 2–5). In brief, unlike for the prehistoric cave dweller and his wall, our ability to produce, store, access, and apply information is massive and nearly instantaneous—truly revolutionary.

In partnership with the microchip, we must also credit innovations in communications technology for the revolutionary changes of the late twentieth century. As we all know, the Internet permits nearly instantaneous communication worldwide between connected computers. In addition, satellites and cellular phone technology have made wireless voice and data communication the norm rather than the

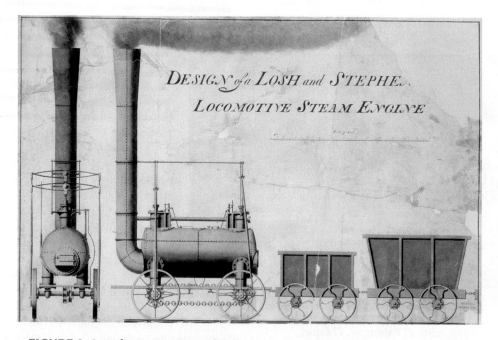

■ **FIGURE 2-4 Early steam-powered train.** This is a drawing of an early nineteenth-century English train powered by a 56-horsepower steam engine. The Industrial Revolution has transformed many parts of the earth but is only now penetrating other regions.

■ **FIGURE 2-5 Auto assembly plant in Shanghai, China.**
The General Motors assembly plant is characterized by the latest robotic assembly techniques.

exception. Geography has not disappeared, but it certainly has been modified by the ability to contact a significant proportion of the rest of the world by simply punching in a number.

Where is this revolution headed? Perhaps a clue can be found in the Elon University/Pew Internet and American Life Project (Fall 2004), in which a survey was conducted of 1,286 Internet stakeholders/experts on the impact of the Internet over the following 10 years. The survey said that news organizations and publishing could expect the most radical changes in the next decade, followed by education, the workplace, and the health industry. Participants were asked to write individual predictions, which included the following: a move toward greater individualism, perhaps "networked individualism," in which people connect increasingly with other like-minded individuals across the globe, and decreasingly with local neighborhoods and groups; an explosion of alternative media for entertainment and communication; an unprecedented reorganization and diffusion of health information, with an accompanying change in the way patients and providers interact with each other; a possible incentive for governments in other countries to open up and become more democratic; and heretofore unheard of changes to the workplace and the ways that businesses operate. One big change has already begun: as workplaces become paperless, it is decreasingly necessary for workers to commute to the office to handle paperwork. On a more ominous note, there is increasing concern among scholars and observers everywhere that the Information Revolution is empowering extremist and fundamentalist religious groups to recruit, manage, and disseminate terrorist activities. There is a chilling awareness that small groups of fanatics can harm large groups of people, thanks in part to the Information Revolution.

The developmental questions surrounding the Information Revolution are profound (see the *Geography in Action* boxed feature, How Does Development Happen in the Information Age?). Will the diffusion of information technol-

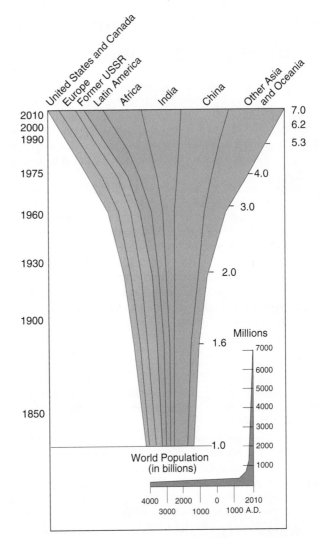

Region	Year 2010 Population (in Millions)	World Share
World	7,020	
United States and Canada	334	4.8%
Europe	519	7.3%
Former USSR	306	4.3%
Latin America	584	8.3%
Africa	1,078	15.4%
India	1,163	16.6%
China	1,376	19.6%
Other Asia and Oceania	1,660	23.7%

■ **FIGURE 2-6 World population growth.** For most of human history, the world held relatively few people. In the last 400 years, however, the total number of people has expanded greatly and at an increasingly rapid rate. The term *population explosion* is often used to describe such rapid growth.

GEOGRAPHY IN ACTION
How Does Development Happen in the Information Age? Is It the Result of a "Creative Class"?

What is really behind economic development? Is it simply finding investment dollars and creating jobs? Or is there something else? For some time, scholars have argued that development is not simply the result of building a new factory or creating more office space; development is the outcome of creative people doing innovative things.

Recent work by Professor Richard Florida has popularized the notion that future development of regions and communities in the United States will depend on their ability to attract creative people who are part of what he labels a "creative class." These people are scientists, engineers, professors, poets, artists, musicians, designers, and architects. They tend to be independent thinkers who are young and highly educated. In many cases they do not have spouses and children, which lets them immerse themselves in their work, occasionally to fanatical excess. They include the eccentric "eat-pizza-for-breakfast" types who like relaxed dress codes, flexible work hours, cultural diversity, "real" experiences in the real world as opposed to cookie-cutter culture, outdoor recreation, and an active night life. And yes, they jog. With these characteristics in mind, Florida makes the interesting and somewhat counterintuitive argument that cities without gays and rock bands have only limited developmental futures.

Which cities have this creative class? The leading large cities are San Francisco, Austin, San Diego, Boston, Seattle, Houston, Washington, D.C., New York, Dallas, and Minneapolis. Smaller cities include Chapel Hill, North Carolina; Madison, Wisconsin; Des Moines, Iowa; Santa Barbara, California; Melbourne, Florida; and Huntsville, Alabama. Notice how many of these cities we associate with large universities and so-called "high-tech" America. On the other end of the scale, those cities without a large creative class include Providence, Rhode Island; Greensboro, North Carolina; New Orleans; Oklahoma City; Grand Rapids, Michigan; Louisville, Kentucky; and Buffalo, New York. Many of these cities are dominated by more traditional working class populations. Chicago stands out as a city that is able to accommodate both a large creative class and a large working class, perhaps because this creative class is viewed as just another "ethnic group" to be worked into the city's already complex cultural fabric (Figure A).

Cities attract the creative class if they are "plug-and-play" communities, which is to say communities into which people fit easily—all sorts of people: eccentrics, gays, and even people who wear ties. These cities provide the types of jobs, often high-tech, that appeal to the creative class. They accommodate artists, writers, performers; they offer authentic downtown experiences and outdoor recreation. And, of course, they have bike paths. Failing cities tend not to create the environment that appeals to the creative class. They stake their developmental hopes on the wrong things—for instance, big-box retailers, subsidized downtown malls, and worst of all, according to Florida, extravagant sports stadium complexes. Planners and public officials in these communities may find it very difficult to discard the developmental attitudes of the past—attitudes, by the way, that may have worked quite well at a different time in history—to embrace the values attractive to the creative class. In other words, there may be too many institutional and cultural barriers to overcome.

It is interesting to note that traditional high-tech centers in the United States are not necessarily included on Florida's list of attractive destinations. Silicon Valley, for example, is too bland, too congested, and too full of strip malls and boring housing developments to attract the creative class. It is more a "nerdistan," according to some observers, than a creative class Mecca. North Carolina's Research Triangle is even worse: it is Silicon Valley without the city.

Are non-creative class cities condemned to irrelevance and decline? Not necessarily, according to Florida. It is largely a matter of embracing the correct development culture—one that cultivates and attracts people who will produce the innovations that will take the economy to the next level. Even Pittsburgh, an archetype of Industrial Revolution values, is showing signs of thriving in the creative age.

Not everyone agrees with Florida's development model. The broad categories and emphasis on class will not appeal to some economic-development professionals. Others argue that his economic logic is flawed and that there is little empirical basis to support his thinking. But few deny the sudden popularity of Florida's ideas. In addition, few will disagree with the proposition that the success of the Information Age can be traced to creative innovations coming from creative people.

Sources: Adapted from Richard Florida, "The Rise of the Creative Class: Why Cities without Gays and Rock Bands Are Losing the Economic Development Race," *Washington Monthly* (May 2002), available at *http://www.washingtonmonthly .com/features/2001/0205.florida.html*. This article provides an abbreviated and simplified version of Florida's book, *The Rise of the Creative Class and How It's Transforming Work* (New York: Basic Books, 2002). For a somewhat different take on creative regions, see E. J. Malecki, "Soft Variables in Regional Science," *The Review of Regional Studies* 30 (2000), 61–69.

■ **FIGURE A Chicago sidewalk café.** One focus of artistic life in Chicago, Illinois, is the Art Institute on Michigan Avenue. Visitors enjoy a well-deserved break and good conversation in the café across the street after savoring the museum's extensive collections.

ogy ultimately be complete, or will there always be a "digital divide," as some call it, that creates a world of digital haves and have-nots? Will the cost advantages of digital technology to corporations help or hurt employees in developing countries; in other words, will a robot or other production technology replace assembly-line workers who have nothing to fall back on if the job is lost? Conversely, will jobs in one country be outsourced to a lower-cost labor environment in another country—outsourcing technology jobs to India is an example of this process—potentially hurting the donor country, but helping the recipient country? There are other questions that could be asked. Many of these questions are wrapped up in larger issues related to globalization that we will examine below.

By the way, as an aside, it is worth noting that we have really been looking at the second great information revolution. The first such revolution we can trace back to Johann Gutenberg of Mainz, Germany, in 1450, who built the first printing press. It would be hard to overstate the fundamental changes that occurred because of the printing press (including being able to read a book like this!).

Where Does Population Change Fit In?

Revolutions create all types of fundamental changes, but some of the most pronounced and enduring transformations occur in the dynamic behavior of human populations. Consequently, we will briefly review how global population counts have changed over time, and then examine selected population concepts and growth models that will help us

understand the pivotal role of population in understanding problems associated with economic change and development.

Global Population Patterns. At the dawn of the Agricultural Revolution, when animals and crops were first domesticated, the world's population was extremely low (Figure 2–6, page 25). Population clusters, often associated with areas of agricultural surplus, later spread throughout the world, along with the diffusion of crops and animals. In a few places, such as Australia, the diffusion process was delayed until the coming of European colonists. Today, only in polar zones, remote dry lands, and other harsh physical environments do small and dwindling numbers of people still live by the age-old occupations of hunting, fishing, and gathering.

At the beginning of the Christian era, world population totaled little more than 260 million people, most of whom were located in the Old World. The majority of those people lived within three great empires: the Roman Empire, around the fringe of the Mediterranean Sea and northward into Europe; the lands of the Han dynasty of China, which extended into Southeast Asia; and the Mauryan Empire of northern India. In those empires, the simpler political, economic, and social organizations of agricultural villages vied with the more complex, integrating structures of the empires and the newly created cities. Urbanism became a way of life. By 1650, the world's population had grown to more than 500 million, despite interruptions of famine, plague, and warfare. Most of the growth was in and around the preexisting

Table 2-1 The World's 20 Largest Urban Agglomerations

Agglomeration	Country	Population (in millions)		
		1800	1900	2005
Tokyo	Japan	1.0	4.5	35.3
New York	United States	.1	3.4	21.9
Mexico City	Mexico	.1	.3	19.0
São Paulo	Brazil	—	—	18.3
Mumbai (Bombay)	India	.2	.8	16.0
Dilli (Delhi)	India	—	—	15.3
Kolkata (Calcutta)	India	.6	.8	14.3
Buenos Aires	Argentina	—	—	13.4
Jakarta	Indonesia	—	.8	13.2
Shanghai	China	.3	.9	12.7
Dhaka	Bangladesh	—	—	12.7
Los Angeles	United States	—	—	12.1
Karachi	Pakistan	—	—	11.8
Rio de Janeiro	Brazil	—	.8	11.5
Osaka	Japan	.4	1.3	11.3
Cairo	Egypt	.3	.6	11.2
Lagos	Nigeria	—	—	11.1
Beijing	China	.7	1.0	10.8
Manila	Philippines	—	—	10.7
Moscow	Russia	.3	1.0	10.7

Sources: *Infoplease,* http://www.infoplease.com/ipa/A0884418.html, *The World Almanac and Book of Facts 2006* (New York: World Almanac Books, 2006); W.S. Woytinski and E.S. Woytinski, *World Population and Production: Trends and Outlook* (New York: Twentieth Century Fund, 1953).

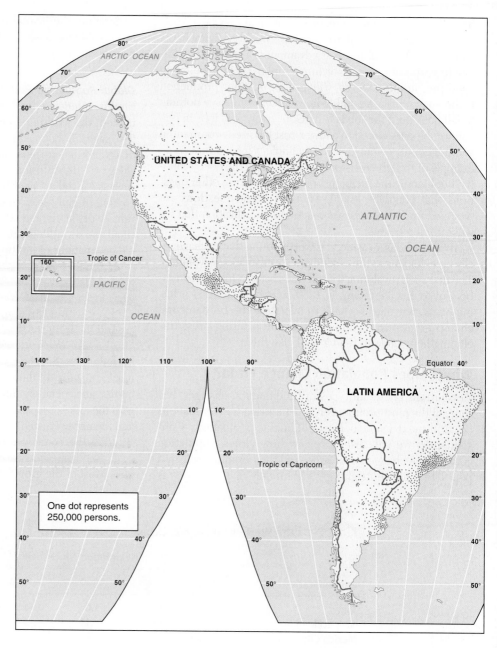

■ **FIGURE 2-7 World population distribution.** The world's population is unevenly distributed. Three large areas of dense population are China, the Indian subcontinent, and Europe. Most of the sparsely populated areas have environmental impediments such as aridity, extreme cold, or mountainous terrain.

One dot represents 250,000 persons.

centers, with gradual expansion of the populace into areas that had been sparsely settled. Productive capacity expanded with improved technology and new resources. Urbanization became more pronounced, though agriculture remained the primary livelihood for most people.

Since 1650, the world's population has increased more and more rapidly. It took an estimated 1,650 years for the population to double from 260 million to 500 million. By 1850, just 200 years later, the population had more than doubled again to approximately 1.175 billion. Within the next 100 years the population nearly doubled a third time, reaching 2 billion. By 1975, it had doubled yet again, to 4 billion. The world now has more than 6.2 billion people and will continue to experience large increases in the foreseeable future. The world's population is increasing at the rate of nearly 1 billion per decade, and by 2025 it may exceed 7.9 billion. An increasingly large proportion of the population is living

in cities, some of which have the populations of medium-sized countries (Table 2–1, page 27).

Distribution and Density of Population As Figure 2–7 shows, the world's **population distribution** (the spatial arrangement of people) and **population density** (the number of people per unit area) show strong ties with the past. Three principal centers of dense population are readily apparent: the Indian subcontinent, eastern China and adjacent areas, and Europe. China and India represent old areas of large populations, stemming both from an early start in the Agricultural Revolution and from empire building. Today at least half of the world's population lives in southern and eastern Asia, where agriculture and village life remain important facets of society. Yet modern cities, with their service and manufacturing functions, are also present. The population density in India and China varies considerably, usually

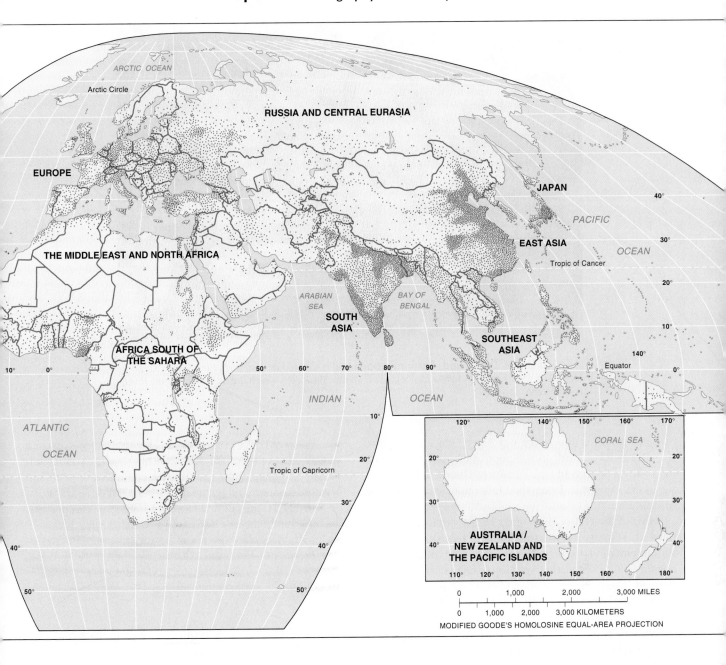

in association with the relative productivity of the land. **Physiologic density**, the number of people per square mile of arable (farmable) land, is a useful expression of the density relationship in agricultural societies. On the coastal and river plains, where alluvial soils are rich and water is abundant, rural densities of 2,000 people per square mile (772 per square kilometer) are not uncommon. Away from well-watered lowlands, such as the Huang He or Chiang Jiang (Yangtze) valleys, densities diminish but may still be in the range of 250 to 750 people per square mile (97 to 290 per square kilometer).

Europe's high population density can be traced back to technological developments originating in the Middle East, which were adopted by the Greeks and Romans and then expanded by the Industrial Revolution. Further increase in the European population is readily associated with developments in technology. Many of the high-density areas in Western Europe are urban regions associated with coalfields or advantageous water transportation, indicating the importance of those assets in the Industrial Revolution (Figure 2–8). Even though Europe's population density is high, it is significantly less than that of comparable Indian and Chinese areas. In addition, agricultural villages and agriculture itself are overshadowed in Europe by modern metropolises and manufacturing.

Secondary centers of high population density are more numerous worldwide. The northeast quadrant of the United States and the adjacent parts of Canada are considered by some a principal cluster, though total population numbers are smaller than those of Europe and far less than those of East Asia. High densities also occur in Africa along the Guinea Coast and the Nile River and in the eastern highlands, but the total number of people involved in each cluster is relatively small. Similarly, around major urban centers

■ **FIGURE 2-8 Dortmund, Germany.** Dortmund is situated in Germany's Ruhr River Valley, one of the most heavily industrialized regions on earth. The valley, which drains into the Rhine River, is rich in coal deposits.

of Latin America and in the old Aztec, Mayan, and Inca realms, small but locally dense population centers are common. Other pockets of high density are found in Indonesia, the Malay Peninsula, Japan, the Philippines, and parts of the Middle East.

Most of the rest of the world's land surface (roughly 80 percent) is sparsely inhabited. Many of those areas present serious environmental challenges—coldness, aridity, rugged terrain—that have made them less attractive to human settlement. The sparse population of other areas, such as some of the humid tropics of South America and Africa, can be attributed to fragile soils of low natural fertility. It is tempting to relate population density to the broad physical patterns described in Chapter 3, and indeed some writers have done so. Yet a correlation between population and physical environment is an oversimplification. Technology and political organization are additional factors to be considered, as are other aspects of culture, such as desired family size and economic organization. As world population continues to increase and levels of technology continue to advance, it is possible that many of these less densely settled regions will become more populous.

Models and Theories of Population Change

Overall, the **population growth rate** for the world is 1.3 percent a year, but that growth is by no means uniform. One explanation for the varied growth pattern is the theory of **demographic transformation**, which is based on four population stages (Figure 2–9). Stage I postulates an agrarian society in which **birthrates** and **death rates** are high, creating a stable or very slowly growing population. Productivity per person is limited. Consequently, large families (or many births) are an

economic asset, particularly since life expectancy is low and security depends on family members, including young children. Employment opportunities outside of agriculture are few in this stage, and technology is stagnant or nearly so.

In Stage II, the cultural custom of large families persists, and the birthrate remains high. The death rate, however, drops dramatically because of better sanitation and medical treatment and because of greater productivity. Productivity may increase in the agricultural sector, but more important is the appearance of alternative economic activity

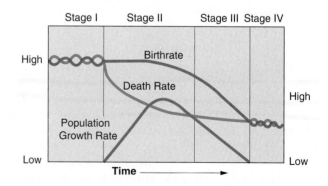

■ **FIGURE 2-9 Model of demographic transformation.** The demographic transformation model is based on the European experience. It may not represent what will happen elsewhere, especially in areas of non-European cultures. The model assumes an initial period of traditional rural life with high birthrates and death rates and no population growth. Then, with increased production and improved sanitation, the death rate declines, creating a condition of rapid population growth. As the population becomes more urbanized, the birthrate also declines until low birthrates and death rates finally prevail, and the population no longer grows.

GEOGRAPHY IN ACTION
Demographic Transformation for More Industrialized and Less Industrialized Countries

The demographic transformation shown in Figure 2–9 is a model that illustrates well the experience to date of most European countries. The model in Figure A distinguishes between the experience of today's economically developed and developing countries. Several observations are appropriate.

The death rate in today's developed countries began a slow, significant decline during the late eighteenth century, but that decline has accelerated during the last century. We should be aware of two important facts. First, the period of time required to lower the death rate was longer for developed than for developing countries. Second, by the time the death rate began its rapid decline in the developed countries, the birthrate had also begun its decline; the same was not true for today's developing countries. The result is that the developed countries overall never had the explosive growth rates of today's developing countries (rates of 1.75 to 2.50 percent a year versus 3 to 4 percent a year, respectively). In addition, the European countries had a safety valve in the form of large-scale migration opportunities, which are not as readily available in today's high-growth countries. Another significant factor is that the death rate declined in the developed countries as a function of the need to discover, invent, and diffuse medical technology—a slow process. In today's world, reducing the death rate is less a matter of discovery or invention and much more a matter of diffusion of medical technology.

We should also understand that today's developed countries experienced a long Stage II and III of their demographic trans-formation, during which major economic development was occurring. By comparison, today's developing countries have been thrown rapidly into high population growth (sometimes in a decade or two in response to a rapidly declining death rate). Many have not experienced economic growth at a similar rate; but they have experienced population growth rates in excess of anything known before. Furthermore, the base population to which those high growth rates are applied is larger; it includes more than 80 percent of the world's population.

The long-term economic consequences of the demographic transformation of both the industrialized and less industrialized countries are unclear. In many of today's more industrialized countries, especially those of Europe, the birthrate has dropped below the death rate, leading to negative growth. In 2002, for instance, Germany, Croatia, and Bulgaria had natural increase rates of −0.1, −0.2, and −0.5, respectively. Will this negative population growth eventually characterize all of the more industrialized countries? Will it ultimately lead to declining economic output and/or large-scale immigration that fundamentally alters the historic, ethnic, and cultural composition of these nations? Will the ongoing decline in the population growth rates of today's less industrialized countries lead to increased levels of personal income? These and many other questions regarding the consequences of the demographic transformation of the more and less industrialized countries remain to be answered.

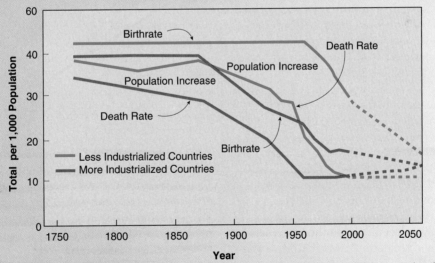

Rate of Population Increase = Birthrate − Death Rate, assuming no immigration or emigration

Note: These are crude birth- and death rates. The projected increase in death rate in developed countries after about 1980 reflects the rising proportion of older people in the population.

■ **FIGURE A** **The demographic transformation for more industrialized and less industrialized countries.** Clearly, the demographic experience has not been the same for all countries. Striking differences are evident for developed and developing countries.
Source: Adapted by permission from *The World Development Report 1982,* New York: Oxford University Press, 1982, p. 26.

■ **FIGURE 2-10 World population growth rates.** There are great differences in population growth from one part of the world to another. Areas of the world that are highly urbanized and have more industrialized economies tend to have low birthrates and death rates and consequently low rates of population growth. Areas that are less urbanized and less industrialized have high birthrates, declining death rates, and moderately to rapidly growing populations.
Source: *World Population Data Sheet,* Washington, D.C.: Population Reference Bureau, 2002.

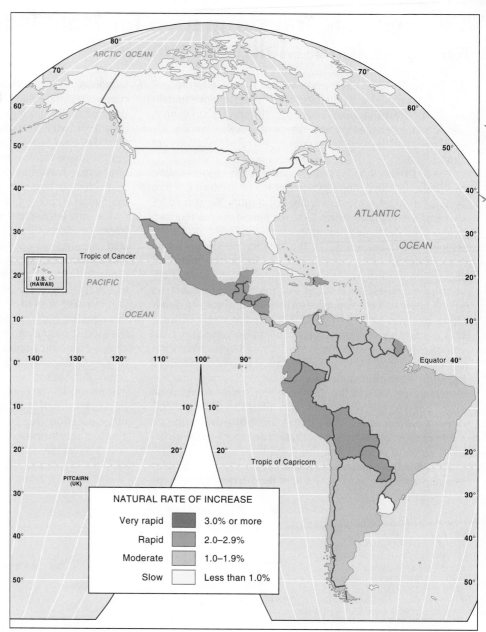

resulting from industrialization. With industry come urbanization and labor specialization. The principal feature of Stage II is rapid population increase.

Numerous countries remained in Stage I until World War II. In the postwar years, however, death rates all over the world have been greatly reduced through the introduction of improved sanitation and health care. Several African countries moved into Stage II only in recent decades. We must remember that the lower death rates, which have contributed in these countries to faster population growth, are likely to decline further before they bottom out, so that even higher rates of population growth are likely. Uganda, Mali, and Niger illustrate precisely such a situation (see the *Geography in Action* boxed feature, Demographic Transformation for More Industrialized and Less Industrialized Countries, page 31).

Stage III is characterized by continued urbanization, industrialization, and other economic trends begun in previous stages. Demographic conditions, however, show a significant change. The birthrate begins to drop rapidly, as smaller families become more prevalent. The shift toward smaller families may be related to the fact that children in an urban environment are generally economic liabilities rather than assets. In Stage III, the population continues to grow, but at an ever-slowing pace.

Stage IV finds the population growth rate stable or increasing very slowly. Both birthrates and death rates are low. The population is now urbanized, and birth control is widely practiced. Population density may be quite high. Countries such as Argentina and Chile may be considered to be in Stage III; the time of their complete transformation to Stage IV remains an open question. Virtually all of the countries in Europe are in Stage IV, as are the United States, Canada, and Japan, countries that exhibit a European style of industrialization and urbanization.

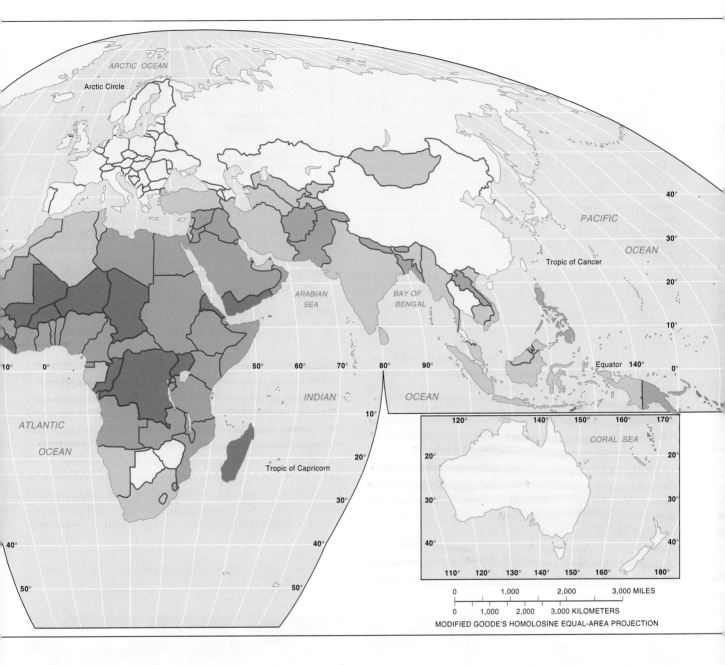

MODIFIED GOODE'S HOMOLOSINE EQUAL-AREA PROJECTION

The difficulty of predicting exactly when a country will enter a particular stage is illustrated by the United States. Birthrates reached a very low level during the 1930s (comparable to current rates) but then accelerated sharply during the postwar years, creating growth conditions equivalent to those in Stage III. Birthrates are sociological phenomena that are strongly influenced by cultural value systems, unlike death rates, which can be reduced by control of disease, proper diet, and sanitation efforts.

The model of demographic transformation is based on analysis of Europe's experience with urbanization and industrialization. Other areas with different cultures may not follow the European example precisely (see the *Geography in Action* boxed feature, Demographic Transformation for More Industrialized and Less Industrialized Countries). If the model is valid, however, population growth in many parts of the world will slow down as Stages III and IV are reached.

Whether and when world population growth will be reduced or stabilized is, of course, unknown (Figure 2–10). Meanwhile, the differences experienced in the rate of population growth, though seemingly minute, have enormous impacts on the absolute growth of population from region to region and on the utilization of resources (Table 2–2).

Malthusian theory is another explanation of population change that has received widespread attention and has numerous advocates. Thomas Malthus, an Englishman, first presented his theory in 1798, basing it on two premises:

1. Humans tend to reproduce prolifically, that is, geometrically—2, 4, 8, 16, 32.

2. The capacity to produce food and fiber expands more slowly, that is, arithmetically—2, 3, 4, 5, 6.

Thus, the population will eventually exceed the food supply unless population growth is checked by society. If growth

Table 2-2 Population Doubling Time by Region

Region	Population (in millions)	Growth Rate	Years to Double
Africa	906	2.3	30
United States and Canada	329	.6	115
Asia	3,921	1.3	53
Europe	730	−0.1	—
Latin America	559	1.6	43
Oceania	33	1.0	70
World	6,477	1.2	58

Source: *2005 World Population Data Sheet* (Washington, D.C.: Population Reference Bureau, 2005).
Note: The number of years required for a population to double is based on the assumption that the current rate of increase will remain constant.

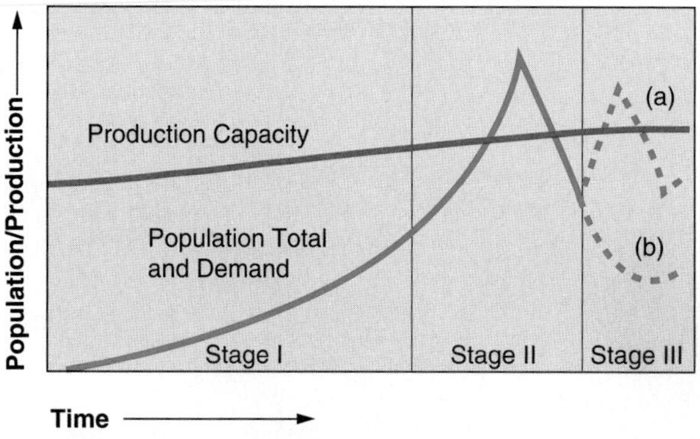

■ **FIGURE 2-11 Malthusian theory.** The Malthusian theory can be illustrated by a three-part diagram. In Stage I, the needs of the population are less than production capacity. In Stage II, however, the increase in population is so great that needs soon exceed production capacity. For a while, the population continues to grow by using surpluses accumulated from the past and by overexploiting resources. Eventually, the pressure on the resource base is too great, and the population begins to die off. Stage III may be a continual repetition of Stage II (a) or a return to Stage I (b).

continues, surplus populations will be reduced by war, disease, and famine.

If we plot Malthus's idea on a graph, we can identify three different stages and outcomes flowing from the relationship between population and production (Figure 2–11). In Stage I, human needs are not as great as production capacity. By Stage II, production capacity and increased human needs are roughly equal. In Stage III, population has grown to the point where its needs can no longer be met.

When Stage III occurs, the population dies off, and we cannot be sure what follows. One idea is that Stage III is simply a repetition of Stage II, with alternating periods of growth and die-off, represented by line (a) in Figure 2–11. Another idea is that the die-off is so great that Stage I is reproduced, as represented by line (b). Malthus's theoretical die-off has not occurred, owing in part to the enormous increase in production associated with the Industrial Revolu-

tion. Malthus assumed that people would reject birth control on moral grounds, and he could not, of course, foresee the impact of the Industrial Revolution. For the past half-century, however, **neo-Malthusians** (present-day advocates of Malthus) have argued that the population/production crisis has merely been delayed, and disaster may yet strike.

Others believe that population growth is beneficial. Ester Boserup, for example, has argued that the stress of increasing population and the corresponding need for more food stimulates intensification of traditional agricultural systems, which in turn leads to greater food production.[1] As long as

---
[1]Ester Bosrup, *The Conditions of Agricultural Growth* (Chicago: Aldine, 1968), 11–27.

■ **FIGURE 2-12 Lower-class housing in Jakarta, Indonesia.** As the income gap steadily widens between and within the rich and poor nations, an ever-growing number of persons are found living in abject poverty. These stilt homes have been built above a garbage-filled canal.

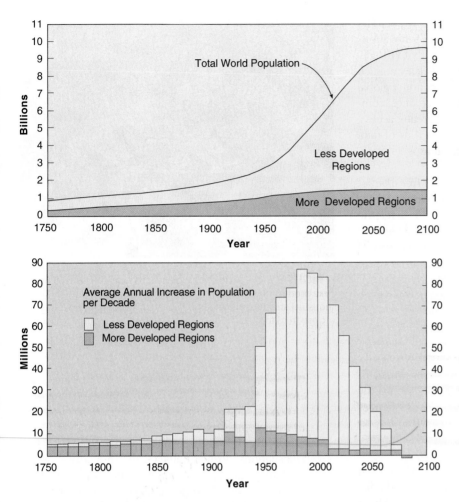

■ **FIGURE 2-13 Population growth in economically developed and developing regions: 1750–2100.** An increasingly large share of the world's population is found in developing regions, a situation that can be expected to continue well into the next century.
Source: Adapted with permission from Thomas W. Merrick, with PRB staff, "World Population in Transition," *Population Bulletin* 41, no. 2, Washington, D.C.: Population Reference Bureau, April 1986, p. 4.

the rate of population growth is slow, the increase in available labor, as well as new technical innovations, sparks a gradual intensification process that sustains the increased population. Then, once the transformation is well under way, the rate of population growth declines in response to modernization. Boserup's thesis is more optimistic than the neo-Malthusian view of population growth, suggesting that food supplies will always stay a step ahead of demand.

All theories aside, the fact remains that the population of the world is large and that increasing numbers of people live in the poor world (Figure 2–12) or the developing nations. Figure 2–13 dramatically illustrates the changing population proportions and projects their continuation to the end of the twenty-first century.

Globalization as a Human Transition?

Up to this point, we have focused on several of the great moments in the history of human change—the Agricultural, Industrial, and Information Revolutions—and how these transitions contributed to the types of improvement that we associate with the development process. We have not examined the latest transition, **globalization**, which for many observers has emerged as the defining social and economic force of the late twentieth and twenty-first centuries. While

it is easy to overstate the significance of globalization—far too many people invoke the term to explain everything that ails and benefits the world as well as ignore the role that globalization has played throughout human history—this transition process is having an impact of some sort on about everybody, everywhere.

What Is Globalization?

Defining globalization can be difficult, given its many dimensions; but in its broadest sense, globalization refers to a growing integration and interdependence of world communities through a vast network of trade and communication links. This process of integration and interdependence can be associated with a wide range of technological, cultural, and economic outcomes that affect our daily lives: communication is virtually instantaneous and worldwide; a visit to the mall is a visit to the factories of the world; a job gained or lost may reflect a decision made by a global corporation; a car assembled in the United States will contain parts from around the world; the music and fashion decisions of an American teenager may be shared by a teenager in south Asia; and perhaps most obvious of all, a trip abroad does not mean giving up a fast-food hamburger.

The increased integration and interdependence of a globalizing world constitute a response to two major forces: technology change and global capitalism. The technological fruits of the Information Age described above are everywhere. The Internet, satellite communications, and other innovations have shrunk time and space to such an extent that virtually no point in the world is farther than a mouse click away. Such virtual proximity has made it much easier to conduct global commerce and to spread ideas. Along the same lines, computing technologies have made it possible to create, access, store, and process extraordinarily large quantities of data with breathtaking speed, thereby making all aspects of society and business more productive.

The embrace of capitalist, or "neoliberal," or free-market policies in their various forms is equally important to understanding the dynamics of a globalizing world. Since the early 1980s, government leaders have joined their business counterparts in expressing a renewed faith in the power of freely operating markets to create prosperity. This expression has been worldwide in scope. Even such bastions of socialist conviction as the former Soviet Union have become converts to the capitalist creed (ignoring for the moment the somewhat rickety and occasionally corrupt market mechanisms that characterize the Russian marketplace). While the free-market lovefest has helped generate extraordinary pockets of wealth, critics contend that the costs of global capitalism outweigh the benefits—costs that include highly uneven economic development, the instability of businesses and worker livelihoods, the breakdown of social fabrics, and the sudden and disruptive movements of production and capital. Time will tell who is right.

Actually, globalization is not new. As Europe awoke from its Dark-Age slumber, it embarked on a journey of world discovery that led to increased global interdependence. These discoveries led to huge Spanish, French, and British colonial empires that created global trade connections and a worldwide diffusion of European culture in its various forms. Along the same lines, the explosive growth of mass production during the Industrial Revolution led to a global search for industrial raw materials (for instance, cotton for British textile factories) and global markets for manufactured goods (often in overseas colonies).

On the other hand, it is possible to argue that today's globalization is different in a qualitative sense from past globalizing events—it is more complex and deeply rooted. Global economic interaction in the twenty-first century involves huge transnational corporations and elaborate production networks made possible by Information Age technologies. Much global trade may in fact occur within the worldwide internal network of an industry or corporation, as opposed to trade among independent firms.

Globalization's Actors

So who is leading the charge in the globalization of societies and economies? Several major players can be identified:

- **Transnational corporations**: The **transnational corporation** (TNC) has a starring role (Figure 2–14). As the name suggests, these corporations have offices, production facilities, and other activities in multiple coun-

■ **FIGURE 2-14 Athletic shoe factory in Vietnam.** This Nike production line outside Ho Chi Minh City is one piece in the industrial structure of Vietnam's largest private employer. Vietnam is a close third to China and Indonesia as a supplier to Nike's global operations.

tries. They are geographically mobile and can take advantage of lower labor costs in one country or a more lenient regulatory environment in another country to minimize production costs and/or to maximize revenues. By the way, it is incorrect to assume that all transnational corporations are the same; they vary in size and shape, and in corporate and national culture. Japanese and U.S. corporations, for example, typically have very different organizational and cultural styles, but they both are players in a globalizing world.

- **Countries**: It is easy to argue that highly mobile global corporations should be able to circumvent the power of national governments, and they can to some degree. Some observers have commented that this mobility and power have led to the "hollowing out" of nation-states, making countries and their governments less relevant to world affairs than in the past. But in reality, countries can still limit access to their spaces, impose regulations on corporations operating in their spaces, and come together (as in the European Union) to create multistate authorities. Countries still have power in a globalizing world.

- **Labor**: Labor is more or less fixed in space and has limited bargaining ability in the global economy. If unskilled labor becomes too expensive in one location, the corporation will simply move production to a less-expensive place. It is possible to argue that such has been the fate of many assembly plants in the United States that have been closed as parent corporations seek lower wage levels in Latin America and Asia. In theory, successful global organization of labor would counter this trend, but as a practical matter such organization is unlikely in the near future.

- **Consumers**: In a real sense, consumers call the shots. If they stop buying, corporations will start dying.

Furthermore, we now live in a world in which products can be marketed to global consumers. While this may appear to be an advantage to corporations, cultural differences can make marketing a single product across cultural boundaries frustratingly difficult (see below).

- **Regulatory organizations and civil movements**: These institutions and movements influence the way in which global society and economy function. They include economic institutions such as the International Monetary Fund, the World Bank, and the World Trade Organization, which were created to provide at least limited multinational oversight of, and support for, global economic activities. They also include other organizations and groups that pressure corporations and governments to act in certain ways. In some instances, transnational social movements have appeared to challenge the perceived excesses of globalization, usually countered by equally vocal and partisan defenders of the globalization faith. The demonstrations of the late 1990s in Seattle, Quebec City, and at other economic conference sites provide cases in point (see the *Geography in Action* boxed feature, Globalization Vignettes).

▪ **FIGURE 2-15 Tradition and modernity in contemporary China.** Traditional curbside market activities in Yangshou, China, combine seamlessly with the use of modern technology.

Globalization's Geographic and Developmental Footprints

Now that we have defined globalization and have identified the major players, it is important to understand the geographic and developmental footprints of the processes behind globalization. In other words, from geographical and developmental standpoints, what does globalization really mean?

Globalization Does Not Necessarily Mean "Homogenization"

There is an increasingly popular assumption that globalization is leading to a worldwide unity of tastes and values, and a concomitant destruction of local cultures (Figure 2–15). It is true that a McDonald's hamburger can be bought virtually anywhere in the world; so can a Coca-Cola or Pepsi. And we all know how important jeans are as a global style statement. Even North American standards related to music, television programming, journalism, and some aspects of political campaigning are spreading to the rest of the world.

In reality, however, this assumption is easily exaggerated. What appears to be homogenization often is little more than a veneer—a penetration of local cultures by outside influences that is only skin deep. Why is this? First, local cultures tend to "domesticate," "indigenize," or "tame" imported consumer culture by giving it a local flavor. South Asians, for example, eat pizza but it may be a "Punjabi Pizza" that includes local toppings and spices; these same Indians may add garlic and chili to McDonald's hamburgers. Speaking of McDonald's, in the Netherlands mayonnaise is the condiment of choice for French fries; in Peru, fries come with a special local sauce called "aji"; in Taiwan, rice patties can be ordered in place of hamburger buns; and in Canada, some stores offer "poutine" to consumers, which is a mix of fries, cheese curds, and gravy. And speaking of fries, in Britain they are called chips and may be free of salt but doused in vinegar; Canadians have some of the same habits. What we call chips in America (as in potato chips) are called crisps in Britain—and yes, they may still be soaked in vinegar. A fast-food outlet in Spain may include beer on the menu, along with their own version of a hamburger and French fries. And in European countries where planning standards are strict, rigid rules that control signage in order to preserve traditional streetscapes often force a global corporation such as McDonald's to minimize its visual impact, diminish its golden arches logo to very modest scale, and adapt its operations to fit within existing buildings rather than construct stand-alone facilities (Figure 2–16, page 40).

Second, many countries promote a consumer nationalism that encourages local over "foreign" goods. For instance, while the Chinese appear to have welcomed globalization wholeheartedly—and there are enough fast-food outlets and Western chain stores to support this conclusion—there remains a deep-seated reluctance to embrace imported consumer cultures too closely. Part of this reluctance stems from a troubled history of contact with European cultures, going back to the nineteenth century when foreign governments indirectly but effectively controlled Chinese affairs. Regardless of the cause, some Chinese entrepreneurs go so far as to advertise their goods as the genuine local alternative to foreign offerings, as in the case of the local fried chicken company that boasted about how its chicken was so much more authentic than the big American brand named after a colonel, because the local chicken did a much better job of taking into account the peculiarities of the Chinese palate.

GEOGRAPHY IN ACTION
Globalization Vignettes

G lobalization has many everyday manifestations. Several examples of how the processes of globalization can be made real are described in the following vignettes:

Where Do Computer Chips Come from?

The transistor replaced the vacuum tube in 1947 and became the basis of the micro-electronics technology that forms the heart of the computer industry today (and enables in some way about every modern industry on the planet). The development of the transistor was closely followed by innovations in semiconductors and integrated circuits. As of 1980, the United States produced about 60 percent of the world's semiconductors. For a while, the United States was overtaken by Japan, South Korea, and Taiwan as they rapidly increased their production of memory chips; but by the mid-1990s, the United States had regained its global leadership, thanks to a shift into the more high-tech, design-intensive microprocessors.

The semiconductor industry was among the first manufacturing activities to have a distinct global geographical separation of functions. The more routine assembly operations were located in developing countries where female labor was abundant and wages were low; and the higher-level design, research, and development operations were located in the United States and other developed countries. A computer purchased in the United States probably had components manufactured in Asia. The first United States overseas assembly plant was located in Hong Kong in 1962; by the late 1960s, plants had been opened in Taiwan, South Korea, Singapore, Malaysia, and northern Mexico. In each case, firms were looking for low wages to keep production costs down.

Over time, however, Asian producers became more sophisticated. A production network appeared that was geographically specific and less oriented toward simple component assembly: Bangalore, India, focused on software design, Singapore on process engineering, South Korea on semiconductor memory, and Taiwan on digital design. Assembly remained important everywhere, but most particularly in Malaysia and coastal China. The global network of semiconductor and related firms had become far more complex, but it was still very much a global operation.

Is Wal-Mart Everywhere?

One of the consummate global players is Wal-Mart. Surprisingly, Wal-Mart did not go global until 1991, when it established its first outlet in Mexico City. Since then, Wal-Mart has opened hundreds of units (although not all are Wal-Mart stores) in China, Argentina, Canada, Germany, Japan, Mexico, and the United Kingdom, often through the acquisition of existing store chains such as the 1994 purchase of Canada's Woolco stores.

It is Wal-Mart's involvement with China that is particularly relevant to our understanding of globalization (Figure A). Much of the merchandise that appears on Wal-Mart store shelves is made in China, creating a huge stream of commerce from China to the United States. Indeed, the Wal-Mart corporation is China's sixth largest export market (and Wal-Mart is not even a country!). Much of Wal-Mart's success as a leader in "everyday low prices" can be traced to its effective use of volume buying power to keep supplier prices low. If a supplier wants a lucrative Wal-Mart contract—which is an invitation to be a part of one of the

■ FIGURE A Wal-Mart in China. Wal-Mart's first superstore in Shanghai was opened in the Pudong New Area in 2005. The company has aggressive plans to expand its Chinese outlets from 48 to 90 stores by the end of 2006.

world's dominant retailing operations—the supplier's prices must be low. To keep prices low, the supplier must keep production costs low, including wages. China has an abundance of low-cost labor that can help the supplier with this task. Critics contend that Wal-Mart's gargantuan size among world retailers and its obsession with controlling costs have depressed already low wages and have encouraged suppliers to ignore harsh working conditions. They also contend that Wal-Mart is using foreign labor at the expense of United States jobs, and is contributing to a huge American trade deficit. Wal-Mart's defenders counter by noting that Wal-Mart has strict labor standards that it imposes on supplier firms, that the use of Chinese labor is more likely to be at the expense of workers in other Asian countries rather the United States, and that Americans are benefiting in a major way from this trade (in spite of pesky trade deficits with China). Interestingly, most of Wal-Mart's suppliers are not Chinese companies; the supplier's factories are in China, but the corporations are based in other countries, including the United States, and operate in China to make use of—you guessed it—China's low-cost labor. One other comment about Wal-Mart and China: in addition to serving as an export platform, China itself has become a major retailing destination for Wal-Mart. Stores and supercenters are sprouting everywhere, and in the next couple of years Wal-Mart hopes to have almost a hundred stores throughout China.

Is There a World Music?

Global flows of culture are very much a part of the processes associated with globalization. These flows have led some to speculate whether a type of "world music" is taking shape that transcends national tastes and traditions.

There has always been a world of music (as opposed to "world music"), which has been embraced by the entertainment cultures of North America and Europe. Witness the popularity of Latin music, along with calypso, reggae, Hawaiian, and other traditions. The United States has a long history of involvement with African music, much of it modified by the slave cultures that populated the antebellum South; the banjo, for example, has an African origin even though it is a very "American" instrument in most peoples' thinking.

World music, though, suggests something more than the transfer of traditions from one region to another, especially from "exotic" tropical origins to North American or European destinations. World music brings to mind a type of music that blends the various traditions into a new sound. Some observers trace its origin to 1986 when Paul Simon recorded *Graceland,* which used English lyrics set over black South African music to produce a single melody reflecting influences of multiple cultural origins. It was not long until the likes of Peter Gabriel, Ry Cooder, and others followed suit in finding their own mixes of global melodies. Suddenly the sounds of Senegalese, Zairean/Congolese, Ghanians, Brazilian Indians, and New Guinea tribesmen were finding their way into popular music. "World" music was born as a music store category.

Of course, with popularity came a marketing blitz that eventually produced a commercial commodification of these new sounds. Some will argue that commercial success led to a stereotyping of, and lack of authenticity in, the sounds of the local cultures being integrated into the new world music. But the rise of world music, if nothing else, constitutes an attempt at global cultural interconnectedness built upon the cultural distinctiveness of places, very much a feature of globalization.

Everybody Speaks English, Right?

There is a humorously outrageous old saying that if the person with whom you are speaking cannot speak English, then speak louder. Nowadays, it is increasingly likely that an English speaker will not have to raise his/her voice. Whether desirable or not, English is a global language. Perhaps 380 million people speak English in countries in which it is the native language; another potential 300 million people speak English in countries in which the language is not native, but was introduced by English-speaking colonizers; and maybe as many as a billion more people worldwide speak English as a second language.

The earliest spread of English was from England to the adjoining territories of Wales (1530s), Scotland (early 1600s), and parts of Ireland (early 1700s). The first true diaspora of English speakers occurred when they colonized North America, Australia, and New Zealand. A second diaspora resulted when non-English-speaking parts of the world were brought under the control of English speakers (e.g., India, Pakistan, and British-controlled Africa). The latest spread of English, as suggested above, has been to parts of the world never controlled by English-speaking countries.

Why is English so widespread? There are several factors at work, of which two are especially noteworthy. First, the reach of British colonization was extensive. There was a time when, quite literally, the sun never set on the British Empire, and hundreds of millions of people were subject to British rule (and the English language). Once these many colonies became independent, they found it difficult to shed their colonial legacies, including English. In some cases, this legacy provided a help-ful unifying function, as in the case of English as a link language among speakers of the many different languages in India.

Second, English has become the worldwide language of business. To some degree, this began with the Industrial Revolution, in which the United States, Canada, and Britain were the early leaders and from which the benefits of the revolution were spread—in English, of course. This trend has been accelerated by the Information Revolution, both in terms of the diffusion of technology and the spread of English-language business communication. Along the same lines, English-language movies, television programming, and media coverage has become worldwide in scale.

Is global English the same everywhere? The answer is "no." We all know about the different accents—British versus American versus Australian versus Canadian—and the different forms of expression (British cars have "bonnets," whereas American cars have "hoods"). In addition, we need to understand that as English has spread to non-English parts of the world, the structure of the language has taken on changes reflective of local cultures. In India, for example, the word "may" is used to express obligation in a polite way—whereas Americans would simply say, "you will do it tomorrow," Indians are more likely to state firmly but politely "you may do it tomorrow." Both expressions intend the same meaning.

Of great concern to many non-English speakers is the fate of local languages as English continues its seemingly inexorable global spread. As evidence of this concern, many languages now have English expressions embedded within them. It is estimated that after World War II, the Japanese incorporated several thousand English words into their language—words such as *man-shon* (mansion) and *aisu-kurimu* (ice cream). But this phenomenon has not been limited to Japan, as illustrated by *dzhazz-saission* (jazz session) in Russia, *baj baj* (bye bye) and *tajt jeans* (tight jeans) in Sweden, and *le weekend* and *le drugstore* in France. French speakers have been particularly offended by the corrupting influences of English in Quebec, Canada, where the French population constitutes an island of Gallic culture surrounded by a sea of English speakers. In the mid-1970s, the Quebec government outlawed the use of English in business and education, except in a few narrowly defined circumstances, to safeguard the long-term linguistic purity of the French language in Quebec. (It is ironic that many French speakers in France show disdain for the Quebec version of French as itself lacking in purity.)

In brief, English has become the language of globalization, for good or for ill. It is the language of business, technology, and much of the world's mass culture. On the other hand, if a billion and a half people speak English, about 5 billion do not. Alas, you may actually find yourself in a situation where you still have to speak louder to be understood.

Sources: Adapted from P. Dicken, *Global Shift: Reshaping the Global Economic Map in the 21st Century,* 4th ed. (New York: The Guilford Press, 2003).
J. Connell and C. Gibson, "World Music: Deterritorializing Place and Identity," *Progress in Human Geography* 28 (2004), 342–61.
D. Elliott and B. Powell, "Wal-Mart Nation," *Time* (June 27, 2005), 36–39.
R. M. Bhatt, "World Englishes," *Annual Review of Anthropology* 30 (2001), 527–50.
R. McCrum, et al., *The Story of English* (New York: Viking Penguin, 1986).

■ **FIGURE 2-16 McDonald's in Munich, Germany.** This fast-food outlet with understated, multilingual signage is seamlessly integrated into a high-style building on the historic Karlsplatz in Munich.

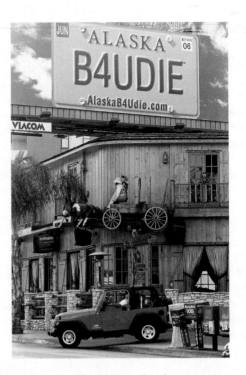

■ **FIGURE 2-17 Alaska Travel Industry promotes tourism.** Tourism is another economic sector in which places compete for income. This oversized billboard on Sunset Boulevard in Los Angeles encourages visitors to see Alaska "Before you die" and is based on marketing studies that indicate many people would like to experience Alaska at least once in their life.

To reiterate, a certain amount of homogenization has occurred—some may call it a transnational, postmodern, or postnationalist culture—but it does not mean that the essential cultural characteristics that distinguish people have been obliterated. It is true that Russians wear blue jeans and eat fast food, as do Japanese, Costa Ricans, and maybe even a few Mongolians; but wearing blue jeans and eating the Kentucky colonel's fried chicken does not destroy the essential "Russian-ness" of the Russian consumer. The fundamental cultural differences that distinguish one place from another remain, and these differences do not disappear the moment that an American fast-food store is built on the corner.

Place Still Matters If homogenization has been overstated, and if differences from place to place still exist, then geography still matters. While the technological foundations of globalization may have helped to annihilate space and time in a functional sense, technology has not destroyed the importance of place in a geographic sense. For example, corporate strategies may be global in reach, but they rely on close associations with distinctive localities to succeed. The corporation that sets up its assembly plants in a country with low wages is clearly establishing a close association with a specific locality, in this case a low-wage locality, and is artfully acknowledging the importance of place in its corporate decision-making. The importance of place is also revealed by the way in which a community will compete with other communities to attract an automobile assembly plant or some other large employer, touting local geographic benefits or differences in the campaign to win the corporation's affections. Competition among communities can be intense, almost ruthless, in pursuit of the prize (Figure 2–17). In brief, the geographic questions posed above—where? why? and so what?—still count in an era of globalization.

By the way, while place still matters, technology has changed the meaning of distance between places. Thanks to the ease of contacting different places across the world through elaborate communications and transportation networks, a person may be "closer" to a chat-room partner living a continent away than his/her next-door neighbor. Functional proximity may be more important to people than physical proximity.

Globalization Has Winners and Losers Contemporary geographic footprints of globalization include those of both winners and losers. In other words, there is an unevenness in the geography of development flowing from the forces of globalization.

Winners include **world cities**, or those centers of global finance, corporate decision-making, and creativity that have worldwide reach, such as New York City, Hong Kong, and London. Winners also include communities that have secured a piece of global commerce (e.g., an automobile assembly plant that provides jobs for the local community). Winners also include consumers who pay less for goods coming from low-cost production abroad (see the Geography in Action story about Wal-Mart, page 38); and they include workers who owe their positions and success to the global expansion of corporate activities. These workers, by the way, may be part of what is increasingly referred to as a **transnational capitalist class**, made up of globe-trotting executives and professionals, who view their work as part of a globally competitive process, who share similarly upscale lifestyles, and who see themselves as citizens of the world as well as citizens of their own countries. Finally, winners include those countries that have been able to transform their low-wage economies into destinations for industries that are higher wage and more technologically sophisticated, a process that is at least partly due to trade ties developed as a part of the globalization process. Good examples include some of the Asian countries that have emerged from low-wage assembly economies to become powerhouse players on the world economic scene. South Korea clearly fits into this category; China and India may be following closely. We should keep this thought in mind as we contemplate the purchase of, for example, a Hyundai car.

Losers include people who have lost jobs due to wage competition elsewhere, including in some cases skilled professionals; people who are too impoverished to take advantage of the consumer benefits of globalization; people who are affected by the pollution and other harmful environmental outcomes of industrial expansion; and people who emigrate because of poverty at home, but become economically marginalized in their destination countries. Also included are many cities that are just "ordinary"—not world class—and countries that suffer resource depletion to support the needs of global corporations. An especially destructive outcome is an expectations gap that is created when impoverished people in developing countries are exposed to advertising and other messages that broadcast lifestyle ideals that are hopelessly out of reach. The participation of these people in a globalizing world takes the form of a vicarious consumption of all that globalization is supposed to offer.

On the other hand, winners and losers are not always easy to identify, and one person's winner may be another person's loser. For example, the teenage girl working in an assembly plant in a developing country for very low wages and in an unsafe environment may be viewed by U.S. counterparts as clearly a loser in the globalization process. The teenage girl, on the other hand, may see her situation as a marked improvement over what she faces at home in her village or elsewhere in the city; she may see herself as a winner. In addition, winners and losers may appear in the same country or locality. Southeastern China, for example, with its explosive economic growth and vast network of connections, epitomizes all that globalization promises. Western China, on the other hand, largely remains a globalization backwater.

Scale Is Important   Most globalization debates focus on two levels: the global and the local. In reality, there are all sorts of middle levels that should be taken into account in identifying the geographic footprints of globalization. For example, and as discussed above, the country or the state constitutes a sort of middle level involved in the processes defining globalization, and sometimes joins in partnership with a global corporation to market the country's products or secure some other economic advantage. Scholars have also identified geographic super-regions such as global cores and peripheries.

In terms of developmental footprints, it is important to recognize that globalization does not necessarily lead to development, and whether it ever will is still subject to debate. Proponents of the capitalistic assumptions driving the globalization process argue that eventually the benefits will, speaking metaphorically, lift all boats and benefit everyone; critics counter that globalization will lift all yachts, swamping the rest of the boats. There is no question that the developmental landscape shows peaks of great prosperity that owe their existence to globalization. At the same time, this landscape reveals regions of deprivation and poverty that may or may not find their ultimate salvation in globalizing societies and economies.

Thus far our story about globalization has focused on cultural, technological, and economic issues, but the interdependence that defines globalization contains other dimensions that should be acknowledged. Increasingly, for example, there is concern about the globalization of environmental problems. Atmospheric warming offers a classic illustration. The greenhouse gases that are the primary culprits behind global warming are produced everywhere fossil fuels are used, making countries dependent on each other's motivations and policies regarding fossil fuels in order to attack the global warming problem (Figure 2–18). In other words, the solution to global warming will have to be a global solution. Another problem is the disposal of wastes, especially hazardous materials. In some cases, developing countries have found it desirable to accept and dispose of other countries' wastes as a source of revenue. In the mid-1990s, for example, India was receiving large quantities of toxic wastes from European and North American countries for purposes of "recycling." Unfortunately, many developing countries do not have the resources to recycle or dispose of hazardous materials without creating environmental damage, thereby creating a new set of environmental hazards with potentially global ramifications.

■ **FIGURE 2-18 Air pollution in Jakarta, Indonesia.** Air pollution in Indonesia's capital contributes to both local health problems and global warming. Even when engaged in recreational activities at the ocean's edge, Jakarta's citizens find it difficult to escape the problems of pollution.

A still darker side of globalization takes the form of globalized terror. As the world has discovered in the last decade, certainly since 2001, religious and other fanatics are able to use Information Age technologies to launch terror campaigns in very different parts of the world. Herein lies a paradox: religious fundamentalists who have sworn to destroy modern societies end up using the technologies and networks of the modern world in their effort to destroy it. If in doubt, just check the web sites of extremist religious groups about their campaigns against modernity. The world has become globalized in ways unforeseen and unwanted.

Sustainability Issues

One of the ways in which the world has changed profoundly in the last two centuries is that sustainability of the biosphere increasingly is threatened by human activities. Globalization contributes to the threat of environmental destabilization by tying distant places closer together, by diffusing new technologies faster to distant lands, and by encouraging changes that may not fit harmoniously into local cultures or may produce unanticipated effects. In this section we examine some of the salient issues associated with the search for sustainable uses of the earth's resources in a development process that often seems prone to producing counterintuitive, unexpected, and undesirable impacts.

Land Degradation

Foremost among these undesired impacts is land degradation. Change is a fundamental part of the natural world. Tectonic forces lift up blocks of land to create mountains; countervailing exogenic (erosive) forces of wind, wave, and water work to lower land levels. If we leave humans out of this equation for a moment, the complex interplay of these forces sculpts the landscapes that we observe today. Blocks of the earth's crust move over very long periods to assume new positions, banging into each other or pulling apart to create zones of earthquakes, mineral formation, and deposition. Climates appear stable only when one considers the last century's "average" conditions. In reality a great deal of variability occurs as do shifts from one set of conditions to another set of dominant characteristics. Ice ages reappear irregularly over cycles of hundreds of thousands of years, droughts periodically afflict even the most humid areas, and the plant and animal communities dependent on one set of parameters may over long periods give way to others. The gradual replacement of the grasslands of the Sahara by drier conditions beginning more than 10,000 years ago is an example. We usually do not regard these natural changes as degradation, although from the perspective of well-adapted plants and animals driven to extinction by major changes in the sustainability status of a habitat it hardly matters whether people or a natural process constitute the driving force.

Actually, humans cannot be separated from nature for analytical purposes as easily as one might think. Humankind lives in and is an integral part of the natural world; our species is both nurtured by nature and an influential driving force in altering the natural environment. Over hundreds of thousands of years people have gradually developed technologies that possess an increasingly powerful ability to change nature and direct energy flows away from other species and into channels that support human life and activity. The pace of this technological change is steadily accelerating and as a consequence the spatial extent of human influence on the natural world is growing and the speed with which signs of such impacts appear is increasing. Until quite recently, changes introduced by people took place relatively slowly. Nature and humankind had an opportunity to adjust to each other and develop along parallel paths as part of coevolved ecosystems. Many of the changes introduced by humans now take place at a pace too rapid for nature to adapt readily to the new and more intense pressures placed upon its adaptive capacities. One result of this excessive pressure is **land degradation**.

Land degradation is a relatively new term, and only entered the United States Library of Congress catalogue system near the end of the twentieth century. Fundamentally, land degradation is a product of human actions that lower a region's biological productivity or diminish its usefulness to humans. Land degradation is not a new phenomenon; it has taken place for a very long time, and certainly has been part of the earth's environmental history for the last 500,000 years, a period in which humankind first learned to control fire. The domestication of fire has allowed our species to cook food (and reduce illness), penetrate colder habitats from which our ancestors were originally barred, and transform ecotypes. Fire controlled and applied to human purposes together with increasingly sophisticated tools were the basic building blocks for the first great transformation of humankind's relationship to nature. Agricultural and industrial revolutions followed and human impacts on their surroundings were successively intensified. By the late nineteenth century these pressures were so great that George Perkins Marsh[2] could describe graphically both the bad historical transformations of nature perpetrated by the Greeks and Romans in the Mediterranean basin as well as the environmental malfeasances of farmers in his native Vermont. In northern New England, clearance of forest had so transformed the landscape that severe soil erosion was a regional problem and habitat change and direct predation on many animals (wolves, bears, mountain lions, deer, and beaver, among others) was so intense that many species had been driven to the edge of extinction. Marsh described this process of environmental change in great detail (Figure 2–19).

The concern with drastic declines in land productivity in Marsh's day has found dramatic echoes in the contemporary world where large dams change the downstream ecology of rivers; wetlands are drained on a massive scale; soil salinization threatens irrigated lands; dust storms increase due to dryland farming practices that expose soil to wind erosion; chemicals washed off agricultural lands pollute rivers, lakes, and

---

[2]George Perkins Marsh, *Man and Nature: Or, Physical Geography as Modified by Human Action*, edited version of the 1864 original (Cambridge, Mass.: Belknap Press of Harvard University Press, 1965).

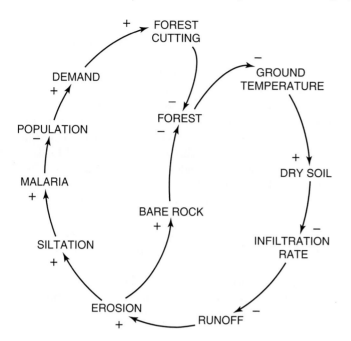

FIGURE 2-19 A nineteenth-century concept of soil erosion. Direct relationships (more = more; less = less) are expressed by plus signs, inverse relationships (less = more; more = less) by minus signs. Thus, more erosion = more siltation (+) = more malaria habitat (+) = less people (−) = less demand for wood (+) = less forest cutting (+).

coastal environments; carbon dioxide emissions from industry and automobiles cause global warming; rising sea levels due to climate change erode coastal habitats; and overzealous groundwater pumping threatens crucial aquifers that sustain life and economy in arid lands. All of these and many other negative consequences of poorly designed and implemented human development activity occur regardless of political, cultural, or social system and are part of the unsuccessful and unsustainable actions that constitute land degradation.

Responding to the Spirit of Place

Human use of resources does not necessarily result in land degradation. There are many examples of people developing land use systems that do not destroy their environment but rather use resources in a sustainable way. Shifting agricultural practices in tropical forest environments use soils of limited fertility by shifting their field sites every three or four years. Burning the plant material cut down to break the forest canopy and allow sunlight to reach the ground also provides a short-term fertilizer boost to the soil. Fallow cycles are long in this "slash and burn" system, since forest regeneration is needed to capture enough nutrients for the next cycle of use. As long as the intervals of use are separated by long periods of rest, a shifting agricultural system works well and sustains its practitioners. But any pressures that encourage shortening of the fallow cycle can produce progressive land degradation and a shift into less productive environmental conditions.

Resource use systems that achieve long-term sustainability do so because they avoid periods of rapid population growth and because they are sensitive to the spirit of place. This **genius loci principle** reflects long-term familiarity with the variations in nature that are part of a location's normal rhythms as well as the potential resources represented by a region's natural resource endowment. When a society has lived in a place for many generations, its members understand how to cope with wet or dry conditions that are part of the annual cycle of the seasons as well as the extreme events that occur less frequently. A rich experiential knowledge is built up over generations as the experiences, experiments, and lessons of living in place are transmitted from generation to generation. Neither the technology nor the theoretical expectations of development agencies and practitioners are an adequate substitute for this locally accumulated knowledge. Techniques and ideas developed in other places can be introduced selectively to augment local knowledge, but the practices of other cultures developed in other places are seldom a panacea for local problems and difficulties.

A case in point is the argan tree (*Argania spinosa*), a plant found naturally only in Morocco, principally in the drier southwestern parts of the country in and around the Sous Valley. The argan tree grows naturally in an open woodland with many other species. It is a valuable tree because its fruit can be pressed into an edible oil much used in local cooking; argan oil is also an important base product in the indigenous cosmetics industry. Moreover, argan leaves provide a high-protein fodder for domesticated livestock, particularly goats. So appreciated are these leaves by both goats and their owners that shepherds assist the goats in their efforts to climb into the tree and consume the leaves. Goats climbing in trees is a signature visual image of rural southern Morocco (Figure 2–20) to such an extent that the image features

FIGURE 2-20 Goats in argan tree in southern Morocco. Although at first glance it would appear the goats are damaging the tree, in fact they are essential to its survival. Only if the seeds contained in the argan's fruits pass through the goat's intestinal tract is germination possible!

prominently in postcards available to tourists to send to friends and family about typical exotic sights encountered in their travels. Goats are regarded by Western-trained foresters as destroyers of trees and inhibitors of forest regeneration, a perception quite at odds with the experience of local resource managers. Over centuries, goats, argan trees, and people have gradually coevolved in a mutually beneficial agroecosystem. The intimacy of this interdependence is revealed by the fact that the seeds of the argan tree germinate only with great difficulty unless they first pass through the intestinal tract of the goat! This relationship (Figure 2–21) results in a gradual process of landscape change. As people exploit their habitat for food and useful products, trees other than argan are removed to provide construction material, fencing, and fuel. In the open spaces this selective removal creates, opportunities exist for argan to expand as goats browse in and around the trees for fodder and fruit. Excreted seeds, neatly packaged in a ready-made fertilizer package, are able to germinate and grow (if protected from goat predation while the seedlings are young) into mature, fruit-producing trees. The result is more fodder and fruit for people and goats! Open spaces can also be used to grow wheat and other field crops that are important dietary staples. Thus the human population can gradually increase as the number of argan trees increases, since, with the survival of more trees, more fodder is produced for goats and more argan oil is available for human use. Larger goat herds generate greater supplies of milk, meat, and fiber, and more open spaces between the argan trees provide opportunities for wheat cultivation for human consumption. The local habitat changes over time, but the result of the change, if change progresses gradually, is a stable set of relationships between people and a selected number of plants and animals that humans define as useful and desirable.

The Doctrine of Creative Destruction

This process is called **creative destruction**, an active transformation in which people alter their habitat in order to produce a new, human-dominated system that meets human needs. Something must be destroyed to create the conditions that promote benefits for humankind. Humans do this

every time they remove one plant community in an area—for example, a forest, or grassland, or suburban lawn—in order to plant a cornfield, grow wheat or soy beans, or cultivate a vegetable garden. The plants that we want to eat require sunlight to grow successfully. We destroy the competition to create the conditions that favor crop growth. We work hard to remove weeds and other invaders that threaten to "steal" nutrients from the plants that we want. We destroy in order to create! The Chinese have destroyed lowland deltaic and estuarian wetlands to create elaborate pond-dike farmlands that combine fish farming and crop production. The Dutch have created farmland from freshwater wetlands and the former Zuider Zee, once an arm of the North Sea. The Nabateans, inhabitants of the northern Negev Desert 1,500 years ago, encouraged soil erosion on hill slopes to transfer soil and water to terraced fields created in nearby valleys. The critical feature of all of these examples of creative destruction is that the transformed agroecosystem must be sustainable and enduring for a long time.

The Precautionary Principle and the Sacrifice Zone

All too often the change that humankind introduces fails to endure. The time perspective employed is much too short and the changes introduced are too rapid and too poorly attuned to the local natural rhythms for the new production activities to survive. The effort and intention is to create, but destruction is the major outcome. Expansion of agricultural settlement into the Great Plains of the United States immediately before and after the Civil War was based on the belief that small farms of 60 acres and crops such as corn that were productive east of the Mississippi would work in the drier landscapes farther west. In the long term this scale of settlement proved overly optimistic and the "Dust Bowl" was created. The severe wind erosion and dust storms that characterized this area in eastern Colorado and New Mexico, northwestern Texas, and western Kansas and Oklahoma in the 1930s was controlled only when land holdings were consolidated into larger parcels, crops better suited to semi-arid conditions were grown, many farmers moved out of the region, and marginal land was taken out of production. Subsequent efforts to intensify production in the same region by tapping the groundwater resources of the Ogallala Aquifer for irrigation have led to severe depletion of the basic water resource and produced an uncertain future for many rural communities.

In the effort to create sustainable livelihoods, people are often willing to sacrifice the future use of potentially sustainable resources or the present productivity of distant areas. These **sacrifice zones** are numerous. Large-scale dams that trap sediments starve rich, downstream delta habitats and coastal beaches of the material they need to maintain themselves against the erosive actions of waves and coastal currents. Coastal lagoons in the Nile Delta are disappearing and the shrimp and fishing industries dependent upon them are declining because of coastal erosion accelerated by reduced river sediment transport. The resources that sustain fishing

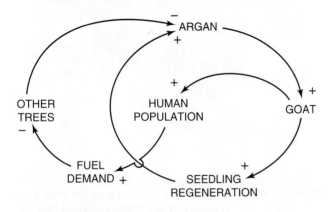

■ **FIGURE 2-21 The coevolution of the goat-argan-human agroecosystem.**

are sacrificed to agricultural and industrial gains elsewhere. Irrigation schemes that convert seasonal floodplain agricultural land to year-round crop production exclude herders from areas that were extremely important to them in the dry season when river levels were low and farming could not be practiced. Forced to stay year-round in what was once only seasonally exploited rangeland, pastoralists overuse grazing resources. The result is a type of land degradation specific to drylands called desertification. Areas degraded in this fashion are sacrificed to promote agricultural production and sustain peoples in other, often distant, regions. To create sacrifice zones is not inherently wrong if the change is understood, accounted for, and essential to the generation of sustainable productivity in another facet of the local system.

To minimize disruption and maximize sustainability requires two essential practices. The first is the application of the **precautionary principle**. This rule mandates that whenever significant change is likely to occur as a result of a proposed development or whenever the long-term implications of a proposed change are obscure, implementation of the anticipated project must proceed slowly, with proper examination of likely impacts and available remedial measures, and that thorough and transparent public discussion of the issues involved must take place. In practice, the precautionary principle would oppose the introduction of novel technologies until their implications are carefully examined and are better known. Deployment of nuclear power technology, for example, would be extremely difficult to justify under the precautionary principle because the wastes from nuclear plants are difficult to dispose of safely, secure long-term storage is almost impossible to guarantee, and, while the benefits of nuclear power are enjoyed by present generations, the costs of dealing with its radioactive wastes are deferred to future generations. The introduction of genetically modified crops without clear understanding of how these new plant strains will interact with other organisms in the natural world, including humans, is another example of leaping into the use of a new technology before looking carefully at the consequences.

Wise use of the precautionary principle requires a second practice: **holistic planning**. To plan in a comprehensive fashion requires an honest attempt to consider all of the implications of development activities. To accomplish this goal, the boundaries of a planning activity cannot be set too narrowly. It is easy to demonstrate that the construction of a particular development project will generate benefits, but it is much more difficult to assess the costs. A new dam will produce a certain amount of hydroelectricity; a dollar benefit from the sale of this energy can be assigned. Water from the dam can irrigate a known number of acres and the value of the crops grown on that land can be estimated. Flood damage can be reduced and the absence of rehabilitation and recovery costs can be calculated and the savings applied as a benefit. It is all too easy to ignore the likely future problems of soil salinization, waterlogging, and siltation that are certain to be encountered in the operation of the irrigation project because the dimensions of those problems are not evident when the project begins. It is even easier to over-look the ecological changes that will occur downstream as natural ecosystems are disrupted while water is diverted to crops and human sustenance. Few people are willing to speak for the plants and animals that lose habitat and even fewer are prepared to assign a cost value to those losses. Holistic planning would require not only that costs be reasonably calculated on both sides of the benefit-cost ledger, but also that alternatives to the construction of large dams be taken into account in the planning process. But all too often the degraded districts sacrificed to development are peripheral to the core areas of concern to policy planners, and, in the words of the old adage, what is out of sight is frequently out of mind.

Leveling the Playing Field: Identifying Gainers and Compensating Losers

This does not have to be the case! Ignoring the adverse consequences of development by planning on a sector-by-sector basis can condemn development efforts to the equivalent of jogging on a treadmill. You work hard and sweat a lot, but in the end you end up pretty much where you began the exercise program. In many real-world situations, gains in one sector of the economy occur at the expense of losses elsewhere. This happens because frequently it is hard to connect the dots that tie cause and effect together. Some of these unconnected consequences are a product of spatial distance. This often occurs when industrial pollutants exit the plant in one place but the consequences appear in distant locations. Tall smokestacks were intended to lift aerosol contaminants out of the local environment into the atmosphere where they were expected to be dispersed by moving air. And they were! But little thought was directed to where those pollutants might come down or how they would impact the sites upon which they descended. Acid rain that fell on distant places was a consequence that changed the water quality of lakes, killed forests, and corroded historical monuments. Another example is the heavy metals from electroplating and other industrial processes that enter terrestrial streams and are transported to the ocean where they enter aquatic food chains. Concentrated increasingly in the bodies of animals at the top of those food chains, they come back to consumers as heightened mercury levels in swordfish. Increased health costs or the economic losses that accompany premature death seldom appear as deductions from the benefits and profits that are earned by factory workers and plant managers whose operations produced the heavy metals in the first place.

Other consequences are temporal in nature and impact generations yet unborn. Factories along rivers in New England were powered by water stored behind a large array of dams. In 1850 no one imagined that the dams and the ponds they created would be a problem for their ancestors in the first decade of the twenty-first century. But they are! The dams are in increasingly poor condition because maintenance has declined as deindustrialization has occurred in many places. Everyone is reluctant to assume responsibility for the task (and cost) of preserving the local dam. State agencies whose mandate includes periodic examination of the structural integrity of dams are underfunded, understaffed, and undermotivated. Warnings about the need to correct poor maintenance are frequently ignored and enforcement of repair orders is generally lax. For many New England towns the threat became very real when seven inches of rain (one-sixth of the region's average annual precipitation) fell in one day, an unusually large rainfall event that was embedded in a week of intermittently heavy rain in the second week of October, 2005. In Taunton, Massachusetts, the Whittenton Mill Dam, a 173-year-old wooden structure, threatened to collapse and flood a three-square-mile area in the central part of the city. Two thousand residents were forced to evacuate to avoid the threat of a six-foot or more flood surge while construction workers attempted to make emergency repairs until a stone dam could be built to replace the wooden edifice.

The issue is much more serious than just the possibility of a dam breaking and flooding businesses and houses downstream from the dam site. Over decades of industrial use, heavy metals have built up in the sediments that also accumulated behind the dams. Any dam rupture would activate the sediments and transport them downstream where they would be a health threat to both terrestrial and marine ecosystems. Scouring out the sediments would be extremely costly; and there are few holes available that are big enough and secure enough to absorb the toxic sediments safely. Not having directly benefited from the wages and profits generated by the now defunct industries that produced the toxic waste products, local and state governments are understandably reluctant to absorb the cost now. And there are other players on the scene! Mill ponds are a prime seasonal habitat for migratory birds, and the ponds (and dams) are defended by organizations such as Ducks Unlimited, whose goals include protecting wildlife habitat so hunters will have something to hunt. So the dams and ponds remain, poorly maintained, as an increasingly serious hazard that will have to be dealt with sooner or later.

Other barriers are conceptual, as when analytical boundaries are created that blot out consideration of peoples and places that are distant. Yet it is critically important to identify who gains and who loses in the process of development, and to provide adequate compensation for those bypassed by the course of economic progress. Displaced refugees from a development project can migrate to urban areas where they may constitute a volatile underclass that undermines the stability of economic and governmental institutions. Dispossessed farmers pushed onto poorer land with steeper slopes may generate sediment-laden runoff that clogs the canals that feed water to lowland irrigation districts. Creating mechanisms that compensate populations bypassed by the development process or developing viable alternatives for such populations is an essential part of long-term sustainability.

Loops, Link, and Feedbacks

The connections between gains and losses are often counterintuitive and difficult to trace, but they are important to identify, follow, and counteract. Investing in holistic planning is one strategy for identifying emerging environmental problems and disadvantaged populations while there is still time to prevent serious and possibly permanent damage. Capturing wastes and monitoring are two ways in which gains and losses can be identified and remedied.

- **Capturing Wastes**: Human activity produces huge amounts of waste material both directly and indirectly. Chinese farmers learned very quickly that the organic wastes produced by people and animals could be turned into a major fertilizer input for fish ponds. Decomposition of these wastes provides the nutrients that fuel pond food chains that ultimately sustain the fish that people wish to eat. Aquatic vegetation and fish waste impregnated mud from the pond bottom provides fertilizer inputs for crops grown on the land. The tight linkage between pond and field in which the wastes from one part of the farm are the valued inputs for another part is ideal. Little waste escapes from the farm to the larger environment. Examples abound of converting unwanted wastes into productive use. Water hyacinth is a dangerous invasive plant in many irrigation systems because its growth clogs canals and wastes precious water. It is very difficult to control and until recently had no viable use. Now people have learned how to use this aquatic plant as the raw material for wicker furniture, converting an unwanted waste by-product of irrigation system maintenance into an income generating finished product. Closer to home, scrap metal from discarded automobiles has become a major input into steel production, a development that also required the development of new technology in the form of the electric arc furnace. Similarly, worn-out car and truck tires are a bulky waste output that is difficult to dispose of unless you are a revolutionary needing a noxious, smoky product to fuel street fires. Increasingly this material is shredded and reconstituted as a building material for boardwalks and high-traffic pedestrian pathways in wetlands. The rubber's impermeability to wet conditions is superior to most other natural products and provides a low-cost alternative construction material with a very long use life. Determining how to use wastes creatively in order to convert them into resource inputs has a bright future if demands for alternative materials can be identified and technologies innovated to facilitate the new products.

- **Monitoring**: We need to know what is happening and what changes are taking place in the world in order to identify feedback loops that promise to produce problems. Monitoring systems can help identify emergent problems before they become so serious that human life and ecosystem health are threatened. These can be as elaborate as the systematic analysis of imagery of the earth photographed from space to monitoring reports by informed local school teachers in rural areas. Without institutional structures in place that can collect appropriate information about environmental and social status and feed these data into planning and decision-making bodies, it will be difficult to use the powerful technologies that people have created in ways that are ultimately sustainable rather than disruptive.

be carelessly degraded in the interest of short-term gain or planned in favor of long-term benefit.

▶▶ Summary

In Chapter 1, we examined what geographers do and investigated the nature of social and economic change. In this chapter, we applied the geographer's approach to understanding what development involves; to the revolutionary kinds of change that reshape societies and economies, with special emphasis on globalization; and to the growing need to guide change so that it is sustainable and beneficial to the earth's inhabitants. We learned that development is a process of change that leads to improved well-being in people's lives, that takes into account the needs of future generations, and is compatible with local cultural and environmental contexts. Globalization plays an increasingly significant role in the development process, although the homogenizing, culturally transforming effects of globalization should not be overstated. Finally, we examined how long-term development requires a commitment by people to understand the environments of which they are a part, and to act creatively in directing the impulses that determine whether human habitats will

▶▶ Key Terms

Agricultural Revolution
agricultural surplus
birthrate
creative destruction
death rate
demographic
  transformation
friction of distance
genius loci principle
globalization
holistic planning
Industrial Revolution
Information Revolution
land degradation
less-developed country
  (LDC)

Malthusian theory
neo-Malthusians
newly industrial country
  (NIC)
old industrial country (OIC)
physiologic density
population density
population distribution
population growth rate
precautionary principle
sacrifice zone
Third World
transnational capitalist class
transnational corporation
  (TNC)
world cities

■ **Deforestation creates spaces for agriculture.** Cutting and burning forest vegetation prepares the land for crop planting in Mindanao, Philippines. Only through long fallow cycles is soil fertility restored by forest regrowth and soil erosion prevented.

Chapter 3

Nature, Society, and Development

- ● **How Places Are Classified Environmentally**
- ● **Use of the Environment**
- ● **Centers of Use and Habitation Modification**

- ● **The Geographic Dimensions of Development**
- ● **Development Explained (at Least in Theory)**

I n some parts of the world, people live under prosperous conditions; in others, many struggle to secure the necessities of life. Two principal factors affect the level of living in an area: the physical environment and its natural resources; and the political, educational, economic, and social systems in place. This chapter focuses on the role that physical environment and human cultural systems play in development and on the geographic patterns of different levels of economic development found throughout the world. With this role in mind, we will also examine some of the measures used to quantify development and survey some of the theories intended to help us understand developmental processes.

The first concern of all peoples is the provision of food, shelter, and other economic necessities through some form of production that necessarily involves a relationship with the natural environment. The nature of that relationship depends on the skills and resources that a society accumulates and on the value system that motivates it. For example, the way people in the United States use a desert environment differs substantially from the way the Bushmen of Africa do.

Using the environment means modifying it, and modification can bring problems by upsetting natural balances and relationships. Our recent experiences with water and air pollution and solid waste disposal sometimes encourage us to think that landscape modification and environmental degradation are new phenomena, but that notion is inaccurate. Ancient inhabitants of the Middle East had culturally induced environmental problems. For example, the increased soil salinity (high salt content) that many areas, particularly Mesopotamia, experienced was a result of repeatedly irrigating the same lands without providing adequate drainage and without allowing for regular fallow periods. Ultimately, this practice reduced grain yields to the point where the worst affected areas became salt encrusted and useless for agricultural production.

Thus, **landscape modification** is as old as humankind. Today few landscapes are truly natural but rather are largely cultural, and have been shaped and formed by the societies that have occupied and used them. Culture is a part of the environment, just as the physical world is, and recognizing this relationship is essential to understanding the condition of the human race in various regions of the world.

How Places Are Classified Environmentally

The numerous elements of the physical environment—rocks, soils, landforms, climate vegetation, animal life, minerals and water—are interrelated. For instance, climate is partially responsible for variations in vegetative patterns, soil formations, and landforms; and organic matter from decaying vegetation is essential for soil development. Environmental elements combine to form intricately interconnected **ecosystems** in which changes in one or more system components produce corresponding variations in other components. Recognizing the relationships among natural processes is vital, as is recognizing that human activities frequently impact those processes. Although natural forces continue to play a major role in shaping the world's ecosystems, the single most active agent of environmental change on the surface of the earth is humankind. Understanding the structure and dynamic forces operating in the natural world is extremely important not only because this is the foundation upon which human livelihood and survival depend, but also because humankind's pressure on the natural world is intensified whenever economic development works to enhance and accelerate natural processes.

Climate

Climate directly affects our efforts to produce the food and industrial crops required by the human race. All plants have specific requirements for optimal growth. Some, such as rice, need substantial amounts of moisture, whereas others, such

49

as wheat, are more tolerant of drought. Coffee requires a year-round growing season; it grows best in the tropics at elevations that provide cooler temperatures. Even though humans have modified the character of many plants, climatic differences on the earth influence which crops are grown in specific geographical areas (Figure 3–1).

The most important climatic elements are precipitation and temperature. Precipitation is water vapor that condenses in the atmosphere and then drops onto the surface of the earth. Rainfall is the most common type of precipitation, but the solid forms—snow, sleet, and hail—are also significant. Melting snow is an important source of water for streams

or as soil moisture that can be used in later seasons. Sleet and hail are normally localized, but they can cause enormous damage and economic losses.

As the map of **average annual precipitation** shows, the rainy areas of the world can be found in parts of the tropics and the middle latitudes (Figure 3–2). In the middle latitudes, where prevailing air movement is from west to east, precipitation is particularly heavy on the western (windward) sides of mountain ranges and continents between 40° and 60° latitude. The eastern (leeward) sides of mid-latitude mountains and continents are often less rainy but may still receive sufficient precipitation to be considered humid regions, particularly if exposure to major water bodies is favorable. Notice in Figure 3–2 that the eastern United States and Canada, as well as East Asia, illustrate such a condition. So too does the Pacific coast of Latin America, where extremely rainy conditions pertain but the Patagonian Desert dominates areas east of the Andes Mountains.

In contrast, the interiors of continents such as Asia, northern Africa, central Australia, or interior North America and the areas on leeward, or downwind, sides of mountains experience moisture deficiency. These areas are located so far from sources of moisture that rainfall is an episodic event (Figure 3–3). Some subtropical areas—northwestern and southwestern coastal Africa, northern coastal Chile and Peru, northwestern Mexico, and the southwestern United States—have only meager amounts of rain. Subsiding air and divergent wind patterns, along with cold offshore waters, reduce the likelihood of precipitation in those areas.

Seasonal rainfall patterns are as important to land use as yearly rainfall totals. Some equatorial areas, including the tropical rain-forest zones, receive a significant amount of rainfall in every season. Much of the tropics, however, is characterized by wet summers and dry winters. Although temperatures in the wet and dry tropics are high enough for the growing season to be yearlong, that advantage is partly offset by the seasonality of the rainfall. Some subtropical climates, particularly those areas around the Mediterranean and their analogs in other regions such as southern California or the Cape of Good Hope in South Africa, also experience distinct seasonal variations, with dry summers and autumns and wet winters and springs.

Variability of precipitation is the percentage of departure from the annual average, which is derived from a 30- or 50-year record of precipitation. The greatest variability is experienced in areas of minimal rainfall, where settlement is limited unless a special source of water is available. In contrast, transitional areas (steppe or savanna grasslands) between humid and dry regions are frequently important settlement zones with significant production of grains. Under normal conditions, the Great Plains of the United States, the North China Plain, the Sahel (on the southern edge of the Sahara), and the Black Earth Region of Russia, Ukraine, and Kazakhstan typify such regions. Unfortunately, these and similar areas are prone to periodic droughts and can be risky habitats if unsustainable land-use practices are used.

■ **FIGURE 3-1 Climatic adaptations of food crops.** All plants, including those used as food by humans, are adapted to specific environmental conditions. Shown here are two that thrive under very different climatic and soil conditions. Cacao (top), or chocolate, trees thrive as understory plants, growing in the deep shade of other trees in the humid rain-forest regions of the earth. The chocolate is obtained by grinding up the large seeds that grow inside the pods. Date palms (bottom), on the other hand, prefer dry, sunny conditions and are widely cultivated in the desert regions of North Africa and the Middle East.

Temperature and Plant Growth Data on the annual amount, seasonality, and variability of rainfall reveal much about the utility of an environment. But the moisture actually available also depends on temperature conditions. High temperatures provide great potential for evaporation and plant transpiration, or a high **evapotranspiration rate**. An unfavorable relationship between precipitation and evapotranspiration means that plant growth will be limited. Deserts represent the extreme cases of that moisture-temperature relationship.

Perhaps the most important aspect of temperature is the length of the **frost-free period**, which relates primarily to latitude but also to altitude and the location of large bodies of water. The modifying influence of water on temperature can be seen in marine locations, where growing seasons are longer than would normally be expected at those latitudes. A comparison of northwestern Europe with similar latitudes in Russia or eastern North America offers an example (Figure 3–4, page 54). Winter temperatures are significant, also. Many mid-latitude fruit trees require a specific dormancy period with temperatures below a certain level. Cold temperatures are necessary to ensure the activation of a new flowering cycle.

In addition, the photoperiod (the length of the day, or the active period of photosynthesis) and the daily temperature range during the growing season affect plant growth. Some plants, such as barley, require a long daily photoperiod to flower or set seed. Others, such as soybeans or rice, benefit from shorter photoperiods. Each plant variety has a specific range of daily low and high temperatures, called *cardinal temperatures*, within which plant growth can occur. Cardinal temperatures are known for most plants. The numerous varieties of wheat, for example, have cardinal temperatures that range from a low of 32° to 41°F (0° to 5°C) to a high of 87° to 98°F (31° to 37°C). Within that range, the optimum temperatures for most varieties of wheat are between 77° and 88°F (25° and 31°C). Although less-than-optimum conditions do not mean that a particular crop cannot be cultivated, departures from the optimum reduce the efficiency with which that crop is produced. And efficiency, translated into cost per unit of production, is a major concern in commercial agriculture.

Climatic Classification Climatic classification is based on both temperature and precipitation conditions (Table 3–1, page 56). Although conditions may vary within a climatic region, a generalized classification system is useful for comparing the characteristics of different areas. Figure 3–4 shows the global distribution of climates and reveals a close relationship among latitude, continental position, and climate. The locational similarity of humid subtropical climates in the United States, China, and Argentina is just one example.

The distinction between dry and humid climates is fundamental to classification, but no single specific precipitation limit can be used to separate the two. For example, areas are classified as desert or steppe if the potential for evaporation exceeds actual precipitation. In the mid-latitudes, deserts normally receive less than 10 inches (250 millimeters) of precipitation; steppes or semiarid regions receive between 10 and 20 inches (250 and 500 millimeters). For classification as a dry climate, the limits of precipitation are lower in the higher latitudes and higher in the lower latitudes. For instance, 25 inches (650 millimeters) of precipitation in the higher mid-latitudes may provide a humid climate and forest growth. That same amount in tropical areas may result in a semiarid, treeless environment. The reason for this is rooted in both the higher temperatures of the tropics, which place greater transpiration stress on plants, and the timing of the bulk of rainfall in subtropical areas, which generally occurs during the high sun period of the year when evapotranspiration is greatest.

Along with the other controls of climate—such as latitude, marine exposure, prevailing winds, and atmospheric pressure systems—elevation is another major factor. The highland climates shown in Figure 3–4 feature variable temperature and rainfall conditions according to specific elevation and position within mountains. Even though many mountainous areas are sparsely settled and provide a meager resource base for farmers, highland settlement is important in Latin America, East Africa, and South and Southeast Asia.

Vegetation

Vegetation patterns are closely associated with climate, as can be seen by comparing world regions with similar locational attributes (Figures 3–4 and 3–5, page 58). For example, the Southeastern United States and Southeastern China exhibit clear similarities in both climatic and vegetational patterns. Less obvious on maps of this scale is the locational distinction between grasslands (herbaceous plants) and forests (woody plants). Grasslands exist as part of mid-latitude prairies, mediterranean woodland shrub vegetation, tropical savannas, and steppes. They are directly attributable to lower moisture totals as well as seasonal and interannual variation in precipitation. A very cold climate (for example tundra) has little woody vegetation, because the growing season is short and because subsoil is permanently frozen. The more humid climates are capable of supporting forest vegetation, which usually requires a minimum annual rainfall of 15 to 20 inches (375 to 500 millimeters), depending on evapotranspiration rates.

Natural vegetation is what would be expected in an area if vegetation succession were allowed to proceed over a long period without human interference. In earlier millennia, natural vegetation permitted our ancestors to move easily across grass-covered plains and mingle with one another, whereas dense forests functioned as barriers, isolating culture groups and providing refuge for those who wanted to remain apart. The grasslands south of the Sahara represent one such cultural transition zone, with features of both African and Arab cultures. The equatorial rain forests of Central Africa, on the other hand, have served to keep peoples apart, primarily because travel through them is difficult, and the soils are generally poor.

■ **FIGURE 3-2 World mean annual precipitation.** Precipitation varies greatly from one part of the world to another. Moreover, there is considerable variability in precipitation from one year to the next. Variability is usually greatest in areas of limited precipitation.

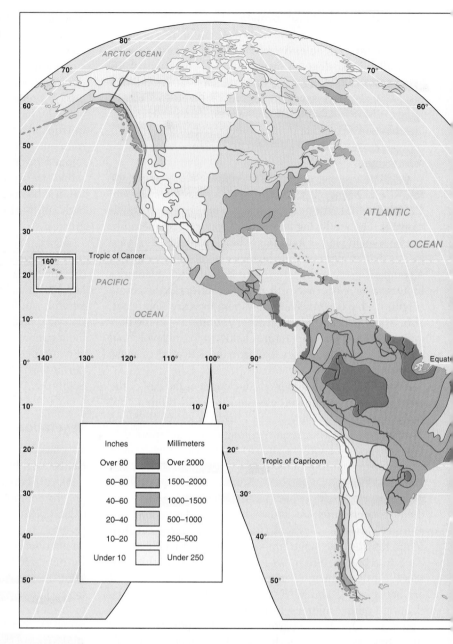

■ **FIGURE 3-3 Atacama Desert.** The Atacama Desert of northern Chile and the coastal regions of southern Peru are some of the driest places on earth, with no measurable precipitation occurring for years on end in some areas.

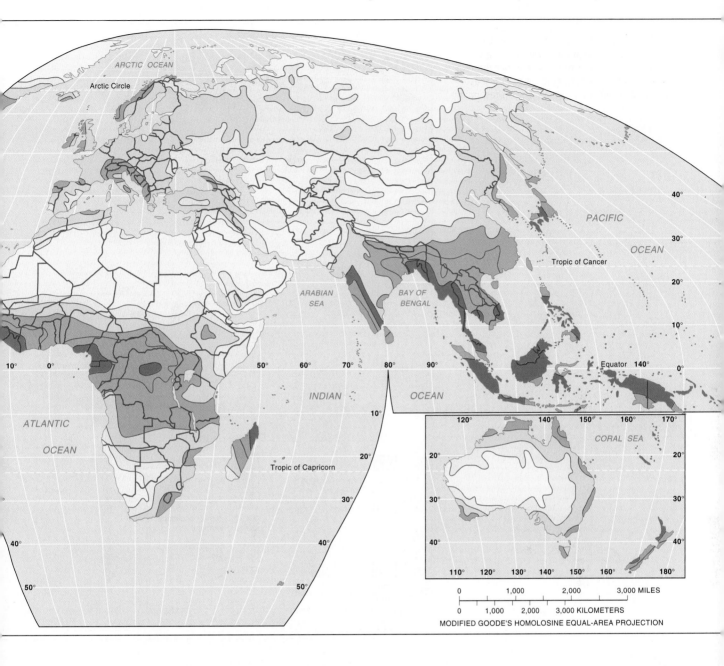

Over the centuries, humankind has greatly altered the world's natural vegetation. In fact, the phrase "natural vegetation" must be used quite carefully today, because few areas of truly natural vegetation remain. Most vegetative cover is a reflection of human activity. In mid-latitude Asia, North America, and Europe, for example, increases in population and changes in agricultural technology have prompted the removal of broadleaf and mixed broadleaf and coniferous forests (see Figure 3–5) from vast areas so that field agriculture could expand. High-latitude coniferous forests and tropical rain forests still exist because they do not lend themselves as readily to permanent settlement or to large-scale modern agriculture. But in many countries situated within the Latin American, African, and Asian tropics, rapid population growth, the need for space, and government economic policy have recently contributed to accelerated rain forest clearing. Some people view the expansion

of Brazilian settlements in the Amazon rain forest, for instance, as an expression of national pride and unity for Brazilians, a means of easing population pressure, and a source of new economic wealth. Others fear that these short-term gains may bring long-term damage to the forest, the soils, the streams, and the atmosphere. History is replete with examples in which short-term or immediate need has clouded our vision in matters of environmental stewardship.

Our attitudes toward natural vegetation have begun to change dramatically. Increasingly we are recognizing that vegetation is significant in many aspects of life and is related to other components of our environment, such as soil and air. Forest vegetation is becoming more critical as our numbers grow and as we consume ever-greater amounts of lumber and paper (Figure 3–6, page 58). Sustained-yield forestry—harvesting no more than the annual growth rate of trees can replace—is a more common practice as we

■ **FIGURE 3-4 Climatic regions of the world.** Climate is the long-term condition of the atmosphere. Although there are many elements of climate, most classifications use only the two most important: temperature (level and seasonality) and precipitation (amount and seasonality).

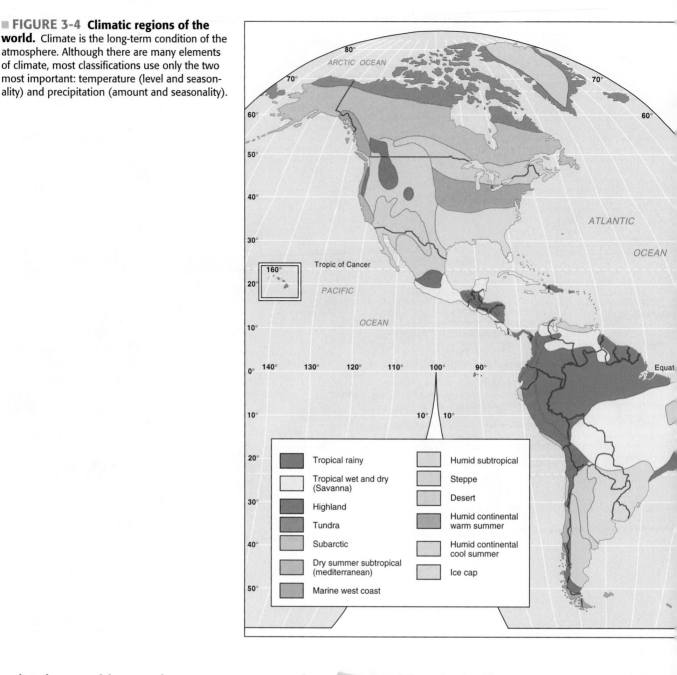

Legend	
Tropical rainy	Humid subtropical
Tropical wet and dry (Savanna)	Steppe
Highland	Desert
Tundra	Humid continental warm summer
Subarctic	Humid continental cool summer
Dry summer subtropical (mediterranean)	Ice cap
Marine west coast	

realize that wasteful use results in greater resource problems. In addition, forests are valued for more than their wood. Not only do they represent an important botanical gene pool, but they also help reduce erosion, a process that can destroy land and lead to the silting of rivers, streams, and reservoirs. Moreover, forests contribute to flood prevention by reducing water runoff. In more affluent countries they are prized recreation areas.

Soils

In order to grow, plants need nutrients as well as moisture and energy from the sun. Nutrients are derived both from minerals in the earth and from humus, organic materials added to the soil by vegetation. The nutrients that a soil contains depend on the kind of rock lying beneath it, the slope of the land, the vegetative cover, the microorganisms within

the soil, and the soil's age. Three processes are particularly important in the formation of soil and greatly affect its supply of nutrients and therefore its fertility: laterization, podzolization, and calcification.

Laterization is a process by which infertile soils are formed in the humid tropics. The plentiful rainfall leaches the soil; that is, it dissolves the soluble minerals in the soil and carries them away. Unfortunately, the soluble minerals that are removed—calcium, phosphorus, potash, and nitrogen—are among the most important plant nutrients. Insoluble compounds of aluminum and iron remain, but those elements alone produce infertile soils. In addition, decomposed organic material, which supplies nutrients, is available only as long as trees and other plants remain to drop their leaves and branches on the forest floor. Once the rain forest vegetation is removed in order to open new lands to farming or animal grazing, the soils in areas of high rainfall and

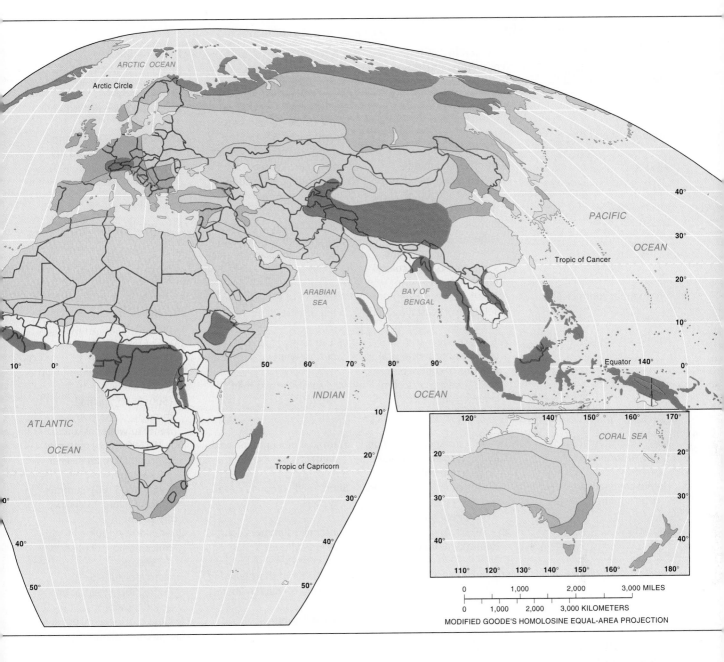

MODIFIED GOODE'S HOMOLOSINE EQUAL-AREA PROJECTION

high temperature—the humid tropical and subtropical regions—show a marked decrease in fertility.

Podzolization occurs in high latitudes or at high altitudes characterized by cold, humid climates where seasonal temperature variations are distinctive. In those locations, normal leaching is restricted, owing to modest rainfall or to the ground being frozen during part of the year. But poorly decomposed organic material, such as that from pine needles, combines with water to form weak acidic solutions that remove aluminum and iron from the soil along with soluble minerals. What remains is a large amount of silica, which is low in fertility.

Calcification occurs on the drier margins of humid regions and in arid zones. In those locations, reduced leaching, in response to limited precipitation, leads to greater accumulations of humus and calcium carbonate. Soils formed under those conditions, including aridisols, alfisols, and mol-

lisols, are generally very fertile but, paradoxically, often remain less productive because of the very moisture deficiency that favored their formation.

Residual soils are those that have formed where they are found. They reflect local environmental conditions and are strongly influenced by laterization, podzolization, or calcification. Some soils, however, are transported and thus have characteristics that bear little relationship to local environmental conditions. **Alluvium** is soil that is transported and deposited by water; **loess**, by wind. Many alluvial soils, loess, and soils formed from some volcanic materials are exceptionally fertile. In humid areas, they are often the most preferred for cultivation; but in excessively dry areas, they may require great amounts of labor or capital for irrigation.

Humans have modified soils in many ways, sometimes improving inherently poor soils, and at other times harming fertile ones. Soils damaged by depletion, erosion, or other

Table 3-1 Characteristics of World Climate Types

Climate	Location, by Latitude (continental position, if distinctive)	Temperature	Precipitation, in Inches Per Year (millimeters per year)
Tropical rainy	Equatorial	Warm, range[a] less than 5°F (3°C); no cold season	60+ (1500); no distinct dry season
Tropical wet and dry	5°–20°	Warm, range[a] 5°–15°F (3°–8°C); no cold season	25–60 (650–1500); summer rainy, low-sun period dry
Steppe	Subtropics and middle latitudes (sheltered and interior continental positions)	Hot and cold seasons; dependent on latitude	Normally 10–20 (250–500)
Desert	Subtropics and middle latitudes (sheltered and interior continental and some cold current coastal positions)	Hot and cold seasons in interior areas; constantly cool in subtropical coastal zones	Normally less than 10 (250)
Dry summer subtropical	30°–40° (western and subtropical coastal portions of continents)	Warm to hot summers; mild but distinct winters	20–30 (500–750); dry summer; maximum precipitation in winter
Humid subtropical	20°–35° (eastern and southeastern subtropical portions of continents)	Hot summers; mild but distinct winters	30–65 (750–1650); rainy throughout the year; occasional dry winter (Asia)
Marine west coast	40°–60° (west coasts of mid-latitude continents)	Mild summers and mild winters	Highly variable, 20–100 (500–2500); rainfall throughout the year; tendency to winter maximum
Humid continental (warm summer)	35°–45° (continental interiors and east coasts, Northern Hemisphere only)	Warm to hot summers; cold winters	20–45 (500–1150); summer concentration; no distinct dry season
Humid continental (cool summer)	45°–60° (continental interiors and east coasts, Northern Hemisphere only)	Short, mild summers; severe winters	20–45 (500–1150); summer concentration; no distinct dry season
Subarctic	50°–70° (Northern Hemisphere only)	Short, mild summers; long, severe winters	20–45 (500–1150); summer concentration; no distinct dry season
Tundra	60° and poleward	Frost anytime; short growing season, vegetation limited	Limited moisture, 5–10 (125–250), except at exposed locations
Ice cap	Polar areas	Constant winter	Limited precipitation, but surface accumulation
Undifferentiated highland (See Figure 3–4)			

[a]The difference in average daily temperature between the warmest month and the coldest month.

forms of misuse may be restored by wise management practices. The addition of inorganic chemical fertilizers can overcome the declining fertility that results from prolonged use but may harm soil structure. The addition of lime reduces acidity, which can be detrimental to many plants, but may also kill beneficial soil organisms such as earthworms. Terraces can prevent erosion or help distribute irrigation water, and in many parts of the world are used to cultivate steep slopes into productive agricultural surfaces. Crops that place a serious drain on soil nutrients can be rotated with less-demanding plants or with leguminous plants, which have the capability of adding nitrogen. Fields themselves can be rotated, with some plots allowed to lie fallow, or rest, for several years between periods of cultivation so that fertility

builds up again through natural processes. Farmers in many areas of the United States have even installed tile drains to remove excess moisture from poorly drained areas, and other drainage techniques have been used in Asia and Europe, and the humid tropics, with soils that otherwise would be unproductive.

Efforts to reduce soil destruction or improve fertility involve expenditures of labor and capital, but the long-term needs of society require that we accept the costs of maintaining the environment. Unfortunately, people are not always willing or able to bear such costs, particularly when short-term profits or immediate survival is the main concern. An exploitative approach may lead to maximum return over a short period, but it often results in the rapid destruction of one of our most basic resources (see the *Geography in Action* boxed feature, Soil Degradation, page 60).

The experience of the United States illustrates the environmental damage and human hardship that can result from less-than-cautious use of resources. The years between 1910 and 1914 are often referred to as the golden years of American agriculture because of the high prices that farmers received for agricultural commodities, prices that continued to be favorable through World War I and into the 1920s. High prices and above-average rainfall encouraged farmers on the western margin of the Great Plains to convert their grasslands into wheat fields. When several years of severe drought occurred, coinciding with the Depression of the 1930s, unprotected cultivated lands were heavily damaged by wind erosion, and the farmers suffered terribly (Figure 3–7, page 61). Many of them moved away from the region that came to be known as the "Dust Bowl" abandoning their land. Banks failed, businesses declared bankruptcy, and small towns declined.[1] But the Dust Bowl did not become a permanent landscape feature because farmers and government institutions made adjustments to the reality of living in a region that episodically experienced severe drought. Farm sizes became larger, marginal land was shifted into less intensive uses, better land management practices reduced erosion, and government programs that paid farmers to take land out of production are some of the adjustments that were put in place. These improvements in agricultural practice and increased institutional support stabilized the situation, although little changed in basic attitudes toward land management. When central pivot irrigation emerged as a technology that could exploit the groundwater resources of the Ogallala aquifer (Figure 3–8, page 61), a new round of intensive development began. Heavy exploitation of the Ogallala's groundwater reserves now threatens a new round of land degradation and diminished intensification in exploitation of the region as people must shift from irrigation to herding or to cultivation of crops nourished by rainfall. The experience of trying (and often failing) to match land-use system and available technology to an unpredictable, non-equilibrium dryland ecosystem with long-term sustainable results is a generic, worldwide problem. The U.S. Dust Bowl has its parallels in the overextension of farming characteristic of the former Soviet Union's Virgin Lands program, the land degradation that is accompanying intensified development of China's Inner Mongolia region, and the Sahelian drought that afflicted the semi-arid fringe of Sub-Saharan Africa in the 1970s.

---

[1]In the novel *Centennial* (New York: Random House, 1974), James A. Michener vividly describes the occupation of the western Great Plains and the problems that farmers faced when they attempted to use the land. John Steinbeck's novel, *The Grapes of Wrath* (New York: Viking Press, 1939), portrays what happened to the farmers of the western Great Plains during the Depression and drought years of the 1930s. Donald Worster's environmental history of the *Dust Bowl: The Southern Plains in the 1930s* (New York: Oxford University Press, 1979) explores the impact of the drought on rural agricultural communities and elucidates the institutional and economic factors that promoted farming in an environmentally risky setting.

Landforms

The surface of the earth is usually divided into four categories of landforms: (1) plains, with little slope or local relief (that is, few variations in elevation); (2) plateaus, or level land at high elevations; (3) hills, with moderate to steep slopes and moderate local relief; and (4) mountains, with steep slopes and great local relief.

Plains are the landform most widely used for settlement and production when other environmental characteristics permit. Large areas of land with little slope or relief are well suited to agriculture. In addition, the ease of movement over plains, particularly those with grassland vegetation, has facilitated exchange with other societies. Not all exchange is peaceful, of course, and the features of plains that contribute to their utility in peacetime can be handicaps in wartime because they possess few natural barriers to afford protection. Nonetheless, great densities of population frequently occur in plains areas where intensive agriculture is practiced or where industrial and commercial activity is concentrated. Consequently, many early high-density agricultural populations were associated with plains. Those plains that exhibit limited settlement or utilization are generally less desirable for climatic reasons.

Hills and mountains offer quite a different habitat. In mountainous regions, small basins and valleys become the focus of settlement and, because they are difficult to penetrate, may lead to the formation of distinct cultures. Although such areas provide security from attack, they may also exhibit economic variation from more favored regions.

Major differences and even conflicts between highland and lowland inhabitants are common and are a part of regional history in many areas of the world. Separatist movements are an expression of those differences, and isolated highland areas provide excellent bases for guerrilla movements. The mountain-dwelling Kurds of Iran, Iraq, and Turkey have been politically at odds with the governments of all three states; and in Myanmar, the Karen of the Shan Plateau have raised communist and indigenous insurrections against a government that is controlled by the lowland river plain majority. Unifying a country that incorporates such contrasting environments and cultures remains difficult.

Minerals

Today's politically and economically powerful countries have built up their industrial structures by using huge amounts of **fossil fuels** (coal, petroleum, and natural gas) to process large quantities of minerals. Any country that wants to be considered a technologically advanced, industrialized nation must either possess such resources or acquire them. To date, neither nuclear energy nor hydroelectricity has surpassed fossil fuels in importance. Nuclear energy as yet contributes relatively small quantities of power in most nations. Hydroelectricity is of considerable importance in some countries but contributes little in many others.

The use of petroleum and natural gas has increased rapidly in the world's leading industrialized areas since World War II. That trend has given those less industrialized

■ **FIGURE 3-5 World vegetation regions.**
The distribution of vegetation closely corresponds
to climatic patterns. A map of vegetation can be
used to determine an area's agricultural potential, since crops and natural vegetation use the
same environmental elements for growth.

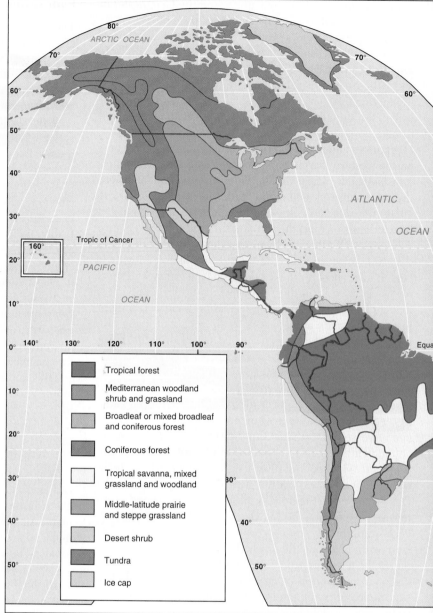

Tropical forest

Mediterranean woodland
shrub and grassland

Broadleaf or mixed broadleaf
and coniferous forest

Coniferous forest

Tropical savanna, mixed
grassland and woodland

Middle-latitude prairie
and steppe grassland

Desert shrub

Tundra

Ice cap

■ **FIGURE 3-6 Clear cuts and
forest regrowth in the Olympia
National Forest of Washington.**
Trees are a valuable resource for many
products, but harvesting of forests may
also destroy wildlife habitats and generate other environmental change, inevitably leading to conflicting notions
about proper use of forests.

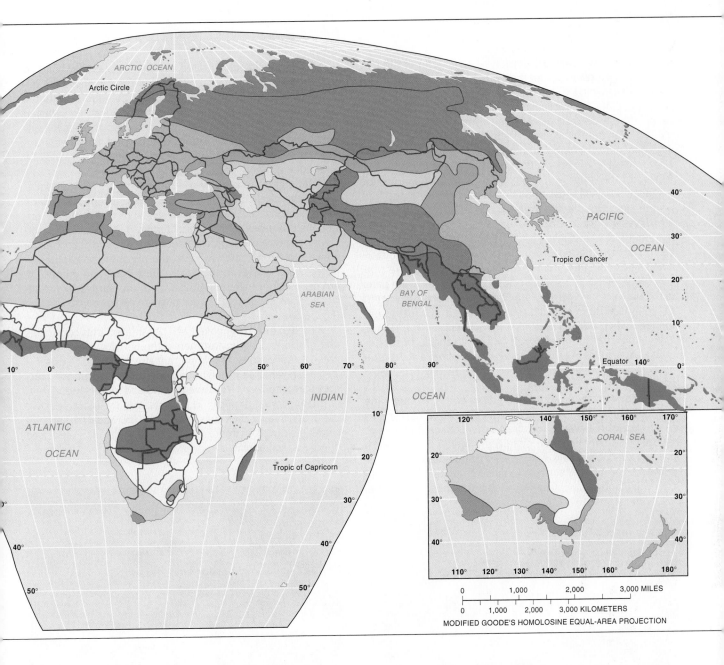

countries with oil wealth a special importance in international politics, an importance far greater than their size or military strength could have provided. Although huge quantities of coal are available, especially in the United States, Russia, and China, in the second half of the twentieth century petroleum was increasingly used as a substitute for coal because petroleum and natural gas are cleaner and contribute less to habitat destruction and air pollution. The increasing cost of petroleum has begun to generate a reverse trend—that is, a move back to coal—and considerable political pressure in the United States to relax environmental standards in order to keep energy costs as low as possible. The impact of this trend is uncertain, but it is unlikely to represent good news for the environment, particularly in terms of atmospheric pollution and an acceleration of the global warming that promotes climate change.

Iron, aluminum, and copper are the most important metallic minerals used in industry. Others include chromium, zinc, lead, gold, and silver. In addition, nonmetallic minerals such as nitrogen, calcium, potash, and phosphate are used for chemical fertilizers. Salt, building stones, lime, sulfur, and sand are other commonly used nonmetallic minerals.

Future discoveries of significant mineral deposits in parts of the world that are currently nonproducing may greatly affect our evaluation of the industrial potential of such areas. At present, however, the United States and Russia appear to possess the most abundant mineral supplies of the world's industrialized countries. The power and position of those two nations in world society are based on their effective use of industrial and power resources. Western Europe is a close third but is somewhat more vulnerable because its supplies

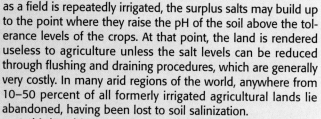

GEOGRAPHY IN ACTION
Soil Degradation

One component of our physical environment that is often taken for granted but which is crucial to the sustaining of life on the earth is the soil. Although the depth of the topsoil is measured in inches rather than feet or miles, it supports all plant life that grows on the land and, directly or indirectly, all animal and human life as well. It is thus difficult to overstate the importance of protecting and, where possible, enhancing the quality of the soil.

One of the most widely recognized forms of soil loss or degradation is soil erosion. Soil erosion is closely associated with the loss of protective vegetative cover through deforestation or desertification. Once the topsoil is exposed to the forces of running water and wind, it can be carried away in very short periods of time. Both slash and burn and other forms of traditional agriculture and modern mechanized farming have contributed significantly to soil erosion (Figure A). In many parts of the upper midwestern United States, for instance, the depth of the topsoil is now less than half of what it was just a century ago when tractors were first introduced into the region. This demonstrates that although nature is capable of replenishing lost topsoils over long periods of time, from a short-term human use perspective, soils are not a readily renewable resource.

A second form of soil degradation occurs through the buildup of excess minerals and salts in the soil, a process known as **salinization**. Because even freshwater surface streams and groundwater contain some dissolved minerals or salts, each time an agricultural field is irrigated, a minute deposit of salts in the soil will occur if surplus water is left standing in the field after the plants have consumed all that they need. Over time,

as a field is repeatedly irrigated, the surplus salts may build up to the point where they raise the pH of the soil above the tolerance levels of the crops. At that point, the land is rendered useless to agriculture unless the salt levels can be reduced through flushing and draining procedures, which are generally very costly. In many arid regions of the world, anywhere from 10–50 percent of all formerly irrigated agricultural lands lie abandoned, having been lost to soil salinization.

A third and increasingly widespread form of soil degradation is chemical contamination. Sources of chemicals include agricultural insecticides and herbicides, sprays designed to control plant diseases, and chemical fertilizers. Another rapidly increasing source of chemical contamination is acid precipitation, which has been known to lower the pH of the soil and associated water bodies to the point where plant and animal life are seriously threatened. Almost all of the earth's industrialized regions are currently at risk of chemical soil contamination (Figure B).

Although often overlooked, soil degradation thus constitutes one of the leading environmental threats of our age. Having noted its seriousness, it is important also to recognize that it is largely preventable through the adoption of agricultural practices that reduce soil exposure to wind and running water, better management of irrigation projects, and the more selective use of agricultural chemicals. In addition, individual farmers and gardeners can often improve their soils through increased used of organic soil additives.

■ FIGURE A **Severe soil erosion in the Andes Mountains of Ecuador.** Advanced soil erosion greatly limits agricultural productivity in many areas of the world.

■ FIGURE B **Agricultural chemical dispensing unit.** Chemicals, including herbicides, insecticides, and chemical fertilizers, are an increasing source of soil contamination in both more and less industrialized nations. This equipment is used on a farm in east-central Texas.

■ **FIGURE 3-7 Wind-ravaged and drought-stricken farmstead of the Dust Bowl period.** The term "Dust Bowl" is used in reference to a portion of the southern American Great Plains that centered on the community of Dalhart in the Texas Panhandle. Cyclic drought combined with unwise farming practices during the 1930s led to the loss of great amounts of topsoil that were transported hundreds of miles by the wind.

■ **FIGURE 3-8 Central pivot irrigation in south Texas.** Rotating around a central well, this overhead sprinkler boom delivers groundwater to a field of cabbage. High rates of groundwater extraction threaten the sustainability of this type of agriculture.

of petroleum are limited. Japan is clearly vulnerable because its industrial structure depends heavily on imported resources.

Use of the Environment

We can conclude that environmental context plays a significant role in the development process. Three considerations are of the utmost importance in our use of the environment. First, our challenge is to avoid environmental deterioration at a time when increasing population and expanding expectations are placing ever greater demands on environmental systems. The problems of human need, drought, famine,

overgrazing, and environmental degradation in the Sahel illustrate the dangerous cycle. Nonuse is not the solution, nor is environmental destruction. What is necessary is intelligent use of the environment, which will be costly as well as elusive, complicated by the varied nature of the environment, about which we still have so much to learn.

Second, we must understand that the environment cannot be separated from the culture of the society that uses it, and societies differ. Motivations, values, attitudes, accumulated technology, and economic and political organization—all of these cultural considerations greatly affect the use of the environment.

Third, we must have long-range plans for using our physical environment. Our attitudes and immediate concerns seldom encourage us to formulate plans that consider the needs of humankind for more than two or three generations to come. Such a limited time perspective is not surprising, but it is ultimately unacceptable if the environment is to sustain humanity far into the future.

Places That Sustain Concentrations of People

With the preceding brief introduction into the nature of environmental systems as context, we can now identify the types of settings that tend to sustain concentrations of people and those that do not. We need to reiterate, however, that environmental conditions do not determine developmental outcomes; at best, the environment provides a significant influence.

If one looks at a map of where the world's population is located (see Figure 2–7), the unevenness of that population is striking. Only about one-fifth of the earth's land surface is densely occupied by humankind. There are good reasons why people live where they do, reasons that at least in part are rooted in the geography and the resources of specific areas. Many of these dense concentrations of population are in "plus places"—sites that have great advantages from an agricultural standpoint. It is no accident that the dense population pattern of Egypt and northern Sudan resembles a giant snake, since the people must live close to the water of the Nile to survive. In Egypt, only 3 percent of the land surface of the country can be cultivated and about 99 percent of Egypt's 74 million people live in this relatively small space. Here they can find good soils, a year-round growing season, and, since the Aswan High Dam has provided storage of Nile floodwater from season to season and year to year, a dependable water supply with which to grow their crops. Similar dense population concentrations are found in the Ganges-Brahmaprutra floodplain of India, the Indus River valley in Pakistan, the Huang He (Yellow) and Ch'ang (Yangtze) river valleys in China, and the Rhine and Po rivers in Europe. The dense populations of arid central Mexico reflect in part the concentrations of water resources from surrounding highlands in interior valleys where intensive, raised-bed cultivation was possible at inland lakes. It did not hurt that the Aztecs' political dominance of surrounding areas funneled resources to sustain the central valley populations, just as Moscow's dominance of the Russian political system throughout multiple regime changes has made the area a plus place. And the rich volcanic soil habitats of Java in Indonesia and Luzon in the Philippines support very large populations despite the risk of eruption that lurks beneath the surface.

"Edge places" are also very positive locations. Dense populations are found in many places where water and land meet. The people who live there are able to exploit the fish of the sea as well as the crops of the land. In addition, these populations have important economic opportunities that are a result of their location. They can engage in trade with distant places, and historically have harnessed the wind to carry cargo across the low-resistance surface of the water. Coastal concentrations in North America, India, eastern China, Japan, Australia, and South America are all based on the richness of multiple resources that a coastal location provides. Those population centers surrounding the Sahara both in the Mediterranean and in the Sahelian zone commanded the caravan routes that linked their two "coastal" habitats across the vastness of the desert "sea." The semi-arid habitats of the Sahel in Africa, the pampas in Argentina, the Eurasian steppes, and the Great Plains of the United States and Canada occupy rich agricultural areas on the edge of much drier, less densely inhabited districts. Like their ocean coastal cousins, who are exposed to the unbridled fury of cyclonic storms, storm surges, and tsunami, denizens of the semi-arid edge environments occupy risky as well as rewarding places. The sufferings of those who experienced the Dust Bowl conditions of the southern Great Plains of the United States or the more recent droughts in the African Sahel, for example, illustrate the danger of dependence on such transitional areas that are as much a part of such spaces as are their opportunities.

On a world-scale map of population, concentrations of people in "rich places" seldom are large enough to appear. Often these places are associated with rich mineral deposits. Butte, Montana, Sudbury, Ontario, and Broken Hill, Australia, are examples of towns that grew up at the site of rich mineral deposits (Figure 3–9). Prosperous for as long as the special resource that called them into existence persists, their survival is threatened whenever the ore runs out or the cost of extraction exceeds the value of the mineral or other, richer sites are discovered. There is a boom or bust character to such rich places, a cycle that is compounded by the post-mining problems of landscape disturbance, residual waste materials (tailings), polluted groundwater, and land subsidence that typify many mining operations. Even densely settled agricultural oases, particularly when dependent on groundwater, experience serious difficulties when technology makes over-extraction of a limited resource possible or water is diverted to other uses. Once the basic resource is removed, it is often difficult to find ways to bolster the local economy, since the base of exploitation was extremely narrow during the rich phase and little remains to keep people in place.

■ **FIGURE 3-9 Nickel mine in Sudbury, Ontario, Canada.** Both tailings (waste) from processing nickel ore and acid rain damage the landscape and vegetation in the vicinity of major mining operations.

Places (Relatively) Devoid of People

Just as dense concentrations of people are present for a reason, so too are the earth's relatively empty places lacking in inhabitants. Nearly 80 percent of the earth's land surface falls into this category. These voids are usually too steep, too wet, too dry, or too cold to support many people. Steep places are generally not the best sites for the agricultural activities that are needed to support people at a basic level. Mountain areas have thin soils, most of the precipitation that falls runs off to other locations, and people have to expend a lot of energy to make these steep areas productive. Flat areas are much easier to farm. But if a population has a reason for wanting to live amidst steep slopes, perhaps a desire to be protected from enemies, then it is possible to develop sustainable agricultural systems as well as to exploit the timber resources often found there. Soil erosion can be reduced by constructing terraces, which have the added benefit of retaining not only soil but also runoff. On the flat areas created crops can be grown. Animals can be grazed in agriculturally marginal areas too difficult to farm. But enormous effort must be invested over very long periods to develop such terraced agricultural systems, and they remain vulnerable to decay. If the labor force necessary to sustain the terraces is invested elsewhere, in more lucrative jobs in the flatland or coastal areas or in cities, what took centuries to construct can waste away in decades as walls fall down and the steep mountain slopes return to their normal gradient.

Lands that are too wet for normal cultivation are also likely to be relative population voids. Because at best only specialized crops can be grown in such places, wetlands are likely to be occupied by few people. The Everglades of south Florida are one such environment, traditionally inhabited only by a sparse population of Native Americans. Despite their value as wildlife habitat, as flood-inhibiting "sponges," as reservoirs of biodiversity, and as seasonal loci for migratory birds, from the human standpoint wetlands are low-productivity environments unable to support many people. Commonly they are valued not for what they are but for what they might become. Massachusetts farmers in the eighteenth and nineteenth centuries may have viewed bogs as useful sources of fodder for their farm animals, but it was not until recognizing that wetlands could be converted to the monoculture of cranberries that the presence of wetland on a farm became a potential major asset. Developers in south Florida regard the Everglades as valuable primarily as potential sites for suburban growth, and constantly pressure local municipalities to promote zoning and land-use policies that will permit wetlands to grow houses rather than grasses and crocodile habitat. Ba'ath Party development planners in Iraq viewed the country's extensive southern marshlands not as a setting for wildlife and the unique local culture of the Marsh Arabs, but as a wasteland to be drained and converted into irrigated fields. Wetlands will remain largely unoccupied unless human intervention makes them into something other than wetlands.

By far the largest "empty" terrestrial areas are the world's drylands. The Atacama Desert of northern Chile, the Sonoran Desert of northern México and the southwestern United States, Patagonia in southern Argentina, the Sahara, large parts of the Arabian Peninsula, the Tarim Basin of western China, and the vast majority of Australia's land surface are too dry for substantial human settlement. Only where exotic streams, which derive their water from sources outside the desert, bring water to limited parts of an arid region do large numbers of people exist. Although small numbers of humans have managed to use arid areas seasonally (pastoral nomads), or in transit (caravans), or as sites for mineral exploitation, dense populations need water to support crops and by definition this is a resource that is in short supply in arid places.

In contrast, some places are too wet to attract dense populations. Indeed, the combination of very high rainfall and high temperatures may contribute to environmental conditions that are difficult to overcome—namely, the infertile soils found in some tropical regions. For this reason the rain-forest areas of the central Congo and Amazon basins traditionally have supported extensive (as opposed to intensive) agricultural systems, and small and widely scattered human populations.

Places that are too cold are also not ideal locations for great numbers of humans. Low temperatures, early frosts in the fall and late frosts in the spring, and a long winter season result in a short growing season and sharp limits on what people can do in cold places. Although some crops, such as quinoa (Figure 3–10), can adjust to adverse conditions, they are seldom productive enough to sustain dense populations. Crops adapted to harsh conditions and short growing seasons at the margins of cultivation are usually the food supply of peoples who are at the periphery of their society's productive resource base. Such marginal areas support limited permanent populations, although they often are utilized seasonally. Herders frequently take their animals to high mountain pastures that are available for only a couple of months of the year; hunters of seals and fish along the Arctic coasts collect berries from tundra plants that grow during the short high-latitude summer. But

■ **FIGURE 3-10 Quinoa field near Cuzco, Peru.** Quinoa is a lesser-known grain native to the high, arid Andes of western South America. Its tolerance of cold temperatures enables it to be cultivated at elevations as high as 12,000–13,000 feet (3,600–3,900 meters) above sea level in the American tropics.

all retreat to more permanent base settlements when summer ends. And in lands of perpetual snow, such as mountain tops, glaciers, or Antarctica, "settlement" is left to scientific researchers and penguins!

Centers of Use and Habitat Modification

The physical characteristics of each place on earth are unique. The course of development that each group of people living in each place has charted is equally different from other places. Styles of dress, forms of houses, food preferences and prejudices, settlement patterns, religious beliefs, and much more vary from place to place. Sometimes these differences are subtle, at other times the variations are dramatic. Each pattern in each place is the product of choices that people make about the life they value, the resources they recognize and exploit, and the contacts they have with and the ideas and technology they learn from the inhabitants of other places.

Humans believe they are unique among living creatures because they can accumulate learned behavior and transmit it to successive generations. As we learn more about the behavior of other animals, this human claim may not be as distinctive as once thought. But what is certain is that humankind's learned behavior patterns, its manifold cultural heritages, have helped to ensure survival, sustenance, and preservation of the social order. With that accumulation of learned behavior, or culture, humans make decisions and create ways of life. Some behavior is based on **inherited**

culture, a society's own earlier experiences; other actions are based on **diffused culture**, the experiences of other societies with which a society has had contact. The entire set of elements that identifies a society's way of life—values, language, technology—also constitutes its culture.

Another way of viewing culture is as a hierarchy of traits, complexes, and realms. A **culture trait** is the way a society deals with a single activity—for example, how people plant seeds. A **culture complex** is a group of traits that are employed together in a more general activity, such as agricultural production (Figure 3–11). A **culture realm** is a region in which most of the population adheres to similar culture complexes. Not all world regions have easily definable culture realms. Some have transition zones between realms, zones in which numerous distinct cultures have met and clashed. Even within a single political unit, cultural pluralism is frequently evident and may complicate achieving national unity.

A **culture hearth** is a source area in which a culture complex has become so well established and advanced that its attributes are passed to future generations within and outside the immediate hearth area. No single hearth exists, as was once thought. Rather, a number of different hearths have contributed various culture traits and complexes to cultures distant from their source area.

Societies advance technologically at uneven rates and along different paths. For example, the Bantu culture of western Africa advanced more rapidly than did that of the Bushmen or Hottentots of southern Africa. The culture complexes of food production in China and Europe during the nineteenth century involved different traits. Cultural distinctiveness is not necessarily a difference in level of achievement; it is most frequently a difference in kind.

The cumulative nature of cultural evolution, its dynamic quality, the unevenness with which it occurs, and its different orientations all contribute to fundamental and intriguing variations among the more than 6 billion people of our contemporary world.

Growth of Culture and Technology

The world's present cultural patterns are a reflection of the major human accomplishments of the past. Early cultural accomplishments common to all human societies included the use of fire, the making of tools, and the construction of shelters. They also included the domestication of numerous food plants and animals that, today, constitute an irreplaceable biological heritage of all humankind.

Hearths of Innovation

The Middle East sustained one of the world's earliest and most influential culture hearths (Figure 3–12). Known as the Fertile Crescent, the region actually consisted of several hearths that were close to one another. The earliest domestication of plants seems to have taken place in nearby hill lands, and agricultural villages appeared in the Mesopotamian lowlands associated with the Tigris and Euphrates river valleys. From the fourth millennium B.C. on, civilizations, city-states, and empires flourished in that area (Figure 3–13). Among other major achievements, those civilizations codified laws, used metals, put the wheel to work, established mathematics, and contributed several of the world's great religions—Zoroastrianism, Judaism, Christianity, and Islam.

Another of the world's early culture hearths developed in the Indus River Valley (in present-day Pakistan), where a mature civilization existed by 2500 B.C. An exchange of ideas and materials with Mesopotamia began early and continued over a long period, from 3000 to 1000 B.C. Much of that exchange of culture was by way of ancient Persia (now Iran), and the two hearths may even have attracted migrant peoples from the same areas. The Indus Valley experienced invasions and migrations of people from northwestern and central Asia, who brought an infusion of new traits with each invasion. The Indus, and the adjacent Ganges River Valley, became the source area for cultural traits that eventually spread throughout India. This particular culture hearth made significant contributions in literature, architecture, metalworking, and city planning. Philosophy also evolved and later contributed to Hinduism and Buddhism.

The valley of the Huang He (Yellow River) and its tributaries were the location for the evolution of the Chinese culture hearth. Wheat and oxen, part of the North China agricultural complex, may have had Middle Eastern origins, and the rice, pigs, poultry, and water buffalo found in China may have originated in Southeast Asia, but the assemblage of cultural traits and their subsequent development are distinctive and unique Chinese achievements. Crop domestication, including soybeans, bamboo, peaches, and tea; village settlement; distinctive architecture; early manufacturing; and metalworking were in evidence at the time of the Shang dynasty (1700 B.C.). That Chinese culture eventually spread into South China and northeastward into Manchuria, Korea, and, later, Japan.

Each of the earliest culture hearths had independent accomplishments, yet each was also a recipient of ideas and

■ **FIGURE 3-11 Chuño.** One little-known component of highland Andean Indian potato use is the production of *chuño,* or sun-dried slices of potatoes. This ingenious form of naturally dehydrated food enables the farmers to extend the "shelf life" of one of their staple crops. Notice the different colors of *chuño* shown in this image. The potato is native to western South America and the smallholder farmers there continue to cultivate literally thousands of varieties, each differing in physical and environmental attributes.

commodities from other areas. Each hearth was able to maintain a distinct identity, which was transferred to succeeding generations or to invading or conquered peoples who were eventually first acculturated and then assimilated.

Two areas in the Americas that began to develop around 3000 B.C. also served as culture hearths for major civilizations. The first appeared in northern Central America and southern México (later becoming the Mayan civilization) and extended to central Mexico (later becoming the Aztec civilization). This large Mesoamerican region supported a sizable sedentary population and was characterized by large cities, monumental architecture, political-religious hierarchies, the use of numeric systems, and the domestication of maize, beans, squash, and cotton (Figure 3–14).

The Middle Andean area of Peru and Bolivia was the site of the other major American culture hearth, which gave rise to the Inca civilization. It advanced more slowly than its Middle American counterpart, but by the sixteenth century A.D., it had developed irrigation, worked metals, domesticated the white potato, established complex political and social systems, set up a transportation network, and built an empire.

Several other areas also functioned as culture hearths with more limited regional influence. The Bantu language family seems to have had its source area in West Africa. The Ethiopian highlands were a domestication center for wheat, millets, and sorghums, and Great Zimbabwe in southern Africa achieved high levels of mining and metallurgy and the manufacturing of clay and wood products and clothing. Central Asia was a domestication center for grains, too, and it was certainly a trade route along which ideas and commodities were exchanged among major civilizations, such as the Middle East and China.

The Spread of Innovations: Early Forms of Globalization?

Contact between individuals and the culture groups to which they belong is not a new phenomenon. It is a process as old as humankind itself. Whether these contacts are peaceful, promoted by trade or by the movement of small groups from one area to perceived new opportunities in another place, or involve force and are the result of conquest and coercion, contact between people inevitably involves the interaction of ideas, practices, and techniques. Good ideas and inventions are readily recognized by people, adopted, and improved on. Thus the history of innovations developed in one culture hearth is their spread to other places.

When Islam spread rapidly in the first two centuries after its emergence, the exchange of ideas and technology was facilitated. As but one example, sugarcane (*Saccharum officinarum*), a tall tropical grass originating in Southeast Asia, was brought westward by Muslim traders and agricultural innovators. Spaniards became familiar with sugarcane through centuries of conflict and culture contact between Muslims and Christians in Spain and North Africa. Where Columbus successfully sailed in 1492, sugarcane followed quickly (Figure 3–15, page 68). And that is not all that Spaniards carried with them to the New World! Cattle,

horses, pigs, wheat, chickens, and much more were imported both to fill unoccupied niches in the Western Hemisphere and to provide the conquerors with the food with which they were familiar. Tomato, corn, beans, cacti, tobacco, potatoes, and other New World domesticates moved eastward and enriched the culinary traditions of Europe. Important and accelerating cultural and technical exchanges and adoptions between different cultures was the result.

In this last great human integration of the earth's ecosystems and cultures, Europe played an extremely important role. Europe had long been a peripheral outlier of the Middle Eastern culture realm. The increasingly global maritime dominance of European countries after A.D. 1500, and their initiating role in the development of industrial technology after A.D. 1700, particularly in England, touched off a scramble for access to and control of resources in virtually all parts of the earth. Many of our contemporary global conflicts have origins in the confrontation between modern European culture, represented not only by individual nation-states but also their colonial offspring, and the traditional systems found elsewhere. That cultural and colonial confrontation did not necessarily represent a deliberate attempt to erase traditional ways of life, but the adjustments involved in the transition to the European system have frequently been difficult.

Between the fifteenth and the twentieth centuries, the Europeans extended their influence and culture around the world, spurred by internal competition and a new interest in science, exploration, and trade. They explored, traded, conquered, and claimed new territories in the name of their homelands. Modern European states emerged with the capability of extending their power over other areas, and many European people resettled in newly discovered lands in the Americas, southern Africa, Australia, and New Zealand.

Complex patterns of **acculturation**, the process by which a group takes on some of the cultural attributes of another society, do not occur painlessly. The European explosion outward into the world has imposed alien structures on people in many parts of the world, encouraging a form of **cultural convergence** as elements of Western culture (fast food, clothing styles, architecture, political systems) have appeared in local landscapes. Many small, isolated non-Western cultures may have disappeared as an unwanted and undesirable consequence. In certain respects political organization, food production, industrialization using inanimate sources of power, and consumption patterns in the now independent former colonies are becoming more similar to those of their former colonial masters. In the former centers of colonial power, migrant populations, often several generations removed from their ancestral homeland, play a prominent, if at times uneasy, role in the economic and social institutions of their "new" homeland. One can predict that this blending of peoples and cultures will continue, with the technological aspects of Western culture being the most readily accepted. Languages, religions, and values adjust more slowly and frequently become sources of friction.

This argument does not assert that traditional non-European cultures are inferior to European cultures or that the non-European cultures will totally disappear. Rather, it suggests

■ **FIGURE 3-12 Early culture hearths of the world.** Three major culture hearths are recognized as the principal contributors to modern societies throughout the world. From those hearths, plants, animals, ideas, religions, and other cultural characteristics have spread. Minor hearths, although locally important, have not had much impact outside their source areas.

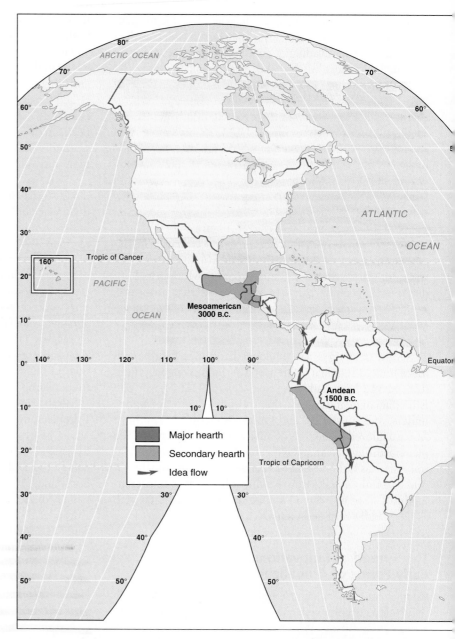

■ **FIGURE 3-13 Baked-brick foundation of the Ishtar Gate.** The Babylonian Empire was one of many ancient Middle Eastern civilizations that rose and fell in the Tigris and Euphrates river valleys. The Ishtar Gate was built by King Nebuchadnezzar II in the sixth century B.C.

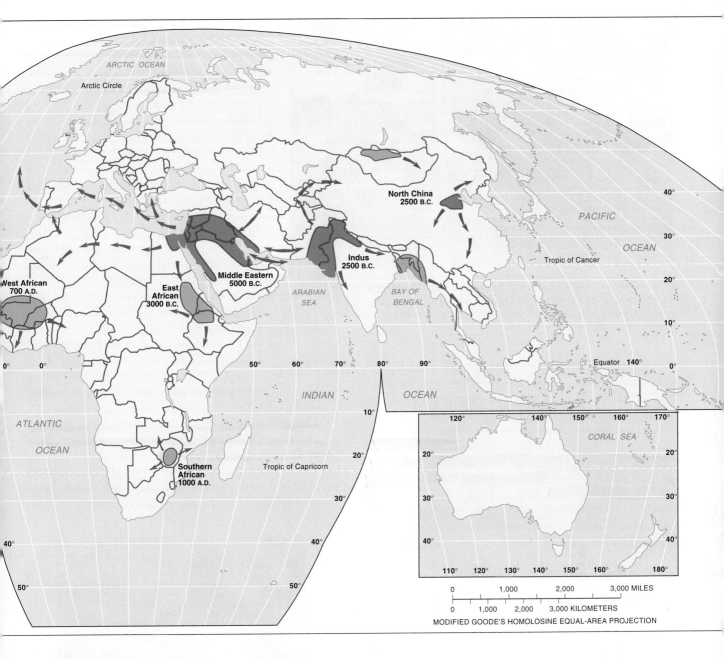

■ **FIGURE 3-14 Mayan stelae.** These stone monuments date back thousands of years and are found in the Mayan ruins of Copán, northern Honduras. At its height, Copán is believed to have supported a population of 20,000–40,000 and served as an educational and commercial center of the region.

■ **FIGURE 3-15 Ecuadorian children use sugarcane as a candy substitute.** Sugarcane is a tropical grass, native to southeast Asia, that is used in its unprocessed form as a sweet in both the Old World and the New.

that large and viable culture complexes will be modified by cultural blending. Japan provides one example of the way in which European traits have been accepted in modified form. Other nations have and likely will continue to experience European acculturation to greater or lesser degrees in a manner that reflects their own cultural heritage. In addition, the argument for convergence recognizes that **regional disparity**, or locational differences in economic, social, and biological conditions, will continue. Full or modified acceptance of the European way of doing things means that particular areas will have great value because of their location and their material and human resources; other areas will not be so favored.

Special Elements of Culture

The patterns of acculturation and convergence that emerge when culture realms come into contact are affected by a number of factors. Arguably the most important, and often the least susceptible to change, are language, religion, and political ideology. These cultural elements are deeply rooted, provide a sense of identity to their practitioners, and express primary differences in values and thought processes that are difficult to bridge.

Language   Language gives a set of meanings to various sounds used in common by a number of people. It is the basic means by which culture is transferred from one generation to the next. Because of the relative isolation in which societies evolved, a great number of languages formed, frequently with common origins but without mutual intelligibility.

Linguistic differences can function as barriers to the exchange of ideas, the acceptance of common goals, and the achievement of national unity and allegiance. Most members of most societies are not bilingual and do not speak the language of a neighboring society if it is different from their own. Sometimes linguistic differences are overcome through the use of a **lingua franca**, a language that is used throughout a wide area for commercial or political purposes by people with different native tongues. Swahili is the lingua franca of eastern Africa; English, of India; and Urdu, of Pakistan (Figure 3–16). In many countries that acquired their independence after World War II, the political leadership has found national unity and stability difficult to achieve. Internal problems and conflict frequently stem, in part, from cultural differences, one of which is often linguistic variation. The situation in East Pakistan (now Bangladesh) and West Pakistan (now Pakistan) is a case in point. From 1947 to 1972, those two regions functioned as one political unit. Political leaders, seeking to promote a common culture and common goals, deemed a single national language necessary, even though several mother tongues were in use. Bengali, the language of East Pakistan, was derived centuries ago from Sanskrit; it provided East Pakistanis with a unifying cultural element. West Pakistan, on the other hand, encompassed a number of languages: Baluchi, Pashto, Punjabi, Sindhi, and Urdu, the lingua franca. That linguistic diversity was just one of the many challenges that eventually led to the establishment of Bangladesh as a separate and independent state.

Pluralistic societies often lack a common language within a country, and that pluralism frequently produces political instability or hinders development. Belgium and Switzerland are noteworthy because they have partially overcome the problem of linguistic pluralism. Others, such as Canada and the former Yugoslavia, have been less successful.

■ **FIGURE 3-16 Newspaper vendor in New Delhi, India.** English language newspapers in India bridge the barriers of communication between many different languages and dialects in this vast country.

Religion   Religions and belief systems have their origins in the universal human desire to comprehend life's purposes and to give and receive support from others. Animism, the worship of natural objects believed to have souls or spirits, was one early form of religion. It includes rituals and sacrifices to appease or pacify spirits, but it usually lacks complex organization. Most animistic religions were probably localized; certainly, the few that have survived are found within small, isolated culture groups. The more important contemporary religions are codified, organized in hierarchical fashion, and institutionalized to ensure the transfer of basic principles and beliefs to other people, including succeeding generations.

Modern religions can be classified as either ethnic or universalizing. Ethnic religions originate in a particular area and involve people with common customs, language, and social views. Examples of ethnic religions include Shintoism (in Japan), Judaism, and Hinduism (in India). Universalizing religions are those considered by their adherents to be appropriate—indeed, desirable—for all humankind. Buddhism, Christianity, and Islam all had ethnic origins but have become universal over the centuries, as they lost their association with a single ethnic group. Religions spread from one people to another through proselytizing, not only by missionaries but also by traders, migrants, and military personnel (Figure 3–17). Proselytizing is often considered a responsibility of those who practice a universalizing religion.

Religious ideology exerts a great impact on culture. Sacred structures contribute to the morphology, or form and structure, of rural and urban landscapes. In addition, religions have shaped many of the routine aspects of daily life. Centuries ago, the spread of citrus growing throughout Mediterranean lands was directly related to Jewish observances in which citrus was required for the Feast of Booths. Religious food restrictions also account for the absence of swine in the agricultural systems of Jewish and Muslim peoples in the Middle East. The Hindus' taboo on eating beef is a response to their respect for the cow, which is considered to be an example of the ideals of selfless service and generosity. This practice has resulted in an overabundance of cattle that require space and feed while returning only limited material benefits (manure, milk, or draft power) (Figure 3–18). In the United States, the economic impact of religion is seen in taxation policies (for example, organized churches are usually exempt from taxation), institutional ownership of land and resources, work taboos on specified days (the so-called "blue" laws), and attitudes toward materialism and work.

One implied function of religion is the promotion of cultural norms, the results of which may be positive. Societies benefit from the stability that cohesiveness and unity of purpose bring. Unfortunately, however, conflict also arises frequently from religion-based differences, evoking intolerance, suppression of minorities, or simply incompatibility among different peoples. Examples include the Crusades of the Middle Ages, in which Christians attempted to wrest the Holy Land from Islamic rule; the partition of the Indian subcontinent in 1947 in response to differences between Hindus and Muslims; and the conflict between Catholics and Protestants in Northern Ireland.

■ **FIGURE 3-18 Sacred cow.** Cattle are allowed to roam freely in Hindu-dominated India. This scene is from Madhya Pradesh.

Political Ideology   Political ideologies also have major implications for societal unity, stability, and the use of land and resources. Such ideologies need not be common to the majority of a population; people may be apathetic, or they may be unable to resist an imposed system of rules and decision making. In some instances, oligarchies, in which control is exercised by small groups, or dictators have made their particular philosophies basic to the functioning of their societies.

Most modern governments assume some responsibility for the well-being of the people in their states. Some socialist governments assume complete control over the allocation of resources, the investment of capital, and even the use of labor, thereby severely limiting individual decision making. In other societies, governments assume responsibility for providing an environment in which individuals or corporations may own and determine the use of resources. National and economic

Table 3-2   Per Capita Gross National Income of Selected Countries, as a Percentage of United States GNI (1977 and 2004)[a]

	Per Capita Income (in U.S. dollars)		Percentage of U.S. Gross National Income	
	1977	**2004**	**1977**	**2004**
Industrialized Economies				
United States	7,060	39,710	100.0	100.0
Norway	6,540	38,550	92.6	97.1
Switzerland	8,050	35,370	114.0	89.1
United Kingdom	3,840	31,460	54.7	79.2
Canada	6,650	30,660	94.2	77.2
Japan	4,460	30,040	63.2	75.6
Sweden	7,880	29,770	111.6	74.9
France	5,760	29,320	81.6	73.8
Germany[b]	6,610	27,950	93.6	70.4
Italy	6,050	27,860	86.7	70.2
Transitional Economies				
Hungary	2,480	15,620	35.1	39.3
Poland	2,910	12,640	41.2	31.8
Russia[c]	2,620	9,620	37.1	24.2
Romania	1,300	8,190	18.4	20.6
Bulgaria	2,040	7,870	28.9	19.8
Newly Industrializing Economies				
China, Hong Kong SAR[d]	1,720	31,510	24.4	79.4
Singapore	2,510	26,590	35.6	66.9
South Korea	2,520	20,400	35.7	51.4
Malaysia	720	9,630	10.2	24.3
Mexico	1,190	9,590	16.9	24.2
Brazil	1,010	8,020	14.3	20.2
Less Industrialized Economies				
China	350	5,530	5.0	13.9
Peru	810	4,890	11.5	13.5
Philippines	370	4,640	5.2	12.3
Jordan	460	4,640	6.5	11.7
Guatemala	650	4,140	9.2	10.4
Jamaica	1,290	3,630	18.3	9.1
India	150	3,100	2.1	7.8
Vietnam	160	2,700	2.3	6.8
Mozambique	310	1,160	4.4	2.9
Mali	90	980	1.3	2.5
Nigeria	310	930	4.4	2.3
Ethiopia	100	660	1.4	2.1

Source: Data compiled from *World Population Data Sheet* (Washington, D.C.: Population Reference Bureau, 1977 and 2004).
[a]2004 data are calculated using the GNI PPP method; 1977 data are calculated using the GNP method.
[b]1977 data are for the former West Germany.
[c]1977 data are for the former Soviet Union.
[d]Special Administrative Region.

development programs reflect the differences in political systems. The approach of socialistic governments, for example, is quite different from that of capitalistic countries.

The Geographic Dimensions of Development

In the previous sections, we defined development as an improvement of the human condition in both material and nonmaterial ways that involves a variety of sustainable cultural and economic changes. In this section, we want to examine the specific criteria that are used to identify levels of development in different parts of the world. We will then use these criteria to construct various maps that show us the geography of development worldwide, and to describe the characteristics of the developed and less developed regions of the world. Finally, we will briefly review some of the theories employed to explain differences in development.

Measures of Wealth and Their Unequal Geographies

Several measures are traditionally used to quantify differences between the developed and less developed countries of the world. While useful in its own way, each measure has weaknesses that prevent it from serving as an exclusive indicator of well-being.

Per Capita Income A popular measure of development is the amount of income earned per person. In general, high levels of economic development are associated with high per capita incomes. Since the end of World War II, many less industrialized countries have shown a significant increase in per capita income. In combination with other indicators—such as death rates, infant mortality rates, and dietary consumption—these increases would initially appear to suggest considerable progress for the residents of those countries. Because of inflation, however, that gain has not always been real. Not only has it not resulted in greater purchasing power in all cases, but in some nations, effective buying power has even decreased. More than 3 billion people—one-half of the world's population—remain in extreme poverty, surviving by traditional economic systems in countries where the per capita GNI PPP (see explanation below) is less than $4,000 per year.

Table 3–2 and Figure 3–19 graphically demonstrate the disparity between more developed countries and Ethiopia, Mali, Mozambique, and Nigeria, which are representative of the poorest African nations and which have per capita GNIs of less than $1,000 per year. Vietnam and India are

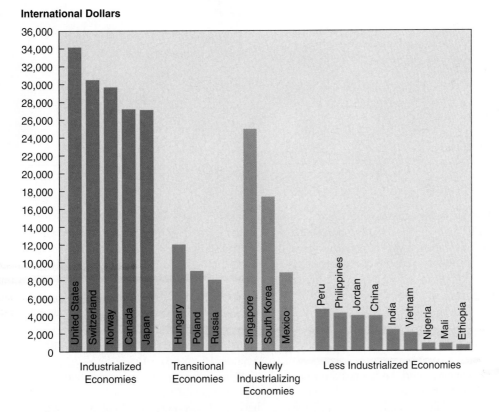

International Dollars

■ **FIGURE 3-19 Per capita GNI for selected more and less developed countries.** Most of the world's countries are experiencing an increase in their per capita GNIs, but great differences remain in income levels.
Note: Data are calculated using the GNI PPP method.
Source: Population Reference Bureau, *World Population Data Sheet.* Washington, D.C. 2002.

■ **FIGURE 3-20 World per capita GNI PPP.** Per capita Gross National Income is considered the best single measure of economic well-being. High per capita GNIs are closely associated with areas of high levels of technological achievement and with oil-exporting countries. Low per capita GNIs are associated with southern, southeastern, and eastern Asia and much of Africa.
Source: Data compiled from *World Population Data Sheet* (Washington, D.C.: Population Reference Bureau, 2002).

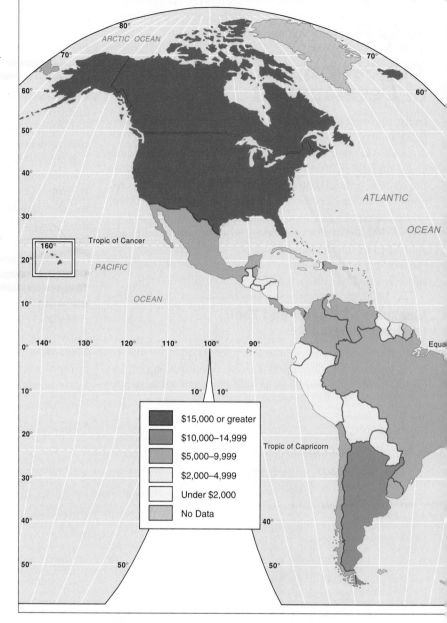

only slightly better off, with average GNIs of less than $3,000 per person per year. For many of the world's poorest peoples, caught in a seemingly endless cycle of poverty, "development" is a relatively meaningless concept. In contrast, the citizens of the United States, Switzerland, Norway, and most other technologically advanced nations have annual GNIs upward of $27,000 per person.

This inequality is also evident when we compute a country's per capita GNI as a percentage of the per capita GNI of the United States; the lower the percentage, the greater the existing **GNI gap**. We have used the United States as a base measure because of its standing as one of the world's most prosperous countries, but we could just as well have used Canada, Japan, or a wealthy European country.

The countries included in Table 3–2 were selected to represent the various world regions and different development experiences. These experiences include the long-

industrialized nations of western Europe, the United States, Canada, and Japan; and the transitional, formerly centrally planned Marxist countries, including Hungary, Poland, and Russia. A third group of nations consists of those that were once clearly in the less developed group but which have shown such marked economic growth that they have come to be called the **newly industrializing countries (NICs)**. The fourth group consists of the less industrialized nations, most of which are situated in Africa, Asia, and Latin America.

When we turn to the current pattern of per capita GNI on a world scale (Figure 3–20), the geography of income haves and have-nots is striking. Areas of high per capita GNI ($10,000 or more) include the United States and Canada, the western European nations, Israel, several of the major oil-exporting countries, Japan, South Korea, Australia, and New Zealand. Russia and the other former states of the

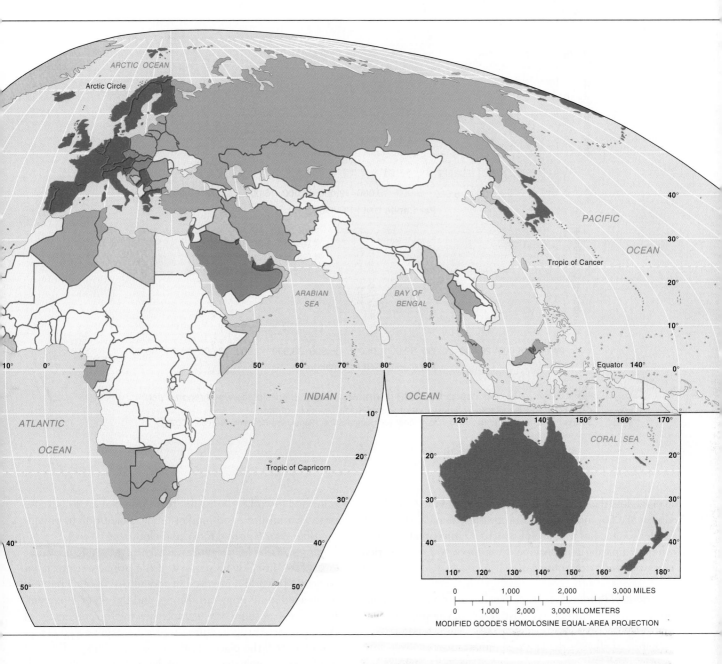

ARCTIC OCEAN

Arctic Circle

PACIFIC

40°

30°

OCEAN

Tropic of Cancer

20°

ARABIAN
SEA

BAY OF
BENGAL

10°

10° 0° 50° 60° 70° 80° 90° Equator 140° 0°

INDIAN OCEAN

ATLANTIC

OCEAN

Tropic of Capricorn

10°

20°

CORAL SEA

120° 140° 150° 160° 170°

20° 20°

30°

30° 30°

40° 40°

40° 40°

110° 120° 130° 140° 150° 160° 180°

50° 50°

0 1,000 2,000 3,000 MILES

0 1,000 2,000 3,000 KILOMETERS

MODIFIED GOODE'S HOMOLOSINE EQUAL-AREA PROJECTION

Soviet Union, while generally industrialized, do not belong to the high-income group. Much of Eastern Europe, Russia, Latin America, and parts of the Middle East have per capita GNIs between $2,000 and $10,000. Countries with a per capita GNI of less than $2,000 are dispersed throughout Africa, Asia, and the poorest areas of Latin America. That group of countries suffers from persistent poverty. They have been identified by the United Nations as part of the growing dichotomy between developing countries that are progressing and those that are stagnating. They are frequently constrained by foreign debt, ethnic conflict, high levels of graft and corruption, and ineffective economic and political systems.

A major weakness of the income measure is that it is an average figure and does not account for income distribution. For example, a few extraordinarily wealthy people or industries can skew upward a per capita income amount, mak-

ing an area appear wealthier (and hence more developed) than it really is. A second weakness relates to the true value of income in a given country. In other words, how much can a person's income really buy? We all know that some countries are prohibitively expensive to live in and other countries are extraordinarily cheap. Consequently, we increasingly see a measure used that is called **Gross National Income in Purchasing Power Parity (GNI PPP)**, divided by population. This measure indicates national income on the basis of "international" dollars, which are dollars that take into account a country's purchasing power relative to other countries. In Figure 3–21, each colored mark indicates a country's per capita GNI PPP. Although countries and regions are positioned all along the continuum, most are clustered at the lower end of the scale. A few regions have achieved the middle range; even fewer are found at the upper end of the scale. Countries in Africa, for example,

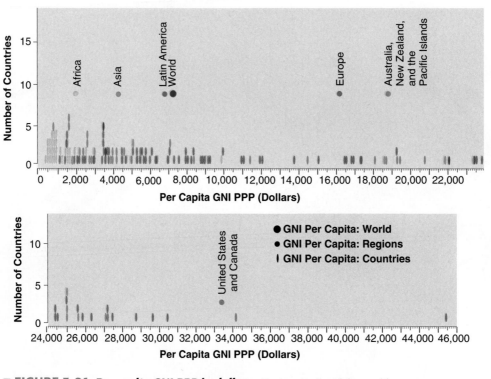

■ **FIGURE 3-21 Per capita GNI PPP in dollars.** Most countries of the world report a per capita GNI PPP substantially below that of the relatively few rich countries.
Source: Data from *World Population Data Sheet* (Washington, D.C.: Population Reference Bureau, 2002).

average less than $2,000; Latin America and the Caribbean almost $7,000; and Western Europe more than $25,000. The United States has a figure of $34,100. Unfortunately, even when comparing purchasing power, the world still has rich countries and poor countries.

Agricultural Production   Employment in primary economic activities is a common yardstick of development. The **primary level of economic activity** focuses on extractive activities, such as agriculture, mining, forestry, and fishing; the **secondary level** includes activities that transform raw materials into usable goods, such as manufacturing; and the tertiary and quaternary levels embrace about everything else. Countries in which a large part of the labor force is engaged in primary activities do not produce much income and use relatively small amounts of power per capita. Conversely, countries with strong secondary and tertiary components usually have greater per capita GNIs and consume more energy. Use of this indicator—percentage of the labor force in primary activities (mainly agriculture)—is based on those relationships.

Countries in which primary production is dominant offer limited opportunities for labor specialization, especially if the economy depends on subsistence agriculture, and the vast majority of effort must be dedicated to meeting local food needs. In theory at least, labor specialization and production diversity are basic to economic growth. Thus, prospects for high levels of individual production are di-

minished if workers must not only grow the crops but also process, transport, and market them, in addition to providing their own housing, tools, and clothing.

Of all the primary activities, agriculture is by far the most important. Roughly 98 percent of the primary-sector labor force in the world is involved in agriculture; about 1 percent is engaged in hunting and fishing, and the other 1 percent, in mining. Consequently, a graph showing the percentage of the labor force in agriculture (Figure 3–22) gives a good representation of the place of primary occupations at different levels of economic development. Note that the pattern is almost the inverse of Figure 3–19, meaning that the higher the proportion of the labor force in agriculture, the lower the per capita GNI is likely to be.

Part of the explanation for the gap between more developed and less developed countries lies in different levels of agricultural productivity. Look, for example, at the greatly contrasting yields of maize (corn) shown in Table 3–3. In the more developed nations, technological innovations have been more widely applied; high-yielding hybrid seed is planted; and chemical fertilizers, pesticides, and irrigation are used more extensively. Those technologies have not been applied as uniformly in the less developed nations, many of which for economic and cultural reasons have been unable to afford and/or are reluctant to embrace the technologies associated with industrialized agriculture (see the *Geography in Action* boxed feature, The Green Revolution).

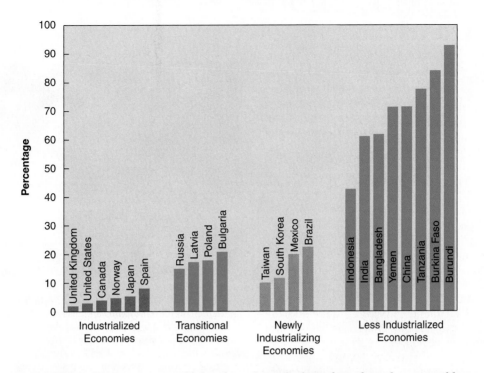

■ **FIGURE 3-22 Percentage of labor force in agriculture for selected more and less developed countries.** The percentage of the labor force employed in the primary sector, almost all of which is in farming, shows the degree of economic diversity of a nation. If a large percentage of the population is engaged in agricultural pursuits, manufacturing and services are limited in their development. Conversely, if only a small percentage of the labor force is in agriculture, manufacturing and services are well staffed.

Source: Data compiled from *Britannica Book of the Year* (Chicago: Encyclopedia Britannica, Inc., 2002), 814–818.

Table 3-3 Maize (Corn) Yields in Selected Industrialized and Less Industrialized Nations in 2004	
Nation	**Yields (hectograms[a] per hectare)**
Industrialized	
Israel	160,000
United States	100,650
Spain	98,946
Italy	91,604
France	89,832
Canada	82,400
Less Industrialized	
Brazil	33,733
Pakistan	30,971
Cambodia	27,612
Mexico	25,000
Bolivia	22,416
India	20,000
Tanzania	17,722
Honduras	15,034
Ecuador	13,851
Malawi	11,181

Source: Data compiled from Food and Agriculture Organization, *http:// faostat.fao.org/faostat/form?collection=Production.Crops.Primary&Domain =Production&servlet=1&hasbulk=0&version=ext&language=EN.*
[a]A hectogram is a metric measure of weight, equal to 100 grams (3.527 ounces).
Note: Yields are from 2004 data.

GEOGRAPHY IN ACTION
The Green Revolution

The success of some countries in significantly improving the wheat, rice, maize, and sorghum yields in selected agricultural regions can be directly attributed to the **Green Revolution**. That term encompasses a number of international research initiatives that serve as models for cooperative efforts but also illustrate the difficulty of solving complex food problems.

The Green Revolution grew out of a joint research effort between Mexican and U.S. scientists during the 1940s, which established the Centro International de Mejoramiento de Maiz y Trigo (CIMMYT) in Mexico. Supported by the Mexican government and the Rockefeller Foundation, those scientists first focused on wheat and maize. In 1960, with additional support from the Ford Foundation, a second international research center, the International Rice Research Institute (IRRI), was established at Los Baños in the Philippines. Now a number of other international agricultural research centers exist, but none has been more successful than the CIMMYT and the IRRI.

The basic goal of those research efforts has been to improve food output and reduce hunger. Simply stated, they have developed high-yield varieties of grains (HYVs). Success has not come easily, but increased food output has been significant, even if localized. Over many years, plant-breeding efforts by the research institutes have been repeated and expanded to develop seed varieties that would resist numerous diseases and pests. Consideration has even been given to taste preferences.

The limited but important achievements of the Green Revolution illustrate several significant points. First, agricultural enterprise, even if largely subsistence, is part of a larger system composed of cultural, economic, and environmental elements. Changing one component of that system may require or generate numerous additional changes. For example, HYVs have required additional irrigation, more fertilizers, better land preparation, and improved tending of crops (Figure A). Some HYVs have also been more susceptible to certain plant diseases. Consequently, the countries that have adopted the higher yielding seeds have found it necessary to improve extension services, transportation, food storage, and marketing procedures—all infrastructure considerations. In addition, the farmers involved need more capital and greater labor input.

FIGURE A Native Philippine rice varieties. Traditional peasant farmers in the Philippines and elsewhere in Southeast and East Asia cultivate simultaneously in their fields multiple varieties of rice. While often lower yielding than the newer Green Revolution hybrid varieties, the native open-pollinated varieties vary in color, taste, length of growing season, and resistance to environmental stresses such as insects and diseases. For this reason, the adoption of the higher-yielding hybrid varieties generally requires that the local people modify their diets and embrace new production techniques and levels of social organization. The rate of acceptance of these changes has varied greatly from region to region and nation to nation. Shown in this image are three native varieties. The purple-colored grains at the top center are *Malagkit;* the white-colored grains at the lower left are *Wagwag;* and the yellow-colored grains at the lower right are *Makan.*

A second point to remember is that rapid innovation and change may not benefit everyone. Some farmers find it difficult to embrace the new products, so differences increase as change occurs. In the case of the HYVs, the larger landowners and more prosperous farmers have generally adapted more easily, owing to their greater ability to afford economic risk, leading to the charge that the Green Revolution, though successful, has contributed to greater income disparity among farmers and even

---

Figure 3–23 charts recent changes in per capita food production. For most regions, as well as for the world as a whole, per capita food production has increased, which is no small feat when we consider the earth's continuing population growth. The gains achieved have been highly variable from place to place, with the greatest increases occurring in selected Asian and South American regions. Much of Africa continues to lag in its struggle to attain higher food productivity per capita. Europe's continuing decline may be attributable to the shrinking of its agricultural sector generally, rather than to decreasing yields per unit of farmland.

Industrial Production   Traditionally, as employment in primary production declined, the percentage of the workforce in manufacturing increased. Consequently, the industrial workforce of a country could be used as an indicator of development. By the twenty-first century, the industrial sector had lost ground relative to the services sector. In addition, technology had eliminated the need for many industrial workers. Employment in manufacturing, consequently, is less valuable as an indicator of development than in the past. However, the strength of the industrial sector as a whole is still relevant.

Most of the less developed countries have a limited industrial sector, the growth of which has often been driven in recent times by foreign investment. In the last several decades, however, at least some less developed countries have begun to distinguish themselves industrially, producing higher

Table A Nitrogenous Fertilizer Consumption of Selected Countries	
Country	**Thousand Metric Tons**
Developed	
Spain	1,180.0
Italy	866.0
Poland	861.3
Less Developed	
Zambia	29.7
Cameroon	26.3
Chad	11.0
Jamaica	9.2
Fiji	6.4
Ghana	5.1
Haiti	5.0
Afghanistan	5.0
Laos	4.5
Botswana	4.1
Congo	2.0
Belize	1.2
Gambia	1.0
Bolivia	0.9

Source: Data compiled from *Statistical Yearbook* (New York: United Nations, 2002), 394–413.

Table B Insecticide Consumption of Selected Countries	
Country	**Metric Tons**
Developed	
Greece	2,864
France	2,590
Germany	1,288
Less Developed	
Dominican Republic	309
Iraq	190
Cameroon	86
Niger	62
Jordan	61
Qatar	60
Suriname	58
Ghana	47
Malta	47
Ethiopia	38
Rwanda	36
Haiti	7
Laos	1
Samoa	1

Source: Data compiled from Food and Agriculture Organization, *http://apps1.fao.org/servlet/XteServlet.jrun?Areas*.
Note: Data are for the year 2000.

increased rural-to-urban migration, both of which are common phenomena in developing countries.

Third, targeting increased output signals a different approach to food problems, one that has been emerging in many parts of the world during the past several decades. In the nineteenth and early to mid-twentieth centuries, output was usually increased by expanding the amount of land in production. In the Green Revolution, we are witnessing a much greater emphasis on improved technology and more intensive use of the land.

Yet it is important to recognize that, in a great number of countries, the Green Revolution has had minimal impact on agricultural production. This is especially true of the poorer developing nations whose citizens have not been able to afford the costs of the associated industrial technologies. This fact is

illustrated in Tables A and B, which show the amounts of nitrogenous fertilizer and insecticides used in selected countries in recent years. Note that in both tables, the combined consumption in many of the developing nations of these basic agricultural inputs is far less than that of a single mid-sized nation that has embraced Green Revolution agriculture.

The Green Revolution has not solved all the world's food supply concerns. Hundreds of millions of rural poor in the developing nations continue to battle malnourishment and undernourishment and their urban counterparts face equally uncertain food supplies. Nevertheless, the Green Revolution has not only enabled the earth's population to grow to record numbers but also, overall, to be better fed and nourished than perhaps at any time in human history.

growth percentages than in already industrialized countries (Figure 3–24). Much of the industrial progress of the less developed countries is concentrated in the small group of NICs that form the upper tier of developing nations. No specific definition exists for inclusion as an NIC, but the usual list includes Hong Kong, Singapore, the Republic of Korea (South Korea), and Taiwan in Asia. These are sometimes referred to as the "four tigers" of Asia because of their dramatic industrial and income growth. Indeed, the question can be asked whether these countries should still be considered "developing." Other countries that may be included are Mexico, Brazil, and Chile in Latin America and other Asian countries such as Malaysia and Thailand. We should also recognize that the absolute increase in production is often higher

in the more developed countries because they have a much larger industrial base to begin with.

Per Capita Inanimate Energy Consumption One of the characteristics of the Industrial Revolution was a shift from animate power (human or beast) to inanimate energy—initially mineral fuels and hydroelectricity and, more recently, nuclear power (Figure 3–25). The degree to which a country is able to supply inanimate energy from internal sources or to import it is an important indicator of applied modern technology and, consequently, of productivity. Just as per capita GNI measures productivity in terms of value, per capita inanimate energy use measures production in terms of power expended.

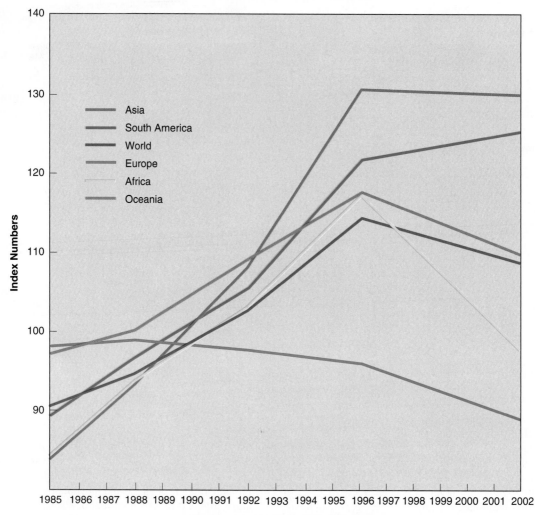

■ **FIGURE 3-23 Per capita food production for selected regions.** An index number relates the production of a specific year to that of the base period. If the index number is 110, it means that production (in this case, per capita production) is 10 percent greater than it was in the base period (in this case, 1989–1991). According to the data depicted here, per capita food production has increased most in Asia and South America, two of the regions most impacted by the Green Revolution. There has been an increase in food supplies for the world as a whole, despite continuing population growth.
Source: Data compiled from *Food and Agriculture Organization, http://appsl.fao.org/servlet/XteServlet.jrun?Areas.*

Thus, energy consumption is closely related to economic activity. Low per capita energy consumption is associated with subsistence and other nonmechanized agricultural economies; high per capita energy use is associated with industrialized societies. Intermediate levels of energy use characterize regions that have both industrialized urban centers and more traditional nonmechanized rural areas. The distribution of per capita inanimate energy consumption is shown in Figure 3–26. The similarity of that distribution to the distribution of per capita GNI (see Figure 3–20) can be seen easily by comparing the two maps.

Other Measures Although GNI, energy use, and agricultural labor force are most often used to identify more developed and less developed countries, other measures are

occasionally employed. Two that indicate quality of life are life expectancy and food supply.

The life-expectancy measure would seem to be the ultimate indicator of development. Because some cultures are less materialistic than others, the standard measures of wealth may mask some important cultural attitudes of a society. All cultures, however, value the preservation of life, and, at least in part, life expectancy is a measure of the end result of economic activity. It tells how well a system functions to provide life support through the provision of medical care and improved sanitation. A great range of average life expectancies characterizes the nations of the earth today. Owing to a number of factors, including the ravages of the AIDS epidemic and continuing poverty, the average life expectancies in some of the southern African nations have now

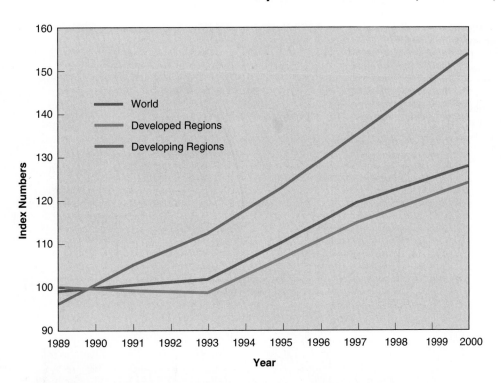

■ **FIGURE 3-24 Industrial production for developed and developing countries.**
Industrial production index values have increased more rapidly in the developing nations than in the developed countries. There are three reasons for this: (1) industrial production in many developing countries is limited, so increases often loom large relative to the base; (2) selected developing countries have been immensely successful in promoting industrial growth; and (3) many developed countries have experienced an economic restructuring in which industrial activity is of decreasing significance relative to other modern economic sectors. Please note that the index numbers are based on a 1990 value of 100.0.
Source: Data compiled from *Statistical Yearbook* (New York: United Nations, 2002).

fallen back to less than 40 years—a shocking and extremely troubling development that none would have predicted just a generation ago (Figure 3–27, page 82). On the other hand, human life expectancy for the entire world has climbed to 67 years, and in a number of countries has now reached the upper 70s and low 80s. Perhaps no single measure better expresses the growing gap between the more and less developed nations.

Two measures of food supply are also fundamental: the number of calories available is an indicator of dietary quantity, and protein supply is an indicator of dietary quality. Adequate quantity is at least 2,400 available calories per person per day. Adequate protein supply is attained if at least 60 grams of protein are available per person per day. Figure 3–28 (page 84) depicts the caloric intake in large areas of the world. A similar pattern prevails with respect to protein supply.

■ **FIGURE 3-25 Traditional and contemporary transport technologies.** Many cities in the less developed nations are characterized by crowded, narrow streets exhibiting multiple forms of human transportation. This scene is from Dhaka, the capital of Bangladesh.

■ **FIGURE 3-26 World per capita consumption of inanimate energy, expressed in oil equivalents.** Per capita energy consumption is a measure of development that indicates the use of technology. Because most forms of modern technology use large amounts of inanimate energy, countries that use small amounts of that type of energy must rely principally on human or animal power.
Source: Data compiled from The World Bank, *World Development Report 2000–2001* (New York: Oxford University Press, 2001).

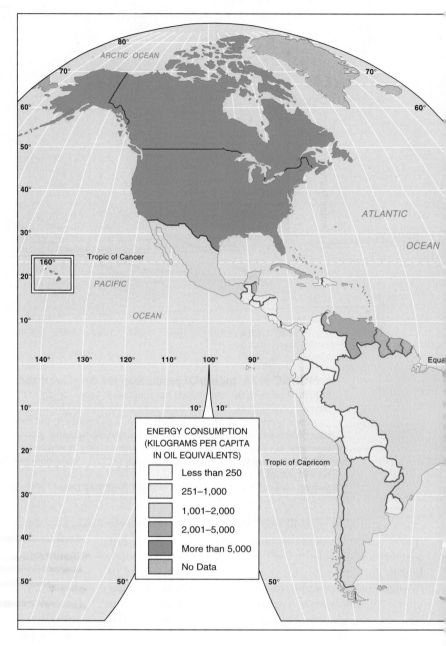

Combined Measures Of course, no single measure tells the full story of development. Many scholars have attempted to combine measures to provide aggregate indices of development that, it is hoped, will be more informative.

One such quantitative index that is used widely to measure and compare the overall development levels of the nations and regions of the world is called the **Human Development Index (HDI)**. The index is derived from three measurable variables: life expectancy at birth, educational attainment, and income, with the score for the highest possible level of development being 1.000 and the lowest 0.000. Table 3–4 (page 84) presents HDI scores for the less and more developed regions of the world. Among the less developed regions of the world, Africa South of the Sahara is ranked lowest and Latin America the highest. While these figures will change slightly from

year to year, they are nevertheless a useful indicator of the comparative levels of development of the regions we will be studying in our text.

A second, relatively new approach by Hoeschele[2] attempts to classify the wealth and poverty of regions based on the proportions of the population not working in agriculture and the types of exports produced. In simplified form, he first divided the world into core and non-core countries. Core countries are those that have a large share of corporate headquarters and from which a significant amount of foreign direct investment originates. Basically, core

[2]W. Hoeschele, "The Wealth of Nations at the Turn of the Millennium: A Classification System Based on the International Division of Labor," *Economic Geography* 78 (2002): 221–244.

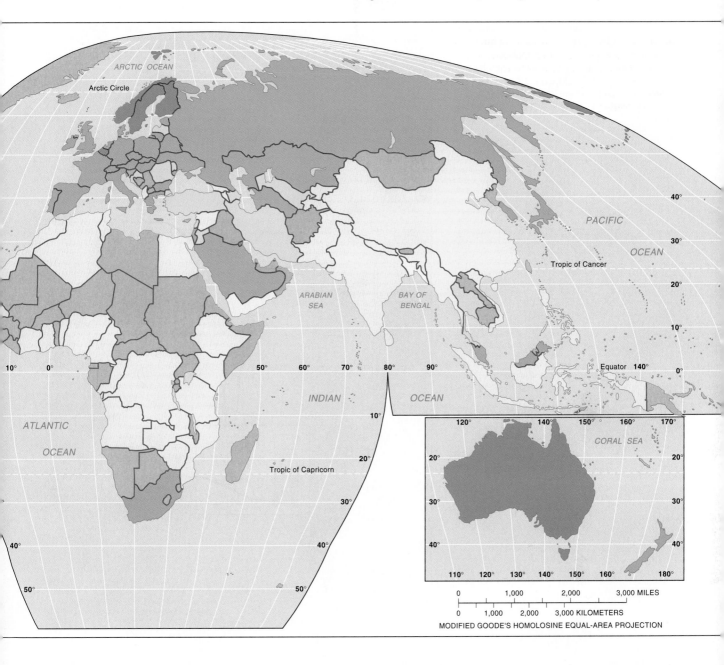

ARCTIC OCEAN

Arctic Circle

PACIFIC

OCEAN

Tropic of Cancer

40°

30°

20°

ARABIAN
SEA

BAY OF
BENGAL

10°

10°   0°   Equator   140°   0°

INDIAN   OCEAN

ATLANTIC

OCEAN

CORAL SEA

120°   140°   150°   160°   170°

20°   20°

Tropic of Capricorn

30°   30°

40°   40°   40°   40°

110°   120°   130°   140°   150°   160°   180°

50°   50°

0   1,000   2,000   3,000 MILES

0   1,000   2,000   3,000 KILOMETERS

MODIFIED GOODE'S HOMOLOSINE EQUAL-AREA PROJECTION

countries are those that we associate with the developed world, including countries such as Switzerland, the Netherlands, Japan, Sweden, Canada, and the United States; non-core countries are all the rest. Second, he then introduced variables related to types of exports, which produced a classification system yielding five types of countries:

- **Industrialized countries that focus on the export of core manufacturing products.** Typical products are textiles, metals, manufactures, chemicals, and paper. Core countries include United States, Canada, most of the western European countries, Japan, and South Korea. Non-core countries include Portugal, Estonia, Greece, and other countries that we think of as only recently having achieved economic prominence as industrial and commercial powers.

- **Industrialized countries that focus on export of natural resources**, including agricultural products, minerals, basic metals, and fuels. Australia, New Zealand, and Norway belong to the core; but most of the countries embraced by this category have achieved wealth through the export of precious commodities such as oil, with Saudi Arabia and Venezuela serving as examples. Indeed, in some cases, it would not be unfair to question the industrial capacity of some of these countries.

- **Partially industrialized countries that emphasize the export of core manufacturing products.** Examples are all found in the non-core part of the world and include Mexico, Malaysia, Philippines, Turkey, and similar countries. To a significant degree, the industrial goods that these countries export

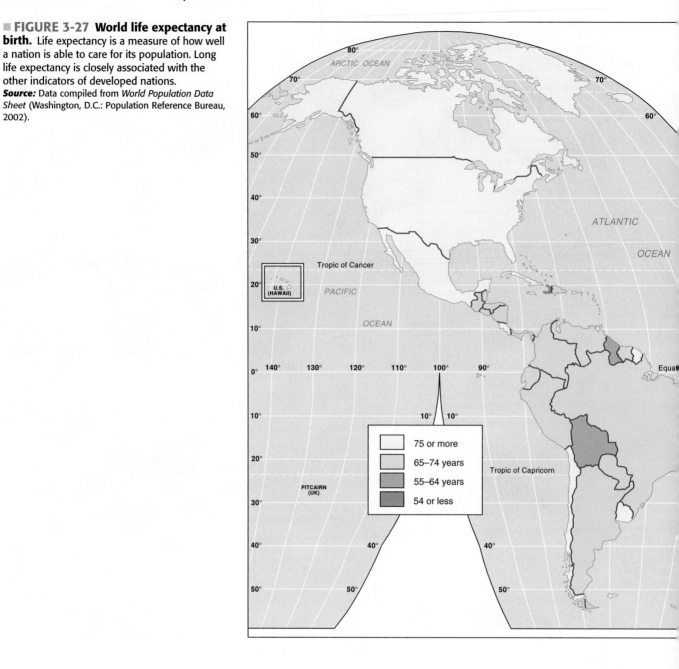

■ **FIGURE 3-27 World life expectancy at birth.** Life expectancy is a measure of how well a nation is able to care for its population. Long life expectancy is closely associated with the other indicators of developed nations. **Source:** Data compiled from *World Population Data Sheet* (Washington, D.C.: Population Reference Bureau, 2002).

are from assembly plants using relatively low-wage labor.

- **Partially industrialized countries that emphasize the export of natural resources.** Examples include countries such as El Salvador, Zimbabwe, Moldova, Morocco, and Syria.

- **Least industrial countries that export natural resources.** This is definitely a non-core part of the world and embraces some of the least-developed economies in existence, including Bhutan, Mozambique, Kenya, Senegal and others. Indeed, a large proportion of African countries fit into this category.

As is the case with all classification schemes, Hoeschele's model has its strengths and weaknesses. For one thing, he

is more interested in wealth specifically than in development generally. He does show, however, just how complex a hands-on definition of developmental regions can become.

Development Mapped

With reference to the measures and indices that we just described, we can construct a highly generalized map of the more developed and less developed countries in the world (Figure 3–29, page 86). These patterns show the United States and Canada, Europe, Russia and the Eurasian states of the former Soviet Union, Australia and New Zealand, and Japan as those regions that most fully exhibit the attributes of technological development that we have discussed. Conversely, those world regions that

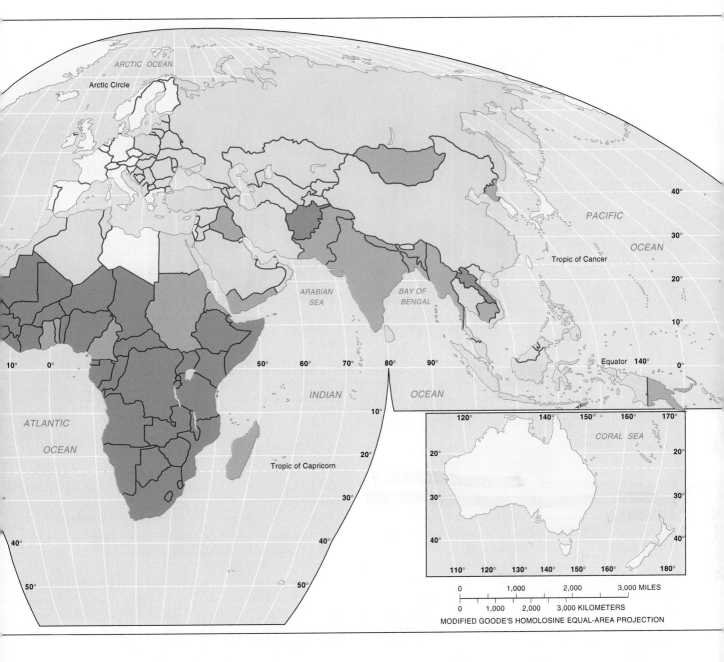

MODIFIED GOODE'S HOMOLOSINE EQUAL-AREA PROJECTION

collectively are characterized by lower levels of the attributes of technological development can be said to compose the less developed world. Included in this realm are East Asia, Southeast Asia, South Asia, the Middle East and North Africa, Africa South of the Sahara, and Latin America and the Caribbean.

As mentioned above, such a map is highly generalized, so that some more developed nations—Israel, South Korea, and Taiwan, for example—are lumped with less developed countries; and some less developed nations—such as Albania, Uzbekistan, and Kyrgyzstan—are included in the more developed regions. Some countries are classified as more developed by all measures, and many are classified as less developed in all ways. A number of countries fall among the more developed in some categories and among the less developed in others.

Summary Characteristics of More Developed Regions

We have already noted several characteristics of more developed countries: high per capita GNI and energy use; a small part of the labor force in primary activities and a consequent emphasis on secondary and tertiary occupations; a longer life expectancy; and a better and more abundant food supply. More developed countries also tend to have a lower rate of population growth, which is primarily the result of having a highly urbanized population for many generations. To a large degree, those countries are found within Stage IV of the demographic transformation (discussed in Chapter 2). In addition, more developed countries share certain other economic and cultural characteristics.

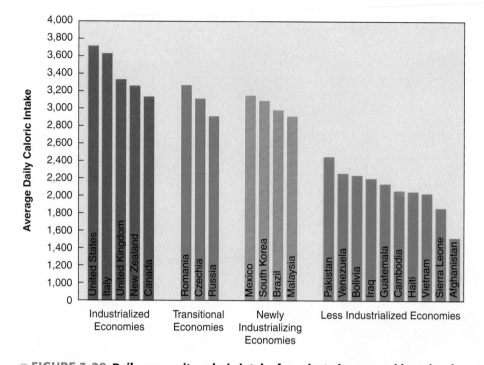

■ **FIGURE 3-28 Daily per capita caloric intake for selected more and less developed countries.** Food supply is a measure of well-being in the most basic sense. Daily caloric intake varies greatly according to level of national economic development.
Source: Data compiled from Food and Agriculture Organization, *http://apps.fao.org/lim500/wrap.pl?* Food Balance Sheet & Domaine. Data are for the year 2000.

Economic Characteristics The most basic economic characteristic of the more developed world is a widespread use of technology (Figure 3–30). In those regions, the fruits of the Agricultural, Industrial, and Information Revolutions are widely applied, and new techniques are quickly adopted and diffused. New technologies create new resources and expand existing ones, thus increasing opportunity for still greater generation of wealth. Advanced technology also leads to increased labor productivity and the creation of an improved infrastructure, including roads, communications, energy and water supply, sewage disposal, credit institutions, and even schools, housing, and medical services. Those support facilities are necessary for accelerated economic activity and for specialization of production, which further enhances productivity.

The heavy dependence of the more developed countries on minerals further differentiates them from the less developed countries. Not only are industrialized countries in an iron-and-steel age, but they are also in a fossil-fuel age, a cement age, a copper-and-aluminum age, a silicon age—the list is almost endless. To a greater and greater degree, more developed countries are importing those minerals from less developed countries and then exporting manufactured goods and, in some cases, food. That practice has led to trade surpluses for most of the more developed countries and trade deficits for many of the less developed countries because the price of raw materials (with the exception of petroleum) has increased more slowly than the price of manufactured goods. The more developed countries gain even more revenues by investing in less developed countries. This outside removal of wealth has encouraged some less developed countries to nationalize or restrict foreign investments.

New technologies, high productivity, and a favorable trade balance result in higher personal and corporate incomes. Consequently, many individuals in more developed countries need to spend only a part of their income for food and shelter; other income can be used for services, products, and savings—where it stimulates further economic growth. For entrepreneurs (that is, organizers who gather together labor, financing, and all those other things necessary to construct an effective economic activity), more developed countries offer numerous infrastructural advantages.

Table 3-4 Human Development Index Values by Region

Region	HDI
Latin America and Caribbean	0.797
East Asia and the Pacific	0.768
Arab States	0.679
South Asia	0.628
Africa South of the Sahara	0.515
All developing countries	0.694
Industrialized countries	0.892
World	0.722

Source: Data compiled from United Nations Development Program, *Human Development Report 2005* (New York: Oxford University Press, 2005).

Table 3-5 Adult Literacy Rates by Region

Region	Adult Literacy Rate
South Asia	58.9
Africa South of the Sahara	61.3
Arab States	64.1
East Asia and the Pacific	90.4
Latin America and Caribbean	89.6
All developing countries	76.6
Industrialized countries	99.2
World	60.8

Source: Data compiled from United Nations Development Program, *Human Development Report 2005* (New York: Oxford University Press, 2005).

Cultural Characteristics   One cultural attribute of the more developed nations is a high level of educational achievement (Table 3–5). Education is enhanced by advanced communications technologies that support the rapid diffusion of new ideas. In addition, education is geared partly toward economic advancement; disciplines such as engineering, geography, economics, agronomy, and chemistry have obvious and direct applicability to resource development. An educated populace is an extremely important resource, for it more readily accepts change. And acceptance of change means that technology is more easily adopted, new products and services are welcomed, and the mobile population can more readily take advantage of opportunities in other areas of the country. Other cultural attributes common to the more developed countries include the presence of a large and economically strong middle class and an expectation that government officials will be honest and accountable to the public for their actions.

Summary Characteristics of Less Developed Regions

We have already learned that less developed nations are characterized by low per capita GNI and energy use, a high proportion of the labor force in primary pursuits, a comparatively short life span, and a diet often deficient in quantity and quality. In addition, less developed countries commonly display a variety of other characteristics.

Population Characteristics   Most less developed countries have a relatively high rate of population growth, the result of a continuing high birthrate and a declining death rate (see Figure 2–9). As a consequence, the age structure of the population of a less developed country is different from that of a more developed country. Figure 3–31 (page 88) illustrates this difference graphically. In less developed countries, a substantial portion of the population is youthful and only a small segment is elderly. In more developed nations, the opposite is true. In a sense, then, per capita comparisons are unfair because all people are counted equally, yet in the less developed nations, a smaller proportion of the population is made up of mature laborers. A large youthful population also foretells continued population growth.

The status of women in the diverse cultures of less developed countries often excludes them from many occupations (see the *Geography in Action* boxed feature, Women, Development, and the Gender-related Development Index, page 90). In the rural sector, women are often expected to work with men in the fields during periods of peak labor requirement while still maintaining the home and caring for the children. While at home, women may engage in some craft industry, such as weaving for household use and for sale in the local market. In addition, many rural women in the less developed world play a pivotal role in marketing the family's surplus on a daily or weekly basis (Figure 3–32, page 88). In the cities, women find employment as domestics, secretaries, teachers, nurses, and, more recently, industrial workers. For many illiterate women, however, non-domestic employment opportunities are not available.

Cultural Characteristics   We have previously noted that the literacy rate in less developed countries is generally lower than in more developed nations. Many of the older generation of rural inhabitants of the developing nations cannot read or write or can do so only at minimal levels. Although literacy campaigns have made great strides in cities and among the younger generations in recent decades, large numbers of urban poor still lack the ability to read effectively and remain functionally illiterate. That inability limits great numbers of people from learning new technologies and from engaging in more remunerative occupations.

The cultures of many of the less developed nations are also often more conservative and resistant to change, thereby presenting a fundamental paradox. Those cultures, by and large, wish to preserve their customs and mores, yet they want to partake of the material benefits that Western society enjoys. Clearly, economic development leads to cultural change, to the possible destruction of traditions, and to the acquisition of new behavior patterns. Sometimes the family loses part of its cohesiveness, villages become subservient to larger urban centers, labor specialization and regional specialization lead to commercialization of the economy, and loyalties to the local community give way to national allegiances. These changes and others are inevitable when economic development occurs. Consequently, social and cultural disruption is a common characteristic of less developed countries, as old and new ways of life come into conflict.

Along the same lines, we can associate certain psychological characteristics with extreme levels of underdevelopment. As mentioned above, one characteristic is a distrust of change and the people who promote change. Subsistence agriculturalists, for example, must avoid miscalculations in their crop production for fear of starving, and taking a risk on a new idea or procedure could lead to devastating consequences. A second characteristic is fatalism about the world—lives are controlled by outside forces over which people have no control, such as a vengeful god or nature's wrath. On a related note, the ability to empathize is largely absent—how can the subsistence farmer ever dream of doing something else? Third, thinking tends to be pre-scientific, with little understanding that a physical cause can have a

■ **FIGURE 3-29 More developed and less developed regions of the world.** This map generalizes the patterns of the preceding maps, graphs, and tables by dividing the world into more developed and less developed regions.

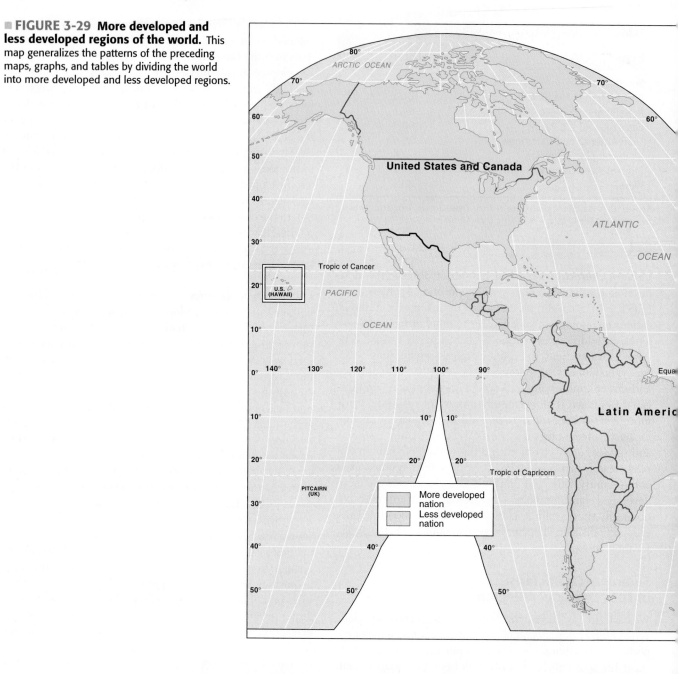

■ **FIGURE 3-30 High-technology manufacturing.** The high technological achievement of Japan is typical of the more developed societies of the world.

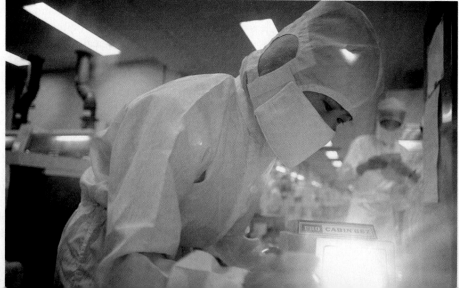

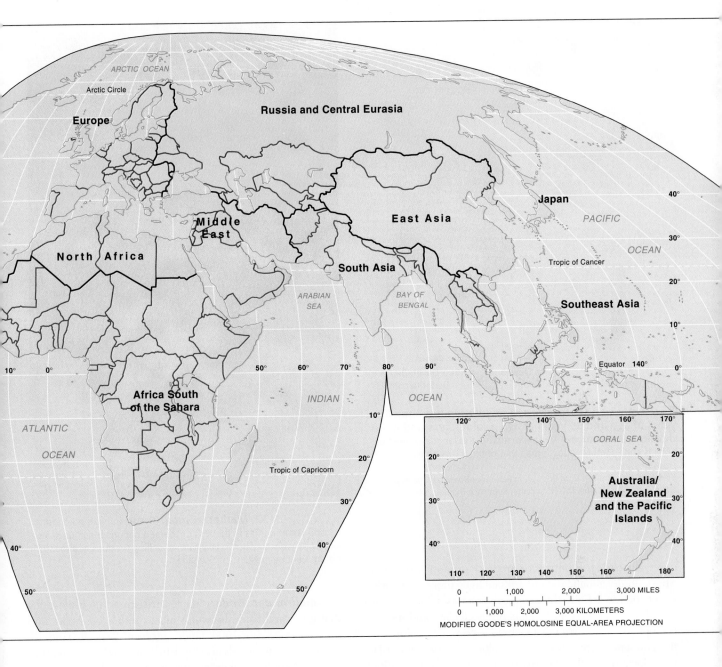

MODIFIED GOODE'S HOMOLOSINE EQUAL-AREA PROJECTION

physical effect. Finally, we can add to the list a restricted worldview, inability to defer gratification, limited aspirations, and a lack of innovativeness. Of course, we are generalizing here and not everybody living in the developing world embraces all, or even some, of these characteristics. But to the extent that they exist, they make even more daunting the challenge of addressing the problems of underdevelopment.

Development Explained (at Least in Theory)

We have dedicated much print to understanding the characteristics and defining the global dimensions of development, but relatively little attention to examining the theories intended to help us understand why development occurs in

some cases and why in other cases underdevelopment can never seem to be overcome. A vast literature exists that attempts to explain underdevelopment, but we will confine ourselves to a general treatment of some of the more prominent theories. Keep in mind that no theory is accepted by everyone and indeed some theories are strongly opposed.

Control One of the earliest theories was **environmental determinism**, which received considerable attention during the 1920s and 1930s but has been largely discredited since then. One of the most effective proponents of that theory was Ellsworth Huntington (1876–1947), who wrote or coauthored twenty-eight books, including *The Pulse of Asia* and *Mainsprings of Civilization*. The premise of environmental determinism is that the physical environment,

Percentage of Total Population

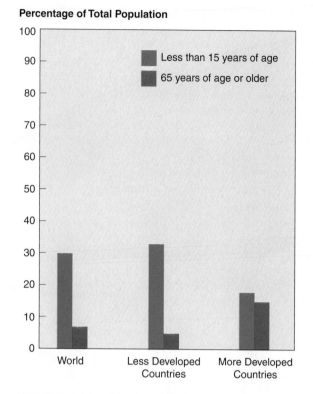

■ **FIGURE 3-31 Percentage of population less than 15 years of age and 65 years of age or older: less and more developed countries.** The population of the less developed nations is considerably younger than that of the more developed countries.
Source: Data compiled from Population Reference Bureau. *World Population Data Sheet.* Washington, D.C., Population Reference Bureau, 2002.

■ **FIGURE 3-32 Market woman.** The selling and purchasing of food commodities in the less developed nations is carried out largely by women. Shown is a woman selling produce at a market on the Caribbean island of Grenada.

especially climate, controls or predestines human behavior. Mid-latitude climatic regions are said to be more stimulating for economic activity than tropical climates, which are believed to make their inhabitants physically lazy and morally corrupt. While it is true that maps of economic development do show a majority of the more developed nations in mid-latitude locations and a majority of the less developed nations in the tropics, we must look beyond climatic differences to the factors discussed previously to account for the spatial patterns of underdevelopment.

Another control theory is **cultural determinism**. According to that theory, which is popular in some quarters, a person's range of action is determined largely by his or her culture. In other words, if the culture emphasizes a work ethic, most members of that society will work. Conversely, if the possession of worldly goods is not among a culture's priorities, economic performance will probably be low.

We could easily combine those two control theories and conclude that harsh environments and cultural barriers frequently work together to ensure a vicious cycle of poverty. But such a simplistic view overlooks the fact that many people may be achieving at low levels merely because they have not had the opportunity to do otherwise.

Colonialism and Trade   **Mercantilism** is the philosophy that governed trade between most mother countries and their colonies after 1600. Under mercantilism, the colonies supplied raw materials and foodstuffs needed by the mother country, and the mother country, in turn, used its colonies as a market for finished products and other surpluses. To ensure that the colonies did not compete with the mother country, they were prohibited from producing goods that the mother country had in abundance. Moreover, the colonies were permitted to trade only with the mother country. That arrangement obviously worked to the advantage of the mother country and severely limited the economic growth of the colonies.

Colonialism in the traditional sense of the word is largely a thing of the past. In the years immediately after World War II, most colonial powers gave up, or were forced to grant independence to, their remaining possessions. Yet many of the trade relationships and patterns of that era have been retained, a condition often called **neocolonialism**. Neocolonialist relationships are cited frequently as an explanation for the continuing uneven distribution of wealth. Many

■ **FIGURE 3-33 Multinational corporate influence in the global economy.** This scene is from the city of Xalapa, Mexico.

less developed countries, it is argued, carry from the colonial era a legacy of economic dependence on the more developed regions of the world. **Dependency theory** advocates believe that present conditions of underdevelopment in the poorer countries are attributable primarily to the ongoing perpetuation of the inequitable trade relationships of the colonial past, with multinational corporations and local elites replacing the former colonial masters as exploiters of the lower classes.

A variation of dependency theory that emphasizes the spatial dimensions of global trade patterns is the **core-periphery model**. This model, which is derived in part from the pioneering work in **world-systems theory** by Wallerstein,[3] argues that the nations of the world can be divided into two groupings: a "core", which consisted originally of the countries of western Europe (not unlike the core described above in our classification system), and a "periphery" comprised initially of Africa, Asia, and Latin America. As with dependency theory, the core-periphery model holds that the economic development of the core has been, and continues to be, achieved through trade relationships that work to the disadvantage of the less-industrialized regions of the periphery (Figure 3–33).

Underdevelopment is not as simple as any single theory. The disadvantage of less developed countries in the traditional trade relationships cannot be denied. However, some less developed countries were never totally controlled as colonies; others have been independent for centuries and yet have not advanced economically as much as neighboring nations that experienced a similar colonial past. Furthermore, it must be recognized that exploitation can come from internal as well as external sources. If we are to understand an individual country's place in today's interdependent world, we must examine its precolonial, colonial, and postcolonial experiences.

Circular Causation Another economic concept known as **circular causation** results in either a downward or an upward spiral. An example of the downward spiral is a farm family that produces barely enough to feed itself. Because the family has little or no savings, it has nothing to fall back on if a minor crop failure reduces the harvest. Its members simply have less to eat, so they work less, produce less, and then have even less to eat. The upward spiral can also be illustrated by a farm family. Perhaps the family produces only enough to feed itself but, by good fortune, obtains some capital to buy fertilizer, which increases crop yields. As a result, the family eats better, works harder to produce more, and sells the surplus, thus increasing its capacity to buy more fertilizer as well as improved seed. That theory is equally applicable to groups and nations.

Stages Theory Walt Rostow compared historical economic data and came up with the idea that there are five **stages of economic growth**.[4] In the first stage we find traditional society. Most workers are in agriculture, have limited savings,

[3]I. Wallerstein, *The Modern World-System: Capitalist Agriculture and the Origins of the European World-Economy in the Sixteenth Century* (New York: Academic Press, 1974): *The Capitalist World-Economy* (Cambridge: Cambridge University Press, 1979).

[4]Walt W. Rostow, *The Stages of Economic Growth: A Non-Communist Manifesto*, 2nd ed. (Cambridge: Cambridge University Press, 1971).

Women, Development, and the Gender-related Development Index

Because people are the greatest resource of any nation or region, it is important that the status and treatment of women, who constitute more than half the world's population, be considered in any study of development. Women have long been the objects of masculine discrimination and prejudice in many societies. In ancient times, and in a few cultures even today, they were treated as little more than the personal property of men, to be acquired, used, and in many cases discarded at the whim and pleasure of the husband or father. Some of the more outrageous expressions of mistreatment have and continue to be the systematic abortion of fetuses that are believed to be female and the practice of female infanticide, or the killing of newborn baby girls whose parents are determined to have male heirs. Other forms of violence include various forms of emotional and physical abuse.

Nonviolent attempts to inhibit the full development potential of women are also widespread. In many traditional societies, women are prohibited from holding offices of authority, and at present, females occupy only 15 percent of the seats in national parliaments or congresses. Although much progress has been realized in recent decades, education has often been intentionally withheld from girls by male authorities who feared that knowledge would ultimately remove women from their traditional home-centered lives. Literacy rates among adult women are far lower than those of their male counterparts in many less developed countries, and men continue to be given preference in admissions to trade schools and universities. Because women in many nations are treated legally as minors in the care of their husbands, they are also the objects of legislation that prohibits or limits their ability to file for divorce, travel, inherit land, or secure loans and other financial services. Only 5 percent of multilateral loan funds for Third World rural development projects, for example, are currently channeled to women.

One reason why these figures are so low is that most women in the less industrialized nations have traditionally worked in the **informal economic sector**, which consists of jobs such as street and market vending, small business operation, and domestic work such as cooking, cleaning, and child care that are not covered by national labor and employment laws. Limited in their earning potential, many live out their lives with no hope for economic advancement. Even those who are fortunate enough to receive an education and secure good career positions often find that they are paid less (an average of 75 percent in the industrialized nations and even less in the developing nations) for the same work than their male counterparts. All of these forces have resulted in 70 percent of the world's adults living in poverty being women, with the figure becoming constantly worse—a process widely referred to as the **feminization of poverty**. Needless to say, with women being the primary caregivers to children, these trends have frightening implications worldwide for child health and nutrition. Yet another consequence of these trends is the so-called **female double-day workload** in which women are expected to put in a full day of work outside the home and then to continue to do the normal levels of cooking, laundry, cleaning, and child rearing once they return home. Recent research in Latin America, for instance, found that women working outside the home still averaged 70 hours of work per week at home while their husbands were willing to put in an average of only 5.5 hours per week assisting with household tasks.

For these and other reasons, the United Nations Development Program has now begun to publish annually a Gender-related Development Index (GDI) as a companion to its broader Human Development Index (HDI). The Gender-related

---

and use age-old production methods. Indeed, all of the characteristics of a truly poor society are exhibited.

"Preconditions for takeoff" are established in the second stage, which may be initiated internally by the desire of the people for a higher level of living or externally by forces that intrude into the region. In either case production increases, perhaps only slightly but enough to cause fundamental changes in attitudes, and individual and national goals are altered.

Takeoff occurs in the third stage, when new technologies and capital are applied and production is greatly increased. Manufacturing and tertiary activities become increasingly important and result in migration from rural areas to bustling urban agglomerations. Infrastructure facilities are improved and expanded, and political power is transferred from the landed aristocracy to an urban-based elite.

The fourth stage is the "drive to maturity," a continuation of the processes begun in the previous stage. Urbanization progresses, and manufacturing and service activities become

increasingly important. The rural sector loses much of its population, but those who remain use mechanized equipment and modern technology to produce large quantities.

The final stage is high mass consumption. Personal incomes are high, and abundant goods and services are readily available. Individuals no longer worry about securing the basic necessities of life and can devote more of their energies to noneconomic pursuits.

It is difficult to place a specific country among Rostow's stages. Furthermore, in large countries, different regions often exhibit different levels of development. Nevertheless, we can speculate on the positions of some countries within the model. For example, in the United States, citizens of European descent were never really in the first, or traditional, stage. The country was settled and developed by Europeans whose homelands already exhibited preconditions for takeoff. The United States approached takeoff by the 1840s, and by 1865 it began its drive to maturity, which

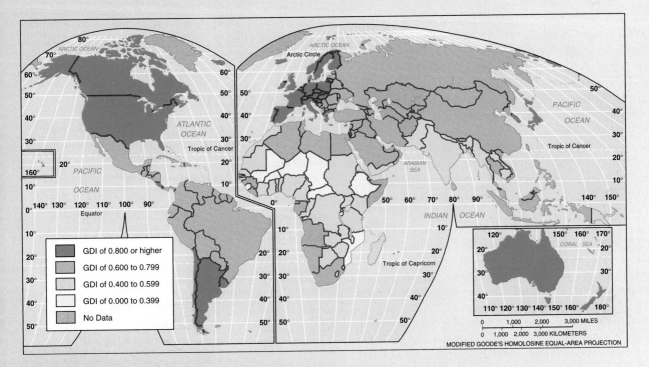

■ FIGURE A A gender-related development index world map.
Source: Data compiled from United Nations Development Program, *Human Development Report,*
http://hdr.undp.org/reports/global/2002.

Development Index takes into account the same development indicators as the Human Development Index but also factors in penalties for gender inequality as it expresses itself in both economic and noneconomic ways. Figure A shows the national GDI rankings with a score of 1.000 representing full gender equality as measured by the GDI and a score of 0.000 representing a total absence of gender equality. While there are differences in the individual GDI rankings and those of other measures of development used in this chapter, it is significant to note that the broad regional patterns are similar, with most of the technologically less developed nations of Africa, Asia, and Latin America also ranking low on the GDI scale. This would suggest that there is, indeed, a correlation between economic and noneconomic development and that the less developed nations will never achieve their full development potential until they remove the barriers to female opportunity and advancement.

was achieved, according to Rostow, by 1900. Even though development was achieved rapidly, great regional variations persisted.

The United States, Japan, Canada, Australia, and many of the nations of northwestern Europe, such as Germany, France, and the United Kingdom, are now in the stage of high mass consumption. Russia and many southern and eastern European countries are in their drive to maturity. Nations in the takeoff stage include Mexico, Brazil, Chile, Argentina, and Venezuela in Latin America and South Korea, Taiwan, Malaysia, and Thailand in East Asia. Countries such as Kenya (in East Africa), Nigeria (in West Africa), and Indonesia (in Southeast Asia) are displaying preconditions for takeoff, while Haiti (in Latin America) and some of the countries of Central Africa and Southeast Asia are still exhibiting the traits of traditional societies. No time frame can logically be specified for passage through any of the stages of the Rostow model, and it is possible

that a nation may actually regress, or slip backward, for a period of time.

The Lacostian Theory   The French geographer Lacoste has offered several cautions for any interpretation of developmental lag.[5] One is to avoid the view that population growth per se causes underdevelopment. According to Lacoste, population growth may be a corollary to lag, along with poverty, but it is not causal. People (population) may be either assets or liabilities in the development process, depending upon whether they are productive or not. Many of the world's most densely populated countries, for example, also rank among the most prosperous nations. Conversely, many of the poorest countries are also very sparsely

[5]The several works of Lacoste are reviewed in H. A. Reitsma and J. M. G. Kleinpenning, *The Third World in Perspective* (Assen/Maastricht, Netherlands: Van Gorcum, 1989): 223–236.

populated. Lacoste further cautions against any view of external forces as the sole impetus for developmental lag. For example, although colonialism has contributed significantly to the problem of underdevelopment, he does not see it as the sole cause. Development problems may also stem from the unwillingness or inability of ruling elites to foster the political, social, and economic infrastructure essential to developmental progress. Lacoste points to their adoption of an "adulterated" capitalist system in conjunction with remnant feudal power relationships. Thus, the **Lacostian view** sees underdevelopment as the result of complex and interacting forces—of both internal and external origin.

In brief, the theories described here are presented in simplified form. Each may have some value in understanding a particular situation, but as a generally acceptable view of development theory, each also has shortcomings. Indeed, of the foregoing notions, all implicitly attribute poverty and related problems to a lack of development. Another viewpoint—arising from the belief that development theory of the past 50 years has failed to provide adequate explanation or solution to global problems—is that development itself, as practiced, has created scarcity and disparity.[6]

Summary

In Chapter 1, we looked at what geography is all about, and the nature of social and economic change. In Chapter 2, our focus was on the meaning of development, globalization, and sustainability. In this chapter, we examined the physical and cultural contexts of development, followed by a survey of development measures and theories. We learned that the combination of climate, vegetation, soils, resource endowments, and landforms create "plus" places that provide resources that favor development; "edge" places where water and land meet that often support dense populations and high development potentials; "rich" places that take advantage of local resource endowments, such as mineral deposits; and of course places that, for a variety of reasons, are largely devoid of people. In most cases, these places reflect millennia of cultural diffusion and inheritance from initial culture hearths; they also reflect a continuing accumulation of technological innovations and cultural practices that shape the developmental settings of countries and regions. Finally, we looked at the value of per capita income, agricultural employment, energy use, and other measures as indicators of development, and we learned that while a number of theories exist to explain the development process, each is limited in its explanatory power and none suffices as a complete explanation of development.

Beginning in the next unit, we will start our analysis of world regions. We will want to keep in mind the principles and explanations provided in these introductory chapters to help us understand the highly variable developmental circumstances of these regions. Indeed, we may discover that

we periodically revisit these introductory chapters for reminders about concepts and processes that enable us better to understand real-world geographical situations.

Key Terms

acculturation
alluvium
average annual precipitation
calcification
circular causation
climate
core-periphery model
cultural convergence
cultural determinism
culture complex
culture hearth
culture realm
culture trait
dependency theory
diffused culture
ecosystems
environmental determinism
evapotranspiration rate
female double-day workload
feminization of poverty
fossil fuels
frost-free period
GNI gap
Green Revolution

Gross National Income in Purchasing Power Parity (GNI PPP)
Human Development Index (HDI)
informal economic sector
inherited culture
Lacostian view
landscape modification
laterization
lingua franca
loess
mercantilism
natural vegetation
neocolonialism
newly industrializing countries (NICs)
podzolization
primary level of economic activity
regional disparity
salinization
secondary level of economic activity
stages of economic growth
world-systems theory

Review Questions

1. Define geography in the context of the following terms: spatial, holistic, interrelationships.
2. Describe the three patterns of change that influence the development process, and explain how geographers would examine these patterns.
3. Define the term *development*. How is development typically measured? Are these measures adequate?
4. Which regions of the world do we normally associate with "less developed"? What are some of the common attributes of the less developed world?
5. What is an economic revolution? Describe how the great revolutions transformed the geography of global societies and economies, giving emphasis to the primary, secondary, and tertiary levels of economic activity. Where does the Green Revolution fit in?
6. What is globalization? Who are globalization's actors? Does globalization mean an end to the importance of localities? Explain.

[6]Lakshman Yapa, "What Causes Poverty? A Postmodern View," *Annals of the Association of American Geographers* 86 (1996): 707–728.

7. How can the concept of sustainability be related to the development process? What kinds of development are NOT sustainable?

8. Physical geography contributes to the development of "plus," "edge," and "rich" places. Explain.

9. What principles might one employ to minimize land degradation?

10. What is the Human Development Index and how is an understanding of it helpful in assessing the current level of development of a nation or region?

Geographer's Craft

1. What is a geographic information system (GIS) and what are some of the potential uses of GIS?

2. You are a budding businessperson and you want to set up a firm that manufactures exotic handcrafted wall decorations made out of a combination of bamboo and mahogany. How would the forces of globalization influence your thinking about where you would locate the different components of your firm?

3. Your government has given you the task of building a dam on a river. Such a dam will clearly contribute to the economic development of the country by providing badly needed hydroelectric power. What factors should you keep in mind as you attempt not only to address short-term electrical needs, but long-term sustainability issues?

4. How would you explain to a class of elementary school students the fundamentals of what geographers do?

5. Explain why the greatest resource of any region or country is its people. Then comment on the development implications of the wasting of human resources through such conditions or practices as rigid social class stratification, poor educational systems, political corruption, and gender inequality.

PART TWO

United States and Canada

Merrill L. Johnson

In the process of settlement and development, both the United States and Canada have marshaled the rich resources of an enormous domain. As a result, the two countries are among the most prosperous and most highly developed nations in the world.

In Chapter 4, we examine the physical, territorial, and demographic contexts in which these two countries evolved. We begin with a review of landform and climate regions. We follow with an examination of the territorial evolution of the two Anglo-American states—the United States and Canada—and an analysis of how, and by whom, these countries were settled. We look with special interest at immigration, since both countries are populated by immigrants who shaped the economies and ethnic composition of their nations.

In Chapter 5, we investigate the geography of economic development in the United States and Canada, with emphasis on the spatial evolution of agricultural, industrial, and urban systems. The highly productive agricultural systems of these two countries emerged from a rich natural resource base, but are experiencing major changes in the twenty-first century. The industrialization of the United States and Canada occurred first in a concentrated core-land but now is dispersed and, indeed, highly international in character. Industrialization promoted widespread urbanization. The challenges faced by these two countries reflect, at least to some degree, adaptations required by economic globalization.

In Chapter 6 we learn that economic development rarely occurs without social and cultural costs. In this chapter, we examine income disparity, regional lag, difficulties experienced by minorities, ethnic change, and relationships between the United States and Canada as economically interdependent neighbors. The challenges are complex, but there is also reason to hope that they can be overcome.

CONTENTS

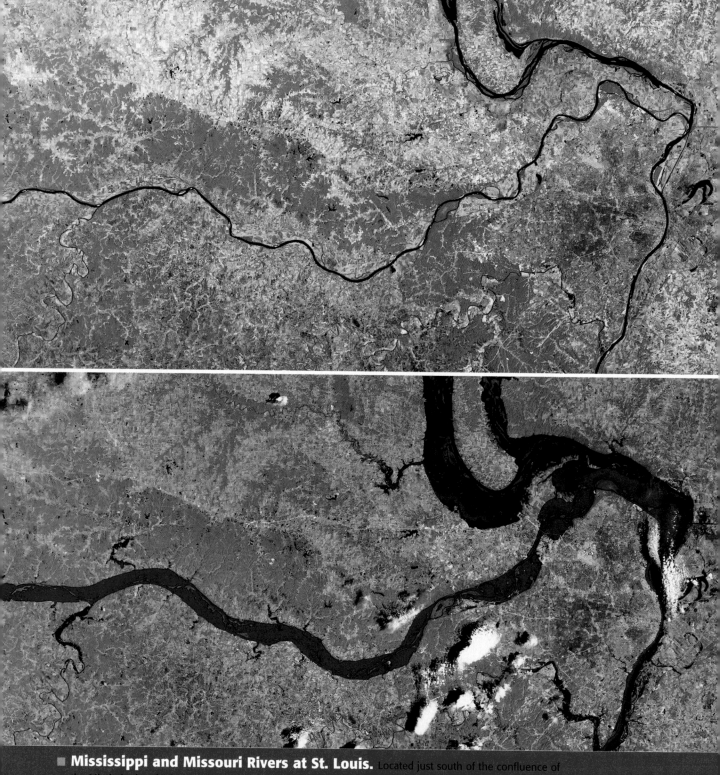

■ **Mississippi and Missouri Rivers at St. Louis.** Located just south of the confluence of the Mississippi and Missouri rivers, St. Louis has long been a major junction of land and water transportation systems. The two images show opposite extremes in the rivers' state: the top image reflects the low flow that occurred in the summer 1988 drought; the bottom image depicts the high water levels encountered during the flood of 1993.

Chapter 4

The United States and Canada:
The Physical and Human Contexts of Development

- **Landform Geography**
- **Climate**
- **Early Settlement: European Culture Cores**
- **Westward Expansion and Receding Settlement Frontiers**
- **Demographic Characteristics**

Together, the United States and Canada cover more than 7.5 million square miles (19.4 million square kilometers), an area that exceeds 14 percent of the world's land surface. Canada is the larger of the two countries by more than 130,000 square miles (336,700 square kilometers), but much of this territory lies in physically harsh, high-latitude environments that are poorly suited for settlement. Canada's population is scarcely more than one-tenth that of the United States. Both countries, however, have highly diverse physical endowments and human legacies that have influenced the political, settlement, and economic patterns of North America, often in profound ways. The purpose of this chapter is to examine these endowments and legacies for their significance in creating the physical and human contexts for development.

Landform Geography

The United States and Canada lie on the North American continent, a crustal mass that, in fact, includes Mexico and Central America. It is for cultural rather than physical reasons that the United States and Canada are treated separately as a world region—that is, the United States and Canada derive mainly from British political origins, whereas Mexico and Central America come from primarily Spanish and Portuguese roots. But many of the generalizations made about the physical structure of the United States and Canada also pertain to Mexico and Central America. The landform geography of North America consists of a shield nucleus, a mountain backbone and adjacent lowlands in the east, a series of mountain backbones and related plateaus in the west, and a large interior lowland dividing the two backbone re-

gions. The continental grain, if you will, runs from north to south.

The Canadian Shield: Continental Nucleus

Our starting point is the great **Canadian Shield**. A shield is a piece of the earth's crust that is very old and geologically very stable—that is, it is not likely to be deformed further. Shields are continental nuclei around which mountain formation tends to occur. All continents, and the tectonic plates to which they belong, have shields or related platforms.

The Canadian Shield extends outward from Hudson Bay to include much of Quebec and Labrador, most of the provinces of Ontario and Manitoba, and a substantial part of Canada's Arctic (Figures 4–1 and 4–2). Exposed pieces of the shield extend into the northern United States, forming the Superior Uplands of Wisconsin, Minnesota, and Michigan, and the Adirondack Mountains of northern New York. Buried portions of the shield continue into the central United States. The economic potential of the shield is mixed. Farming has always been difficult, due to thin and rocky soils. On the other hand, the forests of the shield supported a profitable fur trade in colonial Canada, and in today's economy, the shield is an important source of metallic raw materials and timber products.

The Appalachian Highlands and Adjacent Lowlands

To the east and south of the Canadian Shield lies the eastern mountain backbone of the United States and Canada, the **Appalachian Highlands**. Separating the southern part of this backbone from the ocean is the Gulf-Atlantic Coastal Plain. Since the adjacent lowlands owe their existence, at

97

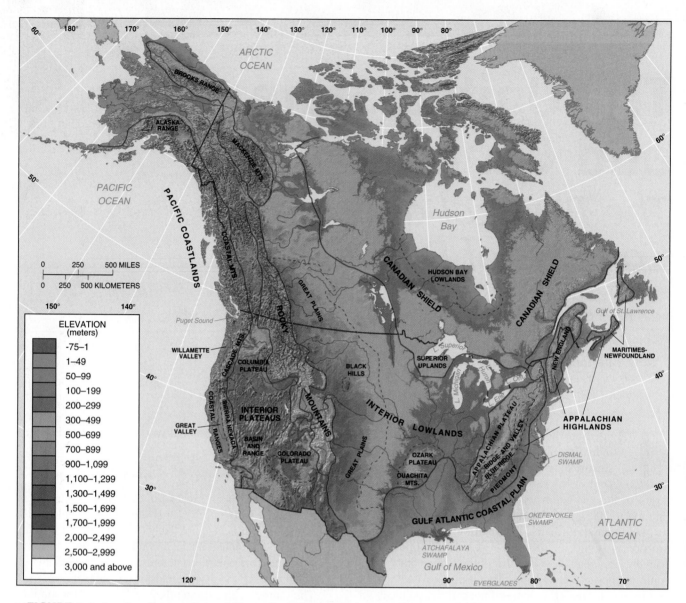

■ **FIGURE 4-1** **Land surface regions of the United States and Canada.** The physiographic diversity of these countries—together with climate, vegetation, and soils—provides highly varied environmental realms.

■ **FIGURE 4-2** **The Canadian Shield.** This region in central Labrador is part of the Canadian Shield, an area dominated by lakes, marshes, bogs, and swamps. Because of repeated continental glaciation, the Canadian Shield has little arable land but provides a wealth of furs, timber, and minerals. The Canadian Shield covers much of Canada and parts of the northern United States and constitutes a physical core area for North America.

least partially, to the degradation of nearby mountains, we will examine the Appalachian Highlands first.

Appalachian Highlands This relatively low-lying, yet imposing, system of mountains extends from Newfoundland to northern Alabama. By global standards, the 6,000-foot (1,800-meter) peaks that mark the highest points are relatively low in elevation. Their low elevations notwithstanding, these mountains created a significant barrier to westward movement in the early years of the United States.

Six distinct landform regions exist within the Appalachian Highlands. The **Piedmont** is an old and highly eroded plateau or upland plain that forms the eastern margin of the Appalachians from Pennsylvania southward to central Georgia. A series of river and stream rapids marks the edge of the Piedmont where it descends onto the coastal plain. These rapids, known as the **Fall Line**, forced early European colonists to transfer their loads to smaller boats upstream or to portage around difficult spots. Many towns were founded along the Fall Line to accommodate the transition from lowland to plateau and many of these towns subsequently grew into important southern cities. The Piedmont itself emerged as an important agricultural and industrial region in the South.

The **Blue Ridge Mountains**, or Great Smokies as they are known in North Carolina, Tennessee, and Georgia, constitute a favored tourist destination today. The **Ridge and Valley Province**, a strikingly folded landscape of long, parallel ridges and valleys, extends from New York to northern Alabama and includes such familiar locations as the Hudson and Shenandoah valleys (Figure 4–3). In terms of geologic age, the Blue Ridge formation is part of the relatively "old" eastern Appalachians and the Ridge and Valley formation the relatively "new" western Appalachians.

Despite the name **Appalachian Plateau**, this western portion of the Appalachian Highlands has been severely eroded into hills and mountains, especially in West Virginia. Other areas, such as the Cumberland Mountains in Tennessee and the Allegheny Mountains in Pennsylvania, retain features that are characteristic of plateaus. This region contains many of the vast coal reserves of North America.

The **New England** section of the Appalachian Highlands is a northward extension of the Piedmont, Blue Ridge, and Ridge and Valley areas but differs from its southern counterparts in having been significantly modified by glaciation. Included in this section are the White Mountains of New Hampshire and Maine and the Green Mountains of Vermont. The Appalachians continue northward into Canada to form a Maritimes–Newfoundland extension.

Gulf-Atlantic Coastal Plain As the name suggests, the **Gulf-Atlantic Coastal Plain** lies along the Gulf of Mexico and Atlantic coasts of the United States. In the east, the plain extends from Cape Cod to Florida, with the Fall Line serving as the western boundary. In the south, the plain embraces Florida, coastal Texas, and much of the lower Mississippi River Valley.

Most of the Atlantic Coastal Plain owes its existence to the erosion of the nearby Appalachian Mountains and con-

■ **FIGURE 4-3 Apple orchards in the Shenandoah Valley.** The Shenandoah Valley is a picturesque segment of the Ridge and Valley region of the Appalachian Highlands.

sists of geologically young rock that inclines gently toward the coast. The Gulf Coast experienced similar sedimentation, although, as in the case of Mississippi River sediments, the sources included much of the interior United States. Sometimes the slopes of the coastal plains are so gentle that it is difficult to discern where the land ends and the sea begins, due to the proliferation of lagoons, sand bars, and barrier islands. Onshore, wetland areas of poor drainage are abundant (and often spectacularly beautiful), as exemplified by Georgia's Okefenokee Swamp, Louisiana's Atchafalaya Swamp, and North Carolina's Dismal Swamp (see Figure 4–4 and the *Geography in Action* boxed feature, The Latest Battle for New Orleans, page 102).

Periodic subsidence of the land and change in sea level have added to the complexity of the coastline. River valleys along the Virginia and Maryland coasts, for example, were flooded as the land dropped below sea level, creating a highly indented coastline. Chesapeake Bay provides an exquisite example of how intricate and jagged such a coastline can become. In addition, periodic sea-level changes exposed the Gulf Coast region to marine deposition, a process that contributed to the formation of oil deposits.

Florida represents something of an exception to the sedimentary processes described above. Too far away to be affected by the eroding Appalachians or Mississippi River deposition, much of the Florida Peninsula comprises limestone formations and coral reefs that are geologically young and related to Cuba, the Bahamas, and Mexico's Yucatán Peninsula. The terrain is relatively flat and low (the highest point in the state is only 325 feet or 99 meters), and is strewn with sinkholes, lakes, and swamps.

The Western Mountains and Plateaus

To the west and southwest of the Canadian Shield lie the mountain ranges and plateaus of North America's western backbone. These features extend southward from Alaska to Central America, and westward to the Pacific Ocean. The look of the western mountains stands in sharp contrast to that of their Appalachian counterparts. The western mountains are, in general, geologically younger and higher than the

■ **FIGURE 4-4 Okefenokee Swamp, Georgia.** The Okefenokee is one of many large wetlands found along the Gulf-Atlantic Coastal Plain. One of the most common tree species found in the southern swamps is bald cypress, a deciduous needleleaf tree that is known for its excellent quality timber that resists rotting and termite infestation.

Appalachians, and are more rugged in appearance. We will divide the West into three parts: (1) the Rocky Mountains; (2) the Interior Plateaus; and (3) the Pacific Coastlands.

The Rocky Mountains When we think of the West, the **Rocky Mountains** and their postcard vistas inevitably come to mind—and they should, since they contain much of the region's scenic grandeur (Figure 4–5). What we should also remember, however, is that the Rockies constitute only a small part of the western ranges. In fact, the Rockies occupy only the eastern third of this region and are known more for their length than their width. While mountains that can be associated with the Rockies are found as far south as eastern Mexico, the major Rocky Mountain ranges begin in northern New Mexico and continue northward through Colorado and Wyoming. A break in the trend appears in central Wyoming, which provided a pass through which nineteenth-century pioneers traveled en route to Oregon and California. In northern Wyoming the full extent of the Rocky Mountains reappears and continues northward into Canada and, ultimately, as the Brooks Range of Alaska.

The present location of the Rocky Mountains was once a vast sedimentary plain. Upward pressure on this plain led to a folding and fracturing of the rock

cover that, along with volcanic activity, produced the mountains that we see today. These sedimentary layers are beautifully exposed as nearly horizontal beds in the jagged peaks of the Grand Tetons of Wyoming and in the mountains of Montana and Canada. Farther to the south is abundant evidence of volcanic activity. Yellowstone National Park, for example, is known for its geysers and hot springs and sits atop a volcanic "hot spot." Ancient volcanic cones and lava flows dot the landscape of northern New Mexico and southern Colorado. Had it not been for ancient volcanic activity in central Colorado, no fortunes would have been made in the gold fields of Cripple Creek.

Interior Plateaus Of course, the Rocky Mountains have no monopoly on beautiful scenery. In fact, many of the vistas that we associate with the "Old West"—the majestic buttes and mesas that serve as backdrops for cavalry charges in old movies—are not in the Rockies at all, but in the **Interior Plateaus** region. This region lies directly to the west of the Rocky Mountains and serves as a transition zone between the Rockies and the Pacific Coastlands.

The Interior Plateaus contain several components. The **Colorado Plateau** lies more than a mile high in southwestern Colorado, eastern Utah, northern Arizona, and New Mexico (Figure 4–6). The plateau is an uplifted piece of the earth's crust comprising sedimentary layers that were downcut by the Colorado River and its tributaries, thereby producing some of America's great canyonlands. The Grand Canyon of northern Arizona, for example, has been cut so deeply by the Colorado River that the canyon's floor lies more

■ **FIGURE 4-5 Pikes Peak, Colorado.** One of the many peaks in Colorado exceeding 14,000 feet (4,000 meters), Pikes Peak rises abruptly on the edge of the Great Plains, signaling the beginning of the Rocky Mountains. Note the barren summit, where harsh, tundra-like conditions prevent the growth of trees and all but the hardiest of plants. En route to the summit, the traveler will pass through various vegetation zones reflecting the changing microclimates produced by increasing elevation. The rock outcrops in the foreground are characteristic of the broken terrain that gives the Rocky Mountain chain its name.

■ **FIGURE 4-6 The Colorado Plateau.** The Colorado Plateau has a rich geologic history and a dramatic landscape characterized by buttes and mesas. This photo was taken in northern Arizona near the Utah State Line.

than a mile below its southern rim. The walls of the canyon expose sedimentary layers that reveal an extended profile of the earth's geological past. Uplift also led to a cracking of the crust, through which lava poured to produce isolated lava flows and volcanic mountain clusters, such as the San Francisco Peaks of northern Arizona.

The **Basin and Range** region lies to the west and south of the Colorado Plateau, occupying much of Nevada and western Utah as the Great Basin and parts of southern California and Arizona. Bordered by the higher Colorado Plateau to the east and the lofty Sierra Nevada Mountains to the west, the Basin and Range region consists of often parallel mountain ridges separated by valleys or basins. Nevada exemplifies this pattern with its rows of north-south-trending mountain ranges. The broken terrain and relative isolation of the Basin and Range region have created drainage conditions characterized by rivers without exits to the sea and isolated saline lakes, such as the Great Salt Lake. Death Valley, the lowest point in the United States at 282 feet (86 meters) below sea level, is situated in the California portion of the Basin and Range region.

To the north of the Basin and Range region lie the **Columbia Plateau** of eastern Oregon and Washington and the Snake River region of southern Idaho. The plateau consists of lava fields often exceeding a mile in thickness. Today, these weathered lava fields are associated with some of the richest agricultural soils in the United States—eastern Washington is a major wheat- and apple-producing region, and Idaho is known for its potatoes and sugar beets. Still farther to the north, the Interior Plateaus region narrows to a series of highly dissected plateaus and isolated mountain ranges in British Columbia and the Yukon, finally ending as the broad Yukon River Valley of central Alaska.

The Pacific Coastlands The Pacific Coastlands constitute a system of mountains and valleys that extends along the western edge of North America. On the eastern side of this region lie two great interior ranges that contain peaks that rival the Rockies in grandeur and elevation. The **Sierra Nevada Mountains** run from north to south in eastern California and have more than 11 peaks in excess of 14,000 feet (4,200 me-

ters). To the north of the Sierra Nevada lie the volcanic **Cascade Mountains** of central Oregon and Washington (Figure 4–7). These interior ranges continue northward as the Coast Mountains of British Columbia and as the Alaska Range. It is within the Alaska Range that we find Mt. McKinley, the highest peak in the United States and Canada at 20,320 feet (6,194 meters).

A smaller mountain chain, the **Coast Ranges**, runs the length of the Pacific coast. In California, these mountains rise gently from the Pacific shore to about 4,000 feet (1,219 meters). By the time they reach the Olympic Peninsula of Washington, they have lost much of their linear form and

■ **FIGURE 4-7 Mt. Rainier, Washington.** Mt. Rainier is one of many beautiful volcanic peaks that comprise the Cascade Mountains of Oregon and Washington. One of the long-term beneficial byproducts of the geological activity of the Cascades is the fertile soils of the Puget Sound Lowland and Willamette Valley, which lie to the west of the Cascades, and of the Columbia Plateau, which extends eastward as far as southern Idaho.

GEOGRAPHY IN ACTION
The Latest Battle for New Orleans

When we think of New Orleans, images come to mind of jazz music, carefree lifestyles, and, of course, a great battle with the British during the War of 1812. What many may not realize, however, is that the city is now fighting for its survival against two interrelated natural threats: soil subsidence and the loss of its coastal wetlands, which increase the hurricane threat.

In a sense, New Orleans should never have been located where it was. The first French settlers were attracted to the relatively high and fertile natural river banks, or levees, along the Mississippi River. Settlers initially avoided the swamps and marshes that lay beyond the levees. As the city expanded, however, the land in the backswamps and marshes was needed for settlement. Drainage of these muck soils, the natural composition of which is roughly 90 percent water by volume, triggered soil subsidence, which continues to the present at a rate of several inches per year. To make matters worse, the remaining 10 percent of the muck soils comprises mostly organic matter that oxidizes when exposed to the atmosphere, thereby compounding the subsidence problem. Subsidence poses especially serious problems for construction, as unstable soils cause foundations to tilt and roads to buckle. The solution is to drive numerous pilings into the ground at the building site, and then to pour the cement slab of the structure over the tops of the pilings. This process results in a building that "floats" on its pilings while the ground beneath and around the structure continues to sink. Thereafter, fill is added to the lawns around the building in a never-ending struggle to keep the "ground" at grade. Subsidence has left much of the bowl-shaped city 5 to 15 feet (2 to 5 meters) below sea level. Man-made levees have been constructed to keep the Mississippi River and Lake Pontchartrain from pouring into the city, and an elaborate pumping system has been built to remove the average 60 inches (1,500 millimeters) of rainfall received annually. For New Orleans residents, the threat of heavy rain presents all sorts of frightening possibilities seldom considered by city dwellers elsewhere.

As if living in a bowl-shaped depression below sea level were not threatening enough, New Orleans is losing its natural hurricane protection through erosion of nearby coastal marshes and barrier islands. Due to a variety of factors, including river control projects, wetland drainage, channel dredging, and the construction of petroleum exploration canals, nature's process of land recharge through deposition of river sediment has been substantially reduced. Coastal Louisiana is losing land to the ocean at the rate of 25–30 square miles (65–78 square kilometers) a year, an area the size of Manhattan. The Louisiana coastline today does not look at all as it did in 1970 (Figure A). In addition, the loss of river sediment has caused barrier islands to shrink as longshore currents no longer carry sediment loads sufficient to replace land as it erodes away. These wetlands and islands historically have helped to protect New Orleans from deadly hurricane-produced surges of water.

The hurricane threat is real. Scientists have long argued that it is a human catastrophe waiting to happen. Researchers at Louisiana State University have attempted to model the effects on New Orleans of a major hurricane tracking from the southwest (Figure B). Massive quantities of rainfall would inundate the city, overwhelming the pumping system. Storm

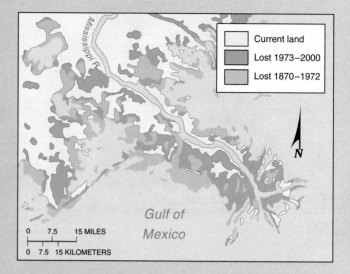

■ FIGURE A **Changes in southeastern Louisiana's coastline, 1870–2001.** Louisiana is shrinking in size as land is continually lost to the Gulf of Mexico. Between 1932 and 1990, the Mississippi River Delta region of Louisiana lost more than 1,000 square miles of land. Part of the cause is the construction of levees that prevent precious sediments from recharging coastal marshes.
Source: D. Foley, Louisiana State University, as presented in M. Fischetti, "Drowning New Orleans," *Scientific American* 285 (October 2001): 6.

surges created by high winds would breach the levees. At the height of the storm, floodwaters in New Orleans would exceed 20 feet (6 meters). Hurricane-force winds would whip up the floodwaters and add wind damage to the mix of horrors already visited upon the city. Evacuation of people during the storm would be impossible. Even evacuation before the storm would be difficult as hundreds of thousands of people would choke the half-dozen bridges and highways that exit the city. Furthermore, substantial numbers of city residents are so poor that they lack reliable transportation and would thus be trapped in the city with no way to escape. The death toll could number in the tens of thousands, or more. Of course, most people regarded such a worst-case scenario as unlikely—it had never happened before. Government agencies, particularly the U.S. Army Corps of Engineers, were attempting to find ways to prevent such a dire scenario from becoming reality. It was convenient for many policy makers simply to ignore or to put off worrying about this next environmental battle for New Orleans.

And then came Hurricane Katrina. Suddenly theory became reality. Early in the morning of August 29, 2005, the eye of Hurricane Katrina passed over St. Bernard Parish, brushed against eastern New Orleans and St. Tammany Parish, and slammed with full force into Waveland, Bay St. Louis, Pass Christian, Biloxi, and other coastal Mississippi towns. Katrina was a category 4 hurricane packing winds in excess of 140 mph (225 kph) and producing up to a 25 foot (8 meter) storm surge. After the passage of the storm's eye, government leaders in New Orleans breathed a sigh of relief: it had not been a direct strike on New Orleans—the feared "Big One" had not happened. But relief quickly turned into horror at the sight of hurricane-

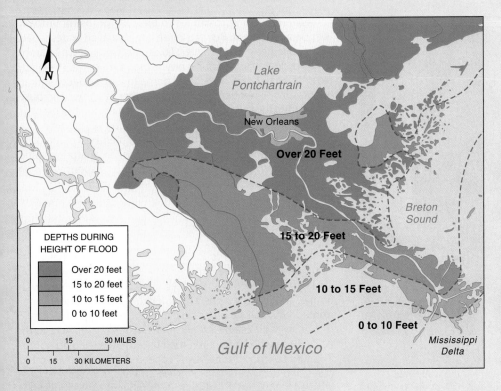

■ **FIGURE B Hurricane flood threat to New Orleans.** Considerable modeling has been done on the hurricane threat to New Orleans. The models suggest that a strong hurricane moving from the southwest would pour more than 20 feet (6 meters) of water into downtown New Orleans. Potential loss of life would be high. Hurricane Katrina did little to dispute the findings of these models.
Source: Computer models by researchers at Louisiana State University, as cited in M. Fischetti, "Drowning New Orleans," *Scientific American* 285 (October 2001): 9.

Labels within figure:
Lake Pontchartrain
New Orleans
Over 20 Feet
Breton Sound
15 to 20 Feet
10 to 15 Feet
0 to 10 Feet
Gulf of Mexico
Mississippi Delta

DEPTHS DURING HEIGHT OF FLOOD
Over 20 feet
15 to 20 feet
10 to 15 feet
0 to 10 feet

0   15   30 MILES
0   15   30 KILOMETERS

protection levees giving way under the pressure of storm surges, funneling millions of gallons of water into the city.

Wind and water damage was extreme throughout the New Orleans and Mississippi Gulf Coast. Perhaps 80 percent of New Orleans was under water, due in large part to the levee breaches (Figure C). The Interstate-10 bridge, a 5-mile concrete twin span connecting New Orleans with the east that daily funneled tens of thousands of commuters to their jobs, was torn apart by storm surge. Some of the Mississippi towns mentioned above largely disappeared from the map because of the combined power of some of the strongest winds and highest surges ever to strike the Gulf Coast. Trees were broken and left lying with roots pointing toward the sky. And most tragically of all, as of this writing, over 1,100 people were killed in Mississippi and Louisiana.

Everybody saw the storm coming and the usual pre-hurricane precautions were taken. Floodgates were closed. Emergency centers were opened. Of particular importance, the mayor of New Orleans ordered a mandatory evacuation of the city, as did neighboring mayors and parish presidents of their communities. Most people heeded the warnings and left, crowding the small number of multilane highways out of the city. Inbound lanes were made available to outbound traffic as part of the region's "contra-flow" exit strategy. Unfortunately, about 20 percent of the population, maybe 100,000 New Orleanians, remained behind. They stayed because they had no cars or other ways of leaving; or they decided to defy the storm, assuming that they could "hunker down" and ride it out (something of a New Orleans tradition, by the way). After the winds and flooding rains of the storm stopped, and particularly after the rising water from levee breaches had its way, many of the people who remained behind found themselves in mortal danger and desperate for help. Tens of thousands of people poured into the Superdome and the New Orleans Convention Center seeking higher ground, and thousands of others congregated on highway overpasses and on rooftops awaiting rescue. Over time, in some cases after several days, these people were taken to shelters elsewhere in Louisiana and in other states. Incredibly, about 10,000 diehards refused to leave, in spite of the the hardships, death, and destruction around them; many of these holdouts were forcibly removed. New Orleanians became scattered throughout Louisiana and across the country forming a New Orleans "diaspora." Nearby Baton Rouge doubled in population. Slightly under 50,000 New Orleans students enrolled in Texas schools. Some New Orleanians even ended up in Utah, where they formed their own small New Orleans community amidst the mountains, deserts, and very non-New Orleans culture.

By mid-September, New Orleans was largely abandoned. The city known for its raucous party life, its cholesterol-charged gastronomic delights, and its "what, me care?" culture was plunged into a surreal silence. There was no electricity. The only night lights left of this town that never slept were the stars and the lamps connected to portable generators. Most TV and radio stations were forced off the air. The city's main newspaper, the *Times-Picayune,* had to locate to other cities to publish. The airport closed. The Orleans Parish public schools suspended operations for the year. The New Orleans Saints football team played "home" games in San Antonio and Baton Rouge. Once the water was pumped out, the mudcovered streets and houses in the formerly flooded parts of the city projected a dusty brown monochrome. New Orleans was very much a ghost city. Thankfully, by late September, New Orleans and the surrounding parishes were able to begin the repopulation process, or at least residents were allowed to inspect the damage to their homes and businesses—many residents knew that they would not be able to live in their dwellings for some time to come.

New Orleans will rebound, but it will be years before anybody knows what the "new" New Orleans will look like. The reconstruction process faces several daunting challenges. First,

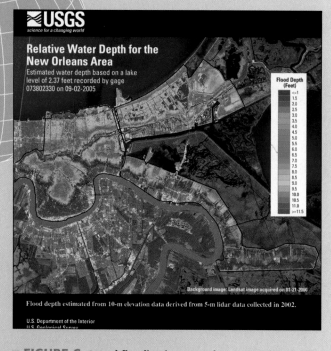

Relative Water Depth for the New Orleans Area

Estimated water depth based on a lake level of 2.37 feet recorded by gage 073802330 on 09-02-2005

Flood Depth (Feet)

<1
1.5
2.0
2.5
3.0
3.5
4.0
4.5
5.0
5.5
6.0
6.5
7.0
7.5
8.0
8.5
9.0
9.5
10.0
10.5
11.0
>=11.5

Background image: Landsat image acquired on 01-21-2000

Flood depth estimated from 10-m elevation data derived from 5-m lidar data collected in 2002.

U.S. Department of the Interior
U.S. Geological Survey

FIGURE C Actual flooding in New Orleans associated with Hurricane Katrina. This image shows where the water was deepest in New Orleans after Hurricane Katrina and the breach of the levees. Note that areas along the Mississippi River, where the natural levees are highest, remained relatively dry. Included here is the French Quarter, the West Bank of the river, and an area called Uptown. Areas adjacent to Lake Pontchartrain also remained relatively dry, but badly windblown. Those parts in the interior "bowl" of the city, and in eastern New Orleans, received in excess of 11 feet (3.5 meters) of water. Only floodwaters within New Orleans are shown; adjacent parishes and outlying areas were also badly flooded.
Sources: United States Geological Service, *http://eros.usgs.gov/ katrina/products.html*, October 5, 2005. The background image is from Landsat 7 and was produced in 2000.

serious environmental problems remained after the storm passed. Water became so contaminated in places that taking showers was dangerous, not to mention drinking. Mold from standing water appeared everywhere, destroying Sheetrock, contaminating food, and staining furniture beyond reclamation; it takes only a little floodwater to create enough mold to make a house unliveable. Some of the dried mud that coated areas after floodwaters receded was contaminated by sewer leaks and chemical spills, and had to be removed before residents could safely return. In addition, many people were haunted by the fear that another hurricane would strike before New Orleans could repair and upgrade its hurricane protection system—Hurricane Rita ravaged western Louisiana only 3 weeks after Katrina—or that a not uncommon monster rainstorm would flood the city before the pumps were fully back in service.

Second, housing continues to be a big problem. Virtually every residence in St. Bernard Parish was damaged or destroyed. Many buildings in south Slidell, Louisiana, were blown apart by winds or floated to new locations by the storm surge. (One lady returned to her damaged Slidell home to find a 40-foot shrimp boat resting in her back yard.) The number of homes and businesses destroyed in the New Orleans area probably exceeds 160,000. A new homeless class has been formed, some members of which were among the wealthiest members of the community before Katrina. Temporary housing in apartments and rental houses is hard to find and extremely expensive when found. Many residents are living in temporary trailer communities funded by the government, little "Katrina towns," embedded within the larger metropolitan area. In other instances, people simply have not come back because they have nothing to come back to.

Third, the damage to the New Orleans economy has been extreme. Repairing the damage will take time and may not be possible in all cases. Businesses were destroyed, along with homes. As business owners returned to the city, they were confronted with the daunting challenge of not only repairing or rebuilding their establishments but also finding workers to replace those who had not returned because they had no homes to return to—"now hiring" signs appeared everywhere. In other cases, workers were laid off. New Orleans, for example, discharged 3,000 city workers in early October of 2005 (thousands of public school teachers had already been let go). Without businesses and without consumers, there was no tax base to support government. New Orleans (and the other affected communites) were faced with a downward spiral of economic hardships. At least initially, there was little government assistance to stop this spiral. The good news is that eventually homes will be rebuilt, tax bases will be restored, and business will be back in operation; but it is unclear what the new economic geography of the region will look like. For example, will there be a regional economic realignment, with the environmentally less endangered areas of Baton Rouge and the north shore of Lake Pontchartrain capturing a share of the economic activity formerly found in New Orleans? Some observers have suggested that the "new" New Orleans may end up with only half of its pre-Katrina population.

Hurricane Katrina did what the British Army could not do in 1815: it brought New Orleans to its knees. This latest environmental and economic battle for New Orleans will in all likelihood last for years and will significantly transform the geography of southeast Louisiana. In addition, it will give leaders of other cities in environmentally precarious settings pause to reflect on how they would respond if their "big one" were to strike. By the way, as hard as it is to believe in view of the horrible damage and loss of life, Katrina was technically not quite the "big one"—the direct strike—that New Orleans had long feared and was described above; it was, however, big enough for the hundreds of thousands of people who suffered loss, and remains a big wake-up call for environmentally endangered cities everywhere.

exceed 7,500 feet (2,286 meters) in elevation. This mountain trend continues into Canada as an island series—the floors of the Coast Ranges now lie below sea level—with Vancouver Island serving as the largest member of the series. California's San Francisco Bay and the Puget Sound of Washington provide two of only a small number of breaks in the mountains and, consequently, two of the limited number of natural harbors on the west coast (the east coast is much better endowed).

Several lowland areas separate the Coast Ranges from the Sierra Nevada and the Cascade Mountains. These include the **Great Valley** of California, a fertile alluvial trough that is home to one of the most productive agricultural regions in the United States. Farther to the north lie the **Willamette Valley** of Oregon and the **Puget Sound Lowland** of Washington. Both lowlands have emerged as important agricultural regions with major population centers—the Seattle metropolitan area, for example, dominates the Puget Sound Lowland.

The western rim of North America stands out as one of the most geologically active regions on the continent. In the south, the infamous San Andreas Fault separates two plates, making much of California susceptible to earthquakes. Farther to the north in Oregon and Washington, the challenge is volcanoes as well as earthquakes. The small Juan de Fuca Plate is still being subducted beneath the North American Plate, a process that has led to the formation of the volcanic peaks of the Cascade Mountains and has created an ongoing risk of volcanic eruption. No better evidence of this risk can be found than the eruption of Mount St. Helens in the early 1980s, an eruption whose fury literally blew off the top of the mountain. Geologic hazards follow the Pacific coast northward into Alaska.

The Interior Lowlands

South and west of the Canadian Shield, and between North America's mountain backbones of the east and west, lies a middle zone of accumulation called the Interior Lowlands. These lowlands reach as far north as the Arctic plains of Alaska, as far south as central Texas, and as far east as the Great Lakes. With several notable exceptions, the terrain is flat to rolling and consists of sediments that have washed out of the eroding Rocky Mountains and Appalachian Highlands. The **Great Plains**, which lie on the eastern side of the Rockies, contain one of the flattest surfaces on earth, interrupted only occasionally by eastward-flowing rivers. The sense of flatness is enhanced by the relative absence of trees, thanks to a semiarid climate. A person's view is seldom obstructed on the Great Plains (Figure 4-8).

Much of the northern interior lowlands region has been glaciated. Minnesota, for instance, is called the "Land of 10,000 Lakes" because of the countless tiny ponds and lakes that formed following the retreat of the ice sheets. The most visible remnants of glaciation are the **Great Lakes** themselves. Prior to the glacial periods, there were no Great Lakes; the lakes that we see today occupy lowlands that were scoured by glaciers and subsequently filled with meltwater.

The generally flat terrain of the Lowlands is interrupted dramatically by the **Ozark Plateau** of Missouri and the

■ **FIGURE 4-8 The Great Plains.** The region of the interior United States known as the Great Plains is one of the world's foremost agricultural settings. Wheat farming and cattle grazing are the principal agricultural activities.

Ouachita Mountains of Arkansas. While these uplands are located in the region that we have identified as the Lowlands, they are in fact remnants of mountain-building processes that produced the Appalachians. The Ozarks are an uplifted western outlier of the Appalachian Plateau, and the Ouachitas are an extension of the Appalachian Ridge and Valley province. Another interruption is the **Black Hills** region of western South Dakota, which has peaks that exceed 7,000 feet (2,100 meters) in elevation. These mountains are part of a very old

■ **FIGURE 4-9 Mt. Rushmore, South Dakota.** The presidential images of Mt. Rushmore are a part of the Black Hills of western South Dakota. From left to right are carvings of George Washington, Thomas Jefferson, Theodore Roosevelt, and Abraham Lincoln.

and highly resistant dome of crystalline rock that protrudes through the land's surface. The sculptors of the presidential faces on Mt. Rushmore were, no doubt, grateful for this resistance to erosion (Figure 4–9).

The eastern and central parts of the Lowlands provide some of the most favorable settings in the world for agriculture. Extreme cold in the north, however, and increasing aridity on the western margin limit agricultural productivity.

Climate

Given the vast size of North America, it is reasonable to expect a variety of climate types. The United States and Canada have climates that range from the subtropics of the Gulf Coast to the tundra of northern Canada (Figure 4–10). Before we examine specific types, however, we should iden-

tify several general factors that influence North American climates. First, most of the United States and Canada lie in the middle and high latitudes, which lead to clearly defined seasonal changes in temperature. Second, for much of the United States and Canada the prevailing wind direction is from the west (thanks to the westerly wind system and its embedded jet streams), which pushes weather systems from west to east. Third, the north-south mountain ranges in the West modify air masses as they move eastward, often drying them out. Fourth, the configuration of mountain systems and the size of the continental landmass create a condition called **continentality** in the interior of the continent in which the atmosphere takes on the more extreme heating and cooling characteristics of land rather than water; in other words, winters tend to be cold and summers hot. Finally, the **Gulf of Mexico** provides an important source of moisture for the Gulf Coast and the Interior Lowlands.

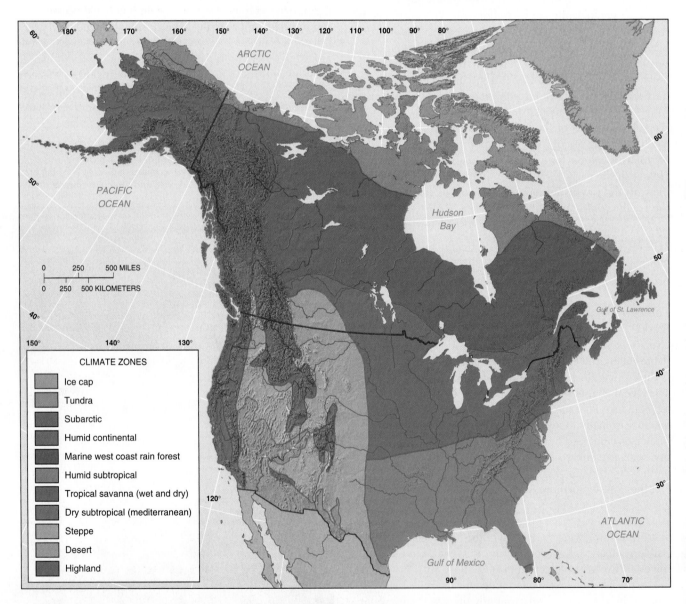

■ **FIGURE 4-10 Climate regions in the United States and Canada.** Note the diversity of climates, from the humid subtropical climate of the Gulf Coast to the Arctic tundra.

The Humid Subtropical Climate

Anyone who has traveled along the Gulf Coast of the United States in the summer has noticed the hot and extraordinarily sticky conditions. This is the humid subtropical climate at its best (or its worst!). This climate extends northward to the southern Middle Atlantic and Midwestern states, and westward to eastern Texas. During the summer, tropical air from the Gulf of Mexico settles over the area, bringing with it tropical warmth and an abundance of moisture that is released in the afternoon as brief but intense convectional thunderstorms. The winter is quite different as cold, dry continental air masses from the north, and cool and moist air masses from the Pacific, compete for space with warm and humid air masses from the south. The result is a roller-coaster ride for residents of the Southeast as the temperatures swing from hot and muggy one day to chilly and dry the next. Winter rainfall is mainly frontal in origin. Snowfalls are relatively infrequent or, in the case of the Gulf Coast and Florida, virtually nonexistent. The vegetation types that we find in this climate include rapidly growing mixed hardwood and softwood forests that yield great quantities of cut timber and pine pulp for paper. The red lateritic soils, however, are acidic and low in nutrients, creating challenges for agriculture.

The Dry Subtropical (Mediterranean) Climate

If we jump across the continent to the Pacific Coast of California, we encounter the dry subtropical or mediterranean climate. It is found as a relatively narrow belt along the California coast from San Diego to north of San Francisco. Summer temperatures are relatively cool and winter temperatures are relatively warm due to the moderating effects of the Pacific Ocean. Precipitation levels are low, with clearly defined wet and dry seasons. Most of the precipitation occurs in the winter and is frontal in origin. Winter storms along the coast can be intense, causing erosion of highways and undermining of homes. On the other hand, snow is virtually unheard of because temperatures rarely dip below freezing. Summer is a time of drought as the fronts shift northward and the adjacent desert climate takes control. The soils in the mediterranean climate are naturally fertile and agricultural yields are high.

The Marine West Coast Climate

North of the dry subtropical climate is the marine west coast climate, which dominates North America's coastal region from northern California to Alaska. Winter temperatures are warmer than expected at such a high latitude, thanks to the moderating effects of the ocean. For the same reason, summer temperatures are relatively cool. Overcast and drippy conditions prevail throughout the year.

The Cascade Mountains exert a blocking effect on climates in the Pacific Northwest. West of the Cascades the climate is maritime, with relatively warm winter temperatures and abundant precipitation. East of the Cascades the maritime effect is lost, leading to drier conditions and greater temperature swings. As the westerly winds are forced up in elevation, copious amounts of precipitation are released (see the *Geography in Action* boxed feature, Landforms and Climate: Orographic Effects in the United States). Huge quantities of snow, measured in feet rather than inches, fall annually above 5,000 feet (1,500 meters) on the windward side of Washington's Cascade Mountains. As the westerly winds descend in elevation on the leeward side of the mountains, the air dries out and little precipitation occurs, which creates a rainshadow effect. The Cascades have endowed the region with fertile volcanic soils that support magnificent forests of Douglas fir and considerable fruit farming and dairying.

The Western Steppes and Deserts

East of the Cascades lies one of the two large areas of mid-latitude steppe climate in North America. This area embraces eastern Oregon and Washington (in the rainshadow of the Cascades) and extends as far south as the Colorado Plateau. The second area lies to the east of the Rocky Mountains (in the rainshadow of the Rockies) and corresponds roughly to the Great Plains. A steppe climate is semiarid, which is to say that the landscape is dry, but not as dry as a desert. Enough precipitation is received to sustain grasses and shrubs. Temperatures tend toward extremes, with frigid winters accompanied by scorching summers. Continentality plays a role in the wide annual temperature swings, especially in the Great Plains. Nineteenth-century settlers initially avoided steppe climates because of their harshness. The Great Plains were known as the "Great American Desert" (although they were not a desert in the true meaning of the word). On the other hand, beneath the prairie grasses of this region lie rich organic soils that sustain one of the world's most productive wheat regions.

The true deserts of North America are concentrated in the Southwest, focusing on western Texas, southern New Mexico, southern Arizona, and southern California. Here the gap between the amount of moisture needed by the environment and the amount of precipitation received is extreme. Phoenix, Arizona, for example, receives an average of only 8 inches (203 millimeters) of precipitation per year. Parts of southern California are fortunate to get 3 inches (76 millimeters) per year. Natural vegetation must accommodate such dryness by storing moisture internally and retarding moisture loss (for example, the cactus with its water storage capability and needles instead of leaves; Figure 4–11). The dessication of the Southwest is the result of lower latitudes producing hotter summers, along with mountain barriers and isolation from moisture sources. Ironically, the southwestern deserts have become popular destinations for city dwellers in search of sunny days and dry air—witness the explosive growth of the great air-conditioned cities of Phoenix and Tucson! Even farming is possible. Irrigation can transform what appears to be a desert wasteland into a highly productive agricultural region.

The Humid Continental Climate

As we move eastward from the steppes and deserts of the West, we enter the humid continental climate, which dominates the northeastern quarter of the United States and

GEOGRAPHY IN ACTION
Landforms and Climate: Orographic Effects in the United States

Mountains can have a profound influence on climate through what is called the **orographic effect**. Mountains create a barrier to a moving air mass that must be overcome, which in the process modifies the temperature and moisture characteristics of the air mass. The result is excess precipitation on the windward side of the mountain and drought on the leeward side of the mountain.

The orographic process is quite straightforward. As an air mass is forced against a mountain's windward side, the air mass rises (Figure A). As it rises, it cools (the stable air lapse rate is 3.5°F per 1,000 feet, or roughly 6°C per 1,000 meters). Because cool air holds less water vapor than warm air, the ascending air loses its capability to hold water vapor to the point that the air becomes saturated and moisture is condensed out. This leads to the formation of clouds and, often, precipitation in the form of rain or snow. Once the air mass passes the summit of the mountain and starts to descend the mountain's leeward side, the reverse process begins. As the air descends, its rising temperature exceeds the condensation point. The drying air thus produces what is called a **rainshadow**—that is, an area that is shielded from rainfall—on the dry side of the mountain.

The United States and Canada contain countless examples of the orographic effect. Especially impressive examples can be found in Hawaii. Many of the major islands have elevations high enough to trap water from the trade winds on the windward sides of mountains, producing rainfall totals that exceed 100 inches (2,540 millimeters) annually, but leaving near-drought conditions on the leeward sides of the same mountains only a few miles away. Another example is found in the northwestern part of the United States (Figure B). Everett, Washington, which lies on the windward side of the Cascade Mountains receives

■ **FIGURE A** **The orographic effect.** As air ascends against the windward side of the mountain, the air cools and condenses, producing precipitation. On the leeward side of the mountain, descending air warms and dries, creating a relatively rainless rainshadow.

38 inches (965 millimeters) of rainfall a year. As a person travels eastward and higher in elevation into the Cascades, precipitation levels increase dramatically. Rainier Paradise Ranger Station, at 5,400 feet (1,646 meters), receives 120 inches (3,048 millimeters) a year; Stevens Pass, at 4,000 feet (1,219 meters), receives 83 inches (2,108 millimeters) a year. In both cases, much of this precipitation occurs as snow. We can only imagine the rain and snow amounts at the higher elevations. By the time that same person travels downslope to Yakima on the leeward side of the Cascades, however, the average annual

southern Canada. Humidity is a hallmark of this environment, especially toward the east where abundant rainfall in the summer and snowfall is the winter can be expected. In terms of temperature, winter conditions are consistently cool to cold, especially toward the western margins. Thanks to continentality, the subzero January temperatures of Winnipeg may be accompanied by 90-degree-plus temperatures in July.

The humid continental climate is home to North America's agricultural heartland, including much of the corn belt and all of the dairy belt. In addition, a substantial proportion of North America's population lives here, including 80–90 percent of Canada's population.

■ **FIGURE 4-11** **Desert vegetation in Arizona.** Desert vegetation must be able to withstand long periods of drought. The saguaro cactus in this picture embodies the adaptations required to survive the desert, as for example, relying on spines instead of leaves to retard moisture loss and a barrel-like trunk to store water. No more plant area than necessary is exposed to the intense evaporative power of the desert sun. Shade is at a premium. The saguaro cactus is a slow-growth plant that Arizonans increasingly attempt to protect as they build out into the desert.

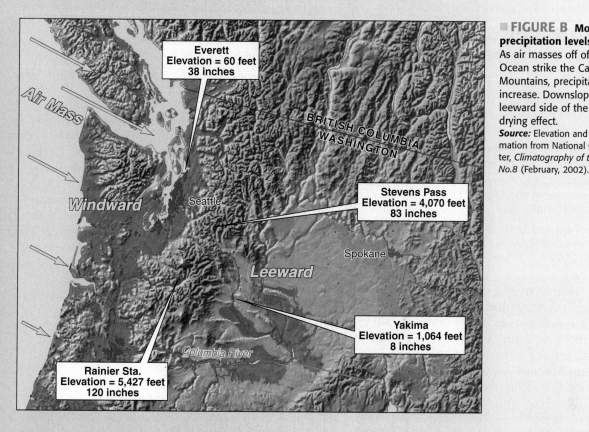

■ **FIGURE B** **Mountains and precipitation levels in Washington.** As air masses off of the Pacific Ocean strike the Cascade Mountains, precipitation levels increase. Downslope winds on the leeward side of the Cascades have a drying effect.
Source: Elevation and precipitation information from National Climatic Data Center, *Climatography of the United States No.8* (February, 2002).

precipitation has plummeted to 8 inches (203 millimeters) a year.

The geographic consequences of orographic effects can be profound. High precipitation levels in the Cascades provide the water that fills the reservoirs that are needed for the region's massive hydroelectric projects. Inexpensive electricity was one reason the aluminum industry developed in Washington, which, in turn, contributed to the growth of the aircraft industry. Farming on the moist windward side of the Cascades focuses on dairy production; on the leeward side, relative dryness has led to greater emphasis on irrigated fruit tree cultivation and wheat farming.

The Subarctic and Polar Climates

Poleward of the humid continental climate lie the subarctic and polar climates. The subarctic climate occupies a wide swath of central Canada and Alaska. It is characterized by intense and prolonged winter cold, with summers that are warm but quite brief. Precipitation levels often are relatively low, but more than enough to meet the needs of a generally heat-starved environment. Contrary to popular belief, most of the precipitation falls as summer rain; the snow that falls during the winter does not thaw until spring and can be whipped around by the wind to create the false impression of constant snowfall. The subarctic is home to large stands of softwood boreal forest and many fur-bearing animals, but few people.

Conditions are even more extreme in the polar, or tundra, climate that is found along the northern edges of Canada and Alaska. Freezing conditions last for most of the year and, in the depths of winter, are accompanied by days without sunlight. Conditions are so harsh that trees are rarely found, replaced by a thin cover of grass, moss, and lichens (Figure 4–12). Very few people other than aboriginal groups live in the far north, and this region's economic potential is limited.

■ **FIGURE 4-12** **The treeless winter tundra near the Arctic Circle in Canada's Northwest Territory.** Daylight hours are short, and winter temperatures can reach −50°F (−46°C).

Early Settlement: European Culture Cores

Europeans in North America were preceded by numerous aboriginal or indigenous groups. At the time of the first European settlement in Canada and the United States, these indigenous groups probably numbered no more than 1 million people. Warfare and exposure to diseases led to drastic depopulation after the arrival of the Europeans. Today's Native American population accounts for less than 0.5 percent of the total population of the United States and Canada. While the historical significance of indigenous populations should not be understated, these groups did not amass populations in core areas sufficient to promote the development and spread of lasting cultural characteristics over large areas of North America.

Of particular interest to geographers is the manner in which groups form cores or **culture hearths**, or centers of cultural innovation, and then extend these innovations across the landscape through diffusion. Often this **diffusion process** takes the form of settlers moving into new territories, or **settlement frontiers**, and establishing their cultures in these territories. In so doing, they gradually create trade and other ties with the core area and surrounding communities, and they eliminate the frontier through a process of **spatial integration.** European core areas on the Atlantic Seaboard provided important early sources of settlers responsible for the spatial integration process in the United States and Canada.

Initial European exploration of the New World was carried out by Spanish, Portuguese, French, and English interests. The Spanish were the first to create a permanent colony at St. Augustine, Florida, in 1565. Somewhat later, a second settlement area was established along the Rio Grande River near what is now Santa Fe, New Mexico. Unlike the English and the French, the Spanish were not able to establish a presence in eastern North America that was sufficiently large to constitute a core area and a source of diffusion into the interior of the United States.

The first permanent non-Spanish settlements founded along the Atlantic Seaboard included the English Jamestown colony in 1607, the French settlement in Quebec in 1608, the English Plymouth colony in 1620, and the Dutch settlement along the Hudson River in 1625 (which soon came under English control). These settlements provided the starting points for core areas that, through the diffusion process, created lasting imprints on the human geography of the United States and Canada (Figure 4–13).

The New England Core

The New England core area, which consisted of parts of modern Massachusetts, Rhode Island, and Connecticut, began as a patchwork of small, subsistence-oriented farms surrounding small villages. While farming was a necessity for settlers in New England, glaciated soils and hilly terrain made cultivation difficult. There was no single crop that provided great wealth or formed a basis for trade, as was the case with tobacco in the southern colonies.

New England was initially a destination for religious and political dissidents who preached thrift, hard work, and piety—a combination of traits that helped the settlers to achieve agricultural self-reliance despite the harsh physical environment. These traits also contributed to New England's later success at realizing the promise of its considerable nonagricultural resource base. White pine from New England forests provided lumber for shipbuilding. Cod fish from offshore banks were abundant and easily harvested. By the late 1700s, the combination of capital and labor from nonagricultural activities, an abundance of streams to turn waterwheels, and a growing maritime trading tradition contributed to New England's early success in the Industrial Revolution. As a consequence, New England became increasingly urbanized and developed a large commercial middle class. The city of Boston figured prominently in this process. New England provided many of the technological innovations, especially firearms, that later accompanied (and were diffused by) westward expansion.

The Southern Core

From the start, the colonies in Virginia and southward differed from those in New England. Like the Spanish in Middle America, the first Virginia settlers harbored the belief that they would become wealthy from gold and silver. Metallic wealth was never found in abundance; instead, wealth was accumulated from the cultivation of subtropical crops that could be exported to consumers elsewhere in the Americas and in Europe. Tobacco became a commercial success almost immediately; indigo, rice, and cotton were added later.

Most of the South's commercial agricultural production relied on the plantation system. Spatially, socially, and operationally, this system could not have been more different from the New England farming system. Indentured and slave labor led to a clear distinction between farm workers and farm managers or owners, with attendant distinctions in wealth, social status, and the built environment of the plantation. The economic emphasis was on the production of one or two crops that could be exported for a profit. While subsistence-oriented activities were present, they were not the plantation's reason for being. Plantations generally were located on navigable rivers or along the coast to make shipment of commodities easy. In the plantation South, few cities other than port cities appeared; not surprisingly, a commercial middle class was largely missing.

The plantation system was established from the Tidewater (lower Chesapeake Bay) area of Virginia to Maryland and southward to Georgia. Inland from the Tidewater and coastal agricultural settlements, and away from convenient water transport routes, were smaller, free-labor farms run by yeoman farmers. This type of farmer was especially prevalent on the North Carolina and Virginia Piedmont. When an improved cotton gin became available after 1800, both the yeoman-farmer culture and the plantation culture spread

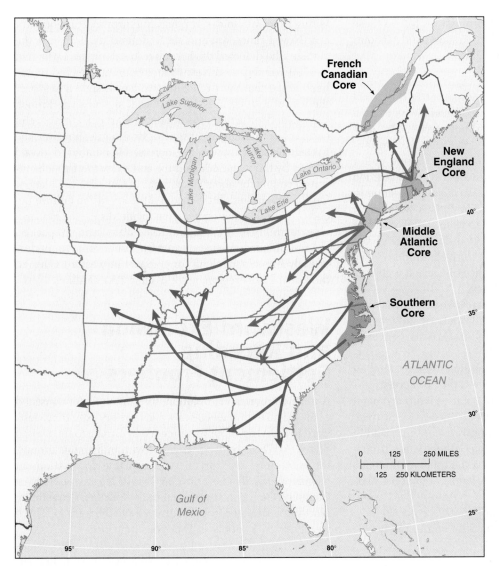

■ **FIGURE 4-13 Early European settlement areas in Anglo America.** Culture traits evolved in these four areas and were later diffused to other parts of Anglo America.
Source: Adapted from W. Zelinsky, *The Cultural Geography of the United States: A Revised Edition* (Englewood Cliffs, NJ: Prentice Hall, 1992), pp. 80–84.

throughout the lower South, with the latter generally securing the best lands. The practices and attitudes of the southern culture hearth, consequently, were diffused well beyond the Tidewater region.

By the mid-1800s, economic differences between North and South were contributing to political conflicts. The issue of tariffs, for example—the duties or customs imposed on imports and exports—came to epitomize fundamentally different regional perspectives. Because the markets for agricultural products were in Europe, southern producers wanted no tariff system that might inhibit the movement of their goods. Northern industrial producers, on the other hand, sought tariffs to protect their young industries. Other differences arose over the socially stratified society that evolved in the South and an economic system based on slave labor.

The Middle Atlantic Core

New York and Pennsylvania, together with portions of New Jersey and Maryland, made up a third settlement core. A variety of peoples settled the middle colonies: English, Dutch, German, Scots-Irish, and Swedish. Neither the cash crops of the South nor the lumbering and fishing activities of New England were significant sources of income. Instead, people in the Middle Atlantic Core developed a mixed agricultural system focused on the fattening of hogs and cattle and on Indian maize, now called corn. These settlers also made tools, guns, and wagons, and they worked the deposits of iron ore found in eastern Pennsylvania. The diffusion of their culture, especially its mixed farming system, had a significant impact on the American Middle West and parts of the Appalachians.

Of special note is the migration of people from the middle colonies into southern Appalachia, including Virginia and West Virginia, and into the Blue Ridge Mountains and the Ridge and Valley region. These settlers came in the early eighteenth century and were descendants of the Scots-Irish, Germans, and English who peopled the middle colonies. They carved out small subsistence farms as slaveless yeoman farmers, and contributed traits (such as a population of small farmers that is almost totally white and Protestant) that still distinguish southern Appalachia from the rest of the lower

South today. Eventually, population pressures forced these people to migrate to the highlands of Arkansas and Missouri and to the hill country of central Texas.

The French Canada Core

The earliest European settlements in Canada were French in origin and, in some instances, did not survive. The Acadian population, for example, settled the valleys of western Nova Scotia under the French flag, only to come under British rule in 1713. Acadian resistance to British rule led the British to transport much of the Acadian population to other parts of North America in 1755. Today's "Cajuns" of southern Louisiana can trace their roots to the Acadians of Nova Scotia—Cajun is an Anglicized corruption of the French word from which Acadian is derived. Modern Nova Scotia is an English-speaking province of Canada with only traces of its Acadian past in evidence.

By far the more enduring French presence in Canada appeared in New France, focusing on Montreal and Quebec City. Immense wealth in the form of furs was extracted from New France's seemingly endless forests. Later, farmers settled the shores of the St. Lawrence River and created an agricultural system that was sufficient, though not extraordinary, in its bounty (Figure 4–14). Farms were arranged in long strips to provide each farmer with access to the river, the principal transportation corridor. French Canadian life was hierarchical in structure, with the Catholic Church, government officials, and landowners at the top, and the mass of tenant farmers at the bottom. A large commercial middle class was absent.

By 1763, there were only about 65,000 people in New France, compared with more than 2 million in the neigh-

boring English colonies. It is hardly a surprise, consequently, that New France was unable to defend itself against the British, who defeated the French colonials on the battlefield in 1759 and acquired New France by treaty in 1763. It has been noted that the French Crown hardly shed a tear over the loss of a "few acres of snow" as it signed away French Canada and left Britain the dominant European presence in eastern North America. In spite of British rule, however, the French Canadian population increased in number to several million by the twentieth century, and today the French culture accounts for about 23 percent of Canada's population. On the other hand, little spatial diffusion of the culture occurred. The French core region, while displaying remarkable durability in Quebec, produced little lasting impact on the rest of North America, apart from occasional reminders of early French-speaking fur trappers in western Canada, the Great Lakes region, and the Ohio River Valley.

Westward Expansion and Receding Settlement Frontiers

After the American Revolution, the United States consisted of the original thirteen colonies and territory extending westward to the Mississippi River (ceded by Britain to the United States in 1783). Barring struggles with indigenous groups, these western territories were available for settlement (Figure 4–15). To the north lay British North America, or Canada, which constituted the bulk of Britain's remaining North American territories. Territories to the west and south were held by a variety of European powers.

It was not long, however, until Americans were looking covetously on foreign-held territories that lay beyond initial U.S. borders. Many in the United States believed that America was entitled to these lands, partly because of their proximity to the United States and partly because of America's growing sense of **manifest destiny**—the conviction that God had willed these lands to the United States to be civilized by Americans and their ennobling institutions.

Territorial Competition in North America

It should come as no surprise that this sense of American destiny was not shared by everyone, particularly those countries that claimed territories lying in the path of American aspirations. Consequently, America's history of westward settlement—its spatial integration process—is largely a history of negotiation and conflict with external powers over the control of western and southern territories.

An early negotiation for land involved the French-controlled Louisiana Territory, which lay immediately to the west of the Mississippi

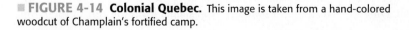

■ **FIGURE 4-14 Colonial Quebec.** This image is taken from a hand-colored woodcut of Champlain's fortified camp.

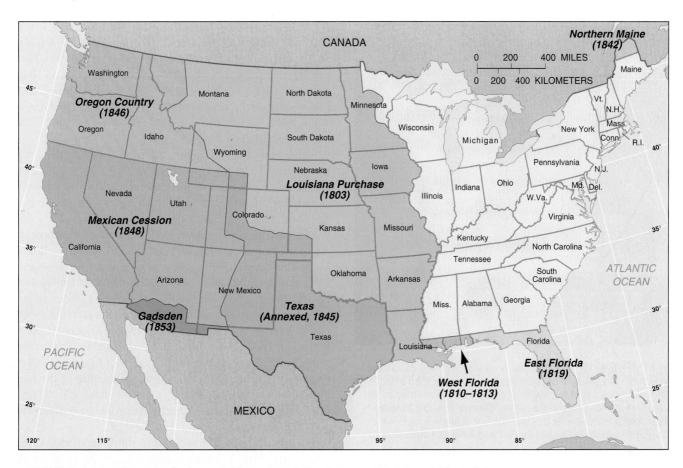

■ FIGURE 4-15 Selected major acquisitions that defined the territorial evolution of the contiguous United States. By 1783, the United States occupied territory westward to the Mississippi River. Over time, however, the United States acquired lands that extended to the Pacific Ocean and, eventually, to Alaska and Hawaii (not shown).
Sources: Adapted from W. Zelinsky, *The Cultural Geography of the United States: A Revised Edition* (Englewood Cliffs, NJ: Prentice Hall, 1992), pp. 80–84; and T. Bailey, *A Diplomatic History of the American People,* 8th ed. (New York: Appleton-Century-Crofts, 1969), selected pages.

River and to the east of the Rocky Mountains. This region included the valuable port of New Orleans near the mouth of the river, the strategic significance of which was clearly apparent to British, French, Spanish, and Americans alike. After a disastrous French military campaign against slave rebellions in Hispaniola, however, and fearful of British intentions in the Americas, Napoleon, Emperor of France, was glad to negotiate a deal that gave the Louisiana Territory to the Americans. While President Thomas Jefferson was interested mainly in New Orleans, his agents consented to an arrangement in 1803 that sold all of the Louisiana Territory to the United States for $15 million—a bargain-basement deal at three cents an acre that nearly doubled the size of U.S. territory.

Adjacent to New Orleans, and posing similar strategic concerns, were the Spanish territories of the Floridas. West Florida was of special interest, since the mouths of many rivers flowing out of Mississippi and Alabama to the Gulf were under Spanish control. After unproductive negotiations with Spain, the United States occupied West Florida between 1810 and 1813 (much to the annoyance of European powers). By 1819, thanks in part to questionable military ma-

neuvers by Andrew Jackson, the Spanish agreed to cede East Florida to the United States for a $5 million settlement.

Farther to the west, American interest in Texas was expressed quite openly, even though official attempts to secure the region supposedly had been discontinued in 1819. The United States tried to purchase Texas from Mexico in the late 1820s, but Mexico refused to sell even though it struggled to populate this outlying province with its own citizens. The shortage of people had long been a problem, to the point that, in 1821, the Spanish government (in one of its last acts as a colonial power in Mexico) granted Moses Austin a large tract of land on which Americans could settle. The Mexican government later made similar grants, and by the mid-1830s, about 30,000 American settlers had moved to Texas. The infusion of Americans, who now saw Texas as the next great frontier to be occupied, created tensions that led Texans to revolt against Mexico in 1835. A short but bloody conflict followed that included battles at the Alamo, Goliad, and San Jacinto and produced historical figures like Antonio de Santa Ana, Sam Houston, and Davy Crockett (Figure 4–16). Mexico lost the conflict and was forced to grant Texas its independence in 1836. Texas subsequently

■ **FIGURE 4-16 The Alamo.** Texas was originally conquered and settled by the Spanish, who lost the territory in the 1830s. Although the battles of the Alamo and elsewhere contributed to the failure of Mexico to retain the region, distance from the Mexican heartland and the ever-increasing number of Anglo settlers were long-term geographic factors that favored American political and economic interests.

requested annexation to the United States, but the issue of slavery prevented the United States from acting promptly on this request, and Texas declared its status as an independent republic in 1838. After a highly contentious debate in Congress, and thanks to the less-than-subtle manipulation of British and French interests by Texans, Congress agreed to annexation in 1845.

The annexation of Texas led to war with Mexico in 1846, a war that the United States entered on the basis of disputes with Mexico over the western border of Texas. The war was prosecuted by the United States with vigor, and by 1848, U.S. troops had invaded Mexico City. The subsequent treaty led not only to a resolution of complaints regarding Texas (that is, a border at the Rio Grande River), but the cession of California and New Mexico. In return, Mexico received $15 million from the United States. When the treaty was approved in 1848, Mexico had lost almost half of its territory to the United States. The U.S. boundary with Mexico was further "tweaked" in 1853 with the Gadsden Purchase, which provided land in southern Arizona for a rail route.

Meanwhile, as the United States was enlarging its southern territory at the expense of Mexico, it was negotiating for clear title to the "Oregon" territory in the Northwest. Perhaps 5,000 Americans had made the difficult trek to the fertile Willamette Valley by 1845, and many more were on the way. Both Britain and the United States claimed the Oregon country, and neither party was initially disposed toward compromise. The United States claimed land as far north as the 54th parallel (well into modern British Columbia), and

British claims extended southward to the Columbia River. Manifest destiny again colored the debate in the United States, and war with Britain was not out of the question. Fortunately, calmer heads prevailed on both sides, and an 1846 agreement set a boundary at the 49th parallel, the current line separating the United States from Canada. The United States had acquired land that would become Oregon, Washington, and Idaho.

Other boundary settlements filled out the domain that we recognize today as the United States. These adjustments included selection of the 49th parallel in 1818 as the line separating the United States and British North America west of the Great Lakes (and, as indicated above, eventually to the Pacific), and the settlement in 1842 delineating a final boundary between Maine and Quebec. The last major territorial acquisitions included the purchase of Alaska from Russia in 1867 and the annexation of Hawaii in 1898. In the case of Hawaii, the manifest destiny that propelled America to become a continental power in the mid-1800s was transformed into an international cause that motivated overseas expansion in the late 1800s. Even Cuba and Canada were targeted by some American politicians. It is important to acknowledge, however, that the international political culture of the late nineteenth century valued such aggressiveness, and the territorial expansion of the United States paled in comparison with that of many European countries.

By 1870 the spatial evolution of the United States was largely complete. Through negotiation, purchase, and war, the United States had acquired territory many times the size of the original thirteen colonies. The settlement frontiers were in place; now they had to be filled.

The Development and Elimination of Settlement Frontiers in the United States

As western territories became populated and tied functionally to the rest of the country, they lost their frontier status. The geography of western settlement in the United States consequently, was one of a spatial integration process leading to the elimination of frontiers.

For many Americans, however, the frontier was more than a territory to be settled—it was a state of mind to be understood. Perhaps no one articulated this position better than Frederick Jackson Turner who, in 1893, presented what became known as the Turner Thesis. He argued that the meaning of much of American history could be discerned only through the prism of the frontier experience. The challenges and opportunities of the frontier molded character, shaped institutions, and provided Americans with a constant source of rebirth. While scholars have generally been reluctant to embrace the Turner Thesis—Turner's perception of the frontier took on almost mythical qualities—most will acknowledge its importance as a reflection of the nineteenth-century American mindset.

Settlement frontiers were populated by a variety of groups, including people moving from, or otherwise influenced by, the core areas on the Atlantic Seaboard. As they migrated to the West, they carried their cultures with them, setting into motion a spatial diffusion process that shaped the cultural landscapes of the frontiers.

The spatial diffusion and integration processes were slow initially. Westward migration meant traveling by land; and before the mid-nineteenth century, overland travel was mainly by foot or animal. Relative to water transport, overland travel was expensive and painfully slow. In 1800, for example, traveling from New York to what is now St. Louis by land could take a month. Overcoming this **friction of distance**—that is, the time and effort required to travel across the continent—required the technical innovations of America's Industrial Revolution.

The nineteenth century was a period of transforming innovations in transport technology, all of which could be applied to the elimination of frontiers. A program of canal building in the early 1800s, which took advantage of the cost advantages of water transport, led to the construction of a series of canals in the east. Particularly prominent was the opening of the Erie Canal in 1825, which connected the Great Lakes and the Hudson River. With the Erie Canal, the cost and time required for shipping goods from Buffalo to New York City plummeted from $100 a ton and 20 days to $5 a ton and 5 days. The Erie Canal facilitated the movement of grains, forest products, and minerals from west of the Appalachians to the East. New York became the center of a new commercial universe, largely at the expense of Boston. While the original explosion of interest in canal building was relatively short-lived, thanks to the promise of other technologies, canals continued to be built, especially in the Great Lakes region. These waterways, many of which were built from the 1850s on, are still in use today. They connect the Great Lakes with each other, with the Mississippi River system, and with the Gulf of St. Lawrence. Today the various waterways in North America penetrate 2,000 miles into the interior (Figure 4–17).

The railroad era began even before the short canal boom ended. Although occasionally competitive with canals, railroads generally complemented water transportation and greatly increased the significance of the Great Lakes as an interior waterway. By the middle of the nineteenth century, new transportation focal points had been established. Rail networks focused on selected coastal ports, such as New York, and converged on interior ports, such as Chicago and St. Louis. By the end of the nineteenth century, both the United States and Canada had transcontinental networks. The integrating effects of the railroad can hardly be overstated. The relative speed of rail transport made it easy to overcome the friction of distance for both passengers and freight, and it was only a matter of time before the railroad, more than anything else, would eliminate the frontier. By the 1860s, the trip from New York to St. Louis took a mere two to three days.

The American Midwest—defined roughly as the Great Lakes states, the Ohio River Valley, and the upper Mississippi River Valley—became a popular frontier destination after the American Revolution. The gently rolling, forest-covered landscape possessed some of the best agricultural soils found anywhere. Settlers from the Middle Atlantic Core region brought their knowledge of mixed farming, which was easily adapted to middle western environmental conditions.

The result of this diffusion process was an agricultural heartland that quickly overshadowed the Middle Atlantic region and produced a surplus of wheat, corn, pork, and beef—commodities that were exported not only to other parts of the United States, but to foreign destinations.

The frontiers of the American South—defined generally as the territory from Alabama to east Texas—were settled at about the same time as the Middle West. Settlers brought with them, or were influenced by, the culture of the southern core area, which included reliance on the plantation system and its focus on the cultivation of commercial export crops. This diffusion process, however, almost did not happen. Prior to 1800, southern commercial agriculture was based on tobacco, a type of cultivation that depleted soils of nutrients and was beginning to erode the economic foundations of the plantation system. Had it not been for Eli Whitney's development of the cotton gin in 1793, it is reasonable to speculate that the plantation system and slavery might have slipped into economic obscurity. The cotton gin permitted the efficient separation of seeds from cotton fibers in short-staple cotton and enabled cotton to be grown in both upland and lowland regions in the South. Driven by a growing demand for cotton fiber in British textile mills, cotton production increased tenfold in the South in the space of 10 years. With cotton, the plantation system was revived and was spread as far west as the 24-inch (600-millimeter) precipitation line in Texas, the line beyond which it was too dry to grow cotton profitably without irrigation.

Continued westward expansion beyond these two initial settlement frontiers was more difficult. The "Great American Desert," what we now call the Great Plains, lay in the path. As mentioned above, this "desert" was not actually a desert, but its extreme temperatures, relative dryness, hard soils, and largely treeless landscape made it unappealing to settlers, especially when more attractive lands lay to the West. By the late 1840s, the Great American Desert was being "leapfrogged" by settlers en route to the farmlands of Oregon's Willamette Valley, and to the gold fields and fertile soils of northern California.

One migration, that of the Mormons, stands out as an exceptional cultural event in the spatial integration of America's settlement frontiers. In effect, this migration produced a new American culture core in Utah from which a separate round of spatial diffusion later occurred. Originating in western New York in the 1820s, the Mormon culture emphasized hard work and thrift, combined with a strong sense of community life and civic obligation. This culture contrasted markedly with the buccaneering individualism that characterized many of the Mormons' gentile counterparts. Public opposition to Mormon beliefs (and Mormon economic successes) led the Mormon community to move progressively farther to the West, eventually settling in Illinois. The murder of their leader, Joseph Smith, and growing harassment by the secular community convinced the Mormons that they needed to leave the more densely populated part of the United States if they were going to practice their religion in peace. Beginning in the late 1840s, tens of thousands of Mormons made the difficult trek to the deserts of the Great Salt Lake Basin in Utah, where they

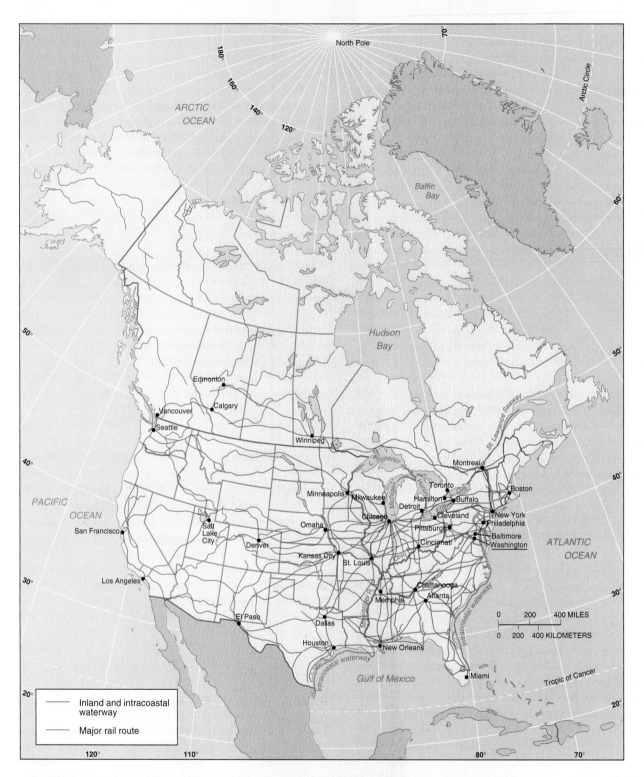

■ **FIGURE 4-17 Major rail routes and inland waterways of Anglo America.** Of great advantage to the United States has been a system of natural waterways, coastal and interior, that have been further linked by rails and have contributed much to the economic integration of a large and rich resource area.

established settlements that were largely isolated from the eastern culture cores. The Mormons developed a thriving agricultural economy based on irrigation (Figure 4–18). To this day, Salt Lake City remains the heart of this culture group.

Eventually, even the Great American Desert yielded to the settler's plow. While hostile environments and persistent conflicts with indigenous groups delayed the early spatial in-

tegration of this settlement frontier, by the 1870s and 1880s the Great Plains were succumbing to the land appetites of both European immigrants and Americans, and to the technologies of industrial America. These technologies included railroads to transport settlers to this frontier, barbed wire to fence in land where wood was unavailable, windmills to pump water out of the ground, and steel plows to break the

■ FIGURE 4-18 Irrigated farmland of the Salt Lake Valley. Mormon settlement of the Salt Lake Valley, like other Anglo settlements of the West, relies heavily on irrigated agriculture. The source of much of the irrigation water is the spring and summer runoff of melting snow, which accumulates during the winter on nearby mountain slopes. The Wasatch Mountains visible in the distance also contain many world-class ski resorts.

heavy prairie sod. By the end of the 1890s, the frontier (excluding Alaska) was officially closed.

Territorial Acquisition and Settlement in Canada

Westward expansion was also important to Canada's nation-building experience, although the Canadian and American approaches were hardly identical. Important differences appeared in the way that Canada became a country and the manner in which Canada opened its western regions.

To begin with, the people of Quebec in the 1770s were largely indifferent to the colonial uprising against the British Crown in the south. In spite of (and perhaps because of) colonial attacks on Montreal and Quebec City, Quebec and the maritime colonies remained loyal to Britain. After the Revolution, what was left of British North America became an important destination for British sympathizers, known as **Loyalists**, who no longer felt welcome in the lower thirteen colonies. Several tens of thousands of Loyalists made the journey northward, many in sea evacuations made possible by the Royal Navy.

The Loyalists established important English-speaking communities in Nova Scotia and New Brunswick, and in the Great Lakes region. In so doing, they created an English-speaking counterweight to French-speaking Quebec and contributed to an Anglicization process that eventually would place French speakers in the minority. In 1791, this linguistic divide was formally recognized with the establishment of French-speaking **Lower Canada** on the lower reaches of the St. Lawrence River and the Gulf of St. Lawrence (modern Quebec), and English-speaking **Upper Canada** on the upper reaches of the St. Lawrence and in the Great Lakes region (modern Ontario). The promise of free or low-cost land in the Great Lakes region led to additional substantial immigration by Irish, Scots, English, and Welsh to Upper Canada. Cultural conflicts and less attractive land inhibited similar immigration into Quebec.

Canada officially came into existence as a self-governing country with the implementation of the British North America Act on July 1, 1867. The new country embraced parts of modern Ontario, Quebec, New Brunswick, and Nova Scotia. As Canadians are quick to emphasize, their country became independent not through a "revolution," as was the case in the United States, but through an "evolution" in the thinking of Canadians and the British Crown. The government that was created combined an American type of federalism—a confederation of provinces—with a British parliamentary form of government—a House of Commons and a Senate.

Unlike in the United States, the western lands into which Canada expanded were already under Canadian or British control. While a sense of manifest destiny may have been implied in the Canadian expansion process, its expression was far more subdued than in the United States—but then, there were no foreign powers controlling territory towards which Canadians may have felt a sense of destiny. By 1870, Manitoba (the Red River area) was brought into the Confederation, as were British Columbia in 1871, Prince Edward Island in 1873, and Alberta and Saskatchewan in 1905. Newfoundland did not join the Confederation until 1949 (Figure 4–19).

As in the United States, Canadian frontiers contained indigenous populations that often stood in the way of European settlement. Some argue that the presence of the scarlet-coated North West Mounted Police, who preceded the settlers, created a civilizing influence that limited the number and severity of conflicts. Armies had to be raised to subdue rebellions only twice, in 1870 and 1885. Others argue that, the relatively peaceful settlement process notwithstanding, the long-term plight of Canadian Indians was only marginally improved over that of their U.S. counterparts.

Immigration into the United States and Canada

The United States and Canada are thus lands largely peopled by immigrants. Immigration is the movement of people into a country of which they are not native residents. Much of the spatial integration process would not have occurred had it not been for immigrants, and without them, the texture of North American culture would have evolved much differently.

U.S. geographer Wilbur Zelinsky observed that immigration into the United States can be divided into two major eras.[1] The first, or "colonial," era occurred between 1607 and 1775. Most of the European migrants were from the British Isles, followed by Germans, Dutch, and Swedes. These migrants generally came with the expectation that they would remain English, or Scottish, or German once they arrived in the New World and that there would be no acculturation process. In addition, large numbers of Africans were brought involuntarily to the United States as slaves, until the slave trade was prohibited in 1808—prohibited as

---

[1]W. Zelinsky, *The Cultural Geography of the United States: A Revised Edition* (Englewood Cliffs, NJ: Prentice Hall, 1992), Chapter 1.

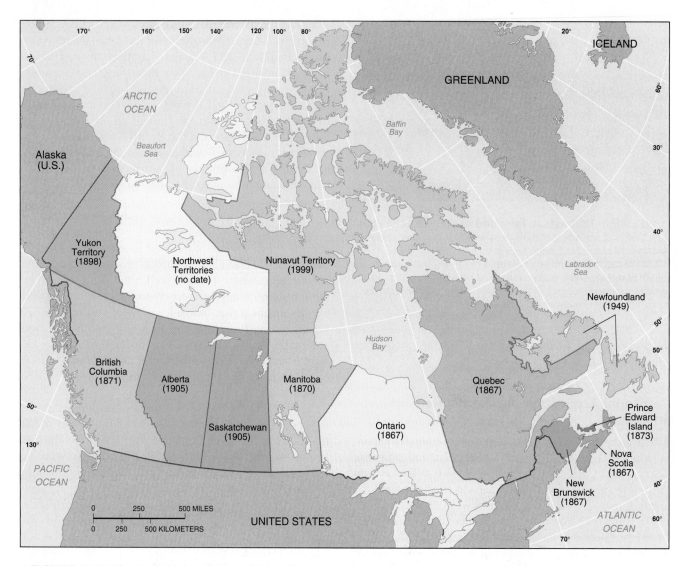

■ **FIGURE 4-19 The territorial evolution of Canada.** By 1912, after final adjustments to the territories of Manitoba, Ontario, and Quebec, the current political geography of Canada was largely in place. Note that a new territory, Nunavut, was carved out of the Northwest Territories in 1999.
Sources: J. Meeker, "Canada: Path to Constitution," *Focus* 33 (1982): 1–16; and Statistics Canada online database, 2000.

a matter of law, it should be added, but not always as a matter of fact. Not surprisingly, the African acculturation process was painful. The second, or "national," era lasted from roughly 1820 to the present. These migrants were faced with a dynamic culture that required major adjustments in terms of language and lifestyle—in other words, they experienced "culture shock." The national era can be divided into three smaller segments, each with its own characteristics: the Northwestern European wave (1820–1870) was still heavily British, Irish, German, and Dutch, but included some of the first migrants from Asia and Latin America; the "Great Deluge" (1870–1920) witnessed the migration of more than 26 million people to the United States, many of whom came from traditional northwestern European sources, but many more of whom traveled from Scandinavia, eastern and southern Europe, China, Japan, and Latin America; and the miscellaneous influx (1920 to the present) made up of a wide variety of origins, especially Asian and Latin American.

Immigrants settling in the United States tended to cluster in regions, many of which are still identifiable (Figure 4–20). This clustering reflected multiple processes, including "environmental affinity," in the words of Zelinksy, or the selection of a familiar physical habitat. Such affinity helps explain the presence of a large number of Scandinavians in the upper Midwest, Dutch farmers in southwestern Michigan, and Italians and Armenians in parts of California's Central Valley (Figure 4–21). In other cases, relative proximity to home led Asian immigrants to focus on the West, Hispanic immigrants the Southwest and south Florida, and European immigrants on Atlantic port cities. In addition, economic circumstances influenced clustering patterns, as in the case of Japanese horticulturists in southern California, Portuguese fishermen in Massachusetts, and Basque shepherds in the arid West. It is noteworthy that most European immigrants during the great waves of the nineteenth and early twentieth centuries avoided the U.S. South.

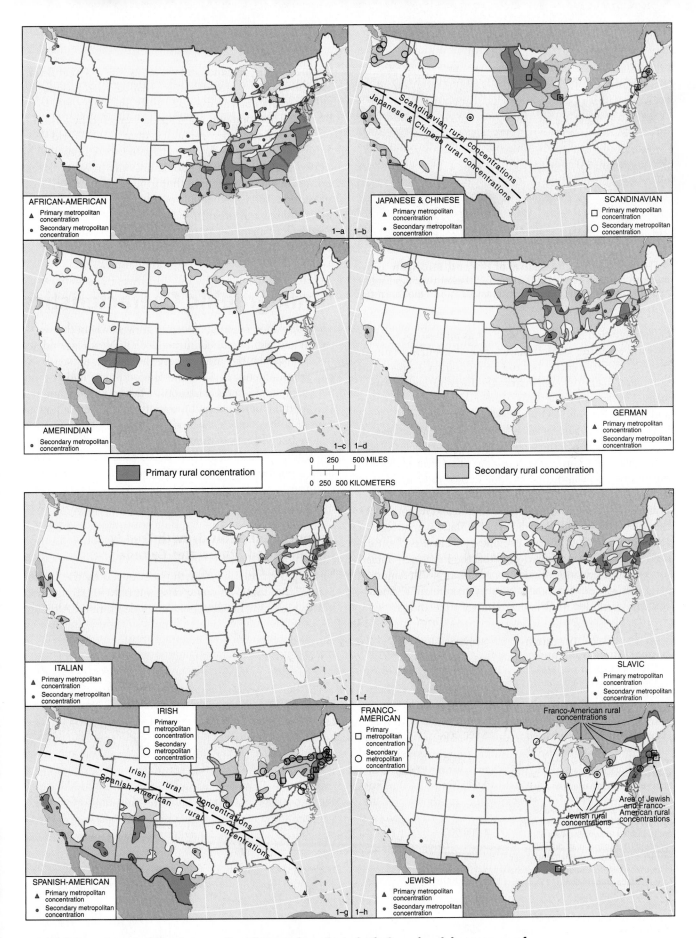

FIGURE 4-20 Generalized areas of settlement for selected ethnic and racial groups, as of the 1960s. While immigrants ultimately settled throughout the United States, specific regions became associated with specific groups of people.

Source: Adapted from W. Zelinsky, *The Cultural Geography of the United States: A Revised Edition* (Englewood Cliffs, NJ: Prentice Hall, 1992), pp. 30–31.

■ **FIGURE 4-21 Tulip festival of Holland, Michigan.** Large numbers of Dutch immigrants settled southwestern Michigan in the mid-nineteenth century. The region continues to bear the Dutch cultural imprint.

Immigration into Canada was just as intense and nation-altering as in the United States, only it occurred later in history (Figure 4–22). Prior to Confederation, Canada was populated mainly by people from the British Isles, by Americans (Loyalists and others who came looking for cheap land), and by the French in Quebec. Most of these people lived in the east. Few Europeans inhabited the great expanse of land that lay between the Great Lakes and the Rocky Mountains, that is the northern extension of the Great Plains and the Interior Lowlands that became known to Canadians as the **Prairies**. As with America's Great American Desert, the Prairies were initially perceived as hostile and remote. With the completion of the Canadian Pacific Railway main line in 1885, however, the Prairies became easily accessible. Furthermore, settlers soon discovered that the Prairies contained the last large expanse of good farmland in North America. Migrants came from everywhere to make claims. Source countries included the Ukraine, Hungary, Iceland, Sweden,

Germany, Italy, Finland, and China, as well as Great Britain and the United States. Between 1896 and 1911, about 2.5 million immigrants arrived in Canada, many en route to the Prairies. The population of Canada increased by a third in the first decade of the twentieth century. After World War I immigration continued, particularly to urban areas, and included large numbers of people not only from traditional European and U.S. sources, but from Commonwealth Asia and Africa. Today, the people of Canada embrace a cultural spectrum of such breadth that Ukrainian, Sri Lankan, and Portuguese surnames (among many others) are almost as common as British and French. Canada, like its neighbor to the south, has become a cultural mosaic.

Demographic Characteristics

The population of Anglo America grew slowly in the seventeenth and eighteenth centuries, even in New England and the Chesapeake Bay areas, where the English presence was most pronounced. In 1776 the United States had perhaps 3 million people, and Canada only a fraction of that number. During the nineteenth and twentieth centuries, the populations of both countries expanded substantially. It is difficult to believe that, from these humble beginnings, by the year 2004 the U.S. population had risen to 294 million and the Canadian population to 32 million (Figure 4–23). By 2020, both countries may see substantial gains in population due partly to natural increases, but probably due more to immigration.

Recent Population Growth Trends in the United States and Canada

The rapid population growth experienced by the United States after 1800 was a response not only to immigration, but to high birthrates and declining death rates. Although

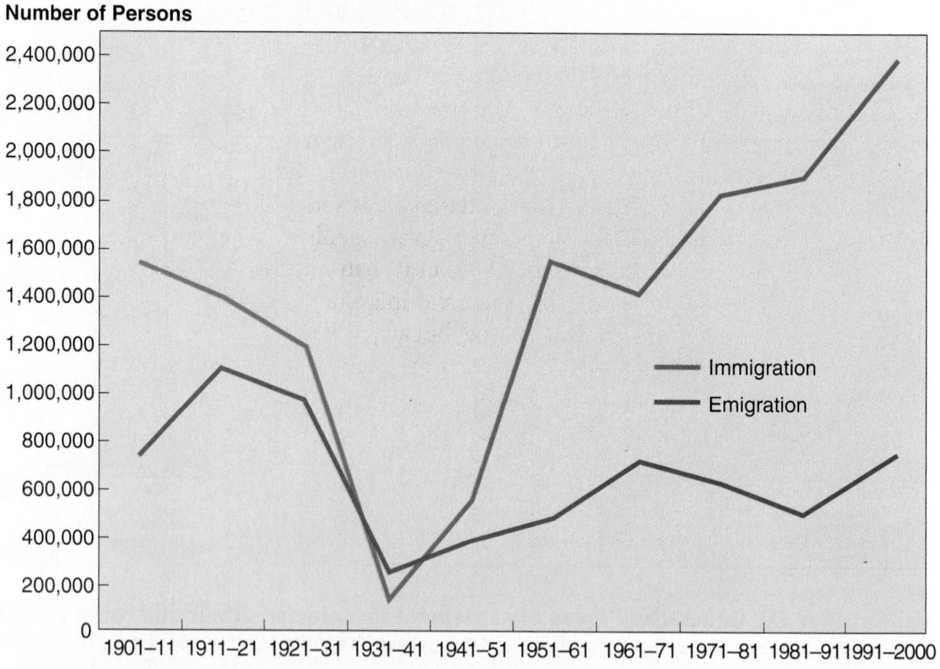

■ **FIGURE 4-22 Twentieth-century immigration and emigration trends in Canada.** Canada, like the United States, is a land of immigrants, which was especially the case during the twentieth century. Unlike the United States, Canada has experienced considerable emigration, or out-migration, often to the United States. **Source:** Adapted from Statistics Canada, *http://www.statcan.ca/english/Pgdb/People/Population/demo03.htm*. Statistics Canada information is used with the permission of Statistics Canada. Users are forbidden to copy the data and redisseminate them, in an original or modified form, for commercial purposes, without the expressed permission of Statistics Canada. Information on the availability of the wide range of data from Statistics Canada can be obtained from Statistics Canada's Regional Offices, its World Wide Web site at *http://www.statcan.ca*, and its toll-free access number, 1-800-263-1136.

Number of Persons

Legend: Immigration, Emigration

X-axis: 1901–11 1911–21 1921–31 1931–41 1941–51 1951–61 1961–71 1971–81 1981–91 1991–2000

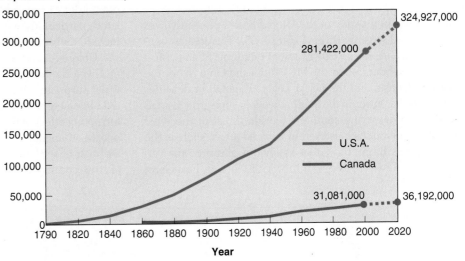

■ **FIGURE 4-23 Population growth trends in the United States and Canada, with projections to 2020.** Note the upward trend in both cases, especially in the United States. The U.S. trend line begins in 1790. Due to data limitations, the Canadian line begins in 1861. The actual year of the Canadian census is one year after the U.S. census—the 1980 Canadian data shown in the table, for example, were actually for 1981.
Sources: U.S. Bureau of the Census, *Statistical Abstract of the United States: 2001* (Washington, D.C.: Government Printing Office, 2001), Numbers 1, 3; Statistics Canada, Population and Growth Components, *http://www.statcan.ca/english/ Pgdb/People/Population/demo03.htm* and *http://www.statcan.ca/english/Pgdb/ People/Population/demo02.htm*.

precise figures are not available, annual birthrates and death rates in the early nineteenth century may have exceeded 5 percent and 2 percent, respectively. A century later (during the Depression of the 1930s) the birthrate had decreased to 1.8 percent per year, but it rose sharply after World War II, possibly to compensate for wartime postponements of child bearing and as a response to postwar economic prosperity. During the baby boom of 1946 to 1965, the birthrate reached a new modern-era high; by 1955, it was 2.5 percent. Subsequently, however, it dipped to a new historic low of 1.46 percent in 1975. When the baby boomers themselves reached childbearing age, a very modest upturn took place, and by the early 1990s the birthrate was slightly above 1.6 percent. Because of the increased population, that rate represents nearly 4.1 million births per year, not so different from the 4.3 million at the peak of the baby boom. Birthrate trends are shown in Figure 4–24. Death rates declined only slightly during the postwar years, down to 8.7 percent in 1999. The relatively small gap between birthrates and death rates produced an overall population growth rate (excluding immigration) of .58 percent by 1999, down from the postwar high of 1.57 percent in the mid-1950s.

The Canadian demographic experience has been generally similar to that of the United States. Canada grew mainly by natural increase between 1867 and 1900; but its natural increase was limited by a low fertility rate. Net migration was negative during most of that early period, but after 1900 a large influx of immigrants arrived from Europe, slowing the decline of birthrates. Like the United States, Canada experienced a low birthrate during the Depression and a sharp rise after World War II. The Canadian baby boom was also followed by a decline in the birthrate during the late 1960s and early 1970s. The birthrate in 2004 was slightly more than 1 percent. Unlike in the United States, emigration, or out-migration, from Canada in the postwar years has been substantial. In 2001, about 3.4 percent of the Canadian population lived abroad.

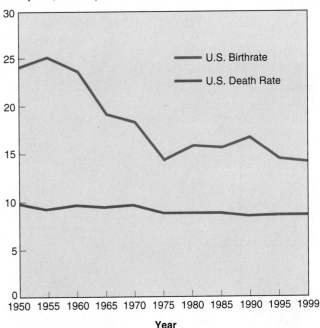

■ **FIGURE 4-24 Post-World War II birth and death rate trends in the United States.** These figures show the number of births and deaths per 1,000 people in the United States since 1950. Note how the death rate line remained relatively stable during this period, but the birthrate line dropped significantly. No doubt the postwar "baby boom" accounts for part of the high birthrate in 1950, although birthrates earlier in the twentieth century were typically higher than today.
Source: U.S. Bureau of the Census, *Statistical Abstract of the United States: 2001.* (Washington, D.C.: Government Printing Office, 2001), Number 68.

Population Distribution

Most of the population of the United States presently lives east of the Mississippi River (Figure 4–25). The greatest concentrations are in the northeastern quadrant of the country, the area bounded by the Mississippi and Ohio rivers, the Atlantic Ocean, and the Great Lakes. Population densities are somewhat lower in the South, except in such growth areas as the Piedmont and southern Florida. Most of the West Coast population is clustered in lowland areas, such as the Los Angeles Basin, the Great Valley of California, the valleys of the Coastal Ranges in the vicinity of San Francisco,

and the Willamette Valley and Puget Sound Lowland. The remainder of the western United States is sparsely populated, particularly west of the 100th meridian. Exceptions include metropolitan areas such as Phoenix, Arizona; Denver, Colorado; and Salt Lake City, Utah.

Given the harsh environments of the Canadian north, it is not surprising that most Canadians live within 200 miles (322 kilometers) of the U.S. border. Were it not for the interrupting effect of the Canadian Shield (which supports few people), Canada's population distribution would appear as a ribbon of settlements stretched across the southern edge of the country. The majority (around 60 percent) of

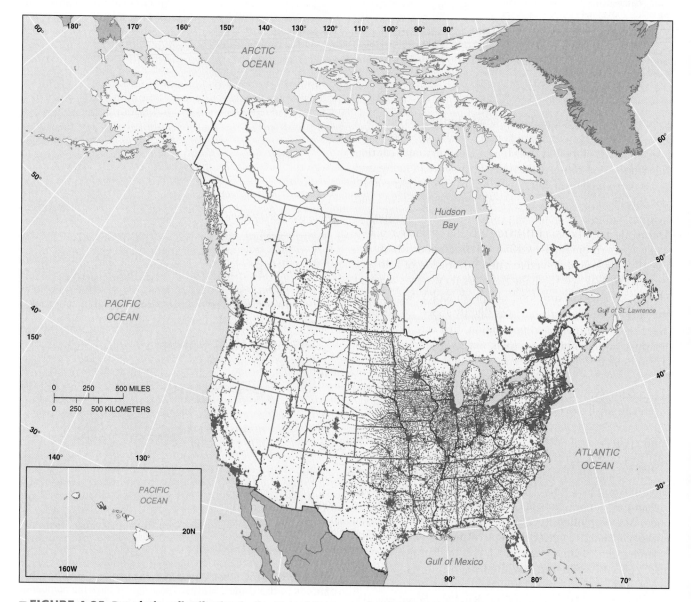

■ **FIGURE 4-25 Population distribution in the United States and Canada as indicated by place density.** The unevenness of population in the United States and Canada reflects numerous influences, including natural environments, early settlement areas, level of urban and industrial growth, and ongoing redistribution (mobility). Notice the relative absence of settlements in the harsh northern climates of Canada and Alaska. *Note:* Each dot indicates one place; larger places such as cities are given slightly larger dots to enhance interpretation, but the intent is to show density patterns rather than proportional circles. Four circle sizes were used for the United States. Data limitations permitted only two circle sizes for Canada and not every settlement is shown.

Source: Environmental Systems Research Institute (ESRI) map databases for 2000, *www.esri.com.*

Number of Persons

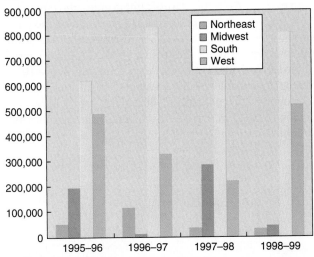

■ **FIGURE 4-26 Net migration by United States region.** In the years shown, more people moved to each region than moved out. Notice that the Northeast and the Midwest had small net gains, while the South and the West received many more people than left.
Source: Adapted from U.S. Bureau of the Census, *Statistical Abstract of the United States: 2001* (Washington, D.C.: Government Printing Office, 2002), Number 25.

Canadians live between Windsor, Ontario, and Quebec City, Quebec. The other major population concentration is on the West Coast around the cities of Vancouver and Victoria, British Columbia. The only exceptions to this southern population ribbon are selected small concentrations in the prairie provinces, most notably Alberta's capital city, Edmonton.

Population Redistribution

Population distribution patterns in the United States and Canada have always been dynamic and, for a long time, focused on westward migration. Many Americans continue to embrace an "Old West" narrative that views the West as a region of boundless opportunity and unlimited new beginnings. This perception has contributed source material for countless novels and movies. But there were other migrations. After the American Revolution, British sympathizers from the United States migrated to Canada. Later in the nineteenth century and into the twentieth century, African Americans from the U.S. South fled difficult social and economic circumstances for the promise of jobs and improved living conditions in northern cities. Also in the twentieth century, Americans rediscovered the West Coast and, by the late twentieth century, the South. Once again job losses and general economic distresses played a role—we remember, for example, the "Okies" who made the trek to the promised land of California during the Depression years in search of opportunities denied them in Depression-weary Oklahoma. In the 1970s and 1980s, the South was perceived to be a good destination for workers displaced by struggling industrial economies in the upper Midwest. Throughout the latter half of the twentieth century, climatic amenities lured retirees to Florida and Arizona, among other southern destinations. In general, this rediscovery of the South and West continued into the 1990s, generally at the expense of the Midwest and Northeast (Figure 4–26). This process is part of a larger Sun Belt migration stream. Canada, on the other hand, has no "Sun Belt," unless the relatively snow-free (but overcast) conditions of Vancouver qualify as such; and indeed British Columbia has been one of Canada's major growth provinces (Table 4–1). Much of Canada's recent

Table 4-1	Canadian Population Data			
	Population, 2005 (estimate)	**Population 2001**	**Population Change, 1996–2001**	**Net Interprovincial Migration, 2000–2001**
Canada	32,078,819	31,081,900	7.7%	
Newfoundland	516,486	533,800	−3.3%	−3,541
Prince Edward Island	137,734	138,500	2.9%	71
Nova Scotia	937,538	942,700	3.7%	−824
New Brunswick	751,257	757,100	2.6%	−81
Quebec	7,568,640	7,410,500	3.8%	−11,782
Ontario	12,449,502	11,874,400	10.4%	17,877
Manitoba	1,174,645	1,150,000	3.2%	−3,094
Saskatchewan	995,280	1,015,800	2.6%	−10,453
Alberta	3,223,415	3,064,200	13.6%	25,748
British Columbia	4,219,968	4,095,900	10.0%	−12,689
Yukon	31,227	29,900	−2.8%	−846
Northwest Territories	42,944	40,900	3.1%	606
Nunavut	29,683	28,200	14.0%	220

Sources: Adapted from *Statistics Canada* population estimates, www.statcan.ca/english/Pgdb/People/Population/demo02.htm; *Statistics Canada* population and dwelling counts by province, http://www12.statcan.ca/english/census01/products/standard/popdwell/Table-CD-P.cfm?PR=10.0; and Statistics Canada components of population growth data. The 2005 estimates were taken from Statistics Canada, www.statcan.ca/Daily/English/050324/d050324c.htm.
Note: Canadian census figures reveal substantial growth in population between 1996 and 2001. Below-average growth occurred in the Maritime provinces, Quebec, Manitoba, Saskatchewan, and the territories (except Nunavut). Above average growth was found in Alberta, Ontario, and British Columbia. The interprovincial migration data indicate the extent to which Canadians moved to or out of a particular province, with negative numbers indicating more people leaving than coming. Even though the data are estimates for only 1 year and hardly establish a trend, they do show that, for the most part, people are leaving the slow-growth provinces. Alberta and Ontario were the most popular destinations.

interprovincial migration, however, has focused on Alberta and Ontario, due in no small measure to favorable economic conditions. Of course, we should not overlook the massive Canadian flocking to Florida's beaches every winter, a holiday tradition among many Canadians in Ontario and Quebec. In effect, Florida is also Canada's "Sun Belt."

An old but continuing process of population redistribution involves North American urban areas and the migration of people from farms to cities and then outward to suburbs. The shift from an agrarian to an industrial society started early in the nineteenth century and continued into the twentieth century. As factories were built in cities, the lure of factory employment attracted people from the countryside, swelling the sizes of cities. Urban growth continued during the twentieth century but was spurred more by expansion of tertiary activities than by industrial growth. By the 1970s, demographers observed a movement of population to smaller metropolitan and nonmetropolitan areas, suggesting a counterflow away from large cities. Whether this countertrend becomes permanent remains to be seen.

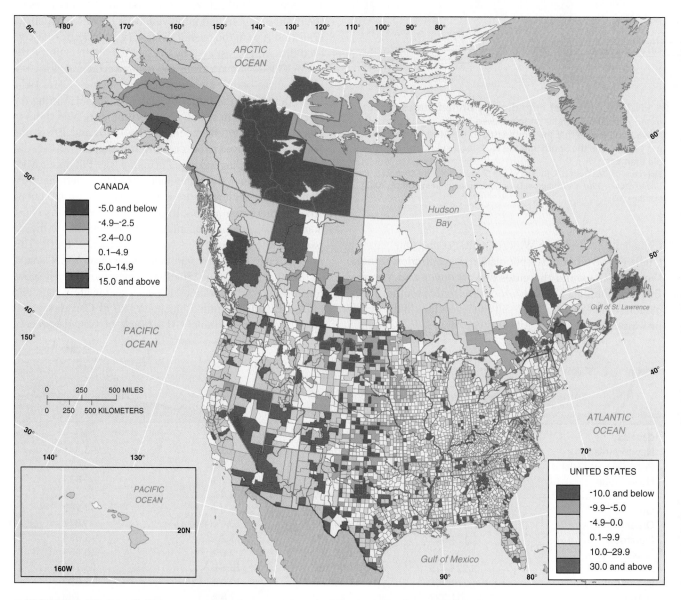

■ **FIGURE 4-27 Population percentage change estimates by county in the United States, 1990–2000, and by county and division in Canada, 1996–2001.** Recent population growth in the United States has focused on the urban areas of the South and the West. Regions of population decline include portions of the Great Plains and the lower Mississippi Delta. Canada also has experienced growth in major metropolitan areas and decline in western farming areas. In addition, substantial population losses can be found in parts of the Maritime provinces and Quebec. Note that percentage population trends in Canada's territories should be interpreted with caution as small changes in number can lead to large percentage changes due to low population levels.
Sources: U.S. data from ESRI county census files. Canadian data from ESRI county and division census files (1996 data) and Statistics Canada population estimates for 2001 by census division, *http://www.statcan.ca/english/census01/products/standard/popdwell/Table-CD-P.cfm?PR=10.0*

County-level data for the 1990s provide clearer insight into the subtleties of U.S. population change (Figure 4–27). First, central cities, while not necessarily losing population, generally did not show meaningful population growth. Counties surrounding central cities, on the other hand, often recorded exceptional growth (the Chicago and Atlanta metropolitan areas provide good illustrations). Second, growth regions tended to focus on the metropolitan South and a variety of counties in the West. Note, for example, the growth in most of Florida, much of the Piedmont from North Carolina to central Alabama, the Dallas-Houston-San Antonio triangle in Texas, most of Arizona, much of the Albuquerque-to-Denver corridor, Nevada, and the Pacific Northwest. Missing from many state-level analyses of population change, but visible here, are the clearly defined pockets of population decline. For example, the Great Plains, from North Dakota to Texas, experienced declines that occasionally exceeded 10 percent of a county's population. This trend reflects dramatically changing agricultural circumstances. In addition, a second pocket of decline shows the other face of the Sun Belt, one of continuing economic distress. Note the negative growth figures in the lower Mississippi Delta and the stagnant growth on much of the coastal plain. The pattern of decline can be extended northward along the western side of the Appalachians from Kentucky to Pennsylvania.

Some of the same trends can be identified for Canada (see Table 4–1 and Figure 4–27). Recent (1996–2001) population change in Canada has focused on its central and western regions. The populous urban-industrial area of Ontario has been experiencing positive net migration and exhibited a population growth rate significantly above the national norm in the late 1990s. Alberta experienced significant growth during the 1970s and early 1980s in response to the development of petroleum resources and remains an attractive destination. British Columbia is experiencing its higher growth as a result of both international and interprovincial migration. The only province that showed a population decline in the 1990s was Newfoundland, although more recent estimates may cause other provinces to be added to the list.

Summary

The United States and Canada present highly diverse physical and human settings. The physical structure is composed of the great Canadian Shield, which anchors the continent, with a north-south trending mountain backbone in the east (the Appalachian Highlands) and a series of north-south backbones in the west (the Rockies, Sierra Nevada/Cascades). Between the backbones lie large areas of relatively flat lands that contain much of the continent's population and economic activity. Climates range from the subtropics of the Gulf Coast to the polar tundra of northern Canada, with the most densely settled climates lying in between. The human setting includes expansion into, and occupation of, North America by a culture that was initially largely Anglo-European in character but ultimately embraced a mosaic of cultural groups that included Africans and Asians, as well as Europeans from outside the British Isles. These people eventually occupied all of the United States and Canada, with distinct clusters along the coasts and in the Great Lakes region. In the next chapter, we will use our understanding of the physical environments and human settlement patterns to examine the economic development of the United States and Canada.

Key Terms

Appalachian Highlands	Interior Plateaus
Appalachian Plateau	Lower Canada
Basin and Range	Loyalists
Black Hills	manifest destiny
Blue Ridge Mountains	New England
Canadian Shield	orographic effect
Cascade Mountains	Ouachita Mountains
Coast Ranges	Ozark Plateau
Colorado Plateau	Piedmont
Columbia Plateau	Prairies
continentality	Puget Sound Lowland
culture hearths	rainshadow
diffusion process	Ridge and Valley Province
Fall Line	Rocky Mountains
friction of distance	settlement frontiers
Great Lakes	Sierra Nevada Mountains
Great Plains	spatial integration
Great Valley	Upper Canada
Gulf of Mexico	Willamette Valley
Gulf-Atlantic Coastal Plain	

■ **Extensive versus intensive animal husbandry.** The more traditional herding of cattle in dry rangeland areas (top) moves the animals to fodder and water. Intensive feedlot operations (bottom) maximize the rate of animal growth by bringing food and water to cattle who are kept in close confinement.

Chapter 5

The United States and Canada:
The Geography of Economic Development

- **Agriculture**
- **Resources for Industrial Growth and Development**
- **Manufacturing in the United States and Canada**
- **Urbanization in the United States and Canada**

Commercial economic activity began in the United States and Canada soon after the first European settlers arrived. Wherever transportation was suitable—which usually meant water transportation—commercialism soon appeared. Agricultural commodities, lumber, furs, and fish were produced or gathered for exchange. Thus, primary production and tertiary activities (that is, trade) were important long before the settlement of North America was complete and before manufacturing activities became significant. Agriculture quickly became the economic mainstay, and it remained so for more than two centuries.

After the United States became independent, a growing population, domestic markets, industrial technology, and transportation stimulated manufacturing, which expanded particularly rapidly from 1860 until the end of World War II. More recently, the most pronounced growth has been in areas concerned with the distribution of goods or the provision of services. A similar progression occurred in Canada, although the timing was slightly different. As a consequence, few people are farmers today. As farm workers were lured to the cities and as agriculture became mechanized, the percentage of the workforce engaged in farming fell to approximately 3 percent by the year 2000 in both countries (see Figure 5–1 for historical trends). These employment changes indicate the transformation of the United States and Canada from an agrarian society, one based on agriculture, to a highly developed urban-industrial society, one in which people live predominantly in urban settings and work in secondary or tertiary occupations.

Agriculture

The story of agriculture in the United States and Canada is one of extraordinary success. This success can be seen in the billions of dollars of farm output each year—over $200 billion in the United States by 2002. This success is visible in the extent to which the United States and Canada supply the food needs of other countries—in 2003, for example, the United States accounted for almost 66 percent of the world's corn exports. Agriculture was an early engine of economic development and remains of critical importance to both countries today (see Table 5–1 for historical productivity comparisons). Several factors have contributed to this success:

- **An abundance of good land.** Approximately one-fifth of all of the land in the United States is classified as cropland, though not all of it is cultivated in a given year. The United States has about 5 acres (2 hectares) per person for the production of agricultural goods, either foodstuffs or industrial raw materials, which is a highly favorable ratio. Canada's northern latitudes give it far less usable land, from 4 to 5 percent of the total land area; but Canada's small population relative to the United States means that the ratio of farmland to people is about the same as in the United States at slightly fewer than 5 acres per person.

- **A high level of mechanization.** The days of the farmer and his mule are over; now it is the farmer and his tractor with the air-conditioned cab and the global

127

■ **FIGURE 5-1 Percentage of economically active population employed in agriculture.** Both the United States and Canada have witnessed major declines in the proportions of their workforces engaged in agriculture since 1890.
Source: B. R. Mitchell, *International Historical Statistics: The Americas 1750–1993*, 4th ed. (London: Macmillan, 1998), Table B1.

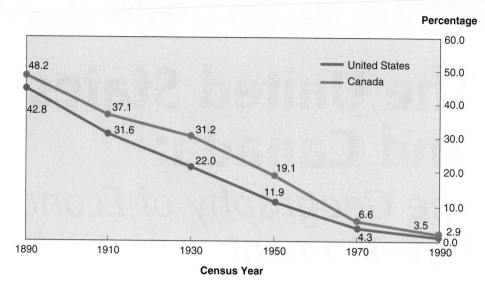

positioning system unit. In 1910, approximately 1,000 tractors were in use on American farms, along with roughly 24 million horses and mules; by the late 1990s, the figure approached 4 million tractors, and not nearly as many horses and mules (Figure 5–2). In addition to machinery, farmers use hybrid seeds, pesticides, herbicides, biotechnology, and scientific farming techniques to increase output per acre. Output per farmer is immeasurably greater than a century ago, thereby requiring fewer people to supply the food and fiber needed to maintain a healthy economy. Today's farmer can produce enough food to support about 70 families.

• **Regional specialization.** Both the United States and Canada occupy large territories with a variety of environmental conditions. Over time, farmers discovered that they were more productive if they concentrated their efforts on what they do best relative to other farmers in other regions. To illustrate this point, assume that the Corn Belt is able to produce more bushels per acre of corn than wheat and produces more per acre of both crops than the Wheat Belt (actually, an accurate assumption); then Corn Belt farmers should focus on corn production and let the Wheat Belt farmers focus on wheat production. In the end, there will be more

corn and wheat for everyone, and the trade process will make up for any regional deficiencies. The principle at work is called **comparative advantage** (at least for the wheat farmers—it would be an "absolute" advantage for the corn farmers) and points out that specialization and trade in the agricultural economy leads to greater productivity overall.

Agricultural Regions

The Corn Belt   The largest expanse of highly productive land in the United States and Canada is the Corn Belt, which extends from central Ohio to eastern Nebraska, to the north into Minnesota and South Dakota, and as far south as eastern Kansas (Figure 5–3). Early pioneers discovered that beneath the forests and prairies of this region lay some of the most fertile soils in the world—soils rich in organic content, and neither too wet nor too dry. In addition, the relatively flat terrain was easy to work, rainfall occurred at the right times of the year (late spring and mid-summer for corn cultivation), and the farmer could expect a frost-free period of at least 150 days.

Historically, corn or maize has been favored by farmers because of its multiple uses—food for people and feed for

Table 5-1   Historical Agricultural Productivity in the United States

	Corn for Grain		Cotton		Wheat	
	Bushels per Acre	Hours Required per 100 Bushels	Pounds per Acre	Hours Required per 500-pound Bale	Bushels per Acre	Hours Required per 100 Bushels
1935–1939	26.1	108.0	226.0	209.0	13.2	67.0
1950–1959	39.4	34.0	296.0	107.0	17.3	27.0
1989–1992	118.7	3.0[a]	650.0	5.0[a]	36.5	7.0[a]

Sources: U.S. Bureau of the Census, *Statistical Abstract of the United States, 1974, 1988, 1993* (Washington D.C.: Government Printing Office, 1974, 1988, 1993).
[a] Average hours, 1982–1986.
Note: All figures are averages for the years indicated.

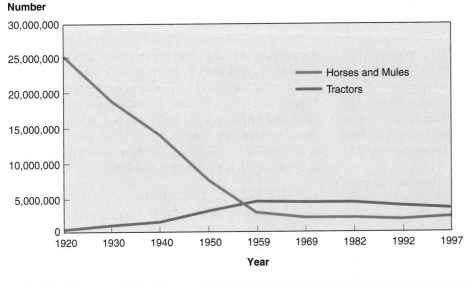

Number

■ **FIGURE 5-2 Horses, mules and tractors used on American farms, 1920–1997.** The number of horses and mules working on American farms has declined substantially since 1920 as farmers have adopted machines. Indeed, by 1997, most of the mules and horses on farms probably were used for purposes other than work.
Source: U.S. Department of Agriculture, National Agricultural Statistics Service, "Trends in U.S. Agriculture," *http://www .usda.gov/nass/pubs/trends/ mechanization.htm.*

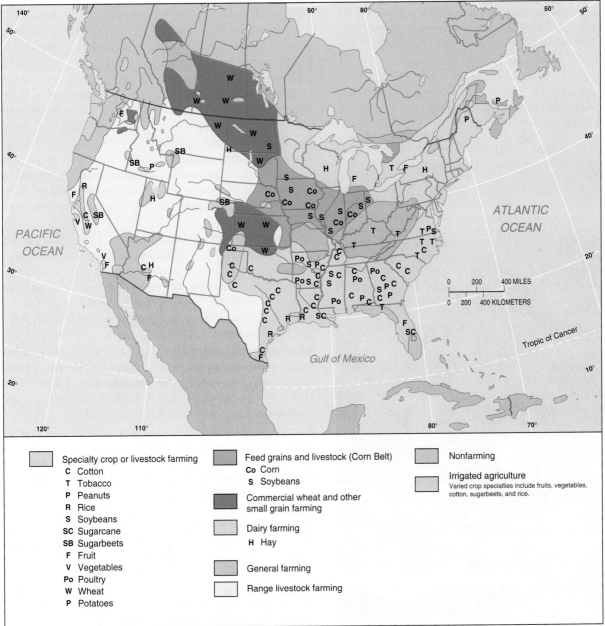

Specialty crop or livestock farming
- **C** Cotton
- **T** Tobacco
- **P** Peanuts
- **R** Rice
- **S** Soybeans
- **SC** Sugarcane
- **SB** Sugarbeets
- **F** Fruit
- **V** Vegetables
- **Po** Poultry
- **W** Wheat
- **P** Potatoes

Feed grains and livestock (Corn Belt)
- **Co** Corn
- **S** Soybeans

Commercial wheat and other small grain farming

Dairy farming
- **H** Hay

General farming

Range livestock farming

Nonfarming

Irrigated agriculture
Varied crop specialties include fruits, vegetables, cotton, sugarbeets, and rice.

■ **FIGURE 5-3 Agricultural regions of the United States and Canada.** A diversity of natural environments has contributed to the great variety in types of agricultural specialties possible in the United States and Canada.

animals—and its high yield per acre. But corn alone no longer supports the Corn Belt farmer. Winter wheat, oats, hay crops, and most especially, soybeans, are grown for additional sources of income, to support livestock operations, and to keep soils fertile. The versatile soybean is now a major source of income. Soybeans can be used for animal feed, as protein-rich human food (tofu, soymilk, soyburgers, cooking oil), and as a component in the manufacturing of everything from paint to plastics (Figure 5–4).

Given the name "Corn Belt," it may come as a surprise that much of the region's income actually comes from the marketing of cattle and hogs. The Corn Belt began as a **mixed farming** region in which equal emphasis was given to crop cultivation and livestock production, with each activity necessary to support the other. Over time, farmers discovered that they could make more money from corn by feeding it to livestock and then selling the livestock for meat—think "corn-fed" steak in this context. Increasingly, the Corn Belt became a "Pork Belt" or a "Beef Belt." Today cattle and hogs are prepared for market in large feedlots, especially in the western part of the Corn Belt. Livestock production accounts for about three-fourths of Corn Belt income, even though most of the farmland is used for cultivation.

■ **FIGURE 5-4 Soybean harvest.** Soybeans are widely cultivated in the eastern United States, with the greatest output coming from the Corn Belt states of the upper Midwest.

The natural advantages of the Corn Belt are enhanced by the location of transportation routes and urban-industrial districts. Various parts of the region lie between or adjacent to the great inland waterways of the United States and Canada, including the Great Lakes, and the St. Lawrence, Ohio, and Mississippi rivers. In addition, one of the densest rail networks anywhere in the world has been constructed in the region. It is easy for Corn Belt farmers to ship their commodities to both domestic and foreign markets.

The Dairy Belt   To the north of the Corn Belt, stretching westward from Nova Scotia and New England to Wisconsin and Minnesota, is the Dairy Belt. While physical conditions are less favorable for agriculture—soils are thinner and less fertile, and growing seasons are shorter—the cool and moist environments promote the growth of hay crops and contribute to high levels of dairy production. Dairy farmers engage in one of two types of production, depending on proximity to markets. Farmers in the east focus on fluid milk production to meet the demand for milk in nearby cities. Fluid milk is heavy and perishable and generally is produced close to the market to avoid excessive transportation costs. The term **milkshed** has been coined to refer to a city's adjacent milk-producing region; cities everywhere, not just in the Dairy Belt, have milksheds. Farther to the west, away from major urban markets, dairy farmers focus on the production of manufactured milk products such as cheese in Wisconsin and butter in Minnesota. Cheese and butter are less perishable than fluid milk and are easier to transport.

Specialty Crop and Livestock Region   This region extends from southern New England to eastern Texas and embraces a wide array of activities. In the Northeast, with its hilly terrain, and its thin and sometimes infertile soils, agriculture was always difficult except in such favored locations as the Connecticut and Hudson river valleys. As much better western lands became available to farmers, the Northeast lost whatever earlier advantages it had, except for one: proximity to large cities. Farmers in the Northeast discovered that city dwellers demanded fruits and vegetables, commodities that often are perishable and difficult to transport and are best grown near consumers. Farmers' efforts, then, became focused on the cultivation of apples, cherries, asparagus, beans, tomatoes, and related crops—commodities that could be trucked to nearby cities. This type of high-value, land-intensive, market-oriented agriculture is known as truck farming. A similar market-gardening region extends from southern Virginia to the Georgia coast. In addition, the Annapolis Valley of Nova Scotia is famous for apples; and parts of Maine, New Brunswick, and most especially, Prince Edward Island, are known for potatoes.

The story in the U.S. South is one of both disappointment and good fortune. On the one hand, southern farmers face big problems: large parts of southern Appalachia are too hilly to farm profitably; the relatively acidic soils of the Piedmont have lost much of the little fertility they once possessed because of repeated row-crop production; and many coastal areas have swamps and sandy **pine barrens** that are not well suited to intensive agriculture. On the other hand, good fortune has come to farmers in the more environmentally favored parts of the South, such as the black-soil belts of Texas and Alabama, the alluvium-rich Mississippi River Valley, and the Nashville and Bluegrass basins. Noteworthy clusters of specialized activities have appeared: tobacco in North Carolina and Kentucky; peanuts in southern Georgia; rice in Louisiana, Arkansas, and coastal Texas; cane sugar in Florida and Louisiana; and citrus in central Florida and south Texas. It is important to note that cotton, which came to define the antebellum South, is no longer dominant except in a few selected locations, particularly the lower Mississippi Valley, and central and west Texas. Disease, insects (the boll weevil), depleted soils, and mechanization were the culprits. In much of the South, cotton has given way to soybeans (although recent attempts have been made to revive cotton in selected areas once abandoned by the cotton economy).

The South also has become a livestock region. While cattle are found throughout the South, it is the rise of poultry broiler production that is most noteworthy. Much of the chicken that is consumed in the United States comes from a series of broiler production areas that begin on the Delmarva Peninsula, continue through north Georgia, and end in Arkansas. Most poultry farms are huge, producing tens of thousands of chickens at a time in a factory-like setting. These farms often are located where other types of farming are no longer profitable and, consequently, have reinvigorated declining agricultural areas. The lay of the land and problems with soil fertility—historic challenges to southern farmers—are largely irrelevant to poultry producers (Figure 5–5).

The Great Wheat Belts
To the west of the Corn Belt lie the wheat belts, characterized by large farms that are highly mechanized. In the **winter wheat belt** of Kansas, Oklahoma, Colorado, and north Texas, wheat is planted in the fall, allowed to lie dormant in the winter, and harvested in the late spring before the onset of summer's scorching heat. Winter wheat is low-gluten soft wheat that is used for spaghetti, crackers, and pastries. In the **spring wheat belt** of the Dakotas, Montana, and Saskatchewan, wheat is planted in the spring and harvested in the late summer. Unlike farther south, winters in the spring wheat belt are too cold for wheat seedlings to survive, but the summers are not so hot as to stress the wheat. The high-gluten hard wheat of this region is excellent for bread. A third wheat-producing region is the Columbia River basin of eastern Washington, which takes advantage of fertile volcanic soils and the relatively dry climates found on the rainshadow side of the Cascades.

When the wheat harvest comes in, a complex set of transportation relationships is put into motion. The wheat is hauled by truck to local grain elevators to await further transport by rail to milling cities such as Wichita, Kansas City, Minneapolis, and Winnipeg. These cities lie on the market side of the wheat belts and constitute convenient stopping points to convert wheat to flour. U.S. and Canadian wheat is exported throughout the world, and it is hardly an over-

■ **FIGURE 5-5 Chicken farm in Lowell, Georgia.** Poultry farming is a common land use in many rural areas of the South.

statement to declare these two countries as the world's breadbaskets.

Wheat is related to the natural grassland vegetation of the prairies and can withstand the relatively dry conditions of the Great Plains (10 to 20 inches, or 250 to 500 millimeters, of annual precipitation). But wheat requires at least some moisture, and many wheat farmers are haunted by the specter of recurring drought. Where possible, particularly on the semiarid western margins of the Plains, irrigation is used to overcome the drought threat.

Farming in the West
Water is the big issue facing western farmers and agriculture is possible only where irrigation is available. Where water is pumped from underground **aquifers** (water-bearing rock strata), or where river water is available, a thriving oasis-type of agriculture is possible. It is important to understand that arid soils generally are not infertile soils, just dry soils. Indeed, some of the most productive and intensively used farmlands in the country are in the steppes and deserts. Numerous examples can be identified: the irrigated cotton lands of west Texas and central Arizona (a new "Cotton Belt"?); the potato lands of the Snake River plains in southern Idaho; the highly diversified vegetable, fruit, and wheat regions of California's interior; the Salt Lake basin of Utah; and the corn and sugar beet areas of eastern Colorado. In the case of Colorado and many other locations, the use of irrigation is clearly apparent from the air by the large green circles of cultivated land produced by center-pivot watering systems (Figure 5–6).

Where irrigation is not possible, little will grow beyond certain types of grasses and xerophytic (drought-resistant) plants like cacti. This type of land is best used for livestock ranching, a land-extensive grazing activity that normally focuses on cattle. Western ranches are very large, in many cases exceeding 100,000 acres (40,000 hectares) in size, because it takes so much land to support a single animal. In the driest parts of the West, a cow or steer may require more than 100 acres (40 hectares) to find sufficient forage to survive.

■ **FIGURE 5-6 Center-pivot irrigation, Colorado.** In much of the western United States, center-pivot irrigation systems draw water from underground aquifers and distribute it through giant, rotating sprinklers, creating great circles of irrigated crops.

Increasingly, ranchers send their cattle to nearby feedlots for finishing. As in the case of the Corn Belt, feedlots concentrate large numbers of animals in small areas for intensive feeding and fattening in preparation for slaughter and transport to market. Because of weight loss, it is more cost-effective to ship processed beef to market rather than live animals. Western feedlots often purchase feed from nearby farms that grow irrigated corn and sugar beets, creating a symbiotic relationship between rancher and farmer. Some of the largest feedlots are found in northern Colorado and in the Texas Panhandle (near towns with names like Bovina and Hereford).

Western agriculture is a study in contrasts. The sparsely populated and relatively unused lands that dominate the West are punctuated by areas of intense activity, both oasis-type agriculture and livestock feeding, that make the West more of an agricultural region than most people realize.

There is one exception to this western pattern. In the well-watered marine west coast climate of the Pacific Northwest, irrigation is not necessary and a productive agricultural system has developed. The Puget Sound Lowland and the Willamette Valley, for example, focus on dairy production, grains, orchard crops, and berries. In the somewhat drier mediterranean climate of coastal California, a variety of crops are grown, including wine grapes in the famous Napa Valley region.

Continuing Adjustments in Agriculture

The U.S. and Canadian agricultural systems, despite their relatively short histories, have experienced many adjustments. While most of these adjustments have been structural in nature, they have produced identifiable geographic

and economic-development ramifications, the most noteworthy and pervasive of which has been a slow depopulation of rural areas. Specifically, these adjustments include:

Fewer Farmers Farm employment as a percentage of total employment in both the United States and Canada declined precipitously in the last century. In 1890, between 40 and 50 percent of the U.S. and Canadian workforces were engaged in agriculture; by 1990, the figures hovered between 3 and 4 percent. People simply left the farm for the city. As discussed above, mechanization played a role in this decline. Perhaps more significant, however, was the attraction of greater wealth and a less demanding workday in the factory or the office.

Growth of Agribusiness/Vertical Integration A trend that has accompanied mechanization has been the growth of corporate farming or agribusiness. The corporate farm is very capital- and energy-intensive, occupies about twice the acreage of the individual farm, and is often part of a food conglomerate. To illustrate the last point, a cereal corporation may own the farms that produce the grain crops that are converted into breakfast food that is then sold to consumers—a "seedling-to-supermarket" concept that is referred to as vertical integration. As evidence of corporate agriculture's growing prominence, these farms accounted for only 4 percent of all U.S. farms in 2002 but 28 percent of the value of farm products sold. The rise of the corporate farm, however, does not necessarily mean the death of the traditional family farm. Indeed, most farms are still individually owned and operated. Even corporate farms have family connections—more than 90 percent of all corporate farms in 2002 were held by families. In some cases, the geographical imprint of the corporate farm is clearly visible, as in the feedlots of the West and the broiler barns of the South; in other cases, the impact is less immediately discernible but still profound, as in the slow but persistent depopulation of rural areas as the agricultural business landscape is restructured.

Fewer, Larger Farms It should come as no surprise that, in an era of declining farm populations and growing agribusiness, the number of farms has declined and their average size has increased. In 1910, the United States had more than 6 million farms; in 2002, the number was down to 2.1 million. Average farm size in 1910 was 138 acres (56 hectares); in 2002, it was 441 acres (178 hectares), with the average corporate farm at 1,469 acres (594 hectares). Rural geography is changing from an intricate mosaic of small farms to a coarse quilt of large corporate holdings.

Greater Increase in Supply Than in Demand Amidst the abundance that defines North American agriculture lies a paradox: success in producing food and fiber has not necessarily led to financial prosperity. Farm output has increased dramatically, but demand has not kept up with this increase. At the same time, the modern farmer spends increasingly larger quantities of money on equipment, fertilizer, insecticides, herbicides, transportation, and other costs of production. The ratio of the prices farmers receive for their products to the costs of inputs has declined precipitously since the 1970s. The result has been a squeeze on profitability that only the most efficient farms can survive, creating yet another reason for the farmer to leave the farm.

Resources for Industrial Growth and Development

Energy and Power

Consumption of energy often is used as a measure of economic development. Because highly industrialized economies have large energy appetites, it should come as no surprise that the United States and Canada are large consumers of energy. Fortunately, both countries have large energy supplies; unfortunately, even these large supplies are sometimes inadequate to meet demand.

Coal   Coal was the initial source of power for the industrial expansion of the United States. It was not long before coal's relative contribution to the nation's energy supply decreased as deposits of petroleum and natural gas were developed. Both petroleum and natural gas are considered cleaner than coal and are preferred for heating and industrial use. Coal now provides less than one-fourth of the country's energy supply (Table 5–2).

The major coal-producing states are Wyoming, Kentucky, West Virginia, and Pennsylvania, with Wyoming accounting for over one-third of U.S. production in 2003 (Figure 5–7). Large quantities of coal are available throughout the western states, but production is handicapped by the small local need and the great distances to major eastern markets. Nonetheless, coal from states such as Wyoming has the advantage of lower sulfur content and can be extracted through relatively low-cost strip-mining techniques. Large quantities of bituminous coal remain, enough to meet the energy needs of the United States for several hundred years. Unfortunately, however, the utility of such coal is limited. Most coal is used in generating electrical power; the rest serves mainly as industrial fuel. Using coal for home heating or transportation would require costly conversion efforts and the application of new and expensive technologies.

Canadian coal reserves are also large; but coal is less important as a source of power there than in the United States. Most of Canada's reserves are in two maritime provinces (New Brunswick and Nova Scotia) and two prairie provinces (Alberta and Saskatchewan), which are far from the major areas of use (the urban-industrial regions of Ontario and Quebec). Consequently, Canadians find it more practical to import coal from the Appalachian region of the United States. Another result of the unfavorable location of Canadian coal is that Canada has placed a greater emphasis on petroleum (which costs less to transport) and water as power sources.

The case of coal is instructive of how environmental concerns can affect the economic feasibility of a specific resource. Any effort to expand the use of coal as a source of power will be affected both by the restrictions and regulations placed on industries and individuals to control air quality, and by concern for the scarred landscapes that result from strip mining of bituminous coal (Figure 5–8). The negative health impacts, especially of black lung disease, associated with the tunnel or shaft mining of coal are especially troubling. Thus, large-scale use of low-sulfur coal likely will be dependent on an economical means of removing sulfurous pollutants from high-sulfur coal or finding an acceptable way to use western coal.

Oil and Gas   Both the United States and Canada are major producers of petroleum and natural gas. In 2000, Canada accounted for about 3.6 percent and the United States about 11.8 percent of the world's production of crude oil and related liquids.[1] In the United States, production is concentrated in Texas, Louisiana, Oklahoma, Kansas, California, and Alaska (Figure 5–9, page 136). Some of America's oldest fields are located in an arc from western Pennsylvania to Illinois, but these fields have been eclipsed by much more productive areas to the south and west. Historically, Canadian production has focused on the province of Alberta, with secondary fields in neighboring Saskatchewan. More recently, however, Canada has begun to drill into potentially large reserves off of the Newfoundland coast in the Jeanne d'Arc Basin—a tough challenge given the winter cold and the high winds. The first fields came into production in 1997.

The United States and Canada are also major consumers of petroleum and natural gas—both countries rely on oil and gas for over three-fifths of their power (see Table 5–2 for U.S. trends). Canadian demand for oil, which accounts for about 2.7 percent of world consumption, can easily be met by domestic production (with a substantial amount left for export). In the United States, on the other hand, demand (about 26 percent of world consumption) cannot be satisfied by domestic production, and the shortfall must be filled by imports. In 2003 the United States imported more than 56 percent of its oil supplies. Saudi Arabia was the largest supplier, followed by Mexico, Canada, Venezuela, and Nigeria.

Evaluating the oil reserves of the United States and Canada is difficult. The term proven reserves refers to oil

Table 5-2   Power Consumption in the United States, by Source

	1960	1980	2000
Petroleum	45.5%	43.6%	38.8%
Natural Gas	28.3%	26.0%	24.2%
Coal	22.5%	19.7%	22.8%
Nuclear Power	0.0%	3.5%	7.9%
Renewable Energy	6.8%	7.3%	6.2%

Source: U.S. Bureau of the Census, *Statistical Abstract of the United States: 2004–2005* (Washington, D.C.: Government Printing Office, 2005), Table 884.

[1]Energy Information Administration, *International Energy Annual 2000.* Tables 1.2, G2.

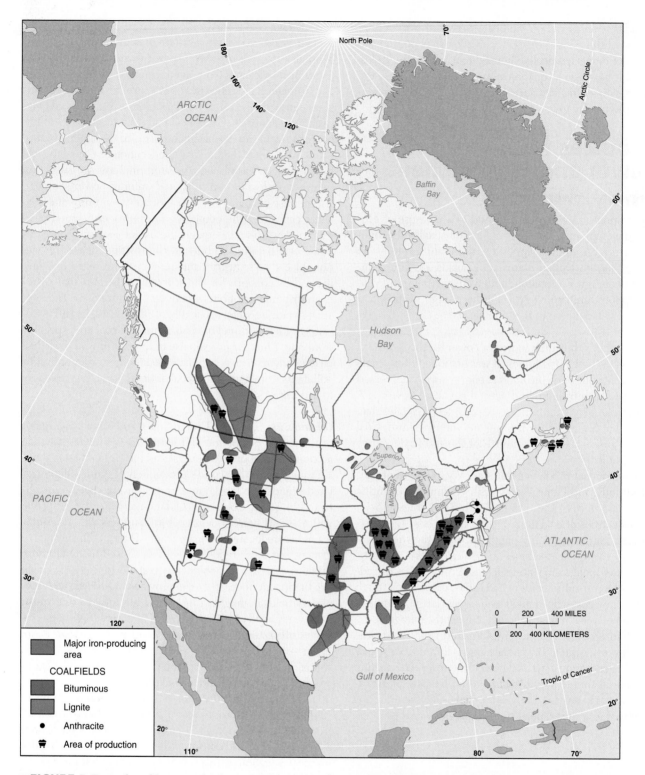

■ **FIGURE 5-7 Coal and iron ore in the United States and Canada.** Major coalfields are found in many regions. The Appalachian fields were major contributors of power for U.S. industrial expansion during the late nineteenth century.

that actual drilling has shown to be available and that can be removed at a given cost with existing technology. Present estimates of proven reserves indicate that more than 21 billion barrels of oil remain in the United States and almost 5 billion barrels in Canada. As exploration continues and new extractive technologies become available, reserves are likely to increase. Thus, depending on the levels of consumption and foreign importation, enough oil may be beneath the surface to last several decades.

The most familiar sources of petroleum are in the reservoirs that lie in sedimentary basins beneath the land or the sea, and these areas are extensive in both countries. But

■ **FIGURE 5-8 Strip mining of coal.** The landscape devastation resulting from the strip mining of coal has raised environmental issues, which in turn affect the feasibility of continued use of these lands.

petroleum is also found in oil shale and in tar sands. The solid organic materials in the shale formations of Utah, Colorado, and Wyoming (that is, the Green River formation) represent one of the world's largest deposits of hydrocarbons, with an immense energy potential. Technological and economic factors will limit the contribution that those resources will make to the energy supply in the near future, but advances in technology or changes in the price of petroleum could increase their role dramatically during the twenty-first century. The technology for extracting oil from shale does exist, although shale oil cannot currently compete in cost with other forms of energy. Moreover, even if shale oil were economically feasible to use, enormous environmental problems would have to be overcome. To produce shale oil, vast quantities of rock have to be processed, and restoration policies for minimizing environmental destruction would be significant factors in any largescale production. In addition, the process of extracting oil from shale requires a great deal of water, a rather scarce commodity in the Green River formation.

Tar sands along the Athabasca River in Alberta also contain enormous quantities of oil. Sophisticated technology and costly capital investments are required to develop the tar sands, but with the right economic conditions the tar sands can become the location of one of Canada's great petroleum production zones.

Water Water can be used to generate electrical power (that is, hydroelectricity). Currently, it provides 60 percent of the electricity in Canada but only 8.5 percent of that in the United States. The reason for such a striking difference is simple. Canada has a large quantity of water compared with coal to use for electricity generation, whereas in the United States there is relatively more coal than water available. Areas of water power generation in North America include the Columbia River Basin, the Tennessee River Valley and southern Piedmont, and the St. Lawrence Valley (Figure 5–10).

Nuclear The use of nuclear power in the United States increased rapidly during the 1970s and 1980s, and by 2002 it accounted for nearly 20 percent of net electricity generation. In contrast, only about 12 percent of Canada's electricity in 2000 came from nuclear sources, with output down by a third since 1994. Once again, with all of its water, Canada has viable non-nuclear alternatives. Whether nuclear power will become a major energy source in the future remains to be seen. Nuclear power plant accidents such as those associated with Three Mile Island, Pennsylvania, and Chernobyl, Ukraine, as well as issues of nuclear waste disposal encourage reassessment of the role of nuclear power. Furthermore, the supply of uranium in the United States may not be enough to meet a high level of demand over the long term.

Metals

Iron Ore Without convenient access to iron ore and coal, the Industrial Revolution that occurred in the United States and Canada would have been, in all probability, substantially diminished in its size and scope. Today, the United States and Canada remain major producers and consumers of iron ore. Canada is in a position to export ore; the United States is a net importer.

More than 90 percent of the domestic iron consumed in the United States comes from the Lake Superior area, especially the great Mesabi Range in northern Minnesota (see Figure 5–7), which has been yielding its iron wealth since the 1890s. The Adirondack Mountains of upstate New York and the area around Birmingham, Alabama, are other locations of historic importance. Numerous iron ore deposits are also scattered throughout western states, notably Texas, Wyoming, California, and Utah. Distance from major domestic markets, however, dictates that those ores be used in the smaller steel centers of California and Utah.

During the 1940s, high-grade iron ores called hematite (that is, ores with an iron content of 60 percent or more), became less readily available in the United States, and the next two decades saw United States dependence on foreign ores grow. Today, about half of the ore imported into the United

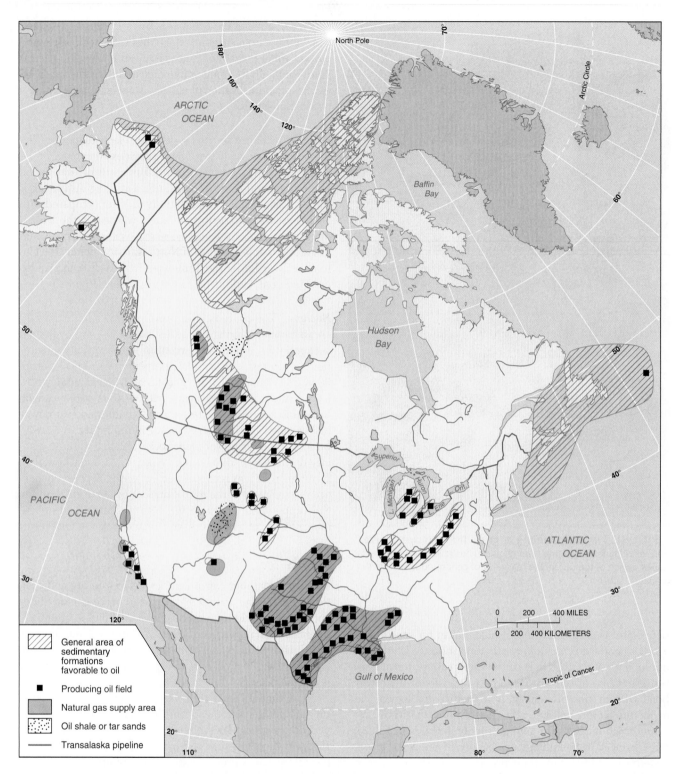

■ **FIGURE 5-9 Petroleum and natural gas in the United States and Canada.** The United States and Canada have had the advantage of large oil and gas supplies. Nevertheless, the level of development, substitution of oil and gas for coal, and high per capita consumption have made the cost and availability of oil a significant issue, particularly for the United States. Power resources remain a long-term concern for many developed countries.

States comes from Canada. Canadian ore is available in the Lake Superior district at Steep Rock Lake, Ontario; other major deposits have been developed in Labrador. In addition, high-grade ores are available from Brazil and Venezuela. Use of those higher-grade foreign ores has been made more feasible by the development and subsequent improvement of the St. Lawrence Seaway and the construction of large ore carriers, which have significantly cut transportation costs.

Almost half of the iron ore used in the United States during the 1970s came from foreign sources, but by the late

■ **FIGURE 5-10** **The Grand Coulee Dam on the Columbia River in Washington.** The dam is located at the northern extremity of the Interior Plateaus. The hydroelectric power generated by this facility has aided the growth of the aluminum industry in the Pacific Northwest.

1980s, the United States was dependent on foreign sources for less than one-fourth of its needs. One explanation for that decrease is that technological advancements now allow the use of ores such as taconite, a very hard rock with low iron content. Accordingly, the taconite industry has expanded rapidly in Minnesota and Michigan. In addition, recent economic problems in the steel industry in the United States have undoubtedly contributed to the reduced dependence on foreign ores.

Other Metals The United States has little domestic production of a number of critical metals. Aluminum is a case in point. Aluminum is used extensively in the transportation and construction industries and for consumer products. Although aluminum is a common earth element, its occurrence in the form of bauxite, the raw material from which aluminum is extracted, is relatively limited. Most bauxite comes from countries such as Jamaica, Suriname, Guyana, and Australia. For some time, the United States produced a small quantity of bauxite in Arkansas; now, however, almost 100 percent of bauxite consumed in the United States is imported. The situation in other metals industries is less critical, but still significant. About 78 percent of U.S. chromium consumption originates as imports, along with about 58 percent of U.S. nickel consumption, 86 percent of tin consumption, and 37 percent of copper consumption. The United States relies heavily on other countries to supply it with metals that are basic to U.S. manufacturing.

Canada is much more of a metals powerhouse. Much of the nickel and copper that the United States lacks, for ex-

ample, can be found near Sudbury, Ontario—indeed, were it not for mining, Sudbury might cease to exist. As mentioned above, iron mines are located nearby in Ontario and farther to the east in Labrador. Note that much of this production occurs on the Canadian Shield, which confirms the importance of shield formations worldwide to the location of metallic resources. Like the United States, Canada has little bauxite and produces large quantities of aluminum using imported bauxite.

Manufacturing in the United States and Canada

Manufacturing activities are basic to modern economies. In the United States more than 16 million people, and in Canada more than 2 million people, are employed in manufacturing. The proportion of the total labor force employed in manufacturing has decreased as relatively faster growth has occurred in tertiary activities, a characteristic of more advanced economies. But manufacturing remains basic to the stability of the American economy. Approximately 14 percent of the 2003 gross domestic product (GDP) of the United States was directly derived from manufacturing, as was 17 percent of Canada's GDP for the same year; but even more important is the great number of employment opportunities provided in the tertiary realm as goods are distributed to consumers and services are provided for manufacturers. These networks form the basis for the extensive urbanization of the United States and Canada.

The Evolution of Manufacturing in the United States

Industry first developed in southeastern New England and the Middle Atlantic area from New York City to Philadelphia and Baltimore, two of the early American settlement cores. Before the railroad era began, access to water offered great transportation advantages, and industrial power was available in the form of mechanical waterpower. In addition, as capital was accumulated from other pursuits—such as lumbering, fishing, and ocean shipping—people who sought financial backing for new industrial activities were able to find entrepreneurs with both available capital and a venturesome spirit. Furthermore, New England had become an agricultural area out of necessity, in spite of its acidic soils and relatively steep slopes. Consequently, the surplus rural populations of New England, suffering from the economic competition of newly opened land in the interior, readily gave up farming for work in the textile mills, leather and shoe factories, and the machine-tool industry, all of which dominated New England industry by 1900. Shipbuilding, food processing, papermaking and printing, furniture, and ironworking were other early industries of the region.

The United States and Canada were largely settled by British and other European peoples whose homelands were already in economic transition. The traditional feudal society was decaying in Europe, and the concept of individual worth was emerging. In addition, capitalism was developing, as those with excess wealth diverted funds into new trade, transportation, power-supply, and manufacturing enterprises. Against this background, the United States reached a take-off stage by the 1840s and Canada by the 1890s.

With takeoff came an era of great industrial expansion, aided by railroads that made possible the movement of materials over great distances. The railroads themselves became a major market for steel, which was available from the 1850s on. By 1865, the United States was experiencing a drive to maturity, which would be achieved by 1900. During those years the improvements in the railroads, continued immigration, and an extremely favorable population-resource balance aided growth. At the same time, the immense growth of industry, primarily in the northern states, was directly responsible for urbanization.

As railroads became the backbone of the transportation system and the dominant market for steel, specific northern locations took on new meaning and value. For example, eastward movement of goods from the interior by way of the Mohawk Valley and the Hudson River strengthened the focus on New York City. In addition, the agricultural goods that moved eastward to market stimulated domestic industry in the market area. The impact of the new and growing steel industries was more complex. As the bituminous coalfields of Appalachia and the iron ore deposits of the Mesabi Range increased in importance, the locations with greatest utility became those between the coal and the iron ore. These locations included Pittsburgh, at the junction of the Monongahela and Allegheny rivers, and Cleveland, Erie, and Chicago, among others, on the Great Lakes. Railroads complemented the lakes by moving coal westward and northward to meet iron ore that was moving eastward by water. At the same time the areas between the Appalachians and the lakes, including along the periphery of the lakes, gained importance in the assembly of materials for production.

The South, remote from the new national transportation routes, continued to produce agricultural goods for external markets. The regional differences discussed in the previous chapter had led the southern colonies to develop a commercial system that was financially and socially rewarding for a select few, who tended to invest their economic surplus in more land, but this system failed to develop fully the talents and skills of African Americans and poor whites. Thus, to some extent, the South was like a colonial appendage. Manufacturing did exist in the antebellum and postbellum South, but it never achieved the rate of growth or the dominance that it did in the North. Following the Civil War (1861–1865), during which the weakness of southern industry was made evident, other conditions made a reversal of the traditional economy even more difficult. And many southern leaders continued to advocate an agrarian philosophy even after the war.

Thus, regional variations were great by the 1930s, when the process of economic development was well into its maturity. The northeastern quarter of the United States had evolved as a major urban-industrial region, with numerous specialized urban-industrial districts interspersed among prosperous agricultural regions. The industrial coreland was an area of relatively high urbanization, high industrial and agricultural output, high income, and immense internal exchange. The South was a region of low urbanization, limited industrial and agricultural growth, and poverty for great numbers of whites and blacks in underdeveloped rural areas.

Manufacturing in the Coreland

The primary manufacturing region in Anglo America is a large area in the northeastern United States and southeastern Canada that consists of numerous urban-industrial districts, within which distinct industrial specialties can be identified (Figure 5–11). That coreland contains a large proportion of the manufacturing capacity of the United States and Canada and the majority of the Anglo American market. The complementary transportation system of waterways, roads, and railroads provides great advantages for both assembling materials and distributing finished products.

Measured by employment, southern New England is one of the most industrialized districts in the United States. The area's prominence is based on the nineteenth-century growth of its textile, leather-working, and machine-tool industries. In the twentieth century, however, the region suffered severe economic problems. Any area that depends on only a few products risks severe economic consequences if the competition comes from more efficient producers. In addition, New England and the significance of its location changed with industrial maturity. Today, New England is a high-tax area, has highly unionized labor markets, and relies on imported resources for power. Moreover, population increases farther west have adversely affected New England's location relative to national markets. Eastern New England is now peripheral in some respects, and the area's once-dominant industries have moved to the South, where labor costs are lower. To compensate for that loss, New England is now emphasizing high-value products—such as electronic equipment, electrical and other machinery, firearms, and tools—that can withstand high transportation, power, and labor costs. A "high-tech" region has developed to the west of Boston, partly due to the presence of highly regarded centers of learning such as the Massachusetts Institute of Technology and Harvard University. The region's success in attracting service activities has also aided its transformation to a new economy. All in all, the New England district illustrates how the industrial structure of a region can change, not always by choice.

Metropolitan New York contains the largest manufacturing complex in the United States. The city's location at the mouth of the Hudson River, its function as the major port for the rich interior, and its own huge population have combined to generate and support up to 5 percent of all manufacturing in the United States. The tendency in that district is toward diversified manufacturing, including printing, publishing, machinery making, food processing, metal fabricating, and petroleum refining. A heavy concentration of garment manufacturing also characterizes industry in greater New York, along with a lack of primary metals processing.

Inertia, immense capital investments, and linkages with other industries once ensured considerable locational

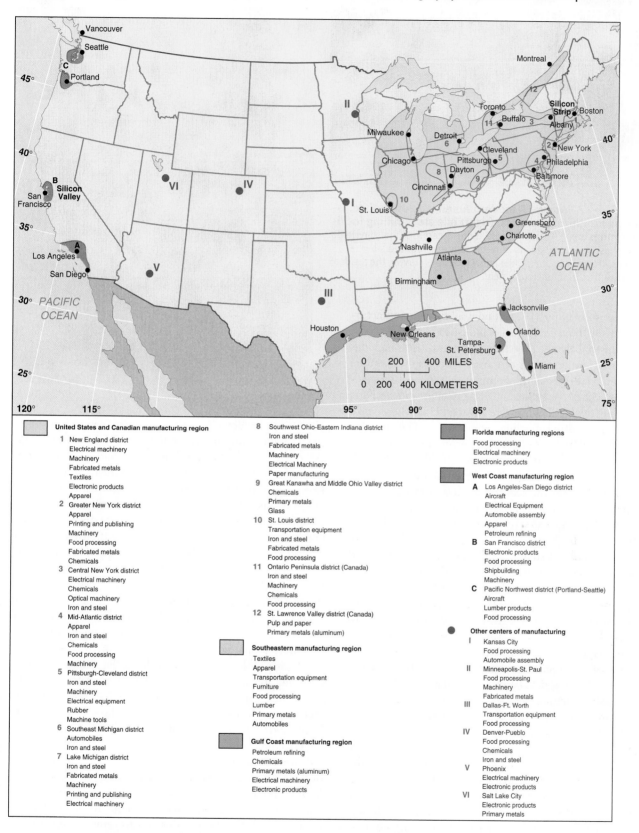

United States and Canadian manufacturing region

1 New England district
  Electrical machinery
  Machinery
  Fabricated metals
  Textiles
  Electronic products
  Apparel
2 Greater New York district
  Apparel
  Printing and publishing
  Machinery
  Food processing
  Fabricated metals
  Chemicals
3 Central New York district
  Electrical machinery
  Chemicals
  Optical machinery
  Iron and steel
4 Mid-Atlantic district
  Apparel
  Iron and steel
  Chemicals
  Food processing
  Machinery
5 Pittsburgh-Cleveland district
  Iron and steel
  Machinery
  Electrical equipment
  Rubber
  Machine tools
6 Southeast Michigan district
  Automobiles
  Iron and steel
7 Lake Michigan district
  Iron and steel
  Fabricated metals
  Machinery
  Printing and publishing
  Electrical machinery

8 Southwest Ohio-Eastern Indiana district
  Iron and steel
  Fabricated metals
  Machinery
  Electrical Machinery
  Paper manufacturing
9 Great Kanawha and Middle Ohio Valley district
  Chemicals
  Primary metals
  Glass
10 St. Louis district
  Transportation equipment
  Iron and steel
  Fabricated metals
  Food processing
11 Ontario Peninsula district (Canada)
  Iron and steel
  Machinery
  Chemicals
  Food processing
12 St. Lawrence Valley district (Canada)
  Pulp and paper
  Primary metals (aluminum)

Southeastern manufacturing region
  Textiles
  Apparel
  Transportation equipment
  Furniture
  Food processing
  Lumber
  Primary metals
  Automobiles

Gulf Coast manufacturing region
  Petroleum refining
  Chemicals
  Primary metals (aluminum)
  Electrical machinery
  Electronic products

Florida manufacturing regions
  Food processing
  Electrical machinery
  Electronic products

West Coast manufacturing region
A Los Angeles-San Diego district
  Aircraft
  Electrical Equipment
  Automobile assembly
  Apparel
  Petroleum refining
B San Francisco district
  Electronic products
  Food processing
  Shipbuilding
  Machinery
C Pacific Northwest district (Portland-Seattle)
  Aircraft
  Lumber products
  Food processing

● **Other centers of manufacturing**
I Kansas City
  Food processing
  Automobile assembly
II Minneapolis-St. Paul
  Food processing
  Machinery
  Fabricated metals
III Dallas-Ft. Worth
  Transportation equipment
  Food processing
IV Denver-Pueblo
  Food processing
  Chemicals
  Iron and steel
V Phoenix
  Electrical machinery
  Electronic products
VI Salt Lake City
  Electronic products
  Primary metals

■ **FIGURE 5-11** **Manufacturing regions and urban-industrial districts of the United States and Canada.** Industrial regions and districts show as much variety in specialties as agriculture does. Industrial specialties reflect the influences of markets, materials, labor, power, and historic forces.

■ **FIGURE 5-12 (Above) A modern American automobile assembly plant utilizing robotics; (right) idle steel milling facilities in Youngstown, Ohio.** Manufacturing enterprises that survive do so by emphasizing capital-intensive technology and reduced labor requirements. The result is that, while many industries survive, the residual labor force suffers from unemployment.

stability for steel industries. But now, all three manufacturing districts with prominent steel industries are undergoing major industrial change. The first of those districts includes Baltimore, Maryland, and the area around Philadelphia, Bethlehem, and Harrisburg, Pennsylvania. A massive steel-producing capacity exists near all of those cities and supports shipbuilding (along the Delaware River and the Chesapeake Bay) and many other machinery industries. The steel industry has expanded there because of proximity to large eastern markets (that is, other manufacturers of fabricated metals and machinery) and accessibility to external waterways. The importance of waterway accessibility has grown as dependence on foreign sources of iron ore has increased.

The second major steel district is a large triangle with points at Pittsburgh and Erie, Pennsylvania, and Toledo, Ohio. It is the oldest steel-producing center in the United States. Its **initial advantage**—that is, early factors that propelled development—was derived from its location between the Appalachian coalfields and the Great Lakes. But the locational advantage of Pittsburgh and its steel-producing suburbs has diminished. Now, South American iron ore is shipped to East Coast works, and Canadian ore comes by way of the St. Lawrence Seaway and the Great Lakes. Thus, the eastern steel district (Baltimore–Philadelphia) is closer to both eastern markets and foreign ores, and Detroit and Chicago are more easily reached by Canadian ore and are closer to midwestern markets.

The third steel area lies around the southwestern shore of Lake Michigan. It includes Gary, Indiana; Chicago, Illinois; and Milwaukee, Wisconsin, which have a vast array of machinery-manufacturing plants that are supplied by the massive steelworks nearby. The steel industries of Chicago and Gary have benefitted from a superb location; ore moving southeastward across the Great Lakes meets coal from Illinois, Indiana, Ohio, Kentucky, and West Virginia. By the late nineteenth century, Chicago had also become a major transportation center; railroads met and complemented freighters, making the southern Lake Michigan area an excellent site for assembling materials and distributing manufactured products. The St. Lawrence Seaway has simply given renewed importance to the location, for now Chicago and other inland cities can function as mid-continent ports from which ships can almost sail in a Great Circle route to Europe (a Great Circle route is the shortest distance between two points on the surface of the earth).

Southern Michigan and neighboring parts of Indiana, Ohio, and Ontario are distinguished by their emphasis on automotive production, both parts and assembly. Those industries are linked not only to the Detroit steel industry but also to steel manufacturers in the Chicago area and along the shores of Lake Erie (cities such as Toledo, Lorain, and Cleveland, Ohio). The automotive industry serves as a huge market for major steel-producing districts on either side of the international border.

Major problems have jeopardized the vitality of manufacturing in the coreland since the 1970s. First, manufacturers of steel and automobiles have faced intense competition from foreign producers, whose labor costs are lower and whose equipment is more modern. The resulting economic and social stress has been felt throughout the industrial community because the steel and automotive industries are integrated into an entire complex of industries. Second, recurring national recessions and economic fluctuations have created economic environments in which it has been difficult for those industries to remain competitive. Third, social problems have plagued the coreland: questions of residential quality, social conflict, air and water pollution, and urban water supply, as well as the need to govern and integrate numerous adjoining political units.

All of those problems have resulted in the restructuring of industry—that is, shifts in emphasis and the spatial adjustment of manufacturing. Some industries have grown and others have declined. Both the steel industry and the automobile industry provide examples of manufacturing activities that have undergone major changes. Both have experienced employment reductions due to reduced market shares (that is, foreign competition), and attempts to modernize and economize production (Figure 5–12). In the process, both have been witness to, and have participated in, significant changes in the location of production.

Restructuring has led to regional changes in the distribution of manufacturing in the United States, with some states gaining, and other states losing, shares of national manufacturing employment. By 2000, many of the states in the industrial core had lower percentages of national manufacturing employment than they had in 1970 (Figure 5–13 and Table 5–3). In contrast, states in the South and West experienced major gains in employment shares, with Texas and California standing out. As pointed out above, however, the overall percentage of the workforce employed in manufacturing has been declining.

The Southern Economic Revolution

The beginning of the economic revolution in the South is difficult to pinpoint. It probably started in the 1880s with the **New South** advocates, who believed that industrialization was necessary for the revitalization of the region. Thus, the recent increase in the southern (and western) share of manufacturing in the United States is really the continuation of a process of industrial dispersion that has been under way for many decades.

The first major manufacturing activity to become distinctly identified with the South was the textile industry, which had evolved as a dominant force in the nineteenth-century industrial growth of New England. By the early twentieth century, however, the industrial maturity of New England was reflected in high wages, unionization, costly fringe-benefit programs, high power costs (imported coal), and obsolete equipment and buildings. Thus, as old New England textile plants closed, new ones were established in the South.

Because the textile industry did not grow on a nationwide basis after 1920, the regional shift benefitted one region at the expense of another. Firms in the South found a major advantage in the quantity of labor that was available at a relatively low cost. The agrarian South had a surplus of landless rural people willing to switch from farming to manufacturing. Other advantages were a better location with respect to raw materials (cotton), lower cost of power, and lower taxes. By 1930, more than half of the nation's textile

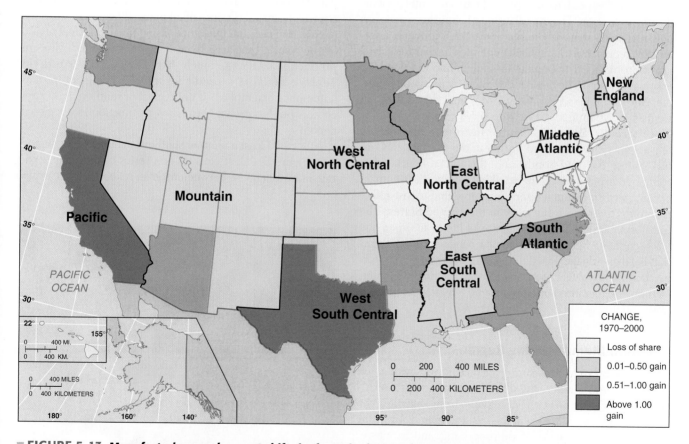

■ **FIGURE 5-13** **Manufacturing employment shifts in the United States between 1970 and 2000.** The percentage of total U.S. manufacturing employment by state shifted considerably during this 30-year period, with most percentage losses in the old industrial core and most gains in the South and West. **Source:** U.S. Bureau of the Census, *Annual Survey of Manufactures,* 1970, 2002 (Washington, D.C.: Government Printing Office).

Table 5-3 Percentage of U.S. Workers in Manufacturing, by U.S. Bureau of Census Division, for Selected Years

Division	1899	1929	1954	1970	2000
New England	18.1	12.4	9.0	7.7	5.6
Middle Atlantic	34.1	29.0	26.6	21.9	11.5
East North Central	22.8	28.8	28.6	26.0	23.2
West North Central	5.6	5.4	6.0	6.3	8.2
South Atlantic	9.7	10.3	11.0	13.5	15.9
East South Central	3.8	4.3	4.5	6.1	7.9
West South Central	2.4	3.4	4.5	6.2	9.3
Mountain	0.9	1.2	1.1	1.9	4.0
Pacific	2.6	5.3	8.8	10.4	14.4

Sources: U.S. Bureau of the Census, *Census of Population and Census of Manufactures* (Washington, D.C.: Government Printing Office, various years); U.S. Bureau of the Census, *Annual Survey of Manufactures,* 1970–1971, Tables 2 and 3, and 2002, Table 1 (Washington, D.C.: Government Printing Office, 1971, 2002)

industry was located in the South. At present more than 90 percent of cotton, 75 percent of synthetic fiber, and 40 percent of woolen textiles are of southern manufacture.

In addition to the labor-oriented textile and apparel industries, material-oriented pulp-and-paper, food-processing, and forest industries have grown rapidly in the South. In Texas and Louisiana, the petroleum-refining and petrochemical industries have contributed much to Gulf Coast industrial expansion. The overall economic transformation has now reached a stage where the South itself provides a significant regional market, which generates further industrial growth to respond to population growth and the needs of existing industry. The automotive industry is a good example. Automobile assembly plants were built decades ago in Louisville, Atlanta, and Dallas to serve regional markets. More recently, after considerable restructuring, automobile assembly has expanded further in the East North Central states, the mid-South, and the lower South. This continued expansion has included both domestic automakers and foreign manufacturers, who now find it cost-effective to do their assembly work in the United States. The transformation from an agrarian way of life to an urban one has brought higher incomes and new consumption patterns that have greatly increased the market importance of a formerly rural populace. Over one-third of the nation's manufacturing, as measured by employment, is now accomplished in the South (see Table 5–3 and Figure 5–14).

Southern Manufacturing Regions

The southeastern manufacturing region coincides with much of the southern Piedmont and the neighboring parts of Alabama (see Figure 5–11). That region—from Danville, Virginia, to Tuscaloosa, Alabama—is dominated by light industry: textiles, apparel, food processing, and furniture. Automobile assembly plants and electronic products manufacturing are also expanding. The chief attraction has undoubtedly been the availability of suitable labor at costs below industry wage scales elsewhere. As an industrial region, the Piedmont is quite unlike the core districts of the

United States and Canadian manufacturing region. Much of its industry is in small cities, towns, and—not infrequently—rural areas. Even where urbanization is occurring, it is a more dispersed form of urbanization than that which occurred with midwestern turn-of-the-century industrialization.

Birmingham and Atlanta are two major exceptions to the general pattern in the Southeastern region. The Birmingham-Gadsden, Alabama, area is known for its steel industry, which started because of an unusual nearby occurrence of coal, iron ore, and limestone. Long the major steel region of the South, that area is now experiencing a decline in production. Atlanta, much like Dallas–Fort Worth in Texas, is noted as a center for aircraft manufacture and automobile assembly. It is not typical of southern industrial centers but has grown because of its function as a regional center.

The Gulf Coast is another southern manufacturing region (see Figure 5–11). It consists of a series of distinctly separate industrial nodes between Mobile, Alabama, and Corpus

■ **FIGURE 5-14 BMW automobile assembly plant in South Carolina.** Heavy manufacturing has expanded rapidly in recent decades in the Appalachian Piedmont from Alabama to North Carolina.

■ **FIGURE 5-15 Petrochemical plant in east Texas.** Scores of petrochemical plants have been established along the Gulf coastal plain from Corpus Christi, Texas, eastward to New Orleans, Louisiana.

Christi, Texas. Petroleum, natural gas, salt, sulfur, and agricultural products provide the base for much of the region's industry and recent growth. Industrial activities include petroleum refining and production of petrochemicals and other chemicals (Figure 5–15); processing of sugar and rice; and steel manufacturing based on local and imported ores. The region's coastal location is a great advantage for both importing raw materials and exporting finished products.

In addition, there has been significant growth in manufacturing in central Florida. Especially important are electronic products and electrical machinery. Those portions of the Coastal Plain falling outside these regions are not without industry. Pulp-and-paper industries and pine plywood industries have expanded greatly since World War II, mainly because the pine forests grow rapidly on the plain.

The Growth of Manufacturing on the West Coast

Approximately 14 percent of all manufacturing in the United States, as measured by employment, is carried out on the Pacific Coast, with the largest single concentration in the greater Los Angeles area. The productivity of California's agriculture and commercial fishing stimulated the food-processing activities that became the state's first major industry and were dominant until the 1940s. World War II then generated the defense industries, aircraft manufacturing, and shipbuilding; and defense has continued to be a major employer in the Los Angeles area during the postwar decades. In addition, automobiles, electronic parts, petrochemicals, and apparel have achieved importance in California's **industrial structure**. Especially in the Los Angeles area, the growing Hispanic population has become a major source of labor for the large apparel industry.

Much of California's early industrial growth was based on local material resources and rapidly growing local markets.

Since the 1950s, however, industries that serve national markets have grown enormously. We need to look no farther than the high-technology clusters of Silicon Valley in the San Francisco Bay region to illustrate this point. For most of us California leaps to mind when we think about computers and technological innovation.

Manufacturing successes in the Pacific Northwest include food processing (dairy, fruit, vegetable, and fish products), forest-products industries, primary metals processing (aluminum), and aircraft factories (Figure 5–16). The emphasis is on the processing of local primary resources and the use of hydroelectric power from the Columbia River system. The region's great distance from eastern markets and the smaller size of local markets have historically inhibited growth. But the economic expansion of many Pacific Rim countries is proving to be a major stimulus to the entire West Coast, including the Portland and Seattle areas.

Canadian Industrial Growth

Secondary activities are important also to Canada. In fact, Canadian manufacturing is closely integrated with that of the United States, as indicated by both trade flows and the high level of U.S. investment in Canadian industry. One difference between the two countries is that industrial development in Canada took off later than in the United States—the 1890s rather than the 1840s. Several factors may have been at work to produce this delay: deferred political independence, late nineteenth-century immigration, harsh physical environments, and a focus on raw materials.

Canadian industry and exports have been closely tied to the production and processing of staples: first, fishing and furs; later, wheat, forest products, and metals. A maturing of the Canadian economy since World War II has reduced the dominance of primary commodities, particularly in Ontario, but that emphasis nevertheless remains an identifying

■ **FIGURE 5-16 Jet aircraft manufacturing at a Boeing Aircraft Company plant in Everett, Washington.** Aircraft manufacturing is one of several manufacturing industries that have experienced highly cyclical employment.

feature. Wheat, primary metals (raw or partially processed), forest products, oil and its refined products, and tourism are still the major dollar earners.

Early twentieth-century growth in Canada was a response to markets and capital in the United States, and Canadians feared that their location would make them little more than a supplier of raw materials for the United States—a kind of economic colony. Consequently, tariffs were adopted to encourage manufacturing and exportation, thereby ensuring that primary production and secondary processing would take place in Canada. Some Canadians have argued that the tariff policy has resulted in higher prices for the commodities they consume and, therefore, in a lower level of living. But tariffs have forced the use of Canadian resources, both human and material, by stimulating manufacturing at home, which has attracted an immense investment of capital in the Canadian economy by U.S. and other foreign companies. In 1987, the United States and Canada agreed to work toward a reduction of trade restraints. That beginning culminated in the North American Free Trade Agreement (NAFTA), which went into effect in 1994. Canada, Mexico, and the United States are now members of a single trade union intended to lead eventually to totally free trade and greater interaction between the member countries. Even now, Canada and the United States are each other's most important trading partners, a relationship not unlike that of the members of the European Union and an indication of the **economic integration**, or interrelatedness, already present in North America. Success may eventually lead to consideration of a larger western hemispheric trade organization.

The Distribution of Canadian Manufacturing

The St. Lawrence Valley and the Ontario Peninsula form the manufacturing heartland of Canada, producing 75 percent of the nation's industrial output. The Canadian area adjoins the industrial core of the United States and may be thought of as its northern edge, specializing in the production and processing of materials from Canadian mines, forests, and farms.

Montreal is the site of a large proportion of Canadian industry. In some ways, it parallels New York City. Both produce a variety of consumer items (chiefly foodstuffs, apparel, and books and magazines) intended for local and national markets. Both cities function as significant ports for international trade.

Outside Montreal, the Canadian industrial structure is more specialized. The immense hydroelectric potential of Quebec is a major source of power for industries along the St. Lawrence River and has provided the basis for Canada's important production of aluminum. With bauxite shipped in from Jamaica and Guyana, the Saguenay and St. Maurice rivers, tributaries of the St. Lawrence, have powered aluminum refining and smelting at Alma, Arvida, Shawinigan Falls, and Beauharnois. Because Canada's production of aluminum far exceeds its consumption, it is a leading exporter. The aluminum industry is an example of Canada's role as a processor and supplier for other countries, using

both national and international resources in that capacity. Other metal-processing industries located near production centers deal with copper and lead at Flin Flon and Noranda, nickel and copper at Sudbury, and magnesium at Haleys.

The valleys of the St. Lawrence and its tributaries also represent Canada's major area of pulp-and-paper manufacturing, with the boreal forests providing the resource base. Canada is the world's leading supplier of newsprint, most of which is sent to the United States and Europe.

Beyond Montreal and its vicinity, the most intense concentration of industry is found in the **golden horseshoe**, a district that extends from Toronto and Hamilton around the western end of Lake Ontario to St. Catharines (Figure 5–17). That district produces most of Canada's steel and a great variety of other industrial goods, such as automobile parts, assembled automobiles, electrical machinery, and agricultural implements. It is one of the most rapidly growing industrial districts in Canada, partly because a significant majority of Canada's market is found along the southern edges of Ontario and Quebec provinces, and industry appears to be increasingly market oriented.

Canada's high protective tariff was another reason for the growth of industry in the golden horseshoe. The tariff was initially important when foreign capital, especially from the United States and Great Britain, was invested in industries that processed Canadian resources for foreign use. It became even more important as a mature Canadian industry sought to serve expanding Canadian markets. Foreign companies found it necessary to locate in Canada to avoid tariffs, and they tended to locate in larger Canadian industrial centers—Windsor, Hamilton, Toronto, and Montreal—or close to the city containing the parent U.S. firm. For example, many Detroit firms with subsidiary operations in Canada located immediately across the river in Windsor.

As in the United States, a drift of manufacturing employment occurred away from established core areas to the West between 1969 and 2002 (Figure 5–18). Quebec suffered the largest loss of employment share for reasons that

■ **FIGURE 5-17 Steel mill of Hamilton, Ontario.** The golden horseshoe district adjoining the western end of Lake Ontario is the source of much of Canada's output of steel, automobiles, and agricultural machinery.

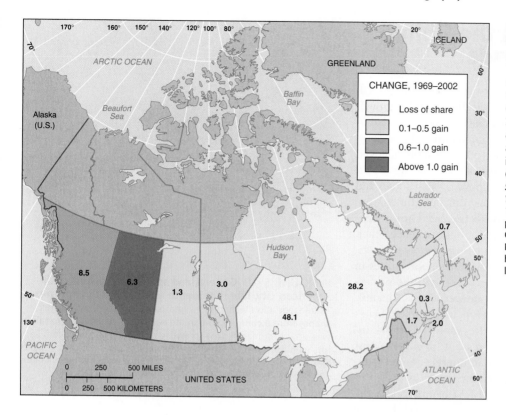

may include uncertainty regarding the province's future in the Confederation. Canada's traditional industrial powerhouse, Ontario, also lost a small percentage. The largest gain was in Alberta, which almost doubled its percentage of Canadian manufacturing employment, no doubt due in large measure to the growth of its resource industries. British Columbia also showed a gain.

A Postindustrial North America

The coal-based Industrial Revolution that reshaped the economic geography of North America is over. What is evolving in its place is often referred to as a postindustrial society with an emphasis on the provision of services and knowledge rather than the manufacturing of goods, a process that began in the 1950s but accelerated in the 1970s (aided by the explosion of new computer technologies). Workers increasingly

are employed in retailing, information and business services, banking, and related activities. In 1950 only 11.9 percent of the U.S. workforce was employed in the service sector, whereas 33.7 percent was in the manufacturing sector; by 2000, this relationship had been reversed, with 36.8 percent in services and only 14.7 percent in manufacturing (Table 5–4). Within urban areas, service providers increasingly are attracted to suburban locations, often to the suburban "downtowns" described below. As with manufacturing, the service industry has become global in its reach; large legal firms, for example, often have offices abroad. Having stressed the importance of services, however, it is important to emphasize that manufacturing has not disappeared and never will disappear. There will always be a need to make goods for consumers. Instead, the manufacturing process has been restructured to take on a more technology-intensive and globally involved form.

Table 5-4   Employment Percentages for Selected United States Sectors, 1930–2000

	1930	1940	1950	1960	1970	1980	1990	2000
Agriculture	22.0	20.0	11.9	8.3	4.3	3.4	2.9	2.4
Mining	3.4	2.9	2.0	1.3	.7	1.0	.6	.4
Manufacturing	32.5	33.9	33.7	31.0	26.4	22.1	18.0	14.7
Transportation/Utilities	12.5	9.4	8.9	7.4	6.7	6.6	6.9	7.2
Services	—	11.3	11.9	13.6	25.9	29.0	33.1	36.8

Sources: Percentages from 1970 to 1990 were computed from the U.S. Bureau of the Census, *Statistical Abstract of the United States: 1999* (Washington, D.C.: Government Printing Office, 1999), Table 578; earlier data were derived from the same source, Table 1432, using slightly different counts; the 2000 data were computed from the *Statistical Abstract of the United States: 2001*, Table 596.
Note: Not all activities are included, and consequently column sums do not equal 100 percent.

Urbanization in the United States and Canada

It should come as no surprise that rapid industrial growth in North America was associated with the explosive growth of cities. From an economic perspective, this association made good sense. Cities offered manufacturers the agglomeration economies that come from location near other, linked activities. In other words, firms discovered that they could reduce costs by sharing services, suppliers, labor forces and other common needs. In addition, many manufacturers were attracted to cities because of the markets created by the large numbers of people living in close proximity to each other. Regardless of the motivation, industrial jobs created multiplier effects that generated employment opportunities in wholesaling, retailing, education, government, the professions, and a host of other urban-oriented activities. Of course, cities and towns had always been present in the United States and Canada, appearing in response to a variety of factors. Industrialization simply created new reasons for cities to exist and to expand, which they did in an unprecedented manner beginning in the early nineteenth century.

By the beginning of the twenty-first century, both the United States and Canada were highly urbanized. Fully 79 percent of the U.S. population was classified as urban by the U.S. Census in 2000, up from 60 percent a hundred years earlier; the equivalent Canadian figure in 2001 was 80 percent. As high as these twenty-first-century percentages are, they in fact still understate the urban nature of much of the rural United States and Canada. The vast majority of rural residents are nonfarmers who lead a largely urban existence. They often work in urban areas. They send their children to school and they shop in towns. Currently, only about 4 million people who live in rural parts of the United States are actually farmers; in Canada, the figure is roughly 700,000 people.

As North American cities grew in response to the Industrial Revolution, the shapes of cities became more complex. Changing transportation technologies and evolving urban cultures led to a sequence of events in the spatial expansion of the North American city that are still visible today, a sequence best described by John Adams in the early 1970s (Figure 5–19). He argued that the form of the modern North American city was the result of four transportation-based stages: (1) a post-Civil War period in which cities were relatively small in area but densely packed, with people living literally on top of each other because the foot and animal transportation of the day did not permit long commutes to work; (2) an electric streetcar stage beginning in the late 1880s in which relatively fast streetcars (moving at a sizzling 15–20 mph) enabled an outbound middle-class migration along streetcar lines, producing spokelike "streetcar suburbs" radiating outward from the city center; (3) a road-building stage for the increasingly popular new form of transportation, cars, by the 1920s, and the resulting filling in of suburban areas between the streetcar lines, further stimulating the middle-class exodus to the edges of the city; and

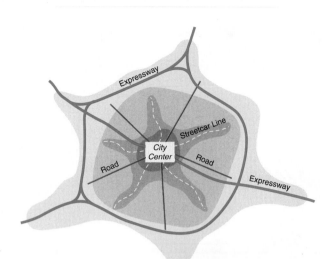

▪ **FIGURE 5-19 Generalized spatial expansion of the American city, late nineteenth and twentieth centuries.** As transportation technologies improved, cities took on distinctive shapes from a compact core in the early years when transportation was slow and difficult, to a sprawling city by the late twentieth century with the construction of expressways.
Source: Adapted from J. Adams, "Residential Structure of Midwestern Cities," *Annals of the Association of American Geographers* 60 (1970): 56.

(4) a post-World War II boom in expressway construction that not only tied the city center with outlying residential areas, but connected suburban areas with each other. Later, suburbs developed their own "downtowns" and their own reasons for being that challenged the prominence of traditional central business districts. Increasingly, service industries supplanted manufacturers as the growth engines of urban economies and as the occupants of urban spaces. The spatial structures of today's cities are complex mosaics of socioeconomic groups and economic activities, reflecting decades of technological and social change.

Regardless of the complexity of today's cities, the one constant that is present throughout urban North America is outward expansion, which is often described as urban sprawl. New suburbs are created ever farther from city cores as people search for the tranquility and convenience of low-density living, all the while maintaining access to urban amenities. Three very geographical consequences have resulted from this expansion process. First, urban sprawl has led to the loss of agricultural land, with old cornfields sprouting new suburbs, thereby removing food-producing resources from production and changing the character of an agricultural community that may have existed for more than a century. Farmers find that new lifestyles, systems of land use, and tax scales invade their domains; they often resent being nudged off good agricultural land in the name of urban progress. On occasion, governments have become involved. New Jersey, for example, has found it necessary to enact legislation to preserve its better agricultural areas.

The second consequence is the coalescence of cities into congested urban corridors. A person driving from Miami to

Jacksonville, for example, may be forgiven for wondering if he/she will ever drive through open country for more than a few miles. As cities grow, they bump into each other—one city's edge becomes the next city's beginning, maybe even to the point of sharing city-limits signs. Jean Gottmann popularized the term **megalopolis**, a Greek word meaning "very large city" (that also refers to a failed planned city in ancient Greece), to describe this coalescence of cities.[2] His initial focus was on the string of northeastern cities that seem to blend together, including Washington, Baltimore, Philadelphia, New York, and Boston. Together with a large collection of nearby cities, they comprise a giant urban region with tens of millions of people. Similar megalopolitan areas can be identified elsewhere: "San-San" (San Francisco to San Diego), "Chi-Pitts" (Chicago to Pittsburgh), and "Ja-Mi" (Jacksonville to Miami), to name only three of potentially many such U.S. regions. Indeed, much has been written about the growth of "spread" cities in the late twentieth and early twenty-first centuries.

■ **FIGURE 5-20 Central city decay, Memphis.** Virtually all cities have at least some central districts that have fallen into economic decline. One of our greatest challenges is to devise ways to revitalize these areas.

Finally, with urban sprawl comes the proliferation of urban governments, and their different resource bases and policy agendas. This fractured political geography complicates regional decision making. Furthermore, relatively wealthy suburban areas may resist the call to help their cash-strapped inner-city neighbors attack problems of poverty, blight, and decaying infrastructure—the problems that suburban commuters pass by en route to their Central Business District (CBD) jobs. Consolidated urban political regions, or political boundaries that coincide with functional cities, could provide a more equitable tax base. Not surprisingly, however, such an idea is often opposed by suburban interests. Among the few cities that have attempted such solutions are Nashville, Tennessee, and Jacksonville, Florida.

Ironically, problems of urban decline must be confronted along with the challenges of growth. An unfortunate corollary of urban expansion has been the abandonment and decay of city centers, resulting not only from middle-class migration to the suburbs but from many businesses frustrated by downtown congestion. Left behind are empty buildings, densely populated low-income residential areas (the "ghettos"), and too often crime and other social pathologies—all potentially within the shadow of a bustling central business district that depends on workers from the suburbs (Figure 5–20). A major economic development goal of many cities has been to revive downtowns, using a variety of investment incentives, tax breaks, and other policies. While central cities are still distressed, signs of renewal are appearing. Many warehouse districts, for example, are experiencing a "gentrification" process as dilapidated buildings are transformed into high-priced condominiums. For some people, accessibility to downtown and all the good things associated with urban culture is important, without having to endure the fist-waving, nail-biting stresses of commuter-choked expressways. Recent news reports have even focused on the attractiveness to retirees of gentrified downtown areas.

Up to now, the focus has been on urban growth in the United States and Canada as a whole, without questioning whether the cities of these two countries reflect any meaningful differences in structure and function. If the frame of reference is the 30 years after World War II, the answer is yes, they do. Canadian cities were more compact than their American counterparts, with especially high population densities in inner-city suburbs. This compactness was a result of greater emphasis placed on the provision of public transportation, and less concern about expressway construction and the role of the car in shaping city form. This emphasis was partly due to less affluence in Canada and to greater public acceptance of government planning—the role of government in day-to-day affairs is generally more pronounced in Canada than in the United States. Beginning in the 1970s, however, the answer to the question is no, as the influence of the car in shaping the Canadian city became more prominent, bringing with it the growth of expressways and the explosion of outer suburbs. It is likely that this convergence in the shapes of North American cities will continue.

[2]J. Gottmann, *Megalopolis: The Urbanized Northeastern Seaboard of the United States* (Cambridge, MA: MIT Press, 1961).

GEOGRAPHY IN ACTION
City Types and the American College Town

All cities and towns are similar in certain ways. They provide goods and services to the local population (so-called "central place" functions), which vary somewhat in scope depending on the size of the community. Cities and towns may also become known for specialization in certain functions. Early towns in the United States and Canada existed mainly to support commercial activities, and to some extent political and military needs. With the Industrial Revolution, large cities appeared that focused on manufacturing and transportation. In the postindustrial world, cities increasingly emphasize service industries. Today in North America, when we think of city types, we associate Washington, D.C., and Ottawa, Ontario, with political functions; New York City, Toronto, Ontario, and Charlotte, North Carolina, with finance and banking; Pittsburgh, Pennsylvania, and Cleveland, Ohio, with manufacturing; Houston, Texas, with the oil industry; and Las Vegas, Nevada, with entertainment. This list is much too short, of course, and oversimplified; but the point is clear.

One type of city that is overlooked in most classifications is the place that many readers of this book call home (at least temporarily): the American college town. This town is a community in which the college or university is the biggest business (bearing in mind that many important universities and colleges are not found in such towns). The college town is a "company" town in every sense.

But is the American college town really "American"? The answer is yes. Few such towns exist elsewhere in the world, even in Canada where Waterloo, Ontario (University of Waterloo), and Wolfville, Nova Scotia (Acadia University) come closest to the American model.

Why did such a town appear in America, but not elsewhere? There are several reasons. The first reason we can label the "Harvard effect": college founding fathers in the United States often embraced the Harvard ideal of setting up colleges away from the contaminating effects of cities, the city being "no place for a college." Specifically, in 1636 Harvard College was established in then remote Cambridge to escape the "evils" of Boston, although time and urban expansion have long since destroyed any physical separation. Other towns received colleges because alcohol was outlawed and/or other urban distractions were missing. Second, the expansive size and rich cultural diversity of the United States led to a proliferation of colleges, often focusing on small towns (the Harvard effect, again), that would serve the educational needs of a highly dispersed population (Figure A). In many cases, religious groups formed the colleges as a part of larger church missions, which tended to embed much of this diversity. Today, most states have a Concordia college (Lutheran), or a Wesleyan university (Methodist), not to mention the myriad Presbyterian, Catholic, and other campuses. Third, many towns in their formative years set out to acquire a college as a way of enhancing the town's prestige and contributing to its economic development. Sometimes bidding wars erupted. In many cases, housing a major college or university was almost as good as having the state capital.

How is the American college town different? All sorts of images and stereotypes come to mind when we think of college towns: ivy-covered walls, eccentric residents, town-gown conflicts, and, thanks to Hollywood, outrageous animal-house behavior. These stereotypes aside, college towns do reflect distinct geographic and socioeconomic patterns (Figure B). They typically have youthful and highly educated populations (no surprises here!), high family incomes, few work opportunities outside the college community, a relatively cosmopolitan but transient population, a disproportionately large

■ **FIGURE A The central campus of Williams College, Williamstown, Massachusetts.** A quintessential college campus in rural western Massachusetts, this view displays the school's immaculate green lawns and stately buildings amidst the emerging multicolored hues of autumn.

▶▶ Summary

This chapter has emphasized the role of agriculture, natural resource extraction, and manufacturing in the development of the United States and Canada. Each of these fundamental economic activities has played an integral part in shaping the geographic landscape of Anglo America, as is evident in the wealth created, the resources used, and the rural and urban landscapes produced. Even as these economic sectors continue to evolve, however, we see their comparative importance diminish as economic restructuring emphasizes the growth of services. The result is a reshaping and expansion of the urban landscape, unemployment in some areas with growth in others, and a redistribution of population. At the same time, many other changes are taking place in the United States and Canada. Their economic systems, like those of other countries, are becoming part of a global economy. These challenges and contemporary changes confirm that economic development is an ongoing process.

FIGURE B Downtown Ithaca, New York in springtime. The presence of Cornell University provides much of the economic support for commercial, cultural, and social activities in this college town.

trict, which may or may not lie adjacent to campus, but in most cases has a subordinate status to the campus core. And then there is the stadium. For large state universities, the stadium is a towering monument to school pride and the sacred alma mater; it is the site of great weekend crusades against visiting infidels from other universities. But of equal importance, the stadium is an economic engine. The weekend pilgrims who fill the stands buy food and drink, T-shirts and jerseys; they stay in local hotels and patronize local shops. Oh yes, and they remember the university when making gifts and writing wills. The number of people involved is impressive: the stadium at the University of Georgia holds almost as many people as reside in Athens; Auburn University's stadium seats nearly twice as many people as live in Auburn, Alabama; the stadium at the University of Michigan in Ann Arbor, one of the largest anywhere, can accommodate more than 100,000 Wolverine fans.

Beyond the educational mission, what may become an equally important legacy of the American college town is just now taking shape, and that is its contribution to the twenty-first-century economy. College towns are centers of innovation and knowledge in an information age; in fact, many universities are now marketing their research capabilities (often to supplement sagging revenues) through university research parks located near the main campus. The hope is that information-age firms will be attracted to the amenity-rich, campus-like setting of the research park, and will find synergies with other park activities, as well as university researchers—in other words, play an economic development role.

As the research park illustrates, it is the setting of the college town that counts. Such a setting increasingly appeals to young professionals who are not related to the university or its research park, but appreciate college-town life, especially if the town is close to a metropolitan area. In addition, the college town has become a retirement destination, to such an extent that some universities are building retirement communities that they can market to senior citizens.

The American college town is a unique urban creation, with its own land uses and culture. And given the dynamics of the twenty-first-century economy, this town is likely to play an increasingly prominent role in the evolution of the American economic landscape.

(Adapted in part from B. Gumprecht, "The American College Town," *Geographical Review* 93 (2003): 51–80.

amount of group housing (apartments and dormitories), a significant number of bicycles, and more than the usual number of people who vote "Green" and listen to National Public Radio (NPR). Distinct land uses develop around the campus, and include: a commercial district heavy in coffeehouses, ethnic restaurants, pizzerias, bookstores, theatres, art galleries, and of course, the ubiquitous bar; a separate fraternity/sorority row sporting stately classical mansions and located where nobody will complain about bad neighbors; a "student ghetto" of high-density and low-cost residences; and in some cases, a faculty neighborhood in which residents work hard to insulate themselves from the late-night parties, front-porch sofas, and other perceived bad manners of nearby student residential areas. The town will have its own central business dis-

▶▶ Key Terms

aquifers

comparative advantage

economic integration

golden horseshoe

industrial structure

initial advantage

megalopolis

milkshed

mixed farming

New South

pine barrens

spring wheat belt

winter wheat belt

Abandoned building, Upper West Side, New York City. Although ranked among the world's most prosperous nations, the United States contains many pockets of poverty and underutilized human resources. These are found in both the urban and rural sectors.

Chapter 6

The United States and Canada:
Challenges in a Developed Realm

- **Income Disparity and Regional Challenges**
- **African Americans**
- **Hispanic Americans**
- **Other Minorities**
- **Canadian Identity and Unity**
- **"Melting Pot," "Stew Pot," or What?**

Even industrialized and prosperous countries such as the United States and Canada have significant and sometimes pressing challenges. We have already discussed the complexities of supplying and consuming enormous quantities of resources. In addition, unbalanced economic growth, ineffective integration of various regions into a national economy, environmental degradation, and the social, economic, and political circumstances of minority groups are other fundamental challenges with which the United States and Canada must deal.

Income Disparity and Regional Challenges

Large numbers of people in the United States and Canada have been unable to acquire the material benefits necessary for an acceptable level of living. Exactly how many poor people live in the United States and Canada depends, of course, on exactly how poverty is defined. In general, **poverty** is a material deprivation that affects biologic and social well-being. Low income levels typically point to poverty, although other factors can be associated with deprivation.

Throughout the United States and Canada, significant **income disparities**—that is, differences in the amount of money people earn—exist among groups and regions. In 2004, about 12.5 percent of the overall U.S. population could be considered poor (Table 6–1). Among African Americans, the figure increased to 24.2 percent nationally, and to a stunning 27.4 percent in the Midwest—in other words, one-in-four African Americans. Poverty levels among Hispanics were only slightly lower than African Americans, with the White and Asian populations showing the lowest figures. It

is interesting to observe, however, that the absolute number of whites in poverty was almost double that of African Americans—15.9 million whites compared with 8.8 million African Americans—confirming that poverty among the white population is still a serious matter, the lower percentages notwithstanding.

At one time, the poor in the United States were almost equally divided between metropolitan and non-metropolitan areas. Currently, however, over 78 percent of all poor people live in metropolitan areas, with most, but not all, of those persons concentrated in central cities. Often poor people represent a smaller proportion of a metropolitan population than they do of a non-metropolitan population, but the metropolitan poor are more geographically concentrated and frequently live in inner-city ethnic communities that social scientists refer to as "ghettos" (Figure 6–1). Poor people in non-metropolitan areas are generally scattered about and are therefore less visible, adding to their problems of securing employment and services.

The map in Figure 6–2 shows clearly defined pockets of poverty. Numerous poor are found throughout the southern Coastal Plain, especially in the lower Mississippi basin; and in the upland South, particularly the Appalachian Plateau and the Ozarks. Rural poverty is not limited to the South, although it is most widespread there. Areas of rural poverty can also be found in New England, the upper Great Lakes (Michigan, Wisconsin, and Minnesota), the northern Great Plains, and the Southwest. In addition, poverty is particularly chronic in the Native American lands of the West and along the border with Mexico.

Why does poverty exist? This age-old question defies easy explanation. Certainly, racial biases and cultural attitudes have created immense barriers for African Americans, Hispanic

Table 6-1   Poverty in the United States, by region and group, 2004

Region	Percentage of Persons Below Poverty Level				
	Total Population	White	African American	Hispanic[a]	Asian
United States	**12.5**	**8.2**	**24.2**	**22.5**	**11.8**
Northeast	11.3	7.5	23.3	25.1	13.6
Midwest	10.7	7.9	27.4	19.4	12.9
South	14.1	8.9	24.4	24.0	11.7
West	12.6	8.0	20.7	21.0	10.9

Source: Bureau of the Census, *Annual Demographic Survey: March Supplement, Table POV41,* as extracted from *www.census.gov/macro/032004/pov/new41_100_01.htm.* Estimated figures.
[a]The White population does not include Hispanics. Hispanics are defined as persons of Puerto Rican, Cuban, Central American, South American, or some other Spanish culture of origin, regardless of race.

Americans, Native Americans, and Appalachian whites in both rural and metropolitan areas. Not only have those biases had a direct effect on employment opportunities, but they have also contributed to unequal education and training.

Perhaps of greater importance is the effect of economic and technological change. As new technologies are introduced and as economies are restructured, new types of production come into existence demanding different resources and workers. The result often is a transformed geography of wealth and poverty, with certain groups and regions benefiting from these changes, and others falling by the wayside. For example, in portions of Appalachia, where agriculture on small, isolated farms occupying poor land has been the basic activity since the area was first settled, the introduction of new technologies and the rise of corporate farming have left these areas impoverished agricultural backwaters. To cite a second southern example, for much of the nineteenth and twentieth centuries, large parts of the South had relatively dense rural populations, where agriculture was labor-intensive and agricultural workers were poorly paid.

In other parts of the country, urban-industrial growth was rapid, and workers developed skills that provided them with higher incomes. The result was a distinct regional variation in income, with greater proportions of the total southern population living in poverty, a pattern that remains visible in the twenty-first century. But Appalachia and the South are not the only regions affected by such changes. New England experienced a decline in its once-dominant textile industry (initially producing a flow to the South to take advantage of the rural labor mentioned above). Miners from Pennsylvania to Kentucky witnessed the disappearance of jobs as mines were automated and the demand for coal failed to increase. High unemployment levels are closely associated with high poverty levels, and the maps showing the two variables often look quite similar (compare Figures 6–2a and 6–2b).

Historically, a typical response to high poverty and unemployment levels has been to move, or to "go where the jobs are." Often this meant leaving the rural areas and moving to the city, or in more drastic cases leaving one region for another. While such migration has improved the material well-being of many people, additional rounds of technological change and economic restructuring have reduced many of these migrants and/or their descendants to lives of poverty in the crowded urban ghettos mentioned above. A lack of skills and an absence of opportunities create cultures of poverty that are persistent and, for too many people, inherited.

In brief, the geography of income disparity in the United States reveals clearly defined pockets of poverty. Any solution to poverty must address a wide variety of geographic circumstances while recognizing that long-term development must be grounded in the strengthening of human resources.

Low Income in Canada

Canada also has regional differences in income and poverty (Figure 6–3). In 2002, approximately 9.5 percent of all persons fell into the "low-income after tax" category, a figure somewhat lower than the U.S. poverty rate discussed above. Other figures suggest slightly higher rates. (It is important to remember that different countries use slightly different measures of poverty, thereby making direct comparisons difficult.) The regional variations are clear. Newfoundland and Quebec had particularly low incomes in 2000. Newfound-

▪ **FIGURE 6-1 Urban poverty.** In sharp contrast to most of the periods of American history, today the majority of poor people live in large urban areas—most often in inner-city communities. This scene is from Boston.

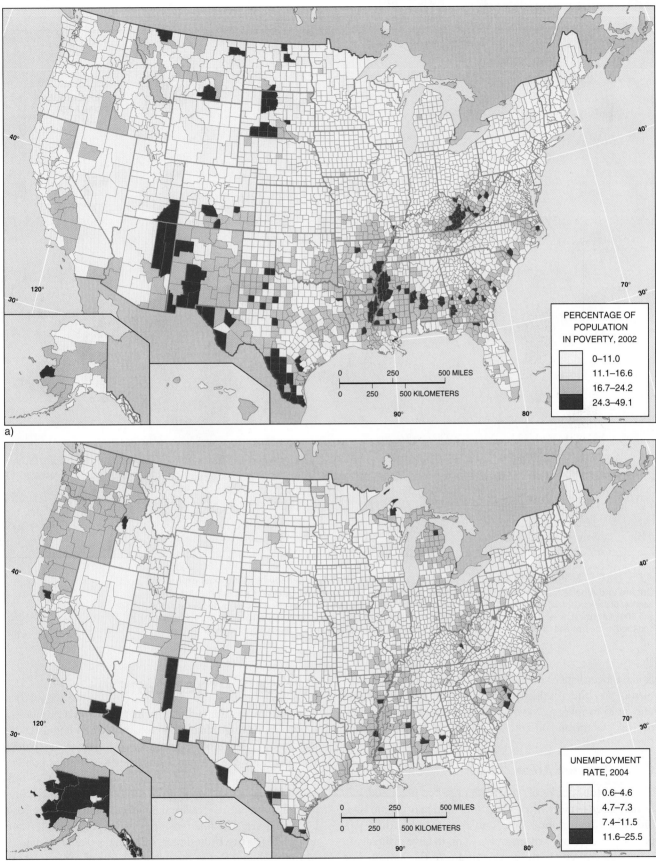

a)

PERCENTAGE OF
POPULATION
IN POVERTY, 2002

- 0–11.0
- 11.1–16.6
- 16.7–24.2
- 24.3–49.1

b)

UNEMPLOYMENT
RATE, 2004

- 0.6–4.6
- 4.7–7.3
- 7.4–11.5
- 11.6–25.5

■ **FIGURE 6-2** **Poverty and unemployment in the United States.** (a) The percentage of county population in poverty varies significantly by region and reflects differences in the development process over many decades. (b) In addition, poverty is closely associated with high unemployment rates. With some exceptions, particularly in the Midwest, the maps show strikingly similar patterns of occurrence of high poverty and high unemployment rates.

Sources: United States Bureau of the Census, Small Area Income and Poverty Estimates, estimates for 2002, *www.census .gov/cgi-bin/saipe*. United States Department of Labor, Bureau of Labor Statistics, Local Area Unemployment Statistics, labor force data by county, 2004 annual averages, via *www.bls.gov/lau/*.

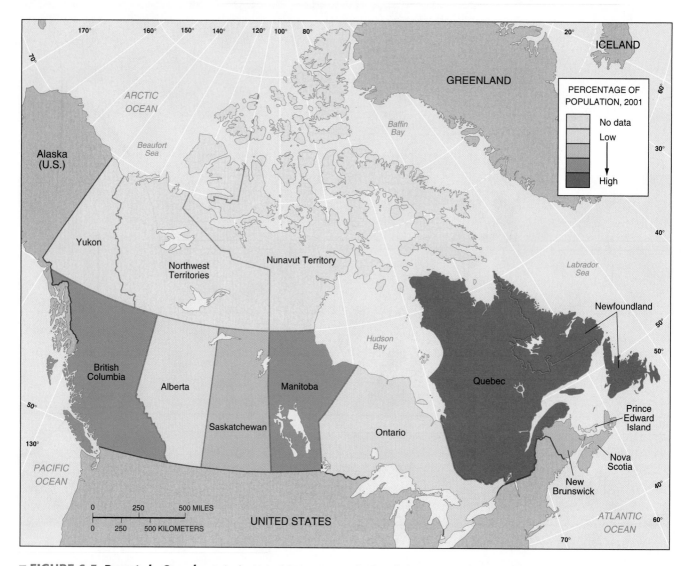

■ **FIGURE 6-3 Poverty in Canada.** As in the United States, poverty in Canada has a regional expression.
Ontario and Alberta have the lowest rates, whereas part of the Maritimes and Quebec have the highest rates.
Source: Statistics Canada, Incidence of Low Income among the Population Living in Private Households, by Provinces (1996
and 2001 Censuses), *www.statcan.ca/english/Pgdb/famil60a.htm.* (Note: this dataset is slightly different in concept from the
one cited in the text and consequently, to avoid confusion, no numbers are used in the Canada map's legend.)

land has been especially hard hit economically with the downturn of its fishing industry (although recent oil discoveries
may help to reverse Newfoundlanders' bad fortunes). Industrial Ontario and oil-rich Alberta had the highest incomes.

Other Problem Areas

The numerous causes of poverty may operate independently
or in concert—more often the latter—and may vary in their
individual importance from one region to another. Appalachia is certainly not the only region to have difficulty in
achieving equality with the remainder of the nation. The
Ozarks, the Four Corners (where Colorado, Utah, New Mexico, and Arizona meet) area, and parts of the Atlantic Coastal
Plain have also experienced economic lag. In addition, the
maritime provinces of Canada have long had income levels
below the Canadian national norms.

African Americans

It is not unusual for political units to contain a number of
subgroups distinguishable by race, ethnic and linguistic differences, and levels of economic achievement. Nonetheless,
such divisions may be major obstacles to a unified political
organization and social and economic satisfaction. The difficulty of integrating diverse groups into a larger society stems
not only from outward cultural differences but also from basic
human nature. The relationship between African Americans
and whites in the United States illustrates that challenge.

In the United States, the initial patterns of African-American residence and African-American–white relationships were
an outgrowth of the diffusion of the plantation system across
the lower South. A small proportion of African Americans,
about one in seven, were freedmen living outside the South
or in southern urban areas, where slightly less rigid social

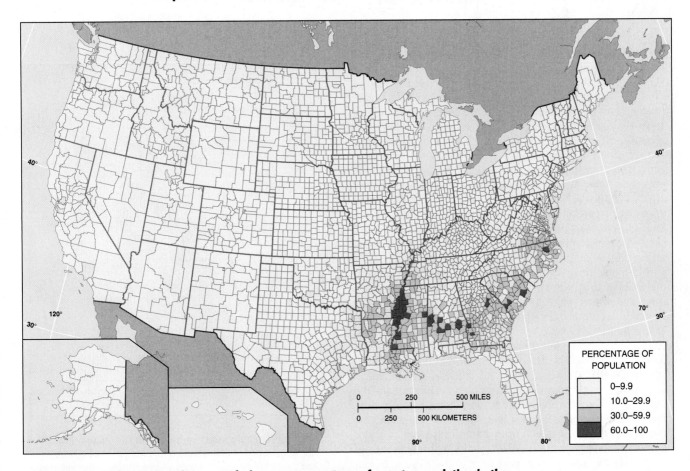

■ **FIGURE 6-4** **African-American population as a percentage of county population in the United States in 2000.** Use caution when interpreting maps based on percentages. Throughout the South, African Americans are found in rural areas as well as urban centers. The African-American population in northern communities (approximately one-half of the U.S. African-American population) is located in the most populous urban areas. Those northern concentrations, therefore, appear underrepresented on the map.
Source: Adapted from U.S. Bureau of the Census, online data resources for the 2000 Census at *www2.census.gov/census_ 2000/datasets/demographic_profile/.*

pressures allowed them to be artisans. The rest of the African-American population formed the backbone of the plantation system.

The southern plantation was a land-based social and economic system, engaged in commercial production using slave labor on relatively large holdings. The original center for that system was Tidewater Virginia and neighboring parts of Maryland and North Carolina. As settlement proceeded southward, slavery accompanied commercial crops, such as rice and indigo, to the coastal cities of Wilmington, North Carolina; Charleston, South Carolina; and Savannah, Georgia. In the more isolated and subsistence-oriented hill lands of the South, as well as northward into the Middle Atlantic and New England colonies, slavery never became important and was eventually declared illegal. The emphasis in these latter areas was usually on less labor-intensive economic activities.

In the areas of the lower South that were considered best for agricultural production, however, the plantation became the basic system. The diffusion of that system went hand in hand with the diffusion of slave labor and thus established the initial pattern of African-American residence across the South. Predominantly African-American populations were found then only in selected areas, including a narrow coastal zone of is-

lands and riverbanks in South Carolina and Georgia, the inner Coastal Plain, the outer Piedmont, the Mississippi Valley (that is, Missouri, Tennessee, Arkansas, Mississippi, and Louisiana), the Tennessee River Valley of northern Alabama, the black-soil belt of Alabama, and—by 1860—portions of Texas. The Nashville and Bluegrass basins also had some plantations and, therefore, some African Americans. But the rest of the South—southern Appalachia, including the inner edge of the Piedmont; the Ozarks; portions of the Gulf Coast; and lower Texas—either had no African Americans or had only a few in proportion to the number of white yeoman farmers.

The distribution of African Americans did not change immediately after the Civil War. The only thing that changed was that they were no longer slaves, but freedmen. They were still agricultural laborers with no land, no money or capital, and little training beyond their farming skills. Consequently, they did not migrate in large numbers immediately after the Civil War, but stayed where they were and entered into a system of tenancy or sharecropping with white landowners who continued to disdain manual labor. With few exceptions, the rural parts of today's South with large numbers of African Americans are the same areas in which the plantation system once flourished.

GEOGRAPHY IN ACTION
Appalachian Poverty

Appalachia provides an excellent illustration of the problems of poverty and the approaches used to resolve these problems (Figure A). For some people, a mention of Appalachia brings to mind beautiful mountain forests, rushing streams, small isolated valleys called hollows, upper valleys with picturesque farmsteads, hillside orchards, and prosperous farms on valley floors favored with soils formed on limestone bedrock. Other people envision homespun mountain folk with little formal education, living a life of isolation and poverty. Indeed, by the 1960s, the Appalachian region was one of several in the United States that lagged far behind the national norm—one of every three persons lived in poverty, per capita income was 23 percent below the national average, and more than 2 million Appalachians had left their homes to find work in other regions.

The problems of Appalachia arose in part from its isolation and inability to participate in the modernization and commercialization of agriculture. The prosperity of the coal-mining era was short-lived, and more wealth ended up in other regions than in Appalachia itself. After 1940, mining rapidly became automated and concentrated on fewer mines, leaving behind unemployed people and scarred landscapes. In much the same way, lumbering operations removed the wealth and beauty of an area in just a decade or two, severely affecting both the physical and the human ecology of the region. Soil erosion, the scars of strip-mining, flooding worsened by poor agricultural techniques, and the removal of forest vegetation from watersheds were all major problems even before the twentieth century. For generations, Appalachia has known poverty, low educational achievement, violence, distrust of central authority, and stagnation as a way of life.

Appalachia consists of uplands that extend from the maritime provinces of Canada to central Georgia and Alabama, including several distinct physiographic regions. The area with the most severe economic and social problems is usually considered to be restricted to the Appalachian Plateau, the Ridge and Valley Province, and the mountain areas south and west of the Mohawk and Hudson River valleys. But the hardcore problem area does not coincide exactly with the physiographic region.

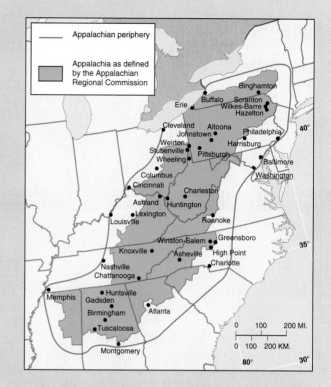

■ **FIGURE A** **Appalachian region of the United States.** The Appalachian region has long been an area of economic and social difficulties. It should be recognized, however, that such a region is not uniform; some portions are areas of growth and economic progress, even as other portions continue to lose population. One of the most persistent and severe problem areas has been the Kentucky and West Virginia portion, where the declining need for coal miners contributed to major economic and social problems after World War II.

By 1900, the South had a large, landless tenant labor force of rural African Americans. By World War I, the great century of European immigration (1814–1914) had ended, and new European immigrants were no longer available as workers. As opportunities for employment outside the South grew, and mechanization of southern agriculture began, African Americans moved in large numbers from rural to urban areas of both the North and South. Although this **African-American migration** slowed somewhat during the 1930s, when economic conditions were as bad in urban areas as they were in rural areas, the northward movement continued into the 1970s. Today, African Americans live in urban areas of the North as well as urban and rural areas of the South (Figure 6–4). Interestingly, the proportion of African Americans living in the South seems to have stabilized and hovers around 53–55 percent (Table 6–2); more African Americans are now moving to the South than are leaving it.

Patterns of migration have implications that go far beyond mere redistribution of population. Many African Americans have clearly improved their economic and social positions in urban areas (Figure 6–5), but the improvement has been a group achievement not easily won and not without failure for many individuals, as evidenced by the poverty-ridden, central-city neighborhoods of many large U.S. cities. Many migrants to the cities have been the better-educated and more-motivated representatives of rural area; but, paradoxically, they have also lacked the skills needed to do well in urban areas. Racial and social biases have further added to the difficulty of obtaining suitable housing, proper education, and access to economic opportunity. Thus, many African Americans have ended up in ghettos, with an unemployment rate that is consistently higher than it is for other segments of society. The seeming inability of the United States to fully absorb this large segment of its population into the national mainstream remains a major challenge.

Beginning in 1964, President Lyndon Johnson spearheaded legislation that resulted in the enactment of the **Appalachian Regional Development Act** of 1965, which was intended to address the dire economic circumstances of Appalachian counties (see Figure A); this act was amended in 2004, but the core mission of eradicating poverty remained the same. As part of this legislation, the Appalachian Regional Commission was created to administer the regional development funds allocated by the federal government. An important emphasis thus far has been on highway development to lessen isolation. One reason that early subsistence farmers in the more remote areas of Appalachia could not adjust to commercial agriculture was their distance from markets. Similarly, although lumber and coal resources were harvested in Appalachia, the profitable markets were elsewhere, and transportation was developed only to the degree necessary to haul the products out of the region. Because outside companies controlled the resources, profits flowed to other regions, not to Appalachia.

Remember the concept of circular causation and growth discussed in an earlier chapter. If advantages can initiate a process of circular causation, disadvantages can do the same, but in reverse. In Appalachia isolation, poverty, low education levels, and limited incentive have fed on each other and have led to further problems. Some believe that improvements in transportation will stop the cycle of poverty by stimulating economic growth. They argue that industry will locate near highways and that Appalachia's potential for recreation and tourism can then more fully be exploited.

Other people view the problem differently. They believe that the interior of Appalachia is disadvantaged by both history and location, and they do not consider transportation development a complete fix. Isolated Appalachian communities have only a small, scattered, unskilled or wrongly skilled labor supply and might easily become overindustrialized relative to their labor supply, even if low-wage, labor-intensive industries were involved. People with this view think that the greatest potential for growth and economic improvement lies in the cities on the edges of the region, where interaction with other cities or population regions is more likely. Most of the urban growth centers in the area are, in fact, peripheral to the Appalachian corridor. The northern edge, in Pennsylvania and New York, is relatively prosperous. So is the peripheral Piedmont, between Virginia and Alabama. Highways may simply help some of the corridor people reach the more advantageous periphery and may not really change conditions within the corridor.

With the dawn of the twenty-first century, the story of Appalachian development contains both bad news and good news. The bad news is that Appalachia remains more impoverished than the United States as a whole. As of 2001, per capita income in Appalachia was still only about 82 percent of the national level; poverty rates in 2000 were 110 percent above the national average, although this figure was down from 114 percent in 1980. The good news, however, is that the 2002 unemployment level was equal to that of the United States as a whole; high school and college graduation rates were up dramatically over 1980; and the region's population actually grew by 9 percent between 1990 and 2000. Earlier successes should also be noted. By 1980, more than 1,700 miles (2,736 kilometers) of highway improvements had been made. Health care had improved to the point that infant mortality was down from 27.9 deaths per 1,000 births in 1963 (not unlike the rates in some less industrialized countries) to 11.4 in 1982.

The struggle continues, nonetheless, as these counties contend with rising competition from imports in basic industries such as textiles and apparel, with chronic distress in the coal industry, with an agricultural system that remains stubbornly dependent on crops such as tobacco, and with a continuing need to address basic services in the most severely distressed areas. Furthermore, economic improvement has not occurred uniformly across the region, with persistent pockets of poverty in eastern Kentucky, West Virginia, and northern Mississippi. For example, poverty levels in the coal-mining counties of eastern Kentucky often exceed 25 percent. These counties are officially classified by the Appalachian Regional Commission as "distressed" and constitute daunting twenty-first century challenges. It remains to be seen what kind of player Appalachia will be in the increasingly technological, service-oriented, and global economy of the twenty-first century, and whether the region will be able to shed its dubious distinction as one of America's impoverished backwaters.

The residential pattern for urban African Americans stems from their economic and social position. Highly concentrated African-American neighborhoods appear in older residential areas, often vacated as economically progressive African Americans and whites flee to the city's periphery or to suburban communities. This process has progressed to the point that many cities have become more African American than white, as evidenced by the increasing number of African Americans elected mayors of large cities.

The 1954 ruling of the U.S. Supreme Court that segregation of schools was unconstitutional is a landmark in race relations in American history. But the process that led to a new social and economic position for African Americans began long before 1954. An important part of that process was the demise of the southern agrarian system as it was structured during the nineteenth and early twentieth centuries. In its place was evolving an urban, industrial, and economically stronger South, complete with a rising African-American consciousness. Fortunately, urbanization has brought economic and educational gains and opportunities for many African Americans. Their new status, hopes, and demands for participation in mainstream society could only have come with a break from the old system; and migration, whether to northern or southern cities, was symptomatic of that break.

Hispanic Americans

Hispanics are persons who trace their backgrounds to a Spanish American culture, regardless of whether it is in the United States or Latin America. The term Hispanic does not refer to a racial group. Indeed, there are established African components to the populations of many Latin American countries, especially in the Caribbean, and these people easily fit into an Hispanic setting. The same can be said for Latin America's Asian populations.

Table 6-2   African-American Population of the Conterminous United States

		Regional Distribution (%)			
	Total (millions)	Northeast	North Central	South	West
1900	8,834	4.4	5.6	89.7	0.3
1910	9,828	4.9	5.5	89.0	0.5
1920	10,463	6.5	7.6	85.2	0.8
1930	11,891	9.6	10.6	78.7	1.0
1940	12,886	10.6	11.0	77.0	1.3
1950	15,042	13.4	14.8	68.0	3.8
1960	18,860	16.1	18.3	60.0	5.7
1970	22,580	19.2	20.2	53.0	7.5
1980	26,505	18.3	20.1	53.0	8.5
1990	29,986	18.7	19.1	52.8	9.4
2000	34,658	17.6	18.8	54.8	8.9

Sources: U.S. Bureau of the Census, Census of Population (Washington, D.C.: Government Printing Office, 1910–1980); U.S. Bureau of the Census, Current Population Reports, Population Estimates and Projections, *Projections of the Population of States by Age, Sex, and Race: 1988 to 2010.* Series P-25, No. 1017; and Bureau of the Census, *Statistical Abstract of the United States, 1996* (Washington, D.C.: Government Printing Office, 1996).
Note: Because these figures have been rounded, the totals by year may not equal 100%.

The Hispanic population in the United States is large and growing at a high rate. As of 2000, the Hispanic population was set by the U.S. Census at 32.4 million people; by 2004, this figure had increased to an estimated 40.4 million people according to the Pew Hispanic Center (acknowledging that different statistical sources may produce different numbers). The majority of Hispanics claims a Mexican origin, perhaps 22 million in 2000; but most Latin American countries are represented (Figure 6–6). Some countries have witnessed large shares of their populations migrate to the United States. The small country of El Salvador, for example, could claim 932,000 Salvadoreans living in the United States in 2000, about 12 percent of the population living in El Salvador itself. By the middle of 2003, the Hispanic population had grown to the point that it had overtaken the African American population as the country's "majority" minority (see the comparisons in Table 6–3).

A surprisingly large proportion of the Hispanic population in the United States is comprised of illegal immigrants. According to the Pew Hispanic Center, the undocumented Hispanic population in the United States reached an estimated 11 million people by 2005, of which roughly 6 million were from Mexico. About a half-million Mexicans have been added to the U.S. population each year since the mid-1990s, with perhaps 80 percent of this migration stream illegal (bearing in mind that this type of information is difficult to obtain). Most of these migrants have been young, with about one in six under 18 years of age. They have tended to

▪ **FIGURE 6-5 African American socioeconomic mobility.** Many African Americans have improved their economic circumstances in recent decades, but others continue to live in poverty.

Table 6-3   Projected Population Percentages by Race and Hispanic Origin

	Percentages			
	1980	1990	2000	2010
White	85.9	83.9	82.2	80.6
African American	11.8	12.3	12.8	13.3
American Indian[a]	.6	.8	.9	.9
Asian or Pacific Islander	1.6	3.0	4.1	5.1
Hispanic Origin[b]	6.4	9.0	11.8	14.6

Sources: Adapted from U.S. Bureau of the Census, *Statistical Abstract of the United States: 2001–Population* (Washington, D.C.: Government Printing Office, 2001), Tables 14 and 15.
Note: The U.S. population in 2010 is expected to exceed 299 million.
[a]American Indian, Eskimo, or Aleut.
[b]Persons of Hispanic origin may be of any race; consequently, the sum of each column exceeds 100.0. Updated Census figures in 2003 indicated that the Hispanic population may now be larger than the African American population.

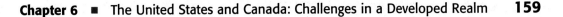

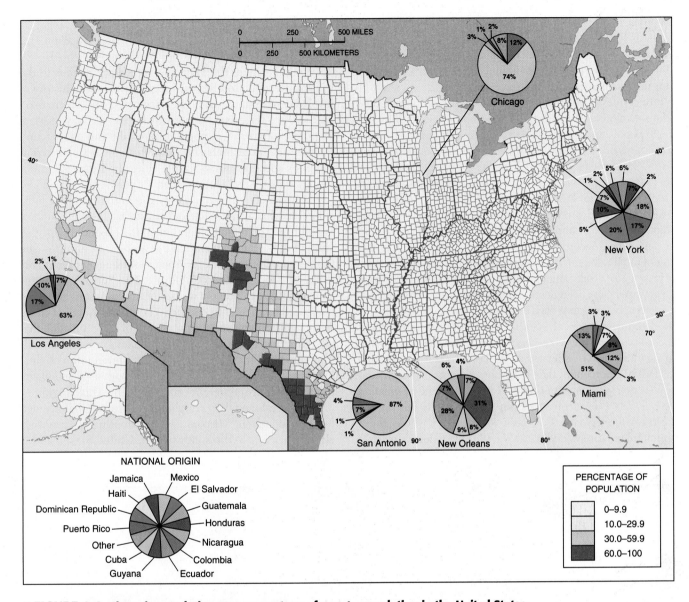

■ **FIGURE 6-6 Hispanic population as a percentage of county population in the United States in 2000.** The distribution of Hispanics exhibits two distinctive patterns. The concentration in the Southwest results from a long tradition of Spanish and Mexican influence. In addition, major cities have received a variety of migrants from Mexico, Puerto Rico, Central America, and the Caribbean, with different cities showing different concentrations. To illustrate the second point, circles are used to show origins of city residents born in Latin America (including Puerto Rico), as of 2000, for selected cities. New York has become a destination for migrants from many origins, but particularly the Dominican Republic and Puerto Rico; Miami has received mainly Cubans and Central Americans; Chicago, Los Angeles, and San Antonio have attracted mainly Mexican migrants. (Note that Latin America includes populations that are not considered Hispanic—such as Jamaican and Haitian. Note also that many people born in the United States consider themselves Hispanic and are not included in these circles.)
Sources: Adapted from U.S. Bureau of the Census, online data resources for the 2000 Census at *www2.census.gov/census_2000/datasets/demographic_profile/*, and Census 2000, Detailed Tables, Summary File 3 (SF 3) Sample Data, P21, PCT19.

settle in eight states that include, in descending order of migrant numbers: California, Texas, Florida, New York, Arizona, Illinois, New Jersey, and North Carolina.

How the United States should respond to this illegal migration has produced intense debate. Advocates of tighter controls argue that illegal migrants have a depressing effect on wage levels and impair the ability of states to provide for the welfare and safety of their citizens. In response to these concerns, some states now require proof of legal residence to collect welfare; other states are attempting to make it illegal for undocumented residents to get drivers licenses. Some advocates would go so far as to argue that the children of illegal residents should be prohibited from attending public schools. On the other side of the issue are those who argue that punishment is not the solution. They contend that illegal aliens do work that legal residents often refuse to do; that these migrants contribute to the U.S. economy through their work; and that sanctions against these people, such as denial of licenses, educational opportunities, and welfare benefits, will only drive undocumented citizens farther underground and deeper into illegal activities. This debate will not go away soon, but in all probability will result in a changed legal landscape for undocumented Hispanic migrants.

Regardless of the legal status of the Hispanic population, their economic circumstances tend to be worse than for the U.S. population as a whole. Hispanic incomes are only about two-thirds that of White non-Hispanics; but even more revealing is the Hispanic household net worth figure, which, at $5,988 in 2002, was only about one-tenth that of white households. Most new jobs for Hispanics in 2004 were in low-skill and low-wage occupations requiring little more than a high school diploma as, for example, building and grounds cleaning, and maintenance. According to the Pew Hispanic Center, employment of Hispanics in these low-wage occupations has been increasing, especially among recent arrivals. At the same time, real wages have been decreasing, suggesting that immigrants are competing against each other in the labor market.

Recent arrivals often send parts of their paychecks back home, creating a business of remittances that has become a big industry. It is estimated by the Pew Hispanic Center that in 2003 14 percent of the population of Ecuador, 18 percent of Mexico, and an astonishing 28 percent of the population of El Salvador received money from friends and family in the United States. For Latin America as a whole, these remittances were valued at roughly $30 billion dollars, a figure

that exceeds the gross national product of all but the largest Latin American countries. The United States is a significant source of income for many Latin American households.

The majority of the Hispanic population resides in the southwestern United States, a region with a long history of involvement in the Spanish-speaking world (see Figure 6–6). As of 2000, Texas was 32 percent Hispanic; New Mexico, 42 percent; Arizona, 25 percent; and California, 32 percent. People with Spanish surnames dominate many smaller communities and even some sizable cities. In Texas, for example, Corpus Christi is more than 50 percent Hispanic and San Antonio almost 60 percent. The initial Hispanic infusion into what is now the southwestern United States resulted from the region's inclusion in the expanding Spanish Empire in the late sixteenth and seventeenth centuries (see Chapter 7). Thinly scattered Spanish settlements eventually extended from Texas to California. The Spanish were unable, however, to prevent a flood of Anglo-American settlement in the nineteenth century. The initial growth of those settlements diminished the proportion of the population that was Hispanic, but it did not erase the long-established cultural imprints. Indeed, these imprints were reinforced during the twentieth century by the tide of both legal and illegal immigration from neighboring Mexico. This migration process, combined with a higher-than-the-national-average fertility rate among Hispanics, is contributing to a renewed Hispanicization of the southwestern United States, and is

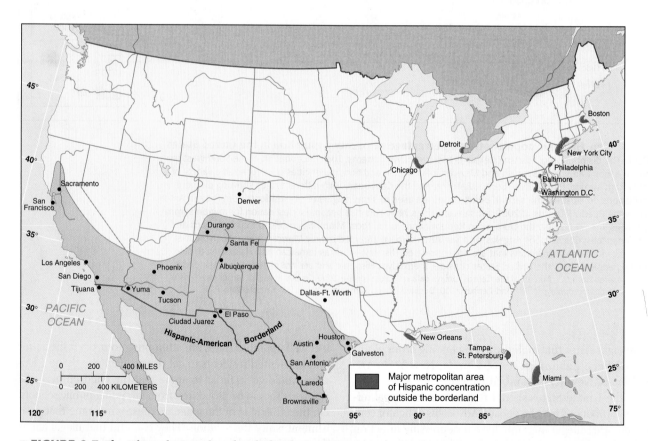

■ **FIGURE 6-7** **The Hispanic-American borderland.** The region designated as the Hispanic-American borderland is an area where Hispanic influence is strong, as reflected in ethnicity, language, and many other dimensions of culture.

creating an Hispanic-American borderland that is at least a partial reprise of an earlier time in history (Figure 6–7).

Several large metropolitan areas outside the Southwest also have significant Hispanic populations (Figure 6–8). In New York and Chicago (27 percent and 26 percent Hispanic, respectively), those populations are largely of Puerto Rican and Mexican descent. In southern Florida the largest Hispanic group is Cuban, a result of the widespread migration of Cuba's middle class after the Cuban Revolution; Miami, for example, is more than half Hispanic. In New Orleans, the Hispanic population is largely of Central American origin, with approximately half of Honduran ancestry.

Other Minorities

The Hispanic-American culture region is also the area that is occupied by almost half of the 2.5 million Native American Indians in the United States (Figure 6–9). Although the Indians of the Southwest have maintained their tribal structure better than other Native Americans have, they endure greater socioeconomic disparities than any other minority (Figure 6–10).

Another rapidly growing segment of the United States includes people of Asian descent, whose numbers have increased

■ FIGURE 6-8 Hispanic cultural imprint. Hispanics now constitute a large and rapidly-growing ethnic group in many cities of the United States and Canada. This business is found in Washington, D.C.

considerably since the 1970s in response to immigration (Figure 6–11, page 164). It is estimated that the 7.5 million Asians in the United States in 1990 increased to almost 12 million in 2003 and will exceed a population of 15 million in 2010. Growth has been particularly noteworthy in the urban centers of the West, although a substantial presence can be found in a number of urban centers outside of the West.

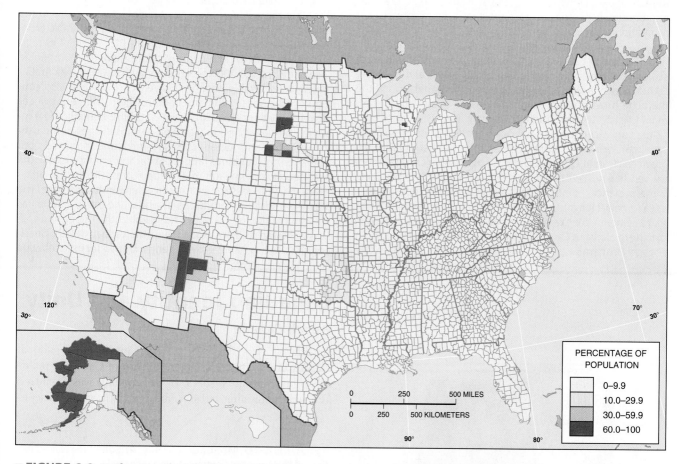

PERCENTAGE OF POPULATION

0–9.9
10.0–29.9
30.0–59.9
60.0–100

■ FIGURE 6-9 Native American Indian population as a percentage of total county population in 2000. The distribution of the Native American Indian population is most strongly associated with the American West.

Source: Adapted from U.S. Bureau of the Census, online data resources for the 2000 Census at *www2.census.gov/ census_2000/datasets/demographic_profile/.*

GEOGRAPHY IN ACTION
Canadian Unity

Canadian unity seems increasingly difficult to achieve. Indeed, political separation (sometimes accompanied by violence) has been promoted by French-Canadian nationalists. In 1980, a referendum in Quebec was conducted by the ruling Parti Quebecois, the separatist party, to determine whether the provincial government should move toward separation from the rest of Canada. The referendum failed, but 42 percent of Quebec voters voted in the affirmative, as did 52 percent of the French-speaking population in Quebec. In 1982, Canada attempted to strengthen its union by revising its constitution to include a bill of rights. The government of Quebec, still under the control of the Parti Quebecois, refused to sign the new document, ostensibly because it would not permit Quebec to secure its cultural identity as a distinct society within the larger framework of Canada. The Meech Lake conference in 1987 produced a document, referred to as the Meech Lake Accord, designed to address Quebec's concerns. The accord was rejected by the governments of Newfoundland and Manitoba, which argued that the accord gave Quebec cultural rights that the other provinces did not have. The accord did not survive its ratification deadline of June 23, 1990. Partly as a consequence, the separatist cause found new strength. In the fall of 1995, a second referendum was held, and the federalist position won by a very slim margin. It is not unreasonable to envision a separatist victory if there is a third referendum, especially as those most inclined to vote against separation move elsewhere.

While the conflict between English and French Canada monopolizes the debate for most Canadians, this conflict often obscures how ethnically diverse English Canada is and how few French speakers live outside of Quebec. The population of the maritime provinces, for example, consists of persons of German, English, Scottish, and Irish descent. New Brunswick constitutes an exception since about one-third of its population speaks French, most of whom live in the northern half of the province close to Quebec. Farther to the west in Canada, there are very few pockets of native French speech (Figure A). Indeed, these provinces are likely to have significant minorities that speak a native language that is neither French nor English. Note that in Ontario, for example, 29 percent of the population (according to Statistics Canada) speak a native language other than English. Furthermore, there are almost as many people in On-

tario who speak Chinese as a mother tongue as speak French. Toronto, Ontario, is one of the most multicultural cities in North America. Of its 4.6 million people, only 58 percent speak English as a mother tongue and a mere 1.2 percent speak French; the remainder speak Chinese (348,000), Italian (195,000), Portuguese (108,000), Punjabi (96,000), Tagalog (Filipino) (77,000), and a wide variety of other languages. In British Columbia, 8 percent of the population speak Chinese as a native language, but only a scant 1 percent speak French; there are twice as many Punjabi speakers from South Asia as there are French speakers. These language distributions reflect shifting immigration patterns. In 2003, the largest source of immigrants to Canada, 16 percent, came from the People's Republic of China, compared with a meager 2.3 percent from Great Britain. There were more immigrants from the Philippines than from the United States. Finally, more than 975,000 Canadians are at least partially of aboriginal origin—largely North American Indian, Metis, Cree, or Eskimo. These "First Nation" populations are in the majority in the Northwest Territories and are significantly represented in the prairie provinces. Indeed, it was the issue of special privileges granted to Quebec in the Meech Lake Accord, but not to the aboriginal population of Manitoba, that led the Manitoba government to oppose ratification.

Even within French Canada, there is considerable cultural diversity. While 81 percent of the population speaks French as a mother tongue, there is a good chance that a visitor to the province will hear English, Italian, Spanish, and even Arabic. Montreal is especially multicultural, with only 67 percent of its population speaking French natively—more than 400,000 people speak English, 120,000 speak Italian, and 70,000 speak Arabic.

The considerable economic diversity of Canada also contributes to a sense of fragmentation. The maritime provinces are relatively poor and dependent on fishing, farming, and mining (all primary activities), as well as subsidies from the central government. Quebec is an industrial province, but slow-growth industries and political uncertainty cloud the province's economic future. Ontario represents the Canadian heartland, with its strong commercial and industrial base. The prairie provinces are producers of wheat and cattle. Alberta has an abundance of oil and the wealth that goes with it; Saskatchewan and Manitoba have little of that wealth. British Columbia is a western growth center known for its lumbering

■ **FIGURE 6-10 Monument Valley, Arizona.** The Navajo Indians of the American Southwest have managed to retain their tribal structure, although they suffer serious economic distress. Settlement on the Navajo Reservation occurs within semiarid regions of great natural beauty.

Canadian Identity and Unity
French Canada

Canada, too, is a pluralistic society. Consequently, national identity and unity are a major concern of Canadian political and community leaders. In fact, Canada was organized as a federation almost 140 years ago because the descen-

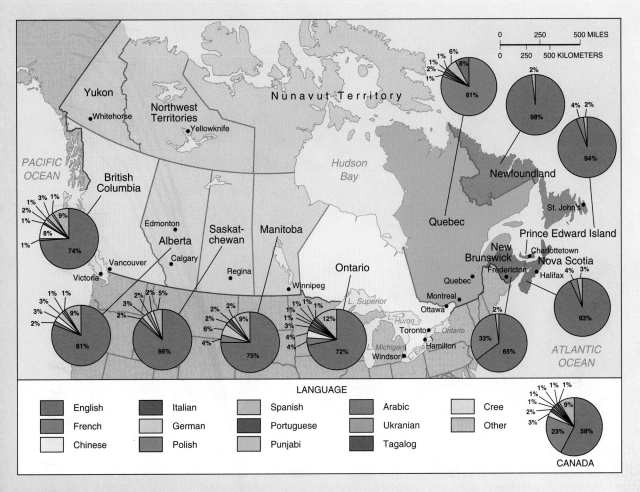

■ FIGURE A The several Canadas. The extraordinary attention given Quebec and its relationship to the remainder of Canada sometimes overshadows the numerous other important differences that exist among the several regions of the vast Canadian territory. Note especially the diversity of languages spoken in Canada in addition to the official languages of English and French as of 2001.
Source: Adapted from Statistics Canada, *Population by Mother Tongue, Provinces and Territories; 2001 Census, www.statcan.ca/english/Pgdb/demo18a.htm.*

and trade, but it is a province far removed from the remainder of Canada both in distance and spirit.

Canada is thus a large and disparate country. Even its geography works against unity and cohesion. When resource en- dowments and economic bases are added to the picture, the various regions seem even more distinct. Achieving unity will remain a full-time job for Canada.

dants of French settlers insisted that any system of union preserve the French identity and influence.

French Canadians have maintained their identity and are intent on continuing to do so. Their distinctiveness is not only linguistic—although that, in itself, is enough to promote a sep- arate cultural identity—but also religious. French Canadians are overwhelmingly Roman Catholic, in contrast to the largely Protestant English Canadians. In other respects, however, the French of Canada have been stereotyped to a misleading de- gree, even by other Canadians. It is a misconception to char- acterize them as a quaint, rural, agrarian, unchanging people with high birthrates. In truth, the province of Quebec is 80 percent urban, part of industrial Canada, and clearly inte- grated into the nation's economic core. Quebec was once largely agrarian and did have high population growth rates, which engendered some fear that French culture would over- whelm English culture. With modernization and an urban lifestyle, however, birthrates in Quebec have fallen below the national average. Yet while the vast majority of French Cana- dians now live in urban areas, their occupational structure still does not match that of English Canadians. They remain over- represented in the primary sector and in unskilled jobs.

Because both French and English are official languages, Canada must forge a national unity in a bilingual frame- work—no easy task. In addition, significant numbers of Poles, Dutch, Germans, Italians, and other national groups have arrived during the twentieth century, sometimes to the consternation of the French, who sometimes oppose

163

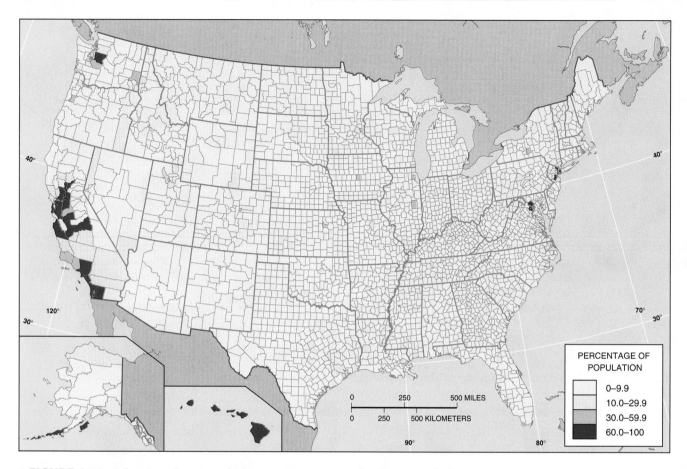

■ **FIGURE 6-11 Asian-American population as a percentage of county population in 2000.**
Asian peoples represent a rapidly increasing population segment in the Pacific Coast states.
Source: U.S. Bureau of the Census, *USA Counties, 1998,* Item AA98P13D.

immigration because it diminishes their numerical strength. Moreover, slightly more than 975,000 Canadians report an Aboriginal or Native American identity (see the *Geography in Action* boxed feature, Canadian Unity, page 162).

Canada and the United States

Canadian identity is further affected by Canada's proximity to the United States. Physically, Canada is one of the largest countries in the world, but its vast area contains only about 32 million people. It is rich in forest, water, and mineral resources, but its value for agricultural production is modest because so much of its environment is harsh. In light of Canada's available resources and its proximity to the industrialized United States, it is not surprising that major economic linkages have evolved between the two countries. Canada and the United States have become the most important trading partners for each other, focusing much of that relationship on the removal and processing of Canada's special material resources. In addition, a high proportion of Canadian industries are controlled by U.S. parent firms.

As mentioned earlier, Canada's fear of a colonial relationship generated its long-standing tariff policy. Paradoxically, the success of that policy has prompted concern over foreign influence and control, which has spurred efforts to "Canadian-

ize" the economy—that is, to increase Canadian investment levels in Canada's economic activities. Nevertheless, Canadian control of some industries has continued to decrease. Economic prosperity in Canada remains significantly tied to prosperity in the United States, thereby contributing to the problem of Canadian identity. Political efforts to redirect trade relationships toward other areas might run counter to the normal economic relationships expected between two well-endowed neighboring countries. The North American Free Trade Agreement embodies recognition of the mutual advantages that may accrue from neighbors benefiting from each other's strengths.

"Melting Pot," "Stew Pot," or What?

A time-worn assumption among Americans is that immigrants eventually lose their separate identities as they blend into a single American **melting pot.** According to this assumption, "foreignness" is lost as immigrants adopt the English language and acquire an American cultural identity. In Canada, with its more pronounced cultural cleavages, the "stew pot" metaphor seems more accurate, with chunks of culture spread across the landscape. Part of the stew-pot phenomenon reflects an immigration process that occurred

GEOGRAPHY IN ACTION

The "Dearborn Effect": Ethnic Clustering in Contemporary American Cities

Dearborn, Michigan, which is located on the southwestern edge of Detroit, is a small city that is known for its prominent role in the history of the U.S. automobile industry. Henry Ford lived and worked in Dearborn, and it continues to house the world headquarters of the Ford Motor Company. One of the great auto plants of all time—the Rouge facility—is located in Dearborn and, at its peak, employed several tens of thousands of workers.

Today Dearborn is known not just for its role in the auto industry, but as the poster child for ethnic clustering in modern America. Out of the 98,000 residents of Dearborn in 2000, more than 17,000 were born in the Middle East—mostly in Lebanon and Iraq. Add to this number the residents of Middle Eastern descent born in the United States, and Dearborn's Middle Eastern population accounts for roughly half of the city's total population. Southeastern Michigan is probably home to 250,000 people from this part of the world. The presence of the auto industry and a Middle Eastern population is not entirely coincidental. Henry Ford initially was reluctant to hire African Americans to work in his car factories (he had equally unenlightened views about other groups, too), but he was willing to hire immigrants from the Middle East. A Middle Eastern core area was thus created—initially populated by Lebanese—that became a magnet for other immigrants. Later immigrants had little or no connection to the auto industry. Civil war in Lebanon and other regional conflicts swelled the number of migrants.

The changes to Dearborn have been profound. Local schools teach in both English and Arabic. The local branch of the University of Michigan offers a curriculum in Arabic and has a Center for Arab-American Studies. The sights and sounds of the city reflect an immigrant culture (Figure A).

The ethnic clustering in Dearborn—what we conveniently call here the "Dearborn Effect"—is only the latest variation on a recurring theme in American history. The United States is a country of immigrants and typically these immigrants have become concentrated in specific locations (see the discussion in Chapter 4 and the maps in Figure 4–20). We are reminded, for example, of the Irish in Boston and the Poles in Chicago. What we see today, however, is a large number of clusters of migrants from origins other than Europe. The Dearborn Effect can be found, for example, in Fremont, California, a city in which a quarter of the city's population of 200,000 was born in Asia, mainly China and India; especially perplexing for Fremont is the large number of different linguistic groups within the Asian community. We find a similar story of clustering in Miami with its large Cuban influx. New Orleans has a large pocket of Vietnamese. New York City has 250,000 people born in China and 115,000 born in the small South American country of Ecuador. Perhaps 35 percent of New York's current population was born abroad.

Ethnic clustering is reshaping the cultural landscapes of American cities, presenting new challenges and offering new opportunities for city leaders. The extent to which these clusters lose their identities as their people become acculturated, and the degree to which these clusters imprint their home cities, remains to be seen. Given present trends, the Dearborn Effect will become even more pronounced in future years.

■ **FIGURE A** **Dearborn, Michigan.** This city is not only a heart of the auto industry, but a core region for Middle Eastern immigrants to the United States.

---

later in history than in the United States and perhaps also reflects a more difficult struggle to identify a truly Canadian identity. Given the dynamics of the global system, however, and their generalizing effects, it is hard to imagine any country retaining distinctive characteristics over the long term in the face of what may be a global melting pot.

Perhaps Wilbur Zelinsky should be given the last word in this conversation.[1] He noted that surprisingly little research has been done on "convergence," or the melting-pot phenomenon. An observer will find, for example, that while acculturation has occurred in the United States, regional differentiation remains. It was Zelinsky's "hunch" that North Americans could never outgrow or dissolve regional "specialness," regardless of the domestic or international pressures. The ultimate reality is that each region has always been in a state of change and, in all probability, will continue to evolve in often disparate directions. In brief, it may be unwise to attach too much weight to a melting-pot argument (see the *Geography in Action* boxed feature, The "Dearborn

[1]W. Zelinsky, *The Cultural Geography of the United States: A Revised Edition* (Englewood Cliffs, NJ: Prentice Hall, 1992), pp. 182–184.

Effect": Ethnic Clustering in Contemporary American Cities).

Summary

Both the United States and Canada are large in size and have immense resources. Each has achieved a high level of living for the majority of its populace and a powerful position in the world, and each will certainly attempt to maintain that level and position.

Out of the American and Canadian development experiences have come technologies and perspectives that can be of great benefit to the remainder of the world. But while the United States and Canada continue to rank among the world's leading countries in many expressions of economic and human development, it is important to recognize that the resources of underdeveloped areas have frequently been used to promote American and Canadian economic activity. In the future, as other countries develop and consume more of their own material resources, it will be important to determine how much the United States and Canada can—or should—gather from these countries, and to understand the social and environmental costs of that gathering process.

The processes of immigration and settlement have affected the two neighbors differently. The larger population of the United States has aided in the accumulation of greater wealth through the processing and use of the nation's resources. The United States, however, is now consuming on such a scale that the cost and availability of basic goods may well encourage or require the limitation of growth, or even stabilization. On the other hand, the economic sluggishness of the early 1980s and again in the early 1990s, and the problems of economically dying communities, suggest that limiting growth may be difficult or unwise. For less populous Canada, greater economic independence and internal industrial growth may require domestic population growth and expansion. The value of population growth for developing and using resources is even now reflected in Canadian immigration policies.

The internal problems of both countries include the need to integrate minority groups into the larger society. As the United States and Canada progressed, entire regions, as well as minority groups, lagged in the acquisition of wealth and position within the system. Effective means are now needed to incorporate the pockets of poverty and isolation into the larger economy.

American society will have to make other adjustments, too. The populations of the United States and Canada are becoming even more urban, though the characteristics of urban areas may differ from those of the past. The large urban formations—the megalopolitan regions discussed in the previous chapter—may require new approaches to government and planning. Certainly such highly integrated urban systems will necessitate development of a regional framework extending far beyond the traditional political boundaries of city governance.

▶▶ Key Terms

African-American migration

Appalachian Regional
   Development Act

Hispanics

income disparities

melting pot

poverty

▶▶ Review Questions

1. The physical structure of the United States and Canada is characterized by several mountain backbones separated by lowlands. Describe where these mountain and lowland features are located.
2. Most of Canada's population lives within 200 miles (320 kilometers) of the U.S. border. Why?
3. Describe the principal influences on North American climate.
4. Identify the major immigrant groups that came to the United States during the nineteenth and early twentieth centuries, and describe where they became concentrated.
5. Where in the United States and Canada has recent population loss been most pronounced?
6. The rural landscape of the United States and Canada has changed dramatically over the last 100 years. What changes have occurred, and why?
7. Where is the historic manufacturing core of the United States and Canada, and why did this core form?
8. Compare and contrast the natural resource endowments of the United States and Canada.
9. Where is the Hispanic-American borderland located? In which other parts of the country are Hispanic concentrations located?
10. In a sense, the city of New Orleans is still fighting a war. Explain.

▶▶ Geographer's Craft

1. Geographers often describe the process of pushing back the frontiers of the United States and Canada as one of spatial evolution. Describe how this evolutionary process occurred.
2. Geographers are interested in the ways that people interact with their physical environments. Explain how people interacted with, and responded to, the presence of the Appalachian Fall Line.
3. You are a consulting geographer for a travel agency. You are asked to identify where the best ocean beaches and the sunniest climates are located. How would you respond to this request?
4. As a geographer, you are asked to explain why Canada faces challenges with ethnic separatism. How would you respond?
5. You are in charge of a committee that has been asked to find a new name for the Corn Belt. What criteria would you use to select the new name and what name would you choose?

PART THREE

ELEVATION IN METERS

	Above 4000
	2000–4000
	500–2000
	200–500
	0–200
	Below sea level

UNITED STATES

Tijuana
Ciudad Juárez
SIERRA MADRE OCCIDENTAL
SIERRA MADRE ORIENTAL
Monterrey
MEXICO
Guadalajara
Mexico City
SIERRA MADRE DEL SUR
YUCATÁN PENINSULA
BELIZE
Belmopan
Guatemala City
HONDURAS
GUATEMALA
Tegucigalpa
San Salvador
EL SALVADOR
NICARAGUA
Managua
San José
COSTA RICA
Panama
PANAMA

BAHAMAS
Nassau
Turks and Caicos (U.K.)
Havana
CUBA
Guantánamo
DOMINICAN REPUBLIC
Kingston
HAITI
Santo Domingo
San Juan
JAMAICA
Port-au-Prince
Puerto Rico (U.S.)
Cayman Is (U.K.)
Caribbean Sea
Curaçao (Neth.)
Aruba (Neth.)
Bonaire (Neth.)
Barranquilla
Maracaibo
Cartagena
Valencia
Caracas

Tropic of Cancer
See Inset Below
ATLANTIC OCEAN

PACIFIC OCEAN

VENEZUELA
Medellín
LLANOS
Ciudad Bolívar
Georgetown
SURINAME
Paramaribo
GUYANA
Cayenne
FRENCH GUIANA
Cali
Bogotá
COLOMBIA
Orinoco R.
Negro R.

Galápagos Is. (ECUADOR)
Equator
Quito
ECUADOR
Guayaquil
Iquitos
Manaus
Belém
AMAZON
São Luís
BASIN
Fortaleza
Natal
Trujillo
PERU
Rio Branco
B R A Z I L
Recife
Callao
Lima
A N D E S
Cuzco
Salvador
Amazon
Tapajós R.
Xingu R.
Araguaia R.
São Francisco R.

Arequipa
La Paz
BOLIVIA
MATO GROSSO PLATEAU
Brasília
Arica
Santa Cruz
Sucre
Goiânia
ALTIPLANO
PANTANAL
Belo Horizonte
CHILE
CHACO
PARAGUAY
São Paulo
SERRA DA MANTIQUEIRA
Asunción
SERRA DO MAR
Rio de Janeiro
PARANÁ BASALT PLATEAU
Paraná R.
ARGENTINA
Pôrto Alegre
Córdoba
Valparaíso
Mt. Aconcagua
Santiago
Buenos Aires
URUGUAY
Concepción
PAMPAS
Montevideo
Bahía Blanca

0   500   1,000 MILES
0   500   1,000 KILOMETERS

Puntas Arenas
Falkland Is. (U.K.)

ATLANTIC OCEAN
N

Puerto Rico (U.S.)
British Virgin Is. (U.K.)
Tortola
Anguilla (U.K.)
Virgin Is. (U.S.)
St. Martin
St. Martin (Fr. and Neth.)
Saba (Neth.)
Barbuda
ANTIGUA AND BARBUDA
St. Eustatius (Neth.)
Basseterre
ST. KITTS AND NEVIS
St. John's
Antigua
Leeward Is.
Montserrat (U.K.)
Plymouth
Lesser Antilles
Guadeloupe (Fr.)
Pointe-à-Pitre
Basse-Terre
Marie-Galante
Caribbean Sea
DOMINICA
Roseau
Ste.-Marie
Martinique (Fr.)
Fort-de-France
Le Diamant
Castries
ST. LUCIA
Soufrière
Vieux Fort
ST. VINCENT AND THE GRENADINES
Kingstown
Bridgetown
Windward Is.
Bequia
BARBADOS
St. George's
GRENADA
Isla de Margarita
Canaan
Tobago
Port of Spain
TRINIDAD AND TOBAGO
Trinidad

0   100   200 MILES
0   100   200 KILOMETERS

Latin America and the Caribbean

David L. Clawson

Although Latin America occupies a major portion of the Western Hemisphere, for many in the United States and Canada, it is one of the least understood culture regions in the world. That lack of understanding and familiarity can be traced in part to the economic and religious rivalries of colonial England and Spain. In modern times, linguistic, political, and technological differences have often masked many of the common needs and aspirations of the Anglo American and Latin American peoples.

Many residents of the United States and Canada think of Latin America as a region of simple peasant farmers living in tiny, isolated villages without benefit of medical care or education. They envision those villagers faithfully attending Catholic Mass in a colonial-era church, while corrupt military leaders and politicians carry out a never-ending series of revolutions in the distant cities. In reality, those stereotypes may have been reflective of selected settings in the past, but they have never been a totally accurate portrayal and are certainly inappropriate images for today. Although thousands of agrarian villages still dot the region, most Latin Americans now live in great cities that are among the largest and fastest-growing in the world.

Urbanization has brought not only better-paying industrial jobs and a growing middle class to Latin America but also increasing pressure for political reform and democratization. With that political change has come fundamental social reform, including the spread of new religious faiths and practices and dramatic improvement in health care and education.

Latin American development has been financed in part through Anglo American, European, and Asian economic assistance. Consequently, the leaders of those countries have often expected Latin Americans to reciprocate by supporting the political and economic interests of the donor nations. Those foreign leaders may even become confused and upset when the Latin American peoples feel threatened by what they perceive to be unwarranted outside intervention in their internal affairs. When emotions run high, the peoples of all regions may lose sight of the many forms of positive interaction that characterize their relationship, including a mutual commitment to freedom and development.

We begin our study of Latin America and the Caribbean by looking first at the physical diversity and historical evolution of Latin American culture and at the common challenges and opportunities that face all Latin American and Caribbean peoples. We then examine the specific regional characteristics and potentials of Mexico and the Central American and Caribbean countries, and of the South American nations. As we proceed, we will come to recognize not only the differences between the Anglo American and Latin American cultures but also the common hopes and needs that are likely to bind them ever more closely in the years ahead.

CONTENTS

El Salvador's national Catholic cathedral. Roman Catholicism was introduced into Latin America and the Caribbean by Spanish, Portuguese, and French settlers at the beginning of the colonial era. The region is now characterized by religious pluralism.

Latin America and the Caribbean:
Physical Diversity and Cultural Heritage

- Physical Diversity
- The Iberian Heritage
- The Colonial Period
- The Era of Independence
- Modern Latin America

Latin America extends from Mexico's border with the United States to the politically divided island of Tierra del Fuego, which faces Antarctic waters in the far south. Latin America is a region of great contrast and rapid change. It includes Brazil, Argentina, and Mexico, three of the largest and most richly endowed countries on earth, as well as Haiti, Nicaragua, and other small and politically unstable nations that rank among the world's poorest countries. The physical environments of Latin America range from the Atacama Desert, where years pass without measurable precipitation, to the luxuriant Amazonian rain forest, and from snow-capped volcanoes that watch over cold and barren plateaus to hot, sandy tourist beaches.

In addition to physical contrasts, Latin America is a region of deep social and economic divisions. Vast mechanized plantations produce export crops of sugarcane, bananas, and other tropical fruits on broad, fertile lowlands, while traditional farmers work tiny plots of extraordinarily steep land on distant mountain slopes. Every year, countless thousands of those subsistence farmers and their families flee the poverty and isolation of the countryside and migrate to the cities, hoping to achieve a better life for themselves and their children. All too often, their dreams are crushed by the realities of urban poverty and unemployment. Unable to afford decent accommodations, many find themselves forced to live in crowded and squalid inner-city tenement housing where dozens of families living in one-room "apartments" may be forced to share a single shower and toilet. Others build tiny, flimsy dwellings from discarded lumber, cardboard, or tar paper—often within sight of the high, guarded outer walls of the great mansions of the elite, where swimming pools, horse stables, and garages for foreign sports cars share space with manicured lawns and tropical flowers (Figure 7–1).

In the cities, old ways and new have been compressed into almost unimaginable combinations (Figure 7–2). Toyotas and donkeys may be double parked on narrow, cobblestone streets lined with historic, colonial-era buildings. Aggressive, well-educated businessmen dressed in Western suits and ties pass barefooted Indian women wearing hats and long, full skirts while carrying babies wrapped in shawls on their backs. Customers in chic specialty shops pretend to ignore the homeless children and peddlers on the sidewalks and streets. Modern supermarkets often stand within view of traditional open-air marketplaces, where housewives and servants go daily to haggle over the price of fresh produce.

In a region of great mineral and agricultural resources, it is ironic that the greatest resource of all—human beings—has often been tragically wasted. Limited educational opportunities, rigid social stratification, graft, and corruption retard present-day economic development just as surely as the colonial exploitation of the past. The Latin American nations' collective burden of wasted development opportunities, both internal and external, has resulted in some of the lowest income levels and heaviest foreign-debt loads in the world.

As mentioned earlier, many of the long-held perceptions of Latin America are no longer accurate. In a land of supposedly universal allegiance to Roman Catholicism, fundamentalist Protestantism is growing at a rapid rate, while the Catholic Church struggles with politically leftist ideology and widespread apathy, both of which tend to undermine church tradition and authority. In a region often described as a land of revolutions, the great majority of the Latin American and

■ **FIGURE 7-1 Squatter housing within a prosperous residential neighborhood.** Latin American urban squatters use any available materials to construct their homes. Assorted discarded items have been placed on the roof of this structure in an attempt to keep the sheet metal from blowing off in a storm.

Caribbean nations are now blessed with freely elected governments. The goal of establishing true and complete democracies remains difficult to achieve, however, owing to the continuing dominance of elite families who have resisted substantive change from the colonial era to the present. Pulled primarily by the perception of economic opportunity abroad, millions of Latin Americans attempt to emigrate to the United States or Canada each year, either legally or illegally.

Although Latin America has a troubled past, it is also a region of great progress and even greater potential. Never before have so many Latin Americans been so healthy; so well fed, housed, and educated; and so free to pursue their lives as they choose. Progress has been greater in some places than in others, however. Future development will depend, in large measure, on the continuing reformation of both internal structures and external relationships.

Physical Diversity

The physical diversity of Latin America is found both in its varied landforms and in its contrasting weather patterns and climatic conditions. In this chapter, we will present a brief overview of the physical geography of Latin America, with more detailed descriptions of the unique natural regions of the nations of Middle and South America occurring in the following chapters.

Landforms

Latin America and the Caribbean consist of three great structural landform zones that have served as the stages or platforms of the varied human activities down through the centuries. The first of these zones, the **Eastern Highlands**, consists of three large upland regions comprised primarily of ancient igneous and metamorphic rocks (Figure 7–3). The largest of these upland areas is found in central Brazil; the second in southern Venezuela, Guyana, Suriname, and French Guiana; and the third in southern Argentina. The Eastern Highlands appear mostly as low, eroded mountains and plateaus averaging 2,000–4,000 feet (610–1,219 meters) above sea level (Figure 7–4). While southern Argentina lacks world-class mineral deposits, the mining of iron ore, bauxite, gold, diamonds, and a host of industrial minerals constitutes a major component of the economies of central Brazil, Venezuela, and the Guiana countries. Owing to the presence in many areas of leached, nutrient-poor soils, commercial food-crop agriculture in South America's Eastern Highlands is generally limited, and vast tracks of cleared land are given over to cattle or, in the case of southern Argentina, sheep grazing.

The second of Latin America's great structural zones consists of three large river basins, which collectively form the **Central Lowlands** of South America. The northernmost of these basins is the Orinoco River Valley of eastern Colombia and southern Venezuela. This region has historically been used for extensive cattle grazing and, beginning in the twentieth century, for petroleum production. Central South America is dominated by the Amazon River, which carries approximately 20 percent of the freshwater discharge of all the earth through an immense network of tributary streams that reach as far inland as the high Andes of southern Peru, more than 4,000 miles (6,437 kilometers) from the Atlantic Ocean (Figure 7–5). Land use in the Amazon Basin has historically been characterized by the hunting and gathering of rain-forest products, such as rubber, and by slash and burn farming. Many of these traditional activities are giving way to large-scale commercial cattle ranching, lumbering, and mining operations, all of which threaten to upset the deli-

■ **FIGURE 7-2 The skyline of La Paz, Bolivia.** The powerful upthrusting buildings of La Paz's central business district stand in sharp contrast to its natural setting and the remnants of older buildings clinging to the edges of the downtown area.

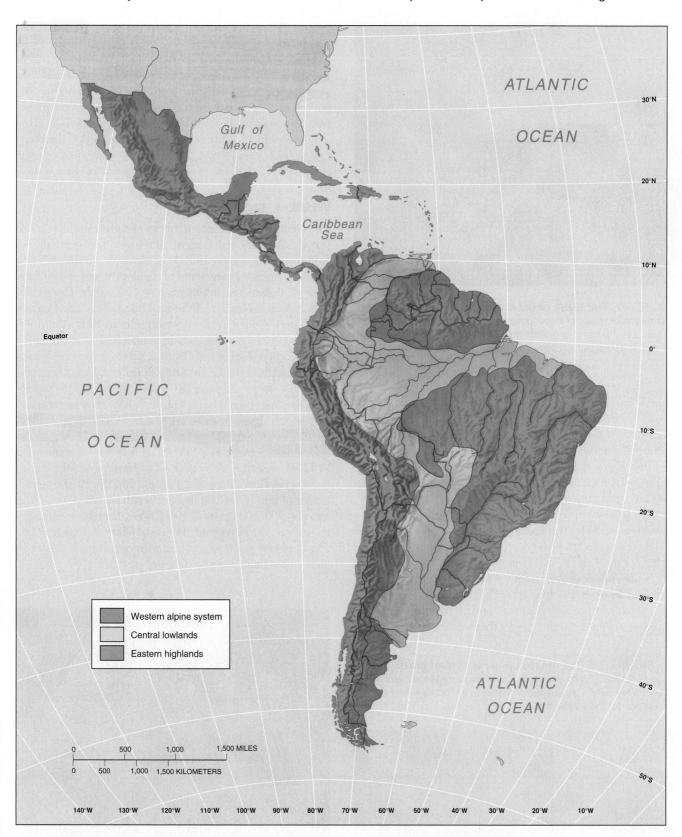

■ **FIGURE 7-3 Landform divisions of Latin America.** Latin America contains three great landform divisions: the eastern highlands, the central lowlands, and the western alpine system. The Caribbean islands are a part of the western alpine system.

■ **FIGURE 7-4 Tepuy, or flat-topped mountain, of southern Venezuela.** Many of these formations, which are common to the Guiana Highlands of northern South America, are islands of biological diversity that support plant species found nowhere else on earth.

cately balanced rain-forest environment with its priceless biological heritage. The southernmost of the great interior river basins is the Paraná–Río-de-la-Plata system of southern Brazil, Paraguay, Uruguay, and northcentral Argentina. Much of this region is blessed with fertile soils and a temperate climate. Agricultural production is characterized by large commercial farms providing abundant harvests of coffee, maize (corn), soybeans, and wheat, destined primarily for overseas markets. The Paraná River is also a leading source of the hydroelectric energy consumed by the large industrial cities of southern Brazil, Uruguay, and central Argentina.

Latin America's third great structural zone is the **Western Alpine System**. Starting in southern Chile and Argentina, the Andes Mountains continue northward through South America in great parallel ranges that average 16,000–19,000

■ **FIGURE 7-5 The Amazon River in central Brazil.** The region's flat topography and heavy rainfall produce broad rivers that are major transportation arteries as well as important sources of fresh fish for the area's inhabitants.

feet (4,877–5,791 meters) above sea level. Between the jagged, folded ranges are high intermontane valleys and plateaus, which for millennia have supported dense Indian populations who raise highland grains and tubers on tiny, steeply sloped fields (Figure 7–6). The Western Alpine System continues northward through the small Central American nations and beyond into Mexico (Figure 7–7). In a physical sense, the Caribbean islands are outliers of the Western Alpine System, whose frequent earthquakes and volcanic eruptions attest to the geological instability of the region.

Weather and Climate

The weather and climate patterns of Latin America are as varied as its landforms. Because air temperatures generally decrease with increasing altitude or elevation above sea level, Latin America contains some of the hottest and some of the coldest inhabited environments on earth (see the *Geography in Action* boxed feature, Altitudinal Life Zones in the Andes). The temperatures on summer afternoons in northern Argentina and Paraguay, for instance, routinely climb above 110°F (43°C), while residents of Potosí, a silver-mining community situated high in the neighboring Bolivian Andes, are accustomed to daytime highs of 50–60°F (10–16°C), with bitterly cold, freezing temperatures at night. Precipitation patterns also vary widely. Much of Latin America experiences alternating wet and dry seasons, with other regions remaining rainy or dry throughout the year. It is interesting to note that some areas of Latin America, such as the Caribbean coastal plain of Nicaragua, Costa Rica, and Panama and the Pacific coastal ranges of northern Colombia and southern Chile, receive as much as 200–300 inches (5,000–7,600 millimeters) of precipitation annually. In contrast, the coastal plains of northern Chile and Peru and much of northern Mexico receive less than 10 inches (250 millimeters) of rain per year.

■ **FIGURE 7-6 Peasant farms, highland Ecuador.** Many parts of Latin America's Western Alpine System are characterized by smallholder agriculture.

■ **FIGURE 7-7** **Irazú Volcano, Costa Rica.** The Central American countries have dozens of active volcanoes and experience frequent earthquakes. Although volcanic lava and ash often destroy homes and crops in the short term, they weather into fertile and productive soils.

Considering the almost countless combinations of temperature and precipitation, and adding to those the impacts of other climatic variables, such as varying latitude (distance from the equator), direction of prevailing winds, and characteristics of ocean currents, it becomes easier to understand why Latin America contains virtually every climate found on earth (Figure 7–8). Not only does Latin America possess extensive areas of tropical climates, including rain forests and savannas, it also contains large areas of humid subtropical, mediterranean, steppe, and desert climates—environments generally associated with midlatitude regions. If one climbs high enough into the mountains, it is even possible to experience the equivalent of subpolar climates, where the temperatures remain extremely cold throughout the year and vegetation is limited to grasses and shrubs capable of surviving the harsh conditions found above the treeline. We will not elaborate further in this introductory overview on the specific attributes of these natural environments, but each will be described in detail within the context of the national regions presented in the following chapters. It is sufficient to reemphasize for now that the economic and cultural development of Latin America continues to occur within widely varying physical environments.

The Iberian Heritage

The Roman Influence

The land use patterns found today throughout Latin America owe much of their existence to practices that were initially instituted by Roman settlers in **Iberia**, the peninsula that comprises the countries of Spain and Portugal. The peoples of Roman society were largely preoccupied with achieving a high social position and an ostentatious lifestyle. Wealth and social standing were achieved primarily through ownership of land, the most basic means of economic production. Ironi-

cally, however, the Romans' participation in the "good life" depended on their residing in Rome or in another major urban center rather than at their rural estates, for it was in the cities that the pleasures of Roman civilization were most available. The Latin word *civilitas,* for example, became the root of two English words—*civilization* and *city*—suggesting that only city dwellers were considered civilized persons in Roman society. Upper-class Romans met the need to be landowners and city residents by acquiring huge estates, called **latifundios**, and practicing absentee ownership, in which the landowner, or *latifundista,* resided in a distant city and entrusted the day-to-day management of the estate to an overseer.

As the Roman, and later the Spanish and Latin American, elites flaunted their wealth by refusing to engage in manual labor of any kind, farming acquired a stigma of inferiority. Simultaneously, huge urban bureaucracies developed in the cities, where countless unnecessary jobs were created and staffed with friends and relatives of aristocratic government officials. It did not matter much that these positions offered limited salaries, since they permitted participation in a broad array of bribes and speculative ventures through which bureaucrats enriched themselves at the expense of the poor.

The *latifundio* system also encouraged use of the best lands for grazing animals, which required relatively little human labor and supervision, rather than for growing food crops, which would have been labor intensive. Land use in Roman Iberia was a frontier version of that in Italy itself. Although some classic Mediterranean crops—wheat, barley, grapes, olives, and vegetables—were cultivated, the dominant land use pattern was one of exploitive animal grazing. Even the fall of the Roman Empire did little to alter the nature of those entrenched patterns.

Roman control also brought a new religious faith to the Iberian peoples. As Roman military conquests had expanded throughout the Mediterranean Basin, Roman society itself had been gradually but steadily transformed by the

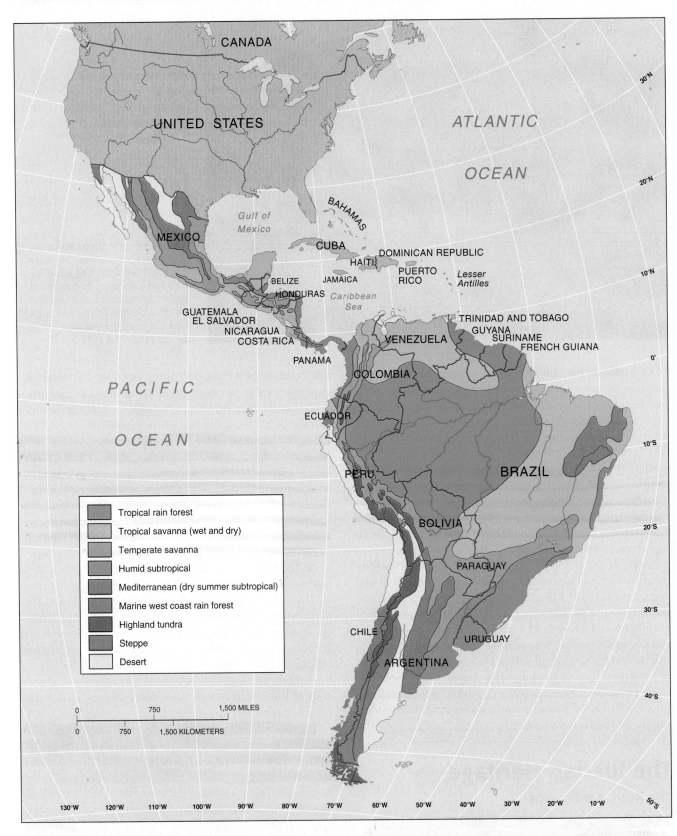

■ **FIGURE 7-8 Latin American climates.** Although most of Latin America is situated close to the equator, variations in altitude, prevailing wind patterns, ocean current temperature, and distance from the sea have resulted in the occurrence of almost every climate found on earth.

GEOGRAPHY IN ACTION
Altitudinal Life Zones in the Andes

To a great degree, the climates of the Latin American tropics are controlled by elevation. Communities located just a few miles from each other often have radically different physical environments as a result of their different altitudes. The temperature of motionless air changes approximately 3.5°F (1.9°C) for every 1,000 feet (305 meters) of altitude. Thus, one could leave the coastal Ecuadorian city of Guayaquil, where the afternoon temperature might be a broiling 96°F (36°C), and arrive by airplane a few minutes later in the Andean city of Quito, elevation 9,350 feet (2,850 meters), only to find the temperature a chilly 63°F (17°C).

A trip by car or train from Guayaquil to Quito would pass through three altitudinal life zones, where natural vegetation and agriculture reflect the changing climates. The lowest zone is the *tierra caliente,* or hot lands (Figure A). The lightly clothed farmers of that zone, which lies below approximately 3,300 feet (1,000 meters), cultivate heat-loving crops, such as coconuts, cacao, rice, bananas, sugarcane, rubber, papayas, mangoes, and

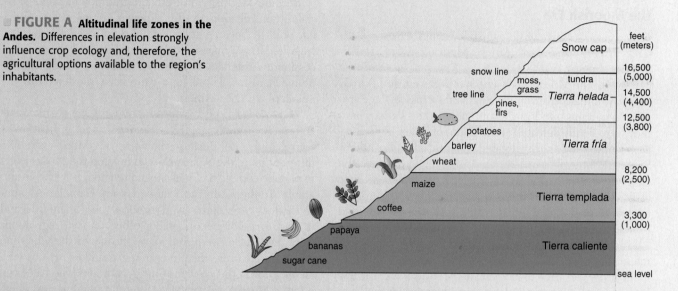

■ **FIGURE A Altitudinal life zones in the Andes.** Differences in elevation strongly influence crop ecology and, therefore, the agricultural options available to the region's inhabitants.

manioc. The second life zone is the *tierra templada,* or temperate land, which extends from the *tierra caliente* to about 8,200 feet (2,500 meters). That zone has year-round mild temperatures conducive to the cultivation of such crops as coffee, citrus, cinchona, maize, and a variety of vegetables, including beans, tomatoes, peppers, broccoli, and onions.

Quito lies in the third altitudinal life zone, the *tierra fría,* or cold land. The inhabitants of those higher elevations must dress warmly throughout the year and must protect themselves as much as possible from the intense solar radiation. Agriculture centers around the cultivation of plums, peaches, apples, and other cold-tolerant fruits, as well as grains and tubers, including wheat, barley, quinoa, and potatoes.

Beyond those zones is a fourth life zone, the *tierra helada,* or frozen land, which begins at around 12,500 feet (3,800 meters) and continues to the glacial snow line, between 16,000 and 17,000 feet (4,880 to 5,180 meters), as seen in Figure B. The *tierra helada* supports agriculture only on its lower slopes; above the tree line, only hardy grasses and lichens can survive.

In reality, it is inaccurate to speak of a single tropical climate; because of vertical zonation, almost every environment on earth, from rain forest to tundra, can be represented in a single tropical region. Thus, travelers in the tropics should dress according to the altitudes included in their itineraries.

■ **FIGURE B Chimborazo volcano, central Ecuador.** Visible in this photograph is the *tierra fría* of the lower slopes and the *tierra helada* of the upper slopes. Notice the tree line where the color tones change and the snow line on the upper slopes of the upper volcano.

177

revolutionary teachings of Christianity. Christianity had first been embraced secretly by the repressed lower-class urban masses but had subsequently gained believers among the ruling elite. It was finally recognized as a state religion in A.D. 313 by Emperor Constantine, whose deathbed conversion in A.D. 337 symbolized the spiritual conquest of the Roman world. Before the end of the fourth century, Christianity became the official state religion of the Roman Empire. Thus, by the time Spain and Portugal were colonized by Rome, all upper-class persons were landowning, urban, and Christian.

The Moorish Era

The outward stability of Roman-style life in Iberia was shattered in 711, when North African **Moors**, bent on spreading the Islamic faith throughout Europe, invaded the peninsula. The Moors, who were of mixed Arab and Berber ancestry, initially overran the Spaniards, crossed the Pyrenees Mountains into France, and marched on Paris, where they were finally defeated at the Battle of Tours. Following that defeat, the Moors retreated into central and southern Spain and brought a golden age to the peninsula over the next 700 years. The Moors introduced advanced hydrologic engineering and metallurgical technology. They also brought numerous crops—rice, citrus, coffee, cotton, sugarcane, and bananas—that were subsequently transported by the Spanish and Portuguese to the New World. In short, the Moors made Iberia a center of education and relative enlightenment at the very time that central and northern Europe had largely receded into the isolation and ignorance of the Dark Ages (Figure 7–9).

■ **FIGURE 7-9 Moorish architecture in Cordoba, Spain.**
This Moorish doorway in Cordoba's former Grand Mosque, now the city's cathedral, reflects the extensive interplay between Islamic and Christian cultures in Spain.

The Moorish occupation influenced the local culture in three additional ways. First, it removed any negative bias that the native Romans might have harbored toward persons of darker skin color. Rather than associating dark skin color with inferiority, as was later to be the attitude among the Puritan and other northern European settlers of the United States, the Spanish and Portuguese peoples came to view racial mixing as socially and politically advantageous. Their willingness to mix with darker-skinned peoples provided the basis for the subsequent development of Latin American race relationships.

Second, Moorish culture had a tradition of political instability and chaotic transfer of power. As time passed, Iberian politics became increasingly dominated by political fragmentation, charismatic personal leadership, and succession to leadership through violent rather than democratic mechanisms. Those traits were later transferred by the Spaniards to Latin America.

The third, and perhaps most profound, impact of Moorish occupation on Spanish culture was the unification of church and state. Because the occupying Moors were of a different faith, the Spaniards who resisted the Moors came to perceive ethnic nationalism and religious belief as inseparable. To them, all true Spaniards were Catholic. All non-Catholics were then, by definition, non-Spaniards and unworthy of holding any major office. In fact, regulations were passed that required an applicant for the Christian priesthood to prove the purity of his bloodline. Anyone with an ancestor in the previous four generations who had been a Jew, a Muslim, or any type of heretic could not gain high social standing.

Catholicism thus became fanatical and militant among the Roman Spaniards who took refuge in northern Spain after their defeat by the Moors. For them, the unification of church and state made religion the overriding issue, which negated earlier racial tolerance.

In the eleventh century, the Spaniards began a long series of wars aimed at driving the Moors from the peninsula. The popes in Rome were so grateful to the Spanish monarchs for reclaiming lands in the name of Christianity that they bestowed the title of Defenders of the Faith on the Spanish rulers. The papacy also ceded to the Spanish monarchs far-reaching privileges that continued in effect during the subsequent conquest and settlement of the Americas. These privileges, which were embodied in authority called the **royal patronage**, gave Spanish kings the power to appoint all clergy and to control clerical assignments and papal communications in conquered lands. In colonial Latin America, that practice resulted in the politicizing of at least the upper clergy, whose decisions and behavior were frequently influenced more by allegiance to the crown than by loyalty to principle. The royal patronage also conferred upon the Spanish monarchs the right to collect the tithe (theoretically, a tenth of a person's income) on behalf of the Church. In colonial Latin America, the tithe came to serve as a government tax. Bound by these extraordinary concessions, church and state became inseparable, first in medieval Spain and later in colonial Latin America.

For young Spanish men coming of age during this uncertain time, wealth and prestige were limited to three primary career options. The preferred choice, that of *latifundista,* was available only to the oldest son in each family. He inherited the entire family estate and, by taking a wife from another prominent family, perpetuated the dominance of the landed aristocracy. The remaining sons were generally forced to choose between a military career, which offered the prospect of glorious conquests and the acquisition of land and an upper-class way of life in conquered territories, and the clergy, which promised high social standing and indirect control of the vast wealth of the Church. Those two great forces, the military and the Church, stood poised, both filled with crusading fervor, as the Moorish era wound down in Spain and Portugal.

It is historically significant that the Battle of Granada, the final defeat of the Moors, was fought only 6 months before Christopher Columbus discovered America. Although the great admiral never accepted the idea that he had opened up a New World, his contemporaries viewed his achievement differently. With Iberia now firmly under Christian control, an entire new continent awaited military and spiritual conquest and, with it, the achievement of personal fame and wealth.

The Colonial Period

The Spanish Conquest

News of the discovery of America spread rapidly through Spain and Portugal. Although the Spanish were the first to settle in the New World, the Portuguese had been actively colonizing Africa and Asia for the preceding 100 years and were anxious not to be excluded from America. In 1493 the pope, hoping to strengthen his ties with the emerging colonial powers and reward them for past devotion, issued a bull (decree) that was ratified by Spain and Portugal in the **Treaty of Tordesillas** (1494). As the earthly representative of Christ, the pope granted half of the non-Christian world to Spain and half to Portugal. The imaginary line separating the two spheres was drawn 370 leagues (approximately 1,100 miles, or 1,800 kilometers) west of the Cape Verde Islands. As a result, Spain received all of the Americas except the easternmost extension of South America, which eventually became Portuguese Brazil (Figure 7–10).

Spanish settlement in the New World was dominated by the pursuit of the three *G*s: gold, God, and glory. Bernal Díaz del Castillo, the chronicler of Hernando Cortés' conquest of Mexico, wrote, "We came to serve God...and to get rich." For the **conquistadores**, or conquerors, the two objectives went hand in hand, and both required the presence of large numbers of native peoples. The natives could be quickly converted to Catholicism and then could be forced to labor on behalf of the Spaniards in the fields and mines.

Native Indian Civilizations   At the time of the conquests, it is likely that some 50 to 100 million Indians inhabited the Latin American and the Caribbean island realms. They were divided into thousands of indigenous groups that differed from one another in dialect, levels of social and political organization, and customs, as well as the physical environments

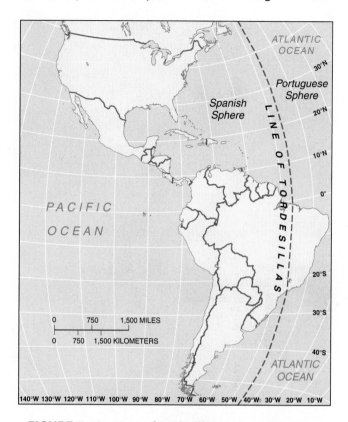

■ **FIGURE 7-10 Treaty of Tordesillas map.** In the Treaty of Tordesillas in 1494, the pope divided non-Christian America between Spain and Portugal.

that they occupied. The largest population clusters were found among four socially and politically advanced groups situated primarily in the highlands of western Latin America (Figure 7–11). The most urbanized of those groups was the **Aztec** Empire, whose capital of Tenochtitlán was one of the largest and most impressive cities on earth, with a population of more than 200,000 residing in the midst of magnificent temples, plazas, and marketplaces, supplied with produce transported from throughout southern Mexico and northern Central America. Tenochtitlán was later razed by the Spanish *conquistadores,* who built Mexico City upon its ruins (Figure 7–12). Prior to being defeated by the Spaniards, the Aztecs had become a very fatalistic and warlike people, offering on special occasions the hearts of thousands of human sacrificial victims to a god named Huitzilopochtli. At the same time, they were also a sensitive and highly educated people who achieved high levels of philosophy, poetry, and the arts.

Of all the indigenous high-culture groups that were present at the time of the Iberian Conquest, none was as ancient as the **Maya**, whose civilization emerged at approximately 3000 B.C. in the highlands of Guatemala and neighboring Mexico. By the seventh century B.C., the focus of Mayan civilization had shifted from the highlands to the surrounding lowlands to the east and north. There, a Classic, or golden, age emerged, characterized by the erection of great monumental cities in the midst of the vast rain forests that extend from the Tabasco lowlands of Mexico's gulf coast to the Copán River valley of northern Honduras

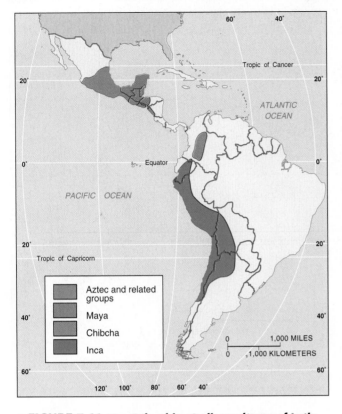

■ **FIGURE 7-11 Pre-Columbian Indian cultures of Latin America.** Before Columbus arrived, most of the American Indians lived in four high-culture clusters, characterized by high population densities. Throughout the remainder of Latin America, the Indian population was widely scattered and socially less complex.

(Figure 7–13). Although the Mayan peoples achieved levels of mathematics and writing equal or superior to those present in Old World civilizations, their world was also one of much uncertainty and violence, as dozens of independent states warred against one another in a never-ending series of shifting political alliances. Then in the eighth and ninth centuries A.D., the cities of the rain forest began to be aban-

■ **FIGURE 7-12 Reconstructed Aztec ruins.** Aztec civilization was thriving in Middle America at the time of the Spanish conquest. Ruins of that civilization are still visible in Mexico today. Shown are recently unearthed preconquest structures found near the present-day main plaza of Mexico City.

■ **FIGURE 7-13 Mayan ruins of Calakmul.** These magnificent, recently excavated ruins lie deep in the rain forest of southeastern Mexico close to the Guatemalan border.

doned, and the population, now much reduced from previous periods, reverted to the practice of shifting, slash and burn agriculture. This period was followed by a brief reflowering of urban life on the limestone plains of the Yucatán Peninsula during the post-Classic period of the thirteenth to fifteenth centuries. The Spanish Conquest brought an end to Mayan political identity but not to the people and their culture, both of which continue to form the basis of rural life throughout the region.

The third of the pre-Columbian high-culture groups of the New World was the **Chibcha**, who at the time of the Spanish Conquest lived in agricultural villages scattered through the eastern Andes of Colombia. Although lacking the large cities and sophisticated agricultural technologies of the Aztec, Maya, and Inca peoples, the Chibcha had developed levels of social and political organization superior to those of the lowland forest dwellers residing within the Amazon and Orinoco river valleys to the east of the Andes.

The greatest of the pre-Columbian empires of South America was that of the **Inca**, which arose in the high mountain valleys of southern Peru and Bolivia in the century prior to the Spanish Conquest. The Inca civilization was, in many respects, inferior to that of the Aztec and Maya. Its capital city of Cuzco never approached Tenochtitlán in size or grandeur, and the Incas never developed a written language. What they excelled at, however, was the imposition of authoritarian social and political control and the construction of remarkable highways and bridges, many of which continue in use to this day. These characteristics enabled the Incas to conquer and subject countless other peoples and to establish an empire that, when the Spaniards arrived, stretched over 2,500 miles (4,000 kilometers), from northern Ecuador to central Chile (Figure 7–14).

Agriculture formed the livelihood base for each of these four high-culture civilizations, as well as the socially and

FIGURE 7-14 Ruins of Machu Picchu. These spectacular Incan ruins are found high in the Andes of southern Peru, close to the city of Cuzco.

politically less-developed peoples of Latin America's central and eastern lowlands. Advanced land management techniques—including land drainage, terracing, fertilization, crop and land rotation, and intercropping—were widely practiced. Among the domesticated crops were maize, beans, squash, chili peppers, tomatoes, manioc (cassava), sweet potatoes, white potatoes, peanuts, pineapples, papayas, avocados, cacao (chocolate), American cotton, and tobacco. As for animals, dogs were everywhere and were sometimes eaten for their meat. In addition, the Incas had guinea pigs as pets and as a food source (Figure 7–15), llamas (a humpless relative of the camel) as beasts of burden, and alpacas and vicuñas (relatives of the llama) as sources of fine wool. However, none of those

FIGURE 7-15 Domesticated Andean guinea pigs. Guinea pigs are raised as a source of food in the Andean highlands. In many homes of the poor, they are allowed to move about freely on the compacted dirt floors, scavenging for food.

animals was used in farming; the Indians cultivated their land entirely by hand, with the aid of simple digging sticks.

To that agricultural complex, the Spaniards added large Old World animals, such as pigs, sheep, goats, donkeys, beef and dairy cattle, and horses. They also introduced animal-drawn plows and a wide array of Old World grains, fruits, and vegetables.

European New World Settlement The initial focus of Spanish settlement in the New World was the Caribbean island of Hispaniola, which is divided today into the Spanish-speaking Dominican Republic and the French-speaking Haiti. Hispaniola served as a testing ground where the Spaniards experimented with crops, animals, and cultural notions they had brought from Iberia. The output of the small gold mines on the island soon diminished, however; and, in what was to become a tragic and common pattern, great numbers of native Indians died from exposure to European diseases, overwork, and undernourishment. With the discovery of the vast wealth and power of the Aztec and Inca empires of Mexico and Peru, respectively, Hispaniola and the other Caribbean islands soon became settlements of only secondary significance to the Spaniards. Most were eventually lost to British, French, Dutch, and American military conquest.

Spain's conquest of Mexico was accomplished under the leadership of Hernando Cortés. The Spaniards were armed with both sword and cross—as well as guns and horses. In addition, an even more powerful weapon was on their side—the Indian belief in the promised return of a great fair-skinned, bearded god named **Quetzalcoatl** (Figure 7–16). As Cortés' ragged party of 400 men steadily advanced toward Tenochtitlán, the Aztec chiefs vacillated, initially engaging the intruders in battle but then, at the final hour, bowing and kissing the earth before Cortés' feet in a futile attempt at appeasement. Unswayed, Cortés abducted and tortured the Aztec chiefs and took possession of the country in the name of God and the Spanish king.

Meanwhile, in Peru and Bolivia, Francisco Pizarro was conquering the internally divided Incan realm in a similar

■ **FIGURE 7-16 Mural depicting the return of Quetzal-coatl.** Cortés' conquest of Mexico was aided by the power of an Indian legend that promised the return of a great fair-skinned, bearded god named Quetzalcoatl. This painting is found on a wall of the government building of the Mexican state of Tlaxcala.

fashion. Within just a few decades, all of Latin America had fallen under European control.

Economic Relationships

Following their military conquests, the Spaniards immediately set about restructuring the economic and social life of the natives. One of their first acts was the establishment of the *encomienda* system, which, in essence was a New World version of the medieval European manorial system. The Indians were informed that they had entrusted themselves to the care of their Spanish masters who, in turn, were required to provide their subjects with physical protection and spiritual salvation. In return for the privilege of being at least superficially indoctrinated into the holy faith and exposed to the alleged civilizing influences of the Spanish language, the Indians were told that they owed their Spanish masters tribute in the form of labor or produce or both. Huge *latifundio*-like land grants were given to the Spanish *encomenderos* who

took up residence in the distant cities, where they could better enjoy the fruits of their Indians' labor.

Ironically, those Spaniards who were given control of the Indians were themselves subjected to unwanted controls by the Crown, which attempted to regulate all economic activity in the new colonies for the benefit of the mother country. The vehicle used to enforce these controls was **mercantilism**. Under the mercantile system, a colony could trade only with its mother country. Trade between colonies was severely restricted. Furthermore, all goods shipped between Spain and its colonies had to be carried by Spanish vessels. Those ships, often laden with gold and silver, were exposed to such great danger from pirates or buccaneers that the Spanish government erected a series of strategically situated fortresses at key cities (Havana, Cuba; Veracruz, Mexico; Cartagena, Colombia; San Juan, Puerto Rico; and St. Augustine, Florida) and instructed the ships to travel only in semiannual convoys, called *flotas* (Figure 7–17). In addition, Spanish mercantile policy curtailed manufacturing-related activities, which meant that even raising sheep and growing wheat, olives, and grapes—all of which competed with Spain's support of its own textile or food-processing industries—were suppressed. And finally, a tax of 20 percent, called the royal fifth, was levied.

The long-term effects of mercantilism on the economic development of the colonies were extremely negative. Economic initiative, diversification, and self-reliance were discouraged and contraband and smuggling became rampant. Graft and dishonesty among government officials became the norm. Competition, rather than unity, was fostered among colonies with a common cultural heritage, and feelings of resentment toward Spain grew ever stronger.

■ **FIGURE 7-17 El Morro Fortress, Havana, Cuba.** El Morro is a remnant of a great series of fortresses erected by the Spaniards in the colonial era to implement mercantilism. The purposes of the fortresses were to protect the key port cities and safeguard the transport of vast quantities of gold and silver from the New World colonies back to Spain.

Social Organization

Society in early colonial Latin America was highly stratified, largely according to race. As the Spaniards and the Portuguese intermingled with the darker-skinned natives, they produced mixed European-Indian offspring called **mestizos**. Then, as great numbers of Indians died and as those who remained proved to be poor workers in the hot and humid lowlands, large numbers of black slaves were introduced, especially into the plantation zones of the Caribbean Basin and northeastern Brazil. The offspring of white-black unions were called **mulattos**, and those of Indian-black unions, **zambos**. Soon, physical race was an inadequate social standard. How, for example, would one even identify, let alone classify, the offspring of a *mestizo-mulatto* union?

Because of the widespread racial mixing, a cultural-racial classification system came to be adopted that ranked individuals by their occupations, languages, and lifestyles. At the top were the Europeans, who held the highest managerial and governmental jobs, spoke Spanish or Portuguese, and lived in cities. Then came the *mestizos* and the *mulattos*, who generally managed the great agricultural estates or performed menial urban tasks. At the bottom were the blacks and the Indians, the latter term becoming synonymous with rural, agricultural laborers. Indians worked with their hands and generally went barefoot or wore sandals. The men wore hats, and the women had long, braided hair and babies in shawls on their backs. Indians spoke in their native tongues and were uneducated in European ways.

The Spanish preoccupation with race and position was a reflection of the social inequalities of colonial society, which in turn were transmitted through the educational system. Members of the wealthy European class, wanting to perpetuate their dominance, paid great sums of money to enroll their children in private, usually Catholic-sponsored schools. Because they then refused to tax themselves to support public education for *mestizo* and Indian youths, the resulting unequal educational system contributed directly to the continued underdevelopment of Latin America. In contrast to the efforts of Spain's and Portugal's northern European rivals, the English and the Dutch, who turned to public education to create a literate and democratic workforce in their colonies, Latin American society remained largely illiterate, totalitarian, and technologically backward.

The Spread of Catholicism

While the *conquistadores* were pursuing the military conquest of the New World, the Catholic clergy were undertaking the spiritual conquest of the native peoples. Spanish and Portuguese conversion efforts, however, emphasized outward compliance with sacramental ordinances, such as baptism, rather than a thorough understanding and acceptance of the teachings and beliefs of the Catholic Church. To facilitate that outward conversion of the natives, the clergy related Catholic practices and beliefs to those of their non-Christian counterparts and permitted indigenous ceremonies and customs to continue.

Not surprisingly, millions of Indian converts were reported annually in the years immediately following the European arrival. Most of the conversions were doctrinally superficial, however. The vast majority of the Indians did not even understand the Spanish language, much less Catholic teachings. What resulted was a fusion, or mixing, of Iberian Catholicism with native American practices.

Three distinct forms of Catholicism evolved in colonial Latin America, and they persist to this day. The first is formal Catholicism, which is practiced by the small, European upper class that resides in the urban centers. Formal Catholicism emphasizes piety, faith, and participation in the Catholic sacraments (Figure 7–18). Women attend mass regularly, and a heavy emphasis is placed on devotional societies, charities, and social clubs. Association with formal Catholicism reinforces a person's upper-class standing within the community.

The second form is nominal Catholicism, which has come to include the majority of the rural peasant population as well as almost all of the urban poor. Nominal Catholics make no financial contributions to the church, except for the rare occasions when they pay a priest to perform a sacrament on behalf of a family member. Priests may be viewed negatively, an attitude known as anticlericalism. Men are unlikely to enter a church more than twice during their lifetimes: once when their parents carry them in as infants to be baptized and once to be married. When asked whether they consider themselves Catholics, their response is often "in name only" or "yes, but in my own way."

The third form is folk Catholicism, which consists of a mixture of European Catholicism and non-Catholic beliefs and practices. Folk Catholicism takes two forms. The first is American Indian, which was developed by isolated tribal peoples. Their beliefs and customs are a mixture of pre-Columbian animistic and medieval Catholic practices, centered in the ancient holy places of mountains and valleys as well as in the local church (Figure 7–19).

■ **FIGURE 7-18 Basilica of Cartago, Costa Rica.** Formal Latin American Catholicism expresses itself in ornate churches patronized by upper-class urban dwellers. Its ceremonies and rituals are only distantly related to those of American Indian and African American folk Catholicism.

■ **FIGURE 7-19 Latin American folk Catholicism.** Among the highland Maya of Guatemala, folk Catholicism may include the worship of ancestral deities.

The second form of folk Catholicism is spiritism, which evolved in the lowland areas of the Caribbean Basin and Brazil among the predominantly black and *mulatto* populations. Its emphasis on close communion with spirit beings is of African derivation; and its doctrines, ceremonies, and celebrations bear only the slightest resemblance to those of the formal Catholic Church. Haitian Voodoo, Cuban and Dominican Santería, and Brazilian Candomblé are some of the many variants of spiritism currently practiced in Latin America and the Caribbean.

Although the colonial church was highly tolerant of diversity in internal belief and practice, it attempted to shield the people, through the office of the church known as the Inquisition, from what it perceived to be corrupting Protestant theology. The Inquisition was originally established in Catholic Europe to preserve the purity of the faith by screening or censoring literature and works of art. As time passed, however, Spain came to dread England and the other Protestant powers so greatly that any book or publication from northern Europe was automatically banned, whether or not its content was religious. Even books that would have aided the diffusion of technology were suppressed. Thus, as the Industrial Revolution gained momentum in northern Europe and in the Anglo American colonies, Catholic Latin America fell further and further behind.

The Era of Independence

As time passed, a major division arose in class-conscious colonial Latin America between those Europeans born in the Iberian peninsula, who were called **peninsulares**, and those born in the American colonies, who were called **criollos**, or creoles. The highest civil and ecclesiastical offices were reserved exclusively for the blue-blood *peninsulares*. In fact, it was so important to be born in Spain that upper-class Spanish women, on finding themselves pregnant, often booked

passage for the mother country. Even a child born at sea was classified as *peninsular* if the mother was en route to Spain.

As long as the *peninsulares* outnumbered the *criollos*, colonial society remained relatively stable. As the centuries passed, however, the increasingly predominant *criollos* grew ever more resentful of their second-class status. The revolutions that swept the continent in the early 1800s were essentially a class war between the two dominant European subclasses. Although freedom and liberty were proclaimed throughout every land by the victorious *criollos*, little meaningful change took place in the lives of the *mestizo, mulatto,* and Indian masses.

Political independence also failed to resolve the fundamental problem of unequal landholdings in a predominantly rural, agrarian society. Indeed, in many areas, access of the lower classes to land actually worsened under the reign of *criollo* and foreign investors. The great estates, or *latifundios*, took three forms, which have survived in varying degrees to the present. The first was the **plantation**, established in the hot, lowland zones, usually near a coast (Figure 7–20). Plantations were, and continue to be, essentially monocultural; **monoculture** identifies agricultural production that centers on one dominant crop, such as sugarcane or bananas. Since the crop is produced primarily for export, plantations are basically commercial ventures aimed at generating cash profits for the owners, typically through the use of industrial Green Revolution technologies and a historically black or *mulatto* workforce.

In contrast, the interior highlands came to be dominated by the **hacienda**, large landholdings given over primarily to extensive cattle grazing and the use of Indian labor (Figure 7–21). The *criollo* owners, though anxious for profits to support their urban way of life, have nevertheless viewed their *haciendas* primarily as a source of prestige rather than as a business. The **estancia**, the third form of *latifundismo*, designated the great cattle ranches manned by European immigrant laborers on the fertile pampas, or prairies, of Argentina and Uruguay.

Both the plantations and *haciendas* used inherited debt to enslave their laborers, thus discouraging the development

■ **FIGURE 7-20 A banana plantation on the north coast of Honduras.** Plantations are examples of monocultural estates that center on a tropical crop produced primarily for export.

■ **FIGURE 7-21** *Hacienda* **of highland Peru.** *Haciendas* continue to be the dominant form of *latifundismo* in highland Latin America. This estate is near the city of Puno.

■ **FIGURE 7-22** **Steet market in La Paz, Bolivia.** Vendors spread their wares on the ground in this street market where the traditional and the modern meet.

of democratic institutions and a middle class. While the huge *latifundios* supported the privileged few in indolent luxury, the countless small farmers of the region continued to work tiny, isolated plots that they often had no hope of ever owning. Toiling under a wide range of local conditions—from steep, rain forest-clad mountain slopes to high interior volcanic basins—those creative, self-reliant small farmers, like their descendants today, sustained themselves and their families through polyculture.

In the humid lowlands of Latin America and the Caribbean, **polyculture** frequently consists of several vertical layers of food crops being grown simultaneously in the same plot of ground (see the *Geography in Action* boxed feature, Tropical Polyculture and the Conservation of Global Biodiversity). A second form of polyculture, common to the interior highlands, consists of cultivating multiple varieties of a single crop in the same field. Today, the Indian farmers of Mexico and Central America, for example, may cultivate four different-color varieties of maize as well as seven or eight varieties of beans of differing shapes, sizes, and colors. Each variety differs from the others in physical appearance, taste, and environmental hardiness. Similarly, the Indian farmers of the high Andes raise twenty to thirty varieties of potatoes, each with distinct physical characteristics and environmental needs. With that genetic security blanket, the traditional farmers of Latin America have been able to feed themselves and their families and to provide for the basic needs of distant urban dwellers.

Marketplaces have functioned historically as the principal interface between the dominant urban culture and the subservient rural culture (Figure 7–22). As housewives and maids haggle daily with peasant vendors and intermediaries over the price of fresh produce and handicrafts, urban dwellers obtain essential food and fuel, and rural folk earn cash with which to purchase commodities that cannot be produced on the farm, such as manufactured appliances and medicines.

Modern Latin America

Latin America today, though strongly linked to the past, is experiencing all the changes and challenges of economic and social modernization. Many of those changes have come so recently and so forcefully that the peoples of the region have had little time to adjust.

Demographic Change

Many parts of Latin America originally supported dense native populations, which were drastically reduced as a consequence of European occupation. Hispaniola's indigenous population, which is estimated to have numbered between 3 and 4 million prior to the Conquest, dropped to approximately 60,000 within two decades of the arrival of the Spaniards. Guatemala's pre-Conquest Maya population of 2 million fell to 128,000 by 1625. Other areas experienced population declines that were similarly radical, not because of any intentional, systematic destruction of the Indians by the Europeans, but as a result of the interplay of three factors.

First, a massive disruption of native food production was brought about by the loss of the best farmlands to Spanish cities and animal grazing. As the Indians were forced to flee to the distant hills and mountain slopes to eke out a meager existence, food shortages became chronic among the native peoples, whose decline in numbers was matched by a corresponding increase in horses and cattle. The second factor that contributed to the great loss of Indian population was the introduction of European diseases, including smallpox, measles, and typhus, to which the natives had no resistance. And third, weakened by both malnutrition and disease, countless Indians died of overwork in the mines and on the plantations. By 1650 the native population of Latin America and the Caribbean was only 5 to 10 percent of what it had been prior to the arrival of the Europeans.

GEOGRAPHY IN ACTION
Tropical Polyculture and the Conservation of Global Biodiversity

The term *biodiversity* is often used to refer to the number of life-forms, plant or animal, that exist in a given area. It is important that we do all that we can to conserve the earth's biological heritage for several reasons. One reason is that each form of life, whether it is prominent or inconspicuous to humans, interacts with other life-forms to create the food webs or chains upon which all life ultimately depends. A second reason why each form of life is worthy of preservation is that many of the earth's most unappreciated plants contain medicinal products that are used to treat serious diseases. More than 3,000 plant species, for example, are known to contain anticancer properties. More than two-thirds of these plants are found only in tropical rain forests. A third reason for wanting to prevent the loss of little-used plants is that many contain genetic qualities that enable them to resist environmental stresses, such as disease, drought, or flooding. Through genetic engineering, scientists are now developing the capacity to transfer the genes that contain these qualities to the most widely consumed commercial crops, thereby creating the possibility of increasing the world's food supplies while simultaneously reducing reliance on expensive and toxic chemicals. A final, more philosophical reason for conserving our biological heritage is that each form of life is irreplaceable, and its loss would constitute a loss of our earthly heritage—in a sense, a loss of part of ourselves.

There currently exist two strategies for conserving the biodiversity of the earth's food crops. The first approach, called *ex-situ* conservation, is to collect as many varieties as possible of wheat, maize (corn), rice, potatoes, beans, cassava, and other stable food crops and to store the seeds or cuttings of these cultivars at very low temperatures in laboratory-like buildings called seed banks. While such measures are clearly useful in the war against "genetic erosion," their shortcomings include the limited storage lives of the seeds and cuttings and an end to positive natural adaptations and mutations that might include resistance to environmental stress.

The second way in which the earth's biodiversity is preserved is through smallholder tropical polyculture. It is not unusual for a farmer in Latin America, Central Africa, or Southeast Asia,

for instance, to cultivate simultaneously twenty to thirty separate crops in the same tiny plot, each occupying distinct vertical and spatial niches or positions (Figure A). The contribution of this same farmer to genetic conservation may be enhanced further through his growing two, or three, or four varieties of each of the twenty to thirty crops, meaning that a single peasant farmer may be preserving 40 to 120 varieties of food crops in a single plot. This strategy, called *in-situ* conservation, has the added advantage of allowing for ongoing natural mutations and adaptations to environmental stress. It is easy to appreciate that the millions of smallholder farmers practicing tropical polyculture in the humid tropics constitute an invaluable resource in our efforts to preserve the earth's biodiversity. It follows that national and international development programs should attempt, among other things, to strengthen traditional agriculture and the village life that sustains it.

■ **FIGURE A Tropical polyculture.** This photograph, taken in the western highlands of Panama, shows bananas, coffee, and other crops growing together in shared space. When skillfully practiced, tropical polyculture is a highly productive and sustainable cropping system that contributes to the conservation of the earth's biodiversity.

Almost 300 years were to pass before the population of the region would begin to grow rapidly again. Indeed, most parts of Latin America did not regain their preconquest population levels until the early- to mid-twentieth century (Figure 7–23). The primary cause of such slow population growth was high death rates, especially among infants, which counterbalanced the traditionally high birthrates. The only exception to that overall pattern was some immigration, principally of southern Europeans to Brazil, Uruguay, and Argentina in the late nineteenth and early twentieth centuries. Overall, however, Latin America has attracted comparatively few immigrants.

The relatively low population levels of Latin America began to change dramatically in the mid-twentieth century. Improved health care and sanitation, including the introduction of an-

tibiotics and mosquito-controlling chemical compounds, made the traditional population centers healthier and opened up vast lowland regions to colonization and economic development. The result was unprecedented population growth, which continues today at a somewhat slower pace.

The magnitude of the change is illustrated by population changes in Mexico. The population of Nealtican, a peasant village in the central Mexican highlands, grew by over 450 percent between 1950 and 2004 (Table 7–1). That pattern, repeated countless times, has also affected the population of Mexico as a whole. In 1950 the population of Mexico only slightly exceeded its preconquest level in 1519; however, since 1950, Mexico's population has almost quadrupled, placing unprecedented pressure on the rural land base.

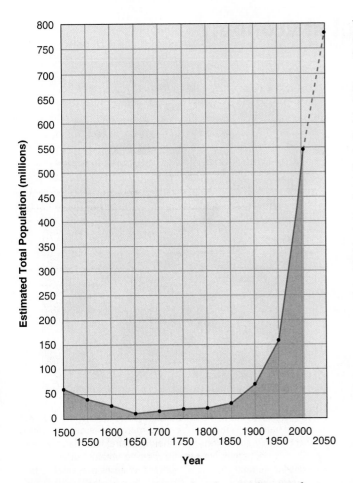

■ **FIGURE 7-23** **Total estimated Latin American and Caribbean population, 1500–2050.**

Urbanization

While opportunities for personal advancement have generally failed to expand in the rural areas of Latin America over the past four or five decades, the cities have increasingly assimilated technological advances that have integrated them into the global economy. The results have been an ever-widening gap in living levels between urban and rural residents and a flood of urban migrants fleeing the poverty of the countryside.

Seemingly overnight, Latin America has been transformed from a predominantly rural region to one in which 75 percent of the people now reside in urban areas. Many Latin American cities that were formerly compact, orderly, and pleasant places in which to live have exploded in population and are rapidly taking their place among the largest and most congested urban centers in the world (Table 7–2). Much of that growth is occurring in huge outer rings around the traditional cores of the cities. As the rural migrants pour in, most settle first in old, rundown inner-city housing stock, which they eventually leave in favor of newer, self-built residences erected within vast **shantytowns**, or squatter settlements, found on the outskirts of the city. Municipal governments—which are hard-pressed to provide basic services such as drinking water, sewerage, electricity, garbage disposal, and police protection to the older, established neighborhoods—find themselves almost totally incapable of servicing the newly settled periphery.

Most of the major urban centers in Latin America are now caught in a self-perpetuating cycle of growth that brings as many problems as benefits. As more and more people flock to the cities, new industries and businesses are established there in response to the growing market. Those industries and businesses, in turn, attract even more migrants. As the accompanying number of automobiles and diesel-powered buses and trucks has also spiraled, traffic has become a nightmare, and levels of air, water, and solid-waste pollution rank among the worst in the world.

Many Latin Americans, dissatisfied with their current economic and political circumstances, attempt either legal or illegal international migration. The crossing of 500,000–600,000 Mexicans into the United States every year, 85 percent of whom are now estimated to be undocumented migrants, has been the most widely publicized of those movements and has prompted a series of new restrictive state and national U.S. immigration policies. Mexico itself, however,

Table 7-1 **Population Growth in Mexico in Selected Years**

	Approximate Population	
	Nealtican	**Mexico**
1519	2,250	25,000,000
1900	1,264	14,000,000
1950	2,448	27,000,000
1960	3,054	35,000,000
1975	7,000	56,000,000
1990	8,380	82,250,000
2004	11,110	106,200,000

Sources: Mexican General Census of Population, various years; Robert C. West and John P. Augelli, *Middle America; Its Lands and Peoples*, 3rd ed. (Englewood Cliffs, NJ: Prentice-Hall, 1989); Population Reference Bureau, *World Population Data Sheet, 2004* (Washington, D.C., 2004); and *The World Gazetteer 2005*: www.world-gazetteer.com.

Table 7-2 **Growth of Major Latin American Urban Centers in Selected Years**

	Estimated Population (in millions)		
	1950	**1970**	**2005**
Mexico City, Mexico	2.0	8.6	22.3
São Paulo, Brazil	2.5	7.8	19.1
Buenos Aires, Argentina	4.5	8.4	13.3
Rio de Janeiro, Brazil	2.9	6.8	11.5
Lima, Peru	1.2	3.3	7.7
Bogotá, Colombia	.6	2.5	7.6
Santiago, Chile	1.4	2.9	4.8

Sources: Charles S. Sargent, "The Latin American City," pp. 201–249, in *Latin America: An Introductory Survey*, edited by Brian W. Blouet and Olwyn M. Blouet (New York: Wiley, 1982); and *The World Gazetteer 2005*, www.world-gazetteer.com.

GEOGRAPHY IN ACTION
Return Migration and the Global Economy

One often overlooked aspect of international migration is the return to the original donor nation of migrants, or the descendants of migrants, who had previously left their homeland in search of greater opportunities for economic advancement abroad. Return migration is thus an expression of changes within the global economy. When economic opportunity in the original country comes to exceed that of the recipient nation(s), it is likely at least some of the migrants will choose to go "home."

An example of return migration currently affecting two Latin American countries is that of Brazilians and Peruvians of Japanese ancestry returning to work in Japan. Japanese peasants emigrated to Brazil, Peru, and other Latin American nations as contract laborers in the early to mid-twentieth century. Once they had fulfilled their contractual obligations, many of the Japanese settlers and their children gained citizenship in their adopted countries and went on to prosperous careers in agriculture and business. Eventually, some 1.2 million ethnic Japanese came to reside in Brazil and approximately 50,000 in Peru. Beginning in the 1980s, however, economic and political conditions in Latin America worsened to the point where tens of thousands of Japanese Brazilians and Peruvians began returning to Japan, which was experiencing labor shortages in some sectors of its economy. By the early twenty-first century, at least 300,000 Japanese Brazilians and 20,000 Japanese Peruvians were residing in Japan (Figure A).

Much to their surprise, however, many of the return migrants have experienced considerable difficulty in acclimating to their ancestral homeland. These challenges are attributable in part to the limited familiarity of many of the migrants with the language, customs, and climate of Japan. The adjustment of the migrants has also been hindered by the traditional Japanese reluctance to assimilate outsiders, even if they are the descendants of their own friends and loved ones. The ethnic Japanese Brazilians and Peruvians who have returned to Japan, for instance, continue to be classified as "guest workers" and often feel discriminated against by the native Japanese. The continued presence of the migrants in Japan, however, is a highly

■ FIGURE A **Brazilian Plaza shopping center in Oizumi, Japan.** Economic hard times in Brazil have led many of the descendants of Japanese immigrants to return to their ancestral homeland, where temporary work papers are easy for them to obtain, but where their divided cultural identity is reflected in stores that cater to their Portuguese heritage.

visible manifestation of the significant role of international labor migration in the global economy.

Source: Adapted from David L. Clawson, *Latin America and the Caribbean: Lands and Peoples.* 4th ed. Boston: McGraw-Hill, 2006, p. 370.

---

is beset with large numbers of Guatemalan refugees who have crossed into the Mexican states of Chiapas and Campeche. Similarly, the Greater Antilles—composed of the Dominican Republic, Puerto Rico, Haiti, and Cuba—have recently sent large numbers of migrants to the United States; yet those islands are viewed as lands of opportunity by the even poorer peoples of the Lesser Antilles. Other significant international migrations affecting Latin America and the Caribbean include blacks from the English- and French-speaking Caribbean to Great Britain and France; Ecuadorians to Spain; Colombians to oil-rich Venezuela; and Salvadorans, Hondurans, and Guatemalans to the United States. Yet another centers on the return migration of Brazilians and Peruvians of Japanese ancestry to Japan (see the *Geography in Action* boxed feature, Return Migration and the Global Economy). Given the ever-increasing income gaps between the wealthy and the poor in much of Latin Amer-

ica, these migration pressures are likely to continue and possibly worsen.

Social and Religious Change

When political independence was achieved by the liberal *criollos* in the early 1800s, the constitutions of the new nations were invariably modeled after those of the United States or France, both of which had recently established new democratic governments. The Latin American constitutions provided for democracy in its fullest sense, including the separation of powers among the legislative, judicial, and executive branches of government as well as extensive guarantees of individual freedoms and rights. The problem, of course, was that the illiterate and socially stratified Latin American societies were inherently incapable of implementing in real life what was prescribed on paper.

Totalitarian governments, often associated with military forces, soon rose to power and lasted only until they were replaced by the next generation of colonels and dictators. Little, if any, real power was exercised by the masses, whom the elite did not consider capable of political responsibility.

That totalitarian political tradition finally began to fade in the post-World War II era, as urbanization and industrialization created pressures for a better-trained workforce. Then, as an urban middle class gradually developed, democratic political practices began to take root. Today, the political trend in Latin America is clearly toward democratic civilian governments. Although it is possible that individual countries will continue to pass through temporary authoritarian regressions, the forces of economic modernization will likely continue to move the region as a whole toward greater levels of individual freedom.

The same forces that are working to increase the levels of individual participation in political matters are also contributing to a transformation of the traditional religious landscape. The institutional Catholic Church in Latin America is being challenged today by three powerful movements. The first is secularism, or a pervasive lack of religious interest and involvement by the general membership. This natural outgrowth of nominal Catholicism has been reinforced by the process of urbanization, which disrupts traditional behavior patterns.

The second challenge, **fundamentalist Protestantism**, originated as a spin-off from Chilean Methodist congregations in 1909 and has since diffused extensively throughout the region at the grassroots level. Appealing first to the repressed masses, its emphasis on education, frugality, and the avoidance of gambling, drinking, and other perceived social vices has resulted in the rise of second- and third-generation members into professional-level positions of influence. The strong sense of community and high levels of mutual support and aid characteristic of most Latin American Pentecostal congregations have proven to be especially attractive to women, who now comprise 65–70 percent of adult converts. Pentecostalism and other non-Catholic fundamentalist Christian faiths—including Latter-day Saints (Mormons), Seventh-day Adventists, and Jehovah's Witnesses—have thus made major inroads in Latin America. All of those faiths are now viewed by the masses as socially upscale, with teachings of benefit in the next life as well as in this one. Fundamentalist Protestantism has reached every corner of Latin America today (Figure 7–24). In many nations, such as Chile, Haiti, the Central American countries, and Brazil, it is possible that more worshipers attend Protestant services on any given Sunday than attend Catholic Mass.

The third challenge to the institutional Catholic Church is leftist ideology, which takes several forms in Latin America. One is overt Marxism or communism. Although some people find it appealing, communism is never likely to freely attract great numbers of Latino followers because of its restrictions on freedom of worship, speech, and movement and because of its emphasis on individuals' placing the interests of the party and state ahead of their own personal ambitions. Those ambitions have traditionally been expressed in Latin

■ **FIGURE 7-24 Latin American fundamentalist Protestantism.** This photo shows members of a Chilean evangelical congregation walking to church while singing and playing instruments.

America through the culturally cherished values of *machismo* (manliness) and *dignidad de la persona* (inherent self-worth of the individual). Many Catholic priests, however, have become intellectually converted to a different ideology known as **liberation theology**, which advocates preferential treatment for the lowest socioeconomic classes. Those priests have organized small self-help study groups, known as basic ecclesial communities (CEBs). CEBs are somewhat similar, in organization and methodology, to many of the small, grassroots Protestant congregations. Viewed collectively, both fundamentalist Protestantism and liberation theology may represent a rejection of the traditional class structure of Latin society and a movement toward increased levels of individual freedom and participation.

Economic Change

Economic development, in its broadest sense, is increasingly associated throughout the world with the diffusion of industrial technologies. Until recently, manufacturing in Latin America was generally limited to small, family-owned-and-operated establishments that focused on food processing, textiles, and other low-technology enterprises. Industrial expansion was hindered by inadequate transportation and communication networks, an untrained workforce, small local markets with limited purchasing power, and little accumulation of money or capital.

Latin American nations have made great strides in overcoming many of those obstacles. New highways and freeways are now crowded with trucks carrying industrial components and products. Telephone, computer, and Internet communications are common (Figure 7–25). Literacy rates now average 85 to 95 percent and are even

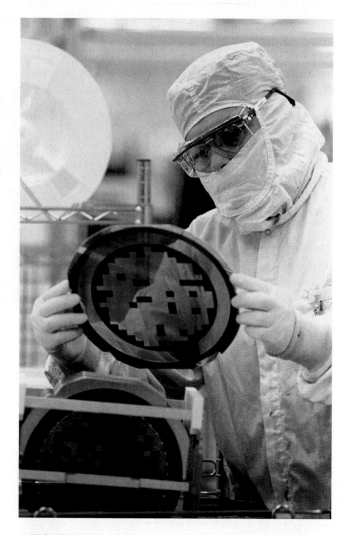

■ **FIGURE 7-25 Intel computer chip manufacturing plant, San José, Costa Rica.** This state-of-the-art plant is emblematic of the great strides currently being made in Latin American communications.

sel, even attempting to do so would imperil the immediate health and the very lives of the family (Figure 7–26). The poor Latin American nations that are struggling to meet the most basic needs of their people—adequate food, health care, sanitation, and education—are hard-pressed to reinvest much of their economic output in industry.

For these reasons, the funding needed for the industrial development of the Latin American nations has usually been obtained from external sources rather than from internal ones. During the 1950s, most external capital came into Latin America through direct foreign investment by multinational corporations. That approach brought much-needed cash into the national economies, but it also increased foreign meddling in the internal affairs of those countries. That situation, often described as **neocolonialism**, was politically unacceptable to Latin Americans because it offended, to varying degrees, their collective pride and freedom to chart their own courses of development. Consequently, in the 1960s, most external funds came through direct foreign-aid grants by friendly nations. But foreign aid also comes with preconditions and expectations for both donor and recipient countries.

The Latin American Debt Crisis

Since the early 1970s, most foreign money flowing into Latin America has come in the form of loans from huge private international banks and their representatives, including the International Monetary Fund and the World Bank. Although the intentions of both the world banking community and the leaders of the Latin American nations have been honorable, the loans have created serious problems.

Many Latin American countries borrowed huge sums of money for development projects. Neither they nor the lending institutions, however, paid sufficient attention to the

higher among the younger generation. The growth of cities is creating an urban middle class, many of whose members are employed in the growing industrial sector of the economy. And yet, despite such progress, Latin America's road to economic development has been marked by serious challenges.

In the minds of many Latin Americans, the greatest obstacle to continued industrial expansion is the accumulation of sufficient capital to establish the types of plants and facilities that will be competitive in the international market. In other words, how can the Latin American nations get enough money to make money?

The classic Western capitalist response is that the Latin American nations should reinvest a substantial portion of their annual economic output in the expansion of the industrial infrastructure by building roads and factories. Although that approach is undeniably correct conceptually, it is somewhat like telling the poverty-ridden parents of eight children that all they have to do to become prosperous is to invest a sizeable share of their annual gross earnings wisely. Regardless of how much they may want to follow that coun-

■ **FIGURE 7-26 A family of the highlands of central Mexico.** For many families, low income and limited education and employment opportunities combine to make economic progress difficult.

countries' long-term ability to repay the loans. Ironically, those nations that showed the greatest potential for economic growth and modernization—including Brazil, Argentina, and Mexico—borrowed most heavily and now find themselves with the greatest debt loads. By 2003 the foreign debts of those three countries stood at $235 billion, $146 billion, and $133 billion, respectively, and other Latin countries were also carrying heavy burdens. Although Latin America's largest nations are presently the most indebted in terms of total

amount of money owed to overseas institutions, many of the smaller countries are carrying a greater debt burden in terms of their capacity to repay the loans (Figure 7–27). Guyana, Nicaragua, and Argentina, for example, each owe more money than the value of all goods and services produced in a given year. When debt reaches those levels, countries find it difficult to make timely payments of the interest due on the loans; repayment of the principal is almost entirely out of the question.

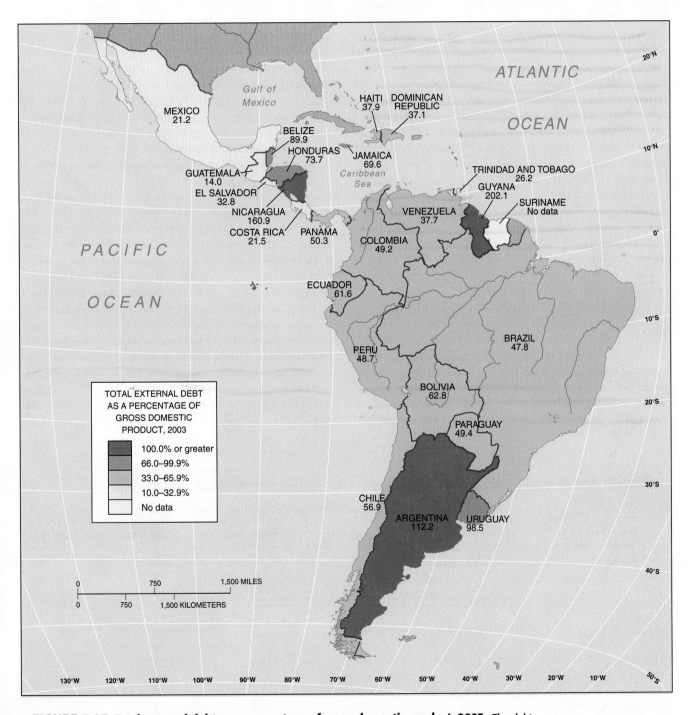

■ FIGURE 7-27 **Total external debt as a percentage of gross domestic product, 2003.** The debt burden of most Latin American and Caribbean nations is extremely high.
Sources: Derived from Economic Commission for Latin America and the Caribbean (ECLAC), 2004: *www.eclac.org*; and World Bank Development Indicators database, 2004: *http://derdata.worldbank.org/data-query/*.

In the 1980s, the initial response of the international banking community to the debt crisis was simply to arrange for "bridge" loans to the Latin American and other Third World nations to enable them to pay the interest due on the previous loans. That strategy, sometimes referred to as the Baker Plan (after then-U.S. Treasury Secretary James Baker), essentially took the questionable position that the solution to unmanageable debt is more debt. Those actions succeeded in buying time for lenders and borrowers alike but failed to address the need for fundamental changes in attitudes and policies among both the international financial community and the debtor countries.

Those long-overdue changes finally began in the late 1980s with the election of new political leadership in the United States and in many key Latin American nations. Faced with the prospect of a collapse of the international financial order, the U.S. administrations implemented new debt-management proposals, which have generally been reluctantly accepted by the debtor countries. Known originally as the Brady Plan, these strategies acknowledge that debtor nations are often financially incapable of repaying all of their loans without severely limiting the domestic investment needed for long-term development. Thus, they propose that a significant part of the debt be forgiven by the commercial banks in exchange for a guarantee by the International Monetary Fund and World Bank of repayment on the remaining debt. The debtor nations are required, in turn, to implement far-reaching internal economic reforms that center on reducing national expenditures.

These reforms, which are currently ongoing in many Latin American countries, have often proven to be very painful. They frequently include the reduction or elimination of long-standing government subsidies for food, fuel, utilities, health care, and other basic services that the lower economic classes have come to expect. These **neoliberal reforms** also entail the **privatization**, or selling to private companies, of countless enterprises formerly owned and operated by the national governments and the opening of the Latin markets to renewed foreign investment. Most of the old government-controlled companies, characterized by bloated bureaucracies and economic inefficiency, were operating constantly in the red, thereby contributing to deficit spending.

As the reforms are implemented, great hardships are often inflicted on the local citizenry. The costs of basic commodities and services may increase so dramatically that the social and political stability of the nation is threatened. In August 1990, for example, the government of newly elected Peruvian President Alberto Fujimori decided, as part of an economic reform package, to remove controls on the exchange rate of the national currency, then called the *inti*. Virtually overnight, gasoline prices went from $.07 to more than $2 a gallon (in U.S. dollars). Private bus lines stopped running, and thousands of commuters in Lima fought one another for room in the backs of pickup trucks. Stores and markets were looted by people desperate for basic foodstuffs, which were disappearing in spite of price increases surpassing 300 percent. Similar scenes have occurred in other countries. Another result of the austerity measures, at least in the short term, is increased unemployment as thousands of workers, formerly employed by government-owned businesses, are discharged, and thousands more are laid off by private companies unable to cope with increased business expenses. Many consider those actions, as bitter as they are, to be necessary remedies for decades of unwise fiscal behavior, characterized by excessive government borrowing and spending. Others consider the changes to be an assault on the poor. In some countries, the failure of the neoliberal reforms to improve the short-term living levels of the masses has led to the election of left-leaning populist presidents who have threatened to dismantle some or all of the reforms.

Regardless of perspective, one of the consequences of the debt crisis was the fueling of inflation (Table 7–3). Official inflation rates of 50 to 400 percent per year became common, with rates of more than 1,000 percent occurring

Table 7-3   Annual Variation in the Consumer Price Index for Selected Latin American Nations

	Percentage of Annual Increase in Consumer Prices				
	1961–1970 Avg.	1971–1980 Avg.	1981–1990 Avg.	1990–1999 Avg.	2000–2003 Avg.
Argentina	21.4	141.6	437.7	10.6	10.6
Bolivia	5.6	19.6	220.0	9.3	2.7
Brazil	46.2	35.6	337.0	253.5	8.9
Chile	27.1	174.3	20.3	9.7	2.8
Colombia	11.1	21.1	23.7	21.7	7.5
Dominican Republic	2.1	10.4	24.6	9.0	16.7
Mexico	2.8	16.6	65.2	19.9	5.8
Nicaragua	1.7	19.6	618.9	35.1	6.3
Peru	9.7	30.3	332.0	31.6	1.9
Suriname	4.2	9.8	12.9	88.8	40.1
Uruguay	47.8	64.0	60.3	38.2	11.2
Venezuela	1.0	8.5	23.3	51.8	21.0

Sources: Inter-American Development Bank, *Annual Report,* various years (Washington, D.C., 1990–1999), various pages; David L. Clawson, *Latin America and the Caribbean: Lands and Peoples,* 4th ed. (Boston, MA: McGraw-Hill, 2006, p. 387).

occasionally. Among the highest official inflation rates recorded in Latin America were Nicaragua's 14,295 percent in 1988, Peru's 7,482 percent in 1990, Argentina's 3,080 percent in 1989, and Brazil's 1,009 percent in 1992. Along with inflation came spiraling devaluations of national currencies, thus reducing even further the purchasing power of the citizenry and pushing ever-increasing numbers of people into poverty. Occasionally, beleaguered national leaders, motivated more by a sense of political expediency than by economic wisdom, attempted to solve their problems by ordering that more money be printed, thereby cheapening the value of the national currency. As a last-gasp measure, some went so far as to start all over by removing three to six zeros from the value of the currency and giving it a new name. But none of those quick-fix solutions were of any real worth in raising the living levels of the masses; they only delayed the implementation of substantive reforms.

The tragic truth is that the **lost decade** of the 1980s brought a significant regression, rather than improvement, in the economic circumstances of most Latin Americans (Figure 7–28). Living levels throughout much of the region fell back to those of the 1960s and 1970s, and when compared to conditions in developed nations, Latin American life seemed to be getting worse rather than better.

However, the countries that persisted in implementing the painful fiscal austerity and privatization policies, and which invested their privatization proceeds wisely, eventually began to experience renewed economic growth in the 1990s. Inflation has dropped in recent years to moderate levels and per capita GDP figures have begun to rise (see Table 7–3 and Figure 7–28). When combined with continued ed-

ucational, social, and political gains, the outlook for improved future personal living conditions throughout the region is now brighter.

There are presently a number of other promising development strategies being pursued by the Latin American nations. None is a panacea, but each is worthy of consideration. One approach that has proven highly beneficial to many Caribbean and Central American nations is the promotion of foreign tourism. Tourism is a major source of foreign exchange and employment in Mexico, Jamaica, Puerto Rico, the Bahamas, the Dominican Republic, and many nations of the Lesser Antilles. Tourism is of relatively minor significance, however, in countries not favored with tropical beaches, Indian ruins, internal political stability, or proximity to the United States.

Perhaps the most basic way to increase a nation's foreign earnings is through trade. Unfortunately, most Latin American countries are handicapped by their overreliance on two or three major exports, usually agricultural products or minerals. The world prices of such commodities are extremely volatile, often rising and falling dramatically in just a few months' time. One example is petroleum, the value of which has fluctuated wildly in recent years. The prices of sugar, coffee, bauxite, and other major Latin American exports have also experienced similar oscillations. In addition, most Latin American nations are overly dependent on a single market, usually the United States or the European Union. Such a narrow focus puts the Latin countries at the mercy of the trade policies and economic cycles of their principal trading partner, which may enact trade restrictions to protect local enterprises.

In an attempt to overcome these limitations, the Latin American and Caribbean nations have given renewed priority in recent years to the strengthening of **multinational economic unions**, similar to the EU. Such regional associations are intended to reduce trade barriers with neighboring countries and more effectively market member products, thereby increasing the volume of trade both within and outside the group. The success of such groups is totally dependent, however, on trust and cooperation among member nations. An example of the difficulties associated with the formation of multinational economic unions is the recently formed North American Free Trade Association (NAFTA), which provides for free trade among Mexico, the United States, and Canada. Although most Mexicans view the treaty as a means of attracting more and higher-paying industrial jobs, U.S. labor and environmental leaders maintain that it has resulted in reduced employment opportunities in the United States, as well as increased pollution and contamination of the Mexican border region. Other economic unions that have recently been formed or strengthened include the Andean Community, the Central American Common Market, the Caribbean Community (Caricom), and the Southern Cone Common Market (Mercosur) (Figure 7–29). Of these, Mercosur has experienced the greatest success to date.

The economic gains of Mercosur led to the establishment in 2004 of a free trade agreement between Mercosur and

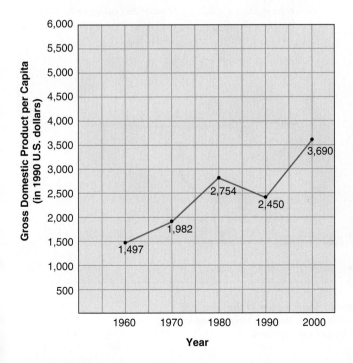

■ **FIGURE 7-28 Latin American economic trends, by decade, 1960–2000.** Note the overall upward trend, with the exception of the "lost decade" of the 1980s.

Legend:
- North American Free Trade Association
- Caribbean Community (Caricom)
- Central American Common Market
- Andean Community
- Southern Cone Common Market (Mercosur)
- No affiliation

Note: Bolivia is an associate member of and Chile and Bolivia have free trade agreements with the Southern Cone Common Market

■ **FIGURE 7-29 Regional economic unions and trading groups.** Many leaders of the Latin and Anglo American nations hope that the day will come when all the countries of the Western Hemisphere will be bound to one another through membership in a proposed Free Trade Area of the Americas.

the Andean Community. Proponents of regional integration hope that the 2004 accord will lead eventually to the formation of a South American Free Trade Area (SAFTA), which might subsequently be merged with the North American Free Trade Association (NAFTA) to create a **Free Trade Area of the Americas** (FTAA). Whether or not this goal is realized, it is clear that there is currently greater momentum toward economic integration than at any other time in Latin America's history.

Summary

Latin America is in a state of accelerated change; conditions, institutions, and values that prevailed for centuries are being challenged and rapidly modified. The difficulties facing a new generation of Latin Americans are possibly the greatest ever, but equally great resources are also available for their solution. In addition, in this age of global interdependence, the challenges of one region are challenges for us all.

As the Latin American nations work to bring about modifications in the international economic order, changes from within are also needed. Additional land reform, meaningful assistance for poor farmers, improved environmental management, and the purging of graft and corruption are all necessary components of lasting economic development. As its human resources are more fully developed, Latin America will begin to realize its full potential.

Key Terms

Aztec
Central Lowlands
Chibcha
conquistadores
criollos
Eastern Highlands
encomienda
estancia
ex-situ conservation
Free Trade Area of the
    Americas
fundamentalist
    Protestantism
hacienda
Iberia
Inca
in-situ conservation
latifundios
liberation theology
lost decade

Maya
mercantilism
mestizos
monoculture
Moors
mulattos
multinational economic
    unions
neocolonialism
neoliberal reforms
peninsulares
plantation
polyculture
privatization
Quetzalcoatl
royal patronage
shantytowns
Treaty of Tordesillas
Western Alpine System
zambos

■ Street vendor in Managua, Nicaragua. The wasting of human resources is one of the leading factors contributing to the underdevelopment of the Latin American and Caribbean nations. Visible in this image is a man confined to a wheelchair who sells newspapers on the busy streets of Managua in order to earn a meager living.

Chapter 8

Mexico, Central America, and the Caribbean:
Lands of Contrasts

- **Mexico: A Nation of Great Development Potential**
- **Central America: Region of Fragile Democracies and Widespread Poverty**
- **The Caribbean: Region of Cultural Diversity and Limited Physical Endowments**

W e now examine more closely the various regions of Latin America in order to appreciate both their differences and their common development challenges and potentials. For our purposes, Latin America is divided into six regions: Mexico, Central America, the Caribbean, Andean South America, southern South America, and Brazil. The focus of this chapter is **Middle America**: Mexico, Central America, and the Caribbean (Figure 8–1).

Mexico: A Nation of Great Development Potential

Mexico, by many indicators, should be among the most prosperous nations on earth. It is a large country, occupying some 72 percent of the land of Middle America and containing 57 percent of the population of that area. It has benefited throughout its history from some of the richest mineral deposits on earth—first its silver in the colonial period and now its petroleum and natural gas. Mexico's proximity to the technologically advanced and wealthy United States is also a potential economic advantage of significance as are its varied agricultural landscapes, which range from irrigated deserts in the north to tropical rain forests in parts of the gulf coastal lowlands.

Yet Mexico, though it has achieved much, has fallen far short of its potential. Some observers attribute that underachievement to physical geography, pointing out that much of Mexico's land is either very dry or mountainous. Other countries, however, including Switzerland and Japan, have demonstrated that well-developed human resources can more than compensate for natural physical limitations. To understand Mexico's limited economic achievement, then,

we must first evaluate the treatment of its people. We will begin by reviewing its quest for political stability, one of the fundamental preconditions of national development.

A Land of Revolutions

From the moment Cortés captured Tenochtitlán, Mexico exceeded almost every other Spanish colony in wealth and grandeur. Not only did the central volcanic highlands possess almost inexhaustible numbers of Indian laborers, but silver and other precious ores were also found in great abundance in Zacatecas, San Luis Potosí, and other parts of the western and eastern Sierra Madres. The quest for wealth and converts led the Spanish explorers ever farther northward, until the colony stretched across the southwestern United States from Texas to California. Had the poor been permitted to own their own family farmsteads, as they were in the United States, and had good public education been established, along with the free exchange of goods and ideas with the non-Spanish world, a vigorous and aggressive middle class would likely have developed and would have placed Mexico in the forefront of the Industrial Revolution and economic modernization.

In reality, colonial Mexico was so strangled by the mercantile system, and its social development was so stunted by class stratification, that when independence came in the early 1800s, the country lapsed into a century of political chaos and despotism. At the center of the storm, initially, was a flamboyant demagogue named Antonio López de Santa Anna. Described as the curse of his native country, Santa Anna managed almost single-handedly to lose Texas in the 1830s and contributed to the loss of California and the remainder of the southwestern United States in the 1840s.

As Santa Anna was pursuing his own personal interests at the expense of the republic, Mexican society became divided

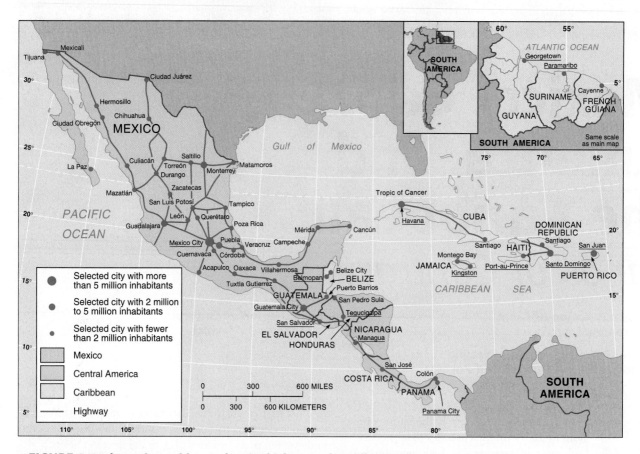

■ **FIGURE 8-1 The nations, cities, and major highways of Middle America.** This designation includes three regions: (1) Mexico, (2) Central America, and (3) the Caribbean, which includes the Greater and Lesser Antilles, Belize, Guyana, Suriname, and French Guiana.

into conservative and liberal factions. The conservatives consisted primarily of upper clergy and landed and monied elements, who favored a return to a European monarchy as the best means of protecting their social and economic privileges. The liberals, though mostly sincere in their Catholic faith, favored a separation of church and state as the best means of achieving democratic reform and economic modernization.

With Santa Anna's final exile in 1855, the liberals rose to power under the leadership of Benito Juárez, an Indian from the southern state of Oaxaca. Almost the opposite of Santa Anna, Juárez ruled until his death in 1872 with a moral rectitude unmatched either before or since in Mexican history.

After that brief flowering of liberalism, Mexico lapsed again into tyranny under the dictator Porfirio Díaz. While paying lip service to the democratic ideals of his predecessor, Díaz proceeded systematically to move the country back to a colonial status. Foreign interests were encouraged to locate in Mexico and, depending on their generosity to the dictator, were given almost unlimited latitude in their operations. Mexico became the mother of foreigners and the stepmother of Mexicans.

By 1910, U.S. investment in Mexico exceeded the total capital owned by the Mexicans themselves. In addition, the English dominated Mexican petroleum, metals, utilities, sugar, and coffee production; the French controlled most of the nation's textile mills; and the Spanish monopolized retail

trade and tobacco. The great *hacendados,* or *hacienda* owners, had grown ever more powerful, and land consolidation had proceeded to the extent that approximately half of the country belonged to fewer than 3,000 families, while 95 percent of the millions of rural peasants owned no land at all.

Revolution finally erupted throughout the country in 1910, and for 7 years Mexico was effectively without a central government as various leaders struggled for power. Then a most remarkable change occurred. A new constitution, drafted in 1917, ushered in a period of political stability that has continued to the present. The constitution provided for a single 6-year presidency with no possibility of reelection, civilian control of the military, limitations on the ownership of property by churches and foreign corporations, universal suffrage and education, a minimum-wage law, and agrarian reform.

National Unity

The achievement of political stability marked the beginnings of economic and social development in modern Mexico. One of the first and most necessary tasks was the building of a new national consciousness that would cause the Mexican people to place allegiance to the nation ahead of old ties to individual charismatic leaders. That objective has been accomplished to an extraordinary degree through three approaches.

The first approach has been to establish good public education as a national priority. Countless schools have been built—everywhere, from the inner cities to the most remote, poverty-stricken hinterlands—and have been staffed with dedicated, capable teachers (Figure 8–2). As a result, Mexico's literacy rate has risen from 56 percent in 1950 to 92 percent at present. In addition, trade schools and universities have broadened their roles and are now turning out increased numbers of skilled technicians and agricultural specialists, practical professions that traditionally were viewed as inferior in Latin American societies. Although private schools still service the wealthy, public education is functioning as the vehicle for the socioeconomic mobility and democratic values that were so seriously lacking in the colonial era.

The second approach to building national allegiance has been the glorification of Mexico's "Indianness." The great Aztec chiefs and other historical Indian figures have become focal points of national pride, to which all Mexicans, regardless of race, can relate. Huge monuments and murals depict the greatness of the Indian forebears, often in exaggerated tones (Figure 8–3). The veneration of Indianness is also carried out in the schools, where the Indian children of today are taught—in Spanish, ironically—to have pride in their Indian heritage.

The third approach to strengthening national allegiance has been the encouragement of political activism by the lower-class masses. This was accomplished initially by the creation of a single dominant political party, the **Institutional Revolutionary Party (PRI)**. From its inception, PRI leaders attempted to portray the party as the embodiment of an ongoing revolution in favor of the poor and repressed. Labor, business, and agriculture were all represented as separate interest groups within the organization. Thus, according to the PRI, Mexico functioned as a

■ FIGURE 8-3 Diego Rivera mural honoring Aztec chiefs. The public veneration of Indianness in Mexico is readily evident in its art.

one-party democracy. Even though minority parties existed on both the far left and the extreme right, PRI candidates swept every presidential and almost all state and municipal elections from 1917 to the mid-1980s, generally claiming more than 90 percent of all votes cast. Charges of electoral fraud were frequently leveled by opposition candidates, but as long as the majority of the people felt that their living levels were improving, few were willing to openly criticize the PRI.

The collapse in the 1980s of Mexico's oil-driven economy, with the attendant decline in personal living levels, opened the floodgates of political pluralism. By the mid 1990s, Mexico was evolving into a multiparty democracy, with the PRI losing a number of key elections. The principal opposition to the PRI has come from a leftist coalition led by the Democratic Revolutionary Party (PRD), that claims to represent the interests of the poor, and a conservative business alliance, led by the right-wing National Action Party (PAN), whose candidate, Vicente Fox Quesada, was elected president of Mexico in 2000. The PRI has also become more open in the way in which it selects its candidates for public office. Viewed historically, Mexico's three-pronged approach to national unity, centering on the strengthening of public education, ethnic allegiance, and political participation, has certainly been successful in promoting a national consciousness and self-respect.

■ FIGURE 8-2 Elementary school classroom in rural Mexico. Education is a national priority in Mexico and a means to achieving national consciousness. Success is largely attributable to the dedication of schoolteachers, since the provision of adequate funding is often difficult in rural areas.

199

■ **FIGURE 8-4 Bean harvest in rural Mexico.** Corn, beans, and squash have formed the basis of traditional diets in rural Mexico. To this day, the crops are cultivated by smallholder peasant farmers who consume what they need and sell the remainder in the urban *mercados,* or marketplaces.

Agricultural Development

Modern Mexico's agricultural development has also followed a threefold approach. The first facet has been agrarian reform or land redistribution. Fundamental to that movement was the reestablishment of the ancient Indian *ejido* system, under which land belonged to villages rather than individuals. Following the revolution, the government expropriated the lands of many *haciendas* and other estates and transferred title and use of the lands to communities of poor peasant farmers. Although some *ejido* lands are now being sold into individual ownership as a result of efforts to privatize the nation's economy, the reestablishment of *ejidos* did much in the twentieth century to promote more efficient and intensive use of the nation's farmlands.

The second thrust of agricultural development in Mexico has been the use of Green Revolution, or industrial, technologies on the farm, including irrigation, mechanized farm equipment, chemical insecticides, hybrid seeds, and inorganic fertilizers. Utilization rates have been high among the large private farmers, whose lands are found mostly in the northern part of the country, but the technologies have not proven to be economically or culturally compatible with the needs of small Indian farmers in central and southern Mexico.

The third agricultural focus has been lowland colonization, particularly along the southeastern gulf coast, from Veracruz to the Yucatán Peninsula. Beginning with the Papaloapan River Basin Development Project of the late 1940s and continuing on several fronts at the present time, integrated development schemes have done much to improve the quality of life in those traditionally neglected regions.

Today Mexico's diverse physical environments facilitate the production of a wide variety of crops. Maize, beans, squash, and chili peppers remain the basic crops of the rural peasantry (Figure 8–4). Commercial agriculture includes irrigated cotton, vegetable oils, and winter vegetables in the northwest; wheat, dairy products, and vegetables in central Mexico; sugarcane, bananas, and mangoes along the south-eastern lowlands; and coffee in the foothills of the Eastern Sierra Madre. Parts of the northeastern lowlands and the Yucatán Peninsula also specialize in the production of citrus and henequen, a hard fiber used in the manufacture of bags, carpeting, and twine (Figure 8–5). In addition, extensive grazing of beef cattle and, in the drier and colder regions, goats and sheep continues in the classic Hispanic tradition.

Mining

Foreign control of mining was almost absolute during the Díaz era. An American named William "Big Bill" Green, for instance, owned the Cananea Consolidated Copper Company, which held mining, grazing, and forest rights to almost 1 million acres (404,700 hectares) of land in northwestern Mexico. But the revolution limited foreign ownership of resources, and in 1938, all U.S. petroleum interests were nationalized (Figure 8–6).

Today Mexico's petroleum reserves, found both onshore and offshore, primarily in the southern Gulf Coast area, are among the greatest in the world (Figure 8–7). Mexico is also of world importance as a producer of silver, copper, lead, zinc, and sulfur. In addition, coal near the city of Sabinas and iron ore deposits from the state of Chihuahua support a growing iron and steel industry centered in Monterrey, Monclova, and Mexico City.

Industrialization and Urbanization

Mexican industry in the colonial and early independence eras was based primarily on textiles, milling, and furniture and tile making. It was centered in Puebla, Orizaba, Mexico City, and Monterrey. Although those activities are still important, more technologically sophisticated forms of manufacturing began to appear around 1940.

At present, Mexico is one of the leading industrial nations in Latin America. Two of its most important industrial regions have evolved around Mexico City and Monterrey; both produce a great array of consumer and producer goods.

■ **FIGURE 8-5 Henequen production in the Yucatán Peninsula.** Henequen is a thorny, fleshy-leaved plant that is well adapted to conditions of heat and drought. Many of the lands previously planted to henequen have recently been converted to citrus and other more profitable crops.

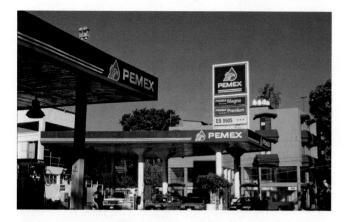

■ **FIGURE 8-6 PEMEX service station, Xalapa, Veracruz.**
The continued presence throughout Mexico of government-owned
PEMEX (*Petroleos Mexicanos*) service stations is a legacy of the nationalization in 1938 of foreign petroleum interests. Mexico remains
one of the world's largest suppliers of petroleum.

Automobiles are assembled in Puebla, Cuernavaca, Saltillo,
and Hermosillo (Figure 8–8); agricultural machinery is made
in León; and seamless steel pipes are produced in Veracruz.

Giant petrochemical complexes have been built along the
gulf coast, in Poza Rica, Minatitlán, Villahermosa, and Tampico.
Additionally, thousands of factories have sprung up in Ciudad
Juárez, Tijuana, and other border cities, where special plants,
called *maquiladoras*, are permitted (see the *Geography in Action* boxed feature, The Globalization of Industry: Mexico's
Maquiladora Program). *Maquiladoras* are foreign-owned
(mostly American) manufacturing plants that operate in Mexico using foreign technology and Mexican labor. The foreign
firms benefit from comparatively low labor costs and the absence of import duties or tariffs on the finished goods that are
shipped abroad. Mexico benefits from the creation of more than
a million relatively high-paying (by Mexican standards) jobs.

In recent years, tourism has become one of Mexico's leading growth industries. Some 20 million international tourists

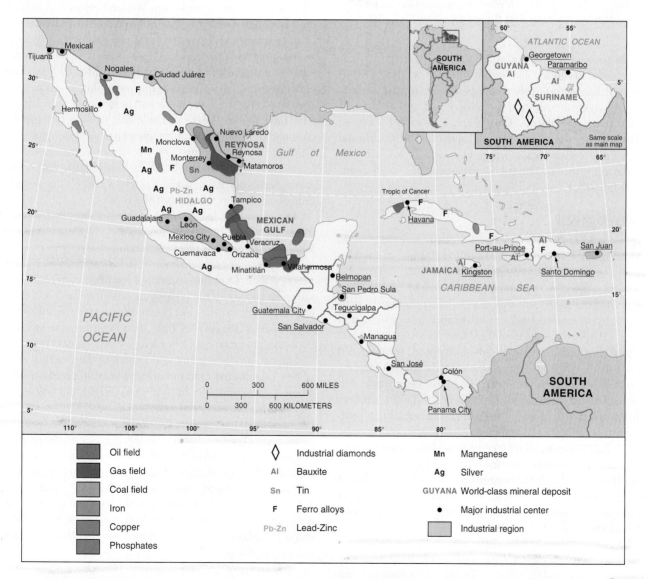

■ **FIGURE 8-7 Principal mineral-producing areas of Middle America.** The major minerals in this region include petroleum and silver in Mexico and bauxite in Jamaica, Guyana, and Suriname. Mexico also contains sufficient coal and iron ore to sustain a steel industry.

■ **FIGURE 8-8 Volkswagen assembly plant in Puebla, Mexico.** Puebla has a well-developed automobile manufacturing tradition, and production in this plant now concentrates on Jettas and modern Beetles rather than the antique variety.

visit Mexico each year, and world-class hotels and facilities now abound in Mexico City and Acapulco. The Mexican government has aggressively developed other tourist centers north of Acapulco along the Pacific Coast and at Cancún on the Yucatán Peninsula. Those travelers who prefer a more relaxed (and less expensive) exposure to the traditional charms of Mexican society frequently seek out Mérida with its Mayan culture or Oaxaca, Cuernavaca, and Guadalajara (see Figure 8–1).

Many Mexicos

Mexico can be divided into seven regions, and they are so distinct that the country has been referred to as "many Mexicos."[1] The first contains the **Border Cities**, where the economy and culture have been greatly altered by proximity to the United States (Figure 8–9). The border economy is dominated by *maquiladoras*, import-export businesses, and tourism. Unfortunately, what most tourists see is a Hispanic culture that has been modified—corrupted, some would say—by U.S. influence. For example, the language of the region is filled with English words that have been made into Mexican verbs, such as *lunchear* ("to eat lunch"), *parkear* ("to park"), and *wachar* ("to watch"). The seemingly rampant mix of smuggling, gambling, quack medicine, and low-quality tourist shops, all set in a dusty, arid environment, can easily mislead visitors into believing that the entire region is a physical and cultural wasteland. This, of course, is not the case, and the Border Cities constitute one of Mexico's most rapidly expanding regions with respect to both industrial activity and population growth.

A second region is Northeastern Mexico, dominated by the giant industrial city of Monterrey and its distant port of Tampico. Supported by extensive reserves of natural gas, private industrialists have developed five large manufacturing

groups that specialize in chemicals, steel, paper, glass, and beer. In addition, vast areas of semiarid land are given over to extensive sheep and cattle grazing on huge ranches dominated by unimproved pasture grasses, mesquite and creosote shrubs, and cacti. In recent years, many of these ranches have been converted to large farms specializing in the production of grain sorghum, which is mixed in poultry feedstocks, whereas others have been transformed into private hunting preserves that cater to foreign tourists.

A third region, one of the most rapidly developing of Mexico, is the **Gulf Coast**, centering on the states of Veracruz and Tabasco. In addition to its great petrochemical and textile industries, the region is one of the leading agricultural zones in all of tropical Latin America (Figure 8–10). An almost continuous stream of cargo trucks laden with sugar, coffee, pineapples, bananas, papayas, cacao, mangoes, citrus, and coconuts passes through Córdoba and Orizaba, as well as the lushly forested escarpment of the Eastern Sierra Madre, on its way to markets in Mexico City and other highland population centers. The rich volcanic and alluvial soils of the southern gulf coast also support Mexico's largest remaining stands of tropical rain forest. However, those forests, which once covered much of southern Mexico, are being converted at an alarming rate to poor, unimproved pasture. Absentee city-dwelling landowners stock the pastureland with mixed Zebu breeds of cattle that can tolerate the disease and periodic flooding characteristic of the region. The Gulf Coast also benefits from the city of Veracruz, the nation's chief port. The traditionally liberal attitudes of Veracruz residents have made the city's fun-loving *jarocho* culture renowned throughout the republic.

A fourth region is the **Yucatán Peninsula**, which consists of a low, flat limestone plateau covered with scrubby savanna forest vegetation that alternates between lush green during the rainy summer months and parched brown during the dry winter season. Over time, rainwater has filtered down through the red topsoils into the underlying limestone bedrock, forming carbonic acid that has begun to dissolve the limestone itself. Subsurface caverns have then been formed and have continued to grow until the surface has collapsed from lack of support, producing sinkholes. Much of the drinking water of the Yucatán has traditionally come from ponds created when high water tables have intersected the sinkholes.

Historically one of the poorest regions in Mexico, the Yucatán is now prospering, as irrigated citrus groves are planted on former henequen lands. Tourism focuses on the beaches of Cancún and the Mayan Indian ruins at Chichén-Itzá. In addition, the capital city of Mérida is attracting foreign and Mexican investment with a courteous, well-trained workforce. Despite foreign influence, the Yucatán has retained a distinctive regional culture characterized by widespread use of the Mayan language, hammocks rather than beds, and colorful clothing (Figure 8–11).

A fifth region, perhaps the poorest and most isolated of Mexico today, is the South, which includes the states of Guerrero, Oaxaca, and Chiapas and consists mostly of vast, rugged mountain ranges. Shortly after the conquest of the Aztec heartland, Cortés took much of the present-day state of Oaxaca as

[1]Lesley Byrd Simpson, *Many Mexicos*, 4th ed. (Berkeley: University of California Press, 1967).

GEOGRAPHY IN ACTION
The Globalization of Industry: Mexico's *Maquiladora* Program

One of the most rapidly industrializing zones in all of Latin America and the Caribbean is Mexico's border region which, since the late 1960s, has gained some 3,000 foreign-owned manufacturing plants that employ more than 1 million Mexican workers. The roots of this unprecedented industrial growth reach back to 1942, when the American and Mexican governments signed an agreement, called the *bracero* program, that allowed Mexican agricultural laborers to work legally in the United States. Although the program initially was designed to be terminated once World War II had ended, when returning American GIs would again be available to work the fields, it proved so attractive to both countries that it was extended until 1964. The following year, the Mexican government established the Border Industrialization Program in the hope of promoting the creation of manufacturing jobs for its displaced farm workers.

The basic idea behind the program was to encourage U.S. and other foreign firms to establish branch assembly plants in Mexico where they could take advantage of inexpensive Mexican labor. In order to entice the so-called *maquiladora* ("milling or processing") plants to come, Mexico agreed to provide subsidized land, utilities, and tax holidays to the foreign firms, as well as to allow the imported U.S.-made parts and exported Mexican-assembled finished products to be transported duty-free across the border. The recently implemented North American Free Trade Association (NAFTA) agreement has not lessened the attractiveness of the border region to many foreign manufacturing interests, which, in recent years, have been joined by an increasing number of Mexican investors.

Most of the products produced during the early years of the Border Industrialization Program were low-technology items such as textiles and toys. As time passed, however, the region evolved into a center for electronics, communications, and transportation manufacturing. Today the Mexican border cities are characterized by large industrial parks that house plants owned by such leading multinational corporations as IBM, Honeywell, GE, Delphi, Hyundai, Sony, Nokia, and Motorola, to name but a few (Figure A).

Although the Border Industrialization Program has proven to be an enormous boon to the Mexican economy, it is not without its problems. The flood of migrants from central and southern Mexico into the border region has led to rapid population growth that has overwhelmed municipal sewerage treatment facilities. This, in turn, has resulted in outbreaks of hepatitis, typhoid, cholera, intestinal parasites, and other serious diseases. The presence of large quantities of inadequately disposed industrial wastes—including solvents, acids, and PCBs—is also raising short- and long-term health concerns. Ironically, many of the Mexican factory workers are young, single women who are recruited for their perceived feminine traits of docility, patience, and dependability. Many of these women subsequently find themselves in the unanticipated role of being the principal wage earners in households headed by their unemployed fathers and grandfathers. Job stress is so great that the typical laborer works for only a few years, despite receiving a comparatively high wage. The program has also been criticized for taking jobs from U.S. workers and for failing to generate spinoff employment through linkages to other sectors of the Mexican economy. Notwithstanding its economic success, then, the program has come under increasing criticism for its attendant social and environmental ills.

■ FIGURE A *Maquiladora* **plant in Nuevo Laredo, Mexico.** This telephone repair plant fixes telephones for AT&T.

his personal *encomienda*. On a later occasion, when asked to describe Mexico, Cortés took a piece of paper and crumpled it tightly, forming a misshapen wad. Holding the contorted object in his hands, he replied, "This, sir, is New Spain."

Outside the administrative and tourist centers of Oaxaca and Acapulco, the South continues to be a region of scattered highland Indian villages with agricultural practices and religious ceremonies that have changed little since preconquest times. Unfortunately, the region is experiencing increasingly severe environmental degradation from a cycle of rural poverty, deforestation, soil erosion, and abandonment of agricultural lands.

The isolation, rugged terrain, and poverty of Mexico's South has also provided refuge, in recent years, to two antigovernment guerilla movements: the publicity-seeking Zapatista National Liberation Army (EZLN), led by its increasingly romanticized, ski-masked leader, "Comandante Marcos," and a far more violent, shadowy collection of leftist extremist groups that call themselves the Popular Revolutionary Army (EPR). This second movement, which first surfaced in 1996, appears to be supporting itself, at least in part, from ransoms received from the release of prominent abduction victims.

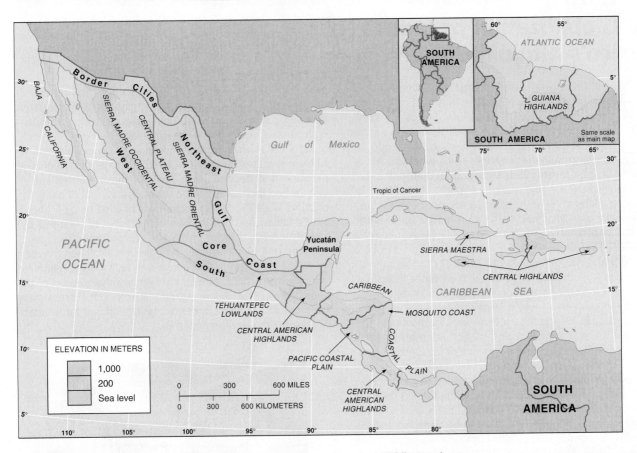

■ **FIGURE 8-9 Regions of Middle America.** The landscape of much of Middle America reveals mountains or mountain basins. Both Mexico and Central America have a long history of earthquake and volcanic activity; along with Andean America, they are part of the circum-Pacific Ring of Fire. Significant tropical lowlands exist, however, as evidenced in Mexico's Yucatán Peninsula and Gulf coastal plains, the Mosquito Coast of Nicaragua and Honduras, and the lowlands of Cuba.

■ **FIGURE 8-10 Tropical agriculture in Veracruz state.** Mexico's Gulf Coast is a lush and highly productive agricultural region. Shown in this scene from the Papaloapan River Basin is a form of commercial intercropping, with cattle grazing under mango and palm trees.

■ **FIGURE 8-11 Traditional rural homestead, Yucatán Peninsula.** In the rural zones of Yucatán, life for the lowland Maya peoples continues much as it has for centuries.

GEOGRAPHY IN ACTION
Environmental Degradation in Mexico City

From its modest beginnings in the early sixteenth century, when the city's *vecino,* or European resident population, numbered only a few thousand, Mexico City has grown to become the world's second most populous metropolitan area, with conservative estimates now placing the number of inhabitants at 22 million or more. With hundreds of rural migrants arriving daily in search of a better life and a 2.1 percent rate of natural increase, Mexico City's population is now growing at hundreds of thousands of persons per year, with no short-term end in sight.

As is true with most Third World governments, Mexico City is hard-pressed financially to provide basic services, such as treated water and garbage pickup and disposal, to the older, more prosperous neighborhoods, much less extend those services to the distant squatter settlements and satellite communities. One result of the poverty and bureaucratic inefficiency is that approximately one-fourth of the estimated 25,000 tons of garbage and other solid wastes generated daily goes uncollected and is left to rot on the streets before being consumed by dogs, cats, rats, and other vermin. The remaining three-fourths of the refuse is transported by the sanitation department to huge open dumps, where it is searched by thousands of poor people who rely on scavenging to provide a meager income. Many of these individuals actually reside in tiny makeshift homes built on top of old garbage mounds.

Compounding the city's solid waste disposal problem is its atmospheric contamination, recognized by knowledgeable observers to be among the most severe in the world. The emissions generated by more than 4 million motor vehicles and some 35,000 factories combine to form 3 to 4 million tons of air pollutants annually. Because the Valley of Mexico is surrounded on three sides by mountains and receives almost all of its annual rainfall during the summer months, atmospheric visibility is often reduced to a block or two during the dry season when pollution levels frequently exceed 400 percent of maximum safe levels established by the World Health Organization (Figure A). Fifty-five percent of school children have unsafe levels of lead in their blood and more than 80 percent of children treated at state-run hospitals suffer from bronchitis, asthma, and other respiratory ailments.

Government attempts to reduce air pollution have centered on the establishment of a network of 33 monitoring stations, whose computerized findings are transmitted hourly throughout the city as the Metropolitan Air Quality Index (IMECA). When the readings reach a certain level, schoolchildren are kept indoors at recess, and the public is advised to refrain from stren-uous physical activity. Other steps have included the closing of eighty of the most high-polluting factories, an urban reforestation initiative, and the adoption of a program that endeavors to restrict the number of cars in use during peak pollution periods. The latter, called *Hoy No Circula* (No Driving Today) uses the last two numbers of a vehicle's license plate to designate which business day of the week car owners are prohibited from driving their vehicles. New cars are also required to have catalytic converters and to use unleaded gasoline. Concurrent with these programs, the government is expanding the subway system, which is one of the cleanest and most efficient in the world, and adding thousands of miles of sewer lines. These small steps, while seemingly ineffective at present, may prove to be the beginning of a long-term betterment of the city's physical environment.

Source: David L. Clawson, *Latin America and the Caribbean: Lands and Peoples,* 4th ed. (Boston: McGraw-Hill, 2006), pp. 415–416.

■ **FIGURE A** **Severe air pollution, Mexico City.** Owing to the interplay of a number of physical and cultural factors, Mexico City has one of the world's highest levels of environmental contamination.

A sixth region—Mexico's Core—consists of a great interlocking series of snow-capped volcanoes and highland valleys extending east to west along the 19th parallel. The largest of those valleys, the **Valley of Mexico**, lies above 7,000 feet (2,134 meters) in altitude and experiences cool temperatures throughout the year. Virtually uncontrolled population growth there has combined with extensive industry and millions of diesel- and gasoline-powered buses, trucks, and automobiles to make Mexico City not only one of the world's most populous cities, with a reported 22 million persons residing in the metropolitan area, but also one of the world's most polluted (see the *Geography in Action* boxed feature, Environmental Degradation in Mexico City). Chronic water shortages, massive traffic congestion, and vast squatter settlements—contrasted with some of the world's greatest museums, finest shopping districts, and most exclusive residential areas—make Mexico City a window to the intense change that is sweeping the nation. Heavy industry and rapid population growth also characterize the other urban centers of the Core region—Puebla, Cuernavaca, Toluca, and León. In the thousands of small peasant villages in the region, however, life goes on much as it has for hundreds of years (Figure 8–12).

■ **FIGURE 8-12 Peasant life in rural Mexico.** In many villages, daily life remains largely unchanged from generation to generation. Here, women wash clothes on stones along the banks of an irrigation ditch.

Mexico's final region is the West, centering on Guadalajara and the oasis cities of Culiacán, Ciudad Obregón, and Hermosillo. This semiarid region is developing rapidly through irrigated commercial agriculture and trade with the United States. It is also the home of the Mexican cowboy tradition.

Mexico is a nation of great potential. If political stability can be maintained and if the government continues to develop its human resources, Mexico will eventually take its place among the leading nations of the world.

Central America: Region of Fragile Democracies and Widespread Poverty

Central America is a troubled region of six small countries, each struggling to raise its people's levels of living in the face of serious economic and political challenges. Unlike Mexico, Central America has limited mineral wealth. The lack of raw minerals, as well as physical isolation, poor transportation networks, and small markets for manufactured goods, has handicapped industrial development. Also, unlike Mexico, which has benefited from the complementary nature of its various regions, Central America has been severely hampered by rivalries and duplication; more than twenty-five attempts at regional economic or political union have failed. Another difference is that Mexico's revolution brought about agrarian reform and the beginnings of a middle class. In Central America, however, social relationships and land tenure patterns have changed little since the colonial era. Political stability has only recently been achieved in this region, which was torn apart during the mid to late 1900s by leftist guerrilla movements, rightist vigilante death squads,

and military intervention by both the former Soviet Union and the United States.

Historical Land Use

Central America was settled simultaneously from two directions. In the south, Vasco Núñez de Balboa first occupied Panama in 1513 on his march to the Pacific Ocean. Panama City was founded 6 years later on the Pacific side of the isthmus, and the area soon established itself as the premier overland, interoceanic transportation route of the Americas (Figure 8–13). Then, from Panama, Spanish colonizers traveled northward into the lowland lake country of Nicaragua. There rivalry between two cities, León and Granada, eventually led to the selection of centrally located Managua as the national capital of Nicaragua.

In the 1560s, rumors of gold prompted a group of Nicaraguans to move southward to present-day Costa Rica. Neither gold nor Indians were found in significant quantities, but the settlers did find rich farmland in the cool, healthful volcanic basins of the central plateau. What evolved was a nation whose population was composed largely of middle-class European farmers, whose traditions of universal education and political stability have made Costa Rica an enclave of relative prosperity and tranquillity.

Meanwhile, settlers pushed into Guatemala, El Salvador, and Honduras from Mexico. Soon all of Central America, with the exception of portions of Costa Rica and Panama, had fallen into the classical colonial land use pattern of great grazing estates and scattered mines, a pattern from which some areas have yet to emerge.

The first commercial crop introduced into Central America was coffee, which had earlier become established on the Caribbean islands as an excellent upland complement to lowland sugarcane. The first coffee plants came from Cuba to Costa Rica in 1796 and, by 1850, small family-operated coffee farms, called *fincas*, or *cafetales*, dominated the central plateau. From Costa Rica, coffee diffused northward, finding ideal growing conditions in the rich soils of the volcanic uplands of El Salvador and western Guatemala (Figure 8–14).

Bananas

Although coffee has had a major impact on the economies of Guatemala, El Salvador, Nicaragua, and Costa Rica, the crop that has most strongly influenced the character and development of Central America as a whole is bananas. The Spaniards introduced the first plants to Hispaniola (present-day Dominican Republic and Haiti) in 1516 and to the floodplains of eastern Central America shortly thereafter. The crop thrived in the hot, moist climate and the fertile, well-drained soils.

Commercial production of such a highly perishable fruit was made possible by the development of refrigerated ships and boxcars in the late nineteenth century. The 1870s brought the first shipments of bananas to Boston and New Orleans, and the fruit quickly assumed a prominent place in American diets. From 1880 to 1930, banana plantations expanded rapidly along the sparsely settled, humid eastern

■ **FIGURE 8-13 Ruins of Panamá Viejo, Panama City.** Panamá Viejo, the first site of Panama City, was sacked in 1671 by the English pirate Henry Morgan and his men, who arrived while the Spanish garrison was away chasing Indians.

lowlands, using imported West Indian wage laborers. As sales soared in the United States, the Standard and United Fruit companies were formed and achieved positions of inordinate influence in many Central American countries. The boom slowed in the 1920s, however, as soil exhaustion and plant diseases made cultivation of the standard Gros Michel variety of banana unprofitable.

The decline of commercial banana cultivation in the Caribbean lowlands prompted a search for alternative plantation sites, which ultimately were found in the Pacific lowlands. This second production phase, which lasted until the early 1960s, was characterized at times by such great dependence on the crop that the countries of the region came to be called "banana republics" by many people. Unfortunately, the deadly plant diseases ultimately established themselves on the Pacific plantations as well, and further changes were necessary.

The third and current phase of banana production has involved a switch to the Giant Cavendish plant variety. Its re-

■ **FIGURE 8-14 A coffee _finca_ of the Central American highlands.** Coffee was the first commercial crop introduced into the region. It continues to be a major export.

sistance to disease has permitted renewed development along both coasts. In addition, because of allegations of political meddling, the multinational fruit companies are increasingly shifting their emphasis from production on company lands to the marketing of fruit purchased from small, individual farmers.

For years to come, scholars will debate the impact of the fruit companies in Central America. On the negative side, many observers are convinced that the companies intervened, often in illegal ways, in the internal affairs of the host countries. The companies are also criticized for monopolizing huge tracts of fertile farmlands that could be given to poor farmers for the production of badly needed domestic foodstuffs. On the positive side, however, it is recognized that those same companies brought high-paying jobs, schools, roads, hospitals, and harbors, all of which benefited local residents. Some scholars have also suggested that the banana plantations have contributed to the development of more progressive, open societies that have often supported industrial as well as agricultural modernization.

Economic Development

Agriculture occupies a prominent place in the Central American economies and is likely to do so for the foreseeable future. Any serious development plan must take into account that one-third of the regional workforce continues to be engaged in agriculture.

Three new agricultural products have joined coffee and bananas in recent decades as important sources of regional revenue. The first is irrigated cotton, which showed rapid expansion from the 1950s into the early 1980s along the dry Pacific lowlands of Guatemala, El Salvador, and Nicaragua. The long-term sustainability of large-scale Central American cotton production is increasingly threatened, however, by its heavy reliance on expensive and highly toxic pesticides and herbicides. The second new product is chilled beef for export. Even though cattle grazing has historically been a major focus of Latin American society, its low output was targeted almost exclusively for local consumption. The past four decades, however, have witnessed the growth of scientific animal husbandry, which places improved animal breeds on imported African pasture grasses. Nonetheless, one of the unfortunate consequences of expanded cattle ranching, both modern and traditional, is an alarming loss of Central America's remaining tropical rain forests. A third category of new commercial agriculture is the growing of cool-weather vegetables, such as broccoli, potatoes, and onions, as well as ornamental plants and cut flowers (Figure 8–15). These are frequently shipped under refrigeration to lucrative Anglo American markets that are only a few hours away by air.

Commercial fishing, especially for shrimp, is also expanding, particularly in the Gulf of Fonseca, an inlet of the Pacific Ocean shared by El Salvador, Honduras, and Nicaragua. Tourism has performed poorly as a rule, with Costa Rica, El Salvador, and Guatemala benefiting the most.

■ **FIGURE 8-15 Central American upland vegetable fields.** The commercial production of cool weather vegetables, such as garlic, has expanded rapidly in Latin America.

Guatemala

Guatemala is the largest of the Central American nations in population and is also one of the poorest and politically most troubled. The northern half, called the Petén, is a sparsely populated, lowland rain-forest region that offers an agricultural frontier. Colonization projects have been established there, and roads have been constructed to connect the area to the rest of the country. However, the highlands that parallel the Pacific Ocean contain the bulk of Guatemala's population. In the western part of the highlands live most of the Indians, who follow traditional ways of life and still maintain many customs of their Mayan ancestors (Figures 8–16 and 8–17). The Indians have long been discriminated against by *mestizos* and whites, who follow European, or **Ladino**, lifestyles in and around Guatemala City, the nation's social, economic, industrial, educational, and political center. Along the southern coast and slopes of the adjacent mountains lies the main commercial agricultural zone.

Sugar, coffee, and cattle are widely produced. In addition to its failure to integrate the Indians, who constitute approximately half the national population, Guatemala's development was seriously retarded by a 36-year civil war that resulted in the deaths or disappearance of more than 200,000 persons and the polarization of society between right- and left-wing sympathizers. A peace accord brought an official end to the war in 1996, but acts of civil violence continue to occur, and the healing process is far from complete.

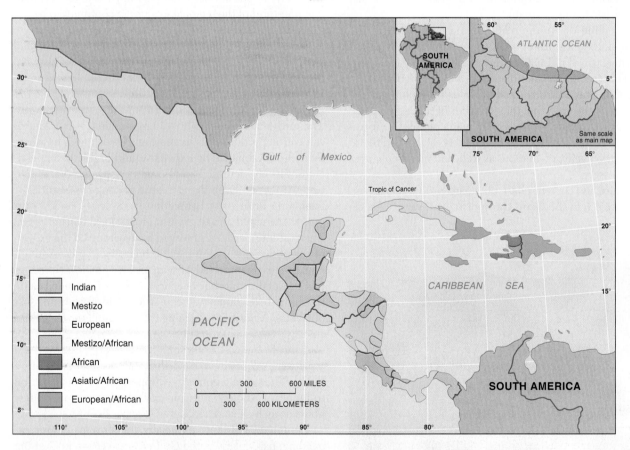

■ **FIGURE 8-16 Dominant ethnic groups of Middle America.** Considerable ethnic and cultural variation exists in Middle America. Mexico, for example, is dominantly *mestizo*, but southern Mexico and Guatemala are populated by Indian peoples, and persons of African descent dominate significant portions of many Caribbean islands. The pre-Columbian Indian population of the Antilles was largely decimated early in the European colonization period.

■ **FIGURE 8-17 Market vendors in the Guatemalan highlands.** Small-scale markets bring together farmers and consumers as they have for centuries.

El Salvador

El Salvador is the most densely populated of all the Central American nations. The effects of rural poverty and landlessness are seen in the flow of rural migrants to shantytowns in the principal cities of San Salvador, the capital, and Santa Ana, and in the migration of Salvadorans to less densely settled areas in neighboring Honduras and Guatemala. In contrast to other Central American nations, El Salvador does not have an agricultural frontier. Thus, increased production must come from higher yields per unit of land rather than from expansion of the area under cultivation. But the traditional landholding patterns have not been completely broken, and little has been done to improve agricultural efficiency. Rather, the emphasis is on industrialization, even though El Salvador does not possess minerals of significance and power supplies are largely imported. The country's greatest economic assets are its relatively skilled and industrious labor force and a strong tradition of business and commerce. With the conclusion in 1992 of the deadly 12-year civil war, which took more than 75,000 lives and displaced approximately 1 million persons, El Salvador has reestablished a democratic government and is endeavoring to become the commercial center of Central America through the promotion of free trade. Impressive progress has been realized in recent years in implementing electoral, judicial, and economic reforms. Despite the devastating earthquakes of 2001, El Salvador is rapidly reasserting its historic leadership role within the region (Figure 8–18).

Honduras

Honduras, with an area of more than five times that of El Salvador, has a population of only 7 million. Thus many parts of Honduras are sparsely settled. Outside the major cities of Tegucigalpa, the capital, and San Pedro Sula, where services and limited manufacturing provide employment opportunities, agriculture is almost the sole occupation. In the highlands surrounding Tegucigalpa, subsistence farming, traditional livestock raising, lumbering, small-scale mining, and commercial coffee growing are the dominant activities. Along the coast around San Pedro Sula, foreign-sponsored banana plantations and apparel-dominated *maquiladora* plants contribute much of the nation's exports (Figure 8–19). The lowland coastal region is more industrialized and open to change than is the interior highland zone. Attempts to promote foreign tourism have met with some success on the English-speaking Bay Islands off the northern coast of the country but have largely failed elsewhere.

Honduras experienced enormous losses of roads, bridges, schools, clinics, and farms when Hurricane Mitch dropped as much as 4 inches (100 millimeters) of rain per hour over a 5-day period in October of 1998. International relief efforts helped avert a calamity in the aftermath of the storm, but reconstruction is not yet complete.

Nicaragua

Nicaragua, like many of its Central American neighbors, has a long history of political instability. For a time in the nineteenth century, the British tried to control part of its Caribbean coast. Later, an American adventurer captured the capital briefly. Then in the early twentieth century, conflict between conservatives and liberals contributed to internal instability, which prompted U.S. military occupation from 1912 to 1933. Finally, in 1937 the Somoza family gained control of the government and held power until it was overthrown in 1979 by the leftist Sandinistas. It is estimated that, at the time of their ouster, the Somoza family controlled half the wealth of the

■ **FIGURE 8-18 Post-2001 El Salvador earthquake reconstruction.** The painting of the word *"Resurgiremos,"* meaning "We will come back" or "We will rebuild," is evidence of the massive reconstruction efforts that have been undertaken in El Salvador since the devastating earthquakes of 2001. In this section of the San Salvador suburb of Santa Tecla, a huge mudslide destroyed homes and businesses. The area has since been converted into a park honoring the victims of the quakes.

■ **FIGURE 8-19 Honduran *maquiladora* plant.** Apparel manufacturing has expanded rapidly in and around the city of San Pedro Sula. Most of the production is exported.

impoverished nation. Thereafter, the Sandinistas attempted to reorient the country toward the Soviet bloc and away from close political and economic ties to the United States. In the process, they achieved a massive military buildup. Although voted out of office in 1990, the Sandinistas remain a potent force in a poor and deeply divided nation.

Nicaragua is basically an agricultural country, with most of its production coming from the lowland area near Managua, the capital. Much of eastern Nicaragua is sparsely populated and little developed. The Mosquito Indians, who dominate this Caribbean lowland region, engage in subsistence farming and fishing and search for gold, which can be found in limited quantities within old streambeds that have since been covered by sandy, unproductive soils. Banana plantations, so common along the Caribbean coast of the isthmus, have never been important in Nicaragua. While receiving less attention than Honduras, Nicaragua also ex-

■ **FIGURE 8-20 Universidad Estatal a Distancia, San José.** This university is one of many excellent institutions of higher learning in Costa Rica, which boasts a 96 percent literacy rate.

perienced massive losses to Hurricane Mitch and will require years to rebuild its infrastructure.

Costa Rica

Costa Rica differs from the other Central American nations in several respects. More than any of its neighbors, Costa Rica has developed a socioeconomic middle class and a spirit of national unity. In addition, 96 percent of the population is literate and excellent universities abound (Figure 8–20). The infrastructure is well developed in the Central Valley but not in the coastal lowlands. The government is democratic and forward-looking, with a history of competence and stability. Many foreign businesses and retired persons have been attracted to the Central Valley by its skilled workforce and temperate climate. Among these is a large Intel microprocessor manufacturing plant that began operations in 1998 in San José, the capital. Large landholdings coexist with small- and medium-sized owner-operated farms, yielding a relatively high productivity. Even though Costa Rica is small in area, it has considerable regional specialization. For example, high-quality coffee, cut flowers, and vegetables are grown in the temperate highlands of the Central Valley around San José. Large-scale cattle ranching dominates in the northwestern province of Guanacaste, and banana plantations are common along the Caribbean coast. Costa Rica has also evolved in recent years as a leader in nature-based ecotourism. Much of the national territory has been set aside as protected national parks, many of them within the tropical rain-forest zones.

Panama

Panama owes its existence and its economic viability to the Panama Canal (Figure 8–21). Built by the United States on land leased from Panama, the canal was completely turned over to the Panamanians on January 1, 2000. The canal divides Panama into two parts: an eastern part, called the Darién, which is a little-developed rain forest; and a western part, which contains numerous banana plantations along the Caribbean coast and beef cattle, rice, and staple food crops in the interior. Tertiary activities in and along the Canal Zone—including retailing, shipping, and banking (some of which is connected to illicit drug trafficking)—contribute significantly to the nation's economic output and represent the most rapidly growing sector of Panama's economy. Panama City, the capital on the Pacific Coast, and Colón, the principal Caribbean port, are the major urban centers. Both cities cater to tourists and transit passengers by offering duty-free goods.

The area in and around the Panama Canal is truly a crossroads of the world. Both English and Spanish are spoken by a majority of the population, and U.S. currency is used everywhere. Since the canal acts as a funnel, finished products and raw materials can be brought together in Panama for transshipment elsewhere, as well as for processing. Manufacturing, however, is only slightly developed. The region's importance to U.S. interests was clearly demonstrated by the

■ **FIGURE 8-21 The Panama Canal, the major focus of Panama's activities.** The canal divides Panama into two parts, of which the western sector is the more developed. The canal has now passed from U.S. to Panamanian jurisdiction.

U.S. military invasion and ouster of former President Manuel Noriega in 1989.

The Caribbean: Region of Cultural Diversity and Limited Physical Endowments

The Caribbean culture realm consists of the **Greater Antilles**: Cuba, Hispaniola, Jamaica, and Puerto Rico; the smaller islands of the **Lesser Antilles**, which are situated in the eastern Caribbean between Puerto Rico and Trinidad; and the continental rimland nations of Belize, Guyana, Suriname, and French Guiana (Table 8–1). Often portrayed in the popular media as tropical paradises, the Caribbean nations face three common challenges in their quest for economic development.

The first obstacle is their small size and relatively harsh physical environment. By world standards, each of the countries is tiny in population, area, or both. Not surprisingly, the island nations with relatively small populations find it difficult to attain economic self-sufficiency. In addition, those small nations face shortages of farmland, as well as occasional shortages of freshwater. For roughly half

of each year, December through April, almost no rain falls. Then, when the tropical afternoon thunderstorms begin, they are frequently accompanied by devastating hurricanes (Figure 8–22).

The second challenge facing the Caribbean nations is the need to overcome a colonial legacy. Although most of the islands are now politically independent, many continue to function as economic colonies, exporting relatively low-value raw materials, such as bananas, sugarcane, or bauxite, and importing relatively expensive processed products. The colonial legacy has also left deep social and racial divisions. Following the almost total extermination of native Indians, African slaves were introduced as a labor supply. Their descendants now constitute the principal population of the Lesser Antilles, Jamaica, and Haiti and a significant minority in Trinidad and Tobago (a single country), Cuba, the Dominican Republic, and Puerto Rico. All of the populations, however, have a European culture base—except Haiti, which has a mixture of African and European cultures. Generally speaking, in divided Caribbean societies the lighter a person's skin color is, the higher that person's socioeconomic standing tends to be. That pattern extends even to those islands that are racially almost entirely black; in those settings lighter-skinned *mulattos* dominate.

The third challenge is to preserve both the local ways of life and the environmental balance in the face of an ever-increasing number of foreign visitors. Along with much-needed jobs for local residents, tourism has often brought drug trafficking, organized crime, and vice.

Cuba

Although Cuba contains well over half the level land of the Caribbean islands, throughout the colonial era it was largely ignored by the Spaniards, who found few Indians and little gold there. By the beginning of the twentieth century, when Spain lost Cuba and Puerto Rico to the United States, Cuba's population was only 1.5 million people, most of whom were centered in or near Havana, and only 3 percent of the land was under cultivation.

Status as a U.S. protectorate brought far-reaching changes to the island. Yellow fever was eradicated, thereby opening up the lowlands to human occupancy. Tourism exploded as the capital Havana became a mecca of nightclubs and gambling. Sugarcane became the grass of Cuba, as American interests oversaw its planting throughout almost the entire island in order to profit from the preferential tariffs authorized by the U.S. Congress. Following the granting of independence, Cuba remained almost totally dependent on U.S. sugar purchases and tourism, and average citizens saw little improvement in their levels of living. Corrupt governments continued.

In 1959 a young Marxist revolutionary named Fidel Castro overthrew the government with the promise of major reforms (Figure 8–23). Castro's first stated objective was to lessen Cuban dependence on foreign nations; his second, to diversify Cuban agriculture; his third, to convert land from private to government ownership; and his fourth, to eliminate the traditional Spanish bias toward farming.

Table 8-1 The Caribbean Culture Realm in 2005

Political Entity	Original Dominant European Culture	Current Dominant Culture	Political Status	Date of Independence or Autonomy
Anguilla	Spanish	British	BOT[a]	
Antigua and Barbuda	Spanish	British	Independent	1981
Aruba	Dutch	Dutch	Autonomous[b]	1986
Bahamas	British	British	Independent	1973
Barbados	British	British	Independent	1966
Belize	British	British	Independent	1981
Cuba	Spanish	Spanish	Independent	1901
Dominica	Spanish	British	Independent	1978
Dominican Republic	Spanish	Spanish	Independent	1844
French Guiana	French	French	FAR[c]	1974
Grenada	French	British	Independent	1974
Guadeloupe	French	French	FAR[c]	1974
Guyana	Dutch	British	Independent	1970
Haiti	Spanish	French	Independent	1804
Jamaica	Spanish	British	Independent	1962
Martinique	French	French	FAR[c]	1974
Montserrat	French	British	BOT[a]	
Netherlands Antilles (Bonaire, Curaçao, St. Eustatius, St. Maarten, Saba)	Various	Dutch	Netherlands	
Puerto Rico	Spanish	American/Spanish	Commonwealth	1952
St. Christopher (St. Kitts) and Nevis	British	British	Independent	1983
St. Lucia	Spanish	British	Independent	1979
St. Vincent and the Grenadines	Various	British	Independent	1979
Suriname	Dutch	Dutch	Independent	1975
Trinidad and Tobago (a single country)	Spanish	British	Independent	1962
Virgin Islands	Danish	American/British	Various	

Source: The Statesman's Yearbook 2005 (New York: St. Martin's Press, 2004); John MacPherson, *Caribbean Lands*, 4th ed. (London: Longman Caribbean, 1980); and *CIA Worldfact Book*, www.cia.gov/cia/publications/factbook/geos/av.html.
[a]British Overseas Territory, considered part of the United Kingdom.
[b]Aruba is an autonomous part of the Kingdom of the Netherlands.
[c]French Administrative Region, considered part of the French nation.

Castro's progress to date in accomplishing his four objectives can be summarized as follows. Cuba's dependence on foreign nations initially shifted from reliance on the United States to subservience to the Soviet Union, whose estimated 6 billion dollars in annual aid compensated partially for the devastating effects of the U.S. economic embargo. The fall of communism and the breakup of the Soviet Union resulted in the loss of Soviet subsidies and a crippling of the Cuban economy, whose output had fallen by 40 percent by 1993. Facing the prospect of even further economic decline, Castro then announced, in a manner reminiscent of former Soviet president Mikhail Gorbachev's actions preceding the fall of Eastern European communism, that temporary compromises would have to be made with the capitalist world in order to save Cuban socialism. These concessions, which have guided Cuban economic policy during the "Special Period in the Time of Peace," as the post-Soviet era is now called, have included legalizing the holding of U.S. dollars by Cuban citizens, which serves to increase the country's foreign currency reserves; encouraging foreign investment, especially in the mining and tourist sectors; and authorizing certain forms of private business and land ownership. Castro also implemented a harsh austerity program, further reducing the already limited quantity of consumer goods available to the Cuban people. Ironically, then, the achievement of economic and political self-sufficiency has been accompanied by a significant deterioration of the living levels of the average Cuban worker. It remains to be seen whether the newfound self-reliance will be temporary or permanent.

Castro's second objective, that of reducing Cuba's dependence on sugar, has yet to be achieved, though nickel mining operations and biomedical-related exports do provide secondary sources of foreign earnings. Castro's third goal, eradicating private property, had largely been accomplished by the early 1990s, but has been compromised during the ongoing "Special Period." Cuba presently has one of the highest levels of collectivized land in the world, but private farms are now permitted. Castro has also succeeded in removing much of the old Hispanic prejudice toward farming and farmers. This goal has been obtained, in part, by allowing a large portion of the country's urban, white-

■ **FIGURE 8-22 Severe hurricane damage.** Hurricanes are frequent visitors to the Caribbean islands and often inflict widespread devastation. This is an aerial photograph of the La Ciénaga neighborhood of Santo Domingo, Dominican Republic, after Hurricane Georges struck in September of 1998.

collar professional class, which was centered in the capital city of Havana, to emigrate to the United States. A second tool used by Castro has been the channeling of most of Cuba's resources into rural development projects, including schools, electrification, and health-care clinics. In so doing, Cuba has considerably improved the social and economic conditions of its rural poor, whose continued sup-

port of Castro has been instrumental in his retaining control of the country.

Cuba thus stands today as a nation of deep ambiguities. Castro himself continues to rely on police-state tactics to quell internal opposition, yet receives considerable support from the rural masses. The nation is now theoretically free of foreign dependence, yet its overall economy and material living levels have seldom been worse. Finally, Cuba ranks among Latin America's leading countries in many social indicators, yet it is governed by an aging dictator whose pride seemingly is preventing him from charting a new course in a changing world.

Puerto Rico

For many years after becoming a dependency of the United States, Puerto Rico progressed little. Unlike Cuba, with its large area of level land and excellent soils for sugarcane, Puerto Rico received U.S. corporate attention only along the ribbon of level land that fringes the island. Interior Puerto Rico is hilly and mountainous.

Following World War II, however, Puerto Rico began a period of continuous economic growth through a program called **Operation Bootstrap**, a three-pronged development plan. The first prong was industrialization. Like other Caribbean islands, Puerto Rico had only a small number of raw materials other than those of agriculture, only a small number of power sources, a small market, and an inexpensive but unskilled labor supply. Nonetheless, the Puerto

■ **FIGURE 8-23 Havana billboard.** Pro-government images continue to dot the Cuban landscape in the Castro era, which has survived for almost half a century. This sign declares that the success of the revolution depends on the degree of participation of women.

Rican government appealed to U.S. industry with attractive tax exemptions, training programs for labor, and a strenuous advertising effort. By 1956 the industrialization movement was so successful that manufacturing produced more revenue than agriculture. Much of that manufacturing is centered in the capital city of San Juan.

Agricultural improvement was the second prong of the program. Experimental stations were established that used the help of soil conservation agents to introduce better land-management practices. Dairying, truck gardening, and poultry farms—new kinds of land use—competed with the traditional crops of coffee in the highlands and sugarcane in the lowlands. In more recent years, rural development has focused on improving housing and bringing water, electricity, and schooling to farm families.

The third prong of the program was expansion of tourism. Again the government played a major role, by sponsoring hotel construction and extensive promotions (Figure 8–24). In addition, tourism benefited greatly from Castro's takeover of Cuba. In 1953, 118,000 persons visited the island; by 1963 tourists numbered almost 500,000; and recently that figure has exceeded 3 million. Much of the tourist trade has come from the United States.

Puerto Rico presently holds the status of a U.S. commonwealth. The people pay no U.S. federal income tax and do not vote in federal presidential elections but are otherwise entitled to all federal programs. The population is fairly evenly divided between those who wish to remain a commonwealth and those who favor statehood.

Hispaniola: Haiti and the Dominican Republic

Haiti occupies the western third of the island of Hispaniola; the Dominican Republic, the eastern two-thirds (see Figure 8–1). Even though the two nations have the same physical environment, the contrasts between them are marked.

The dominant culture of Haiti is a blend of African and French elements, and its racial composition is almost entirely black. It is a country of extreme poverty—statistically the poorest nation of the western hemisphere. Small subsistence farm plots worked with hoes and machetes still support about two-thirds of the population. Decay is seen everywhere: from potholed roads and erratic water and power supplies in the capital city of Port-au-Prince, to widespread deforestation and severe soil erosion in the rural countryside. Criminal and lawless behavior are frequent.

After being run by the Duvalier family as an almost feudal kingdom for some thirty years, Haitian politics have been dominated over the past two decades by a leftist Catholic priest, Jean-Bertrand Aristide, who, since 1990, has twice been elected and deposed as president. Haiti is presently struggling with the need for massive social, political, and economic reforms.

In contrast, the Dominican Republic has a Hispanic flavor, with areas of subsistence agriculture as well as large ranches and mechanized farms. Road transportation is good, and the infrastructure is improving. Tourism is booming, especially along the north coast and in the capital city of Santo Domingo. Numerous U.S.-owned apparel factories have also been established in the country in recent years. Although the Dominican Republic remains a poor country, it has achieved one of the highest rates of economic growth in Latin America during the past three decades.

Jamaica

Jamaica consists of a high, rugged core surrounded by a limestone plateau that grades into some of the most beautiful and popular beaches in the Caribbean. Bauxite, or aluminum ore, is abundant in lowland basins and is the nation's leading export (see Figure 8–7). Industry is limited by an almost total lack of domestic energy sources.

Jamaica today is largely British in culture, having been captured from Spain in 1655. It is also an island of contradictions. Neat and orderly in many aspects, with English town clocks and efficient scheduling, the island is also experiencing severe social turmoil. Fostered by a colonial prohibition against slave women

■ **FIGURE 8-24 Tourist hotels in Puerto Rico.** Tourism is a major industry in the West Indies, where beautiful beaches, warm temperatures, first-class accommodations, and proximity to the United States result in millions of visitors annually, mostly during the drier months of December through April.

marrying, which resulted in many slave "families" without an adult male in residence, a matriarchal society now dominates the island's overwhelmingly black population. Unemployment is chronically high, and crime is increasing, from the shantytowns of Kingston to the Rastafarian marijuana vendors in the northern beach resorts. Although mining and tourism are welcome and needed revenue sources, long-term development will come to the island only as Jamaican society is strengthened.

Belize

Considered a Caribbean rimland nation primarily because of its ethnic makeup, Belize gained its independence from Great Britain in 1981 and was recognized as independent by Guatemala in 1991 (see Figure 8–1). Its population is less than 300,000, and much of its 8,866 square miles (22,963 square kilometers) is swampy and little used. For many years, the nation's principal exports were tropical hardwoods and other forest products. Within the past 30 years, however, sugar and citrus have also become significant exports. The English influence on the predominantly black population is evidenced in its language and many Protestant faiths (Figure 8–25). The city of Belize is the principal urban center, although a new capital city, Belmopan, has been built inland, where hurricane damage is likely to be less severe.

Guyana, Suriname, and French Guiana

These three little-known, culturally Caribbean nations occupy the northeastern coast of South America (see Figure 8–1). The region was settled intermittently during the sixteenth and seventeenth centuries by Spanish, Portuguese, Dutch, English, and French colonists who sought to establish Caribbean-like plantations along the banks of the main rivers, all of which flow northward from their forested headwaters in the Guiana Highlands. Soil exhaustion caused subsequent settlements to diffuse eastward and westward along the marshy coasts. Eventually, the areas that today are Guyana and French Guiana came to be governed by Britain and France, leaving Suriname a Dutch possession.

The cultivation of sugarcane on the coastal estates of Guyana and Suriname led to the importation of large numbers of slaves. When the slaves were freed in the nineteenth century, thousands of workers from the Indian subcontinent were imported as indentured laborers. After 3 to 7 years of farm service, the so-called **East Indians** were considered free men. Many of them have remained in the rural areas as independent farmers. Most blacks, collectively referred to as **Creoles**, have sought residence in the urban centers.

Collectively, the three nations, for all their physical size, have far fewer people than either Jamaica or Puerto Rico, and the population is clustered along the coast. Bauxite and diamonds are mined in the interior highlands. French Guiana, which the French consider to be an integral part of their nation, serves as a principal European spaceport. After the French withdrawal from Indochina, modest numbers of Hmong refugees from Laos were resettled in French Guiana, where many have become prosperous farmers. The infamous penal colony at Devil's Island was closed in 1945. Both Guyana and Suriname are experiencing much social turmoil, as East Indians and blacks vie for power.

▶▶ Summary

Middle America is a region that is composed of one large country, Mexico, and numerous small ones that have been searching for national identity and environmentally balanced economic development. The quest for national identity centered first on gaining political independence from European colonial masters. It has focused subsequently on the need to maintain friendly, yet independent, relationships with the economically and politically powerful United States, and on the challenge of unifying diverse subregions and peoples.

It is ironic that the struggle for national identity has also contributed indirectly to the economic underdevelopment of Middle America. It is unlikely that a highly fragmented area such as Middle America, composed as it is primarily of ministates with populations far smaller than that of the city of Los Angeles, can be economically competitive without some form of regional cooperation. The extent to which political stability and social equity are achieved in the years ahead will influence to a large degree the economic and social development of the Middle American peoples.

▶▶ Key Terms

Border Cities
bracero
Creoles
East Indians
ejido
fincas
Greater Antilles
Gulf Coast
Institutional Revolutionary Party (PRI)

Ladino
Lesser Antilles
maquiladoras
Middle America
Operation Bootstrap
Valley of Mexico
Yucatán Peninsula

■ **FIGURE 8-25 Anglican Episcopal church in Belize City.** English influence in Belize is manifest in its numerous Protestant congregations and widespread use of the English language.

■ **Hillside *favela* housing and high-rise office buildings, Rio de Janeiro, Brazil.** Rural poverty and urban opportunity have contributed to massive population shifts in Latin America. Many of the poorest migrants end up living in makeshift shantytowns, called *favelas* in Brazil, often in sight of luxury apartment and business complexes.

Chapter 9

South America:
Regions of Uneven Development

- **Andean South America: Lands of Physical and Social Fragmentation**
- **Southern South America: Lands of Economic Growth**

- **Brazil: Latin America's Potential Economic Giant**

T hree South American regions (together with the three regions of Middle America) complete the broad realm of Latin America: Andean South America, southern South America, and Brazil, a single nation with an area larger than any of the other five regions.

Andean South America: Lands of Physical and Social Fragmentation

Andean South America includes Venezuela, Colombia, Ecuador, Peru, and Bolivia (Figure 9–1). Each is dominated by a high mountainous core, which gives way on the eastward side to the humid lowlands of the Amazon and Orinoco river basins. With the exception of Bolivia, each of those countries also has a narrow coastal lowland, which facilitates trade with the outside world. Although the Andean nations have been independent since the early nineteenth century, each continues to struggle to overcome internal physical fragmentation and isolation, colonial social relationships, and regional inequalities.

Venezuela

Venezuela has a rich and diverse resource base, consisting of four regions (Figure 9–2). The first is the **Andean Highlands**, which curve eastward from Colombia to run parallel to the Caribbean coastline. Within the mountains are numerous large basins endowed with temperate climates and fertile soils. Finding little gold and few Indians, the Spanish colonists settled in those basins. Today, they house much

of the Venezuelan national population in the urban centers of Caracas (the capital), Maracay, Valencia, Barquisimeto, Mérida, and San Cristóbal.

The northwesternmost part of the Andes divides into two ranges, which almost totally enclose the broad **Maracaibo Lowlands**. Those lowlands are dominated by the brackish Lago de (Lake) Maracaibo, which is 75 by 130 miles (121 by 209 kilometers) in size and opens into the Caribbean Sea. Lake Maracaibo, having one of the hottest and most humid climates in Latin America, was inhabited by Indians who lived in stilt houses above the shallow water at the time of the Spanish conquest. Those houses inspired the name Venezuela, or "Little Venice." Although largely ignored by the Spaniards in colonial times, the Maracaibo Lowlands have prospered since the early twentieth century from some of the largest petroleum deposits on earth, and the city of Maracaibo ranks today as the nation's second largest (Figure 9–3, page 220).

In the southeast, a second mountainous region, called the **Guiana Highlands**, is under development for its vast iron ore and bauxite reserves. Between the Andes and the Guiana Highlands lies Venezuela's fourth region, the great plain of the Orinoco River known as the *Llanos*. The *Llanos* have one of the most colorful and troubled histories of all Latin American regions. Tall, coarse grass dominates the lowlands, which are alternately inundated during the rainy season and baked to a rock-hard consistency during the dry winter months. Hordes of mosquitoes and other noxious insects heighten the human risk of diseases such as yellow fever and malaria (see the *Geography in Action* boxed feature, Health and Development in the Humid Lowlands of Latin America, page 222). Since the Spanish Conquest, the *Llanos* have been used primarily for the grazing of tough *criollo* cattle, whose hides,

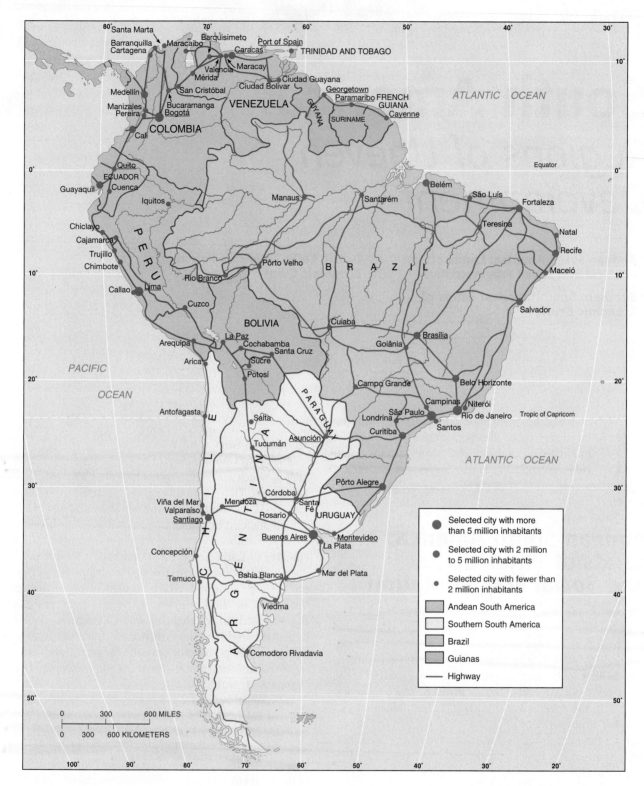

■ **FIGURE 9-1** **The nations, cities, and major highways of South America.** Settlement remains largely on the periphery of the continent.

tallow, and meat are destined for the highland market centers. More recently, extensive deposits of petroleum have been developed in the eastern *Llanos* (Figure 9–4, page 221). Together with that of the Maracaibo Lowlands, this production has helped Venezuela sustain its historic position as one of the world's leading petroleum-producing countries. In addition,

the *Llanos* produced a series of violent, illiterate dictators, who ruled the country off and on until 1935, when a tradition of democratic government was established.

Venezuelan Economy Until recently, Venezuela was widely viewed as one of the most stable and progressive nations in

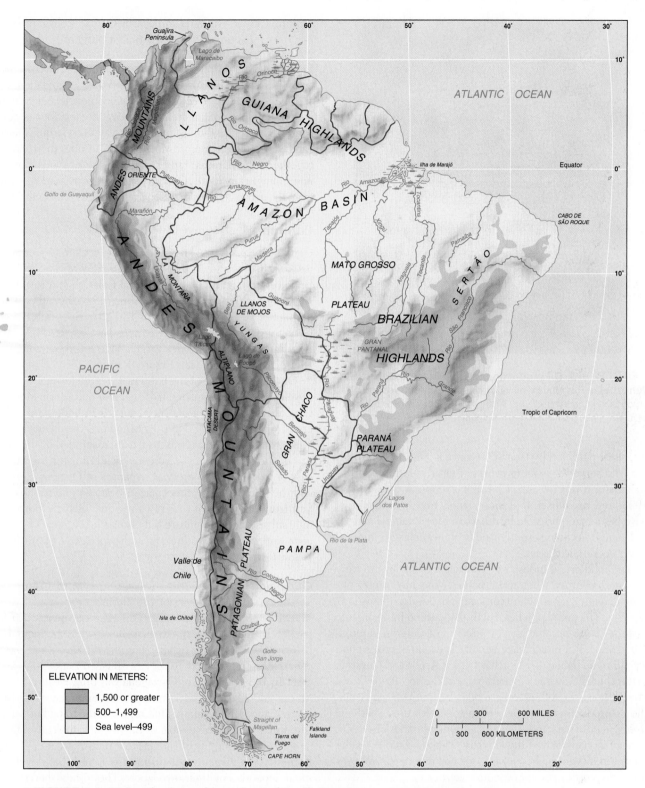

■ **FIGURE 9-2 Major physiographic regions of South America.** South America consists of many different physiographic regions, each of which possesses a distinctive environment with particular opportunities and challenges for development.

Latin America, owing largely to its open governments and to petroleum-derived revenues, which allowed the country's leaders to subsidize food, housing, and transportation costs and to invest great sums in schools and roads. This stability has been undermined in recent years, however, by declining oil revenues, which have forced government leaders to remove the subsidies that long had been taken for granted by the masses. Widespread corruption, at all levels of government, has also wasted much of the nation's wealth. Even the tradition of democratic government is now threatened by deep divisions between the long-established political parties that represent the interests of the traditional elites and a new

■ **FIGURE 9-3   Oil production in Lake Maracaibo, Venezuela.** This modern oil rig continues nearly 100 years of oil production in an extensive but shallow water body, where pollution problems pose a constant threat to habitat stability.

reformist party led by Hugo Chávez, a former military officer who served 2 years in prison after attempting two unsuccessful coups in 1992. Chávez's election as president in 1998, and his subsequent increasingly authoritarian rule, have been made possible by the worsening financial and social conditions being experienced by Venezuela's middle and lower economic classes.

The modern oil industry of Venezuela began in the early twentieth century, when dictator Juan Vincente Gómez encouraged foreign petroleum companies to develop the nation's reserves. The petroleum fields of the Maracaibo Basin and the eastern *Llanos* became major sources of oil entering international trade. For many years, Venezuela exported crude oil to refineries on the nearby Dutch islands of Aruba and Curaçao, to the United States and to northwestern Europe. Since World War II, however, refineries have been built in Venezuela, and the by-products provide the raw materials for tires, synthetic fibers, medicines, and a host of other industries. In 1976, Venezuela nationalized the industry. Nationalization was the final step in Venezuela's move to gain complete control over its single most important economic activity.

Iron ores and bauxite deposits along the northern fringe of the Guiana Highlands have also contributed substantially in recent decades to Venezuelan exports (see Figure 9-4). The ores are located in what was a little-developed part of Venezuela and required large capital investments and an extensive infrastructure. Originally developed by U.S. corporations, operations in the area received additional investments from the Venezuelan government, with the goal of creating a national center of heavy industry. Hydroelectric power and imported coal are now used to make iron, steel, and alu-

minum. Ciudad Guayana and nearby Ciudad Bolívar are the centers of that ambitious development program.

Urban Centers   Venezuela is one of the most urbanized nations in Latin America, with 87 percent of the population residing in cities. Of the country's 26 million people, 3.6 million live in the metropolitan area of Caracas; the next four largest centers are Maracaibo, Valencia, Maracay, and Barquisimeto, each with a million or more inhabitants. The latter three cities, together with Caracas, represent one of several megalopoli now emerging in South America (Figure 9–5, page 224). Away from the urban centers, traditional agriculture is characteristic.

Colombia

Colombia is presently one of Latin America's most troubled nations. Its deep social divisions and history of violence can be traced in part to intense physical isolation and resulting regional rivalries. Western Colombia, where the bulk of the population resides, is divided into two great river valleys, the Cauca and the Magdalena, by three parallel, north-south-trending arms of the Andes, which have severely restricted east-west movement (see Figure 9–2).

One consequence of the nation's fractured terrain was the development of a conservative regional culture in Medellín, Cali, and other cities of the Cauca River Valley and a liberal stronghold centered in Bogotá and the Magdalena River Valley. The historic divisions have been magnified by the failure of the government to effectively administer the lands east of the Andes, where vast expanses of Orinoco *Llanos* and the rain forests of the Amazon Basin extend to the distant frontier. Much of those two regions today harbors not only subsistence agriculture but also Latin America's largest coca and cocaine production and processing operations (Table 9–1, page 225).

Colombia's Past   The Spanish occupation of Colombia began with the founding of the Caribbean port of Santa Marta in 1525. Searching for gold, the ***conquistadores*** pushed up the Magdalena River Valley toward the highland basin of Cundinamarca, where they met significant numbers of Chibcha Indians. There they established the capital, Bogotá. In 1533 the port of Cartagena was built west of Santa Marta, and a second settlement cluster emerged along the lower Cauca River Valley. That region came to be known as Antioquia. Unlike the people of Bogotá, who took advantage of an abundant supply of Indians to establish great highland haciendas, the isolated ***Antioqueños*** found that development depended on their own labors. They established small, independent farms and businesses and produced a conservative, self-reliant middle-class society that, fed by an extremely high birthrate, pushed southward up the Cauca.

Colombia was dominated throughout the colonial era by an oligarchy of church, military, and civil officers. Independence did little to change those patterns, which persisted until 1936, when a new land law was enacted. The intent of the framers of the law was to assist the poverty-stricken small farmers, many of whom illegally worked unimproved lands belonging to *latifundistas*. The law declared that if the

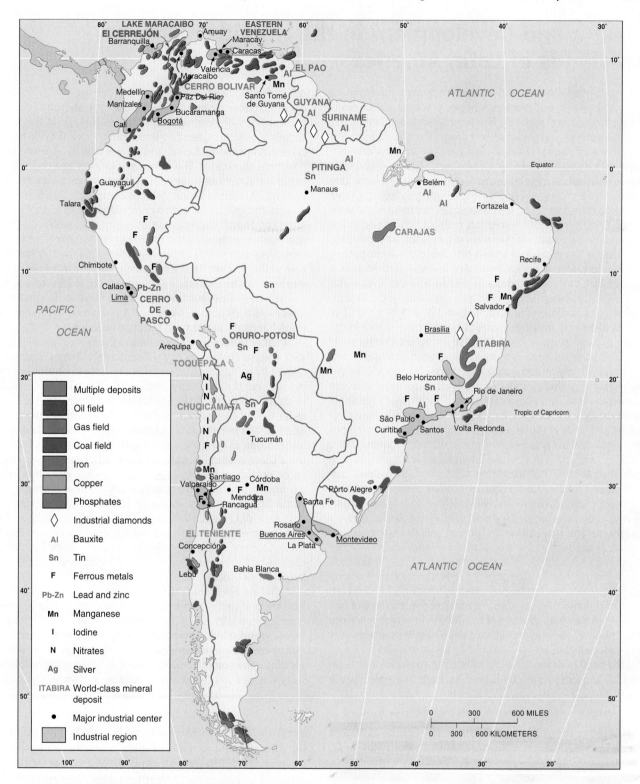

■ **FIGURE 9-4 Principal mineral-producing areas of South America.** Petroleum production in Venezuela, Brazil, and Ecuador is of world importance, as is iron ore production in Brazil and Venezuela, coal in Colombia, tin in Peru, Brazil, and Bolivia, silver in Peru, copper in Chile and Peru, and bauxite in Brazil and Venezuela.

colonos, or small squatter farmers, could show that they had brought certain lands into economic utilization through fencing, cultivation, or the pasturing of cattle, they were entitled to receive those lands as their own.

Although well intentioned, the land law had the unanticipated effect of provoking the complacent land barons into evict-

ing the *colonos* from their subsistence plots, in order to prevent the breakup of the barons' great estates. Deprived of their livelihood, the *colonos* emerged as a vast, homeless subclass, and Colombia lapsed into an era of almost indiscriminate violence. Outwardly, *La Violencia*, as it was called, which lasted into the 1960s, was a war between the **liberals** of Bogotá and

GEOGRAPHY IN ACTION
Health and Development in the Humid Lowlands of Latin America

An often-overlooked dimension of human development is health and disease. A population that is well nourished and relatively free from disease will live longer and be more physically active and productive than one that is not. While all natural environments have certain physical characteristics that make their occupants more or less vulnerable to various diseases, the humid lowlands of Latin America have long been home to a range of serious illnesses that, collectively, have hindered the settlement and economic development of these regions. Contrary to public perception, many of these diseases are continuing to spread, and their rates of incidence are increasing.

Three of the most well-known diseases currently impacting the humid lowlands are transmitted by the mosquito. Malaria is caused by a protozoan parasite or plasmodium that lives in the body of the *Anopheles* mosquito. Because the plasmodium requires summer temperatures of 59–60°F (15–16°C) or higher in order to develop, malaria is of little consequence in the *tierra fría* and *tierra helada* altitudinal life zones. On the other hand, the disease is so widespread in the *tierra caliente,* and its effects so deadly or debilitating, that it is widely acknowledged as having contributed more to the underdevelopment of the humid lowlands than any other physical factor. Many residents of these regions experience the intense pain and alternating fevers and chills of the disease over and over again throughout their lives. Malaria was formerly treated with Chloroquine, but a number of resistant strains have emerged in recent years and incidence rates are now increasing rapidly in most nations. Sulphur-based treatments are effective but prohibitively expensive.

Yellow fever, formerly known widely as the "black vomit," is a viral disease transmitted primarily by the *Aedes aegypti* mosquito. The most common form of yellow fever is urban-based, being passed from person to person through incidental contact. It was this form that resulted in the loss of thousands of lives in the ill-fated attempt of the French to build the Panama Canal. Yellow fever can be prevented by a vaccine, but the high cost of treatment has not only precluded eradication but has permitted the reintroduction of the disease in recent years into a number of urban areas that had been free of its effects over the past half century.

The third, and most rapidly increasing, mosquito-borne illness of the Latin American humid lowlands is a viral disease called dengue fever. The symptoms of the less-severe form include high fever, intense headache, diarrhea, and joint pain. The more serious hemorrhagic form causes leakage in the blood vessels of its victims, sending them into shock if they do not receive prompt medical care, and proves fatal in approximately 5 percent of all cases. The illness is known to have been present in the American tropics for at least two centuries but had been much reduced by the 1970s as a consequence of an *Aedes aegypti* mosquito eradication campaign organized by the Pan American Health Organization. However, the control program was subsequently discontinued owing both to concerns for the environmental effects of the insecticide DDT, and to the high cost of the initiative and its apparent success in lowering the incidence of the disease. This complacency led, in turn, to the reemergence of dengue fever as a major threat to human health in the 1980s (Figure A). More than 1 million cases are now reported annually in the Latin American nations.

Another serious illness endemic to the *tierra caliente* is Chagas' disease (American trypanosomiasis). Chagas' disease is caused by a protozoa that is carried by lice and a number of triatomid insects of the *Reduviidae* family. Wild and domestic animals, including rats, dogs, armadillos, and opossums, can serve as hosts. The disease is most commonly spread by insects that often nest in the air spaces of thatch roofs and the daub-and-wattle walls of dwellings of the rural poor. Once established, the disease passes through three distinct phases, which result in the enlargement of the heart, liver, spleen, and other internal organs, and increase susceptibility to other diseases. Because its victims often succumb first to other maladies, the incidence of Chagas' disease has been, until recently, widely underestimated. Some 16–18 million persons in Latin America are currently believed to be infected, of whom 50,000 die each year. Those too poor to have their homes treated regularly with insecticide are especially vulnerable.

Other life-threatening diseases common to the Latin American lowlands include river blindness, or onchocerciasis, which is associated with several species of the *Simulium* genus of black fly; and schistosomiasis, which is caused by suckerworms that utilize snails as intermediate hosts. The larvae of the worms penetrate the skin of persons bathing or working in shallow, infested waters and enter the circulatory system of the victim,

the **conservatives** of Antioquia, with its regional capital of Medellín. More fundamentally, however, it was a class conflict between the haves and the have-nots, and it resulted in the migration of great numbers of rural dwellers to the more secure cities. For example, Cali, in the heart of the violence-torn middle Cauca Valley, grew from 27,000 inhabitants in 1912 to 200,000 in 1950 and to approximately 2,300,000 in 2004. Now 71 percent of all Colombians live in cities.

Hope for domestic tranquility increased in 1957 when the two dominant political parties, liberal and conservative, agreed to divide national cabinet posts, governorships, and mayoral offices in proportion to the number of votes each obtained—an arrangement that was abandoned in 1974. The elections of 1990 marked another historic turning point on the difficult road to national unity when the leadership of the powerful leftist guerrilla April 19 Movement (M19) announced an end to its 16-year insurgency and campaigned for national office. Unfortunately, peace is still not at hand, as numerous other groups continue to terrorize the citizenry.

Perhaps the greatest threat to Colombian society in recent times has been organized crime syndicates. Armed with both wealth and weapons, the Medellín and Cali drug warlords have hired guerrilla bands and gangs of homeless street children as private armies to effectively control parts of the major cities and large expanses of the rural hinterlands. When opposed by elected national officials, they have been so bold

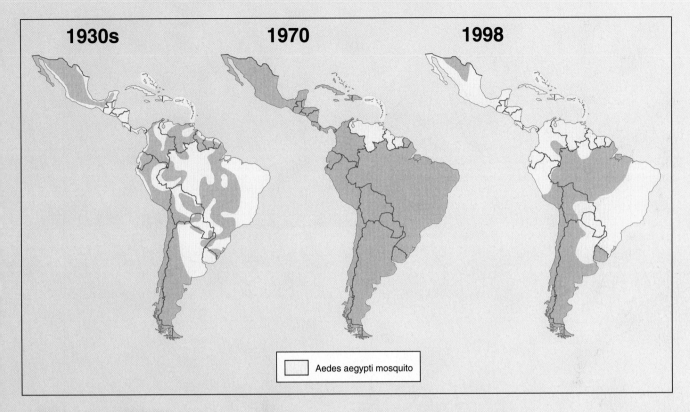

FIGURE A Geographic distribution of the *Aedes aegypti* mosquito in Latin America: 1930s, 1970, and 1998.

Aedes aegypti mosquito

eventually resulting in chronic weakness, diarrhea, anemia, and, in advanced stages, lung and liver failure. At least 14 million persons are currently believed to be infected, primarily in and around the major cities of northeastern Brazil, with secondary concentrations in north-central Venezuela and Suriname. Considerable progress has been achieved recently against the disease in the Lesser Antilles and Hispaniola. Success in limiting the disease ultimately will depend in large measure on the provision of clean potable water and sanitation systems.

While the presence and, in many cases, expansion of these life-threatening illnesses in the Latin American lowlands justifies cause for concern, it is important to recognize that other natural environments, including the highlands, are also associated with different widespread and serious threats to human health. It is increasingly evident that no one environment is more or less "healthy" than another. Rather it is poverty, with its associated inability to secure safe housing, clean drinking water, qualified medical care, adequate food, and host and vector control, that is most responsible for the high levels of disease common to lower socioeconomic classes. It is to be hoped that the human resources of the Latin American nations can be increased significantly in the coming decades through improvements in health and nutrition.

Sources: Adapted from David L. Clawson, *Latin America and the Caribbean: Lands and Peoples,* 4th ed. (Boston: McGraw-Hill, 2006), pp. 410–414; and *Centers for Disease Control and Prevention, www.cdc.gov/ncidod.*

as to kidnap, assassinate, or bribe the highest officers of the land, from supreme court justices to cabinet ministers and presidential candidates. Currently, leftist guerrilla and opposing rightist paramilitary groups are estimated to control almost half of Colombian territory, and the long-term survival of the nation's constitutional government is threatened.

Modern Economy Despite these challenges, Colombia has experienced some economic growth in the past decades. Bogotá has emerged as one of the leading cities of South America. Its cosmopolitan population of 7.6 million prides itself on its culture and produces a broad array of agricultural and industrial products. Antioquia, built on agriculture, continues to produce much of the world's high-grade coffee, providing about half of the nation's legal exports. Barranquilla has become Colombia's leading Caribbean port. Textiles, food processing, and banking sustain the economies of Manizales and Pereira as well as that of Medellín. Perhaps the most rapid growth in recent years has occurred in Cali, where a diversified economy is fueled by both Colombian and foreign interests. In addition, major coal deposits have been worked since the 1970s at El Cerrejón, in the dry Guajira Peninsula northwest of Venezuela's Maracaibo Lowlands, and petroleum production has greatly increased (Figure 9–6).

In recent years major transportation projects have linked formerly isolated areas. Cartagena has been connected by a

■ **FIGURE 9-5 Population distribution in South America.** The distribution of population in South America is oriented toward the fringes of the continent; almost 90 percent of the population is located within 200 miles of the coast. Densely settled areas include the highlands formerly occupied by pre-Columbian cultures, the northeast coast of Brazil, the Rio de Janeiro–São Paulo area, the Argentine *Pampa,* and the Central Valley of Chile.

Table 9-1 Leading Latin American Coca and Cocaine Producing Nations

Country	Hectares of Illicit Coca Cultivation	Coca Leaf Harvest (metric tons)	Cocaine Production (metric tons)
Colombia	86,000	168,000	440
Peru	44,200	50,790	155
Bolivia	23,600	17,100	60

Source: Compiled from the United Nations Office on Drugs and Crime; *World Drug Report 2004:* www.unodc.org/pdfWDR_2004/Chpt3_coca.pc.

waterway called El Dique to the Magdalena River and by a superhighway to Medellín and the Cauca Valley. In addition, a railroad now links Santa Marta to Honda on the middle Magdalena and indirectly to Bogotá beyond. Colombia's portion of the new Simón Bolívar Highway, which is an improved roadway intended to eventually link much of northern South America from Caracas to Quito, is also complete. Air transportation has been improved so that it now ranks among the best in Latin America. These advances will do much to accelerate Colombia's future economic growth if the nation can establish political and civil order.

Ecuador

Ecuador has long been beset by deep social and economic divisions. During the colonial period, it was largely neglected in favor of the wealth and splendor of Peru. The few Spaniards who settled there gained their livelihood from the sweat of Indian workers on large *haciendas* in the high basins of the Andes. The Spaniards themselves lived in Quito, the capital and former Inca center, and in smaller regional centers. For many of the Indians, European conquest had little impact beyond the introduction of new crops and animals. In the **Oriente**, an area east of the Andes, the fierce Amazon Basin Indians and the lack of gold and silver kept the area free from Spanish colonial administrative control. To

the west, the Pacific lowlands were largely neglected because of the prevalence of malaria and yellow fever in the humid north and severe aridity in the southern Guayas Basin.

Today, the Pacific lowlands are Ecuador's most dynamic and progressive region. Guayaquil has surpassed Quito as the nation's largest city and functions as the principal port and manufacturing center. Since 1940, a vigorous road-building program has opened large areas of formerly mosquito-infested lowlands to settlement and the commercial farming of rice and cacao. In addition, Ecuador is a leading exporter of bananas, which are produced by smallholder lowland farmers. Fishing in Pacific waters has contributed substantially to the nation's income as well. Furthermore, natural gas and shrimp farming are being developed near Guayaquil.

Unlike the Pacific lowlands, the Andean highlands of Ecuador have changed little since colonial times. Population is concentrated in the **Callejón Andino**, which consists of a series of intermontane basins and the surrounding hillsides. Quito, the urban center of the region, serves as the national capital, the home of the landed aristocracy, and a regional service center. Agriculture, largely for local consumption, is the major economic activity. The best bottomlands are in large farm units that specialize in dairying and maize. The remaining lands are divided into thousands of minuscule, often incredibly steep Indian subsistence plots (Figure 9–7). Wheat and barley are cultivated on dry lands, potatoes are grown at higher elevations, and sheep are pastured in alpine meadows. The Andes of Ecuador present a classic example of both **minifundios**, or small landholdings, and **latifundios**.

The *Oriente* is a sparsely settled rain forest inhabited largely by Amazon Indian groups, such as the Jívaro, who possess few modern techniques of production. In past times, the *Oriente* has received temporary influxes of people from the Andes in search of balsa wood, rubber, and cinchona (quinine used to battle malaria). Each time, however, the gathering of those wild products has lost out to commercial production elsewhere or to synthetic substitutes. Then, in

■ **FIGURE 9-6 Cerrejon open pit coal mine, Riohacha, Columbia.** The fifth largest coal mine in the world, much of El Cerrejón's output is exported overseas.

■ **FIGURE 9-7 Subsistence farms in the Andes.** Thousands of small subsistence farms dot the landscape of the Andean highlands of Ecuador. This scene is from the *tierra fría.*

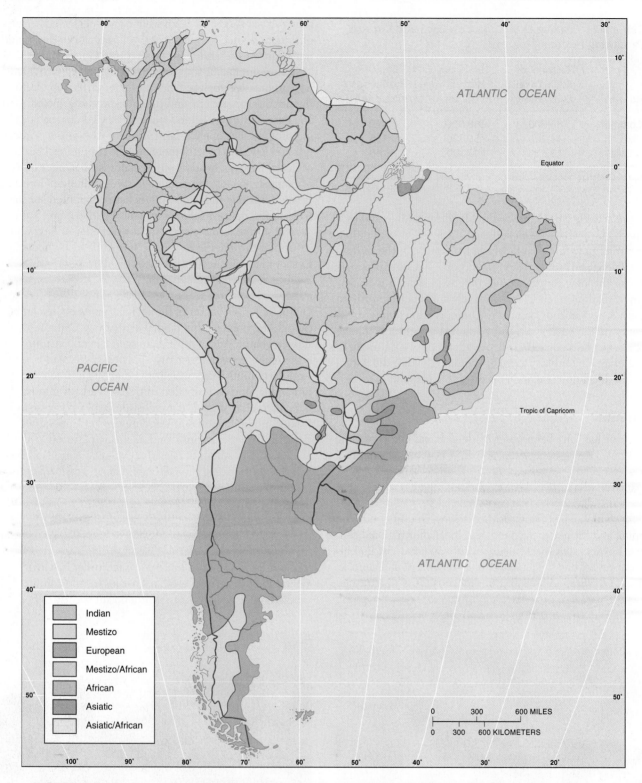

■ **FIGURE 9-8 Dominant ethnic groups of South America.** The most common group in South America is the *mestizos,* but Indians still predominate in the highlands of Ecuador, Peru, and Bolivia. Europeans dominate in Chile, Argentina, Uruguay, and southern Brazil. Africans dominate in many coastal lowland areas of Colombia, Venezuela, and Brazil.

1967 a major oil field was discovered in the *Oriente,* and Ecuador became a leading exporter of unrefined petroleum. Almost 400,000 barrels a day presently can be shipped by pipeline to a Pacific terminal near the port city of Esmeraldas, and a second pipeline currently under construction could increase Ecuador's oil transport capacity to 850,000

barrels per day. Ironically, however, Ecuador must still import some refined oil.

Despite the disparity in their areal distribution, Ecuador's 13 million inhabitants are about evenly divided between Andean Indian farmers and the Spaniards and *mestizos* who live in the cities and the Pacific lowlands (Figure 9–8). The two

■ **FIGURE 9-9 Ecuadorian textile market.** Highland Ecuador was a center of textile production long before the Spanish conquest and has maintained that role to the present. Shown here are wall hangings, blankets, and carpets for sale at the point where the Pan American highway crosses the equator. Also visible in the photo is a large stone globe that serves as a geographic symbol of the equator.

groups differ greatly, and communication between them is minimal. The Indians, some of whom are still illiterate, are oriented toward their local highland communities rather than toward the nation as a whole. Although some have begun to prosper through the manufacture and sale of high-quality textile products (Figure 9–9), most earn a meager existence from manual toil, and their material possessions are few. Their diet consists of tubers (such as potatoes, ulluco, and oca), guinea pigs, and highland grains, including wheat, barley, and quinoa (see the *Geography in Action* boxed feature, High-Altitude Adaptation in the South American Andes). In contrast, the Westernized urban dwellers are highly nationalistic, mostly literate and aspire to white-collar employment. The failure to integrate the European and Indian spheres has contributed greatly to Ecuador's enduring underdevelopment, as has its notoriously unstable political environment. In recent times, for instance, Ecuador has averaged almost one new president per year. Although bananas, petroleum, and fish products have provided Ecuador with valuable exports, the nation's long-term well-being hinges on its willingness to more fully develop its human resources.

Peru

Like Ecuador, Peru emerged from the Spanish conquest as a nation deeply divided between European and Indian values and economic production systems. The Spanish introduced the concept of privately held property; the Indians were accustomed to working communal lands. The Spanish established large-scale commercial agriculture; the Indians continued to practice semisubsistence farming. The Spanish economy was geared toward mining; the Indians were ori-

ented toward agriculture. Spanish loyalty was directed primarily toward the nation-state; Indians were loyal to their local villages. The Spaniards lived in cities and were Westernized in their culture; the Indians resided in the rural highlands and strove to maintain their traditional languages, dress, and diet.

Peru is also similar to Ecuador in physical geography. Each has three major areal units: the coast, the Andes, and the east. Unlike Ecuador, however, Peru has expended much effort in incorporating the Amazon lowlands into the national economy. Peru has also aggressively pursued a policy of national unity, attempting to integrate the Indian societies into a common national culture. Peru's economy is also more diversified than Ecuador's.

Coastal Peru   The Peruvian coast is a narrow ribbon of desert. Its extreme aridity and surprisingly cool temperatures are the result of prevailing westerly winds that blow inland over the cold, offshore Peru Current. As the air is warmed over the coastal plains, it is also dried. When it reaches the lower slopes of the westward-facing Andes, it is forced upward and cooled, producing a thick fog called the *garúa* that sustains grasses and lichens; at higher elevations it produces rain.

The cold, upwelling Pacific waters have also traditionally supported one of the world's richest fishing grounds, with great harvests of anchovy, sardine, mackerel, pilchard, and hake (Figure 9–10). Huge flocks of seabirds feed off the fish, and for years, the bird droppings, called **guano**, constituted Peru's most valuable export, nourishing some of the finest gardens of Anglo America and Europe. In recent years, Peru's coastal fishing industry has suffered from inconsistent management policies and periodic invasions of a warm equatorial current, called ***El Niño***. In addition to disrupting the fragile marine ecosystem, El Niño brings unaccustomed heat, humidity, and flooding to the desert lowlands.

Despite its struggles, the offshore fishing industry remains an important segment of the Peruvian economy, the most prosperous components of which are concentrated along the arid coastal plains. Coastal settlement focuses on the

■ **FIGURE 9-10 Peruvian anchovy harvest.** The cold Peru Current is one of the earth's most productive fishing grounds.

GEOGRAPHY IN ACTION
High-Altitude Adaptation in the South American Andes

The Andes Mountains of South America are the highest mountain chain of the western hemisphere. Appearing on the island of Tierra del Fuego as relatively low, snow-covered mountains rising up out of cold Antarctic waters, the Andes increase rapidly in height as they proceed northward through the glaciated valleys of Chile and neighboring Argentina. They reach an elevation of 23,035 feet (7,021 meters) on the summit of Mt. Aconcagua, a great volcanic peak situated a short distance north of Santiago, Chile. From that point northward, the Andes begin to broaden in width, with two, and sometimes three, roughly parallel ranges extending through Bolivia, Peru, Ecuador, Colombia, and Venezuela. Between these towering ranges lies a series of intermontane plateaus and basins, called *punas,* which contain many of the highest human-occupied regions on earth.

All forms of life found in the high Andes have adapted to the harsh physical environment. Much of the region lies above the treeline and is characterized by tundra-like bunch grasses and herbaceous plants that miraculously cling to life in the cold and stony soil. Native animal life is dominated by four members of the *Camelidae* family. In descending order of size, these include the llama (*Lama glama*), the alpaca (*Lama pacos*), the vicuña (*Lama vicugna*), and the guanaco (*Lama guanicoe*). Growing to 300 pounds (136 kilograms) and heights of 4.5 feet (1.4 meters), the surefooted llamas are marvelous transport animals, capable of carrying loads of 75–110 pounds (34–50 kilograms) for a 3-week trip averaging 10–15 miles (16–24 kilometers) per day over incredibly rugged mountainous terrain. If pushed too hard or forced to carry excessive weight, however, the temperamental animals will sit down on the ground, hiss and spit at their masters, and refuse to take another step. The smaller alpacas and vicuñas are raised primarily for their high-grade wool, and the few remaining wild guanaco are hunted for their meat (Figure A). Another indigenous meat source is the domesticated guinea pig (*Cavia aperea porcellus*), which evolved originally in central and southern Chile and has subsequently diffused throughout the Andes. The animals are raised by the millions, often being allowed to forage freely on the interior dirt floors and outside yards of human dwellings. Their meat is flavorful and fatty and is commonly used as a complement to potatoes and other highland tubers in soups and stews (see Figure 7–15).

Humans living in the high Andes may experience a number of physiological changes. Because atmospheric oxygen levels decrease with increasing altitude, visitors to the region often develop hypoxia, or oxygen deficiency illness. Called *soroche,* or mountain sickness, by the highlanders, its symptoms may include dizziness, shortness of breath, nausea, headaches, and fatigue. The condition usually passes after one or two weeks, but some individuals are never able to adjust and find themselves incapable of functioning normally at the high elevations. Permanent residents have been found to have larger-than-average chest cavities, lungs, and hearts; greater-than-average bone marrow and red blood cell counts; and elevated basal metabolisms. The Andean highlanders also experience higher-than-average incidences of respiratory illnesses, including tuberculosis, bronchitis, and pneumonia.

In noting these adaptations, we should not conclude that the Andean highlands are inherently less healthy and consequently less capable than the adjoining lowlands of economic development. Every physical and cultural environment has certain health risks. Residents of Latin America's humid lowlands may be exposed to a host of serious illnesses, including malaria, yellow fever, dengue fever, and Chagas' disease, maladies that rarely, if ever, occur in the highlands (see the *Geography in Action* boxed feature, Health and Development in the Humid Lowlands of Latin America). Good health is associated more with adequate income, which enables persons of all natural environments to secure adequate food, housing, and health care.

FIGURE A **Alpacas grazing on the southern Peruvian *Altiplano.*** Alpacas are raised in the Andes primarily for their wool, which has excellent insulation qualities.

margins of some forty rivers that rise in the Andes and cross the desert to the sea, and most of the nation's commercial crops are grown in the river oases, with sugarcane, cotton, bananas, and rice being the most important export crops. Those crops used to be grown on large estates, many of which were owned by foreign corporations. In the late 1960s, however, most of the estates were nationalized and turned into worker cooperatives. The oases near the capital city of Lima and the nearby port of Callao (with a joint population exceeding 8 million) are oriented to truck gardening and dairying for the urban market. The southern oases have a subsistence economy, but some grapes and olives are grown for the national market and assorted food crops for the regional center of Arequipa. Since 1950, the government has implemented many projects to provide greater amounts of water in order to expand the irrigated area.

Lima–Callao forms the nation's social, political, and economic focal point. While preserving many of its colonial

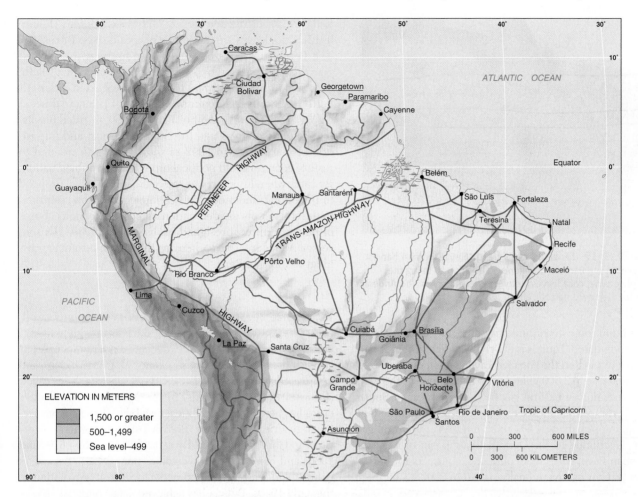

■ **FIGURE 9-11 Proposed Amazon Basin highway system and connections to other regions.**
An extensive highway system is being built as part of a major development effort in the Amazon Basin. The northern Andean nations have also proposed the construction of the Marginal Highway, which may, in time, connect to the Trans-Amazon Highway network.

traditions, Lima has become increasingly industrialized. Historic buildings on narrow streets contrast with modern architecture and broad avenues. Around the city's fringes and on formerly vacant land near the city center are squatter settlements, occupied by the urban poor and migrants from the rural highlands and small towns. The population of the Lima–Callao area is growing at about twice the national rate.

The Peruvian Interior   While the coastal area is the home of Europeans and *mestizos*, the Peruvian Andes are inhabited mainly by peasant Indians of **Quechua** and **Aymara** descent. The official languages of Peru are Spanish, which is spoken by 80 percent of the population, and Quechua, which is still the dominant tongue of 17 percent of the people. Only 3 percent speak Aymara.

Population in the Andes is dense, levels of technology are low, and many peasants migrate temporarily or permanently to the coast and, to a lesser degree, to the east in search of work. Land tenure in the highlands remains much the same as it was in colonial times. Large *haciendas* control the best lands, while poverty-ridden Indians continue to work tiny privately or communally owned hillside plots, much as they did in pre-Columbian times. The Peruvian government has attempted to break the aristocracy's hold on the land and the peasants, and some of the large estates have been broken up through land-reform programs and allotted to the smallholders who have worked them. Those actions are intended to further a larger policy of national unification by integrating the Indians into the national Spanish-*mestizo* culture.

The mountains of Peru are also noted for their extensive mineral deposits (see Figure 9–4). The most famous mining district is Cerro de Pasco, an old colonial silver center that has been redeveloped to produce copper, silver, gold, lead, zinc, and bismuth. In addition, coal is mined in several locations along the western flank of the Andes.

Peru has made repeated attempts to tame its vast eastern lands in the Amazon Basin. For more than 100 years the nation has sponsored colonization projects in the area, many in the region of transitional border valleys called the *montaña* (see Figure 9–2). Those attempts have greatly intensified in the past 50 years. Roads have been built across the Andes into the east, and along each road, highland peasants have come in search of new economic opportunities. Peru's proposed Marginal Highway is projected to extend along the eastern flank of the Andes (Figure 9–11). To date, however, many of the roads have failed to meet their

▪ **FIGURE 9-12   Terraced coca field.** Native women harvest coca leaves in the Peruvian *montaña*. In addition to being the source of cocaine, coca leaves are often chewed raw in the Andean nations for their mild narcotic effect.

development objectives and much of the east remains sparsely populated. Iquitos is the regional urban center and Peru's major port on the Amazon River, which empties into the Atlantic Ocean 2,300 miles (3,700 kilometers) to the east. The giant Camisea natural gas field, the largest in South America with estimated reserves of 8–11 trillion cubic feet of gas and approximately 500 million barrels of natural gas liquids, is situated in the Ucayali River Basin of eastern, Amazonian Peru. After years of delays, contracts for development of Camisea were finally signed in 2000.

Government Policies   In recent years the Peruvian government has been playing an active role in directing change, both economic and social. Many of Peru's policies mirror those that were instituted much earlier in Mexico: seizure of political power from the landed gentry, unification of society, emphasis on education, and agrarian reform.

Unfortunately, the 1980s and early 1990s brought mounting economic and social reversals to the proud nation that once was Spain's greatest colonial prize. During these years, Peru had one of Latin America's highest rates of inflation and heaviest foreign debt loads. Living levels declined dramatically. Much of the economic dislocation was attributable

to the expansion of **coca** farming within the *montaña* (Figure 9–12). Coca leaves are the source of cocaine (see Table 9–1). The Peruvian drug producers joined forces with a Maoist guerrilla group called *Sendero Luminoso,* or Shining Path, to form a formidable threat to internal peace and order. Alberto Fujimori, an agricultural engineer of Japanese descent who was elected president of Peru in 1990, justified his disbanding of Congress and the Supreme Court and imposition of martial law in 1992 as short-term evils needed to restore economic and political order to the troubled nation. By 1995, when Fujimori was reelected, inflation had dropped to 10 percent, and moderate economic growth had returned to the still deeply divided nation. Accused of serious abuses of power, Fujimori resigned in disgrace in 2000. His successor, Alejandro Toledo, became the first Amerindian president of modern Peru.

Bolivia

Bolivia, like Peru, has long been divided between a small, elite, European upper class, who have arrogantly referred to themselves as *gente decente* (decent or civilized people) or *gente de razón* (people capable of reasoning or thinking), and a large underclass of Indians who derisively have been called *naturales* (native life-forms) by the Spaniards. Yet another curse has been chronic political instability flowing from factional rivalries. As a result, Bolivia lost large sections of territory to Chile, Brazil, and Paraguay between the 1870s and 1930s and has long been viewed as one of Latin America's poorest and least-developed nations. Since its revolution of the early 1950s, Bolivia has achieved some progress in correcting the historic neglect of its human resources. However, it continues to be a deeply divided nation and its considerable development potential remains largely unfulfilled.

The Altiplano   The heart of Bolivia is a series of high mountain basins nestled between two branches of the central Andes. Centered on Lake Titicaca—which, at 12,507 feet (3,812 meters), is the world's highest large freshwater lake—the ***Altiplano***, or high plain, is a perpetually cold and arid region dominated by hardy, windswept grasses (Figure 9–13). There, rural Indian farmers in remote villages raise sheep and llamas and cultivate potatoes and other highland tubers

▪ **FIGURE 9-13   Lake Titicaca.** Shared by Bolivia and Peru, the world's highest large freshwater lake, Lake Titicaca, was believed by the ancient Inca to be the point of origin of humankind. Visible in this image are small agricultural plots (*minifundios*) and a coastal settlement.

■ **FIGURE 9-14 Andean tin mine, Bolivia.** High-altitude mining has long sustained the Bolivian economy. In recent years, the country has begun to develop its fossil fuel and agricultural resources, as well.

and grains. Indian dress is a distinctive blend of medieval European serf and pre-Hispanic American styles, with the men often wearing knee-length pants and the women multiple long, ankle-length skirts. Both sexes wear sandals, woolen ponchos, and hats to help fight the effects of the cold and the intense solar radiation. The dress, language, and customs of the rural Indians stand in stark contrast to those of the Westernized, urban elite living in La Paz, Bolivia's largest city and administrative seat of government.

Within the mountains that tower above the *Altiplano*, the Spaniards discovered some of the richest mineral deposits of the New World. At their zenith, the fabled mining centers of Potosí and Oruro produced more than half of the world's silver. Unfortunately for the country, in the early independence era, Bolivian mining interests came to be dominated by three aristocratic families whose members ran the country as an economic colony while they lived in lavish residences in Paris and other overseas locations. Little regard was given to the Aymara- and Quechua-speaking Indian masses, whose leaders became politically radical over time. As pressure for change mounted, the president of the republic announced in 1939 that the great mining companies, which by then were primarily producing tin, must invest part of their profits in Bolivia itself. Shortly thereafter, the president met a violent death. In the early 1950s, the mines were expropriated by the government, and a sweeping agrarian reform law was enacted.

Economic Development To the extent that Bolivia emerges from its legacy of social and political turmoil, the prospects for economic development appear bright. The highlands continue to be one of the greatest mining districts in the world. Tin, zinc, and lead mines at altitudes of 12,000 to 18,000 feet (3,658 to 5,486 meters) are the source of most of the minerals mined in Bolivia and account for almost half of the nation's exports by value (Figure 9–14). But extraction from

deep-shaft mines is difficult, and transportation is costly. Large natural gas and petroleum deposits are also under development in the eastern lowland region known as the ***Chaco***. Much of the nation's natural gas output since 1999 has been exported over a 2,000-mile pipeline to the great industrial regions of São Paulo and Porto Alegre in southern Brazil.

Bolivia is also a nation of great agricultural potential. In the southeastern lowlands around Santa Cruz, considerable recent development has made the nation self-sufficient in sugar, rice, and cotton. In addition, the northern lowland Amazonian region known as the ***Yungas*** is a frontier zone of active colonization and is the source of most of Bolivia's coca. For centuries, the highland Indians have chewed the leaves of the coca tree for their mild narcotic effects, but much of the crop is now processed into cocaine, which is by far Bolivia's most valuable agricultural commodity. As is the case with Peru and Colombia, Bolivia's future development will reflect not only social and technological advancement but also continued progress in overcoming the consequences of dependence on the export of narcotic drugs.

Southern South America: Lands of Economic Growth

Southern South America is composed of Chile, Argentina, Uruguay, and Paraguay, a region known collectively as the Southern Cone (see Figure 9–1). Chile, Argentina, and Uruguay share several features that separate them from many of their Latin American neighbors. Each is relatively prosperous, has a well-defined middle class, and has a high literacy rate. In each, the nation-state idea is well established, and all three populations are culturally unified. Measured by many standard indicators, such as levels of education, food supply, and life expectancy, Chile, Argentina, and Uruguay appear to be among the world's more-developed nations. In addition, each has developed a significant industrial base. Yet despite these attributes, personal income levels are lower than those of the Western European and Anglo American countries, owing partly to periods of extended political and economic turbulence.

Chile

Chile has achieved cultural unity in spite of its unique long, narrow shape. Extending 2,630 miles (4,232 kilometers) north to south but never more than 250 miles (402 kilometers) east to west, Chile is physically almost the inverse of the varied environments of the Pacific Coast of North America, from Baja California to southeastern Alaska. The heart of the nation is the great, alluvium-filled **Central Valley** situated between 30°S and 42°S latitude, where the climate is mediterranean, with hot, dry summers and cool, wet winters that are similar to those of central and southern California. Within that area, a cohesive society has formed (Figure 9–15).

To the north is the bone-dry **Atacama Desert**, where decades can pass without measurable rainfall. The Atacama

■ **FIGURE 9-16 Harvesting grapes in the Central Valley of Chile.** With a mild climate and seasons opposite to the Northern Hemisphere, central Chile is a world leader in producing quality commercial fruits and vegetables.

■ **FIGURE 9-15 Santiago, Chile.** Santiago dominates the Central Valley of Chile. In the background are the Andes Mountains, which rise up to more than 20,000 feet and are the source of large quantities of copper and meltwater for irrigation.

is sparsely populated but possesses nitrates, copper, and iron ore, which contribute significantly to the nation's mineral exports. Southern Chile has a marine west coast environment with abundant rainfall and cool temperatures year-round, much like that of coastal Oregon, Washington, and British Columbia. Occupied by the descendants of nineteenth-century German immigrants, the south is a fjorded, mountainous region endowed with large tracts of temperate rain forests that are coming under increasing development pressures from international timber and paper interests. In the far south, some coal, petroleum, and natural gas are found on and near the island of Tierra del Fuego, and sheep are raised for wool. Fishing is a growing industry all along Chile's extensive coastline.

Central Chile   Most of the population, industry, and commercial agriculture is located in the temperate and fertile lands of the Central Valley. The country's major cities—including the capital, Santiago, Viña del Mar, Valparaíso, and Concepción—all lie within this zone, which is also the agricultural heartland of the nation. In addition to the traditional grain and tuber crops, central Chile has emerged as a major world supplier of high-quality fresh fruits. Table grapes, apples, pears, peaches, plums, and citrus are produced in abundance and have found acceptance overseas, especially in Northern Hemisphere countries, where seasons are the opposite of those in Chile (Figure 9–16).

Political History   Chile's unified society is of recent vintage. In colonial times, the economy revolved around large, inefficient *haciendas,* whose labor supply consisted of poor tenant farmers called *inquilinos.* The War of the Pacific (1879–1884) resulted in the acquisition of the northern Atacama mining region from Bolivia and, together with German settlement in the south, led to a period of rapid industrialization. That shift was reinforced in the early twen-

tieth century, especially during World War I, when goods from the more industrialized nations became difficult to obtain and necessitated local manufacturing. With industrialization, political power passed to reform-minded urban-oriented parties and, by the mid-1900s, Chile had come to be viewed both at home and abroad as a model of democratic socialism patterned after the welfare state societies of Western Europe. The focus of the reform movement centered on greater control of foreign investment, especially the mining industries that provided most of the nation's exports, and land redistribution in the rural sector.

In 1970, Salvador Allende stunned the world by becoming Latin America's first freely elected Marxist head of state. Although supported only by a minority of the electorate, with barely more than one-third of the vote in a three-man race, Allende moved aggresively to reorganize Chile's economic and social structure, nationalizing the mining industry, accelerating the redistribution of land, and taking over most banks and communications media. A true socialist economy was in the offing.

The restructuring was not without problems and serious opposition, however. Inflation and unemployment soared, and homemakers, businessmen, and factory workers demonstrated in the streets. In 1973 the protests were brought to a head and the military revolted, ending 46 consecutive years of constitutional government, the longest among South American nations at that time. Allende was killed, and a military government, led by General Augusto Pinochet, was formed to rule the country. Pinochet's 16-year reign was marked by accusations of massive human rights violations and international isolation. It was also characterized by a return to a market-oriented economy. Pinochet's democratically elected successors have restored respect for civil liberties while maintaining the free-market policies that have sustained Chile's impressive economic growth. Chile is rapidly becoming one of the most prosperous and progressive nations of Latin America.

Argentina

Argentina has the second largest territory in Latin America and one of the richest agricultural bases in all the world. By Latin American standards, Argentina's annual per capita GNI of $10,190 is high, yet the country economically lags far behind the nations of Western Europe and Anglo America. It has failed to achieve its potential primarily because of misguided political and economic policies that have restricted the initiative and energies of its workforce throughout much of history.

Argentina's History   During the colonial period, Spanish interest in southern South America centered on the silver and gold mines of the high Andes in arid northwestern Argentina, bordering Bolivia. Agricultural communities—such as Salta, Tucumán, and Córdoba—were founded along rivers in the adjacent foothills and lowlands to provide mules and other work animals, hides, and food for the highland mining centers (see Figure 9–1). Buenos Aires and other settlements along the humid Atlantic coast languished from Spanish neglect and served primarily to block further Portuguese expansion from Brazil.

The development of Buenos Aires and its rich agricultural hinterland, the ***Pampa***, began in the eighteenth century (see Figure 9–2). Argentine cowboys, called *gauchos,* roamed the *Pampa* in search of hides and tallow, which they obtained from rangy, wild cattle that had evolved from European animals released on the grassy plains two centuries earlier. By the end of the colonial period, the wild herds were mostly exhausted, and *estancias,* or large cattle ranches, began to be established near the cities.

Argentina gained its independence in 1816, but it remained a sparsely settled nation until Great Britain became interested in the country as a potential source of wheat and beef and a market for British industrial products. The turning point in Argentina's economic development came around 1870, with the invention of refrigerated railroad cars and ships, which made possible the exportation of fresh Argentine beef to Europe. The few remaining Indians were either killed or driven southward into the dry, cold Patagonian Plateau, and the *estancias* were fenced, in preparation for British purebred cattle. Unwilling to work on the *estancias* themselves, the landed aristocracy imported between 1860 and 1930 more than 6 million workers from southern Europe, mostly from Italy and Spain, as indentured laborers.

Under the terms of the indenture agreements, passage to South America was provided by the landowners, who also contributed small plots on which the immigrants could have a family garden. In return, the workers were obligated to farm the estate lands for 3 to 7 years. Many of the natural grasslands, formerly used for the inefficient grazing of the scrawny *criollo* cattle, were then planted with wheat. But at the conclusion of the indenture period, much of the land was converted from wheat to improved pasture crops, such as alfalfa. In contrast to the native prairie grasses, those provided a tender pasture suitable for the imported European breeds of cattle.

Once released from indenture, the dispossessed farm workers, called *descamisados,* or shirtless ones, gravitated toward Buenos Aires, Rosario, and other urban centers in search of employment. They eventually came to form an urban underclass whose poverty made them highly vulnerable to the influence of charismatic political or military leaders.

The worldwide depression of the 1930s severely damaged the fragile Argentine economy, as Great Britain cut back on purchases of Argentine agricultural products. Matters were further aggravated by the fall of the constitutional government and the assumption of power by the military. In 1946 a young army colonel named Juan Perón took control in a coup. Eager to build a base of political support, Perón nationalized most of the major urban industries. Countless *descamisados* were given jobs in the new, inefficient, state-controlled factories, which were kept operating through government subsidies derived from heavy taxes on the agricultural sector. Without sufficient money for investment and improvement, the output of the previously productive *Pampa* dropped, and Argentina's economy nose-dived.

Perón was ousted in 1955, but many of his policies continued. Runaway inflation became chronic, reaching thousands of percent in some years. In addition, widespread nepotism in the large industrial firms restricted the opportunities for career advancement among the most capable of the younger generation and led to an out-migration of skilled workers, which further damaged Argentina's chances for recovery.

Frustrated by economic decline, Argentinians experienced political and social polarization. One military regime followed another in the midst of strikes, demonstrations, and urban violence. A nostalgic longing for the stability of the past led to Perón's return from exile in 1973, but he died the following year. He was succeeded by his widow, Isabel, who in turn, was deposed in 1976. Subsequent military and civilian rulers, eager to shift attention away from the nation's economic problems, engaged in a series of questionable activities. In 1982 they declared war on Great Britain over the Falkland (Malvinas) Islands, which resulted in a humiliating defeat for Argentina. Later, they announced a plan to transfer the capital from Buenos Aires to the small Patagonian port town of Viedma. In the meantime, the effects of the trade deficit were temporarily delayed by foreign borrowing, leaving Argentina with one of the highest per capita levels of indebtedness in the world.

The vicious cycle of inflation and declining levels of living was seemingly altered by the election in 1989 of a reform-minded president named Carlos Menem. Menem privatized thousands of inefficient state-owned industrial plants and encouraged foreign investment. He failed, however, to curb the country's rampant corruption or to impose fiscal discipline on the federal and provincial governments. Once the privatization revenues declined, the fiscal deficits grew to unsustainable levels and the nation experienced yet another political and fiscal crisis. In late 2001, Argentina went through three presidents in 15 days and shortly thereafter suspended payments on its international debt. By 2002, unemployment had risen to 21.5 percent, and half the population of the once proud nation was living below the poverty line. Economic and political conditions have improved somewhat in recent years, but the nation continues to feel the effects of its turbulent past.

Argentine Economy Argentina thus remains an enigma, a country that has accomplished much yet fallen far short of its potential. Its population is both literate (97 percent) and urban (89 percent). In addition, some 98 percent of the nation's 38 million people are European; only 2 percent are *mestizo* (see Figure 9–8).

Buenos Aires and the *Pampa* form the country's heartland, by far the nation's most important area. Greater Buenos Aires, with a population of 13.3 million, is one of the largest and most sophisticated cities in the world. A well-integrated road and railway system radiates outward from the city, and goods funnel through the city's port to the world market. From the fertile, subtropical *Pampa* come beef, maize, and wheat, which are largely exported overseas (Figure 9–17). Imports are mainly manufactured goods from Western Europe and the United States.

Most of Argentina's manufacturing and service activities are in Buenos Aires, where much of the nation's industrial labor force is also located. Most of that force, however, is employed in small, inefficient workshops that specialize in metal fabricating, petrochemicals and plastics, meat packing, milling, and textiles. About half of all those who hold jobs in the tertiary sector are also found in the capital. Thus, with approximately one-third of the population of the entire country residing in greater Buenos Aires, the city constitutes a classic example of a primate city, an urban area that totally dominates the economy, culture, and politics of the nation. Although it functions as the center of development and change, Buenos Aires also experiences the greatest levels of congestion, pollution, and poverty.

Several other Argentine regions provide products that complement the economy of the *Pampa*. In the west, the old cities of Córdoba, Mendoza, and Tucumán have become important growth centers. The west, a dry region in the rain shadow of the Andes, has an agricultural base. The arid zones are devoted to raising cattle and the irrigated districts to sugarcane and grapes for the national market. In the Andean highlands of the extreme northwest are a few small mines of asbestos and several metals. In addition, some petroleum is produced from a giant geologic trough just east of the Andes.

To the north of the *Pampa* is a low, humid area between the Paraná and Uruguay rivers. Those two rivers are wide and subject to great variation in water flow, so transportation has been difficult. However, the area, called Entre Rios, is noted for its tea, maize, flax, cattle, and sheep, and the Río de la Plata watershed development program—in which Argentina, Paraguay, Uruguay, and Brazil participate—may make the Argentine "Mesopotamia" even more productive.

In southern Argentina is **Patagonia**, a sparsely populated, arid, windswept plateau (see Figure 9–2). Most of Patagonia is used for raising sheep, but the lower, well-watered valleys have irrigated alfalfa fields, cattle ranches, vineyards, and mid-latitude fruit orchards. In this century, some mineral resources have also been exploited. Small quantities of petroleum have been discovered in the south near Comodoro Rivadavia, and limited amounts of coal and iron ore are shipped to Buenos Aires to be processed into iron and steel.

Argentina's outlying regions are primarily product suppliers for the *Pampa* and its great cosmopolitan capital, Buenos Aires (Figure 9–18). The outlying regions support much of the population of the *Pampa,* and their development is reflected in the improved well-being of the nation's heartland.

Uruguay

Although Uruguay is one of the smallest nations in South America, it has historically enjoyed one of the highest levels of living on the continent. Much of its prosperity can be traced to the unity of Uruguayan society, which has been supported by the country's uniformly rich agricultural base. Geologically, Uruguay is a transition zone. In the north it contains a low-altitude extension of the Brazilian Paraná Plateau, and in the south, a continuation of the Argentine *Pampa* (see Figure 9–2). The central part of the nation is a region of low, rounded hills. All three of those low-lying landform regions are characterized by a humid subtropical climate and fertile soils. As a result, a higher proportion of land

■ **FIGURE 9-17 Cattle herding in Argentina.** Beef from cattle raised on the fertile *Pampa* of Argentina is exported to markets all over the world.

■ **FIGURE 9-18 A view of Buenos Aires showing the obelisk.** Buenos Aires is one of the world's great urban centers and the economic, cultural, and political heart of Argentina.

(90 percent) is used for agriculture in Uruguay than in any other Latin American nation.

The agricultural transformation of Uruguay began in the late 1800s, when high-grade Merino sheep were introduced to the traditional *estancias*. Animal grazing now occupies most of the nation's agricultural lands, with mutton, wool, and beef as the primary products. The remaining farmlands produce rice, sugarcane, wheat, maize, and fruits. Industry is concentrated in the capital, Montevideo, and focuses on processing the nation's agricultural produce such as meat packing, foodstuffs, leather, and textiles. Most of the factories are small, family-controlled operations.

■ **FIGURE 9-19 The Gran Chaco region of Paraguay.** This is a sparsely settled area with one of the harshest environments on the continent and scarce natural resources. The tropical wet and dry climate experiences summer flooding and winter droughts.

Uruguay's long tradition of two-party democratic government ended in 1973 when the military assumed control of the country following years of economic stagnation associated with national welfare economics. Civilian rule was restored in 1985.

Paraguay

Paraguay is a poor, landlocked nation in the heart of South America. The western two-thirds of the country, known as the **Gran Chaco**, is a sparsely settled, semiarid region of intermittent streams lined with quebracho trees, a source of tannin (Figure 9–19). The *Chaco* has one of the harshest environments on the continent. During the rainy season from November to April, the rivers overflow their braided channels and flood vast stretches of land. In May, the waterlogged soils begin to dry out, and by the end of the dry season they are often caked with thick layers of mineral salts left by the evaporating waters. Summer temperatures frequently reach 110°F (43°C), the highest in South America, adding to the discomfort of the scattered inhabitants, many of whom reside in small, dispersed agricultural communities. Although considerable oil exploration has taken place in the *Chaco*, no significant deposits of petroleum or other minerals have been discovered.

Eastern Paraguay is an extension of the fertile volcanic Paraná Plateau of southern Brazil (see Figure 9–2). Situated between the Paraguay and Paraná rivers, that humid region supports the bulk of the population, including the capital city of Asunción. The great majority of the people are *mestizos*, of mixed Spanish–Guaraní Indian stock (see Figure 9–8). The Guaraní were taught sedentary agriculture by Jesuit missionaries, who were very influential until they were expelled in 1767. After that and for most of its independence period, the country was ruled by somewhat moderate, benign dictators, largely content to perpetuate colonial socioeconomic structures. In 1989, however, the aged General Alfredo Stroessner, who had been elected president every 5 years since 1954, was deposed in a military coup led by General Andrés Rodríguez. Although Rodríguez and his successors have since governed as freely elected civilian officials, the nation's military leaders continue to exercise great power behind the scenes.

Much of the little economic development that has occurred in Paraguay has come from external sources. Paraguay is a partner with Brazil in the development of Itaipú, one of the largest hydroelectric dams in the world, which is situated on the Paraná River (Figure 9–20). Paraguay has also participated with Argentina in the development of a second major Paraná River hydroelectric project at Yacyretá. However, the bulk of the power generated from both projects is routed to Brazil and Argentina.

The limited industry that exists within Paraguay focuses on the processing of local agricultural products and textiles. Land is abundant, but most farmers do not own their own

GEOGRAPHY IN ACTION
Latin America's Street Children

Great numbers of Latin American youth, or **street children**, live all or part of their lives away from home. UNICEF, the United Nations Children's Fund, has estimated that there are approximately 100 million street children worldwide, of which about 50 million are found in the Latin American and Caribbean countries. As with any large group of persons, Latin American street children exhibit considerable variation in attributes and behavior. While the largest numbers are found in Brazil, Colombia, and Mexico, every country now has at least some *niños de la calle,* or *gamines,* as they are generally called. A few have been totally abandoned by their parents, but most maintain regular contact with their families. Some are engaged in productive employment and others are not.

Although conditions vary according to locale, an in-depth study by Lewis Aptekar[1] of the street children of Cali, Colombia, provides helpful insights into the circumstances of many Latin American *gamines.* Aptekar found that 95 percent of Cali's street children were males ranging in age from 6 to 16 years, with 43 percent being under 12 years of age. Most kept in regular contact with their families and were free to return home at night, although many rarely did so. Those that lived away from home for extended periods of time generally developed a close relationship with a benefactor and had places where they could go to obtain food and rest. The younger children tended to establish a close personal relationship with a person of the same sex. As they grew older, their social focus shifted first to gang and then to boy-girl associations. Petty thievery was commonly practiced, but drug abuse, prostitution, and homosexuality occurred infrequently. Although Aptekar noted that all of the young experience some trauma, he concluded that most are emotionally resilient, and about three-fourths eventually become law-abiding, contributing adult members of society.

Most researchers, including Aptekar, have come to believe that the phenomenon of Latin American street children is primarily a response to deepening poverty (Figure A). The greatest surge in numbers, for example, occurred during the "Lost Decade" of the 1980s. While virtually every Latin American country has well-intentioned legislation requiring children to attend school and discouraging them from leaving home, the scholars argue that such laws are ineffective because the parents of the youths feel compelled to encourage their children to contribute to their own maintenance. If these conclusions are correct, it may be that the wisest response to the problem would be to concentrate our efforts on assisting the children to secure appropriate jobs, or on devising forms of alternative education that enable the participants to earn a little money while receiving both literacy and vocational training. A successful example of the latter strategy is the *Proyecto Axé* initiative of Salvador, Brazil.

■ **FIGURE A A group of street children at Paulista Avenue, the financial center of São Paulo.** Latin America is estimated to have approximately half of the world's street children.

[1] Lewis Aptekar, *Street Children of Cali* (Durham, NC: Duke University Press, 1988).

---

land; the vast majority work subsistence crops of manioc (cassava) on the estates of the aristocracy or on otherwise unused public lands. Attempts to stimulate commercial agriculture have focused on imported Mennonite agricultural colonists around Filadelfia in the heart of the *Chaco,* as well as on Japanese, Korean, and Brazilian settlers and investment in the eastern section. Three of the leading sources of income are the undocumented smuggling of foreign goods to both Argentina and Brazil, money laundering, and drug trafficking.

Brazil: Latin America's Potential Economic Giant

Brazil is by far the largest of all Latin American nations, only modestly smaller than the United States. Indeed, it is the fifth largest country in the world in terms both of area and population. Brazil's large area is underlain by a diversity of geologic formations, some of which contain rich and extensive ore deposits. These include gold and diamonds as well

■ FIGURE 9-20 Itaipú Dam. Much of the electricity generated by the Itaipú Dam is transmitted to São Paulo, Brazil, the leading industrial region of South America.

as such important industrial minerals as iron ore, bauxite, and ferroalloys. Brazil is also blessed with several different climatic environments, within which a wide array of crops are grown (see Figure 7–8).

Brazil's physical resource base is as abundant as that of any nation on earth. But Brazil has yet to fulfill its great potential. While ranking among the world's leading nations in industrial and agricultural output, it also carries the greatest total debt burden of all Latin American countries and has been ravaged in recent years by some of the region's highest inflation rates. Corruption and crime are rampant at all levels of society, and the income gap between the rich and poor is one of the widest in the world. The often life-threatening plight of its street children, of whom there are a greater number in Brazil than in any other country on earth, is well known (see the *Geography in Action* boxed feature, Latin America's Street Children). Personal living levels of the middle and lower economic classes have stagnated rather than improved over recent decades.

Boom and Bust Cycles

One reason for Brazil's limited development thus far has been the exploitative nature of its economy. Since the days when Brazil was a Portuguese colony, emphasis has been placed on deriving maximum wealth in minimal time, without regard for building a stable long-term economic base. Several cycles of boom-and-bust economic activity have resulted.

Sugarcane along the northeast coast was the basis of Brazil's first economic boom. Planted on large plantations worked by African slaves, sugar yielded great profits during the sixteenth and seventeenth centuries. But the boom eventually collapsed in the face of stiff competition from other parts of Latin America and the Caribbean. Brazil had failed to apply the improved technology that was being developed elsewhere by Dutch, English, and French planters.

The second boom cycle occurred within the Brazilian Highlands north of Rio de Janeiro, where deposits of gold and diamonds were discovered. Exploitation of those minerals led to partial settlement of the interior, and the discoveries encouraged other colonists to come to Brazil. Unfortunately, after the many surface ores were removed, there was an exodus from the area.

The third cycle was an Amazonian rubber boom, centered on the inland port city of Manaus (see Figure 9–1). Trappers extracted the latex, smoked it into balls, and sold the balls to traders for shipment to the United States and Europe. Rubber first gained importance with the development of the vulcanization process, but the demand increased greatly in the latter part of the nineteenth century with the development of the pneumatic bicycle tire and, shortly afterward, the automobile tire. Rubber prices skyrocketed. Such great fortunes were made that many of the wealthy residents of Manaus could afford to send their shirts and dresses out to Europe to be laundered, and a magnificent opera house was constructed of the finest Italian marble. But the boom ended as quickly as it had begun when some Brazilian rubber seeds were smuggled to England, providing the stock for the great rubber plantations of the Far East and West Africa. Once plantation rubber came on the market and petroleum-based substitutes for rubber were developed, world prices of natural rubber collapsed, and wild-rubber gathering in the Amazon dwindled to minimal levels (Figure 9–21).

The fourth cycle took place after 1850 with the search for areas that were suitable for coffee cultivation. Coffee production reached its zenith on the famous fertile volcanic *terra roxa* soils of São Paulo and southern Brazil. Until early in the twentieth century, coffee commanded such a high price that its cultivation expanded into cooler areas that were environmentally marginal for its growth. Then, when the inevitable fall in prices occurred, the peripheral lands were abandoned.

Nonetheless, coffee production did put Brazil on the map, and for many years Brazil exported more coffee than all other nations combined. By around 1910, however, world production exceeded demand, and prices dropped rapidly. The Brazilian government tried to protect its most important export by buying the harvest and holding back portions of it from the world market until an acceptable price was attained. But other Latin American nations, notably Colombia, undercut Brazil and captured a large part of the market. Since then, Brazil has occasionally repeated its attempt to control coffee prices but has lost markets to other countries each time. In recent years, other coffee-producing nations have joined Brazil in trying to limit production and control prices. But such endeavors, called valorization, can work only when producing nations cooperate fully, which rarely occurs.

The production of coffee in Brazil illustrates a method of land development that requires little money and is similar to the tenant-owner agreement used in the Argentine *Pampa*. When coffee production began its great expansion,

■ FIGURE 9-21 Wild rubber tapping in Acre state, within the Brazilian Amazon. The white-colored material visible in the two lowest bark cuts is slowly flowing sap, which is collected in the can secured to the side of the tree. The sap will eventually be heated to the point where only solid rubber remains. The rubber will then be formed into large balls that are transported downriver to market.

thousands of emigrants from Italy, Germany, and other parts of Europe entered Brazil. Most went to the southern half of the country as tenants on large *fazendas* (farms), the owners of which were rich in land but poor in money. Thus, owners and tenants made agreements that involved little or no cash. Tenants leased portions of the *fazendas* (often uncleared land) and planted coffee bushes provided by the owners. For a period of years, the tenants tended the young coffee plants and cultivated their own crops, both subsistence and commercial, between the rows of coffee bushes. After 5 to 7 years, the coffee plants began to bear heavily, and the tenants moved on to other plots. In that manner an entire *fazenda* was planted in coffee with little capital outlay. Once harvesting began, cash-wage employment was available to the tenants and their families.

The Regions of Brazil

Today Brazil is a nation of striking contrasts. The western half of the country is sparsely populated, but the east is densely settled. In addition, Brazil is both traditional and modern. Traditional Brazil survives in parts of the agricultural sector, with landowners of European extraction controlling workers of African, European, and Indian stock. The

workers are poor and often illiterate; they cultivate their small farms or work on large *fazendas* using hand tools. Modern Brazil is urban and industrialized. Living levels are relatively high, literacy is nearly universal, and machines have replaced muscles. In the south, modern Brazil also includes mechanized farming areas.

Brazil is divided into six regions, all of which differ significantly in both population and economic activity (Figure 9–22). Our discussion orders those regions according to a decreasing level of technological development.

São Paulo The state of São Paulo is Brazil's most modern and productive region. Per capita income is far above the national average, and approximately one-third of the nation's economic output originates there. São Paulo state alone accounts for almost two-thirds of the country's industrial output, producing practically all of Brazil's motor vehicles and leading the nation in the manufacture of textiles, cement, shoes, paper products, processed coffee, pharmaceuticals, and electrical goods. Petroleum refining, petrochemicals, and steel production are important as well.

The state of São Paulo also leads the nation in agricultural production. Coffee, soybeans, beef, sugarcane, cotton, peanuts, truck crops, and rice are grown under scientific conditions on large, mechanized farms. Much of the sugarcane is processed into ethyl alcohol (ethanol) for use as an alternative motor vehicle fuel.

Urbanization has proceeded rapidly in this region. The city of São Paulo has grown from a backland trade center to one of the world's largest urban areas. Growth has been especially rapid in the last century, with the city's population

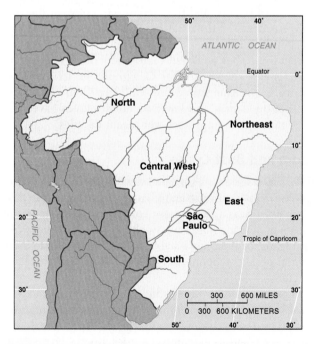

■ FIGURE 9-22 Brazil's six regions. Brazil's coastal regions are largely settled, but the interior regions (Central West and North) are part of Brazil's frontier. Ambitious projects are under way to develop those interior regions.

■ **FIGURE 9-23 São Paulo, Brazil.** This view of the central portion of São Paulo illustrates the degree of urban development that has occurred in some parts of Brazil.

increasing from 600,000 in 1935 to more than 8 million in 1970. The metropolitan area now contains approximately 19 million residents, making it one of the largest cities in the world. It generates 55 percent of the entire nation's manufacturing, with more than 30,000 factories. It is also Brazil's leading financial center, with the largest number of banks and the biggest stock exchange (Figure 9–23).

São Paulo's rapid economic growth has not come without problems, however. Many migrants from the countryside and from other regions have been attracted to the city. Most, illiterate and unskilled, live in great poverty. Many lack access to the city's water supply and sewerage network, and housing, public health, and schools are generally inadequate for them. The city has become so large and its streets so choked with traffic that residents often spend 2 to 3 hours each way commuting to and from work. Air pollution is so severe that the World Health Organization considers breathing the city's air to be dangerous to human health. Water pollution is also severe. Other cities in the region, including the port of Santos, are experiencing similar problems on a smaller scale.

The South Southern Brazil has its own distinctive flavor. In the nineteenth century, large numbers of German and Italian farmers settled in the area, and many have retained their ethnic identity into the twenty-first century (see Figure 9–8). German and Italian architecture is still noticeable.

Agriculture is the basis of livelihood in the area, but several large cities are situated on or near the coast, chief among them Pôrto Alegre and Curitiba. Most of the region is well settled, but the western section is still part of a pioneer fringe. Transport facilities have been greatly expanded and have contributed much to the South's prosperity.

Exploitative agriculture, which focuses on maximum immediate return, is less prevalent in the South, which is char-

acterized by cattle ranching and mechanized wheat and soybean farming. Coffee farming occurs in the northernmost parts of the region but is prevented from reaching farther south by freezing winter temperatures. Rice, maize, and hogs are raised in the traditional manner. Hogs are associated especially with German settlements, whose products include sausage. Many Italian communities are noted for their grapes and wines.

Urban areas in the South are basically regional service centers, but in the larger cities, such as Pôrto Alegre, industrialization based on the processing of farm and ranch products provides employment for a sizable portion of the labor force. Milling, meat packing, tanning, textiles, and breweries are typical manufacturing activities (Figure 9–24).

■ **FIGURE 9-24 Meat processing plant, Pôrto Alegre.** Southern Brazil is a productive cattle ranching region, with much of the beef being processed under modern, sanitary conditions in Pôrto Alegre.

The South's mineral resources are limited. Small deposits of low-grade coal are mined to provide energy in the area and to be shipped to São Paulo and the eastern region of Brazil.

The East Modern Brazil meets traditional Brazil in the East, where large, modern urban centers stand in stark contrast to nearby rural areas that have changed little in the past century. The East is basically the hinterland of Rio de Janeiro, the former capital and the second city of the nation with more than 11 million inhabitants. The region has experienced several waves of exploitation—first sugarcane, then gold and diamonds, followed by coffee, rice, and, most recently, citrus fruits. After each boom has come a time of population decrease and reversion to a grazing economy.

Rio de Janeiro is the focal point of the East (Figure 9–25). Nestled around one of the most beautiful and easily defensible harbors of Brazil's Atlantic coast, Rio grew rapidly throughout the colonial era owing to its proximity to the inland gold and diamond fields of Minas Gerais. When Brazil's capital was moved from the Northeast to Rio de Janeiro, the city's functions increased. As long as Brazil was a coastal nation, Rio continued to be the national focal point. Access to the interior was difficult, however, because a steep escarpment rises just behind the narrow coastal plain. The advent of motor vehicles and paved roads has partly offset that natural disadvantage, but São Paulo and other cities located inland have taken some of Rio's former trade area. Rio itself faces a severe shortage of level land for urban development.

The hinterland of Rio de Janeiro has both an agricultural and a mining base. Most of the land is used for grazing. Although some of the cropland is devoted to the cash crops of sugarcane, coffee, rice, and citrus, much of it is used for subsistence crops. Land rotation is common.

In addition, the East contains one of the most mineralized areas in the world, the **Mineral Triangle**. Gold, diamonds, and a number of precious and semiprecious gems have been mined in this region for centuries. Industrial minerals—such as manganese, chromium, molybdenum, nickel, and tungsten—are of growing significance, but iron ore is the most important mineral. The amount of iron ore reserves is unknown but is certainly one of the largest in the world.

Volta Redonda is Brazil's iron- and steel-making center. The site is almost ideal: it is close to iron ore deposits, limestone, water, and the markets of São Paulo and Rio de Janeiro (see Figure 9–4). Coal, the one vital ingredient that is lacking, is brought in from the South or is imported from the United States. Since the Volta Redonda plants opened in 1946, their production capacity has increased several times over. Today Brazil is the leading steel producer in South America and one of the largest in the world.

The Northeast The Northeast was once the center of Brazilian culture, the location of the capital, and the most developed part of the nation. Most-developed status was based on sugarcane cultivation under the plantation system, which used African slaves, and on the region's location as the part of Brazil closest to Portugal. That proximity was advantageous during the days of sailing ships but lost importance with the advent of steamships. In addition, although sugarcane is still grown in the region, that industry is no longer prosperous. Today much of the Northeast is poverty-stricken; per capita income is less than half the national figure, and feudalistic social structures continue to retard the development of human resources. Great numbers of people in this region have migrated to other parts of Brazil, adding to the number of unskilled laborers searching for work in the cities farther south and in the new capital of Brasília.

Nonetheless, levels of technology remain low in the Northeast, and livelihood is precarious even in the best of times. Over the years, settlement has moved inland, away from the moist coast into a drier environment called the *sertão* (Figure 9–26). In the backlands, hillsides are cultivated with minimum regard for conservation, and pastures are overgrazed. Both practices have resulted in soil destruction and rapid water runoff, causing the naturally semiarid environment to become even more moisture deficient. Prolonged, severe droughts are frequent yet unpredictable. When they occur, large numbers of people move from the interior to the already overcrowded coast and to other parts of the nation. Yet so strong is the tie to the Northeast that many of those persons return when they can.

The Northeast has three cities with more than 2 million people each: Recife, Salvador, and Fortaleza. All three are port cities that function as service centers. Manufacturing is limited.

Since about 1960, a number of government-financed projects have been directed to the Northeast. Dams for hydroelectric power have been built, irrigated agriculture is expanding, and roads are being improved and extended. In addition, industries that locate in the Northeast receive special tax rebates from the national government. But none of those efforts has altered the widespread poverty of the region.

■ **FIGURE 9-25 Rio de Janeiro.** Rio de Janeiro is the second largest city of Brazil and is known for its spectacular physical setting and distinctive *Carioca* culture.

■ **FIGURE 9-26 Cattle grazing in the backlands of Northeast Brazil.** Once the center of Brazilian culture and the most developed part of the nation, the Northeast today is poverty-stricken. Despite significant migration from the region, levels of technology remain low and livelihood is precarious.

The Central West The Central West region occupies a large portion of the **Brazilian Highlands**, a vast, sparsely settled interior upland region of rolling plains and low, flat-topped mountains. Brasília was founded in 1960 on the eastern edge of the Central West region, some 600 miles (960 kilometers) inland from Rio de Janeiro and São Paulo, to symbolize and encourage Brazil's occupation of its interior (Figure 9–27). In addition, the nearby city of Goiânia has developed into a second regional center. Roads have been built to connect Brasília with other parts of the nation and to lead westward to the new settlement frontier. Some commercial agriculture has evolved along the eastern fringe of the region, but

■ **FIGURE 9-27 Brasília, Brazil.** Brasília was founded in 1960 as Brazil's new planned capital. The distinctive architecture of its government buildings is recognized worldwide. The interior location chosen for the city is an expression of Brazil's commitment to settle its vast interior regions, many of which are drained by the Amazon River and its tributaries.

elsewhere subsistence agriculture, grazing, small-scale lumbering, and mining predominate.

The frontier, or pioneer zone, is gradually moving westward as new lands are opened for agriculture. Like the United States in times past, Brazilians believe their country's future lies in the west. Pioneering is at best a risky venture, but the farmers have discovered a relationship between vegetation and the quality of land for cropping. Six vegetative types are recognized. In descending order of quality for agriculture, they are first-class forest, second-class forest, mixed grassland and woodland, scrub woodland, grassland with scattered trees, and grassland.

Forested land is regarded as good and as suitable for cropping. On the other hand, grasslands and scrub woodlands are generally considered less suited to agriculture, though mechanized cropping has occasionally been attempted, usually with disappointing long-term results. Most often, however, those types of vegetation are used only for grazing. Use of mixed grasslands and woodlands holds the key to occupation of the Central West, for that vegetative type, called *campo cerrado*, covers about 75 percent of the region. Unfortunately, many of its soils are low in nutrients and are subject to drought. Extensive mechanized agriculture emphasizing wheat and soybeans has been somewhat successful in the eastern part; alfalfa, a deep-rooted pasture crop, offers another possibility. However, the main use of mixed grasslands and woodlands at this time is for grazing.

The North In recent decades, much effort has been exerted to settle the Amazon Basin. That vast region, however, with its enormous expanses of multilayered tropical rain forest and incredible diversity of plant and animal life, has repeatedly turned back attempts at large-scale commercial agriculture and forest development. Two of the most common mistakes of these failed projects have been their assumption that the soils of the region must be fertile since they support such a lush forest and their reliance on the cultivation of a single dominant crop. In reality, most rain-forest soils are leached, low in nutrients, highly acidic, and consequently ill-suited to the cultivation of many food crops. The great height of the rain forests is attributable to competition for sunlight, a year-round growing season, and the ability of the plants to live off their own fallen leaves and other decayed organic matter. Furthermore, monocultural cropping systems are so vulnerable to disease and insect infestations in the tropics that great quantities of fungicides, insecticides, and other chemicals are necessary to produce even a modest harvest. The costs of these Green Revolution inputs are often so great that the farmer cannot realize a reasonable profit and eventually abandons the enterprise altogether. Once this happens, the land generally reverts to unimproved pasture, which is used to support extensive cattle grazing where the magnificent rain forests once grew. Research is currently being undertaken to study polycultural alternatives to monocultural agriculture in the Amazon Basin. These can include an almost limitless variety of crop combinations, many of which incorporate **agroforestry**, the utilization of tree crops in agricultural cropping systems.

Although the basin covers two-fifths of Brazilian territory and has long been considered a region of great economic potential, it supports relatively few people, most of whom reside in communities situated on the natural levees of the river and its principal tributaries. Manaus, near the center of the basin, and Belém, near the mouth of the Amazon, are the major urban centers. Away from the rivers, an ever-dwindling number of Indians live and raise crops of manioc, maize, and beans, supplemented by hunting, fishing, and gathering.

The Amazon Basin remains one of the world's largest wilderness areas, but change is accelerating throughout the region. Commercial agriculture has been stimulated by immigrant Japanese farmers, who successfully cultivate commercial crops of rice, jute (a hard fiber), and black pepper in some of the more favored parts of the basin. Mining is also playing an increasingly prominent role in the development of the area. The largest mineral district in the world has been developed at **Carajás** (see Figure 9–4). In addition to an estimated 18 billion tons of hematite, or high-grade iron ore, an abundance of manganese, copper, bauxite, and nickel can be found there. Additional Amazonian mining sites include the Pitinga mine, which yields tin ore, and several world-class bauxite and titanium deposits. Other important mineral resources are likely to be discovered in the future. Taken together, the mineral deposits of the Amazon Basin and the Brazilian Highlands have already made Brazil one of the world's leading mining nations.

Recently, the Brazilian government initiated an industrial development program for the basin that focuses on Manaus. Attracted by tax exemptions and other fiscal incentives, more than 400 companies, including many leading foreign high-tech firms, have established industrial plants under the sponsorship of the Manaus Free Trade Zone. Development is also coming from tourism; hundreds of thousands of visitors come every year, many of them for jungle safaris and Amazon River cruises.

In 1970 work began on the **Trans-Amazon Highway** (see Figure 9–11), the start of a massive road system throughout the basin and the connection of that region to other parts of the nation (Figure 9–28). One of the principal purposes of the road system was to encourage underemployed migrants from the Northeast and elsewhere to settle the area. In addition to the building of the highway system itself, plans were drawn up for the construction of hydroelectric dams at locations where the major tributaries of the Amazon drop off the Brazilian and Guiana highlands onto the Amazon plain. Numerous regional development poles centering on either agriculture or mining were also designated, and several reservations were created for the exclusive use of native American groups.

Each of these initiatives has subsequently met with both success and failure. Although new highways continue to be built, many are now in poor states of repair owing both to weather-related deterioration and to the limited funds allocated to road maintenance. Not all of the planned hydroelectric dams have been built, but serious problems, including reservoir siltation, changes in the species compo-

■ **FIGURE 9-28** **A segment of the Trans-Amazon highway system.** Effective use and settlement of the Amazon Basin has long been a dream of many Brazilians. It is now being realized through costly and ambitious governmental development programs, many of which have contributed to the destruction of the tropical rain forest.

sition of fish and other aquatic organisms, and outbreaks of schistosomiasis, yellow fever, malaria, and other diseases associated with stagnant water bodies, have arisen among those that have been constructed. The great mining centers have been criticized for contaminating rivers, lakes, and the land itself and for contributing to the deforestation of the region. Yet another concern is the low level of housing and health care available to the mine workers and their families. The native American reservations have also proven controversial, with some observers blaming them for hindering the technological advancement of the resident Indians and others decrying the lack of enforcement of laws designed to keep out non-Indian interests. Global environmental issues associated with intensified occupation of the Amazon include the potential loss of countless rain forest plant and animal species and concern that the loss of much of the earth's remaining rain forest may contribute to planetary warming. Amazonia thus continues as it has for centuries to be a paradox, immense in size and rich in natural resources, yet not easily developed in a sustainable manner.

▶▶ Summary

South America is one of the most richly endowed regions on earth. Blessed with a wide array of physical environments, it has provided much of the world's greatest mineral production for centuries. It also possesses the largest remaining stand of tropical rain forest and some of the world's most productive agricultural regions. Favored with vast expanses of underutilized land, South America is likely to strengthen its position as one of the world's leading suppliers of raw materials in the decades ahead.

We must recognize, however, that the presence of raw materials, however abundant and valuable, does not constitute an assurance of economic development. The greatest resource of any region will always be its people. South America's steady development of its human resources—as expressed in improved health care and sanitation, housing, literacy and education, and political and social democratization—offers the most compelling evidence of the region's progress. Even though the pace of change has varied from place to place and even though much remains to be done, the progress of recent decades gives hope for the future development of the continent.

▶▶ Key Terms

agroforestry	Guiana Highlands
Altiplano	*latifundios*
Andean Highlands	*La Violencia*
Antioqueños	liberals
Atacama Desert	*Llanos*
Aymara	Maracaibo Lowlands
Brazilian Highlands	Mineral Triangle
Callejón Andino	*minifundios*
Carajás	*montaña*
Central Valley of Chile	*Oriente*
Chaco	*Pampa*
coca	Patagonia
conquistadores	Quechua
conservatives	*sertão*
El Niño	street children
fazendas	*terra roxa*
Gran Chaco	Trans-Amazon Highway
guano	*Yungas*

▶▶ Review Questions

1. Name the three great landform divisions of Latin America and list the principal natural regions of each division.
2. Summarize the general location and seasonal temperature and precipitation characteristics of the major climate regions of Latin America and the Caribbean.
3. Who were the Moors and what were some of their technological and cultural achievements in medieval Spain? How did these ultimately come to influence the cultural geography of Latin America?

4. Name the four great pre-Columbian Indian civilizations of Latin America. Where were they located and what were the principal accomplishments of each?
5. Define the following mixed races of Latin America: *mestizo, mulatto, zambo*. To what degree does culture and lifestyle influence the current definitions of race in modern Latin America?
6. List and describe the three forms of Roman Catholicism currently found in Latin America. How do these forms relate to the cultural races of the region?
7. What are shantytowns or squatter settlements? In Latin America, are they located in the inner-city or on the periphery of the great urban centers? How does the relative spatial distribution of economically prosperous and poor city neighborhoods in Latin America differ from that found in the cities of the United States and Canada?
8. List the regions of Middle America and summarize the physical and economic characteristics of each.
9. What is coca? What are its primary production and processing zones in South America? What have been the economic, social, and political consequences of cocaine production in Colombia?
10. Name the principal regional economic unions and trading groups of Latin America. How has geography influenced the composition of their memberships? What are their primary long-term challenges and potential benefits?

▶▶ Geographer's Craft

1. If you and a friend were planning a trip to Latin America and the Caribbean, what information would you need to know in order to bring the proper wardrobe?
2. Discuss the reasons why Latin America is characterized by almost innumerable physical microenvironments. How can this knowledge guide individuals and government agencies contemplating small-scale polyculture versus large-scale monoculture agricultural development strategies in the New World tropics?
3. Compare the social class structure established in Latin America by the Spanish *conquistadores* to that established by the early settlers of New England. How did these differing values lead in subsequent centuries to differing levels of human resource development?
4. Summarize the demographic history of Latin America over the past five centuries. In comparison with the other major geographic regions of the world, is Latin America more or less densely populated?
5. Compare and contrast the physical, social, and economic conditions found in the Brazilian regions of São Paulo and the Northeast. As a geographer, what recommendations could you make to government officials and planners endeavoring to promote the human and economic development of the Northeast?

PART FOUR

GREENLAND
(DENMARK)

Arctic Circle

ICELAND
Reykjavík

60°

ELEVATION IN METERS

Above 4000
2000–4000
500–2000
200–500
0–200
Below sea level

Norwegian Sea

Faroe Islands
(DENMARK)

Shetland
Islands
(U. K.)

Orkney Islands
(U. K.)

LAPLAND

FINLAND

SWEDEN

NORWAY
Oslo

Helsinki

Tallinn
ESTONIA

Stockholm

Riga
LATVIA

0   200   400 MILES
0   200   400 KILOMETERS

Glasgow

UNITED
Belfast

Dublin
IRELAND   KINGDOM   Sheffield

Birmingham

North
Sea

Goteborg

Copenhagen
DENMARK

LITHUANIA
Vilnius

RUSSIA

Hamburg

Leeds

Baltic Sea

EUROPEAN PLAIN

BELARUS

50°

Thames R.

NETHERLANDS
Amsterdam
Rotterdam

London

English Channel

ATLANTIC

OCEAN

Brest

Paris

Seine R.

Loire R.

FRANCE

Bay
of
Biscay

Bordeaux

Bremen

Hannover

Essen

Brussels
BELGIUM

LUXEMBOURG

Elbe R.

NORTHERN

GERMANY

Frankfurt

Luxembourg

Stuttgart

Rhine R.

Munich
LIECHTENSTEIN

Bern   Zürich
SWITZERLAND

Geneva
Lyon   Mt. Blanc

Poznan
Berlin

Leipzig

Prague
CZECHIA

Vienna

AUSTRIA

A L P S

Milan

Lodz

Oder R.

Warsaw

POLAND

Wroclaw   Krakow

CARPATHIAN MOUNTAINS

SLOVAKIA
Bratislava

Danube R.   Budapest

HUNGARY

SLOVENIA   Zagreb
Ljubljana   CROATIA

UKRAINE

MOLDOV
Kishinev

ROMANIA

Bucharest

Vistula R.

Garonne R.

Toulouse

PYRENEES

ANDORRA

Torino

Genoa

Nice
MONACO

Marseille

IBERIAN

Rhône R.

Po R.

APENNINES

SAN
MARINO

Zaragoza

Madrid

Barcelona

BOSNIA &
HERZEGOVINA
Sarajevo

Adriatic Sea

40°

Porto

PORTUGAL
Lisbon   SPAIN

Ebro R.

PENINSULA

Tagus R.

Sevilla

Malaga

Valencia

Corsica
(FRANCE)

Rome   ITALY

Naples

Sardinia
(ITALY)

Belgrade

YUGOSLAVIA

Danube R.

BULGARIA
Sofia
BALKAN

Skopje
MACEDONIA

Tirana

ALBANIA   PENINSULA

Thessaloniki

Aegean Sea

GREECE

Balearic Islands
(SPAIN)

Mediterranean   Sea

Palermo

Sicily
(ITALY)

Athens

Strait of
Gibraltar

MALTA   Valletta

Crete
(GREECE)

10°

20°

Europe

Bella Bychkova-Jordan
and Terry G. Jordan-Bychkov

O ver the past five centuries, Europeans have played an enormously important role in almost every facet of human life and
culture. They have demonstrated a remarkable ability to in-
novate, discover, migrate, exploit, conquer, and destroy. Europeans
gave the world the Age of Discovery, in which they visited virtually
every foreign land and established intercontinental and intercultural
contacts that permanently bound the world together for the first time.
They created the Industrial Revolution, which forever changed the
way we manufacture goods and live. From Europe, too, came significant advances in technology and modern science, as well as nationalism, imperialism, genocide, democracy, communism, fascism,
individualism, and world war.

European colonial empires, at one time or another, encompassed all but a few far corners of the earth,
and Europeans spilled out as settlers to colonize all or part of five continents. The Slavs colonized Siberia,
Iberians made South and Middle America their new home, and Germanic Europeans settled Anglo America, Australia, and parts of South Africa. In the process, European languages, religions, and customs spread
widely across the world. For better or worse, the European legacy touched literally every land, far and near.

Who are these Europeans? How could the inhabitants of such a small corner of the earth have developed so influential a culture? What, in fact, is Europe? We often find it defined as a continent—that is, a
large landmass standing mostly or entirely apart from the others. But in reality Europe fails to meet that
common definition, for in the east, it is fused solidly with Asia, forming the huge continent of **Eurasia**.
Europe, in fact, occupies only one small part of Eurasia.

If Europe is not physically a conventional continent, why do we regard it as a separate geographical entity? The answer is that Europe is a people and a way of life—a separate culture. Its defining characteristics are human, not environmental. A thousand years ago and more, "Europe" could already be distinguished
by three main human traits—the Christian faith, the Caucasian race, and the family of related languages
called Indo-European. In succeeding centuries, the list of distinctive European traits grew. Europeans came
to be the best educated, healthiest, longest lived, best fed, wealthiest, most industrialized, most urbanized,
and, to many, the most individualistic people in the world. Still more recently, they became the first in the
world to experience zero population growth, with the number of births and deaths balanced each year. In
the process, they have become the most aged population in the world.

We devote three chapters to these remarkable people and the land they inhabit, and explore the characteristics that help explain European uniqueness in world regional development. Our overarching definition of Europe implies a geographical uniformity, but that is, as we will see, an illusion. Europe displays
vivid internal contrasts. Most of these obey one or another of three basic spatial patterns—north versus
south, east versus west, and core versus periphery. That is, northern Europe differs from southern, western from eastern, and center from margins.

CONTENTS

■ **Trollfjord, Norway.** Europe is a region of varied physical and cultural landscapes. The coast of Norway possesses spectacular fjords, drowned U-shaped glacial valleys that penetrate deep into the interior highlands.

Chapter 10

The European Habitat

- Landforms
- Climate Regions
- Hydrogeography
- Mineral Resources

- Environmental Modifications
- Environmental Crises
- European Union Environmental Policies

We include in Europe the forty tiny to medium-sized independent countries—forty-one, if you count the Kaliningrad exclave of Russia on the Baltic—lying in western Eurasia, extending from the Atlantic Ocean on the west to the eastern borders of Finland, the Baltic States, Poland, and the Balkan Peninsula (Figure 10–1). This is an arbitrary border. We could well argue that the eastern Slavic lands—Russia, Ukraine, and Belarus—also belong to Europe. But human entities, including cultural regions, rarely possess sharp boundaries, and Europe is no exception.

While Europe is not a spatially separate continent, it does have a distinctive physical character. We might best call it a **subcontinent**—a westward-projecting peninsular section of Eurasia. Europe is a peninsula of peninsulas, possessing a highly irregular, deeply indented coastline with numerous offshore islands (of which four encompass independent states). Four larger peninsulas, the **Scandinavian**, **Iberian**, **Italian**, and **Balkan**, dominate the coast of Europe; and diverse bordering seas and bays—the Baltic, North, Biscay, Mediterranean, Adriatic, Aegean, and Black—penetrate deeply into Europe (Figure 10–2). As a consequence, fully twenty-eight of Europe's independent states possess a coast, and no point in Europe lies more than about 400 miles (650 kilometers) from the sea.

The human drama of Europe has been acted out on this oddly shaped segment of the physical earth. In this chapter, we consider Europe's terrain, climate, soils, waters, mineral resources, and environmental problems. The habitat has influenced the inhabitants' way of life in diverse ways, and we cannot understand Europe without knowing the nature of its physical framework. At the same time, Europeans have massively altered their environment, for both better and worse, to the extent that the original "natural" environment no longer exists. Environment, people, and region remain abstractions unless considered together.

Landforms

Europe has a great internal variety of landform types. To generalize, we can say that European terrain displays the north-versus-south spatial pattern (Figure 10–3, page 250).

Mountain Ranges

Mountains are confined largely to the southern half of Europe, except for the Kjölen chain in Scandinavia and Iceland. In marked contrast to those of Anglo America, most European mountain ranges are oriented in an east-west direction. The southern concentration of mountain ranges has formed largely as a result of **plate tectonics**, as the African and Eurasian continents slowly collide. The collision has caused folding, fracturing, and uplifting of the earth's surface, building these "young" ranges.

The greatest of the mountain ranges of southern Europe include the **Alps**, **Pyrenees**, **Dinaric**, and **Carpathians**, some towering to elevations over 14,000 or even 15,000 feet (4,300 to 4,600 meters). Most bear the scars of glaciation and remain snowcapped through the summer (Figure 10–4, page 251). In spite of their impressive appearances, however, these ranges historically have not constituted significant barriers to human movement and interaction. Italians, in fact, refer to the Alps as the "magnificent traitor," since invaders have repeatedly found their way through the mountains into Italy. One of the most astounding of the ancient invasions was led by the North African warrior Hannibal, who crossed the Alps with his cavalry and elephants. Most European ranges can be circumvented or crossed via numerous low passes. Tunnels exist on many well-traveled routes.

Instead, the most disruptive aspect of the European mountain ranges lies in their frequent earthquakes and widespread volcanism. Huge losses of life sometimes accompany southern European earthquakes, as when 60,000 inhabitants of the Italian island of Sicily perished in a single early-twentieth-century quake. The most recent earthquake fatalities occurred in Athens, Greece, in 1999. Active volcanoes also present a danger, and Mount Etna, also on Sicily, has claimed an estimated 1 million lives during the past 2,500 years (Figure 10–5, page 251).

In between the various mountain ranges lie fertile valleys, basins, and tablelands, created by the same tectonic forces that built the mountains. The larger among these are the **Hungarian Basin** in the Balkans, the **Padana Valley** of the Po River in northern Italy, and the **Meseta**, a

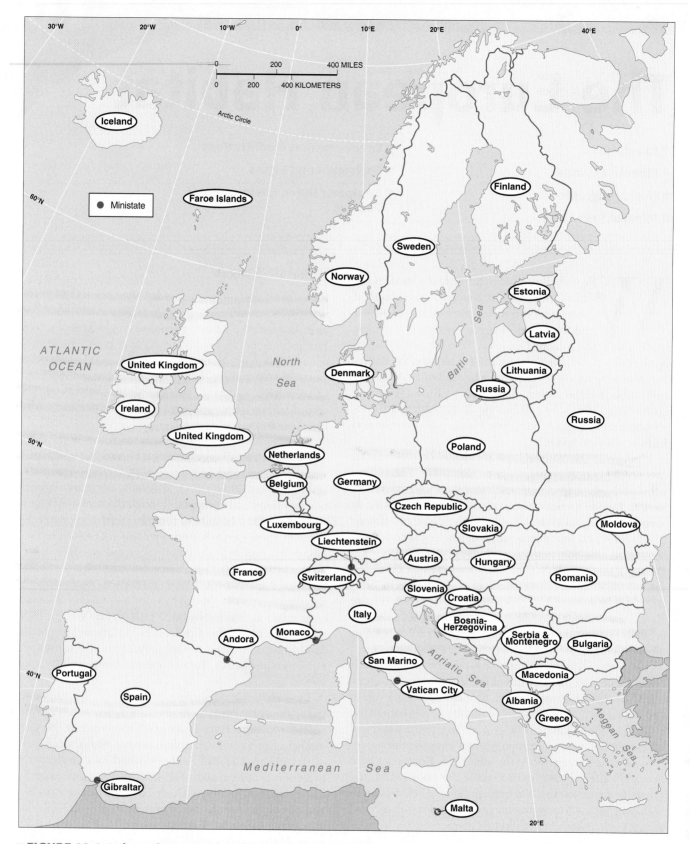

■ **FIGURE 10-1 Independent countries of Europe.** The following abbreviations are internationally accepted and appear on automobiles and other vehicles to identify their origin. A=Austria; AL=Albania; AND=Andorra; B=Belgium; BG=Bulgaria; BiH=Bosnia-Herzegovina; CH=Switzerland; CZ=Czechia; D=Germany; DK=Denmark; E=Spain; EST=Estonia; F=France; FL=Liechtenstein; GB=United Kingdom; GR=Greece; H=Hungary; HR=Croatia; I=Italy; IRL=Ireland; IS=Iceland; L=Luxembourg; LT=Lithuania; LV=Latvia; M=Malta; MC=Monaco; MK=Macedonia; MOL=Moldova; N=Norway; NL=Netherlands; P=Portugal; PL=Poland; RO=Romania; RSM=San Marino; RUS=Russia (including here only the Kaliningrad enclave); S=Sweden; SF=Finland; SiM=Serbia & Montenegro; SK=Slovakia; SLO=Slovenia; V=Vatican City; and YU=Yugoslavia. In addition there are two dependent territories: FR=Faeroe Islands, ruled by Denmark but apparently headed for independence in the next decade, and GBZ=Gibraltar, an autonomous British colony threatened with annexation by Spain.

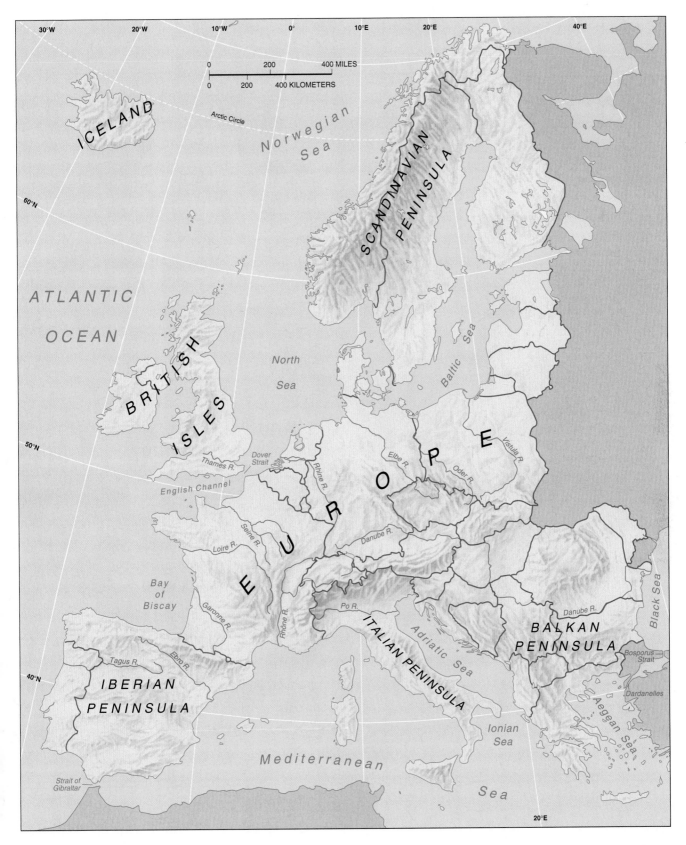

■ **FIGURE 10-2 Some basic locations in Europe.**

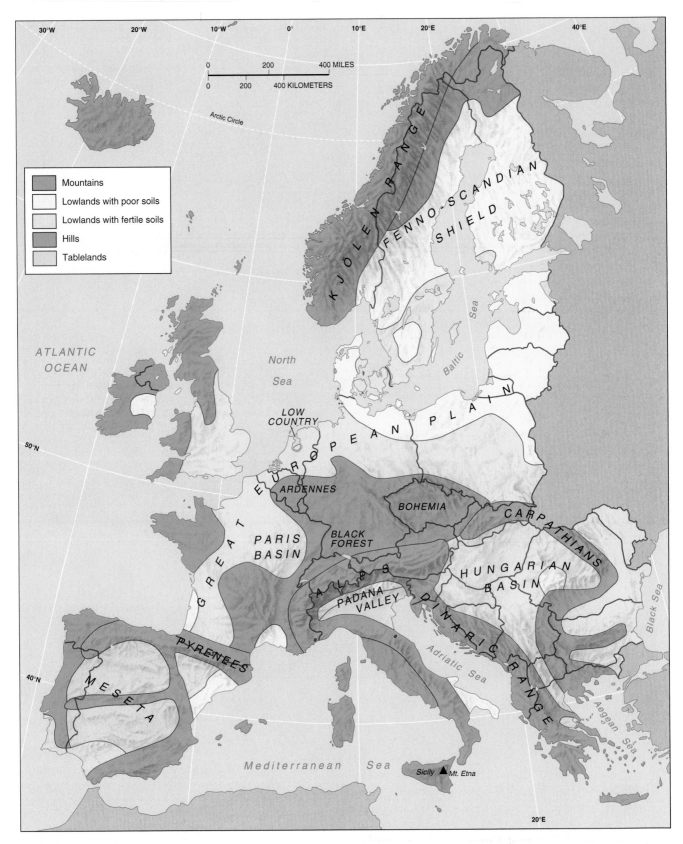

■ **FIGURE 10-3 Landforms and soils of Europe.** Southern Europe is dominated by mountains, northern Europe by plains.

■ **FIGURE 10-4 Snowcapped Alps.** The Alps are located in the heart of Europe and contain some of the most beautiful landscapes on earth. This view from atop Rufikopf Summit overlooks the town of Lech Am Arlberg in the Austrian Alps.

■ **FIGURE 10-5 Mt. Etna, Sicily spilling lava.** The Mediterranean Basin contains a number of active volcanoes. In this November, 2002, photo, bulldozers are constructing a wall to protect a hikers' shelter from the advancing river of lava.

tableland that dominates interior Iberia (see Figure 10–3). These plains support dense agricultural populations and numerous large cities.

The Great European Plain

A much larger area of plains lies beyond the mountains and dominates the northern half of Europe. This vast lowland, collectively named the **Great European Plain**, stretches from the Atlantic coast of France northeastward to the borders of Russia and beyond into the Eurasian heartland. Many local names are attached to this plain, such as the Paris Basin, the Low Country, and the Fenno-Scandian Shield. Many areas within the Great European Plain have highly fertile soils derived from limestone, ancient wind-deposited ma-

terials called loess, or river-borne alluvium. Sizable agricultural populations have concentrated there since time immemorial (Figure 10–6). But in areas once covered by the continental glaciers of the last Ice Age, especially the northern half of the Great European Plain, the potential for agriculture is far poorer, and huge expanses of land are covered with commercial forests.

Hill Lands

The remainder and far smaller part of Europe has hilly terrain. These hill lands occupy two basic locations—a broken belt along the Atlantic coasts of Portugal, Spain, France, Ireland, and Great Britain and a central European zone wedged between the Great European Plain and the mountain ranges

■ **FIGURE 10-6 Agriculture of the Paris Basin.** The Paris Basin, a subregion of the Great European Plain, is a highly productive agricultural region. Visible in the foreground are vineyards and the town of Avize.

■ **FIGURE 10-7 Rolling hills of western Ireland.** The glaciated regions of Europe's northwestern highlands have cool, damp climates and stony soils. Shown here is the town of Cliften, County Galway.

of the south. For more than a thousand years, the coastal hills have served as refuges for Celtic peoples such as the Irish, Scots, Welsh, and Bretons, creating a distinctive regional culture (Figure 10–7). The hills of Central Europe offer some of the most charming landscapes in regions such as the Black Forest, Ardennes, and Bohemia.

Climate Regions

Europe lies largely in the temperate mid-latitude zone, tucked safely north and west of the Eastern Hemisphere's great desert belt. In that fortunate location, it enjoys climates that are generally humid and mild. These moist, mesother-

mal (moderate temperature) conditions favored Europe and helped it become one of the major homelands of humanity. Still, the climate is far from uniform, and we can recognize three major types: marine west coast, humid continental, and mediterranean. Both north-versus-south and east-versus-west patterns can be detected (Figure 10–8).

Marine West Coast Climate

Europe lies in the latitudes of the **westerlies**, prevailing winds that come onshore from the Atlantic Ocean, both at the earth's surface and aloft, as the jet stream. These winds, many of which originate over a warm ocean current called the **North Atlantic Drift**, are marine in character—wet and

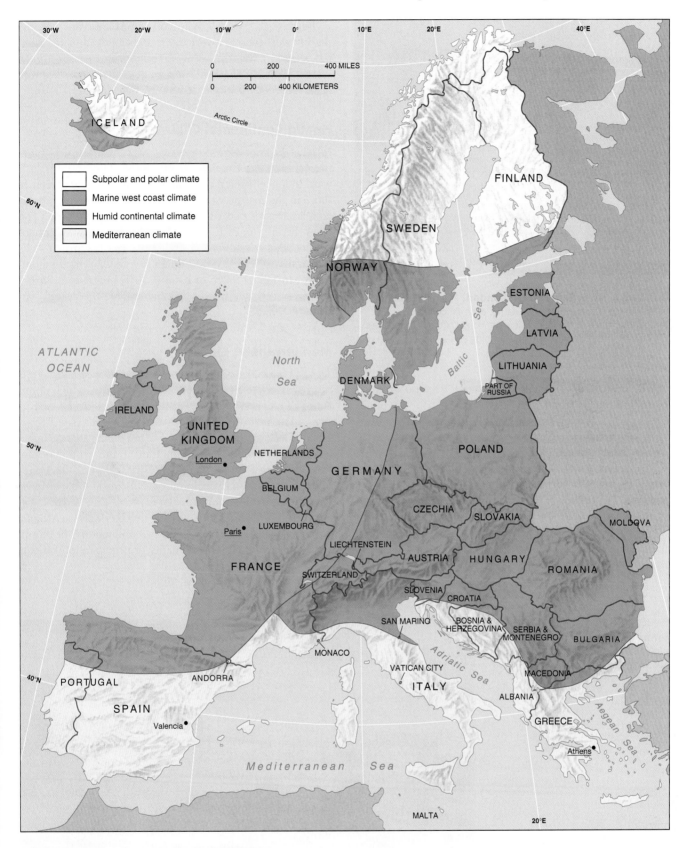

■ **FIGURE 10-8 European climate regions.** Most climates of Europe are temperate. The
polar climates of the far north are atypical of Europe, and that area remains very sparsely inhabited.

Because the winds coming off the Atlantic are wet, the marine west coast is a very moist and cloudy climate. Rainfall occurs frequently during all seasons. In an average year, rain falls on 188 days in Paris, in small amounts. London once recorded 72 consecutive rainy days (Figure 10–9).

Humid Continental Climate

Most of Eastern Europe is dominated by a **humid continental climate**. Enough Atlantic moisture penetrates eastward to make this a moist climate, with summers similar to those in the marine west coast. In winter, however, bitterly cold air masses enter from the frozen northern ocean and from the interior of Eurasia. A durable snowcover perhaps best characterizes the continental winter (Figure 10–10). Rivers, lakes, and bays freeze. In severe winters, temperatures can fall to −20° or even −30°F (−29° to −35°C). Thus, summer and winter differ sharply in the humid continental climate.

Mediterranean Climate

The north-versus-south dimension of European climate is represented by the **mediterranean climate**, which dominates all or parts of the three southern peninsulas of Europe—Iberia, Italy, and the southern Balkans (see Figure 10–8). The distinguishing trait of the mediterranean climate is the concentration of precipitation in the winter season, with exceptionally dry summers. Generally, less than one-tenth of the annual precipitation falls in the quarter-year comprising the summer months of June, July, and August, and the month of July is almost totally rainless. Athens, Greece, averages only 0.3 inches (8 millimeters) of rain in July. This pronounced seasonality of precipitation reflects

▪ **FIGURE 10-9 Rainy day in London.** The marine west coast climate is characterized by damp, overcast conditions throughout much of the year. This view along Oxford Street shows the core of one of London's main shopping areas.

cool in the summer and mild in the winter. Because Europe's mountain ranges trend west-to-east, the marine air penetrates far into the interior. As a result, much of western Europe is characterized by a **marine west coast climate**, which extends eastward about halfway across Europe. Cool summers, in which no month averages above 72°F (22°C), alternate with gentle winters, with no month averaging below freezing. Temperatures do not differ all that much between summer and winter. In London, the majority of January nights have no frost, and a snowcover only very rarely and fleetingly blankets the countryside. Heat waves are also rare.

▪ **FIGURE 10-10 Snow-covered farms of Poland.** Most of Poland lies within the European Plain. Shown is a village near Bialowieza.

the transitional position of the Mediterranean basin between the humid marine west coast climate to the north and the parched Sahara Desert to the south. In winter, the Mediterranean lands come under the influence of the westerlies, which shift southward with the direct rays of the sun. The region consequently experiences the impact of precipitation-producing marine air masses and migrating storm centers. In summer, the region comes more under the influence of a great subtropical high-pressure center, which causes fair weather in North Africa and the Mediterranean peninsulas. Winter precipitation occurs as rainfall in the small lowland plains, but often as snow in the numerous mountain ranges in the mediterranean region. Accumulated snow in the highlands is of crucial importance to farmers because the meltwater runoff in spring and summer provides irrigation water for the drought season.

Summers in the Mediterranean region are hot. Athens has recorded 105°F (40.5°C) and Valencia, Spain, 109°F (43°C). Relative humidity is low in the summer, allowing pleasant nighttime cooling. People in the Mediterranean lands tend to nap during the hottest time of day, then stay up past midnight to enjoy the evening coolness. Most Europeans regard the mediterranean as the perfect climate, gifted with warmth and sunshine. Northern Europeans often vacation in these southern lands or even move there in retirement.

Hydrogeography

The waters of Europe, including its flanking seas, connecting ocean straits, and rivers, possess a great importance to the inhabitants. The seas and straits facilitate the movement of people and goods around the perimeter of Europe, and a desire to control strategic straits, such as the Bosporus-Dardanelles, Gibraltar, and the English Channel-Dover Strait, has often led to war (Figure 10–11).

■ **FIGURE 10-11 Strait of Gibraltar.** The Strait of Gibraltar occupies an extremely strategic site that controls access between the Atlantic Ocean and the Mediterranean Sea at the top of the image. It also constitutes a divide between Europe and Africa. The northern tip of Morocco on the right is only 13 kilometers (8 miles) from southern Spain on the left. The Rock of Gibralter is located on the small peninsula in the left background of the image.

The great navigable rivers of Europe lie disproportionately on the Great European Plain. Flowing in a southeast-to-northwest direction, each has a major port city at its juncture with the sea. The port cities include Rotterdam, Netherlands, at the mouth of the Rhine River (see the *Geography in Action* boxed feature, The Rhine River: Economic and Cultural Significance); Hamburg, Germany, at the mouth of the Elbe River; and Szczecin, Poland, at the mouth of the Oder River. Numerous canals interconnect these rivers, providing a splendid inland waterway system. Most of southeastern Europe drains to the Black Sea, particularly through the great Danube River, which, rising in southern Germany, passes through or borders nine independent states and is known by five different names along its course. The rivers of southern Europe, not being navigable, generally lack port cities at their mouths.

Most rivers of the Great European Plain have been channelized and canalized with locks and small dams to enhance transport, and the great rapids at the Iron Gate on the Danube, where the river cuts through at the Carpathian Mountains, have been blasted away. Riverbanks are often manicured and lined with facing stones, giving them an artificial appearance.

Mineral Resources

The European habitat offers abundant fertile plains, mild climates, and useful waterways, but mineral resources are less abundant. The most notable exception is coal—much of very high quality—which occurs widely and helped fuel the Industrial Revolution (Figure 10–12). Major deposits of petroleum and natural gas occur only in the North Sea and in Romania, and neither of those fields can rival those of the Middle East or Russia. Most of the modest iron ore deposits have been exhausted, and only Sweden remains a major supplier.

We will not find, then, the explanation for Europe's greatness in its mineral resources, because few exist. Even coal mining is in major decline today, employing only a small fraction of its former workforce. Europe must import the far greater part of its mineral needs. Instead, human resources are the key to understanding the economic development of Europe, just as culture has provided the basis for our definition of the region.

Environmental Modifications

For millennia, Europeans have been modifying their habitat on a massive scale. Some of the changes they wrought have been intentional and often beneficial—the clearing of forests for agriculture, the draining of malarial marshes, the linking of rivers with canals, the piercing of mountain ridges with tunnels. Other changes, often catastrophic, have come as unintentional side effects of human activity. Europe, for both better and worse, is a thoroughly humanized habitat (see

GEOGRAPHY IN ACTION
The Rhine River: Economic and Cultural Significance

The Rhine River is the most important waterway of Western Europe (see Figure 10–2), as well as the busiest inland river in the world, carrying millions of tons of cargo between some of the most densely populated conurbations in Europe. From its sources in the Alps, the Rhine becomes navigable near Basel in northern Switzerland and runs north along the French-German border for about a hundred miles before cutting through western Germany and the low-lying broad plains of the Netherlands, where it drains into the North Sea at the port of Rotterdam, the "Gateway to Europe" and the largest port in the world (Figure A). It is also connected through navigable tributaries to the canal network in France and Belgium, and the Rhein-Main-Donau-Canal, which was completed in 1992, allows access to Austria and countries of southeastern Europe via the Danube River. Thus, despite its relatively modest length of about 820 miles (1,320 kilometers), the Rhine's strategic location in the heart of Western Europe makes it a major international shipping route. This fact was recognized as far back as 1868 when the riparian countries signed The Rhine Navigation Act, freeing the parties of all duties and promoting cooperation along the river and in the maintenance of its banks and dams.

Today the Rhine River ties together the most important Western European regions of industrial production, including centers of chemical industry (Basel, Ludwigshafen, Frankfurt-Höchst, Wesseling, Rotterdam), heavy industry (the Ruhr region), motor industry (Rüsselsheim, Köln), and engineering (Mannheim). That also means a high degree of environmental impact. Since the late 1980s, the installation of powerful water filters and the latest technological improvements of the canal and lock systems have greatly improved the environmental quality of the Rhine.

But the Rhine Valley has another value for humanity—it represents one of the most romanticized landscapes in the world. The word "Rhineland" conjures up powerful idyllic images, most famously that of Lorelei—the sad maiden sitting on a rock, luring sailors to their demise with her forlorn songs. The lasting beauty of this image was immortalized by one of Germany's greatest poets—Heinrich Heine. Picturesque landscapes, dotted with the ruins of medieval castles and vineyards on the Upper and Middle Rhine (Figure B) have been a magnet for international tourism from as early as the eighteenth century.

FIGURE A **The port of Rotterdam, in the Netherlands.** The largest port in the world, Rotterdam's harbor is the collection point where cargo is transferred between ocean vessels and the barges and smaller boats that carry products via the Rhine and canals to and from the interior of Europe.

FIGURE B **The Rhine at Bacharach, Germany.** Deeply entrenched in its middle course, the Rhine flows northwards toward the North Sea past castles, small towns, and spectacularly steep vineyards. This view looks southeast from Stahleck Castle, and vineyards cling to those slopes with a southwestern orientation where exposure to sunlight is greatest.

the *Geography in Action* boxed feature, Dutch Coastline Alteration).

Forest Clearance

A great forest once stretched almost unbroken across the northern half of Europe, consisting of oaks, beeches, birches, firs, spruces, and pines. Most of it fell before the axes of farmers and charcoal burners, especially between about 500 and 1300 and again in the 1500s and 1600s. Some countries,

such as Ireland and the Netherlands, are less than 10 percent forested today, and it is fair to say that virtually no virgin woodland survives anywhere (Figure 10–13). The remnant forests are, in effect, artificially planted tree farms.

Similarly, the Mediterranean forests vanished. Once covered by an open woodland dominated by huge live oaks, interspersed with tall grasses, the Mediterranean peninsulas today offer only bare pastureland, densely cultivated valleys, wide expanses of shrub growth, and planted forests of exotic species such as Australian eucalyptus. Soil erosion followed

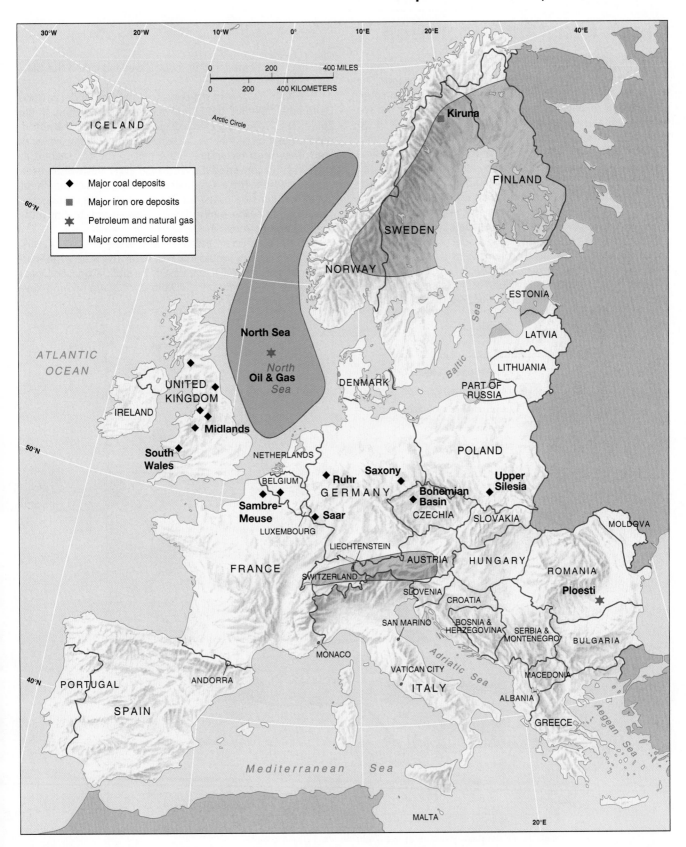

■ **FIGURE 10-12 Mineral and forest resources of Europe.** Europe has relatively few natural resources.

GEOGRAPHY IN ACTION
Dutch Coastline Alteration

Nowhere is the human modification of the European habitat more impressive than in the Netherlands. The huge multiriver delta that forms much of that country has been sinking for thousands of years, allowing the sea to encroach and to threaten the very existence of the Dutch homeland. For the past 2 millennia, the Dutch have battled the sea, creating a wholly artificial land in the process. At first, they merely built mounds of earth on which to live and erected silt traps to extract soil from the rivers. Then they began erecting long earthen embankments called **dikes** to enclose their fields. As the land sank well below sea level, the Dutch first harnessed wind power to pump the water, then later shifted to steam engines and finally to electricity. Beginning in the 1600s the Dutch went on the offensive, draining freshwater lakes to expand their farmland.

Setbacks occurred. In 1421 a flood near Rotterdam drowned seventy-three villages and caused 100,000 deaths. This land was never reclaimed and is today a wetland called the Biesbosch National Park.

Two ambitious twentieth-century efforts—the Zuider Zee and Delta projects—were brought to successful completion. In the Zuider Zee Project, the Dutch diked off the mouth of a notorious arm of the sea that had repeatedly flooded adjacent farmland. They then reclaimed about half of the old seabed as farmland and converted the remainder into a freshwater reservoir. In the Delta Project, at the collective mouths of the Rhine, Maas, and Schelde rivers, the Dutch erected huge protective sea dikes, protecting another vulnerable area and creating more freshwater reservoirs.

Had the Dutch not battled the sea for the last 2,000 years, half of their country would not exist today (Figure A). The "natural" coastline lies well to the east of the artificial one, and great cities such as Amsterdam and Rotterdam stand on land literally made by human hands.

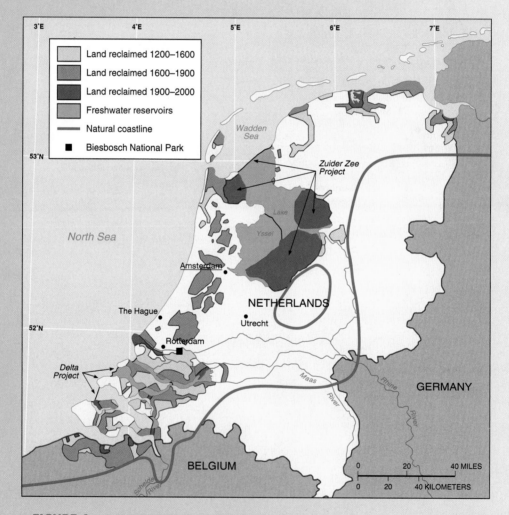

■ **FIGURE A** **Land reclamation in the Netherlands, 1200–present.**

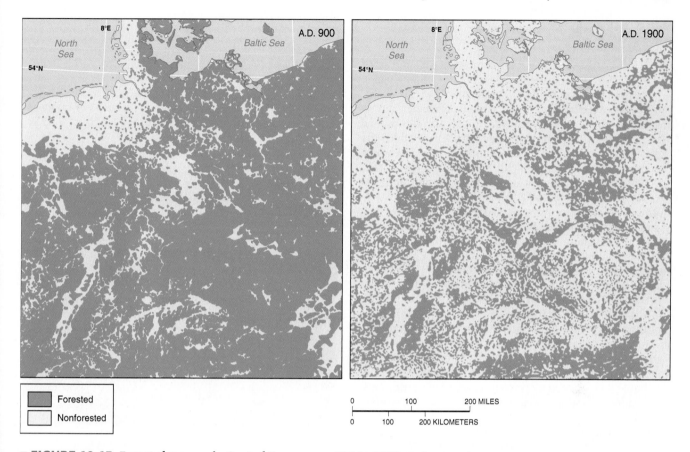

Forested

Nonforested

0   100   200 MILES

0   100   200 KILOMETERS

■ **FIGURE 10-13 Forest clearance in Central Europe, A.D. 900 to 1900.** Today, even the remnant woodlands are perishing, due to acid rain.
Source: T. G. Jordan-Bychkov and M. Domosh, *The Human Mosaic: A Thematic Introduction to Cultural Geography.* New York: W. H. Freeman, 9th ed., 2003, p. 281.

deforestation in the Mediterranean, leaving some landscapes stripped to bare rock and countless harbors silted up by material washed down to the coast.

It is hard to find natural vegetation anywhere in Europe, for all has been modified to one degree or another. The only exception is the Bialowieza National Park in eastern Poland, the last undisturbed forest habitat in Europe, where gigantic oaks, hornbeams, elms, ashes, and linden trees survive. Here the European bison, long extinct in countries like England, France, and Germany, still freely roams in the woods. No forest activity has been allowed in Bialowieza since 1921.

Terrain and Weather Modifications

Not even the "everlasting hills" (to use a Biblical term) have escaped the human hand in Europe. In parts of the Mediterranean, **terracing** has stair-stepped steep hillsides for agricultural purposes, trapping soil behind myriad stone retaining walls (Figure 10–14). From ancient times, Europeans have also thrown up artificial mounds and hills for defensive or ceremonial purposes, and the rubble of cities destroyed in World War II was piled up to form artificial hills beyond many city limits, as in Berlin, where the *Teufelsberg* (Devil's

■ **FIGURE 10-14 Terraced agriculture in Portugal.**
Southern European agriculture is often labor-intensive and characterized by relatively low crop yields.

GEOGRAPHY IN ACTION
The Black Triangle

The European region most catastrophically impacted by ecological disaster is the notorious "Black Triangle," straddling the borders of Germany, Poland, Czechia, and Slovakia (see Figure 10–15). The westerlies bring airborne pollutants from as far away as Great Britain and France, and these mix with pollutants generated locally to produce the highest acid rain and forest death concentrations in Europe. The local industries, especially those producing chemicals, added massive amounts of poisons to the rivers and land until the mid-1990s, when industrial decline began. Poland's Vistula River departs the Black Triangle laced with heavy metals such as lead and mercury, coal mine salts, and organic carcinogens. The local population suffers from diverse pollution-related disorders and has a decreased life expectancy. Children, especially, have many health problems (Figure A). The northwestern part of Czechia is widely acknowledged to be the most devastated area within the Black Triangle.

This situation has somewhat improved in the process of accession of Czechia, Slovakia, and Poland to the European Union. Since 1993, the EU has been monitoring the environments of countries in transition under the aegis of UNECE—United Nations Economic Commission for Europe. A series of regional and international conferences with the motto "Environment for Europe" resulted in the formulation of common environmental legislation and standards in the candidate countries. Cooperation between Czechia, Germany, and Poland led to the adoption of a Joint Air Monitoring System (JAMS). This does not imply a happy end to the story of the Black Triangle, however. As the air pollution from mining and heavy industrial activities has decreased, the air pollution from car exhausts has significantly increased as more citizens of Poland, Czechia, and Slovakia can afford the ultimate symbol of increasing economic prosperity—a family car.

■ **FIGURE A Schoolgirl wearing a protective face mask, Czechia.** The Black Triangle's air pollution is particularly harmful to the health of the young.

Mountain) is visible for miles around and today provides ski slopes and sled runs.

The air pollution generated in Europe's cities and the heat absorbed by the pavement and masonry of the urban infrastructure have created numerous **urban heat islands**, characterized by significantly higher temperatures than those of the surrounding areas. Urban temperatures are often 10° to 15°F (5° to 8°C) warmer than those in neighboring rural regions. Because cities lie close together in many parts of Europe, some urban heat islands have merged, producing elevated temperatures over entire regions. Europeans contribute quite significantly to global warming. Air pollution now envelops virtually the entirety of Europe, and even on sunny days, visibility is usually restricted.

Environmental Crises

In all too many places in Europe, environmental pollution has reached disastrous proportions. Such problems may result from individual accidents such as oil spills, radioactive dumping, chemical spills, and explosions (Figure 10–15).

The western promontories of Europe have repeatedly suffered major oil spills as tanker ships ran aground and ruptured. The worst of these occurred along the coasts of Ireland, Great Britain, France, and Spain—one or more in each of the past four decades. At Seveso, northern Italy, a chemical plant exploded in 1976, spreading toxic dioxin over surrounding areas and causing long-term elevated occurrence of leukemia, liver cancer, and lymphoma. This ranks as one of Europe's worst industrial accidents ever. In 2000, two separate toxic spills from sites in northern Romania severely polluted the Tisza River, a tributary of the Danube. As a result, some species of aquatic fauna and flora in the Tisza and Danube have become extinct.

Acid Rain

Other European environmental disasters have afflicted more sizable regions and result from ongoing, multiple pollution (see the *Geography in Action* boxed feature, The Black Triangle). Such ecological crises usually reveal a core-versus-periphery geographical pattern, since major environmental problems occur much more frequently in the central part of Europe, where the population is concentrated. Perhaps the best example is **acid rain** (see Figure 10–15).

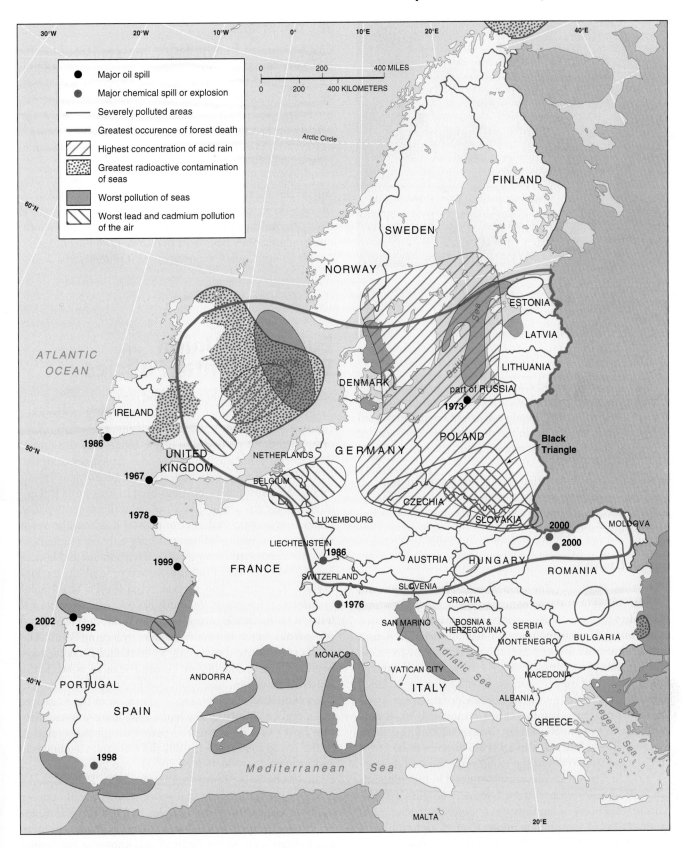

■ FIGURE 10-15 Environmental crises and disasters of Europe. Problems are worst in
the core of Europe and generally less severe in the peripheries.

■ **FIGURE 10-16 Forest death caused by acid rain.** The Krusnehory Mountains, Czechia, are a part of the infamous Black Triangle.

Acid rain (acid precipitation) occurs as various substances, including sulphur and carbon, are released into the air as a consequence of burning huge amounts of fossil fuels—coal and petroleum products—in the densely populated core of Europe. These substances mix with water vapor in the atmosphere and fall to the earth again as mildly acidic precipitation. This precipitation is sufficiently toxic to diminish soil fertility, damage standing crops, poison lakes, kill fish, eat away medieval stone structures, and even ruin paint finishes on automobiles.

Forest Death

The most disturbing effect of acid rain is **forest death**, for the remnant woodlands of central Europe are perishing today. In the mid-1970s, forestry experts began to notice tree damage, especially in needle-leaf species. Then, as a critical threshold was crossed, the percentage of trees exhibiting damage increased rapidly. In western Germany, only 8 percent of the forest showed visible damage in 1982, in the form of leaf or needle loss or discoloration, but a year later the proportion had increased dramatically to 34 percent and by 1984 to 50 percent. Switzerland's forest damage rose from 17 percent in 1984 to 40 percent by 1988. In Czechia, 96 percent of the woodland exhibited damage by 2000, and forest death was widespread in the hill lands along the border with eastern Germany (Figure 10–16). The Black Forest of southwestern Germany became another center of the phenomenon. Even peripheral lands such as Greece, Norway, and Estonia had damage in more than half of their forests by the late 1980s. Twenty percent of the trees throughout Europe showed damage by 1991, and in the decade that followed, "a worsening of the forest condition" continued, according to the European Union.

The Green Vote

One European reaction to the problem of pollution has been the rise of **Green politics**. Characterized by party platforms favoring the protection of the natural environment, skepticism of technology, opposition to the cooperation of industry and government, rejection of excessive consumerism, and a desire to change basic lifestyles, the Green movement arose in the 1980s in environmentally troubled, yet wealthy countries such as Germany.

The Greens emerged as an electoral force to be reckoned with in the elections of 1989, achieving a tenth or even more of the vote in some regions, with their greatest strength in Belgium, northern France, western Germany, and southern England. They have held their own until now but seem unable to approach a majority or even plurality status anywhere.

European Union Environmental Policies

Major environmental initiatives and reforms have been initiated by the European Union. Since 1972 the member countries have been working together on developing coordinated environmental policies. Since that time more than 200 different environmental laws have been adopted. In 1997 this process culminated in the Amsterdam Treaty, proclaiming sustainable development in Europe as a major objective for the union. The essence of this incentive is to balance economic and environmental considerations for development, taking into account not only economic benefits but also the protection of the environment and human health, rational use of natural resources, and close cooperation at regional and international levels. The implementation of these new standards has met with deep-rooted tension between the interests of the more developed and prosperous North and the economically lagging South. The northern countries, especially Sweden and Finland, have been pushing for the highest possible standards, while the southern countries, like Spain and Portugal, have emphasized economic development. To harmonize the situation, the member countries reached a balancing compromise. The best example is the EU policy on global warming. At the Kyoto conference in 1997 the EU took a leading role and proposed a reduction of greenhouse gas emissions of 8 percent by 2012. To achieve this goal, the EU adopted a policy of burden-sharing that reduces costs for the poorer member countries. For example, in 1998 the EU Environmental Council decided on quotas: Germany and Denmark (together with the Netherlands known as the "green troika" for their leadership in the protection of the European habitat) agreed to reduce their emissions by 21 percent, while Greece is allowed to increase its emissions by 25 percent—but the Union as a whole remains responsible for the overall reduction target.

▶▶ Summary

In the physical sense, Europe is unremarkable. Nothing in the European habitat might explain the enormous importance and economic development of this rather small world region. Its mountain ranges, plains, and hills are duplicated on other continents, as are its climate types. Europe's natural resources are, by world standards, not abundant. Moreover, Europe is plagued by serious local and regional environmental problems caused by ecological abuses. For the answer to how this modestly sized and endowed region came to be so important, we must turn to the European people.

▶▶ Key Terms

acid rain
Alps Mountains
Balkan Peninsula
Carpathian Mountains
dikes
Dinaric Range
Eurasia
forest death
Great European Plain
Green politics
humid continental climate
Hungarian Basin
Iberian Peninsula

Italian Peninsula
marine west coast climate
Mediterranean climate
Meseta
North Atlantic Drift
Padana Valley
plate tectonics
Pyrenees Mountains
Scandinavian Peninsula
subcontinent
terracing
urban heat islands
westerlies

Tyn Church, Old Town Square, Prague, Czechia. Many European cities have historic districts that date back hundreds of years and provide cultural continuity in a rapidly changing world.

Chapter 11

Europe:
Culture, Society, Economy

- Religious Regions
- Language Patterns
- Population
- Standard of Living
- Immigration

- Europe's Cities
- Primary and Secondary Industries
- Industrial Rejuvenation
- Service Industries
- Agriculture

Europe can be viewed as a human phenomenon. We can identify twelve basic "European" traits and attach—if arbitrarily—numerical values to each. "Europe" includes those areas that have the following characteristics:

1. 80 percent or more Christian in religious faith or tradition

2. 80 percent or more Indo-European in speech

3. more than 90 percent Caucasian (or Europoid) in race

4. healthy populations, perhaps best exemplified by a low infant mortality rate (fewer than 12 infants per 1,000 die in the first year of life)

5. educated populations (more than 90 percent literacy rate)

6. high urbanization (more than 50 percent of the population resides in cities and towns)

7. wealthy populations (a per capita gross domestic product of $22,000 or more)

8. a densely built transportation network (500 or more miles of highway per 100 square miles of territory—that is, 300 kilometers or more per 100 square kilometers)

9. a small segment of the workforce in agriculture (less than 25 percent)

10. stabilized or declining populations—that is, have zero population growth

11. a well-established tradition of democratic government (no substantial restrictions of personal freedom today)

12. dense population (lie within an area called the "European zone of continuous settlement")

If we map all of these "European" traits, a **core-periphery pattern** emerges (Figure 11–1). The most purely European areas lie in the center, in countries such as Germany and France, while peripheral areas—especially on the east and south—exhibit fewer of the defining traits. Cultural borders are rarely sharp, and Europe has no clearly demarcated limits, though for the purposes of this textbook we have arbitrarily chosen one.

But there is more to understanding Europe as a cultural region than merely these defining traits. Even more striking are its internal cultural geographical contrasts. Europe is Christian, but the faith is fragmented into very different churches and denominations. The Indo-European language family is dominant, but it is divided into numerous, mutually unintelligible tongues. Let us now turn our attention to these and other internal contrasts, all of which obey the previously mentioned east-versus-west, north-versus-south, and core-versus-periphery models.

Religious Regions

Monolithic Christianity never existed in Europe. From the very first, the more ancient contest between Greek and Roman civilizations dictated fragmentation. The subsequent split of the Empire into western and eastern halves, Roman and Greek, foreshadowed a religious schism. In a separation that finally became official in 1054, the western church became **Roman Catholicism**, the Greek church **Eastern Orthodoxy**. The dividing line between Catholicism and Orthodoxy remains the most fundamental religious border in Europe today (Figure 11–2). It has changed little in a thousand years and provides the basis for many of the contrasts between west and east in Europe.

Much of the border between eastern and western Christianity also forms the eastern limit of Europe, as used here, but in the south the religious border cuts westward, placing most

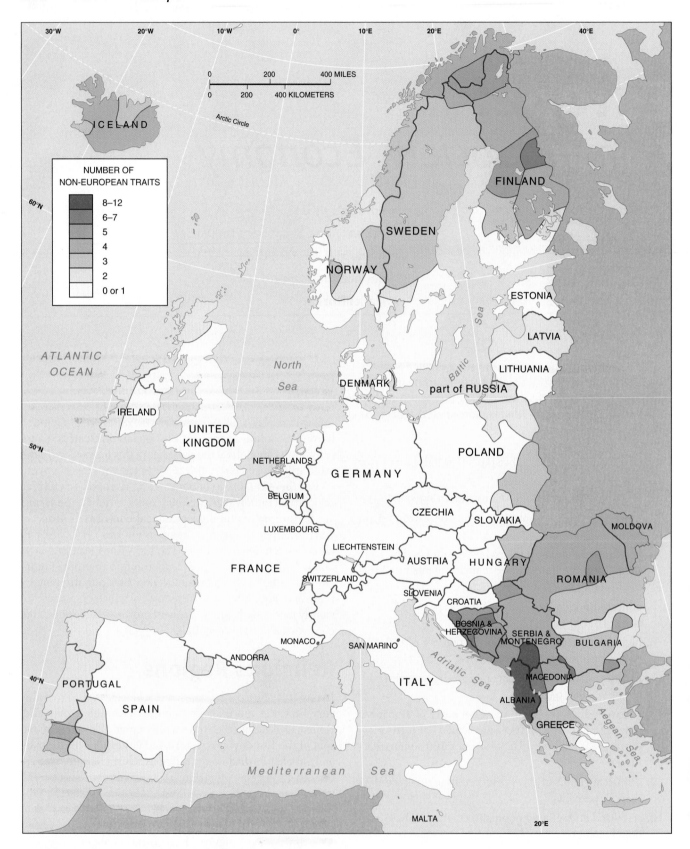

■ **FIGURE 11-1 Europe and non-Europe.** The "European" traits are listed in the first paragraph of the chapter. Europe has few sharp borders in the human/cultural sense.

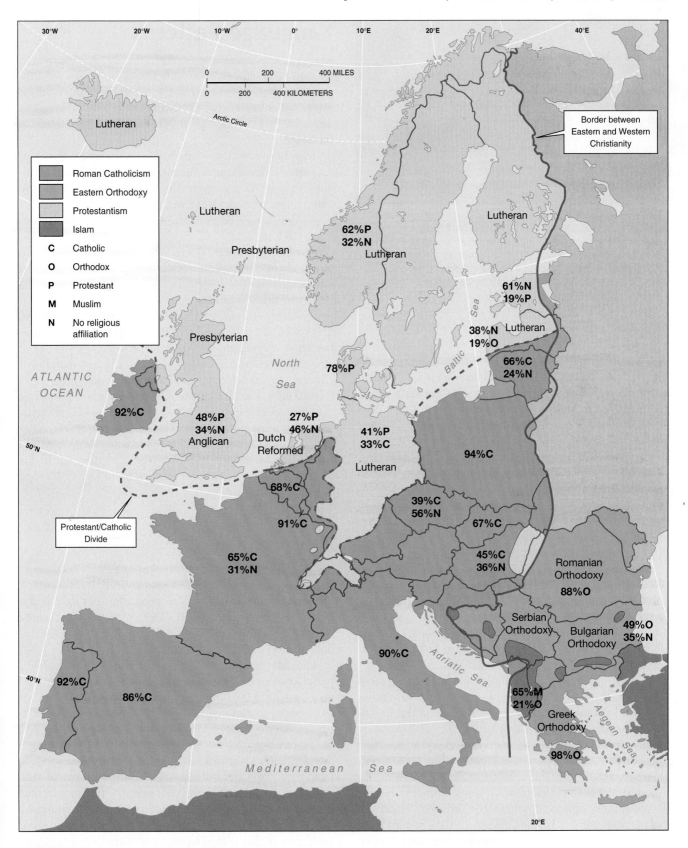

■ **FIGURE 11-2** **Major religions of Europe.** Northern Europe is predominantly Protestant, Western Mediterranean Europe is primarily Roman Catholic, and Eastern Orthodoxy prevails in much of Eastern Europe. Pockets of Islam are found in the Balkans.

■ FIGURE 11-3 Leonardo da Vinci's *Last Supper*. Much of Europe's greatest art, architecture, music, and literature has been inspired by Christian religious belief. This photograph was taken in 1982 during the early phases of a 22-year restoration project that ended in 1999. The painting was begun by da Vinci in 1495 in Milan's Our Lady of Grace church.

of the Balkan Peninsula in the realm of eastern Christendom. The Eastern Church is subdivided into several independent national denominations, including Greek, Romanian, Serbian, and Bulgarian orthodoxy. The distinctiveness of the eastern churches rests not only in their Greek origin but also in their greater emphasis upon mysticism and stoicism.

The second great schism, the Protestant Reformation, which split the western church, added a north-south aspect to the religious map of Europe (see Figure 11–2). **Protestantism** presently prevails in northern Europe, Roman Catholicism in the south. About 250 million Europeans are practicing or nominal Catholics, while about 90 million of diverse denominations claim Protestant affiliation.

About one-quarter of the European population professes no religious faith. Some scholars even describe Europe as a "post-Christian" society. Atheist/agnostic sentiment is strongest in the center, north, and east. The proportion of people claiming no religious faith is 46 percent in the Netherlands, 56 percent in Czechia, 61 percent in Estonia, and 32 percent in Norway (see Figure 11–2).

In spite of religious fragmentation and widespread secularization, the Christian heritage—more than any other trait—still defines Europe today. It remains at the root of the "we-versus-they" mentality of the typical European. Christianity underlies and inspires both the good and bad aspects of Europe: its great art, literature, music, and philosophy as well as its religious wars, genocides, ethnic cleansings, and inquisitions. One cannot imagine European culture devoid of the magnificent cathedrals, altarpieces, crucifixes, and religious statuary. Christianity gave us the *Commentaries* of Saint Thomas Aquinas, Leonardo da Vinci's *Last Supper*, Michelangelo's *David* and Sistine Chapel, Dante's *Inferno* and Milton's *Paradise Lost* (Figure 11–3). For many centuries, the Church was Europe, and Europe was the Church. All Europeans, regardless of their present religious beliefs, bear the permanent stamp of Christianity.

Language Patterns

Europe's language map is rather similar to the religion map (Figure 11–4). Three major subdivisions of the Indo-European family are widespread: the western and southern **Slavic languages** in the east, the **Romance languages** in the south, and the **Germanic languages** (including English) in the north.

Together, these three subdivisions account for the large majority of Europeans. Romance speakers number 186 million, or about 35 percent of Europe's population, while the 190 million, who speak one of the Germanic languages, account for another 36 percent. Some 86 million Slavs form about 16.5 percent of Europe's people. German, the language of 90 million Europeans living in eight different countries, is by far the largest single language. However, English functions as the most important lingua franca of international communication in Europe.

Smaller, peripheral Indo-European groups such as the Celts, Greeks, Latvians, and Lithuanians collectively number about 20 million, and the small remaining minority of languages, also concentrated on the outermost margins of Europe, do not belong to the Indo-European family.

Adding to the already complicated linguistic pattern is the division of Europe into two major alphabet traditions. Most Europeans employ the Latin alphabet, as does English, but in the southern Balkan Peninsula, the Greek and closely related Cyrillic alphabets prevail. Further north, this alphabet divide follows our eastern boundary of Europe (Figure 11–5).

Population

Within the boundaries of Europe, as defined here, 526 million people live, forming one of the most densely populated regions of the world. Four European countries have populations in excess of 50 million: Germany with more than 82 million, France and the United Kingdom each with 60 million, and Italy with 58 million.

The pattern of population density reveals a striking core-periphery contrast. Most of the densely populated districts—those having 500 or more persons per square mile (200 or more per square kilometer)—lie in or near the center of Europe, while regions of sparse settlement—fewer than 130 per square mile (50 per square kilometer)—appear around the perimeter (Figure 11–6, page 271). The thickly populated core is sometimes called the "European zone of continuous settlement." In the past century, this pattern has intensified, with emigration from the peripheries to the core.

The population of Europe, following three centuries of rapid growth, has ceased to increase. Births are almost precisely balanced with deaths, ending natural increase and achieving zero population growth. The small increases evident at present are due entirely to immigration. Natural population decreases are now occurring in Europe's center and east, causing some predictions of **demographic collapse**.

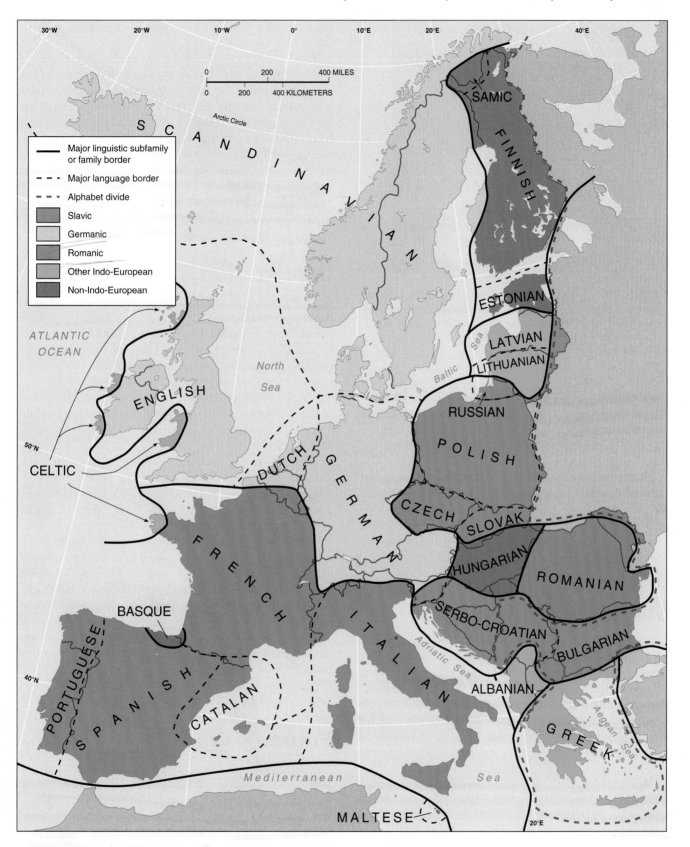

■ **FIGURE 11-4 Principal languages of Europe.** Three main branches of the Indo-European family of languages dominate Europe.

▪ **FIGURE 11-5   Road sign using the Latin and Cyrillic alphabets.** Language has historically been a divisive influence in Europe, yet linguistic differences need not prevent cooperation and trade between regions or nations. This photograph shows a Russian-Finnish border crossing.

Even modest natural increase can be found only in certain peripheral regions of Europe, including Albania, Macedonia, Ireland, and Iceland. Figure 11–7 shows the "total fertility rate" for each country, which is a good predictor of future population change.

Zero population growth, coupled with excellent health conditions, cause Europe's population to be the world's oldest. Of the twenty-one countries in the world with the highest proportion of their population aged 65 years or older, twenty lie in Europe. Monaco and Italy lead the list, each with 23 and 19 percent of the population, respectively, being elderly.

Standard of Living

Europeans, for the most part, are healthy, wealthy, and wise. Many different measures can be applied to reveal the level of living—literacy rate, life expectancy, per capita annual income, and the like. Many demographers have concluded that the single best indicator is the **infant mortality rate**—the number of children per thousand live births who die before reaching the age of one year. From this measure, we can learn much about nutrition, the availability and quality of medical care, and the health of the mother.

Europe enjoys the lowest infant mortality rate of any sizable part of the world—seven—and all of the countries bordering Europe, whether in Africa, Asia, or Russia, have much higher rates. Only the countries of the Balkan Peninsula, excluding Greece, have high infant mortality rates, and even there the rates are not high by world standards (Figure 11–8, page 273). In some places along Europe's outer border, the contrast is startling. Spain, for example, has an infant mortality rate of 4.4, while in adjacent Morocco, across the narrow Strait of Gibraltar in Africa, the rate is 41.6. Greece

has a rate of 5.5, while neighboring Turkey, across the border in Asia, has 41.0.

Does this high level of living mean that Europeans are content? Responses vary. In Denmark, 64 percent of the people report that they are "very satisfied"—Europe's most contented country by far. Only 4 percent of the Danes are "not satisfied" with their lives. By contrast, in Portugal only 7 percent are very satisfied and fully 35 percent report dissatisfaction. In Bulgaria, 3 percent are very satisfied and 67 percent are not. The highest suicide rates in the world occur in Europe, often in countries that rank among the wealthiest, healthiest, and best educated. Paradoxically, Europeans enjoy the world's highest living levels but are far from being the most contented people in the world.

Immigration

Europe's high level of living, coupled with a low birthrate, has attracted millions of immigrants from poorer parts of the world. This great immigration began around 1950 and was, in part, the result of the reluctance of the increasingly wealthy Europeans to engage in manual labor. Increasingly, manual labor and lower-paying jobs in the service sector employ non-European immigrants. For instance, 20 percent of the population in affluent Switzerland now are foreigners, and in Germany the proportion of foreign-born inhabitants is 10 percent.

Several major migration flows occurred. Asian Indians, Pakistanis, and West Indians, all citizens of the British Empire, migrated to the United Kingdom, while Germany drew mainly upon Turkey, and France drew upon its former colonies in North Africa and the Caribbean.

Most of this migration was legal and by invitation. Increasingly, however, Europe is being inundated by waves of illegal immigrants from Africa and the Middle East. Spain and Italy are the two main entry points. Should conditions reach crisis levels in Russia, another wave of illegal immigration might come from the east.

Even if no further immigration were to occur—an unlikely scenario—Europe's population has already been substantially altered. Eight million Muslim immigrants and their children live in Europe, as do perhaps 1.5 million blacks (Figure 11–9, page 274).

Europe's Cities

The overwhelming majority of Europeans, 74 percent, live in cities and towns of 10,000 or more people. In Belgium, Europe's most urbanized country, 97 percent of the people live in cities. European civilization has been closely bound to urbanism, and to belong fully to this vibrant culture, one must reside in a city (Figure 11–10, page 274). Not even the recent phenomenon of **counterurbanization** (see the *Geography in Action* boxed feature, Counterurbanization, page 276)—the movement of people from the larger cities out to more rural settings—has had much effect. Europe, as a culture, resides in its cities and towns.

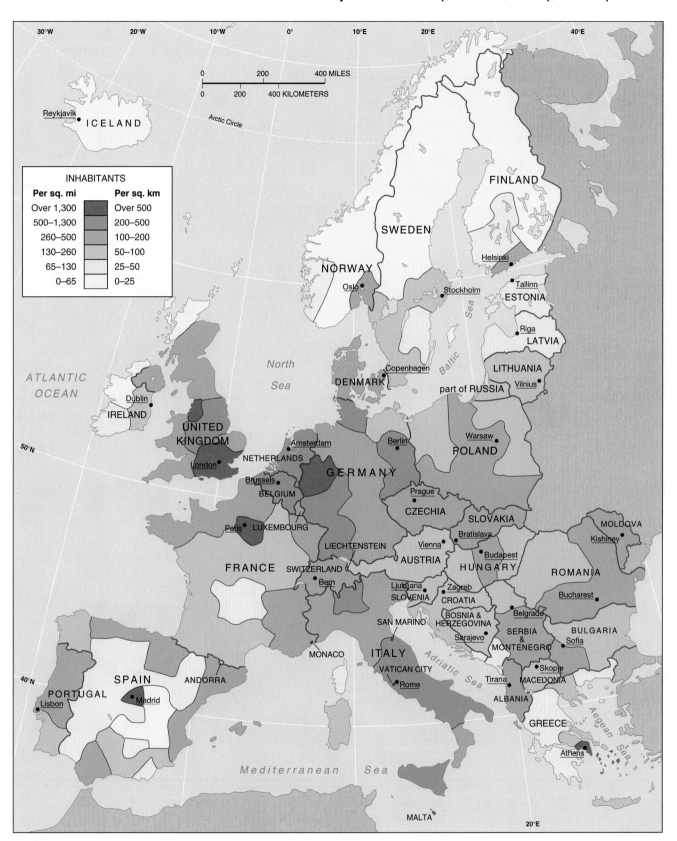

▪ **FIGURE 11-6 Population density of Europe.** A core-periphery pattern is evident.

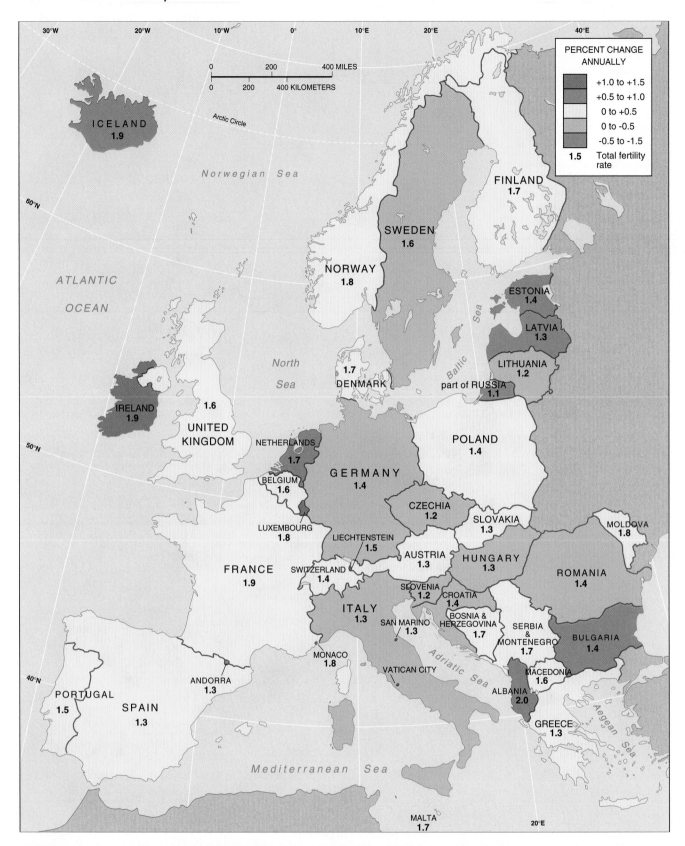

■ **FIGURE 11-7 European natural population annual growth or decline and the total fertility rate (TFR).** The TFR is the number of children, on average, born to each woman during her life; a TFR of 2.1 is required to avoid future population decline. In all of Europe, only Albania has a TFR that high, and as the map shows, population decline has already begun over much of Europe. Immigration and emigration are not considered in these percentages.

■ **FIGURE 11-8** **Infant mortality rate.** This is one of the best indicators of level of living. Even the highest rates, in Romania and Moldova, are low by world standards. Worldwide, the infant mortality rate is 54. An east-versus-west pattern prevails within Europe.

■ **FIGURE 11-9 Muslim of North African descend resid-ing in France.** North Africans constitute one of the largest groups of Western European migrant workers. Owing to historical ties, most of the North African workers have migrated to France.

A core-periphery contrast is evident in the spatial patterns of urbanization (Figure 11–11). The great cities of Europe, of which London and Paris are the largest, are concentrated in the core, as are the highest urban percentages. In the peripheries, especially in northern Europe, the eastern borderlands, and the Dinaric Range in the Balkans, the proportion of the total population residing in cities falls below 50 percent.

In many ways, European cities differ from those in North America. As a rule, the European city is far more compact, occupying half the area or less of an American city of comparable population. There is less suburban sprawl, and the inhabitants are much more likely to be apartment dwellers and to use mass transit than their American counterparts. Commuting substantial distances to work is uncommon.

The center of the European city consists of the **preindustrial core**, a medieval or even ancient nucleus of twisting, narrow streets and marketplaces—many now pedestrian zones—fronted by buildings not more than two or three stories tall (Figure 11–12). Church spires still dominate the skyline. The preindustrial core is usually a prestigious place to live, and housing there is very expensive. No expressways penetrate the core or, for that matter, any part of the European city. Around the perimeter of the preindustrial city you can sometimes find remnants of the old city walls and gates.

Surrounding the preindustrial core in the European city lies the **inner ring**. It began as a lower-class settlement zone outside the city walls—in French the faubourg, or "false city." In time, the hovels gave way to more substantial buildings as the city grew, and some fine neighborhoods developed. In function, the inner ring remains predominantly residential, as opposed to the retail/institutional focus of the city center. This zone is sometimes called the "preindustrial

■ **FIGURE 11-10 Athens, Greece.** Athens is the capital and largest city of Greece and has, for thousands of years, been a living expression of the European love for city life. Surrounded by the modern city, the Acropolis with the Parthenon dominates the traditional core of Athens.

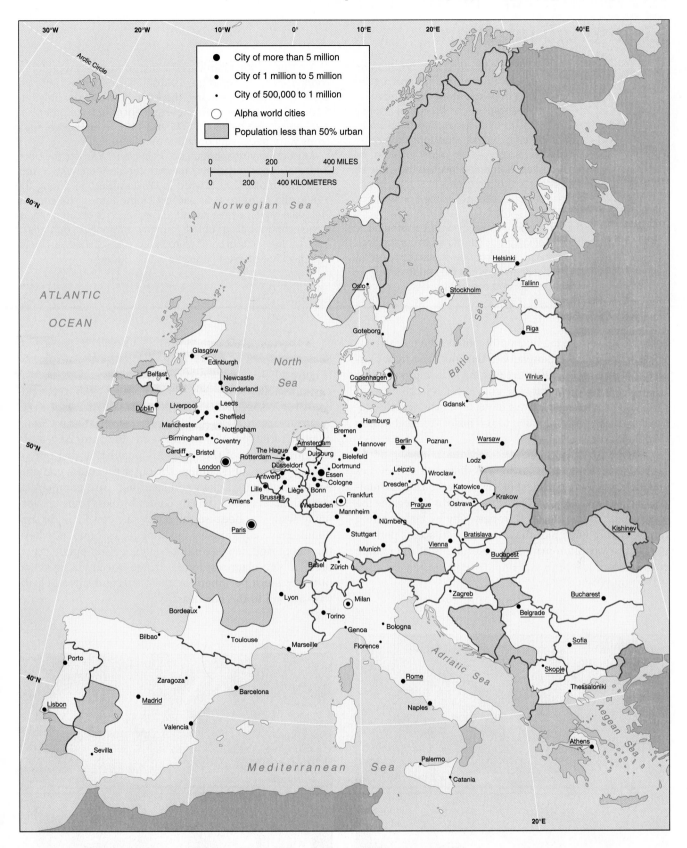

■ **FIGURE 11-11 Europe: distribution of large cities and predominantly rural areas.**
Urban refers to cities and towns of more than 10,000 inhabitants. Again, a core-periphery pattern is
evident. An **alpha world city** is not only large but also plays a major economic role in the world.

GEOGRAPHY IN ACTION
Counterurbanization

Beginning about 1965, a trend called counterurbanization began in Europe. Counterurbanization occurs when sizable numbers of people leave cities and move to rural areas. The larger cities stagnate or decline in population, while villages and small towns grow. Improved transportation and communication networks allow long commutes to work, and with the use of computers, many jobs can be performed at home. Counterurbanization involves the more affluent, professional people and often occurs in stages, beginning with the acquisition of a second home for vacation purposes.

The aging of the European population also contributed to counterurbanization, as affluent people retired to rural places. The sudden, rapid deindustrialization of Europe also contributed. Cities became unattractive to many, while the countryside gave the image of harmony, peace, and civility.

The United Kingdom was the first to experience profound counterurbanization, in the 1960s, followed almost immediately by western Germany, then the Netherlands, Belgium, France, and Denmark. France experienced a large growth in rural population between 1975 and 1982, even though many farms were being abandoned at that time.

Most counterurbanization occurs in areas closest to the cities. In some respects, it simply involves the development of suburbs. The movement slowed in Western Europe during the 1980s and came to a virtual halt by the turn of the century. The main region where counterurbanization is occurring today is the economically troubled, formerly communist countries of the east. The focus of counterurbanization has shifted east, helping magnify east-west contrasts within Europe.

---

suburb." The inner ring frequently contains the railroad stations, which ring the periphery of the core. The tight clustering of venerated old structures made it impossible for railroad lines to penetrate the center. Such large cities as Paris and London have a circle of rail terminals, and the stations typically bear the name of the major city that lies on their rail route.

The Industrial Revolution added a **middle ring** to the European city, mainly in the 1800s. A dingy halo of factories and huge workers' apartment blocks or row houses, the middle ring—often called the "industrial suburbs"—became the largest part of the typical European city by far, dwarfing the older core and inner ring. A controlled-access perimeter expressway often marks the outer limits of the middle ring.

After about 1950, many European cities, especially in the west, added an **outer ring** or "postindustrial suburb" (Figure 11–13). It consists of a loose assemblage of residences dominated by detached single-family houses, modern factories devoted to high-tech industry, and firms specializing in data gathering and processing. Some planned satellite towns also appear in the outer ring, as do the garden plots of the inner-city population. In spite of the low density of buildings, the outer ring often contains American-inspired glass-and-steel high-rise commercial structures. Access to the city center by mass transit is normally available. Because the outer ring developed in segments, it usually remains incomplete. In London, where the outer ring lies beyond a zoned, largely open "green belt," a high-class residential fringe began taking shape in the western periphery by the 1950s and spread clockwise around to the northeast.

■ **FIGURE 11-12 Riga, Latvia.** Like many older European cities, Riga, the capital of Latvia, has a well-preserved preindustrial core.

■ **FIGURE 11-13 Residential suburb of a Western European city.** La Queue-en-Brie, a residential suburb southeast of Paris, typifies the middle-class communities that have grown up within commuting distance of major European urban areas.

Primary and Secondary Industries

Europeans are an urban people because, preponderantly, they find employment in industrial and service-related pursuits (Table 11–1).

Primary industries are those involved in extracting resources from the earth and seas. Mining, forestry, fishing, and agriculture are four principal examples of primary activities. **Secondary industry** is the processing stage, commonly called manufacturing, in which the materials collected by primary industries are converted into finished products. Ore is converted to steel, fibers are made into cloth, and steel becomes automobiles and machines. The Industrial Revolution, which involved the shift to machine-made goods

and inanimate sources of power, began in Europe, and manufacturing remains an important component of the European economy.

By the beginning of the twentieth century, a huge industrial complex had developed in the European core, including twenty-five or so local manufacturing districts, many situated atop coalfields. Heavy industries, such as iron and steelmaking, vehicle manufacture, shipbuilding, and coal mining played prominent roles, as did chemicals, textiles, and food processing. In 1900, Europe accounted for about 90 percent of the world's manufacturing output. The majority of the European labor force worked in these and other industries.

Then, quite abruptly, everything changed, beginning about 1950 (Figure 11–14). The most important mass-production industries, both primary and secondary, went into severe and irreversible decline, prompting the use of the term **deindustrialization**. Once-prosperous industrial districts became derelict, pauperized, and eligible for economic assistance within two decades. Entire working-class communities were devastated and became dependent upon unemployment relief.

The statistical and visual evidence of deindustrialization is both convincing and sobering. In the United Kingdom, birthplace of the Industrial Revolution, the labor force employed in manufacturing plummeted from 9,119,000 in 1966 to 5,172,000 twenty years later. In the Midlands district, where the modern coal and steel industry was born, manufacturing employment fell by 55 percent during that two-decade span. In Wales, with its spirit and wealth broken, employment in mining dropped from an all-time high of 250,000 to only 24,000 as early as 1982. Overall, European employment in steel manufacture plummeted from 800,000 in 1980 to 280,000 by 2002. In just a two-year span, 2000–2002, steel production in Poland declined by 16 percent (Figure 11–15). Following the collapse of communism, Eastern Europe's primary and secondary industries declined even more rapidly in a single decade (see Figure 11–14).

Industrial Rejuvenation

Rejuvenation followed almost immediately, but the new industrial system and its geography differed fundamentally from the vanished older order. The new manufacturing industries specialized in high-quality and luxury goods requiring a skilled labor force, ongoing innovation, and sophisticated technology.

Europeans have successfully moved from an emphasis on mass-produced goods requiring large factories and minimally skilled labor to a focus on labor-intensive operations producing items of high value. They have chosen to depend upon the cardinal virtues of European culture—education, individualism, and innovation.

The most glamorous of these new enterprises can be grouped under the much-used term, "high-tech." These include industries manufacturing high-technology products such as electronic and microelectronic devices, data-processing equipment, robotics, telecommunications apparatus,

Table 11-1 Percentage of Labor Force by Economic Sector in Selected European Countries

Country	Mining, Forestry, Fishing, and Agriculture	Manufacturing	Services (including construction)
Albania	57	20	23
Austria	4	29	67
Belgium	1	25	74
Bulgaria	11	33	56
Croatia	3	32	65
Czechia	5	30	65
Denmark	4	17	79
Estonia	11	20	69
Finland	8	22	70
France	5	22	73
Germany	3	33	64
Greece	12	20	68
Hungary	6	27	67
Iceland	10	18	72
Ireland	8	29	63
Italy	5	23	72
Latvia	16	18	66
Lithuania	18	19	63
Luxembourg	1	13	86
Macedonia	25	24	51
Moldova	51	9	40
Netherlands	3	14	83
Norway	4	13	83
Poland	16	29	55
Portugal	13	20	67
Romania	38	21	41
Serbia & Montenegro	6	27	67
Slovakia	6	30	64
Slovenia	6	40	55
Spain	6	18	75
Sweden	2	24	74
Switzerland	4	17	79
United Kingdom	2	19	80

Sources: *Europa World Year Book 2004* (London: Europa Publications, 2004) and various national statistical yearbooks.

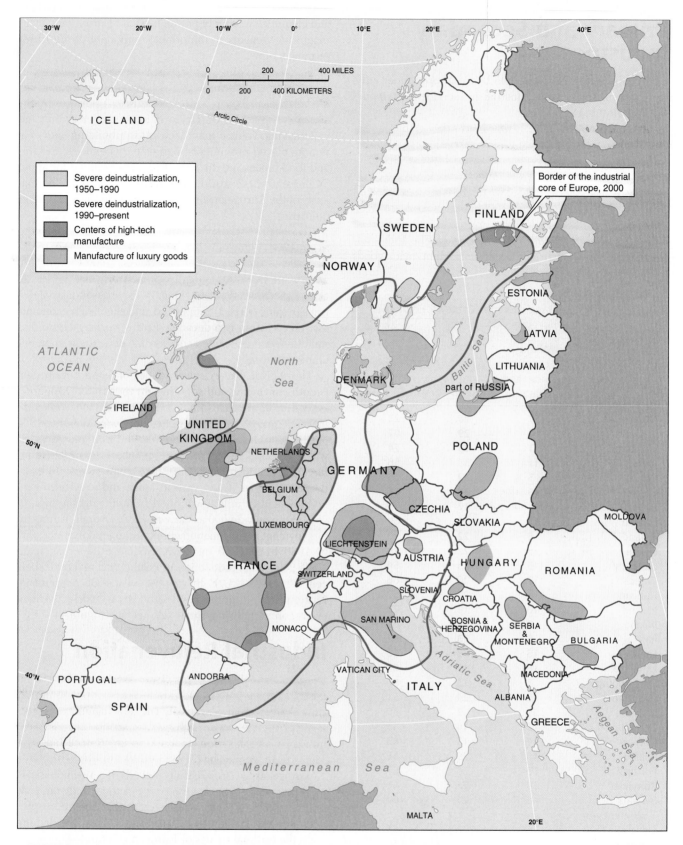

▪ **FIGURE 11-14   Europe: the shifting pattern of primary and secondary industry.** The modern industrial core is donut-shaped. Deindustrialization afflicts many areas of Europe's former manufacturing and mining core.

High-tech manufacturing, computer-based information services, and research-and-development activities tend to be highly concentrated geographically into areas called "technopoles." The United States has its "Silicon Valley" in California, and Europe has comparable districts, or "new industrial spaces." Most lie near major universities and airports, in suburban areas.

Among Europe's high-tech districts are England's "M–4 Corridor," a crescent-shaped area curling around the western and northern edges of London; Scotland's "Silicon Glen" near Edinburgh; the "Scientific City," a district stretching southwestward from Paris; south Germany's Bavarian complex, lying between Munich, Nuremberg, and Augsburg; the Netherlands' "Randstad"; and numerous other, smaller areas (see Figure 11–14).

So far, eastern and southern Europe have failed to develop technopoles. As a result, those regions fell ever further behind the European core and entered the twenty-first century in a disadvantaged position.

---

and the like. In addition, there are firms which make preponderant use of such sophisticated products in the manufacturing process, such as pharmaceutical firms and pesticide makers. Central to the entire high-tech enterprise is the computer (see the *Geography in Action* boxed feature, "Technopoles"). While much is made of high-tech manufacturing, and several European countries have tied their industrial future to such activity, these firms employ far fewer people than the old, collapsed ones.

Finland provides an example. Thirty-one of every 100 cell phones sold in the world in 2004 (and worth more than $25 billion) were made by Nokia, a Finnish company. This single enterprise accounts for one-fifth of Finland's exports by value and two-thirds of the country's stock market value. And yet, in a country of 5.2 million people, Nokia employs only 22,000 workers, with another 20,000 in related jobs.

More promising for Europe's future is the manufacture of high-quality expensive luxury goods, mainly for export. These factories operate as labor and design-intensive (rather than technology-intensive) craft industries. The highly skilled workforce enjoys high wages, job security, and the satisfaction of laboring in "a positive culture of work." Creativity, skill, and craftsman pride are all essential components. Factories tend to be small and the workplace pleasant. Typical products include ceramics, pottery, decorative glassware, fine clothing, jewelry, and quality leather goods such as shoes, well-crafted wood products, and luxury automobiles.

Geographically, deindustrialization and rejuvenation occurred in different regions, with little overlap of either area or workforce (see Figure 11–14). The districts benefiting from high-tech and craft industries lie outside the economically depressed, deindustrialized areas. In effect, a complex of new industrial districts replaced the old, all within the core of Europe. The new industrial heartland is smaller, most notably because formerly communist Eastern Europe has, so far, failed to participate in the new prosperity. Also, the new core is hollow in the center, excluding the British hearth areas of the Industrial Revolution. Yet the basic core-periphery configuration remains unaltered.

Service Industries

Service industries involve neither the extraction of resources nor manufacturing but instead include a broad range of activities such as government, transportation, banking, retailing, and tourism. The term **postindustrial** denotes the rise of dominance of **service industries** in modern Europe, coincident with deindustrialization.

The most outstanding fact concerning service industries in Europe is their disproportionate recent growth. While most service activities have always been present, a marked increase has occurred, especially in the western part of Europe. Well over half of the labor force in many countries and provinces now finds employment in service industries (see Table 11–1). In Austria, for example, jobs in the services sector rose from 30 percent of the workforce in 1961 to 67 percent today. In all parts of Europe, job losses in the traditional primary and secondary sectors of industry have been offset by large gains in the service sector.

No service activities better epitomize Europe than transportation and communication. For centuries, European culture has thrived upon the most complex and complete transportation and communication

■ **FIGURE 11-15 Deindustrialization in Poland.** An abandoned factory near the Gdanski train station in 1999.

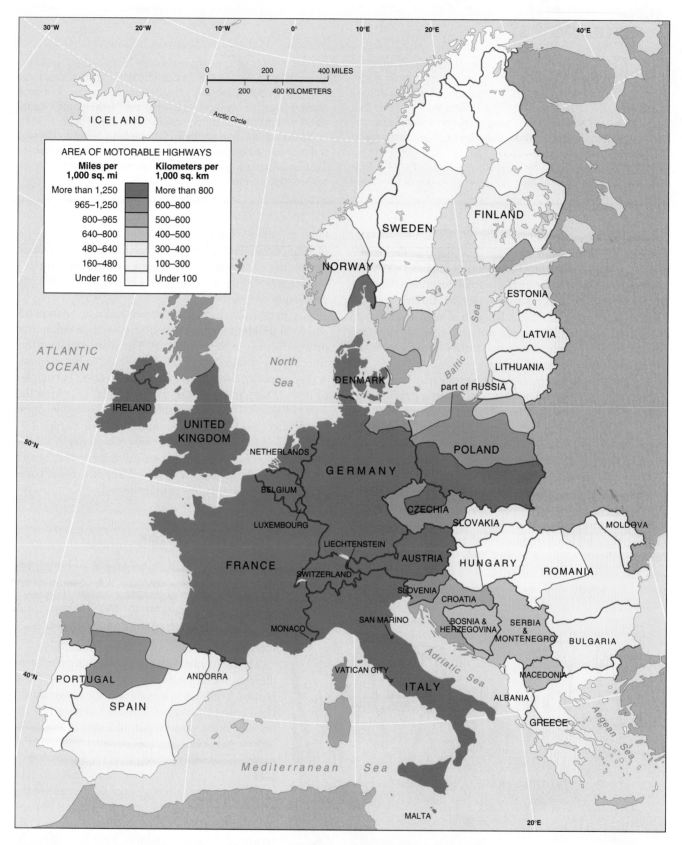

■ **FIGURE 11-16 Europe: density of all-weather, motorable roads.** The core-periphery pattern is vivid. Compare to Figure 11–6.

network of any region of comparable size in the entire world. Since Roman times, Europe has possessed an extraordinary network of all-weather roads. The geographical concentration of modern highways lies in the center of Europe, again revealing a striking core-periphery pattern (Figure 11–16). This distribution also typifies railroads, airline connections, and the network of canals and navigable rivers.

Energy production is another major service industry. Europeans consume vast amounts of energy in sustaining their high level of living. Once dependent upon its own coal mines for most of the energy needs, Europe now relies upon a mixture of petroleum, natural gas, hydroelectric, aeolian, geothermal, and nuclear power (Table 11–2).

Physical geography plays a significant role in the production of electricity. Predominantly mountainous terrain allows countries like Norway, Iceland, Switzerland, and Albania not only to generate inexpensive hydroelectric power but also to contribute minimally to air pollution and global warming by basically avoiding the burning of fossil fuels. Countries like Denmark and the Netherlands are leaders in harnessing wind energy. A growing alternative to coal, oil, and gas as sources of energy is nuclear power. In just five years countries such as France, Belgium, Finland, and Spain have more than doubled their reliance on nuclear fission reaction as the source of energy for commercial purposes. France is an internationally recognized leader in scientific research on the utilization of nuclear energy and has more nuclear energy plants than any other country in Europe. In 2005, France was selected as the site for an experimental nuclear fusion reactor. The hope is that this developing technology will, over the long term, prove to be able to provide large amounts of energy with lower levels of waste. This is an important issue, as the problem of the safe, long-term disposal of radioactive waste has yet to be solved satisfactorily.

An entire sector of postindustrial activity consists of higher-order services catering to large business enterprises. These **producer services** include banking, accounting, advertising, legal services, research and development, insurance, marketing and wholesaling, real estate brokerage, various types of consulting, and the processing and provision

Table 11-2 Percentage Production of Electricity by Source for Selected European Countries

Countries	Nuclear	Hydro	Fossil Fuels	Other	Comment on "Other"
Albania	-	97.1	2.9	-	
Austria	-	12	76	12	combustible wastes
Belgium	59.3	1.8	38.4	<1	
Bosnia, Herzegovina	-	46.5	53.5	-	
Bulgaria	44.1	8.1	47.8	-	
Croatia	-	66	33.6	0.4	
Czechia	20	2.9	76.1	1	
Denmark	-	0.1	82.7	17.3	notable wind generation
Estonia	0.2	0.1	99.8	-	
Finland	30.4	18.7	39	11.8	combustible wastes
France	77.1	14	8.2	0.7	
Germany	29.9	4.2	61.8	4.1	
Greece	-	3.8	94.5	1.7	
Hungary	39	0.5	60.1	0.4	
Iceland	-	82.5	0.1	17.4	geothermal
Ireland	-	2.3	95.9	1.8	
Italy	-	18.4	78.6	3	
Latvia	-	70.9	29.1	-	
Lithuania	77.7	5.7	16.5	-	
Luxembourg	-	25.2	57.3	17.5	combustible wastes
Macedonia	-	16.3	83.7	-	
Moldova	-	9.4	90.6	-	
Netherlands	4.3	0.1	89.9	5.7	wind generation
Norway	-	93.3	0.4	0.4	
Poland	-	1.5	98.1	0.4	
Portugal	-	31.3	64.5	4.1	wind; combustible wastes
Romania	9.9	27.6	62.5	-	
Serbia, Montenegro	-	37.1	62.9	-	
Slovakia	53.6	16	30.4	-	
Slovenia	36.8	27.3	35.2	0.7	
Spain	27.2	18.2	50.4	4.1	wind generation
Sweden	43	50.8	4	2.3	
Switzerland	37.1	59.5	1.3	2	combustible wastes
United Kingdom	23.7	0.9	73.8	1.6	

Source: http://www.odci.gov/cia/publications/factbook/geos/ee.html (2005 est.).

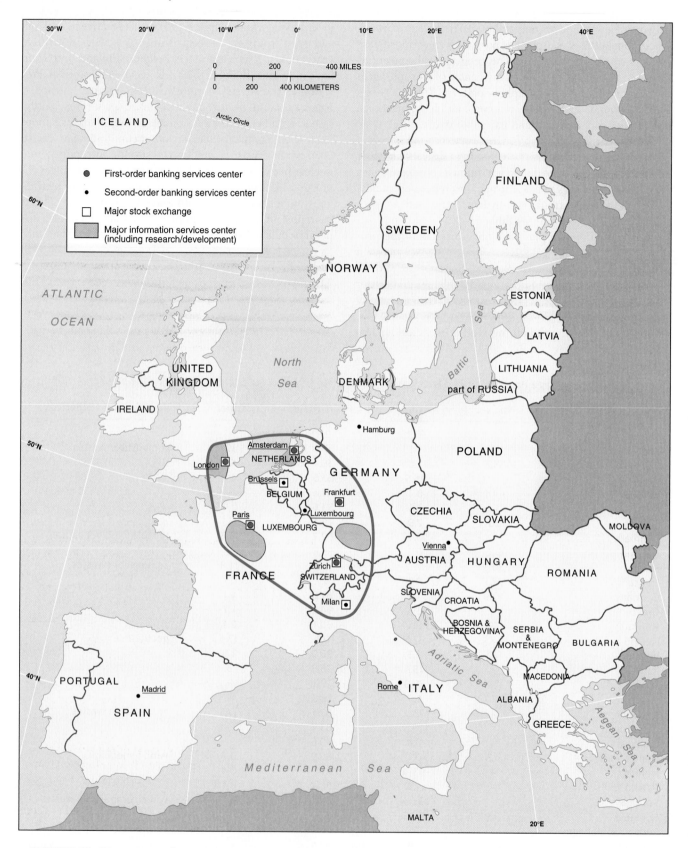

■ **FIGURE 11-17 Centers of European producer service industries.** Banking, insurance, research and development, information processing, stock exchanges, and other such activities are highly concentrated in an inner-European core.

Table 11-3 Number of Tourist Arrivals Annually in Selected European Countries

Country	Tourist Arrivals (thousands)	As % of Country Population
Albania	342	11
Austria	18,180	225
Belgium	6,451	63
Bulgaria	6,189	79
Croatia	6,544	147
Czechia	5,194	51
Estonia	1,231	91
France	77,012	129
Germany	16,977	21
Greece	14,033	127
Iceland	303	105
Ireland	6,450	162
Italy	35,768	62
Latvia	322	14
Lithuania	350	10
Luxembourg	876	195
Macedonia	122	6
Moldova	23	>1
Monaco	262	819
Netherlands	9,499	58
Norway	4,244	93
Portugal	12,167	117
Romania	3,274	15
Serbia & Montenegro	351	3
Slovakia	1,219	23
Slovenia	1,302	65
Spain	50,093	117
Sweden	9,133	102
Switzerland	7,454	102
United Kingdom	24,180	41

Sources: Europa World Year Book 2004 (London: Europa Publication, 2004) and *Yearbook of Tourism Statistics 2003* (Madrid: World Tourism Organization, 2003).

■ **FIGURE 11-18 Seaside resort in Spain.** Many Mediterranean Basin countries rely heavily on tourism to promote economic development. This is a beach on the Spanish island of Mallorca.

of knowledge and information. Such activities, as well as many corporate headquarters, cluster in the inner core of Western Europe, revealing the dominant economic position of that small region (Figure 11–17).

London and Paris quickly became the financial centers of the world. While later surpassed by New York and Tokyo, they remain very powerful, especially within Europe. London still boasts eight of the world's twenty largest insurance firms. Zürich ranks as a world-famous banking center, as does Frankfurt. The service industry devoted to the computer-assisted generation, storing, and processing of diverse types of information is clustered in the same European inner core that houses financial services.

Tourism—the short-term movement to destinations away from the place of permanent residence for reasons unconnected to livelihood—forms an integral part of the European lifestyle and is another service industry. In the western half of Europe, tourism experienced very rapid growth after 1950, especially in the two decades following 1965, during which the number of tourists doubled. Tourism constitutes an important growth sector in the economy and provides one of the main reasons why employment in service industries has grown so large.

Some countries, including France, Spain, and Italy, especially benefit from the growing numbers of international tourists (Table 11–3). A significant proportion of these travelers come on religious pilgrimages. Lourdes, a small town in southwestern France, alone hosts more than 1.5 million people each spring and autumn. In April of 2005 more than 2 million people came to the Vatican to bid their farewells to Pope John Paul II.

There is also a persistent seasonal pattern of tourist migration within Europe. Each summer Europeans from colder climates descend on the Mediterranean coast (Figure 11–18). For southern countries this "climatic pilgrimage" is economically very important. Croatia, for example, heavily relies on tourists from rich countries. With a sunny climate, relatively undisturbed habitat, and pristine beaches on the Adriatic, it is an attractive destination. Millions of Swedes, Germans, and Italians especially patronize Istria, Croatia's northern region. Despite the disruption caused by the civil war, Croatia's tourist industry has been rapidly recovering and already by the late 1990s this vital economic sector was earning billions of dollars.

Europe, then, has evolved into a postindustrial status, in which most people find employment in a diverse array of service industries. This development leads the Europeans into uncharted waters, just as the original Industrial Revolution did a quarter-millennium ago. As always, change remains the hallmark of the European way of life.

Agriculture

Today only a small minority of Europeans work in agriculture. In Belgium, for example, farmers make up less than 3 percent of the workforce. While percentages are higher

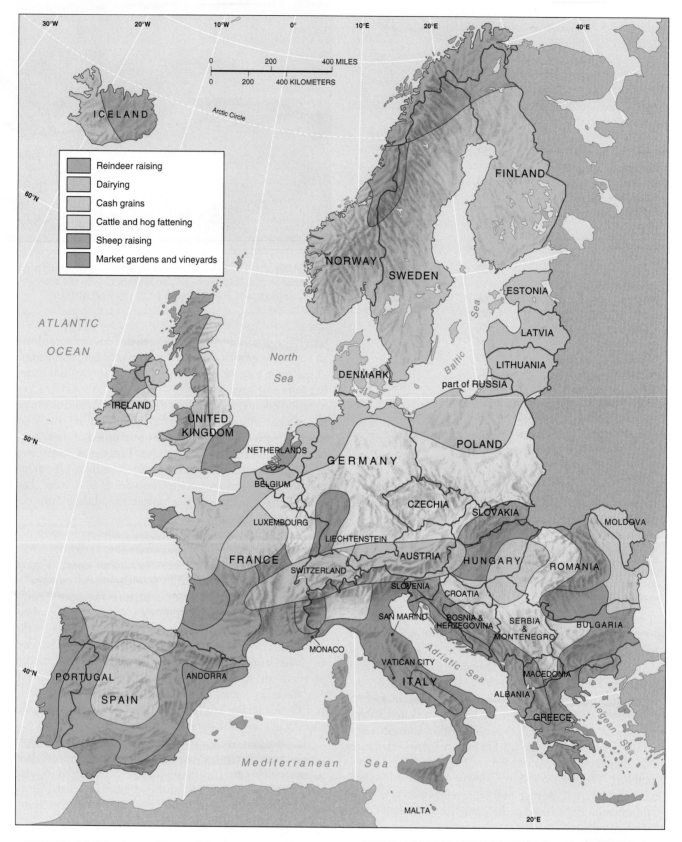

■ **FIGURE 11-19 Types of European agriculture.**

in the southern and eastern peripheries of Europe, the typical European is a city dweller. Rural depopulation is common, and abandoned farms dot many landscapes.

Those who remain on the land pursue some form of commercial agriculture, oriented to supplying urban markets. Highly market-oriented farming dominates, often employing Green Revolution technologies. These types of agriculture vary from one region to another (Figure 11–19). In the sunnier regions, mainly in the south, **market gardening** prevails. In this system, fruits and vegetables are produced for market. The main products are wine, citrus, olives and olive oil, green vegetables, apples, and other tree fruit.

On the Great European Plain, as well as in the Alps, dairying is the dominant type, supplying the enormous northern European appetite for cow's milk, butter, and cheese. Central Europe produces fattened hogs and cattle, and in the far north, the Sami (Lapps) raise reindeer for their meat. Wheat is produced in certain fertile plains areas, from Spain to the Balkans—Europe's main cash grain areas. Hilly, less fertile regions often specialize in raising sheep, both for wool and meat.

Of all European countries France has the most versatile agricultural sector (see Figure 11–19). More than one-third of its land is of good arable quality. The combination of mainly flat plains and gently rolling hills and a relatively mild climate provides a good base for a wide variety of agricultural activities.

In Normandy the wheat fields are interspersed with fruit orchards and dairy farms, while in the south famous French wines are produced (Figure 11–20). But it is not the diversity of French agricultural produce that generates most of the revenues. The French have been able to capitalize on the image of French taste for elite cuisine and have successfully promoted so-called luxury foods worldwide, including select wines and cheeses.

But all of these agricultural activities do not produce nearly enough food for Europe's 526 million people. For at least three centuries, Europe has been a food-importing region. In the 1960s the European Community adopted the Common Agricultural Policy (CAP), with the goal to produce enough food for internal consumption and reduce dependency on food imports. In economic terms it dictated the allocation of a substantial share of the joint budget to farming subsidies. Artificially high prices were maintained, and at one point up to 75 percent of the budget was spent on the CAP. Countries with a relatively strong agricultural sector benefited from this policy. Others, especially highly industrialized countries like the United Kingdom, rebelled against what they saw as a disproportionate and unfair allocation of common financial resources. In the 1970s, then British Prime Minister Margaret Thatcher was able to get a so-called budget rebate for the United Kingdom through a series of rough negotiations in Brussels, especially with France, Germany, and the Netherlands.

At present more than 40 percent of the European Union budget is spent on subsidies for agriculture, with France receiving the largest share. Almost 5 percent of the French

■ **FIGURE 11-20 A vineyard estate in Bordeaux.** Many estates specialize in the production of elite wines, often priced at several hundred dollars a bottle.

labor force is involved in agriculture, the highest percentage for any advanced economy. The French government maintains it is vital to keep people working the land, not so much for economic gains but for the preservation of the national spirit and heritage. There is certainly a great clash of interests between the leading European powers when it comes to expenditures. Most recently British Prime Minister Tony Blair expressed his country's dissatisfaction with the situation, stating that Europe should invest more in technology and advanced education to be able to compete globally, rather than "to give two euros to every cow."

▶▶ Summary

Europe, in spite of its small size, is a culturally, socially, and economically diverse region. A complex mixing of core-periphery, east-west, and north-south patterns seems to dissolve almost into geographical chaos. How simple, in some ways, does Anglo America seem by comparison! This fundamental complexity is further revealed in the political geography of Europe.

▶▶ Key Terms

alpha world city	outer ring
core-periphery pattern	postindustrial service
counterurbanization	industries
deindustrialization	preindustrial core
demographic collapse	primary industries
Eastern Orthodoxy	producer services
Germanic languages	Protestantism
infant mortality rate	Roman Catholicism
inner ring	Romance languages
market gardening	secondary industry
middle ring	Slavic languages

■ **Checkpoint Charlie in Berlin.** Europe's political geography is diverse, complex, and deeply rooted in the past. This split image shows one of the major crossing points between the former American and Soviet zones in post–World War II Berlin. At the height of the Cold War (top) crossings were difficult, dangerous, and infrequent. More than ten years after the fall of the Berlin Wall in 1989, only a replica guardhouse remains in commemoration of that historical past, and traffic now flows freely in a reunified Berlin.

Chapter 12

Europe:
Political Geography

- **Separatism**
- **Germany**
- **France**
- **Switzerland and Belgium**
- **Yugoslavia: A Balkan Tragedy**
- **The United Kingdom**

- **Italy**
- **Spain**
- **Poland**
- **Romania**
- **The European Union**

I n the political geographical sense, Europe presents the paradox of diametrically opposed trends—fragmentation and, simultaneously, federation and union. At the dawn of the twentieth century, Europe (excluding Russia) had twenty-two independent states. As we entered the twenty-first century, that number had increased to forty in an area significantly smaller than the United States. Six **ministates** are included, which are so small you can hardly see them on a map of Europe (see the *Geography in Action* boxed feature, Liechtenstein: A European Ministate).

Separatism

The fragmentation has not necessarily ceased, for nineteen restive provinces exist today where at least part of the population has expressed a desire for independence (Figure 12–1). Violence has occurred in many such regions.

Separatism in Europe is located peripherally within existing countries and is usually cultural in nature. That is, the groups seeking independence most often form an **ethnic minority** on the edge of a larger state. They differ linguistically or religiously from the country's majority. If a country's population is relatively homogeneous linguistically and religiously, as in Greece, Portugal, Germany, and Austria, separatism will not likely arise, but most European states house sizable, regionally concentrated ethnoreligious or ethnolinguistic minorities rooted in ancient homelands. Ethnicity is territorial by nature, and territoriality provides the very essence of **nationalism**—the feeling of belonging to a separate people.

Not all ethnic minorities pursue separatism in Europe. Switzerland successfully joins four linguistic and two religious groups, none of which seeks to depart from the Swiss Confederation. To spark separatism, ethnicity must be linked to some grievance. If ethnic minority status means persecution, attempted ethnocide, forced assimilation, domination, lack of autonomy, or denial of access to the country's power structure, producing second-class citizenship, grievances develop. Europeans possess long memories, and even abuses that occurred centuries ago are rarely forgotten. If the peripheral location of the ethnic homeland also causes it to be economically poorer than the country at large, then class struggle is added to ethnic grievance—a potent combination. Ethnicity then provides the vehicle for class struggle.

Not all secession movements in Europe have an ethnic basis. Other reasons can cause people to identify more closely with their province than with the central state, and the resultant regionalism and sectionalism often become sufficient to fuel separatist sentiment. Geographers recognize that nationality can derive as much from region and place as from cultural affiliation. Attachment and loyalty to regions must be considered in studying nationalism, especially in Europe, where people typically live in areas inhabited by their ancestors for centuries or even millennia. Dramatic and mundane things happen in places over the centuries, giving them nationalistic meaning. We should not be surprised, for example, when the inhabitants of many German-speaking regions seek or desire no connections with Germany, as is true of the German–Swiss, or when many northern inhabitants of Padania, the prosperous Po River valley region of northern Italy, seek to secede from Italy. Here the political party called the Northern League advocates independence in order to break free from the poverty-plagued, crime-ridden Italian south.

With these considerations of political fragmentation and separatism in mind, let us now turn our attention to some representative independent countries in Europe. Perhaps

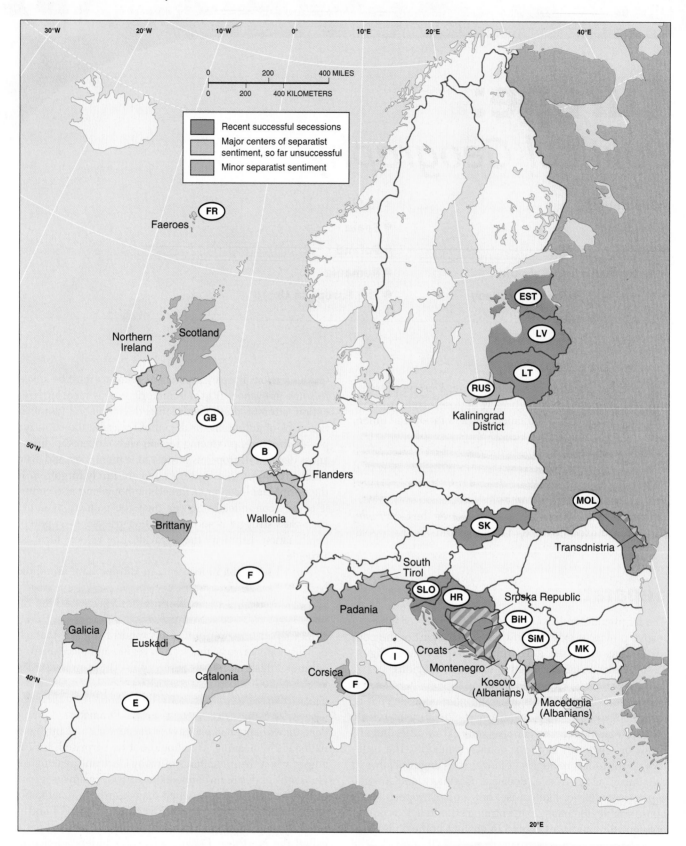

▪ **FIGURE 12-1 Political separatist movements of Europe.** Between 1990 and 1995 alone, nine new independent states emerged, and the end may not be in sight, as secession movements are afoot in numerous countries. For the key to the country symbols used on this map, see Figure 10–1.

Liechtenstein: A European Ministate

Among the independent countries in Europe, six are so small as to be referred to as ministates. If your eyesight is worse than 20/20, you will not be able to see these tiny countries on a map of Europe! How do they survive? The example of tiny Liechtenstein will help us understand.

A wedge-shaped splinter of land lodged between Austria and Switzerland in the Alps, Liechtenstein has only 62 square miles (161 square kilometers) of territory—two-thirds of it rugged mountains—and 33,863 inhabitants (Figure A). By comparison, a typical county in the United States has fifteen times as much territory as Liechtenstein. The words of the national anthem honor their "little home on the young Rhine." Liechtenstein was founded as a principality in 1719, within the First German Reich, and attained full independence in 1866. It has no army.

The local language is German, and most of the people are Roman Catholics. Liechtenstein functions as a constitutional monarchy, ruled by a prince who is the last reigning Hapsburg and lives in a mountaintop palace in Vaduz, the capital. The people are healthy, wealthy, and well educated. Swiss currency is used in Liechtenstein, and the country has a customs union with Switzerland, which also handles most of its foreign affairs.

Prosperity is based largely on the presence of numerous corporations, attracted to establish headquarters there by Liechtenstein's low corporate taxes. Liechtenstein also gets income from tourism, from a banking establishment that offers cus-

tomers considerable anonymity, and from the sale of its national postage stamps, printed since 1912, to collectors. If it were to join Switzerland or Austria, these advantages would be lost. On the negative side, in 2000 Liechtenstein was placed on a blacklist of notorious international tax havens by the Organization for Economic Cooperation and Development. It joined fellow European ministates Andorra and Monaco as countries where ill-gotten wealth could be hidden and income taxes avoided by foreign depositors.

■ FIGURE A Flag of Liechtenstein.

---

the most appropriate one to look at first is Germany, which dominates the core of Europe.

Germany

Germany is both an ancient and a relatively new country. The German "First Reich" (Empire) arose over a thousand years ago, only to fragment gradually into numerous small principalities and duchies. In 1871, under the leadership of Prussia, a northern German kingdom, Germany was united again, forming the Second Reich, which in time became the enlarged, totalitarian Third Reich under Nazi rule. Defeated and occupied by foreign armies in 1945, Germany again fell into disunity.

The modern Germany emerged only in 1990, when the western Federal Republic, which had arisen in the military occupation zones of the Americans, French, and British, reunited with the eastern German Democratic Republic, the old Russian occupation zone. In this way, a country over a millennium old was once again united.

The essential problem of the German state—a problem that helps explain its repeated fragmentation—is that while it occupies the core of Europe, it also straddles the east-versus-west and north-versus-south divides that characterize Europe (Figure 12–2).

The north-south divide is ancient, going back 2000 years to the time when southern Germany was part of the Roman

Empire, while the north was controlled by various Teutonic tribes. Rome placed its cultural imprint on the south, and such imprints never completely disappear, even after many centuries. Thus, the German south remains predominantly Roman Catholic, while the north is largely Protestant. Dialects of the German language are also arranged in belts separating south from north. Add to this the fact that the Second Reich was formed when the Prussian North annexed southern Germany, and you can fully appreciate the depth of the north-south divide.

The east-west division within Germany is equally ancient and deeply rooted. A thousand years ago and more, German peoples lived only in lands west of the Elbe and Saale rivers (see the "German/Slav border, A.D. 850" in Figure 12–2).

Slavic tribes possessed eastern Germany, and the Teutonic peoples very gradually spread east, absorbing the Slavs. In the process, they became considerably Slavicized. In that eastern area, the state of Prussia arose, centered at Berlin and in the surrounding Mark Brandenburg, and the ultimate origins of the Second Reich are to be found there. Significantly, Berlin was restored as the country's capital city in 1999, acknowledging its ancient role. As late as 1795, Prussia remained confined to the east, as did residual feudalism, in which landless peasants worked on manorial estates. The distinctiveness of eastern Germany was powerfully reinforced between 1945 and 1990, when it was a communist country under Russian domination while West Germany

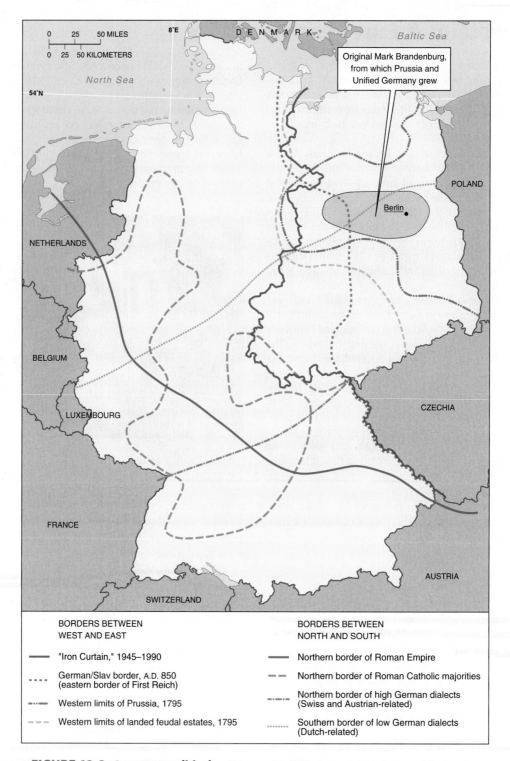

■ **FIGURE 12-2 Germany: political patterns.** The challenge to the German state has always been to unite contrasted east and west, north and south.

belonged to the free, democratic, capitalist part of Europe. In spite of the dismantling of the Berlin Wall in 1989 and reunification in 1990 (Figure 12–3), the old "**Iron Curtain**" border has not entirely disappeared. *Ossis* ("easterners") and *Wessis* ("westerners") still differ sharply in standard of living and political ideology within Germany.

The modern Germany has sought to overcome the east-west and north-south splits by structuring itself as a **federal**

state, in which considerable powers are vested in the individual provinces, and the central government is relatively weak. In earlier incarnations—the Second and Third Reichs—Germany responded to these internal divisions with a strong central government, a **unitary state**.

Today, fifteen years after reunification, Germany continues to struggle with the enormous challenges of integrating its disparate parts into a harmonious whole, overcoming the

economic, social, and political divide that was reinforced over so many decades. While infrastructure in the east—roads, railway lines, airports and the like—is often the most modern in the land due to large subsidies, the employment situation is extremely difficult and creates much social unrest. The unemployment rate in the eastern states of the federal republic is around 20 percent, twice the size of unemployment in the western states. It has proven far more difficult than anticipated to revive and modernize the economy of the eastern part of the country and make it sufficiently competitive. Far from being self-sustaining, it continues to devour enormous subsidies, which constitute a big burden on the federal budget and the taxpayer community. While eastern Germany, given its western German support system, has fared better than other countries of the former communist bloc, it will take decades to create economic equality between the two disparate parts. Change is a slow and painful process that tests the patience of all Germans in varying ways. The new government that has emerged from the most recent federal election and has propelled forward Germany's first female Chancellor, has the formidable task of reducing the colossal budget deficit by reducing spending and increasing tax revenue.

France

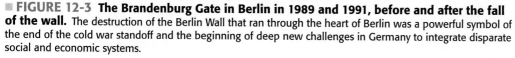

An old and stable country, France, in many ways, offers a model for the evolution of a viable, successful state. It grew from a nucleus, the *Île de France,* located in the center of the Paris Basin (Figure 12–4). The Franks, a Germanic tribe that had invaded the collapsing Roman Empire and adopted Latin speech, built the French state and gave their tribal name to it.

The Paris Basin, forming a deep embayment of the Great European Plain, provided a splendid natural framework for the infant France. From that fertile base, the French expanded, moving westward to dislodge the English from the Atlantic coastal margins, southward to reach the Mediter-

ranean Sea and the Pyrenees Mountains, and eastward, at the expense of Germany, to the banks of the Rhine River. The city of Paris, the *Île de France,* and the Paris Basin never lost their early dominance. A unitary state with an all-powerful central government based in Paris has always characterized France (Figure 12–5).

Two challenges, both involving ancient cultural regionalism, have faced the French state. The more profound of these was the task of uniting a Germanized north with a Mediterranean south. The northern French, occupying the Paris Basin, descended in large part from Germanic tribes, not only the Franks, but also Normans (Vikings), Burgundians, and others. While these tribes abandoned their ancestral languages and adopted French, they remained Germanic in many other ways. Their houses, villages, and farm field patterns are Teutonic in appearance and origin; their Germanic diet relies heavily on bovine dairy products, including butter for cooking; and their dialect of French is distinctive (see Figure 12–4).

The lands in the south retain a much stronger Roman and Mediterranean imprint, with minimal Germanic influence. We can detect this in the cultivation of grapes for wine, the use of olive oil in cooking, and the separate dialect (or language) known as *Occitan* (or Langue d'oc). These are just a few expressions of a deeply rooted, distinctive southern regional culture. France's essential difficulty, then, has been in welding a Germanized north to a Latinized south.

The other challenge to the French state results from the annexation, centuries ago, of peripheral lands inhabited by ethnic minorities. The outer margins of France are peopled by speakers of the German, Italian, Dutch, Catalan, Basque, and Celtic languages. These minorities have sought independence or annexation to neighboring countries, where their languages are spoken. Alsace-Lorraine, the German-speaking part of France, was periodically lost in wars with Germany.

The French reaction to these challenges has been to develop an all-powerful unitary state, based in Paris, and to

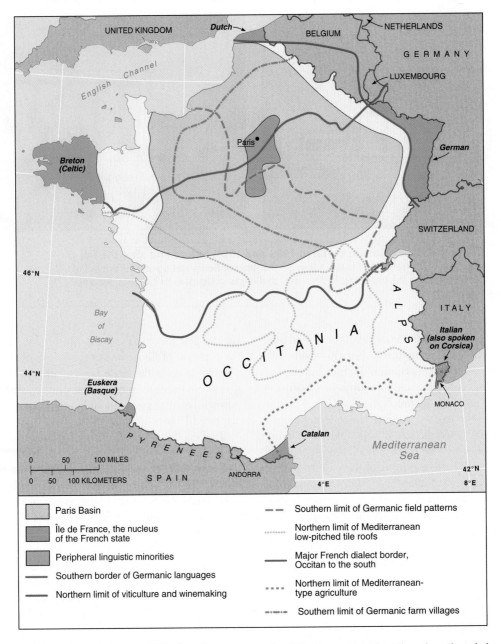

■ **FIGURE 12-4** **France: political patterns.** France has joined a contrasted north and south and also dominated ethnic peripheries through a strong central government that denies power to the provinces.

impose the will and culture of the north on both the south and the restive peripheries. This strategy has succeeded. France is a stable, firmly unified country. No province possesses enough power to challenge Paris.

Switzerland and Belgium

The most important language frontier in Western Europe separates the Romance languages in the south from the Germanic languages of the north. Switzerland and Belgium both straddle this cultural border and are similar in certain other respects, but only one of the two countries has achieved stability and relative internal harmony.

Switzerland, one of the most viable European states, has enjoyed considerable success in joining different linguistic

and religious groups in a multinational country. About 73 percent of the Swiss speak German, 21 percent French, and 4.3 percent Italian, in addition to a tiny Romansh minority. Religiously, the country is almost evenly divided between Catholics and Protestants (Figure 12–6). The Alps provide additional challenges as they cut directly through the state, dividing much of Switzerland into separate valleys, each with its own identity and sense of autonomy. In spite of these problems, Switzerland recently celebrated its 700th anniversary as an independent state.

The origins of the Swiss state can be traced to a core area in the 1200s around the shores of Lake Lucerne. Feudal lords there banded together in 1291 for mutual defense, and over the centuries, they gradually increased their autonomy within the loose-knit German First Reich,

■**FIGURE 12-5 Paris, France.** Paris has long dominated the rest of France and continues to function as the nation's primate city. The Seine River is visible to the right.

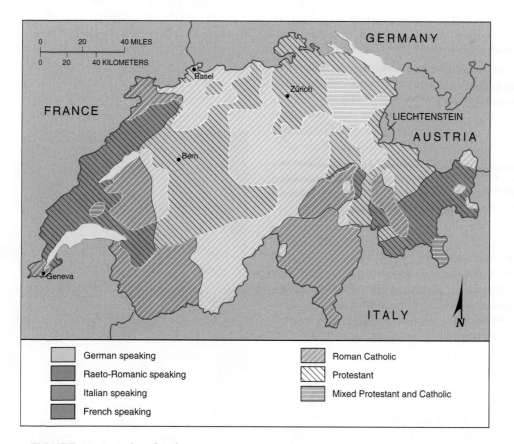

■**FIGURE 12-6 Switzerland.** An old successful confederation.

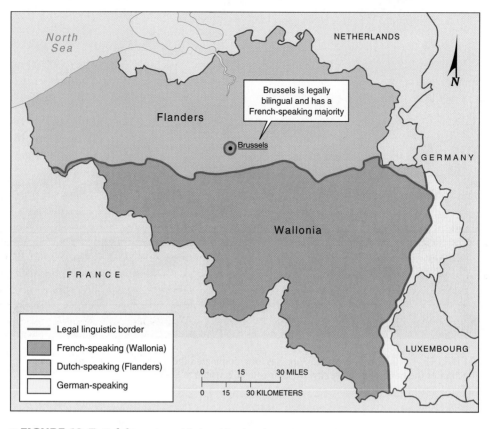

■ **FIGURE 12-7 Belgium.** A troubled multinational country.

strengthened by their control of a strategic alpine pass. As time passed, neighboring cantons (districts) and towns were annexed by conquest, purchase, or voluntary association. From the first, Switzerland formed a very loose federal state, or **confederation**, a framework that successfully accommodated subsequent linguistic and religious diversity as the country grew from its tiny core area. In recent times, Switzerland achieved prosperity and demonstrated an isolationist ability to avoid wars, accomplishments that greatly enhanced the value of Swiss citizenship and fostered nationalism.

Belgium, a much newer state than Switzerland, dates only from 1830, when its Catholic population, with approval and encouragement from the Great Powers of Europe, broke away from the Protestant-dominated Netherlands. The crucial mistake came when the newly independent Belgium was constituted as a unitary state under the domination of the French **Walloons**, in effect excluding the Dutch-speaking people from the power structure. Only very gradually did the Dutch **Flemings** attain equal cultural rights, as in 1891, when their language first appeared alongside French on the country's postage stamps and currency, or in 1898, when a regulation requiring that laws be published in Dutch first took effect. As a result, the two groups developed a deep-seated mutual antagonism. Virtual partition came in 1963, with the drawing of an official language border within Belgium, north of which the Dutch tongue enjoyed preference (Figure 12–7).

The tension flared into violence in the 1960s with street fighting in the capital city, Brussels, which has a French majority but lies north of the linguistic line (Figure 12–8). Students at the traditionally French University of Leuven, in Dutch-speaking territory, rioted in successful support of demands that instruction be in Dutch and that French faculty be dismissed. Tension persists between the two groups, though in recent years, the violence has ceased. Reflecting this tension, the government has not enumerated linguistic groups in censuses since 1947.

■ **FIGURE 12-8 A Brussels street corner.** In Brussels, Belgium's capital and a city with a French majority, dual-language street signs are legally mandated in a country with a sharp division between Francophones and speakers of Flemish.

Yugoslavia: A Balkan Tragedy

Belgium's problems astride a major cultural divide pale in comparison to the fate of Yugoslavia, the "land of the South Slavs." Formed following World War I when peacemakers appended various territories to previously independent Serbia, the new Yugoslavia straddled profound human fault lines. It joined shards of diverse former empires—Roman, Byzantine, Turkish, Austrian, and Hungarian—each of which had left indelible cultural residues. While largely Serbo-Croatian in speech, Yugoslavia inherited three major mutually antagonistic faiths—Catholicism, Orthodoxy, and Islam; two alphabets—Latin and Cyrillic; diverse ancient legacies of hatred; and, physically, an awkward spanning of the rugged Dinaric Range with the Hungarian Basin and Adriatic coast on either side.

Making matters worse, the **Serbs** dominated Yugoslavia from the first, effectively regarding it as their empire, a unitary Greater Serbia. Belgrade, the Serb capital, became the national capital, and the street signs there always remained exclusively in the Cyrillic alphabet. Serbs saw themselves both as defenders of the Orthodox faith and of European civilization at large.

Also, major economic disparities developed among the various provinces of Yugoslavia, with Slovenia and Croatia becoming relatively prosperous, while Macedonia and parts of Serbia fell behind. On top of the economic imbalance, both Croatia and Bosnia-Herzegovina harbored memories of medieval independence. The rise to power of an uncompromising Serbian nationalist leader in Yugoslavia sealed the doom of the state. In 1991 and 1992, four provinces seceded from Yugoslavia and achieved independence—Slovenia, Croatia, Macedonia, and Bosnia-Herzegovina. In 2002 the remnant Yugoslavia, consisting of only two republics, adopted a new name—Serbia & Montenegro (Figure 12–9).

The new international boundaries trapped hundreds of thousands of people on the "wrong" side of the lines, leaving Serbs in Bosnia and Croatia and Croatians in Bosnia. In the war that followed, mass murder and **ethnic cleansing**—the forced removal of minority groups—were common. The Serbs of Bosnia created the Srpska Republic in the areas they control, and they hope for eventual reunion with Yugoslavia. In **Kosovo**, a southern province of Yugoslavia, conflict broke out between Serbs and Albanians—a largely Muslim, non-Slavic group that formed nine-tenths of the province's population. Kosovo, today under NATO military occupation, will not likely rejoin Serbia, and Montenegro may also secede.

▪ **FIGURE 12-9** **Yugoslavia.** The death of a state and a model for turmoil. Yugoslavia no longer exists and has been replaced by five independent countries.

The United Kingdom

Yugoslavia was not the only European country to fragment in the twentieth century. This also happened to the United Kingdom, one of the oldest European countries and widely regarded as one of the most politically stable. The Republic of Ireland broke away in the 1920s, culminating a revolt that had endured for centuries. The Irish secession occurred along one of several natural and cultural dividing lines that cut across the country's territory.

The United Kingdom was the creation of Germanic tribes, particularly the Saxons and Angles, and Normans, who came as waves of invaders from across the North Sea and English Channel, eventually mixing to become the English people. Their new state, England, arose in the lowland plains that form the southeastern part of the island of Great Britain (Figure 12–10). London soon became the capital.

Once the English had unified the lowlands, they turned to the second great task—conquering and annexing the hilly lands on the west and north, inhabited by the **Celts**. One by one, these Celtic refuges—Cornwall, Wales, and Scotland—fell to the English.

Even before these conquests were completed, the English began their third, and final, major expansion into Ireland, another Celtic land, lying across a narrow sea to the west. The two British Isles, in this way, came to be a single country, the United Kingdom, ruled by the English from their base in the lowlands and London.

While outwardly successful, the expanded United Kingdom never fully absorbed the conquered Celts. Cultural and natural borders snaked through the country's territory, always threatening a rift. These borders included the environmental divide between fertile lowlands and barren hills and the separation into two major islands, as well as the borders between Anglo-Saxons and Celts; Protestants and

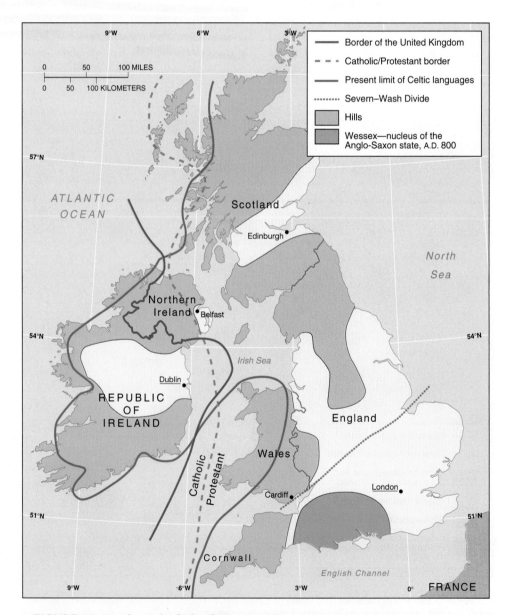

■ **FIGURE 12-10 The United Kingdom.** An old but endangered state.

Catholics; rich and poor; industrial and rural; prosperous postindustrial and decayed deindustrialized.

It was along the Protestant/Catholic divide that the United Kingdom fragmented, though some Catholic counties in Northern Ireland remained under English control. The "troubles" that have plagued Northern Ireland for decades suggest that the borders of independent Ireland were not wisely drawn. Irish separatists want the entire island to be part of the Irish Republic.

On the larger island, Great Britain, other divides continue to be troublesome to the United Kingdom. An independence movement in Scotland has many supporters, and the Scots have established their own separate parliament, as has Wales.

Another fissure opened up as deindustrialization struck the United Kingdom after 1960. The older coal-and-steel districts—the "Black Country"—fell into decay as factories and mines closed. As the United Kingdom emerged from this episode into the prosperity of the high-tech, postindustrial, service-dominated era, the older, decayed industrial districts were left behind. The **Severn–Wash Divide** developed, separating prosperous southeastern England from the rest of the country. Voting patterns now typically reveal this north-south division, and socioeconomic tensions exist.

The future of the United Kingdom now hangs in the balance. Will Northern Ireland be lost? Will Scotland opt for independence? Can the Severn–Wash Divide be closed?

Italy

Italy exemplifies another type of European state, one resting upon the memory of a vanished ancient greatness. The Roman Empire collapsed a millennium and a half ago, and Italy fell into political fragmentation, but the memory of Roman glory and unity persisted. In the nineteenth century, an Italian unification movement arose, which quickly reassembled Italy from an array of weak, small states ruled by feudal lords, princes, and popes (Figure 12–11). Although the unification movement originated in the north, in

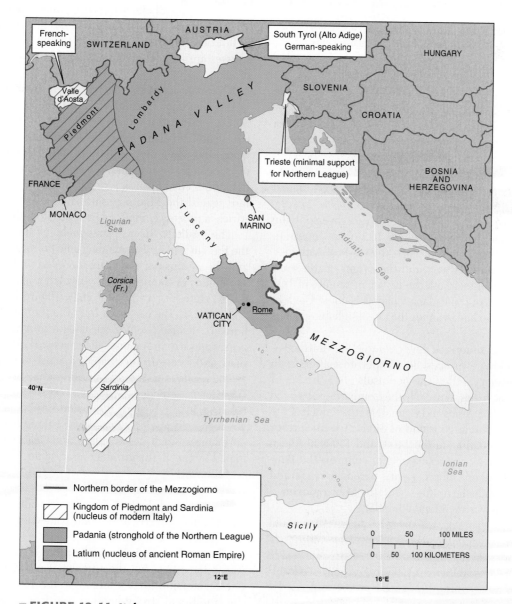

■ **FIGURE 12-11 Italy.** An improbable union in a peninsular location.

■ **FIGURE 12-12 Rome, Italy.** Visible in this evening scene is the Tiber River with St. Peter's Basilica and the Vatican in the distance.

Piedmont, the ancient legacy demanded that Rome be the capital (Figure 12–12).

The major unifying factor for the new Italy, aside from the Italian language (fragmented into many dialects) and the ancient Roman legacy, was the country's natural framework—a large freestanding peninsula fringed by great mountains, the Alps, in the north. Italy, it seems, was designed by natural forces to be a single country.

In truth, however, this peninsula was very diverse in the cultural sense and had been so even in Roman times. To unify Italy more than 2,000 years ago, the tribes of Latium—a district surrounding Rome and the origin of the word "Latin"—had to conquer both the Etruscan peoples of Tuscany and the Celts of the Padana Valley (Cisalpine Gaul) in the north. They also had to annex the Greek-inhabited lands of south Italy.

These ancient minorities never completely vanished—in Europe, nothing ever does. To complicate matters even more, new invaders settled the peninsula after the fall of Rome. Germanic tribes, especially the Lombards, seized the Padana Valley, while Arabs and Crusaders ruled much of the south, including the large island of Sicily. To still further complicate the situation, in the 1400s and 1500s northern Italy gave birth to a great awakening of culture—the Renaissance—and an era of prosperity based in trade and manufacturing, while the south—called the **Mezzogiorno**—remained in impoverished feudal bondage.

The modern Italian state has struggled perpetually to hold the north and south together. We can sum up northern Italy today as "European"—prosperous, secularized, democratic (and occasionally fascist), cosmopolitan, innovative, and capitalistic. The Mezzogiorno, by contrast, remains poor, devout, feudalistic (the Mafia being a classic feudal survival),

traditional, corrupt, and provincial. Some of the resultant internal pressure was relieved by the mass migration of poor southerners to factory jobs in the north after 1950. In addition, the central government transfers enormous amounts of northern wealth to the south as subsidies, relief, and developmental funds.

Still, the tensions persist. Northerners have disdain for southern Italians, and some even speak, as we saw earlier, of Padania, an independent northern Italy. And if these problems were not enough, a German-speaking part of Alpine Italy, the South Tyrol, has agitated for a half-century for autonomy or even independence (see Figure 12–11).

Spain

Spain, like Italy, Germany, Belgium, and the United Kingdom, has historically struggled to hold the north and the south together, but its internal stresses are far more complicated than this simple twofold division.

The Spanish state had its beginnings in far northern Spain in the Christian resistance against the Muslim Arabs (or Moors), who had invaded from Africa in 711. In particular, two small Christian states, Castile and Aragon, led the resistance, and when they united in 1469 through the marriage of Isabella and Ferdinand, the Spanish state was born. The defining deed of Spanish national identity was successful warfare against the Moors, pushing them southward back into Africa, a process completed in 1492, after almost 800 years.

The resultant Spanish state included the greater part of the Iberian Peninsula, excluding only Portugal in the far west; and the peninsular structure, coupled with the Pyrenees Mountains in the northeast, provided a natural framework for the state, just as it had in Italy (Figure 12–13).

But as was the case with Italy, the Iberian Peninsula was home to diverse peoples, speaking different languages and possessing strong regional allegiances. Portugal was one of these and succeeded in escaping Spanish rule. Other regions were incorporated, more or less against their will. The **Castilians**—the people of the central Meseta—built and ruled Spain, establishing the capital city, Madrid, in the center of the elevated interior plains. Ethnic minorities in the peripheries—Basques, Catalans, and Galicians—were dominated by the Castilians in a unitary state.

In addition, the cultural legacy of the Moors and Africa remained strong in southern Spain, particularly in the province of **Andalusia**. Southern regionalism today finds its expression in poverty, aristocratic landed estates, secularism, and socialist political leanings.

These cultural tensions almost tore Spain apart in a civil war in the 1930s, but the outcome was a Castilian-run fascist dictatorship that endured into the 1970s, perpetuating the traditional strong-handed rule from Madrid and the

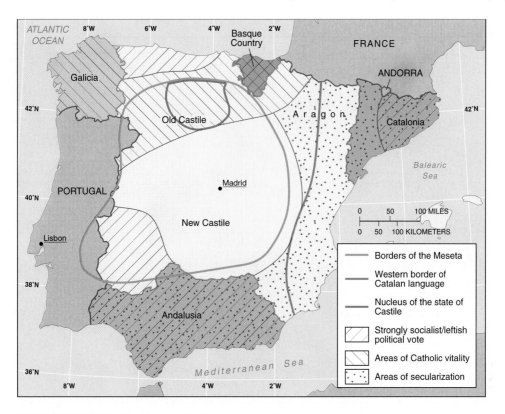

■ **FIGURE 12-13 Spain.** In Spain, the central Castilian Meseta regional core is attempting to retain control of the restive peripheries.

Meseta. In a radical shift, Spain adopted a federalist democracy and extended sweeping freedoms to the provinces and ethnic minorities. This has not proven entirely successful, as both the Basques and, to a lesser extent, Catalans now exhibit strong separatist sentiments (Figure 12–14). The Spanish state is core versus periphery in miniature.

■ **FIGURE 12-14 Separatist graffiti in the Basque-dominated region of Spain.** The Basque view themselves as culturally separate from all other peoples and have resisted efforts to integrate them politically into Castilian-controlled Spain.

Poland

Poland's task, by comparison, has been to try to join both east and west and core and periphery in Europe. The task has more than once led to tragedy for the country. More than 1,000 years ago, about 960, several Slavic tribes banded together to form the nucleus of the Polish state, located in the district called Wielkopolska, or "Great Poland" (Figure 12–15). The largest of these tribes, the Polians, gave their name to the new country. In less than a century, Poland grew from this core area to reach the Carpathian Mountains in the south, north to the Baltic Sea shore, west beyond the Oder River, and east to the Bug River—almost precisely the territory of Poland today.

It expanded further, creating an empire that spanned the breadth of Eastern Europe, between the Baltic and Black seas. But Poland could not hold together this large geographic entity. More powerful empires arose on all sides—Germany, Russia, Austria, and Sweden. The first three of these rivals eventually seized all the territory of Poland, and the country ceased to exist in 1795, although its people rose repeatedly in rebellion against the invaders.

Reborn in the aftermath of World War I, when Russia, Germany, and Austria all suffered defeat, the new Poland had its capital in Warsaw (see Figure 12–15). Only two decades later, in 1939, Poland vanished from the map again, partitioned between Germany and the Soviet Union (Russia's successor). When it was reborn again in 1945, Poland reassumed its ancient borders, but fell under Soviet

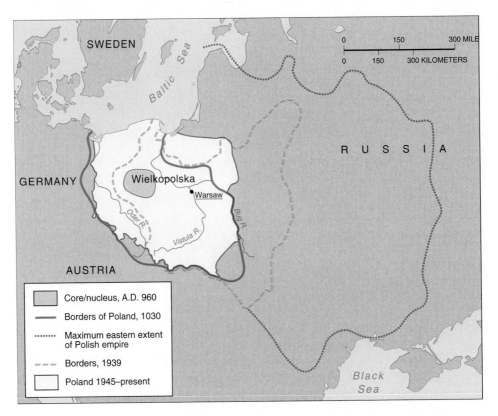

▪ **FIGURE 12-15 Poland.** An old and troubled state joining core and periphery, east and west in Europe. It has historically been trapped between powerful neighbors, but its people have struggled valiantly for their independence and freedom.

domination, with an imposed communist dictatorship. Only in 1989 did Poland emerge as a free, democratic country.

Though still troubled by its communist legacy, the modern Poland has notable strengths. Its geographical shape approximates an ideal, compact hexagon—the best shape, according to political geographers, because it minimizes the length of borders and internal transport routes. France enjoys this same advantage of shape (see Figure 12–4). Only a few small ethnic minorities exist, and the large majority of the people speaks Polish and is Roman Catholic. A plains country, Poland possesses much fertile land. Baltic seaports provide access to the world's shipping lanes (Figure 12–16). National pride remains strong, and in a 2001 survey, 89 percent of the people said they were proud to be Polish.

But the east-west and core-periphery problems persist. Western Poland has long been more prosperous, urbanized, industrialized, and "European," while the eastern provinces remain deeply rural and impoverished, on the very margins of Europe. Additionally, the old problem of its geographical location between powerful neighbors has not gone away. Poland's neighbor to the west, Germany, dwarfs it economically, while to the east lies another ancient enemy, Russia—slumbering for now. Poland could again become the victim of its geographic location.

Romania

Unlike Poland, Romania never enjoyed an earlier period of independence and imperial glory. Hardly had a proto-

Romanian principality emerged in Valachia in the 1300s than it was conquered and annexed by the Turkish Empire, Europe's ancient southeastern enemy. Turkey ruled the Romanians for centuries. A freedom movement began to achieve autonomy in the 1850s, followed by full indepen-

▪ **FIGURE 12-16 Port of Gdansk, Poland.** An important port on the Baltic since medieval times, the modern harbor facilities handle a large share of Poland's foreign trade. It was in the shipyards of Gdansk that the Solidarity labor movement emerged to challenge successfully the Communist regime and transform the Polish state.

dence in 1878. But Romania had more problems than Turkey alone. The empires of the Austro-Hungarians and Russians also intruded upon its territory. The Romania of 1878, with its capital at Bucharest, consisted only of Valachia and western Moldavia (Figure 12–17). Dobruja was added in a war. With the collapse of the Russian and Austro-Hungarian empires in 1918, Romania annexed Bessarabia (eastern Moldavia), Banat, Bukovina, and the huge province of Transylvania, doubling the size of the country.

Romania could not retain control of all of these new lands, even though they were Romanian-speaking. Its most serious loss was eastern Moldavia to Russia. More recently, that province became the independent country of Moldova.

Romania's strengths include its language, a survivor of the Latin dialect spoken here two thousand years ago during the time of the Roman Empire. Hence the name Romania. More than 90 percent of the population speaks this ancient language. The only sizable linguistic minority is Hungarian, spoken by the Szekeli people, 6.6 percent of the total population. They live in the very heart of Romania, away from the border of Hungary, lessening the chance of secession. Romania also has about 2.5 million Gypsies, in a population of 22 million, but nearly all of them can speak Romanian. The formerly large Jewish population today numbers only 12,000. Most Romanians belong to their own independent branch of the Eastern Orthodox Church, adding further unity.

The country has abundant fertile plains, but its agriculture remains unmodernized and underproductive. Like France and Poland, Romania has a roughly hexagonal shape, another advantage. It also has mineral resources, including petroleum.

Problems include widespread poverty and low living standards, chronic conditions only made worse by four decades of communist dictatorship that ended in a revolution in 1989. The country is cut through the middle by a major mountain range that serves as a divide, especially since an imperial border followed its crest for centuries before 1919. Add the problem of Romania's location at the very edge of Europe, a disadvantage only partially offset by the country's position on the greatest navigable river of Europe, the Danube. And there is the problem of what to do about neighboring independent Moldova, culturally Romanian and part of the country between 1919 and 1940. Moldova is even more impoverished than Romania.

The European Union

Faced with this bewildering fragmentation into numerous independent states and separatist provinces, a half-century ago the Europeans began a countermovement toward unification. This decision, which would eventually produce the **European Union (EU)**, was also prompted by fear of

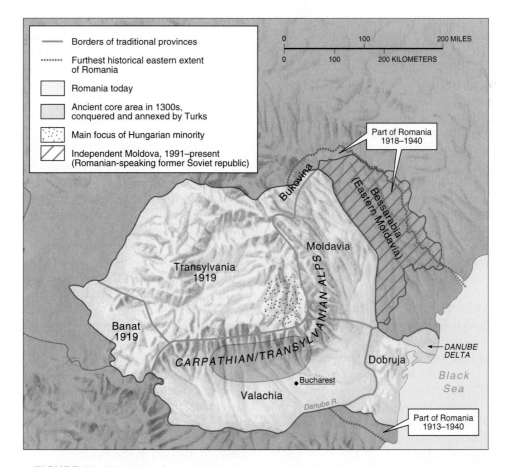

■ FIGURE 12-17 Romania. A Balkan state pieced together from the remnants of three empires, but today the independent nation of an ancient and proud people descended from the Romans.

Russia, the powerful empire to the east. Indeed, the EU was a Western European movement and is the most potent expression today of an east-versus-west division within Europe.

In the 1950s, Western Europe began moving toward a supranational federalism. This initiative began in the economic sphere with agreements reducing coal and steel tariffs, and the greatest success so far has been in the realm of free trade. From the very beginning, the member countries introduced the "four freedoms of movement" of labor, capital, goods, and services. Creation of a single market and determination to protect the member countries from external competition allowed them to revive their economies from the devastation of the last world war in record time. The rapid economic growth characteristic of the 1960s and 1970s contributed to the success of the European Economic Community. The multinational organization began with six states and has since expanded to twenty-five (Figure 12–18). In 1993, it changed its name from European Economic Community to European Union. This change indicated a major shift in the ideology of the organization as it emphasized an unprecedented institutional transformation. What began as a basically economic enterprise over time evolved into a sophisticated process of political, judicial, and cultural integration. In that sense the European Union is a unique entity. One of the most important factors contributing to this unique arrangement is the new, supranational identity of the Europeans. The founding fathers of the Common Market notably proclaimed that the union "was created not for the countries, but for the people." Today most important policies and decisions are made at the supranational level.

The EU major policymaking institutions are represented by the Brussels-based European Commission, the Parliament that meets in Strasbourg, and the Council of Ministers. The Commission is independent from national governments and responsible for the formulation and implementation of common policies, especially in the economic sphere. The Council of Ministers serves as a counterpart to the Commission and advocates the national interests of the member states. The European Parliament is a forum directly elected by the European citizens for a term of five years. Each country has a quota for the number of representatives based on population size, but once in Strasbourg the Parliament members unite according to their political affiliations. The Christian Democrats and Social Democrats form the two largest factions in the Parliament, while the number of Green Party votes is increasing each year. The European Court of Justice (ECJ), located in Luxembourg, is an important safeguard of EU law, guaranteeing its precedence over national interests.

The EU spends a significant share of its budget on its bureaucratic apparatus. Its codification of laws and standards alone is about 80,000 pages long and has to be translated into every member country language. In 2005 Gaelic, the native tongue for 4 million Irish became the twenty-first official language of the organization. While the preservation of national languages and dialects is a noble cause, it means that EU citizens will have to pay an additional 4.15 million euros more in taxes each year. In addition to the broad institutional network, the EU has 127 overseas delegations and permanent diplomatic missions in 164 states, though it is premature to speak of a single and clearly defined common foreign policy. When it comes to international policymaking, the national elites exercise unrestricted power. Traditionally, so-called Atlanticist states such as the United Kingdom and the Netherlands tend to support American international projects, while Germany and particularly France emphasize the "continental European" perspective.

Once the concept of the united Europe proved to be viable and successful, the European Community became a club by invitation only. By 1995 all the willing Western European states were in. The collapse of the eastern socialist bloc opened a door for ten more countries from the margins of Europe in 2004. The addition of 75 million people with different social and economic histories and demographic patterns represents both new opportunities and challenges. The Westerners are especially concerned with the perceived dangers to their living standards. They assume that lower paid Easterners will take over lots of jobs. In reality this does not happen, as newly admitted citizens of the EU are not allowed to seek jobs outside their countries for seven years. But politicians have often used the image of the "Polish plumber" or the "Czech nanny" as threatening symbols of cheap labor import. Also, the richer countries now have an obligation to subsidize heavily the economically less rapidly growing partners in the East and South. The plans to admit impoverished countries such as Bulgaria and Romania in 2007 will put an even heavier financial burden on the EU. Between 2000 and 2006 the EU will have spent more than 6 billion dollars in aid to the Balkan countries, which is more than it spends anywhere else in the world per capita.

The latest indication that membership in the EU does not automatically bring about unanimity came with a deep political crisis over the ratification of a new European Constitution. After a decisive majority of French and Dutch citizens rejected this new document, the idea of European evolution and integration suffered a significant blow. Some political analysts point out that the fear of Turkey's accession to the "Christian club" played a decisive role in the process. Many Western Europeans are threatened by the idea of a country of 70 million people with a profoundly different culture and history being their peer. For the average European, the concept of the European bloc sharing a common frontier with Iraq, Iran, and Syria is quite alien.

To add to the many divisions in Europe, there is also a strong dichotomy between the Atlanticist social and economic vision of development, represented by the United Kingdom, and the continental European social model advocated by major players like Germany and France. The Anglo-Saxon Atlanticist concept of development emphasizes a laissez-fair economy and argues for liberalization of markets, in order to be able to compete on a global scale. This perspective sees the more protective and rigid system in continental Europe as a major hindrance to the European ability to improve technologically and innovatively. Unemployment rates over 10 percent have plagued both Germany and France throughout the last decade, yet the ideology of

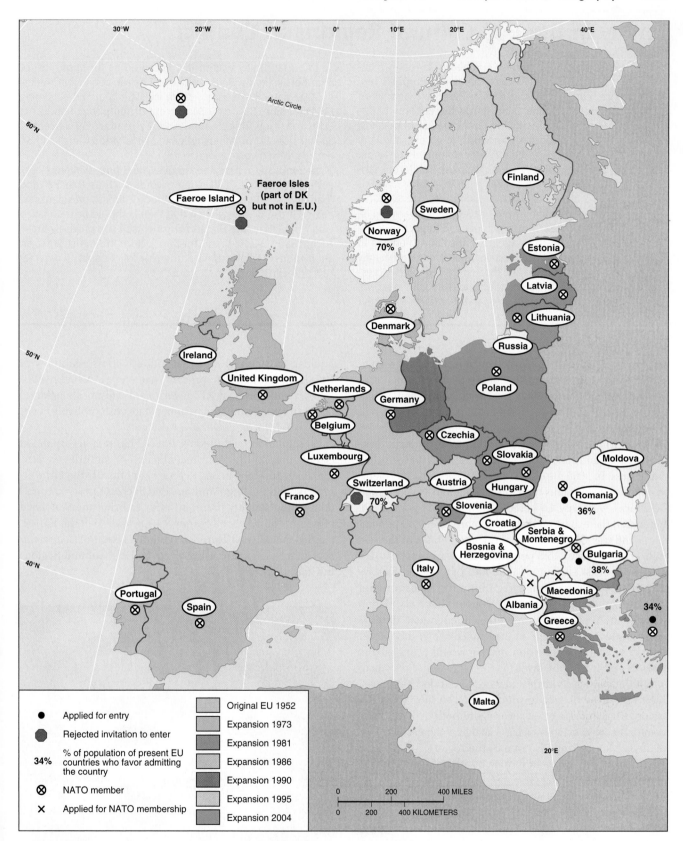

■ **FIGURE 12-18 The growth of the European Union.** The expansion of the EU to the east has profoundly changed the economical and geopolitical situation of the European continent; all new Eastern European members have also joined NATO.

GEOGRAPHY IN ACTION
Turkey in the EU? Cultural Regions in Conflict

Is the European Union only for Europeans? Is Turkey a European country? The Turks once threatened Europe's very existence, conquering the Balkan Peninsula and three times laying siege to Vienna, a city that is the epitome of Europe. They destroyed the Greek Empire, annexed all of its lands, and built a mosque atop the Parthenon in Athens, the very symbol of classical Greece. Can Europeans forget these old memories? Can they forget that Turks speak a different language and adhere to an unfamiliar religion—Islam? Can Europe forget the centuries of warfare required to drive the Turks back to the small foothold they now have in Europe around Istanbul, the former Greek Constantinople and forget the genocide committed by the Turks against Christian Armenians in 1915? Can they set aside the Turkish invasion of the dominantly Greek island nation of Cyprus and their continued military presence there? And, finally, is Europe willing to overlook the recent, brutal suppression of the Kurds, a separatist minority in Turkey?

In 1999 Turkey was at last accepted as an applicant for membership in the EU. But will it ever be allowed to join? Turkey in 2002 passed extensive reform legislation, aimed at gaining EU admission. Still, only 34 percent of the population in the EU today favors Turkish admission. By contrast, 59 percent of all Turks say EU membership would be a "good thing" and 68 percent would vote to join.

In books still in print, the famous Nobel-prize-winning Greek author, Nikos Kazantzakis, uses words such as "bestial" and "barbaric" in reference to Turkey. A recent Turkish president castigated those who "maintain that the EU should be a Christian club," but in fact the EU, and Europe itself, are precisely that— a Christian (or is it post-Christian?) club. In August 2002 an anonymous EU diplomat said, "there's a lot of doubt . . . whether Turkey really belongs in the club."

---

protection of the internal market and governmental regulation of major economic spheres of production has deep social and political roots among the first founding states.

Despite a slowed rate of economic growth and a high rate of unemployment (over 9 percent for the EU as a whole), however, Europe is one of the world's most powerful economic actors, and economists evaluate the EU's performance as a success. In 2005 the euro (Figure 12–19) area (EU members before 2004 minus the United Kingdom, Sweden, and Denmark), with only 4.9 percent of the world's population, dominated the export markets with more than a 30 percent share, and was second only to the United States in the size of its economy, with 15.3 percent of the world's GDP.

The strongest challenge to the idea of a unified Europe might come from political and cultural regionalism rather than diverse, at times conflicting, economic interests. Within the EU, the power of the member independent states is being eroded or bypassed on two fronts. While the master plan envisioned that decision making would occur only at national and federation levels, many regions and provinces within the various countries now exert considerable power, due to the issue of separatism discussed earlier. Spain, for example, has lost aspects of sovereignty, both to the EU and to autonomous provinces such as Catalonia. Increasingly, one hears of a federalized "Europe of the regions," implying a federal structure linking a mosaic of ethnic homelands, in which the formerly independent states have been doubly demoted. The homelands would presumably offset the excess and sheer size of EU integration, allowing Europe's diverse regional cultures to survive and flourish. The cultural impetus within the EU runs counter to the economic thrust, favoring regionalism, separatism, and nationalism. Quite substantial proportions of Europeans, both within and outside the union, never think of themselves as "Europeans" rather than as Dutch, Scots, Spaniards, or Catalans.

The eventual level of federalism and sacrifice of national sovereignties that will occur, even in the inner tier, remains unclear, particularly now that military confrontation with Russia, at least for the present, has ended. A highly centralized and powerful EU government will not be created, but at the same time, the member states will probably never regain their former level of sovereignty.

In spite of those problems, the EU has steadily expanded (see Figure 12–18). Many recent member countries are economically troubled in one degree or another. Is the EU willing to spend the huge sums needed to create prosperity and democratic stability there, in order to bring the beneficial effects of the EU to all corners of Europe? Will the EU expand to include non-European countries (see the *Geography in Action* boxed feature, Turkey in the EU? Cultural Regions in Conflict)?

■ **FIGURE 12-19 Euro notes and coins.** The unprecedented degree of European integration has led to the creation of a common currency, named the **euro**. To date, not all member countries of the European Union have adopted the euro.

To a degree, the **North Atlantic Treaty Organization (NATO)** serves as the military arm of the EU, with the qualifications that it is dominated by the United States and also has Canada as a member. NATO's role is more a result of the EU not developing its own military capacity, opting instead to rely upon the United States, as in Cold War times. Most of the same Eastern European countries that applied for EU membership also seek to join NATO, and the 2004 expansion countries have already been admitted. This is partly because the EU cannot and does not have the will to spend more on defense, allocating a mere $4,000 per soldier as compared to the hefty U.S. amount of $26,800. Its military technology lags behind the American. But after the collapse of the Eastern bloc of the Warsaw Treaty countries and the expansion of NATO to the border of Russia (see Figure 12–18), the EU was faced with a new geopolitical situation and had to reconsider its relationship with NATO. Two opposite attitudes, linked to different perspectives on common foreign policy, influenced this process. The United Kingdom insisted that EU military decisions should comply with NATO directives, while France and Germany advocated more independence from NATO. Continuous negotiations and summits have not changed the status quo. Moreover, the civil war in former Yugoslavia underscored the inability of the EU to organize any significant military campaign without the backing of NATO and especially the United States. At present, the relationship between the EU and NATO is tentative. It remains to be seen if the EU will be able to develop its own feasible program for defense and security.

Summary

Europe, then, is an ongoing economic, political, and cultural experiment that is experiencing two diametrically opposed trends—separatism and union. If both movements succeed, producing an array of seventy or eighty autonomous homelands bound loosely together in the framework of the EU, then the future of the various present independent countries would seem clouded. But if the independent states survive, old violent conflicts could continue to surface, and the dream of a Europe at peace could be elusive.

Nonetheless, the European Union has come to be perceived as the embodiment of a new idea of Europe—a Europe of cooperation without borders, a Europe with a new identity, and a powerhouse in economic and political terms. So is Europe moving logically to what some have labeled the "United States of Europe"? European scholar Göran Therbon is optimistic about the future of Europe in the globalizing world and emphasizes that "a rare composition of a relatively balanced internal composition of power, intensive socioeconomic internal exchange, and considerable economic strength on a global scale has meant that 'Europe' (i.e., the EU) is widely perceived in the world as one 'bloc'."[1] It is an image and a reality in the making.

---

[1]G. Therbon, "Europe—Superpower or a Scandinavia of the World?" In *European Union New Regionalism,* edited by M. Teló (Aldershot, England: Ashgate, 2001), pp. 227–243.

Key Terms

Andalusia	Kosovo
Castilians	Mezzogiorno
Celts	ministates
confederation	nationalism
ethnic minority	North Atlantic Treaty
ethnic cleansing	Organization (NATO)
euro	separatism
European Union (EU)	Serbs
federal state	Severn–Wash Divide
Flemings	unitary state
Iron Curtain	Walloons

Review Questions

1. What is Europe? That is, how do we define it?
2. What are the main problems of habitat damage facing Europe?
3. What is meant by "demographic collapse" and why is it a problem for Europe?
4. What are the internal zones of European cities?
5. What is "deindustrialization" and which part of Europe has suffered most as a result of it?
6. Which features of Europe's geography reveal a core-periphery pattern?
7. What have been the internal divisions that plague Germany and the United Kingdom and that destroyed Yugoslavia?
8. What is the European Union and what major challenges does it face?
9. What are the three major Christian subdivisions of Europe and where is each located?
10. What are the three major Indo-European linguistic subdivisions of Europe and where is each situated?

Geographer's Craft

1. If you were a geographer and wanted additional defining traits for Europe, besides the twelve listed earlier, how might you select them?
2. What geographical expertise might you draw upon to address the question of whether mass immigration of non-Europeans from poorer countries will greatly change or even destroy European culture?
3. The European Union expansion to new member countries in eastern and southeastern Europe includes great regional differences in levels of prosperity and standards of living. How might the expertise of the geographer be employed to help solve these inherently geographical problems?
4. The European way of life and prosperity seem to be destroying the habitat. How might geographers help solve this very difficult problem?
5. How might a geographer go about helping a country like Spain, Macedonia, or Serbia and Montenegro solve the problem of separatist minorities in their peripheral homelands?

PART FIVE

ELEVATION IN METERS

Above 4000
2000–4000
500–2000
200–500
0–200
Below sea level

0   500   1,000 MILES

0   500   1,000 KILOMETERS

Russia and Central Eurasia

Robert Argenbright and William C. Rowe

The northern Eurasian region consists of eleven former Soviet countries together with Afghanistan. The Soviet experience remains the common thread for these countries, unlike the former Soviet lands that have become distinctly European in terms of identity and political orientation. In contrast, Armenia, Azerbaijan, Belarus, Georgia, Russia, and Ukraine have yet to fully redefine or redevelop themselves; they remain "in-between" in several important respects. The other countries in the region—Afghanistan, Kazakhstan, Kyrgyzstan, Tajikistan, Turkmenistan, Uzbekistan—might also be seen as in-between in certain ways. However, what is more important for these countries today is their reemergence in world affairs after many decades spent in obscurity in the Soviet periphery.

Formally, the Soviet Union disappeared with a few strokes of a pen, yet the communist empire's legacies continue to haunt the region. Russia inherited the USSR's (Union of Soviet Socialist Republics) permanent seat on the UN Security Council, as well as the Soviet nuclear arsenal, thus guaranteeing that Russia would be treated as a great power, if no longer a "superpower" comparable to the United States. Although the Russian economy no longer rates in the top rank globally, income from oil and gas exports has provided relative stability and enabled President Vladimir Putin to reassert the Kremlin's authority both at home and abroad. From Russia's standpoint, the other former Soviet countries constitute the "near abroad," a realm where Russia seeks to have unparalleled influence.

However, not all of the near abroad countries, especially Ukraine, welcome the Russian embrace. Georgia and Armenia seek to get beyond merely struggling to survive, but they must contend with an unfavorable geopolitical situation. The former Soviet Central Asian republics, although maintaining cordial relations with Moscow, now face new problems, including the need to deal with the revival of Islam. In the post-9/11 context, the specter of Islamist terrorism haunts the new states, which are mostly led by former Soviet-elite figures, and guarantees that the United States, Europe, and China will take more of an interest in the region today than in the past. The substantial oil reserves of Azerbaijan, Kazakhstan, and Turkmenistan also boost the region's global profile. Meanwhile, Afghanistan is being reinvented, with the aim of eliminating the terrorist threat and developing the economic basis for stability.

Whatever their individual issues, without exception all of the countries in the Northern Eurasian region face the same pair of challenges. All of them need to restructure their economies to make them more diverse and robust so that they can profit from globalization while simultaneously providing for citizens' needs. Closely related to this challenge is the need to firmly establish and enforce the rule of law. Without the rule of law, economic development is distorted by corruption and the growth of civil society is stunted by repression. So long as these countries lack economic stability and healthy civil societies, they will remain vulnerable to political extremism.

CONTENTS

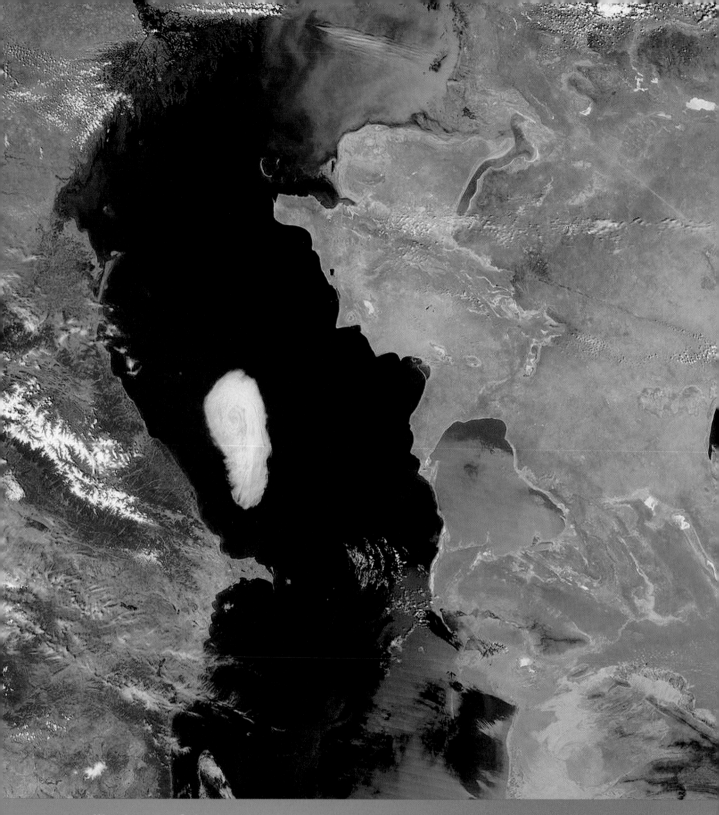

■ **Satellite image of the Caspian Sea.** In this satellite image of the Caspian Sea, taken in March 2003, the deeper waters appear as dark blue and the more shallow portions of the sea are light blue to turquoise in color. Extensive cloud cover can be seen to the northeast and southwest. The Caspian Sea is the focus of major new petroleum exploration activities and is coming under increasing environmental stress.

Chapter 13

The In-Between Countries of Eurasia:
Physical Geography and Historical Context

- **Physical Geography**
- **Historical Overview**

There are many differences to be found in the physical and cultural environments of Belarus, Ukraine, Georgia, Armenia, Azerbaijan, and Russia. But all six countries share the quality of being "in-between" in two senses. In terms of political and cultural geography, they are not strictly European. Culturally, the region has been Eastern in many ways, such as the dominance of the Eastern Orthodox and Muslim religions and the traditional lack of democratic institutions. Today these countries lie to the east of NATO and the EU, and probably will remain "outsiders" for more than a few years to come. Yet all six countries have known westernizing currents, which today mesh with forces of globalization. They also are all "in-between" countries in the historical sense. All six countries were constituents of the Soviet Union. They have all been in transition away from this anti-capitalist one-party dictatorship since 1991. Today all six have economies that have only partially been converted to capitalism, and in none of them is democracy deeply rooted.

The Russian Federation is the successor to the Soviet Union and the Russian Empire (Figure 13–1). In the two chapters devoted to the in-between countries, Russia will receive the most attention for a number of reasons. With respect to territory, Russia is the largest country in the world. It inherited the Soviet Union's permanent seat on the UN Security Council and its nuclear weapons, so it remains a great power, if no longer a superpower. Russia's gravitational pull on the other five in-between countries is still considerable, more so perhaps than on the other former Soviet countries. From Russia's perspective, the former Soviet countries form the "near abroad," where Russia intends its influence to outweigh the great powers of the West and East, and even the UN. Yet Russia no longer can simply dictate terms to its neighbors, and in fact must struggle against centrifugal forces within the federation. All six in-between countries feel the contradictory pulls of West and East, the challenges of globalization, and the tension between embedded practices and new developments.

Physical Geography

In the west the immense Eurasian landmass presents no obstacles to the eastward movement of air masses or to the intrusion of cold air from the Arctic and Siberia in winter. Even the **Ural Mountains**, conventionally considered to divide Europe and Asia, reach a maximum of 6,250 feet (1,905 meters) in the remote north, but rarely exceed 5,000 feet (1,524 meters) in the settled area of the country. As a result, there are immense regions that are characterized by relative uniformity of climatic conditions and vegetation patterns. In contrast, the **Caucasus Mountains** and the Transcaucasian lands to the south are highly diverse, while the southern end of the Crimean peninsula in the Black Sea and western Ukraine's slice of the Carpathian Mountains also stand out as exceptional. Eastern Siberia consists of rugged, eroded plateaus bounded on the east and south by substantial mountain ranges. Narrow strips along the Pacific shore and in coastal river valleys benefit from the moderating influence of the sea, but most of the immense area east of the **Yenisey River** is isolated and inhospitable (Figure 13–2).

Continental Climate

Three main factors combine to give most of the region a vigorously **continental climate**, which is marked by a long, relatively dry, and very cold winter and a short but surprisingly warm summer. First, formidable mountain systems rise to the south in Central Asia and in the east, blocking the Pacific's influence from all but a small area of the Russian Far East. The one ocean where Russia has vast frontage is

FIGURE 13-1 Russia and the Eurasian states of the former Soviet Union. These recently independent countries, which together once constituted the Union of Soviet Socialist Republics, cover a large area of the world. About a quarter of that area is considered part of Europe; the remaining three-fourths lies in Asia.

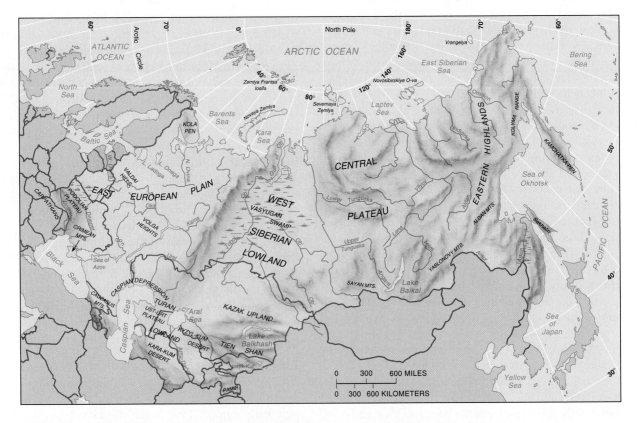

FIGURE 13-2 Landform regions of northern Eurasia. The surface of this part of Eurasia is extremely varied. Most of the European part of the region is characterized by rolling plains with small, scattered uplands and mountainous areas. In Siberia and Central Asia, vast lowlands, uplands, and extensive mountain landscapes create a complex landform map, which, as it does in the European area, helps shape the maps of population and economic geography.

the Arctic, which is frozen much of the year and generally contributes little moisture. Second, because of its high-latitude location, the region receives little insolation in winter but has long days in summer. North of the Arctic Circle, the "White Nights" of summer are followed six months later by the dark days of winter, when the sun does not appear above the horizon.

The last factor is the great size of the landmass. The territory of the in-between countries stretches over 6,200 miles (10,000 kilometers) west to east and as much as 1,200 miles (2,000 kilometers) north to south, and the rest of the vast Eurasian landmass must be included in the consideration of climate as well. When sending an e-mail to Vladivostok from St. Petersburg, one must take into account an 11-hour time difference—St. Petersburg, is almost as close to Boston as it is to Vladivostok! Between Russia's great port cities lies a huge territory that enjoys little or none of the moderating influence of the ocean, except near the Black Sea, the Sea of Azov—the shallow northern arm of the Black Sea—and, to a lesser extent, the Caspian Sea.

Winter is the longest season throughout most of the region, and it is brutally cold. Only Antarctica gets colder than

East Siberia. Surprisingly, it is also rather dry, especially to the east and southeast of the **Volga River**. Whereas Moscow receives 24 inches (610 millimeters) of precipitation on average, the Siberian center of Irkutsk receives 15 inches (380 millimeters), while Astrakhan on the Caspian's north shore receives just 8 inches (200 millimeters). The crucial factor is the relationship between precipitation and evapotranspiration. A strong moisture surplus in the north shifts to a pronounced moisture deficit in the south and southeast.

Natural Regions

The result of the northern location, continentality, and topography is a pattern of exceptionally large bands of essentially uniform vegetation. These are the natural regions: tundra, taiga, mixed forest, deciduous forest, forest steppe, steppe, semidesert, and desert (Figure 13–3). Starting in the north, the **tundra** region stretches all across Russia's Arctic shore, and in places extends southward for hundreds of kilometers in Siberia. Although summer days are long, the intensity of the insolation is weak—no month averages 50°F (10°C). Only a thin "active" layer on the surface thaws;

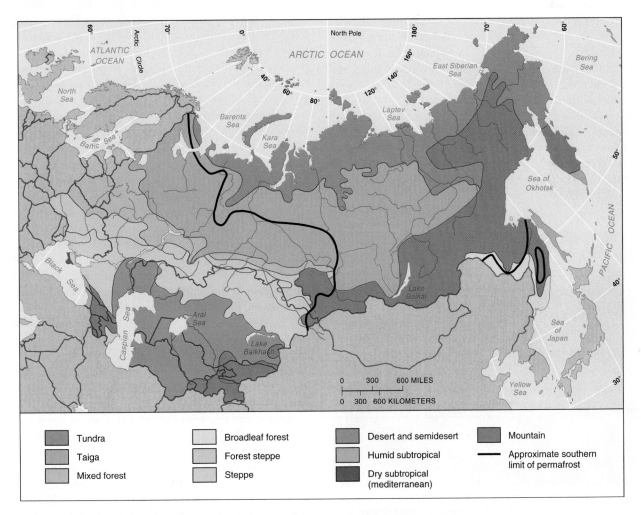

■ **FIGURE 13-3 Natural regions of northern Eurasia.** Many different physical environments are found on this large landmass. That diversity provides an opportunity to grow a wide range of crops, and the large area enhances the possibility of a varied mineral resource base. Much of the area, however, is at high latitudes and is little used.

■ **FIGURE 13-4 Tundra summer vegetation.** Although the Russian tundra is cold and bleak in appearance throughout the long northern winter, it possesses a quiet, solitary beauty during the brief summer season.

beneath it lies **permafrost**, permanently frozen earth. No trees grow in the tundra because of the short growing season, infertile soil, and the shallow active layer that is insufficient for tree roots (Figure 13–4). Lichens and moss, and sometimes shrubs, feed herds of reindeer, which in turn have supported indigenous peoples of the north for ages.

Taiga is the Russian word for boreal forest—northern forest dominated by conifers—and Russia contains more of it than any other country. Taiga covers much of northern Russia west of the Urals as well as most of Siberia (Figure 13–5). West of the Urals, where permafrost is not as extensive and there is more moisture available, a variety of conifers flourish. In the West Siberian lowland, much of the ground

■ **FIGURE 13-5 Summer in the taiga.** The Russian taiga is characterized by areas of close-growing coniferous trees, interspersed with swamps and meadows in the great river valleys. Traditional land use has consisted of hunting fur-bearing animals, mining gold and other high-value ores, and reindeer herding.

is poorly drained and the forest is interspersed with bogs and meadows. In spring snow begins to melt in the south, and the tributaries of the **Ob River** thaw out. The ice in the lower Ob acts like a giant plug, sometimes until late in the summer, and this causes extensive flooding. Taiga continues to the east, but its diversity diminishes to the point that great swaths of forest consist of just one species, the larch, which is uniquely adapted to the bitter winter, the short growing season, and the constraints imposed by the underlying permafrost. Taiga soils are acidic and not very fertile. Because of the great ice sheets of past glaciations, drainage systems in the taiga region west of the Urals are chaotic. Bogs and marshes abound—and so do ravenous mosquitoes in summer! Where the drainage is passable, the ground is likely to be very rocky, which also makes agriculture difficult. In the Soviet era, heroic efforts were made to farm in some taiga areas, but today the market rewards productivity, not heroism (see the *Geography in Action* boxed feature, Russia's Natural Resources: Are They Worth It?).

While conifers tolerate winter better than deciduous trees, from about 60°N and increasingly to the south deciduous trees begin to appear, creating a band of mixed forest. This band, like the forest-steppe and steppe further to the south, is much broader in the west than the east. This tapering reflects the decreasing length of the growing season, as one moves toward the "cold pole" of East Siberia, and the reduction in available moisture with distance from the sea. Historically, these three bands have been best suited for human settlement. At the western extreme of the region, agriculture in some form or another is possible generally from the St. Petersburg area in the north to the Black Sea coast. In Ukraine, roughly 58 percent of the land is arable, but only 7–8 percent of Russia is. Increasingly as one moves eastward into the depths of the immense continent, the northern area with a growing season that is too short and the southern area with a growing season too dry put the squeeze on agriculture. Mixed forest provides a much greater array of resources than does taiga, and the leaf litter builds up humus in the soil. But drainage can still be a problem, as it is in much of Belarus. Another type of forest, mostly Asiatic deciduous trees, grows in the Amur River basin and along the coast in the Russian Far East. With commercially valuable species and good accessibility, these forests are extremely vulnerable to excessive logging, given the burgeoning demand for wood products in Japan and China.

To the south of the mixed forest the growing season lengthens but the moisture surplus becomes a deficit. Grasslands grow more extensive, first in the mixed area of the forest steppe, and take over entirely in the **steppe** proper. The more grass, the better the soil. The Tatars who once held sway in the steppe are said to have been able to mount their horses and still be concealed by the tall grass! The soil is so rich with organic matter that it appears as dark as compost—this is **chernozem**, "black earth." Soil fertility and the relatively long growing season, up to months in southern Ukraine, would make the steppe region perfect for farming, if only precipitation were more plentiful and reliable (Figure 13–6).

GEOGRAPHY IN ACTION
Russia's Natural Resources: Are They Worth It?

Russia is a vast storehouse of natural resources. Its petroleum and natural gas are especially important—in 2004 the country was the world's number one producer of both fuels. And there are extensive coal and iron ore deposits. Russia has most of the strategic-metal ores, such as cobalt, nickel, platinum, and tungsten, as well as gold. So why is Russia not a wealthy country? Perhaps geography has something to do with it.

Much of Russia, in the north and Siberia, is inhospitable to human settlement and far from global markets. In the past, the state overrode economic rationality in order to exploit strategic resources. Today globalization has opened up a world of resources to exploitation by private capital, which seeks to minimize costs of production and transportation. If the market is to rule, then Russia's economic geography will be transformed. The rate of return will not cover the costs of permanent settlements in most remote and inhospitable regions.

Accessibility is poor in much of Russia. In 2004 President Putin ordered the construction of several hard-surface roads in Siberia, which, when completed, would make it possible for the first time to drive from one end of the country to the other. In winter Siberia's frozen rivers still are the best roads. The traditional name of spring, however, is "the roadless season," given the pervasive mud created by the spring thaw (Figure A). Even in western Russia many villages still lack hard-surface roads.

The cold climate also adds to the price of development. Consider just the cost of heating for Moscow's 10 million people—nowhere else in the world is such a large city located away from the sea at such high latitude. Now consider Yakutsk, which with 200,000 people is the largest center in East Siberia. January temperatures frequently fall below −58°F (−50°C). Massive energy consumption is a given. Moreover, ordinary steel and rubber tires will shatter at −30°C, so all equipment and machinery must be specially made to endure the Siberian winter.

The permafrost that underlies part of European Russia and nearly all of Siberia presents costly engineering challenges (Figure B). Since the active layer is saturated and rests on frozen ground, it is not very stable, especially on slopes. But even in winter, care must be taken with permafrost to make sure no source of heat causes melting. The ground may simply give way or buckle unpredictably. These risks are compounded in eastern Siberia by frequent earthquakes, which add to engineering costs and may cause massive destruction.

Shipping suffers because all of Russia's and Ukraine's ports are hampered by ice in winter, with the exception of Murmansk, on the Kola peninsula's northern coast. The Gulf Stream keeps the port open, but Murmansk's eccentric location adds to the cost of transport. All the major Siberian rivers flow northward, which exacerbates flooding in the early summer and means that their mouths are frozen much of the year. Atomic-powered icebreakers can extend the shipping season, but they are costly. Development in this region generally incurs extraordinary costs, especially in the north and eastern Siberia.

■ FIGURE A Ob River at flood stage. The spring and summer flooding of vast stretches of the West Siberian Lowlands by the Ob River and its tributaries greatly limits human settlement and economic development in this region.

■ FIGURE B Siberia, Russia. This vast region of frozen tundra, forest, and rich mineral resources includes more than two-thirds of Russia. Despite the problem of accessibility, this region served as an industrial frontier for the Soviet Union and still has potential for Russia's development.

Mountain areas hold a virtually infinite array of microenvironments. The mountains also create conditions for two exceptional natural areas. The southern tip of the Crimean peninsula, shielded from the cold winter air masses, enjoys a mediterranean climate. Beyond the Caucasus there is much variety, but all the lower elevations enjoy longer growing seasons and milder winters than the lands to the north. The eastern shore of the Black Sea has a humid subtropical climate, the only area of this type in the whole "in-between" region. Moving to the east, away from the Black Sea, the climate becomes more arid. A host of diverse crops can be grown, including some of the best varieties of wine grapes in the world, but irrigation may be required.

■ **FIGURE 13-6 Wheat cultivation in the Russian steppe grasslands.** In the better-watered western steppe grasslands of Ukraine and Russia, cultivation has largely replaced the natural grassland with wheat and other cereal crops.

Historical Overview

In-betweenness is nothing new in this part of the world. As Christian outliers in the Caucasus since the fourth century, Georgia and Armenia have always persevered in between more powerful neighbors. As for the rest of this vast land, its story begins with the movement of Slavic tribes through forested river valleys out into the steppe. There the Slavic farmers encountered nomadic, herding peoples with whom they traded and often warred. By the ninth century, the Slavs had developed a complex society based on towns and trade, as well as agriculture.

From Kievan Rus to Tsarist Russia

Ukraine, Belarus, and Russia share a common ancestor, Kievan Rus. The **Dnieper River** was the axis of an extensive trade network that linked the Baltic and Black Sea regions. Kievan Rus's close relationship with Byzantium, based in Constantinople (today Istanbul), brought Rus the Cyrillic alphabet and the Orthodox faith. Kievan Rus from the tenth to the thirteenth century was not at all "backward" compared to other European realms. Although the state was a monarchy, the towns usually managed their own affairs. The merchants of Novgorod, for example, set up an elected assembly to rule their town, consigning to their prince the role of military leader.

Kievan Rus disintegrated due to three main causes. (1) Because of the size of the state, competition among the princes, and a lack of centralizing institutions, Kievan Rus began to fragment from within. (2) The European Crusades shifted trade away from the Dnieper network, causing the Kievan Rus economy to decline. In the second half of the twelfth century the frontier region of Vladimir-Suzdal in the northeast emerged as the main power center. (3) Attacks in the early thirteenth century by the Golden Horde of Ghengis Khan proved unstoppable. In 1240 Kiev itself was sacked.

Kievan Rus's demise was swift and violent. Of 300 towns before the invasion, only 80 survived. History offers few better examples of "backwardness" imposed by defeat in war.

Over time Galicia (now western Ukraine) came under the influence of Poland, while the lands to the north (including most of present-day Belarus) came under Lithuanian rule.

In Vladimir-Suzdal the town of Moscow gradually came to dominate the region. Moscow's princes sought to gather the lands of Rus together again, but they remained subservient to the Tatars[1] until the 1470s, when the horde split three ways and Russia was able to begin expanding to the east. During this period, the foundations for a society much different from Kievan Rus' were established. The new state's leaders were determined to overcome the political fragmentation that had undermined Kievan Rus. Steadily Moscow's Grand Princes increased their power over lesser princes and other aristocrats. In conquered areas, the Grand Prince installed new elites composed of landlords who swore to serve the state. To support loyal servitors, peasants were bound to the land and required to serve the landowners. Over time this condition of **serfdom** turned into slavery. Russian peasants were losing freedoms at the same time that feudalism was breaking down in Europe. In the West, the emergence of capitalism and the cultural ferment of the Renaissance led to a focus on the individual. But in Russia there was a renewed emphasis on the greater community, as embodied by the state and the Orthodox Church (Figure 13–7). It fell to Russia to uphold the Orthodox faith after the fall of Byzantium. As leader of the "Third Rome," Russia's Grand Prince became Tsar, or "Caesar." Thus emerged Russian society's fundamental principle from the rise of Moscow until the collapse of the USSR, that is the duty of all people, of whatever rank, to serve the state.

In the mid-sixteenth century, Ivan IV ("the Terrible") took this principle to extremes when he attacked the nobility, seeking to replace them with hand-picked servitors. This effort failed in the end, but Ivan did greatly expand state territory. Russian fur hunters pushed into the East, and would reach the Pacific a century later. Ivan defeated the Tatars of Kazan and Astrakhan to acquire the Volga region. He also tried to gain a foothold on the Baltic, which would have completed a vast trade network under Moscow's control, but 25 years of war gained nothing. When Ivan died the exhausted state collapsed and chaos reigned in the "Time of Troubles." Finally in 1613 an assembly of nobles elected a new Tsar, Mikhail I, the first of the Romanov dynasty. The centralized autocracy was restored and territorial expansion resumed. Russia reached the Dnieper and won Kiev in 1654.

The Russian Empire

Russia in 1700 was big, but backward in terms of trade, technology, and modern culture. Tsar Peter I ("the Great") transformed the country by introducing new ideas and technology from the West. But he did not reject the fundamental principle of service to the state. To the contrary, he pushed it to

---

[1]The Golden Horde was led by ethnic Mongols, but most of the troops were Turkic-speaking warriors, whom the Russians knew as "Tatars." Their descendents today form the largest non-Russian nationality in the Russian Federation.

an extreme not seen again until the reign of Stalin. Peter even subordinated the church to the state. Having modernized the army and created the Russian navy from scratch, Peter defeated Sweden and gained access to the Baltic. Now "Emperor," he chose a marshy, flood-prone site on the Gulf of Finland for St. Petersburg, Russia's "Window on the West." A beautiful European city was built by serfs, thousands of whom died due to deplorable conditions on the site. Peter I also established a new industrial zone in the Urals to exploit the plentiful iron ore deposits. He built canals to link St. Petersburg with the Volga and to gain access to the Urals' metals. Transportation remained difficult but the state would not have to depend on foreign suppliers for cannons or bayonets.

Peter the Great cast a long shadow. But Catherine II ("the Great") in the latter eighteenth century made her own mark. Highly intelligent, German-born Catherine II was attracted by the Enlightenment, and she nurtured the arts and education in Russia. But she did not allow new ideas and sensibilities from the West to radically affect her governance of the empire. The position of the serfs in society continued to deteriorate as the power of the landlords grew (Figure 13–8). In the West, American and French national revolutions plus "revolutions" in industry and agriculture were ushering in the modern age, but in Russia the state-society regime remained fundamentally the same. Autocratic, state-centered Russia relied on expansion, not innovation.

Catherine II acquired an area the size of Spain that would later allow Russia to become an agricultural giant and an industrial power. First to fall were the Crimean Tatars, which gave the empire control over the north shore of the Black Sea. Thus Russia acquired the black earth of the Ukrainian steppe, with its longer growing season, as well as maritime access to world markets. In time, even more valuable resources would be found under the ground, especially hard coal and iron ore. Catherine had the beautiful port of Odessa built to celebrate the success of her reign and to support the grain trade. Russia also played a principal role in the partitioning of Poland, which caused that country to disappear from the map. Thus, Russia acquired the territory that would become Belarus, as well as the rest of Ukraine, except for Galicia, which went to the Austro-Hungarian Empire.

Early in the nineteenth century Russia acquired Finland and Moldova, and began to move into the Caucasus (Figure 13–9). By 1828 Russia controlled Dagestan, Azerbaijan, Georgia, and eastern Armenia, but Chechens and other Caucasian peoples resisted for decades. Georgia was an ancient country, Christian from

■ **FIGURE 13-7 St. Basil's Cathedral, Moscow.** The Russian Orthodox Church is a branch of Eastern Orthodox Christianity. As is true of many religious faiths, it has provided cultural continuity and spiritual refuge through centuries of political turmoil and economic uncertainty.

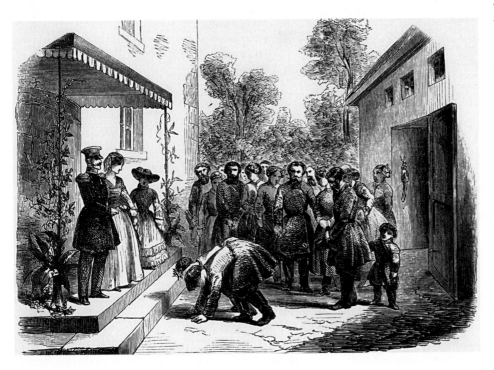

■ **FIGURE 13-8 Tsarist Russia.** Landlords were absolute rulers over their lands and people. This couple seeks their lord's blessing on their wedding day.

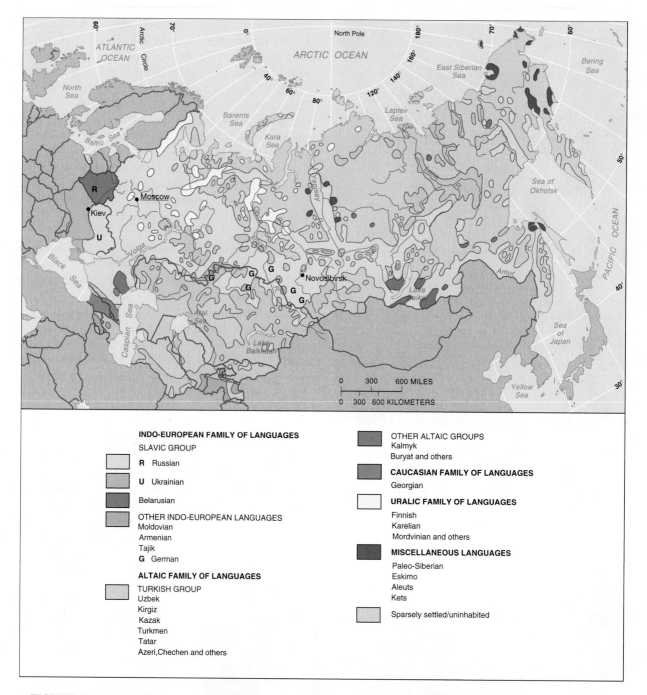

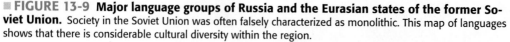

INDO-EUROPEAN FAMILY OF LANGUAGES
SLAVIC GROUP

R Russian

U Ukrainian

Belarusian

OTHER INDO-EUROPEAN LANGUAGES
Moldovian
Armenian
Tajik
G German

ALTAIC FAMILY OF LANGUAGES
TURKISH GROUP
Uzbek
Kirgiz
Kazak
Turkmen
Tatar
Azeri,Chechen and others

OTHER ALTAIC GROUPS
Kalmyk
Buryat and others

CAUCASIAN FAMILY OF LANGUAGES
Georgian

URALIC FAMILY OF LANGUAGES
Finnish
Karelian
Mordvinian and others

MISCELLANEOUS LANGUAGES
Paleo-Siberian
Eskimo
Aleuts
Kets

Sparsely settled/uninhabited

■ **FIGURE 13-9 Major language groups of Russia and the Eurasian states of the former Soviet Union.** Society in the Soviet Union was often falsely characterized as monolithic. This map of languages shows that there is considerable cultural diversity within the region.

the fourth century. Stuck between the Iranians and the Turks, Georgia may have found Russia the lesser evil. The Armenians, also Christians from the fourth century, were in the same predicament. In 1800 Armenia had long been divided into western and eastern parts, held respectively by Turkey and Iran. Azerbaijanis were Turks who settled west of the Caspian in the eleventh century. For a long time they too were ruled by Iran, from where they acquired the Shi'ite faith.

In 1800 Russia produced more iron and steel than any other country in the world. But soon the Industrial Revolution left Russia behind. In the Crimean War of 1853 Russia

was defeated on its own territory due largely to the technological edge enjoyed by Britain and France. This shock energized the reform movement. The serfs were formally emancipated in 1861, although most remained bound to their villages by law. But some escaped to the cities to find jobs in industry. The government began to industrialize in order to catch up with the West. The focus was on steel, railroads, and textiles, while grain exports paid for technology imports.

The development program was a mixture of government control and capitalist enterprise, the latter mostly foreign. The Urals languished because modern steel production required coal, which the region lacked. Instead, two linked industrial districts emerged in the south—the coal-rich Donbas, in the Donetsk River basin, and the mid-Dnieper River region with its plentiful iron deposits. Railroads linked the complementary regions and consumed much of their steel output. Azerbaijan's leading city, Baku, became an oil boom town. St. Petersburg was the country's largest industrial center and most important port. Moscow grew on the basis of textiles and its centrality in the growing railroad network. Kiev, Kharkov, and Odessa in Ukraine all flourished thanks to industry and trade.

The rural population more than doubled in 50 years. But agriculture remained backward. The peasants paid ten times as much tax per unit of land as the landlords. Desperate peasants plowed marginal lands, which caused erosion and exacerbated poverty. They thought the solution was to obtain more land, but the landlords and the state stood in their way.

For industrial workers, working and living conditions often were very harsh. Life expectancy was lower in urban slums than in primitive villages. But the city offered possibilities for remaking oneself. New urban workers encountered a range of ideologies, some radical, that advocated progress by means of reforming or overthrowing the regime.

Russia may have had a developing country's economy, but it had the ambitions of a superpower. In the 1860s the empire expanded into Central Asia, soon building a railroad to bring the region's cotton to Russia's textile centers. The **Trans-Siberian Railway** was completed to link west and east, and to propel Russian expansion. As originally constructed it cut across Chinese Manchuria, which Russia sought to develop and dominate. Subsequent efforts to project Russian influence into Korea led to a disastrous war with Japan in 1904–1905.

The regime ruled by tradition and force—the people were subjects not citizens. When the war went badly a massive uprising broke out. Peasants seized landlords' estates and workers began to organize their own governments—"soviets" (councils). The regime overcame the 1905 Revolution with a mixture of violent repression and compromise. Many workers and peasants were executed or imprisoned. But the Tsar did agree to a representative body, the **Duma**, which reformers hoped would lead to a democracy.

Industrial development resumed, with strong government involvement and dependence on foreign capital. The government attempted to replace traditional communal agriculture with detached family farms, hoping to create a conservative class of rural property owners. Perhaps Russia would have successfully modernized, but World War I intervened.

The Soviet Union

Russia's economy broke down in World War I and breadlines turned into riots. In 1917, the Tsar was forced to step down. The Provisional Government was democratic, but it continued fighting the war. The people wanted what **Vladimir Lenin**, leader of the **Bolsheviks** (the majority) promised to give them: "Bread, Peace, and Land." Lenin charted a radical course, extreme even by Marxist standards. Karl Marx argued that capitalism would have to be com-

pletely developed and the world's workers become the majority before the revolution could succeed. Lenin felt that a dedicated vanguard party could free the masses, which would inspire the workers in more advanced countries to rebel and come to the aid of socialist Russia. The 1917 October revolution soon led to chaos. Lenin resolved to build a new type of state, one that would restore and maintain order while at the same time transforming society. In theory, governing was the job of a hierarchy of soviets—hence the name Soviet Union—while the party, soon renamed the **Communist Party**, was to advance the social revolution. In practice, the party took control over the soviets.

The new regime's first great challenge was to survive civil war and foreign intervention. Although there were pitched battles, probably more people were executed in captivity. But the worst scourges were hunger, cold, and disease, which killed about 10 million people. In the civil war the state tried to control all economic activity. But in 1921, with the economy in shambles, Lenin announced the New Economic Policy (NEP), which allowed some marketing of agricultural produce and consumer goods. Political opposition was not tolerated, but culturally this was a period of relative freedom, except with respect to religion. Throughout the Soviet period, no one could rise in society who was not an atheist.

Lenin died before he could flesh out the NEP or anoint a successor. In the end, a Russified Georgian, **Josef Stalin**, overcame all rivals. Stalin argued that Russia would be conquered if it did not rapidly develop. He aimed to create "socialism in one country." To start, Stalin launched collectivization to directly take charge of all farming. Peasants who resisted were killed and many more were exiled to remote labor camps—the Gulag labor camp system. The government exported much of the harvest to pay for imports of advanced machinery. The result was a massive famine in Ukraine, along the middle Volga River, and other regions. Again, millions died.

The other thrust of Stalin's strategy was rapid industrialization under the banner of the **Five Year Plan**. In reality, Soviet planning meant micromanagement down to the level of specifying how many loaves of bread would be baked. The plan was backed by force from the start—economists and engineers who voiced doubts were shot—but the industrialization drive also generated tremendous enthusiasm, especially among younger workers. They believed they were building a new world. Unprecedented avenues of social mobility opened up for those who could achieve results. At the same time, the government provided education, health care, pensions, and other benefits. Under Stalin, everybody served the state, while the state promised to take care of them. However, millions were identified as enemies of the people and sent to the Gulag.

"Socialism in one country" aimed for self-sufficiency. The regime expanded existing industrial centers, such as the Donbas, the mid-Dnieper region, Moscow, and St. Petersburg (Leningrad). But the Soviets also invested massively in the development of the East for self-sufficiency and to base industry beyond the reach of invaders. One key project found coal for the Urals in the Kuznetsk Basin, 1,500 miles (2,400 kilometers) to the east (Figure 13–10). They built two huge

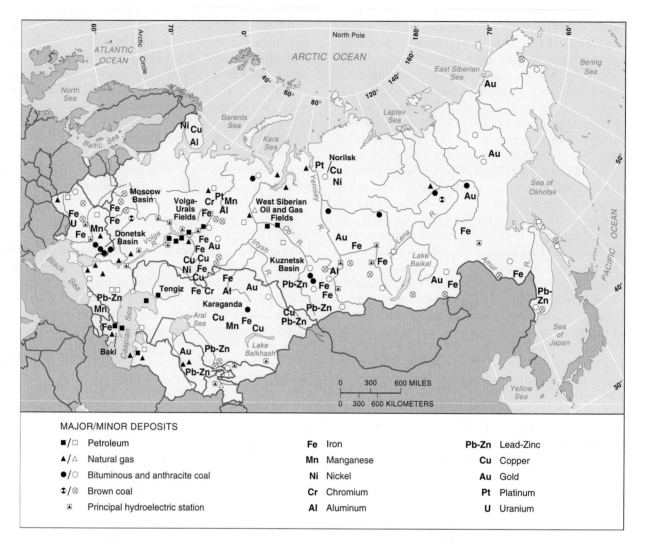

■ **FIGURE 13-10 Natural resources.** Russia, Ukraine, and Kazakhstan, in particular, are rich in minerals. High-quality coal is found in the Ukrainian Donetsk and Russian Kuznetsk basins and in the Kazak Karaganda deposits. Oil and gas are produced in Azerbaijan, Kazakhstan, Turkmenistan, and the Russian West Siberian and Volga–Urals fields. The Volga River provides hydroelectric power. Most of the area's metals are found in the west and along the southern fringe of the region.

metallurgical centers, sending Siberian coal to the Urals and the latter's iron ore to the Kuzbas by rail. Only a state that took no account of transport costs could contemplate such a project! Labor costs also were held down on difficult projects—such as the Pechora coal mines, the Kolyma gold fields, and the smelting center of Noril'sk—thanks to the forced labor of the Gulag.

The Soviet industrialization campaign was a huge success in the heavy-industry sector, but light industries such as textiles and food processing lagged far behind—in the USSR there was no consumerism (Figure 13–11). Heavy industry supported a military buildup in the expectation of war, which came when Nazi Germany invaded in 1941. Soviet losses were astronomical, but the country managed to evacuate 1,200 facilities and over a million industrial personnel to the East. The people rallied in support of the war effort. Millions of women moved into industry, some worked double shifts in freezing conditions, even sleeping on the shop floor until the next shift.

Victory in World War II confirmed for Stalin the superiority of his repressive, closed system. The "liberated" countries became Soviet puppets. The Soviet military's strength lay not just in sheer numbers, but also in technology. The USSR soon joined the nuclear "club" and also began the space race when they were first to put a human into outer space in 1961. The Soviet system excelled at focusing science and industry on the regime's top priorities. Yet there was hunger in much of the country.

Stalin's successors were able to enforce political control without resorting to massive terror campaigns (Figure 13–12). The Soviet bloc competed with America and its allies for influence throughout the world. In this **cold war**, the Soviets developed maximum self-sufficiency. Huge projects transformed remote regions, especially in the East—they were depicted in a heroic and patriotic light to attract the participation of young communists, the leaders of the future. The greatest of these was also the last, the Baikal-Amur Mainline (BAM) railroad.

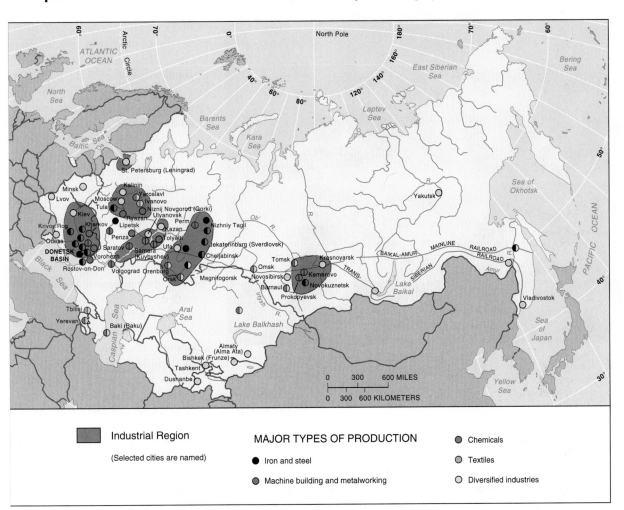

■ FIGURE 13-11 Industrial regions and selected cities. Industry in the former Soviet republics is concentrated in six regions. Except for the Moscow and St. Petersburg areas, these regions possess abundant local energy supplies. Soviet industry was oriented toward capital goods. In recent years, production of consumer goods has increased.

The Disintegration of the USSR

What caused the downfall of the USSR? Several factors stand out from a geographical perspective. Scale heads the list—that is, in many respects the scope of Soviet ambitions exceeded the capabilities to manage. Take the military for example. Soviet troops were based en masse in Eastern Europe and an even larger force covered the long border with China. The entire length of the Soviet border had to be patrolled because it was a closed country—airplanes that accidentally strayed into Soviet airspace were shot down. And the Soviet military was fighting in Afghanistan for 10 years, while trying to keep up in the military-technology race with the United States.

The **Kremlin**, seat of the Russian government in Moscow, was trying to maintain control over everything and everybody from the heart of Europe all the way to the Bering Strait. Although some people, mostly Russians of the World-War-II generation, saw themselves primarily as Soviet citizens, national identities grew increasingly more important, not just in the countries of the Eastern European bloc, such

as Poland, but inside the Soviet Union itself, especially in the Baltic lands, Ukraine, Georgia, and Armenia.

Following the Chernobyl nuclear catastrophe in 1986, outrage over Soviet despoilation of the environment swept the country, often fueling the flames of nationalism. Chernobyl contaminated much of Belarus, including devastating the health of thousands of children. Ukraine had suffered from pollution dating back well over a century. None of the fifteen "soviet socialist republics" that constituted the USSR was exempt from the ecological destruction of Soviet forced-march development. In Russia itself, every day seemed to bring another depressing revelation of ecological distress (see the *Geography in Action* boxed feature, Environmental Problems and Prospects).

People were tired of having to stand in line for a limited supply of low-quality products that lacked diversity and style. As the populace came into contact with more and more foreigners and foreign goods, some Western products—Levis, Marlboros—became fetish items. The West seemed the land of plenty, while Soviets put up with rationing.

■ FIGURE 13-12 Symbol of the Soviet economy. A 1967 Soviet poster urges the Russian people toward united action to achieve greater economic effort.

work out a coherent plan for *perestroika*. He allowed cooperatives to operate on a market basis, hoping they would be efficient providers of services to the population, but he was unable to protect them either against the bureaucrats or the extortionists who formed the first Russian mafias. Meanwhile, industry and agriculture remained under bureaucratic control.

In fall 1991 Gorbachev was kidnapped by conspirators who sought to restore the old regime. The coup was poorly organized and proved incapable of overcoming the opposition of the Russian government, led by Boris Yeltsin. Gorbachev was freed, but he no longer mattered. The Soviet regime had lost its legitimacy. While Gorbachev struggled to hold the Soviet Union together, the leaders of Russia, Ukraine, and Belarus met and agreed to declare their countries' independence. No referendum was held to ratify their decision, but neither was there much protest. The immense anti-capitalist empire that changed the course of history was finished because three men had the courage to say so and 280 million people followed their lead.

Putting an end to the political regime was surprisingly simple, but breaking up the tremendous economic system, "USSR, Inc.," was much more problematic. The Soviet economy to a large extent was a unified closed system. To separate Belarus, for example, from Russia was to intervene in chains of production for many key products. Most important in this regard were the unified networks for the provision of natural gas, petroleum, and electricity, as well as the railroad system. Nationalist politicians, especially in Ukraine, overestimated both the market value of their countries' products and the economic benefits of separation from Russia. There was also the expectation of massive aid from the West, which proved elusive.

Other problems from the Soviet system carried over into the post-Soviet period. Territorially, the USSR was an ethnic patchwork. From Kyrgyzstan around to Estonia, the Russian Federation was bordered by fourteen "Autonomous Soviet Socialist Republics" (ASSRs) that were based on non-Russian nationalities and formally granted the right of secession by the Soviet Constitution. Nobody ever expected that any would be allowed to secede—the laws of the Soviet Constitution had rarely been observed in the regime's practices. Yet when the country did disintegrate, the fissures followed the lines of the Soviet map. But this radical transformation did not put an end to all ethnic tension. In all fourteen of the former ASSRs there were sizable Russian populations that were made into minorities overnight. Where they were viewed as former colonizers, for instance in Estonia and Latvia, their status became that of second-class citizens. In Ukraine and Kazakhstan, the large concentration of Russians in industrial districts and other key areas

When **Mikhail Gorbachev** became General Secretary of the Communist Party in 1985 he was the youngest leader the country had known since Stalin consolidated his grip on power in the late twenties. At first he tried a moderate course of reform based on the historical precedent of the NEP, but soon he found that more radical measures were required because the party itself was the main obstacle to progress. The reforms were deployed under the slogan: *Glasnost, Perestroika, Demokratiia* (Openness, Restructuring, and Democracy). Chernobyl supplied a test of *glasnost* that the regime failed at first when it denied that anything serious had happened. But soon the floodgates of information were forced open and not only all the current environmental and political issues were brought out into the open, but also all the bloody secrets from the past.

Demokratiia made considerable progress. When it became clear that the East European nations would eagerly leave the Soviet bloc if given the chance, Gorbachev let them go without firing a shot. At home, Gorbachev sponsored the First Congress of People's Deputies, an elected representative body that was beyond party control. *Glasnost* became reality when the Congress's sessions were broadcast nonstop to the entire country. Nobody could remain unaware of human rights issues, national minorities' grievances, economic shortcomings, and environmental problems.

The result was traumatic. In the past the Soviet media had not even reported traffic accidents or fires. Now everything bad that had happened over 80 years plus all the current problems of the immense multiethnic empire came flooding in all at once. It did not help that Gorbachev, who became the first and only elected President of the USSR, never managed to

GEOGRAPHY IN ACTION
Environmental Problems and Prospects

Perhaps the world's worst environmental horror stories emerged in the USSR—the **Chernobyl** nuclear-reactor meltdown (Figure A) and the decline of the Aral Sea. Countless other environmental problems are known to exist in the region, involving air, water, and ground pollution, erosion, threats to biodiversity, and other issues. All the major rivers in the European part of the area suffer from diverse and numerous sources of pollution. The Black Sea and Sea of Azov are heavily polluted. Forests have been clear-cut in cold areas where it takes many decades for them to regenerate. The Ural Mountains include a broad array of environmental hotspots, including the site of another nuclear-radiation disaster near Chelyabinsk. Oil pipelines from western Siberia leak into fragile tundra and taiga landscapes, harming both wildlife and indigenous people. The Arctic city of Noril'sk, with its massive smelters for nickel and other ores, has destroyed a forest the size of Britain with acid precipitation. Nobody in cities drinks tap water—St. Petersburg's water is infamous for giardia, a tenacious intestinal parasite.

As bad as the environmental problems are, the situation today is better than it was in the USSR. The economic crisis that followed the breakup proved to be a blessing for the environment—industrial production declined by as much as 50 percent and so pollution decreased. But the transition economy prevents further advance in environmental protection. Many industries survive, thanks to non-monetary survival tactics, notably barter and payment in kind. Firms can get by without much money, but they cannot invest in technology to clean up production processes. Most cannot even afford to repair their Soviet-era pollution-abatement equipment. Environmental protection agencies are weak and most places are desperate for economic growth, whatever the ecological damage.

Although there are far too many environmental hotspots, vast tracts of generally unpolluted wilderness remain. The Russian government and local authorities have worked with various international agencies and NGOs to protect many unique environments. *Russian Conservation News* counts ninety-five national natural reserves and thirty-two national parks that have been established so far in Russia. They include Wrangel Island (Vrangelya) with its polar bears and walruses, a central Siber-ian reserve in the heart of the taiga, the Samara Bend National Park on the Volga River, and the Dagestan Marine Reserve on the Caspian Sea. Wrangel Island and nine other natural areas in the Russian Federation have been designated World Heritage Areas by UNESCO.[2] Infrastructure and the types of services needed by most tourists are lacking, but the growing popularity of ecotourism offers hope that the reserves and parks can increasingly pay for their own protection.

Of all these areas, one in particular stands out as a mecca for nature lovers. **Lake Baikal** is one of the greatest natural wonders in the world (Figure B). Situated in a geologic rift zone, its maximum depth of more than a mile (1.6 kilometers) easily makes it the world's deepest lake. It holds approximately 20 percent of the world's freshwater, comparable to all of the Great Lakes combined. Naturally isolated by mountain ranges from its surroundings in East Siberia, the ecosystem of Baikal is unique. Perhaps the most attractive of Baikal's native species is the *nerpa,* a freshwater seal. Baikal was the focus of the first environmental protection activity in the USSR and, although there are some sources of pollution in the area, it remains relatively undisturbed.

The other in-between countries also have potential ecotourism assets. Belarus shares with Poland the Belovezhskaya Forest, one of the last undisturbed representatives of Eastern European mixed forest, which is home to lynx, wolves, and European bison. Ukraine stretches to the Carpathians in the west and the Black Sea in the south. But nothing surpasses the Crimean peninsula, particularly its southern mediterranean area. Georgia also boasts of a mild climate, as well as outstanding vineyards. Azerbaijan offers access to mountains and the Caspian Sea. And Armenia's Lake Sevan is the world's largest alpine lake. These and other natural attractions could become economic assets in the future, if stability can be achieved.

FIGURE A **Contaminated village of Chernobyl, Ukraine.** The radioactive contamination of sizable areas of Ukraine and Belarus from the Chernobyl nuclear disaster was one of the greatest environmental failings of the former Soviet Union. The effects of the disaster on human health and regional land use will likely be felt for decades to come.

FIGURE B **Lake Baikal in Siberia.** A unique freshwater ecosystem, Lake Baikal is known for its clear water and unique animal species. Recent efforts to reduce effluent flow into the lake have diminished the danger of pollution from Siberian industrial development.

---

[2]To learn more about Russia's natural areas that have received UNESCO World Heritage designation, consult *http://whc.unesco.org/pg.cfm?cid =31.* Details and up-to-date information on Russia's many natural areas and reserves can be found at *http://www.russianconservation.org/.*

GEOGRAPHY IN ACTION
War in Chechnya

It would take twenty-five Chechnyas to equal the territory of Iraq, and Chechen guerillas are outnumbered 50 to 1 by Russian troops and police. Yet they not only endure but continue to inflict serious damage on their foes, much as their ancestors did. In the nineteenth century, Chechen warriors fought the Imperial Army for 50 years. In the Russian Civil War, Chechens again fought for independence. For this reason Stalin distrusted the Chechens and had them exiled en masse to Central Asia in World War II.

In 1994 the Chechen government declared independence, which led to a Russian invasion. In the fighting, Groznyi, the capital, was virtually destroyed (Figure A). But the Russian army was bedeviled by low morale and corruption. In 1996 the two sides agreed to disagree, but peacefully. The Chechens again proclaimed independence, which Russia declined to acknowledge, but Russian troops withdrew in ignominy. Yet independent Chechnya, landlocked and small, was no paradise. Crime flourished, as did radical Islamist ideology. In 1999 fundamentalist fighters swept into neighboring Dagestan. Again Russia intervened, chasing the guerillas back into Chechnya and reigniting the war.

Russian forces control the towns and most of the infrastructure, at least in the daytime. The guerillas hide in the mountains and some of the villages. They set ambushes, plant mines, shoot down Russian aircraft, and dispatch suicide bombers. Es-

■ **FIGURE B The North Caucasus.** Ethnic warfare has resulted in the loss of tens of thousands of lives and physical and economic ruin. It has also divided the Russian people, whose armies, for all their supposed military might, have been unable to subdue Islamic rebels.

■ **FIGURE A War damage in Chechnya.** Some 300,000 Chechens, out of a million, have fled their homeland to escape the violence. Most of those who remain seek only to preserve their lives and property.

timates of Russian losses vary from about 3,400 to about 15,000. Russian forces have been accused of frequent, serious human-rights violations. Today the Russians are attempting to build up loyal Chechen forces; however, the latter are grouped around different clan chiefs instead of forming a unified force.

Chechen terrorists have had highly dramatic successes, including a series of suicide bombings and mass hostage takings, such as the bloody slaughter at a school in Beslan, North Ossetia (Figure B). These acts have outraged the Russian people, who overwhelmingly support President Vladimir Putin's hardline approach to the war. Now the unrest is spreading. Corrupt local administrations breed resentment, and many of the small nations have grievances with Russia or with each other. Political violence is mounting, especially in Dagestan, one of the most ethnically diverse areas on earth. For the rest of the world, the war in Chechnya presents a dilemma. The Chechen separatists are ruthless terrorists. Yet from their perspective, they are fighting for independence from an empire that has no right to rule over their homeland.

---

complicated domestic politics and gave cause to worry that Russia might one day seek to regain these territories.

Within Russia itself the main issue was the degree of autonomy allowed the remaining jurisdictions that represented non-Russians. At first, Yeltsin followed a policy of devolution of power. But when the government of Chechnya declared independence (see the *Geography in Action* boxed feature, War in Chechnya), Yeltsin changed course and tried to compel submission. The conflict in Chechnya has yet to be resolved, while the situation in many of the other non-

Russian territories has deteriorated. President Putin has the final say concerning the composition of regional governments and demands total loyalty. Yet the local elites have been allowed great latitude in the repression and exploitation of the people under their rule, which sparks bitter discontent.

Unhappiness takes various forms. Of all the in-between countries, only Azerbaijan and Armenia have positive natural population increase rates, and their rates are very low— 1.0 and 0.3 percent, respectively (Table 13–1). To a degree this is because they all have gone through the demographic

Table 13-1 The In-Between Countries of Eurasia

	Capital	Area Sq. Miles (km.) in thousands	Natural Increase Rate (%)	Population, 1999 (in millions)	Population, 2005 (in millions)	Largest Ethnic Group (% of total)	Second Largest Group (% of total)	Principal Religions
Russia	Moscow	6,593 (17,076)	−0.6	146.5	143	Russian (82)	Tatar (5)	Eastern Orthodox
Ukraine	Kiev	233 (604)	−0.7	49.9	47.1	Ukrainian (78)	Russian (17)	Eastern Orthodox
Belarus (Belorussia)	Minsk	80 (208)	−0.6	10.2	9.8	Belarussian (81)	Russian (11)	Eastern Orthodox
Azerbaijan	Baki (Baku)	33 (86)	1.0	7.7	8.4	Azeri (91)	Russian (2)	Islam
Georgia	Tbilisi	27 (70)	0.0	5.4	4.5	Georgian (84)	Azerbaijani (6.5)	Georgian Orthodox
Armenia	Yerevan	12 (31)	0.3	3.8	3	Armenian (98)	Kurds (1)	Armenian Orthodox

Source: USSR Central Statistical Committee, Narodnoye Khozyaystro SSSR za 70 let (Moscow: Finansy i Statistika, 1987), 373; *1995 CIA World Fact Book; World Population Data Sheet* (Washington, D.C.: Population Reference Bureau, 1999 and 2005); Wikipedia, *http://en.wikipedia.org/wiki/Main_Page.*

transition—all are primarily urban countries in which women have, at least in principle, the same rights as men. Everywhere in the world these conditions have resulted in lower birthrates. But in Russia, Ukraine, and Belarus there is the unprecedented twist of rising death rates, especially for men (Figure 13-13). Currently in Russia male life expectancy is just 58, which is 3 years lower than the average for developing countries. Males fare as well in Ghana as in the heart of the former superpower! Many explanations have been put forward, such as high rates of smoking and vodka consumption, decline in public-health services, high-cholesterol diets, environmental pollution, accidents, and stress from a perceived loss of status. While there is some truth to each of them, taken altogether they do not entirely explain why so many men still die prematurely 15 years after the end of the Soviet system.

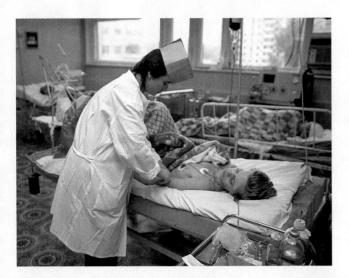

■ **FIGURE 13-13 Health care in Russia.** Since the end of the Soviet period, health care in Russia has deteriorated, and life expectancy for Russian men and women has declined.

▶▶ Summary

Geography does not determine countries' destinies. And history is what was, not necessarily what is going to be. However, we deal with geography at various scales every waking moment of our lives. Geography is an especially important force in economic activity. The formidable physical conditions found in Russia, and the remoteness of most locations, inevitably add to the costs of development. Just how this development will proceed, in this age of globalization, remains to be seen. History also matters, especially with respect to political culture. How people understand history affects what they expect from their leaders, how they see the role of the state, and how they judge their institutions. Democracy played little part in the history of this region, so the cultural environment is not particularly hospitable for a transition to democracy.

▶▶ Key Terms

Bolsheviks
Caucasus Mountains
Chernobyl
chernozem
cold war
Communist Party
continental climate
Dnieper River
Duma
Five Year Plan
Josef Stalin
Kremlin
Lake Baikal

Mikhail Gorbachev
Ob River
permafrost
serfdom
steppe
taiga
Trans-Siberian Railway
tundra
Ural Mountains
Vladimir Lenin
Volga River
Yenisey River

Trans-Siberian Railway. An engineering marvel that stretches from Moscow to Vladivostok, the world's longest rail line passes through a variety of landscapes including desert, steppe, forest, and mountain habitats in its 9,300-kilometer- (5,800-mile) long route from the Russian heartland to the Pacific Ocean.

Chapter 14

The In-Between Countries of Eurasia:
The Geography of Their Political Economies

- **The Russian Federation: Economic Policy and Society**
- **Ukraine: A New Start**
- **Belarus: Stuck in Transition**

Since 1991 Western observers have spoken of the in-between countries as being in transition. Usually it is simply assumed that after the transition the countries will resemble the United States or Western Europe in having democratic states and market economies. But overuse of the "transition" label may obscure some unpalatable realities. None of the in-between countries has an economy that is primarily governed by the market. Armenia, Georgia, and Ukraine have moved toward democracy, but these gains are far from secure. Azerbaijan and Belarus have authoritarian governments, while Russia is clearly moving in that direction. This chapter explores current realities in the "in-between" countries from a geographical perspective and considers the gap between what is there today and what is needed for them to succeed in this era of globalization.

The Russian Federation: Economic Policy and Society

In late 1991 there was much excitement in the new Russian Federation. For the first time in their lives people were free to voice their opinions in public, and many did. It appeared that all of the old, repressive institutions would be replaced by new democratic agencies of the people's will. Russian President Boris Yeltsin even launched a fundamental reform of the dreaded KGB, the secret police. However, despite some initial successes, euphoria faded and the institutional bases for democracy were slow to materialize. The KGB, for instance, fought off major reform and regrouped as the Federal Security Service (FSB). Many people who had been part of the old elite found ways to enter the new one—communists became "democrats" overnight,

by donning an imported suit and having a photo-op at a church.

President Yeltsin himself had been a Communist Party boss for decades. He was skilled in the arts of dealmaking and political maneuvering, but he knew little about economics. Russian economists and some Western advisors were convinced that the quickest transition to capitalism would be the least painful, but hyperinflation was a grave threat, in large part because Yeltsin had pursued a very liberal fiscal policy to bolster his popularity in the context of his struggle with Soviet President Gorbachev. There were too many rubles in circulation, in savings accounts, and in mattresses. Yeltsin's advisors were afraid to just eliminate state controls and let the market determine prices and wages because there might be "too much money chasing too few goods," the classic recipe for inflation. Instead, the government launched "shock therapy," the first stage of which was a government-controlled price increase on goods, including basic foodstuffs—as much as 800 percent overnight. People called the policy "shock with no therapy," as they spent their life savings to put food on the table. They felt they had been robbed.

Worse yet, inflation was not averted. Trade barriers were knocked down and the global economy surged in, in the form of consumer goods that were often inferior by European standards but marked up in price in Russia, where there was a huge repressed demand for imports. Few Russian industries were competitive in world markets. Consumption was globalized, but production declined radically.

Demoralization was not the only negative effect of "shock with no therapy." Corruption existed before, but now it was an epidemic. Police, judges, medical personnel, teachers, and others who provide vital services to the public all struggled against instant impoverishment as best as they could.

Some former professionals, and more than a few college students, helped form the "mafias" that proliferated in the cities.

Had this been the last such reform, perhaps "democrat" would not be a dirty word for most of the public today. But it was not. A currency reform followed that again penalized people who had saved rubles. Even more damaging was the persistent practice of failing to give people their pay. Some employees, and pensioners, went many months on end without receiving their rightful income. What better way to prevent "too many rubles" than to refrain from paying salaries! But perhaps the most controversial stage of the transition was **privatization of industry**. Under the Soviet system, the country's economy in theory was the property of the people, although in practice it belonged to the state. The privatization reform promised to make all citizens shareholders in corporations, but in reality elite figures concentrated control over all valuable assets. A social-Darwinian struggle of survival between the new entrepreneurs led to the emergence of the **oligarchs**, a few individuals who controlled vast economic empires (Figure 14–1). Ordinary people watched aghast as dollar billionaires emerged when just a few years previously nobody amassed personal fortunes—there were no checking accounts, credit cards, or ATMs.

"Shock therapy" received strong support from the International Monetary Fund (IMF), which helped make Russia attractive for foreign investment. Most successful were investors in the booming retail sector and real estate, especially in Moscow and St. Petersburg. After the initial crash the economy slowly began to recover until the 1998 financial crisis struck. Foreign investment nearly dried up and the government was forced to devalue the ruble. The period was painful for many, but the ruble devaluation made foreign goods more expensive, which helped Russian producers recapture domestic markets. Growth resumed, and now it was more diversified. Foreign investment returned—not just in retail, but also in producing for the Russian market, particularly in the vicinity of Moscow and St. Petersburg, which

■ **FIGURE 14-1   Lukoil headquarters building, Moscow.**
The new corporate office building of Lukoil, Russia's largest oil company, reflects the prosperity, and even opulent lifestyle, that has come to characterize the elite entrepreneurial class in the post-communist era.

are the largest markets with the best accessibility. In his first term, President **Vladimir Putin** called for reforms to put business on a more stable footing. Several of the oligarchs, including oil billionaire Mikhail Khodorkovskii, began improving the transparency of their companies' operations and urged development of appropriate institutions to regulate the economy legally.

Unfortunately, foreign investors have grown wary again recently due to the failure to build institutions that would protect property and guarantee the rule of law. The government has imprisoned Khodorkovskii, ostensibly for tax evasion. Interpretations vary, but it appears that Putin's government does not welcome the oligarchs' involvement in politics. Also, the regime appears to want more direct control over the crucial oil and gas industries. Natural gas in fact is virtually monopolized by the government-controlled company **Gazprom**. Gazprom is also moving into the oil industry. Finally, it appears likely that some of Putin's appointees, called *siloviki*, men from the "power ministries," the FSB or the military, are taking slices of the economic pie for themselves. The government still calls for economic innovation—especially in the area of information technology, in which Russia does possess some important advantages. But overall, the economy suffers from a lack of strict rules concerning government intervention, property rights, financial regulation, and transparency. The result is *rent-seeking*, the use of the resources of the state for the benefit of private interests, including the "private interests" of many public officials. In general, corruption is pervasive. Excessive reliance on income from the export of oil and gas makes the entire economy highly vulnerable to price declines in either industry.

Post-Soviet Changes in Society

State policy may be stuck in transition, but on the street the explosive growth of consumerism has changed the Russian way of life. Although most people earn paltry salaries by Western standards, Russia still heavily subsidizes housing and utilities for most of the population. Now people can buy new TVs or laptops, or even cars, such as they could only have dreamt of in the Soviet past. In major cities shopping malls and megastores are proliferating, while even in the poorest neighborhoods small groceries offer a much greater range of goods than in the past (Figure 14–2). The message is clear and most urban people, at least, have gotten it—make money and you can buy all kinds of things.

Some people have prospered. Although only a few became billionaire oligarchs, the so-called **New Russians** emerged as the first class of wealthy Russians since 1917. Their conspicuous consumption and uncultured ways make them the target of scorn—few believe that such riches could be accumulated overnight by ethical means. Indeed, many did use connections and insider information to corner the market on something in high demand. However, another group, a **new middle class**, has also appeared on the scene, at least in the larger urban areas. Unlike the New Russians, new middle-class people are professionals who draw salaries in fields

■ **FIGURE 14-2 New shopping complex in Moscow.** The retail sector is booming in Moscow and other cities. This "Ramstore" was built on open space next to an apartment building, despite the residents' protests. There is no zoning in Russian cities, and citizens are not invited to public hearings before projects are undertaken.

valued by the global economy. They earn more than the Russian average, but they work long, intense hours, often in fields that did not exist in Soviet times. New middle-class people are making globalization work for them in terms of consumption. But few have time for politics or civic engagement, which slows the formation of civil society in Russia.

For older professionals—the remnants of the Soviet middle class—the picture is mixed. Some have managed to exploit the possibilities of their situation to maintain a comfortable life, especially in Moscow or St. Petersburg. In other locations, the opportunities are much more limited—educators, engineers, and other professionals have sunk into poverty. For workers, the picture is bleak, because most industries need to shed labor and boost productivity. Agricultural workers in general have the grimmest prospects of all, although there are some bright spots. Overall, most new jobs are in producer and consumer services. For men a huge sector has opened up in private security services. Criminal activities such as prostitution and extortion occupy many more people than in the Soviet past.

With respect to Russian society today, three types of difference deserve special attention: old/young, male/female, and Russian/non-Russian. In the first case, there clearly is a generation gap in Russia today, a profound difference in perspectives, which has resulted from the old and young growing up in different worlds. Middle-aged people and seniors are uneasy at best with the continuing erosion of social-security institutions. The only significant protest movement during Putin's presidency was in opposition to the monetization of in-kind benefits such as discounted food and transport for the elderly. Many older people react on principle to what they see as a pervasive dog-eat-dog mentality. Young people, on the other hand, are more comfortable with a rapidly changing world in which their own individual merits and efforts determine their success. They tend to be very confident about the future, both for themselves and their country.

Communism was a patriarchal system, but the state did provide substantial maternal and child-care benefits. Women were not exploited as sexual objects in the mass media. And quotas were set in legislative and other bodies to ensure significant female representation. The quotas are entirely gone now, and the family-welfare benefits are negligible. Sexual exploitation of women is rampant, and not just in the media (Figure 14–3). In the 1990s job ads in major newspapers specified "girl with no complexes wanted," which meant that she would be expected to provide sexual favors. Prostitution has become a major industry, not just domestically. Many have been lured by offers of legitimate employment in exotic locations, only to end up trapped, when their documents are held by their "agents," and forced to work as sex slaves.

On the other hand, life for women under the Soviet system was stifling in many respects, not least because the regime's ideology was antithetical to style—female tractor operators and "heroic mothers" with 8–12 children got their pictures in the news. By and large, Russian women are indifferent to feminism, but keenly interested in expressing and displaying femininity. Therefore, the flood of fashionable clothes and beauty products has been extremely popular. Although industrial jobs for women have declined drastically, young professional women find diverse and challenging opportunities in new fields such as information technology and real estate.

The USSR became fifteen independent countries, of which only Russia has attempted to create a federation. By the government's count there are some 150 nationalities and ethnic groups in the Russian Federation. Twenty-one of these nationalities inhabit "autonomous republics," which have certain rights with respect to the preservation of their cultures and economic development of their territories

■ **FIGURE 14-3 Billboard fantasy in Moscow.** Despite the government warning against smoking on the bottom of this billboard, the caption at the top promises that "everything is possible" for a man who smokes "West" cigarettes. Blatant exploitation of women as sex objects appeared after the downfall of the Soviet Union.

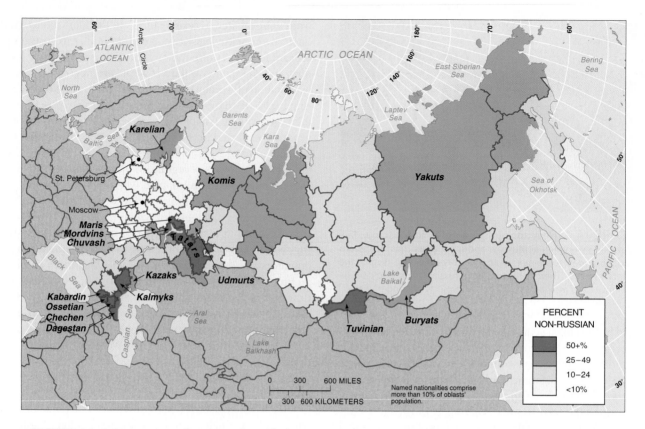

■ FIGURE 14-4 Percentage of non-Russian ethnic groups in Russia. These ethnic groups, most of which are found in the Russian Caucasus, the middle Volga region, and Siberia, may prove problematic for the future of Russia.

(Figure 14–4). Another 12 nationalities inhabit lesser-order territorial units based on ethnic criteria. Relations between the non-Russian nationalities and Russians vary considerably. Some nationalities, including the Chechens and Ingush, suffered historically and still have bitter grievances today. Others, such as the Udmurts, appear satisfied with their situation. Sometimes economic matters complicate ethnic politics. For example, Tatarstan has highly profitable oil wells and substantial reserves. Tatars compose the largest non-Russian nationality, with about 5.6 million people.

Russian attitudes have hardened against many of the non-Russian nationalities, particularly the Caucasians or others who resemble them in Russian eyes. Preexisting prejudices have intensified in the wake of terrorist attacks in Moscow and the North Caucasus. Denigrated as "blacks," Caucasians are frequently attacked by skinheads and detained by the police. Thus, Chechen extremists' terrorism has contributed to a situation in which ordinary Caucasians dwelling outside their homelands live in fear of fellow federation citizens.

Russia's Economic Geography

The more the Russian economy is subjected to market forces the more important become typical economic-geographic factors, such as costs of production and transportation costs. Hardly anything is produced more cheaply in Russia than anywhere else because of harsh climatic conditions, distance

to market, inadequate transportation and communications infrastructures, obsolescent technology, corruption, and other factors (see Figure 13–10 and Table 14–1). Places where such costs can be minimized may do relatively well under market conditions. But there are other centers of production in Russia that may not survive in a competitive economic environment.

Industry

The brightest spots in the economic landscape are Moscow and St. Petersburg, along with their surrounding regions (see the *Geography in Action* boxed feature, Moscow: Emerging Global City?). The economic structure of even the "winner" regions is changing. Textiles are in decline everywhere, even in their traditional core area of the Moscow region. Machine building and metalworking industries are not as hard hit as textiles, but the transition to competitive conditions has been painful. However, because St. Petersburg and Moscow have the largest concentrations of population and the best transportation facilities, they have attracted industries that produce consumer goods for the Russian market, including automobiles and parts, household appliances, and a variety of food products. In general, with the transition to consumerism urban markets have become crucial in determining what is produced and where production is located (Figure 14–5).

Recently, the government resolved to promote the development of information technology parks. The Soviet Union

Table 14-1   Economic Resources of the In-Between Countries of Eurasia

	Per Capita GNI, 2004	Major Resources	Principal Industrial Cities	Principal Manufactured Goods	Major Agricultural Products
Russia	9,620	Oil, natural gas, coal, iron ore, hydroelectric power, gold, polymetallic ores, aluminum ore, timber, diamonds, platinum, fertile soils	Moscow, St. Petersburg, Niznyi Novgorod (Gorki), Yekaterinburg, Novosibirsk, Perm, Chelyabinsk	Steel, machinery, metal working, textiles, chemicals, armaments, paper, food products, transport equipment, electronics [del woodworking]	Wheat, rye, potatoes, dairy products, beef, swine, corn, barley, oats
Belarus	6,900	Peat, oil	Minsk	Machinery, food, woodworking, electrical goods	Potatoes, dairy products, flax, beef, swine, wheat, rye
Ukraine	6,250	Coal, oil, iron ore, manganese, natural gas, uranium, fertile soils	Kiev, Kharkov, Dnipropetrovsk, Donetsk, Odessa	Steel, machinery, metal working, chemicals, food products, agricultural equipment	Wheat, sugar beets, corn, barley, beef, swine, sunflowers
Armenia	4,270	Aluminum ore	Yerevan	Chemicals, jewelry, food, machinery, light metallurgy	Wine, cognac, grains, fruits, vegetables, tobacco
Azerbaijan	3,830	Oil, natural gas	Baki (Baku)	Petrochemicals, food, machinery	Cotton, tea, subtropical fruits
Georgia	2,930	Manganese, coal, oil	Tbilisi	Machinery, food, building materials, steel, chemicals	Tea, subtropical fruits, wine, cognac, tobacco

Source: *World Population Data Sheet* (Washington, D.C.: Population Reference Bureau, 2005). Data are calculated using the GNI PPP method.

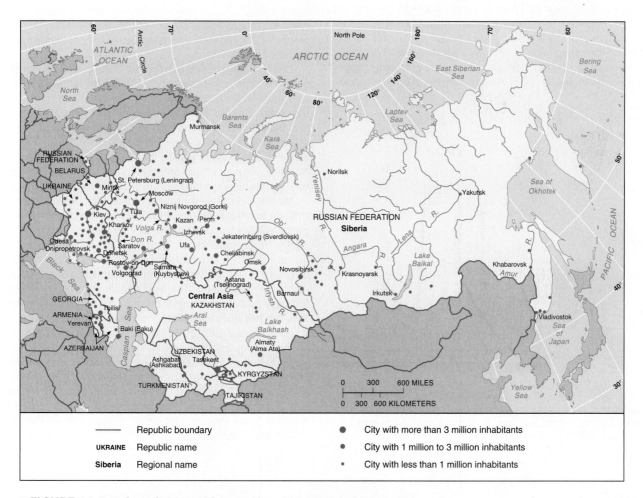

■ **FIGURE 14-5 Selected cities with more than 100,000 inhabitants.** Most of the area's major cities are found in the western fourth of the region.

GEOGRAPHY IN ACTION
Moscow: Emerging Global City?

Moscow's Mayor Yurii Luzhkov has dedicated himself to making Moscow a global city, one of the world's leading centers of trade, finance, and entertainment. The economic and cultural pressures of globalization, together with currents of power that shape the national, regional, and urban contexts all affect Moscow's transformation. However, the crucial agent is the city government, because it owns most of the city's land and controls what will be done with it. Not only is the city to be global, but the mayor has also sworn to make it "comfortable." These aims dovetail, since the "comfort" campaign is dedicated primarily to the new middle class (Figure A), which contributes to globally oriented sectors of the economy while also forming the vanguard of cosmopolitan consumerism.

The mushrooming proliferation of residential and office towers along with the flashy new superstores all contribute to Moscow's image makeover. Most prominent among the new developments is the Moscow City complex, which is now under construction on the banks of the Moskva River. The name "City" evokes London's famous financial center and indicates Moscow's high ambitions. The first stage of construction includes a building for city government agencies concerned with promoting investment, a trade center, parking, and a metro station. A pedestrian bridge that includes a mall has already been completed. The plan for the future includes hotels and a water park, as well as new buildings to house the legislative and executive organs of the city government (Figure B).

Top city officials are sometimes called city builders in the media. The designation is deserved, because the city is being reconstructed. One of the most important geographical changes is the "avtomobilizatsiia" (automobilization) of the metropolis.

■ **FIGURE B Amenities for the new middle class.** Little unused space remains in the Russian capital, yet room was found for this nine-hole golf course. Now New Russians can enjoy a middle-class lifestyle in the heart of Moscow, a city that is going global at the same time that it rediscovers its cultural heritage.

The first project was the Outer Ring Road, which bounds the city, turning it into a modern limited-access, divided highway. Subsequently, the city began the Third Transport Ring (TTR), the most expensive engineering project since the end of the USSR. TTR had to be retrofitted into the built landscape between the Garden Ring, which bounds the historic center, and the Outer Ring. The multibillion-dollar project has transformed the movement of traffic in the city and attracted new shopping complexes and residential towers.

However, city building in a limited, built-up territory inevitably requires some "city destroying" as well. This is particularly the case in the historic center, where space is at a premium. Many restoration projects have retained no more than the facade of the original structure. Some buildings that were landmarks in the cultural landscape have been completly destroyed, such as the Hotel Moskva just to the north of Red Square. Such "creative destruction" has provoked protests by some residents, who feel some of the changes are ill-considered or in poor taste.

Nevertheless, the extensive reconstruction of Moscow maintains its breathtaking pace. Investment in the city continues to grow, as does the city's economy—led by the construction industry. Urban space is changing from industrial to post-industrial at a gallop as old industries are relocated, or simply closed down, to make more spaces for residential, office, and commercial developments. The new middle class is prominent in Moscow and the city's global connections are proliferating. Moscow has not yet reached the top rank of global cities, but it has charted a clear path. What does this mean for Russia? St. Petersburg is following a similar trajectory and the region around Moscow is also developing briskly. But countless towns and villages in Russia languish in poverty and despair. Ultimately Moscow's economic health depends on it becoming the gateway between the world economy and all of Russia. The big question is: will Moscow's globalization approach "trickle down" to the rest of the country?

■ **FIGURE A New business center in Moscow.** Global cities have agglomerations of producer services. This center is well located on the Garden Ring next to metro and rail connections, with train service to an airport.

■ **FIGURE 14-6 Novosibirsk, Siberia.** Novosibirsk is Siberia's largest city and a major transportation and industrial center. Shown is the Gorki Cultural Center.

invested heavily in science and technology; whole towns were established as closed "Science Cities." Today their original consumer, the Soviet military complex, is gone and they lack the resources to diversify research and develop marketable products. Considerable human resources are still there, although many professionals emigrated as part of a brain drain. The Russian Federation needs to invest and attract foreign investment in centers of innovation. Moscow, St. Petersburg, and Novosibirsk, with their concentrations of scientific institutions and personnel, already have concentrations of scientists and technical specialists (Figure 14–6).

Other areas with potential include the Volga region. Although some industries have failed, the automotive sector appears to be recovering. The general diversity of the region's industrial base is an asset as well (Figure 14–7). Also

Rostov-on-the-Don shows promise. Long called the Gateway to the Caucasus, post-Soviet Rostov is also Russia's southern gateway to the sea and it serves Russia's most fertile agricultural region. Don River traffic is important, but also a canal connects Rostov with the Volga and the Caspian Basin. Finally, beyond the Urals centers of oil and gas production are prospering, such as Tiumen in western Siberia (Figure 14–8).

The big losers in the transition are regions that relied on a narrow base of heavy industry, especially those in remote areas. The coal mines of the Pechora region—above the Arctic Circle west of the Urals—appear to have no prospects, yet the miners soldier on, hoping for a miracle. The Urals, once the USSR's industrial heartland, has slumped due to obsolescent technology, world competition, and difficulty accessing world markets. The Kuzbas region, dependent on coal and steel, has the same problems, with even higher transportation costs. In the 1970s and early 1980s, Soviet media championed the construction of the Baikal-Amur Mainline (BAM), which was going to serve as a land bridge for East-West trade and be the springboard of development in resource-rich Yakutia. Operational since 1991, this civil-engineering marvel is too expensive to operate on the basis of ordinary cargo, as current revenues cover less than 50 percent of maintenance expenses. The 300,000 people who settled along BAM's route expecting prosperity now endure a grim struggle to survive.

Agriculture, Changing in Two Opposite Directions

As can be seen from Figure 14–9, some large-scale crop production continues in Russia, but much of the land most suitable for grain, sugar-beet, and sunflower production now

■ **FIGURE 14-8 Oil and natural gas development in Siberia.** Despite its harsh physical environments, Siberia has become Russia's most important source region for petroleum and natural gas. Almost all of Siberia's production is transported by pipeline to the more densely settled European areas of Russia or beyond to European export markets.

■ **FIGURE 14-7 Tolyatti automobile plant.** Russia's Tolyatti automobile manufacturing plant is one of the largest industrial enterprises in the country. It is located in the mid-Volga region.

■ FIGURE 14-9 Commercial sunflower farm. Sunflowers are grown extensively in the western portions of Zone II (see Figure 14-10). They are raised for their seeds, which are a source of much of the cooking oil used in Russia and adjoining countries. This farm is near Barnaul, Russia.

belongs to Ukraine and Kazakhstan (Figure 14–10). In the area devoted to diversified agriculture, soil quality is inferior and poor drainage is often a problem north of the Smolensk-Moscow axis. And of course the growing season is short. Because self-sufficiency was vital for the Soviet regime, the government invested heavily in developing agriculture in difficult areas.

The Soviet ideal was **state farms**, large-scale rural enterprises that were operated as much as possible on industrial principles. In the early 1930s, the regime set up **collective farms**, also very large scale, which were seen as precursors of state farms. The farmers themselves were supposed to be the owner-operators, but in practice the state maintained strategic control over them. Stalin tried unsuccessfully to abolish farm families' home gardens, their private plots. As it turned out, Soviet agriculture depended heavily on **private-plot production**, especially for fruits and vegetables.

In the 1990s, Russian agriculture was privatized. Western advisors expected that enterprising families would withdraw from the huge state and collective farms and set up family farms on a commercial basis. There are more than a quarter million family farms in Russia today, but their contribution to overall output is only about 4 percent. Families have been reluctant to leave the state and collective farms, which are now independent corporations. Dividing the land raises issues, but a greater problem is the distribution of the farms' other assets, such as the machinery. Potential family farmers often lack the assets, especially capital, to get started in agribusiness. Credit is weakly developed in Russia generally and especially in the countryside, where family farmers have little collateral and no credit histories. Farmers who stay with the corporation have access to its machinery, fertilizer, and other assets, which they often use on their private plots. In these cases, the corporation may fare

badly while the farm families survive thanks to the private plots. However, this does not seem viable in the long run. Over half of the corporate farms run at a loss. Outside interests have gained shares steadily, to the point now that the farmers themselves typically form a minority of the shareholders. Bankruptcies and liquidations are becoming more frequent—especially where physical conditions are unfavorable or where farms are remote from markets (Figure 14–11).

Current trends are contradictory. On the one hand, new types of commercial operations are emerging. These are not relabeled collectives; their only aim is to make a profit. These new agribusinesses are found where climate, soil, and market accessibility are most favorable. This is normal under market conditions and so may be considered progress, even if it causes hardship for farmers in ex-collectives that fail. On the other hand, the area of arable land devoted to private plots at least doubled in the 1990s. More than 13 million people are employed on household plots; some 10 million of them are subsistence farmers. About half of the world's countries have populations of 10 million or less, so Russia has a "nation" just surviving through subsistence gardening. This is not a sign of progress for a developed country.

Political Geography and Political Economy

The focus of this section is on how the state works throughout the country, but first a sketch of the central government is in order. Vladimir Putin, previously head of the Federal Security Bureau, has been elected President twice. There were some irregularities, but most believe he would have won easily had the elections been flawless. So does that make Russia a democracy? Perhaps, but if so, it is one with serious problems. President Putin did not run as the candidate of a party; instead a new party formed around him, and now it dominates the Duma, the legislative branch. In the judiciary there undoubtedly is corruption, but worse is the ingrained habit of serving as an agency of the executive branch rather than upholding the law impartially. The tax authorities also act as enforcer of the regime's will rather than a law-based institution, targeting certain alleged evaders while ignoring the crimes of others. The major television stations and most of the radio stations are government-controlled and rarely mention anything that would be unflattering to the government. In recent years the government has taken steps to prevent public protest and threatened to make NGOs get official approval. Laws are selectively enforced to punish perceived opponents. On the whole, the drift is toward a more authoritarian government.

But what goes on outside Moscow? When the USSR broke up, local elites sought to control the resources of their territories (see the *Geography in Action* boxed feature, The Caspian Sea: Geopolitics of Oil). Former communists, for whom ethnicity had been insignificant and religion suspect, suddenly embraced new flags and faiths. President Yeltsin, because of his precarious political situation, cultivated relations with regional, especially non-Russian, leaders. Some called this an overdue decentralization, which promised to

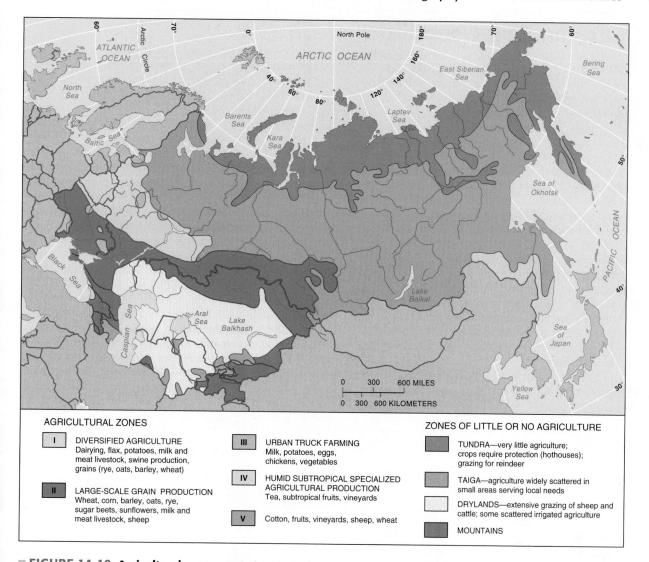

AGRICULTURAL ZONES

| I | DIVERSIFIED AGRICULTURE |
Dairying, flax, potatoes, milk and meat livestock, swine production, grains (rye, oats, barley, wheat)

| II | LARGE-SCALE GRAIN PRODUCTION |
Wheat, corn, barley, oats, rye, sugar beets, sunflowers, milk and meat livestock, sheep

| III | URBAN TRUCK FARMING |
Milk, potatoes, eggs, chickens, vegetables

| IV | HUMID SUBTROPICAL SPECIALIZED AGRICULTURAL PRODUCTION |
Tea, subtropical fruits, vineyards

| V | Cotton, fruits, vineyards, sheep, wheat |

ZONES OF LITTLE OR NO AGRICULTURE

TUNDRA—very little agriculture; crops require protection (hothouses); grazing for reindeer

TAIGA—agriculture widely scattered in small areas serving local needs

DRYLANDS—extensive grazing of sheep and cattle; some scattered irrigated agriculture

MOUNTAINS

■ **FIGURE 14-10 Agricultural zones.** Agriculture in the former Soviet Union was confined largely to the southwest quarter of the region. After World War II, attempts were made to expand the area of crop production into the dry lands east of the Caspian Sea and into western Siberia.

stop the Kremlin from stifling local initiative and monopolizing the country's resources. However, all too often regional leaders actually were building up their own networks for monopolizing political power and maximizing the "rent" taken from local economic activity. With their people in most of the important positions, with the local media under control, and with financial resources available through back channels, these leaders had little to fear from elections.

President Putin set out to re-centralize control over the country by establishing a layer of super-governors, who were hand-picked by the President himself, basically to oversee the elected governors. Putin's selections were all *siloviki*. This move was not popular in the West, but it occasioned little protest in Russia. Many people were disgusted with corrupt officials and economic stagnation, and so believed a firm hand was necessary.

Immediately after the Beslan massacre, when hundreds of innocents died after Islamist fanatics seized a school full of children, Putin declared that henceforth he would appoint governors. Political commentators argued that the President

■ **FIGURE 14-11 Farming in eastern Russia.** Farm workers sift grain 900 kilometers (560 miles) east of Moscow. Distance to large markets, high transportation costs, and low profits make a transition from subsidized state farming to a privatized, market-based economy a painful and slow process.

GEOGRAPHY IN ACTION
The Caspian Sea: Geopolitics of Oil

In May 2005 the Baku–Tbilisi–Ceyhan oil pipeline opened, forging a new link between the Caspian and the Mediterranean, 1,100 miles (1,770 kilometers) away. But long before a single mile of pipe was laid, petroleum geopolitics made the project a burning issue. The chosen route reflects the preference of the United States, as well as Azerbaijan, Georgia, and Turkey. The United States opposed alternatives that would have made Russia or Iran partners in the deal, probably because both already are major players in the oil business and chart independent courses in foreign policy. Also, the alternative North Caucasus route was made impractical by the war in Chechnya and the spread of violence in the region (Figure A).

Azerbaijan and Georgia now assume geopolitical roles at the global level, as well as bolstering their revenues. Turkey is glad to divert oil transport from the crowded Bosporus and Dardanelles Straits, which may be an ecological time bomb due to the high volume of tanker traffic in the narrow channels. And Ceyhan becomes a vital connection with the West just as Turkey seeks to enter the European Union. Also, Kazakhstan will be using the pipeline to move its oil in the near term, and Turkmenistan stands to benefit in the future.

When the controversy was at its peak, some argued that the Caspian region would one day rival the Persian Gulf in petroleum production and, therefore, in geopolitical significance. The rhetoric has since dissipated, but the question has not been settled. Estimates of the ultimate size of Caspian reserves range from 32 to 220 billion barrels. If the larger figure is proven, the Caspian region becomes another Saudi Arabia. If the lowest estimate proves correct, then the region's reserves will compare with Mexico's, but they will be split among as many as five countries.

Today the pipeline is important mainly because of where the oil comes from, rather than its quantity. When the pipeline becomes fully operational in 2009, it will carry 1 million barrels per day, about 1 percent of the world's current needs. But it will be non-OPEC oil, unaffected by Persian Gulf instability or Russian influence. The pipeline will pay significant revenues to Turkey and Georgia in transit fees—this is especially important for impoverished Georgia. But the big winner is Azerbaijan. Experts predicted that Azerbaijan would reap $50 billion over the next 20 years, but they assumed a $25/barrel price for oil. In 2005 the price reached $60/barrel, a huge windfall for Azerbaijan.

Unfortunately, the Transcaucasian region is far from stable. Georgia's 2003 "**Rose Revolution**," which got its name because demonstrators carried roses to emphasize non-violence, ousted Eduard Shevardnadze, the former Gorbachev ally whose corrupt regime failed to revive the economy or keep the country intact. President Mikheil Saakashvili has been assertive territorially, notably by regaining control over the breakaway region

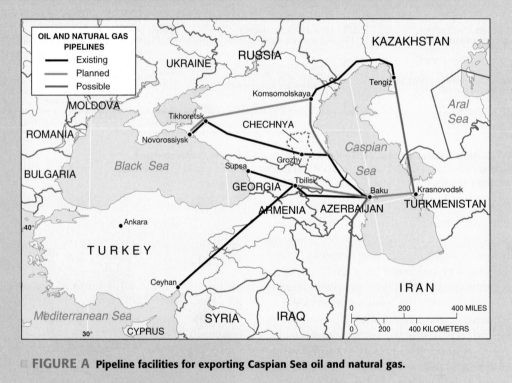

FIGURE A Pipeline facilities for exporting Caspian Sea oil and natural gas.

cynically had used the tragedy as an excuse to grab more power. Again, however, ordinary Russians raised no protest. Dissatisfaction with local leadership remains high. Moreover, people were all too aware that at Beslan, as well as in other terrorist incidents, some citizens must have accepted bribes to look the other way. That people took money to allow the slaughter of innocents was profoundly shocking. If

Putin—sober and serious—believed he could enhance security by appointing governors, most people were willing to let him try. Yet almost immediately the word spread that Putin soon would also assert the authority to appoint mayors for all of Russia's cities. While as of June 2005 that policy had not been enacted, many mayors of Russia's cities were publicly asking the President for his blessing, evidently

of Ajaria. He has insisted, so far unsuccessfully, that Russia remove military bases dating back to the Soviet period. But South Ossetia remains largely beyond the Tbilisi government's control, as it gravitates toward North Ossetia, which is part of the Russian Federation. The Georgian government also does not recognize the independence of Abkhazia, from where 250,000 Georgians were expelled in the 1993 war. Bitter feelings persist on both sides. Finally, Saakashvili has not eliminated corruption or boosted the economy, leaving many people discontented. Georgia remains highly unstable.

Azerbaijan and Armenia have their own internal problems, but they pale compared to the conflict between the two. In the late 1980s the Armenian majority in the **Nagorno-Karabakh region** began agitating for independence from Azerbaijan (Figure B). The Soviet government failed to resolve the conflict, which contributed to its downfall. With the breakup of the USSR, the fighting spread to Azerbaijan, where Armenians were attacked, and to Armenia, where the same fate befell Azerbaijanis. Both minorities fled. Then full-scale war began between the two new states. Armenia managed to seize a corridor between itself and Karabakh, while both countries endured economic decline and political instability. Fighting ceased in 1994, but the fundamental issues have yet to be resolved.

Armenia, by in-between standards, is a relatively democratic country. However, this factor does not necessarily work to-ward peace because most Armenians want Karabakh united with Armenia. President Robert Korcharyan was born in Karabakh and is unlikely to compromise. But Azerbaijan's growing income, thanks to the Ceyhan pipeline, may change the balance of power.

When Azerbaijan's government essentially collapsed during the Karabakh war, leadership passed to Heydar Aliyev, formerly a very high-ranking Soviet official. He established an authoritarian regime and ruled until 2003 when, like a king, he transferred power to his son Ilham. Ilham Aliyev's position later was bolstered by a questionable election, but he lacks the legitimacy of his long-serving father. There are many oppositions in Azerbaijan, and some are radical Islamist. In that regard, Azeris occupy an interesting position in the Muslim world, as they are a Turkic people of Shi'ite heritage, with a large minority in neighboring Iran.

The Aliyev regime has a history of violently suppressing protests. Perhaps the pipeline will bring prosperity and reform. Yet Azerbaijan's regime is as corrupt as any in the in-between region, so the "trickle down" effect may be weak. Continued corruption and poverty could well lead to unrest. If so, a patriotic war to recover Karabakh may appeal as a means of rallying, or distracting, the populace. Given Azerbaijan's new importance in petro-politics, this region deserves close attention in the West.

■ **FIGURE B** **Armenians in Karabakh in Azerbaijan.** The Armenians represent a minority in Azerbaijan and would prefer a political affiliation with neighboring Armenia. This territory is now under Armenian control after a bloody war in the early 1990s.

considering the approval of the President more important than the votes of their constituents.

To most people, only Putin appears capable of dealing with corruption, terrorism, and other sources of insecurity. There also is the vital question of Russia's national economy. Because of the country's size, relative inaccessibility, and harsh climate, some argue that a **state-capitalist economy** would be better than a globalized market economy. Many Russians believe that privatization was a huge swindle and that the country's natural wealth should benefit all the people, not just oligarchs and New Russians. They would applaud if the state interceded on behalf of the people. A recent study shows that only 8.2 percent believe that individuals should be primarily responsible for improving their own

■ **FIGURE 14-12 Russian President meets industrialists in the Kremlin.** In happier days of cooperation in 2001, President Putin and Mikhail Khodorkovskii meet to discuss the state of Russia's economy. By 2003, the billionaire chairman of the YUKOS oil conglomerate was no longer in favor and a judicial probe of Khodorkovskii's business affairs resulted in the oligarch's arrest.

living standards, while 60 percent want increased state spending on old-age pensions, student stipends, and state-employees' wages.[1]

However, the state-capitalist approach assumes that the public's welfare will be the government's top priority. In reality, public officials privatize their government functions by seeking rent in exchange for allowing development to proceed. Since officials may own and run for-profit corporations, firms may be used for political purposes, without regard for economic factors, while government positions become sources of wealth (Figure 14–12). This topsy-turvy system is somewhere between state socialism and capitalism and incorporates much of the worst of both worlds.

In sum, Russia has a **rent-seeking state** from the Kremlin down to the local level. Today the income from oil and natural gas exports makes Russia's economic data look impressive. But this is a dangerously narrow basis for development. Funds that could be used to encourage innovation or diversification are lost as rent. Locally there are bright spots where entrepreneurial activity flourishes and tangible development takes place, but generally rent taking exacts its toll, stifling initiative and discouraging investment.

Ukraine: A New Start

Although dwarfed by Russia, Ukraine is almost as big as Texas and larger than any country in Europe. Mostly it consists of rolling or flat steppe land with extremely fertile chernozem soil. Except for Crimea's southern strip, Ukraine has a continental climate similar to the upper Midwest of the United

States. Ukraine has tremendous agricultural potential—nearly 60 percent of the land is arable. A century ago it was the "Breadbasket of Europe." Ukraine is better suited to grow wheat than most of Russia since its climate is sufficiently mild to permit fall planting (Figure 14–13). This winter wheat sprouts in the fall, goes dormant for winter, and then resumes growth in the spring, giving it a head start over wheat planted in the spring, which has lower yields. Also, maize can be grown to maturity, as well as soybeans, along with the traditional crops of sunflowers and sugar beets.

Ukrainian industry was also world-renowned. Ukraine has coal, iron ore, and manganese—the basic ingredients for steel production. The metallurgical centers of the Donbas and mid-Dnieper areas were the mainstays of Tsarist industrialization and leaders in the Soviet period as well. Machine-building and chemical industries were also developed extensively. The giant industrial complexes, sizable hard coal reserves, and rich agricultural potential led many people to feel that Ukraine would prosper immediately upon attaining its independence (Figure 14–14).

But post–Soviet Ukrainian industry and agriculture have declined and stagnated. Ukraine attempted to avoid shock therapy, but the gradualist approach to reforms made little progress. Agriculture has been privatized in theory, but most farms continue to function as in Soviet times, although not as productively because essential inputs are too expensive. Ukrainian industry was undergoing privatization piecemeal via insider deals that outraged the population. The market has but a small toehold in Ukraine as individuals, enterprises, and institutions have all relied on non-market survival tactics. Vegetable gardening and barter are pervasive, as are many activities that would be illegal in law-governed societies. The Ukrainian state was occupied with rent seeking even more than its Russian counterpart, until the public finally rebelled in 2004. The **Orange Revolution**, which took

■ **FIGURE 14-13 Farm fields in central Ukraine.** This patchwork of seasonally inactive fields (purple and brown) shows the intensity of mechanized agriculture near the Dnieper River. Hedges separating the fields, and forests along the river, appear yellow in this colored radar satellite image.

---

[1]Natalya Alyakrinskaya, "Forty Percent of Population Believe They Are Poor," *The Moscow News*, No/24 (4178) 29 June–5 July, 2005, pp. 1, 4.

■ **FIGURE 14-14 Heavy manufacturing plant in Ukraine.** Ukraine has an abundant natural resource base and a skilled industrial workforce.

■ **FIGURE 14-15 Coal mine, Donetsk, Ukraine.** The Donetsk region in eastern Ukraine is one of the leading coal-producing zones of the former Soviet Union. Shown here is the Skochinski mine.

its name from the opposition party's color, led to a relatively fair presidential election which was won by Viktor Yushchenko, who has promised to root out corruption and restructure both state and economy.

Yushchenko's victory was seen by many in the West and Russia as a blow against the latter, and as an expression of the Ukrainian people's desire to move closer to Europe. However, Ukraine cannot literally move away from Russia. Ukraine is indebted to Russia, it benefits from Russian investments, and it depends on Russia for oil and natural gas. Moreover, although Ukrainian nationalism thrives in western Ukraine, most Ukrainians are little concerned about separating themselves from Russians. Russians make up the largest minority in Ukraine, with local majorities in the Crimea and some parts of eastern Ukraine. More Ukrainians consider Russian rather than Ukrainian to be their native tongue, while many people in rural areas speak a mixture of the two. Also, intermarriage is very common.

However, many Ukrainians do support Yushchenko's effort to improve relations with Europe and the rest of the developed world, not least because it will require rooting out corruption and making institutions more transparent. Then, perhaps, increased investment will help the Ukrainian economy so that ordinary people benefit, not just the "New Ukrainians," who profit from insider deals and rent seeking. This may be Ukraine's best strategy, if it can simultaneously mollify the wary Russian government.

There are practical difficulties. For example, consider Ukrainian coal, which is vital for metallurgy and potentially for export. Coal also could be substituted for petroleum and natural gas in electricity production, thus reducing Ukraine's dependence on Russia (Figure 14–15). But the Ukrainian coal industry is in wretched shape. Ukraine, with 450,000 coal miners, produces less coal than nearby Poland, with just 140,000 miners. About two-thirds of Ukraine's coal mines lose money, while the government doles out billions in subsidies. Ukraine needs fewer mines, but more productive ones that rely on advanced technology rather than muscle power. However, it is not easy for any government, especially a fledgling democracy, to deprive hundreds of thousands of workers of their livelihoods.

Similar dilemmas persist in other industrial sectors and in agriculture. Difficult decisions must be made that will be painful for many. Meanwhile, the West awaits signs of progress, which would likely irritate the Russian government. The Orange Revolution could not change everything overnight. There is no easy way, but a government that earns the people's trust and support is more likely to find some way to make progress.

Belarus: Stuck in Transition

Belarus is about two-thirds as large as neighboring Poland, or about the size of Kansas. With just over 10 million people, there should be plenty of room, but good quality land

■ **FIGURE 14-16 Harvesting potatoes in Belarus.** Potatoes are grown on private plots throughout Russia, Ukraine, and Belarus. Hopefully these Belarusan gardeners are not in an area contaminated by the Chernobyl nuclear reactor accident.

has always been scarce in Belarus. The glaciers left Belarus 11,000 lakes and a chaotic drainage pattern. The Soviets' strenuous efforts left almost 30 percent of the land arable, but very little compares to Ukraine's chernozem. Belarus contains few natural resources, apart from its forests. Moreover, the Chernobyl nuclear accident contaminated about one-fourth of the country's territory (Figure 14–16).

After World War II the Soviets systematically developed industry in Belarus. But with no resource base, Belarus was bound to be dependent on the USSR Inc. Energy dependence is extreme. Belarus does produce some machinery, vehicles, and appliances, but they are not in demand in the West, so Russia remains the main market, another form of dependence.

Chernobyl aroused nationalist feelings in Belarus, but this moment was atypical. For decades Belarusans had gradually been assimilating into Russian culture. Belarus' nationalist first government accomplished little. The in-between economy is still much closer to the Soviet model than to capitalism. Also, being landlocked is a disadvantage in this era of globalization. Yet Belarus's economy is isolated institutionally and politically as well, largely because of its government.

First elected in 1994, President Aleksandr Lukashenko has consolidated power in his own hands. Supporters laud him for eschewing shock therapy and maintaining a social-security safety net to keep ordinary people out of abject poverty. Detractors fault him for curtailing civil liberties. Several outspoken opponents have disappeared or died under mysterious circumstances. Lukashenko is hostile to the West and deals with countries deemed rogue nations

by the United States. Foreign investment in Belarus is insignificant.

In 1999, Belarus and Russia agreed in principle to a union of the two countries. Russia has neither repudiated the agreement nor moved to implement it. Russians and Belarusans are not foreigners to each other, so unification seems natural. But Putin has little reason to proceed. Russia already dominates Belarus so there would be nothing to gain. Lukashenko may be a useful loose cannon where he is, to show the West a contrast to Putin, but having him as some sort of co-leader inside the federation would be intolerable.

▶▶ Summary

The in-between countries have yet to complete the dual transition that has been expected of them—to become democracies with market economies. All six countries left the Five Year Plan behind, yet they have not restructured their economies so that the market can function as it does in developed countries. All six countries have held elections, some of which have been relatively fair. But they have not developed the necessary institutions to make the rule of law supreme. They lack sufficient separation and balance of powers within their governments. They need territorial administrations that honestly serve their diverse populations. They have just begun to create civil societies and democratic political cultures.

What does the future hold for the in-between countries? A few trends seem fairly clear. Georgia and Armenia will have a long, hard climb to achieve minimal prosperity, and both must avoid violent conflicts to succeed. Azerbaijan has been dealt a wild card—how the forthcoming oil money will be used is crucial. Ukraine has a chance to expand its connections with Europe in a productive way, perhaps someday to leave in-between status behind. Belarus is not only landlocked but apparently "time-locked" as well, unlikely to change unless Russia provides some leadership in that direction.

In many ways Russia's course will be crucial for the whole region. The oil-export windfall poses opportunities and dangers. Currently the government is bolstering its financial position, which should help attract investment. But Russia needs systematic reform of the rent-seeking state from top to bottom. All areas of public life and economic activity would benefit from more transparency and even-handed enforcement of the law. Then the government and people would have the opportunity to decide the best strategy for developing the country's vast resources. However, the danger is that the oil money may allow more entrenchment of the rent-seeking state and permit an increasingly authoritarian government to assert itself more forcefully abroad.

▶▶ Key Terms

collective farms
Gazprom
Nagorno-Karabakh region
new middle class
New Russians
oligarchs
Orange Revolution
private-plot production

privatization of industry
rent-seeking state
Rose Revolution
siloviki
state-capitalist economy
state farms
Vladimir Putin

Bukhara, Uzbekistan Bukhara is one of the most ancient cities in central Asia and was an important trade center on the Silk Road connecting China and the Mediterranean. This is a view from the Kalyan minaret of the Poi-Kalyan Mosque complex in the heart of the city..

Chapter 15

Central Asia and Afghanistan:
Historical and Geographical Context of Economy and Environment

- Natural Regions
- Central Asian Heritage
- Society and Culture
- "De-Modernization"

- Modern Afghanistan
- Post 9/11 Impacts in Central Asia and Afghanistan

From 1920 until 1991, the Central Asian countries of Kazakhstan, Kyrgyzstan, Tajikistan, Turkmenistan, and Uzbekistan were part of the Union of Soviet Socialist Republics (USSR). The USSR was a major international political player in the 1900s and the counterpart to the United States during the cold war in the second half of the twentieth century. As such, the area was viewed as a part of a superpower and thus grouped with the "developed" world along with the other republics of the USSR, like Russia and the Ukraine. That Central Asia was not as advanced as the Slavic areas of the country was not widely acknowledged until after the fall of the USSR and the independence of the five countries in 1991 (see Figure 13–1, page 310). Since then, Central Asia has had a more difficult time adjusting to independent status and to the difficult economic changes than any other region of the former USSR.

Prior to the Soviet period, the region was more closely tied to areas to the south and west because of the Islamic heritage it shares with the Middle East and western South Asia. Afghanistan has been an independent country since the nineteenth century, but had close trade and cultural ties with the region the USSR absorbed. Much of that region was of Turkic origin and at the time known as "Turkistan." The etymology of the ending "-istan" is Persian, the other major culture group in the region, and means "place of"; therefore, Tajikistan, for example, means "place of the Tajiks." These ties ranged across many groups and across the borders that would eventually separate Afghanistan from the other five

countries. Half of the Tajiks, for example, live in Afghanistan and the other half in Tajikistan and Uzbekistan. Also, significant minorities of Uzbeks and Turkmen live in Afghanistan. For this reason, Afghanistan is included in this chapter.

Central Asia and Afghanistan connect four of the great "realms" of the world: Russia, and through it Europe, the Middle East, South Asia, and East Asia. It is this very centrality that gives Central Asia both its name and its place in history today. Central Asia was at the crossroads of some of the most important ancient empires and regions, such as Mesopotamia (in modern Iraq), the Indus River valley, the Iranian Plateau, Greece, Rome, and China. Because of its centrality and strategic importance, the region has remained important and pivotal to many political and economic issues the world faces today.

Natural Regions

When looking at Central Asia and Afghanistan, three major natural regions dominate (see Figure 13–3, page 311). This might seem surprising when one realizes that the area of Central Asia and Afghanistan crosses latitudes equal to the distance between Austin, Texas, and Edmonton, Alberta, in Canada and roughly three-quarters the longitudinal distance of the continental United States. Because of its position at the heart of the Eurasian landmass and the massive

■ **FIGURE 15-1 A young Tajik herder in the Fan Moun-tains.** Mountains are a distinctive feature of Afghanistan, Kyrgyzstan, and Tajikistan. This herder practices transhumance, a way of life whereby he drives his cattle to mountain pastures in the summer time, where melting snow and ice create lush vegetation, then takes them back to the river valleys in the winter when the snow falls in the mountains.

mountains that cut off the monsoonal flows from the south, the region is dominated by complex mountain climates in the southeast and deserts and mid-latitude steppes in the lower-lying areas to the north and west.

■ **FIGURE 15-2 Steppe grasslands.** The steppe regions of Central Asia are natural grasslands, whose primary historical use has been animal grazing.

Mountains

The climate, soils, and vegetation of the complex mountain areas are diverse, reflecting the location of the mountains, their local relief, and, most importantly, their altitude (see Figure 13–3, page 311). Like the Himalayas to the east, the tectonic upthrust caused by the Indian plate sliding under the Eurasian plate created the mountainous region of Central Asia. Due to continuing tectonic activity in the region, Central Asia and Afghanistan are prone to frequent earthquakes. The major mountain chains are the **Hindu Kush** (in Afghanistan), the **Pamirs** (in Tajikistan and Afghanistan), the Fan (in Tajikistan and Uzbekistan; Figure 15–1), and the Tien Shan (in Kyrgyzstan). In Afghanistan, Kyrgyzstan, and Tajikistan particularly, the majority of the population lives in river valleys, as well as foothills and meadows located in and at the base of these mountain chains, and the streams on which they depend are fed by spring and summer runoff from these mountains. Summers are mild in the mountains, but extreme cold and abundant snowfall characterize the winters.

Steppe

This natural grassland region dominates much of Kazakhstan in the northernmost part of Central Asia (Figure 15–2). Grasslands that extend both north and west from Kazakhstan into Russia with only a few trees located in river valleys characterize the region. Although the soils are generally rich, much like those of the Great Plains in the United States, the generally low level and high variability of precipitation makes agriculture in this region more precarious and susceptible to drought than that in either the Great Plains or the forest steppe in southern Russia.

Desert

The desert areas of Central Asia and Afghanistan comprise the largest portion of the region and are arid, that is potential evaporation is greater than potential precipitation. Precipitation is generally less than 10 inches (250 millimeters) a year with very hot summers and cold winters. Xerophytic plants—that is, species tolerant to aridity—predominate, although there are scattered grasses. Most major economic activities within the desert regions are concentrated in the major river valleys that flow from the mountains to the east and provide extensive irrigated agriculture in particular to the populations of southern Kazakhstan, Uzbekistan, and Turkmenistan. The other major economic activity is herding, which historically was the main livelihood of the Turkmen that lived in the deserts (Figure 15–3). The most prominent deserts in the region are the **Kyzl Kum Desert** (east of the **Aral Sea** in Uzbekistan and Kazakhstan), the **Kara Kum Desert** in Turkmenistan, and the Registan and Dasht-i Margo in western Afghanistan. To the west of this region lies the Caspian Sea, the

■ FIGURE 15-3 Desert land use. Aridity and great temperature fluctuations between summer and winter characterize the desert regions of Central Asia and Afghanistan. Where irrigated agriculture is not present, the deserts are often used for extensive animal grazing. This scene is from Turkmenistan.

surface of which lies at 92 feet (28 meters) below sea level, and within the region (west and north of the Kyzl Kum and Kara Kum deserts respectively) lies the Aral Sea, which is fed by the **Syr Darya** and **Amu Darya rivers**. The other major rivers in the region are the **Zerafshan River**, which flows from Tajikistan into Uzbekistan and disappears into the Kyzl Kum Desert, and the Harirud, Helmand, Kabul, and Kunduz rivers in Afghanistan.

Central Asian Heritage

Within Central Asia and Afghanistan, two major groups predominate: Turkic ethnic groups and Persian ethnic groups (see Figure 13–9, page 316). The Persians were the earliest known settlers in Central Asia. The two most important settled Persian populations were the Bactrians and the Sogdians. Alexander the Great conquered this area in the fourth century and chose as his wife a Bactrian princess. The Greeks did not rule long and the region reverted to Persian rule until the Arab conquest in the eighth century, which brought Islam to the region. Unlike Iran to the west, most Central Asians are Sunni Muslims, the exceptions being the Shi'ite Hazaras in central Afghanistan and the Ismailis in the Pamir Mountains. Persians continued to dominate Central Asia and Afghanistan until the conquest of the main city of Bukhara by Turkic tribes in 999. The Turkic peoples came from what are now Siberia and Mongolia. While the Persians speak an Indo-European language (the Indo-European family of languages also includes English, French, Russian, and most other European languages), the various Turkic people speak languages that are from the Altaic family, which some linguists believe to be related to Mongolian and Finnish.

Successive waves of Turkic peoples swept through Central Asia. Some like the Turkmen stayed; others like the Seljuks moved on to other places. But they dominated Central Asia from the eleventh to the thirteenth centuries until they were conquered by the Mongol Empire under Genghis Khan. The Mongols sacked and destroyed the major cities of Bukhara, Samarqand, Khiva (all three in modern Uzbekistan), and Balkh (in modern Afghanistan) before moving to-

wards Iran. The worst damage of this campaign was the destruction of the irrigation systems, which meant the destruction of both agriculture and the water supply. Balkh remains a ruin to this day, but Bukhara, Samarqand, and Khiva were all rebuilt over the next centuries and flourished on the **Silk Road** (Figure 15–4). This caravan route, stretching from China to the Middle East and ultimately on to Europe, was the main procurement corridor for goods from the east, carried mostly on camel caravans through the Central Asian river valleys and oases and on to the markets of the Mediterranean. Although called the Silk Road, silk was not the only good to be traded in this fashion. Paper, for example, was first used outside of China in Samarqand and later made its way to the west.

Turkic groups continued to move into the region after the Mongol interlude; by the sixteenth century the Uzbeks came to dominate the river valleys, which were mostly inhabited by eastern Persians known as Tajiks. Later still the Kazakh-Kyrgyz moved into the steppe and northern mountainous regions and a related group, the Karakalpaks, moved into the region east of the Aral Sea. Although Turkic people lived in the area we know as Afghanistan, they did not dominate it in the same way as the rest of Central Asia. Rather, a linguistically Persian group—the Pashtun—rose to dominance in Afghanistan except for two short interludes in the twentieth century when Tajiks led the country. From about the time of the arrival of the Uzbeks, the Silk Road began to be eclipsed by the sea route pioneered by the Portuguese, which bypassed Central Asia and the Middle East and allowed Europeans to trade directly with India and China and carry their goods back around Africa to Europe.

By the nineteenth century, Central Asia had stagnated to the point that it became a pawn in the imperial land grabs made by Russia and Great Britain. This period became known as the Great Game, as Great Britain, fearing Russia was a danger to its Indian colony, engaged in three wars in Afghanistan and extensive exploration in Central Asia and Afghanistan. By the later half of the nineteenth century a

■ FIGURE 15-4 The Registan of Samarqand. This image shows the famous *madrasah* (Quranic school) and mosque complex that make up the Registan in Samarqand. The descendants of Tamerlane built this complex, and it has become the symbol of the renaissance of Central Asia after the destructive Mongol interlude.

truce was declared and Afghanistan was considered to be part of the British sphere of influence and Central Asia part of Russia's. The two governments set the boundaries of modern-day Afghanistan; and to separate the powers, a narrow strip of land, called the **Wakhan Corridor**, was created between the Russian controlled Pamir Mountains and British India. This feature is still on the map today and connects Afghanistan to China.

Although the British never colonized Afghanistan, the Russians did colonize much of Central Asia, specifically the areas that would become the republics of Kazakhstan, Kyrgyzstan, and Turkmenistan, along with Samarqand, the Ferghana Valley, and the Pamir Mountains. Russia directly administered these areas and the rest of the region remained independent, but vassals of the Russian Tsar. The Russian Revolution of 1917 did not have an immediate impact on most of Central Asia. Initially, an attempt was made to create a new country allied with Soviet Russia, one that was independent and Muslim, called Turkistan; however, the area was too important agriculturally to the new USSR to permit only indirect control, and by 1925 the USSR had annexed all areas of Central Asia. The government was afraid that the region would try to unite again as Turkistan and carved the region into five different republics, each with a titular majority, but they made sure that small populations of each group would be in other republics. Hence, there are many Uzbeks in all five republics. Likewise there are many Kazakhs, Turkmen, Kyrgyz, and Tajiks in Uzbekistan.

Three important developmental and economic processes characterized the Soviet period: centralization, collectivization of agriculture, and Russianization. **Centralization** is important to the development of Central Asia because the government in the USSR's capital, Moscow, decided all political, economic, and cultural questions. Decisions were then put into practice throughout the USSR. The economy was a **command economy**, and the state planned nearly all economic activities. This manifested itself in industry through the process of distributing specific industrial activities throughout the various republics. Because Tajikistan had ready supplies of hydroelectric power, a major aluminum smelter was located there. The alumina (the main component of aluminum) came from the Ukraine and much of the workforce came from Russia and Belarus. The aluminum was then sent out to Russia, where workers would use it as a component in many items. This type of industry, however, was unusual in Central Asia. The government primarily utilized the region to harvest its natural resources such as agricultural products, most importantly cotton, and natural gas and oil (from Turkmenistan and Kazakhstan), as well as gold and uranium (from Kyrgyzstan and Tajikistan) (see Figure 13–10, page 318). The USSR also centralized transportation, communications, and other decision making. All highways, railroads, and services made their way to Russia from each of the countries. Goods traveled to Russia to be manufactured into products that would then be shipped back to the republics. Issues as diverse as education curricula, production decisions, and even architectural decisions were decided in Moscow. For purposes of education, this theoretically

meant that everybody got the same education and access to education, without geographic discrimination. But this centralized approach could also have disastrous results when, for example, the government decided how to construct apartment buildings based on models for Moscow. Moscow is not a tectonically active area, but Central Asia and the Caucasus are. This has caused many buildings to be constructed that could hold up to a Moscow winter but not to an earthquake.

Collectivization of agriculture is an important hallmark of communist regimes. This is the process whereby all land, goods, equipment, and crops are appropriated by the government. Instead of numerous medium-sized and small farms, these are consolidated into large farms that the former owners work collectively to produce a certain crop or a variety of crops. The farmers are paid in produce from the farm, a periodic monetary stipend, or a combination of the two. In the USSR, these were known as collective farms or, in some instances, state farms. Prior to collectivization, the major crop in Central Asia was wheat, followed by other food crops, orchards, vineyards, and—if the farm was large enough—market crops like cotton or tobacco, depending on the terrain and environment (see Figure 14-10, page 333). After collectivization, the most important crop in the lower river valleys became cotton for use in the Soviet Union (Figure 15–5). Central Asia therefore became a colony, not unlike India or Egypt, where as much land as possible was put into cotton production and the government shipped in food items that the Ukraine or northern Kazakhstan could grow more efficiently. In the mountains the major crop was tobacco in the narrowest and highest of the river valleys (Figure 15–6). The USSR also collectivized herds, although many herders, rather than turning over their animals to the government, either slaughtered them or drove them to Afghanistan. In this way, the numbers of Turkmen and Uzbeks grew in that country and a small population of Kyrgyz was established. The herders who stayed faced a situation similar to that of the farmers. They continued to look after herds, but now the state owned those herds.

▪ **FIGURE 15-5 Women and children loading cotton.**
Scenes like this are prevalent throughout Kyrgyzstan, Tajikistan, Turkmenistan, and Uzbekistan. Both during and after the Soviet period, Central Asian children were taken from school during the cotton harvest and sent to the collective farms to work on the harvest along with the women of the farm.

FIGURE 15-6 Upper Zerafshan River valley. Farmers along the upper reaches of the Zerafshan River were forced by the Soviet government to put all available land in this narrow valley into tobacco, as the climate and features were favorable to this crop.

The USSR also tried to control culture as much as industry (Figure 15–7). There was a generic ideal of a *Homo sovieticus*, which was visualized for everybody in the USSR. The main instrument in this transformation was **Russification**. Russian culture and language along with European ideas of society and culture were expected to become the norm for all citizens. Teachers taught Russian along with local languages, yet they increasingly emphasized the Russian language, Russian literature, and Russian culture as the decades wore on. The government also mandated that universities emphasize Russian and for most students to get a college degree this meant that they had to be fluent in Russian. To get a good job, either in industry or government, one had to be at least proficient in Russian. The idea behind this initiative was that a manager or government official would have to speak to people from all over the Soviet Union and he or she should not be responsible for learning all the languages of the country, rather all people should learn Russian. This has created a mostly bilingual culture, in which local languages, literature, culture, and art were all seen as second class to Russian.

Society and Culture

The population of Central Asia and Afghanistan is not large when you compare the region to South Asia or China, but considering that much of the land is desert or mountains, it is nonetheless densely settled in the arable parts of the region (Table 15–1). Tajikistan and Bangladesh, for example, are roughly the same area, both just over 55,000 square miles (142,000 square kilometers). But Tajikistan has only about 6.6 million people, while Bangladesh has roughly 141.3 million people. When one looks at the terrain, however, Bangladesh lies in the highly fertile plains of the Ganges and Brahmaputra River valleys, whereas Tajikistan is 93 percent mountains, with most of the Pamir Plateau over 12,000 feet (3,600 meters). Afghanistan and Uzbekistan have by far the largest populations, but these are densely settled in river valleys with only minor populations living in the extensive desert regions.

The USSR always considered Central Asia as its most backward area. Much of this had to do with a typical European colonial mentality toward Asia, but also with Islam and the fact that the area had only very limited industry and relied mostly on agriculture. However, when looking at late Soviet demographic data, the government felt there was much for which they could be proud. Literacy rates had skyrocketed to 98 percent from less than 5 percent; they had built extensive transportation and communication links between the Central Asian republics and Russia; many goods and services were now available (even if people had to stand in line for them); they provided universal and free health and child care; everyone had a job; and together they formed a superpower. That these statistics masked a very different reality became extensively apparent after independence. Demographically, Central Asia was the fastest growing area in the former USSR. Statistics today still show a major growth rate for all five countries, although they are behind Afghanistan in these numbers (Table 15–2, Figure 15–8).

FIGURE 15-7 Opera and ballet theater in Central Asia. The Soviet government determined that all republics should have the same access to cultural events. Above is the distinctly European-looking opera house of Dushanbe, Tajikistan.

FIGURE 15-8 Village family in the Fan Mountains. In the countryside particularly, large families are common. Note the womens' distinctive dress patterns that were maintained throughout the Soviet period as indicative of their culture.

Table 15-1   The Countries of Central Asia

Country	Capital	Area	Population 2004 est.*	Largest Ethnic Group (% of Total)
Afghanistan	Kabul	251,772	28.5	Pashtuns (38%)
Kazakhstan	Astana	1,049,151	15.0	Kazakhs (46%)
Kyrgyzstan	Bishkek	76,641	5.1	Kyrgyz (52%)
Tajikistan	Dushanbe	55,251	6.6	Tajiks (67%)
Turkmenistan	Ashkhabad	188,456	5.7	Turkmen (77%)
Uzbekistan	Tashkent	172,741	26.4	Uzbeks (80%)

Source: *World Population Data Sheet* (Washington D.C.: Population Reference Bureau, 2004).

Compare these numbers to Russia (with a birthrate of 11 and a death rate of 16) or the Ukraine (with a birthrate of 9 and a death rate of 16). Infant mortality (an indicator of the development of a country) was higher in these countries, most of the population still lived in rural areas (with the exception of Kazakhstan where nearly half the population is Russian), and per-person purchasing power was lower than in the European republics. When a woman in the Soviet Union had ten children or more, she was presented with a van and a medal and declared a "heroine of the Soviet Union." Most of these "heroines" were in Central Asia. The social benefits of large families disappeared with the USSR, so today it is unusual to see very large families among younger people. Most young families state they want two or at most four children because now they must pay for education, health care, and other services formerly provided by the USSR.

With the fall of the Soviet Union, the population became faced with other issues, particularly ones to do with culture. What parts of their culture were intrinsically "Soviet" and did they want to embrace them or renounce them? Where would they look for ethnic or cultural examples to supersede Soviet examples? For each of the five countries, the answers are different and show the growing divide between them. Unlike the Baltic nations of Estonia, Latvia, and Lithuania, Central Asia did not wish to break away from the Soviet Union. Independence more or less caught them by surprise and in all but one country (Kyrgyzstan), the ruling Soviet elites stepped directly into power. Since then, they have tried to underscore the uniqueness of their cultures and ethnici-

ties by creating new architecture, art, heroes, and even history.

In Uzbekistan and Turkmenistan, the approach was to distance themselves as much as possible from Soviet culture. In Kazakhstan, Kyrgyzstan, and Tajikistan, the strategy was to integrate the Soviet past with an often government-controlled sense of ethnic sensibility. They also dealt differently with Russification. Kazakhstan and Kyrgyzstan still try to maintain a bilingual society. Turkmenistan and Uzbekistan have de-emphasized Russian, mostly for cultural purposes. Tajikistan declared Tajiki the first language of the country, with Russian an important second language, but due to the hardships caused by their post-independence history, they have de facto de-emphasized Russian. Although Lenin still proudly stands in the main square of Kyrgyzstan's capital, Bishkek, many iconographic Soviet statues have been replaced by historical figures, like Ismail Somoni, the founding emperor of the last Persian (or Tajik) empire in Central Asia, in Tajikistan and Tamerlane in Uzbekistan. Tamerlane, a descendant on his mother's side of Genghis Khan, and his descendants ruled from Samarqand in the fourteenth and fifteenth centuries until the coming of the Uzbeks, so it is an interesting cultural point that the Uzbeks have built some of their identity around this better known historical figure (Figure 15–9). All countries have also changed the names of streets, cities, provinces, and even mountains, although most people on the street still refer to them by their Soviet names. The highest point in the former Soviet Union is now in Tajikistan and was called Stalin Peak until the 1960s, when it

Table 15-2   Demographic Data for the Countries of Central Asia

Country	Birth Rate*	Death Rate*	Infant Mortality Rate*	Percent Urban
Afghanistan	48	21	165	22
Kazakhstan	17	11	52	57
Kyrgyzstan	21	8	42	35
Tajikistan	25	6	50	27
Turkmenistan	25	9	74	47
Uzbekistan	24	8	62	37

Note: * = per 1000
Source: *World Population Data Sheet* (Washington D.C.: Population Reference Bureau, 2004).

■ **FIGURE 15-9 Statue of Tamerlane in Tashkent.**
Although Tamerlane was of Turko-Mongol descent (he claimed a relation to Genghis Khan through his mother) and not Uzbek, the Uzbek government regards him as a national hero and the one who created an amazing renaissance in architecture and learning in Central Asia.

was changed to Communism Peak until independence; now it is Somoni Peak.

"De-Modernization"

It is difficult to know exactly how to characterize the economic reality that affects Central Asia in its post-independence years. Some authors have used the term de-industrialization, but this is to imply that Central Asia was industrialized. Central Asia was part of an industrialized country, the USSR. But the main economic activities in Central Asia were resource extraction and agriculture. De-modernization is another term that has been suggested and although it also has problems, specifically when related to well-educated societies, economically it does conjure up the idea that output significantly slowed down as equipment failed, infrastructure fell apart, and each government was faced with an immediate cessation of central governmental funding and subsidies for most economic activities. Bureaucrats suddenly had to stop concentrating on maximizing subsidies from Moscow and figure out from their meager economies and limited transportation routes how to forge their economic future.

All five former Soviet countries are landlocked (as is Afghanistan) and all transportation routes went to Russia. This meant at the onset that trade must continue with other former Soviet countries because none of these countries has the resources to build railroads or highways to other countries. In the context of Kazakhstan, Kyrgyzstan, and Tajikistan, the extremely high mountains separating them from China and South Asia compounded this connectivity problem. In the case of Turkmenistan and Uzbekistan, the neighbors were war-torn Afghanistan and the theocratic government of Iran, both of which worried those governments. The country that first stepped into this trade void was Turkey, a country that is linguistically and culturally related to the Turkic nations of Kazakhstan, Kyrgyzstan, Turk-

menistan, and Uzbekistan and has positioned itself as the model of the secular Muslim nation to emulate. Unfortunately, Turkish planning was much more expansive and audacious than was their economic capability and many of the planned ventures collapsed because of poor planning and not enough money. This left some bitterness in Central Asia, particularly in Uzbekistan and Kyrgyzstan, where the investment was badly needed and land and resources had been dedicated to these purposes. Another problem was the attitude of many early entrepreneurs toward Central Asia. They viewed the areas as "developing," a term that, economically, is true. Hand in hand with this attitude usually comes a patronizing attitude that did not sit well with Central Asian bureaucrats and businessmen, many of whom are very well educated and did not like these foreigners telling them how to run their affairs.

In terms of development, post-independence Central Asia displays a wide disparity of developmental issues, and none of them can any longer be classified as economically developed. Kazakhstan has perhaps the greatest potential, both in resources and potential productivity. It has the strongest economy in the region and a large agricultural base, in addition to one of the world's largest oil and natural gas deposits. This has already attracted many multinational firms into the country to help exploit these resources, although they have not been developed at a pace hoped for in the West because of continued bureaucratic wrangling with the government. Turkmenistan also has sizeable oil and gas deposits and a considerable agricultural base, considering its desert climate. This is because extensive irrigation projects promulgated by the Soviets diverted large amounts of water from the Amu Darya River that normally would have gone to the Aral Sea (see the *Geography in Action* boxed feature, The Aral Sea). The country is hampered by its severely authoritarian government that remains adamantly state run and largely discourages foreign investment and the application of Western development concepts.

Privatization has been slow to nonexistent in Central Asia; however, during the Soviet period, families were allowed to have a small amount of land adjacent to their homes (what would be considered yards in the United States) as private land. On this land, people could grow whatever they wanted either for their own consumption or for trade. Families often had a number of fruit trees, depending on the size of their plot, around which they would plant vegetables and herbs. Over time in the Soviet Union, these **kitchen gardens** became the basis for some of the most productive agriculture in the republics. These gardens constituted as much as 23 percent of land use in some provinces and, because of personal incentive, were some of the most productive land in Central Asia. After the fall of the USSR, the people retained these kitchen gardens and used them initially to augment their families' food supply. Over time, they have maximized these gardens (where possible) to grow fruits and vegetables for sale in local markets to augment family income.

Uzbekistan has the largest population and the largest market-driven agricultural base. The diversity of the country in terms of resources and people has resulted in less of

GEOGRAPHY IN ACTION
The Aral Sea

The issues and developments of the Aral Sea represent one of the saddest intersections of economic, environmental, cultural, demographic, medical, and physical geography in the world. The economic decisions made by Soviet planners in the 1960s have created a situation that has devastated an environment, its population, and, in the findings of some researchers, its climate. In the 1960s, there was a huge push throughout the USSR to rapidly expand agriculture. In Central Asia, the goal was to produce more of the all-important cotton crop. Land was shifted from the production of food staples and farmers planted more cotton. But this conversion still was not enough for the needs of the USSR. So more marginal areas (land that was near river valleys, but never was irrigated in the past) were brought into cultivation. Production went from 6.7 million acres (2.7 million hectares) in 1965 to 10.4 million acres (4.2 million hectares) in 1989. Massive new irrigation canals were dug, including the 1,000-mile- (1,600-kilometer) long Kara Kum Canal, the longest diversionary channel in the world, which did not feed drainage water back into the river system but took water away from the Aral Sea drainage system toward the republic of Turkmenistan.

According to Soviet archives from that time, planners knew that this additional irrigation and diversion would affect the Aral Sea, but they contended that the impact would affect the Aral Sea only slightly. What happened, however, is one of the worst ecological disasters in modern history. It has badly affected the people who live around the Aral Sea, the Karakalpaks, who are linguistically and culturally related to the Kazakhs and Kyrgyz but settled around the Aral Sea both as herders and fisherfolk. The fishing was good through 1960, there were over twenty different species of fish living in the Aral Sea, and several fishing ports grew during the early period of the USSR to service the fishing fleet. The Soviets also built fish-canning factories to increase jobs and produce more food for the local republics.

In the wake of the construction of the new irrigation works and the opening of the Kara Kum Canal, the Aral Sea almost immediately started to shrink. At first the government tried to dredge a canal for fishing ships to get to the Aral from the ports, but before long this was no longer feasible. Currently, the port of Muynak in Uzbekistan is more than 30 miles from the Sea (Figure A). By 2000, the Aral Sea's surface level had fallen by more than 48 feet (15 meters) and its volume by 75 percent. From twenty fish species in 1960, it went to two species in the early 1990s, one of which was introduced because it was more

■ **FIGURE A** **Abandoned ship at a former port of the Aral Sea.** These ships now constitute part of the graveyard of ships around the former port of Muynak in Uzbekistan. Scenes of abandonment like this are common around the former shoreline of the Aral Sea.

salt-tolerant. Neither of these species has been caught for a long time. Because of the shrinkage of the sea, the concentration of salt has intensified and now the salinity of the Aral Sea is second only to the Dead Sea in Israel/Palestine and Jordan. The canning factories continued to work, but from 1981 to 1998 the fish were imported from the Baltic Republics to keep the jobs going. Since 1998, the canneries have been closed and between fishing and canning more than 60,000 jobs have literally evaporated. The vegetation has changed radically and the native hydrophytic (water-loving) plants that have historically grown there are all gone and have been replaced by halophytic (salt-tolerant) plants. The reason for this is that more than 7.4 million acres (3 million hectares) of former seabed land have been exposed; the residual salt left from evaporation of the seawater readily became airborne from the newly exposed land, was transported considerable distances from its origin, and settled over thousands of square miles of irrigated land. Along with a poor drainage system that allows too much water to evaporate, this has caused cotton productivity in Uzbekistan to decline drastically. As one moves closer to the Aral Sea, the plants become stunted and many cannot grow because of the high salinity of

an economic slide than in the other non–oil-producing countries. The Soviets invested more into industry (relatively speaking) because of the size of the population, although cotton remains the largest sector of the economy. Uzbekistan does not have the large oil deposits of its northern and western neighbors and has only small reserves of natural gas. In the long term, this puts the country in a more precarious position because its industry is rapidly aging and it has to rely on a crop that continuously erodes the country's environment and, ultimately, its productivity (see the *Geography in Action* boxed feature, The Aral Sea). Another

potential conflict is driven by Uzbekistan's location as a downstream country. Only 9 percent of the water used in Uzbekistan's extensive agricultural systems originates within the country. The water flows from Kyrgyzstan, Tajikistan, and Afghanistan; and all of these countries have signaled that they would like to use more of what they consider their own water. This has caused a war of words with the considerably more powerful Uzbek government insisting that the water-sharing agreements signed during the Soviet period—which overwhelmingly benefit Uzbekistan, Kazakhstan, and Turkmenistan—remain in place.

■ FIGURE B **Salt deposits on stunted cotton fields.** Over-irrigation and airborne salt particles have contributed to the salinization of the soil to the extent that it has become very difficult for cotton to grow.

the soil (Figure B). The crop that caused the disaster in the first place, cotton, is now in danger of not being able to grow at all near the Aral Sea because of the impact airborne salt imposes on future agricultural operations.

The airborne particles (around 150 million tons annually) from the seabed also include many other items, including fallout from nuclear and biological testing carried out during the cold war on Vozrozhdenie Island in the sea. This has caused a catastrophic drop in health amongst the Karakalpaks who live nearest to the former seashore. Infant mortality in this area is 110 (nearly twice the country average), kidney diseases are up 60 percent, strontium (a by-product of nuclear testing) has been found in the thyroids of 23 percent of Karakalpaks tested, and incidences of anemia, respiratory diseases, and cancer are up 600 percent since 1980. In a health survey in the late 1990s, every single expectant mother was found to have traces of DDT (an insecticide) in her milk, 70 percent of residents in the port cities have cancerous or precancerous growths in their throats, and birth defects have skyrocketed. **Karakalpakstan**, as it has been known, is now sardonically being called "Malthustan." This is in reference to Thomas Malthus, the eighteenth-century political economist who predicted that eventually the planet would reach its human carrying capacity, resulting in a subsistence crisis and eventually an environmental collapse.

Because the Karakalpaks have little power in Uzbekistan, virtually nothing has been done for them except by a handful of international agencies. The five Central Asian countries

have had many meetings and signed many agreements, but no real progress has been made in returning more water into the sea from the Amu Darya River. As the sea has divided into two parts (Figure C), the Kazakhs have built a dam to protect the northern segment and there are early signs that it might be stabilizing. But the culture, environment, and health of the residents have been too effectively destroyed for this development to have much effect in the short term on the regional environment. Perhaps in the long term, it could be cause for some hope in the region, but for now the world is seeing the destruction of a culture through poor economic choices that did not take potential environmental issues into account.

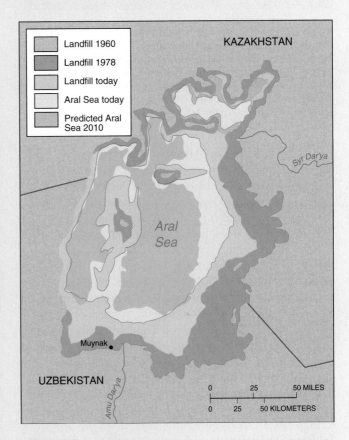

Landfill 1960
Landfill 1978
Landfill today
Aral Sea today
Predicted Aral Sea 2010

KAZAKHSTAN

Syr Darya

Aral Sea

Muynak

UZBEKISTAN

Amu Darya

0   25   50 MILES
0   25   50 KILOMETERS

■ FIGURE C **Projected area of the Aral Sea in 2010, if loss of water intake is not controlled.** If current trends are not arrested, the Aral Sea will soon be reduced to a small fraction of its former size.

Kyrgyzstan has taken a more Western approach to its development and has privatized more land and industry than any of the other countries. Because of its mountainous terrain, transhumance herding is prevalent (see Figure 15–1) and agriculture is limited to river valleys and upland meadows, although water is available for irrigation and the terrain is well suited to the dams built there, which produce much needed hydroelectric power. Nonetheless, Kyrgyzstan's economy has shrunk considerably since independence and, as witnessed by the recent coup against the government, the population has grown weary of economic hardships and the

obvious income disparity between government officials and the majority of the population.

As difficult as the post-independence years have been in the four Turkic countries, nowhere were these problems more apparent than in Tajikistan. Tajikistan, which was the poorest country during the time of the Soviet Union, fell even further behind in the post-independence years as the country immediately devolved into a civil war (Figure 15–10). The war was based not on ethnic differences but on regionalism and tribalism. During the Stalin administration, many people from overpopulated river valleys east of the

■ **FIGURE 15-10 Bus top turned fence.** Throughout Central Asia, there was an extensive transportation network; however, during the civil war in Tajikistan, because of insecurity and a lack of spare parts and gasoline, many of these buses used in the countryside sat idle and rusted. The former overseer of transportation to and from the villages west of the capital of Dushanbe decided that a better use could be had for these rusted bus tops and used them as a goat pen.

capital of Dushanbe were forcibly moved to another valley where the USSR had recently expanded irrigation and the agricultural base. This caused conflict between the immigrants and the people who already inhabited that region, most notably around the city of Kulob. The effects of the civil war were the worst in this agriculturally important area, but the impacts were felt widely as fighting broke out all over the southern part of the country. A peace accord was reached by the victorious Kulobis, their allies, and their opponents in 1997, although the peace was only tentative until 2001, when the government cracked down on many of the **warlords** and gained better control. These warlords gained financing for their weapons and soldiers through the **opium** drug trade (see the *Geography in Action* boxed feature, Warlords and Opium). Since then the economy has grown enormously from a near standstill. The country relies on cotton, a growing mining sector that includes gold and uranium, and foreign aid. Until recently, all goods coming to and from Tajikistan had to go through Uzbekistan. Considering the fact that relations between the two countries have been severely strained over the civil war, Islamic rebels operating in the mountains of Tajikistan, and issues pertaining to water distribution, this transportation reality has put the entire Tajik economy at the mercy of relations with Uzbekistan. This has begun to change in the past couple of years as a highway was recently opened to China, which has already had a positive economic effect on the whole country. With aid from the United States government, another road is planned as a link to Afghanistan and ultimately Pakistan's ports to overcome the constraints of a landlocked location.

Modern Afghanistan

Historically, the Hindu Kush Mountains naturally divided Afghanistan (Figure 15–11). The areas north and west were allied closely with Iran and Central Asia, and the areas to the south and east with India. When the competing Russians and British set the borders of Afghanistan at the end of the Great Game, a very disparate and fiercely independent population was thrown together to attempt to make common cause. They have famously gotten together to fight off invaders, both British and Russian, but when left alone, they have more often than not fought amongst themselves along ethnic and tribal lines. The composition of the country (Table 15–3) means that the Pashtun have enjoyed near continuous power running the country, a situation that has caused strife with the other ethnic groups. Only twice has the Tajik minority run the government, once in the early twentieth century for a brief time and, most disastrously, during the traumatic civil war years before the rise of the Taliban.

An attempt at economic liberalization occurred during the 1960s and 1970s, but a communist coup in 1978 turned the economy and government firmly toward the USSR. When many in the society rebelled, the Soviets occupied the country from 1979 to 1989, when they were driven out in defeat. The Soviets, both before and during the occupation, did set up some industry and expanded transportation and communication; but they also caused extensive damage in rural areas, both economic and environmental, in their attempts to pacify the country. The most egregious act was the distribution of more than one million land mines (a figure enlarged further by land mines placed by the Mujahideen, or insurgents, fighting the occupation) throughout the countryside. This has caused untold deaths and maimed more than a million people, many of them children, who stumble across the mines while doing chores or playing. An international campaign to mark the minefields and especially de-mine the agricultural land is currently underway.

With the retreat of the Soviets, Afghanistan continued under its communist government until 1992, when the rebels, led by the Tajik, Ahmed Shah Masud, captured Kabul. The presidency of the country was supposed to revolve between the various ethnic groups, but the Tajik, Burhanuddin Rabbani, refused to relinquish the presidency and the country was plunged into civil war (Figure 15–12, page 353). It is extremely difficult to understand and describe the ensuing catastrophe because different groups continuously changed sides, if they felt they would get a better deal. The Uzbek warlord, Rachid Dostum, was especially famous for this as he changed sides on what seemed like a weekly basis. Masud was rumored to have inquired each morning whether he was fighting with or against Dostum on that particular day. This all changed with the rise of a group that became infamously known as the Taliban.

The **Taliban movement** began in the *madrasahs*, or Quranic schools, financed by the Saudi Arabian government and located on the border of Pakistan and Afghanistan. The students were mostly Pashtun refugees whose families were extremely poor and many sent their sons to the schools for no other reason than they would get at least one meal a day. The *madrasahs* followed a very strict interpretation of Islam known as Wahhabism, named after Muhammad ibn 'abd al-Wahhab (1703–1787). His main goal was to get rid of all innovations that occurred later than the third century of Islam (900s); he was particularly vehemently opposed to Shi'a

GEOGRAPHY IN ACTION
Warlords and Opium

As a crop, poppies—and the opium they produce—do very well on marginal soils and need very little space to render a crop that will give a good payback for the resources expended (Figure A), something that farmers in better endowed habitats do not have to consider (Figure B). As such, the poppy is the ideal cash crop for many parts of Afghanistan. The mountainous core is suited only for subsistence agriculture and transhumance herding. Although a living could be made from the herding, it is virtually impossible to grow enough crops in the narrow river valleys that pour from the Hindu Kush Mountains. Farmers typically plant a variety of crops (including poppies) on very small plots of land (some as small as eight feet by five feet) in a checkerboard pattern alongside sources of water. This makes eradication by aerial spraying, especially of the poppy, extremely difficult, as it will destroy all the other crops and leave the farmer with nothing. Another characteristic conducive to poppy production is the region's periodic droughts. Vegetables and wheat require a great deal of water to grow; poppies require less. With poppies, farmers not only are better assured of a return on their crop and labor, but also with a good crop they can suddenly afford to buy food and animals rather than trade their meager amounts of wheat and vegetables for what they could not grow or raise. The United Nations Office on Drugs and Crime estimated that the average farmer conservatively earned around $475 for his crop annually between 1994 and 2000. Considering the absence of viable alternatives, this is a fortune for as many as 3 million people who are thought to depend on poppies in Afghanistan.

During the early Taliban era, economically the country was stagnating. The only major export product from Afghanistan by this time was opium and its major refined form, heroin. The Taliban, although they did not encourage the trade, did tax it heavily for much needed hard currency for their continuing war with the Northern Alliance. When they had succeeded in taking over 90 percent of the country by 2000, to gain recognition by the international community they bowed to pressure and enforced a ban on growing poppies. Given the amount of production at this time, they were incredibly successful and the amount of

◼ FIGURE B The lower Kunduz River Valley. Compare Figure A with this view over the extensive lower Kunduz Valley where farmers have considerably more land available and can grow market crops such as rice (pictured), cotton, and melons below their villages.

opium coming out of the area held by the Taliban fell by over 90 percent, causing prices to spike internationally in the heroin market. Some of this shortfall was made up by the Northern Alliance and primary production shifted to the north. This changed many things as the product was carried overland to Pakistan to be refined into heroin in mobile labs, then carried across northern Afghanistan into Tajikistan, where warlords trafficked the heroin to the Russian Mafia, who then sold it on to Europe, thus further destabilizing Tajikistan and threatening the other Central Asian countries. Thus the route through Central Asia supplanted the historic Balkan Route through the Black Sea countries to Europe. After 9/11, however, farmers throughout the country began growing poppies again and the production levels have been the highest ever. The Afghan government, with help from the British and American governments, has an ongoing eradication program; but eradicating the crop means that the worst affected are the poorest farmers, who many times lose not only their poppy crop but also all crops because of the piecemeal planting they use in the mountain valleys. This puts them into debtor situations with warlords to provide seeds for conventional crops. Often the only way out of this situation is a form of "debt slavery," whereby farmers must grow poppies to repay the warlords or, in some extreme cases, according to the Revolutionary Association of the Women of Afghanistan, turn over a daughter for marriage or servitude to the warlord or dealer.

Although many farmers benefit, most grow the crop for sheer survival, and most of the money that stays in Afghanistan goes into the pockets of local warlords, who finance and arm the militias that keep the countryside of Afghanistan unstable and out of the reach of the elected government. Although there are stories of warlords forcing farmers to grow poppies, it is more usual for farmers to elect to grow it either because of debt, basic survival, or as the only viable source of income in the region. They are aware of the ultimate use of the crop, and know that the number of Afghan heroin addicts has grown sharply since the days of the civil war, but plead, not without justification, that they have no alternative and if they want to feed their families and provide for their children in a drought-prone and barren landscape, this is their one choice.

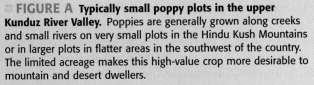

◼ FIGURE A Typically small poppy plots in the upper Kunduz River Valley. Poppies are generally grown along creeks and small rivers on very small plots in the Hindu Kush Mountains or in larger plots in flatter areas in the southwest of the country. The limited acreage makes this high-value crop more desirable to mountain and desert dwellers.

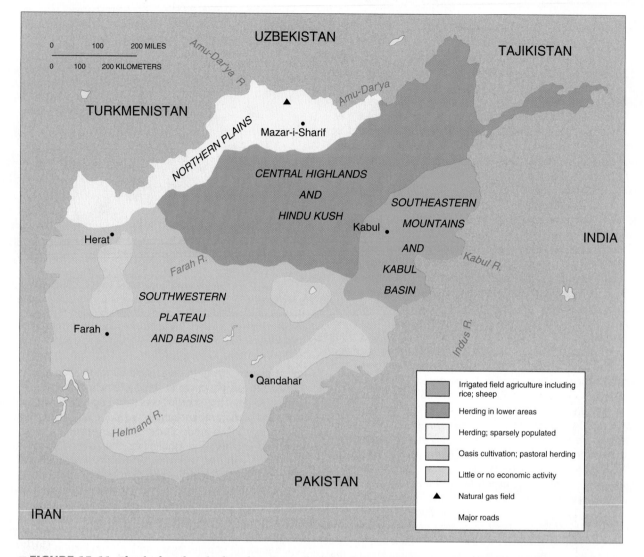

▪ **FIGURE 15-11 Physical and agricultural regions of Afghanistan.** Afghanistan exhibits many Middle Eastern features, such as aridity and adherence to the Islamic faith. However, the country's contact with the outside world has been predominantly through South Asia. The Afghan population is generally conservative, and regional loyalties are often more important than allegiance to the nation as a whole.

Muslims, whom he considered heretics. These students (*talib*, plural *taliban*) formed the core of the followers of *mullahs*, or religious leaders, who led the movement. Their financing mainly came from Saudi Arabia and Pakistan (although the latter vehemently denied it) and they began to militarize around 1994.

The Taliban got their break when a warlord in southern Afghanistan held for ransom a convoy of trucks owned by the powerful Pakistani trucking union. The Taliban managed to defeat the warlord and release the convoy to continue on to Central Asia. The union, along with the Pakistani government, began to funnel money to the Taliban in order to keep the roads open for transportation of goods. In this way, the Taliban began their conquest of the south and west of the country and subsequently turned their attention toward Kabul. The various other ethnic groups then joined together to defeat this "outside group," a characterization they made because the Taliban were financed by Pakistan, although the movement was mostly made up of displaced local Pashtun. In 1996, the Taliban took Kabul and began their assault on the warlords in the north. They were particularly brutal to the Hazara in the center because they are Shi'a, and many of the atrocities of this period were perpetrated by both sides. The major city in the north, Mazar-i Sharif, controlled by Dostum, fell in 1998. It was assumed that the Tajik resistance in the northeast would soon fall, but as late as 2001, the Northern Alliance under Masud held.

Table 15-3   Ethnic Composition of Afghanistan

Ethnicity	Percentage
Pashtun	38
Tajik	25
Hazara	18
Uzbek	6
Baluchi	1
Turkmen	1

FIGURE 15-12 Destruction in Kabul, Afghanistan. Note the juxtaposition between the ruined residential area to the left and the mosque rebuilt by the Taliban to the right. Scenes of destruction such as this are very typical throughout Afghanistan from the time of the civil war.

Although the north continued to live under a constant state of war, the rest of the country, especially the Pashtun regions, mostly rallied around the Taliban at the beginning. In interviews, many people expressed happiness that a single group had gained power, that all the bickering and infighting had stopped, and that they could get on with their lives. When asked about their views, most people either sympathized with the movement or else thought that the extreme positions held by the group were just rhetoric. When the war began to bog down in the north, however, the Taliban showed that their movement was not just rhetoric. They unleashed a steady terror on the population that was particularly aimed at women and minorities. Many restrictions pertaining to women had existed in Afghanistan, but now they were mandatory. Women were not allowed an education or an occupation. Women could not go out unless they were dressed in the heavy, confining *burka* mandated by the Taliban and unless a male relative accompanied them. Men were required to wear beards and closely cut their hair. Any infractions were dealt with swiftly and publicly. All sports were stopped and the soccer fields were converted to public disciplinary centers where people accused of thieving had their hands chopped off, women accused of prostitution were whipped, and all executions took place. Music was banned, as were dancing, kite flying, gambling, and any images that included human or animal figures. Some of the Taliban's greatest acts of destruction occurred at Bamiyan, where they destroyed the famous standing Buddhas, and at the Kabul Archaeological Museum, which was considered one of the important repositories of human history in the world (Figure 15–13). Members of the Taliban destroyed anything that predated Islam as well as anything representative of another religion or which displayed a human or animal form.

Post 9/11 Impacts in Central Asia and Afghanistan

On September 9, 2001, the al-Qaeda network headed by Osama bin Laden, in coordination with their hosts, the Taliban, assassinated Ahmed Shah Masud. The coordinated attack on New York and Washington, D.C., using pirated commercial jets on regular scheduled flights, happened two days later.

That these events were connected soon became apparent. In looking for allies, the government of the United States soon saw the Northern Alliance along with Uzbekistan, with its nearby army bases and airports, as the best potential partners in the area in its war on terror that began in Afghanistan. Prior to 9/11, the Northern Alliance held only about 10 percent of the country; however, with coordinated air assaults and help from American special forces, they managed to retake the strategic northern cities of Kunduz and Mazar-i Sharif before capturing Kabul and eventually driving the Taliban and al-Qaeda into the eastern mountains along the Pakistan border.

In this struggle, the Americans and their allies supported various warlords, who had ruled different parts of Afghanistan during the civil war, in their efforts to retake "their" regions and establish order. This trend has unfortunately continued until today and has created various ethnic fiefdoms throughout the country, which have only nominal allegiance to the government of the new president, Hamid Karzai. These warlords were the main reason the civil war continued throughout the 1990s until the Taliban defeated most of them. The international community has questioned putting power back in their hands, but the American government and its allies no longer have the will or the means to break their power or to at least force them to acknowledge the authority of the elected president given the overwhelming attention now paid to Iraq. This leaves Afghanistan

FIGURE 15-13 Great Stone Statues of Buddha in the Bamiyan Valley. Afghanistan's historic role as a geographical crossroads between south, central, and southwestern Asia was long evident in the great stone statues of Buddha in this overwhelmingly Islamic nation. The statues, which were carved some sixteen to nineteen centuries ago out of a sandstone cliff in the Bamiyan Valley 90 miles west of present-day Kabul, were considered one of the world's great archaeological treasures and functioned historically as a sacred Buddhist pilgrimage site. In March 2001, despite worldwide pleas that they be spared, they were destroyed by the Taliban government, which then ruled Afghanistan. The left image shows one of the statues before it was destroyed, and the right photo shows the cliff after the statues were destroyed.

in a very precarious position. The Karzai government must depend on the warlords to keep order in the countryside, but it cannot control them and can only hope for their continued goodwill into the future. Given their pre-Taliban history, this is extremely unlikely, and there have already been worrying signs of violence with the warlord of Herat in the west of the country and with Taliban sympathizers in the south and east. Many analysts predict that it is only a matter of time before a major showdown between the government and the warlords takes place. How that confrontation develops will decide the future of this precarious democracy.

In Central Asia, the U.S. administration has forged closer ties with the government of Uzbekistan in particular, as well as with Kyrgyzstan and Tajikistan. Kyrgyzstan is perceived as stable, and Tajikistan and Uzbekistan are geostrategically placed along the northern border of Afghanistan and are therefore the front line in the war on terror. This has placed some strain on U.S.-Russian relations because Russia still views this area, along with the Baltic and the Caucasus, as the "near abroad" and within its sphere of influence, both politically and economically. The Central Asian administrations, however, see greater diversity in international contacts and relationships beyond those with the former Soviet countries as a positive move. But these new relationships often are not smooth. Tajikistan still has a major Russian army presence in the country and along the border with Afghanistan and cannot afford to offend Russia or its hard-line president, Vladimir Putin. Although Kyrgyzstan was initially viewed as the "Switzerland of Central Asia" because of its mountains and stability, the recent elections in early 2005 showed that the government was very nearly as authoritarian as the administrations in Kazakhstan, Turkmenistan, and Uzbekistan.

In an election marred by widespread irregularities, the cronies and family members of President Askar Akaev won every seat for which they ran. This sparked a revolution in the volatile southern half of the country that spread throughout Kyrgyzstan and eventually caused the president and his family to flee. Although the reason Kyrgyz citizens initially protested the results of the election and joined protest marches was for a less authoritarian government and less fraud, the issue at the root of the people's problems was economic. The government had continually promised better economic conditions and had never delivered on these promises. The economy continued to stagnate, and the obvious wealth of the president and his family fueled the feelings of economic desperation amongst many in the population. A similar scenario seemed likely in Uzbekistan soon afterwards in the very river valley just across the border from where the Kyrgyz revolution occurred. However, unlike Akaev, Islam Karimov, the President of Uzbekistan, called for a violent and bloody crackdown on the protesters that left numerous dead and caused extreme concern amongst the United States and its allies over the future of the country. Uzbekistan, as has often been the case in the past, used the war on terror and the fight against Islamism as the justification for the crackdown. The government is already deeply unpopular in many areas of the country and the scene seems set for future violence and disruption in this country.

■ **FIGURE 15-14 Muslims in Central Asia.** Islam, which diffused from the Middle East, is the dominant faith of Central Asia. Although most Central Asians never abandoned Islam during the Soviet period, they are now witnessing a renaissance of the religion in many places and the practice of Islam is becoming more public. The four men in the image wear the headcovering, beard, and clothing traditional for Muslims in Central Asia.

One of the most important causes for the popularity these authoritarian governments have is their continued fight against what they call the excesses of religious fundamentalism. Fundamentalism as a concept is strictly a Christian one. But the most common local acknowledgment of this phenomenon employs the term Islamism. For most people in Central Asia, the term and the idea is one of religious renaissance (Figure 15–14). During the time of the USSR, the government persecuted obvious displays of all religions and particularly Islam. After independence, many people saw this as the chance to reawaken religious feelings in the region. On no level does this imply that they became terrorists or that they were planning violence or advocating a forcible renewal of the religion. However, it became clear that the former communists who had taken over the government saw them as a threat to their power and began to persecute them anew. Some of the Islamists turned to violence, especially during the Tajik Civil War and after crackdowns in the late 1990s in Uzbekistan. The governments, particularly after 9/11, use the cloak of the fight against Islamic terrorism to perpetuate their authoritarian regimes against this threat to their power. This turned especially violent in the city of Andijon in the Ferghana Valley of Uzbekistan where eyewitnesses reported hundreds of people were killed by government forces while they were protesting Uzbekistan's actions against Islam and attempting to start a Kyrgyz-style revolution. While many Uzbek citizens remain secular and support any crackdown against a potential Islamic state, those who wish to explore their religiosity in a more public way have become increasingly frustrated with the government and fearful of its tactics against Islamists.

▶▶ Summary

Central Asia and Afghanistan stand in a very precarious place. In Central Asia, the revolutionary fever has already hit Kyrgyzstan and the leaders of the other countries fear that they could be next. This has caused them to be increasingly authoritarian in their actions and to be suspicious of any overtly religious people. Economically, the countries were devastated after independence and only Kazakhstan has made any real strides toward economic development. Turkmenistan and Uzbekistan have the ability to move in this direction, but the economic and political policies of both have caused the economic potential of their countries to stagnate. Tajikistan and Kyrgyzstan are economically in the most precarious positions, and it is difficult to imagine ways that they can alleviate the poverty that afflicts much of the countryside and even large parts of their urban populations. Except in Kyrgyzstan, privatization of land continues to proceed at an extremely slow pace, which further impoverishes the large rural populations while increasingly outdated industries make even blue collar jobs scarce. Positive trends include the recent opening of many of the economies, especially those bordering China, to international trade and competition to Russian and other former Soviet countries' goods, and the negotiations to build oil pipelines that bypass Russia. The latter has the potential to set up an economic situation like that of the Middle East, where countries with large oil reserves (Kazakhstan and Turkmenistan) manage to develop, while the other countries (Afghanistan, Kyrgyzstan, Tajikistan, and Uzbekistan) remain poor and continue to struggle to develop. There is still a long and difficult path ahead for all of these countries, but especially for Afghanistan. Its devastated countryside teeters on the edge of becoming a narco-economy controlled by warlords unless something dramatic can be done quite rapidly. This very scenario adversely affected the entire area throughout the 1990s and has the potential to do so again. Historically, Central Asia was at the very center of trade in Eurasia. Unfortunately, the main trade today is in narcotics. However, the prospects of a better future for the region lies in the ability of the region's various governments to address a range of economic, political, and environmental issues that will allow Central Asia to once again become a center for transcontinental trade and permit the opening of their economies to world markets, which in turn will hopefully promote investment in the future of these people.

▶▶ Key Terms

Amu Darya River	Kyzl Kum Desert
Aral Sea	opium
centralization	Pamir Mountains
collectivization of agriculture	Russification
command economy	Silk Road
Hindu Kush Mountains	Syr Darya River
Kara Kum Desert	Taliban movement
Karakalpakstan	Wakhan Corridor
kitchen gardens	warlords
	Zerafshan River

▶▶ Review Questions

1. What are the obstacles to economic development in the Yakutsk region?
2. Discuss the current developmental prospects of Belarus.
3. Describe Ukraine's agricultural potential. To what extent is that potential realized? Explain.
4. Identify winners and losers among Russian regions in the economic transition to capitalism.
5. Explain the significance of the Baku–Tbilisi–Ceyhan pipeline.
6. Describe current trends in Russian agriculture. What do they say about the transition in the country?
7. Imagine traveling from Murmansk to Yalta in the southern Crimea in June. Describe the natural regions through which you would pass.
8. What are the major obstacles to trade and economic development in Central Asia and Afghanistan?
9. Why are questions about religion so heated in Central Asia and Afghanistan?
10. Why do so many farmers in Afghanistan grow poppies?

▶▶ Geographer's Craft

1. Consider Russia's environment and resource base. Which economic-development strategies are being debated? Given the current situation (explain), what strategy would you suggest?
2. Explain the geographical dimensions of the downfall of the Soviet Union. To what extent has the current situation in the in-between countries solved these problems?
3. The recent rise in the price of petroleum has been a windfall for Russia. Combined with earnings from the export of natural gas, the Russian government has begun paying off debts early and saving money in a stabilization fund. If it were your job to spend this windfall to develop the economy, given your knowledge of Russia's economic geography, where would you invest?
4. What was the fundamental principle underlying the relationship of state and society from the rise of Muscovite Russia to the end of the USSR? Use examples to draw contrasts with the Western experience. What remains of this political culture today?
5. In some parts of Central Asia, cotton is the most important economic resource; in other parts oil and natural gas play the same role. Contrast the impact of cotton as opposed to oil dependence, and discuss how dependence on and development of one major resource have cultural and political ramifications in addition to economic implications.

PART SIX

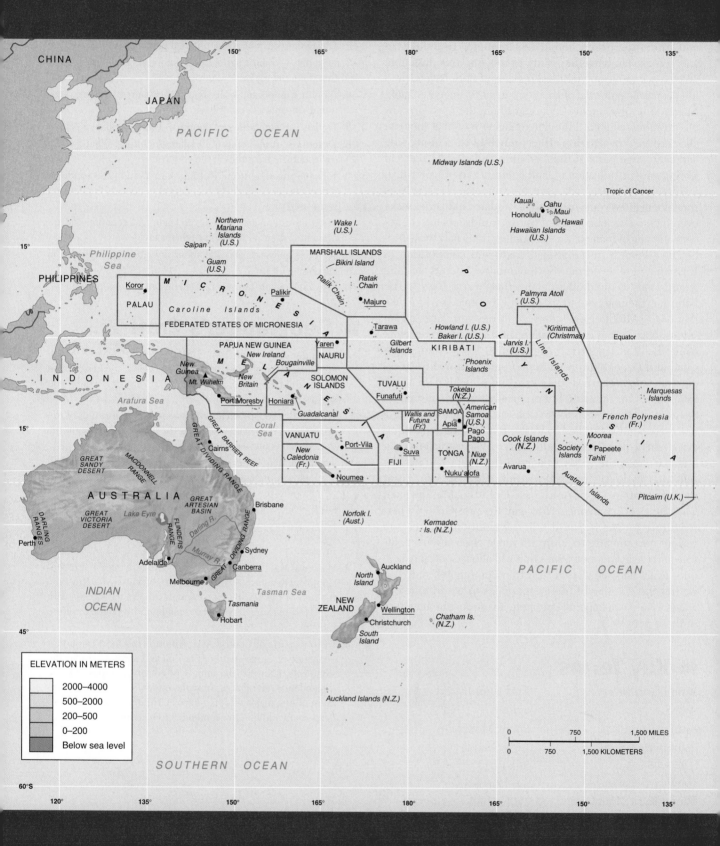

CHINA

JAPAN

PACIFIC OCEAN

Philippine Sea

PHILIPPINES

Midway Islands (U.S.)

Tropic of Cancer

Kauai
Honolulu • Oahu
Maui
Hawaii
Hawaiian Islands (U.S.)

Northern Mariana Islands (U.S.)
Saipan

Wake I. (U.S.)

MARSHALL ISLANDS
Bikini Island

Guam (U.S.)

M I C R O N E S I A
Koror
PALAU
Palikir

Caroline Islands

FEDERATED STATES OF MICRONESIA

Ralik Chain

Ratak Chain

Majuro

Tarawa

Palmyra Atoll (U.S.)

P
O
L
Y
N
E
S
I
A

Kiritimati (Christmas)

Equator

INDONESIA

PAPUA NEW GUINEA

New Ireland

Bougainville

Yaren
NAURU

Howland I. (U.S.)
Baker I. (U.S.)

KIRIBATI

Jarvis I. (U.S.)

Line Islands

New Guinea
Mt. Wilhelm

M E L A N E S I A

New Britain

Gilbert Islands

Phoenix Islands

Marquesas Islands

Arafura Sea

Port Moresby

Honiara

SOLOMON ISLANDS

TUVALU
Funafuti

Tokelau (N.Z.)

French Polynesia (Fr.)

Coral Sea

Guadalcanal

Wallis and Futuna (Fr.)

SAMOA
Apia

American Samoa (U.S.)
Pago Pago

Cook Islands (N.Z.)

Moorea
Papeete
Tahiti

Society Islands

VANUATU

Port-Vila

Suva

New Caledonia (Fr.)

FIJI

TONGA

Niue (N.Z.)

Avarua

Austral Islands

Cairns

Noumea

Nuku'alofa

Pitcairn (U.K.)

GREAT BARRIER REEF

GREAT DIVIDING RANGE

GREAT SANDY DESERT

MACDONNELL RANGE

GREAT ARTESIAN BASIN

Brisbane

Norfolk I. (Aust.)

Kermadec Is. (N.Z.)

PACIFIC OCEAN

AUSTRALIA

GREAT VICTORIA DESERT

Lake Eyre

FLINDERS RANGE

Darling R.

Murray R.

GREAT DIVIDING RANGE

Sydney

Canberra

Auckland

North Island

DARLING RANGES

Perth

Adelaide

Melbourne

Hobart

Tasmania

INDIAN OCEAN

Tasman Sea

NEW ZEALAND

Wellington

Christchurch

South Island

Chatham Is. (N.Z.)

ELEVATION IN METERS

	2000–4000
	500–2000
	200–500
	0–200
	Below sea level

Auckland Islands (N.Z.)

SOUTHERN OCEAN

0   750   1,500 MILES

0   750   1,500 KILOMETERS

Australia, New Zealand, and the Pacific Islands

Jack F. Williams

Australia, New Zealand, and the Pacific Islands are unique among the world's subregions. Collectively, they contain less than .5 percent of the earth's population (roughly 35 million in all), yet these nations and islands occupy the largest geographical realm on the planet, albeit one composed mostly of water. If we think of Australia as an island, the region covering much of the Pacific Basin can be described as a collection of islands, which are incredibly diverse in size, physical geography, population, culture, ethnicity, levels of economic development, and political structure. Australia is the dominant nation of the region, and, culturally and technologically, it has much in common with New Zealand. The smaller islands of Melanesia, Micronesia, and Polynesia differ greatly from Australia and New Zealand and face enormous challenges in their quests for human and economic development. One of the most basic, yet difficult, tasks faced by the peoples of many of the smaller Pacific Islands is the creation of a clearly defined national identity. Despite the differences among the Pacific Basin countries, it is clear that the economic development of all the nations of the region is becoming increasingly linked to the powerful countries that face them from either side of the Pacific—China, Japan, and others in Asia to the west, and the United States and Canada in North America to the east—and less dependent upon Great Britain, France, and other European nations. The strategic location of this region gives it a prominent role to play in world affairs, despite its small population and limited natural resource base.

CONTENTS

357

■ **Waterfront and opera house in Sydney, Australia.** Sydney is Australia's largest city, industrial center, and major seaport. More than 40 percent of Australia's highly urbanized population lives in the two most prominent cities—Sydney and Melbourne. Both are beautiful cities occupying magnificent harbor sites.

Chapter 16

Australia, New Zealand, and the Pacific Islands:
Isolation and Space

- **Australia**
- **New Zealand**
- **The Pacific Islands**

Australia and New Zealand are both blessed with abundant space and natural resources and physical landscapes of great variety. They have also long had reputations for high levels of material development, which are associated with the transplantation of Western society and economy in territories that, by accidents of geography and history, had been largely unknown and untouched by the peoples of Asia. Although Australia and New Zealand differ greatly from each other in size and physical environment, they share many characteristics of historical development and economic circumstances. Both were established in the late eighteenth century as British colonies for white settlers. Both have large land areas in proportion to their populations and high levels of living. In addition, both developed as supermarkets for Britain prior to World War II; that is, they provided many of the foodstuffs that Britain and the British Empire needed. That function is still important today, although it is no longer dominant.

New Zealand remains the more pastoral and agriculturally based of the two countries; its high level of living depends on abundant production of dairy products, meat, wool, and other animal and forest products. Australia's wealth is more diversified, with rich deposits of metals, coal, and natural gas; a bountiful agricultural assortment of meat, dairy products, wool, wheat, and sugar; and, increasingly, industrial manufactures.

Australia and New Zealand both depend on trade with industrialized nations to maintain their high levels of living. For many decades, that trade was directed primarily toward Great Britain and the British Commonwealth. Because of preferential tariff treatment and other trade privileges, as well as exclusion elsewhere and a lack of real markets in nearby Asia, it was profitable for Australia and New Zealand to market their products thousands of miles away. Since World War II, however, and particularly since Britain entered the European Common Market, the overseas relations of Australia and New Zealand have undergone a metamorphosis. Ties with the British have gradually weakened, whereas those with the United States, Japan, East Asia, and, to a lesser extent, other Asian countries have assumed new importance.

The Pacific Islands are similar to Australia and New Zealand in certain physical and historical aspects, yet differ greatly in their cultural and economic attributes. The Pacific Islands also are characterized by great natural beauty set in the context of extreme physical isolation. A second common feature is a shared legacy of European or American colonialism, which resulted over time in the loss or dilution of many components of indigenous culture. Perhaps the most obvious difference between the Pacific Islands and Australia and New Zealand is the much smaller size and more limited natural resource base of the former. Taken collectively, the human resource base of the Pacific Islands, as measured in health and education, is also less developed than that of Australia and New Zealand. Finding it increasingly difficult to successfully compete with the larger and more technologically advanced nations of the world, the Pacific Island peoples must now consider multinational economic cooperation and the filling of specialty niches in the global economy as development strategies.

Australia
A Vast and Arid Continent

Much of Australia's development experience is related to the continent's physical environment, particularly its isolation, vastness, aridity, and topography. With almost 3 million

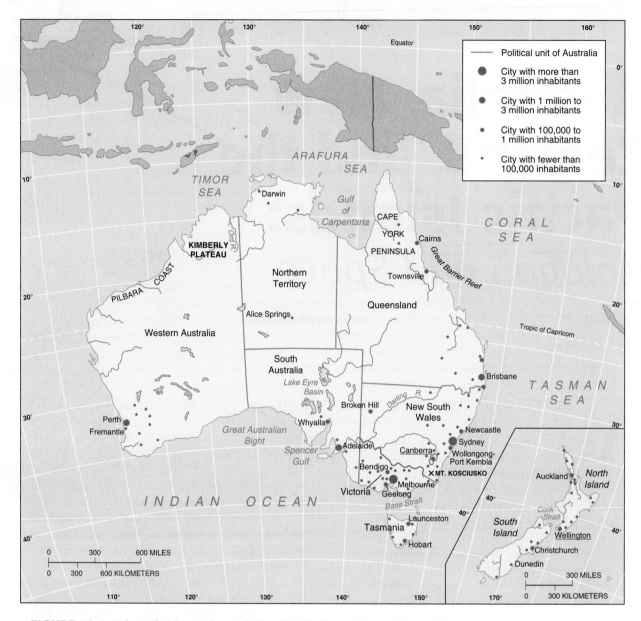

■ **FIGURE 16-1 Selected cities and population distribution of Australia and New Zealand.** Australia and New Zealand are two distant outposts of European culture. Both are sparsely populated, with the major cities situated along the coastlines.

square miles (7.8 million square kilometers), including the offshore island of Tasmania, Australia extends for 2,400 miles (3,862 kilometers) from Cape York in the north at 11°S latitude to the southern tip of Tasmania at 44°S latitude; it also extends about 2,500 miles (4,023 kilometers) from east to west. Although that land area is approximately equal to the area of the lower forty-eight states of the United States, Australia's population is far smaller at about 20 million. On the basis of population, Australia is actually one of the smaller countries of the world. Although Australia's average population density is low, its people are concentrated in a relatively small portion of the continent (Figure 16–1).

Only 11 percent of Australia gets more than 40 inches (1000 millimeters) of rain a year; two-thirds of the country receives less than 20 inches (500 millimeters) (Figure 16–2). On the basis of both climate and relief, five major natural regions can

be distinguished. The core region is the humid highlands, which extend in a belt 400 to 600 miles wide (644 to 966 kilometers) along the east coast. In addition to Tasmania, the narrow and fragmented coastal plains along the base of the highlands are the only part of Australia that is not subject to recurrent drought, and most of the nation's population, major cities, agriculture, and modern industrial economy are concentrated in that coastal fringe. In the southwestern corner of Australia and in a band along the eastern portion of the southern coast, the climate is mediterranean, or dry summer subtropical. Those two areas have the second major concentration of population, particularly around the cities of Perth and Adelaide, but the total population is still sparse.

The three other natural regions have various disadvantages for human settlement, and land use is confined largely to mining and livestock raising, which results in a very low

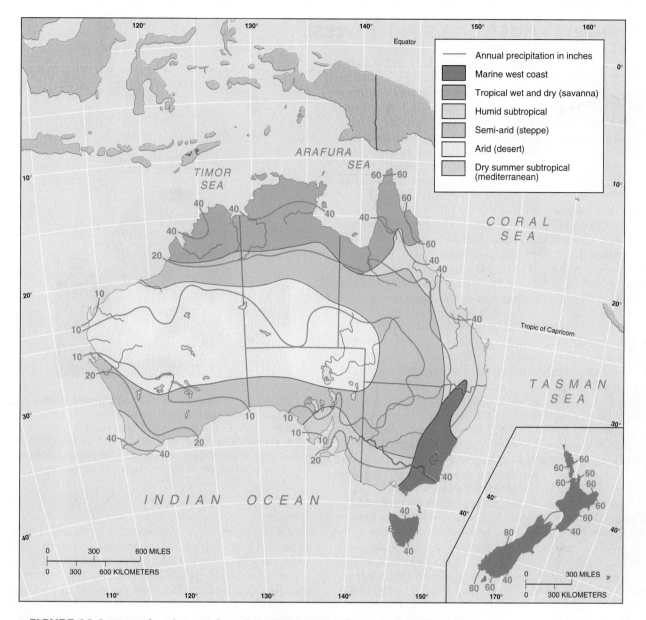

■ FIGURE 16-2 Natural regions and precipitation in Australia and New Zealand.
Much of Australia is little used because water is scarce; a large area of Australia receives less than 20 inches (500 millimeters) of precipitation annually. New Zealand has a much wetter climate.

population density (see the *Geography in Action* boxed feature, Frontiers: Australia's Northwest). Along the northern fringe of Australia are the tropical savannas, where the climate—3 to 4 months of heavy rain followed by 8 to 9 months of almost total dryness—has made commercial agriculture more difficult. The huge interior of Australia is desert surrounded by a broad fringe of semiarid grassland (steppe), which is transitional to the more humid coastal areas. These two dry areas cover nearly two-thirds of the continent.

The western half of Australia is a vast plateau of ancient rocks with a general elevation of only 1,000 to 1,600 feet (305 to 488 meters) (Figure 16–3). The few isolated mountain ranges are too low to influence the climate significantly or to supply perennial streams for irrigation. The continent as a whole was once joined to Antarctica, as the easternmost part of **Gondwana**, which broke up into the present conti-

nents in the distant geologic past. Australia moved northward, carrying with it unique wildlife, especially kangaroos and other marsupials (Figure 16–4).

The Shortage of Arable Land

For such a large landmass, Australia has a remarkably small amount of land that can be farmed without irrigation. The arid interior region, which makes up fully one-third of the continent, has not yet been developed for any agricultural purposes, even for livestock raising. Another 40 to 42 percent of the continent—to the north, east, and west of the arid interior—receives only enough rain to support cattle and sheep. The remaining land area, about 25 percent of the total, receives sufficient rainfall to support agriculture, but rough terrain and poor soils further reduce the presently arable area to about 10 percent.

GEOGRAPHY IN ACTION
Frontiers: Australia's Northwest

One of the few true frontiers remaining on the earth's surface is Australia's northwest, which is composed of the Northern Territory and the Kimberley Plateau and Pilbara Coast regions of Western Australia. That vast area, which covers about one-fourth of Australia, illustrates the challenges associated with the development of many frontier regions: small population, isolation and poor transportation linkages, a harsh physical environment, difficulties in resource exploitation, and neglect by the country's power center (urban-industrial core). Much the same can be said of other frontier regions in the world, such as Brazil's Amazonia.

Australia's northwest region runs in a band roughly between 12°S and 25°S latitude. Tropical wet and dry savanna climate dominates in the north, gradually blending into steppe and eventually becoming desert in the southern part of the region. Most settlement and development have taken place in the savanna, a region where heavy rains combine with temperatures as high as 113°F (45°C) during the summer months. An eight-month dry period follows, with daytime high temperatures cooling down only to about 86°F (30°C) in the winter. Less than half a million people live in the entire region today, most in widely scattered towns and settlements, the largest being Darwin (just over 100,000).

Lack of access to good transportation greatly hindered regional development here until just recently. There was no rail connection to the south; the nearest railhead was at Alice Springs in the interior. Darwin's ties with the rest of Australia were long, straight roads running south to Alice Springs and east to Queensland, plus limited air links. Port facilities were poor, limiting regular shipping service out of Darwin.

Now Darwin (and the region) appears on the brink of a renaissance, due to completion (finally) of a new railway line running south through the heart of Australia all the way to the southern port city of Adelaide, and also due to Darwin's proximity to Asia (Figure A). Darwin is closer to Jakarta in Indonesia than it is to Sydney, hence it made sense for the northwest to orient itself more toward Asia than toward the rest of Australia. The northwest historically has been more tolerant of Asians than has the rest of the country. Darwin has even had ethnic Chinese mayors, and intermarriage with Asians is common. Now, with the new rail link, Darwin has the potential to become a conduit for trade and other linkages with Asia.

The new railway line will have a number of benefits. One of these has to do with oil and natural-gas fields under the waters between the new nation of East Timor and Australia, which is cooperating with East Timor in development of these energy resources. A newly built LNG plant in Darwin will process the natural gas when it starts flowing in 2006 through a 175-mile (280-kilometer) pipeline from under the Timor Sea to Darwin. The project operator, ConocoPhilips, will market the LNG to Japan and possibly China. In addition, the new railway is expected to give a big boost to Australia's mining economy in the Northwest, making it economically feasible to exploit several major mineral deposits (such as magnesium), which were previously too remote and regarded as "stranded deposits." The government also hopes that cheap energy and improved rail links will

■ **FIGURE A Australia's north-south transcontinental railroad.** The first train to travel Australia's newly completed Adelaide to Darwin railroad reached Darwin on February 3, 2004. The Ghan train is named after the Afghan camel drivers who were instrumental in opening up the Australian Outback for explorers and early settlers of the interior in the nineteenth century.

stimulate a manufacturing base in the Darwin area. In the 1990s the government established a **trade development zone** (the Australian equivalent of a special economic zone or SEZ, as it is known in China and other parts of Asia). Aiming for capital-intensive finishing activities, the development zone had already attracted a few Asian manufacturers, mostly from Taiwan and Hong Kong. Now, with cheaper energy and transport costs on the horizon, the attractiveness of the development zone to other foreign investors may become much greater.

Darwin's economy is already benefiting from an increased role in Australia's defense needs. The Defense Department made the decision to relocate substantial elements of the country's Defense Force from Victoria to forward bases in the Northern Territory in the 1990s. Some 10 percent of Darwin's population are now defense personnel and their families. The new railway line will make military shipments to the region much easier. Even agriculture, long problematic in this unfriendly agroclimatic environment, may now benefit from cheaper transport costs to markets in the south and east.

While Darwin and the northwest are likely to benefit greatly from these new developments, the inherent handicaps of this isolated frontier region can never be totally overcome. Hence, this will never be a region of significant population concentration, and likely will continue to play a minor, albeit enhanced, role in the national economy.

■ **FIGURE 16-3 Arid landscape of western Australia.**
Much of interior and western Australia experiences a desert climate. These regions are used primarily for extensive animal grazing. This image shows a cattle station in the Outback.

Presently, about 5 percent of Australia's agricultural land is used for food crops (Figure 16–5), with a further 5 percent sown to improved pastures and grasses. Most of this land is irrigated in some fashion. These figures have held fairly constant since the 1980s, before which the area of land cropped or sown to pastures and grasses had been expanding rapidly, owing to increased use of chemical fertilizers, improved water supplies, and success in reduction of the rabbit population. Crop production is heavily concentrated in the states of New South Wales, Victoria, and Queensland. By contrast, the vast state of Western Australia has less than 2 percent of the total, while the Northern Territories scarcely register. The reverse side of this, thus, is that vast areas of the country are used for livestock raising on natural grasslands.

Settlement and Population Growth

The long delay in the European discovery and settlement of Australia was caused by many factors, including the vast-

ness of the Pacific Ocean, the direction of the prevailing winds and currents, and the lack of any sign that the continent possessed worthwhile resources. Until 1788, Australia was inhabited only by **aborigines**, who some estimate may have numbered as many as 1 million, members of some 300 distinct "nations" (Figure 16–6). These native people of complex origin had been in Australia for some 50,000 years, leading a traditional existence in close harmony with nature as hunters and gatherers. In 1770, Captain James Cook became the first European known to reach the east coast of Australia, the part of the continent that appeared most suitable for settlement. However, the first British ships did not disembark, at Sydney Cove, until 1788, marking the formal beginning of white settlement. For the early British settlers, who were primarily convicts, Australia served as a remote penal colony.

Exploration and settlement by adventurers, emancipists (convicts who had served out their sentences), and others continued into the nineteenth century. Immigration was encouraged by Britain through land grants in the developing continent, but the total population remained small. A great stimulus to development and immigration was the gold rush of the 1850s, which brought large numbers of prospectors and settlers. In 1901 the six Australian colonies—Queensland, New South Wales, Victoria, South Australia, Western Australia, and Tasmania—were federated into the Commonwealth of Australia, with the planned city of Canberra as the national capital.

One of the most important developments of the nineteenth century was the implementation of the **White Australia policy,** officially termed the Restricted Immigration

■ **FIGURE 16-4 Marsupial wildlife in Australia.** Gray kangaroos give a distinctively Australian flavor to the landscape of New South Wales.

■ **FIGURE 16-5 Land use in Australia.** Only about 10 percent of Australia's total land surface is arable. Of this area, about half is devoted to cultivation and the other half is used as improved pasture for animal grazing. The mediterranean climate of southeastern Australia encourages the cultivation of vines, and Australia has become an increasingly important producer of wine.

■ **FIGURE 16-6 Aborigine potters in Northern Territory, Australia.** The native inhabitants of Australia greatly declined in number after white settlement. These contemporary potters, members of the Hermannsburg Aboriginal Community, decorate pots with motifs derived from their local habitat.

Policy, after the first nonwhite settlers set foot on the continent. Successive Australian governments recognized the vulnerability of a relatively small white population controlling such a large land area so close to densely populated regions of Asia. Immigration into Australia was thus a major concern. Government policy was characterized by alternating support for large-scale immigration during periods of domestic prosperity and opposition to immigration during recessions; strong preference for people of British origin; and the exclusion of nonwhites, with only a few exceptions, from the late nineteenth century until the 1970s.

Britons predominated in the immigration pattern until World War II and were aided by the Australian government. After World War II, the government changed its policy and accepted other European and Anglo American settlers as long as they were white. Since then, more than 3 million

■ **FIGURE 16-7 Australian immigration trends.** Since 1975, Asian-born settlers have formed an increasing proportion of immigrants to Australia.

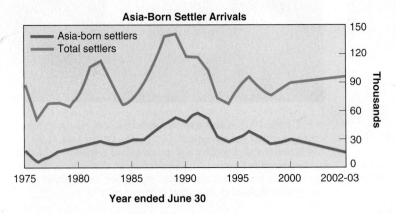

Asia-Born Settler Arrivals

— Asia-born settlers
— Total settlers

150
120
90
60
30
0

Thousands

1975   1980   1985   1990   1995   2000   2002-03

Year ended June 30

people—about one-sixth of the present total population—have moved to Australia from various other parts of the world. These new residents have helped to create an increasingly distinct Australian character, to replace the dominantly British type of Australian society. Australia continues to cultivate immigration at a carefully controlled rate.

The White Australia policy was quietly shelved in the early 1970s, as Australia's focus shifted away from Europe. Instead, the nation now selects immigrants on the basis of education, job skills needed by Australia, and potential for adapting to life in Australia (primarily, an English-language ability), as well as for family reunion.

Figure 16–7 shows the pattern of immigration since the 1970s. The peak of immigration was actually in the 1980s, but has since settled at a more modest level under the government's carefully controlled immigration program. In general, those from English-speaking countries are diminishing in numbers, while those from Asia are increasing. More than a third of each year's intake comes from Asian countries. A quarter of Australia's population was born abroad, and another quarter is made up of first-generation natives. About two-thirds of immigrants are admitted on the basis of their skills; the remaining third to reunite families. Critics complain that Australia is contributing to the "Brain Drain" in less developed countries, while Australia gains mostly well-educated and skilled people to help support a booming economy with a serious labor shortage. Fortunately for Australia, illegal immigration is a much less significant problem than in the United States.

How many immigrants could Australia effectively absorb each year, if the doors were opened wider? This is a much debated question in Australia; estimates run as high as 250,000 a year (a higher proportion in relation to population than in the United States). It should also be noted that immigration is countered to some extent by emigration, which averages more than 100,000 a year. This out-migration is due to many factors, but includes the perception of greater opportunities outside of Australia, and the relative isolation of the country.

Australia's Asian minority includes a sizable number of Indochinese refugees, as well as people from Malaysia, Singapore, the Philippines, and Hong Kong. The swelling tide of Asian immigration has renewed concern among some Australians. One camp argues for immigration to be cut back to 50,000–60,000 a year and for reduced numbers of family reunions, which tend to put a strain on welfare and social security systems. An extremist camp, rallying around the minority One Nation Party, had a brief period of prominence in the late 1990s, wanting to see a complete end to nonwhite immigration as well as less state support for the aborigines. The pro-immigration camp argues that Australia can easily absorb at least 160,000 settlers a year (under 1 percent of the total population). The latter group appears dominant at present, with most Australians voicing approval of the current immigration policy and the increasing multicultural character of Australian society. They also argue that Asian immigration is in Australia's long-range interest, given that more than 60 percent of Australia's exports now go to Asia, whereas only 11 percent go to Europe.

The other major nonwhite minority remains the original native population, the aborigines, who total around 450,000 today or about 2 percent of Australia's population. They are heavily concentrated in the Northern Territory, where they account for about a quarter of that region's population of roughly 200,000. The aborigines can be found today throughout the country, though, including the big cities. As a group, they are at the bottom of the socioeconomic ladder in Australia. Their life expectancy is 21 years less than the Australian average. They are three times as likely to be unemployed, 16 times as likely to be in prison. Those in the cities have fared the worst, but even those who stayed on their ancestral land and worked for white farm owners have done poorly and alcoholism is widespread. Unlike the Native Americans in the United States, the aborigines in Australia do not have a casino economy to help them. Wherever they live, they tend to be on the margins of society, ill-prepared to survive in modern Australia, rural or urban. Like Native Americans, the aborigines rely heavily on welfare. The Australian government has yet to apologize for the "stolen generation," that is aboriginal children forcibly taken from their mothers to be brought up by whites (something that also happened to Native Americans in the United States), a practice not stopped until the 1970s.

A few promising signs offer some hope, however. The aborigines' distinctive art has caught on in Australia and abroad, and some have been able to profit from this trendy line in world art markets and the tourist trade. A handful of aborigines have also succeeded and made it into political office, opening a niche for justice and reconciliation. Recent court decisions have opened the door to "native title" that, while not restoring ownership to land, at least is starting to give the aborigines some rights to use of the land, including possibly royalties from mining. Some land has been returned, but only to collective "Land Councils," not individuals, and the land is almost always remote and unfarmable. In short, the aborigines have a very different perspective on the so-called

"Lucky Country" they live in today. Their experience is not unique, however, in that minority natives in many countries around the world have shared similar fates.

How many people can Australia ultimately support in a sustainable manner? This question has intrigued geographers and others for over 200 years. Dreamers in the nineteenth century once speculated on an eventual population of 100 million. Pessimists, seeing the current challenges presented by severe shortages of water, have estimated the optimum population as small as 8 million. Those in between, who might be called realists, think that 50 million may be a reasonable figure by mid-century, assuming that the government and people develop more efficient use of water in all sectors. Regardless, a population of that size would require a much stepped-up rate of immigration in the decades ahead.

An Urbanized Society

For a country with so much land and so few people, Australia's high degree of urbanization is unusual. More than 85 percent of the population lives in urban areas. Around 40 percent of the people live in the two great metropolitan areas of Sydney and Melbourne, with populations of 4.2 and 3.6 million, respectively (Figure 16–8). One reason for this urban pattern is that Australia's agricultural production is extensive and employs few people. In addition, especially since World War II, Australia has encouraged industrialization to provide more home-produced armaments for defense and to secure greater economic stability through a more diversified and self-sufficient economy.

All five of Australia's largest cities—Sydney, Melbourne, Brisbane (1.7 million), Perth-Freemantle (1.4 million), and Adelaide (1.1 million)—are seaports, and each is the capital of one of the five mainland states of the commonwealth. Much of the country's production is exported by sea, and much internal trade is also conducted by coastal steamers. Before federation in 1901, each state had built its own rail system linking the hinterland to the chief port and thus to international markets. Ultimately, however, the

■ **FIGURE 16-8 Melbourne, Australia.** Melbourne, Australia's second largest city with over 3 million inhabitants, is located on Port Phillip Bay at the mouth of the Yarra River.

varied railroad gauges that were used hampered national integration.

Australia thus reflects a strong degree of urban primacy, with each of the Australian states having a single large city totally dominating. Moreover, many small towns, and the outback in general, are losing population, in spite of the mining boom. As in the United States, the trend is toward suburbanization, with urban sprawl around the central cities. Australia as a whole is becoming one of the most highly urbanized countries in the world, with 91 percent of the population already living in urban places today.

The Bases of Australia's Economy

Australia's high level of living can be attributed to its small population and its reasonably well-developed and diversified export economy, which depends on the production of

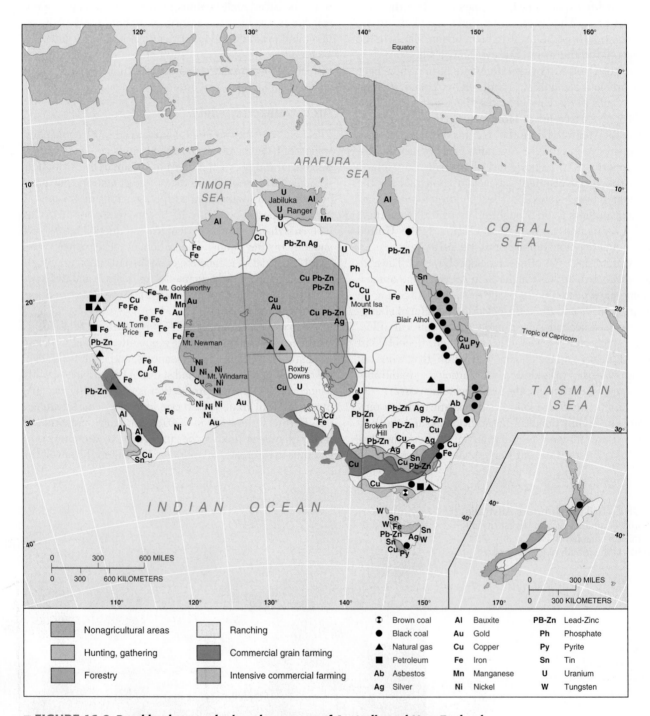

■ **FIGURE 16-9 Rural land use and mineral resources of Australia and New Zealand.**
Australia and New Zealand have extensive forms of agriculture, with ranching dominating much of the area of each country. Other extensive activities include mechanized commercial grain farming in Australia and forestry in New Zealand.

agricultural, mineral, and industrial goods. That production trilogy provides a solid base for a prosperous economy.

At the time of federation (1900–1901), Australia's largest single industry was agriculture (including forestry and fishing), which contributed about 20 percent to GDP. Manufacturing and mining each accounted for roughly 10–12 percent then, and services accounted for the rest. Industrialization and growth stimulated by World Wars I and II, plus the Korean War, brought significant changes. By 1950–1951, agriculture accounted for 30 percent of GDP, manufacturing had doubled to 22 percent, while mining had slipped to under 3 percent. Today services provide nearly half of the GDP, reflecting a highly urbanized society and maturing economy, while manufacturing is down to about 12 percent, mining stands at roughly 6 percent, and agriculture at 3 percent.

Agriculture is dominated today by sheep and cattle ranching and by wheat farming—extensive forms of agriculture that are well suited to Australia's environment (Figure 16–9). Sheep and cattle ranches, or **stations**, are usually quite large, often encompassing thousands of acres and making motor vehicles and airplanes important equipment. For example, Anna Creek cattle station in South Australia encompasses 12,000 square miles (31,080 square kilometers), an area about equal in size to Maryland and Delaware combined. Sheep ranching became a mainstay of the economy in the nineteenth century, when it provided wool for Britain's textile industry. By 1850, Australia was the world's largest exporter of wool, and it still produces a major share of the world's wool, though wool now accounts for only a tiny portion of total exports. In addition, since World War II, mutton and lamb have been exported in increasing quantities.

In terms of overall production value and importance to agriculture, beef cattle have become number one, followed by wool, wheat, and dairying. Sheep and lambs peaked at more than 175 million head in the early 1970s and have declined ever since, to about 100 million head. Beef and dairy cattle numbers have gone upward, by contrast, to around 27 million. Beef cattle are concentrated in Queensland and New South Wales, while dairy farming is largely confined to the eastern and southeastern coastal fringes and

to the Murray Valley north of Melbourne. Development of refrigerated shipping after 1880 enabled Australia to supply European markets with both meat and dairy products. In recent years, much of the increased demand for wool, beef, and dairy products has come from the countries of East and Southeast Asia, as well as the Middle East, where Australian meat and dairy products are common in supermarkets and food stores.

Australia's wheat production has also benefited from modern technology. The introduction of mechanization in the twentieth century permitted wheat to be extensively cultivated and about 50 percent of Australia's total cropland is now devoted to wheat, by far the largest area under any one crop (Figure 16–10). Like Canada and the United States, Australia has become one of the great breadbaskets of the world (although wheat accounts for only 3.1 percent of Australia's total exports, by value). The twentieth century witnessed a dramatic expansion in wheat cultivation and production. In 1900, just 1 million tons of wheat were produced on 2 million hectares of land (half a ton per hectare). By the early 1950s wheat cultivation had spread to more than 4 million hectares, with an average yield of 1.2 tons/hectare. Today wheat is grown on 11 million hectares, producing 22 million tons (more than 1.8 tons/hectare).

Australia also produces many other crops and is self-sufficient in foodstuffs. Sugarcane, one of the more important crops, is grown along the northeastern coastal fringe. Annually, Australia produces 3–4 million tons of sugar, most of which is exported to Japan and other Asian markets. Thus, Australia is the world's fourth largest sugar exporter, behind Brazil, the EU, and Thailand. Other important crops include a wide variety of temperate and tropical fruits for both domestic consumption and export markets.

Further growth in agricultural exports will be constrained by the tariffs levied by the European Union, which limit the amounts of Australian meat, butter, grain, fruit, and sugar that can be sent to Britain and other European markets. In addition, the European Union and the United States send their own agricultural surpluses to other markets in which Australia is competing. Consequently, the growth potential is not as great for Australia's agriculture as it is for its mining.

■ **FIGURE 16-10 Commercial grain farming.** With 50 percent of its cropland devoted to wheat, Australia has become one of the world's largest breadbaskets.

In terms of mineral resources, Australia is a veritable cornucopia (see Figure 16–9); it is among the most favorably endowed of all major world regions. Australia has enormous reserves of most of the key minerals needed in today's global economy, being at or near the top. Australia is first in production of bauxite and alumina, diamonds, lead, and uranium (with 40 percent of the world's uranium deposits); second in production of gold, nickel, and zinc; third in iron ore, and manganese; fourth in coal, copper, and silver. No other country in the world can boast such rich reserves of so many important minerals. Many countries buy these mineral products, but in the latter decades of the twentieth century, Japan came to be the leading player, and a key investor in development of Australia's mineral deposits. Now, since the economic take-off of China in the last 20 years, that country, with its skyrocketing demand for raw materials, is starting to play a commanding role as a key purchaser of Australia's mineral commodities. Hence, Japan and China together have become the most important export markets for Australia. The one commodity that remains in insufficient supply is petroleum. Although production of oil and natural gas has improved dramatically over recent decades, and the country exports some, nonetheless Australia also is forced to import some petroleum.

Manufacturing is the relatively weaker link in Australia's economy. So far the facilities that have developed are primarily **import-substitution industries**, which are geared to consumer goods, as well as to the partial processing of mineral and agricultural products, and the modest beginnings of some heavy industry, such as iron and steel and automobiles. Manufacturing is still protected by tariff barriers because the relatively small and scattered domestic market does not yet prompt production that is competitive with foreign imports.

Manufacturing in Australia is concentrated in the state capitals, where three-fourths of all factory workers reside and where industry also has access to available markets, fuel, business and government contracts, and both overseas and internal transportation systems. Much of the remaining industrial activity is located in a few large provincial centers on or near the coast, such as Wollongong and Newcastle. The leading industrial state is New South Wales, centered in Sydney; Victoria is second, focused in Melbourne.

In the 1980s, tourism began to assume new importance as a fourth leg in the Australian economy. Although hampered by the long distances international visitors must travel, Australia now attracts nearly 5 million foreign visitors a year, having doubled in the last decade, with the greatest increases coming from Asia. In 2003, New Zealand was the largest source of international visitors (18 percent), followed by the United Kingdom (14 percent) and Japan (13 percent). Other Asian countries contributing significant numbers of tourists included China, South Korea, Singapore, Indonesia, Taiwan, Hong Kong, Malaysia, and Thailand. Altogether, 40 percent

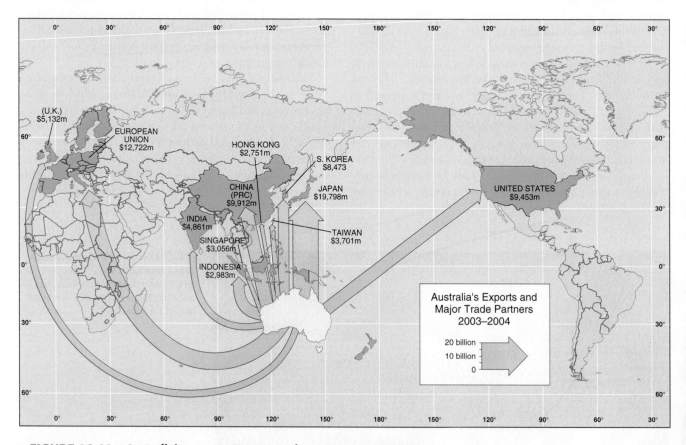

■ **FIGURE 16-11a** **Australia's exports: Top ten trade partners, 2003–2004.**
Source: Australian Government, Bureau of Statistics.

of all visitors come from Asia. These figures help to reinforce the pro-immigration lobbyists, who argue that Australia's economy would be much weaker were it not for Asia's proximity. In addition to traditional draws such as Sydney and the **outback** (interior), other attractions include resort developments along the **Gold Coast** in Queensland, one of the fastest growing parts of the country. Northward from there is the famous 1,250-mile (2,000-kilometer) long **Great Barrier Reef**, one of the world's greatest natural wonders. Cairns in northern Queensland has seen phenomenal growth, partly as a result of adding an international airport, as the jumping-off point for visitors seeking tropical recreation on the Reef. Tourism generates 4 to 5 percent of GDP and provides about 6 percent of total employment in the country. The prospects for tourism are very good, although Australia still accounts for less than 1 percent of total international visitor arrivals in all countries. Australia's relative isolation remains a hindrance to achieving its full potential.

Although there have been significant shifts in Australia's economic sectors in recent decades, the country still depends heavily on exporting primary or semiprocessed primary products in exchange for manufactured goods from the more developed world (Figures 16–11a and 16–11b). The gradual shift toward Asia and the Pacific Rim countries is very apparent from the trade data. The most dramatic recent change has been the rapid increase in trade with China, which replaced Japan as the number two market for exports and be-

came the number three source of imports, further evidence of the fierce competition Japan can expect from China in this century. Besides the big two, though, a number of other Asian countries have become key trading partners, especially South Korea, Taiwan, Singapore, and now up-and-coming India, which is following China's example in globalizing its economy. Australia's near neighbor Indonesia remains much below its potential as an economic partner because of ongoing internal economic and political problems. The United States remains the number one supplier to Australia (for the time being), but ranks third in purchases of exports, making Australia one of the few countries in the world with which the United States has a favorable trade balance. The greatly diminished economic ties with the United Kingdom are also evident, in both exports and imports.

Early in the twenty-first century, Australia is at a critical juncture in its national development. On the one hand, it has enjoyed remarkable economic growth for the past decade, bringing a high standard of living to the country. At the same time, the old ways that made life so comfortable for so many years—central wage management, protective tariffs, high tax rates to sustain a relatively egalitarian society—have been under attack. From the 1970s on, Australia actually declined in its relative economic standing among developed countries, but has since recovered to where it was then and still has untapped potential. While admired for the high quality of individual life, the country

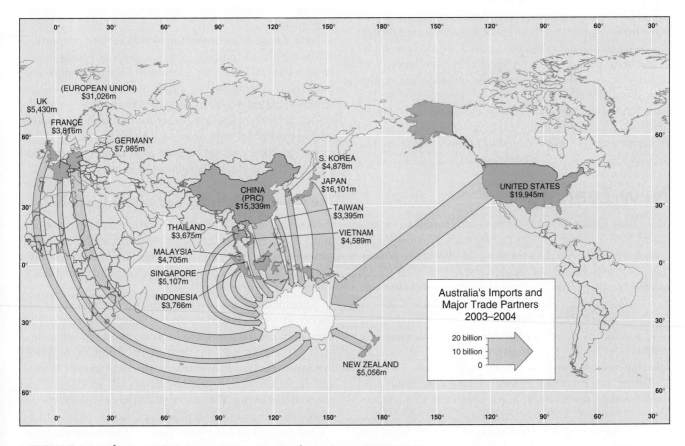

■ **FIGURE 16-11b Australia's imports: Top ten trade partners, 2003–2004.**
Source: Australian Government, Bureau of Statistics.

acquired a reputation for being more concerned with how to spend time off the job than with increasing productivity and national competitiveness, especially given the dramatic rise of Japan, China, and other Asian countries. Asia clearly is both the challenge and the future of Australia.

Some see the future as lying overwhelmingly with Asia, which will require making Australia more attractive than at present for Asian capital investment, producing manufactured products that have greater appeal and marketability in Asia, and stamping out vestiges of anti-Asian racism that still lurk beneath the surface of Australian white society, perhaps including increased immigration from Asia. None of these developments is likely to occur quickly. Australia must still contend with the problems of too small a population in relation to land area and resources and its relative physical isolation, even in the age of jet planes and modern telecommunications.

An important, albeit symbolic, step in the right direction, in the minds of an increasing number of Australians, would be for Australia to declare itself a republic, thereby ending its recognition of the British crown as the nation's titular head of state. Proponents of this measure point to the economic realities of Australia's trade, tourism, and investment, which are now dominated by Asia, not to mention the changing patterns of immigration. They argue, further, that becoming a republic would also signal to the world that Australia has come of age and accepted its new, emerging role in the Asian-Pacific realm. A referendum on becoming a republic was held in late 1999. Had it been approved, it would have called for the changeover to take place in 2001, exactly 100 years from the date that the original British colonies formed the Australian Commonwealth in 1901. The referendum's defeat at the polls has been interpreted by many analysts not as a vote against becoming a republic, but rather an expression of public concern over a lack of agreement regarding what kind of government the republic would have. Proponents are expected to work hard to develop a consensus on this issue and then call for another election. The Republic of Australia may be only a question of when, not if.

New Zealand

More than 1,000 miles (1,600 kilometers) southeast of Australia, New Zealand consists of two main islands—North Island, with a smaller area but three-fourths of the population, and South Island—as well as a number of lesser islands (see Figure 16–1). The country is located entirely in the temperate zone, from about 34°S to 47°S latitude. Like the Japanese islands, New Zealand was formed from the Ring of Fire, a section of the orogenic zone, or unstable crust belt, rimming the Pacific. The islands are the crest of a giant earth fold that rises sharply from the ocean floor.

Almost three-fourths of South Island is mountainous, dominated by the Southern Alps, which rise to elevations above 12,000 feet (3,600 meters). North Island, which is volcanic, is less rugged, but many peaks still exceed 5,000 feet (1,500 meters). New Zealand has a humid temperate cli-

mate, commonly known as marine west coast, with mild summers and winters; but in the highlands of South Island, weather conditions are severe enough for glaciers to form. The country's great north-south extent means that average temperatures in the north are at least 10°F (4.5°C) warmer than those in the south.

A Pastoral Economy

Settlement of New Zealand has been confined largely to the fringe lowlands around the periphery of North Island and along the drier east and south coasts of South Island. A large section of the country is relatively unproductive, though a tourist industry is being developed in the mountains, capitalizing on the attractive scenery (magnificently shown in the film trilogy, *Lord of the Rings*, 2001–2003).

In 1642 the Dutch explorer Abel Janszoon Tasman became the first European known to have sighted the islands, but it was not until Captain Cook arrived in 1769 that exploration began, and settlement really did not get under way until the 1840s. Hence, the development of New Zealand closely parallels that of Australia in time. Because the climate is ideal for growing grass and raising livestock, New Zealand, like Australia, has specialized in pastoral farming from the very founding of the country. Some 200 years later, New Zealand remains dependent on its **pastoral economy** and the production of livestock and livestock products (see Figure 16–9). Even the 3 percent of the land area that is cropped is devoted in large part to animal feeds. Horticultural crops, wood, and wood products are also of some importance.

New Zealand has one of the world's highest proportions of livestock (cattle and sheep) to human population—a ratio of 14:1. Pastoral industries completely dominate exports and, because of the country's small population, make New Zealand a world leader in per capita trade. It is among the world's top two or three exporters of mutton, lamb, butter, cheese, preserved milk, wool, and beef (Figure 16–12). In exchange, New Zealand imports most of its manufactured goods and considerable quantities of food.

The Need for Industry and Diversification

With such a heavy dependence on trade and a narrow economic base, New Zealand is far more vulnerable to the vagaries of world economic conditions than is Australia. As a result, New Zealand's per capita income is only about three-fourths that of Australia's.

Attempts at diversification, primarily through industrialization, met with limited success. Although New Zealand has coal, gold, natural gas, some iron ore, and a few other minerals, it does not have the rich menu of mineral resources that Australia has. Furthermore, the local market is small and dispersed, thus restraining large-scale production and efficient marketing. The cost of skilled labor is also high, and competition from overseas producers, such as Japan and the United States, can be severe. Most of the current manufacturing industries in New Zealand are high-cost producers

■ **FIGURE 16-12 New Zealand sheep being moved to pasture.** Australia and New Zealand have both been major exporters of meat, dairy products, and wool. New Zealand currently retains the more pastoral economy, whereas Australia is developing a manufacturing economy in addition to its traditional agricultural sector.

that survive because they are protected by tariffs. Overall, manufacturing contributes more than 25 percent of the national income in New Zealand and employs about the same proportion of the labor force.

The economy of New Zealand has benefited from a free-trade agreement known as Closer Economic Relations, which was signed with Australia in 1983. That agreement opened Australia's larger domestic market to New Zealand's products and gave a much-needed stimulus to New Zealand's industry, making Australia the country's largest trading partner in both exports and imports.

In spite of the predominantly agricultural economy, most of New Zealand's population of about 4.1 million lives in cities, similar to the population distribution in Australia. However, the cities of New Zealand are generally much smaller; the largest is Auckland on North Island, with just over 1 million people. Other major cities are Wellington, the capital, which is also on North Island, and Christchurch and Dunedin on South Island (see Figure 16–1).

About 80 percent of New Zealand's population is of European origin, mostly British, but including people from the Netherlands, Yugoslavia, Germany, and other nations. The indigenous **Maori** population, a Polynesian group who have lived there for at least 1,000 years, are the largest minority, at about 14–15 percent. The remaining minority population is made up of other Pacific Island peoples (mostly from Niue, Samoa, Tokelau, and Tonga), who migrated here after 1960. The Maori went through a long period of decline under European settlement after the Treaty of Waitangi signed with the British in 1840 until about the 1970s, but now are on an upward growth curve and expected to account for 20 percent of the population by 2050. Economically and socially, however, the Maori have much catching up to do and remain mired in the lower rungs of the socioeconomic lad-

der, not unlike the aborigines of Australia. More than a quarter of all Maori aged 18–64 are on welfare, and per capita income of the Maori is well below the national average. In recent years, the Maori have been exercising their political muscle in an effort to stand up for what they see as their land rights and other privileges. They are in a relatively stronger position in New Zealand, compared to the aborigines in Australia, because of their larger share of total population. Resolving the "Maori problem" remains one of New Zealand's key national issues.

New Zealand also faces increasing concerns about how to deal with Asia. **Kiwis** (New Zealanders) know that New Zealand must integrate more fully with Asia if the economy is to be turned around. At the same time, many are worried about Asian immigration and the social and economic impact of wealthy Asians (mostly Chinese and Indians) buying up real estate and changing the social composition, which is still much more British than in Australia. Asians number about 130,000 in all, just over 3 percent of the population, but a minority anti-Asian group lobbies against the Asians, paralleling Australia's experience. In some ways even more isolated than Australia, geographically as well as psychologically, New Zealand faces difficult choices if it hopes to maintain long-term prosperity in the twenty-first century.

The Pacific Islands

The vast realm of the Pacific Islands, or **Oceania** as it is frequently called, is often ignored in world geography textbooks. One reason for this neglect is that the total population of this huge area, which extends over several million square miles of Pacific Ocean, is only about 11 million (using the broadest interpretation of the region's boundaries), and the economic importance of the area in the modern global economy is slight. Yet such disregard does injustice to the region and its peoples, and restricts our understanding of a number of important geographical concepts and issues. The study of the Pacific Islands brings new insights into the challenges of isolation and distance, scarcity of land and other physical resources, small and scattered populations, limited economic opportunities, and social and political dysfunction derived in part from colonialism.

Regional Groupings

The Pacific Islands, some 30,000 in all, consist of several levels of regional groupings. At the first level, there are the three realms of Melanesia, Micronesia, and Polynesia (Figure 16–13 and Table 16–1). While geographically necessary, we should recognize that these are Western names of convenience that do not fully reflect the complexities of race, history, and physical environment. **Melanesia** consists of islands stretching along the northern perimeter of Australia, from New Guinea eastward to Fiji. The name derives from *melanin*, the pigment in human skin, and refers to the dark-skinned, Papuan-speaking peoples who predominate in this area. New Guinea, with more than 7 million people (counting both halves of the island; the western half is part of Indonesia), has the largest population by far in the Pa-

cific Island realm and gives Melanesia the largest total population of the three subregions at about 9 million. **Micronesia**, with more than half a million people, comprises the groups of islands just north of Melanesia, from Guam and the Marianas on the west to Kiribati on the east. This region's name refers to the "small" islands that predominate here. **Polynesia**, whose name means "many islands," is the largest in size of the three Pacific subregions, stretching in a huge triangular area from Midway and Hawaii on the north to New Zealand on the south, and eastward as far as Pitcairn, of *Mutiny on the Bounty* fame. Some of the native peoples of both Micronesia and Polynesia may be of Austronesian ancestry who migrated northward in the distant past. If one includes Hawaii's 1.2 million people, Polynesia's total population is slightly under 2 million. It is the best known of the Pacific Island subregions to most Westerners, who long have associated romance and tropical beauty with Hawaii and Tahiti (Figure 16–14). These two island groups receive, by far, the greatest number of tourists, with the majority coming from North America. Historically, New Zealand was also part of Polynesia because of its native Maori people. However, European settlement has effectively transformed New Zealand into a European country in which the Maori are a highly disadvantaged minority. Thus, New Zealand really belongs to the Polynesian realm only in a historical context.

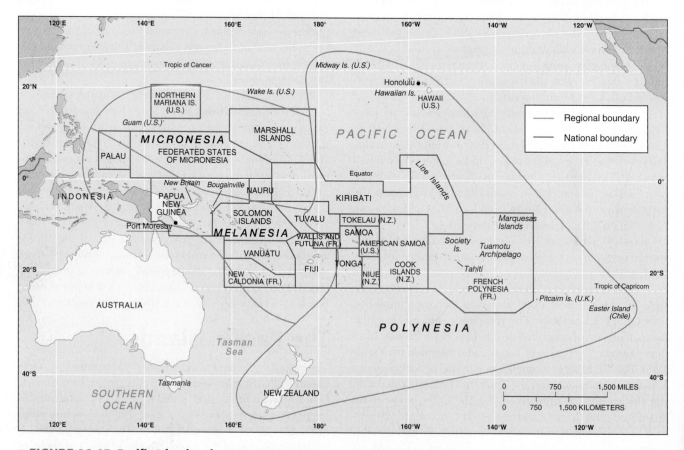

■ FIGURE 16-13 Pacific Island realm. The Pacific Islands can be divided into three subregions: Micronesia, Melanesia, and Polynesia.
Source: Dean W. Collingswood, "Japan and the Pacific Rim," in *Global Studies*, 5th ed. Guilford, CT: Dushkin Press, 1999, pp. 14–15.

Table 16-1   The Pacific Islands

Name	Political Structure	2005 Population
Melanesia		
Papua New Guinea	Independent	5,545,000
Bougainville	Part of PNG	(160,000)
New Britain	Part of PNG	(315,000)
Irian Jaya	(Province of Indonesia)	(2,200,000)
Solomon Islands	Free Assoc.–UK	538,000
Vanuatu	Independent	205,000
New Caledonia	French Terr.	216,000
Fiji	Independent	893,000
	Subtotal	7,397,000
		(10,072,000)
Micronesia		
Palau	Independent	20,000
Fed. States–Micronesia	Independent	108,000
Guam	U.S. Terr.	168,000
Northern Mariana Islands	Free Assoc.–U.S.	80,000
Marshall Islands	Independent	59,000
Nauru	Independent	13,000
Kiribati (incl. Line Is.)	Independent	103,000
	Subtotal	551,000
Polynesia		
Hawaiian Islands (Incl. Midway Is.)	U.S. State	1,258,000
		(2003 est.)
Tuvalu	Independent	11,600
Wallis & Futuna	French Terr.	16,000
Tokelau Islands	N.Z. Terr.	1,400
Samoa	Independent	177,000
American Samoa	U.S. Terr.	57,900
Tonga	Independent	112,000
Niue	Free Assoc.–NZ	2,100
Cook Islands	Free Assoc.–NZ	21,400
French Polynesia	French Terr.	270,000
Pitcairn	U.K. Terr.	(46)
	Subtotal	1,926,000
	Grand Total	**9,874,000**
		(12,549,000)

Source: CIA Factbook, 2005.

■ **FIGURE 16-14 Tahiti.** Many of the Pacific Islands possess great natural beauty.

The second level of groupings in the Pacific Islands consists of the various political arrangements and governmental structures that characterize the region. Altogether, there are twenty-three political entities: eleven independent nations (Papua New Guinea, Vanuatu, Fiji, Palau, Federated States of Micronesia, Marshall Islands, Nauru, Kiribati, Samoa, Tuvalu, and Tonga); four self-governing entities in free association with former colonial rulers (Solomon Islands, Northern Mariana Islands, Niue, and Cook Islands); seven continuing dependencies of France (New Caledonia, French Polynesia, Wallis, and Futuna), Britain (Pitcairn), the United States (Guam, American Samoa), and New Zealand (Tokelau); and one of the fifty U.S. states (Hawaii). This mosaic of political structures is the result of the region's complex colonial history and post-independence struggles. None of these structures has been a guarantee to success in terms of modern economic development.

There is a third subregional pattern that consists of extreme variations in levels of development and population distribution within each of the political entities noted above. This regional imbalance, common to larger nations of the world, may be found both on single islands, such as Papua New Guinea, or within island chains such as the Marshalls and Hawaii.

The Challenges of a Vast Island Realm

Even had this realm not been invaded and colonized by various outside powers, the Pacific Islands still would have faced daunting challenges in the current global economy. Traditional island societies were typically hierarchical and community based, dependent on fishing and subsistence agriculture, which centered on the cultivation of coconuts, taro, breadfruit, and other fruits and tubers. To outsiders, the lifestyle of the islanders often appeared to be an idyllic blend of a tropical climate, beautiful environments, and seemingly stable populations and societies. As long as the outside world remained outside, the island peoples might have been able to maintain their traditional, isolated lifestyles, ignorant of the turbulent world about them. Geography and history, however, were not in their favor, and the region fell piece by piece into the hands of foreign powers during the colonial era. By the late nineteenth and early twentieth centuries, foreign culture and technology had severely disrupted traditional societal and economic patterns, and the island populations had been decimated through the inadvertent introduction of exotic diseases. The islands were governed through artificial colonial administrative groupings that often disregarded historic cultural and resource utilization patterns. Plantation agriculture was introduced, focusing on the production of sugarcane, pineapples, coffee, tea, and cacao; and large-scale mining of gold, copper, nickel, manganese, petroleum, and natural gas was initiated. Over time, the native societies were further disrupted by the colonial rulers' systematic importation of alien laborers, including Indians to Fiji; Chinese to French Polynesia; Koreans to Guam and American Samoa; Filipinos to Guam, Palau, and

the Northern Marianas; Vietnamese to New Caledonia; and Japanese to various parts of Micronesia. Hawaii became perhaps the most extreme case, with the native Hawaiian population eventually reduced to a neglected minority.

The physical geography of the Pacific Islands made economic development difficult. Many are of volcanic origin with extremely rugged interior cores. Others are relatively flat atolls that have formed on the tops of coral reefs. A few combine volcanic and atoll landscapes. The vast majority of the islands are relatively small and limited in their population carrying capacities. Individual islands within archipelagoes are often situated great distances apart, making communication difficult. As the subsistence economy was destroyed by the colonizers, the islanders became dependent upon imported foodstuffs and the industrial manufactures that had become part of island lifestyles. Destroyed also were traditional diets, social patterns, and mores.

Today many of the islands face a host of serious challenges: low levels of income and health, social inequalities, weak governments, and uncertain national identities. As a result, political instability, separatist movements, political/military coups, and other troubles have afflicted numerous of the island groups at various times, including Bougainville, the Solomon Islands, Fiji, New Caledonia, Papua New Guinea, and others. The Solomons remain currently one of the most troubled states in the Pacific. There have been sharp declines in agricultural occupations, with corresponding increases in migration to urban centers, both domestic and foreign. Many rural areas and outer islands are experiencing depopulation, thus exacerbating imbalances in regional development within archipelagoes. In some cases, entire regions have been abandoned as the remnant native peoples seek better lives in the current or former colonial nations. Of those who remain, most are employed in the service sector, including the tourist industry, which has a spotty record throughout the region both as a generator of jobs for local residents and as a contributor to long-term economic development. Quality of education also varies widely, but literacy levels and years of schooling are generally low.

Pacific Islanders today are thus reappraising their options and trying as best they can to adapt to the global economy and integrated world into which they have been thrust. Many of the island peoples should be credited for their creativeness in attempting to generate international revenues in the context of severely limited natural resources. Some have sold fishing licenses to fleets from Japan, Taiwan, South Korea, and the United States. Tonga and Kiribati have sold passports to Hong Kong Chinese. Kiribati has sold satellite launch and tracking services. Now a number of the islands are trying to market themselves as loosely regulated global "financialservice centers," in the hope of emulating the experience of some Caribbean island nations with similarly limited economic options. Unfortunately, some of these operations are developing into money laundering opportunities for organized crime groups. Vanuatu appears to be leading the charge down this dubious road. Most of the Pacific Islands have joined to form a supranational

■ FIGURE 16-15 Panguna copper mine. Although one of the world's largest, the Panguna copper mine was caught up in a bitter civil war that wracked Papua New Guinea for over a decade.

organization, the **South Pacific Forum** (or Pacific Islands Forum), which was founded in 1971 and ostensibly is designed to promote the region's collective interests and build strength from unity.

In reality, it is unlikely that many of the island entities will achieve accelerated near-term economic growth. Hope for long-term development appears to hinge largely on strengthening education, increasing regional cooperation, and creating specialty niches in the global economy.

Three Examples of Challenges

Three representative entities within the region illustrate the development prospects: Papua New Guinea, Hawaii, and the Marshall Islands.

Papua New Guinea Papua New Guinea (PNG) consists of the eastern half of the island of New Guinea (the western half is the Indonesian province of Irian Jaya or West Papua), plus the islands of Bougainville and New Britain. Gaining independence from Australia in 1975, this tenuous union has had a troubled history. Papua New Guinea is a nation in search of a collective identity. The New Guinea part of PNG is an extremely rugged, mountainous land with densely forested peaks reaching more than 14,000 feet (4,267 meters). Roads and highways are almost nonexistent away from Port Moresby, the capital—at less than 200,000 people the third largest city in the Pacific realm after Honolulu and Auckland—and several small towns that dot the coastline. The country was the site of bitter fighting between Allied and Japanese forces during World War II. A polyglot collection of tribes, speaking more than 700 languages, make up the country's disunited population.

Papua New Guinea is the largest and the best endowed of the developing Pacific Island states. Rich forest and mineral resources (especially gold, copper, and petroleum which account for 72 percent of export earnings) offer some real potential for economic advancement. One of the largest copper mines in the world is at Panguna on the island of Bougainville (Figure 16–15). Unfortunately, that island was wracked for 10 years by a bitter civil war, led by secessionists who protested the dominance of Port Moresby and New Guinea over their island. As a result, PNG's export earnings from copper plummeted, and the last decade was filled with economic and political strife.

The struggle over Bougainville appeared to be moving toward settlement by an agreement signed in January 2001 giving Bougainville a referendum on independence 10–15 years after an autonomous government was to take over in 2002. The permanency of this arrangement has yet to be proven. Worrisome, also, is the position of the natives of Irian Jaya, who are more sympathetic and culturally aligned with the peoples of Papua New Guinea than with the nation of Indonesia. Secessionist tendencies continue to simmer in Irian Jaya. Australia is in the uncomfortable position of trying to assist in the stabilization of PNG, including investing money each year to sustain the economy of its former colony and having to deal with a neighboring unstable Indonesia, as well as trying to assist in stabilizing the nearby Solomon Islands.

Papua New Guinea illustrates the challenges of developing a culturally diverse nation in a volatile region even when that country possesses a rich endowment of natural resources.

Hawaii Hawaii, the fiftieth U.S. State (1959), consists of a string of mountainous volcanic islands stretched along the upper perimeter of Polynesia, on the northern margin of the tropics. This was the farthest north that the Polynesian sailors reached in their epic voyages, some of which may have originated centuries ago from today's French Polynesia. The total population of more than 1.2 million relies upon Western technology for its comparatively high levels of living. In some ways, Hawaii is the most changed of the Pacific Island groups today and, like New Zealand, might properly be described as no longer belonging culturally to Polynesia. The native population has declined to about 200,000 as successive waves of Chinese, Japanese, Filipino, white North American, and other immigrants, settled on the islands. Asians now account for 42 percent of the population, whites for 24 percent, Hawaiians and other Pacific Islanders for a mere 9 percent. However, there is a great deal of racial mixing through marriage.

Some 900,000 persons, or approximately three-fourths of the population of Hawaii, are concentrated in the urban center of Honolulu and its suburbs, which occupy a good share of the island of Oahu. Honolulu is the second largest city of the Pacific realm, with a population exceeded only by

■ **FIGURE 16-16 Marshall Islands.** The tiny coral atolls that make up the Marshall Islands contain few natural resources, despite their tropical landscapes.

Auckland, New Zealand. The city is highly dependent upon trade with the U.S. mainland and parts of Asia (especially Japan) for food and virtually all the industrial goods needed to sustain its economy. Spending by some 7 million tourists each year has cushioned the effects of the decline of the sugarcane and pineapple plantations and sharp cutbacks in military spending associated with Pearl Harbor, the headquarters of the U.S. Pacific Fleet. Although Hawaii's multicultural Asian-Pacific-Anglo population appears to have created a harmonious society, serious undertones of conflict exist between socioeconomic classes closely tied to different ethnic groups and state politics. Struggling for cultural survival in the midst of accelerated change are the native Hawaiians, who seem largely relegated to tourist-sector service jobs while fighting for their ancestral rights.

Even with its relative prosperity derived largely from U.S. statehood, Hawaii faces the same obstacles that confront the other Pacific Islands in the quest for economic success. Hawaii did not reap the benefits of the booming U.S. economy of the 1990s and fell behind in various socioeconomic indicators. Severe recession in the 1990s led thus to efforts at revitalization of the economy, focused on improving the tourist environment and trying to diversify the economy. Some 70 percent of the tourist business was concentrated in the grossly overbuilt, often tacky, Waikiki Beach area, prompting efforts to promote tourism elsewhere in the islands, particularly through emphasizing native Hawaiian culture, eco-tourism, self-contained beach resorts, and other measures. Plans to develop a high-technology industrial sector have been disappointing, as Hawaii's physical isolation, even in the age of jet travel, and relatively small population remain formidable obstacles to attracting industry. Even maintaining tourism is a challenge, as other, more affordable and attractive sites in the Pacific (such as Tahiti), the Caribbean, and others lure Japanese and mainland U.S. tourists. Nonetheless, since 2000, the tourism industry has rebounded to some extent, and the economy is improving. Compared to most of the other Pacific Island groups, the majority of the people of Hawaii still live quite well.

The Marshall Islands In stark contrast to the physical size of Papua New Guinea and the prosperity of Hawaii, the Marshall Islands are very small and very poor. The islands consist of a tiny string of coral atolls with a total population of 59,000 (Figure 16–16). Most of the residents are crowded into two towns: one on the island of Majuro and the other on the island of Ebeye. Granted a United Nations trusteeship in 1947, the United States chose to use the islands as a site for nuclear weapons testing (Bikini) in the 1940s and 1950s and still maintains a strategic Army missile range at Kwajalein Atoll. Although the Marshall Islands were granted independence in 1986, this group of islands could hardly be described as a viable nation state today. Propped up by continued U.S. subsidies and military expenditures, the traditional culture and lifestyle of the islanders has been almost totally destroyed, and the islands exhibit all of the environmental and socioeconomic challenges common to Oceania, compounded by health problems related to nuclear weapons testing. The Marshall Islands are a case study of the cultural dilution and economic dependence that so often have followed foreign involvement in the Pacific Island realm.

▶▶ Summary

Australia, New Zealand, and the Pacific Islands occupy a remote but strategically significant part of the earth. Although far from the Western world in location, Australia and New Zealand are Western in culture and in their approach to economic development. Production in both countries has been oriented toward utilizing agricultural resources (and minerals, in the case of Australia) for export to the developed world.

Past trade relationships with the United Kingdom were particularly strong. Both Australia and New Zealand, however, are now in the process of reorienting their economic relationships, largely toward the Pacific rim, with Japan and the United States currently playing dominant roles. That reorientation also includes increased attention to diversification of economic activity, a goal that will be more easily attained by Australia than New Zealand because of differences in resource endowment. Economic changes, along with the significant change in immigration policy during the 1970s, signal that Australia and New Zealand recognize the realities of their location in the economically developing world.

Papua New Guinea is endowed with a rich natural resource base. The other Pacific Islands, however, are challenged in their economic development by the scarcity of land. Many of the island states of Oceania are also characterized by cultural disunity and social inequality. Future economic growth of Oceania will hinge on the achievement of greater regional cooperation, filling specialty niches in the global economy, and the fuller development of the region's human resources, including the remnant indigenous populations whose societies have been seriously disrupted by Western conquest and settlement.

Key Terms

aborigines
Gold Coast
Gondwana
Great Barrier Reef
import-substitution
  industries
Kiwis
Maori
Melanesia

Micronesia
Oceania
outback
pastoral economy
Polynesia
South Pacific Forum
stations
trade development zone
White Australia policy

Review Questions

1. Summarize the physical attributes of the five natural regions of Australia. How have their physical geographies influenced their economic development?
2. Why has it been difficult for Australia, as well as New Zealand, to develop large manufacturing sectors in their economies?
3. Why is Australia so highly urbanized, given its extremely low population density?
4. Describe the urban geography of Australia in terms of city sizes, locations, and functions.
5. How can one explain the small population of both Australia and New Zealand?
6. What was the White Australia policy and what has happened to it?
7. Where is the core region of Australia and why is it located where it is?
8. Distinguish between the three great Pacific Island regions of Melanesia, Micronesia, and Polynesia.
9. In what ways have Australia's overseas linkages been changing, and why?
10. What are the physical challenges inherent to the development of the small island nations of the Pacific?

Geographer's Craft

1. How would you advise the Australian government to develop the northern part of the country?
2. How might Australia become a major industrial exporter?
3. How might Australia overcome its water shortages?
4. What policies or initiatives might improve the conditions for Australia's aborigines and New Zealand's Maori?
5. If you could redesign the national boundaries of the Pacific Island states, how would you do it, and why?

PART SEVEN

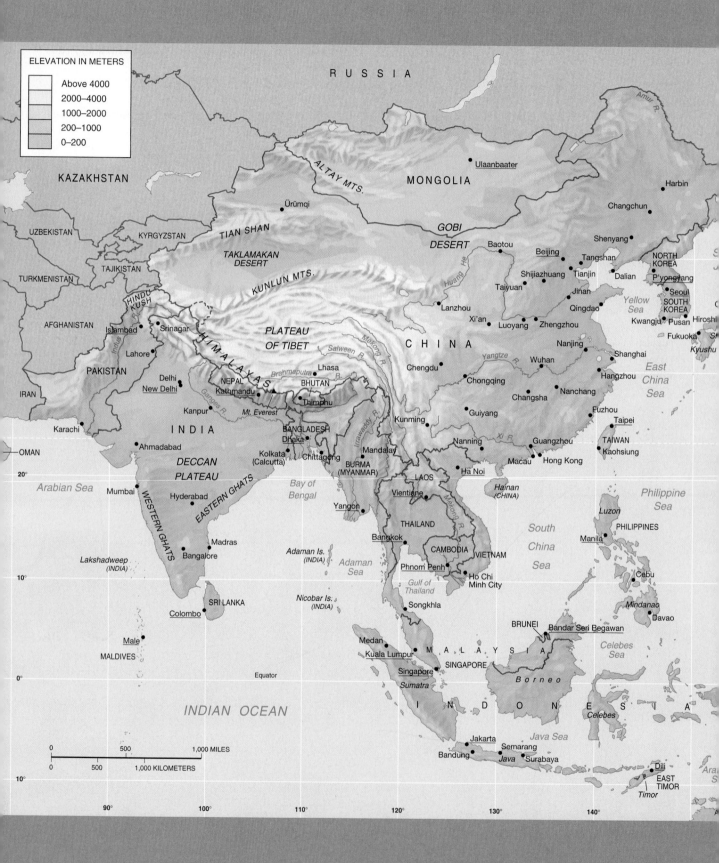

Asia

Christopher A. Airriess and Jack F. Williams

We should not confuse the human geographic term "Asia" as a major world region with the more traditional physical geographic term "Asia" that describes a continent. The human geographic term "Asia" refers to the southeastern quarter of the huge Asian landmass composed of twenty-four independent states. Once referred to as the "Far East" or the "Orient" by the western world, this human geographic region of Asia stretches east to west from Japan to Pakistan and north to south from Mongolia to Indonesia. Although Asia may appear at first glance to be too large to qualify as a world region based on homogeneous characteristics, it possesses numerous physical and human characteristics anchored in either unity or diversity. Nevertheless, the region differs significantly from the neighboring world regions of Russia; Central Asia; the Middle East; and Australia, New Zealand, and the Pacific Islands.

Asia is certainly homogeneous based on its physical characteristics. The most prominent of these is the presence of monsoonal winds, which bring annual alternating wet and dry seasons. Together with fertile alluvial soils, the rains associated with the wet monsoon have, for millennia, sustained many of the densest populations on earth. This in part is because Asia's world share of great rivers is unmatched. In addition, Asia exhibits the greatest cultural diversity of any major world region in terms of both religion and language. Perhaps with the exception of Sub-Saharan Africa, no other world region was as directly impacted by such a diverse group of colonial powers as Asia was.

Finally, when compared to other developing world regions, Asia stands at the forefront of the many economic and social changes associated with the globalization process. In part related to globalization, Asia is thus characterized by the growth of some of the world's largest urban agglomerations, many of which are confronted with very serious environmental problems. Although many of the nations of the region in recent times have relied on aggressive government policies promoting export-led industrialization as a development strategy, they exhibit widely varying levels of personal and collective well-being. Japan and Singapore, for example, rank among the most prosperous nations on earth, while Cambodia, Vietnam, Bangladesh, and India are among the poorest. Unfortunately, much of the recent economic growth that has come to the **Newly Industrializing Economies** or **NIEs** of the Pacific Rim nations of South Korea, Taiwan, Malaysia, and China has been restricted to particular socioeconomic classes as well as to specific subregions. The key to the long-term sustainable economic development of the nations of the region will be the provision of opportunity to all their citizens. The unity and diversity found among the peoples of Asia thus make it an ideal region to examine the complex issues of the developing world.

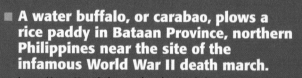

■ **A water buffalo, or carabao, plows a rice paddy in Bataan Province, northern Philippines near the site of the infamous World War II death march.**

Inset: Hong Kong's intensely urbanized character contrasts dramatically with the agricultural landscapes of much of rural Asia.

Chapter 17

Asia:
A Physical and Human Overview

- **The Monsoon Climate: Wet and Dry Seasons**
- **Population and Favored Environments**

- **Cultural Diversity: Past and Present**
- **Colonial and Modern Economies**
- **Development Indicators**

Asia may be divided into four distinct subregions based upon cultural differences within the larger region (Figure 17–1). The first subregion is **Japan**, the only subregion defined by a single country. Japan's inclusion within Asia is based on its geographical location and its intimate economic relationship with the rest of this broader region. Because of the country's advanced level of technological development and cultural distinctiveness, Japan is deserving of a separate chapter. The second subregion is **East Asia**, consisting of territorially dominant China, with Mongolia, North and South Korea, and the island state of Taiwan on its margins. The various countries of **Southeast Asia** include the mainland states of Myanmar, Thailand, Cambodia, Laos, and Vietnam and the maritime states of Malaysia, Singapore, Indonesia, Brunei, the Philippines, and East Timor. The fourth subregion, **South Asia**, consists of India, Pakistan, and the smaller states of Nepal, Bhutan, Bangladesh, Sri Lanka, and the Maldives.

The Monsoon Climate: Wet and Dry Seasons

The term **monsoon** is derived from the Arab word *mausin*, meaning season. More specifically, monsoon refers to the prevailing winds that occur during particular seasons of the year, bringing pronounced wet and dry periods. Human activity has long been influenced by these dramatic seasonal wind shifts, which are caused by the differential heating of ocean and land during the summer and winter, coupled with such additional factors as the jet stream position and strength. In a sense, these shifts in wind direction are simply a gigantic version of the more localized land and sea breezes that many coastal residents experience daily. During the Northern Hemisphere summer, a huge low-pressure cell develops over

southwestern Asia in response to the gradual heating of the landmass (Figure 17–2). As the air over the landmass is heated, it rises, drawing in warm, moist air from the Indian and Pacific oceans to the south and east. These moisture-laden winds persist for months at a time, bringing substantial amounts of precipitation to fill rivers and saturate dry soils. During the Northern Hemisphere winter, however, the Asian land surfaces become much colder than the adjacent oceans, leading to the formation of a large, strong high-pressure cell over east-central Siberia (Figure 17–3, page 384). A dry northerly or easterly wind then spreads over much of southeastern continental Asia. This dramatic seasonal shift of wind and precipitation produces stark landscape changes in agricultural areas (Figure 17–4, page 384).

The monsoon expresses itself differently in each of the four subregions of Asia. In South Asia, the summer, or southwestern, monsoon is divided into two branches, one originating in the Arabian Sea and the other in the Bay of Bengal to the east. From the Arabian Sea, southwestern winds begin to form in May. As they strike the southwestern coast of India, they release between 60 to 100 inches (1,500 to 2,500 millimeters) of rain over a 6-month period ending in October. The mountain wall of the Western Ghats captures much of this moisture for the narrow coastal plain and thus reduces the rainfall of the Deccan Plateau. By June, the eastern flank of the southwestern monsoon is sweeping across the Bengal lowland on its course to the Himalayan mountains, where tremendous tropical downpours drench the southern slopes. **Orographic** (mountain-induced) **precipitation** is greatest in the Cherrapunji region of the Assam Plateau, where total annual precipitation averages 420 inches (10,668 millimeters). The Himalayan Mountains then direct the wet air-masses westward into the Ganges River Valley and eastward into the Brahmaputra River Valley. The arrival of the rains is welcomed—both for the life-giving moisture and for the

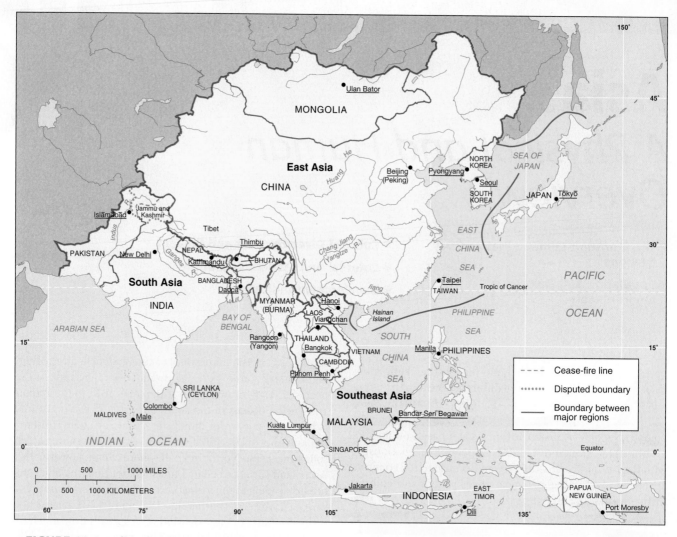

■ **FIGURE 17-1 Political units and capital cities of Asia.** The region consists of East Asia, Southeast Asia, and South Asia.

associated cooling, which relieve the 90°F–100°F (32°–38°C) temperatures characteristic of much of interior India at the end of the dry season. During the late southwestern monsoon season, **cyclones** (as Indian ocean hurricanes are called) frequently devastate the lowland coastal areas of the Bay of Bengal. Rain and occasionally tidal waves (more than wind) are the most destructive forces associated with the storms that lash the densely settled portions of southern Bangladesh. The dry and cooler winter monsoon, which lasts from October through March, brings less than 10 inches (250 millimeters) of rain to all but the extreme tip of India and Sri Lanka, with temperatures averaging 60°F (15.5°C) in northern India.

In East Asia, the summer, or southeast, monsoon envelops southernmost China by April and lasts approximately 6 months, but it does not reach northeastern China and Korea until June. By October, the summer monsoon is retreating in advance of the northerly winds associated with the winter monsoon. Rainfall amounts are greatest in China where the annual average exceeds 50 inches (1,250 millimeters). As a result, the lowlands of eastern China south of the Chang

Jiang (Yangtze) River experience warm and sultry summers similar to those of the southeastern United States. Interior regions receive substantially less precipitation because of the shorter summer monsoon and the greater distance from the sea. Annual rainfall totals for Beijing, for example, only average 25 inches (635 millimeters), which is similar to the rainfall levels on the American Great Plains. The coast of East Asia, stretching from northern Vietnam to Japan, is also subjected to hurricane-like **typhoons** (Figure 17–5, page 385). Originating in the Pacific Ocean and lashing the Philippines in their eastward tracks, typhoons most frequently make landfall between Shanghai and Hong Kong but occasionally loop northward in the East China Sea, bringing high winds and heavy rain to the Korean Peninsula and Japan. When compared to Atlantic hurricanes, Pacific typhoons are more numerous and powerful because the Pacific Ocean is larger and warmer.

The dryness of the winter monsoon contributes to the relative aridity of north-central China, which experiences cold, dry winds with clear and bright days, broken by occasional snowfalls. Because of the Qin Ling mountain

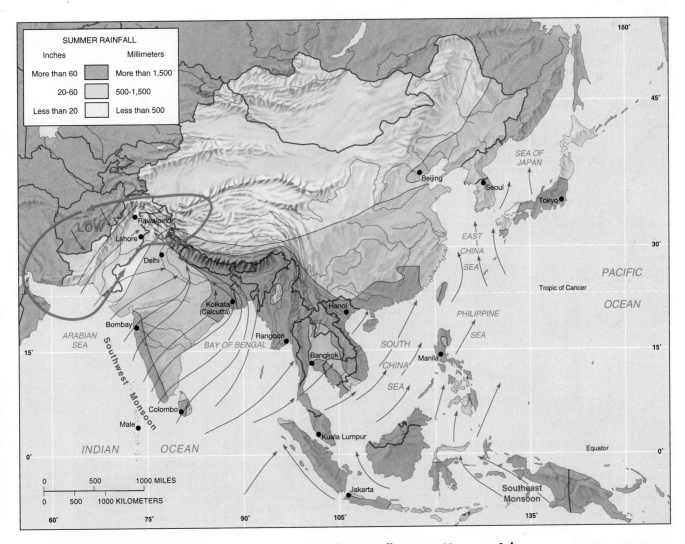

■ FIGURE 17-2 Annual rainfall and dominant atmospheric streamlines over Monsoon Asia during the summer. Asia's summer monsoon rains are fed by southerly wind flows.

barrier, the middle Chang Jiang (Yangtze) River Valley rarely experiences hard and long freezes; the average January temperature in the city of Wuhan, for example, is only about 40°F (4.5°C). Western China is less affected by the monsoons because of its sheltered interior location north of the Himalayan wall. During the winter monsoon, only dry and frigid winds sweep across the grassland and desert basins of this area.

In Japan, the monsoon climate is modified because of strong maritime influences. The summer monsoon arrives in the southern part of the country by June and slowly spreads northward into Hokkaido by July. Unlike China, Japan receives copious amounts of precipitation during the winter monsoon. When cold winds from continental Asia sweep across the warmer Sea of Japan, they pick up moisture, which then precipitates in the form of snow, particularly in western Japan.

The seasonal wind patterns of mainland Southeast Asia resemble those of South Asia. In the maritime states, however, the wet-dry pattern is less pronounced because of prox-

imity to the equator and to the Indian and Pacific oceans. Rainfall is distributed more evenly throughout the year, with the greatest amounts often occurring during the winter monsoon period as the continental winds pick up moisture while crossing the seas. The southwestern monsoon brings sufficient rains as well, except for the eastern half of Indonesia, where a pronounced dry season is common. Persistent convectional thunderstorms are also common in the equatorial region.

Population and Favored Environments

World population distributions reflect the interaction of complex physical and human forces. Climates, soils, and landforms, for example, may interact to form regions capable of supporting dense human settlement. This is true for Asia, much of which has been characterized by high population levels for thousands of years (see the *Geography in Action*

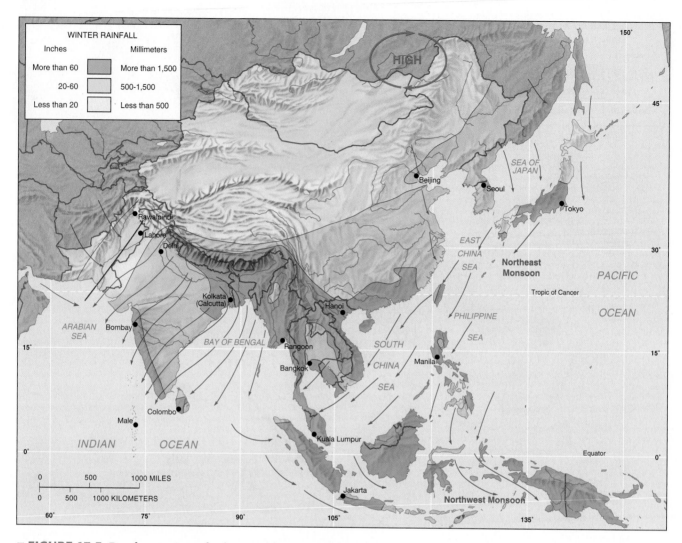

■ **FIGURE 17-3 Dominant atmospheric streamlines over Monsoon Asia during the winter.**
The dry winter season is associated with northerly continental wind flows.

■ **FIGURE 17-4 Monsoon contrasts.** (a) Lush agricultural scene of peasant farms in southern India during the wet summer growing season. (b) A dry season view of the same farms shown in (a).

■ **FIGURE 17-5 Typhoon Rananim in August 2004 moves to the northwest to make landfall along China's densely populated central coast.** The northwest Pacific experiences three times the number of such storms when compared to the northwest Atlantic.

boxed feature, The Economy and Culture of Rice). Significant population concentrations and agricultural production have long characterized Asia, as evidenced in its early culture hearths. Both the Huang He (Yellow River) in China and the Indus River Valley of Pakistan supported early civilizations on par with those of Mesopotamia and Egypt.

Figure 17–6 (page 388) illustrates the association between the physical environment and population distribution in Asia. In South Asia, notice the dense population concentrations in the fertile, level plains of the Indus and Ganges river valleys, as well as along the coastal plains of western and southern India. Less abundant precipitation coupled with a more dissected land surface have resulted in less dense populations in the interior Deccan Plateau. In East Asia, the valleys of the Huang He and Chang Jiang rivers and their tributaries support ribbons of dense population concentration (Figure 17–7, page 388). Under extreme population pressure, flat spaces have been created from slopes through the laborious work of terracing. Much of Western China has traditionally supported pastoral nomadism because it is arid and thus less favorable to intensive agriculture (Figure 17–8, page 389). Due to the absence of substantial level surfaces in Japan, the majority of the population occupies discontiguous coastal plains.

In mainland Southeast Asia, the relationship between dense human populations and intensive agriculture is pronounced. The tongues of high population density paralleling the Irrawady River in Myanmar, the Chao Praya in Thailand, and the Mekong and Red rivers in Vietnam provide a vivid contrast to the lightly populated surrounding uplands that enclose these linear river valleys. Maritime Southeast Asia presents a different ecological situation because of the scarcity of wide riverine valleys and plains. Here, the coastal plains are frequently narrow and, where they are wide, extensive swamps tend to cover the level land until the tropical rain forest of the nearby mountains is reached. These waterlogged environments characterize the eastern third of Sumatera and virtually the entire coastal plain of the island of Borneo. The only significant population cluster in the region is the island of Jawa, where the fertile volcanic soils of relatively small upland valleys support dense population concentrations.

Such favorable agricultural environments, coupled with historically high population growth rates in some countries, have produced a region that accounts for approximately 55 percent of the world's population. Indeed, six of the ten most populous countries of the world are found in Asia. The countries of China, India, Indonesia, Pakistan, Bangladesh, and Japan are respectively the first, second, fourth, sixth, eighth, and tenth most populous countries of the world. Just the two countries of China and India account for 37 percent of the world total. Having some of the world's most populous countries also means that Asia is dotted by some of the most populated cities in the world. Of the world's ten most populated cities, four are located in Asia. Because of the cumulative impacts of rapid urban population growth caused by high rates of rural to urban migration, coupled with increased industrialization and automobile use associated with the greater globalization of national economies, Asia's "megacities" have become some of the most polluted in the world. Urban areas aside, the density of some of the more populated rural regions of Asia exceeds the density of many suburban areas in Western countries.

Cultural Diversity: Past and Present

Asia is characterized by considerable religious and linguistic diversity. Two of the great world religions, Hinduism and Buddhism, originated within the region (Figure 17–9, page 389). Hinduism remains the primary religion of India. While Buddhism first emerged in what is now India, it was absorbed into Hinduism in South Asia at an early date. It subsequently spread throughout mainland Southeast Asia and China, Mongolia, and into the Korean Peninsula and Japan. Both religions have left a clearly visible imprint on the cultural landscape in the form of temples, shrines, and pilgrimage sites, and they indirectly influence social and economic life today (Figure 17–10, page 390). Diffused from the Arab World by way of India, Islam has also become a major religious belief system in Asia. Indeed, the number of adherents to the Islamic faith in South Asia and Maritime Southeast Asia outnumber the number of believers in the Arab World (Figure 17–11, page 390). One of the local responses to the perceived threat of increased secularization and cultural

GEOGRAPHY IN ACTION
The Economy and Culture of Rice

Although rice is grown on every inhabited continent, its cultivation is ideally suited to the seasonal water conditions of Asia. With the exception of taro (a broad-leafed tropical tuber), rice is the only major food crop capable of growing in standing water. Rice is the major food crop in southern India and China and most of Southeast Asia. In the poor and rural countries of Cambodia and Myanmar, the share of rice in the total calorie supply is an astounding 80 percent and 77 percent, respectively. The countries of China, India, Indonesia, Bangladesh, and Vietnam account for 75 percent of world production. The place of rice as a dominant food crop of Asia is attributable, in part, to the successful research efforts of the International Rice Research Institute (IRRI) in the Philippines. The IRRI has been responsible for the development of numerous hybrids or "miracle rice" varieties, which have boosted yields under a variety of environmental conditions. The increased output of the IRRI hybrid rice varieties has enabled rice yields to increase faster than population growth in East and Southeast Asia and to keep pace with population growth in South Asia. A handful of countries have even become net exporters of rice. Despite a decrease in rice consumption in the more-developed Asian countries due to changing food habits associated with increased income, projections indicate a 69 percent increase in rice demand over the next 30 years.

Rice was domesticated in mainland Southeast Asia at the dawn of human history. Of the many species of rice genus *Oryza*, *Oryza sativa* is the most common, with the subspecies, or varieties, of *japonica* and *indica* being the most widely cultivated. Like maize, wheat, and other grain crops, rice is a member of the grass family, which produces edible seeds.

There are three primary types of rice culture, or ways of cultivating rice. The oldest but least common method is dry or upland rice farming, in which rice seeds are thrown, or broadcast, into the air and allowed to sprout and grow in rain-fed fields. The two most widespread cropping systems today are rain-fed paddy and irrigation paddy farming. In these methods, the dyked field is flooded, either from rainfall or water from a nearby river or irrigation canal, and then plowed to the consistency of a thick sludge. Young seedlings from a nearby seedbed are transplanted one by one into the mud (Figure A).

■ **FIGURE A** **Young transplanted rice plants growing in a southern Philippine rice paddy.** Traditional rice farming is very labor-intensive, often requiring as much as 1,000 hours of human labor per hectare per season.

---

diversity associated with globalization is the rise of rigid forms of Hinduism and Islam in India, Malaysia, and Indonesia.

Other belief systems add to Asia's complex religious mosaic. The older native belief systems of Taoism and Confucianism have blended with Buddhism to create a composite spirituality in China, Korea, and Vietnam. The same process of blending took place in Tibet, where Buddhism was grafted onto a Tibetan animism to create a large Buddhist-monastic population before Chinese Communist control in the early 1950s. Buddhism has blended with Shintoism in Japan and Sikhism and Jainism have added to the religious diversity of western India. Roman Catholicism is dominant in the Philippines and East Timor, and a significant Protestant minority exists in South Korea, with additional pockets of Protestantism in Indonesia, Burma, the Philippines, and northeastern India.

Compared with religion, language presents an even more complex geographic picture in Asia, where some fifty-six distinctly different languages are spoken (Figure 17–12, page 391). Unlike Latin America and the Arab World, where a common spoken tongue assists in creating a map of cultural homogeneity, each country of Asia possesses its own language. The only exception are the states of Malaysia, Brunei, and Indonesia, where a variety of Malay language dialects are spoken. Some individual countries such as India and China are characterized by many languages, which heighten domestic cultural diversity. However, this religious and linguistic diversity sometimes exacerbates political tensions within and between the countries of Asia.

Colonial and Modern Economies

Three of the four subregions of Asia experienced the effects of **colonialism**. Japan was never a colonial possession but became a regional colonial power itself through the projection of its military and economic reach into various modern East and Southeast Asian nations. While the social and economic impact of colonialism varied from place to place, each colonial possession experienced a loss of political freedom and economic independence to a distant European, American, or Japanese power. The objectives of the foreign governments were to secure sources of raw

As the plants grow, the field remains flooded during the long rainy season until the seeds or grains begin to mature (Figure B). The field is then drained to prevent the heavy grains from falling into the water and rotting. Paddy rice is an extremely labor-intensive form of agriculture, with field preparation, dyke mending, transplanting, maintaining sufficient water height, weeding, and harvesting done almost exclusively by hand in traditional cropping systems. Except for plowing, rice cultivation is very much a gendered activity, because it is primarily women that engage in the arduous tasks of transplanting, weeding, and harvesting.

As a dominant staple food crop, rice provides the starchy carbohydrate component of meals to complement the vegetables and fish protein in a reasonably well-balanced diet. Unfortunately, much like white bread, the important vitamins of thiamine, riboflavin, and niacin, which are found in the brown outer bran and germ, are lost in the milling process to produce the familiar white polished rice. Used much like bread or potatoes in the West, rice possesses the qualities of enduring palatability and is thus often consumed for every daily meal among the middle and lower classes. In southern China and to some extent in the other traditional rice growing areas of Monsoon Asia, rice flour is used in noodles, cakes, and candies. The harvested plant, or biomass, is an important source of animal fodder and house thatching straw as well.

The economic importance of rice is matched by a strong social and cultural meaning. In southern China, for example, the phrase *chi fan,* meaning "to eat rice," also means "to eat." The word *fan,* meaning "cooked rice or grain," also means "food." In Hindu Bali, one of the most beloved deities is Dewi Sri, who is frequently honored in small temples located in the rice fields. Villagers in Thailand think of rice in more mythic proportions. In addition to providing physical sustenance, the Thai people believe that rice renews the body in a more spiritual sense and that human "tissue is made of rice because it derives from rice. Mother's milk is blood purified to a whiteness. Just as mothers give their food and bodies to nourish children, *Mae Phosop,* the Rice Mother, gives her body and soul to make the body of mankind." While the cultural significance of rice in modernized Japan has diminished, it lives today in the word "Toyota," which means "bountiful rice fields."

Sources: N. Anderson, *The Food of China* (New Haven: Yale University Press, 1988).
M. Hossain, "Rice Research for Food Security and Sustainable Agricultural Development in Asia: Achievements and Future Challenges," *GeoJournal* 35 (1995): 286–298.
R. Huggin, "Co-Evolution of Rice and Humans," *GeoJournal* 35 (1995): 362–265.

■ **FIGURE B** **Mature rice fields being harvested in a rural county near Shanghai, China.**

materials for the benefit of industry and trade in the home country and to expand their political empires. Thus, colonial powers transformed the indigenous economic systems to suit the needs of the home economies. An example of such a transformation was the substantial number of native peoples who were forced to cultivate commercial crops or labor on plantations rather than being allowed to grow food or fiber for themselves. This resulted in the disruption of local and regional food production systems, which, in turn, led to the importation of food and the progressive underdevelopment of the agricultural economy. Indigenous trading patterns were also disrupted as foreign trade and transport concerns replaced local interests that had for centuries offered an efficient system of interregional exchange. Much of the modernization accompanying colonial rule was not spatially uniform; it was restricted to a handful of coastal cities that functioned as foreign administrative and commercial centers. It was during the colonial period that a distinct **core-periphery relationship** evolved in which the more modern economic structures of mines, plantations, or colonial urban centers developed in contrast to the surrounding traditional subsistence agricultural economies.

The great differences in living conditions between these relatively well-developed urban centers and the surrounding impoverished agrarian countryside have led to a continual stream of rural-to-urban migrants searching for increased social and economic opportunities for themselves and their children during the postindependence or modern period. Although the present pace of industrialization is rapid in many Asian countries, employment opportunities in industry are only sufficient to absorb a small fraction of those seeking these livelihoods. As a result, millions of urban dwellers struggle to support their families through unskilled jobs, such as street sweepers, domestic servants, and market vendors (Figure 17–13, page 391).

During many years of colonial rule, the indigenous economic and political systems were subjugated to the will and might of the distant foreign powers. In recent decades, the nations of Asia have embraced a number of Western economic development and political philosophies in forms modified to suit the cultural, political, and economic circumstances of the newly independent countries (see the *Geography in Action* boxed feature, Asian Values, Development, and Globalization, page 392).

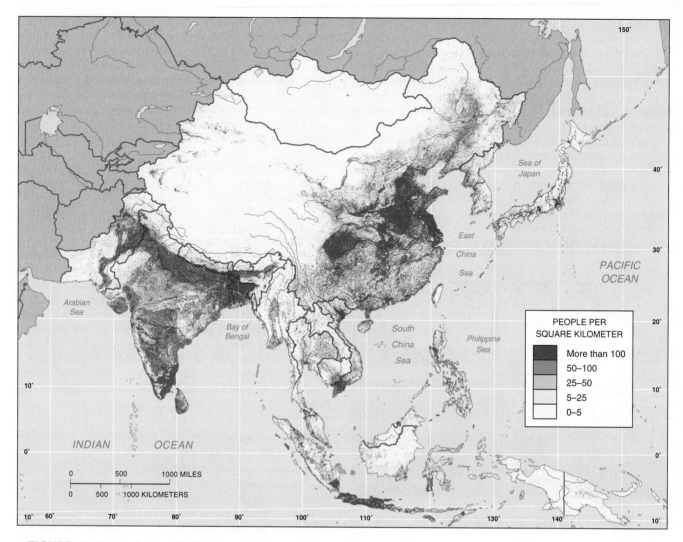

■ **FIGURE 17-6 Population distribution of Asia.** Monsoon Asia's population is concentrated in coastal areas and in interior river valleys.

■ **FIGURE 17-7 Densely settled Chang Jiang floodplain.** Canals that cross the Chang Jiang floodplain provide a focus for dense, linear settlements, deliver water for agricultural activities, and serve as highways for the movement of bulky goods.

■ **FIGURE 17-8 Goats on the Pamir Plateau, Tashikurgan, Xingjiang, China.** The aridity and cold temperatures that characterize parts of western China result in much of the land being used to support the seasonal grazing of animals.

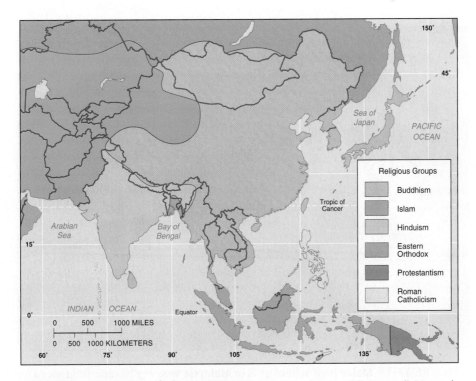

■ **FIGURE 17-9 Major religions.** Asia is home to many of the world's great religions and belief systems.

■ **FIGURE 17-10 A delicately styled Buddhist temple in Chiang Mai, Thailand.** Throughout much of Mainland Southeast Asia, Buddhist temples and shrines are ubiquitous in both rural and urban landscapes and are evidence of the central role the faith has played in the culture of the region.

In economic development, the range of philosophies has been great. Aside from Japan, which possesses the second largest economy in the world, the more prosperous countries of Singapore, Taiwan, and South Korea have stressed export-oriented industrialization, the cultivation and control of an educated workforce, and the ability to adjust their domestic output to the demands of the global economy through constant technological upgrading. Similar export-oriented economic strategies have been adopted by Malaysia, Indonesia, Thailand, and more recently China (Figure 17–14). These outward-looking economic models required reducing or eliminating many of the barriers imposed during the early postindependence period that were intended to protect the domestic economy from foreign competition. While the economies of the South Asian nations are becoming more export-oriented, they have nevertheless retained protectionist economic policies that prevent greater competition or increased participation in the global market. At the opposite end of the philosophical spectrum are the Marxist and inward-looking economies of North Korea and, until recently, China. In the political realm, one finds authoritarian regimes such as those of North Korea, Myanmar, and China and, more commonly, democracies in countries such as Japan, the Philippines, Malaysia, and India. It is important to note, however, that the Western-oriented free market and democratic nations continue to be characterized by high levels of central government economic control and limitations on opposition political parties.

Central government planning and control has often taken the form of selected state-run industries operating under commercial principles. Western-style political freedoms are tempered by the fact that Asian-style democracy often entails the partial sacrifice of certain personal and political

■ **FIGURE 17-11 Malay high school girls in Malaysia wearing Islamic-influenced uniforms.** Islam is the dominant faith of Malaysia and Indonesia, and its cultural imprint is evident in architecture, dress, and social customs.

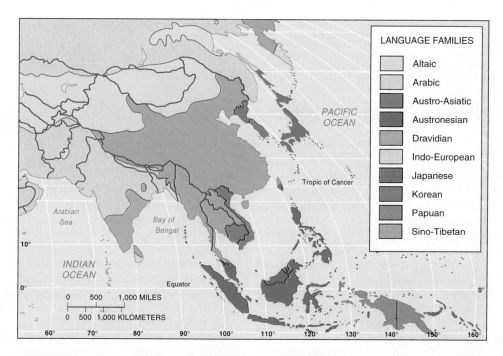

■ **FIGURE 17-12 Linguistic patterns.** Asia exhibits great linguistic diversity.

freedoms. With the exception of India and the Philippines, where opposition parties have had free reign, Asian democracies generally hold free elections, but they often permit the dominant political party to curtail the activities of the opposition political parties through government action. From the perspective of the national governments, these modifications of the Western political ideal are seen as suiting national circumstances and necessary to secure the greatest possible benefits for all.

Development Indicators

The philosophies discussed above have led to widely ranging levels of economic development in the nations of Asia (Table 17–1). Because GNI PPP per capita is a reasonably good surrogate measure of the integration of individual national economies with the global economy, the nations of Asia with high GNI PPP per capita tend to be characterized by a vigorous export-oriented focus. The value of

■ **FIGURE 17-13 Indian street vendor.** Tens of millions of urban laborers in the cities of Asia subsist on low-paying jobs that require only unskilled manual labor.

■ **FIGURE 17-14 High technology industry in Southeast Asia.** In contrast to the unskilled employment represented in Figure 17–13, an ever-increasing number of well-compensated positions in the high technology sector are becoming available in Asia to persons possessing the necessary education.

GEOGRAPHY IN ACTION
Asian Values, Development, and Globalization

The global emergence of Japan in the 1960s and 1970s; South Korea, Taiwan, and a handful of Southeast Asian countries in the 1980s; and, of course, China in the 1990s has made Asia one of the most economically dynamic regions of the world. In their study of the causes of the growth of these **Pacific Rim nations**, Western and Asian scholars alike have frequently concluded that **Asian values** have contributed significantly to the region's economic success. These values include the importance of collective action, loyalty, respect for authority, education, hard work, and frugality. The power of these values to influence human behavior in Asian societies contrasts sharply with what many view as their diminishing roles in Western societies. This argument for a distinctive set of Asian values challenges the long-held Western belief that Asia could not become politically and economically "modern" without adopting notions of Western individuality. That logic established a false dichotomy of the West versus the Rest. The absence of Western-style capitalism is perhaps why some Western development experts have often used the term "Asian Miracle" when describing the dynamic nature of economic growth on Asia's Pacific Rim. This growth of course has not been a "miracle," but simply a regionally variant form of capitalism. In addition, this so-called miracle has had its flaws, as evidenced by the Asian Financial Crisis of 1997, from which some national economies are only now recovering. If we claim Asian values for economic success, then would we not also have to blame these same values for the region's economic failures?

While values may indeed account in part for Asia's recent economic growth, the explanation fails to consider the cultural diversity of the western Pacific Rim nations, many of which, including China, Indonesia, and Malaysia, lack a common set of national values, much less a shared set of regional beliefs and behaviors. In addition, because these perceived traditional values arose in widely varied historical contexts, one must inquire whether they remain intact today under conditions of economic and cultural globalization.

If the Asian values and behaviors argument is harnessed to explain the region's economic trajectory, it may best be conceptualized as a tool used by Asian national governments wishing to confront the perceived external social and cultural threats posed by globalization. Is it possible that the process of globalization based upon Western values threatens the political hold of long-governing Asian regimes through its association with notions of increased individuality and choice, expanded labor union activity, a more independent media, and, of course, greater individual involvement in the political process? If such is the case, is it possible that the leaders of the Pacific Rim nations continue to promote these perceived Asian values in an attempt to keep their citizens and workers docile or submissive and thus maintain the political status quo? Under such circumstances, it would not be surprising that multinational corporations with investments in the western Pacific Rim countries would also support these "ideologies of the state," in the hope of minimizing the increased social and economic instability that is so often associated with globalization.

Source: Derived in part from R. Pertierra. "Introduction (to Asian Ways: Asian Values Revisited)," *Sojourn* 14 (1999): 275–294.

Table 17-1	Development Indicators, Asia			
	GNI PPP Per Capita (U.S. dollars) 2003	**Life Expectancy at Birth (years) 2002**	**Ratio of Girls to Boys in Secondary School (%) 2003**	**Human Development Index (HDI) 2002**
Japan	28,620	82	97	.94
Singapore	24,100	78	88	.90
South Korea	17,930	74	91	.86
Malaysia	8,940	73	103	.79
Thailand	7,450	71	93	.77
China	4,990	71	71	.74
Indonesia	3,210	68	80	.69
India	2,880	63	55	.59
Pakistan	2,060	64	45	.50
Bangladesh	1,870	62	49	.51

Source: Population Reference Bureau, *World Population Data Sheet 2004* (Washington, D.C.: 2004); and *World Development Report, 2005* (New York: Oxford University Press, 2002); ESCAP Population Data Sheet, 2004 (ESCAP: Bangkok, 2004); Human Development Report, 2004 (New York: Oxford, 2004).

exports is critical here because without exports to obtain hard currency, imports are unaffordable. The East and Southeast Asian countries of Japan, Singapore, South Korea, and Malaysia export higher-value manufactured goods when compared to the South Asian nations of India, Pakistan, and Bangladesh. The latter rely upon earning hard currency through lower-value agricultural products or lower-value manufactured goods such as textiles.

The range of human development attained among the Asian countries is quite broad as well. When we refer to basic development indicators, life expectancy is arguably the ultimate measure. The range of life expectancy is substantial between Japan and Singapore, which average 82 and 78 years at birth, respectively, and Bangladesh, which averages only 62. The ratio of girls to boys enrolled in secondary schools provides additional insights into human resource development and the quality of the labor force, particularly within the context of gender equality. Access to secondary education in the five richest countries is relatively equal for both males and females. While the gender ratios for China and Indonesia are not as equal, they are relatively equal considering the low GNI PPP for the two countries. This is an expression of the government desire to provide basic needs regardless of gender. The much poorer South Asian countries, however, provide far less equal access to educational opportunities for females. While it is true that secondary school enrollments for males are low as well, the ratio of girls to boys attending secondary school is on average 50 percent lower. What this means, of course, is South Asian governments are not fully developing the human resources of half their populations. The Human Development Index or HDI provides a general and useful indicator of development levels. Based on the combined variables of life expectancy, educational attainment, and income, South Asia is second only to Sub-Saharan Africa in possessing the lowest HDI among developing world regions. Again, despite its relatively low GNI PPP, China possesses a relatively high HDI. While many poor countries suffer from scarce financial resources, it must be recognized that governmental will and determination to address human resource problems may be the key to determining a country's long-term development.

Summary

The countries of Asia, like other countries throughout the developing world, manifest a wide range of development levels. As in most of the developing world, agriculture remains central to the livelihoods of many of this region's people. Despite technological innovations such as irrigation, the majority of peasants continue to rely on the arrival of the wet monsoon as a source of life-giving water. The monsoon rains, together with fertile soils, explain the region's extensive agricultural base. This productive agricultural base, in part, explains the region's huge population base.

The colonial inheritance of most Asian countries continues to influence the structure of their modern-day economies. These countries have adopted a wide variety of development philosophies as guides to reduce the lingering effects of colonialism. In some countries, these development philosophies have proven successful, while in others attempts at reconstructing a national economy to better serve domestic concerns have resulted in stagnant or even lower levels of development. Last, the cultural diversity of the region is vast. In Asia we encounter the wide range of experiences characteristic of the developing nations of the world.

Key Terms

Asian values
colonialism
core-periphery relationship
cyclones
East Asia
Japan
monsoon

Newly Industrializing
  Economies (NIEs)
orographic precipitation
Pacific Rim nations
South Asia
Southeast Asia
typhoons

Tokyo Auto Salon 2006.
The production and sale of high quality manufactured goods have become
significant to Japan's economy. Here
enthusiastic visitors to Tokyo's annual
auto show capture images of their favorite new models to carry away and
share. Inset: Spring in the Japanese
countryside.

Chapter 18

Japan

- Resources: Compensating for Scarcity
- Human Resources: The Hybrid Culture
- The Japan Model
- The Consequences of the Japan Model
- Japan's Third Transformation: Charting a New Course

Japan has always been something of an enigma to outsiders and the decade of the 1990s and the first years of the twenty-first century have only intensified the puzzlement of many who try to understand this economic giant of Asia and the world. By the 1980s Japan had succeeded so well in its economic development that unflattering expressions, such as "Japan, Inc." and "economic imperialism," were often applied to the country, as Japan played an economic role far out of proportion to its population, geographic size, and natural resource base. It was a stunning reversal after the country's abject defeat in World War II. Now outsiders are just as likely to look upon Japan with wonderment at how the nation's economy became so troubled at the close of the twentieth century. If nothing else, Japan's experience dramatically demonstrates that nation-states undergo constant change, and in today's highly competitive world economy it is not easy to stay on top.

The dilemma facing Japan is how to come out of what has been dubbed the **Great Stagnation** that has gripped the nation since the economic bubble burst at the end of the 1980s. The 1990s have also been called Japan's "lost decade" and the "10-year recession," except that the period of stagnation has now gone on well past 10 years with no clear end in sight. Pessimists contend that Japan's days of glory may be over permanently, as China rises to dominance in Asia in the twenty-first century. Others argue that Japan still has underlying strengths that will eventually overcome the problems holding back the country from advancement and that outsiders would be unwise to dismiss Japan too quickly. Who is right? Obviously, any assessment of Japan and its role in Asia and the world today needs to take careful stock of both the country's strengths and weaknesses, in light of what has happened in the decades since World War II.

Historically, Japan was the first non-Western country to be counted among the economically developed nations. To be sure, Japan has since been joined by several other states in Asia, notably Singapore, Hong Kong, Taiwan, and South Korea, with others moving briskly in that direction. Japan is still the economic kingpin, but not yet the political leader, of a group of **Pacific Rim nations** that collectively have become the world's third major economic center, rivaling the two older centers of Europe and North America. The economic dominance of the United States, while still very evident, is facing increasing competition from these up-and-coming Asian powers.

In spite of its recent struggles, Japan is still a very wealthy nation. Among the world's countries, Japan and its key economic partner, the United States, still collectively account for at least one-third of the world's economic output. Within Asia, Japan is still the richest nation in per capita income, but China moved ahead of Japan in total output of goods and services (measured in terms of PPP) with the new century (China's economy is now about half the size of the U.S. economy). Thus, Japan now ranks third in Asia. Japan's per capita income has slipped well below that of the United States (it used to be ahead), and is about the same as that for the key European nations (especially Germany), not even taking into account housing, infrastructure, social services, and so forth, which have always been stronger in the United States. Japan's share of Asia's economic output dwindles steadily each year, as the booming economies of China, India, and others make inroads. Japanese brand names, especially in automobiles and electronics, fill the marketplaces of the world and still are renowned for style, quality, and dependability, but other nations are giving Japan stiff competition. Japan remains a key member of important international financial and economic organizations, and many nations, including the United States, depend on Japan for manufactured goods and investment capital. Japan is Southeast Asia's most important economic partner and foreign aid donor. Japanese firms consistently account for about 80 percent of the largest Asian enterprises, but that margin is also decreasing. Hence there still is ample statistical evidence to argue against discounting Japan too greatly, in spite of its troubles.

What, then, are Japan's strengths and weaknesses, and how have the Japanese people maximized the former and minimized the latter to create this still highly productive country?

Resources: Compensating for Scarcity

In striving to understand Japan's economic success, it is important to recognize that it has occurred in spite of its limited physical resource base and sometimes hazardous natural environment. The Japanese have learned to make do with little and to acquire abroad what they do not have at home. This approach was the essence of the national development strategy after the Meiji Restoration of 1868, which is normally used to mark the beginning of Japan's modernization effort.

Location and Insularity

Japan's unique role in East Asia, past and present, can be attributed in part to the nation's relative isolation off the east coast of Asia (Figure 18–1). The Japanese call their country *Nihon* (often anglicized as *Nippon*), meaning literally "Source of the Sun," reflecting both the fact that the emperors claim direct descent from the sun goddess as well as that the Japanese archipelago is the first part of Asia south of Siberia that sees the morning sun. Hence, the popular term, "Rising Sun," in reference to Japan and the design of the Japanese national flag, although that design was not formally adopted until 1870.

The country consists of four main islands—Hokkaido, Honshu, Shikoku, and Kyushu—as well as many lesser ones. Together they form an arc about 1,400 miles (2,250 kilo-

meters) long that reaches from approximately 31° to 45° north latitude. Japan also owns the Ryukyu Islands, with Okinawa being the most important, which continue southward from Kyushu in another arc to almost 24° north latitude. Running due south of central Honshu are the Nampo Shoto, Ogasawara (Bonin) Islands, and Kazan Retto (Volcano Islands), all of which also belong to Japan. The latter includes Iwo Jima of World War II fame. Japan also claims the southernmost four of the Kurile Islands, currently occupied by Russia, that stretch northeast from Hokkaido toward the Kamchatka Peninsula of Siberia. So far, Russia's refusal to accept this claim has meant that Russia and Japan still have not signed a formal peace treaty ending World War II. Other contested islands, reflecting unresolved territorial disputes from the past, are the Senkaku (Diaoyu) Islands off the northeast coast of Taiwan (claimed by China, Japan, and Taiwan) and Takeshima (Tokdo) Island in the Sea of Japan (claimed by Japan and South Korea).

The impact of Japan's location off the northeast coast of Asia has been significant and sometimes is compared to the position of Great Britain off the coast of continental Europe, except that Japan's historical isolation was more profound. Unlike Britain, Japan had never been successfully invaded or occupied by foreign powers until the end of World War II. Rather, it had developed over some two millennia, adopting and adapting what it wanted from foreign cultures, particularly from China via Korea, and then much later from Europe and the United States. This selective process and relative isolation, within an environment of resource scarcity,

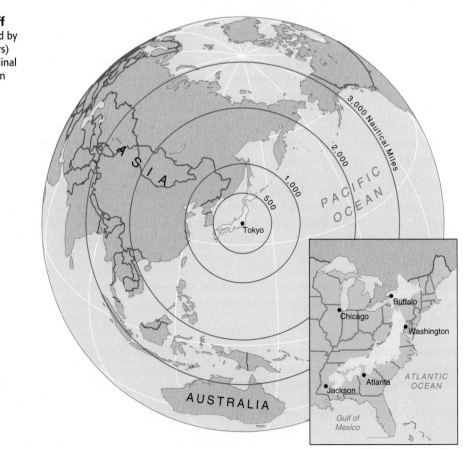

■ **FIGURE 18-1 Japan's location off the east coast of Asia.** Japan is formed by an island arc 1,400 miles (2,250 kilometers) long. The inset map depicts Japan's latitudinal position as compared to that of the eastern United States.

strongly shaped Japanese attitudes toward themselves and others. Moreover, the Japanese also acquired an acute awareness of their vulnerability and a determination to survive.

A Crowded and Temperate Land

Japan's long latitudinal sweep within Asia's mid-latitude zone, combined with its being surrounded by water, have resulted in the country having a humid continental climate in the north and a humid subtropical climate in the south, comparable in many ways to the eastern United States from New York to Georgia (see Figure 18–1). Asia's monsoonal patterns greatly affect Japan as well (see previous chapter), with the warm summer monsoon from the south bringing heavy rainfall and sometimes typhoons, and winds from the northwest in the winter carrying heavy snows to the Sea of Japan side and northern parts of Japan. One of the most significant boundaries is a line dividing the country roughly along the thirty-seventh parallel, which marks the northern limit of double-cropping of rice (Figure 18–2).

Japan's total land area is relatively small in comparison to its population. Japan has about 2.3 percent of the world's people living on a mere 0.3 percent of the land area. Smaller than France but larger than Germany or Great Britain, Japan has a population of 127 million, which is much larger than that of any European country and makes Japan the tenth most populous nation in the world. Moreover, Japan's people live in a geologically youthful, tectonically unstable landscape carved out of giant mountain ranges thrust up from the bottom of the Pacific Ocean as part of the **Ring of Fire** mountain-building zone that encircles the Pacific, the result of the earth's shifting continental plates. This location gives Japan some challenging physical characteristics, including a mountainous environment in which only 25 percent of the total land area has level-to-moderate slopes, at low elevations, while 75 percent consists of hills and mountains too steep to be easily cultivated or settled (Figure 18–3). Thus, most of Japan's population is actually crowded into an area slightly smaller than the state of Indiana, resulting in one of the world's highest population densities

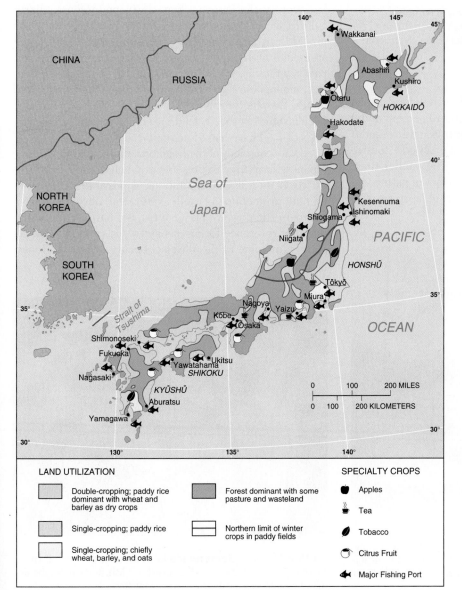

■ **FIGURE 18-2 Land utilization in Japan.** Nearly 75 percent of Japan is too steep to be easily cultivated. Farms on Honshu, Shikoku, and Kyushu remain small and are intensively worked, with rice the dominant grain crop. Agriculture in less densely populated Hokkaido, a middle-latitude environment, is less intensive, and farms are commonly larger than they are in the remainder of Japan. Self-sufficiency in agriculture is steadily declining in Japan.

LAND UTILIZATION

- Double-cropping; paddy rice dominant with wheat and barley as dry crops
- Single-cropping; paddy rice
- Single-cropping; chiefly wheat, barley, and oats
- Forest dominant with some pasture and wasteland
- Northern limit of winter crops in paddy fields

SPECIALTY CROPS

- Apples
- Tea
- Tobacco
- Citrus Fruit
- Major Fishing Port

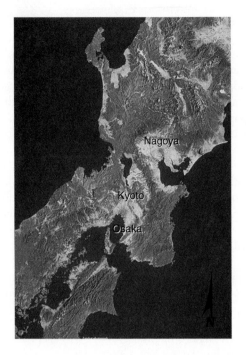

■ FIGURE 18-3 Landsat photo of Central Honshu.
Japan's landscape is mountainous, with limited lowlands and a rugged coastline. Visible are the major cities of Osaka, Kyoto, and Nagoya.

(comparable to that of the Netherlands or Belgium). Japan also has hundreds of volcanoes (including 10 percent of the world's active ones), numerous hot springs, and frequent earthquakes. In 2004, for example, a quake caused extensive damage in central Japan (Figure 18–4). Little wonder, then, that the Japanese feel vulnerable in the hands of nature and that disaster themes have long been a part of Japanese popular culture.

■ FIGURE 18-4 Earthquakes in Japan. Major earthquakes are common in Japan and other Pacific Rim nations and often bring widespread destruction of buildings and infrastructure. A 6.8 magnitude earthquake struck central Japan in 2004 and derailed the "Toki No. 325" bullet train in Nagaoka.

A Maritime Nation

Given its size and location, Japan has always relied heavily on the sea for food and transportation, much like Great Britain. Most of the people have lived close to the sea, in narrow river valleys, coastal plains, and scattered pockets of level land. The **Seto Naikai**, or **Great Inland Sea**, between Honshu and Kyushu/Shikoku, is akin to the Great Lakes of North America, albeit much smaller, as a vital waterway for shipping. Japan is one of the top fishing nations of the world and one of the few, and controversial, whaling countries left today (Figure 18–5). Japan also has a huge merchant marine to handle the immense volumes of goods moving into and out of the country.

Agricultural and Mineral Resources

Providing the raw materials needed to sustain Japan's economy has always been a challenge to the Japanese. Early on, the Japanese developed a form of intensive irrigated agriculture, much of it adopted from China and Korea, that focused on the cultivation of rice on tiny plots of land. In the last century, the Japanese have developed intensive horticulture, including hothouse cultivation in cooler months, to provide a rich and varied diet with some of the highest unit area yields in the world (Figure 18–6). Japanese agricultural patterns thus are very different from the large-scale, extensive, monoculture systems common to much of the United States. Despite this progress, domestic cultivation has long been insufficient to fully feed the population and large quantities of food have had to be imported.

Always conscious of the fragility of their environment, the Japanese have tried to preserve as much of the natural forest cover on hills and mountains as possible for environmental protection as well as for the development of hydroelectric power. The Japanese also have a cultural tradition of reverence for forests and wood products that accounts for some of the physical beauty of Japan. Hence,

■ FIGURE 18-5 Seafood stalls at an indoor market in Japan. The Japanese diet relies heavily on fish, and creates the demand that drives an extensive fishing industry.

■ **FIGURE 18-6 Kagawa vineyards.** Owing in part to its scarcity, agricultural land in Japan is worked very intensively.

natural forests still cover 67 percent of the land, an astonishingly high figure for such a heavily industrialized country (Figure 18–7).

On the debit side, Japan is severely lacking in mineral resources and must import nearly everything it needs for energy production (except hydropower) and industrial development. Japan has the most limited natural resource endowment of any of the world's major economic powers, especially when compared to the United States. As a consequence, Japan must shop throughout the globe to obtain the critical raw materials it needs, the most important being crude oil, wood, coal, iron ore, nonferrous metals, and wood pulp (Table 18–1). Adding all food and raw materials together, Japan is one of the largest consumers of such imports in the world today.

Japan's energy resources illustrate especially well the country's dependence and vulnerability. Until the 1950s, the primary energy source was coal. After that, the government tried to steer Japan toward greater reliance on petroleum, which peaked at 73 percent of energy needs in 1975 (Figure 18–8). The oil crises of the 1970s made Japan increasingly aware of the political and economic insecurity of world oil supplies. Since then, the government has been pushing expansion of natural gas and nuclear power; the goal is to bring dependence on oil down to less than 50 percent of energy needs by 2010. Nuclear power already accounts for 30 percent of electric power production. Currently, the Persian Gulf states provide a whopping 85 percent of Japan's oil imports, with Southeast Asia (mainly Indonesia) providing an additional 9 percent. Japan once had high hopes that China could become a major supplier, but that hope has withered as China finds itself increasingly short of oil for its own needs. Indeed, China and Japan are now competing for access to pipelines that will bring oil from the Central Asian states and the Caspian Sea area.

It was Japan's very real shortage of raw materials and limited food production capabilities that contributed to the decision in the late nineteenth century to embark on an imperialist expansion into Asia, which culminated in the catastrophe of World War II. Other factors played a role, of course, but one can understand the sense of vulnerability that led Japan's leaders to take such a tragic and ultimately self-defeating path. Since World War II, Japan has chosen peaceful means to acquire the foods and raw materials it needs from abroad, but still sometimes with controversy, as, for example, in the case of logging operations in tropical forests of Southeast Asia.

■ **FIGURE 18-7 Forest scene.** The Japanese have long protected their natural forests, which continue to cover two-thirds of the country, and which lend great beauty to this highly industrialized nation.

Table 18-1   Origins of Japanese Imports, 2002

	Source Countries (% of imports)						Share of Top 3 (%)
	1		2		3		
Coal	Australia	58.2	China	17.1	Indonesia	10.2	85.5
Crude oil	Saudi Arabia	26.2	U.A.E.	24.3	Iran	12.4	62.9
Liquid natural gas	Indonesia	32.7	Malaysia	20.0	Australia	13.3	66.0
Iron ore	Australia	51.9	Brazil	22.3	India	11.7	85.9
Nonferrous metal ores	Indonesia	28.6	Chile	20.4	Australia	16.0	65.0
Copper ore	Chile	29.8	Indonesia	27.0	Australia	11.8	68.6
Soybeans	U.S.	74.5	Brazil	14.8	Canada	4.5	93.8
Wood	Canada	26.4	U.S.	16.8	Russia	11.4	54.6
Pulp	Canada	36.7	U.S.	32.4	Brazil	9.8	78.9
Wool	Australia	31.6	Taiwan	27.5	Thailand	9.5	68.6
Cotton	U.S.	41.0	Australia	37.0	India	3.9	81.9
Wheat	U.S.	53.6	Canada	26.9	Australia	19.4	99.9
Corn (for feed)	U.S.	96.1	Brazil	1.4	China	1.4	98.9

Source: *Japan 2004: An International Comparison,* Keizai Koho Center.

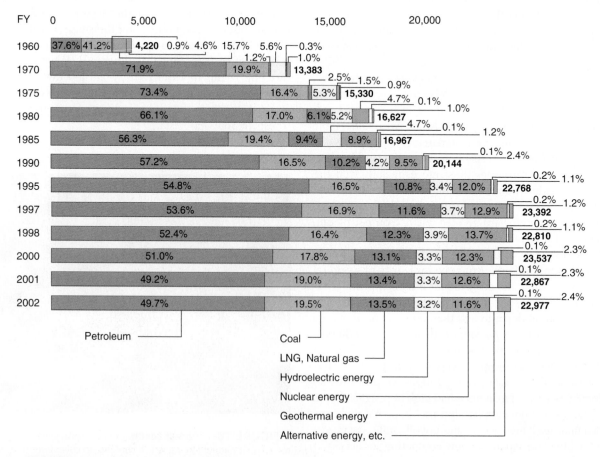

■ **FIGURE 18-8 Primary energy supply.**
Source: Japan Almanac 2001. (Tokyo: Asahi Shimbun, 2000), p. 169, and *Statistical Handbook of Japan, 2005.*

Human Resources: The Hybrid Culture

Much is made of the homogeneity of Japan's people. The Japanese themselves are fond of emphasizing this as one of their attributes. Certainly, the Japanese are a far more homogeneous people than the pluralistic societies of most other countries of the world today. Without question, racial and cultural homogeneity helped to foster a sense of national identity and unity, which was critical to Japan's successful modernization in the past century. Unfortunately, that homogeneity also bred ethnocentrism (a belief that one's race is superior to all others), xenophobia (fear of foreigners), and chauvinism (aggressive patriotism)—faults certainly not unique to Japan but which contributed to Japan's difficulties in the twentieth century, including the harsh occupation of much of Asia up to 1945. These are still important factors in Japanese society today.

Japan's homogeneity, however, is far from complete and has been achieved by forcing conformity on all who deviate from socially accepted norms. Japan does, in fact, have ethnic and cultural minorities who generally have not fared well within Japan (see the *Geography in Action* boxed feature, The Outsiders: Minorities in Japan), and efforts by these groups to achieve justice have been an important trend in postwar Japan.

The distinctive linguistic and physical identity of the Japanese emerged 2,000 years ago, when China was already a flourishing civilization some 2,000 years old. In physical form, the Japanese are the product of the blending of various ethnic groups from different parts of Asia over a long period of time (Figure 18–9), the principal source being the Mongoloid people, the major Asian racial stock, who migrated from the mainland via Korea starting with the Han dynasty in China (206 B.C. to A.D. 220). The main settlement patterns first emerged in the western parts of Honshu and Kyushu and gradually spread eastward to encompass what

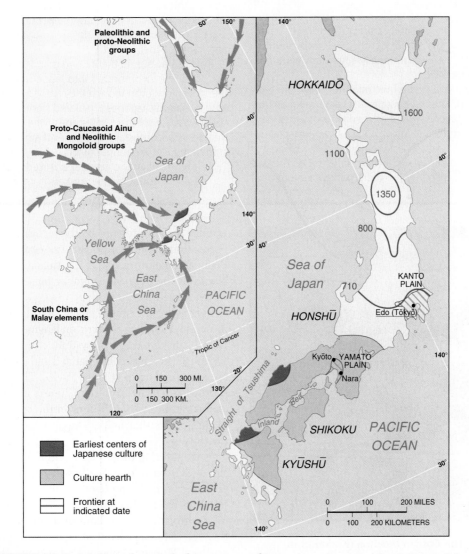

▪ **FIGURE 18-9 Origin and spread of Japanese culture.** Many of Japan's cultural characteristics originated in China, but Japan's insular location ultimately fostered the development of a distinctive culture. The Japanese culture hearth began to form along the Sea of Japan and the Strait of Tsushima; it then spread to the Inland Sea area, especially the Yamato Plain, and eventually northward to the Kanto Plain (Tokyo).

GEOGRAPHY IN ACTION
The Outsiders: Minorities in Japan

Japan's human resources and culture are not as homogeneous as popular perception would suggest, for ethnic and social minorities make up approximately 4 percent of the population, or roughly 5 million people in all. The ethnic minorities consist of Koreans, Chinese, Okinawans, Ainu, and foreign residents. The social minorities are composed of the *burakumin,* persons with disabilities, and children of interracial ancestry.

Historically, the cultural homogeneity of Japan has been obtained through policies and practices that limit opportunities of persons who do not belong to the cultural and social mainstream. The Japanese regard themselves as a unique people, completely different from all other races. They sometimes refer to themselves as the **Yamato people,** in reference to the Yamato Plain around Kyoto, where the Japanese culture and state developed in centuries past and from which the ancestry of the imperial family is derived. There is a strong current in Japanese society to preserve the purity of the majority, or Yamato, strain; anyone else is automatically an outsider and can never hope to be fully accepted into the mainstream. The 1947 constitution expressly prohibits discrimination in political, economic, or social relations because of race, creed, sex, social status, or family origin. But that U.S.-imposed provision has been unable to fundamentally alter many centuries-old attitudes and practices.

The Koreans are the largest ethnic minority, numbering officially more than 630,000. The Koreans first came to Japan during the colonial occupation (1910–1945), when thousands were forcibly brought or enticed to move to Japan as low-cost laborers. By the end of World War II, some 2.5 million Koreans were living in Japan. Those who chose to remain after the war were deprived of citizenship when the Japanese government declared them aliens in the 1952 peace treaty with the United States. Birth in Japan does not guarantee citizenship even today, and the government makes it very difficult for Koreans to obtain citizenship, although most of them have Japanese names, speak the language fluently, and have attempted to integrate into Japanese society. The Koreans remain mired at the lower end of the economic ladder, victims of social and economic discrimination, and tend to live in ghettos in the larger cities.

The Chinese number more than 270,000 today, about one-fifth of whom are descendants of residents who were in Japan before World War II. The remainder are more recent temporary residents, guest workers or migrant laborers. The Chinese tend to be especially visible in "Chinatowns" in various cities, the largest being in Yokohama. The Chinese fare somewhat better in Japanese society than the Koreans, although they still are discriminated against.

The Ainu were among the earliest inhabitants of Japan. Racially different, they also were treated as aliens by the Yamato Japanese. The Ainu have been reduced to only about 25,000 pureblood persons, mostly in a few locations in Hokkaido. The Ainu have been heavily assimilated into Japanese culture, but a few, like Native Americans, struggle to maintain some of their heritage.

The Okinawans, in the Ryukyu Islands south of Kyushu, were not politically incorporated into Japan until early in the seventeenth century, even though they are of basically the same stock as the majority Japanese. Isolated from the main islands and speaking a variant form of Japanese, the Okinawans have been treated as second-class citizens ever since their incorporation, something like Japanese "hillbillies."

Foreign residents consist of a mixed bag. The majority are Asians. Aside from Chinese workers from the People's Republic

■ FIGURE 18-10 The Golden Pavilion (Kinkaku) in Kyoto. This structure is one of the most famous and treasured relics in Japan, illustrative of the influence of Chinese culture on Japan in previous centuries.

is today the core region. By the eighth century, the focus of political and cultural power was centered in the Yamato Plain, at the eastern end of the Inland Sea. For a short time Nara was the political capital of Japan, followed by Kyoto, which served for more than 1,000 years. Japan continued to borrow from Chinese civilization those features it found desirable, including the written language, Confucian values, Buddhism, the emperor system, city design, architecture, art, music, and more, all changed and adapted to suit Japanese needs and tastes (Figure 18–10). The Japanese language, borrowing heavily from the Chinese in the form of written characters but very different in the spoken form, was one of the factors that helped Japan preserve its cultural distinctiveness.

Encountering the cooler environment of northern Honshu and Hokkaido, as well as the native **Ainu,** a minority people who had long lived in that part of the islands, the Japanese tended to concentrate their efforts south of about 37° north latitude, the limit of double-cropping of rice. To this day, only 12 percent of Japan's total population lives in northern Honshu (Tohoku District) and Hokkaido. During the **Tokugawa period** (1615–1868), the last feudal or premodern era of Japan, the focus of power shifted further east-of China, the largest group is from the Philippines, with smaller numbers from several other countries, such as Thailand, Indonesia, and Vietnam. They have all been drawn to Japan for the same reason—economic opportunity, found in low-wage jobs at the bottom of the ladder, which ordinary Japanese do not want to touch. These migrant workers typically come from densely populated, poorer nations that rely on repatriation of wages earned by their emigrants living in Japan. Another group of foreign residents is difficult to classify. These are the **Nikkei**, foreigners of Japanese ancestry. There are about half a million in Japan now, mostly from Brazil, Peru, and other parts of South America, former emigrants to that continent in the twentieth century who have returned "home" to Japan to seek work. They look Japanese but often cannot speak Japanese and have Latin American cultural outlooks. Hence, they are very much outsiders.

The **burakumin** are the largest and most abused social minority (Figure A). They are physically indistinguishable from other Japanese because they have the same racial and cultural origins. However, somewhat like India's untouchables, Japan's *burakumin* have been discriminated against for centuries because of their past association with the slaughtering of animals and similar occupations. In Buddhism and Shintoism, the two major religions of Japan, those activities are regarded as polluting and defiling. Thus, a subclass of Japanese was branded forever as unfit for association with the "pure" majority. That discrimination was formalized and legalized during the Tokugawa period and is still entrenched in Japanese society. Most of the *burakumin,* who number some 2–3 million, live in ghettos scattered throughout the country. Denied access to better-paying jobs, housing, and other benefits, they eke out a living at the bottom of the socioeconomic ladder. Many *burakumin* try to hide their origin and quietly integrate into the mainstream of society, but they are usually found out when background checks are made for marriage or employment.

■ **FIGURE A** *Burakumin. Burakumin* occupy the lowest socioeconomic levels of Japanese society and, despite official government policy, continue to be the object of widespread discrimination.

Resistance by minorities through such actions as lawsuits, sometimes with the support of progressive Japanese individuals and organizations, is beginning to crack the system and bring a certain amount of improvement. The younger generation seems more tolerant and open-minded. However, most observers believe that this kind of discrimination is not likely to disappear soon, given the context of Japanese society. Yet how can Japan overcome the threat of depopulation without coming to terms with pluralism in Japanese society? This is a critical question for the nation in the twenty-first century.

---

ward to Tokyo (then called *Edo*) in the Kanto plain. Today Japan's core area remains in a belt between approximately 34° and 36° north latitude, extending from Tokyo on the east to Shimonoseki at the western end of Honshu and encompassing the Inland Sea. The major share of Japan's population, cities, industry, and modern economy is still concentrated in that core region (Figure 18–11).

The Japan Model

Japan's transformation from a small feudal Asian state into the economic giant of today began in 1868 with the **Meiji Restoration**. This was a process of redesigning Japan from top to bottom by a group of young Japanese responding to challenges posed by the forced opening of the country by the United States, with the arrival of Commodore Perry's fleet off the coast of Japan in 1853. From that date onward, the United States and Japan forged one of the most important bilateral relationships in the world. Now, after 150 years of learning to deal with each other, the two countries have an interdependency that cannot be broken. During those 150 years, Japanese leaders dealt with successive challenges

that forced Japan to constantly change and adapt to new ideas and new ways of national development, as Japan struggled to become and then remain a major power. Out of that struggle emerged what could be called the **Japan Model**, a unique adaptation of Western methods with indigenous (Japanese) culture and values. This model can be summarized as follows (not in any order of priority):

1. Government guidance, not control
2. Competent bureaucracy
3. Proper sequencing of the development process
4. Focus on comparative advantage and regional specialization
5. Wise investment of surplus capital
6. Development of infrastructure
7. Emphasis on education and upgrading of the labor force
8. Population planning
9. Long-range perspective

Other factors were involved, of course, but these nine certainly have been vital to the success of Japan. This model

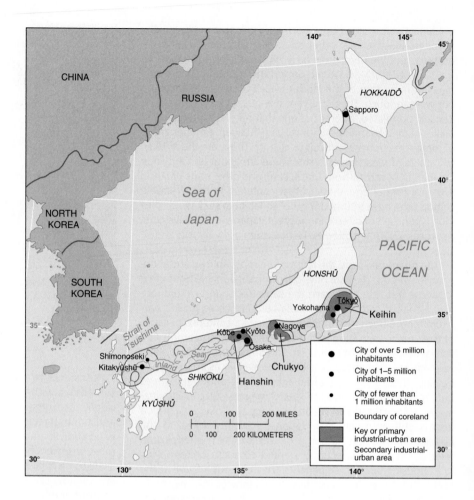

■ **FIGURE 18-11 Japan's coreland and selected cities.** Japan's coreland, stretching from Tokyo to Kitakyushu, contains the vast majority of Japan's cities, population, and modern economy.

was a powerful force in the twentieth century as other Asian states attempted to emulate Japan's success, especially after Western colonialism diminished in the region in the postwar era. The newly industrializing economies of South Korea, Taiwan, Hong Kong, and Singapore, followed later by China and others, have adopted many of these practices or policies developed by Japan, albeit with adaptations to fit each country's unique conditions. Some of these factors are derived from the Confucian tradition that permeates many of Asia's cultures. Hence, today it might be more appropriate to label the above as the **East Asian Model**. Regardless, Japan now finds its system no longer working very well, under rapidly changing circumstances, and is trying to reinvent itself once again.

To understand the model, it is important to note that Japan is now in the third of a series of transformations that began with the Meiji era. The first transformation consisted of modernization and industrialization through the Meiji Period (1868–1914), with a shift toward heavy industry, foreign imperialism, and militarization that reached its peak with World War II. The second transformation was the reconstruction and return to international power in the postwar period. The third transformation, now in its formative stage, involves the search for new directions in a world of increasing competition, economic stagnation at home, and discontent with past strategies. Through the evolution of these three transformations, one can see the various com-

ponents of the Japan Model as it took shape over a long period of time.

Japan's First Transformation: Rise to Power

A modern Japan was actually born in the Meiji Period when a coalition was formed between government bureaucracy, banks, and big business that provided the leadership and policies to steer Japan through its transformation into a world economic power. **Zaibatsu**, large industrial and financial cliques that provided an effective means of marshaling private capital for investment, came into being. The *zaibatsu* worked through vertical and horizontal integration of the economy; thus, a single *zaibatsu* might have control of an entire operation, from obtaining raw materials to retailing the final product. The nearest equivalent in the West is a giant conglomerate or multinational corporation. By the 1920s, the *zaibatsu*—particularly the big three of Mitsui, Mitsubishi, and Sumitomo—controlled a large part of the nation's economic power. The *zaibatsu* were efficient and provided the entrepreneurial strength that led to modernization.

The Japan Model involved a pragmatic approach to development, characterized by efficient government and bureaucracy, a sound currency and banking system, development of education, and effective harnessing of the abilities of the Japanese people, who responded favorably to the new opportunities. Unfortunately, in the pre-World War II era, democracy and human rights took somewhat of a back seat in the process.

Japan passed from the traditional society stage in the Rostow model to preconditions for takeoff at the beginning of the Meiji period. A critical aspect was the emphasis first on developing the traditional areas of the economy: agriculture, commerce, and home or cottage industry. The modern industries that were deemed important by the government were those on which military power depended, because it was to be military power that would enable Japan to achieve equality with the Western powers and avoid the disaster unfolding in China under colonial intrusion. Hence the government led the way in developing shipbuilding, munitions, iron and steel, and communications. The first railroad was built between Tokyo and Yokohama in 1872. At the same time, as the need for importing raw materials grew, export industries were encouraged, particularly silk making (which the Japanese learned from China and eventually came to dominate) and textiles.

At the end of the nineteenth century, the country still had a small industrial base, but it was sufficient to boost Japan into economic take-off after the Russo-Japanese War of 1905 and lasted until the start of World War I in 1914. Japanese manufacturing production increased enormously between 1900 and the late 1930s, particularly for inexpensive light industrial and consumer goods. Foreign markets, while important, were secondary to the growing domestic market within Japan.

Two decisive events that shaped the course of Japan's development were the victories over China in 1895 and Russia in 1905. Those military successes encouraged Japan to pursue a course of territorial expansion, motivated by the need for secure sources of raw materials and the desire to be accepted as an equal by the world powers. Japan thereby lost the chance to be recognized as the political and moral leader of the Asian nations struggling to throw off the bonds of colonialism and imperialism. Instead, Taiwan, in 1895, and Korea, in 1911, became the first pieces of a growing, but short-lived, Japanese Empire that stretched eventually all the way to Southeast Asia and halfway across the Pacific by the early 1940s. By the end of World War I, Japan was a fully recognized—and increasingly feared—world power.

A significant change that accompanied the modernization of Japan after 1868 was an upsurge in population growth. By late Tokugawa times, Japan had a relatively stable population of about 27 million. Between 1868 and 1940, however, Japan provided a classic illustration of the interaction of economic and demographic factors, as industrialization and urbanization proceeded. Population increased to just over 73 million by 1940, as Japan went through the second stage of the demographic transformation.

At the same time, the Japanese people gradually shifted from reliance on agriculture and fishing to industry and services, with somewhat less than half the population living in cities by the start of World War II. Much of the agricultural land that is now farmed in northern Honshu and Hokkaido—including the terraced hillsides so common in those regions—was brought into cultivation during the period after World War I. In spite of significant advances in agricultural technology and much higher unit-area yields, Japan needed the food production of its colonies in Taiwan and Korea to help sustain the homeland. In addition, many Japanese functioned as tenant farmers, not owning the land they worked. Thus, the gap between urban and rural levels of living tended to increase, causing rural people to seek part-time employment in industrial activities, a trend that continues to this day.

From World War I to the late 1930s, Japan went through its drive-to-maturity stage of development. The transfer of workers from agriculture to industry, hastened by the depression of the 1930s, kept industrial wages down. Textile and food-processing industries gradually gave way to heavy industry, especially as Japan further militarized in the 1930s.

The war years, 1937–1945, saw major reversals even as the empire expanded. Personal consumption declined in the face of increased austerity measures, chronic shortages, and eventually the destruction of Japan's cities and economy. Japan's extreme vulnerability—its lack of domestic raw materials—contributed greatly to its defeat by the United States. Japan's attack on Pearl Harbor demonstrated not only its profound lack of real understanding of the United States, but also how insular and inward-looking the nation still was.

Japan's Second Transformation: The Quest to Be Number One

When Japan surrendered in August 1945, the nation was prostrate. Destruction from the war had been catastrophic, especially in urban/industrial areas. The bombing of Hiroshima and Nagasaki had turned Japan into a test case for the era of modern warfare. Stripped of its empire, the nation was shrunk back to just the main archipelago (the Ryukyu Islands were not returned until the 1970s). The future of the Japanese people seemed bleak. The United States, in the initial phase of its occupation, had the intention of not allowing Japan to reindustrialize.

Yet within a decade Japan was free of foreign occupation and well on the road to recovery, going through what amounted to a speeded-up repeat of the Rostow model, with takeoff in the 1950s, drive to maturity in the 1960s, and high mass consumption from the 1970s onward. The postwar recovery is attributable in substantial part to the resilient fiber of the Japanese people. War does not destroy the inherent qualities of a strong society, and Japan, like Germany at the same time, still had its greatest resource—that of a well-educated, technically proficient people. Its able administrators and entrepreneurs were eager to seize the reins and rebuild the nation, just as fast as U.S. authorities would allow. Japan was also greatly helped by the expansion of communism in Europe in the late 1940s and the communist victory in China in 1949. Fearing further communist expansion, the United States, in effect, reversed its occupation policy and decided that a reindustrialized, strong Japan could be a key ally. Thus, independence was granted to Japan in a peace treaty in 1952, much earlier than might otherwise have happened.

The Role of the United States

Japan would not be where it is today without the vital role played by the United States at a critical juncture in Japanese history. American assistance took several forms. Financial aid was especially needed immediately after the war. The Korean War (1950–1953) was a blessing in disguise because Japan was used as a procurement site for war material; thus, billions were funneled into Japan's reindustrialization to fight the Korean War. A similar pattern occurred during the Vietnam War (1965–1975). To help Japanese exports, the United States gave Japan open access to the huge American consumer market and sold American technology at bargain prices to Japanese companies. One of the most famous cases was semiconductors, which spawned a revolution in consumer electronics and propelled Japan to world dominance in that industry. Since the 1960s, the United States has absorbed 25–30 percent of Japan's total exports annually. Japan was also spared most of the cost of defending itself by being placed under the U.S. nuclear umbrella, although Japan does pay a major share of the cost of maintaining American forces and bases in Japan. The United States imposed a U.S.-designed constitution on Japan that still is in effect and brought great changes to government, politics, and society. That constitution also forbids Japan from redeveloping an offensive military capability, permitting only "Self-Defense Forces." Japan thus could divert virtually all its energies into peaceful economic growth. In effect, the United States bought a loyal ally but unwittingly created a formidable future competitor, as Japan reasserted itself, reborn in a new postwar form.

The Japan Model Reinvented

Many of the features of the Japan Model reemerged after World War II as Japan was helped to reindustrialize; in addition, new features were added that reflected the changed conditions of the postwar era. For one thing, close cooperation between government and business grew. That cooperation was particularly important in financing the modernization process. Banking credit was backed by the government, which made heavy capital investment possible. The Japanese economy became geared to a high growth rate strategy that relied on large increases in productivity to provide the surpluses to pay back capital debts. High rates of personal savings gave the banks more money to loan. Japanese firms were able to concentrate on long-range strategies and research and to be less concerned about short-term, even quarterly, profits or comparisons (as is typical in American business). In addition, the breakup of the *zaibatsu* during the Occupation was unsuccessful, as they returned in a new form, known as the **keiretsu**, after the Americans left. These *keiretsu* played a critical role in Japan's dramatic postwar growth. The strategy, espoused by Japan's leaders, was to make Japan No. 1 in the world in economic strength and production, and it was the seemingly broad-based support of the Japanese people for this ultimate goal that gave rise to the stereotyping of the country as a kind of all-encompassing national corporation, dubbed "Japan, Inc."

Economic growth was indeed impressive, increasing at an average rate of almost 9 percent per year in the 1950s, more than 11 percent in the 1960s, and more than 10 percent in the 1970s. The maturing of the economy by the 1980s caused the growth rate to slow to about 4 percent per annum, but that was still higher than in most industrialized countries. In fact, Japan passed the United States in per capita income (on an exchange rate basis) by the late 1980s, which was in many ways the best decade for postwar Japan.

The Japanese government guided the economy, yet the system was far from a centrally planned, command-type economy, as in communist countries. The Japanese seemed to have found a happy middle ground between the two extremes of communism and laissez-faire capitalism that was highly successful, at least in the Japanese context. Through such key government organs as the Ministry of International Trade and Industry (MITI), since renamed the **Ministry of Economy, Trade, and Industry (METI)**, growth industries were targeted and then supported with generous assistance of many kinds, to maximize the country's comparative advantage. Textiles and food processing, which had dominated until the war years, shrank rapidly in importance after 1950. At that time, the government stimulus shifted to iron and steel, petrochemicals, machinery, automobiles and other transportation equipment, precision tools, and consumer electronics. Automobiles and electronics are probably the products most familiar to consumers around the world (see the *Geography in Action* boxed feature, From Dattogo to Toyota). Heavy industry's share of total industrial production passed the 50 percent mark around 1965. In iron and steel alone, Japan surpassed France, Great Britain, and West Germany in the early 1960s to become the world's third largest steel producer. It moved ahead of the United States by the late 1970s and, following the collapse of the Soviet Union in 1991, became the world's largest producer until China caught up and surpassed Japan in the late 1990s. Now Japan is second only to China (Figure 18–12).

During the 1950s and 1960s, especially, Japan followed a protectionist policy by raising tariff barriers against foreign products. Foreign investment in Japan was also restricted. The rationale for that policy was that Japan's economy was too weak to withstand uncontrolled imports and foreign investment. By the mid-1970s, however, the barriers began to fall, and since then Japan has become much more open economically. Much of the frustration of foreign companies trying to do business in Japan has stemmed from their difficulty in penetrating the complex marketing system and their ignorance of the Japanese and their culture.

Another feature of Japan's economic system has long been its double structure, which consists of a pyramid of a relatively small number of modern, giant companies at the top; a greater number of medium-sized firms in the middle; and thousands of tiny workshops and family establishments at the bottom. That structure had fully emerged by the 1930s. Large firms tend to dominate such industries as transportation equipment, electrical machinery, steel, precision machinery, and chemicals, where economies of scale are

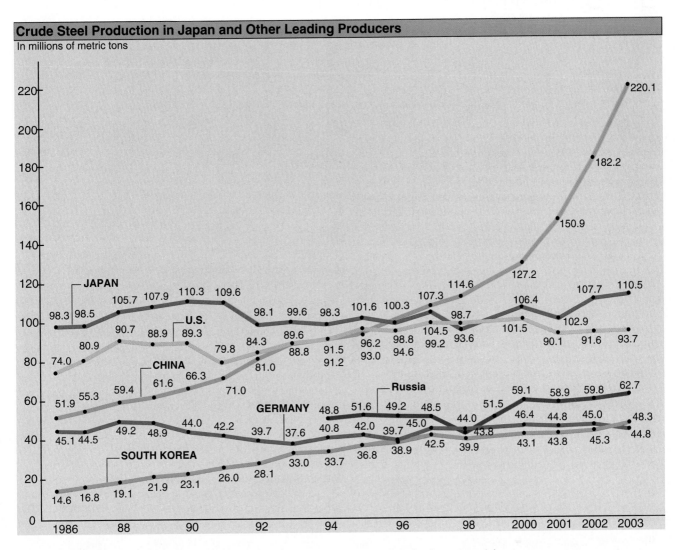

Crude Steel Production in Japan and Other Leading Producers

In millions of metric tons

■ **FIGURE 18-12 Top six steel-producing countries.** Since the 1960s, Japan has been one of the world's leading producers of steel.

Source: *Japan 2001: An International Comparison* (Keizai Koho Center), p. 45, and International Iron and Steel Institute website.

needed; the smaller firms, conversely, are concentrated in consumer goods areas such as leather products, textiles and apparel, and foodstuffs. The larger companies contract out substantial parts of their production to smaller firms. The larger companies also tend to have the famous lifetime employment system developed in the postwar years as a way of placating labor unions. This paternalistic relationship between employer and employees has been much envied and studied by other countries, but is now proving increasingly difficult for many companies to sustain.

Population Stabilization

Another aspect of Japan's model was stabilizing population growth. In 1945, Japan faced the specter of becoming so densely populated that the quality of life might seriously deteriorate without a corresponding increase in

economic output. To control population growth, especially in the grim years immediately after the war, in 1948 the government passed the Eugenics Protection Law, which legalized abortion. In addition, efforts were made to disseminate information on birth control practices, and public education stressed the raising of small families. Growing levels of urbanization also contributed to a steady decline in the rate of population growth, a decline made more absolute by the nearly total absence of legal immigration into Japan.

Japan currently has an annual growth rate of 0.2 percent, one of the lowest in the world, and is near the end of stage three in the demographic transformation. Thus, as would be expected, the age structure of Japan's population has changed significantly since 1950, with a sharp decline in the youthful age group and a large increase in the over-65 age group (Figure 18–13, page 411).

GEOGRAPHY IN ACTION
From Dattogo to Toyota

No product more visibly typifies the extraordinary success of Japan's postwar transformation than the automobile. Probably few people have heard of the Dattogo, Japan's first all-Japanese car from the 1920s. Most people have heard of Toyota, Japan's "General Motors" or largest corporation and auto producer. In just about 20 years, from 1960 to 1980, Japan emerged as the world's second largest producer of motor vehicles, and for a time in the late 1980s and early 1990s it soared to first place, ahead of even the United States. No other industry more effectively illustrates the principles of comparative advantage and international division of labor.

Japan began experimenting with cars around 1902, about the same time as Henry Ford and others in the United States. The industry had a slow start in Japan, however, because of an underdeveloped machine-building industry, a small domestic market, and competition from the United States. For a short while in the 1920s and early 1930s, Ford and General Motors even had subsidiary plants in Japan to assemble trucks and passenger cars, using imported American parts! All this stopped abruptly with World War II. After the war, truck production resumed first because of procurements for the Korean War.

Then, in the 1950s, the Japanese government targeted automobiles (and motor vehicles in general) as an industry with a huge growth potential and, assisted by the classic collaboration between government and business, the motor vehicle industry took off. The government's income-doubling plan, which aimed to make 1970 per capita income twice that of 1960, was hugely successful and led to a surge in domestic demand for autos. Foreign markets also grew rapidly, especially in the United States and Europe. By 1970, Japanese car manufacturers shrewdly saw the potential for small, well-designed, fuel-efficient cars, which proved a boon in the oil crises of that decade. An industrial legend thus was born. By the 1990s, a mystique had developed around Japanese vehicles, noted for their high quality in relation to price, even though the competition had spurred U.S. auto manufacturers to turn their declining business around and become much more competitive.

The growth in Japanese production was phenomenal, as shown in Figure A. Production of motorcycles (not shown) peaked at more than 7.4 million in 1981, because of competition from Taiwan and other lower-cost producers, but is now down to less than 2 million. The "Big Five" (Toyota, Honda, Nissan, Mitsubishi, and Mazda, in that order) account for about 80 percent of the 10 million or so vehicles produced each year (Figure B). Peak production was 13.5 million in 1990. Japan's lead over the United States as the number one producer disappeared, with Japan slipping back to second place by the late 1990s as Japan's economy stagnated, but Japan is still far ahead of number three Germany (Figure C). In total exports, however, Japan is still by far the leader worldwide, with nearly 5 million vehicles sold abroad in 2004, about half of total production. Motor vehicles constitute Japan's largest export product, accounting for 16 percent of total exports (although down from about 18 percent in the 1980s). Thus, the motor vehicle industry has acquired the same vital role in Japan's economy that it has in the United States. In fact, about 5 million workers in Japan, more than 8 percent of the workforce, are engaged in some aspect of the industry. Most of these work in small and medium-sized firms that subcontract with the big auto producers. It is estimated that 70 percent of the cost of making a car consists of these subcontracts.

■ **FIGURE A** **Japanese motor vehicle production.** Japan's motor vehicle production peaked in 1990 but continues to be one of the world's highest.
Source: JAMA (Japanese Automobile Manufacturers Association), 2005: The Motor Industry of Japan (www.jama.org)

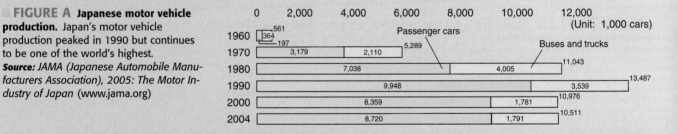

| | 0 | 2,000 | 4,000 | 6,000 | 8,000 | 10,000 | 12,000 (Unit: 1,000 cars) |

Passenger cars — Buses and trucks

Year	Passenger cars	Buses and trucks	Total
1960	561 / 364	197	
1970	3,179	2,110	5,289
1980	7,038	4,005	11,043
1990	9,948	3,539	13,487
2000	8,359	1,781	10,976
2004	8,720	1,791	10,511

☐ **FIGURE B** **The Mazda automobile plant in Hiroshima.** This factory is typical of many built on reclaimed shoreline in postwar Japan.

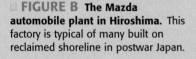

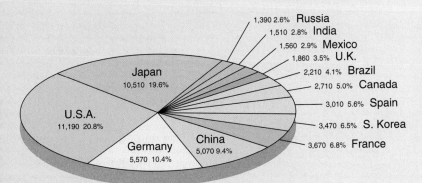

■ **FIGURE C** **World car production, 2004**
(Numbers in thousands of vehicles). By 2000, Japan had slipped back to second place in motor vehicle production (the United States produced 12,778,000 vehicles to Japan's 10,145,000); the gap narrowed significantly by 2004.
Source: *JAMA (Japanese Automobile Manufacturers Association), 2005: The Motor Industry of Japan* (www.jama.org)

The industry faces ceaseless competition from producers in other countries. Newer and lower-cost producers, such as South Korea, are making some of the smaller Japanese autos increasingly uncompetitive. The rebound of the U.S. auto industry, together with the rise in the value of the yen in recent years, caused a turnaround in the volume of Japanese cars sold in the United States, from a peak market share of about 30 percent by the late 1980s to about 25 percent in recent years. The fears of U.S. auto workers in the 1980s that Japan would take over the American domestic market proved exaggerated.

The basic problem facing the entire world auto industry is that there is a huge surplus production capacity, with more than 55 million vehicles produced worldwide each year. Automakers have been forced to merge or go out of business, and thus today six big groups dominate the world industry: General Motors; Ford; Daimler-Chrysler/Mitsubishi; Toyota; Renault/Nissan; and Volkswagen. The trend is toward greater integration of national car production, making motor vehicles truly international products (Figure D). The main reason for the overcapacity is that auto production is a very attractive industry for many countries. It can reduce expenditures that would otherwise go to purchase foreign-made vehicles while simultaneously generating much-needed export earnings, and there are strong positive multiplier effects on other sectors of the economy.

One has only to look at China, for example, to see this bandwagon effect. Currently, only 1 in 100 Chinese has a car. Automakers dream of this figure someday becoming 1 in 2 of China's 1.3 billion people. China has already moved into fourth place and is likely to soon replace Germany as the world's third largest producer. With relatively stagnant domestic demand for cars in Japan, automakers there see no choice but to expand production and sales in other countries. Production of Japanese cars in other countries is now much greater than production in Japan itself (Figure E).

Part of this overseas investment and production stems from sensitivity to Japan's auto exports, especially in the United States and Europe. To offset the threat of protectionist measures, as well as cut costs, Japan began opening assembly plants overseas in the late 1980s. Figure F shows the plants in the United States. The major concentration is in the Midwest, along what is called "Toyota Road," Interstate 64, running from West Virginia to Missouri. The popular Camry, for example, is produced in Kentucky. Another concentration is in California.

Japan is likely to remain at or near the top of world auto production—certainly in the category of exports—for the foreseeable future. But sustaining the country's prominent role will require the utmost attention to styling, quality, and price. Some believe that the motor vehicle industry's ability to continue operating in Japan will be a test of how the nation withstands the forces of deindustrialization. Right now, the future for automobiles, at least, looks good.

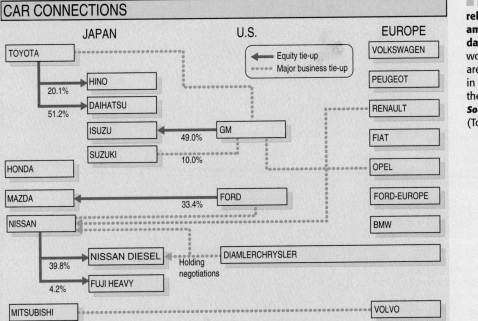

■ **FIGURE D** **Major equity relationships and business ties among the world's carmakers at the dawn of the twenty-first century.** The world's major motor vehicle producers are increasingly linked to one another in the manufacture and marketing of their products.
Source: *Japan Economic Almanac 1999.* (Tokyo: Nikkei Weekly) p. 97.

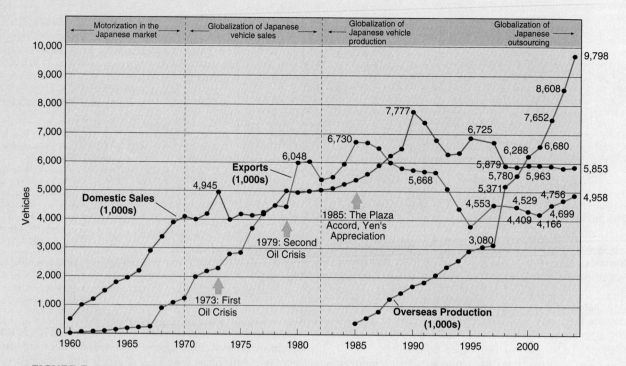

■ **FIGURE E** **Trends in Japan's domestic sales, exports, and overseas production of motor vehicles.**
One of the most dramatic changes in the manufacture of Japanese motor vehicles has been the rapid
increase in overseas production since 1985.

Source: *JAMA (Japanese Automobile Manufacturers Association), 2005: The Motor Industry of Japan*
(www.jama.org)

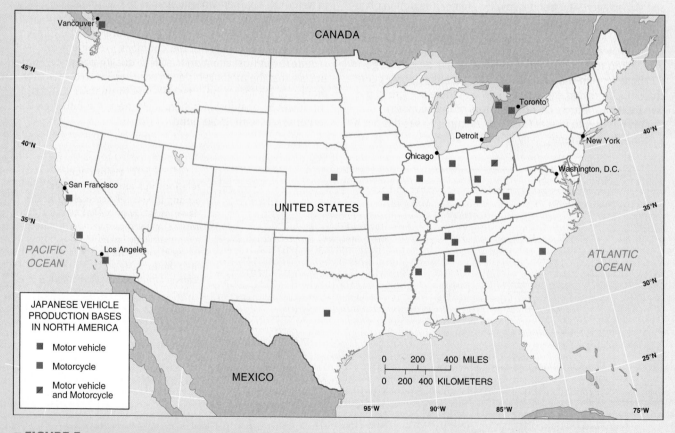

■ **FIGURE F** **Japanese auto manufacturing in the U.S.**
Source: *JAMA (Japanese Automobile Manufacturers Association), 2005: The Motor Industry of Japan*
(www.jama.org)

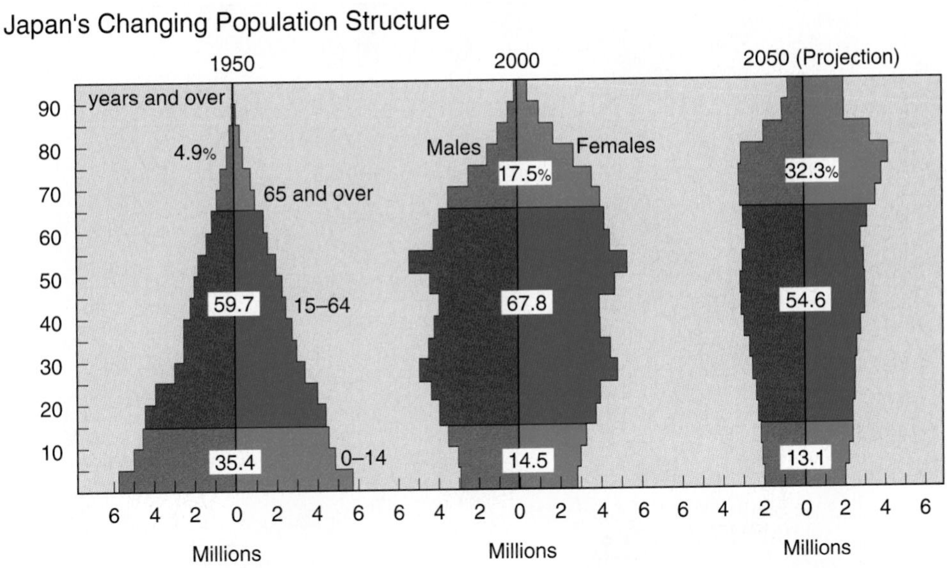

Japan's Changing Population Structure

■ FIGURE 18-13 **Japan's changing population structure.** Japan's population has aged greatly since 1950 and is expected to continue to do so in the decades ahead. **Source:** *Statistical Handbook of Japan 2004* (www.stat.go/jp).

Growth of Cities and Industry

One of the most dramatic developments of the postwar era, and one with profound consequences for Japan, has been the rapid increase in urbanization. Farm population as a percentage of total population declined from 85 percent early in the Meiji period to about 50 percent in 1945 and to less than 3 percent by the end of the twentieth century. Urban population, in turn, increased. As late as 1960, the urban population accounted for less than half the total population; today that figure is 78 percent. Thus, Japan's transformation to a predominantly urbanized nation has taken place relatively recently; by contrast, the United States passed the 50 percent urbanization rate before the 1920s.

A striking characteristic of the urbanization of Japan is the concentration of that population in a small portion of the country, in cities scattered through the core area (Figure 18–14). Altogether, Japan has twelve cities with 1,000,000 or more population, nine with 500,000–1,000,000, and forty-five with 300,000–500,000. The whole country has a total of 664 cities, as officially defined by the government.

The urban concentration is most intense in three huge urban nodes, which are gradually coalescing into what is called the **Tokaido Megalopolis**, named after a road that ran through the region in pre-Meiji times. The three nodes are Tokyo–Yokohama, Nagoya, and Osaka–Kobe–Kyoto, also known as the *Keihin, Chukyo,* and *Hanshin* industrial regions, respectively. (The fourth largest industrial region, Kitakyushu, is outside the Tokaido Megalopolis.) Today 57 million, or about 45 percent of Japan's total population, live in those three nodes. The Tokyo agglomeration alone has more than 34 million (still the largest urban concentration in the world), the Nagoya region has 5 million, and the Osaka area has just under 18 million (Figure 18–15). This latter is also one of the largest megalopolises in the world.

The remaining 70 million Japanese are found predominantly in other cities scattered throughout the core region.

The few significant cities outside the core include Sapporo and Hakodate on Hokkaido; Sendai in northern Honshu; and Toyama, Fukui, and Niigata on the Sea of Japan side of Honshu. There are no major cities in southern Shikoku or southern Kyushu. The growth rate in the central cities of the Tokaido Megalopolis has slowed almost to a standstill. The trend most recently has been for the fastest growth to occur in the suburbs and satellite cities of the major metropolitan centers, a pattern analogous to that of the United States in recent years.

In Japan, as in most developed countries, industrialization provided the major stimulus for urbanization. In Japan's case, however, an important additional factor was the desire of Japan's business and government leaders to concentrate industry, especially heavy industry, in a few areas, most of them near the coast. Concentration was useful to take advantage of economies of scale, and location near the seashore made it cheaper to handle large quantities of imported raw materials, such as iron ore, coal, and oil. Much of the postwar development of industry occurred on reclaimed land built along the shoreline of the Inland Sea and the **Pacific Belt** of the core region (see Figure 18–14), a process dating back as far as the expansion of Edo's shoreline in the Tokugawa era of the seventeenth century. All the major cities, but especially the big three, now have largely artificial, expanded coastlines built up over the decades. The most notable recent addition is the new Kansai International Airport, built on an artificial island just off of Osaka (Figure 18–16).

The pattern for light and small-scale industry is somewhat different. Because access to overseas sources of raw materials and shipping is not as important for them, those types of industry are more widely distributed in both urban and rural areas. Indeed, the incidence of part-time farming in the agricultural sector was made possible, in part, by the distribution of industry in rural areas, where farmers could commute daily between farm home and nonfarm job. As a result, many small towns and nonurban centers around the

■ **FIGURE 18-14 Urban and industrial patterns of Japan.** Much of Japan's industrial activity is concentrated along the southern coast of Honshu, particularly in the three regions of Tokyo–Yokohama, Nagoya, and Osaka–Kobe–Kyoto. Those three conurbations constitute the Tokaido Megalopolis.

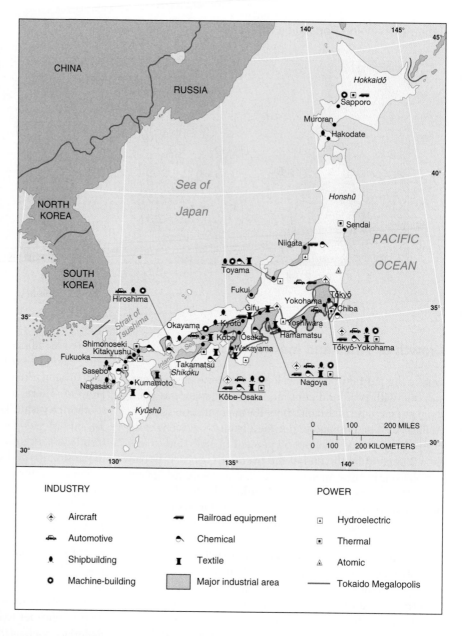

country have industrial plants that form important bases for local economies.

The Consequences of the Japan Model

By the 1970s, Japan began to realize the price it was paying for its development strategy. National attention had been focused on economic growth at the expense of social welfare, the environment, and relations with other nations. The costs of Japan's "economic miracle" have been high.

Urban Challenges

As population became ever more concentrated in the core, a number of urban ills have surfaced or worsened. To accommodate huge urban populations, the Japanese government developed some of the best public transportation

■ **FIGURE 18-15 The Dotambori section of Osaka.** The Dotambori section of Osaka illustrates the modernity and glitter of Japan's second most important city and rival of Tokyo.

■ FIGURE 18-16 **An example of postwar development.** The new Kansai International Airport, just off the coast of Osaka, is one of the latest reclaimed or manmade sites built in recent years, a long-standing practice to provide room for urban/industrial growth.

systems in the world. These complex systems, marvels of modern engineering and design, function with amazing efficiency and punctuality, carrying millions of intra- and interurban commuters daily. Nonetheless, the crush of people during rush hour is incredible. Moreover, since the 1960s Japan has moved aggressively toward an automobile society. Today there are more than 74 million motor vehicles in Japan, a ratio of one for every 1.7 persons. This ratio is similar to that of the United States, yet in Japan there is much less space and fewer roads, in spite of massive freeway and viaduct construction in recent decades. Japan has discovered that the private motor vehicle is a mixed blessing, opening a Pandora's box of problems, including traffic congestion, parking shortages, air pollution, noise, strip development, and urban sprawl.

Aside from the traffic problems, most foreigners are also struck by the seemingly unplanned sprawl of Japan's cities, especially Tokyo, which seem characterized by dense siting of buildings, lack of architectural unity with sometimes bizarre adaptations from the West, and the relative lack of green space and recreation areas. On the positive side, though, Japan's cities are renowned for their low crime rates and general civility of life.

Housing remains one of the most critical urban problems. Escalating land costs through the postwar decades led to housing becoming less and less affordable to middle-class families, especially in Tokyo. Typical housing space has always been less in Japan than in the United States, or even compared with most European countries, averaging just 33 square meters by the end of the 1990s, compared with 38 in Germany or the United Kingdom, and 60 in the United States (Figure 18–17). The cost differential is far greater. The cost of a home in Tokyo is roughly thirteen times the mean annual income, compared with ten times in Osaka, but only three times in New York City. Housing costs in Japan likely will never be comparable to the American average,

although some of the priciest locations in the United States, such as parts of Hawaii and California, are comparable to some of the less-expensive Japanese cities. The sewerage access rate, although increased to 58 percent of all homes by 1999, nonetheless still lagged significantly behind all the major developed countries.

Pollution of the Environment

Another clear consequence of Japan's development path has been intensified environmental pollution. Until the late 1960s, Japan had done less than any other major industrial country to protect its natural environment from the effects of uncontrolled industrial and urban development. Air and water pollution became acute. Beloved Mt. Fuji could rarely be seen

■ FIGURE 18-17 **Japanese family members in their small home.** The average Japanese dwelling is only about half the average size of a home in the United States.

from Tokyo, which was shrouded in a brownish smog most of the time. Fishing was halted in Tokyo Bay. Poisonous chemicals, such as organic mercury and cadmium, got into the food chain in certain localities, producing horrible deformities and suffering, which received international notoriety.

Spurred by growing public protest, by the early 1970s the Japanese government finally recognized the seriousness of the problem and created the national Environment Agency (similar to the Environmental Protection Agency established in the United States at about the same time). Billions of yen have since been invested in environmental cleanup, education, and protection, and the results are beginning to show. Japan, with its immense wealth and human resources, has managed to turn the tide and the country is unquestionably much cleaner today than 30 years ago, but there is still much to be done.

Regional Imbalances

Every country in the world has regional imbalances in its distribution of population and levels of development, but especially sharp contrasts have emerged in Japan. The population and modern economy became increasingly concentrated on the outward, or Pacific, side of the nation, at the expense of the inner side, which borders the Sea of Japan. Hokkaido, northern Honshu, the part of Honshu facing the Sea of Japan, southern Shikoku, and southern Kyushu all lost population as massive movement toward the core took place. This regional imbalance has bedeviled the government for decades, which has tried, with limited success, to alleviate the problem.

Since the 1960s, Japan has developed five National Comprehensive Development Plans, plus other measures, to promote industrialization in a more regionally planned manner. Tunnels and bridges now link all four main islands and are tied in with the *Shinkansen* (high-speed) railway system, aimed at facilitating dispersion of people and industry (Figure 18–18). Various industrial zones have been built outside the core region with government assistance, for the same purpose. There has been some dispersion of industrial production. Comparing 1960 with the late 1990s, the four major industrial areas collectively declined from having 61 percent of Japan's total manufacturing production to 44 percent. Keihin fell from 25 percent to 15 percent, Hanshin from 21 percent to 12 percent, Chukyo increased its share from 11 percent to 14 percent, and Kitakyushu fell from 4 percent to 2.5 percent.

In spite of this trend, over the decades Tokyo became even more the control center for Japan and its modern economy, even while manufacturing was dispersing. Tokyo dominates Japan as the nation's primate city much the way that Paris does in France. Hence, an ongoing issue in Japan is that of decentralizing the Japanese government and moving the national capital, or at least some central government functions, out of Tokyo to other sites. Frequently mentioned are Sendai (in northern Honshu), Nagoya, and Kyoto, among many others. The immense financial cost of relocating the capital, not to mention the political hurdles to overcome, make this option very unlikely for the foreseeable future, at least until Japan's fortunes make a comeback. At the same

■ FIGURE 18-18 **High-speed Japanese railway system.** Japan's *Shinkansen* railway system is among the most efficient in the world and does much to facilitate the movement of goods and people between the major urban centers.

time, it has been recommended that the existing forty-seven prefectures, laid out during the early Meiji era, be consolidated and reduced in number to give them more autonomy and power and hence reduce Tokyo's control. This politically charged initiative also faces a tough uphill road.

Rural Challenges

For political and social reasons, farming has been slow to change in Japan. The farmers' natural conservatism was reinforced during the U.S. occupation by the land reform program that transformed tenant farmers into independent smallholders. Today the average farm size is still a mere 1.6 hectares (about 4 acres); farms are largest on Hokkaido, at more than 16 hectares (40 acres). In many prefectures in the core region, the average size is well under a hectare. No one can make a full-time living from such small plots of land. Thus, through the decades the government has lavishly subsidized farmers with price supports and stiff tariffs on imported food, in exchange for loyal political support of the postwar conservative governments under the Liberal Democratic Party (LDP). Those tariff barriers started to fall in the 1990s under pressure from the United States and other countries, but the wall has not been completely breached. As a consequence, food prices for consumers in Japan remain among the highest in the world.

Ironically, in spite of such subsidies, most Japanese farmers still cannot make a living from farming only and must work part-time in industry or services (Figure 18–19). Proponents of the support system argue that Japan cannot afford to totally

▪ **FIGURE 18-19 Combining of rice in Japan.** Miniature mechanization is common in the form of small tillers, tractors, and threshers. In spite of mechanization and high government subsidies, agriculture remains unprofitable as a full-time occupation.

lose its food-producing capability, which would be the likely outcome if tariff barriers completely disappeared. The root issue is the sense of vulnerability felt by the Japanese people and government. Although Japan could easily afford to buy all the food it needs from abroad, such dependency would be terribly risky in the eyes of many. As it is, Japan's food self-sufficiency ratio has been steadily slipping year by year. Overall self-sufficiency had fallen below 40 percent by the end of the 1990s. In the meantime, the farm population continues to dwindle and become increasingly older as young people leave the farms for the cities. In some respects, agriculture has become almost superfluous to Japan, insofar as its contribution to the national economy is concerned.

Japan's International Relations: Trade, Aid, and Investment

The most visible consequence of the Japan Model to outsiders is in Japan's relations with the rest of the world, through trade, aid, and foreign investment. These activities have been alternately cheered and jeered at by other nations.

The United States, Japan, and Germany are consistently the top three world trading nations by far, with Japan and Germany jockeying for second place behind the United States. The big difference is that Japan has consistently had the highest trade surplus since the 1980s, in spite of the stagnation that set in during the 1990s. The surplus of exports over imports is one of the root problems Japan has had with a number of trading partners, but especially the United States, which still accounts for about half of Japan's trade surplus (Figure 18–20).

The overall trend in recent years, however, has been for Japan's trade and investment to gradually shift more and more to Asia, at the expense of North America and Europe (Figure 18–21). In 2003 the United States bought 25 percent of Japan's exports but supplied only 15 percent of its imports. Europe's shares were 17 percent and 16 percent, respectively. Asia accounted for about 49 percent of exports and 58 percent of imports. China alone accounted for 12 percent of exports and 20 percent of imports. Nonetheless, the United States continues to dominate as the single largest market for Japan's products, and Japan has also been a major investor in American real estate and government securities, especially in the "golden years" of the 1980s. The net result is that both sides have come to depend heavily on each other for their economic well-being (Figure 18–22). Although the United States still has a large trade deficit with Japan, the anti-Japanese rhetoric of the 1970s and 1980s has dissipated significantly, as the perceived "threat" from Japan's economy receded in the 1990s and the U.S. auto industry rebounded to first place. Unhappiness in the European Union over trade and investment relations with Japan also has receded in the last 10 years, although it has not disappeared.

The growing concern with Japan now is in Asia. Japan has a large trade surplus with most of the Asian countries, except China, Australia, and Indonesia, which have large raw material exports that Japan needs. More than just trade imbalances create friction with Japan's Asian neighbors, of course. Bitter memories of World War II Japanese occupation and frustration over Japan's reluctance to make amends for its actions during the war continue to rankle many Asians, and not just the older generation who experienced the war

■ **FIGURE 18-20 Japan's balance of trade.** Since 1980, Japan has generated one of the world's largest national trade surpluses.
Source: *Japan Almanac 2001* (Tokyo: Asahi Shimbun, 2001), p. 113, and *Japan 2001, an International Comparison* (Tokyo: Japan Inst. for Social and Econ. Affairs, 2001) p. 74, and *Statistical Yearbook of Japan 2005.*

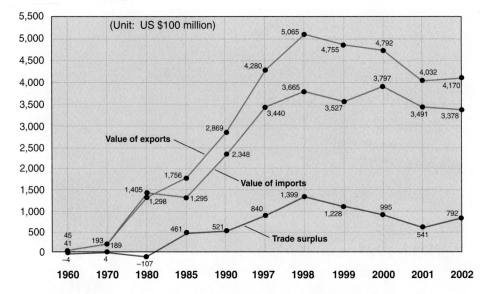

firsthand. Even Japanese investment in Asia, as well as foreign aid (larger as a percentage of economic output than U.S. foreign aid), does not sit well with many. Critics have spoken of the emergence of a neo-**Greater East Asia Co-Prosperity Sphere** (which Japan tried to create militarily in World War II) of economic dependence on Japan for machinery, high-value industrial goods, and capital in exchange for raw materials and low-value consumer goods, and of aid tied too closely to Japanese investment and trade. Such an argument is overly simplistic, of course, because a number of other wealthy states in Asia are growing sources of investment for manufacturing in the rest of Asia and beyond and doing some of the same things Japan has long been doing.

But in the eyes of many Japanese, the growing nemesis is China. As China's economy boomed in the 1980s and 1990s, just as Japan's was weakening, Japan began to view China as a growing rival and potential threat to its economic hegemony in Asia. China and Japan have a natural economic complementarity, given their relative sizes, resource bases, and levels of development, but a very long and troubled relationship. While China has become a major trade partner, Japanese investment in China's booming economy lags behind that of overseas Chinese and Western businesses. China's fast-growing domestic companies, such as in consumer electronics and household appliances, are moving aggressively into international markets long dominated by Japan, including Southeast Asia. China is now

■ **FIGURE 18-21 Japanese exports and imports by area.** Japanese trade extends to every region of the globe.
Source: *Statistical Yearbook of Japan 2005* (website).

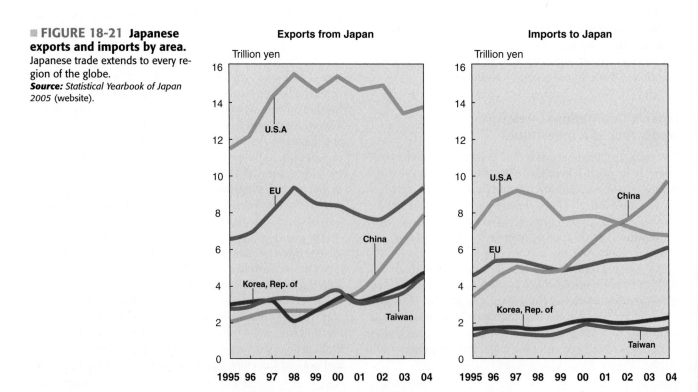

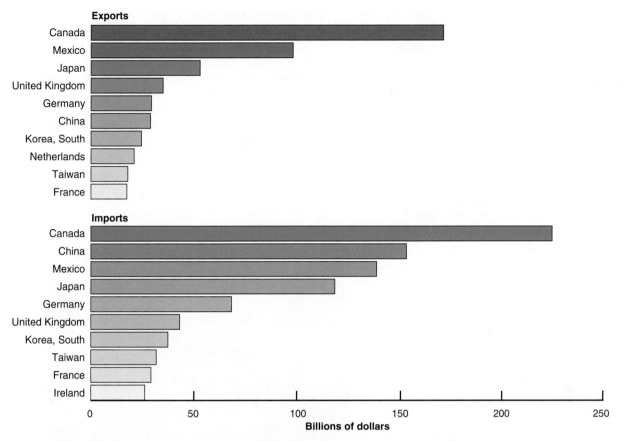

■ **FIGURE 18-22 Top purchasers of U.S. exports and suppliers of U.S. general imports: 2003.**
Source: *Statistical Abstract of the United States.* 2004–2005, Figure 28.2.

second only to Japan in output of electrical and electronic products in the world. China has proposed a Free Trade Agreement with the member states of ASEAN. Diplomatically, China views Japan with some distrust, especially over the perceived threat of renewed Japanese militarism and any hint of Japanese interference in the festering issue of Taiwan, Japan's former colony, for which most Japanese have affection and maintain close economic and cultural ties. The contest thus is on. Japan's dilemma is how to respond to the rise of China and to what some predict as a coming "Chinese century."

Japan's Third Transformation: Charting a New Course

In the first decade of the twenty-first century, internal and external pressures, as well as the natural evolution of the economy and society, are slowly nudging Japan toward new paths of national development, a task made more urgent by the continuing stagnation of the economy. A growing body of literature has emerged trying to explain and offer solutions to Japan's economic malaise, but nothing has worked so far. This is because Japan's economic troubles stem from serious structural defects in the political and economic system. The first of these defects is that the triumvirate of business, bureaucracy, and banks that led Japan's economy to such stunning success in the past may have outlived its usefulness, having bred rampant corruption, cronyism, and sometimes bad decisions behind a wall of inadequate and incorrect public information. Successive weak prime ministers have been unable to change the system, even those who came into office as reformists. A second defect is the banking system, which is in dire straits, having made vast numbers of bad loans over the years, a substantial portion involving the *yakuza*, or Japanese mafia. Some have even called this the "*yakuza* recession." The cost of redressing this problem is dragging the economy downward, as bankruptcies and unemployment increase, while investment in new plants and equipment lags. Third, the government has relied increasingly on massive deficit spending on public works projects to try to jump-start the economy, without success, and the country is now saddled with innumerable white-elephant projects in roads, railroads, bridges, and land reclamation that are not really needed.

Consequently, consumer confidence is at a historic low. Yet although Japan is experiencing its most serious economic difficulties since World War II, the general public still is quite well off and does not sense the full dimensions of the problem. The outside visitor to Japan gets little impression of any crisis, especially if he or she does not know the language. As one Japanese leader put it, "This has almost become a Japanese disease: the slow strangulation of an affluent and egalitarian society." Because the government has high savings, it can continue to issue debt and thus postpone the day of reckoning. Critics argue, however, that nothing short of a drastic shakeup of government, and emergence of a true reformist leader, can reverse this dangerous trend.

Deindustrialization

In terms of specific changes Japan may experience in this current round of reinvention, one of the most significant is **deindustrialization**, a process that is likely to eventually affect all of the more industrialized nations. The hollowing out of Japanese industry began in the 1970s, as the costs of production in Japan escalated. The only way to hold down production costs, besides adopting more advanced technologies, is to seek lower-cost labor sites. By 2000, almost 15 percent of Japanese manufacturing had been shifted to other countries, primarily East and Southeast Asia, North America, and Europe. Some in Japan argue that the country should accept the inevitable and learn to live with the status quo, with reduced personal consumption, as production shifts inexorably overseas. Increasing numbers have begun to question if the current lifestyle, with the extreme emphasis on high mass consumption and all the negative environmental consequences associated with it, is the best that Japan can do. It is a concern raised in many highly developed industrial countries, including the United States, and one that is likely to grow in urgency for Japan and other nations in this century. It is argued that **sunset industries**, those losing their international competitiveness, should be allowed to wither and die. In Japan, these industries include shipbuilding, textiles, some chemicals (especially petrochemicals), and aluminum smelting (Figure 18–23). Other industries that are fully mature but still have strength must retrench, diversify, or shift more operations overseas. The most important of these are iron and steel, electronics, and motor vehicles (see the *Geography in Action* boxed feature, From Dattogo to Toyota). Others argue that Japan's economy is based primarily on manufacturing and the country cannot afford to let too much slip abroad, since the nation does not yet have a strong information technology industry to replace smokestack industry, as does the United States.

An Aging Society and Shrinking Population

Part of the urgency for Japan to adapt to changing global conditions stems from the country's evolving social and demographic patterns. Japan is on the verge of entering stage four of the demographic transformation, with a population of just under 128 million people. After that the population will decline slowly, since the growth rate will be below replacement level. Some projections suggest that the population could sink as low as 67 million by 2100, a decrease of nearly 50 percent. There are only two ways to reverse that trend: boost the birthrate, which would likely

prove extremely difficult in Japan's social context, and allow increased legal immigration. Adoption of the latter policy would have profound implications for Japan's much vaunted "homogeneous" society. As it is, the existing level of legal and illegal residents has been causing anxiety among some Japanese. Currently, there are more than 1.5 million legal foreign residents, led by (in descending order): Koreans, Chinese, Brazilians, and Filipinos. The Koreans and most of the Chinese date from the pre-war colonial era. The Brazilians are mostly some 230,000 *Nikkei,* or foreigners of Japanese descent. The others have come as "temporary" migrant workers in the postwar boom economy. The number of il-

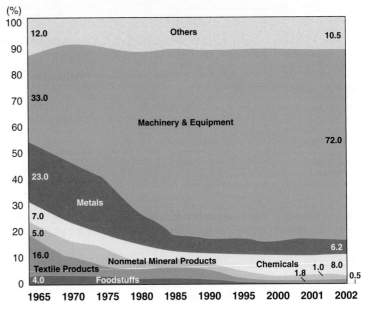

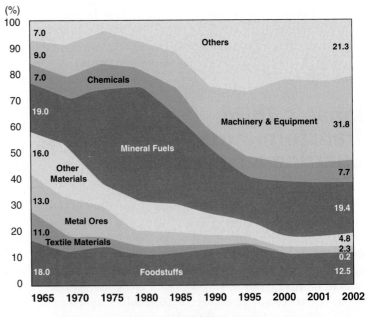

■ **FIGURE 18-23 Japan's major exports and imports.**
Source: Japan 2004: An International Comparison (Tokyo: Japan Institute for Social and Economic Affairs) website.

legal residents is about 250,000 and increasingly consists of Chinese from the People's Republic of China, plus others, who come on tourist or other visas and then overstay. The economic stagnation of recent years has slowed but not stopped the numbers of foreigners who want to get into Japan, legally or illegally, to work.

Aside from the cultural and social problems associated with allowing greater numbers of non-Japanese to live and work in Japan, the economic issue is about the shortage of labor, as the population ages. Already, 18 percent of the population is older than 65, one of the highest rates in the world. That elderly group is projected to reach one-third of the population by 2050. In addition to figuring out how to take care of so many elderly people, Japan also has to find a way to provide adequate labor for the economy and to maintain the entrepreneurial vigor of the population. Complicating the problem have been signs of the younger generation questioning the wisdom of their elders and not wanting to be part of the Japan Model. This generation gap is a growing worry for Japan's leaders, not to mention parents.

All sorts of ideas are being discussed to solve the labor problem, including raising the current mandatory retirement age of 60, encouraging more elderly people to continue working after retiring, and greatly improving the employment opportunities for females (Figure 18–24). Those measures will only go so far, however. Many argue it is inevitable that Japan will have to open its closed doors and internationalize its population much more if the country is to survive. The United Nations estimates Japan would have to admit at least 600,000 immigrants annually to stabilize the population and maintain the workforce at current levels. If the political courage to accept this reality is not forthcoming, or if the domestic birthrate does not increase, Japan will likely experience significant depopulation in the twenty-first century.

▪ **FIGURE 18-24 Elderly Japanese women in productive pursuit.** Twin 107-year-old sisters demonstrate how to make an old-style kimono in a Yokohama kimono shop. They report that they worked at sewing jobs until they were 80 years old.

▶▶ Summary

Japan is an economically wealthy but troubled nation, rich in accomplishments, but very poor in natural resources with which to further develop and grow. Japan produced a spectacular economy, and became a world economic leader, by making the most of its remarkable human resources and by acquiring abroad the raw materials it lacks at home. Although homogeneous by most international standards, the Japanese population has significant ethnic and social minorities whose roles in Japan's national life have been marginal to the mainstream culture, which stresses conformity as well as devotion to national and group goals at the expense of individual aspirations.

Emerging out of a long period of feudalism by the mid-1800s, Japan developed its own unique model of development, which other Asian states have emulated to varying degrees. That model evolved out of three major transformations the country has gone through since the Meiji Restoration of 1868. The first transformation was the initial modernization and industrialization, followed by militarization, that led to empire, world power status, and defeat in 1945. The second transformation was the rebuilding after the war and the attainment of new heights as the world's second most powerful economy, closely allied politically and economically with the United States. The third transformation is just beginning. It is a result of the natural evolution of the economy; the emergence of Asia as a major world power center; a stabilizing and aging population; changing social and economic values in the younger generation; evolving relations with its key international partners, especially the United States; and China's ongoing emergence as a serious economic competitor to Japan. The current transformation is made more urgent by the economic stagnation that has beset Japan since the start of the 1990s, a lasting solution to which has yet to be found.

▶▶ Key Terms

Ainu
burakumin
deindustrialization
East Asian Model
Greater East Asia Co-Prosperity Sphere
Great Inland Sea (*Seto Naikai*)
Great Stagnation
Japan Model
keiretsu
Meiji Restoration
Ministry of Economy, Trade, and Industry (METI)

Nikkei
Pacific Belt
Pacific Rim nations
Ring of Fire
sunset industries
Tokaido Megalopolis
Tokugawa period
yakuza
Yamato people
zaibatsu

Pudong New Area, Shanghai. One of the world's largest seaports, Shanghai, like much of coastal China, has experienced historically unprecedented growth and technological modernization since the late twentieth century. Much of interior China, however, continues to be associated with traditional economies and is increasingly lagging behind the dynamic coastal regions.

Chapter 19

China and Its Pacific Rim Neighbors

- Environmental Diversity
- Spatial Evolution of Chinese Culture
- East Meets West
- Transformation Under Communism
- Transformation of Agricultural Production
- Industry and Regional Economic Growth

- Urbanization and Migration
- Population Contours
- Environmental Sustainability
- Challenges to China's Future
- China's Pacific Rim Neighbors

Anchoring East Asia is the large state of China (People's Republic of China). To the north of China is medium-sized and landlocked Mongolia, to the northeast are the small-sized peninsular states of North Korea and South Korea, and to the southeast, the island state of Taiwan (Republic of China), to which China lays claim politically (Figure 19–1). Chinese civilization stretches back some 4,000 years, and most of its present cultural and economic attributes are of ancient origin. The durability of Chinese-centered culture in East Asia is explained, in part, by its relative geographic isolation throughout history from other centers of global change. Compared to the rest of the non-Western world, for example, East Asia was spatially marginal to the process of European-based colonialism. This is not to suggest, however, that East Asian societies have always been static or unchanging. Change has come, but it has generally been gradual and linked strongly to past beliefs and practices.

Since World War II, however, the peoples of the East Asian nations have experienced social and economic change at an unprecedented pace. During the cold war, South Korea and Taiwan became allied with the United States, while China and North Korea chose to align themselves with the Soviet Union. South Korea and Taiwan have subsequently experienced rapid economic growth by following an export-based industrial development philosophy that has engaged their national economies with the global economy. China has also become a major global political and military power while following socialist economic principles that have been more inward-looking and self-reliant. Over the past 25 years, however, the Beijing government has partially sacrificed its avowed socialist ideology to principles of economic efficiency,

and the country has embarked upon a new economic and social development path. Indeed, China is experiencing a radical transformation as the world's most populous country confronts the forces of capitalist-based globalization. It is important to remember, however, that China has not adopted a form of capitalism common in the West. Beijing has adopted capitalism as the primary engine of economic growth, but within the context of strong government control and planning. China's economy cannot be characterized as being either socialist or capitalist in nature, but as a hybrid of both. The implications of the globalization of China's economy and society are wide-ranging and include changing gender relationships, growing regional economic disparities, increased rural-to-urban migration, and the transformation of agriculture and food habits, to name just a few. With the fastest-growing economy in the world during the 1990s, China's globalizing economy has reconfigured the economies of the Pacific Rim and beyond.

Environmental Diversity

Because the distribution of China's natural environments is similar to those of the United States, a comparison of the two countries helps one to better appreciate China's environmental diversity (Figure 19–2). China and the United States are, respectively, the third and fourth largest countries in the world, with China's area slightly exceeding that of the United States. If only the lower forty-eight states of the United States are considered, the two countries also share a similar longitudinal, or east-west, dimension. The north-south, or latitudinal, dimension exhibits slight differences in

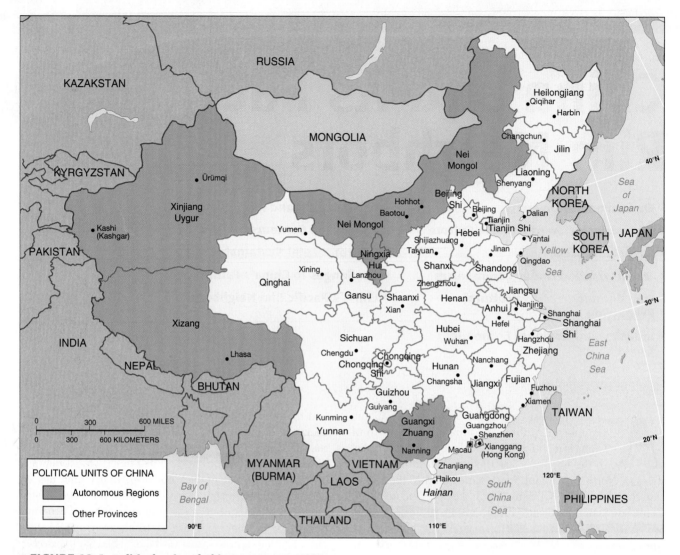

■ **FIGURE 19-1 Political units of China.** As the largest nation of East Asia, China is divided administratively into provinces, urban municipalities administered directly by the central government, special administrative regions, and autonomous regions serving areas of cultural minorities.

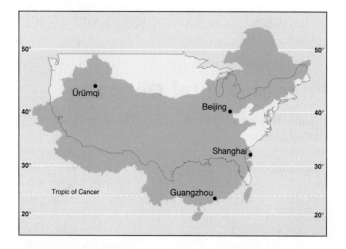

■ **FIGURE 19-2 Size and locational comparison of China and the conterminous United States.** China occupies an area of about the same size and latitude as the United States.

that a small portion of China extends as far south as Cuba and its northeasternmost territory lies at the same latitude as Canada's Quebec Province.

The spatial distribution of landforms and climates within the two countries is similar as well. Both China and the United States have low plains and mountains in their eastern zones and higher mountains in the west; moreover, the eastern portions of both nations have humid climates, with more arid conditions dominating in the western areas. China's northeast lowlands possess shorter summers and drier, colder winters than those in the United States because of the dominance of the interior high-pressure cell of Asia that reduces the influence of the moderating maritime air masses to the south and east. The western half of China is generally drier than the American West, simply because China does not receive the moist, westerly maritime air masses that bring rain or snow to the upper elevations of the western United States. Nonetheless, it is this shared east-west climatic divide that in part produces a population pattern of a more densely settled east and a more sparsely populated interior west in both countries. The popular Western perception of China as a rice paddy and bamboo-grovedominated landscape applies only to its southeastern quarter and thus belies the realities of an environmentally diverse country.

Environmental Regions

We are able to classify China's varied environments into a number of generalized regions based primarily on their importance to human development (Figure 19–3). We will begin with an overview of the landscapes of the more densely populated eastern basins and uplands and then progress to the drier basins and highlands of western China. Perhaps the most specialized environmental zone is that of the Loess Plateau, an elevated tableland 4,000 to 5,000 feet (1,200 to 1,500 meters) above sea level situated between the Ordos Desert and the North China Plain. Loess is a term used to

describe a fine, yellow dustlike soil deposited thousands of years ago by winds originating in the Mongolian grasslands to the north of China. This windblown material was deposited in mountain valleys up to depths of 600 feet (200 meters), but most areas contain deposits 60 to 90 feet (20 to 30 meters) thick. Loess is very fertile but at the same time highly susceptible to erosion. Infrequent but strong summer thunderstorms in this otherwise dry climate region have caused gullying on the slopes of hills and mountains that has resulted in the loss of soil to streams flowing into the Huang He for centuries. This river is appropriately translated as the "Yellow River" because of the yellow hue of its loess sediments. Farmers have for centuries terraced the hillsides of the Loess Plateau in an effort to limit soil erosion (Fig-ure 19–4).

Downstream from the Loess Plateau is the North China Plain, an extensive riverine surface built up from the silt deposited by the Huang He on an ancient shallow continental shelf. Through the centuries, the loess-derived silt has been deposited on the riverbed, gradually elevating

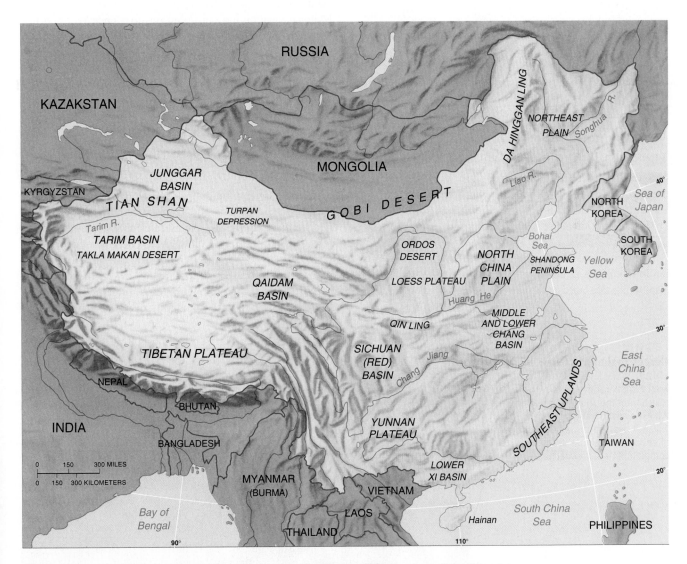

■ **FIGURE 19-3 Physiography of China and its Pacific Rim neighbors.** The nations of East Asia contain a wide variety of landform regions, ranging from some of the world's highest mountains to great coastal plains and inland plateaus.

■ **FIGURE 19-4 China's Loess Plateau.** Loess is a very fertile soil, but it is very prone to erosion. In order to cultivate on steep slopes, terracing is essential if erosion is to be controlled. Where terracing is absent, heavily dissected slopes often are the result.

it above the surrounding plain and necessitating the construction and continual raising of levees or dikes to hold the river within its channel. Periodic levee breaks, as well as river channel changes, have brought devastating floods to this densely settled region for thousands of years. Between 602 B.C. and 1950, there were 1,573 serious flood-induced levee breaks and 26 channel changes, the most recent of which occurred over the period 1938–1949. The disastrous 1938 flood resulted in 890,000 deaths and left some 12 million people homeless. It is no wonder that the river's nickname is the "river of sorrow." Carrying a heavy silt load and extending its delta length some 20 miles during the 1975–1991 period alone, no other major world river is as silt charged as the Huang He. While the river has allowed the North China Plain to support one of the highest rural population densities of any major world landform region, it has also been associated with many of history's greatest natural disasters.

The Sichuan Basin, one of the largest interior basins of China, is located in the south-central area of China. Surrounded by high mountains and plateaus, the Sichuan Basin is densely inhabited by an agricultural population situated primarily on the Chengdu Plain. Because of the protection of the surrounding mountains, Sichuan experiences relatively cool but humid winters and warm and humid summers. With high year-round humidity levels and foggy days averaging 300 days per year, extended sunny periods in Sichuan are rare. Between the eastern edge of the Sichuan Basin and the flatter riverine plains of south-central China is a section of the Chang Jiang, or "Long River," called the **Three Gorges**. Here, river water is forced to flow through a narrow, 150-mile (240-kilometer), steep-walled valley, which is no greater than 350 feet (107 meters) in width (Figure 19–5). Much like sand in an hourglass, water from spring snowmelt upstream and wet summer monsoon rains can fill this narrow passageway with such great fluid pressure that river levels may easily rise 20 feet (6 meters) over a 24-hour period. Motor navigation has only been possible since World War II and then only at times of lower water levels when near-surface rocks or shoals are exposed. To improve navigation, and more importantly, to harness the potential hydropower of the Three Gorges, the government is currently constructing the largest dam in the world. Experiencing staggering cost overruns and bad press from foreign environmental organizations, the government pushes on with this project that symbolizes national pride (see the *Geography in Action* boxed feature, The Three Gorges Project: Dams and Development).

After leaving the constricted confines of the Three Gorges region, the Chang Jiang meanders sluggishly across the flat surface of the Middle and Lower Chang Jiang Plain. The middle plain is encircled by numerous low mountains and hills, dotted by many shallow lakes. These lakes act as flood reservoirs for the Chang Jiang during the high-water summer monsoon season. Many have silted up, however, and been reclaimed for agriculture, thus reducing the ability of south-central China to manage floods. The lower plain of

■ **FIGURE 19-5 The Three Gorges region of south central China.** Water levels of the Chang Jiang will be significantly raised with the construction of the Three Gorges Dam. While navigation and power generation will improve, there will be a deep sense of environmental loss as the projected completion date approaches.

GEOGRAPHY IN ACTION
The Three Gorges Project: Dams and Development

When completed in 2009, the Three Gorges Dam will become the largest dam in the world and China's single largest construction project since the Great Wall. Understanding the project is important because it not only represents a microcosm of China's future environmental, economic, and political challenges but it also raises larger issues concerning the dam fetish of many developing-country governments. Although feasibility studies were first carried out in the 1920s, it was not until 1989 that a Canadian study recommended the dam construction. This 600-foot-high and 1.2-mile-wide dam will create a 360-mile-long and 500-foot-deep reservoir (the length of Lake Superior) that extends to the city of Chongqing Shi. While official Chinese government cost estimates are between $24 billion and $34 billion, less biased projections put the cost at $77 billion. Because of the size, the cost, and the potential environmental impact of the dam, few large-scale infrastructure projects have engendered such great global-scale debate.

The government believes the Three Gorges Dam to be critical to the country's future development. With a poor land transport system, for example, greater control of water flow will allow large cargo vessels to penetrate 1,500 miles into China's interior, a boon to Sichuan, the third most populous province. The project would also make Chongqing Shi the busiest inland port in the world. Because 75 percent of China's energy needs are met by burning high-sulphur coal, the resulting air pollution has made for very unhealthy living conditions in many areas. The hydroelectricity generated by the dam will serve the purposes of diversifying energy production, providing 10 percent of the country's energy needs at a reasonable price, improving urban air quality, and reducing the already high levels of acid rain. The government also claims that the periodic floods that ravage the middle and lower Chang Jiang will be greatly reduced, benefiting those that farm the 35 million acres of land downstream. Floods in 1998, for example, left more than 2,000 dead and required approximately 14 million people to be evacuated from their homes.

In an effort to terminate the project, both domestic and international critics have challenged almost every claim made by the government. Because of technological advances in energy generation, megahydroelectric projects are now seen as uncompetitive dinosaurs. Coal-fired plants with pollution controls, for example, are estimated to be twice as cost-effective. Natural gas, an energy source that China has yet to fully exploit, is even more cost-effective. Critics have called for a more flexible system of small dams that could generate substantial amounts of energy at half the cost. Critics also question the flood-control capabilities of the Three Gorges Dam. As experienced by other dam projects in China, sediment buildup will drastically reduce the water storage capabilities of the dam and thus the flood-control purpose of the dam. The constant dredging of sediment from the reservoir will inflate the cost of energy even more.

Leaving aside the cost-benefit shortcomings, critics have been extremely vocal concerning the dam's tragic environmental and human impact. The rising reservoir level will submerge habitat for many endangered animal species, including migratory cranes and river dolphins, and it will destroy some 1,000 culturally important sites as well. The direct human impact involves the relocation of approximately 1.5 million people because rising waters will submerge 13 cities, 140 towns, 1,352 villages, and 115,000 acres of valuable agricultural land. While the national government has promised to financially compensate inhabitants, too many of them are victims of corrupt local officials who have pocketed resettlement funds. Too often, land provided for resettlement is of lower quality.

Because of widespread opposition by dam opponents, most of the major foreign financing has been withdrawn, forcing the government to invest funds from domestic sources. Facing the prospect of a dam that is not economically viable, the criticism of international environmental groups, and the growing social unrest of a resettled population, why does the central government continue to politically support the project when the governments of many other nations no longer view huge dam projects in a favorable light? Part of the answer lies in the fact that China's government has always perceived dam building as a way of taming the environment. Indeed, when the government came to power in 1949, there were only 22 large dams. By 2000, the total number of large dams had increased to 368, more than any other country in the world. Secondly, as in any country, a monumental dam project is a crowning symbol of nationalism, prestige, and modernization. In part, the Three Gorges Dam is a huge vanity project.

Sources: P. Adams and G. Ryder, "China's Great Leap Backward," *International Journal* (Autumn, 1998): 686–704.
P. Berkman, "The Three Gorges Dam," *Education About Asia* 3 (1998): 27–35.
David Murphy, "Dam the Consequences," *Far Eastern Economic Review* (May 23, 2002): 28–30.

---

the Chang Jiang sits less than 10 feet (3 meters) above sea level, and this wetland environment is characterized by a patchwork of rice paddies and fishponds, laced by a dense network of streams and canals (Figure 19–6). The discharge volume of the Chang Jiang is approximately seventeen times greater than that of the Huang He. With high summer humidities, cities along the river's course are often referred to as the "ovens of China." Lacking the protection that the Qin Ling Mountains afford to Sichuan, the Lower Chang Jiang Plain experiences winters that are much colder, with periodic heavy frosts and some light snow.

One of the most physically stunning of China's eastern regions is the Yunnan Plateau, which occupies the environmental transition zone between the cold Tibetan Plateau and Eastern Monsoon China. With elevations ranging from 5,000 to 9,000 feet (1,500 to 2,700 meters), this dissected upland is laced by mountainous spurs of the Tibetan Plateau. Between the mountain ridges are deep river gorges and small upland valleys dotted with agricultural settlements. Influenced by both the Indian and Pacific ocean monsoons, the region is quite wet, but because of its intermediate elevations, it experiences moderate summer temperatures. Each year this magical environment is

■ **FIGURE 19-6 Lower Chang Jiang Plain.** Agriculture in this area features a mixture of rice paddies, fish ponds, and field or tree crops. In the foreground rice paddies are separated by narrow footpaths. The more rectangular shape of fish ponds is observable in the middle ground. Raised dikes separate fish ponds and are the site of vegetable, grain, and fruit cultivation. The haze reflects pollution generated by nearby urban areas.

the favored region of millions of domestic tourists seeking eco-tourism experiences. Stretching along China's southeast coast are the Southeast Uplands, which average 3,000 to 4,000 feet (900 to 1,200 meters). With rugged hills and low mountains reaching down to the narrow coastal plain, the region possesses most of China's deepwater harbors. Farming tiny plots of land, many peasants have looked to the sea as a supplemental source of income. Not surprisingly, this is the source region of many

Overseas Chinese. The last environmental region of eastern China is the Northeast Plain, an extensive rolling hill surface that has become an important grain farming region. Although presently threatened by deforestation, the surrounding uplands are China's most important source of timber.

Two major environmental zones occupy the vast area of western China. The first and largest is the Tibetan Plateau, an area that encompasses approximately 25 percent of China's territory. It is the largest, most elevated plateau in the world, and it is sometimes referred to as the "rooftop of the world" or the "third pole" of the world because of its cold and inhospitable environment (Figure 19–7). Averaging 13,200 feet (4,000 meters) in elevation, the Tibetan Plateau is affected by snow year-round, except in the lower forested valleys. Most of the population is found on the wetter and warmer southern edge, where the warmth and precipitation of the summer monsoons are felt. North of the Tibetan Plateau is the Tarim Basin, centered on the Taklamakan Desert. It is a region of internal drainage where rivers flowing from the surrounding mountains disappear into the desert sands, sometimes forming salt lakes such as at Lop Nor or Turpan, the latter lying at 500 feet (150 meters) below sea level. Pockets of agricultural settlement are found along a belt of river-fed oases located at the mountain-desert edge (Figure 19–8).

Spatial Evolution of Chinese Culture

Chinese cultural, economic, and political institutions have been greatly modified over time, but their long-term persistence explains why Chinese civilization is often referred

■ **FIGURE 19-7 Tibetan Plateau.** The Tibetan plateau is a remote and sparsely settled region of southwestern China. Summer nomadic movement of herders with their sheep and yaks to high altitude pastures is a traditional way to access seasonal grazing; animal herding is one of the few ways in which many of Tibet's people can exploit their limited resource base.

■ **FIGURE 19-8 Oasis settlements in Xinjiang.** Farmers capture water from mountain-derived subsurface streams to make the soil bloom at the mountain–valley edge. Many settlements date to the Han period, when they were critical travel links along the Great Silk Road.

GEOGRAPHY IN ACTION
Language and Development

While written Chinese is consistent in form throughout China, spoken Chinese varies widely in the sense that a given character, while having the same meaning, is often pronounced differently from one region to another. In many southern areas, the peoples of one region often cannot understand those of another because spoken Chinese is tonal in nature, and a single word possesses different meanings, depending upon the tone or inflection. It is these tonal differences that have traditionally made it difficult for Chinese from different regions to converse with one another.

Throughout much of China's history, the written language has served to promote cultural, political, and economic integration. While the vast majority of people were illiterate, government bureaucrats and scholars who held the empire together administratively were able to communicate through writing. In the modern period, however, both the complexity of the uniform written language and the variety of spoken languages have posed serious problems to national integration. To increase literacy, the communist government simplified the characters in common usage by reducing the number of strokes composing a character. The government also introduced **pinyin**, a standard system of spelling words in the Latin alphabet, to replace the older Wade–Giles transliteration system. Implementing pinyin has increased literacy, and it has also afforded China the opportunity to rid itself of a colonially imposed transliteration system that misspelled and mispronounced Chinese words. The spelling, for example, of Peking, Fukien, Shantung, and Sinkiang under the Wade–Giles system was changed to Beijing, Fujian, Shandong, and Xinjiang, respectively, under the pinyin system.

The spatial standardization of the spoken language required even greater fortitude on the part of the government. While the Beijing dialect of spoken Mandarin had long been in wide use throughout northern and central China, and some movement toward a national or common language referred to as **Putonghua** was made in the 1920s and 1930s, it was not until 1956 that the government officially designated **Mandarin** as the national spoken language. Although southerners may view the forced imposition of Mandarin as a case of cultural imperialism originating from the political core, most regard it as necessary to forge national unity. Today there are more speakers of Mandarin in the world than people who use English as their first language. Mandarin is the language of government, school instruction, radio, television, and movies, and it is almost exclusively the language of public space. The regional languages, particularly those of the south, continue to be used, but mostly in the more remote and isolated areas and in private places such as the home.

---

to as the world's oldest surviving culture. This distinct culture was perceived by the Chinese themselves as being indigenously derived, without substantial external influences. It is understandable how the Chinese, hemmed in by mountains to the north, west, and south and the Pacific Ocean to the east, came to think of themselves as the bearers of a superior civilization that possessed a deep sense of cultural unity. The Chinese thought of themselves as inhabitants of the **Middle Kingdom** or *Zhongguo*.

Out of the Northern Culture Hearth

The Chinese culture hearth emerged on the great bend of the Huang He that includes both the Wei River Valley and adjacent loess lands. From 1766 B.C. to A.D. 1912, fifteen major dynasties ruled China, each important in its own right. During certain dynasties, however, important cultural, political, and economic traits emerged, and it is these dynasties that are given attention here (Figure 19–9).

The first authentic Chinese dynasty was the Shang (1766–1122 B.C.), centered at the great bend of the Huang He where the river enters the North China Plain. The Shang contributions to Chinese cultural identity included the development of metal working and the distinctive character-based written language (see the *Geography in Action* boxed feature, Language and Development). Elaborate bronze ceremonial vessels, dating to the Shang dynasty, were unearthed at many archeological sites during the early decades of the twentieth century. Most importantly, the Shang period marked the transformation of Chinese society from one based on egalitarian agricultural communities to one oriented toward socially and occupationally stratified urban centers. Large walled cities supported artisans producing goods for local consumption and trade and an aristocracy whose primary function was to assist the emperor in the administration of his feudal domains. Government officials possessed the power to forcibly enlist peasants for "corvée" duty during times of war or when government construction projects required large inputs of labor.

Replacing the Shang were the **Zhou** (1027–256 B.C.), who originated farther upstream on the Huang He and established their capital at a site close to modern-day Xian. The Zhou were more nomadic than the Chinese of the Loess Plateau or North China Plain, but they eventually experienced a cultural transformation toward Chinese norms. The militaristic Zhou extended the imperial domains beyond the Chang Jiang, farther west up the Huang He and Wei River valleys, as well as to the northeast. A number of cultural and political traditions make this dynasty central to the evolution of the traditional Chinese way of life. These traditions emerged at a time of political and social discord, as the Zhou political center lost control of many vassal states during the "Warring States" (403–221 B.C.) period.

The most enduring of these traditions was the ethical philosophy of **Confucianism**, based on the writings and teachings of Confucius. Confucianism emphasized that humans were moral by nature and that immoral behavior, such as that experienced during the Warring States period, was a

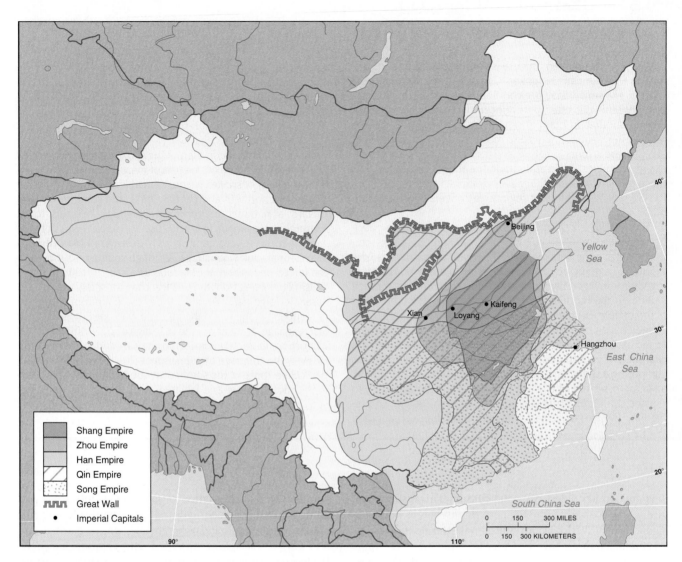

■ **FIGURE 19-9 Spatial evolution of Chinese political territory.** Chinese political control and accompanying Han culture spread out over thousands of years from its core at the great bend of the Huang He River.

product of the loss of virtue by all. Confucius argued that authority within the government or family should not be based on formalized law and brute force punishment but rather on living a virtuous life characterized by obligations to others. Much like the philosophy of Socrates and Jesus, both of whom preached "man's moral duty to man," the sayings of Confucius constituted a new ethical system that has endured in East Asia even to the present (see the *Geography in Action* boxed feature, Confucianism and Local-Global Processes). The opposing philosophy of **legalism** advocated the idea that humans are essentially selfish and that a system of strictly imposed uniform laws is required to ensure acceptable behavior. Compliance was sought through the rule of the emperor or, at the family scale, the rule of the father without questioning the morality of actions. Most Chinese dynastic governments ruled by applying concepts from both philosophies.

The next dynasty to bring in a new imperial "order" was the Qin dynasty (221–207 B.C.). Although it was short-lived,

the Qin dynasty was the period during which China became a single state and culture; in fact, the name China probably originated from the word Qin. The totalitarian Qin emperor Shih-huang-di forged a unified and spatially integrated state by dividing imperial territory into forty military regions, each administered by a staff of officials who were appointed by the central government based on merit rather than birth. The central government unified writing systems, standardized weights, measures, currency, and even axle lengths to provide uniformity on the hundreds of miles of imperial highways and canals that bound the empire together. The uniformity imposed by the Qin was characterized by a dominant legalist vein and bore a strong resemblance to later communist rule. Another aspect of Qin rule that defined Chinese culture and space was the completion of the Great Wall, which served to spatially separate even more of China Proper from the grasslands of Inner Asia (Figure 19–10). The drier northwest frontier, unlike the southern frontier, posed a barrier to Chinese settlement; decreased rainfall prevented

GEOGRAPHY IN ACTION
Confucianism and Local-Global Processes

When the Chinese Communist government came to power in 1949, Confucianism was attacked as being old-fashioned and a barrier to modernization. In the globalized economy and society of China today, however, where increasing wealth promotes materialism and a perceived decline in moral values, Confucianism is being reintroduced into elementary school curriculums; approximately 2 million students have enrolled in formal Confucian morality classes. This reintroduction of Confucian classics in schools is in a sense a local reaction to the forces of globalization, or to what geographers refer to as local-global processes. In any country experiencing the economic and social change associated with globalization, there is a perceived threat to traditional or nation-based cultural values. In China,

where Marxist ideology is being replaced by capitalist-based market reforms, the government is wary that the country's rising middle class is losing its sense of "Chineseness" that emphasized courtesy and loyalty to the family and state. Aside from Confucianism promoting a more civil society in a time of rapid economic and social change, the government perceives Confucianism as a philosophy capable of promoting loyalty to the communist political system that is under threat by the democratic values that have accompanied the process of globalization. Much like Confucianist learning, the revival of traditional school subjects such as martial arts and the art of character writing known as calligraphy is simply viewed as a localized response to the perceived negative effects of the homogenizing forces of globalization.

---

sedentary agriculture, which was a defining characteristic for being Chinese. Symbolically a defensive structure to contain pastoral nomads to the north, the Great Wall also became an agroecological boundary, separating pastoral nomadic from sedentary ways of life. The Qin dynasty is remembered for forging the idea of a single unified state and culture "within the walls."

The Qin was followed by the **Han dynasty** (206 B.C.–A.D. 220), a period so important that the Chinese today still refer to themselves as "people of the Han." As organizers of the first truly large-scale empire in East Asia, the militarily powerful Han expanded their zone of influence by territorially digesting much of what encompasses present-day China. The Han wrestled a wide corridor of far western lands in Inner Asia from pastoral nomadic kingdoms and brought these new territories within the Chinese orbit. The Great Silk Road was then established, indirectly connecting the Chinese and

Roman empires through traders who traveled Inner Asia as merchant intermediaries. It was the opportunity gained from Inner Asian trade, but also the potential nomadic threat to that trade, that led the Han to establish their imperial capital at Ch'ang-an (Xi'an). As a forward capital, Ch'ang-an was located in the northwest, the direction in which Chinese expansion was pointed and from which both opportunity and threat originated.

With the northwestern frontier secured, the imperial capital was then moved to the North China Plain site of Loyang, in the direction of the southern frontier, where opportunities eventually proved greatest. Southern expansion beyond the Chang Jiang increased the prestige of Han rulers and generals but, most importantly, enabled the northern peoples to secure an abundant source of food. The south-central region then became a key economic region upon which the economic and political power of the northern-centered

■ FIGURE 19-10 **Great Wall of China.** Completed during the third century B.C., this 3,000-mile-long structure was constructed to separate the pastoral nomadic life of the northwest from the sedentary agricultural life of the south. Today many sections of the Great Wall are crumbling because of neglect. This particular section remains intact and one of a handful open to tourist excursions.

■ **FIGURE 19-11 Grand Canal in the city of Wuxi.** This canal connected many regions of the Empire, improving transport. Today, it continues to provide the means for low-cost movement of bulk goods.

The Song dynasty provides us with two very important lessons concerning economic development. First, China and its people are not static or constant (as is often perceived), but, in fact, they have experienced substantial economic and social change over time, based primarily on internal forces. Second, evidence from China, including the Song period, demonstrates that Europe has not always been technologically superior to the rest of the world and that many civilizations have fostered high levels of social and economic development. European colonialism did much to create both the advantaged and disadvantaged worlds of today.[1] China in the thirteenth century resembled eighteenth-century Europe.

government largely depended. The south came to be viewed as a granary for the north, which, over time, became increasingly parasitic and exploitative. Although constructed during the Sui dynasty that followed the Han, the 1,400-mile-long **Grand Canal**, between modern-day Hangzhou (just south of Shanghai) and the heart of the North China Plain, facilitated the much-increased trading of commodities among the regions of a now more-integrated empire (Figure 19–11).

The **Song** or Sung, **dynasty** (A.D. 960–1279) was a distinctive period of Chinese history. Many of the social and economic development patterns that emerged then may be likened to similar revolutionary changes that China is currently experiencing. Having been forced south by Mongol invasions, the new Song capital was Hangzhou, a city of huge physical proportions that easily outsized the largest urban settlements in Europe at that time. Although Hangzhou was an administrative capital, its lifeblood was commerce, and along with the handful of other highly populated coastal and coastal plain cities within Song domains, it strongly reinforced the pattern of urban dominance of the countryside.

The Song dynasty was also characterized by the expanded use of early-ripening rice varieties, improvements in irrigation, and better marketing and distribution systems. These advancements made the Chinese agricultural economy one of the most sophisticated in the world. The government printed money and issued letters of credit that facilitated interregional trade throughout the Chang Jiang basin. For the first time in China's history, long-distance maritime trade developed, bringing Arab and Indian merchants and products to the coastal cities and encouraging Chinese merchants to establish trading networks in Southeast Asia. Great private wealth was accumulated and invested by many in art, music, and literature for the merchant class and commoners alike. To the legalist Qin as well as the later communists, the development of a merchant class in competition with the state was a sign of disorder and decadence.

East Meets West

By the eighteenth century, China had, for all intents and purposes, led an isolated and self-sufficient existence for 4,000 years. The foreign religion of Buddhism had taken hold but had become transformed and assimilated into Chinese culture. The arrival of Western traders during the early eighteenth century, however, signaled the beginning of the collapse of the world's oldest culture. Fundamentally challenged by Western technologies, sciences, and economic systems, China's traditional view of the world with the Middle Kingdom as its center was turned upside down. The meeting of the East and the West did not involve a mutually interactive environment in which China could reject things foreign or, as in the past, transform them to suit spiritual and intellectual traditions.

The East India Company, under the authority of the British government, was in competition with other European colonial powers for the lucrative China trade. With great hesitation, the Qing government restricted foreign trade contacts to Guangzhou (Canton) in 1702. The British bought tea, silks, porcelains, and medicines but had to pay in silver because, except for cotton from India, the Chinese government thought English goods were inferior. Insistence on this requirement seriously drained British silver reserves and created a deficit-payment problem. The English solution was to sell Indian **opium** (a dangerously addictive drug from which heroin is derived) to the Chinese in exchange for silver bullion. The opium trade became extremely profitable, and by the end of the nineteenth century, some 40 million Chinese (roughly 10 percent of the population) were estimated to be addicted. China experienced a dramatic outflow of silver and, in response, attempted to stamp out opium use by confiscating opium chests from British ships docked in Guangzhou. The British declared war, resulting in the First Opium War (1839–1842), in which the British soundly defeated Qing government forces. The opium trade contributed in two ways to the process of creating a global

---

[1]James M. Blaut, *The Colonizer's Model of the World* (New York: Guilford Press, 1993).

colonial economy. The first significant issue was the use of resources from one region (cotton and opium from India) to purposely gain economic advantage in another, more potentially profitable region, such as China. The second issue revolved around the use, by supposedly more civilized and developed nations, of a dangerous narcotic drug to gain economic advantage.

Treaty Ports

China's defeat in the First Opium War resulted in further spatial penetration by Western interests. In 1842 the Chinese were forced to sign the humiliating Treaty of Nanking, which opened up five coastal ports to Western interests. Although many more ports were opened by 1920, only about fifteen ports experienced substantial economic activity (Figure 19–12). These enclaves essentially became foreign-owned territories where alien, not Chinese, laws governed human behavior. Essentially, many Chinese became foreigners in their own land. Treaty ports became the primary centers of foreign trade and large-scale industry, to which great numbers of Chinese people were attracted. Shanghai was the most important of the treaty ports. It was known as the "Paris of the East" and was one of the busiest ports in the world during the early twentieth century (Figure 19–13).

Foreigners viewed China as a market for manufactured goods at a time when the mass-production innovations of the Industrial Revolution were already beginning to saturate markets in Europe. Treaty ports also provided the opportunity for Western banks and shipping companies to expand their business through dominating both the import and export trades. Western-owned factories could produce low-cost manufactured goods at treaty ports because of inexpensive domestic labor and then market those goods in China or export them to other colonial territories. The treaty ports also became symbols of what the West wanted China to become. With the appearance of railroads, modern banking, Western medicine and Victorian morality among other Western imports, the treaty ports came to be viewed by

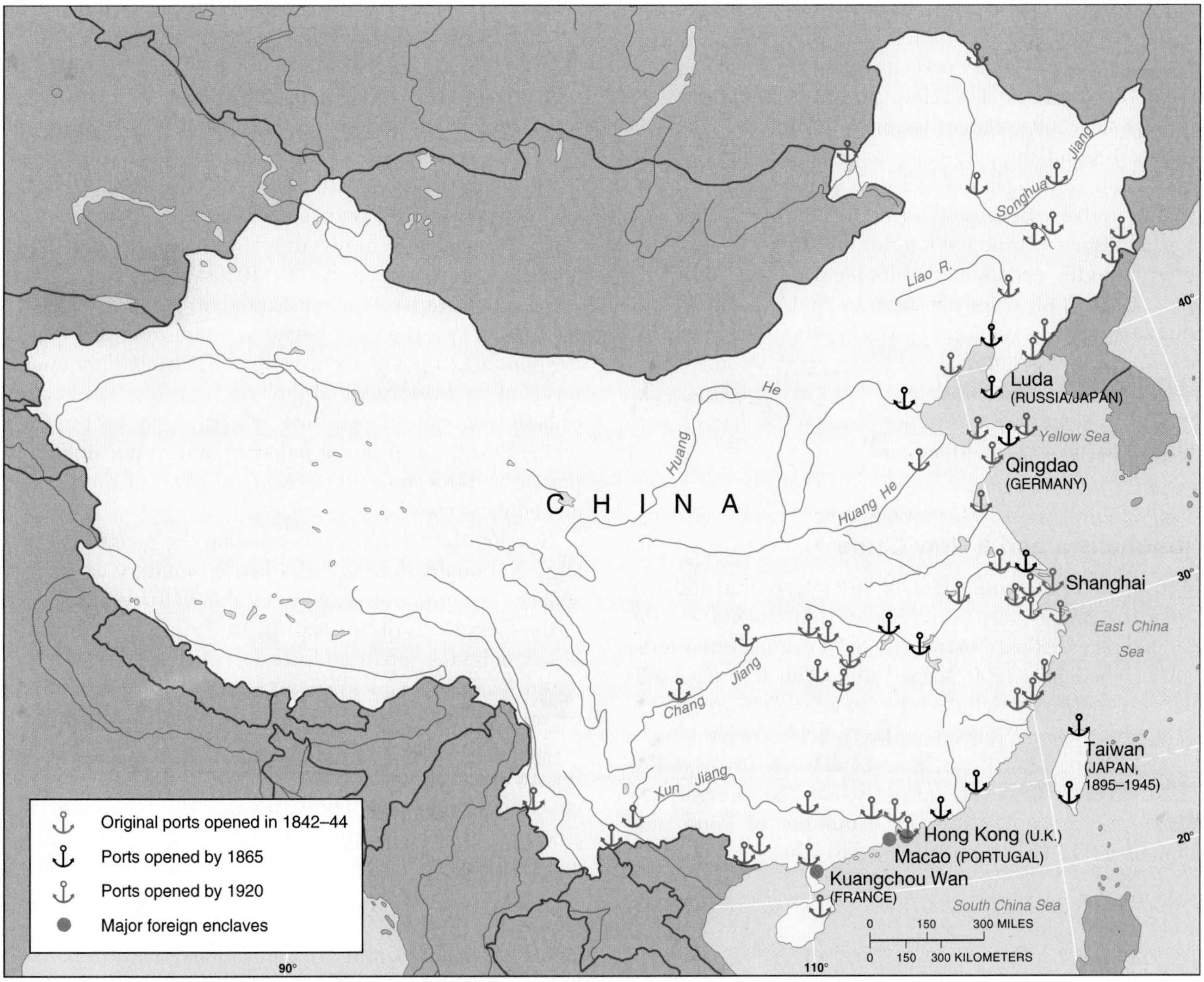

■ **FIGURE 19-12 Treaty ports of coastal and riverine China.** With the European occupation of major port cities, China was forcibly opened to Western influences.

▪ **FIGURE 19-13 Shanghai Bund.** The bund is the old European colonial sector of Shanghai, associated with the treaty port era. Facing the Huangpu River, it is now surrounded by modern skyscrapers but continues to function as a popular recreation and tourist district.

Europeans as "model settlements" capable of enabling "China to follow the Western path toward wealth and power, rejecting its own clearly inferior and 'backward' tradition."[2] The impact of the treaty ports on China's economic development was mixed. While a class of treaty-port Chinese came to own a substantial share of factories and founded Western-style banks and shipping companies, the vast majority of the Chinese residing in the enclaves were poor and exploited factory workers. Because the treaty ports were all located at great distances from the interior of China, they did little to promote the economic development of the interior regions. Indeed, the colonial experience in China did not resemble similar colonial processes in other regions of the non-Western world, in that Western powers colonized specific locations along China's coast but did not colonize the country as a whole. Geographer Carolyn Cartier refers to this process as **semi-colonialism**.[3]

Nationalism and a New China

In response to the humiliation of losing territorial and economic sovereignty, various nationalistic movements soon developed. The first outward expression of nationalism was the Boxer Rebellion in 1900. Named after an ancient art of self-defense, the Boxer Rebellion began as an antidynastic village-based peasant movement but quickly evolved into a movement opposing the presence of all foreigners as well as Chinese connected even indirectly with the foreign presence. After sweeping through a number of flood- and drought-stricken rural areas during the early months of 1900,

the Boxers stormed foreign embassy compounds in Beijing. Because the imperial government failed to provide protection, an international military force was dispatched from the nearby coastal treaty port of Tianjin (Tientsin) and easily defeated the Boxers, taking revenge through widespread looting and raping.

In 1911 the Qing government collapsed, and numerous nationalistic movements emerged in treaty-port cities, where foreign influence was greatest. One of these movements was led by **Sun Yat-sen**, a medical doctor trained in the West and sometimes referred to as the "father of the republic." After several attempts to establish a sustainable political party, Sun revived his earlier Nationalist Party, Guomindang, in 1919 and established its headquarters in Guangzhou. Its platform centered on the "Three Principles of the People": nationalism, democracy, and livelihood. Sun died in 1925 and was replaced by a general named Chiang Kai-shek, who took the party in a very different direction. Enlisting the financial help of treaty-port Chinese capitalists and conservative rural landlords, as well as the United States government, those closest to Chiang became corrupt, while at the same time amassing great personal fortunes.

Founded in Shanghai in 1921, the **Chinese Communist Party (CCP)** was also a nationalistic movement. **Mao Ze-dong** emerged as its leader in 1935. Receiving both financial and moral support from the Soviet Union, the CCP's initial base of power was among the progressive intellectuals and poor treaty-port factory workers. Although the two parties were once united to replace warlord rule, the Nationalists brutally drove the urban-based CCP into the countryside in the late 1920s. It was from among the rural peasantry who constituted 80 percent of China's population that the CCP began to mobilize mass support. Responding to peasant needs, the CCP embarked upon a series of land reforms that included outright confiscation of land from greedy landlords who disregarded the Confucian traditions of moral behavior and responsibility. In the early 1900s, only 10 percent (at most) of the farmers owned their own land.

Sensing the CCP's success among the peasantry, Chiang's Nationalist forces embarked on military campaigns against the communists and were able to force them north to the loess lands of Shaanxi. In 1937, the brutal invasion of China by the Japanese forced a retreat of the Nanjing-based Nationalist government to Sichuan, thus preventing any further assault on Communist forces. The civil war between the nationalists and communists resumed after the end of World War II, with the communists eventually emerging victorious. The communists' ascendancy was aided by the fact that the Nationalist government had become increasingly identified with foreigners and totalitarianism and was supported by a poorly paid and poorly equipped army, which possessed little ideological passion. By the end of 1949, much of the Nationalist army had been killed or captured, and Chiang and his followers were forced to flee to the island of Taiwan, leaving the communists in total control of the mainland.

[2]Rhoads Murphey, "City as Mirror of Society," in *The Chinese Adapting the Past, Building the Future,* edited by R. F. Dernberger. K. J. DeWoskin, and S. M. Goldstein (Ann Arbor: University of Michigan, Center for Chinese Studies, 1986).
[3]Carolyn Cartier, *Globalizing South China* (Oxford: Blackwell, 2001).

Transformation Under Communism

The new Communist Party of China (CPC) faced the daunting problem of rebuilding both a society and an economy that, since the 1920s, had experienced civil war, the Japanese occupation, and civil war again. The CPC chose socialism as its primary development philosophy and soon came to monopolize political power, reserving the right to make all decisions pertaining to economic and social policy. In agriculture and industry, the government has controlled all capital investment through state-owned industrial and financial institutions and has restricted the influence of foreign capital and external economic forces in the classic socialist pursuit of economic self-reliance. State control of social development has been equally pervasive. Political dissent has been ruthlessly repressed, and in the 1970s, population-control measures were introduced in an attempt to manage the reproductive lives of hundreds of millions of women.

In the late 1970s to early 1980s, the government began to realize the limitations of these guiding economic principles in building a modern state and thus embarked upon a road of economic reform that was as revolutionary as the events of 1949. With the death of Mao in 1976 and the rise of the more pragmatic leader **Deng Xiaoping**, China began to emulate the capitalist economies of many of its East Asian neighbors. Testing the waters of capitalism entailed decentralizing government economic power and promoting greater interdependence within the global economy, both of which signaled the loss of economic self-determinism, a philosophy central to communist rule. This process of decentralizing financial and decision-making power engendered substantial regional disparities of wealth, which tended to favor coastal provinces. Naturally, then, there exists much caution on the part of the Chinese government about moving too quickly toward a capitalist and market-based economy because of the uncertainty such a dramatic shift entails. An appropriate metaphor to describe this gradual approach and vigilant attitude on the part of the national government is "crossing the river by feeling the stones."[4]

Transformation of Agricultural Production

With approximately 60 percent of the labor force directly or indirectly engaged in agricultural activity, China remained a predominantly rural country until the 1990s. The achievements of Chinese governments, both past and present, must be measured largely on the basis of agricultural productivity, because approximately one-fifth of the world's population must be maintained on only 7 percent of the world's arable land—only half of which is considered of good quality.

---

[4]Godfrey Linge, "Toward 2020: Crossing the River by Feeling the Stones," in *China's New Spatial Economy: Heading Towards 2020,* edited by Godfrey Linge (Hong Kong: Oxford University Press, 1997).

Agricultural Regions and Dominant Crops

In contrast to agriculture in Europe or North America, China's agricultural energies are, for the most part, focused on the production of food crops rather than the growing of plants for industrial and animal feedstock purposes. China may be divided into three primary agricultural regions based upon both farming systems and dominant food crops (Figure 19–14). The first and most obvious regionalization of agriculture is between the wetter eastern and drier western halves of the country. In western China, pastoral nomadism has long been dominant, but during the past century, Han Chinese have settled oases and river valley bottoms to cultivate wheat, barley, and maize. Even within the eastern region of China, however, substantial differences exist between the south, which is moist with mild and relatively short winters, and the north (except the northeast), which suffers from precipitation shortages and great seasonal variations between long, harsh winters and short, mild summers. In the far south and southeast, paddy rice is the dominant grain crop. Because of an almost year-round growing season, many farmers are able to engage in the double-cropping of rice, whereby the first rice crop is planted in May and harvested in August and the second crop is planted in August and harvested in November. In the southwest and the Sichuan Basin, rice is also the dominant grain crop, but it is often rotated with maize, wheat, and potatoes. The middle and lower-eastern Chang Jiang basin represents the agricultural transition zone between north and south. The crop pattern south of the river is dominated by the double-cropping of rice, with tea as a dominant nonfood crop, while the common mix of crops north of the river is the double-cropping of summer paddy rice and winter wheat.

The dominant crop of the North China Plain is winter wheat, which is planted in the fall and harvested in the spring. Because of shorter summers, double-cropping with other crops is difficult. When double-cropping takes place, however, it is usually with maize or with *gaoliang*, which is a variety of millet. Cotton requires a long summer and is unable to be double-cropped with other food crops. On the terraced slopes of the Loess Plateau where rainfall is sparse, one growing season is the rule; winter wheat is the dominant crop, joined by the dry-climate grains of millet and maize. The northeast is a productive and highly mechanized agricultural region, which has only been settled (largely during this century) by pioneering Han Chinese. In the river plains, spring wheat is grown along with maize and potatoes, while the elevated slopes support wheat and millet farming.

Superimposed on these agricultural regions, which are based largely upon the types of grain produced, is the widespread process of agricultural intensification, whereby a broad range of supplementary crops are cultivated for commercial purposes and for adding variety to the diet (Figure 19–15). Vegetables and various types of melons are grown everywhere, as are soybeans, to make traditional bean curd—an important source of protein in a meat-scarce diet. Fruit orchards are important; in the north,

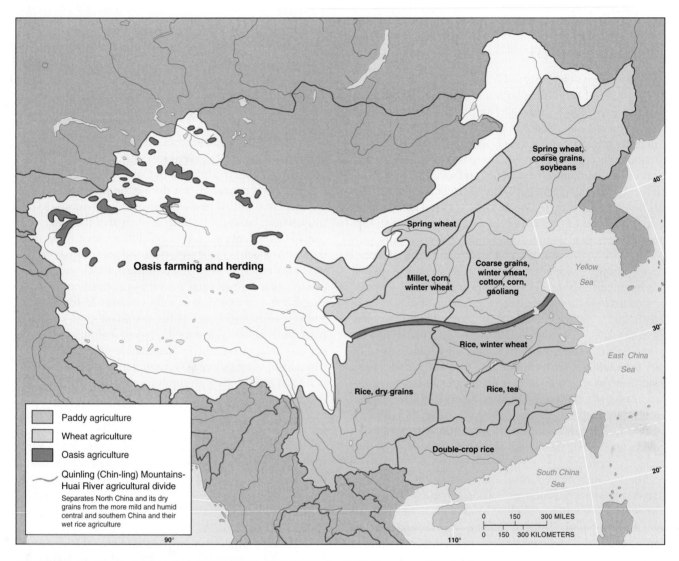

■ **FIGURE 19-14 Agricultural regions of China.** China can be divided into three broad agricultural regions: rice in the Southeast, wheat in the Northeast, and oasis agriculture in the West.

■ **FIGURE 19-15 High-value urban gardening.** On the urban periphery, access to efficient road transport coupled with agricultural reforms has produced greater prosperity for China's vegetable farmers.

apples and pears are common, while in the south, citrus dominates. Also in the south, fishponds are an important secondary source of agricultural income. Supplementary crops increase opportunities for peasant farmers to earn additional income.

Agriculture During the Pre-Reform Years

In the years directly following the 1949 revolution, the Communist Party embarked upon a program to radically reconstruct agricultural production. During the 1950s land ownership was abolished and collectives were established to be followed by the creation of People's Communes, which attempted to control every aspect of agricultural production. The state determined which crops were to be grown, the proportion of various crops to be grown, and the distribution of those crops to state agencies. Peasants, in turn, were remunerated through food ration coupons based upon time spent in the fields. Household vegetable plots and small animals were officially discouraged. Adopting the revolutionary phrase of "take grain as the key link," the official state policy was to produce grain to the general exclusion of other crops. Results were certainly mixed. While supplies of subsistence grains were adequate, and few went hungry, the diet of the average peasant became less diverse, and there existed little regional specialization based on comparative advantage. Favored in part by nature, the south continued to be relatively prosperous, while villages in the north remained miserably poor. The inability of the government to effectively control production and distribution became evident during the period of the **Great Leap Forward** (1958–1961) when the spirit of communalization was greatest. Bad weather and poorly conceived incentives for communes to maximize output resulted in widespread famine, which led to the death of some 14 million to 26 million people from famine-related illness and starvation. Environmental degradation, particularly soil erosion resulting from the use of formerly nonarable lands, was severe.

Post-Mao Agricultural Reforms

In an attempt to boost both financial incentives and enthusiasm for agricultural production, Bejing introduced the **household responsibility system** in 1978. This policy is based on a production contract in which the peasant household is allotted a piece of "responsibility" land and is obliged to produce a specific amount of grain or cotton to be sold to the state at a regulated price. Once that contract is fulfilled, the household is free to produce cash crops to be sold privately, at local markets, or to the state at above-market prices. Therefore, peasants became more empowered through being able to plan production based on local environmental resources, their level of skill, and perceived financial returns. By 1991 the production of major crops had increased dramatically (Table 19–1). Most increases occurred, despite decreased grain acreage, through greater use of fertilizers and green-revolution hybrid varieties. Indeed, cropland as a percentage of total land area dropped from 11 to 10 percent during the 1980–1995 period. Like so many other Asian countries, the application of fertilizer was critical to increased agricultural production. In China, for example, fertilizer consumption increased almost tenfold from 38 kilograms per hectare between 1968–1970 to

339 kilograms per hectare between 1993–1995. Required to import a substantial quantity of fertilizer to meet domestic demand, the central government partially subsidizes fertilizer to provide farmers incentives to increase agricultural production.

While the responsibility system has boosted agricultural production, resulting in greater general prosperity in rural areas and a parallel reduction of urban-rural income disparity, there remain some serious drawbacks. Seeking the greater profits available from the more specialized and higher-value crops such as vegetables, fruit, and livestock, Chinese peasants have tended to underproduce most cereal grains and cotton (see Table 19–1). During the 1991–2004 period, the growth rate of cereal-grain production declined relative to the 1981–1991 period, while the production of vegetables and fruit continued to increase dramatically. Several reasons explain this divergence in production. First, unlike cereal grains, government price controls on vegetables and fruits were less stringent, thus producing greater profits for peasant households. Second, the free markets where vegetables and fruits are sold increased dramatically since agricultural reforms were first implemented in the late 1970s; between 1978 and 1995, the number of free markets increased from 33,302 to 82,892. Third, the food consumption habits of the growing number of more affluent urban consumers demanded greater variety in the diet, prompting both suburban and distant producers to cultivate higher-quality vegetables and fruits, which were then sold through a more efficient marketing system. The opportunity to specialize and diversify crop production is not geographically uniform because those households located close to roads or, better yet, close to an urban area are more easily able to specialize (owing to better transport access). Because urban incomes are higher in the coastal provinces, greater opportunities to specialize and increase rural incomes characterize the more eastern provinces (see the *Geography in Action* boxed feature, Chinese Apple Exports and Globalization). More efficient agricultural production has also created a huge surplus of rural labor, forcing an exodus of displaced people to the cities. By 2001 the surplus of rural labor was estimated to be 200 million!

Industry and Regional Economic Growth

As with agriculture, the primary objective of Chinese government industrial policy has been to promote regional self-sufficiency through the spatially equitable distribution of, in this case, manufacturing activity. The government was relatively successful in meeting that goal until the reforms of the late 1970s. Since then, the influence of the global economy and the parallel rise of foreign investment in manufacturing have made China a much richer nation. Indeed, second only to Vietnam, China is the fastest-growing economy in the world. At the same time, however, some regions have benefited far more than others.

Table 19-1 Output of Selected Chinese Agricultural Products, 1981, 1991, and 2004

Agricultural Product	Metric Tons (millions)		
	1981	**1991**	**2004**
All cereals	286.4	398.5	422.6
Rice	143.9	185.7	186.7
Wheat	59.6	95.9	91.3
Soybeans	9.3	9.7	17.7
Cotton	2.9	5.6	24.0
Vegetables	63.3	130.7	423.3
Fruits	17.5	38.6	168.4

Source: Food and Agriculture Organization, Statistical Databases.

GEOGRAPHY IN ACTION
Chinese Apple Exports and Globalization

All of us are very well aware of the flood of imported manufactured goods from China. Few would have predicted, however, that Chinese farmers, too, are exporting a wide variety of higher-value farm products such as apples, red peppers, garlic, honey, flower bulbs, and even crayfish. Indeed, one year after China's 1997 entry into the World Trade Organization, China became a net exporter of agricultural products to the tune of $2.26 billion. Many Chinese farmers have thus joined the global supermarket.

In 2004 geographer Gregory Veeck and a Chinese colleague studied the rise of apple product exports outside the city of Qingdao, the capital of Shandong Province. Shandong is the leading apple producer in China, a country that produces four times the volume of apples when compared to the United States. Like so many farmers in the region, rural families entered apple production because prices for wheat, which is the primary grain crop, are substantially lower. In response, Shandong apple farmers have seized the opportunity to cultivate this higher-value crop, a portion of which enters global trade flows to European Union countries, the United States, and Japan and Korea in the form of table fruit, but especially dried slices, canned pie slices, and juice concentrate; China currently accounts for an amazing 43 percent of global apple juice exports (Figure A).

Much like the case of manufactured products, lower production costs or factors of production account for China's agricultural exports success, despite claims from apple importing countries that China is dumping its apple products on the world market. For example, inputs of fertilizer, pesticides, and all important fungicides are much cheaper in China. In addition, much animal manure applied to apple orchard soils reduces the cost of chemical inputs. Total input costs were $413.20 per acre when compared to $2,754 per acre in Washington state. Labor costs, too, are lower. In Shandong, labor is almost exclusively family labor, while in Washington or Michigan, migrant labor

■ **FIGURE A Apple production in Shandong Province.** China produces four times as many apples as the United States, and Shandong is the leading Chinese apple producer.

costs the orchard owner $10.27 per hour and $7.50 per hour, respectively. As a result, the wholesale prices for major apple varieties in the United States are ten times higher than in Qingdao, despite the lack of government programs common in the United States. The concept of comparative advantage, which is the driving philosophy of capitalism and globalization, has arrived to some Chinese farming households.

Source: G. Veeck and Z. Liu, "Fruit Farmers in Shandong, China: A Survey of Farmer's Strategies and Views." Paper presented at the Association of American Geographers Annual Meeting, Denver, CO, April 5–9, 2005.

Mineral Resource Endowment and Distribution

China possesses almost a full complement of mineral resources to further its industrial goals (Figure 19–16). For example, since the mid-1990s, China has ranked as the world's largest coal producer, with almost 35 percent of total global production. Much of the coal is found in the Loess Plateau, the North China Plain, and the western mountains of the northeast. China also was ranked as sixth in oil production, with much of its production located in the northeast, the far west, and the southwest. With its rapidly growing economy, however, China has become the second largest consumer of oil in the world and must import huge quantities of fuel oil. The recent spike in global oil prices is in part explained by China's exploding demand for oil and the resulting squeeze on global supplies. China ranks third in the global production of iron ore, but because most of its reserves are of low grade, a substantial amount must be imported. In addition, China ranks first in the world in tin and third in lead and zinc production. In terms of the production of hydroelectric power, China ranks fifth in the world

and will improve its ranking with thirty new hydroelectric dams in the planning stages. In general spatial terms, China's rich endowment of mineral resources favors the interior regions.

The Modernization of Industry

The communist government of 1949 inherited an industrial sector that had employed only 1 percent of the workforce. In addition, most of that industrial activity was spatially concentrated in regions that had experienced past foreign investment, namely, the former coastal and riverine treaty ports and the northeast, where Russian and Japanese colonial interests developed heavy industries based on iron and coal resources. The government's goal was not only to increase the contribution of industry to the national economy but also to spread the effects of industrial growth more evenly over the country. All of these industries, of course, were state-owned enterprises (SOEs).

Following the centrally planned Soviet pattern, much of the early industrial development was in the heavy industries, such as iron and steel, chemicals, electricity generation, and

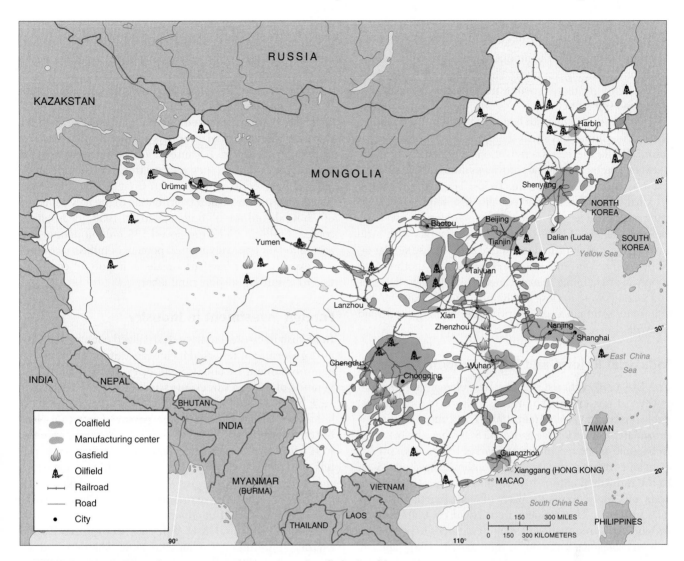

■ **FIGURE 19-16 Mineral resources and transportation links in China.** China is richly endowed with mineral resources, but economic development of interior regions has been limited by inadequate ground transportation systems.

textiles. New industrial development, based on large individual plants, was concentrated in a few locations, particularly in inland areas, where mineral resources were more abundant (Figure 19–17). During the 1950s, slightly more than half of all capital invested was allocated to SOEs situated in more isolated inland provinces. Because much of this industry was military-related, an interior location was important because sites distant from the coast were perceived as less vulnerable to external attack during the threatening cold war period of antagonism with the United States. During the 1970s, at least half of total government investment was consumed by eleven strategically located interior provinces. Apart from the megascale industrial plants, smaller-scale industrial enterprises were also established by provincial and city governments. The Communist Party officials managing these medium- and small-sized facilities sought to achieve regional self-sufficiency through captive markets and the use of local resources. The smallest

■ **FIGURE 19-17 Industrialization in China.** This aging industrial facility is a product of China's 1960s industrial drive. While many state-owned plants have been closed, those that remain continue to be inefficient and a source of heavy pollution.

industries promoted by the central government were the rural, commune-based initiatives that focused on such areas as agroprocessing and the production of agricultural equipment and building supplies. These areas of focus were intended to promote greater peasant productivity and commune self-sufficiency.

The report card on both large-scale, urban-based industry and medium- and small-scale, rural-based industry is mixed. While there was a reduction of regional inequality between coastal and interior provinces during the 1960s, the gap became wider during the 1970s. The avowed goal of regional self-sufficiency and equality resulted in a substantial waste of scarce government investment funds as well because managers of industries were bureaucrats with little training in production and marketing and because similar facilities in the same area competed against one another, with supply overwhelming demand. Because both raw materials and markets were in close proximity, China also failed to develop any semblance of a nationwide or integrated road network.

Industrial Reform and Rapid Growth

The late 1970s ushered in a dramatic change in industrial policy that was anchored by increasing levels of financial and decision-making decentralization. Thus, most of the approximately 380,000 large and small SOEs have been forced to become more responsive to market forces. Essentially, state and even provincial planning authority has been given to the individual enterprises, so that government bureaucrats and local communist party officials function as capitalist managers. This transition from a rigid, centrally planned economy to a free-market economy continues to pose monumental challenges. Among the thousand or so large and key industrial SOEs, such as iron and steel, the process of decentralization has been fraught with many difficulties. In an effort to reduce losses, the central government demanded that factories reduce their workforce. The result is that between 30 to 40 million factory workers have been laid off. The shedding of surplus industrial labor through the closing of less-efficient factories will put serious strains on China's social fabric. The region most impacted by SOE closures has been the northeast, comprised of Heilongjiang, Jilin, and Liaoning provinces. Although accounting for 8 percent of the national population, the region possesses 25 percent of the laid off SOE workers in the country as well as 25 percent of the nation's urban poor. When workers lose their jobs, they also lose their pensions and health care benefits. While there are a handful of successful and globally competitive SOEs, state sector industrial production in the northeast still accounts for 60 percent of industrial output compared to 37 percent for the nation as a whole. Because the central government is unable to allow such inefficient, but strategic, industries across the country to fold because they still contributed to approximately 40 percent of GDP in 2003, it must continuously make bank loans to the underachieving SOEs in the hopes that loans will afford SOE managers sufficient time to successfully restructure production. Given the fact that SOEs account for 80 percent of all loans from government banks, the opportunities for the fledgling and energetic private sector to borrow money from state banks are much reduced.

The process of decentralization has been accompanied by the rapid growth of town and village enterprises (TVEs), which are collectives owned by towns and villages that produce a wide variety of products for both domestic consumption and export. Because such enterprises are owned by local communities, they are free from the decision-making and fiscal constraints imposed from provincial and national governments. Their contribution to the new economy of China should not be understated. In 1996, for instance, some 23 million TVEs employed 135 million people and accounted for approximately 30 percent of industrial output and exports. TVEs have also made an important contribution to absorbing surplus rural labor.

Foreign Investment in Industry

In its rush to create a more industrialized country, the Chinese government reversed its hostile attitudes toward foreign investment by adopting an **open door policy**. The benefits of foreign direct investment (FDI) to both sides have been substantial. Foreign investors are provided tax holidays as well as access to inexpensive labor to assemble consumer goods for export and the domestic market. FDI has also been used to construct highways, ports, hotels, and shopping centers. In theory, China gains exposure to foreign technologies and managerial skills as well as expanded employment opportunities for Chinese workers. The tide of FDI commenced in 1979 when the government established four **Special Economic Zones (SEZs)**, centered on Zhuhai, Shenzhen, Shantou, and Xiamen, and in 1988 a fifth SEZ was designated on Hainan Island (Figure 19–18). In 1984 the government also established fourteen **open coastal cities**, which have operated much like SEZs but with lower levels of government funding for site improvement. A year later, three open economic regions were established—two around SEZs and one encompassing the immediate environs of the port city of Shanghai. The location of SEZs, which function in many respects as modern-day treaty ports, is not surprising in the sense that they are viewed as social and economic laboratories, geographically restricted to the country's margins. This experiment has been more than successful as FDI accounted for 40 percent of GDP and about half of export value in 2003. In addition to FDI investments in lower-value consumer products, China is expected in the near future to be a global center of electronic product exports. The country will produce half of all DVD players and digital cameras, one-third of DVD-ROM drives and personal computers, and one-quarter of mobile phones, color televisions, and car stereos.

The two most successful Special Economic Zones are Shenzhen, located in Guangdong Province just north of Hong Kong, and Xiamen, in Fujian, across from Taiwan (Figure 19–19). As the first experiment with market

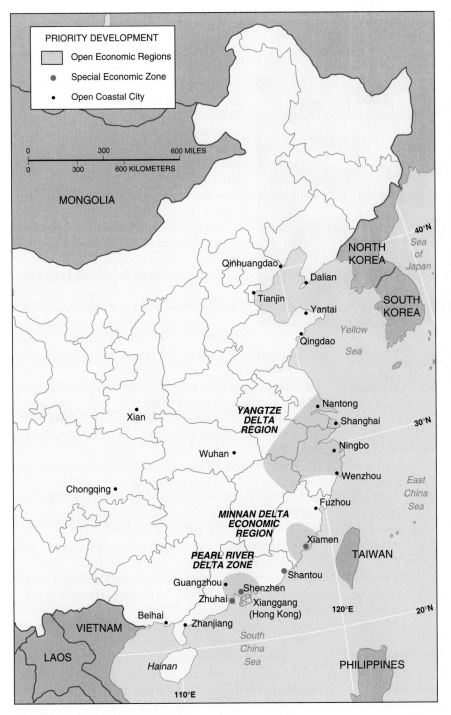

■ FIGURE 19-18 Special Economic Zones, open cities, and open economic regions. This map shows the locations of areas designated by the Chinese government for accelerated industrial and economic development.

zen's industrial and hi-tech parks now attract investments, both domestic and foreign, in electronics, biotechnology, and producer services. Another substantial recipient of foreign investment is southern Jiangsu Province, which is part of the open economic region centered on the lower Chang Jiang basin. Although the government provided FDI incentives to an additional 30 noncoastal cities in 1992, foreign investors continue to favor coastal and near-coastal locations. Between 1992 and 1995, approximately 87 percent of FDI was located in Beijing, Tianjin, Shanghai, and the 9 coastal provinces. By 2005 these regions continued to attract identical volumes of FDI. While investment from Japan, the United States, and the EU countries is significant at 25 percent, it is important to recognize that in 2001 58 percent originated from **Overseas Chinese** living in Hong Kong, Taiwan, and Southeast Asia, where land and labor costs are substantially higher (see the *Geography in Action* boxed feature, Hong Kong's Past, Present, and Future). Also important to locational decisions are kin and ethnic ties between Overseas Chinese, many of whom fled immediately after the 1949 revolution, and their home regions. These groups include the Cantonese from south central Guangdong, the Teochiu from northeastern Guangdong, and the Hokkien from southern Fujian. This is particularly true for the provinces of Guangdong and Fujian, which accounted for 28 percent of total FDI in 2003. The targeting of Guangdong and Fujian provinces by Overseas Chinese investment is not only based upon economic considerations, but upon historical, social, and cultural variables as well.

Growing Spatial Inequalities

Based on FDI, private domestic investment, and the economic impact of TVEs, interprovincial economic inequalities as measured by per capita GDP have emerged (Figure 19–20, page 442). Especially evident is the dramatic growth of wealth in those coastal provinces that are home to the majority of foreign-owned manufacturing facilities, domestically owned modern industries favored by government investment, and prosperous specialty crop farmers catering to more affluent urban markets. Coastal economic growth has far outstripped that of the central and western interior provinces, whose relative share of national wealth has

economies, Shenzen is the most dynamic of SEZs. Just a small fishing town of 30,000 in 1980, it experienced a population increase to 4.7 million by 2002. Taking advantage of decentralization policies granting greater local autonomy as well as competing with other SEZs for FDI, Shenzen has become an "entrepreneurial city" by attracting increasingly higher-value FDI. No longer promoting labor-intensive FDI in the form of textiles and cheaper consumer products, Shen-

■ **FIGURE 19-19 Billboard at the Hong Kong–Shenzhen border.** Shenzhen, the most successful of China's Special Economic Zones, has been the site of much foreign direct investment. The billboard is an example of the Shenzen government's entrepreneurial or boosterism development philosophy.

declined. The per capita GDP of Guangdong, for example, is more than triple that of the interior provinces of Gansu and Guizhou. Through time, the income gap between eastern and western provinces has in fact widened. In 1990, the average income of residents in the eastern provinces was 2.1 times greater than residents of western provinces. By 2003, however, the income gap between eastern and western provinces was 3.2 greater. With economies anchored in primary sector agriculture and mining, a core-periphery relationship between interior and coastal provinces is emerging whereby coastal provinces extract surplus value from products of interior provinces. Of course, China's leaders hope that a spatial "trickle-down" effect will occur and eventually enable interior provinces to economically benefit from coastal investment. Nonetheless, the majority of China's 16 million people who live below the poverty line are rural inhabitants of interior provinces. In an attempt to promote economic development in western provinces, the government in 2000 launched its "Go West" program, which entailed substantial investments in basic transport infrastructure. The program, however, has met with little success in attracting investments from other regions of China. To address economic disparities in China's first tier inland provinces, Beijing will begin making transport investments beginning in 2006 in the hope that better land communications with the coast will attract export processing industries.

Let us briefly examine some of the economic contours of selective, provincially based regions of eastern China to better understand the nature of economic inequality since the 1978 reforms. The Southeast, comprising the provinces of Guangdong and Fujian, is by far the most changed region. In 1978 the two provinces accounted for only 7 percent of national GDP, but that figure doubled by 2003. In Guangdong, the majority of economic growth is centered on the Zhujiang or Pearl River Delta, anchored by the provincial capital of Guangzhou (formerly Canton), but also including the SEZs of Shenzhen and Zhuhai. In 2003, Guangdong accounted for 21 percent of national FDI and Southern Fujian, centered on the SEZ of Xiamen, is the second most dynamic region of the Southeast. Even more than the Zhujiang Delta, the economy of Fujian is very enclave in nature because most of its products are exported. As a result, the two subregional economies of the Southeast lack internal linkages.

The dominant economic region of eastern China is that of the Chang Jiang River Valley, sometimes envisioned as the head of a dragon in the form of Shanghai, and the body and tail composing the rest of the river. As a region possessing a quarter of the country's population, it accounts for 30 percent of FDI and 20 percent of the country's exports by value.

Although lagging in economic reform until the early 1990s, Shanghai was chosen by the central government as a counterweight to the rapidly reforming Guangdong. Since central government control has lessened, the city-region has attracted many major foreign multinational corporations producing goods for both export and domestic consumption. Economic growth has spatially spread outward from Shanghai to include the nearby provinces of Zhejiang and Jiangsu.

In 2003 these two provinces plus Shanghai accounted for 21 percent of national GDP and 33 percent of FDI. Attracted to newly constructed expressway connections, other cities that have prospered in Shanghai's shadow include Nanjing, Suzhou, Wuxi, and Ningbo. While traditionally supporting heavy industry, the city of Wuhan is eager to become the premier upstream growth center, based on an economic development zone that has attracted substantial foreign investment, an interior finance and information technology center, and a transport intermediary between eastern and western China.

The region of the Bohai Sea Rim comprises the two municipalities of Beijing and Tianjin and the provinces of Shandong and Hebei. Encompassing almost the entire North China Plain, this region supports 14 percent of the national population and accounted for 19 percent of national GDP in 2003. Despite being the country's historic core, capital city region, and an important source of industrial production, the Bohai Sea Rim is perceived as not reaching its potential. Reasons for its economic shortcomings include land pressure because of high population densities, a shortage of water for industrial growth, a dependence upon imported coal to fuel its inefficient heavy industries, an inefficient rail system, and the most polluted air and water environment of any region in the country. Regional economic fortunes must be anchored in the technological upgrading of industry so that the Beijing and Tianjin regions can assume the roles of economic gateways between northern China and the Pacific Rim.

GEOGRAPHY IN ACTION
Hong Kong's Past, Present, and Future

Hong Kong Island was first ceded to Britain in 1842 as part of the spoils of the First Opium War, and the adjoining mainland areas of Kowloon and the New Territories were added by 1898. On July 1, 1997, the British Crown Colony of Hong Kong (population 6.8 million) that was often referred to as a "borrowed place on borrowed time" was returned to China. The nearby Portuguese colony of Macao was returned in December 1999.

Hong Kong's economic functions have changed over time. For much of the period before 1949, Hong Kong served as an entrepôt, or transshipment point, for much of south China's external trade. After the Communists gained power, Hong Kong lost most of its commercial links to mainland China, and the colony was forced to shift its economic energies to the manufacturing of inexpensive consumer goods for the export market. Cheap labor was abundant, with many refugees continuing to flee the mainland even decades after the 1949 revolution. While some sweatshop industries remain, exporting goods primarily to the West, Hong Kong's economy has become well grounded in the high-value service sector, which in 2004 represented 87 percent of GDP. Hong Kong has become a major global banking and finance center and a leading maritime shipping center (Figure A). Tourism is its second most important income generator.

The critical question is how long will this prosperity last? Hong Kong has been designated as a Special Administrative Region by Beijing and has been promised that it will be allowed to retain its economic and social system, with Beijing controlling the functions of foreign affairs and defense for the next 50 years. Essentially, the arrangement proclaimed is a one-country, two-systems relationship. Despite these guarantees, tens of thousands of wealthier Hong Kong inhabitants, lacking confidence in the assurances of China's communist leaders, have emigrated to Singapore, Australia, Great Britain, and North America. Because China stands to reap such substantial economic benefits from its relationship with Hong Kong, it would appear unlikely that Beijing will significantly alter the character of the region. What is more probable, however, is that the concentration of wealth in Hong Kong will not be so great as in the past. Beijing will likely distribute growth opportunities more widely and will continue to promote Shanghai for a Hong Kong-like role in northern and central China. Hong Kong can no longer rely on its traditional role as China's window to the world but must economically remake itself. While continuing its role as a global financial, maritime transport, and tourist center, cosmopolitan Hong Kong must promote an even higher-value, knowledge-based economy anchored in information technology and communications.

Resigned to a reduced role in China's national economy, many Hong Kong residents are increasingly concerned about the erosion of democratic institutions in the aftermath of the colony's return to China. The system of checks and balances between the Beijing-appointed chief executive, the locally elected legislature, and the judiciary has weakened in the past few years in favor of the chief executive. This means that Beijing has increased power to influence the internal political and legal environment of Hong Kong.

■ **FIGURE A** **Victoria Harbor and the skyline of Hong Kong.** Hong Kong has emerged in recent decades as one of the world's leading financial and shipping centers. Its return to China in 1997 greatly strengthened that country's global economic position.

The Northeast region has fared the worst during the reform process. Once the heart of heavy industry, the Northeast's share of national GDP had declined to only 9 percent in 2003. Several reasons explain the Northeast's inability to conform to China's economic transition. First, the substantial natural resources that fueled the region's energy-intensive heavy industries are declining; both oil and coal must now be imported from other provinces or from overseas. Second, the economy of the Northeast was traditionally based on a few mega-sized SOEs, with a long tradition of central planning. Under such an institutionalized economic environment, the regional economy has become inefficient and inflexible under the post-1978 reform regime. Third, reduced central government investment has meant substantial debt and the resulting inability to dedicate finances to technological upgrading. One subregional bright spot is the port city of Dalian, where Japanese and South Korean corporations have invested in plants utilizing the cheap domestic labor.

Gender Impacts of Economic Reforms

Male bias in the development process in China is not restricted to the post-reform development period but stretches back to the time when the Communist government was established in 1949. While gender equality improved dramatically when the Communists came to power (new marriage laws, for instance, legally allowed women to choose their marriage partners and granted rights of divorce), Maoist policies did not totally transform the traditional male-dominated gender roles. In fact, the government used these traditional gender roles to

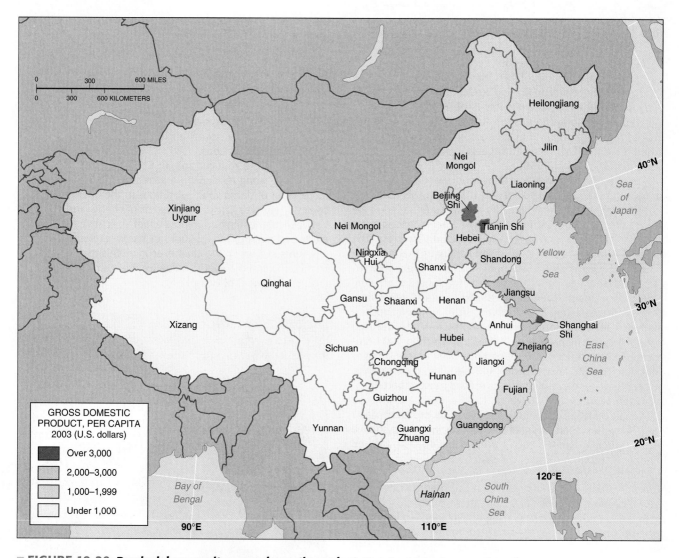

■ **FIGURE 19-20 Provincial per capita gross domestic product.** Despite economic prosperity at the national level, China's interior provinces lag seriously behind the coastal provinces.

promote implicit political or economic purposes rather than allowing women a more self-determining role. During the Great Leap Forward, for example, women's status was elevated to that of men because women were needed to engage in heavy manual field labor. Women were liberated during the Cultural Revolution based upon the freedom to dress like men in an "androgynous" or "unisex" manner.[5]

During the current period of economic reform or "transition" in which the government has adopted capitalism to drive economic growth but retains many facets of government control and planning, policies have also harnessed traditional Confucian philosophy to gender the development process. This hybrid philosophy directly impacts the different employment opportunities of male and female who engage in rural to urban migration. Rural migrants comprise the majority of labor for urban growth and export-led manufacturing, as the native urban population view these occu-

pations as low status. Of these migrants, however, males comprise approximately two-thirds of the flow while females account for one-third. This large flow of males from villages has created a "feminization of agriculture" so that women now are responsible for 60 percent of agricultural work. Farming females are perceived by the government as being "virtuous" because by remaining in the kitchens and fields, they are allowing their husbands' money-earning opportunities in urban areas. In this sense, traditional social construction of gender roles is reinforced through Confucian ideals of women's position within the family, and females lack opportunities to economically empower themselves through wage income generating opportunities.

These traditional female roles are further reproduced among females who do migrate to urban places in search of wage income work. The most common work for females is in manufacturing that requires detailed assembly work (Figure 19–21) such as garment workers or seamstresses, or as domestic servants, restaurant servers, and hotel workers. In contrast, males are highly represented in more physically demanding factory work, and as constructing laborers,

---

[5]Xiaorong Li, "Gender Inequality in China and Cultural Relativism," in *Women, Culture and Development,* edited by Martha C. Nussbaum and Jonathan Glover (Oxford: Clarendon Press, 1995).

FIGURE 19-21 Female factory workers in the Pearl River Delta in Guangdong Province. Most workers are rural-to-urban immigrants and assemble finished products for export.

Urbanization and Migration

In the world's most populous country, issues of urbanization and migration become central to the processes of economic planning and social development. The agricultural and industrial reforms of the modern period have resulted in increased rural-to-urban migration and accelerated social change.

Controlled Urbanization Under Mao

During the opening decades of Communist rule, the government limited peasant migration from the countryside through food rationing and restrictions on urban industrial development. The impact of these policies in reducing the unplanned growth of urban communities was so substantial that the percentage of the population classified as urban had reached only 12.8 percent by 1978 (Table 19–2). This extremely low level of urbanization was significant because it enabled China to avoid the rapid growth of peripheral squatter settlements so common in most non-Western countries.

Certainly, the Chinese variety of Marxist philosophy indirectly provided arguments in favor of restricting rapid urban growth. Urban places were viewed with contempt and distrust because cities were inhabited by more educated and commercially oriented capitalist urban classes. Cities were also viewed as being "parasitic" or "nonproductive" because the urban elites were perceived to be living off the products of rural labor. A bold and disruptive policy expressing this antiurban philosophy materialized during the **Cultural Revolution** (1966–1969), which was marked by a period of deurbanization. Millions were sent to the countryside to receive "correct political thinking" from the rural masses. In reality, however, government policies were actually anchored in the more practical concern of a potential inability to feed

porters, and manual laborers. This gendered division of the migrant urban labor force reflects the situation whereby migrants do not choose jobs, but the jobs choose the migrants. Time spent by migrant laborers and their age is also gendered. Female workers are primarily younger, single, and the majority are between 15–24 years old. Urban employers operating under the capitalist model of production efficiency prefer these "maiden workers" because they are perceived to be more docile and better able to engage in detail work. Married women are viewed as less desirable because they demand higher wages and time off for family reasons. Migrant males, however are able to engage in demanding physical work well into their 40s, and family responsibilities are less disruptive for husbands when compared to wives.

Traditional marriage norms restrict the time spent by female work migrants in urban places. Not only are young females no longer desirable as workers by their late 20s, but finding a husband in the city is difficult because female migrants do not possess an urban residency permit. In addition, migrant females believe they have little to offer more sophisticated, urban men. By the time a woman is in her late 20s, there exists substantial pressure to get married, and she is pushed back to her village of origin to find a spouse. When compared to male migrant workers then, work in urban places is far more an "episode" in women's lives, as marriage generally means returning to a rural place of residence. Once a woman marries, there also exist socioeconomic pressures to remain in the village. Despite cutting short opportunities for greater socioeconomic mobility, migrant work allows women to send remittances to their families, and greater empowerment through greater exposure to worlds outside the village. Nonetheless, it is appropriate to make the observation that poorly paid migrant work is an extension of women's family roles because migrants have simply replaced the constraints that they left at home for a different form of subordination in the city.[6]

---

[6]Cindy C. Fan, "Rural-Urban Migration and Gender Division of Labor in Transitional China," *International Journal of Urban and Regional Research* 27:1 (2003): 24–27; Fan Dai, "The Feminization of Migration in the Pearl River Delta," *The Journal of Chinese Geography* 8:2 (1998): 101–115; Xiaorong Li, "Gender Inequality in China and Cultural Relativism," in *Women, Culture and Development*, edited by Martha C. Nussbaum and Jonathan Glover (Oxford: Clarendon Press, 1995).

Table 19-2 Urban Growth in China, 1949–2004

Year	Total Urban Population (millions)	Percentage of Total Population Classified As Urban
1949	49.00	9.1
1953	64.64	11.0
1957	82.18	12.7
1960	109.55	16.5
1964	98.85	14.0
1969	100.65	12.5
1978	122.78	12.8
1982	152.91	15.1
1986	200.90	18.9
1994	345.39	29.0
1998	372.75	30.0
2002	486.67	38.0
2004	533.04	41.0

Sources: T. Cannon and A. Jenkins, *A Geography of Contemporary China* (New York; Routledge, 1992), p. 210; *World Development Report, 1998/1999* (New York: Oxford University Press, 1998/1999); and *World Population Data Sheet 2004* (Washington, D.C.: Population Reference Bureau, 2004).

a rapidly expanding national population if rural-to-urban migration were left unchecked. The administrative tool to legally control the movement of people between rural and urban areas was a place-of-residence or *hukou,* registration system, wherein rural and urban dwellers were issued registration booklets that restricted their access to food, employment, housing, and health services to their home area. Without being registered as urban inhabitants, rural-to-urban migrants found it very difficult to live permanently in the city.

Economic Reform and Rapid Urban Growth

With the loosening of the place-of-residence registration system, the surplus rural labor created by the agricultural reforms was free to move from villages to small, medium, or large urban places, thus swelling the urban proportion of the country's population. Between 1978 and 2004, the proportion of China's population living in urban areas more than tripled, from 12.8 percent to 41.0 percent. China's urban population of 533 million is almost twice the total population of the United States.

By 2002, more than 120 million migrant workers had found their way to China's rapidly growing cities. Because of the large population base, this was the single largest movement of people within the borders of any country during the twentieth century. Interestingly, the majority of the migrants have moved to urban places within their home county; some 28 percent have left their home county but remain within their home province. As a result, a dramatic change in the overall geographical distribution of population and rankings of urban places has not occurred. Longer-distance interprovincial migration has favored the more economically dynamic coastal provinces. The five largest cities not including Hong Kong are, in order, Shanghai, Beijing, Tianjin, Wuhan, and Guangzhou. Of the 174 urban places in 2003 with greater than 1 million population, 94, or 54 percent, are located in coastal provinces. Just the provinces of Guangdong and Jiangsu (the province neighboring Shanghai on the north), possess 22 and 20 cities with populations over 1 million. In addition, 7 of the 11 urban places with 4 million or more people are located in coastal provinces.

Urban Housing

Problems associated with inadequate urban housing and pollution were pervasive during the prereform era. With most of the government's investments directed toward building "productive" heavy industry, urban living levels remained low. A land-use philosophy that kept industrial workers housed in apartments adjacent to factories (thereby reducing travel costs to work, shopping, and school) resulted in extremely high levels of noise, water, and atmospheric pollution. Traditionally controlled by the work unit, housing was subsidized by the government and was thus officially viewed as a social good rather than an economic good. In a sense, low wages translated into low rent. The postreform period, however, has been marked by the government slowly but gradually dispossessing itself of its landlord role through the

introduction of market principles. This includes all new construction being sold on the private market and the sale or renting of public work-unit housing to existing tenants. To promote home ownership, the government has introduced a savings fund whereby employees dedicate a small percentage of their annual wages to a future housing purchase. With the average living space per urban resident increasing from 3.9 square meters (42 square feet) to 8.5 square meters (91.5 square feet) over the past decades, it is obvious that market reforms have benefited many. Nevertheless, private housing ownership rates remain relatively low because of two interlocking factors. Because rents in state-owned housing are often less than 5 percent of annual income, many households continue to rent rather than purchase their own home or apartment on the private housing market. The second barrier to home ownership is the high cost of private housing. Despite high personal savings rates, purchasing a home on the private market is beyond the financial capabilities of the vast majority of urban residents. It is a foregone conclusion, however, that the living conditions of the poorest households will become worse as the pool of rental housing shrinks.

The new housing programs, unfortunately, have had little impact on the flood of rural migrants, who are sometimes referred to as the floating population. In Shanghai, for example, the rural migrant population has increased from 1 million to 3.5 million. In many cities, rural to urban migrants compose as much as 20 percent of the urban workforce. Lacking the economic means to purchase or rent adequate housing, many migrants who are not provided housing by their employer have illegally settled on land on the urban periphery, where they live in makeshift shanty dwellings that lack clean water as well as sewage and garbage facilities. Social disruption as a consequence of the poor living conditions and government laws restricting access to education and health services by the rural migrants are growing problems that must be addressed by the urban authorities. The problem will only get worse, as government experts predict an additional influx of 400 million rural migrants by 2020, with about half of this total moving to economically dynamic coastal provinces.

Contrasting Beijing and Shanghai

Urban places are often mirrors of social and economic change. Think, for example, of the differences between New York and Los Angeles, Moscow and St. Petersburg, or Munich and Berlin, and how these city pairs symbolize competing visions within their own respective countries. Throughout the last 100 years, the same distinction can be drawn between Beijing and Shanghai, the two largest cities in China, with 13.8 million and 16.7 million inhabitants, respectively (Figures 19–22 and 19–23). Beijing's origins date back to the thirteenth century when the newly installed Mongol government established its capital at the northern edge of the North China Plain. Until the modern period, Beijing remained one of the world's grand imperial centers. Shanghai's origins are less impressive. Until the sixteenth century,

■ **FIGURE 19-22 Beijing skyline.** The urban landscape of Beijing has both a traditional and contemporary flavor. The walled Forbidden City in the foreground reflects the traditional and inward-looking nature of Chinese political culture. Economic liberalization is reflected in the contemporary skyline.

it was a fishing village, after which it grew into a minor walled administrative center; its meteoric growth only commenced with its designation as one of the handful of treaty ports in the mid-nineteenth century.

Beijing has traditionally typified the conservative, orderly, and inward-looking nature of Chinese culture. Centered on the Forbidden City and fronted by Tiananmen Square, the older parts of the city are crisscrossed with broad avenues that invoke certain principles of traditional Chinese cosmology and that are punctuated by monumental architec-

ture centered on large open spaces. As the political center of the Middle Kingdom, Beijing's primary function was as a cultural and administrative center. Shanghai, on the other hand, represented the outward-looking and commercial nature of Chinese culture that was characteristic of the southern part of the country. The symbolic center of the city was the "bund," a wall of Victorian-period commercial establishments fronting the Huangpu River. Situated at the delta of the Chang Jiang, the city had command over a vast and productive economic riverine hinterland. Unlike the Huang He, which is too shallow to accommodate seagoing vessels, the Chang Jiang is navigable by large vessels some 600 miles (960 kilometers) upstream. By the 1930s, Shanghai had already become the eighth busiest port in the world.

Following the rise of the Communists, the functions of the two cities changed little. Although Beijing possesses some light and heavy industry, much of its potential for industrial growth was diverted to the nearby former treaty port of Tianjin. Shanghai, on the other hand, exploded with economic activity as access to its riverine hinterland was fully exploited. It became the leading industrial center of China, focusing primarily on the chemical, textile, metal, and food-processing industries. The reforms of the late 1970s, however, brought substantial changes to the landscapes of both cities. Beijing has become a far more cosmopolitan city, with high-rise hotels and office buildings, separation of industrial and residential land uses, and high-technology industries using a well-educated workforce trained at the many world-class universities located in the city. The modernization of Beijing's landscape means that many of the traditional, compact, and narrow-laned urban neighborhoods are rapidly disappearing. A partial reason for the emergence of a more modern and globalized urban landscape is to prepare Beijing for hosting the 2008 Summer Olympics.

Shanghai's urban landscape makeover has been even more dramatic. Many of the older state-run factories are being torn down or relocated to the outskirts of the city. The huge Pudong development area was established in 1990, just across the Huangpu River, to attract foreign investment in the hopes that China will become the Pacific Rim's next "economic tiger" (Figure 19–24). Located on the south side of the Huangpu River, Pudong supports the new international airport, the urban region's largest container port, nine development zones focusing primarily on high value manufacturing such as autos and electronics, as well as financial services. While capturing only 7 percent of FDI for the entire Shanghai urban region in 1990, Pudong by 2002 accounted for 37 percent of total FDI. Much like the Pearl River Delta, Hong Kong provides the single largest share of Pudong's FDI. As a symbol of China's commitment to economic growth and globalization, Pudong supports the eighty-eight story Jin Mao Building. The futuristic ninety-five story Shanghai World Financial Center

■ **FIGURE 19-23 Shanghai skyline.** From its treaty port period in the nineteenth century, reflected in the low-rise buildings of the colonial-era Bund along the Huangpu, to its contemporary efflorescence as a modern commercial center of more than 16 million inhabitants, Shanghai has become one of China's most important commercial and manufacturing centers.

■ FIGURE 19-24 The Pudong Financial District. Rising from rice paddies and aging factories, Pudong symbolizes the "new" Shanghai beginning in the early 1990s.

is scheduled for completion in the near future. Government planners have not only set their sights on Shanghai replacing Hong Kong as the financial and trade center of the country, but also they intend for Shanghai to assume the rank of a "world city."

Population Contours

The magnitude of China's huge population of more than 1.3 billion persons can be appreciated through the simple observation that the number of people added to China's population since 1975 is slightly greater than the entire population of the United States. With only 7 percent of the world's arable land, China faces great challenges in raising its citizens' levels of living. Indeed, while growth rates have been substantially lowered over the past 30 years, the population is projected to increase to approximately 1.5 billion by 2025 before beginning to decline.

Population Distribution

The spatial distribution of China's population is highly uneven (Figure 19–25), generally reflecting the climatic patterns that we have already discussed. Some 94 percent of China's population resides in the humid eastern regions, which comprise only 43 percent of China's land area. Closer examination reveals that about 40 percent of the population occupies some 10 percent of the land, mostly the alluvial valleys of the lower Huang He and Chang Jiang. Population densities in some of these more crowded rural regions may easily reach 150 persons per square mile (58 persons per square kilometer). As a general rule, population densities progressively decrease with increasing distance from the coast. Some provinces are far more populated than others as well. For example, in 1997, Henan supported some 92 million people, which—were it an independent country—would rank Henan as the world's twelfth most populous nation. In contrast, Qinghai, the province encompassing the huge Tibetan Plateau, possesses a population that is only one-third the number of persons residing in metropolitan Shanghai.

Spatial variations also exist in the distribution of cultural minority groups. The Han are China's majority culture group,

accounting for some 93 percent of the population. The government officially recognizes fifty-five different minority groups, of which fifteen possess populations greater than 1 million persons. The largest ethnic group is the Zhuang (13.3 million), who are scattered across several provinces in the southwest. The second largest minority group is the Hui (8.6 million), who are Muslim and primarily inhabit Ningxia, with secondary concentrations throughout the arid west. The third largest group is the Muslim Uygur (7.2 million), who inhabit much of Xinjiang. With the exception of Xinjiang and Tibet, every province or autonomous region possesses a Han Chinese majority.

Population Growth

China's large population reaches back many centuries (Figure 19–26). By the mid-1700s the population was already approaching 200 million, up from only 20 million during the mid-1600s. China was, therefore, the most populated country in the world in the mid-1700s. This massive increase in population is attributable to a century of relatively peaceful conditions and improved intensive farming technologies. Equally important was the introduction of New World crops such as maize and sweet potato in the 1600s, which allowed marginal land to be cultivated, thus increasing the capacity to feed the people. By the time the Communists assumed power in 1949, the population had ballooned to approximately 548 million people, and during the next 30 years the number of people nearly doubled again, to 1 billion.

Population growth increased dramatically during the first two decades of communist rule. While population-control programs were instigated during the early years of the communist era, they were short-lived because of the lack of political support. In general, Mao Ze-dong held "pronatalist" views based on the Marxist notion of more workers producing more economic goods. This relationship between production and population size was perhaps best captured in Mao's popular saying that "every stomach comes with two hands attached." Rapid population growth was also a byproduct of the very successful health-care delivery system that was implemented. A healthier population meant a drop in the death rate, but because birthrates remained high, the population growth rate increased. It was not until the early 1970s, when the government came to view people as consumers rather than producers, that a systematic national birth-control program was implemented.

Policies to Reduce Population Growth

The policies first introduced to reduce population growth were extremely successful. Between 1970 and 1978, the total fertility rate was more than halved, from 5.8 to 2.7, and the natural rate of increase declined from 2.6 percent to 1.2 percent. The family planning programs, which took on a revolutionary flavor, were able to convince people to marry later, to extend the interval between births, and to have fewer children. Birth quotas were expected to be adhered to, and educational programs specifically designed to modify traditional attitudes toward children became widespread. Two specific

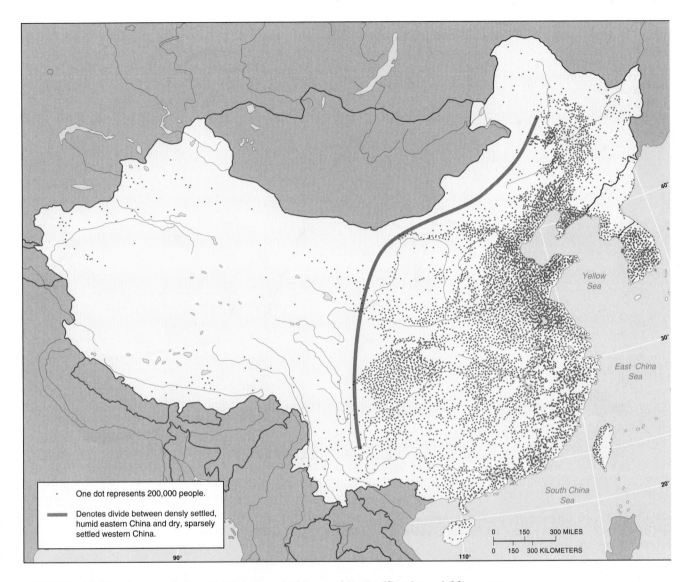

■ **FIGURE 19-25 The population distribution of China and its Pacific Rim neighbors.** China, the earth's most populous country, is most densely settled along the eastern coastal regions and the interior river valleys. South Korea and Taiwan also support high population densities.

programs were very important. First, the distribution of contraceptives became more extensive, so that by the end of the decade, some 70 percent of married couples of childbearing age were using some form of contraception. Second, improving educational opportunities for women proved successful in reducing fertility rates, as many of the more educated women chose to delay marriage. By 1980, the proportion of women married by age twenty was only 24.9 percent, down from the 1958 level of 70.9 percent.

While these population-control policies were very effective, the government inaugurated an even stricter program in 1979, the goal of which was to reduce the population.

The suggested two-children-recommended policy throughout much of the 1970s was replaced by a mandatory one-child policy. Attempting to reach the one-child goal has required a complex system of economic rewards for those who conform and penalties for couples who have two or more offspring. The economic rewards remain in place through the child's teenage years and include better living accom-

modations, extra grain rations for the parents, and improved education and employment opportunities for the single child. Parents producing two or more children are fined and, in many cases, the mother is forced to have an abortion. There are exceptions to the one-child policy, predominantly among the country's non-Han minority groups, who are exempted. Under conditions of the physical disability of the first child, special needs to preserve the family line, and sometimes in cases where the only child is a girl, the one-child restriction may also be waived. The number of exceptions to the policy increased in 2001 when Beijing allowed provinces to enact their own policies to meet local circumstances.

While the one-child policy has substantially reduced population growth, it has not been completely effective because of the relaxation of strict enforcement in rural areas where two children are allowed, particularly when the firstborn is a girl. Despite the obvious surplus of labor in the countryside, families still desire more than one child, due to tradition and also family labor needs. This slight upward trend

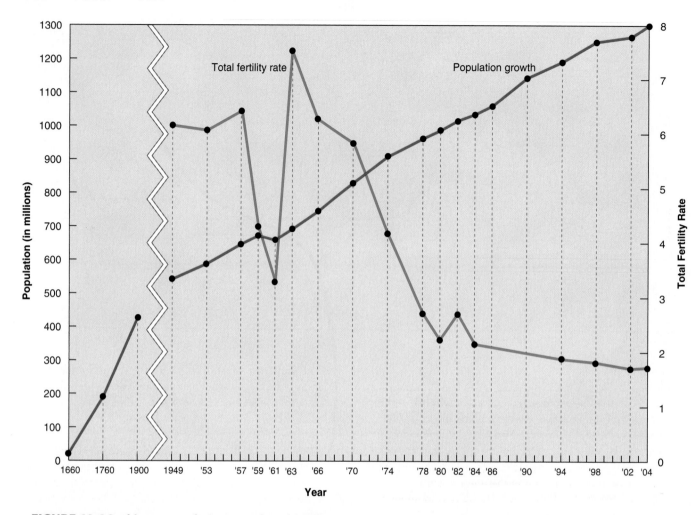

■ **FIGURE 19-26 Chinese population growth and fertility trends.** Although China's total fertility rate has dropped dramatically in the last decades, population growth continues, owing to large numbers of people reaching their reproductive years.

in annual natural increase, however, may significantly threaten long-term population-reduction goals.

While many rightly believe that the fertility reduction of the one-child policy resulted in improved living conditions for the majority of Chinese citizens, they also acknowledge negative moral, social, and economic consequences. Chinese, like a handful of other Asian cultures, have always shown favoritism toward male offspring for a variety of reasons. In China, female infanticide (the systematic and intentional killing of infants) was a problem until the Communist revolution, when the practice all but disappeared because of the government's ideology of male-female equality. After implementation of the one-child policy, the deficit of females skyrocketed through not only infanticide, but also sex-selective abortions, which were aided by prenatal sex detection technologies such as ultrasounds. As a result, some sources suggest that since the one-child policy went into effect, some 150 million females are missing from the population. The consequences are that China's sex ratio stands at 117 males for every 100 females, compared to the world average of 105 to 100.

The geography of missing females in China indicates that it is in Han-dominated provinces where the problem is most

acute. Nonetheless, there are some Han-dominated provinces, many in the south, with far more abnormal sex ratios than others. Economic factors are often suggested as a reason for fewer females. While poorer provinces do indeed exhibit a missing-girl problem, some rich provinces possess more skewed sex ratios. Indeed, higher than normal rates of female deaths occur in all types of households, regardless of economic conditions. Oddly, the same is true for educational attainment, in that education does not possess strong explanatory power for the decision to abort a female fetus. While it is recognized that the missing-girl phenomenon is higher in rural areas, abnormal urban sex ratios were still much higher than normal. While economic status, place of residence, the one-child policy, and technology all play a role in the shortage of females, the most important explanation is traditional son preference, despite the government's many campaigns to convince families that a single female offspring family is virtuous (Figure 19–27). All things equal then, female abortions and infanticide are most common in families that possess only one or more daughters or more daughters than sons.

The direct consequences of single-child families as well as a shortage of females are many. The shortage of females

■ FIGURE 19-27 A government billboard advertising the belief that a single female offspring family is a happy family. The nuts or seeds of the lychee fruit symbolize close family ties.

has led to a "marriage squeeze" in which there are too few females for men wanting to marry. While this situation can sometimes lead to an increase in female status because more women can now exercise greater choice in mate selection, economically poor men are ultimately disadvantaged. Coupled with the impact of globalization and capitalism, the shortage of females has fueled the illegal trafficking of hundreds of thousands of girls and women to support a growing sex work industry. This is not a sign of development, when the basic human rights of females are violated. Another economic and social consequence of the one-child policy is future high dependency ratios. In a one-child society, the number of elderly dependents increases dramatically relative to the number of children, and the ability of sons and daughters to fulfill their traditional social safety-net obligations to their parents is threatened. This problem is exacerbated by a deficit of females because in any culture in the world, it is generally daughters who are more dedicated to taking care of aging parents. The capacity of the government to care for the elderly will also be threatened, since it is predicted that by the year 2030, the elderly proportion of the population will reach a staggering 25 percent.[7]

Environmental Sustainability

A substantially large population coupled with rapid economic growth begs the question of whether China's environment has suffered as a result. In general, the answer is that the environment has been sacrificed in the government's drive to modernize, whether during the earlier communist period or the present era of reform. Much like the experience of the former Soviet Union and its European satellite states, the race to modernize industry and agriculture in order to prove

to the capitalist world that the socialist system was superior, left the environment much degraded. The bulk of industries, for example, have been dominated by high energy and resource consumption, where quantity rather than quality and efficiency remained paramount. During the reform period, environmental degradation has only worsened, simply because there now exist many more and different types of polluters. Although environmental protection officially became a government priority in the early 1980s period of reform, legislation and enforcement has left much to be desired. Many current and future environmental problems face the government, but only two of the most serious are discussed here.

Perhaps the most basic environmental challenge facing China is the provision of clean freshwater. Because of demand levels coupled with wastage, half of the country's medium-to-large cities regularly experience water shortages. In Beijing, for example, increased demand between the 1950s and early 1990s lowered the water table from 5 meters to 50 meters below sea level. The dramatic rise in subsurface freshwater extraction in coastal cities has resulted in substantial saltwater intrusion during the dry season. Water wastage is substantial because low water prices to the user provide few financial incentives to repair distribution systems or reduce consumption. Factories in China recycle only 20 to 30 percent of water as compared to 70 to 80 percent in developed countries. Mismanagement is also causing severe shortages in some regions. Because so much water is being diverted for irrigation, hydropower generation, and industry use in the upstream stretches of the Huang He, water shortages are frequent in the North China Plain. In fact, the lower stretches of the river dried up for months in 1995 and 1996! These water problems dramatically impact agricultural production in provinces north of the Chang Jiang, which supports approximately 64 percent of cultivated land but possesses only 17 percent of exploitable national water resources. The planned

[7]Judith Banister, "Shortage of Girls in China Today." *Journal of Population Research* 21:1 (2004): 19–45.

construction of a north-south water transfer project transporting water from the Chang Jiang to the dry north further complicates the issue. With respect to water pollution, it is estimated that 80 percent of China's sewage and other dirty waste enters rivers and lakes untreated. Some three-quarters of the 500 rivers monitored by the Ministry of Water Resources are badly polluted. The many township and village enterprises that have become critical to China's economic growth drive are the sources of much of this pollution.

The other major environmental problem is air pollution in both older urban areas and regions, as well as those experiencing dramatic postreform growth. Indeed, a 1998 World Health Organization report listed seven of the ten most polluted cities in the world as being in China. Beijing is a good example of the former. Beijing residents are literally choking in a city that annually experiences an average of 265 days of smog-filled skies. The sources of the pollution are high-sulfur coal and leaded gasoline. The city's residents and factories are highly dependent on coal as a heating and energy source. It is estimated that 70 percent of coal used does not meet the government's stated environmental standards. Consumed in automobiles, trucks, and buses, cheaper leaded gas and diesel fuel also contribute to Beijing's dim, hazy skies. Adding to the problem are slow-moving vehicles, which attempt to negotiate streets that simply were not designed for such a large volume of traffic. While the Beijing government has implemented tougher pollution legislation, compliance is weak at best. It is no surprise that respiratory problems are widespread, particularly among children.

In the once agriculturally lush landscape of the Pearl River Delta of Guangdong, air pollution is also serious as the thousands of small, export-oriented enterprises choose to use outdated and highly polluting coal-burning technologies because they cost less. Because of the decentralized nature of production, the spread of pollution is substantial; a blanket of polluted air hangs over much of the delta region for months each year and even negatively impacts the air quality of Hong Kong during the winter months.

Challenges to China's Future

While few would disapprove of China's choice to pursue a more market-oriented economy and to promote a more open society, the break with the past has been so substantial that numerous development challenges have emerged. Foremost is the growing regional inequality in economic development between the increasingly prosperous coastal and riverine zones and the poorer interior regions. This problem exacerbates regional antagonisms, and it places great economic and social strains on coastal governments who are being forced to support millions of poor migrants. Although not a potential problem in interior Han China, the continued economic impoverishment of non-Han territories in China's interior far west might generate feelings for greater political autonomy; minorities have already developed stronger economic and social ties with their ethnic cousins in central Asia. This regionally based economic inequality is matched by a widening gap in income inequality as well. The rich have gradually come to control an increasingly larger slice of the economy; approximately 10 percent of the population, for example, now controls 60 percent of bank deposits.

Because of the shedding of surplus labor in the agricultural sector and the shift to more profitable crops at the expense of grain, some scholars question whether China can continue providing for itself nutritionally from domestic sources. Grain production has already leveled off, and some speculate that the demand for livestock feed will increase dramatically as dietary habits reflecting increased prosperity switch to greater meat consumption. The conversion of prime agricultural lands to industrial development within the immediate areas around cities only increases the difficulty of China feeding itself in the future. Likewise, the prospect of shedding millions of SOE workers also threatens social and political stability because the government will no longer be able to rely on their loyalty to the Communist Party; beginning in 1999, cities across China witnessed an increasing number of demonstrations and strikes. Equally threatening to sustained development is the high level of corruption among local government officials. For example, local officials often overtax farmers, or directors of failing SOEs siphon off profits to line their own pockets. Additionally, the Communist government must confront the specter of increased political discontent as the economy expands. As symbolized by the **Tiananmen Square** uprising in 1989, economic growth begs for parallel reforms that have not yet been forthcoming in the political sphere of development. However, China is an example that economic growth is not necessarily dependent upon more democratic forms of government.

The rapid globalization of China's economy has and will continue to cause, at least in the near term, concern among its trading partners. North America and Europe possess high trade deficits with China, prompting calls for a variety of trade reforms through increased import tariffs on Chinese imports. We must remember, however, that a substantial slice of imports originate from Western-owned plants in China, or Chinese-owned plants with production contracts with Western retailers such as Wal-Mart. The West has also asked Beijing to revalue its currency downward so that Chinese imports become less costly, which in turn would reduce trade deficits. The West, and particularly the United States, is also concerned about China's growing military strength and the potential threat this poses to the larger geopolitical stability in the western Pacific Rim region. Such concerns, however, should be cast aside as the Chinese government believes that military confrontation would jeopardize its current rapid economic growth trajectory and engagement with the global economy.

China's Pacific Rim Neighbors

The three independent countries which are China's Pacific Rim neighbors, are North Korea, South Korea, and Taiwan. While North Korea remains a hermit-like communist

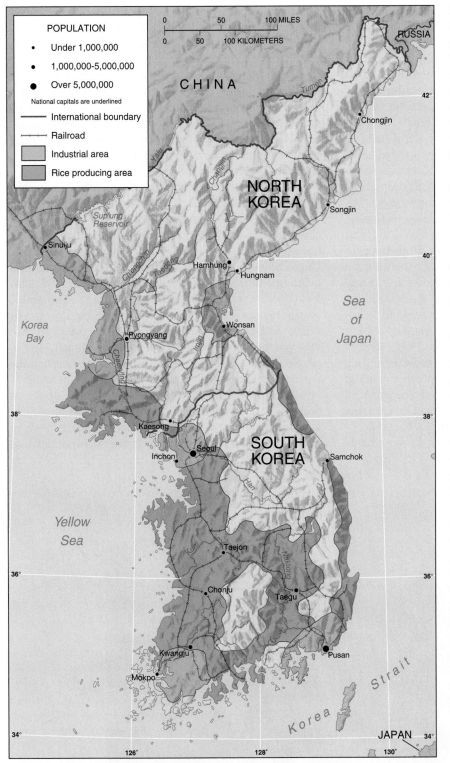

■ **FIGURE 19-28 The two Koreas.** The Korean peninsula is divided into the countries of North Korea and South Korea, which vary in levels of agricultural and industrial development.

in which the state either partially owns corporations that operate under commercial principles or directs the structure and orientation of the national economy through various economic policies. Although Pacific Rim economies are classified as capitalist, the role of the state is more pervasive than in the free-market capitalism of the West. Two explanations for the emergence of this "managerial capitalism" must be considered. First, because private capital was so scarce at the onset of industrialization, the national governments largely raised and coordinated the allocation of investment capital. Second, strong state control was deemed necessary to survive in an intensely competitive global economy.

The Two Koreas

Between Japan and China lies a rugged and mountainous peninsula that is a little larger than the state of Utah (Figure 19–28). Although native kingdoms such as the *Silla* and *Koryo* emerged during the first millennium, the Japanese and particularly the Chinese exerted significant influence on the development of Korean culture. The peninsula received a Chinese cultural package that included a system of writing, art, Confucian principles, Buddhism, and political norms. Japan's influence has been more recent and primarily economic in nature. From 1905 to the end of World War II, the peninsula became an integral part of the Japanese Empire, functioning as a source of minerals for resource-poor Japan, as well as a market for Japanese industries. In serving Japanese colonial interests, roads, railroads, ports, factories, shipyards, and oil refineries were constructed, which dramatically altered the economic landscape of the peninsula. The Japanese influence also included learned technological and business skills that prepared the south for later entry into the global economy. Nonetheless, the often cruel treatment of Koreans by Japanese colonial masters continues to be a nagging issue in present-day political relations.

Today the peninsula is politically divided between North Korea (Democratic People's Republic of Korea, with 23 million people) and South Korea (Republic of Korea, with 48 million). As a result of World War II, postwar North Korea gravitated toward the Soviet orbit, and South Korea became part of the United States–led anticommunist alliance. In

dictatorship with an economy that is practically lifeless, South Korea and Taiwan, anchored by export-oriented manufacturing sectors, possess two of the most robust economies in the world. Like Japan and the handful of emerging Pacific Rim economies, both South Korea and Taiwan adhere to a development philosophy of **state-led industrialization**,

1950, North Korea attempted to reunify the peninsula by force, but after 3 years of war, a ceasefire line was drawn at the 38th parallel to function as a demilitarized zone as well as a new international boundary. Despite the events of the Korean War and the continuing political subdivision of the peninsula, most Koreans on both sides of the international boundary continue to feel a deep sense of shared cultural unity.

North Korea's economic and political systems have resembled the closed world of China during the Mao years. A bleak and dictatorial communist regime controls every aspect of economic production and social relations in the north. Based upon a relatively good mineral resource base, state-owned heavy industries remain tied to the domestic market, although a limited amount of trade is carried on with China and the states of the former Soviet Union. Unfortunately, much of North Korea's productive energies have been spent building an impressive military capability, particularly its missile and nuclear programs. Depressingly symbolic of these misplaced priorities are the severe food shortages and, sometimes, outright famine experienced by North Korea since 1994. The death toll from a famine between 1995 and 1998 is estimated between 2.5 and 3.0 million people, or slightly more than 10 percent of the country's population, and up to 62 percent of children under the age of seven experienced stunted growth as a result of malnutrition. The situation has become so serious that some 100,000 to 400,000 people have temporarily crossed the unguarded international border with northeastern China in search of food from ethnic-Korean Chinese. As a result, North Korea is one of the few counties in the world where peacetime mortality rates have increased and life expectancy has decreased.

South Korea, on the other hand, has followed an outward-oriented path of economic development. South Korea's transformation from a war-ravaged and impoverished country in 1953 to the proud host of the 1988 Summer Olympics was expressive of one of the most radical national economic makeovers in the twentieth century. Its gleaming skyscraper capital of Seoul is now one of the world's most populated cities and the primate city for the world's fourteenth largest economy (Figure 19–29).

The GNI PPP per capita, which was in the hundreds of dollars in 1963, skyrocketed to $18,000 by 2003. This remarkable economic transformation to become a high-income country was based upon both external and internal economic forces. Foreign grants and loans—from the United States immediately following the war and from Japan beginning in the 1960s—enabled South Korea to finance its economic transformation. The opening of the American and Japanese markets to South Korean exports was crucial as well. Internal factors were equally important. In a development approach reminiscent of Meiji Japan, the government transformed the agricultural sector by dismantling the large feudal estates and returning the land to the farmers. Then the financially compensated landlords were recruited to become the country's business leaders. The mechanization of agriculture also reduced the need for farm workers, enabling many peasants to migrate to urban areas and secure employment in the rapidly expanding industrial sector.

■ **FIGURE 19-29 Seoul, South Korea.** South Korea's prosperity is expressed in the contemporary skyline of Seoul. This primate city has grown to occupy much of the limited coastal plain between the Yellow Sea and the peninsula's mountainous backbone.

South Korea's initial industrial development was based on an import-substitution strategy that protected industries producing for the domestic market from foreign competition by imposing high import duties on foreign products. As with Japan, industrial development eventually became focused on export-oriented manufacturing. Labor-intensive industries, such as textiles, were developed first, but by the 1990s, the focus had changed to more capital-intensive industries such as shipbuilding, petrochemicals, heavy machinery, electronics, and automobiles. The export of automobiles to the United States has become very aggressive as evidenced by the increasing number of Hyundai and Kia cars now on American roadways. The role of the government as general manager of the national economy was instrumental in coordinating this successful industrial transformation. For example, the government controls the issuing of business licenses that determine the nature and scope of a company's activity, and it funnels investments into industries deemed most likely to help the country achieve its development goals. The state also controls the spatial distribution of industry through detailed regional planning, and many formerly neglected regions have benefited from this approach. In fact, regional development has benefited all regions so serious income inequalities do not exist between regions. While we would expect the primate city of Seoul to monopolize much of the country's export-led manufacturing, urban areas in the far southern provinces are centers of prosperous export-led manufacturing as well. In fact, one of the five busiest container ports in the world is in the southern city of Pusan.

From the Western perspective, South Korea's prosperity also has a negative side. Although Korean laborers rank among the best educated in the world, their quality of life leaves much to be desired. Working conditions are relatively poor, and workers have only recently been able to organize collectively. Workweeks are among the longest in the world, and pay is relatively low when compared to corporate profits. As a result, the work stoppages and strikes that are characteristic of South Korea's economy contrast sharply with other globalized economies of the western Pacific Rim, where the labor force is relatively docile.

Government control over business activity has also engendered substantial corruption through bribes to obtain contracts. In 1996, two of South Korea's recent presidents were found guilty of accepting hundreds of millions of dollars in corporate bribes. In addition, the intimate company-government relationships have promoted the economic fortunes of a few select family-owned megacorporations called *chaebol*, or family-owned conglomerates. In 2000 the top four *chaebol*, which included the familiar names of Hyundai and Samsung, accounted for 40 to 45 percent of South Korea's economic output. In numerous respects *chaebols* are much like the Japanese *keiretsu*.

The *chaebol*-centered economy caused the Asian financial crisis to be especially severe in South Korea in many ways. Having unfettered access to bank loans and diversifying their business holdings to include many unprofitable subsidiaries, even the largest of the *chaebol* incurred serious debt. When the crisis hit in 1997, the South Korean currency was devalued by 50 percent, many subsidiaries went bankrupt, per capita economic output declined by 35 percent, and the International Monetary Fund rescued the national economy with the largest loan ever made to a single country. The direct human impact was substantial, with 1.6 million people becoming unemployed. The economic crisis was also a harbinger for positive change. In the late 1990s the government began to institute reforms that will reduce the power of the *chaebol*, privatize state-owned industries, and extend loans to medium- and small-sized companies. Ultimately, it is to be hoped that the reforms will also lead to a more democratic system of government representation.

The first decade of the twenty-first century will likely prove to be a critical watershed for the Korean Peninsula. In the post-cold war "new world order," North Korea has become increasingly isolated and is now considered by some to be a "rogue" country. North Korea's inclusion on the U.S. government's list of "axis of evil" countries is because of its secret development of nuclear weapons. In 2002, the North Korean government expelled International Atomic Energy Agency inspectors and withdrew from the Treaty on the Non-Proliferation of Nuclear Weapons. While not yet possessing advanced missile delivery systems, many experts agree that the government will use the nuclear weapons issue to extract economic concessions from its much richer neighbors. Although the reunification of north and south seems highly unlikely in the frothy political atmosphere of the present and near future, perhaps a clearing of the mist is close at hand. Indeed, South Korean President Kim Dae Jung in 2000 reduced tensions between the two countries by instituting the **Sunshine Policy** to promote reconciliation and peace on the peninsula. The policy has already led to the reconnection of rail lines between the two countries, the reuniting of some cross-border families, and closer economic cooperation. In the long run, reunification is a win-win proposition offering distinct economic advantages to both sides in the form of regional complementarity. The north possesses mineral resources such as coal that the south must import, and the south possesses the financial and technological ability needed to rebuild the north's economy. As in Germany, however, unification will likely come at high financial cost to the South Korean government.

Taiwan

The island country of Taiwan (Republic of China) is another East Asian economic tiger (Figure 19–30). Lightly populated when it became an administrative unit of China's Fujian Province in 1683 and a separate province in 1888, Taiwan was conquered by the Japanese in 1895 and remained a part of their empire until 1945. Much like Korea, the island's function within the empire was primarily as a supplier of raw materials and a consumer of Japanese finished products. When

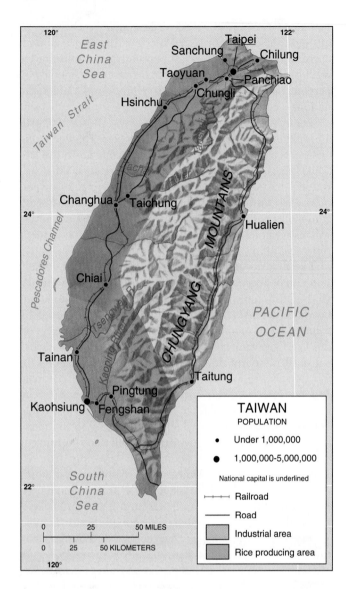

■ **FIGURE 19-30 Island of Taiwan.** Taiwan was occupied by Chinese Nationalist forces following the communist conquest of the mainland. The nation has developed one of the world's most rapidly expanding economies.

Nationalist forces fled from the Chinese mainland to the island on the heels of the communist victory in 1949, Chiang Kai-shek and his followers claimed to be the government of China in exile. Taiwan was recognized by the United Nations as the true representative of China until 1971, when its seat was given to the People's Republic of China. A severe blow to Taiwan's political identity came in 1978 when the United States, its staunchest ally, recognized the People's Republic of China as the legitimate government of China and transferred its embassy to Beijing. China continues to view Taiwan as a renegade province that will be "liberated" sometime in the future.

Taiwan's early economic growth was spurred by investment aid from the United States and open markets for Taiwanese exports. Coupled with the foreign aid was a large pool of Nationalist entrepreneurs and a skilled labor supply. Much like South Korea, the development of domestic human resources through government guidance has been a key to sustained growth. Because of land reform, Taiwan supports an efficient and productive agricultural base, although an increasing amount of food must be imported. Unlike South Korea, however, whose export-led economy is primarily based on heavy industry which is controlled by megacorporations, Taiwan's economy is more diversified, with a strong presence of competitive small and medium-sized firms. In addition to the basic heavy industries of shipbuilding, iron and steel, textiles, and chemicals, the country is an exporter of precision instruments, telecommunications equipment, electronic parts, and computer-related products such as monitors, motherboards, scanners, and software. The country's high-tech firms do not simply assemble products but also engage in their own advanced research and design, utilizing a highly educated workforce. It is this export-driven economy that has produced the fourth largest currency reserves in the world and a rapidly growing economy. The presence of small- and medium-sized firms, coupled with substantial foreign reserves, shielded Taiwan's economy from Asia's economic crisis because smaller firms tended to borrow from savings clubs or mutual aid associations rather than from more formalized institutions such as banks.

Much like Japan, Taiwan relies heavily on importing most of the raw materials used to produce finished exports. Consequently, the vast majority of its industries are located along the broad alluvial lowlands of the west coast, where raw materials are unloaded. The adjacent industrial corridor is anchored on the north by the capital of Taipei and its outport of Keelung and to the far south by the port city of Kaohsiung. In 1998 plans were formulated to construct a 211-mile (340-kilometer) high-speed train between Taipei and Kaohsiung. This west coast region that faces the Chinese mainland is the key to Taiwan's economic future. Much of the growth of China's Special Economic Zone located in Xiamen in Fujian Province is the result of Taiwanese investment capital taking advantage of inexpensive mainland labor. A significant slice of Taiwan's world-class electronics production has migrated to Fujian and elsewhere in China. Indeed, Taiwan's trade with the mainland over the past decade has increased dramatically and likely will continue to expand, so long as political stumbling blocks between communist China and democratic China are reduced. Those stumbling blocks are formidable because the mainland government has threatened Taiwan with invasion if any serious move is made toward claiming political sovereignty.

▶▶ Summary

The past decades have witnessed a fundamental structural rearrangement of the world's economy, owing in part to the emergence of many East Asian countries as central players in the new global production system. Fifty years ago, few could have envisioned the remarkable economic transformation of the region that has occurred. The entrance of

South Korea and Taiwan onto the global economic stage has been achieved largely through state intervention. In China, too, the state has been instrumental in rousing the most populous country in the world from 30 years of economic isolation. China has yet to incorporate its interior regions into the new global production system. Given its huge workforce and enormous market, however, the country has assumed the role of a major global economic power.

▶▶ Key Terms

chaebol

Chinese Communist Party (CCP)

Confucianism

Cultural Revolution

Deng Xiaoping

Grand Canal

Great Leap Forward

Han dynasty

household responsibility system

legalism

Mandarin

Mao Ze-dong

Middle Kingdom

open coastal cities

open door policy

opium

Overseas Chinese

pinyin

Putonghua

semi-colonialism

Song dynasty

Special Economic Zones (SEZs)

state-led industrialization

Sunshine Policy

Sun Yat-sen

Three Gorges

Tiananmen Square

Zhou dynasty

Vietnamese girls. Southeast Asian populations are youthful and growing rapidly. The lives of these young girls are likely to be very different from those of their parents and grandparents.

Chapter 20

Southeast Asia:
Development Diversity

- **Areal Organization and Environmental Patterns**
- **Pre-European Culture and Economy**
- **Colonialism and Development**
- **Modern Economic Growth and Stagnation**

S outheast Asia is a region of large and small peninsulas and islands, sandwiched amidst East Asia, South Asia, Australia, and the Pacific and Indian oceans. While the regional label Southeast Asia might give the impression that this area of the world is "what is left over" from East Asia and South Asia or perhaps a watered-down version of these two culture realms, in truth it is a distinct region deserving to stand apart from its larger neighbors. Unlike China and India, which culturally are relatively homogeneous, Southeast Asia is a fragmented transition zone, or **shatterbelt**, in both the physical and cultural sense. The thousands of mountains, basins, and islands that make up the region have produced a politically and culturally complex mosaic which, at the outset, seems to defy generalization.

The physical fragmentation of Southeast Asia may partly account for its control by so many distant Western powers during the colonial period. The multitude of colonial influences further increased the cultural diversity of the region. After the colonial era, many of the independent states of Southeast Asia, with the obvious exception of Vietnam, played relatively minor geopolitical roles during the cold war era, rarely attracting international attention. During the past three decades, however, a number of the Southeast Asian countries have achieved increased economic importance through their integration into the emerging global economy. Today Southeast Asia and East Asia share the distinction of possessing many of the highest rates of national economic growth in the world. Interestingly, much of this growth has been achieved through East Asian rather than Western foreign investment.

Areal Organization and Environmental Patterns

At its widest dimensions, Southeast Asia stretches more than 3,000 miles (4,800 kilometers) from Myanmar (Burma) in the west to Irian Jaya, the Indonesian half of the island of New Guinea, in the east (Figure 20–1). When the oceans and seas surrounding Southeast Asia are included, its dimensions are approximately equal to those of India and the neighboring states of South Asia. Southeast Asia is situated almost entirely within the tropics and, with the exception of northernmost Myanmar, is characterized by tropical forests and monsoon climates.

Mountains, Basins, and Islands

In the north, Southeast Asia is separated from the rest of Asia by the mountain spurs of the Himalayas. Further south, on both the eastern and western margins, Southeast Asia's physical boundaries are marked by sweeping volcanic arcs that have pushed up at the edges of the Indian and Pacific plates (see Figure 20–1). In the west, the arc encompasses the Indonesian islands of Sumatera, Jawa, Bali, Lombok, and Sumbawa. The eastern margin, which constitutes one small segment of the volcanic belt known as the Pacific Ring of Fire, follows a line extending from the Philippines to New Guinea. While no single volcano or mountain in Southeast Asia is extremely lofty, many peaks reach 10,000 feet (3,048 meters) or higher (Figure 20–2). At the edge of both volcanic arcs are deep oceanic trenches marking tectonic plate boundaries. These tectonic plate boundaries have produced some of the most destructive earthquakes and volcanic eruptions in the region (see the *Geography in Action* boxed feature, The Asian Tsunami, page 460). Between the western and eastern volcanic arcs is a shallow platform called the Sunda Shelf. Measuring no deeper than 600 feet (180 meters) in most locations, much of the Sunda Shelf was a swampy exposed surface connecting much of Southeast Asia until recent geological times, when sea levels rose as a consequence of post-Ice Age warming.

Physical Environments and Human Activity

The broadest examination of Southeast Asia's physical environments requires division of the region into mainland (continental) and insular (maritime) subregions. The mainland

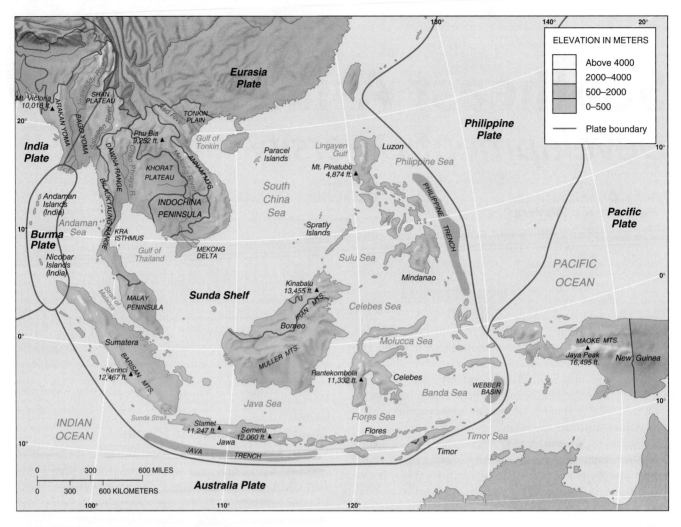

■ **FIGURE 20-1 Physiography of Southeast Asia.** Southeast Asia is a physically fractured region consisting of thousands of islands, and the mountains and valleys of the continental realm.

■ **FIGURE 20-2 Mt. Mayon, Philippines.** Southeast Asia forms a part of the volcanic Ring of Fire that encircles the Pacific Ocean. Many of the region's beautiful volcanoes, including Mt. Mayon, are geologically active.

states include Myanmar, Thailand, Laos, Cambodia, and Vietnam, while the insular states are Malaysia, Singapore, Indonesia, Brunei, the Philippines, and East Timor. In mainland Southeast Asia, broad alluvial river valleys are separated from one another by north-south trending mountain chains. The lowland river valleys today, as in the past, are centers of substantial population concentrations and agricultural production sustained by soil-enriching floodwaters. Insular Southeast Asia is an island and sea environment with most of the population clusters located along the coastal plains and traditional economic activity focusing on rice cultivation, fishing, and maritime trade.

Despite these environmental differences, the mainland and insular realms share their dependence on forests and water bodies. Because movement by land has always been difficult in this heavily forested region, water has historically proven to be a more efficient medium of travel. In mainland Southeast Asia, rivers and canals play vital roles in those heavily populated regions. Indeed, the floodplains and deltas of the Irrawady, Chao Praya, and Mekong rivers are laced with canals (Figure 20–3). In insular Southeast Asia, monsoon winds and calm, shallow waters make coastal boat travel

■ FIGURE 20-3 A canal or *klong* in Bangkok, Thailand.
Although many canals have been filled in, waterways are still used for transporting all manner of products or selling consumer goods such as food to residents living along their banks.

relatively safe and efficient. Southeast Asia, then, is a region oriented toward the sea. Forests have also traditionally played prominent roles in the livelihoods of the Southeast Asian peoples. Much of the region was originally covered by tropical forest vegetation, owing to the year-round warmth and humidity. During much of Southeast Asia's premodern period, the forests provided materials for building homes and boats and food and medicinal substances for domestic consumption or long-distance trade. As will be noted throughout the remainder of this chapter, the mainland-insular distinction, coupled with the human dependence on the aquatic and forest environments, has greatly influenced the cultural and economic contours of Southeast Asia, past and present.

Pre-European Culture and Economy

Southeast Asia's forest and water environments have historically protected the region from long-distance migration and conquest originating in interior Asia. Contact with India and China, from which many culture norms and economic opportunities originated, was generally of a small-scale and selective nature, involving adventurers and merchants rather than large-scale military action.

The emergence of economic, political, and cultural cores in Southeast Asia commenced with the introduction of Hindu and Buddhist systems of belief and political organization from India. As a result, one of the older European names for Southeast Asia was "Further India." In mainland Southeast Asia, we can identify several early centers of development (Figure 20–4, page 462).[1] In Myanmar, Indianized kingdoms emerged

in both the delta region of the Irrawady during the early centuries A.D., as well as the drier middle reaches of the river by A.D. 850. In Thailand, downstream migration of successive capitals to the lower Chao Praya began in the twelfth century, until near coastal Bangkok was established as the capital of Siam in 1767. In the lower valley and delta regions of the Mekong, the powerful and Indianized kingdom of Funan was established during the early centuries A.D. By the tenth century, however, the focus of power moved inland to the great lake of Tonle Sap, where the powerfully commercial Khmer Empire ruled until the fifteenth century. The monumental ruins at Angkor Thom and **Angkor Wat** stand as testimony to the agricultural productivity and trade capabilities of this most powerful of all pre-European Southeast Asian empires (Figure 20–5, page 462). In Vietnam, two cores emerged during the early centuries A.D. One was the Chinese-influenced Annamite region centered on the Red River delta, and the other was the Indianized culture of Champa in the south, which practiced rice agriculture on the narrow coastal plain. It was not until the fifteenth century that the northern Annamite peoples completely subjugated Champa and thus opened the way for the southward migration of what we now call the Vietnamese people.

With the exception of Champa, all of these early political, economic, and cultural cores were centered on riverine environments, particularly in their lower stretches, where alluvial lands could be transformed from forest and freshwater swamp environments to productive rice-growing areas. The power of the state rested on taxing rural production to finance defense, trade, industry, and irrigation. With the exception of Annam, all the mainland Southeast Asian cores emerged only after adopting Indian models of state organization based on kinship, the court, secular and religious bureaucracies, and a royal and sacred capital. These lowland-dominant cultures contrasted sharply with upland-subordinate cultures, the latter characterized by village life, oral traditions, folk religions, and slash and burn agriculture coupled with forest-gathering activities. Even today, there are substantial cultural distinctions between the more populous and powerful lowland peoples and the upland minority groups, such as the Karen of Myanmar and Thailand and the Hmong of Vietnam and Laos.

In insular Southeast Asia, many small kingdoms dotted coastal regions, but only three large-scale pre-European cores are recognized (see Figure 20–4). The first was the Indianized and Buddhist kingdom of Srivijaya, which flourished from the seventh to the fourteenth centuries, with its capital of Palembang in the freshwater swamp region of southeastern Sumatera (in present-day Indonesia). Srivijaya was a **thalassocracy**—that is, a sea-based state. Like many other smaller maritime kingdoms of the region, its power was derived from the long-distance maritime trade that flowed into and out of its capital of Palembang. Little reliance was placed on agriculture. The second pre-European insular core area was a group of successive Indianized states that first emerged during the early centuries A.D. on the island of Jawa. The earliest of these Hindu-Buddhist states occupied small fertile valleys of central and eastern Jawa

---

[1]Many of the precolonial observations are extracted from William Kirk, "Southeast Asia in the Colonial Period: Cores and Peripheries in Development Processes," in *Southeast Asian Development,* edited by Donald Dwyer (New York: John Wiley, 1990), pp. 15–17.

GEOGRAPHY IN ACTION
The Asian Tsunami

While most of Southeast Asia is not threatened by typhoons or cyclones, one natural hazard that is not predictable and can potentially produce greater infrastructural destruction and loss of life are tsunamis, meaning "harbor wave" in Japanese. **Tsunamis** are seismic sea waves triggered by the energy released from deep earthquakes, but also from massive landslides and volcanic eruptions. The wave is barely visible at the ocean surface, but from the ocean floor to the surface, waves can travel up to 1,000 kilometers an hour across wide expanses of ocean. Tsunamis really only become a serious hazard when approaching more shallow coastal waters when wave energy is compressed into a now visible wave that can reach 40 meters in height.

Although occurring in all oceans, the most destructive have occurred in the Pacific and Indian Oceans, where geologic convergent tectonic plates are especially active. In 1883, for example, the volcanic cone of Krakatoa between the islands of Jawa and Sumatera explosively erupted and the following tsunami killed 36,000 inhabiting the surrounding costal lowlands. In 1896, an earthquake-triggered tsunami killed 26,000 along the

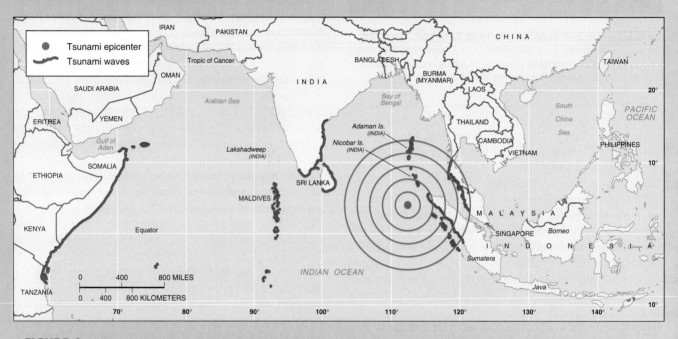

■ FIGURE A 2004 tsunami epicenter in the Indian Ocean. With an epicenter off the coast of Sumatera, the December 26, 2004, "megathrust" earthquake produced destructive above-the-surface tsunami waves along many Indian Ocean coastlines.

where, much like the mainland riverine states, they gained political control through the taxation of rural economic activity. With the demise of Srivijaya, however, their energies became more dedicated to maritime trade. The archetypical Jawanese state during this later period was the Majapahit, which loosely controlled many of the eastern islands, Borneo, Sumatera, and the Malay Peninsula.

During the fifteenth and sixteenth centuries, the arrival of Arab, Indian, and Chinese traders dramatically transformed the development of insular Southeast Asia. Arab and Indian merchants introduced Islam, which was adopted by local royalty and subsequently spread rapidly throughout the local population. Islam eventually became the dominant religion of the entire archipelago, the main exceptions being the isolated interior peoples, who clung to animism, and Bali, which remained Hindu (see the *Geography in Action* boxed feature, Islam in Insular Southeast Asia). These small coastal and river-mouth trading sultanates procured forest products

from their respective riverine hinterlands and shipped them to Malacca, the successor state to Majapahit, which was located on the southwestern coast of the Malay Peninsula. Like Srivijava's Palembang, the entrepôt of Malacca was a cosmopolitan thalassocracy visited by traders from throughout much of Asia and for a short period of time, it was the busiest port in the world. Mainland Southeast Asia is mostly Buddhist, while Islam prevails in the insular realm, with the exception of the predominantly Christian Philippines.

Colonialism and Development

Although colonialism radically changed the traditional economic and social contours of Southeast Asia, its impact varied. The impact of the West may be divided into two time periods. The first period, from about 1500 to 1800, was

■ FIGURE B **Banda Aceh, Indonesia.** This aerial photograph shows the destruction caused by the tsunami of 26 December 2004, which struck Banda Aceh with devastating force.

coast of central Honshu island in Japan. Of course the most destructive and deadly tsunami in the world since Krakatoa was the December 26, 2004, tsunami off the northwest coast of Sumatera when a magnitude 9 "megathrust" earthquake resulted in the 1,200-kilometer section of the Indian plate being thrust up some 20 meters over the Burma Plate (Figure A). The energy released from this plate slippage was equivalent to 23,000 Hiroshima-sized atomic bombs. Within hours of the earthquake, tsunami waves ranging from 10 meters in Sumatera and 4 meters elsewhere made landfall.

Confirmed deaths are approximately 300,000 persons in the thirteen countries impacted, with Indonesia (243,530), Sri Lanka (30,957), and India (16,389) being the most affected. In the city Banda Aceh at the northern tip of Sumatera, approximately 15 percent of the city's population of 400,000 perished (Figure B). The wave that washed over this coastal city penetrated 3 kilometers inland. For the region as a whole, some half million were injured and 1.6 million were displaced from their homes. While the Pacific coast countries possess a tsunami early warning system, the Indian Ocean countries do not.

The economic, social, and environmental impact of the Asian tsunami on individuals, families, communities, and countries is substantial over both the short and long term. Because of salt water entering the water table, village wells have become unusable. The soils of coastal padi fields and fruit orchards too have become polluted with salt. Successive wet monsoon seasons, however, will flush much of the salt deposits out of the soil. The fishing industry too was devastated, as many fishing villages lost their crews and boats. The negative impact of the tsunami on coastal tourism is obvious, particularly in Sri Lanka and Thailand. The big question is whether Western tourists will want to return to the same stretches of beach that witnessed such tragedy. Along Thailand's southwestern coast that has witnessed a boom in resort construction during the 1990s, some half of the 5,393 deaths were tourists.

Often unrecognized in the media's portrayal of the disaster is that women's death rates were often triple that of males and that the spatial expressions of traditional gender roles in part explains this disparity. In India, for example, male fisherfolk were at sea while women waited on shore or at home for their return. Throughout the region, men often work away from home while women remain at home taking care of domestic duties. Indeed, having to gather children together when the wave struck cost many women their lives. In addition, women and children, when compared to men, did not possess the physical strength or the experience to climb trees or tread water for extended periods of time and simply drowned. This tsunami-induced gender imbalance in many coastal communities begs questions with reference to future gender and development issues. For example, will the female workload now increase or will the traditional dominant gender roles be transformed as widowers or adult male orphans take on additional domestic responsibilities? Will younger adult women now marry earlier, and thus lessen their opportunities for education and socioeconomic mobility? These new brides, for example, might be prodded by their husbands to have more children at shorter intervals to make up for tsunami-induced losses.

characterized by **mercantile colonialism**, which consisted of Western governments trading with local elites for native luxury goods. The impact of the trading on indigenous societies was generally minimal. The exceptions were in localized areas in the Philippines and the Indonesian island of Jawa, where local political structures and economies were dramatically disrupted by forced agricultural production of crops for export back to Europe. This period was followed by an era of **industrial colonialism**, which lasted from 1800 to 1945 and which brought Western political control of colonial territory and the involvement of Western private or corporate economic interests.

Creation of Core-Periphery Exchange

Although short in duration, industrial colonialism had a deep and widespread impact on Southeast Asia. Europe and the United States became dependent upon tropical lands for raw materials to fuel industrialization as well as to establish new markets overseas. The array of raw materials obtained from Southeast Asia was vast and included oil, rubber, tin, hemp, sugar, palm oil, tea, and tobacco. On the global scale, the impact of industrial colonialism was to create a core-periphery exchange relationship between the Western core countries and their peripheral Southeast Asian colonies. Within the colonial possessions themselves, however, regional cores and peripheries emerged that resulted in substantial differences in local levels of economic development.

The principal focus of the colonial economy was commercial agriculture, which supplied raw materials to the controlling Western nation and which financed colonial administration. One form of commercial agriculture was the large-scale corporate plantation. Rubber and oil palm were the primary plantation crops of the west coast of British Peninsular Malaya and the northeastern coast of Sumatera in Indonesia (Figure 20–6). Sugarcane dominated in the

461

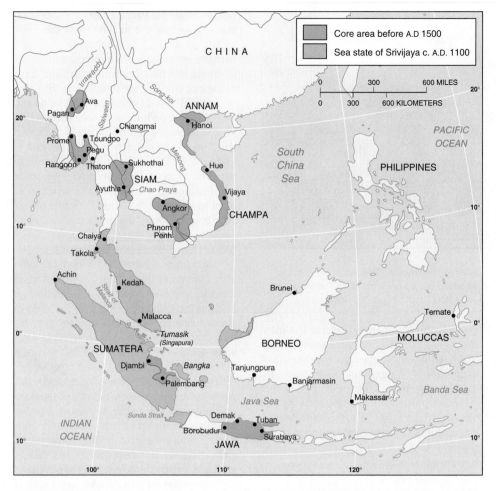

■ **FIGURE 20-4 Precolonial states of Southeast Asia.** Southeast Asia supported a number of advanced civilizations prior to European conquest.

■ **FIGURE 20-5 Angkor Wat ruins of Cambodia.** At Angkor Wat, a series of temple cities stands as testimony to advanced urban development in precolonial Southeast Asia. Damaged from years of civil war during the 1970s and 1980s, United Nations funds are being used to restore these 900-year-old cultural heritage sites.

GEOGRAPHY IN ACTION
Islam in Insular Southeast Asia

Central to Islam is the concept that Muslims around the world constitute a single community of faith or *ummah*. While this core belief binds Muslims in Insular Southeast Asia to the wider world of Islam, many of the social, political, and economic expressions of Islam in Southeast Asia differ from those of the Middle East. Several historical and cultural processes explain these differences. The first is that Islam did not diffuse directly to Southeast Asia from the Middle East, but rather from South Asia, particularly India during the fourteenth and fifteenth centuries. As Islam spread into Southeast Asia, its impact was modified by the previously established Hindu-Buddhist religious culture. Southeast Asian Islam thus became a hybrid religion. A second factor that contributed to the nature of Southeast Asian Islam was the pre-Islamic role of women in society, particularly their prominent status in household and village economic exchange. This has tended to reduce in Southeast Asia the patriarchal or male-dominated nature of Middle Eastern Islam. While the arrival of Islam certainly did stress the traditional roles of women with regard to morals, the wearing of veils by Muslim women in Southeast Asia, for example, is a matter of choice. In fact, until the 1970s, when Islam began playing a much greater role in national cultures, the head coverings for women that make the cultural landscape in some religiously conservative Middle Eastern countries so uniform were not common in Insular Southeast Asia. Finally, the colonial impact in Insular Southeast Asia was far deeper than in most Middle Eastern nations, resulting in the formation of more secular governments in Indonesia and Malaysia. For example, Islamic law or *sharia* in Southeast Asia is less pervasive when compared to most Arab countries, in part because of the inheritance of European legal systems.

The subject of Islam in Insular Southeast Asia has attracted greater interest in the wake of increased levels of world terrorism, and the growing politicization of Islam in Malaysia and Indonesia. Many are now wondering whether these countries will experience a significant rise of a stricter form of Islam that may lead, for example, to the establishment of an Islamic state. In all likelihood, the answer to this hypothetical question is no, for two reasons. The first is that the governments of Indonesia and Malaysia possess democratic institutions, allowing for some dissent, and thus making a politically radical form of Islam less likely to emerge. Indeed, in the 1999 national elections of Indonesia, only 25 percent of the political parties taking part in the election called for the establishment of an Islamic state, and these same parties garnered only 16 percent of the vote. Second, both countries are ethnically diverse, and their governments do not want to alienate non-Muslim minority groups, particularly the economically powerful Chinese, by allowing for the establishment of an all-pervasive Islamic state. Because Islamic revivalism in Insular Southeast Asia is of a social rather than of a political nature, an Islamic religious revival in the context of a secular grounded government may actually have the effect of strengthening democratic institutions.

Some radical Islamic groups have emerged in Southeast Asia, but these are found primarily in regions where grinding poverty and corrupt local politicians are endemic, thus providing fertile ground for enlisting recruits. For example, in very poor Muslim regions of the southern Philippines long neglected by the national government, the extremist Abu Sayyaf group, which has kidnapped foreigners for ransom, is essentially a band of criminals that endeavors to use Islam to justify its violent actions. Muslim-on-Christian violence on the island of Sulawesi in Indonesia was provoked by the radical Laskar Jihad, who hired unemployed males in Jakarta to do their dirty work. Similarly, a radical form of Islam has emerged in the last few years in the three Muslim majority provinces of southern Thailand. Only becoming politically part of Thailand in 1902, the rise of an Islamic insurgency movement is complex, but primarily revolves around corrupt regional Thai government officials and poverty because of central government neglect. It seems that long-term solutions to the growth of radical Islam in Southeast Asia, much like the Middle East, must address its basic sources of economic and political inequalities.

Sources: John Gershman, "Is Southeast Asia the Second Front?," *Foreign Affairs* 81, no. 4 (2002): 60–74.
Anthony Shih, "The Roots and Societal Impact of Islam in Southeast Asia: An Interview with Professor Mark Mancall," *Stanford Journal of East Asian Affairs* 2 (2002): 114–117.

---

central Philippine islands and Jawa, and rubber was the chief plantation crop in southern French Indochina. Because plantations were associated with industrial production technologies, they evolved as economic enclaves, possessing stronger and more extensive economic linkages with the outside Western world than with their surrounding hinterlands. There thus emerged a **dual economic system**, consisting of a modern plantation sector operating in the midst of traditional peasant cropping systems.

In addition to promoting plantation agriculture, the Western colonial powers frequently forced the peasant farmers to dedicate a portion of their traditional paddy land to the cultivation of export crops. This **cash cropping system**, as it was called, was especially common in the Philippine Islands and Jawa. Its impact was to greatly lower grain production, which, in combination with continued population growth, eventually necessitated the importation of vast quantities of rice, much of it from Thailand, Burma, and southern Vietnam. Using local labor, the British in Burma and the French in Indochina purposefully drained much of the Irrawady and Mekong deltas, respectively, for the purpose of creating "rice bowls" for Southeast Asia. The ongoing conversion of land from subsistence to export crop production is one of the principal reasons why both Malaysia and the Philippines, as well as other plantation areas throughout the region, continue to rely on food imports.

The concentration of colonial activity in the core region or selected enclaves of the colonial periphery also tended to marginalize minority peoples who did not participate in the economies of the core. One reason for neglecting the minority populations was the high cost of connecting the mountainous home regions to the core. A second reason

■ **FIGURE 20-6 Harvesting rubber on a Malaysian plantation.** Rubber has been a leading commercial crop of Southeast Asia since its introduction from South America's Amazon Basin during the colonial era.

for not exposing these minorities to technological advances was the paternalistic philosophy of colonial officials, who decided that these "backward" societies should be preserved. This policy, whether right or wrong, created an historical legacy of marginalizing cultural minorities that has continued to the present day.

The concentration of economic activity in the colonial cores also contributed to ever-increasing differences between the more Westernized urban centers and the indigenous rural hinterlands. While modernization proceeded at a rapid pace in the larger cities, the rural peripheries attracted little investment in industry, education, or medical care. This domestic core-periphery relationship remains much the same today. In the British possession of Malaya, for example, the small island of Singapore functioned as the primate city for a vast rubber, oil palm, and tin mining periphery on the Malay Peninsula. In the Philippines, Manila was both a regional primate city and a gateway to the rest of world; it was where raw materials from the southern archipelago were processed and exported to the United States. In Burma, the British established Rangoon in the Irrawady delta as their colonial capital, reviving the pre-European core that had been replaced by the upstream core centered on Mandalay. In Vietnam, the French reinforced the two core areas of the north and south that dominated the country prior to colonial conquest.

Chinese Immigrant Middlemen

Intimately linked to the formation of colonial cores was the presence of non-Western foreigners—in particular, Indians and **Overseas Chinese**. Chinese settlers became a significant presence in many Southeast Asian port cities as early as the fourteenth century. Their populations increased further when trading, agricultural, and mining activities expanded during the early European colonial era. The civil wars in China were an added inducement for many Chinese to migrate and seek opportunities in *Nanyang,* or "southern seas," a term the Chinese use to describe coastal Southeast Asia. It was during the late colonial period, however, that the flow of Chinese reached its highest levels. By the end of World War II, there were approximately 10 million Chinese residing in Southeast Asia, with the greatest concentrations in Thailand and Malaya. Most arrived as desperately poor plantation contract laborers. Over time, however, many opened small retail shops and trading concerns, while others engaged in interisland shipping or worked as clerks in Chinese- and European-owned companies. Overseas Chinese thus came to be indispensable to the success of industrial colonialism. Eventually, some became captains of industrial banking, insurance, shipping, and agro-processing corporate empires that competed with European firms. Excluding Dutch Indonesia and Burma, Overseas Chinese accounted for some 38 percent of Southeast Asian capital investment during the late 1930s (see the *Geography in Action* boxed feature, Overseas Chinese).

The colonial empires constructed by the West rapidly crumbled when Japanese forces conquered most of Southeast Asia during World War II. The Japanese coveted the same natural resources that had fueled the factories of Europe and the United States. Despite Japanese propaganda that promised a Southeast Asian economic development model based on Asian values and needs, Japan proved to be an even more paternalistic (if not dictatorial) colonial master than the Europeans and Americans. At the conclusion of the war, each Southeast Asian territory was reoccupied by its former European colonial power, the latter hoping, in most cases, to resume the former colonial relationship. However, the Western superiority myth had been shattered by the Japanese and, inspired by the global call for decolonialization, many nationalistic movements soon surfaced demanding independence. The struggle for independence was relatively peaceful except in Indonesia and Vietnam, where military conflict ensued. The Philippines (1946) were the first to gain independence, followed by the mainland Southeast Asian possessions between 1947 and 1954. Indonesia achieved independence in 1949, but it was not until 1963 that Malaya became a sovereign state. Singapore split from the Malayan Federation that also included Sarawak and Sabah, on the island of Borneo, in 1965. Because of its fear of annexation by the much larger Malaysia, the tiny oil-rich British protectorate of Brunei was not granted independence until 1983.

GEOGRAPHY IN ACTION
Overseas Chinese

Between the mid-1800s and World War II, an estimated 30 million persons left China, with approximately 75 percent of this total migrating to Southeast Asia. Like the early migrations of Chinese to the United States, most of the **Chinese diaspora** originated from the two southern Chinese provinces of Guangdong and Fujian. Indeed, many "Chinatowns" around the world might better be called "Guangzhoutowns," named after the largest city in the Pearl River Delta. Today, the Overseas Chinese continue to play a prominent economic role in many Southeast Asian countries (Figure A). Their economic impact is strongest in Singapore, Malaysia, and Thailand, where ethnic Chinese constitute 78.0, 30.0, and 10.8 percent of the respective populations. Even in countries with smaller ethnic Chinese populations, such as Laos (3.8 percent), Cambodia (3.7), Myanmar (3.7), Vietnam (3.0), Indonesia (2.0), and the Philippines (1.3), their numbers belie their economic strength. In Indonesia, for example, ethnic Chinese control roughly half of the country's total private assets.

The economic success of the ethnic Chinese can be traced to both internal and external factors. The commonly held view that Chinese are "good at business" relative to indigenous Southeast Asian populations is certainly true. Confucian traditions that promote the family as the basic social unit have led the Chinese to use their extensive familial connections to engage in a wide range of business activities. The previous experience of some ethnic Chinese in the highly commercial environments of South China is a second internal factor. External factors have been equally important. For example, while colonial governments exploited Chinese labor for their own purposes, they never accorded the Chinese the same social status as Europeans and never allowed them to assume positions in government. It was incumbent, then, for the Chinese to focus their energies on business, where fewer barriers were laid in their path. Because both colonial and modern governments discriminated against the immigrants when it came to schooling, Chinese families often were forced to send their children to superior schools abroad. The dominance of Chinese doctors, accountants, engineers, and lawyers in many Southeast Asian countries is thus, in part, a by-product of local educational discrimination.

The nature and extent of the discrimination against ethnic Chinese minorities in Southeast Asia vary from country to country. In the Philippines, for example, discrimination has been insignificant, because most of the Chinese have long inter-

■ **FIGURE A Chinese influence in Kuala Lumpur, Malaysia.** Over the past centuries, tens of millions of Chinese have migrated to Southeast Asia. In many countries, they are a highly visible and economically powerful presence.

married with the local population, converted to Catholicism, and adopted indigenous names. Many of the Filipino political elite, in fact, possess some measure of Chinese ancestry. In Thailand, a common Buddhist religion, similar physical appearance, and the adoption of Thai names have reduced the potential for systematic discrimination. In Vietnam, however, overt discrimination is common. During the early and mid-1980s, for example, the Communist regime expelled tens of thousands of ethnic Chinese as boat people for their capitalist tendencies. In Islamic Malaysia and Indonesia, the vast majority of lower- and middle-class ethnic Chinese are legally discriminated against in a number of ways, although the financially and politically connected elite suffer little from official prejudice. The civil strife experienced in Indonesia in response to the 1997 Asian financial crisis was especially hard on the middle class; thousands of Chinese businesses were burned or looted as scapegoats of the economic meltdown. Despite latent anti-Chinese feelings, many Southeast Asian countries continue to depend upon this minority group as a source of capital and entrepreneurial talent, particularly as their national economies become further intertwined with global markets.

Modern Economic Growth and Stagnation

A new economic orientation that better suited the needs of these newly independent countries was in order. Two of the most urgent needs were to diversify economic production and to reduce dependence on exports of raw materials to the West. Delivering modernization to the rural population of the economic and cultural peripheries has been another daunting task. Yet another challenge has been that of nation-building, that is, the creation of unified populations from a heritage of multicultural societies.

How each nation responded to these social and economic challenges will be addressed shortly, in the context of individual country descriptions. While there exists a great number of economic indicators to measure economic performance and growth, two of the most important are the growth of export-led manufacturing, which in turn is often tied to levels of foreign direct investment. With reference to these development indicators, geographer Jonathan Rigg has divided Southeast Asian countries into the two categories of "modernizers" and "reformers." Countries classified as modernizers are those countries that together make up the original core of the **Association of Southeast Asian Nations (ASEAN)**, which was established in 1967 to promote regional political stability. These states were the first to adopt economic policies that promoted manufacturing as an engine of economic growth. Because much of this growth in export-led manufacturing was tied to Foreign Direct Investment (FDI), these countries were the first to globalize their economies. These modernizing countries include Singapore, Malaysia, Thailand, Indonesia, and the Philippines.

Countries classified as reformers were those whose governments during the 1970s and 1980s were tied to economic policies anchored in strong socialist principles, and thus were adverse to foreign direct investment. War and recovery also obviously had an impact on economic growth. Only in the 1990s did some governments begin to open their national economies to globalization. These reformer countries are Myanmar, Laos, Cambodia, and Vietnam. As an indicator of openness, these countries joined ASEAN during the 1990s. While Brunei is a relatively rich country, its wealth is derived almost exclusively from oil and natural gas rather than manufacturing exports. The newly independent, but desperately poor, country of East Timor does not fit into this modernizer and reformer economic growth model either. It is interesting to observe that with the exception of Thailand, all modernizer countries are those in Insular Southeast Asia, while all those countries classified as reformer countries are in Mainland Southeast Asia.

For those modernizer economies, the principal sources of FDI in the 1960s and 1970s were corporations from Western Europe and the United States. Since the 1980s, however, firms from Japan, South Korea, Taiwan, and Hong Kong have accounted for most of the FDI. These East Asian firms have been attracted by the region's diverse natural resource base, relatively abundant and inexpensive labor, and government tax breaks. Another extremely important, but often overlooked, development factor in most countries has been the high level of political stability. Only with these factors in place have the modernizer countries been able to attract foreign-owned manufacturing facilities and thereby become integrated into the global production system. Only recently have reformer countries attracted FDI in manufacturing, and this has been primarily from East Asian firms, but also Singaporean capital. Economic openness at the regional scale was enhanced by the establishment of the **Asean Free Trade Area** or **AFTA** in 1992 to reduce tariffs on trade between member countries. AFTA was established in part to compete against the ascendancy of China's highly competitive export economy in a more globalized economy.

The differences in economic growth trajectories in terms of export-led manufacturing and FDI between modernizer and reformer economies tells much about the level of globalization experienced by these countries during the 1980–2003 period (Table 20–1). With the obvious exception of the city-state of Singapore, the transformation of resource and agricultural-based to manufacturing-based national economies between 1980 and 2003 has been dramatic. The value of merchandise exports between 1990 and 2003 has doubled if not tripled for most countries. Based on both merchandise exports as a percentage of exports plus the value of exports, both Malaysia and Thailand have joined the ranks of **Newly Industrializing Economies** or **NIEs**. Singapore's high merchandise export values are in part explained by exports from other Southeast Asian countries being shipped through Singapore and counted as exports from this city-state. Singapore and Malaysia are particularly reliant upon FDI for economic growth. While all reformer countries have opened their economies to global forces and experienced an increase in manufacturing exports and FDI, only Vietnam made a serious attempt to join the ranks of modernizer countries. In terms of FDI, Vietnam accounted for 87 percent of FDI

Table 20-1   Economic Growth Trajectories of Southeast Asian Countries

	Merchandise Exports as % of Total Exports			Merchandise Exports in $ (U.S. Millions)		Gross FDI as % of GDP	
	1980	**1990**	**2003**	**1990**	**2003**	**1990**	**2003**
Modernizers							
Indonesia	2	35	52	25,675	60,955	1.0	1.7
Malaysia	19	54	77	29,452	99,369	5.3	5.7
Philippines	37	38	90	8,117	36,502	1.2	0.6
Singapore	50	72	85	52,730	144,127	20.6	18.6
Thailand	28	63	75	23,068	80,522	3.0	1.7
Reformers							
Cambodia	–	–	1	86	1,690	1.7	2.8
Laos	–	–	–	185	524	0.7	1.4
Myanmar	–	10	–	325	2,600	–	–
Vietnam	–	–	50	2,404	20,176	–	4.0

Sources: *World Development Indicators 2005*, http://www.worldbank.org/data/wd2005/wditext; Rigg, Jonathan. *Southeast Asia: The Human Landscape of Modernization and Development*, 2nd ed. New York: Routledge, 2003.

in all reformer countries in 2003. It is important to recognize, however, that Southeast Asia's share of global FDI decreased over the past 20 years. For example, ASEAN accounted for 23 percent of global FDI during the 1980–1996 period, but decreased to 11 percent during the 1997–1999 period. In 2000, its share was only 4 percent. This decline in FDI is explained in large part by the Asian Financial Crisis of 1997, which hit Southeast Asian countries particularly hard, but also by the attraction to the large market and inexpensive labor of China by FDI. China's global share of FDI was less than 6 percent during the 1980–1985 period, but dramatically increased to 22 percent between 1997–1999.

Urban and Rural Transformations

Compared to other major world regions, Southeast Asia has traditionally been described as possessing relatively low levels of urbanization. In 2000, for example, the region's total population was only 37 percent urbanized; only Sub-Saharan Africa possesses lower levels of urbanization. As expected, levels of urbanization are much higher in insular Southeast Asia, with Indonesia, the Philippines, and Malaysia averaging 50 percent and the mainland Southeast Asia countries of Thailand, Vietnam, Laos, and Myanmar much lower at 25 percent. Nevertheless, Southeast Asia possesses the three mega-urban regions of Jakarta, Manila, and Bangkok. These mega-urban regions can be conceptualized as **extended metropolitan regions** or **EMRs** that include a core, inner, and outer zones comprising the larger urban area. Inclusion of the outer zone as part of the EMR that often reaches 50 miles (80 kilometers) from the urban core is justified because while land use is primarily agricultural, it is characterized by factories dotting the landscape, a significant number of inhabitants working in non-agricultural occupations, and good transport networks allowing for high mobility. Partially explaining the growth of EMRs is FDI in manufacturing as these cities become "production platforms" functionally integrated into the global economy. Coupled with rural to urban migration, the 2000 EMR populations of Jakarta and Manila exceeded 15 million, and Bangkok 11 million.

The question of whether the EMR outer zone is classified as urban, begs the larger question of whether it is rural or urban spaces that are being transformed. The rural spaces of Southeast Asia beyond the EMRs are witnessing a dramatic transformation as both globalization and urbanization proceeds at a rapid pace. The region is experiencing a "deagrarianization" because young rural people prefer urban occupations as opposed to the drudgery of farm work. In Indonesia and Thailand, for example, many farmers have diversified their income-generating activities by either seasonal work in cities or by engaging in nonagricultural pursuits such as piecemeal handicrafts. Research informs us that 30 to 50 percent of rural household income is derived from non-farm labor. In older and poorer households, for example, younger adult members often work in urban factories, on construction sites, or in public transport, while older adult members work the land in a survival strategy that incorporates both the paddy field and an urban place. For richer village

families, the income earned by family members in the city is remitted and used to accumulate even more wealth in the countryside through the purchase of more modern agricultural technologies. So while national censuses enumerate individuals based on their permanent place of residence, rural household income-generating activities cross rural-urban spatial boundaries.

While it is true that many seasonal migrants to the city are underpaid, exploited, and live in cramped conditions, and that rural spaces are increasingly characterized by a "graying" population coupled with a landscape of abandoned agricultural fields, these conditions are expressions of the larger modernization process associated with globalization. It is a process in which urban development is dependent upon changes or innovation in rural space. In turn, innovations in agriculture are dependent upon remittances from urban wage relatives. It is difficult then to conceptualize rural development or urban development as mutually exclusive processes. Much like villages within the outer zone of EMRs, but only more so, more distant villages across Southeast Asia might be described as spatially isolated in the physical sense, but less so with reference to economic and social isolation.

Singapore

Singapore is a state within Southeast Asia that is distinctive for a variety of reasons. With a population of 4.2 million inhabitants, it is often referred to as a **city-state** because it measures only 239 square miles (619 square kilometers), making it the smallest and most urbanized of all Southeast Asian countries (Figure 20–7). Singapore is also distinctive because it is the only developed nation in Southeast Asia where ethnic Chinese constitute the majority of the population. Because of economic and ethnic differences with Malay-majority Malaysia, this ethnic Chinese enclave became a separate independent country in 1965.

Much like Hong Kong, Singapore functioned as an entrepôt for regional maritime trade, becoming the "crown jewel" of Britain's Southeast Asian empire. Blessed with a strategic location at the eastern entrance to the Strait of Malacca and with superior port facilities, Singapore came to function as the leading transshipment and processing center of the Malay Peninsula and of a substantial portion of Dutch Indonesia. Today Singapore continues to function as a leading regional maritime trade center. During the past decade, Singapore has been second only to Hong Kong as the second busiest container port in the world. While the island generates substantial manufacturing exports to fill those containers, approximately 60 percent of the containers are transshipped from neighboring countries using small feeder vessels and are then loaded onto enormous oceanic container vessels (Figure 20–8). In addition to Singapore's being a regional and global shipping hub, the government has promoted it as a "global maritime center," offering financially rewarding **producer services** such as banking, insurance, communications, and consulting for maritime transport-related activities.

While its maritime transport-related businesses have helped Singapore achieve the twenty-second rank in GNI

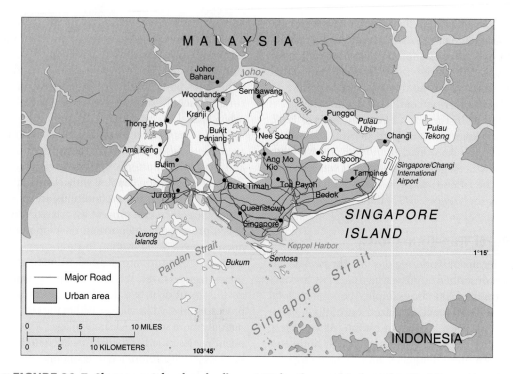

■ **FIGURE 20-7** **Singapore Island and adjacent Malaysian and Indonesian territory.** Although one of the world's smallest countries, Singapore possesses one of the highest levels of economic prosperity.

■ **FIGURE 20-8** **Singapore harbor.** Singapore's many container terminals are constructed on reclaimed land and are considered the best operated in the world. Collecting container cargo from as far away as the Persian Gulf, value-added transport services generate substantial income.

PPP per capita income among the countries of the world and third to Japan and Hong Kong in Asia, its wealth is grounded in a diversified economic base. Singapore serves as the regional headquarters for many multinational corporations that generate substantial numbers of high-paying jobs. More importantly, the government has attracted many foreign high-technology firms to its state-planned industrial parks. During the 1970s and 1980s, Singapore attracted firms engaged in the assembly of lower-technology consumer products, such as stereo equipment, televisions, and video recorders. Because land and labor costs were more than double those in neighboring countries, these multinational corporations eventually began moving their manufacturing or assembly operations to Malaysia and later to Thailand and Indonesia. Replacement industries were found, however. Foremost among these was the global electronics industry, which in 2001 accounted for approximately 50 percent of manufacturing output and 60 percent of exports. In recent years, for example, Singapore has been the world's largest producer of hard drives for personal computers, whether marketed under Japanese, Taiwanese, or American brand names.

As land and labor costs have risen relative to levels prevailing in surrounding countries, the Singapore government, much like its economic-tiger cousins of South Korea and Taiwan, has been forced to promote higher-technology and capital-intensive industries domestically and to entice lower-technology industries to move abroad. Overseas relocation has been either long-distance—to Thailand, Vietnam, or southern China—or to neighboring Malaysia and Indonesia. This example of a "borderless world" associated with globalization has led to the emergence of a **growth triangle** centered on Singapore, the nearby southern Malaysian state of Johor, and the Indonesian islands of Batam and Bintan, situated just south of Singapore. This industrial twinning phenomenon, much like the relationship between Hong Kong and China's Pearl River Delta, enables both foreign and Singaporean investors to base the labor-intensive portions of production in Johor and Batam before completing further value-added activities in Singapore. Singapore's pursuit of higher-value economic activities stems from its highly trained workforce, as well as aggressive government promotion of an information-technology culture that has virtually become a national obsession. The critical role of information technology and higher value economic activities such as producer services is critical to globalizing the city-state's economy and providing a competitive edge in the region. In this respect, government development policies have been successful as Singapore accounted for 44 percent of Southeast Asia's total value of commercial services exports in 2003.

The pervasive role of government in economic and social planning has been central to Singapore's success. The state controls many profitable domestic industries and actively seeks to attract foreign investment. Social planning policies have assured social stability in an ethnically diverse population, adequate housing for all, a relatively crime-free environment, high personal savings rates, and an immaculately clean urban environment. Despite being perceived by some as an overly regulated state, Singapore possesses a government that is attracting continued foreign investment because it is virtually free of corruption. Although on paper Singapore is organized as a constitutional democracy, these accomplishments are the product of a "soft-authoritarian" government dominated by a single political party that has handily won every election. Because the government views its primary responsibility to be the promotion of economic growth, political freedoms in the Western sense become less important in legitimizing political power.

Malaysia

Malaysia is a spatially fragmented country consisting of the more densely populated Peninsular portion, which is called West Malaysia, and the sparsely populated states of Sarawak and Sabah (East Malaysia), situated on the northern coast of the island of Borneo (Figure 20–9). Since the 1970s, Malaysia has evolved from a traditional developing nation, whose economy was dependent on exports of tin ore (cassiterite) and rubber and other plantation crops, to one of the newly industrializing economies, whose leading exports are now manufactured products. Guiding this economic transformation has been the goal, laid out in a 1970–1990 master plan called the **New Economic Policy (NEP)**, of increasing the economic contributions of the majority ethnic Malays, called *bumiputras,* at the expense of both Chinese and Western economic interests. State-owned industrial enterprises were established to provide nonagricultural jobs for Malays and, by the mid-1980s, when manufacturing replaced agriculture and mining as the most important economic sector, the Malay share of the manufacturing workforce had increased to 50 percent.

A second means adopted to raise the incomes of the predominantly Malay rural workforce was to increase government funding of roads and other infrastructural improvements needed to expand the commercial cultivation of rubber, oil palm, and coffee by native smallholder farmers in frontier areas. Anchored by small urban places supporting agroprocessing industries, these schemes have often been touted as the most successful rural land development programs in the developing world. A significant negative consequence, however, has been the large-scale destruction of the natural rain-forest vegetation. The proportion of Peninsular Malaysia classified as forested plummeted from 74 percent to 42 percent between the late 1950s and 1991 (see the *Geography in Action* boxed feature, Forest Clearance and Underdevelopment).

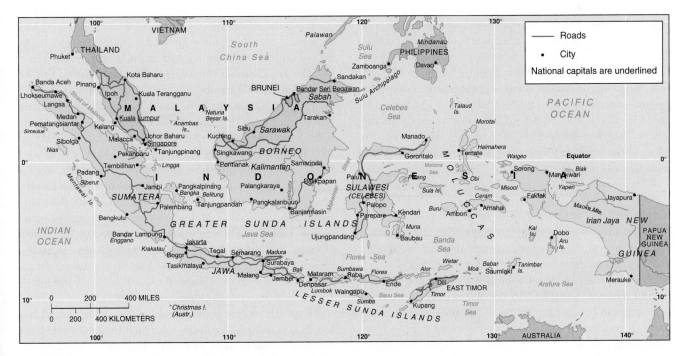

■ **FIGURE 20-9 Malaysia and Indonesia.** Malaysia and Indonesia are insular nations, which rely extensively on maritime shipping for their economic development.

GEOGRAPHY IN ACTION
Forest Clearance and Underdevelopment

The depletion of natural forest cover in Peninsular Malaysia is only one example of the increased pressure on natural resources that often occurs among the economically less developed countries in their quest for national development. In all the Southeast Asian nations, the loss of tropical forests has been rapid. The countries with the greatest losses between 1990 and 2000 were the Philippines, Myanmar, Indonesia, and Malaysia, each of which lost 12 percent or more of its forest area during the decade. Thailand, Cambodia, and Laos also experienced significant loss. In all of these countries, forests are found today only in small patches, often on the highest mountain slopes.

The causes of the alarming rates of deforestation are numerous and can be traced to both domestic and foreign economic forces. One of the earliest sources of long-term deforestation was the foreign demand for raw logs that were exported and made into finished wood products in the importing countries. Interestingly, the first systematic programs designed to increase foreign exchange earnings through the sale of Southeast Asian forest products were promoted in the 1960s by the World Bank. The largest foreign consumer of tropical timber during this early period was Japan. Realizing that greater incomes could be earned by processing logs into plywood and pulp, the Southeast Asian governments next began to promote the use of logs for domestic wood-based industries, with the goal of then exporting the finished products. Domestic timber corporations were thus able to gain access to forest concessions. Tragically, this access has frequently been realized through various monetary incentives offered to corrupt government officials. The loss of forest cover may also be at-

tributed in part to land-hungry peasants, who have been clearing land for years in order to cultivate subsistence crops. Another important source of deforestation is government-connected oil palm plantation corporations which clear vast swaths of forest to make way for this industrial form of agriculture (Figure A). This source of deforestation is especially common in East Malaysia and in Indonesian Borneo and Sumatera. Frequently, the oil palm companies will purchase a tract of forest and let peasant smallholders clear the land and grow subsistence crops for a few years before planting the oil palms.

■ FIGURE A Oil palm plantation, state of Pahang, Malaysia. The clearing of lands in Southeast Asia for the planting of oil palm plantations is a major source of regional deforestation.

Malaysia's per capita GNI PPP of $8,940 has been achieved largely through foreign investment in export-based manufacturing. Between 1980 and 2003, manufactured goods as a percentage of export value increased from 19 percent to 77 percent. While many of the export goods have traditionally been lower-technology products, Malaysia has increasingly attracted higher-technology computer-related manufacturing, a substantial portion of which is located on the ethnic Chinese-dominated island of Pinang, which has been dubbed "Silicon Island" (Figure 20–10). Most of the workers there—employed by such companies as Intel, National Semiconductor, Hewlett-Packard, Sony, and Panasonic (to name a few)—are young rural Malay females, who are referred to as "children of the NEP." Although still part of the west coast economic core, the electronics industry of Pinang, coupled with the southern (Singapore-centered) growth triangle in Johor state, contributes to a healthy geographic deconcentration of industrial development away from the already saturated national capital and emerging EMR region of Kuala Lumpur. In an attempt to become less economically reliant on revenues earned from foreign-owned assembly plants whose long-term competitiveness is challenged by lower labor costs in neighboring countries, and motivated by a desire to propel the country into the "post-industrial" age, the Malaysian

government announced in the mid-1990s a new development program, cleverly called "Vision 20/20," that sets the stage for the country to develop a globally connected and regionally dominant information technology-based economy. The first spark for such an economic transformation was the establishment in 1996 of the **Multimedia Super Corridor (MSC)**, stretching for 31 miles (50 kilometers) south from Kuala Lumpur. Anchored by "Cyberjaya" or "Intelligent City," the corridor is envisioned as a "technopole,"—an information technology research center whose inhabitants reside in "telecommunities." MSC's goal is to foster indigenous creativity and expertise, and it is hoped that the presence of large multinational firms such as Sun Microsystems, Oracle, and Intel will aid in the development of a national or Malaysian-based information technology culture. Although obviously linked to increased globalization, the MSC is perceived as a means of balancing the large presence of foreign multinational firms through the creation of a new economic sector intended to serve Malaysian development aspirations.

Despite its recent economic successes, Malaysia still has much to accomplish. Although originally established to increase the Malay share of wealth, the state-owned enterprises are a huge financial burden. The government has been wary of rapidly reducing the state-owned sector because

■ **FIGURE B** **Burned forest lands of Indonesia.** The burning of tropical rain forests has become a major environmental issue in Southeast Asia. Much of the region's atmospheric pollution has been tied to the practice.

Whether through illegally logging the forest or through clearing the land for eventual plantation use, forest fires have become an associated environmental hazard (Figure B). Occurring mostly in Indonesian Borneo and Sumatera, the smoke haze layer of the 1997 fire season blanketed more than 1,158,000 square miles (3 million square kilometers). On Borneo, some 32 days were characterized by "unhealthy to hazardous" air conditions, while in Malaysia and Singapore, 40 and 12 days, respectively, possessed similar air pollution levels. Aside from the cost of poor health conditions, indirect economic losses asso-

ciated with the forest fires are substantial. Many persons must stay indoors because of the high cost of respiratory masks. Schools are sometimes forced to close, and the reduction of visibility negatively impacts the transport, construction, and tourism industries. Although 1997 was the worst fire year in recent memory, the fires continue; approximately 3,000 fires in Indonesian Borneo alone were burning out of control in the summer of 2002.

There is growing evidence that tropical deforestation ultimately contributes to the underdevelopment of the formerly forested regions. One reason why forest loss has been so rapid is that, on average, only 20 percent of the trees felled are of commercial value. Felling the forest also leads to the loss of nontree plants possessing subsistence and commercial value, pollutes streams that provide fish and drinking water, makes river travel hazardous because of logging debris, and increases the chance of flash floods that destroy agricultural lands in downstream communities. This "commodification of nature" has thus placed the state and timber interests in direct conflict with indigenous forest dwellers who have long practiced sustainable agroforestry.

Frequently, governmental and commercial interests have attempted to make scapegoats of tribal groups, blaming traditional slash and burn agricultural practices for the loss of the forests. Environmental groups, however, have noted correctly that slash and burn agriculture has been a part of traditional sustainable forestry in the tropics from time immemorial and thus cannot possibly be the primary cause of deforestation. It is easy to understand why the loss of forest cover is one of the most important development issues in Southeast Asia.

it provides a source of employment for the politically powerful ethnic-Malays. Recent developments have attempted to address these perceived problems. Many state-owned enterprises have been partially privatized, and many government policies favoring ethnic-Malays were addressed in the New Development Policy (1991–present). Also, the long-term goal of spatial equity in development has not been met. Of Malaysia's fourteen political units, only five possess incomes above the national average, and four of these are located in the west coast core region. As part of the national economic periphery still tied to natural resource exploitation, three of the four peninsular east coast states, as well as Sarawak and Sabah in East Malaysia, remain relatively poor. The colonial inheritance of uneven spatial development persists to the present day.

Indonesia

Measuring some 741,100 square miles (1,920,000 square kilometers), the archipelagic state of Indonesia is by far the largest in Southeast Asia (see Figure 20–9). Its original territory was enlarged considerably when the newly independent state annexed the western half of the island of New Guinea in 1963. With 219 million citizens, Indonesia is Southeast Asia's most populous country as well. Although

five islands support the vast majority of Indonesia's population, the problem of physical fragmentation is pervasive and lies at the heart of most of its development challenges.

The overarching goal of the newly independent government was to forge national unity from a diverse array of ethnic groups that inhabited the far-flung island realm. The achievement of unity was particularly difficult because of the political, economic, and cultural domination of Jawa, the most populous island. In pursuit of the goal of national unity, Malay was successfully promoted as the national language.

Although Indonesia is the most populous Islamic country in the world, religious freedom has generally been respected. While the Indonesian population is 87 percent Muslim, there exist Hindus on Bali, substantial pockets of Protestants on Sumatera and Sulawesi, Roman Catholics in West Timor, animist groups in Kalimantan and Irian Jaya, and, of course, Chinese Buddhists in urban areas.

Much of Indonesia's early development planning was focused on two population-related issues: growth and distribution. With a population approaching 100 million during the late 1960s, the government implemented a noncoercive volunteer family planning program promoting the two-child family. The program has been recognized around the world as quite successful, considering the relative poverty of the country. Indeed, average annual population growth rates

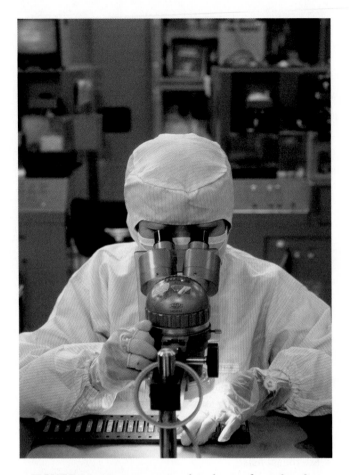

■ **FIGURE 20-10 Computer-related manufacturing, island of Pinang, Malaysia.** Pinang has become a center of high-technology industrial activity in Southeast Asia.

declined from 2.9 percent in 1970 to 1.6 percent in 2004. Reflecting other government initiatives, the proportion of the population living below the poverty line before the Asian financial crisis in 1997 dropped to under 15 percent. The poverty rate dramatically increased to 23.5 in 1999 in response to the financial crisis, but has declined to 16.6 in 2004, slightly higher than pre-crisis levels.

Linked to poverty eradication has been a program aimed at raising agricultural yields through the widespread adoption of hybrid rice seed and fertilizers. This approach has been so successful that Indonesia reached rice self-sufficiency by the mid-1980s, after having been the world's largest importer just 15 years before. However, rice self-sufficiency ended in the mid-1990s as a result of the cumulative effects of El Niño and La Niña, the financial crisis, and especially because the government dramatically reduced import tariffs on rice as part of a larger strategy of trade liberalization associated with the globalization process. Because the crowded island of Jawa supports 58 percent of the country's total population, the government has also sponsored migration programs designed to resettle landless Jawanese peasants on the outer islands, where they can engage in subsistence and cash crop agriculture. By 1994 approximately 1.7 million peasants had been relocated through government financial assistance for transport, settlement, land, and social services. Although the numbers are staggering, they represent just a small portion of Jawa's total population of more than 113 million persons. Nor have the transmigration programs slowed rural-to-urban migration—four of Indonesia's five most populous cities are located on Jawa. With an estimated population of more than 17 million, Jakarta has become a prototypical EMR and primate city, consuming huge quantities of food, fuel, and other raw materials while monopolizing the nation's wealth (Figure 20–11).

During the 1960s and 1970s, Indonesia's economy was inward-looking and dependent on a narrow manufacturing base supported by exports of petroleum and a variety of plantation crops. Depressed global petroleum prices during the early 1980s forced the Indonesian government to liberalize the economy, transforming it over the past two decades into one that is now dominated by industry. Although the economy is still largely based on inefficient state-owned companies, the national capital region of Jakarta has attracted a large number of East Asian and other foreign-owned export-based manufacturing operations focused on the production of lower-value electronic parts, footwear, textiles, and household appliances. Diversifying the country's manufacturing base has also involved promoting resource-based industry on the outer islands. Adding value to the well-endowed outer island resource base is important both to the economic development process and to Indonesia's political stability. With more than 80 percent of national industrial employment concentrated in Jawa, deep-seated resentment exists among the outer island political and economic interests.

■ **FIGURE 20-11 The skyline of Jakarta, Indonesia's primate city.** The Istiqlal Mosque is set in a garden landscape that contrasts sharply with the dense, sprawling city and the thrusting towers of its international-style high-rise buildings. The more discrete twin steeples of St. Mary's Cathedral, reflective of Indonesia's colonial past, are visible just beyond the Istiqlal Mosque's minaret.

The provinces of Irian Jaya and Aceh provide ideal examples of outer island resentment against Jawanese interests. Irian Jaya province (western New Guinea) is home to one of the world's largest gold and copper mining operations, which has been developed by Freeport-McMoRan, a U.S.-based multinational firm (Figure 20–12). The Jawa-based Indonesian national army has been used to silence the substantial environmental concerns of the many indigenous tribal groups. In addition, the Indonesian army, or its surrogates in the form of "special police," have killed or imprisoned various members of the Free Papua Movement who seek independence. At the other end of the archipelago in the northern Sumatera province of Aceh, the nativist movement called the Free Aceh Movement has waged a war of independence since 1975, which has resulted in the deaths of more than 10,000 innocent civilians. Having resisted Dutch rule during the colonial period and having always been resentful of the central government in Jakarta, the Acehnese desire to keep more of their natural resource revenues at home, especially those from the huge Exxon-Mobil natural gas facility. While Aceh has been given "special autonomy" status, the killings of innocent civilians continues. Antagonism between Jakarta and Aceh has only increased since the 2004 Asian tsunami because of the central government's slow and inefficient delivery of financial aid and materials with which to reconstruct the province's basic infrastructure.

The unequal economic relationship between the outer islands and Jawa, coupled with outer island resource extraction by foreign multinationals, is a classic example of an increasingly developed core benefiting at the expense of a less developed periphery. Coupled with the pervasive influence of Jawanese culture, those in the outer islands often feel colonized by the government on Jawa, which they perceive as a form of **internal colonialism**.

The regional tensions associated with the economic disparities between the core and periphery were brought to the forefront as a result of the Asian financial crisis. The meltdown of Indonesia's economy forced the 1998 resignation of President Suharto who had guided the country since 1965. The issue of corrupt government practices became magnified by the economic crisis and forced the new civilian-led government to grant greater political representation to both the common person and the twenty-seven provincial governments. To diffuse political tensions, a popular national election was held in 1999 and, in that same year, Jakarta began allowing the provincial governments to keep a greater share of revenues derived from natural resource exploitation. These measures failed to prevent further geographic disintegration of the state, however. In 1999 the tiny province of East Timor, which was a Portuguese colony until 1975 when it was ruthlessly annexed by the Indonesian army, gained independence after 25 years. The desire for independence is certainly understandable given the fact that some 200,000 East Timorese (who are culturally more similar to Melanesian Irian Jayans and Catholic) lost their lives during Indonesia's brutal occupation of the island. Economic reasons are equally important, however, as the non-Timorese population that constituted only 5 percent of the province controlled 75 percent of the economy. East Timor's economic recovery will be difficult. The Indonesian Army and its militias destroyed 70 percent of the region's economic infrastructure during the 1999 war of independence. Some 42 percent of East Timor's population are below the poverty line and its only important export is coffee. A 2005 agreement between the Australian

■ **FIGURE 20-12 Papua province gold mine.** This mine, situated high in the mountains of New Guinea, has been the object of prolonged and serious conflict between the Indonesian government and a coalition of international environmental and indigenous tribal interests.

and East Timorese governments to develop rich oil and gas fields in the Timor Sea is the only short-term opportunity for lifting this ravaged country out of poverty.

In an odd twist of reasoning, the Asian financial crisis signaled a number of positive developmental changes for Indonesia. First, the institution of "crony capitalism," in which the rich and politically well-connected derive huge financial windfalls, may now be less of an endemic feature of national economic growth. Second, revenue-sharing reforms can only lead to a more spatially even distribution of the national wealth as provinces are able to formulate their own development agendas. Third, a more democratic and open election system, no longer controlled by an all-powerful military, provides the people of Indonesia with hope that they, too, can be part of the development process. The multitude of socieconomic problems that currently beset Indonesia, however, will not be solved overnight.

The Philippines

Composed of some 7,000 islands and encompassing an area of 115,831 square miles (300,000 square kilometers), the Philippines is perhaps Southeast Asia's most distinctive country because of its early colonization by Spain (Figure 20–13). After an initial settlement in Cebu in the central Philippine Visayas region, the Spanish plundered Manila and transferred their base of operations to this small Malay-Muslim settlement on the island of Luzon in 1574. Much as they did in Latin America, the Spanish granted large tracts of land to colonial officers and the Catholic Church. Most of the population eventually converted to Catholicism, with the exception of Muslims living on or near the southernmost large island of Mindanao, where the colonial government never gained strong administrative control. With a powerful and exploitative regional landlord base and Chinese merchants who often intermarried with the Spanish and converted to Catholicism, a highly rigid socioeconomic class structure developed that persists to the present day.

The Philippines was ceded to the United States in 1898 as part of the spoils of the Spanish–American War, and the United States proceeded to modernize the almost feudal Philippine economy. However, although the transportation network was upgraded, agroprocessing industries were established in the provinces, popular education was instituted, and health and social services were upgraded, the fundamental sources of income disparities—namely, social class inequity and corruption—remained. Furthermore, the overwhelming presence of American culture (English is the lingua franca) contributed in various ways to the failure to develop a distinct national identity, an essential ingredient in its own right for long-term economic development. Owing, then, to both Spanish and American influences, the

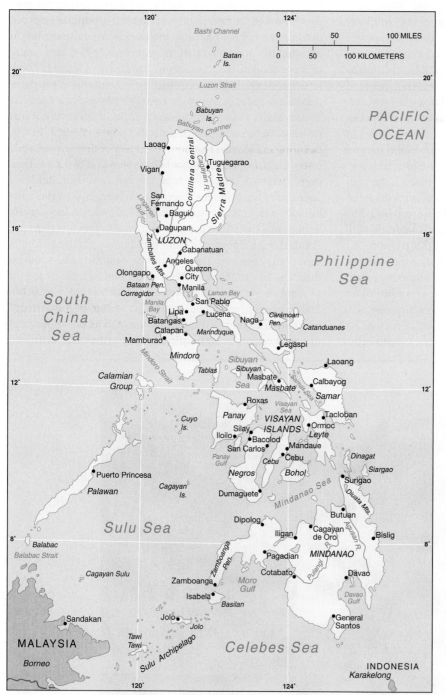

■ **FIGURE 20-13 The Philippines.** The country of the Philippines includes more than 7,000 islands, many of them uninhabited. The Philippines has experienced the greatest degree of Western cultural influences of Southeast Asia.

archipelago has long been regarded as an Asian outpost of Western thought and practice, and in an odd sense shares many social and economic commonalities with Latin America.

Following the Filipinos' independence in 1946, Philippine economic growth remained tied to agriculture, whether of the subsistence, cash cropping, or industrial agroprocessing variety. Land reform to benefit the vast numbers of poor peasants has been a part of the agenda of every recent government, but progress has been minimal owing to widespread corruption and to the fact that most of the political decision makers are largeholders themselves (Figure 20–14). A strong manufacturing base has evolved, but this sector of the economy is restricted to the Manila metropolitan region and a handful of second-tier regional centers. With the exception of textiles, which are produced by women in thousands of home-based shops, these industries tend to be capital intensive rather than labor intensive. The country's limited economic growth has kept per capita income levels among the lowest in Southeast Asia. The country's economic growth was also negatively affected during the 1970s and 1980s, when martial law was imposed to strengthen President Ferdinand Marcos' grip on political power and to contain the country's growing communist insurgency movement. The latter fed upon widespread social unrest arising from deepening economic disparities. The results were the flight of domestic capital overseas, a decline in foreign investment because of political instability, and a large-scale outflow of workers, both educated and undereducated, seeking better economic opportunities overseas (see the *Geography in Action* boxed feature, Migrant Labor and Economic Development). In 2003 there were some 8 million Filipinos working abroad, equal to approximately 10 percent of the country's population. In fact, the government has established the Philippine Overseas Employment Administration as part of its economic development strategy because overseas workers relieve unemployment at home and money sent back home (US$7.6 billion in 2003) helps offset the country's large trade deficit.

The national economy began to rebound with the fall of Marcos and the election of Corazón Aquino as president in 1986. The return of funds that had fled in the previous decade, coupled with migrant worker remittances, injected huge amounts of investment capital. Foreign investment has also returned, particularly at newly established export processing zones where electronics and textiles are commonly produced for export. The former U.S. naval facility at Subic Bay has attracted numerous offshore manufacturing facilities, as have zones on the southern and northern periphery of Manila and Mactan Island near Cebu. Although much of this recent economic growth has been restricted to the central Luzon region around Manila, industrial investment in the outer-province cities of Cebu, in the central Visayas, and Cagayan de Oro and Davao in northern and southern Mindanao attempts to provide a geographic counterweight to Manila's economic dominance. Unlike Indonesia, the more evenly spaced distribution of urban centers in the Philippines has allowed for the emergence of distinct regional economies. Nevertheless, manufacturing still accounts for only 10 percent of the national labor force, and the greater Manila region accounts for 50 percent of that total. As the national capital region and an EMR, rural migrants have flooded into Metro Manila seeking a better life. With limited economic opportunities, many reside in squatter settlements or other forms of unimproved housing with little access to services such as clean water and utilities. Some 35 percent of the population in Metro Manila reside in squatter settlements (Figure 20–15).

The Philippines possess a number of interconnected problems that are very difficult to remedy. High unemployment and underemployment rates are in part caused by high population growth rates. In other words, not enough jobs are being created for those entering the workforce. This in part explains the 30 percent of the population that is below the poverty line. Closing the income gap between the rich few

■ **FIGURE 20-14 The widening quality of life gap.** Often residing in larger regional cities or Manila, absentee landlords possess country homes like this one in southern Luzon.

■ **FIGURE 20-15 Squatter settlement in Manila.** Squalor and sophistication reside in close proximity, yet remain worlds apart, as Manila's downtown towers over the congestion of nearby impoverished neighborhoods.

GEOGRAPHY IN ACTION
Migrant Labor and Economic Development

Different rates of economic growth among states of the larger Pacific Rim region have created a substantial flow of migrant labor from poorer labor-surplus countries such as Indonesia, Thailand, and the Philippines to richer labor-shortage countries such as Singapore, Malaysia, and Taiwan. Migrant laborers are willing to engage in low-paid work that is deemed undesirable by local populations. Because of the infrastructural needs associated with rapid economic growth, the greatest labor demand has been in the construction industry, with secondary demand in domestic services, production workers in smaller firms, and plantation agriculture. In turn, countries contributing the migrant laborers rely on them to reduce unemployment at home as well as to provide needed foreign exchange through remittances. The pool of Southeast Asian migrant laborers numbers in the millions. With approximately 8 million migrant laborers, the Philippines is the largest source country. The Malaysian construction industry employs about half a million Indonesians, and plantation operations would be dealt a serious blow without them. Thailand supplies close to 430,000 migrant workers, primarily to Singapore, Hong Kong, and Malaysia. Japan and Taiwan have become the semipermanent homes to one million and a quarter million migrant workers, respectively.

The push to seek employment overseas is driven by both internal and external forces. Source countries are generally characterized by high unemployment levels. Receiving countries with higher economic growth rates often suffer labor shortages despite the fact that both the husband and the wife are part of the labor force. Under these circumstances, Indonesian construction workers in Malaysia, for example, earn ten times the Indonesian minimum wage. The governments of sending countries often see labor migration as a vehicle to diffuse social and political protest. The flip side is true as well in that receiving countries are provided with a relatively inexpensive source of labor for their booming infrastructural projects, which are tied to the fast-growing Pacific Rim economy. A Hong Kong construction company, for example, is able to get away with paying a migrant worker only one-quarter of the salary usually paid to native workers. Not covered by local employment laws and lacking legal residency status, migrant workers are often financially cheated by job placement agencies or their employers. Fees charged by these private agencies are often equal to half a year's wages in the receiving country. The "maid trade" of well-educated Filipina, who are unable to find employment at home, is perhaps the most visible form of labor migration. In 2003, there were approximately 150,000 and 140,000 Filipina maids in Hong Kong and Singapore, respectively, because under conditions of the tight labor markets associated with economic growth, many married women in these two cities sought high-wage employment. The physical abuse of Filipina maids in the Persian Gulf states, as well as in various Southeast Asian countries, is well documented.

On the positive side, remittances have contributed substantially to the economic betterment of many of the villages from which the migrant laborers come. Filipina maids in Hong Kong, for example, are able to earn $500 per month, far more than what they might expect to earn back home. The money is certainly welcomed in the stagnant economy of the poor villages, which benefit immensely from the funds sent by the migrant workers to family members. On the negative side, migrant labor also exacts a serious sacrifice on the part of family members because the spouse left behind must try to do the work of two, and children grow up without having both parents at home. The emergence of a migrant-labor culture as a by-product of the global economy is an issue that most contributing and receiving governments would prefer to ignore because of the uneasy position of treating human beings as commodities and the sometimes blatant ethnocentric or racist attitudes on the part of some in the host countries.

Sources: H. Jones and A. Findlay, "Regional Economic Integration and the Emergence of the East Asian International Migration System," *Geoforum* 29, no. 1 (1998): 87–104. Murray Hiebert, "Give and Take," *Far Eastern Economic Review* (May 25, 1995): 54–59.

---

and sea of poor is thus critical to the country's future economic prospects. In addition, this most democratic of Southeast Asian countries experiences endemic political corruption and instability, which in turn drives potential foreign investments elsewhere. After Spanish and American domination, followed by a half century of rule by domestic elites, the people of the Philippines deserve better.

Thailand

Occupying the geographic heart of mainland Southeast Asia, Thailand is distinctive among its Southeast Asian neighbors because it never was a Western colonial possession (Figure 20–16). Changing its name from Siam to Thailand in 1939, the country has been ruled by successive military governments under a system of constitutional monarchy in which the king possesses little political power but is much revered and carries substantial moral influence. The country has experienced many military coups, but most have been relatively peaceful in nature. Until the late 1970s, Thailand's economy was primarily based on a very healthy agricultural export sector that included large quantities of rice, tropical fruits, seafood, and canned foods. Throughout much of the 1960s and 1970s, Thailand also gained economic benefit through servicing the various Western military needs associated with the Vietnam War, as well as from foreign aid provided in the expectation that the country would serve as a bulwark against the perceived spread of communism.

Beginning in the early 1980s, the government established economic policies promoting export-based manufacturing activity on the part of both foreign and domestic investors. Since then, manufacturing activity has increased rapidly. Aside from the usual factors, such as economic incentives for foreign industrial concerns, reduction of government regulations, political stability, and inexpensive labor, Thailand offered a friendly business culture to East Asian investors

■ **FIGURE 20-16 Myanmar (Burma), Thailand, Cambodia, Laos, and Vietnam.** The nations of mainland Southeast Asia consist largely of north-south trending mountain ranges and broad intervening lowland river valleys, which support the majority of the population.

who shared the Buddhist religion. The country was also friendly to ethnic Chinese, a factor important to Hong Kong and Taiwanese investors. Equally important, particularly for the future, is Thailand's central location within mainland Southeast Asia that allows investors access to the resource-rich markets of neighboring countries.

Although primary commodity exports remain important, manufacturing has become the engine of economic growth. As a result, Thailand's GNI PPP per capita in 2003 was $7,450, and manufactured goods comprised 75 percent of exports by value. Unlike Malaysia, Singapore, and even the Philippines, where FDI in manufacturing exports is anchored in electronics, Thailand has become the center of Southeast Asia's auto industry. While most ASEAN countries produce auto

parts or assemble vehicles, Thailand is home to more than twelve foreign corporations assembling autos as well as 300 parts suppliers. Japanese manufacturers arrived in the 1980s, and later firms such as General Motors, Ford, BMW, and even Porsche have established assembly operations. While some exports are to Japan and Europe, the primary markets are in Asia, and primarily Southeast Asia, where AFTA allows for low tariffs on imports of auto parts or vehicles between member states. With rising middle class affluence in the region, Southeast Asia is set to become the fifth largest automobile market in the world. Also important to Thailand's economy is tourism. With one of the best developed tourism industries in the world, tourists are attracted to the sun and surf locations in the country's south, cultural heritage sites based on Buddhist temples in the Bangkok region, and eco-adventure destinations in the mountainous north.

The economic growth associated with export-oriented manufacturing, however, has been almost totally restricted to the capital city region and EMR of Bangkok. With only 16 percent of the national population, Bangkok now accounts for slightly more than 50 percent of economic output and half of all the industrial establishments in the country. In contrast, the territorially large northeast region supports approximately 40 percent of the country's total population but accounts for only 12 percent of economic output. While Bangkok has prospered, much of Thailand remains poor and agriculturally based. With an EMR population surpassing 11 million, Bangkok is perhaps the best example of a primate city in Southeast Asia. Accounting for 80 percent of Thailand's urban population, Bangkok is approximately thirty-four times as large as the second largest city of Chiang Mai. With the disparities in development between Bangkok and the rest of the country, it is fair to ask the question whether Thailand is a newly industrializing country or whether Bangkok is simply a newly industrializing urban region.

The high concentration of economic activity has generated severe environmental consequences in this once pleasant city of canals and Buddhist temples (Figure 20–17). Industrial development, coupled with greatly increased private automobile use, has created almost intolerable levels of air pollution. The levels of carbon monoxide, lead, and dust pollution, caused in large part by the widespread use of low-grade and cheaper leaded fuels, are often three times the international safety levels. "Rush-hour" traffic persists throughout the day, with the new elevated expressway system unable to keep pace with traffic growth. As a result of road construction not keeping pace with the rise in the number of vehicles on the road, traffic speed averages 6 miles (9.6 kilometers) an hour. Water-borne pollution, originating from industries and households, is a serious problem as well. The Chao Praya River south of Bangkok, as well as the spider-web of canals that made Bangkok the "Venice of the East," are dangerously polluted. Another source of human-induced water pollution is the saltwater intrusion of the city's freshwater table, which is facilitated by the overextraction of freshwater. With the freshwater table dropping, flooding problems are increasing as the city experiences subsidence averaging 4 inches (100 millimeters) a year. Those urban residents living in the city's burgeoning squatter settlements bear the brunt of the floodwaters.

Vietnam

Vietnam was conquered by Viets from China in the second century B.C. and for most of the following 2,000 years was dominated, both politically and culturally, by China, before being captured by France in the mid-to-late-nineteenth century. A coalition of nationalist groups emerged in the 1940s that, under the leadership of a Communist guerrilla named Ho Chi Minh, succeeded in driving out both Japanese and French colonial forces (see Figure 20–16). The 1954 Geneva Agreement, which formally separated Vietnam into northern and southern regions, was reflective of historic regional divisions. Over the next two decades, northern forces waged a war of unification against the South and its U.S.-led allies. The establishment of an independent, unified Vietnam occurred in 1975, following the hasty withdrawal of the last U.S. troops after a physically and emotionally devastating war. Known by the North Vietnamese as the "American War," it resulted in 58,000 American war deaths, but 2 million Vietnamese casualties. There was, however, no reconciliation with the South, a region that, from the northern communist perspective, had been fully "corrupted" by long-term exposure to Western capitalist influences. Similar to the experience in early Maoist China, farms, factories, and businesses were confiscated by the state, reeducation was required of those tainted with Western ideas, and tribal highlanders were forced to settle and engage in the more "civilized" Vietnamese form of intensive agriculture. More than 1 million persons were forced from southern cities into new economic development zones, where the government believed they would be more productive. The 2 million or so refugees that fled Vietnam over the following 15 years were a testament to the harsh conditions of conformity imposed upon Vietnam's population.

By the early 1980s, the hard-line Marxist policies of the national government had fashioned a country that, within just a few years, had become one of the poorest in the world. Agricultural stagnation, lack of investment capital, a virtual absence of consumer goods, and declining support from the Soviet Union all contributed to the economic malaise. Recognizing its decline, and aware of

■ **FIGURE 20-17 Bangkok, Thailand.** Islands of traditional Buddhist architecture such as this temple co-exist with the new international-style buildings of central Bangkok.

communism's growing economic challenge in the Soviet Union, Vietnam's leaders began to loosen the strings of the centrally planned economy and adopted a program of **doi moi**, or "economic renovation." This policy shift entailed dismantling agricultural communes, opening the country to foreign investment, reforming the financial system, attracting foreign investment, and reducing subsidies for state-owned companies. The economy as a whole has responded well to these new doi moi economic opportunities integrating Vietnam into the global economy. Although starting at a low base, Vietnam experienced the highest rate of GDP growth as well as the highest export growth rate in the world during the 1990s (surpassing even China). Share of exports in GDP rose from 24.9 percent in 1994 to 47.5 percent in 2002. The impact of economic growth on poverty reduction was substantial as poverty rates of 58 percent in 1993 declined precipitously to 29 percent in 2002. Poverty reduction was not geographically even, however, as the southeast experienced greater declines when compared to the Central Highlands and Northern Uplands. Even in the traditional rural economy where rice anchors production, increased property rights by rural households assisted in greater levels of production so that Vietnam shifted from being an importer of rice in the mid-1980s to the second largest exporter of rice in 1996.

The engine of Vietnam's economic turnaround, however, has been FDI that is dominated by East Asian (70 percent) and other ASEAN members, particularly Singapore, Malaysia, and Thailand (20 percent). Aside from investments in hotel construction, tourism, and oil development, FDI has specifically targeted garments and footwear for export. Garments, footwear, electronics, and seafood in 2005 accounted for approximately 43 percent of export value. Its export market is primarily Asia, but with the lifting of the U.S. trade embargo in 1995, the United States now consumes approximately one-quarter of Vietnam's exports. The country is also an important tourist destination, particularly for sun and surf and eco-adventure recreation. Unfortunately, much like in other Southeast Asian countries, the openness that accompanies globalization, tourism, and increased domestic wealth has dramatically increased the pool of sex workers and HIV infection rates as well. For economic growth to continue, the government must continue to liberalize the national economy. Although the number of state-owned enterprises has been halved, confusing foreign investment laws, corrupt government officials, and continued suspicion of foreign intentions on the part of the government prevents an even more rapid economic transformation (Figure 20–18). The potential for economic and social change is great, however, because some 60 percent of Vietnam's 81 million were born after the war, and this younger generation lacks the emotional ties to the revolution symbolized their aging government leaders.

In considering the possible future development geography of Vietnam, it is important to recognize the historic differences that have divided the northern and southern regions. Centered on the densely populated Red River delta, northern Vietnam is the traditional core of Vietnamese culture. It was here, among millions of poverty-stricken peasants, that the communist revolution took root, and it was from the North that Marxism spread to the South. The South, centered on the Mekong River delta, was historically more prosperous and commercially oriented, and it became even more outwardly focused as a result of French and later U.S. influences. It was the Cholon district of Saigon, for instance, that supported one of the largest single concentrations of Overseas Chinese in the world. In a pattern similar to that in China, southern Vietnamese culture is a variant of the original culture from the North, and it is likely that Vietnam's South will ultimately assume a leadership role as the country becomes more fully integrated into the global economy. Much as in China, this process of integration is not without great social cost. In a country where 76 percent of the population live in the countryside, and urban incomes are four times those of rural zones, it is not difficult to predict the increasing regional income gap that will flow from an export-led development strategy that favors urban areas over rural. Over the past decade, for example, the number of street children in Vietnam has increased tenfold; most of these children migrated from the countryside in an attempt to earn income for their families residing in the peasant villages.

Cambodia (Kampuchea)

Cambodia's past has been even more troubled than Vietnam's. Called Kampuchea by its own people, Cambodia was once a proud empire centered on the magnificent twelfth-

■ **FIGURE 20-18 Ho Chi Minh City (Saigon), Vietnam.** Increasing contact with the global economy has made a flood of products available to the Vietnamese consumer.

century Hindu temple complex at Angkor Wat and sustained by productive rice agriculture nurtured by the Mekong River (see Figure 20–16). Since the fall of the Khmer Empire in the fourteenth century, however, this culturally homogeneous land has been periodically occupied by the neighboring states of Vietnam and Thailand. The French ruled Cambodia from 1863 until World War II, but, following independence, political events once again drew Cambodia into a wider regional conflict.

The war in Vietnam spilled over into eastern Cambodia as the United States attempted to prevent North Vietnamese troops from using this region to launch operations into South Vietnam. In the early 1970s, the monarchy of Prince Sihanouk was overthrown by military leaders who, in turn, lost control of the country to the communist **Khmer Rouge** in 1975. The Khmer Rouge isolated the country from the outside world and launched an unspeakable reign of terror by emptying cities of their inhabitants, murdering most of the educated elite, and suspending formal education in a process of "reconstruction" that the world has now come to know as the "killing fields." The result was widespread starvation and the loss of some 2 million lives. Once North Vietnam conquered the South, it invaded Cambodia, ousted the Khmer Rouge, and installed its own Communist government in 1979. The various factions entered into a coalition government in 1982, and by 1988, Vietnamese troops had withdrawn. In 1992, the United Nations sent in peacekeeping troops, and this was followed by parliamentary elections in 1993. The Khmer Rouge, however, engaged until the late 1990s in limited insurgency activity in the far western regions of the country.

The legacy of war and genocide has obviously hampered Cambodia's ability to reach greater levels of economic growth and development. With its GNI PPP per capita at $1,970 and some 30 percent of its 13.1 million population below the poverty line, Cambodia remains a desperately poor country. The government continues to rely on foreign aid as a critical life support; in 2002, for example, foreign aid assistance accounted for more than half of the central government's budget. Manufacturing accounts for only 20.2 percent of GDP, and FDI remains at relatively low levels, with textiles and garments comprising Cambodia's major source of exports. Government corruption, less than clear FDI investment policies, and an unskilled workforce remain impediments to further foreign investment. Angkor Wat-based tourism is also an important source of foreign earnings. Agricultural productivity is a barometer of Cambodia's poor economic health. While some 50 percent of the total labor force is engaged in rice agriculture, this sector accounts for only 9 percent of GDP. The lack of irrigation during the dry season is the primary problem. In Vietnam, for example, some 54 percent of total harvested area is irrigated, while in Cambodia it is only 8 percent. As a result, rice production per hectare in Vietnam is more than double that of Cambodia.

Another barrier that has slowed the pace of Cambodia's economic recovery is the millions of land mines and unexploded bombs used by Vietnamese, American, and Khmer rouge military forces that continue to maim or kill an average of ten people every day. Cambodia is one of three countries in the world that account for 85 percent of unexploded land mines. Because the land mines are most often planted in locations that prevent farmers' access to their paddy fields, agricultural productivity is negatively affected. Land mines also negatively impact other aspects of life not directly related to agriculture. Children, for example, must forego an education to care for maimed family members. This, coupled with the financial burden of disarming the land mines that could be dedicated to more productive activity, greatly hinders Cambodia's attempts at economic recovery.

Myanmar (Burma)

Surrounded by mountains and centered on the large Irrawady River basin, Myanmar possesses an ecological framework similar to Thailand's (see Figure 20–16). While its eastern neighbor has taken advantage of its physical environment, the development policies of this most western mainland Southeast Asian state have, for decades, spelled economic decline. Myanmar is the one reformer country most resistant to the openness associated with globalization. Led by an authoritarian and isolationist regime that has embraced an indigenous brand of socialism, Myanmar has seen its economy steadily crumble despite a well-endowed resource base of timber, rice, gems, gold, tin, and petroleum. In 1962 the single-party military dictatorship of Ne Win was installed in the midst of ethnic rebellion on its national frontiers and a growing communist insurgency (see the *Geography in Action* boxed feature, The Golden Triangle, Drugs, and AIDS). After the implementation of a new constitution in 1974, the government's isolationist policies moderated to allow increased foreign exposure. The result was a gradual rise in prosperity, although by the mid-1980s, economic growth had stagnated.

The future economic development of Myanmar will surely hinge on the more complete development of its human resources, which can only be realized through the granting of greater levels of political freedom. Political liberalization appears unlikely to occur soon, however, because it would threaten the continued existence of the military-led government, whose state-owned corporations control almost every aspect of the country's economy. A serious threat to military authority arose in the late 1980s with massive popular uprisings. From these uprisings emerged a heroic young woman named Aung San Suu Kyi, the daughter of a national hero, who has challenged the government to democratize the political system; in 1991 she received the Nobel peace prize. ASEAN, through a quiet process known as "constructive engagement," is also endeavoring to promote political change. While external pressure is certainly helpful in encouraging long-term political liberalization, sustainable substantive change will only be realized in any country through the determined efforts of its own citizens.

GEOGRAPHY IN ACTION
The Golden Triangle, Drugs, and AIDS

A region of Mainland Southeast Asia called the **Golden Triangle** encompasses the shared borders of Myanmar, Thailand, Laos, and southwestern China. It is an area of mountainous terrain, poverty, ethnic hill minorities, and frontier lawlessness associated with drug production. In 2002, the Golden Triangle accounted for approximately 21 percent of the world's opium (the source material for heroin) production, with Myanmar accounting for 87 percent of this output (Figure A). More recently, the region has become a major global producer of methamphetamines that have become quite popular among the newly affluent of Southeast Asia, especially in Thailand. Both drugs are controlled by a handful of drug lords, some of whom are protected by powerful elements in Myanmar's military government.

Because of the huge sums of money involved, drug production and trafficking activity in the Golden Triangle possesses a number of negative side effects. Young and impoverished women in Golden Triangle countries, for example, have migrated to the many booming border towns to become sex workers in karaoke bars and massage parlors catering to truck drivers, businessmen, soldiers, policemen, and international tourists. As a result, HIV has spread dramatically into the general population of the region. Myanmar's 2003 HIV positive population between the ages of 15–49 accounts for 22.5 percent of the Southeast Asian total of 1,425,700, and is ranked second behind Thailand (39.2 percent), with Vietnam (14.6 percent) ranked third. In addition, HIV-positive sex workers in the border region are lured to other cities in Myanmar and Thailand only to spread the virus. Even the border regions of China, Bangladesh, and India's northeast are now impacted. While

■ **FIGURE A Myanmar and the Golden Triangle.** Despite efforts by the central government to suppress poppy production, Myanmar's ethnic farmers in the eastern and northern highlands still produce over 80 percent of the Golden Triangle's output of opium.

international development agencies have promoted alternate crops to opium and have been relatively successful, methamphetamine production is in no way affected by such alternate crop development strategies. Saddled by an authoritarian government and poverty-stricken population, Myanmar's development hopes are now burdened by a rising HIV population.

▶▶ Summary

The precolonial political and economic systems of Southeast Asia were forever changed by the effects of colonialism. In attempting to restructure their economies to better suit the domestic circumstances of the postcolonial period, each newly independent country embraced a different development path. Each of these paths, however, has been characterized by state-directed economic growth. The insular Southeast Asian states and Thailand have chosen to temper state-directed development with strong injections of private sector involvement. In mainland Southeast Asia, the development paths have been overwhelmingly centered upon the state. After decades of economic stagnation, the mainland nations are slowly integrating into the private-sector-driven global economy. It is important to recognize, however, that the recipe for economic growth in other developing world regions may not be the same as Southeast Asia's, owing to differing physical and cultural circumstances. The rapid pace of growth in Southeast Asia has been overly dependent upon economic forces emanating from the more industrialized regional neighbors of East Asia. Because relying on foreign investment as a source of long-term economic growth is problematic, it is critical that domestically generated investment be harnessed by the Southeast Asian nations to reach the next plateau of development.

▶▶ Key Terms

Angkor Wat
Asean Free Trade Area (AFTA)
Association of Southeast Asian Nations (ASEAN)
cash cropping system
Chinese diaspora
city-state
doi moi
dual economic system
extended metropolitan regions (EMRs)
Golden Triangle
growth triangle
industrial colonialism

internal colonialism
Khmer Rouge
mercantile colonialism
Multimedia Super Corridor (MSC)
New Economic Policy (NEP)
Newly Industrializing Economies (NIEs)
Overseas Chinese
producer services
shatterbelt
thalassocracy
tsunamis

■ **Cattle walking the streets of New Delhi.**
As unperturbed by the flow of traffic around them as mo-
torists are unfazed by their presence, cows, regarded by Hindus as sacred animals, are allowed to roam freely even when
causing minor obstructions to traffic. India, the world's second most populous country, exhibits both traditional and
modern lifestyles and its rapidly growing cities have become
centers of cultural and technological change.

Chapter 21

South Asia:
India and Its Neighbors

- Environmental Contrasts
- The Precolonial Heritage
- Colonial Transformation
- Independence and Nation-State Building
- Population Contours of India
- Agricultural Development in India

- India's Industrial Economy
- Urban India
- Pakistan
- Nepal
- Bangladesh
- Sri Lanka

South Asia is often referred to as the "Indian subcontinent" because of the territorial dominance of India. On India's periphery are the smaller nations of Pakistan, Nepal, Bhutan, Bangladesh, Sri Lanka, and the archipelagic country of the Maldives. One of the obvious demographic characteristics of South Asia is its substantial population density. India, Pakistan, and Bangladesh respectively rank as the second, sixth, and eighth most populous nations in the world.

High population concentrations do not of themselves cause poverty or underdevelopment, for people can and do function as both producers and consumers of resources. When levels of human productivity are low, however, high population densities frequently are associated with widespread poverty. Such has been the case with the South Asian countries, each of which ranks among the world's poorest nations. Because South Asia is primarily a rural region where the majority of the people are smallholder subsistence farmers, levels of poverty are greatest in the countryside, despite the dramatic increases in agricultural output achieved through the Green Revolution. The South Asian tradition of state control of industry has not provided the required competitive environment and economic growth to absorb the huge pool of surplus labor. South Asian economies thus possess little of the dynamism characteristic of their Pacific Rim neighbors. In the past decade, however, India has made significant strides by reducing state participation in the economy and engaging the forces of globalization. An additional factor that distinguishes the South Asian realm from the rest of the Asian region is the divisive role of ethnicity, religion, and politics in the economic development process. Cultural diversity does not necessarily constitute a barrier to the forging of national unity, but when such differences are exploited by competing political groups whose interests do not address the well-being of the larger community, the development process is substantially impeded.

Environmental Contrasts

Much like China, the triangular-shaped region of South Asia possesses a wide range of physical environments because of its substantial latitudinal and longitudinal reach (Figure 21–1). In their west-east extent, South Asian environments range from bone-dry deserts in the west to some of the wettest rain forests in the world along India's eastern border with Myanmar. We will now study the principal natural regions of South Asia.

The Northern Mountain Rim

No elevated surface of the world provides such a sharp boundary between world regions as the Northern Mountain Rim of South Asia. Its origins reach back to ancient geologic times, with the subduction of the northeast-migrating Indian plate under the Asian plate. The edge of the Asian plate consequently became upthrusted and folded to resemble an imposing mountain wall. From west to east, this loftiest elevated surface in the world encompasses the Karakorum in northern Pakistan and India and the even loftier Himalayas that separate lowland India from the Tibetan Plateau (Figure 21–2). The moisture of the southwest monsoon is captured by this mountain barrier, thus providing water for many of the rivers that flow from this highland realm into the lowlands to the south.

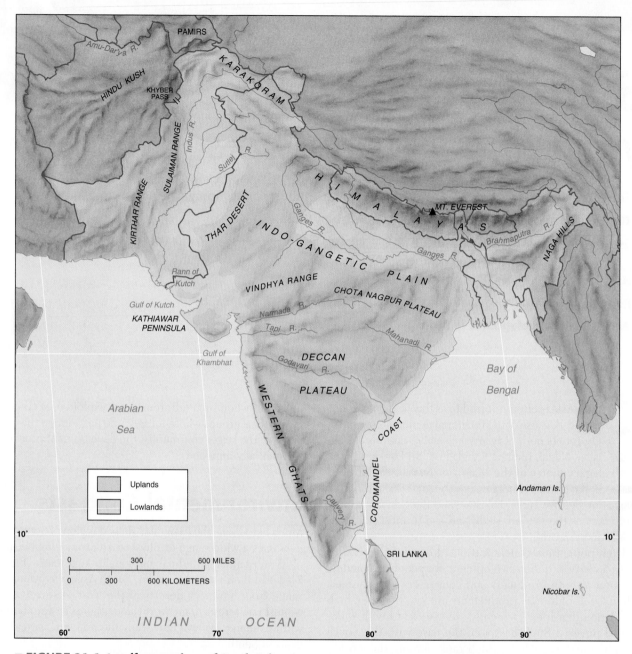

■ **FIGURE 21-1 Landform regions of South Asia.** Three major physiographic regions compose the subcontinent: the southern plateau, the northern and western mountain wall, and the intervening Indo-Gangetic Plain. There are also narrow coastal plains on the eastern and western coasts.

The Indo-Gangetic Plain

This physical region encompasses the alluvial plains of the Indus, Ganges, and Brahmaputra rivers. The Indus and Ganges rivers are separated by the uplifted Thar Desert. Measuring some 300 miles (480 kilometers) wide, the Ganges is the largest of the three river plains. Due to the path of the southwest monsoon, annual precipitation totals increase from west to east so that much of the Indus plain is arid, whereas humid Bangladesh is characterized by semideciduous and tropical rain-forest vegetation. The nutrient-rich Indo-Gangetic alluvial soils support approximately half of South Asia's population. The Gangetic Plain itself supports approximately 10 percent of the world's population, sustained in part by one of the world's largest groundwater aquifers. The water supply is, however, very seasonal because the summer or wet monsoon provides approximately 85 percent of annual rainfall totals. Unlike developed countries, where water resources are primarily used for urban-industrial purposes, 94 percent of India's water is consumed by agriculture.

The Peninsular Plateau

Much of the southern half of India consists of an elevated plateau surface underlain by a core of ancient igneous and metamorphic rocks. The most extensive portion of the plateau is called the Deccan, meaning "south," where elevations average between 2,000 to 3,000 feet (600 to 900

■ **FIGURE 21-2 Machhapuchare and Pokhara peaks, Nepal.** There is a stark environmental divide between the Indian lowlands and the Himalayas. Perceived as religiously sacred, no other mountain chain in the world possesses as many peaks exceeding 20,000 feet as the Himalayas.

meters) above sea level. The western margin is the most elevated portion, with a general downward tilt evident to the north and the east. Except in the northwest, where rivers flow westward into the Arabian Sea, the predominant stream flow is eastward. While the Deccan possesses the largest share of India's minerals, agriculture is hampered by the relative absence of water. Much of the natural vegetation is tropical dry deciduous forest. The edges of the Deccan Plateau abut the Western and Eastern Ghats, meaning

"steps." The Western Ghats are a steep mountain range with average elevations of 5,000 feet (1,500 meters), below which is a narrow, lushly vegetated coastal plain (Figure 21–3). The Eastern Ghats are actually a line of disconnected hills, and the eastern coastal plain is substantially wider than the western.

The Precolonial Heritage

South Asia possesses far greater cultural diversity than China. The region was subjected to numerous external influences from the west, which penetrated through Afghanistan's famed Kyber Pass. These successive waves of cultural infusions were accomplished primarily through invasions rather than through peaceful means, and they came to form the basis of the major linguistic and religious spheres of present-day South Asia.

Empires, Culture, and Society

South Asia's earliest civilization was centered on the Indus River, from which the word India derives. Dating to approximately 2350 B.C. the Indus Valley empire, commonly referred to as the **Harappan culture** was centered on the three cities of Harappa, Ganweriwala, and Mohenjo-Daro (Figure 21–4). Each city was well planned, dominated by a citadel, surrounded by massive walls, and laid out using a grid street pattern. Farmers working the rich alluvial soils cultivated wheat, barley, legumes, dates, and cotton and supplemented their diets with meat from domesticated sheep and goats. Substantial long-distance trade was carried out with lands further west through various Arabian Sea trading

■ **FIGURE 21-3 Western Ghats of India.** Southern India is dominated by the elevated Deccan Plateau. The Western Ghats are the mountainous edge, or escarpment, of the plateau.

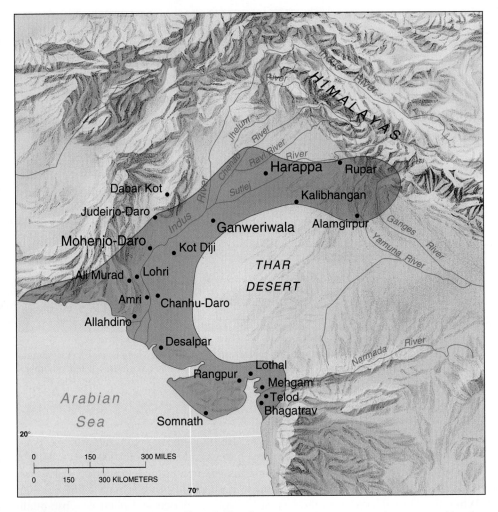

■ **FIGURE 21-4   Indus River Valley civilizations.** The early Harappan culture centered on the lower Indus River Valley.

ports. Harappan culture began to decline about 1900 B.C. Its decline may have been triggered by waves of Indo-Aryan invaders from the west or from human- or naturally induced environmental changes in this delicately balanced desert environment.

Beginning about 2000 B.C. a new culture group called the Aryans invaded from the west to produce a mixed Indo-Aryan civilization, which at first was primarily nomadic in nature but which adopted sedentary agriculture over time. The Indo-Aryan culture subsequently spread eastward into the forested regions of the Ganges River Valley, where large urban settlements were established. Indo-Aryans also introduced the religion of **Hinduism**, the world's oldest major religion, which now spatially dominates South Asia (Figure 21–5). Integral to Hinduism is the **caste system**, wherein individuals are believed to be preordained at birth to remain throughout their lives as members of one of four major socioeconomic groupings, called jati. In Hindu belief, just as each individual possesses a duty to the larger society, so too do entire jati and the hundreds of sub-jati. Only when individuals fulfill their own duty within their jati will an idealized Hindu world function harmoniously

(Figure 21–6). In a general sense, the caste system institutionalized both social status and economic function within the larger society, and only through cyclical rebirth, or reincarnation, is mobility to a higher caste believed possible. To achieve this upward spiritual mobility, one's soul requires the accumulation of good karma, or good deeds, over many generations. At the apex of this highly stratified socioeconomic system is the religious Brahman caste. One group, known as the untouchables or *Dalit* (meaning broken or depressed), engages in economic activities deemed dirty or polluting and stands altogether outside the caste system.

Indo-Aryan culture did not completely transform southern peninsular India. There Hinduism and the caste system were adopted, but the languages of the native **Dravidian** peoples were not displaced (Figure 21–7). In the many centuries following the establishment of Indo-Aryan civilization, a succession of empires ruled over parts of South Asia. After the brief occupation of the upper Indus and Ganges valleys by Alexander the Great in the third century B.C., the **Mauryan Empire** became the first political state to control most of South Asia. The greatest Mauryan ruler

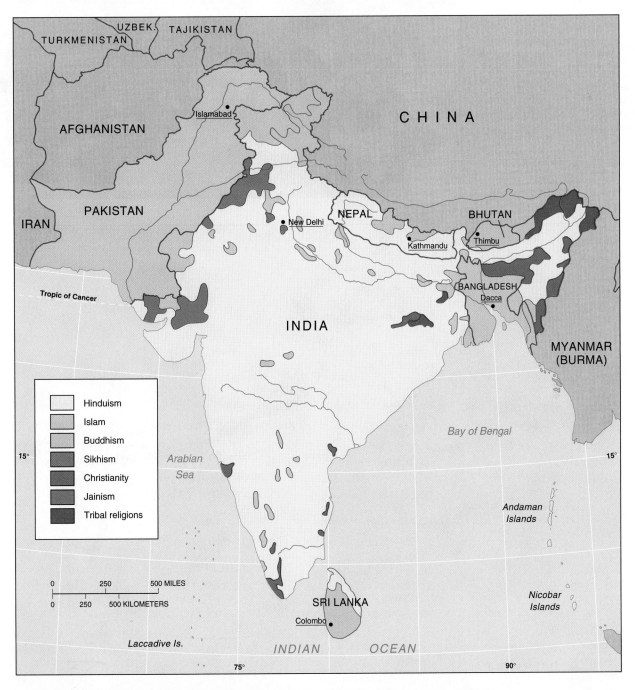

■ **FIGURE 21-5 Major religions of South Asia.** Hinduism dominates in most of India, while Islam is the principal religion of Pakistan and Bangladesh. Buddhism prevails in most of Sri Lanka and in portions of the northern Himalayan regions. Sikhism, Christianity, and Jainism form scattered enclaves.

was Asoka, who converted to Buddhism and assisted in its diffusion through India and beyond into Southeast Asia.

The Impact of Islam

As with the Aryans earlier, Islam entered South Asia through the mountain passes of the Hindu Kush. Converting virtually all the population of the Indus River Valley, Islam next spread into the Gangetic Plain, where some of the resident Hindus were converted but others were not. Islam was em-

braced more widely further east in the lower Ganges River Plain that today makes up much of Bangladesh. In southern India, conversions were limited primarily to the narrow Malabar coastal plain, where Islam was introduced through Arab maritime traders.

The most powerful of all the Islamic empires was the **Mughal Empire** of the sixteenth and seventeenth centuries (Figure 21–8). After having conquered the Gangetic Plain, the Mughal Empire shifted its seat of power to Lahore, in the upper Indus Basin, which subsequently became one of

Dasaswameda bathing ghat in Varanasi on the Ganges River in northern India. Thousands of Hindus come to Varanasi each year to bathe in the sacred waters of the Ganges. Originating in the Himalayas, the river's waters symbolically represent the purity of heaven and thus wash away the worldly sins of religious pilgrims. Varanasi is also a spiritual center for Buddhists because Buddha formulated his principles in this vicinity around 500 B.C.

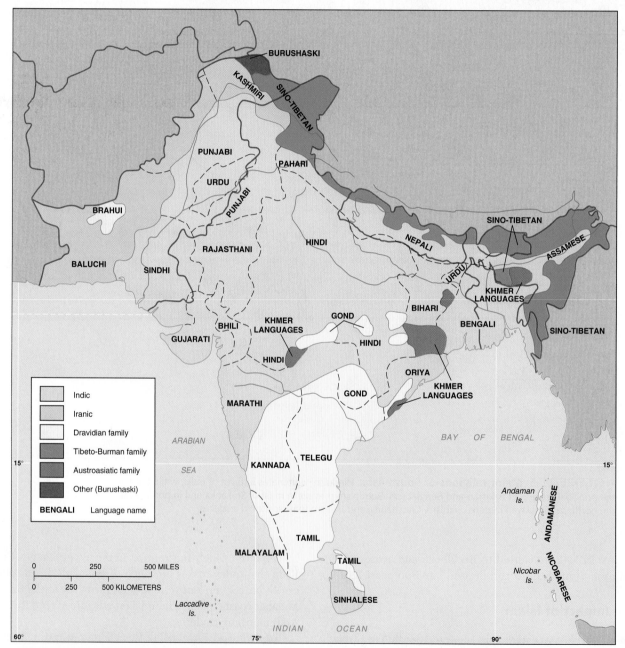

■ **FIGURE 21-7 Languages of South Asia.** The languages of the Indian subcontinent derive from two major groups: the Indo-European languages, which are dominant in the central and northern regions, and the Dravidian languages, dominant in the south. The diversity of languages and cultures on the subcontinent contributes to the difficulty in forging unified national states.

■ **FIGURE 21-8 Taj Mahal, Agra, India.** The Taj Mahal, meaning "crown palace" in Persian, was completed in 1647 as a royal mausoleum for the wife of the king. Although an Islamic structure in a Hindu majority state, the Taj Mahal is India's most popular visitor attraction.

the most celebrated Islamic cities in the world. Apart from its distinction as a seat of Islamic learning, Lahore's function was very much as a **forward capital** to better control the threat of Afghan tribes disrupting the caravan trade between South and central Asia. The Mughals then conquered the balance of India, with the exception of the far southern tip of the peninsula.

Islam was especially attractive to untouchables and Hindus of lower caste because it freed them from the rigid social order of Hinduism, in which there existed little opportunity for individual betterment. Among the rural middle class of the strategic Punjab region in the northwest, rejection of the caste system produced the blended Hindu and Islamic religion of **Sikhism**. While conversion to Islam brought marginal improvement in their lives, converts continued to be viewed by the higher castes as a convenient pool of exploitable labor; a less severe caste system remained in place, even among later Christian converts. Unlike the experience with Buddhism, Jainism, and Sikhism, the diffusion of Islam brought sharp conflict with established Hindu society. Islam was highly doctrinaire and closely associated with a nonnative conquering culture. These considerations suggest why Islam, despite the political and economic power of the Mughal Empire, failed to convert most South Asians. Fewer than 12 percent of Indians today profess the Islamic faith.

Colonial Transformation

In the competitive environment that characterized the early European trading forays into Asia during the sixteenth and seventeenth centuries, India provided an ideal intermediate

trading location between Europe and Southeast Asia and China. The Portuguese arrived in the early sixteenth century, followed by the British, French, and Dutch. By the beginning of the eighteenth century, European trading stations dotted the coast of South Asia. The British eventually gained political and economic dominance (Figure 21–9).

Early British Influence

The sustained economic exploitation of South Asia was left to the quasiprivate British East India Company. Gradually replacing Mughal rule and usurping the political power of regional states, the company indirectly came to control approximately two-thirds of South Asia. After annexing much of the east coast by 1760, the East India Company was able to gain control of the prized lower Ganges Plain, or "Bengal," by 1765. Slowly annexing coastal Indian states up the west coast during the ensuing years, eventually the company came to control the Bombay region in 1818. The large state of Sind at the lower Indus Valley was annexed in 1843, as was the Punjab in 1849. The Company replaced native administration by giving land deeds to local higher-caste "zamindar," who had functioned as tax collectors during the Mughal period. Through this new landlord–tax collector class, the British were able to exact high taxes from peasants to pay for the cost of colonial administration. Because taxes had to be paid in cash rather than in kind, the peasants were forced to cultivate a wide variety of cash crops, such as cotton, peanuts, indigo, jute, and opium, in order to pay their taxes. Unable to pay both high land rents and high taxes, many peasants lost their land. The dramatic reduction of subsistence cultivation in the British-administered regions

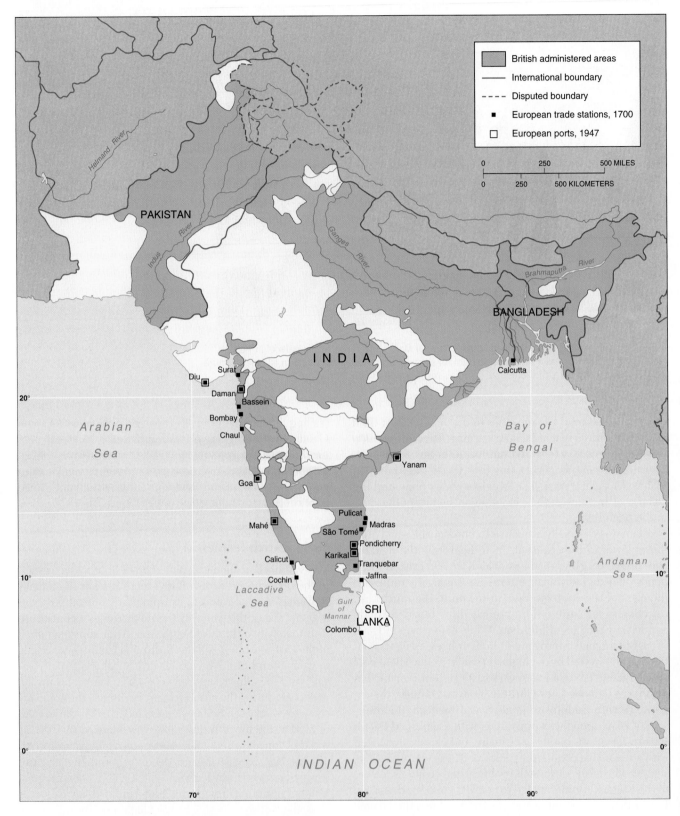

■ **FIGURE 21-9 The spatial evolution of the British Empire in India.** British occupation of South Asia focused initially on coastal trade stations, with later penetration into the interior regions.

resulted in recurring famines, particularly during the first half of the 1800s. Early British rule thus created a huge class of landless peasants that contributed to the region's future underdevelopment.

Beginning in the nineteenth century, the British also took actions that decimated South Asia's textile industry. No longer able to compete with inexpensive British machine-made cotton goods, and having Indian textile exports banned from the British domestic market, the once thriving domestic handloom industry disappeared almost overnight. By the mid-1800s, when South Asia became a colonial possession, this newly "deindustrialized" region had begun taking on the attributes of an underdeveloped country. In reorienting the economy to service the needs of British capitalism, a new spatial organization of the region's productive capacity emerged. Cash crops were grown in the interior hinterlands and exported through colonial trading ports. Cotton was the primary export commodity of Karachi and Bombay (Mumbai), while rice and jute were shipped through Madras (Chennai) and Calcutta (Kolkata), respectively. As emerging colonial economic cores, these trading centers gradually assumed greater economic power, leaving interior and indigenous settlements to stagnate.

South Asia's Economic Contours as a British Colonial Possession

The imposition of direct rule through a formal colonial administration was, in part, a response to the Sepoy Rebellion in 1857. Sepoys were the Indian troops used by the British. Their rebellion was an expression of the intense resentment toward the British that had emerged in all quarters of Indian society. British commanders and their loyal Indian troops swiftly crushed the Sepoys and, realizing that India might be lost, created a colonial possession in 1858. To directly administer colonial territory, Britain developed a huge civil-service system consisting of Englishmen in the top leadership roles and higher-caste Anglo-Indians at upper- and middle-level positions.

As a colonial possession, South Asia's economic contours were profoundly altered. The cultivation of cash crops was expanded with the addition of coffee and tea. The peasantry was taxed even more in order to pay the costs of the road and railroad construction needed to transport increasing quantities of cash crops to ports and to allow for the more rapid deployment of British troops for perceived security needs. Increased taxation resulted, of course, in the further impoverishment of the rural population. By the 1940s, for example, some 40 percent of the farming population in India was landless, approximately triple the percentage at the turn of the century. Famines became more frequent and widespread, with the second half of the nineteenth century witnessing a famine or severe food shortage about every 2.5 years. Despite the food shortages, wheat was still exported to England from the Punjab.

Industrial development was slow paced; by 1940, less than 1 percent of India's population was employed in the industrial sector, which was limited to forms of manufacturing that did not compete against British factories at home. Industry continued to be confined almost exclusively to the port cities, in part because railroad pricing policies discriminated against manufacturing located at a distance from the coasts. The relative absence of industry meant that the population remained predominantly rural. Much of the limited urban growth took place in the few coastal cities that had become true metropolitan centers. The administrative capital of colonial India was not relocated from Calcutta to New Delhi until 1911. Delhi, sometimes referred to as Old Delhi, had been the site of seventeen former historical imperial capitals. Having transformed the economy of India to favor the coastal periphery, the British administration's move from coastal Calcutta to New Delhi in the upper Gangetic Plain was viewed as a deliberate historical imitation of Mughal greatness.

Independence and Nation-State Building

It was not in the interests of the British to cultivate cultural and political homogeneity. Rather, they engaged in a systematic policy of divide-and-rule so that an indigenous unified front would not emerge to threaten British rule. The list of cultural and political entities in colonial India was almost endless. There existed some 113 "princely states" that never came under direct British rule. Most were very small, but a handful, such as Hyderabad and Ladakh, were as large as Greece or Great Britain. The caste system remained rigid, and **communalism** (an uncompromising allegiance to a particular ethnic or religious group) persisted. Despite such divisions, the desire for independence grew steadily and, indeed, was strongly called for during the early 1930s. In 1947 colonial rule ended, and a number of independent states emerged from the former British India (Figure 21–10).

Carving Up Political Space

At the largest scale, territory was divided into predominantly Muslim and Hindu regions. Owing to the historic clustering of Islam at the western and eastern margins of South Asia, when Islamic Pakistan was created in 1947, a West and East Pakistan were formed, separated by more than 900 miles (1,500 kilometers) of Indian (Hindu) territory.

The union of West and East Pakistan was tenuous from the beginning. The people of West Pakistan were largely oriented toward the West. The population of East Pakistan, on the other hand, was more culturally oriented toward Southeast Asia. East Pakistan achieved independence from West Pakistan in 1971, following a short (2-week) civil war, and was renamed Bangladesh.

The creation of a separate India and Pakistan entailed the massive movement of some 14 million persons; Muslims and Hindus, in about equal proportions, crossed the newly demarcated national boundaries. The greatest movement was from border areas with no clear communal majority. For various reasons, many Muslims remained behind in India. Today, India's 130 million Muslims constitute the largest

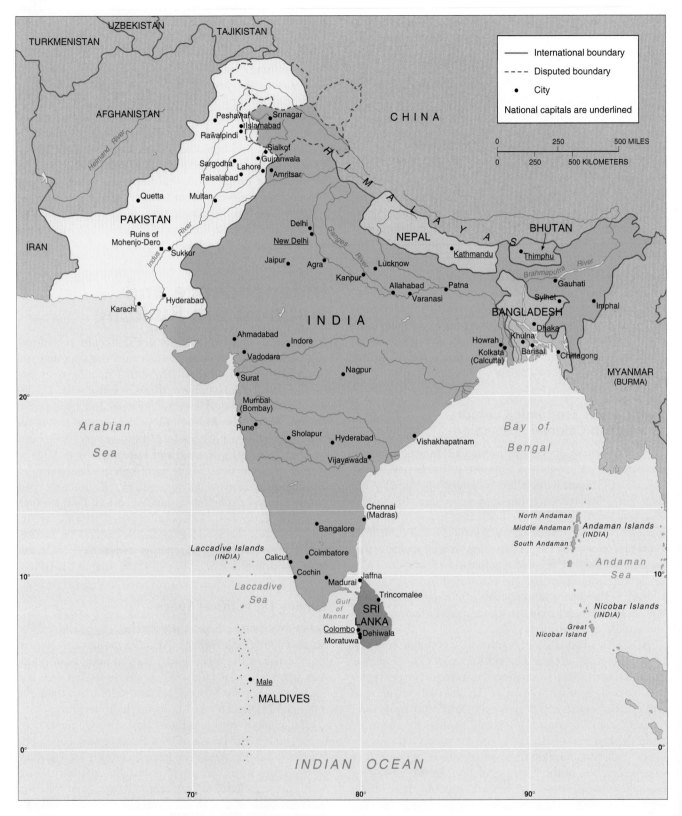

■ **FIGURE 21-10 Political map of South Asia.** South Asia consists of one large country (India), one medium-sized country (Pakistan), and the smaller nations of Bangladesh, Nepal, Sri Lanka, and Bhutan.

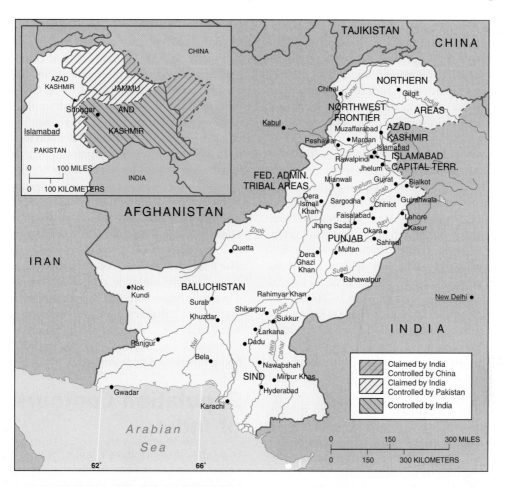

▪ **FIGURE 21-11 Jammu and Kashmir.** The rugged mountainous region of Jammu and Kashmir is subject to conflicting territorial claims from India, Pakistan, and China.

Muslim national minority in the world. Hindu-Muslim strife continues to be endemic in the northernmost Indian state of **Jammu and Kashmir**, where a Hindu-dominated government rules a Muslim majority population (Figure 21–11). Although the United Nations established a cease-fire line in 1949, Pakistani-backed Islamic insurgents have waged a recurrent war against the Indian army, which has taken some 20,000 lives. It is unclear whether Pakistan desires to annex Kashmir or whether its motive is to gain exclusive control of the headwaters of the Indus River, which is vital to Pakistan's irrigated agriculture. The most recent major conflict took place in 1999. Another region that has experienced considerable religious turmoil is the agriculturally rich Punjab, where the historically anti-Islamic Sikh homeland was partitioned by the new India-Pakistan boundary. Characterized by a violent separatist movement against the Indian government throughout the 1980s, the Sikh goal of creating an autonomous state named Khalistan has fizzled because of the lack of unity within the movement. Sikhs, Hindus, and Muslims have thus found themselves in new states that are potentially antagonistic to their respective religions. Although the members of each religious community have traditionally expressed tolerance toward the others, some 1 million persons have died as a result of the continuing religious conflicts.

Accommodating Diversity in India

The cultural diversity of India has demanded that the government accommodate the aspirations of both ethnic and religious groups, as well as socioeconomic classes, in this most populous democracy in the world. Because cultural and economic differences are often religious-based, India was created as a secular state containing constitutional guarantees that the government will not seek to promote nor interfere with any particular religion. While secularism has been instrumental in reducing communal antagonisms, it has not been able to eliminate discrimination toward the lower castes nor the increasing incidence of Hindu-Muslim and Hindu-Sikh confrontations. Religious identity has become the preeminent political issue, owing in part to fear, on the part of Brahmans and other upper castes, that reform would only bring a loss of their political and economic power.

To improve economic opportunities for Hindu and Sikh untouchables, the government has designated them as **"scheduled castes"** eligible for special consideration in university admissions and government-sector employment as clerks, railroad laborers, postal workers, and sometimes middle-level administrators. But with overall unemployment high and untouchables viewed as being competitors for prized government jobs, higher-caste hostility has indirectly resulted in violence in urban areas. Far more subtle and pervasive forms of discrimination are common in the office workplace. While government policies have softened the edges of the caste system, class-based status has not yet replaced caste-based status as a primary social determinant. Constituting about 17 percent of India's population (approximately 180 million persons), the untouchables remain desperately poor and represent an enormous wasting of human resources.

Communal unrest in India also continues to fester between Hindus and Muslims. Political parties seeking greater power are generally inclined to promise increased economic opportunities to Muslims, the poorest of the major religious communities. Political appeasement of Muslims, however, often leads to aggressive and sometimes violent reactions by other religious groups, primarily Hindus. Such reactions occur most often where Muslims have achieved some degree of political or economic advancement. In this sense, the destruction of a Muslim house of worship by Hindus becomes a political and economic, rather than a religious, act. The recent rise of the conservative Hindu-based **Bharatiya Janata Party (BJP)** does not bode well for near-term communal peace. The worst sectarian violence in India, for instance, has centered on the western state of Gujarat, which is a BJP political stronghold. Fortunately, the violence has been concentrated in a handful of cities. In the villages, where most of the people live, Hindus and Muslims generally co-exist peacefully.[1] The fracturing of society along religious and political lines only deprives the country of energies that might best be focused upon many constructive tasks of economic development.

The government has chosen to accommodate the many regional cultures and economies through the administrative framework of **federalism**, which provides the thirty states and seven union territories a measure of economic and political autonomy (Figure 21–12). Because language is central to cultural identity—and, indeed, most of the state and territorial boundaries coincide with linguistic boundaries—the government recognizes fourteen different official languages that account for some 90 percent of Indian speakers. The Indo-European **Hindi** language is the most widely spoken and is the principal official tongue. Hindi is spoken by approximately 43 percent of the population and is particularly widespread in the northern states. The colonial language of English continues to function as an informal official language. Most adult Indians are multilingual, speaking their regional language plus some Hindi or English.

Greater linguistic autonomy is only one small part of the efforts of the federal government to reduce the sources of separatism, or the desire of ethnic or religious groups to gain territorial independence from the larger country. With the obvious exception of Jammu and Kashmir, the strongest challenge to national unity has originated in Punjab, where Sikh separatists, seeking an independent "Khalistan" in the 1960s, forced New Delhi to redraw boundaries so that Punjab became a Sikh majority state. New Delhi also granted state-

hood to the handful of tribal regions of the northeast after separatist movements emerged during the 1960s and 1970s.

Finally, in 2000, regional dissent prompted the federal government to create three new states. Jharkhand was created from the southern half of Bihar, Chhatisgarh from the eastern third of Madhya Pradesh, and Uttaranchal from northwestern Uttar Pradesh. Perhaps the most celebrated and long desired of the three statehood movements was that of Jharkhand, which is populated primarily by forest-based tribal peoples who have a distinctive cultural identity and who desired greater economic self-determination. Southern Bihar is mineral-rich, but its mines and processing plants were controlled by interests in northern Bihar, part of the Hindi-speaking heartland. In fact, these mining related industries generated 70 percent of Bihar's annual tax revenue. The Jharkhand people, many of whom worked as poorly paid mine laborers, saw little economic benefit from being part of a larger Bihar state. The Jharkhand movement is just one example of tribal resistence to larger hegemonic forces at work in India and other developing nations where the economies and environments of traditional societies are being threatened or destroyed.

Population Contours of India

Inhabited by more than 1 billion people, India is the second most populous country in the world. With an annual population growth rate double that of China, India is expected to surpass China in population by 2050. Because Pakistan and Bangladesh have even higher population growth rates, the population of South Asia likely will surpass that of East Asia within the next few years. In India, as well as the rest of South Asia, such rapid population growth requires increased economic productivity if levels of development are to improve.

Population Distribution

As discussed in the Asian introductory chapter, India's population is not distributed evenly. With almost three-fourths of the people living in rural villages, the most densely settled regions tend to coincide with the most favorable agricultural environments, where soils are fertile and where water supplies are relatively abundant and dependable. The single largest population concentration is in the modern core area of the Gangetic Plain, with Uttar Pradesh being the most populated state. The next most densely populated regions are the coastal plains of the Western and Eastern Ghats, where abundant precipitation allows for population densities that exceed those of Gangetic Plain states in some locations.

Population Growth

Over the past 85 years, India's population has more than quadrupled (Figure 21–13). Population growth was relatively low until the 1920s, when it commenced a steady rise. Annual

---

[1]Ashutosh Varney, *Ethnic Conflict and Civic Life: Hindus and Muslims in India* (New Haven: Yale University Press, 2002).

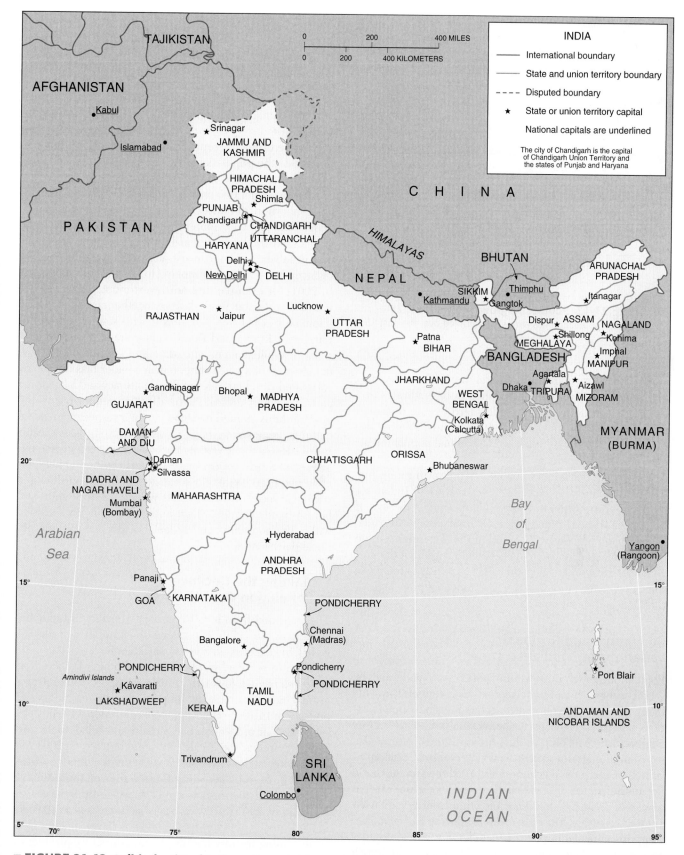

■ **FIGURE 21-12 Political units of India.** India, one of the world's largest countries, is divided into thirty states and seven union territories.

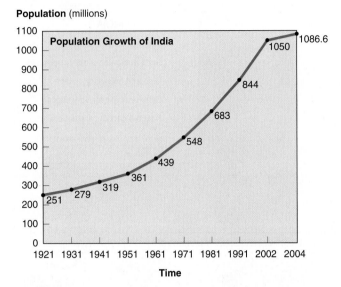

Population (millions)

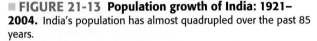

▪ **FIGURE 21-13 Population growth of India: 1921–2004.** India's population has almost quadrupled over the past 85 years.

growth rates rose above 2 percent in the 1960s and continued unabated until the 1980s, when both birthrates and death rates began to decline. India's fertility rate remained around six children per woman until the mid-1960s but declined to 3.1 by 2004. India's average annual population increase is now 1.7 percent.

The recent decline in the population growth rate is further illustrated by the changing age structure of Indian society. The proportion of the population under the age of fifteen fell from 41 percent in 1971 to 36 percent in 2004. The responding downward trend in the dependency ratio may be expected to lessen the economic burden of the next generation of working-age population as well as to reduce to some degree the burden on the government of providing educational and social services.

Communal and Regional Variations in Population Growth

Population growth rates within India vary considerably by ethnic, religious, and caste groups. Although culture certainly influences fertility rates, economic status better explains these differences. In general, Hindus have fewer children than Muslims. In the southern state of Karnataka, for example, Hindu women have 2.7 children, compared with an average of 3.9 children for Muslim women. In the far southern state of Tamil Nadu, "scheduled caste" Hindu women average 2.8 children while other Indian women average 2.4 children. The differences in religious and economic conditions are the basis for variations in regional population patterns. Those states possessing the most rapid growth rates during the 1991–2001 period, for example, were the poorer tribal northeast states (Figure 21–14). This region contrasts sharply with the southern states of Tamil Nadu and Kerala, where growth rates are significantly below the national average. Above-average growth rates also character-

ize the highly populated northern "Hindi belt" states of Bihar, Uttar Pradesh, Madhya Pradesh, and Rajasthan. As an indicator of relative levels of poverty in the early 1990s, urban fertility rates stood at 2.7 per woman, while in rural regions, 4.4 was the norm.

These statistics describing regional variation in population growth mask the regional **gender bias** of India's population structure (Figure 21–15). Gender bias exists when one or the other sex represents an abnormally large percentage of the population. In 2001, for example, India exhibited a gender bias in favor of males because there were 933 females per 1,000 males. Whether for reasons of infanticide, abortion, or nutritional or medical neglect, India possesses, like many developing world countries, a deficit in females. At the basic regional level, the degree of gender bias is substantial; northwest and northern India exhibit stronger gender bias as compared to the southern provinces. In 2001, for example, the northwestern states of Punjab, Haryana and Uttar Pradesh were together characterized by a female ratio of 878, whereas the far southern states of Karnataka, Kerala, and Tamil Nadu averaged a female ratio of 1003. While male in- and out-migration rates impact the female ratio of a given state, female gender bias is strongly correlated with high rates of female literacy and labor-force participation. A critical factor affecting gender bias is the general status granted to women in their respective regional cultures. In southern states, women tend to possess greater social and economic freedoms, such as owning land and engaging in a wide variety of economic activities, which empower them to be greater social actors. In the northwest, a patrilocal social structure exists whereby a bride moves to her husband's parents' village and is consequently secluded within the household and is denied access to land and political participation.[2]

Explaining the Decline in Population Growth Rates

The transition from high to moderate fertility in India, beginning in the 1970s, is explained by a number of factors. As in China, increased educational attainment of women has resulted in reduced fertility levels. Although female literacy rates remain lower than male rates, government programs have dramatically improved women's access to education. In 1961 only 13 percent of women were literate; by 2003, that proportion had more than tripled, to 48 percent. By comparison, the literacy rate for males in 2003 was 70 percent. Increased educational attainment tends to delay marriage, thereby reducing the number of reproductive years. In 1961 the average age of marriage for females was 16.3, but by 2001 the average age had increased to 20 years of age.

Declining fertility has also been a direct by-product of the central government's family planning programs. Although

[2]S. Bhutani. "Spatial Patterns of Change in Indian Sex Ratio: 1981–1991." *Asian Profile* 27, no. 2 (1999): 157–168; M. Murthi et al. "Mortality, Fertility, and Gender Bias in India. A District-Level Analysis." *Population and Development Review* 21, no. 4 (1995): 745–772.

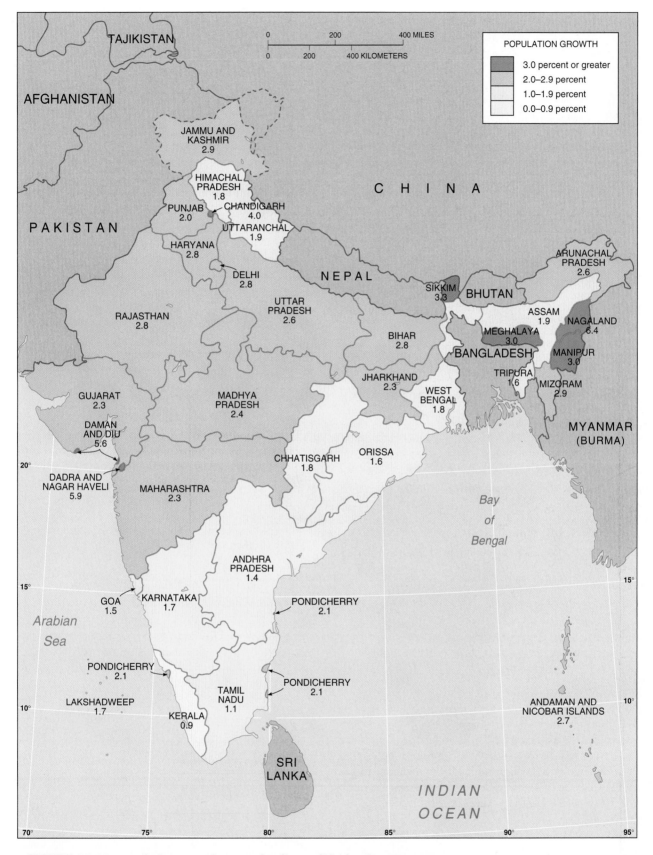

■ **FIGURE 21-14 Population growth rates of Indian political units, 1991–2001.** The highest rates of population growth are found in northeastern India and the lowest in the southern states.

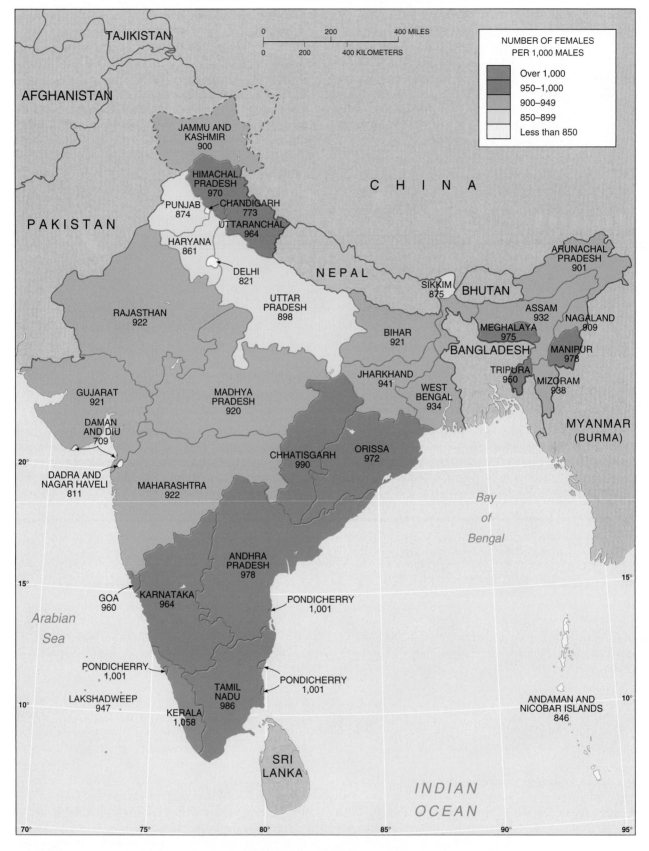

■ **FIGURE 21-15 Gender bias of Indian political units, 2001.** Owing to a variety of factors, much of India possesses a severe deficit in females. The figures shown in this map represent the number of females per 1,000 males.

GEOGRAPHY IN ACTION
Being Poor and Female in South Asia

One of the primary barriers to economic development in South Asia has historically been the undervaluation and, in some cases, total disregard for the importance of women's contributions to the development process. The "invisibility" of South Asian females characteristically begins at birth. Because a young girl will ultimately move to her husband's household upon marriage, she is viewed as an expendable and temporary member of her birth family. This traditional economic view of young females has led to a deficit of women in India. The 2001 gender ratio of 93 females to 100 males was the result of abortions of female fetuses, female infanticide, and females receiving less food and health care than their male siblings. It is not uncommon in Pakistani hospitals for males to account for more than 70 percent of the children admitted.

Young females have also been denied equal educational opportunities because they are more likely to be withdrawn from school by their parents to engage in household work. Much like their mothers, most young South Asian girls can expect to spend 70 percent of their waking hours at various domestic chores. In Bangladesh, for example, the national literacy rate stands at 62 percent; but female literacy is only 51 percent, while male literacy is 74 percent. With the vast majority of females marrying by fifteen years of age, thus renewing the cycle of **gender subordination**, it is not surprising that the South Asian nations score very poorly on the Human Development and Gender Development Index rankings.

Within the subsistence agricultural economy, female access to land, credit, and other resources is restricted because their labor is judged to possess little monetary value. While males engage in money-earning activities deemed as productive, women's subsistence agricultural work, together with cooking, washing, child rearing, and gathering fuelwood and water, is traditionally viewed as nonproductive because little income is generated. South Asian women, like many others in the developing nations, shoulder a double workload and yet are widely viewed as contributing little to the larger development process. In fact, many well-intentioned efforts to bring development to rural South Asia have led to the further economic marginalization of women. The expansion of the amount of land dedicated to cash crops, for instance, has resulted in less land being available to produce the subsistence crops women use to feed their families. Mechanization associated with commercial agriculture has caused many peasant women to lose their farm jobs, making them even more dependent on wage labor. As the pool of women in wage labor has increased, wages have declined, increasing poverty among rural women. If notions of work, value, and production are not redefined, rural development efforts will continue to exact increasing hardships on women.

Improvement of women's lives is possible through a variety of institutional mechanisms that allow for greater mobilization of the poorer segments of society. In the southern Indian state of Kerala, for example, a basic-needs approach that implicitly benefits females has been heralded as an outstanding development model for the developing world. Female literacy rates, for example, stand above 85 percent—almost equal to those of males. Despite widespread local poverty, health care access in Kerala is the highest in India. Indicative of the heightened status of females, Kerala possesses the highest female-to-male ratios in India. Females in Kerala, as well as throughout India and Bangladesh, are also being empowered through the formation of women's cooperatives in both rural and urban areas. Many of these programs are designed to achieve greater female access to financial credit and vocational training, while others seek to improve the financial bargaining power of women who are engaged primarily in the informal economic sector. The overarching goals of all of these programs are to instill self-reliance and self-confidence, to increase access to resources, and to enhance women's capacity to generate and allocate income.

---

limited by serious logistical constraints, low levels of financing, and a past history of unpopular mandatory sterilization programs, various initiatives aimed at lowering birthrates have contributed to the slowing birthrate. In 1970, for example, only 13 percent of women of childbearing age used some form of contraception, but the use rate had increased to 48 percent by 2004. This increase has come from offering voluntary birth-control techniques at a modest cost. Substantial nonparticipation still exists, however, among the poorer segments of the population, including Muslims and the scheduled castes.

Another factor that often influences population growth rates is unequal gender relationships (see the *Geography in Action* boxed feature, Being Poor and Female in South Asia). Women who have no viable economic alternatives to marriage, and whose husbands control key economic resources such as land, dowries, income, and savings, for instance, are likely to desire to have sons in order for some of their offspring to have access to wealth. Fertility rates among these women are likely to be relatively high because they and their husbands will continue to produce children until they have at least two surviving sons.[3]

Based on its recent decline in fertility rates, India appears to be approaching stage three of the demographic transformation model. This does not mean that India's past rapid population growth will soon end, however, because huge numbers of people under age fifteen remain who have not yet reached their reproductive years. Even if every couple were to produce only two offspring, for example, India's population would still not level off for another 60 to 70 years. Assuming that replacement levels are reached by the year 2020, India's population will exceed 1.4 billion by 2025. Without improved rates of economic growth and, more importantly, basic socioeconomic development, India's government

---

[3]Sonalde Desai, "Engendering Population Policy," in *Gender, Population and Development*, edited by Maithreyi Krishnaraj et al. (Delhi: Oxford University Press, 1998): 11–69.

may experience great difficulty in providing sufficient food and shelter for its people in the years ahead.

Agricultural Development in India

With some 64 percent of the population engaged in agricultural pursuits and 72 percent residing in villages and small towns, India is an overwhelmingly rural country. Contrasts with China's agricultural landscape are noteworthy; while the percentage of the population engaged in agricultural pursuits is roughly the same in both countries, some 46 percent of India's land area is classified as productive cropland, compared to only 10 percent of China's. With approximately 1,300 and 3,200 people per square mile of cultivated land in India and China, respectively, India's agriculture is far less intensive overall than China's (Figure 21–16).

Agricultural Regions

The spatial distribution of agricultural systems and crop types in India is very much influenced by the availability of water (Figure 21–17). In drier regions, precipitation is supplemented by the use of water from wells, canals, and small impounded ponds called "tanks." Along the Western and Eastern Ghat coastal plains, the eastern Deccan, the middle and lower Gangetic Plain, and the Brahmaputra lowlands, rice is the dominant crop. More land is dedicated to rice production than to any other crop. In a broad south-to-north swath of territory from the western Deccan to the edge of the Thar Desert, drought-tolerant millet is the primary grain crop, with pockets of cotton and peanuts assuming local im-

portance. In the drier north-central and northwest regions, wheat and chickpeas are the primary staple food crops, with the politically strategic northwestern state of Punjab functioning as India's "breadbasket." The primary plantation region is in the humid northeast, where tea was first introduced on a grand scale during the colonial period (Figure 21–18). Maize-based shifting cultivation agriculture is also practiced by the tribal hill peoples.

As in China, the vast majority of India's farmers supplement their primary crops with other crops or livestock, whether for subsistence or commercial purposes. Along the rice-dominated coasts, for example, coconuts are extensively grown. Various types of beans, fruits, and vegetables are cultivated in dooryard gardens throughout much of India. The spread of irrigation has also promoted double-cropping in the more arid regions during the dry monsoon season.

In India, unlike the situation in land-scarce China, villagers rely heavily on livestock. The most important is cattle, which are rarely consumed for their meat because of the traditional Hindu taboo against consuming beef. Dairy products in the form of milk, yogurt, and "ghee" (clarified butter) are important sources of animal protein. Cattle dung is gathered for cooking fuel in many regions where fuelwood is scarce (Figure 21–19). Cattle are the primary sources of power for plowing and are widely used for short-distance transport. Cattle hides sustain India's thriving leather and tanning industries.

Agricultural Productivity and Change

Many scholars believe that the modernization of Indian agriculture is the single most important step needed to achieve sustained economic development. Compared to China and other more industrialized countries, the productivity of Indian agriculture is relatively low, particularly in the case of rice (Figure 21–20). The dramatic yield differences between China and India are attributable in part to China's having a greater proportion of its land under irrigation, applying greater amounts of fertilizer, and practicing double-cropping more widely. Despite its relatively low yields per unit area, however, high-yielding hybrid-grain varieties associated with the Green Revolution have dramatically boosted Indian agricultural productivity since the 1970s. Between 1973 and 2004 for example, the production of cereal grains increased from 119 million to 226 million metric tons. As a result, India's food production has kept pace with population growth and, in fact, India is now an exporter of many agricultural commodities. India's agricultural population, however, remains relatively poor; in 2002 some 48 percent of the population under the age of five were classified as malnourished.

The sources of continued rural poverty are many. Although the government has dedicated much energy and money to improving village education, electrification, health

■ **FIGURE 21-16 Extensive resource use on the Deccan Plateau.** Drier than the coastal mountain ranges that surround it, the Deccan Plateau is used less intensively as a result. Here goats and cattle graze on fallow land and post-harvest residues on the fields that surround village centers.

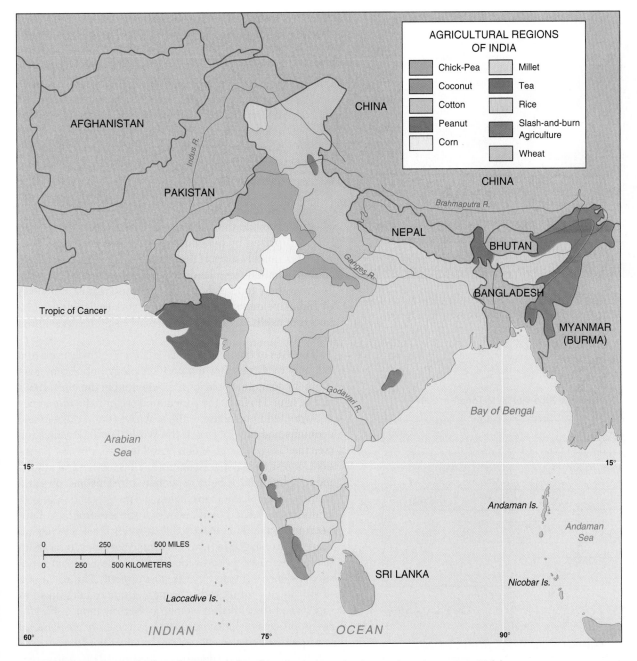

■ **FIGURE 21-17 Agricultural regions of India.** Rice farming dominates along the western Malabar coast and in humid eastern India. The more drought-tolerant grains of wheat and millet prevail in the arid central and western zones.

■ **FIGURE 21-18 Harvesting tea leaves, state of West Bengal.** Tea is the primary plantation crop of the humid Himalayan foothill regions of northeastern India. This scene is from near Darjeeling.

■ **FIGURE 21-19 Dried cattle dung.** In much of India, cattle dung is collected and dried to be used as cooking fuel.

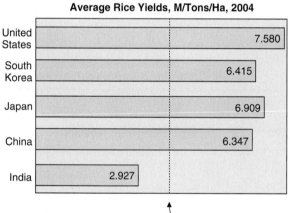

Average Rice Yields, M/Tons/Ha, 2004

United States — 7.580
South Korea — 6.415
Japan — 6.909
China — 6.347
India — 2.927

World Average 3.970

Average Wheat Yields, M/Tons/Ha, 2004

Mexico — 4.464
China — 4.202
India — 2.639
United States — 2.898

World Average 2.868

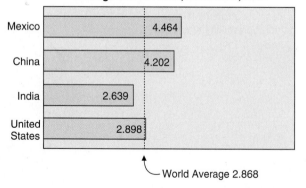

■ **FIGURE 21-20 Grain yields for selected countries.** Despite the expenditure of many hours of work per land unit, rice yields in India are significantly below the world average. South Korea and Japan, on the other hand, have rice yields almost twice that of the world average.
Source: From Food and Agriculture Organization, *http://apps.fao.org/page/form?/collection=Production.Crops.Primary.*

access, and other social services, meaningful land reform has hardly begun. Because of inheritance laws, most farm plots are small, thereby discouraging mechanization and the cooperative use of irrigation. The average size of a household plot, for instance, is only 6.5 acres (2.6 hectares), but some 34 percent of households cultivate plots measuring only .5 acre (.2 hectare). Cash cropping and the Green Revolution have only aggravated the land ownership disparities between the rural rich and poor. The hybrid seeds, chemical fertilizers and insecticides, and mechanized irrigation that might improve yields are either too expensive or fail to provide the harvest security and self-reliance so prized by peasant farmers. Thus unable to adequately support their families, many peasants have leased their lands to Green Revolution farmers, only to lose their self-employed status. Fully half of India's peasants are landless and work as agricultural laborers or as sharecroppers on the margins of the agricultural economy.

The lives of the rural poor have been negatively impacted not only by Green Revolution technology but also by government economic policies implemented in the early 1980s and designed to force the agricultural sector to be more market-oriented in response to the globalization of the national economy and the policies of the World Trade Organization (see the *Geography in Action* boxed feature, Peasant Farmers Contest the Global Economy). These structural adjustment policies have been especially burdensome on rural women, many of whom live below the poverty line. The government-subsidized food and social service programs have been reduced. This means that women must now spend more time engaged in income-earning activities to make up for the increased price of foodstuffs and lost critical social services once provided by the government. The reduction of subsidized rural credit through commercial banks has threatened the ability of women to borrow money. Finally, as agroprocessing industries have become more mechanized in response to greater international competition, female agricultural laborers, who depend on this critical off-season employment opportunity, are losing their jobs. While some degree of market liberalization is required in the agricultural sector, the overwhelming burden of the structural adjustment policies has fallen disproportionately on the shoulders of women.[4]

To date, technology has not solved the problems of rural poverty but, in fact, has created even greater economic inequalities. One might argue that the process of machinery displacing agricultural labor is simply part of the modernization process because Western countries experienced the same phenomenon during the Industrial Revolution. A critical difference, however, is that the rural population in the West responded by migrating to urban areas to seek

[4]Nata Duvvury, "Women and Agriculture in the New Economic Regime," in *Gender, Population and Development,* edited by Maithreyi Krishmaraj et al. (Delhi: Oxford University Press, 1998): 223–244.

GEOGRAPHY IN ACTION
Peasant Farmers Contest the Global Economy

In the early 1990s, the Indian government enacted a variety of policies intended to increase the global competitiveness of the agricultural sector. Peasant responses to these development strategies were confrontational and focused on the presence of U.S. fast-food chains such as Kentucky Fried Chicken and agribusiness firms. The farmers' outrage was directed not only toward the growing foreign presence in the food industry but, more importantly, toward the changing policies of the Indian government that had marginalized many peasant producers.

During the first two decades of India's independence, the government invested far more money in industry than in agriculture, believing that industrial growth would eventually benefit the rural masses by generating increased demand for agricultural commodities. Because India's growing industrial base was located in the larger cities, a strong antiurban and antigovernment bias developed among the rural population. As a result of the investment disparity between industry and agriculture, food production was unable to keep pace with population growth during the 1950s and 1960s, resulting in massive grain imports. In response, a multitude of independent political organizations emerged calling for increased governmental allocations to the rural sector. It was this political climate that led to the election of Indira Gandhi, who campaigned on a platform of strengthening the "real" India, which comprised peasant farmers and agricultural laborers. Green Revolution technologies, supported by a host of government subsidies, were subsequently introduced in support of the goal of achieving national economic self-sufficiency.

The 1990s package of reforms that opened India's economy to the forces of globalization was seen by India's farmers as a betrayal of Gandhi's pro-agrarian policies. Lower-class farmers argued that national food security was being sacrificed for the sake of profits from agricultural exports. Those middle-class farmers who had benefited from the Green Revolution protested cuts in government subsidies for fertilizers, insecticides, hybrid seeds, and other agricultural inputs. Critics of the new policies further argued that the increased use of grain for livestock feed amounted to supplying foreign fast-food restaurants in the urban areas while tens of millions of rural dwellers went hungry. As one farmer-friendly politician put it; "In a country where ordinary people cannot get one square meal a day, we are inviting multinationals to prepare fast foods which only the affluent can afford" (Gupta, p. 333). In a sense, small pockets of prosperous urban Indians who are connected to the global economy and culture have outbid the poor for grain. Many Indians also view the growth of the urban-based agroindustries as a symbol of the increasing income gaps between the wealthy and the poor, and thus a return to the deprivation experienced by so many during the 1950s and 1960s.

Source: Akhil Gupta, *Postcolonial Developments: Agriculture in the Making of Modern India* (London: Duke University Press, 1998).

---

wage-based industrial employment. As we will learn in the following pages, such urban-based opportunities are not available to the dispossessed Indian farmer.

India's Industrial Economy

Despite the fact that India's most important development challenge lies in the countryside, the country's political leaders have long chosen to dedicate scarce financial resources to industrial growth. This development strategy is understandable given India's huge population and unequal distribution of land. Although inefficient, India's aging rail transport system continues to be a valuable asset in its drive to become an industrialized country (Figure 21–21). Progress, however, has been slow. In 2003, for example, industry accounted for only 26 percent of GDP, compared to 53 percent in China.

The Industrial Resource Base

In addition to its railroad system, another industrial asset possessed by India is its relatively well-endowed resource base (Figure 21–22). With few exceptions, India's energy resources are found in the southern two-thirds of the country, and most of these in the Deccan Plateau. Although consisting primarily of lower-quality bituminous coal, India's

fossil-fuel endowment is adequate to power the country's industrial base. The majority of coal production takes place in the Chota Nagpur Plateau. India is also a major producer of iron ore mined on the Chota Nagpur Plateau and on the

■ **FIGURE 21-21 Commuters atop an overcrowded train.** From the British colonial era to the present, trains have played a major role in India's economic development. This scene is from New Delhi.

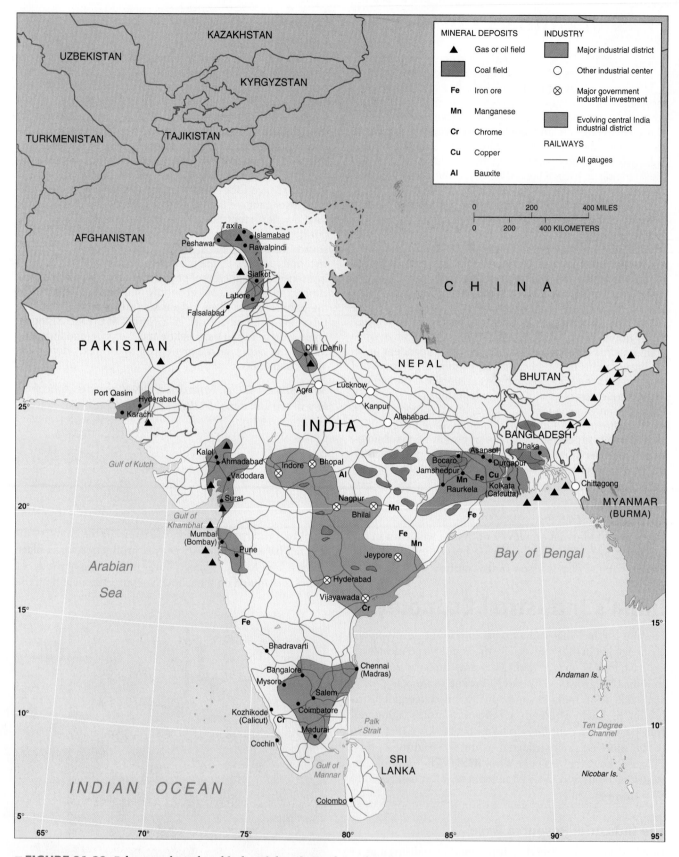

■ **FIGURE 21-22 Primary mineral and industrial regions of South Asia.** Although poor, the subcontinent has a well-developed surface transportation network, inherited in part from British colonial investments. The major industrial regions are centered on the principal cities.

edges of the Deccan. India's iron ore deposits account for 6 percent of world production and 5 percent of the world reserves. Unfortunately, India's oil and natural gas endowment is not substantial, despite the presence of older fields in Assam, Gujarat, and Punjab, plus newly developed offshore fields around Mumbai (formerly Bombay). To partially make up for the shortages of commercial energy, as well as to promote irrigation and flood control, the government has constructed numerous hydroelectric facilities, which are sustained by the summer monsoons.

A Centrally Planned Industrial Economy

Because Great Britain left independent India only a rudimentary industrial infrastructure, the first Indian governments attempted to develop a number of key industries in order to attain greater self-sufficiency. Adopting the Soviet model of industrial development anchored by 5-year plans, but within the context of democratic institutions, India channeled huge investments into many state-operated firms engaged in the processing of cotton, jute, sugar, and other agricultural raw materials. Emphasis was also placed on the development of heavy industries, including iron and steel, chemicals, shipbuilding, and automobile manufacturing. While a handful of large private-sector corporations and small-scale manufacturing and commercial enterprises thrived, the state, rather than the market, determined what and how much was produced. Because the government regulated the structure of the industrial economy through the issuing of business licenses and the implementation of import tariffs to protect state-owned firms, the industrial sector was effectively isolated from the more competitive forces of the global economy.

By the late 1990s, there existed approximately 300 central government-owned corporations and thousands of corporations owned by state and municipal governments. Using outdated technology and possessing a huge pool of surplus labor, the Indian economy was stagnating at a time when the economic growth of its Pacific Rim neighbors was skyrocketing. Most exports remained tied to lower-value textiles and leather, as well as to gems.

In response to its declining global economic standing, the government embarked upon a series of sweeping reforms in 1991 designed to increase the involvement of both domestic and foreign firms. Much of this new investment has been aimed at the domestic rather than the export market and often takes the form of joint ventures between domestic and foreign corporations. These reforms have shown positive signs, but the government's industrial policies continue to be mired in the past. Unlike China and many other Pacific Rim economies, India has not been able to attract a substantial amount of foreign investment. This is attributable both to India's continued reliance on import substitution industries that are protected by high tariffs and also to the small-scale nature of India's low-technology and labor-intensive industries. The government has traditionally reserved the production of low-technology products such as clothing, shoes, and toys for small-scale and labor-intensive companies because it was thought that large-scale and higher-technology firms would reduce employment opportunities. In a global economic environment, where multinational firms such as Nike or Mattel mass-produce identical products, the small-scale Indian firms simply cannot compete. It is hoped that further deregulation of industry might attract larger-scale investment. For that to happen, however, an improved infrastructure is needed to support industrial growth. This is why the private sector is investing heavily in transport, telecommunications, and power generation.

Industrial Regions

Although the spatial distribution of India's industrial base continues in part to reflect colonial influences, a more diversified industrial sector is emerging. Foremost among the industrial regions is the **Damodar Valley**, west of Kolkata (formerly Calcutta), where the presence of coal and iron ore have promoted the growth of the metallurgical, machine, fertilizer, and chemical industries. The Damodar Valley is similar in its products and national significance to the Ruhr Valley of Germany, and in fact it is one of the single largest industrial regions in the world.

India's second most important industrial region is centered on the city of Mumbai, which supports, along with Ahmadabad to the north, one of the largest global concentrations of cotton textile manufacturing. Mumbai also possesses a diverse array of other industries that engage in the production of automobiles, aircraft, pharmaceuticals, plastics, and chemicals. Pune, a city of some 4 million inhabitants to the southeast of Mumbai has attracted a large number of multinational and domestic manufacturing plants as well. Chennai (formerly Madras) and its hinterland have become another major industrial region specializing in the production of textiles and an array of light manufacturing. To the west of Chennai is the city of Bangalore, which has been dubbed the "Silicon Valley of India" because of the growing presence of domestic and foreign electronics manufacturers in this more temperate and higher-elevation garden city. Bangalore has attracted the likes of Texas Instruments, IBM, and Compaq, as well as domestic computer software firms. In 2004–2005, India's software and service exports were valued at $12 billion, up from $2.5 billion in 1999. This spatial concentration of a highly educated workforce was an important reason for U.S.-based Dell Computers to locate its global Tech Support Center in Bangalore. The contrast between Bangalore and much of the Indian economy provides a perfect illustration of the concept of "dual economies," so common in the developing world (see the *Geography in Action* boxed feature, "IT for All" and the Digital Divide). As is the case with China, India's national capital of New Delhi is not a national industrial center.

The deregulation of industry in India, if sustained, will likely usher in radical changes to the national economy. A

"IT for All" and the Digital Divide

India's recent and seemingly meteoric rise as a global center of IT or information technology research and service in Bangalore, Mumbai, and Hyderabad has prompted the government and media to promote an "informational economy" as a solution to the country's endemic development problems through improving economic growth, productivity, and competitiveness. Indeed, the government now supports a Ministry of Information and Technology. While it is now self-evident that managing knowledge and information is critical to prosperity in the global economy, it is also based on the unfounded belief or faith that technology leads directly to human progress. The question asked in the Indian context then is whether the Indian government's promotion of an "IT for All" (primarily via the Internet) program as a development tool to empower even the poor is an illusion. The success of the program is dependent upon the adoption of Internet technologies in rural areas where most poor people live.

The first problem is educational attainment. While the Internet provides greater economic opportunities for those with college and high school educations, the technology possesses little use to those with only primary schooling or less. Though the quality of a college education in India is quite good, elementary school education, particularly in rural areas, is poor, in part because of the low quality of teachers. In addition, real per capita expenditures on education has gradually declined since the 1970s. With 50 percent of students dropping out before completing the 8th grade, the minimum educational attainment threshold required for the perceived positive benefits of technology to be effective is unable to materialize.

A second barrier to the implementation of an "IT for All" program is inadequate teledensity (the number of telephones per person), particularly in rural areas. In 2001, approximately 376,000 of India's 660,000 villages possessed telephones, but more than half use outdated technology. Being very capital intensive, it is estimated that only 8 percent of the population could afford a minimum annual expenditure on telephone and Internet connections. While the small farmer might be able to afford Internet access, the farm labor class and all those below the poverty line would remain "out of touch." A third problem for a more deeper spread of the "IT for All" program is a common language. While India possesses 100 million English speakers, they account for only 10 percent of the country's population. Although Internet users have increased dramatically over the past decade, there is little growth in the use of Internet-based regional languages, with the exception of e-zines or a handful of localized entertainment sites. Knowledge of English, then, seems to be the present criterion for "membership in the Internet club" in India. The digital divide between urban and rural areas, between rich and poor, and between English and non-English speakers will continue for perhaps a generation or more. The most basic structural barrier to the spread of the Internet in India is minimal educational attainment, that most basic of all development goals.

Source: Joyojeet Pal, "The Developmental Promise of Information and Communications Technology in India," *Contemporary South Asia* 12, no. 1 (2002): 103–119.

---

more market-driven economy that is subject to foreign competition will almost certainly increase the rate of industrial growth and promote long-term technological modernization. Despite low overall literacy rates, India possesses a core of well-educated university graduates. Offering rewarding career opportunities to this group is essential if the **brain drain** of highly educated Indians migrating overseas in search of improved employment opportunities is to be stemmed (see the *Geography in Action* boxed feature, Globalization and India's Diaspora). Although not as formidable a task as in China, the process of industrial transition will also likely require that millions of workers at state-owned enterprises obtain employment within the private sector. The macrogeography of industrial location in India may thus be altered substantially as a result of government deregulation. Many scholars predict that the western half of the country will benefit the most from new private-sector investment, while the older heavy industrial region centered in Kolkata and the lower Gangetic Plain may gradually decline and assume a "rust-belt" economic position. The more dynamic western region is expected to geographically embrace a triangular area whose imaginary lines connect the cities of New Delhi, Chennai, and Mumbai (see Figure 21–12).

Urban India

India has always been characterized by stark contrasts between its urban and rural worlds. In the colonial period, this contrast became even sharper as the European-controlled centers of trade and administration became islands of modernization. In the postcolonial period, however, as the prosperity gap between cities and rural regions has widened, India has experienced greater levels of rural-to-urban migration. In recent years, the stream of urban migrants has become so great that it has exceeded the ability of the urban economy to provide adequate employment and housing.

Urban Growth

Levels of urbanization in India, like those of other Third World countries, remained relatively low into the middle of the twentieth century. As recently as 1941, for example, only 13.9 percent of the population lived in cities. Since then, the proportion of the population residing in urban places has doubled (Table 21–1). Although these figures are substantially lower than those of most devel-

GEOGRAPHY IN ACTION
Globalization and India's Diaspora

India has long been an important source of emigrants worldwide, as the 20 million Indians living overseas attest. During the colonial period, for example, Indian workers provided the backbone of plantation and railroad labor in many British colonial possessions. Today, two types of Indian emigrants predominate. The first comprises unskilled and semi-skilled laborers working primarily in the Persian Gulf states. The second comprises Indian professionals who migrate to primarily English-speaking countries, especially the United States and Great Britain. Whether historic or modern-day, Indians living abroad make up the **Indian diaspora**. The term *diaspora* is traditionally used to describe the dispersal or permanent migration of people from their home regions or countries. Today, however, many diaspora Indians are able to maintain close contact with their relatives, friends, and business contacts in India through the time- and space-collapsing technologies of globalization. These multidirectional technologies include the Internet, e-mail, low-cost telephone connections, more frequent and less expensive air passenger service, and entertainment videos. In an era of globalized interconnections and linkages, the elite segment of non-Resident Indians (NRI) plays a significant role in the economic, social, and political development of India.

The economic impact of the Indian diaspora has been substantial because of the flow of NRI wealth back home. For example, the Indian government promotes NRI investments in stock shares of Indian companies. Many successful software firms in Bangalore are financially linked to NRI software engineers in America's Silicon Valley. Developers of exclusive residential complexes in India explicitly advertise to potential NRI investors. Economic change at a smaller scale is also common. Villages in Punjab, Gujarat, and Kerala states, where emigration rates are high, have sections of improved housing as a result of NRI remittances. The great demand for a particular variety of Gujarati mango by NRIs in Britain has driven up the price for this fruit in India; the flow of migrants and the ability to cheaply air-freight cases of fruit creates a global uniformity of prices that in India works against the lower-income population but favors the fruit trader.

The social exchange between the diaspora population and India is profound as well. NRIs are able to remain immersed in Indian popular culture via videos and satellite television programing. In fact, Indian popular media caters to the NRI population; Bollywood, the term used to describe the Indian film industry, which is the largest in the world, derives a third of its annual revenue from Indian diaspora customers. Bollywood's movies, however, increasingly feature the Westernized and affluent lifestyles of NRIs to moviegoers in India through name-brand designer fashions and youthful fun rather than the standard or traditional singing and dancing common to most Indian films. Many NRI-influenced films are also responsible for promoting "Hinglish" (a combination of Hindi and English words in the same language) among educated young Indians.

Politics in India is also ideologically influenced by the Indian diaspora. The diaspora elite tend to be politically conservative because many NRIs were members of the conservative upper classes in India. Conservative politico-religious organizations in India, such as the Vishwa Hindu Parishad (World Hindu Council) and the Rashtriya Swayamsevak Sangh (RSS), as well as the Bharatiya Janata Party (BJP), operate web sites that solicit funds from NRIs, recruit members, and lobby for right-wing issues. In 1998 the BJP encouraged NRIs to e-mail their friends and relatives in India to vote for the BJP party. Although much smaller than their right-wing counterparts, some North American-based left-wing diaspora groups attempt to influence awareness of social movements back in India as well.

Long-distance migration, diasporas, and space-collapsing technologies are all characteristics of globalization. No longer are we able to conceptualize impenetrable national boundaries, closed national economies, nor cultures that are exclusively tied to a particular place. For the Indian as well as other ethnic-based diasporas, technology allows people in a sense to possess multiple "identities" and "homes."

Sources: Pamela Shurmer-Smith, *India Globalization and Change* (New York: Oxford University Press, 2000); Claire Dwyer, "Migrations and Diasporas" in *Introducing Human Geographies,* edited by Paul Cloke, Philip Crang, and Mark Goodwin (New York: Oxford University Press, 1999), pp. 287–295.

---

oping countries, they nonetheless represent massive increases in real terms, given the country's enormous population base. It is significant to note also that the Indian government requires that a settlement possess 5,000 or more residents before being classified as urban. This requirement eliminates thousands of communities that, in many countries, would be enumerated as small urban places. These particular urbanization statistics, therefore, do not provide a clear picture of the extent of urban population growth.

The growth of India's urban population is attributable both to rural-to-urban migration and to the natural increase of the urban population. In recent years, approximately 60 percent of urban population growth has been realized through natural increase. Much of this population growth has taken place in larger cities. India now possesses thirty-five cities with 1 million or more people, accounting for almost 40 percent of India's total urban population. The three largest cities of Delhi (Dilli) (17.3 million), Mumbai (17.3 million), and Kolkata (14.3 million) are three of the ten most populated cities in Asia and have joined the ranks of global megacities. Unlike Tokyo, Hong Kong, and Singapore, however, India's three largest cities are not classified as "world cities" in the economic sense because of insufficient finance, transport, and telecommunications linkages with the global economy.

507

Table 21-1 Urban India, 1901–2004

Year	Total Population (millions)	Urban Population (millions)	Percentage Urban of Total Population
1901	236.3	25.7	10.9
1911	252.1	26.6	10.6
1921	251.4	28.6	11.4
1931	279.0	33.8	12.1
1941	318.7	44.3	13.9
1951	361.1	62.6	17.3
1961	439.2	78.8	18.0
1971	545.5	108.8	19.9
1981	688.6	163.2	23.7
1991	838.5	214.7	25.6
2004	1,086.6	304	28.0

Source: Government of India; Population Reference Bureau, *World Population Data Sheet* (Washington, D.C.: 2004).

The Urban Poor

Because of the endemic poverty and limited cultural and educational resources of the countryside, first-and second-generation rural-to-urban migrants perceive the city as a place of greater opportunity. Members of the lower castes and untouchables are particularly attracted to urban areas because the greater anonymity of urban life frees them from the socioeconomic constraints of the village culture. A severe shortage of wage-based opportunities, however, has forced the vast majority of the migrants to engage in a wide array of unskilled tasks. Transport and construction work, food vending, domestic services, and even small-scale cottage industries are common sources of income generation. Underemployment rather than unemployment is the norm.

Lacking adequate income to secure durable housing, tens of millions of India's urban poor are forced to live in substandard dwellings that vary from single-room rental structures to makeshift shelters built of assorted discarded materials. Because many of the migrants are squatters living on land at the outskirts of the city that does not

■ FIGURE 21-23 **Kolkata shanty home.** A woman washes clothes in the contaminated Hooghly River. As in cities in other developing nations, the urban poor of India live under conditions of extreme poverty.

belong to them, their settlements are said to resemble a "village in the city," or **bustee**. Indian squatter dwellings commonly have dirt floors. Public services such as electricity, sewerage, and clean water are rare (Figure 21–23). It has been estimated that more than 60 percent of the urban population is not connected to municipal sewerage and water systems. The absence of such basic needs infrastructure leads to high levels of intestinal diseases, particularly among children, because water extracted from near-surface sources with hand pumps is often contaminated. At the bottom of the poverty chain is a large pool of sidewalk or pavement dwellers who do not even possess a roof over their heads. Mumbai, for example, is home to 250,000 sidewalk or pavement dwellers. Unfortunately, India's recent economic advances have produced even wider disparities between the urban rich and poor. Many scholars are concerned that if the urban poor are not allowed to benefit from the economic growth, a potentially combustible situation could arise from the social and economic inequality that now prevails.

Contrasting Kolkata and Mumbai

A comparison of two of India's most populated cities provides an illustration of the country's varied urban fortunes, especially with reference to engagement with the global economy. In the Western mind, at least, Kolkata (Calcutta) epitomizes the urban problems of the developing world.

▪ **FIGURE 21-24 A section of the skyline of Mumbai (Bombay).** Members of India's professional and middle classes enjoy the sunset from Mumbai's oceanfront corniche, the Marine Drive.

Situated on the banks of the Hooghly River (a distributary of the Ganges), Kolkata has been depicted since colonial days as a hopelessly poor tropical port, infested with malaria and cholera. The late Mother Teresa spent most of her adult life caring for the destitute of Kolkata. The prestige of being the colonial capital of British India was lost when this administrative function was transferred to New Delhi in 1911. The city became the recipient of a wave of Hindu refugees as a result of the 1947 partition that created East Pakistan, thus taxing an already overburdened urban infrastructure. Kolkata remains the capital of West Bengal state and the largest city within India's traditional manufacturing core; but these establishments, many of which are state-owned, are part of the country's aging industrial past.

Although characterized by similar rates of poverty, Mumbai (Bombay) has come to symbolize the opposite face of urban India (Figure 21–24). During the colonial period, this peninsula-sited city, which is exposed to the open ocean, attracted a diverse array of South Asian ethnic entrepreneurs. Owing largely to its dynamic, outward-looking commercial and financial sectors, Mumbai has grown to become India's most cosmopolitan city, symbolizing the new era of national economic growth linked to the global economy. Modern skyscrapers, hotels, and apartment buildings dot the cityscape, evidence of a highly trained workforce of professionals and entrepreneurs flush with new investment capital. The city is India's busiest international seaport as well as the country's primary air transport gateway. Mumbai is the capital of Maharashtra state, the third most populated and, territorially, the fourth largest federal state. The state government has taken the lead in injecting greater competition into the remaining state-owned industries as well as in attracting foreign investors. Mumbai has thus become the showcase of the new and progressive India.

Pakistan

With a GNI PPP of $2,060, Pakistan continues to rank as one of the world's poorest countries. Industry accounts for only 23 percent of GDP value added, and the growth rate of its primary exports of textiles, leather products, rice, and carpets declined during the 1990s. The agricultural sector, based primarily on wheat, rice, and cotton cultivated within the narrow confines of the Indus River Basin, has not fared well over the past 20 years either (Figure 21–25). The average annual growth rate of agricultural production has declined, and Pakistan is forced to import a significant amount of foodstuffs. The explanation for agricultural underdevelopment and food insecurity is the power of the conservative feudal-landlord class. A mere 2 percent of Pakistan's richest landlords control a quarter of the country's land and millions of serflike peasant farmers who are dependent upon landlords for survival. Agricultural productivity is less of a concern for the landlords than the social prestige and political power that land ownership yields; more than half of the members of the National Assembly are feudal landlords and

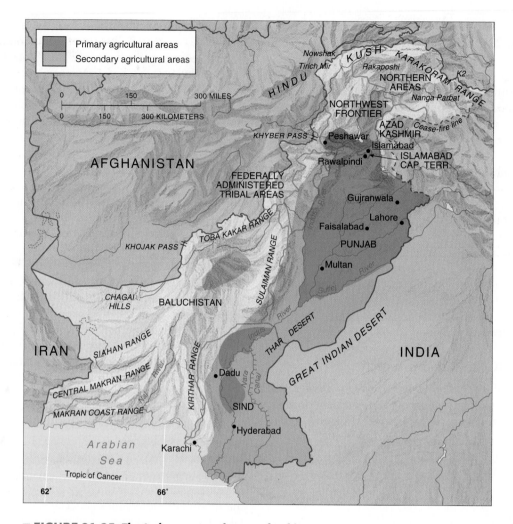

■ **FIGURE 21-25 The Indus-centered state of Pakistan.** As illustrated by the location of major urban places, the most productive economic regions are those paralleling the Indus River valley.

tribal leaders. Except at election time when votes are needed, landlords simply do not view peasants, many of whom are female and child bonded laborers because of debt owed to the landlord, as human resources, but rather as commodities.

Pakistan is one of the larger debtor nations in the world, with its 2003 foreign debt equal to approximately 45 percent of its GNI. Much of the debt is attributable to widespread corruption and chronic economic mismanagement of inefficient state-owned companies. Little serious effort has been made to reduce the role of government in industry, as has been done in India. Without increased agricultural and industrial output brought about through a strengthening of the private sector, Pakistan's economic fortunes are unlikely to improve significantly in the foreseeable future.

An additional obstacle to economic growth is the substantial ethnic and regional tension that has arisen from the country's transitional location between the Islamic Middle East and Hindu India. Cultural diversity does not of itself preclude the achievement of national unity, but in Pakistan efforts to forge a national identity have been largely symbolic. Chief among these efforts was the move-

ment of the national capital from the port city of Karachi to the interior Islamabad, directly after the 1947 partition. The transfer was partly a political and historical statement because Pakistan's cultural hearth is in the north. Islamabad also functions as a forward capital, designed to confront the perceived threats from Kashmir and the northwest frontier region. In addition to these attempts to create an identifiable cultural core, Pakistani nationhood has been strengthened by its outward religious homogeneity. Some 96 percent of the population professes the Islamic faith, and with more than 159 million people, Pakistan is the second most populous Muslim country in the world. Islam is the state religion (the country's official name is the Islamic Republic of Pakistan), and although not a true theocracy, Pakistan has incorporated elements of Islamic law, or *shariah,* into its legal system.

Despite these two cementing bonds of nationhood, ethnolinguistic fragmentation, based primarily on an east-west divide, has frustrated the goal of national unity. Three Indo-European languages are spoken in eastern Pakistan. Urdu is the official national language, but is spoken only by some

10 percent of the population, residing primarily in the north-central region. Punjabi is spoken by about 50 percent of the population, particularly in the densely settled north-eastern region. To the far south, Sindhi is the most commonly spoken tongue. Several Iranic languages are used throughout much of the more arid and mountainous western half of the country. As in India, English is commonly used for government administration and commerce.

One is not able to explain Pakistan's current social, economic, and political problems without discussing the war in Afghanistan, and the unintended role that the United States played in these problems. Pakistan's northwest frontier has traditionally been inhabited by ethnic **Pushtuns**, a group that also occupies portions of Afghanistan. Islamabad has historically been wary of Pushtuns, and this wariness only increased with the flow of millions of refugees into Pakistan's north-west and Baluchistan after the Soviet invasion of Afghanistan 1979. The refugees brought some 2.5 million head of livestock with them. The effect on the environment in these two territories was devastating as refugees and their animals caused serious overgrazing problems in fragile grass-land ecosystems, soil erosion as well as deforestation in response to the need for firewood for cooking and heating. Trans-border smuggling has also created economic problems. Imported goods in transit to Afghanistan via the port of Karachi were not taxed. Traders would cross the Afghani border and turn around and smuggle their goods back into Pakistan. The Pakistani government then lost millions in import revenues, as these goods were illegally sold in every Pakistani city.

Perhaps more importantly, the Afghan War has caused serious political problems in Pakistan. Before the invasion of Afghanistan by the former Soviet Union in 1979, Pakistan's relations with the United States were poor because of Pakistan's dismal human rights record, dictatorial governments, and its desire to become a nuclear power. After the 1979 invasion, however, this relationship took a very positive turn as the United States wished to cultivate friendly relations with Pakistan in order to provide a bulwark against the spread of communism. As a result, the U.S. government lifted its economic embargo against the country, and Pakistan became the recipient of advanced military hardware and economic aid. Pakistan became the fourth largest recipient of U.S. military aid and the second largest recipient of economic aid in the world during the 1980s. Because the Zia government became a new bulwark against communism, the U.S. government looked the other way to continuing human rights abuses by the dictatorial regime.

In an effort to combat communism, both the United States and Pakistani governments promoted Islam. This program resulted in the increased establishment of Islamic schools, or **madrasahs**, many corrupted their purpose to promote hatred toward the United States and to create distrust between the various sects of Islam. Because much of the foreign economic aid was spent on consumer goods rather than investments in education, health, and rural development, the growing poverty in the country only fueled this orientation to a radical form of Islam. In addition,

much of the military hardware provided by the United States to help the **Mujahideen** or freedom fighters in Afghanistan was stolen to create one of the largest illegal arms markets in the world. The United States inadvertently then created the present conditions for extreme lawlessness or a "Kalashnikov culture" among militants, criminals, and ethnic and sectarian groups that challenge the authority of the government. As part of U.S. strategy, the CIA encouraged the Mujahideen groups to cultivate poppy production for the heroin trade to finance the war against the Soviets. As a result, Pakistan became one of the largest heroin producers and trafficking countries in the world. In the late 1980s, Pakistan earned more from this heroin trade than all other exports combined, and today the country has more than 3 million drug addicts.

By most measures, Pakistan has become a dysfunctional country with little evidence of a civil society. At present, few foreign or domestic investors possess sufficient confidence in the nation's political and social environment to make any substantial long-term financial commitments. The continued economic stagnation has created a pool of more than 2 million Pakistani migrant workers, who have sought employment from Japan to Saudi Arabia.[5]

Nepal

Nepal, much like neighboring Bhutan to the east, is a land-locked country that occupies a physical and cultural transition zone between Tibet to the north and India to the south (see Figure 21–10). This Himalayan state is divided into three east-west-oriented landform regions, each characterized by distinctive culture groups. The southern region, or Terai, is part of the subtropical Gangetic Plain and is populated primarily by Hindi-speaking Hindus. Paralleling the lowland is the central foothill region, where some two-thirds of the national population resides in the more temperate mountain valleys and adjoining slopes. The most populated portion of the central foothills region is the Kathmandu Valley, which contains Kathmandu, the nation's capital and largest city, as well as the second and third largest cities. Although ethnically diverse, the peoples of the central foothills practice a Hindu-Buddhist syncretic belief system and speak the national language of Nepali, which is a fusion of Hindi and several Tibeto-Burmese languages. The cultural mix in the central foothill region produces colorfully textured scenes of religious structures that represent some of the most recognizable tourist landscapes in the world (Figure 21–26). To the north is the sparsely populated and spectacular high-mountain Himalayan zone, inhabited by peoples of Tibetan descent.

Nepal has experienced worldwide success in marketing its natural and cultural heritage. Adventure tourism, as it is often termed, has become a key component of the

---

[5]A.Z. Hilali, "The Costs and Benefits of the Afghan War for Pakistan," *Contemporary South Asia* 11, no. 3 (2002): 291–310.

■ **FIGURE 21-26 Buddhist monastery in Nepal.** The Tengpoche Monastery is both a spiritual retreat and a destination for visits from a growing number of tourists.

country's development plans and is a leading generator of foreign exchange. The tourist industry also creates substantial domestic employment opportunities. The long-term consequences of adventure tourism in Nepal still remain to be assessed. Environmental degradation from trekking activities includes soil erosion, trailside litter, and deforestation associated with the fuel needs of tourists and the construction of lodges. Loss of the social and cultural integrity of the most heavily affected local populations is also an issue of concern.

As prominent as tourism has become, its significance pales in comparison to the host of problems facing underdeveloped Nepal. Political instability has long been the norm. In 1990, Nepal became a constitutional monarchy, but this more democratic system of government has not reduced the historic regional rivalries. With a GNI PPP of only $1,420, Nepal is the second poorest South Asian country. Almost 90 percent of its 24 million people are engaged in subsistence agriculture, but even this most basic economic activity is threatened by severe soil erosion associated with deforestation. With virtually no industry to gainfully employ the population, which is growing by 2.4 percent annually, land-hungry peasant farmers are cutting the remaining forests at alarming rates. Under such conditions, the female gender burden increases dramatically on two fronts; women are forced to spend more of their daily lives walking farther to obtain water and firewood, and they are forced to work longer because eroded upland soils have forced many males to migrate to the plains in search of work. Approximately 42 percent of Nepalese live below the poverty line and literacy rates are extremely low. Industry is so poorly developed that the value of imports is more than double the value of exports. Nepal's landlocked location also results in in-

creased transport costs associated with external trade. With a high dependence on foreign aid, Nepal's development challenges, compared to those of other South Asian states, at times seem overwhelming.

Nepal's severe economic problems explain in part the rise of an agrarian-based Maoist Communist movement in 1996. With subsistence farmers living in near feudal conditions, the Communists find a willing support base for their cause. The constitutional monarchy only helps the Communist cause as the king dissolved the parliament in 2005, shut down the independent media, and jailed opponents. The Communists control more than half of the countryside in a struggle with government forces that have left more than 11,000 dead. Needless to say, the important contribution that tourism makes to the national economy has suffered tremendously.

Bangladesh

Few countries in the world have symbolized the developing world's poverty as much as Bangladesh. The country emerged from its 1971 civil war with West Pakistan as a desperately poor nation with few natural resources (Figure 21–27). The war devastated the country's roads, bridges, ports, and electrical power plants. To compound the misery, the destructive cyclone of 1971 resulted in a great loss of life and sharply reduced agricultural production. The international community responded with substantial amounts of aid to avert even greater human suffering. After a quarter century as an independent country, Bangladesh continues to rely heavily on foreign aid, and its GNI PPI of $1,870 remains one of the lowest in the world. Almost half of the population live below the poverty line.

The economy of Bangladesh rests almost exclusively upon agriculture, which absorbs three-fourths of the labor force and land area. The primary crops are rice, jute, wheat, and barley. Occupying, for the most part, portions of the Ganges and Brahmaputra river deltas, Bangladesh possesses some of the most fertile soils in the world, but opportunities to exploit this rich agricultural environment are matched by great liabilities. Because much of Bangladesh lies barely above sea level, annual **cyclones** (hurricanes) and their associated tidal surges occasionally cause great loss of crops and animal and human life. In 1991, for example, an unusually strong cyclone resulted in more than 100,000 deaths (Figure 21–28). Although not resulting in such great loss of life, the 1998 cyclone so damaged agricultural production that food imports doubled. The threat to human life by this recurring natural hazard rests upon two interdependent factors. The first is the fact that Bangladesh's 138 million people occupy a territory slightly smaller than the U.S. state of Georgia, making it one of the most densely settled countries in the world. The second source of vulnerability is the country's poorly

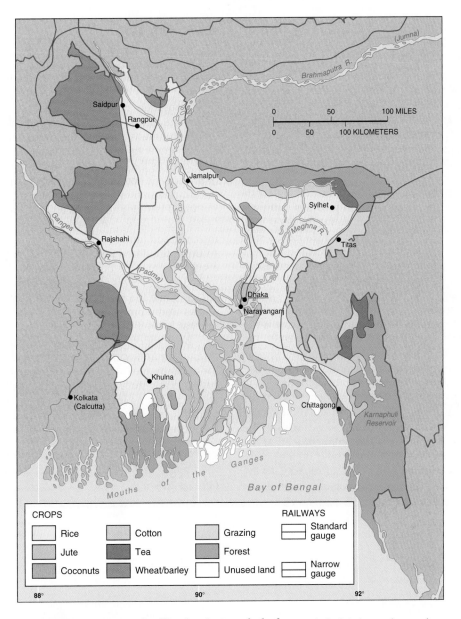

■ **FIGURE 21-27 Land utilization in Bangladesh.** Bangladesh is focused around the main drainage channels of the Ganges and Brahmaputra rivers. Almost the entire country is a large deltaic plain, much of which is given over to rice production to feed the country's large and growing population.

developed transportation and communications network, which makes large-scale evacuation ahead of a storm's landfall virtually impossible (see the *Geography in Action* boxed feature, Water Resource Security and Development).

The industrial economy of Bangladesh is poorly developed, with industry contributing only 25 percent of GDP value. Manufacturing is anchored in the low-value agroprocessing of jute and cotton. Like Pakistan and India, Bangladesh has become an exporter of inexpensive clothing and apparel products made from locally grown cotton. The total value of exports accounts for only 12 percent of economic output. With little industrial production, Bangladesh is mostly a rural country; only 23 percent of the population resides in urban areas. Only the national capital of Dhaka and the country's major port of Chittagong possess 1 million or more inhabitants.

Unlike Pakistan, where the potential for ethnic strife makes foreign investors wary, Bangladesh has a relatively stable government. Some 87 percent of the population is Muslim, with much of the balance being Hindu. Both religious communities, however, share a common ethnic heritage and the Bengali language. Unfortunately, the economic future of Bangladesh does not look bright. The scarcity of industry and the predominance of subsistence agriculture limit opportunities for individual economic advancement. Child labor is widespread, and half the rural population is landless.

GEOGRAPHY IN ACTION
Water Resource Security and Development

The concept of national security has traditionally been viewed in South Asia through the lens of military or geostrategic power. In Bangladesh, however, where the Ganges and Brahmaputra rivers join in the geographic center of the country to create one of the largest river deltas in the world, there is a growing concern that declining water resources might threaten national security and agricultural development. The problem stems from the fact that India, which almost totally surrounds Bangladesh, controls 90 percent of the catchment area of the rivers that flow through Bangladesh. Although the wet or summer monsoon provides sufficient water resources during the rainy season, during the dry or winter monsoon, the flow of water through Bangladesh is substantially reduced. Water shortages are particularly acute in the southwestern and northwestern regions of the country—areas that constitute 37 percent of the national territory and that support a population of more than 30 million.

The causes and consequences of the threatened water shortages are many and complex. One contributing factor is the Farakka Barrage Project, which was completed in 1960. The project diverts water from the Ganges to the Hooghly River in India for the purposes of increasing water flow to the Indian states of Uttar Pradesh and Bihar and improving navigation at the great downstream port of Kolkata. A second cause of the water shortages is the ongoing deforestation and resultant soil erosion of the Himalayan slopes, where the rivers originate. The effect of the erosion is to increase the sediment load of the rivers, which has caused them to gradually build up their beds, leading in turn to a decreased water carrying capacity. As the amount of water carried by the rivers has diminished, the Bangladesh farmers have responded by drilling more wells in order to supplement the river water with groundwater. This, however, is now depleting regional groundwater reserves. Another consequence of the reduction in river flows is that less freshwater is now being discharged into the Ganges delta, which, in turn, is allowing the more dense salt water of the Bay of Bengal to invade the coastal wetlands during the dry monsoon. As the salt water penetrates inland, it alters the pH of the aquatic ecosystems, thereby threatening the survival of the Sundarban forest, the largest mangrove forest in the world. The forest has long served as an important source of raw materials for pulp wood used in the manufacture of paper, matches, and furniture.

Unlike the Chang Jiang in China, which flows entirely through a single country, the problems associated with the Brahmaputra and Ganges rivers are magnified because of their binational nature. While the respective governments have proposed solutions to the problems of water control and access, resolution has been difficult to achieve. Such political problems will only be compounded if China and Nepal choose to build planned hydroelectric power plants and associated dams on the upper courses of the rivers. Although the specific issues discussed in this study are unique to South Asia, water resources are becoming an increasingly prominent security concern in many other regions of the world and are only now beginning to receive the scholarly attention they merit.

Source: Nahid Islam, "Indo-Bangladesh Common Rivers: The Impact on Bangladesh," *Contemporary South Asia* 1 (1992): 203–226.

■ FIGURE 21-28 **Flooding in the Ganges River Delta.** The farming opportunities in the Ganges delta are matched by the equally great liabilities associated with flooding.

Sixty-eight percent of children under the age of five experience some degree of malnutrition. Some 60 percent of the population are illiterate, and only 19 percent of children aged twelve to seventeen are enrolled in school. Although Bangladesh is led by a female Prime Minister, and women have just recently been allowed to run for government and labor-union offices, male-constructed barriers continue to withhold important decision-making power from women concerning development policies. The underutilization of the nation's human resources thus poses substantial barriers to future economic development.

Sri Lanka

Gaining independence from British colonial rule in 1948 and changing its name from Ceylon in 1972, this teardrop-shaped country, separated from southern India by the 21-mile-wide (34-kilometer-wide) Palk Strait, is almost equal in size to West Virginia (Figure 21–29). Economically, Sri Lanka stands apart from its South Asian neighbors. While still considered a poor country, it has a GNI PPP of $3,730, substantially higher than that of India. Government programs have significantly bettered the quality of life of the Sri Lankan

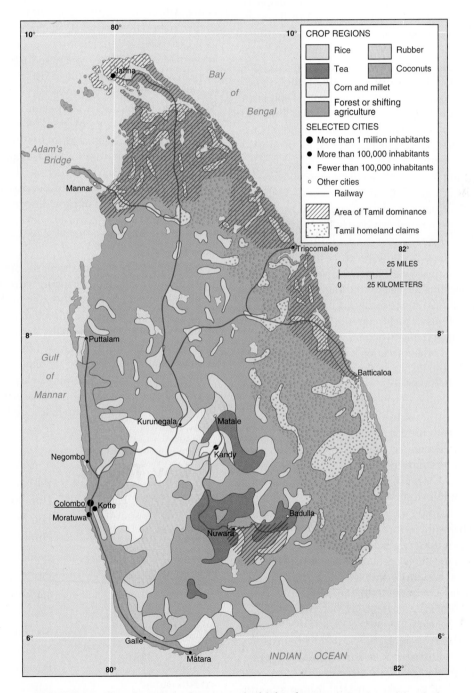

■ **FIGURE 21-29 Sri Lanka, a large tropical island.** Sri Lanka's substantial agricultural economy is based on both estate (cash) crops and subsistence farming. Much of the island's interior remains underdeveloped, supporting either shifting cultivation or forest vegetation.

population: the average life expectancy of 72 years matches that of some developed countries. Exports account for some 34 percent of economic output, almost triple the level of India. The higher quality-of-life indicators are the result of a basic-needs approach to development adopted by a moderate socialist government. It is not surprising then that Sri Lanka's Human and Gender development indices are the highest in South Asia, are higher than half of the Southeast Asian countries, and approximate the world median.

The cultural contours of Sri Lanka are also significantly different from those of its South Asian neighbors. Of the 19 million Sri Lankans, some 74 percent are Sinhalese-speaking Buddhists, with much of the balance being **Tamil**-speaking Hindus of Dravidian descent. Migrating from northern India during the first millennium B.C., these Buddhist Indo-European-speaking peoples established an advanced agricultural state in the drier northern half of the island. Successive Hindu migrations and the extension of southern Indian political influence by A.D. 1000 forced the Sinhalese to migrate into the forested central and southern mountains, where they established successive kingdoms until the arrival of the British in the early 1800s.

Upon establishing rubber and tea plantations in the central and southwestern areas of the island, the British imported hundreds of thousands of Tamil-speaking Indians as plantation workers, thereby increasing the Tamil share of the island's population. Owing in part, perhaps, to their having experienced discrimination at the hands of the British, the newly independent Sinhalese-dominated government began to systematically discriminate against the Tamil minority. The policy of discrimination was outwardly justified through Sinhalese-Buddhist notions of racial superiority but, in reality, centered on the fear of being overrun by Dravidian Tamils. The Tamil reaction was the formation of a guerrilla army demanding a separate Tamil state in the northern and eastern regions of the island. The 22-year old civil war between the "Tamil Tigers of Eelam" and Sri Lankan government forces has taken 62,000 lives, most of which were innocent civilians. Despite good-faith proposals from the government for greater political and economic autonomy for the northern and eastern Tamils, the most powerful Tamil militia group continued to resist any form of political accommodation for fear of losing its political power. In early 2002, the most promising cease fire to date was signed, which will hopefully end the bloody civil war.

The tragedy of cultural conflict in Sri Lanka has been a serious impediment to national economic progress. Sri Lanka continues to rely on the traditional exports of rubber, tea, coconut-based products, gems and, more recently, inexpensive apparel. Modern industry is limited to the national capital of Colombo. The government perceives the economic potential of Sri Lanka to be similar to that of the Pacific Rim's economic tigers and has created an export-processing zone that has attracted a handful of foreign investors. The government has also been successful in establishing a South Asian container-transport load center; the gross inefficiency of South Asian continental ports has led to an important feeder and transshipment role for Colombo. The resurrection of the once thriving tourist industry based on the spectacularly beautiful highlands and beaches would certainly be a welcome source of renewed revenue. A precondition to Sri Lanka's assuming the status of a middle-income country, however, is the resolution of its deep-seated ethnic problems.

▶▶ Summary

The challenges to South Asian economic development are many. Although progress is being made to further industrialize national economies, the nature of industrial activity is extremely varied. High-tech manufacturing plants may be situated adjacent to grimy industries exploiting child labor. A similar dual economy exists in the agricultural sector where large-scale commercial farmers utilizing Green Revolution technologies are becoming prosperous in part at the expense of a sea of landless and desperately poor subsistence cultivators. It is increasingly obvious that the top-down economic-development models conceived in the West do not necessarily best serve the needs of the non-Western developing countries. Because development in its broadest sense consists of providing increased opportunity for all, a basic-needs, bottom-up development strategy may be the most effective in the densely populated South Asian nations.

Although the persistent threat of communal violence appears to be intractable, a broader grassroots development philosophy would help resolve this problem. Ethnicity or religion is not the ultimate source of communal strife, but rather unequal access to human advancement. Each ethnic or religious community possesses its economic elite, middle class, and poor. Lasting development will occur only through significant reduction of class inequalities within individual communities rather than by treating any particular ethnic or religious group as a completely homogeneous entity.

▶▶ Key Terms

Bharatiya Janata Party (BJP)	Harappan culture
brain drain	Hindi
bustee	Hinduism
caste system	Indian diaspora
communalism	Jammu and Kashmir
cyclones	*madrasahs*
Damodar Valley	Mauryan Empire
Dravidian	Mughal Empire
federalism	Mujahideen
forward capital	Pushtuns
gender bias	scheduled castes
gender subordination	Sikhism
	Tamil

▶▶ Review Questions

1. What is the "Japan Model" and how has it influenced the economic development of the other nations of Asia?

2. Why has Japan suffered recurring recession since the early 1990s, and what can be done to reverse this trend?

3. How has ethnic and linguistic diversity challenged the development of many of the nations of Asia?

4. Compare the differing levels of globalization in the national economies of the mainland and insular nations of Southeast Asia.

5. Identify some of the environmental challenges facing China's emerging industrial cities.

6. How likely is it that the Chinese communist party will be able to maintain its political monopoly if the country's economy continues to become more open to international influences in the years to come? Is it possible over the long term to have economic freedom without political freedom?

7. How have the cultural landscapes of the countries of Asia been impacted by the great belief systems of the region—Buddhism, Hinduism, and Islam?

8. Compare and contrast the different levels of globalization in the Indian cities of Mumbai, Bangalore, and Kolkata.

9. How has the emergence of primate cities in many of the nations of Southeast Asia impacted the economic development of those countries?

10. What are the physical factors that contribute to the monsoon climates of the countries of Asia? How does the monsoon influence the agricultural and settlement geography of the region?

▶▶ Geographer's Craft

1. What will be the likely consequences of the continuing diffusion of Green Revolution agricultural technologies on both the rural and urban sectors of the nations of Asia? How might those consequences impact the long-term economic and human development of the region?

2. How has Japan compensated for its limited physical resource base to achieve a high level of economic development?

3. How would a geographer explain the importance of Singapore to the larger regional economy of Insular Southeast Asia?

4. What geographic considerations have influenced the recent economic growth of coastal southern China? Why have interior and northern China lagged behind southern China in their integration into the global economy?

5. What are the economic consequences of India's marginalization of females? How might a geographer contribute to overcoming this human development deficit?

PART EIGHT

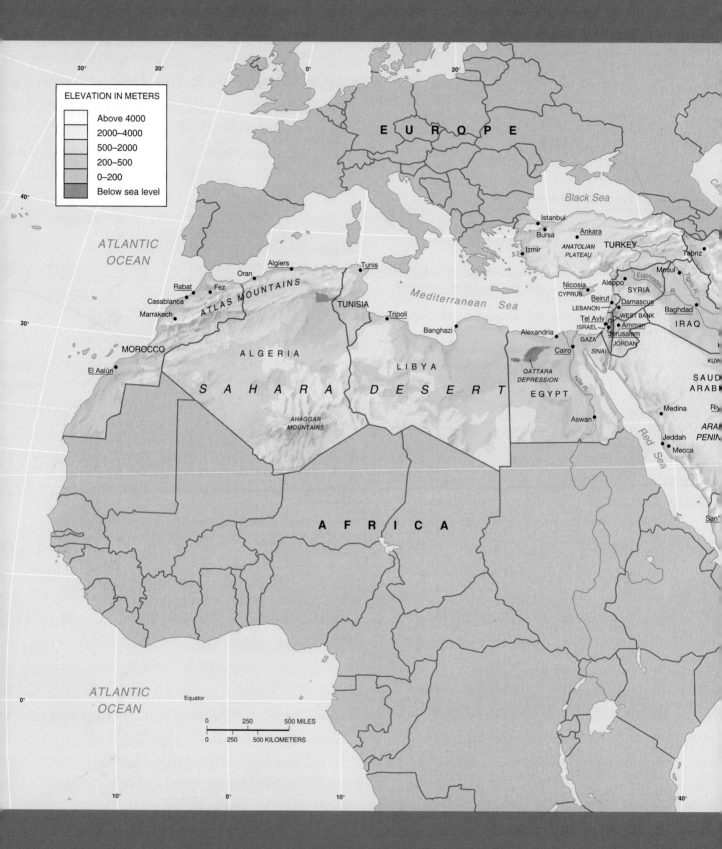

ELEVATION IN METERS

Above 4000
2000–4000
500–2000
200–500
0–200
Below sea level

EUROPE

ATLANTIC OCEAN

Black Sea

Istanbul
Bursa
Izmir
ANATOLIAN PLATEAU
Ankara
TURKEY
Tabriz
Mosul

Mediterranean Sea

Algiers
Oran
Tunis
Rabat
Fez
Casablanca
Marrakech
ATLAS MOUNTAINS

TUNISIA
Tripoli
Banghazi

Nicosia
CYPRUS
Aleppo
SYRIA
Beirut
LEBANON
Damascus
Baghdad
IRAQ
Tel Aviv
WEST BANK
ISRAEL
Amman
Jerusalem
GAZA
JORDAN
Alexandria
Cairo
SINAI
KUW

MOROCCO

ALGERIA

SAHARA

LIBYA

DESERT

EGYPT

QATTARA
DEPRESSION

SAUD
ARAB

El Aaiún

AHAGGAR
MOUNTAINS

Aswan

Medina
Riy

Jeddah
Mecca

ARAB
PENIN

Red Sea

A F R I C A

ATLANTIC OCEAN

Equator

San'

0   250   500 MILES
0   250   500 KILOMETERS

The Middle East and North Africa

Douglas L. Johnson and Viola Haarmann

The Middle East and North Africa constitute a large and distinctive region composed of Mediterranean and Persian Gulf countries characterized as much by the unifying bonds of Islam and Arab culture as by deep internal contrasts and severe conflicts. The contrasts are evident in levels of technological development, the extent of dependence on oil for national revenues, cultural differences, religious divisions, the influence of militant Islam, ethnic diversity, and uneven progress toward stable societies.

The intertwining of these contrasting elements has created major areas of tension and conflict within the region. The most notable is the struggle between Israel and the Palestinian Arabs, who are supported by most of the region's Arab states. Even when neighboring states have recognized Israel's existence and signed formal peace treaties, substantial elements of their population do not accept their government's official policy. Unresolved, this conflict impacts the domestic politics of North African and Middle Eastern countries as well as regional and international relations and adversely influences economic investment strategies, development opportunities, and human rights practices.

Major ethnic divides within many states in the region further magnify these issues. In Turkey and its neighbors to the south and east, Kurdish aspirations for independence or autonomy have produced serious internal tensions. In Morocco the desire of Berbers for greater linguistic and cultural self-expression has recently met with a less hostile attitude on the part of the state, a model that might offer hope for diversity within unity if it were applied more widely.

Interstate rivalries are another area of conflict and sparked the inconclusive eight-year war in the 1980s between Iraq and Iran over resources, territorial claims, and regional leadership. Regional rivalries as much as principles governed which Arab states participated in the military opposition in 1991 to Iraq's invasion of Kuwait. Weak involvement by Arab countries in the American- and British-led invasion of Iraq in 2003 and its aftermath of violence reflects their different assessment of the regional political realities and extreme discomfort with the future impact of events in Iraq on their own internal stability.

The geography and future of this complex region are also influenced by various elements of the Middle Eastern and North African physical and cultural environments. Aridity dominates the region in a way that renders much land of limited utility for agriculture. The uneven distribution of oil reserves explains many of the differences in extent of development, levels of living, and political influence. The Islamic faith remains a powerful force in structuring society and state, despite the presence of nationalist forces and non-Islamic minorities. Finally, the Middle East and North Africa long has functioned as a crossroads for Europe, Asia, and Africa, enabling an exchange of goods and ideas throughout history. The colonial influence of the late nineteenth century, and the continuing American and European economic and military presence in Israel, Iraq, and a number of other Arab states, have contributed to widespread resentment of Western ideas and technology, which remains a significant issue in the politics of many of the region's countries.

■ **Hassan II Mosque in Casablanca, Morocco.** Religion is an extremely important part of self-identity for many people in North Africa and the Middle East. Completed in
1993, this imposing center of Islamic worship is the most spectacularly sited mosque on the
Atlantic coast of North Africa and a monument to traditional craftsmanship.

Chapter 22

The Middle East and North Africa:
Physical and Cultural Environments

- **Physical Geography of the Region**
- **Land-use Patterns**
- **Historical Overview**

Physical Geography of the Region

The Middle East and North Africa form a vast region that stretches from the Atlantic Ocean eastward to the Persian Gulf and the lands bordering Monsoon Asia (Figure 22–1). Although in many ways diverse, it has become one of the earth's most distinctive geographical realms, bound together for the most part by circumstances of aridity and a common Islamic cultural heritage. We will begin our study of this pivotal region with an overview of its varied physical environments.

The Role of Aridity

Aridity, or dryness, is the dominant climatic characteristic of North Africa and the Middle East. Because much of the region is so dry that it can support little or no vegetation, we find extensive areas of barren rock, gravel, or sand that support very few people. Where water is present, people are found, often in large numbers. These more densely populated settings include upland areas that intercept storms, coastal lowlands watered by runoff from the uplands, inland plateaus exposed to seasonal storms, seasonal streams and exotic rivers, and oases.

Much of the Middle East and North Africa, particularly the portion bordering the Mediterranean Sea, experiences a dry summer subtropical climate with less than 20 inches (500 millimeters) of annual rainfall (Figure 22–2). This means that precipitation falls mainly in the winter and early spring. In contrast, during the summer and autumn seasons, typically there is very little rain. Winter rainfall throughout the greater Mediterranean Basin is produced by frontal or cyclonic storms that move across the region from west to east. Whenever these storms encounter a mountain barrier, the air is lifted and cooled, often leading to condensation

and precipitation. The Atlas Mountains in northern Africa, the Jabal al-Akhdar in eastern Libya, Mount Lebanon and the Judean hill country in the lands bordering the eastern Mediterranean, and the Zagros Mountains of Iran are all associated with localized **orographic**, or mountain-induced, **precipitation**. Where mountains are high, this precipitation falls as snow and remains on the ground throughout the winter. Some of the mountain ranges close to the sea, such as the Pontic Mountains along the Black Sea coast of Turkey or the Elburz Mountains on the Caspian coast of Iran, generate huge quantities of rain or snowfall. Even the ranges deep in the Sahara, such as the Ahaggar Mountains in southern Algeria or the Tibesti Mountains along the border of Libya and Chad, serve as water catchment areas for the surrounding regions and support "islands" of vegetation that are very different in type and productivity from the plants of the surrounding lowlands. These mountain barriers trap moisture on their windward sides, while their interior or leeward sides are much drier, owing to the **rainshadow** effect produced by descending, drying air masses. These local rainshadow drylands, including the High Plateau of Algeria, the central Turkish or Anatolian Plateau, and the Syrian Steppe, often are important cereal and livestock production zones. Only where the surrounding mountain ranges are too high and extensive, as in the mountain rimlands of Iran, do these interior environments become extensive deserts.

The moisture that nourishes the upland mountains and plateaus also sustains the adjacent lowlands. While some of this moisture falls directly onto the lowlands as the frontal storms cross them, much of the water comes from runoff from the nearby uplands. The streams that flow to the sea, such as the Umm er Rhbia, Bou Regreg, and Sebou in Morocco; the Medjerda in Tunisia; the Orontes in Syria; and the Derbent, Gediz, Mederes, and Seyhan in Turkey all support rich agricultural environments in the coastal lowlands.

521

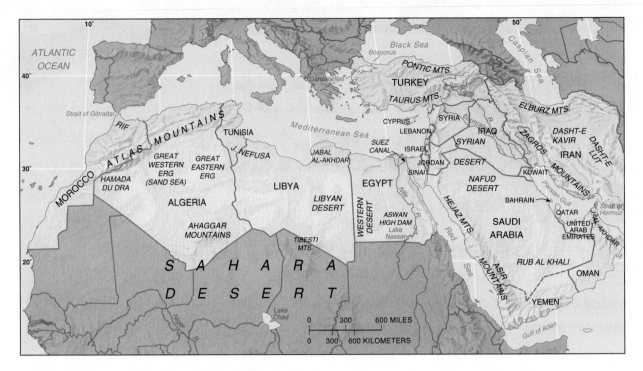

■ **FIGURE 22-1 North Africa and the Middle East.** North Africa and the Middle East extend physically from Morocco on the shores of the Atlantic Ocean to Iran on the Persian Gulf. Most of its peoples are bound together by allegiance to Islam and by the need to cope with water scarcity.

In some cases, **rivers** are **exotic**, or foreign, to the area to which they supply water. This means that they derive their water supply from outside the region and cross dry areas on their journey to the sea, collecting little if any additional runoff along their lower channels. The Nile is the best-known example. Rising in the mountains of Ethiopia and East Africa, the Nile flows northward across the Sahara Desert before emptying into the eastern Mediterranean. The Nile receives almost no water northward of the point where the Atbara River joins it in northern Sudan. In effect, the river is a linear oasis with narrow banks and floodplain along which vegetation grows. Move a short distance away from the banks

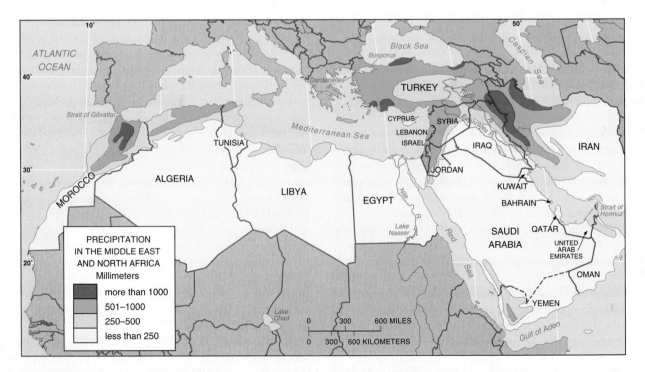

■ **FIGURE 22-2 Precipitation patterns.** Most of the Middle East and North Africa receives less than 250 millimeters (10 inches) of precipitation annually. Areas receiving more than 1000 millimeters (40 inches) annually are centered on the high mountain zones.

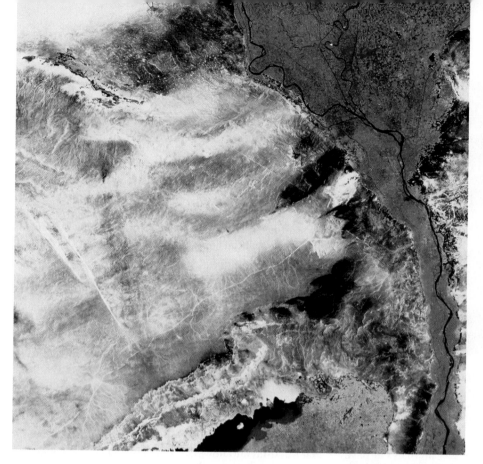

■ **FIGURE 22-3 Landsat photo of the Nile Delta.** Vividly shown in this satellite photo, the Nile River forms a linear oasis with human settlement along its banks and none away from the river. The large blue spot at the base of the delta is the city of Cairo and its suburbs.

of the river, and vegetative growth is impossible (Figure 22–3). The Euphrates and Tigris rivers, rising in the highlands of eastern Turkey, are also exotic streams as they flow through the deserts of Mesopotamia. South of Baghdad, practically no water is added to either stream as they proceed southward to the Persian Gulf.

Similar "islands" of life also exist deep within the desert in **oases** (singular, *oasis*), where erosion has lowered the land surface and relatively high groundwater levels result in water being found close to the surface. In some cases, saline lakes, nourished by groundwater from below and diminished by evaporation from above, are found in the lowest central portion of the oasis. Near-surface groundwater is often recharged by rainfall entering regional **aquifers** far outside the desert. Lateral movement of subterranean recharge water is always very slow and withdrawal of groundwater at rates greater than is possible using traditional animal-powered technology invariably leads to drawdown and eventual depletion of the aquifer. Deeper aquifer layers are composed of fossil water trapped underground during wetter epochs thousands of years in the past. This fossil water is a finite resource, and its use constitutes a form of mining. Oases historically served as staging points for the caravan trade that linked more productive regions outside the desert across the generally barren arid wastelands (Figure 22–4).

Living in the desert requires special adaptation on the part of animals, plants, and people. No animal is better suited to the desert than the camel, whose endurance, capacity to travel long distances with heavy loads, and ability to survive without drinking more than once every 5 to 8 days in the dry season have made it the animal of choice for desert dwellers

(Figure 22–5). To a lesser degree, donkeys and goats are also tolerant of water shortages and poor fodder conditions. Only in places where local water and grazing conditions improve do cattle, sheep, and horses appear as important livestock. Plants adapt to aridity by finding ways to evade dry conditions or by adjusting to particular local conditions. Wheat, barley, and rye are domesticated grasses native to this region that avoid drought by completing their life cycle quickly and by producing seeds that can survive for many years, if need be, until the next rainy period produces good growing

■ **FIGURE 22-4 Al-Jawf oasis, Kufra, Libya.** Amid the dryness that is characteristic of North Africa and the Middle East are desert oases like this one in Libya. These "islands of life" exist where surface erosion and relatively high groundwater levels result in water being found close to the surface.

■ **FIGURE 22-5 Nomads at a Moroccan oasis.** Oases are an essential component of desert pastoral livelihood. Here camels, the animal of choice for desert herders, can find abundant water, and their owners can obtain dates and other products.

conditions. Most of the wild grasses that animals eat have adapted in the same way. Other plants sink deep taproots into the ground where they can find water or, like the olive or the almond tree, withstand drought by limiting their activity to the wet season or to wetter years. Yet other plants thrive under conditions of extreme environmental adversity. The date palm is able to use much saltier water than most plants and thus is a major component of oasis agriculture. Its fruit is a vital source of food for desert dwellers, and its palm fronds and woody trunk are important fencing and building materials.

The peoples of North Africa and the Middle East have also adapted to dry conditions. The flowing robes worn traditionally in the Middle East reduce moisture loss through sweating by insulating the wearer from the intense desert sun and by allowing free air flow to cool the skin. Their light color reflects rather than absorbs sunlight, thus keeping body temperature lower, and the use of head coverings reduces the risk of sunstroke. Housing also reflects climatic adaptation (Figure 22–6). Vaulted roofs, high ceilings, and the use of tiles promote cooler conditions by allowing warm air to rise or by absorbing less heat. Building houses around a central courtyard provides both shade and private outdoor space. In some places, houses are constructed partially in the ground or extending back into caves in order to moderate the extremes of daytime temperatures. Cooling towers promote air movement that produces cooling breezes. Nomads make tents that can be shifted seasonally from place to place. Often a heat-absorbing black tent is used in winter, when concern for warmth is greatest, while a light-colored, more reflective tent is typical of summer, when cooling is the chief concern. People also adapt economically to the seasonal or perpetual aridity, with nomads moving their herds between wet and dry season pastures in order to find grass and water for their animals. Many farmers compensate for limited local water supplies by constructing irrigation systems that bring water from distant sources in order to grow their crops. Today many farmers use glass or plastic greenhouse-like structures in order to control temperature and water loss and create a suitable environment for the growth of vegetables, fruit, and flowers. These high-value specialty products are then sold in the region's major urban markets or exported to Europe as "out-of-season" crops, which can command a high price to compensate for the costs of transport.

Landforms

We often think of sand dunes as being the most common landforms in the deserts of North Africa and the Middle East. While there are very large sand areas (*erg*) composed of mobile sand dunes, these spectacular environments are the exception rather than the rule. Dominated by crescent-shaped *barchan* dunes or long, linear *seif* dunes, such areas are largely devoid of vegetation (Figure 22–7). Far more common, however, are the stony *hammada* and *serir* surfaces that are covered with rocks and pebbles, respectively.

■ **FIGURE 22-6 Typical housing in El Oued, Algeria.** These barrel-vaulted and domed structures, set in a landscape of linked barchan sand dunes, allow more interior space in which hot air can rise, thus helping to maintain indoor temperatures at a more reasonable level.

■ **FIGURE 22-7 Desert land-scapes of North Africa.** These sand dunes dramatize some of the most inhospitable desert landscapes of North Africa and the Middle East. These areas are largely devoid of vegetation. While large areas of sand dunes, such as the Grand Ergs of Algeria or the Nafud of Saudi Arabia, are spectacular landscapes, they are only a small portion of the total desert scene.

These areas also lack the resources of soil and water with which to sustain vegetation and human life.

Of much greater importance for human settlement are the mountain ranges and upland plateaus that provide a basis for livelihood and economy. These highlands stand across the paths of the frontal or cyclonic storms that pass through the region and force these moisture-bearing air masses to rise, cool, and produce precipitation. The amounts of precipitation produced are often substantial. At Ifrane in the central Middle Atlas Mountains of Morocco, as much as 59 inches (1500 millimeters) of precipitation can fall in a year. Much of this moisture occurs in the form of snow since the mountains are high, and the rainy season comes in the cooler part of the year. Inland from Ifrane, in the area of Midelt, less than 100 kilometers away, fewer than 10 inches (250 millimeters) fall, on average, each year. This sharp contrast between wet and dry habitats is a classic rainshadow effect. A similar contrast can be observed on a journey between Beirut on the Mediterranean coast and Damascus, only 50 miles (80 kilometers) to the east. At Beirut there is an average of 37 inches (940 millimeters) of rainfall per year, while Damascus is functionally an oasis that receives less than 9 inches (230 millimeters) annually. The intervening Lebanon and Anti-Lebanon mountain ranges absorb much of the moisture produced by the winter storms passing through the Mediterranean, leaving the interior regions dry but creating humid zones along the coast and in the uplands.

These uplands are of tremendous importance to the nearby lowlands. Originally the uplands supported large forests that provided building materials for the cities and fleets of the coastal states. Most of these forests have been removed, and only remnants survive in the higher and more inaccessible places or in protected stands near sacred sites.

The rivers that drain the uplands transfer large amounts of water as runoff to the nearby lowlands. These river flows provide water for irrigated agriculture, the lifeblood of the settled populations of the lowlands. The seasonal contrasts in temperature and precipitation between the uplands and lowlands have historically constituted the basis for important nomadic movements of livestock. Although much reduced today in many countries, this upland-lowland form of migratory animal husbandry can still be encountered much as it existed in former times when it helped support the great lowland civilizations by supplying meat and other important animal products.

The highlands themselves have also supported substantial sedentary agriculture populations throughout recorded history. In many respects these uplands are good places to live. In a region where aridity is widespread, uplands offer reasonably secure water supplies. Here minority religious and ethnic groups have often concentrated. The same mountain environments that provided food also provided protection from the central governments of the region, typically located in lowland or coastal sites. Maronites in Lebanon, Druze in southwest Syria, Berbers in North Africa, and Zaidis in Yemen all took advantage of the defensive features of their mountain homelands. With only limited ability to expand spatially outside their folk fortresses, these groups often invested in labor-intensive agricultural improvements such as terracing. But when better employment opportunities appeared in adjacent lowland areas during the last half century, the loss of laborers has often led to a collapse of this kind of agricultural infrastructure. As terrace walls crumble due to the lack of maintenance, the soil accumulations of centuries run the risk of washing downslope to the sea.

Land-use Patterns

As a region of great spatial extent and environmental diversity, the Middle East and North Africa support a wide variety of land-use activities (Figure 22–8). These sustain the region's ecological trilogy of farmer, herder, and city dweller. Land use can be grouped into four major categories: agricultural, pastoral, mining, and urban.

Agriculture: Traditional and Modern

Despite the region's aridity, a great deal of agriculture in the Middle East and North Africa consists of **dry farming**, wherein farmers rely exclusively on rainfall to produce their crops. Concentrated in the region's better-watered semiarid environments, dry farming typically centers on the cultivation of cereal crops. Wheat and barley are the most widely sown, with tree crops such as almonds, olives, and grapevines also being important, particularly in the coastal districts of the Mediterranean. Cereal cultivation is an extensive activity and, in recent decades, it has been aided greatly by the use of tractors and other mechanized farm equipment. This

has often allowed farmers to expand into rangeland areas that were once the exclusive domain of nomadic herders. The major zones of dry farmed cereals are the coastal plain south of Rabat in Morocco, the High Plateau in Algeria, the Anatolian Plateau in central Turkey, the steppe areas of Syria and other portions of the Fertile Crescent, and the semiarid zones in interior Iran between the mountain rimlands and the interior deserts. Cereal cultivation in these areas, however, remains very vulnerable to drought. **Run-on farming** techniques, which collect water from a larger area and concentrate it in valley bottoms and on terraced slopes, permit agriculture in regions that otherwise would be too dry for crop cultivation. These run-on systems are best known in the northern Negev zone of southern Israel, where they were developed by the ancient Nabateans, but they are found widely distributed throughout North Africa and the Middle East. Terraced agriculture in mountain districts, such as the High and Anti-Atlas mountains of Morocco or the highlands of Yemen or Mount Lebanon, is another important traditional form of dry farming.

In contrast to dry farming, **irrigated agriculture** has long been concentrated along the region's major rivers and

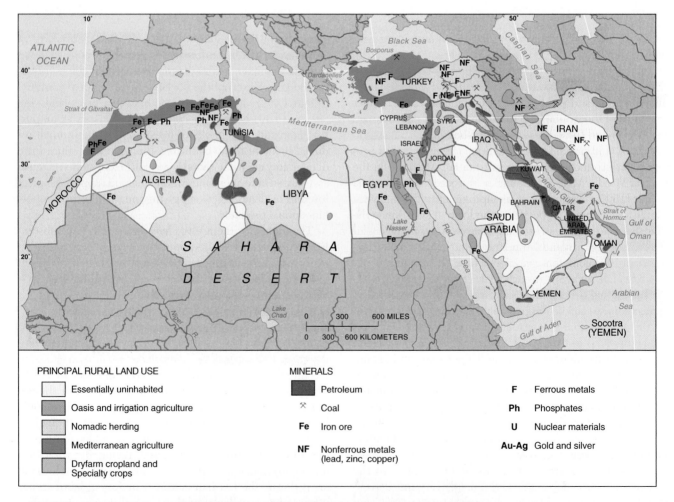

PRINCIPAL RURAL LAND USE

Essentially uninhabited

Oasis and irrigation agriculture

Nomadic herding

Mediterranean agriculture

Dryfarm cropland and Specialty crops

MINERALS

Petroleum

✕ Coal

Fe Iron ore

NF Nonferrous metals (lead, zinc, copper)

F Ferrous metals

Ph Phosphates

U Nuclear materials

Au-Ag Gold and silver

■ **FIGURE 22-8 Land use in the Middle East and North Africa.** Drier parts of the Middle East and North Africa are used for herding, but oases based on groundwater, and Mediterranean agriculture in higher rainfall areas play the most prominent roles. Oil is the most important mineral resource.

in oases. Oasis irrigation is always of limited extent because the amount of available water is relatively small, and the effort required to raise that water to the surface is relatively great. Large amounts of human and animal power must be invested in the system in order to make it function. Irrigation based on river water is a different situation. Large amounts of water are usually available, at least seasonally, but, although temperatures permit year-round cropping in many places, until recently no way existed to store water on a significant scale. Consequently, most irrigation systems have historically relied on retaining the annual flood in small-scale basins maintained with local labor. This was the system typified by the Nile. Water-lifting devices, such as the *shaduf* (a counterweighted, lever-mounted bucket), the *noria* (waterwheel), and *tambur* or *Archimedes' screw* (a wide, hand-turned, threaded screw enclosed in a cylinder that is used to raise water a few feet from one level to another), made enough water available to permit limited double-cropping in the dry season near the river (Figure 22–9). Canal construction along the Tigris and Euphrates rivers permitted more extensive and integrated irrigation to take place in ancient Mesopotamia (contemporary Iraq). These developments, however, never allowed for over-season water storage and were always constrained by low levels of dry season stream flow and the need to practice alternate year fallow in order to avoid waterlogging and salinization.

Only the **qanat** system in Iran (*falaj* in the Arabian Peninsula; *foggara* in North Africa) permitted year-round irrigation. Qanats tap groundwater, but are fundamentally different from wells. Because available groundwater and rich agricultural soils are often not located close together, the problem is how to get the water out of the ground and move it to a site with good soil that farmers can cultivate. When fields and groundwater sources are distant, qanat systems sometimes dozens of kilometers in length have been constructed. A qanat is an underground tunnel that moves water from source to field. A great deal of labor and skill is invested to construct the tunnel, which starts by digging a vertical shaft into the ground until water is encountered. Usually this takes place near the base of an upland at the upper edge of an alluvial fan where water collects between bedrock and looser, eroded material. From this point the tunnel is extended horizontally at low gradient so the water can move laterally by gravity along the tunnel floor. Vertical shafts are dug at intervals to make removal of the tunnel's excavated material easier. Gradually the tunnel's floor approaches the surface as each successive shaft is somewhat shallower than its predecessor. Eventually the tunnel emerges at the surface and groundwater has been brought to the surface by gravity under its own power. Creation of the tunnel avoids the very expensive and energy-consuming animal and human labor needed to raise water from a well before steam or gasoline pumps were invented. Only occasional maintenance is needed to clear debris from the tunnel and keep the water flowing. When the water reaches the surface, it can be moved by a ditch, which often is covered to reduce evaporation loss, to the field where the farmer can use it. Although the vol-

■ **FIGURE 22-9** **Donkey power in North Africa.** In the Dukkala region of Morocco, a donkey-powered *noria,* or waterwheel, irrigates the fields and permits limited double-cropping in the dry season. This traditional lift device is rapidly being replaced by gasoline-powered pumps.

ume of flow in a qanat shows some seasonal variation, in most cases a remarkably constant rate of flow is produced. In most areas of the Middle East, until the mid-nineteenth century, year-round irrigated agriculture was unknown unless the area could be commanded by a qanat or a traditional lift device.

Everything changed with the construction of man-made barriers (barrages), first on the Nile and later on the region's other perennial streams. In that part of the year when a river's water level is low, it is difficult to remove the water from the stream without expending enormous amounts of energy in the form of animal and human labor. A barrage is a low structure placed across the stream channel that blocks the flow of water in this low-flow period. The goal is to raise the water level to a height that is much closer to the level attained when the stream is in flood. With the water level artificially raised, water can then be diverted into canals where gravity flow carries it to nearby fields. This innovation made possible the irrigation of much larger areas than was possible by the use of lift devices or qanats. Gradual expansion of the area commanded by barrages enabled year-round cultivation of more and more of the Nile's floodplain and delta. In these areas of perennial cultivation, a succession of crops could be grown throughout the year and the yields of food and fiber per unit of land could be dramatically increased. This pattern of development culminated in the construction of the High Dam at Aswan in southern Egypt (completed in 1971). From that point on, most of the Nile's floodwaters, which formerly had flowed into the eastern Mediterranean, were stored in a large lake, to be released as needed throughout the year. Two and three cultivated crops are now possible, and the primary problem has become how to drain the excess irrigation water in order to avoid waterlogging and salinization of the soil (see the *Geography in Action* boxed feature, Land Degradation in North Africa and the Middle East). The large-scale perennial irrigation system of the Nile has been emulated throughout the Middle East. Most notable are the large-scale dams on the tributaries of the Tigris constructed by the Iraqis; the major dam built by the Syrians at Tabqa on the Euphrates; the series of dams built or under construction on the Turkish

GEOGRAPHY IN ACTION
Land Degradation in North Africa and the Middle East

The physical environment of North Africa and the Middle East has changed in major ways over thousands of years of human occupancy. Perhaps the oldest pressure on Middle Eastern and North African environments has been deforestation. Because the region is quite dry, forested habitats of the upland valley bottoms, where water and soil resources are concentrated, have been the most vulnerable to **deforestation** owing to the expansion of agriculture. This pressure on woody vegetation began on a small scale when humankind first used fire to clear land for cultivation. Many of the woody species in the Mediterranean flora possessed or developed adaptations to this pressure, for they either sprouted from the base when their woody stems were cut or burned or, like the cork oak, they developed a thick, insulating bark in order to resist the heat of ground fires.

The development of cities more than 5,000 years ago in the lowland river basins and coastal districts of the Middle East and North Africa did not lessen the pressure on the forest. On the contrary, one of the earliest epic tales of the Sumerians, inhabitants of what is now Iraq, recounts how Gilgamish, a legendary hero, organized expeditions to the mountains, presumed to be Lebanon, to acquire lumber for building projects. Rot-resistant timbers, such as cedar, were much in demand for construction in ancient Iraq, as well as elsewhere in the region. The dry plains of the Tigris and Euphrates rivers, while they had date palms, reeds, and mud aplenty, were devoid of trees.

Thus, both agricultural clearing and lowland cities' need for timber—for doors, roof beams, decorative elements, city gates, and shipbuilding—placed heavy demands on the forest resources of the uplands. Only the least accessible places escaped the attentions of the lumberman. Many upland habitats attracted agricultural settlement, partly because the water resources were relatively abundant and partly because the mountains provided defensible sites for persecuted minority groups. The result was the gradual impoverishment of the upland forest cover, as well as the exposure of the hillslopes to accelerated erosion. Recent population growth and agricultural expansion have increased the pressure on forest resources. In Lebanon, for example, only a dozen stands of the famous Lebanon cedar still survive. These clusters are preserved either because they are hard to reach or because they are located at sacred sites.

Another land degradation issue is a set of three linked problems—siltation, waterlogging, and salinization—all of which affect irrigation systems. Anything that affects irrigation strikes at the basis for survival. Irrigation is essential to the maintenance of food supplies for the large urban civilizations that have developed in the river floodplains and coastal districts of the re-

gion, particularly in Iraq and Egypt, where the bulk of the population lives in riverine oases threading through the desert zones. The deposit of waterborne silt is a particular threat to irrigation. Clogged canals carry less water, which means that less land can be cultivated. Much of the silt comes from the denuded slopes of the uplands, the forests of which were removed to support the urban centers of the adjacent lowlands. Also seriously affecting irrigation systems in the region are waterlogging and **salinization**. Many farmers have a tendency to use too much water to irrigate their fields. When fields are over-watered, the groundwater table rises. As the soil moisture content increases, salts accumulate in the root zone of crops. Irrigation no longer washes the salts down through the soil to below the root zone. Instead, surface evaporation draws salts toward the surface, where they can accumulate as a crust and reduce crop yields. This is an ancient problem, noted in the fields of Iraq as long as 5,000 years ago.

As more large-scale dams are built on the region's rivers, and over-season storage capacity is expanded, keeping agricultural fields irrigated all year long is much easier. The problems caused by this greater water supply, however, make the

FIGURE A **Salt-encrusted soils.** These fields show the result of w and salinization, which have resulted in the considerable loss of produ agricultural acreage.

portion of the Euphrates; Sudanese development of the Gezira, Rahad, and Khashm el Girba schemes based on diverted water from the Nile and the Atbara; and the Moroccan development of the Oued Sebou in the Gharb Plain.

Animal Husbandry

Although many animals such as cattle, oxen, donkeys, and mules are kept by settled people in the Middle East for food, plowing, and transportation, the bulk of the animal stock tra-

ditionally has been held by pastoral nomads. **Nomadic herding** of sheep, goats, and camels covers large areas, a system encouraged by the spatial extent of aridity that characterizes the region (see Figure 22–8). These nomadic pastoralists are really practitioners of rotational grazing on a grand scale. Often moving hundreds of miles in an annual migration cycle, nomads bring their animals to grass and water that is only available on a seasonal basis in each district. Thus, movement is essential to their existence (Figure 22–10).

balance between irrigation and drainage ever more delicate and harder to maintain. At the present time in the region, every new hectare of land brought into production by the extension of irrigation is matched by the loss of a hectare to salinization or related problems (Figure A).

Other land degradation problems take place when large amounts of sediment are trapped behind new dams, such as the Aswan High Dam in Egypt or the Euphrates dam at Tabqa in Syria. The entrapment of sediments behind the dam not only reduces the dam's storage capacity but also diminishes the development of coastal deltas. The deltas begin to erode, endangering coastal lagoons and tidal marshes that are important breeding grounds for fish. The changed river turbidity downstream of these dams alters the river environment for fish, leading to the death of many species, and also initiates new patterns of stream erosion. Having more irrigation water present year-round rather than just seasonally increases the incidence of such diseases as schistosomiasis and malaria. The occurrence of both of these diseases is increased by perennial irrigation because the new, wetter environment favors the expansion of animal populations that help transmit the disease. In the case of schistosomiasis, it is the snail population that benefits from the new conditions produced by irrigation. Year round irrigation means that there is abundant food throughout the year for the snails. As a consequence, the snail population no longer is limited by a restricted food supply during part of the year and instead can increase to extremely high levels. The snail serves as host to a parasite that invades the bodies of workers who are in prolonged and unprotected contact with irrigation water. In the case of malaria, swamps and small bodies of standing water serve as breeding grounds for mosquitoes that transmit the disease.

Middle Eastern and North African rangeland environments also face serious problems of degradation such as the expansion of dry farming into zones formerly the exclusive domain of pastoral nomads. Aided by mechanization, farmers are able to plow formerly inaccessible tracts. Once the grass and shrub cover is removed by plowing, the field surfaces are very susceptible to wind erosion. Moreover, as nomadic herders are pressed into smaller and less productive rangeland spaces, they are forced to use these areas more intensively. Reduced access to river floodplains, now converted to year-long cultivation, also denies herders access to critical dry season pasture areas. The result is **overgrazing** in the remaining rangeland areas, reducing the quality of available fodder. Efforts to improve the productivity of rangelands by introducing wells and surface water collection projects often backfire. These water provision facilities lead to concentrations of animals whose fodder needs exceed the available local resources. The result-

ing vegetation deterioration, in the absence of control over the number of animals, can lead to serious losses of animals when droughts reduce naturally occurring fodder.

The drawdown of groundwater reserves is another serious problem. In the Jefara Plain of western Libya, overzealous exploitation of groundwater for irrigation purposes has led to a dangerous decline in the groundwater table and the imposition of severe constraints on agriculture. In many oases, the introduction of diesel-powered pumps lowers the groundwater to depths that cannot be exploited by more traditional lift technologies. Often traditional producers as well as formerly productive areas are driven out of production as a result. In coastal districts where urban expansion has heavily increased the extraction of groundwater, the invasion of saltwater into the lowered groundwater table has reduced the quality of water available for drinking purposes.

Tourism (see the *Geography in Action* boxed feature, Tourism in the Middle East and North Africa, in the following chapter), often promoted in many areas as a boost to local economies, can have undesirable impacts on local resources as well. Particularly threatened are Red Sea coral reefs, which are subject to physical damage from diving and which are swamped with sediments and effluents produced by nearby residential and industrial areas.

Also seriously affected by development pressures are the offshore fishing resources of the region's oceans. Dams constructed on rivers cut the flow of organic materials that form the basis for marine food chains, resulting in the decline of fish stocks. Heavy fishing in the region's few prime marine fishing areas is placing fin fish stocks under severe pressure, as well. The capture of floodwaters for irrigation purposes also has an important side effect on adjacent oceans. As less freshwater reaches the sea, the salinity content of nearly enclosed water bodies such as the Mediterranean Sea, the Red Sea, and the Persian Gulf increases, thereby reducing fish stocks. Unfortunately, efforts to establish fishing in the lakes formed by the region's dams have not yet compensated for the decline in offshore fishing. But in recent years signs that fish stocks in the eastern Mediterranean may be recovering somewhat have appeared. Speculation that lavish use of fertilizers in Egyptian agriculture may restore via runoff some of the nutrients needed to feed the base of aquatic food chains is a counterintuitive development that shows how complex management of environmental problems can be.

The story of human impact on the Middle Eastern and North African environment is one of increasing pressure on the region's ability to sustain its populations. The remarkable feature of this environment is its great resilience, its ability to absorb heavy pressure without a complete collapse. In many places, however, this capability is increasingly at risk.

Two major types of nomadic herders exist: vertical and horizontal. Vertical nomadism alternates between winter pastures in the lower elevations and summer pastures in the uplands. These herders follow regular routes through the mountain valleys and passes between their major grazing areas, and they often plant cereal crops in their winter grazing areas. Wealthier nomads frequently buy agricultural land along their route, while less prosperous herders sell their labor to farmers during the harvest season. Nomadic herds bring manure to help

fertilize the fields in autumn, while the stubble on those fields provides fodder for the animals. These same herds are a major source of meat and animal products for the settled population. At the same time, much of the food and equipment used by herders comes from farmers and urban merchants. Thus, many linkages exist between farmers and herders.

Horizontal nomadism uses variations in the availability of grass and water and generally operates in relatively flat areas. Such nomads move from wells that provide water to their

■ **FIGURE 22-10 Kurdish sheepherders, Suleymaniya, northern Iraq.** Nomadic herders move their animals from one pasture to another on a seasonal basis.

herds in the dry season to distant regions where rainfall produces surface water and grass. This movement pattern rests the grazing areas around the dry season wells and ensures herd survival except under the most severe drought conditions. Unlike vertical nomads, who concentrate more on sheep, a species popular in urban markets, horizontal nomads tend to keep camels as their major animal, with goats and smaller numbers of sheep being maintained for subsistence purposes. Camels, not particularly valued as a source of meat, once were important beasts of burden in the caravan trade. Now largely replaced by trucks, camels are still valued as racing animals.

Never encompassing the largest segment of the North African and Middle Eastern population, except in specific areas such as the Red Sea Hills of Egypt and Sudan or the Empty Quarter of Saudi Arabia, nomadism recently has declined in importance throughout the region. Much land once reserved for grazing has been converted to agricultural use. The borders of nation-states now cut across the routes of many nomadic groups, making annual migrations more difficult. The reluctance of many countries to see large groups of nomads moving from one place to another has put a lot of pressure on pastoralists to settle. Thus, many changes have taken place in the nomadic way of life, and many herders have become sedentary (Figure 22–11).

Mining

While a large number of minerals are found in the Middle East and North Africa, they generally occur in small deposits (see Figure 22–8). These deposits were previously important to local populations, but they are quite insignificant today on a global scale. Some that are still mined include chromite in Turkey and mercury in Algeria. Moroccan phosphate exports are significant and command a substantial share of the world market. Turkey is the only country in the region with sufficient domestic coal and iron ore reserves

to support a broadly based industrial economy. In other countries of the region where steel mills have been constructed, such as Egypt and Saudi Arabia, the raw materials are imported. This has not discouraged countries from looking for new deposits in the Sahara Desert and on the Arabian Peninsula. Saudi Arabia, Egypt, Libya, and Algeria have all found rich iron ore deposits in ancient pre-Cambrian shield formations, but their exploitation has been hindered by the extremely arid climatic conditions where they are found and the great distances from the deposits to the coastal locations where they can be utilized. On the whole, the region is not well endowed with most minerals.

The big exception is the widespread occurrence of petroleum. Many parts of the Middle East and North Africa have abundant oil and natural gas deposits, although the distribution of this "black gold" is very uneven (see Figure 22–8). The richest deposits are found in and around the Persian Gulf, and in the Saharan territories of Libya and Algeria (Figure 22–12). From these areas, exports to the rest of the world are made, and enormous revenues are earned for the producing countries. A few other countries have found modest supplies of oil that are sufficient for domestic consumption but insufficient for large-scale export; Egypt, Syria, and Yemen are in this category. The remaining countries in the region, including Jordan, Lebanon, Morocco, Tunisia, and Turkey, are basically "have-not" countries, outsiders to the region's major foreign exchange-earning resource. A more extensive treatment of the significance of petroleum in North Africa and the Middle East is found in the following chapters.

Urbanization

More than 50 percent of the population in the Middle East and North Africa now live in urban areas, and the tradition of urban living has deep roots in the region. This gives the Middle East and North Africa a higher level of urbanization than most other Third World regions, including China, South Asia, and Sub-Saharan Africa. These urban areas grew rapidly in the second half of the twentieth century, and some of the

■ **FIGURE 22-11 Saudi Bedouins.** Across the Middle East and North Africa, many formerly nomadic pastoral peoples have adopted sedentary lifestyles.

■ **FIGURE 22-12 Shaybah oil field in the Rub al-Khali Desert in Saudi Arabia.** Oil and gas often occur together. Here an industrial plant separates natural gas from the crude oil before it is moved via pipeline to coastal refineries. Petroleum is by far the most valuable mineral resource of the Middle East and North Africa, but not all countries within the region share equally in its wealth.

of valuable farmland. Other land is lost when it is mined for sand, gravel, and other materials for building construction. As a consequence, urban growth has absorbed some of the best farmland in a region where prime farmland is at a premium.

Since World War II, migration to the cities has also brought large numbers of recent migrants into the *madina* as many of the *madina's* wealthier professionals, merchants, and entrepreneurs have left the old central district for the nearby quarters once occupied by foreign colonial officials. In countries that did not experience colonialism directly, such as Saudi Arabia, many people have also left the central city for more spacious urban and suburban districts with larger and more modern housing. The vacated inner-city properties have been filled by large numbers of rural migrants, often turning ancient palaces and other structures into multifamily dwellings. The resulting increase in population densities has produced extremely crowded living conditions, often

region's cities, most notably Cairo, Egypt, Tehran, Iran, and Istanbul, Turkey, now rank among the world's largest urban areas (see the *Geography in Action* boxed feature, Cairo: Dryland Metropolis in Distress). Rapid urban growth in the Middle East is the result of high rates of natural increase and large numbers of rural dwellers shifting their residence to the city.

This rural-urban migration has had several consequences. One is the development of squatter settlements on the outskirts of urban areas. These structures are generally illegal in the sense that they are usually situated on underused private or state land without the approval of the landowner(s). Few if any municipal services, such as electricity, water, sewerage, health-care facilities, or schools, are provided, at least in the initial stages. Occasionally governments try to remove such settlements, but the long-term pattern is for these communities to become stable and permanent. From an initial stage of transitional construction with flimsy materials, most houses are rapidly transformed into stone or concrete structures. The more durable the structure, the more likely that urban authorities will recognize the reality of occupancy and grant legal tenure to the inhabitants. In the oil states, national governments have been able to afford the costs of providing subsidized housing for all who seek it. The non-oil states have fewer resources and often large populations to care for and have been unable to provide sufficient housing, infrastructure, and other services. Many of the rural migrants possess considerable resources and, if given tenure to the land that they occupy, are able to erect their own housing. When recognized by the state, such squatter communities are often able to build neighborhoods of decent living levels.

In this way, many Middle Eastern cities have expanded considerably around their ancient *madina*, or urban core (Figures 22–13 and 22–14, page 534). Often this growth has extended into nearby agricultural areas, resulting in the loss

■ **FIGURE 22-13 Khan al-Khalili Souk (market), Cairo, Egypt.** Many cities of the Middle East and North Africa are centered around a traditional urban core called the *madina.* One of the most distinct elements of the *madina* is the *souk,* or marketplace, where a great variety of foods, handicrafts, and manufactured items are bought and sold.

Cairo: Dryland Metropolis in Distress

Fast approaching an estimated 15 million inhabitants or more than 20 percent of Egypt's total population, Cairo is by far the largest urban agglomeration in North Africa and the Middle East and a prime example of the difficulty of maintaining a sustainable metropolis in a dryland country with limited resources.

Steeped in thousands of years of history, Egyptians affectionately call their capital *um ed-dunya,* "Mother of the World," much like the way Bostonians refer to their home town as the "hub of the universe." Cairo was founded at the prime location where the Nile's wide delta—an area about twice the size of Egypt's Nile Valley—fans out east and west and the many-armed waters of the Nile begin their final 125-mile (200-kilometer) journey to the Mediterranean Sea. This is Cairo's and Egypt's prime agricultural hinterland that has played such a major role as the country's breadbasket over the centuries. These days, however, more than 40 percent of Egypt's inhabitants live here, a population that is still fast growing at an annual rate of 2 percent.

Between metropolitan Cairo, the thickly settled delta, and the inability to expand into the mostly waterless expanses of the desert that make up most of the country, it is thus not surprising that urban sprawl; housing, infrastructure, and industrial development encroaching on prime agricultural land; traffic congestion from the daily swells of urban commuters only partly alleviated by the construction of an underground Metro light rail system; environmental pollution; and waste disposal problems loom in much magnified form (Figure A). It is increasingly difficult for the "Mother of the World" to maintain a reasonable quality of life for its residents who are steadily increasing in numbers. Infrastructural and social services are not keeping pace.

Affordable housing is one of the major problems. Like in so many other metropolitan areas of the world, the gap is widening between the "haves" and "have nots," those who can afford gentrified, high-quality residences and those who cannot find adequate housing. Greater Cairo is dotted with more than a hundred informal settlements consisting of illegally built accommodations that house 80 percent of the population without benefit of running water, proper sewage disposal, and electricity. Many of the poor live illegally in tombs in Cairo's vast City of the Dead. At the same time many better quality apartments in the suburbs and satellite towns around Cairo remain empty as entrepreneurial landlords are waiting for more affluent renters.

With all manner of land uses concentrated in the Cairo and delta area and competing for prime locational space—40 percent of Egypt's industry and 32 percent of the country's vehicles are accounted for in metropolitan Cairo alone—pollution of all kinds is creating major health issues. Cairo is among the most polluted cities in the world, though some strides are being made in relocating noxious industrial facilities away from the city and applying tighter emissions regulations.

Source: F. and B. Ibrahim. 2003. *Egypt: An Economic Geography* (London/New York: I. B. Tauris).

in buildings that are rapidly deteriorating (Figure 22–15). Lax inspection practices and poor maintenance by absentee landlords have occasionally resulted in spectacular structural failures. In the older urban areas, streets not designed for automobiles and trucks have become increasingly congested, producing both poor safety conditions for drivers and pedestrians and serious air pollution problems.

Historical Overview

An Economic and Cultural Crossroads

The Middle East and North Africa (MENA) constitute a common meeting ground linking Europe, Sub-Saharan Africa, and Asia. The use of "middle east" to designate the eastern portions of this larger region is something of a historical accident. In the nineteenth century Europeans divided the "orient," the world east of Europe, into near, middle, and far east. In this division, the near east comprised the area now generally regarded as the Middle East and North Africa. Early in World War II, difficulties sending troops quickly from Britain to defend Egypt from attacks by Italian and German forces resulted in shifting troops and command staff from India to Cairo. Dispatches from this Middle East command retained their traditional by-line, as did the articles filed by journalists, and over time popular identification of the Middle East with the eastern Mediterranean took deep root. Today this is reflected in the widespread use of the term Middle East in popular literature, by academic programs engaged in area studies, aid organizations, and most international organizations. Because the region is in an intermediate location between three major cultural and continental-scale areas, use of the term "middle east" retains a certain serendipitous geographic logic.

This interactive position is reflected in the connections the Middle East and North Africa have historically had with the cultures of adjacent land masses. With Europe, the region shares many common scientific and literary traditions; much of the ancient cultures of Greece and Rome were preserved by Islamic scholars. In addition, there was trade across the Sahara, as well as overland trade between the Mediterranean Sea, the Red Sea, and the Persian Gulf that connected European economies with those of Africa and Asia. The expansion of Islam from its origin in western Saudi Arabia into

■ **FIGURE A** **Traffic congestion in Cairo, Egypt.** Cars, trucks, handcarts, and pedestrians compete for space in an effort to make progress through the congested streets of Cairo's neighborhoods. In the last decade, attempts to reduce gridlock in Cairo's streets have encouraged the construction of subway lines. This metro train compartment is reserved exclusively for women.

adjoining regions also served to link the surrounding areas together in patterns of conflict and accommodation. Islam's expansion eastward into Pakistan, India, Bangladesh, Indonesia, and the Philippines promoted trade contacts between the Middle East and Monsoon Asia and provided a network of connections maintained by pilgrimage and a sense of religious brotherhood. The contemporary expansion of Islam into Sub-Saharan Africa maintains the connections established by the trans-Saharan caravan trade and Omani mercantile contacts with East Africa many centuries earlier. The legacy of Muslim connections with Europe is found not only in the architectural and cultural survivals in Spain and Portugal, but also the Muslim communities in Albania, the former Yugoslavia, and Bulgaria. These latter products of crossroads contact historically have generated serious political and cultural problems within the Balkans, which resurfaced in the 1990s. Similarly, European colonization in the nineteenth and twentieth centuries had a highly disruptive effect on North African and Middle Eastern societies. The conflicts sparked by that contact continue to disturb the region despite the ending of the colonial era. The decline of Beirut, Lebanon, as a regional banking and commercial cen-

ter linking Europe and the Middle East is just one illustration of the difficulties associated with a **crossroads location**, where contrasting cultures meet and often clash.

Ethnicity and Identity

Much of the world views the Middle East and North Africa as a region inhabited primarily by **Arabs**. The Arabs are indeed the largest single ethnic group in the region, accounting for more than half of its approximately 400 million inhabitants. Grouped politically into the Arab League, speaking a common Semitic language, and possessing a shared history and culture, the Arabs certainly are a readily identifiable group. But this sense of unity is somewhat misleading. While Arabic is a shared language, many regional dialects, often mutually incomprehensible, exist. Despite the existence of an umbrella political organization in the form of the Arab League, Arab politics are highly variable. Monarchies (Morocco, Saudi Arabia, Oman, Jordan), one-party states (Algeria, Tunisia, Egypt, Syria, until recently Iraq), military dictatorships (Yemen, Libya), religiously based states (Iran), secular states (Lebanon), and the relatively powerless proponents of

533

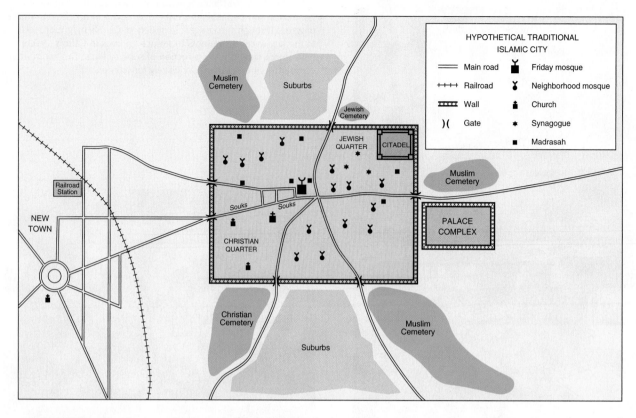

■ **FIGURE 22-14 Diagram of a hypothetical traditional Islamic City.** The major components of the traditional Islamic city include the main residential quarters, markets (*souks*) located along a complex of streets near the main mosque; Islamic schools (*madrasahs*); a citadel; suburbs and cemeteries outside the main gates; and a palace complex where the ruler resides. Attached to the *madina,* or urban core, is a modern town organized on the European pattern.

multiparty democracy all vie for attention and compete for influence. Believing that they share a common culture and history, most Arabs long for a political unity that reflects their perceived oneness. Yet this pan-Arab nationalist goal has been very hard to achieve. Attempts to form a meaningful union among the Arab countries have consistently failed.

Many other ethnic groups are also found in North Africa and the Middle East (Figure 22–16). Where the Arabs occupy the central core of the region, having spread out from their ancestral homeland in the Arabian Peninsula, the major minorities of Turks, Persians, Kurds, Berbers, and Nilo-Hamitic groups are found in more peripheral locations.

■ **FIGURE 22-15 Cihangir Quarter of Istanbul.** Recent migration to the cities has resulted in extremely high urban population densities and has produced crowded living conditions.

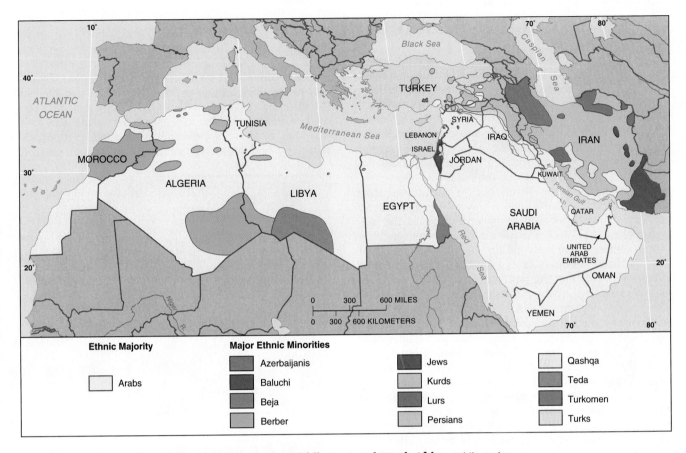

■ **FIGURE 22-16 Major ethnic minorities in the Middle East and North Africa.** While Arabs are the largest ethnic group of North Africa and the Middle East, the region contains numerous ethnic minorities. Most of these are found in the higher mountainous areas where physical isolation has protected them from cultural assimilation.

Turks and **Persians** are the two largest non-Arab groups, dominating the countries of Turkey and Iran, respectively. A substantial majority within their own country, Turks are important minorities elsewhere in the region and in Europe. The presence of Turks in the Balkans, for example, reflects the former extent of the Ottoman Empire, the predecessor state of modern Turkey. The widespread presence of Turks in German urban areas is a product of post-World War II labor migration. While Turks and Persians are the largest ethnic groups in Turkey and Iran and dominate the local political and economic scenes, each country contains sizable linguistic and ethnic minorities.

In the mountainous regions of southeastern Turkey, large numbers of **Kurds** (about 20 percent of the Turkish population) are found, a pastoral and agricultural people split among several adjoining countries. Smaller numbers of Arabs (mostly around Adana on the Mediterranean coast near Syria), Greeks (in and around Istanbul), Circassians and Georgians (in the northeast), and Armenians (in the southeast) also are part of Turkey's ethnic mix. Kurds also are concentrated in northeastern Iraq (17 percent of the Iraqi population) and in the northwestern border zone of Iran (7 percent of its population), where they occupy mountainous areas that are difficult to reach from the outside (Figure 22–17). Smaller numbers of Kurds are found in Syria, Armenia, and Azer-

baijan. Possessing an Indo-European language as do the Persian speakers of Iran, Kurds have long agitated—sometimes violently, but always unsuccessfully—for autonomy within Iraq and Iran and for independence from Turkey. At present, Iraq's Kurdish political parties seem prepared to work within the structure of a unified Iraqi state, but it is unclear what degree of internal autonomy might be needed to satisfy their longer-term aspirations. Since both substantial autonomy within Iraq and outright independence for Iraqi Kurds are likely to be viewed as a potential long-term threat to the internal unity of both Turkey and Iran, the bounds of what is possible and desirable to Kurds or would be permitted by their larger and more powerful neighbors remain obscure.

Iran, although dominated by speakers of Farsi and related Persian dialects (60 percent of the total), has an equally diverse population. Azerbaijani Turks, concentrated in the northwest corner of the country, are the largest minority (24 percent), while Turkish-speaking Turkoman (2 percent) and Qashqa (1 percent) communities are regionally prominent in the northeast and southwest, respectively. Baluchi (2 percent) are widespread in the arid southeast of Iran, while Lurs (2 percent) and other smaller ethnic groups are scattered through the rugged Zagros Mountains of western and southwestern Iran. Arabs, largely concentrated in the lowlands east of the Tigris River, make up 3 percent of the Iranian

GEOGRAPHY IN ACTION
The Reassertion of Islamic Values

Perhaps the most significant development influencing the Middle East is the reassertion of traditional values by fundamentalist Islamic groups. That powerful force, which has come into sharp view in many countries and which the Western public and popular press are increasingly recognizing, has deep roots. The Islamic resurgence represents a reaction to centuries of increasing contact with and pressure from Western European and North American values and institutions.

That contact took many forms, but the initial consequence was to erode the self-confidence of Islamic states and societies and to undermine their ability to defend themselves. The rapid industrialization of Western societies in the eighteenth and nineteenth centuries gave them military superiority over their traditional Middle Eastern rivals. As a result, the Ottoman Empire gradually lost its outlying provinces and tributary states as the caliphate—the secular and religious leadership of the Islamic community—proved incapable of defending the House of Islam against the encroachments of Austria, France, and Russia. Christian populations reemerged to form non-Islamic states (Lebanon, Armenia, Greece); foreign powers asserted themselves as the protectors of Christian minorities within the boundaries of the shrinking Ottoman empire; missionaries, both Protestant and Catholic, attempted to win converts; and French and Italian colonists arrived to settle the best agricultural lands in Algeria and Libya. Both France and Italy considered those territories to be integral parts of the mother country, rather than colonies, and had limited respect for indigenous Islamic institutions and values.

Where military power led, economic penetration followed. By the twentieth century, most of the Middle East and North Africa was divided into political and economic spheres of influence or control, and the modern sectors of the economy, particularly the oil industry, were dominated by Western interests. Most Muslims (believers in Islam) who sought a modern education were forced to find it in schools permeated with Western values, emphasizing individualism; separation of church and state; equality of the sexes; material and technological progress;

■ **FIGURE A** **Men praying in a mosque in Riyadh, Saudi Arabia.** Islam remains a powerful force throughout the Arab world. Fundamentalists in many countries seek the elimination of secular values, especially those associated with the West, and a return to traditional values and practices.

population. The result is a core of cultural homogeneity surrounded by very substantial regional diversity.

A smaller but regionally prominent ethnic minority in the Middle East is the Jewish populace. Concentrated in Israel, **Jews** constitute 76 percent of that country's population.

■ **FIGURE 22-17** **Kurdish bazaar, Erbil, Iraq.** Kurds constitute a sizable minority of the populations of Turkey, Iraq, and Iran, where they have long sought greater political autonomy.

This population contains both Jews of European origin and those from North Africa and the Middle East who migrated to Israel after it gained independence in 1948, a process of return that continues today. Small Jewish populations are still found in most of the other countries of the region. Most of the non-Israeli Jews live in urban areas, the remnants of much larger and commercially more important populations predating Israeli independence (Figure 22–18). The remainder of Israel's population is composed of Arabs (20 percent) of different religious faiths and an immigrant population (4 percent) whose ethnic and religious origins are not known.

Other small minority communities are scattered through the region. Armenians, once an important element in the population of eastern Turkey, are now only a residual community. Large numbers of Armenians perished in a series of genocidal "ethnic cleansing" attacks in World War I. Survivors dispersed throughout the region, where they are prominent in urban areas, particularly in Lebanon and Syria. Circassians, nineteenth-century Muslim refugees from the Caucasus Mountains between the Black and Caspian seas, are found in Syria and Jordan. Many of the gulf states, including Kuwait, Qatar, and the United Arab Emirates, have transient populations of non-Arab foreign workers (Iranians,

rationalism; and secular nationalism. Traditional Islamic knowledge and understanding of state and society were set aside, despite protest and resistance.

Traditional voices were never completely silenced. With the emergence after World War II of movements advocating political independence, proponents of more traditional Islamic values also have gained increasing influence (Figure A). Successful leaders of that time—such as Muhammad V in Morocco, Habib Bourguiba in Tunisia, or Gamal Abdel Nasser in Egypt—were able to use traditional values in a nationalist context to defy the Western powers. Increasingly, though, fundamentalist Islamic forces now demand a return of the Sharia, the traditional legal code based on the Quran, in place of legal systems derived from European models. They also demand intensified resistance to Israel and increased support for the Palestine Liberation Organization (PLO), the widely recognized political organization representing most Arab Palestinians. The more fundamentalist the Islamic group or nation-state, the more likely it is to oppose the prospect of peace and compromise with Israel, for to Islamic purists Israel constitutes an alien entity in the Islamic body politic. Israeli control of Islamic holy places in Jerusalem is a constant source of humiliation and a reminder of past political, technological, and military inferiority.

In each country, the particular local situation has determined which forces have dominated the fundamentalist revival. In Iran in the 1970s, the Shi'ite clergy, led by the Ayatollah Ruhollah Khomeini, spearheaded resistance to the secular policies of the shah, or king, and opposed the inequitable distribution of wealth generated by the oil economy. In Egypt, a fundamentalist group on the fringes of the Muslim Brotherhood assassinated President Anwar Sadat in 1981, in part because he signed a peace treaty with Israel. In Tunisia, the primary effective opposition to President Bourguiba was confined to religious groups, until an unexpected coup within the ruling

elite removed the ailing, aging president. In Syria, resistance to the authority of the secular Ba'ath party, which contains a significant proportion of members of the minority Alawite sect among its leaders, is maintained by orthodox Sunni movements. And in Algeria, fundamentalist political groupings have posed a serious challenge to the current one-party socialist state structure.

The influence of the Islamic revival also extends to regions and countries outside those normally regarded as Islamic. Muslim missionaries often have a fundamentalist perspective and increasingly act to radicalize Muslim minorities. Several violent confrontations between radical Muslims and government forces in northern Nigeria are an example of that process. A variant of the theme is the fighting in southern Sudan, where Christian and animist rebels feel threatened by the government in the north, which is influenced by the Muslim Brotherhood. In Iraq, the removal of the authoritarian and secular Ba'ath party of Saddam Hussein has given increased power to Shi'ite clerics, and it has unleashed violent opposition to the U.S. military presence. This resistance is more likely to come from Sunni communities than from Shi'ite. An element in the mix are Mujahideen (Islamic warriors) attracted to Iraq by the opportunity to confront the United States directly. These fundamentalist Muslim fighters are prepared to die for their beliefs and thus they often provide the human resources for suicide bombing attacks. They also are receiving invaluable on-the-job training in urban combat skills, which can readily be transferred to other locations in the Middle East and beyond in the future.

In short, no country in the Middle East is without the tension generated by an increasingly intense struggle between secular and religious forces. The outcome of that struggle will determine the nature of both internal politics and international relationships.

Pakistanis, Bengalis, Filipinos, and Indians), as well as Palestinians, Egyptians, and Yemenis living among them.

In North Africa, the ethnic scene is somewhat simpler. Much of the indigenous population has, over many centuries, been assimilated through the adoption of the Arabic language and culture. Still, Berber speakers remain a significant minority, numbering somewhat more than 35 percent of the total population of Morocco. **Berbers**, concentrated in mountain districts or in the central Sahara, are a less prominent ethnic group in Algeria (approximately 25 percent), Tunisia (2 percent), and Libya (1 percent). In south-central Libya, the nomadic Teda, who speak a Hamitic language, dominate the northern slopes and foreland of the Tibesti massif.

Religion and Identity

Religion is an extremely important part of self-identity for many people. Just as not all Middle Eastern and North African people are Arabs, so too not all are Muslims. Nonetheless, **Islam** is the dominant religion in the region. Established by the prophet Mohammed, who began to have divine visions in A.D. 610, the community of believers practicing Islam (which literally means submission to the will of

God, Allah) formally began in 622 on the Arabian Peninsula, when the Hegira (flight) occurred. This event was the forced departure of Mohammed and his followers from Mecca to Medina to the northwest. From his new base in Medina, Mohammed eventually conquered Mecca and expanded the area of Islamic control to include, at the time of his death in 632, most of the western half of the Arabian Peninsula. His successors, the Caliphs, rapidly extended the area of

■ **FIGURE 22-18 Jewish synagogue in Tunisia.** In addition to forming the majority population of Israel, small Jewish communities are still found in many cities of the Middle East and North Africa.

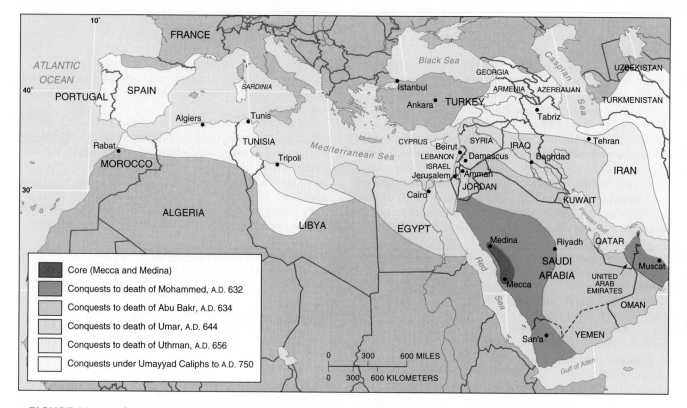

■ **FIGURE 22-19** **The expansion of Islam from A.D. 622–750.** Islam expanded rapidly from its core in the western Arabian Peninsula to dominate most of North Africa and the Middle East by the mid-eighth century.

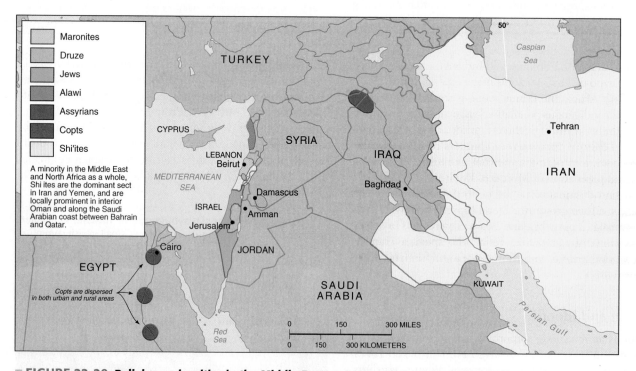

■ **FIGURE 22-20** **Religious minorities in the Middle East.** Only the spatially most prominent religious minorities appear on this map. The most prominent are the Shi'ites of Iran (90 percent of the population) and southern Iraq. Coptic Christians are a significant minority in Egypt (around 10 percent of the population). Jews are a minority in the region, largely concentrated in cities, but constitute a majority (76 percent) of the population of Israel.

Islamic political control to the east and west until, little more than a century after the Prophet's death, the Umayyad Caliphate stretched from Spain to beyond Afghanistan (Figure 22–19).

The faith established by Mohammed was and is appealingly simple. At its root are the **Five Pillars of Islam**—the creed, prayer, charitable giving, fasting, and pilgrimage—that govern the behavior and belief of the faithful. The fundamental tenet of faith is the *shahada*, the creed, which states that "There is no God but God, and Mohammed is his prophet." **Muslims** believe that Allah (God) is a transcendent, divine being, surrounded by angels and prophets, and Mohammed was the culmination of a tradition of prophetic ministry that began with the Jewish Old Testament prophets, continued through the teachings of Jesus of Nazareth, and reached its fullest form in the *sura* (books or chapters) revealed to Mohammed and collected in a sacred text, the **Quran**. The second pillar of faith is the offering of regular prayer, which takes place five times a day. *Zakat* (alms) is the third pillar, and expresses the responsibility of the more prosperous Muslim to aid those less fortunate. Fasting during the month of Ramadan, the ninth month of the lunar calendar, is the fourth pillar. During this period, the practicing Muslim avoids eating, drinking, and smoking between sunrise and sunset in order to exercise command over bodily cravings and to subject them to spiritual control. Finally, at least once, if physically and financially able, the Muslim is expected to perform the *hajj*, the pilgrimage to Mecca, where visits to sacred sites and religious devotions predominate. From the Quran is derived the **Sharia**, the rules and laws that are to govern the society in which Muslims live. To devout Muslims there is no distinction between the religious and the secular, between church and state; rather, both spheres are part of one integrated whole, which should be governed by the Sharia.

It is this belief in the religious state that is at the root of the conflict between secular and fundamentalist forces within the region (see the *Geography in Action* boxed feature, The Reassertion of Islamic Values, page 536), just as in earlier times when a dispute over power and authority within the Muslim community led to a division into **Sunni** (the orthodox, who believe that potentially any practicing Muslim can exercise power) and **Shi'ite** (those who believe that only descendants of Mohammed should hold political power). Today Sunnis constitute about 90 percent of all Muslims, although Shi'ites are particularly important in Iran, where nine out of ten Muslims are Shi'ites, southern Iraq (more than 50 percent of the population), southern Lebanon (30 percent of the population), Yemen (60 percent of the population) and the interior of Oman, and the gulf coast districts of Saudi Arabia (approximately 6 percent of the population) (Figure 22–20). Most of the Shi'ites in Iran are Persian; in the rest of the Middle East they are Arabs. Other Arabs include the Alawi, an Islamic sect living in the coastal mountains of Syria and totaling 11 percent of that country's citizens, and the Druze, an Islamic offshoot constituting 6 percent of the Lebanese, 3 percent of the Syrian, and 1.7 percent of the Israeli, populations. Similarly, most North African and Mid-

■ **FIGURE 22-21 Coptic Christian service, Wadi Natrun, Egypt.** The Coptic Christian liturgy is based on the Greek rite. The faith is of great antiquity in Egypt and has a strong monastic tradition.

dle Eastern Christian communities, regardless of the rite practiced, are Arabs. The most important of these are the Maronites (30 percent of the population) of Lebanon and the Copts (10 percent) of Egypt (Figure 22–21).

▶▶ Summary

North Africa and the Middle East together are a complex region. Superficially, the region appears to be largely a desert wasteland inhabited by one people, the Arabs, who are uniformly Muslim in religion. In reality, the region is characterized by diverse physical environments, many with considerable productive potential that contradict its desert image. In these more productive zones, great population densities are found. The cultural diversity of the region's human population is as great as its environmental variation. More than one-third of the region's population belongs to ethnic, linguistic, or religious minorities that are either not Arab or not Muslim. The two chapters that follow examine these diverse environments, cultures, and economies in greater detail.

▶▶ Key Terms

aquifer	nomadic herding
Arabs	oases
aridity	orographic precipitation
Berbers	overgrazing
crossroads location	Persians
deforestation	*qanat*
dry farming	Quran
exotic rivers	rainshadow
Five Pillars of Islam	run-on farming
irrigated agriculture	salinization
Islam	Sharia
Jews	Shi'ites
Kurds	Sunnis
madina	Turks
Muslims	

■ Mediterranean agriculture in Turkey. Citrus orchards and olive groves in the foreground complement the wheat fields and small vineyard in the distance, a typical combination of crops in the Mediterranean basin.

Chapter 23

The Mediterranean Crescent:
Maximizing Limited Resources

- **Large States, Integrated Economies**
- **Small States, Unique Economies**

The Middle East and North Africa together constitute a distinctive region within which we recognize two subregions: the countries of the Mediterranean Crescent along the Mediterranean coast and the Gulf States around the Persian Gulf (Figure 23–1). North Africa and the countries along the eastern end of the Mediterranean Sea share a long history of cultural and economic interaction with Europe, whereas the Arabian Peninsula and the countries along the shores of the Persian Gulf became more prominent in the world with the discovery of oil.

The Mediterranean Crescent countries encompass the North African states of the **Maghreb** (Morocco, Algeria, Tunisia) as well as Libya and Egypt. The subregion also includes the smaller states of the eastern Mediterranean, the **Levant**, as well as Turkey, a unique Middle Eastern country in that it possesses a physical foothold in Europe and has associated membership status in the European Union. All of the countries in this broad spatial and political grouping share strong historic ties to Europe. They were once part of the Roman Empire and shared in a regionwide culture and political economy. During the Ottoman Empire, most of North Africa was grouped with the Balkan lands of southeastern Europe and the eastern Mediterranean into one political unit. These historical ties are symbolized in the contemporary landscape not only by extensive Roman monuments and ruins but also by similarities in house types, traditional clothing styles, irrigation and other agricultural production technologies, legal arrangements, and patterns of trading and raiding over centuries. Today many of the citizens of these countries, most notably the Maghreb states and Turkey, are migrants to Europe in order to find employment. They constitute an important source of laborers in France, Germany, and other countries, and the remittances that they send home to families in their homeland are a significant source of income to the recipient countries. These patterns of interaction and movement also work in the reverse direction, for many Europeans travel to Morocco, Tunisia, Egypt, and Turkey for vacation. Attracted by the warmth and sun of the Mediterranean winter season, these tourists contribute in important ways to local economic prosperity. The Mediterranean Crescent is also linked to southern Europe by a shared climatic and agricultural regime. Dry in the summer and autumn, its precipitation is concentrated in winter and spring. Throughout the subregion, mountains close to the coast trap this moisture and make it available for the traditional crops of wheat, olives, and grapes. Historically, these coastal districts have been staging points for trade across the desert to Sub-Saharan Africa and South Asia. Today a rich variety of contrasts and similarities link the Mediterranean countries one to another. This diversity of opportunity and experience is the focus of our chapter on the Mediterranean coastal countries.

Large States, Integrated Economies

We will now discuss three of the largest countries in the Mediterranean Crescent—Algeria, Egypt, and Turkey—which have many basic similarities. Thereafter, we examine four of the smaller remaining states—Israel, Lebanon, Jordan, and Syria—as well as the possibility of a future Palestinian state. Morocco, Tunisia, and Libya share numerous similarities with these other Crescent states and are not discussed independently.

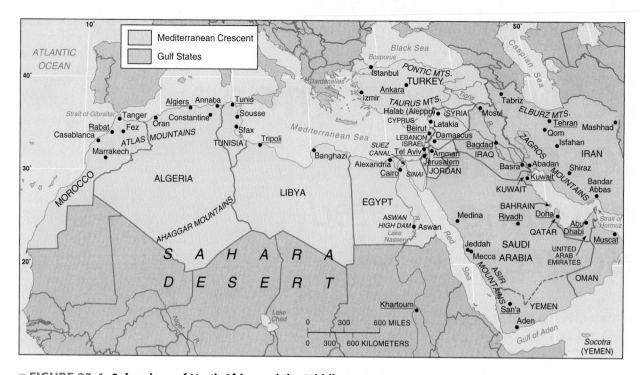

■ **FIGURE 23-1 Subregions of North Africa and the Middle East.** The Mediterranean Crescent sub-region extends around the Mediterranean Sea. The Gulf States comprise Iran, Iraq, and the countries of the Arabian Peninsula.

Features of Similarity

One important common feature of Algeria, Egypt, and Turkey is that they are relatively self-supporting, as a result of three factors: an important industrial sector that processes raw materials; a diversified and productive agricultural base; and a developed service sector. Together with a large population that includes a significant proportion of well-educated individuals, such a diverse natural and human resource base provides the foundation for a balanced, integrated economy.

A second common feature is that a high proportion of the populace of each country lives in urban areas, the product of a long, rich history of urban life. These three countries contain several of the region's largest cities: Cairo, Egypt; Istanbul, Turkey; and Algiers, Algeria. These primate cities dominate their national landscapes. Massive immigration from the countryside has swelled those urban populations far beyond the capacity of urban planners to calculate accurately and beyond the ability of urban systems to absorb satisfactorily. Uncounted, underprivileged people have ballooned the population of Cairo's metropolitan area to an estimated 15 million inhabitants and Istanbul's to 11.2 million. In addition, other urban centers have attained major size and importance: Ankara, Izmir, and Adana in Turkey; and Alexandria and Giza in Egypt.

Because Algeria, Egypt, and Turkey all have large populations—32.3 million, 73.4 million, and 71.3 million, respectively—their cities serve complex functions. They are administrative, industrial, cultural, trading, and often religious centers. Their economic role is enhanced by their countries' large, well-integrated, and relatively powerful economies.

A third common feature is that each country has a rich agricultural base, which is essential both to feed the local population and to earn foreign exchange in overseas markets. Egypt's cotton, Turkey's citrus fruits, and Algeria's wines, dates, and olives are important export commodities; and the domestic food-processing industries are a significant focus of development. Nonetheless, in all three countries (particularly in Egypt, where the pressure of population is especially severe), maintaining and increasing agricultural productivity is a continuous struggle.

A fourth significant common feature is each country's appreciable mineral wealth. Algerian oil and natural gas are important world resources. The coal, iron ore, and chrome of Turkey are less abundant but provide the base for the largest heavy-industry complex in the Middle East. Egypt's oil and phosphates meet internal needs and, together with hydroelectric power from the Aswan Dam, are important bases for economic growth.

Finally, each country is characterized by a large area of semi-arid and arid territory of little value to the agricultural economy. Thus, agriculture tends to be concentrated in limited areas and to be more intensive than it is in many other countries of Africa and Asia. Egypt—with only the narrow, fertile strip of the Nile Valley and its delta—is the extreme example.

Algeria: A Maturing Economy

More than a century of cultural conflict and discontinuity, terminated by a lengthy struggle for independence, severely interrupted the indigenous patterns of life and development in Algeria. Algeria's major cultural conflict began when the country became a French protectorate in 1830. Large-scale migration of Europeans to Algeria followed, and tremendous pressures were placed on local cultural institutions. Today, despite freedom from French control since 1962, Gallic (French) civilization maintains a very visible presence in Algeria. The resulting French-Arab conflict mirrors the long opposition of the Berber culture, protected by its rugged mountain retreats, to a succession of Phoenician, Roman, and Arab invaders. Ironically, that guerrilla opposition to the French has done much to reduce the differences between Arabs and Berbers in Algeria and to speed the assimilation of Berbers into the mainstream of Arab culture (Figure 23–2).

In some ways Algeria is similar to other oil states. It has enjoyed a large income from oil exports and has tried to use that wealth as the basis for economic growth. But Algeria has gained a position in the world far above that which might be suggested by its size and economic status. One reason for that position is Algeria's long struggle for independence from French domination. As a result of that struggle, Algeria sympathizes with and is respected by many Third World revolutionary groups. Another reason for Algeria's position is the strong and expanding economy that Algeria is building, made possible by the good mixture of resources available to the nation.

Algeria has the most diverse and integrated economy of the North African countries. With Libya, Algeria shares a wealth of petrochemical resources (Figure 23–3). Like its near neighbors, Morocco and Tunisia, Algeria possesses rich agricultural resources in the northern and coastal sections of the country (Table 23–1). The combination of minerals (iron ore, phosphates, lead, zinc, and mercury), coupled with petroleum and farming, provides particular breadth and richness to the Algerian resource base in North Africa (Figure 23–4).

About 80 percent of the country is classified as too dry or too steep to be productive for crops or pasture. Nevertheless, the desert portions of Algeria are vital to the national economy. Of some importance are the high-quality dates produced in small desert oases for export, but the major resources of the desert are oil and natural gas (Table 23–2). Those two commodities account for most of the value of the country's exports. Algeria has been a pioneer in the production of liquefied natural gas (LNG) for export to industrialized states, with much of its LNG being shipped to the United States. To reduce that market dependence on the

■ **FIGURE 23-3 Oil rig in the Algerian Sahara.** Much of Algeria's interior, except for a few oases, is too dry to support agriculture. But substantial oil and natural gas deposits are found in precisely such desert areas, and the extraction of these resources makes an important contribution to the Algerian economy.

Table 23-1   Land Use (in Thousand Hectares) in Algeria, Egypt, and Turkey

| | Labor Force in Agriculture | | | | |
| | Total (000) | | % | | Land Use |
Country	1990	2000	1990	2000	2000
Algeria	1,827	2,545	26.1	24.3	
Total Land Area					238,174
Arable Land					7,525
Permanent Cropland					520
Other[a]					230,129
Egypt	7,899	8,591	40.3	33.1	
Total Land Area					100,145
Arable Land					2,825
Permanent Cropland					466
Other[a]					96,854
Turkey	13,012	12,426	53.6	39.8	
Total Land Area					77,482
Arable Land					24,138
Permanent Cropland					2,583
Other[a]					50,761

Source: Food and Agriculture Organization of the United Nations, *www.fao.org/statistical databases/faostat.*
[a]Other—Seasonal pasture, forest, and wasteland

United States, a pipeline to Italy was completed in 1981. Most of the revenues earned by the sale of oil and natural gas are invested in the northern part of the country, where the bulk of the population and the productive agricultural land are located. Declining oil reserves and inefficient, government-controlled industries have contributed to much slower industrial growth for the last 15 years and have resulted in economic stagnation.

The northern region of Algeria is relatively fertile and well watered. Because soils are often thin and slopes are steep, it is difficult to farm this potentially fertile area without encountering environmental problems. Once the granary of the Roman Empire, the area today grows wheat and other cereal crops, especially in the drier plateau country between the Saharan Atlas and the coastal mountain ranges. The coastal climate is ideal for citrus fruits, grapes, olives, and other fruits and vegetables, and European markets are close at hand. But penetrating those markets is a problem; the preferential treatment given by France during the colonial era has disappeared, and the membership of Spain and Portugal in the European Union has put Morocco, Algeria, and Tunisia at an increasing disadvantage.

Outside the commercial agricultural zone, many peasants till poor soils for a low subsistence level of life. As a result, many people have left those areas and moved to the coastal cities, where economic and social opportunities are greater. Also, many Algerians, as well as Moroccans and Tunisians, have migrated to France. Their labor is important to the French economy, and the money they send back home sustains the family members left behind. However, recent restrictions on migration to France place this source of income increasingly at risk.

The basis for industrial growth clearly exists in Algeria. Oil and gas resources are backed by phosphates and major iron ore deposits; and, together with lesser quantities of coal, lead, and zinc, those resources form a solid base for the development of heavy industry and metal fabrication. In addition, worker skills and infrastructure are sufficient to allow the major development initiative to remain firmly in Algerian hands. But in recent years, the involvement of the central government in the economy and an excessively large bureaucracy have held back industrial change. This stagnation has intensified as a result of the rise of antigovernment terrorism, which not only targets government officials, military figures, and prominent supporters of the military regime but also attacks foreigners. This violence discourages foreign investment and makes it difficult to attract outside managerial expertise when needed. The political turmoil also undermines the desire of tourists to visit and contributes to stagnation in business activities that depend on tourist income, such as hotels, resorts, local crafts, and transportation.

A distinctive feature of Algeria in its immediate postindependence years was the single-minded political strength, wisdom, and insight of its leaders. Those characteristics enabled the government to demand and receive major internal sacrifices in return for the promise of long-term benefits. Accordingly, some 40 percent of Algeria's economic output was reinvested annually in the development of heavy industry. But the country's yearly industrial growth rate of 7 to 10 percent during the 1970s proved difficult to duplicate in the 1980s because a world oversupply of both oil and gas reduced prices and diminished the funds available for investment. Moreover, Algeria's oil and gas resources are not

■ **FIGURE 23-4 The commercial/market center of Ghardaia in the Algerian Sahara.** This busy open-air marketplace, surrounded by shops under the arcades, reflects the prominent business and trade roles that merchants and shopkeepers play in the major and minor urban areas of the countries of the Mediterranean Crescent.

unlimited, and concern has developed that they may have only a relatively short life span.

Today Algeria is finding that its economic growth is insufficient to provide jobs for many of its citizens. The unrest and dissatisfaction that such a situation provokes has promoted simultaneously a fundamentalist Islamic revival, street violence, a decline in the authority of the single-party socialist government, and substantial gains by increasingly powerful opposition forces in Algeria's initial experiments with a more democratic political process. In the early 1990s, when opposition forces appeared poised to win the first free elections in the country's history, the army stepped in to cancel the elections. The consequence of this action has been a heightened level of political violence, an increased scale of repression by the government, and uncertainty and stagnation in the economy.

Traditionally, Algeria has been able to take a leadership role in the Arab League, the **Organization of Petroleum Exporting Countries (OPEC)** (a group of thirteen oil-producing nations), and the Third World in general. That position of influence is accompanied by respect in both capitalist and socialist countries. As a result, Algeria was able to serve as intermediary between the United States and Iran to help settle the hostage crisis that destroyed relations between those two countries in 1979 and 1980. At the same time, Algeria supports those revolutions and wars of national liberation that it views as legitimate, a consequence of its own struggle for national independence against French colonialism. Algeria's history also explains its active assistance to the **Palestine Liberation Organization (PLO)**, the Palestinian umbrella organization formed to coordinate opposition to Israel. There is reason to believe that continued international political in-

Table 23-2 Population and Petroleum Production in the Mediterranean Crescent Countries

Country	Population (millions)	Petroleum Production (in thousand barrels per day)			
		1980	1990	2000	2004 (est.)
Algeria	32.3	1,000	797	800.0	1,205.0
Egypt	73.4	585	873	810.5	712.0
Israel	6.8	0.7	0.3	0.1	0.1
Jordan	5.6	—	0.4	—	—
Libya	5.6	1,780	1,369	1,407.5	1,550.0
Morocco	30.6	0.2	0.3	0.3	0.2
Syria	18.0	165	385	510.2	504.0
Tunisia	10.0	100	93	74.5	70.0
Turkey	71.3	42	70	56.0	42.0

Sources: *Oil and Gas Journal*, vol. 78, no. 52 (December 29, 1980); vol. 88, no. 52 (December 24, 1990); vol. 98, no. 51 (December 18, 2000); vol. 102, no.47 (December 20, 2004); and Population Reference Bureau, *World Population Data Sheet* (Washington, D.C.: Population Reference Bureau, 2004).

volvement will be characteristic of Algeria, but that its influence will be directly dependent upon its ability to solve its internal political, social, and economic problems.

Egypt: Gift of the Nile

The richness of Egyptian civilization was a wonder even in Greco-Roman times, and the splendor of its scholarly institutions increased further after the Arab conquest in A.D. 640. At al-Azhar University and other religious centers in Cairo, Arab scholars were largely responsible for preserving the writings of the classical world and transmitting them to the Christian West. When the Dark Ages enveloped Western Europe, the peak of scholarship and invention was reached in the universities of the Middle East. Even when political independence was lost in the region, a spirit of national independence survived and fueled the drive for complete independence after World War II.

Today Egypt is a country of contrasts. Cairo and Alexandria are two of the largest cities in Africa, but many peasants in Egypt live under relatively the same conditions that their ancestors experienced a thousand years ago. The Egyptian economy is diversified, with a wide range of basic and processing industries and many large and small consumer-oriented enterprises; but its population is increasing at a rate of 2 percent per year, so constant gains in productivity are necessary just to stay abreast of population growth. Currently, Egypt has the largest population of any Arab state, yet less than 5 percent of its land surface is arable (see Table 23–1).

Egypt is both a center of Muslim culture and tradition and a powerful political force, but employment opportunities for college-educated Egyptians are limited, and many must seek work elsewhere. For example, Egypt manufactures most of the world's Arabic-language films, and its colleges and universities enroll more than 600,000 students, many of whom come from other Arab countries. Many Egyptians of all skill levels are employed in other Arab countries, and the money they send home is an important contribution to the Egyptian economy.

Egypt's employment pattern has increased its influence outside Egyptian borders. However, when oil prices decline, causing opportunities for migration to diminish throughout the region, that influence also diminishes, along with the flow of income back to Egypt. In addition, both the decline of the pan-Arab nationalism of the 1950s and 1960s and Egypt's support for the 1979 Camp David agreements, which established peaceful relations with Israel, decreased Egyptian influence by isolating Egypt politically from other Arab states. In the wake of the two decades of political and military conflicts in the Gulf, Egypt, which was readmitted to the Arab League in 1989, has regained its regional power status, and most Arab countries have reestablished diplomatic ties with Egypt. This improvement in its regional political influence made it possible throughout the 1990s for Egypt to play an important role in the Israeli/Palestinian peace process. Participation in the coalition against Iraq in the 1991 Gulf War had the additional benefit for Egypt of ensuring a favored position in receiving U.S. civilian and military aid.

Egypt has built its history on a limited resource base. Its agricultural resources, especially, are constrained by a lack of water and suitable soils. Because very little rain falls in Egypt, except over a thin strip of land along the Mediterranean coast, cultivation is concentrated in the valley and delta of the Nile River, where both adequate water and good soils coincide. Outside the Nile Valley, significant cultivation is possible only in a few small oases. Alfalfa, cotton, rice, maize, and wheat are the main crops in Egypt. Cotton is the most valuable export crop and is also the basis for a substantial textile industry. Land use is already intense in Egypt, and without significant changes, major breakthroughs to higher levels of production are unlikely.

Nonetheless, one attempt to achieve such a breakthrough was made when a high dam on the Nile was constructed in the 1960s at Aswan, creating a huge lake (Figure 23–5). That water represents the flood that previously inundated Egypt each year, providing a natural supply of irrigation water and soil-enriching silt to the fields. The Egyptians expected to increase their agricultural production in two ways. First, releasing the stored water slowly makes it possible to raise two, and sometimes three, crops a year on the same land. In addition, surplus water can be channeled into the desert to develop areas that are without rainfall—an effort called the **New Valley Project**.

The economic and ecological effects of the **Aswan High Dam** have been both good and bad (see the *Geography in Action* boxed feature, Land Degradation in North Africa and the Middle East, in the previous chapter). For agriculture, water stored behind the dam has provided larger and more reliable water supplies for development. By capturing the

■ **FIGURE 23-5 Aswan High Dam.** Visible in the center of this 6-meter resolution satellite image taken from the International Space Station are the dam and the lake (Lake Nassar) that has formed upstream behind the dam. The Nile River is also visible as it resumes its downstream course.

Nile's floodwater behind the dam, agricultural security is increased dramatically. Destructive floods no longer occur. Droughts in the Nile's headwaters, where most of the river's water originates, no longer translate into irrigation water deficit, crop failure, and famine in Egypt. Farmers now potentially operate in a dependable environment. But desert soils, devoid of organic matter and difficult to manage, have often been afflicted with **salinization**, the process in which salts accumulate in the upper soil horizon and inhibit crop growth. In addition, the steady expansion of permanent irrigation to allow year-round cropping creates a year-round moist environment that nourishes wild as well as cultivated vegetation, which has led to a parallel expansion in the occurrence of schistosomiasis, a debilitating parasitic disease. Snails are the intermediate hosts for the parasite that causes the disease. In the one crop, flood-dependent basin irrigation system practiced for centuries in Egypt, the snail population shrank by 90 percent when irrigation stopped in the dry season. With water now abundant all year long, snail populations flourish, the parasites find abundant intermediate hosts, and an increased percentage of farm workers become infected. Malaria also has begun to make a comeback. Moreover, even with the water stored by the dam, Egypt faces serious water shortages in the foreseeable future. This is because all of the Nile's water is already allocated between the countries along its course; population growth requires more water for domestic, agricultural, and industrial purposes; and the present supply appears inadequate to meet all of the growing demands likely to be made on it over the next 25 years. Thus, Egypt's situation calls for conservation efforts and closer political and economic ties with Sudan and Ethiopia, Egypt's neighbors to the south, where new water sources might be developed.

The major action envisaged to make new water available is the **Jonglei Canal Project**. This project called for the construction of a canal in southern Sudan that would bypass the Sudd swamps. Much water is "lost" to evaporation and the sustenance of this wetland, and the canal was intended to divert White Nile water before it could reach the swamp. The diversion would result eventually in an increase of 4 billion cubic meters of water available at Aswan. However, before it could be completed, the civil war in southern Sudan interrupted work on the Jonglei Canal. Similar water conservation projects in the Machar marshes along the Sobat River, a tributary of the White Nile, are also planned but have not yet been undertaken. Although unsettled political conditions in Sudan have made realization of these plans impossible, the prospect of an end to Sudan's civil war means that reactivation of water development projects once again becomes feasible. Augmenting existing water supplies is important to both Egypt and Sudan because both countries are rapidly approaching full utilization of presently available water. If Ethiopia, which was shortchanged by the original agreements dividing up the Nile's water despite being the source for 80 percent of the Nile flood, were to press its claim to a larger share, the pressure on Egypt and Sudan to manage existing available water supplies as efficiently as possible would increase substantially.

On the other hand, hydroelectricity produced at the Aswan High Dam has promoted industrial growth, and a fertilizer industry has emerged to restore fertility to the soils now deprived of Nile silt. Equally important, the dam has been an important symbol of Egypt's determination to develop. Thus, the dam has had a psychological effect that may be as significant as its economic impact.

Despite efforts to increase agricultural productivity, Egypt's food production situation is weaker today than it was in the early 1970s. Thirty-five years ago Egypt was largely self-sufficient in food and some agricultural products were exported. In the intervening decades, food consumption has increased dramatically, but agricultural output has risen at a much slower rate. The decline is particularly striking in specific staple items in the Egyptian diet such as wheat (>40 percent imported), corn and sugar (>45 percent imported), and edible oils and milk (>65 percent imported). Among major cereals, Egypt is self-sufficient only in rice. High levels of self-sufficiency continue to characterize production of other important foods such as eggs, fruit, meat, poultry, and vegetables. But a measure of agriculture's declining relative importance in the overall economy is its drop from a 30 percent share of national output three decades ago to less than 14 percent today. Strenuous efforts have been made, often accompanied by vigorous official propaganda hyping development achievements, to extend agriculture to new areas. Land reclamation is undeniably the present and future hope for Egyptian agriculture. The reason for this is simple. Today little more than 3 percent of the total land area of Egypt is cultivated. Even if areas traditionally cultivated with Nile water can be double or triple cropped using water reserves captured by the High Dam at Aswan, the effect is only as if a mere 10 percent of the total national territory was devoted to food production.

Thus bringing new areas into production is an imperative national goal. Major plans exist to bring Nile water westward along the Mediterranean coast and eastward to districts along the Suez canal and in the Sinai as well as to drain coastal lagoons in the northern edge of the Nile Delta. In theory hundreds of thousands of hectares of new land have been converted from waste or rough grazing to cultivated fields. In practice achievements have fallen far short of the official line extolling development success. In the Western Desert, after 40 years of effort to enhance productivity of the well-established oases in the "New Valley" region west of the Nile, the areas reclaimed only slightly exceed the total area once cultivated by traditional methods in the oases of Bahariya, Dakhla, Kharga, and Farafra. There are many reasons for this state of affairs: soils in newly developed areas are innately less productive than in traditional zones close to the Nile; management of the newly exploited soils has proven more difficult than anticipated; as a result, salinization and waterlogging plague many districts in new and old agricultural areas; where development depended on groundwater development, drawdown of the water table by diesel-powered pumps dried up traditional shallow wells and artesian springs; many of the occupiers of farm concessions in new agricultural areas are inexperienced farmers or are members of the Egyptian social and military elite who are more interested in land speculation than serious farming; and rapid urban expansion in old agricultural districts is consuming alluvial soils in the Nile Valley and Delta that once were extremely productive. In consequence, increases in agricultural productivity have proven difficult to achieve.

Egyptian industry, which is quite diversified, contains both well established textile and food-processing industries and modern chemical, electrical, and metalworking industries that emerged after 1950. In these modern industries, government financial investment was substantial. This was part of a concerted strategic plan in the socialist era of President Nassar to reduce dependence on Western industries and the import of their products. Despite a record of substantial growth into the late 1970s, as well as domestic self-sufficiency in oil requirements, the rapid expansion of industry and the labor force employed in the industrial sector is constrained by a limited domestic resource base deficient in major deposits of minerals. These limitations combined with a recession in the international economy to produce economic stagnation in Egypt by the late 1980s. Food processing, textiles, and fabricated metal products remain the major industries—reflecting the important role of agriculture in the overall economy, the central position that cotton occupies with respect to the textile industry, and the significance of imported materials. Not reflected in standard statistics is small-scale manufacturing of craft products organized on a traditional, household basis. Frequently engaging too few employees to be included in official statistics, those craft enterprises nonetheless make an important contribution to the national economy.

Another component of Egypt's economy is international tourism, which generated U.S. $4.3 billion in foreign exchange in 2000 from visitors who came to view the country's historic monuments and enjoy the recreational amenities. The pyramids, temples, monuments, and art and history museums complement such modern fare as markets and craft products and are major tourist lures. Also appealing to tourists are the coral reefs and snorkeling opportunities of the Ras Mohammed National Park, the beach resorts at Sharm-el-Sheikh at the southern tip of the Sinai Peninsula, and the Hurghada resort area on the African side of the Gulf of Suez. Both international and domestic recreational tourism have boomed in this region, spreading southward along the western shore of the Red Sea in a sprawling complex of villa and resort village developments. But these sources of tourist revenue are associated with potential problems. Traditional tourist destinations in the Nile Valley are adversely impacted whenever attacks by Islamic terrorists occur; concern about the security of Nile River boat tours between Cairo and Luxor, which must sail past a center of Islamic fundamentalism at Asyut, has almost completely stopped this branch of the tourist industry north of Luxor. Sporadic terrorist attacks, such as the July 23, 2005, car bombings at Naama Bay in the Sharm el-Sheikh resort complex that killed more than 85 Egyptians and international tourists, have a serious dampening impact on tourist visits for some time after each attack. Revenues decline as a result of tourist decisions to postpone visits due to security

fears, and analogous declines take place whenever periods of economic stagnation occur in tourist origin countries.

The Red Sea and Sinai coastal resorts share the economic volatility endemic to the tourist industry as a whole, but they experience an additional set of problems. These difficulties are environmental and stem from the rapid growth of tourist facilities and the increased volume of tourists in the past decade, which has exceeded all expectations. Planners hoped to attract 1.2 million tourists per year to Egypt by the end of the twentieth century, but by 2002 more than four times that number of tourists entered the country. The intensity of aquatic recreational use in the coastal tourist districts, as well as the construction of housing and beach enhancement facilities, is having a detrimental impact on the coral reefs, which are extremely sensitive to sedimentation, eutrophication, sewage seepage, and improper rubbish disposal. All of these environmental problems are a particularly significant problem along the African and Sinai coasts of the Gulf of Suez. The irony of these pressures generated by the coastal tourist industry is that they impact that part of the natural resource base—the corals and the clear, clean ocean water—that represents one of the prime attractions for tourists in general, and snorkelers and divers in particular, in the first place.

Turkey: Unrealized Potential

As heirs to the Byzantine (Eastern Roman) Empire, the Ottoman Turks assumed the mantle of power in the eastern Mediterranean during the mid-fifteenth century. Only when it was outstripped by the industrial development of the West did the Ottoman Empire lose its internal cohesion. Even then, Turkey retained enough vitality and creative drive to renew itself and escape colonial rule.

The end of World War I witnessed the demise of Ottoman Turkey and the start of a new, smaller, and more homogeneous Turkish state. After a revolution led by Kemal Atatürk in the 1920s, the forces of modernization cut the ties to the past. Throughout the whole process, the army played a crucial role, serving as an integrating institution of national regeneration to replace the discredited Ottoman bureaucracy. In the end, many old customs were abandoned, a secular state was created, and great impetus was given to industrial and agricultural development.

Turkey is unusual among Middle Eastern states in two ways. First, it possesses sufficient coal and iron resources to develop its own heavy industry without having to import basic materials (see Figure 22–8). Second, it has a greater percentage of usable land than any other country in the region (see Table 23–1). Those advantages, combined with virtual self-sufficiency in petroleum, have helped to create the most powerful and integrated economy in the region (Figure 23–6).

Much of Turkey's development has been centered on tourism and on industry, which consume a large part of the capital invested in overall national development (see the *Geography in Action* boxed feature, Tourism in the Middle East and North Africa). Until the global economic recession of the early 1980s, Turkey's industrialization program progressed well. In addition, concentration on basic metal-

▪ FIGURE 23-6 Istanbul, Turkey. Historically significant for thousands of years, Istanbul, like its predecessors, has a preeminent position in Turkish cultural and economic life. The Blue Mosque and the Hagia Sophia basilica/mosque/museum are two of Istanbul's prominent cultural symbols. They appear in this view across the Golden Horn looking toward the densely occupied neighborhoods of Galata at the southern entrance to the Bosporus Strait.

Even though agriculture remains important in Turkey, contributing 16 percent of economic output (see Table 23–3) and continuing to be the primary occupation for 40 percent of the labor force, agriculture's share of total exports has declined. Today the value of manufactured goods is twice the value of agricultural products, and manufactured commodities represent 78 percent of total merchandise exports, although high-tech exports constitute only a small share of the total (4 percent). Tobacco and cotton, along with fruits and nuts, continue to be important agricultural products. Nonetheless, the bulk of Turkey's significant industrial output continues to be consumed in its domestic market, and only rare minerals—such as chromite, meerschaum, and manganese—enter world trade.

Much of Turkey's commercial crop production is concentrated in coastal districts that have a mild, dry summer subtropical climate, rich soils, gentle slopes, adequate moisture, good potential for irrigation, and relatively easy access to good transportation facilities. Since 1970 agricultural activity has intensified considerably in those coastal districts. Because most of Turkey's farmland is privately owned, continued agricultural growth depends on farmer recognition of the positive economic benefits accompanying a change from subsistence to commercial activities. That change in orientation has occurred in many areas as dry farm operations are converted to irrigation, more fertilizers are employed, and more high-yielding fruit and vegetable crops are planted. Such changes are assisted by liberal credit policies, government extension services, and support from cooperative societies, in which Turkish farmers are participating more frequently.

Until recently, the progressive trends in the coastal districts were not matched elsewhere in the country. Although much of interior Turkey is suitable for cereal cultivation and animal husbandry, there are serious constraints. High elevations and steep slopes, cool temperatures, short growing seasons, and semiarid conditions limit the productivity of much of the more traditional agricultural zones. In addition, gains in permanent cropland are made at the expense of permanent pastureland and the dry farmland included in arable

lurgy and textiles stimulated the development of the country's primary resource base. Because of the economic slowdown in the 1980s, and the general economic doldrums that afflicted Turkey's economy in the aftermath of the 1991 Gulf War, Turkey's manufacturing sector stagnated (Table 23–3). In 2000, Turkish industrial activity generated only 25 percent of the country's economic output, while service activities accounted for 59 percent.

Despite economic growth, rapid population increases throughout the 1960s and 1970s outpaced the ability of Turkish industry to absorb new workers. As a result, large numbers of Turks sought employment abroad, replicating the pattern of labor migration experienced by the North African countries. The bulk of Turkish migrant labor flowed to Germany, Turkey's traditional European ally, and the remittances of that labor force continue to be important contributors to the homeland's economy.

Table 23-3 Manufacturing and Development in Algeria, Egypt, and Turkey: Year 2000

Country	Exports of Manufactured Products as a Percentage of Total Merchandise Exports	GNI PPP ($/capita)	Population below $1/day (%)	Value Added as % of GDP			Trade (millions $)	
				Agriculture	Industry	Services	Exports	Imports
Algeria	3	5,040	<2	10	69	21	19,550	9,200
Egypt	37	3,670	20	17	33	50	4,700	13,600
Turkey	78	7,030	2	16	25	59	27,324	53,983

Source: The World Bank, *The World Development Report 2002: Building Institutions for Markets* (Washington, D.C.: The International Bank for Reconstruction and Development/The World Bank, 2002).

GEOGRAPHY IN ACTION
Tourism in the Middle East and North Africa

Although on a global scale the countries of the Middle East and North Africa are not major players in the world's tourism industry, tourism is very important in the economy of most countries. Tourism is a service industry that contributes in significant ways to local employment opportunities, supports household-based craft industries, and provides major foreign exchange inputs to the economies of Mediterranean coastal countries. Every year millions of visitors, particularly from Europe, seek out the warm climate, abundant beaches, and spectacular historical monuments of the southern and eastern Mediterranean. In terms of popularity as a tourist destination, Turkey ranks highest in the region as measured by tourist arrivals and in terms of money earned from tourism. For the countries of the Middle East and North Africa with significant tourist industries, the revenue earned from tourist visitors is an important part of their national economy given that in many other ways their resource base is limited. But sunshine, warm weather, and attractive coastal locations they have in abundance, and tourism is an appealing way to transform the negative constraint of aridity into a positive generator of cash income.

In addition to Turkey, the major earners of income from tourism are Egypt, Israel, Morocco, and Tunisia. All of the Mediterranean Crescent tourist destination countries have strong linkages historically and commercially to Europe, so they have a known quality to them without sacrificing a sense of the exotic. It is, perhaps, ironic that they all lack significant oil reserves, so tourist development is viewed as a substitute earner of foreign exchange. Egypt, for example, generates nearly one-fifth of its foreign exchange earnings from the tourism industry. This is in sharp contrast to the oil-rich Persian Gulf countries, for whom tourism is an insignificant factor in their total economic structure. The more than two million Muslim pilgrims who perform the *hajj* each year to Mecca and Medina, despite bringing significant economic impact to the region, come for religious reasons and do not fall into the same category as tourists.

In each of the five leading tourist destination countries, European-origin travelers are the largest component of the international visitor stream. Two factors in particular contribute to the appeal of these destinations. One is the mystery and romance of Middle Eastern culture, which has long fascinated Europeans. Practicing customs that are rich in diversity, producing high-quality, sophisticated craft goods in wood, leather, metal, ceramics, textiles, and precious metals and stones, the primary tourist destinations offer aesthetic and creative options that are simultaneously exotic yet familiar. The sense of the familiar is based on a long history of circum-Mediterranean contact and on the substantial efforts made by governments to develop a modern infrastructure in support of tourism. From tourist bureaus to modern beachfront hotels, from restaurants and cafes to good quality transportation systems, the Mediterranean countries have the basic support systems in place to cater to tourists at both the high-price and the budget ends of the income continuum. Advertisements that tout the Egyptian Red Sea Coast as the "Egyptian Riviera," for instance, suggest both the known and a patina of luxury. Second, Mediterranean Crescent countries share a historical connection to the cultures of Greece and Rome. Some of the most impressive ruins of antiquity are found in North Africa and the eastern Mediterranean (Figure A). These sites are well preserved and presented and are very attractive to tourists, as are the Pharaonic monuments along the Nile. Israel benefits from the fact that many sites of significance to Christians and Jews can be found within its borders. Because more than 85 percent of Israel's tourists come from Western Europe and North America, they spend significant sums during their visits and make Israel the third largest tourist revenue earner in the region.

One of the most significant developments that has taken place in the destination of tourists heading to the Mediterranean Crescent is the rise of Turkey to the status of most popular destination in the region. In the mid-1980s Turkey's tourist industry was one of potential rather than reality, and it was outdistanced by the North African countries of Tunisia, Morocco, and Egypt. By the late 1990s, Turkey's number of tourist arrivals increased tenfold while its North African competitors managed only to double the number of tourist visits. But all countries with active tourist industries are vulnerable to fluctuations in tourist interest. Any economic downturn in the economy of the tourist-origin country rapidly is reflected in a drop in tourist arrival figures in the Mediterranean Crescent. Even more likely to generate a rapid decline in tourist arrivals is a terrorist attack that targets tourists. While 35 to 40 percent of the tourist visitors to Egypt come from other Arab countries and historically have not been dissuaded from their travel plans by terrorist attacks, interna-

land, the only categories of agricultural land use to decline in past years. That shift results in both overgrazing on the remaining pastureland and encroachment on and degradation of remaining forest land.

Although the productivity of Turkish agriculture has kept pace with Turkey's slowing population growth (1.4 percent per year), the government's continued emphasis on commercial and industrial crops such as cotton and tobacco, at the expense of food and livestock production, could result in serious future difficulties. A critical challenge in Turkey's development is to extend the modernization and development that characterize its urban, industrial, and coastal districts into the still largely rural and traditional interior areas.

One effort to do this is the U.S. $32 billion suite of continuing development projects undertaken by the Turkish government over the last 15 years in southeastern Turkey. The GAP (Guneydogu Anatolu Projesi), or Southeast Anatolia Project, is intended to control the flow of the Euphrates and Tigris rivers in Turkey for irrigation use, flood protection, and hydroelectric power production. If the project reaches its goals, 22 dams ultimately will permit irrigation of 1.7 million hectares. The vision of turning large tracts of semi-arid grazing or unproductive land into intensive food and fodder producing areas motivates the project. That these same developments may cause problems in the quantity and quality of water supplied to downstream irrigation farmers in

tional terrorists are more likely to feel directly threatened. The July 2005 attacks on popular tourist areas along the Mediterranean coast of Turkey southeast of Izmir and at Sharm el-Sheik in Egypt will certainly be reflected in a drop in tourist income in both countries. Because tourist facilities are by their very nature dispersed and because tourists wish to move freely and casually in public spaces, security for the tourist industry is difficult to provide. No one wants to vacation behind barbed wire in a high-security environment or be discouraged from visiting local markets or monuments. Thus opposition groups with political, social, and religious agendas, willing to resort to violence in an effort to press their claims, find tourists a convenient target. If an industry rule-of-thumb is correct, and every million tourists creates 200,000 jobs, then Egypt's 8 million tourists in 2004 provided jobs to more than 1.5 million Egyptians. Any substantial decline in tourist visits, then, has major economic repercussions.

If most Mediterranean Crescent countries are open to international tourists and encourage their presence, Persian Gulf Coast countries tend to be much less receptive or congenial. Iran has virtually disappeared from the charts of international tourist destinations, and Iraq, already a less-than-desirable destination for Western tourists for many years, is now a war zone. Although Saudi Arabia earns a substantial sum from visitors, this income is almost entirely derived from Muslim pilgrims making the *hajj* as part of their religious obligation. Rather closed societies, with stricter controls on dress code, public behavior, and the types of entertainment popular elsewhere (mixed gender beaches, nightclubs, casual dress, consumption of alcohol, and so on), and the current high level of violent conflict in the area make the Persian Gulf countries unlikely destinations for casual vacationers. Bahrain is a remarkable contrast to this generalization, because for Saudi Arabians it has become a reasonable alternative to vacations in more distant, non-Islamic destinations. For Bahrain, tourism represents a way to diversify its oil-dependent economy, especially in light of its limited ability to expand its petroleum reserves.

Syria and Iraq has not loomed large in Turkish political debate or practical planning.

A hard-working and disciplined people, a strong military, and a growing agricultural and industrial base have made Turkey a powerful state. Yet that political and economic power is seldom applied outside its own borders for three reasons, all of which relate to issues that were never fully resolved by Atatürk's revolution. First, Western notions of democracy are imperfectly transplanted in Turkey. The nation was founded by a strong military leader, and the military perceives itself as both a guarantor of the ideals of the revolution and a preserver of national unity. As a result, the army has intervened several times in the national government and has consistently maintained an internal focus except for its intervention in Cyprus in support of that country's Turkish minority.

Second, Atatürk created a secular state, even though 98 percent of the population is Muslim. With one-third of the population still living in rural areas, the influence of the religious leadership remains strong, but the government remains secular for the time being. However, Turkey is affected by the fundamentalist current now surging through the Middle East, which appeals to conservative Turks. Sensitivity to the strong feelings of solidarity felt by this segment of the population to Iraqi Muslims was a factor in the Turkish government's unwillingness to permit American military forces

551

to use Turkish bases in support of the U.S. invasion of Iraq in 2004. Despite Turkey's interest in joining the European Union, the country may no longer have the luxury of remaining aloof from its Islamic heritage.

Third, Turkey's ambivalent location between East and West causes uncertainty over the best foreign policy to follow. Although Turkish leaders remember traditional quarrels with Russia, a vocal minority continues to agitate for closer ties to that nation. In light of recent changes in East-West relations, it is likely that concerns over past problems will exert less influence on Turkey's future relations with countries of the former Soviet Union. In addition, even though it was Turkey's strategic location at the outlet of the Black Sea that made it a valuable member of the North Atlantic Treaty Organization (NATO), that military role may be less important now. As Europe achieves greater political union, attention will inevitably turn to economic concerns. Can Turkey, with only a foothold in Europe, achieve entry into the European Union? With the bulk of its territory and population rooted in the Middle East, does Turkey want to? Ironically, Turkey has achieved associated status with the European Union at the same time that a pro-Islamic party, with the support of a quarter of the Turkish electorate, has become a prominent player on the political scene. Turkey's in-between position is complicated further by a series of historical hostilities and contemporary economic conflicts with its fellow NATO member Greece over the island of Cyprus, control of the Aegean

Sea, and offshore mineral- and oil-exploration rights. In addition, a large Kurdish minority exists in eastern Turkey. With aspirations for recognition, autonomy, and independence that go largely ignored by the Turks and the international community—but with an increasingly active underground liberation movement—the Kurdish minority is certain to cause future problems for development in the dry and mountainous southeastern portion of Turkey.

Caught between Europe and the Middle East, between modernization and tradition, between freedom and authority, Turkey struggles to establish a sense of identity and purpose. If the country achieves agreement on a course of action, its regional effect could be enormous.

Small States, Unique Economies

Four countries on the eastern margin of the Mediterranean Sea, together with one increasingly recognized movement that aspires to full state status, present a different set of issues. In those countries agricultural and mineral resources, basic factors for many countries, are of secondary importance. They are superseded by questions of human resources, service functions, and various kinds of inflowing support for the nations concerned. Israel, Lebanon, Jordan, and Syria are the states in this group, with the PLO representing the

Table 23-4   Land Use (in Thousand Hectares) in Israel, Jordan, Lebanon, and Syria

| Country | Labor Force in Agriculture | | | | Land Use |
| | Total (000) | | % | | |
	1990	2000	1990	2000	2000
Israel	73	70	4.1	<1	
Total Land Area					2,106
Arable Land					333
Permanent Cropland					85
Other[a]					1,688
Jordan	123	179	15.1	11.4	
Total Land Area					8,921
Arable Land					244
Permanent Cropland					157
Other[a]					8,520
Lebanon	62	47	7.3	3.7	
Total Land Area					1,040
Arable Land					190
Permanent Cropland					142
Other[a]					708
Syria	1,147	1,434	33.1	27.7	
Total Land Area					18,518
Arable Land					4,552
Permanent Cropland					810
Other[a]					13,156

Source: Food and Agriculture Organization of the United Nations, *www.fao.org/satistical databases.faostat.*
[a]Other seasonal pasture, forest, and wasteland

national aspirations of the bulk of Palestinians. Syria is significantly larger than the other states (Table 23–4). But its occupation of parts of Lebanon in the 1980s, its similar problems and potentials, and its involvement in the affairs of other eastern Mediterranean states make it reasonable to include in this group.

The eastern end of the Mediterranean has a reputation as a strategic area that controls important nodes of communications and trade. Because routes linking Africa with Asia and the Persian Gulf with Europe pass through the area, control of those routes has long been important. Beirut, for example, was a vital air link between Europe and the East for a time, although jet travel and the political crisis of the late 1970s and early 1980s have encouraged travelers to route their flights through other intermediate points.

Focal point of three major world religions, the religious significance of this area is also remarkable. It is the seedbed of the Jewish religion, out of which grew the various Christian denominations (Figures 23–7 and 23–8), as well as being significant to Muslims. Although Mecca, located in Saudi Arabia near the Red Sea, is the central holy place of Islam, Jerusalem is the site of the Prophet Muhammad's ascension into heaven and is a major Muslim pilgrimage center (Figure 23–9). Medieval maps show Jerusalem as the center of the world, with Asia, Africa, and Europe all focused toward it. Present power politics have greatly modified but not totally destroyed that view. From a religious and a geopolitical viewpoint, the area remains vital to the world at large.

Israel: Resurrected Homeland

Israel is unique among countries in that it became established territorially on the basis of 2,000-year-old claims and in response to Jewish persecution around the world. That establishment on disputed land long occupied by indigenous

■ **FIGURE 23-8 Sunday morning Christian worship service overlooking the Sea of Galilee.** Christianity has strong historical ties to this region.

peoples unsympathetic to the Israeli national cause was bound to be contested.

The small state of Israel, established in 1948 from the British mandate of Palestine amid considerable conflict, was difficult to defend against surrounding Arab pressures. As a result of the 1967 War, Israel was able to rationalize its frontiers and make its borders more defensible, but it did so at the territorial expense of Egypt, Jordan, and Syria. At the same time the Palestinians, who had lost part of their lands in the 1948 hostilities, found all of their traditional territory in Israeli hands after the 1967 War (Figure 23–10). That war sparked a new wave of Palestinian migration into neighboring states, and subsequent hostilities culminated in Israel's 1982 invasion of southern Lebanon, which pushed refugees from the 1948 conflict into a new round of forced migration. Thus, Israel's search for a secure and peaceful existence has involved permanent insecurity for Palestinians.

Apart from this continuing struggle to maintain and protect Israeli territory, statistics can provide some clues as to the kind of state Israel is. In an area where per capita GNIs are generally low, Israel's (adjusted for Purchasing Power Parity, or PPP), is $19,000, nearly six times those of Jordan and Egypt. Thus, in spite of a high rate of inflation, Israel is a comparatively rich country in a poor part of the world. In addition, Israel is basically an urban country; 92 percent of the population lives in urban areas. Industry is an important part of the economy, with diamond cutting, the manufacture of textiles and woolen goods, and many other small industries providing a livelihood for the bulk of the population.

Part of Israel's wealth comes from the industriousness of its people. The farming cooperatives of Israel are especially well known; the **kibbutz** (a collective farm) and the **moshav**

■ **FIGURE 23-7 Tomb site of Abraham.** The prophet Abraham is viewed by Jews and Muslims alike as their progenitor. This building sits upon the site where his remains and those of his son Isaac and grandson Jacob are believed to have been buried.

■ **FIGURE 23-9 Muslim worshipers in Jerusalem.**
Jerusalem is a holy city to Muslims. Visible is the al-Aqsa Mosque
compound with the Dome of the Rock Mosque seen in the back.

derived population, Israel is actually more European in its values and orientation than it is Middle Eastern. Thus, Israel is doubly at odds with its neighbors; their political conflicts are reinforced by a clash of cultural values. In addition, that cultural conflict is mirrored within Israeli society, as Jews who migrated to Israel from other Middle Eastern countries (**Sephardim**) find themselves at odds with Jews whose origins are European (**Ashkenazim**).

Even with such rich human resources, Israel would not be a self-sufficient state without the large flow of capital received from Jews around the world. And that support is augmented by massive assistance from governments that are Israel's allies or friends. Israel receives more foreign aid from the United States than does any other country, and both private and public assistance flows to Israel from Western European countries. The high burden of defense costs is one reason for that capital inflow, but current levels of living would probably be difficult to maintain on internal resources alone, even without the defense costs.

(a smallholders' village) are the two main types of production and marketing cooperatives that help Israel maximize production on land that is fertile only in the narrow coastal plain. Water that is moved southward from the Sea of Galilee along the national water carrier assists in the intense cultivation of citrus fruits, vegetables, and some grain crops. High-quality farm exports make a significant contribution to the economy, although less than 1 percent of the population works in agriculture.

Extremely efficient use of water is absolutely necessary for Israeli agriculture and urban life. Almost all of Israel's available water is currently being exploited, and desalinization of seawater is still too expensive to be a practical alternative. One-half of Israel's area is unusable (essentially, the area of the Negev Desert), and aridity and drought reduce the productivity of the bulk of the country's unirrigated land. In order to increase agricultural yields, Israelis have become among the most efficient managers of water in the world, developing innovative techniques in irrigation, hothouse agriculture, and rainfall and runoff management (Figure 23–11).

Another striking feature of the country is its high level of education. It has four main universities, and many people who immigrate to Israel from the United States, the countries of the former Soviet Union, and other countries are highly qualified academically and technically. Israel is one of the few countries that has difficulty because of the high level of its human resources; providing satisfying employment that matches existing skills is not always easy. As a result of those resources, however, the quality of Israel's technical and academic achievement is unsurpassed in some fields. For example, Israel's work on both hydrology and horticulture as well as in some electronic fields is particularly important and has enabled Israel to provide significant aid to developing countries.

With its high living levels, technological sophistication, large and well-educated elite, and substantial European-

Lebanon: A Crisis of Identity

Lebanon, which shares the eastern Mediterranean coast with Israel, is also small in both area and population, but Lebanon is culturally complex, with a society highly fragmented along religious and ethnic lines. The political parties and armed militias that proliferated during the 1970s and 1980s were generally based on a Christian or Muslim community or sect, many of whose adherents were historically attracted as refugees or traders to Lebanon's rugged mountain interior and strategic economic location. But the very mountains that promised protection pose serious challenges to prospects for economic growth in the agricultural sector.

As a result, the Lebanese have for many years moved out from the poor agricultural lands of their own country to become merchants and entrepreneurs in other parts of the world. Until recently, most of the traders in many West African capitals were Lebanese, as were the bankers of the Arab world and parts of Africa. Large numbers of Lebanese also moved to Latin America and the Caribbean Islands and important Lebanese communities can be found in many U.S. cities. Remittances from the overseas Lebanese population have supported many of the country's rural villages at living standards well above the traditional subsistence level.

Until the mid-1970s that overseas activity was balanced by the steady growth of Beirut as the financial center of the Arab world, with more and more traders returning to their homeland. Beirut also played a major educational role in the Middle East; at the American University of Beirut, 30 percent of the student population was composed of non-Lebanese Arabs. However, because its curriculum emphasized a liberal arts orientation and because new universities were being founded elsewhere in the Middle East, American University and other Beirut institutions of higher learning lost influence to schools that emphasized engineering,

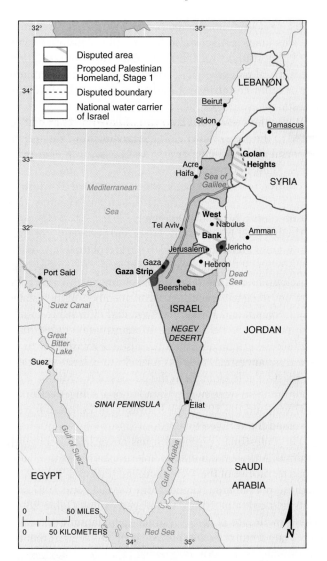

■ **FIGURE 23-10** **Small eastern Mediterranean states.**
Serious conflict continues along the eastern coast of the Mediterranean Sea, largely as a consequence of the Arab-Israeli confrontation. Recognition in the Oslo Accords of an autonomous zone of Palestinian authority in the Gaza Strip and around the city of Jericho seemed to establish the basis for a larger Palestinian homeland that would incorporate the West Bank. Constant disagreements over the pace and direction of greater Palestinian autonomy have produced escalating cycles of violence. Failure to settle the Palestinian statehood issue makes a more general peace between most Arab states and Israel exceedingly difficult to achieve, undermines the prospect for a return of the Golan Heights to Syria, and promotes continued instability along Israel's northern frontier with Lebanon.

■ **FIGURE 23-11** **Irrigated agriculture in Israel.** Israel has become a world leader in irrigation technology. Shown here is a field of zucchini squash at a kibbutz in the Judean desert.

science, and technology. Nonetheless, the Christian and Muslim mixture that characterizes Lebanon, together with a considerable veneer of French culture from the colonial era, made Beirut the most cosmopolitan city of the Middle East, albeit with deep social, cultural, political, and religious antagonisms that eventually flared into serious conflict.

Lebanon has no firm agricultural base, although bananas and citrus fruits grown in the coastal plain and deciduous fruits from the upland area form the basis of an export trade with surrounding Arab states. Industry is only now beginning to expand, with metal goods, processed foods, textiles, and pharmaceuticals the leading components. Before the late 1970s, Lebanon was a major center for Arab services, which accounted for 70 percent of the Lebanese national income and provided jobs for 55 percent of the active population. Moreover, traditionally, oil sheikhs, who were reluctant to invest in non-Arab countries, invested in Lebanese real estate much of the capital that could not be employed in development projects at home. Thus, Lebanon was a classic crossroads economy; its location resulted in a good living standard and a steadily expanding economy as the Arab hinterland grew in wealth.

Unfortunately, the prosperity generated was not evenly distributed. Urban residents with service sector skills, both Muslim and Christian, gained wealth from the growing economy, but recent migrants from rural areas were less fortunate. Christians who migrated overseas sent money back to benefit their villages, but Muslims, especially Shi'ites from southern Lebanon, did not have access to similar support.

The status of the entire country changed in 1975, when tensions burst into the open. The pact that accompanied independence from France in 1943 had guaranteed a dominant political and economic role to Christians, particularly the Maronites, the largest Christian group in Lebanon. The president of Lebanon was required to be a Maronite, whereas the prime minister had to be a Sunni Muslim. Other important political and administrative posts were also distributed on the basis of religious and ethnic affiliation. Over time, however, the numerical balance between Christian and Muslim populations changed, without any shift in the political structure. In addition, although jobs were distributed according to a community formula, the nation's wealth was not equitably shared. Adding to the unrest was a large Palestinian population that lived in Lebanon but often operated outside state control. In 1975, a civil war erupted in which many were killed, and effectively the central government collapsed. By 1983, Lebanon had a rather low GNI of $3,750. Since the restoration of peace and economic confidence, material conditions have improved and modest economic growth has occurred as reconstruction has been undertaken. But the fragility of this somewhat artificial burst of prosperity is reflected in the inability of the Lebanese economy as a whole to translate the reconstruction bubble into sustained growth; the country's current GNI (PPP) of U.S. $4,600 reflects this lamentably slow recovery rate.

Today Lebanon's long-term political and economic future remains uncertain. In 2005 a strong anti-Syrian people-power, pro-democracy movement emerged in Lebanon. It was motivated by many factors: unhappiness with the continued Syrian military presence long after the end of the Lebanese civil war; frustration with Syrian interference in internal Lebanese politics; and anger at the assassination of several critics of Syria, which included a recent prime minister, Rafiq Hariri. Combined with international support, the movement ultimately resulted in the withdrawal of the Syrian army from Lebanon. Free elections to the Lebanese parliament followed in which the anti-Syrian political groupings associated with Saad Hariri (a Sunni) and Michel Aoun (a Maronite) emerged as the majority force. This is a reappearance of the traditional power axis in Lebanon. But if a more balanced formula for sharing internal political power can be developed and a solution to the Palestinian issue, hardly within Lebanon's power to resolve, can emerge, then there is a prospect that Lebanon can rebuild its economy and infrastructure and recapture something of its former crossroads function (Figure 23–12). In this way, Lebanon, thoroughly Arab, yet neither fully Christian nor wholly Muslim, neither completely capitalist nor totally socialist, neither entirely pro- nor anti-Palestinian, may yet again turn its internal contradictions into a positive advantage.

Palestine: A People Struggling for a Home

The **Palestinians**—or, more specifically, the Arab population of the former British mandate of Palestine, represented largely by the Palestine Liberation Organization (PLO)—are unusual. They are a people who view themselves as a nation but have not had an internationally recognized national territory. Areas in which Palestinian Arabs are a majority of the population, such as the **West Bank** and the **Gaza Strip**,

■ **FIGURE 23-12 Beirut, Lebanon.** Beirut has historically functioned as an international banking, transportation, and educational center.

have long been controlled by Israel. But significant numbers of Palestinian refugees also live in other countries, while many Palestinians have become citizens of other nation-states. Until recently, Israel, the United States, and a number of other countries refused to recognize the PLO and Palestinian aspirations.

It was the creation of the state of Israel in 1948 that resulted in the suppression of Palestinian nationalism and the dispersal of many Palestinian Arabs among neighboring states. Many fled to the Gaza Strip, which fell under Egyptian control; to the West Bank, which was annexed by Jordan; or to Syria and Lebanon. Grouped into refugee centers, they subsisted on UN relief handouts.

Unabsorbed into the structures and economies of their fellow Arabs and unreconciled to the loss of their homes and property, the Palestinian Arabs developed a distinct national consciousness. At the same time, neighboring governments were unwilling to accept them as full members of their states; only Lebanon and Jordan have granted citizenship to limited numbers of Palestinians. In many cases, Palestinians have refused to be absorbed because they wish only to regain their ancestral lands. As a result, they have long been the most volatile element in the politics of the Middle East. Their numbers were swelled by additional refugees after the 1967 war, and the 1982 Israeli invasion of Lebanon further increased the refugee total.

The Palestinian population has never been accurately counted. Although 3.5 million Palestinians live in the Palestinian territories of the West Bank and the Gaza Strip, many of these people are refugees from the 1948 and 1967 conflicts and continue to live in refugee camps (Table 23–5). The growth rate of the Palestinian population is the highest of any country in the region. Estimates suggest that the Palestinian population is growing at a rate of 3.5 percent per year and that its fertility rate is very high (5.7 children per female). Statistics on GNI (PPP) do not exist, but are likely to be among the lowest in the region. This combination of rapid population growth and low income is a potentially highly volatile mix. Nearly 3 million Palestinians live in neighboring Arab states. In Jordan, Palestinians categorized as refugees constitute nearly one-third of the Jordanian population. Through intermarriage with Jordanians and acquisition of Jordanian citizenship, many Palestinians have reached at least a nominal accommodation with the Jordanian state. If this category of individuals were to be counted as Palestinians rather than Jordanians, Palestinians would constitute somewhat more than half of the total Jordanian population. Counting only those individuals listed in official records as refugees, and excluding those refugees living in the West Bank or Gaza Strip, the majority of all Palestinians live outside present or prospective future Palestinian territories.

In late 1993, Israel signed a peace agreement (the Oslo Accords) with the PLO that granted autonomy to the Gaza Strip and to the Jericho region of the West Bank (Figure 23–13). Palestinians expected that this increased level of control would extend over the next 5 years to the remaining parts of the West Bank and at least some part of Jerusalem. When

Table 23-5   Estimated Palestinian Refugee Populations

Country/Region	Total Host Country Population (millions)	Palestinian Refugee Population	Percentage of Total Palestinian Refugee Population	Percentage of Host Country Population
Egypt	71.2	50,000	1.1	<0.1
Iraq	23.6	90,000	2.1	0.4
Jordan	5.3	1,639,700	39.0	30.9
Kuwait	2.3	35,000	0.8	1.5
Lebanon	4.3	383,000	9.1	8.9
Libya	5.4	30,000	0.7	0.6
Palestinian Territory	3.5	1,460,400	36.3	41.7
Saudi Arabia	24.0	123,000	2.9	0.5
Syria	17.2	391,600	9.3	2.3
Yemen	18.6	150	<0.1	<0.1
Total		4,202,850		

Sources: United Nations High Commissioner for Refugees, *The State of the World's Refugees 2000* (Oxford: University Press, 2000); Population Reference Bureau, *World Population Data Sheet 2002* (Washington, D.C., 2002); United States Committee for Refugees, *World Refugee Survey 2002* (Washington, D.C., 2002).

this did not happen—and the resolution of issues such as the status of Israeli settlements in the West Bank, the recognition of a part of Jerusalem as the Palestinian capital, withdrawal of the Israel Defense Force from the West Bank, and the reconciliation of Israel's security concerns with the Palestinian desire for a sovereign state with a modern army did not occur—resistance to Israel (the ***intifadah***) grew and acts of violence increased. The PLO and its then leader, Yasir Arafat, and his successor, Mahmoud Abbas, were unable or unwilling to suppress organizations such as the armed wing of Hamas, whose members and supporters engaged in street violence, bombings, and armed attacks on Israeli civilians and military personnel.

Unable to bludgeon the resistance into oblivion or elicit a serious parallel suppression effort from the PLO and the Palestinian Authority, whose leaders fear they would initiate a civil war, Israel has decided on a unilateral policy of disengagement. This involves evacuating Israeli settlers from the Gaza Strip, where defense of the settlements is difficult and the Israeli presence constitutes a constant irritant to the Palestinians of Gaza. Despite vocal opposition from the religious and political supporters of the settlers, this strategy appears to have the support of the majority of Israelis. A second part of the disengagement strategy involves constructing a security barrier to separate the West Bank Palestinian areas from Israel. The course of this barrier—a barbed wire fence in places, a concrete wall in others—does not follow the border of Israel. Rather its somewhat unpredictable course seems designed to protect Israeli settlements close to the border, incorporate locally strategic areas into the protected area, and disrupt Palestinian livelihood activities in the region of the fence. The ostensible goal of the barrier, to reduce terrorist movements into Israel in order to eliminate or reduce to manageable levels Palestinian suicide attacks on Israelis, has attained considerable success. The longer-term implications and consequences of the barrier fence/wall remain unresolved.

Solving the enormous difficulties associated with a resolution of the Arab-Israeli problem would only be the first step in erecting a viable state with a functioning economy. But it would be not only a giant step in resolution of local conflict but also a major factor in reducing tensions that complicate and impede regional development. If such a Palestinian state does emerge to share the space of the former Palestine mandate with Israel, its national profile will show strong similarities to those of Israel, Lebanon, and Jordan.

■ **FIGURE 23-13 Welcome to Jericho!** One of the world's oldest cities, with a settlement history of at least 8,000 years, contemporary Jericho is an important regional center on the West Bank. The current welcome sign near the oasis city's grand mosque is an allusion to one translation of its name: city of the moon god.

Jordan: A Precarious Kingdom

Jordan and the future of the Palestinian people are closely linked. Jordan's population is composed of three elements: Palestinians; the pre-1948 settled Arab population of the Transjordan (the area east of the Jordan River); and the **Bedouin**, a people who have traditionally been predominantly mobile animal herders. Jordan was originally created after World War I as a British-mandated territory with a member of the Hashemite family as its ruler. It possessed few agricultural and mineral resources and had a sparse population of largely pastoral nomads. Those former nomadic herders were loyal supporters of the Hashemite dynasty and today contribute a disproportionate share of soldiers in the Jordanian army. Between 1948 and 1967, Jordan occupied large sections of the West Bank and attempted to carry out programs of industrial and agricultural development throughout its territory. But these programs were hampered by Jordan's limited resource base (see Figure 22–8 and Table 23–4). Other obstacles to development included significant land-use problems associated with deforestation, overgrazing, and soil erosion, as well as military conflicts with both Israel and supporters of the PLO.

In a sense, Jordan is a homeland in search of a people. Over time, Palestinian refugees and West Bank Arabs with Jordanian citizenship have drifted eastward across the Jordan River. But the loyalty of that Palestinian population to Jordan is uncertain. Numerically, they outnumber the Bedouin, who dominate both the government and the army. Israel might prefer to see Jordan designated as the Palestinian homeland, since that solution would undercut PLO demands for a national territory on land now occupied by Israel, but this idea has found no international support. However, Jordan's resource-base limitations do place severe constraints on economic development opportunities. Without linkage of some sort to the richer and better-developed West Bank territories, an economically viable Jordanian state is hard to envisage. Thus, movement toward settlement of the Palestinian issue presents Jordan with new opportunities for growth and development.

Syria: An Expanding Agriculture

Unlike its smaller neighbors, Syria is endowed with natural resources, primarily agricultural but also including small-scale mineral deposits and sufficient petroleum to meet immediate domestic requirements. Although much of the country lies in the rainshadow of coastal mountain ranges, sufficient rainfall occurs in the central and northern areas to support dry farming. There cereal grains and Mediterranean tree crops (olives and grapes) are the important products. To the east, rainfall becomes too small in amount and too variable in occurrence to make cultivation secure or successful.

Historically, the areas east of the cities of Homs and Halab (Aleppo) have been important zones of nomadic animal husbandry, and they continue to be significant producers of animal products for the urban market. Now largely organized on a cooperative basis, those pastoral districts are increasingly integrated into the cereal-based agriculture of the fertile semiarid zones along the Homs-Halab axis. Efforts to intensify production and reduce the impact of drought in rangeland and cereal farming districts are numerous. They include programs to protect and improve range fodder resources, integrate fodder crops into the fallow cycle of grain farming operations, utilize mechanization more effectively, increase the use of fertilizers, and introduce improved crop varieties (especially of strategic crops such as cotton and wheat) in order to increase yields.

Major capital investments in agricultural intensification are concentrated in irrigation, which is an ancient technology in Syria. Lever-operated buckets, underground water-collection tunnels (*qanat*), and waterwheels have long been used to bring water to agricultural fields. After World War II, however, major efforts were made to develop the irrigation potential of the Orontes River by using modern technology. Recently, more attention has been directed to improving rain-fed farming and pastoral activities, which utilize 80 percent of the country's productive land.

The gains from such development projects, though important, have been minor when compared with Syria's total arable area (see Table 23–4). Consequently, attention turned to the Euphrates Valley, which represented the last major source of water available for irrigation. The giant **Tabqah Dam** completed in 1978 and located 20 miles (32 kilometers) south of Meskene, created an artificial lake, Assad, that is 50 miles (80 kilometers) long and has the potential to provide water to irrigate 1.5 million acres (607,000 hectares) (Figure 23–14). If developed successfully, the Tabqah Dam project would more than double Syria's irrigated area, dramatically increase crop yields, and produce more cotton for export. But many of the soils slated for intensive development naturally contain high levels of gypsum salts, and are notoriously difficult to manage without experiencing salinization or waterlogging problems. So the planned contributions of Euphrates Valley irrigation development to the Syrian economy were probably overly optimistic. Even when salts are successfully washed from the soil without inducing waterlogging, this often only transfers the problem downstream

■ **FIGURE 23-14 Tabqah Dam, Syria.** The Tabqah Dam on the Euphrates River has helped Syria greatly increase its irrigated farmland.

to Iraq for whom wastewater from upstream Syrian irrigation is a prime input for agriculture. A series of three smaller dams planned for the Khabur River valley in the northeast is intended to increase that region's irrigated acreage tenfold, matching the productive potential of the Euphrates project. Unfortunately, poor management of existing irrigated areas has led to salinization, which has meant that irrigated cropland has increased less rapidly than expected.

The Tabqah Dam could improve Syrian levels of living considerably. Its hydroelectric power capacity, which exceeds that of the Aswan High Dam in Egypt, was intended to compensate for Syria's lack of coal and extensive oil resources. But environmental and political problems have made it difficult for the dam to reach its full power potential. Drought, combined with Turkey's increasing withdrawal of Euphrates water for its own irrigation and power projects, has made it difficult for the Tabqah Dam to solve Syria's energy needs in the ways that were originally intended. But in the late 1970s, oil deposits discovered at Qara Shuk in northeast Syria began to help fill this energy gap. Today a pipeline links these oil fields to the Mediterranean coast and Syria now generates 85 percent of its foreign exchange from oil exports. In addition to energy generation difficulties, the Tabqah Dam has encountered other problems.

Construction of the Tabqah dam displaced more than 60,000 farmers from floodplain areas that are now inundated by Lake Assad. Relocating those people and establishing new farms and villages have involved serious social costs, which—together with possible ecological side effects of the project—constitute the major hidden costs of this vast, technologically sophisticated scheme. The expectation is that local problems will be compensated by substantial gains at the national level and that potential conflicts with Iraq and Turkey over the distribution of Euphrates River water can be avoided.

Syria's major regional significance resides in its role as a frontline state confronting Israel politically and militarily. That role solidified when Gen. Hafez al-Assad emerged from a power struggle in Damascus in November 1970 as the dominating force in the Syrian branch of the Ba'ath Party. Quickly solidifying power through his election to the presidency in the following year, al-Assad derived much of his support from members of the Alawi minority community, which today constitutes about 11 percent of the Syrian population. Heavily recruited into the military during the French colonial regime, the Alawi community provided an increasingly cohesive core of support for Arab nationalist ideas and confrontation with Israel. That role has involved Syria in several unsuccessful military confrontations with Israel, a protracted and costly intervention in Lebanon, the support of resistance groups such as the PLO and Hezbollah that has earned Syria a place on the United States' list of terrorist-supporting states, and the loss of the Golan Heights, an upland area in southwest Syria. Under Israeli military control since 1967, and effectively (although not officially) annexed in 1981, the return of the Golan Heights to Syrian authority has been a central foreign policy goal for nearly four decades. When Hafez al-Assad died in June, 2000, he was succeeded as head of state by his son, Bashar al-Assad. Despite promises of reform and liberalization, little has happened to suggest a loosening of one-party control of Syria's political and economic life. A distinct downside of the highly centralized, authoritarian regime created by the Ba'ath Party has been the stultifying effect this has had on economic growth. Syria's GNI (PPP) of U.S. $3470 would be the lowest in the Middle East and North Africa were it not for Yemen's U.S. $800 figure. The high cost of the struggle with Israel includes a diversion of energy and resources from development, and that struggle also partially masks the political and religious conflict within Syria itself between orthodox Sunni Muslims and the exponents of secular pan-Arab socialist nationalism.

▶▶ Summary

The countries of the Mediterranean Crescent have attained a significant level of economic development; and, except for Algeria and Libya, that development has been achieved without the aid of massive oil revenues. Instead, Egypt, Israel, Turkey, and other Middle Eastern states have made progress by utilizing their agricultural, industrial, and environmental/cultural resources. That feature of development distinguishes the Crescent countries from the Gulf States, which are discussed in the next chapter.

Despite economic progress, however, the stability and future of the Mediterranean Crescent countries are jeopardized by severe political, ethnic, and religious tensions. The fascinating cultural mosaic that resulted from the interactions of the Middle East, North Africa, and Europe is also a source of conflict and instability. The civil war in Lebanon pitted different religious, ethnic, and political forces against one another and nearly wrecked the Lebanese state and economy. The Israeli-Palestinian dilemma is underlain by historical claims that are fundamentally opposed and require major compromise on both sides if the parties are to reach a mutually satisfying resolution. In addition, the fundamentalist Islamic revival movement, which has inspired profound change in and continues to dominate the government of Iran, has also had a major impact on the contemporary political scene in Algeria, Egypt, Lebanon, and the Palestinian Territory. The resurgence of more traditional Islamic values appears certain to contribute in major ways to political debate and conflict in additional Middle Eastern countries in the future.

▶▶ Key Terms

Ashkenazim

Aswan High Dam

Bedouin

Gaza Strip

intifadah

Jonglei Canal Project

kibbutz

Levant

Maghreb

moshav

New Valley Project

Organization of Petroleum Exporting Countries (OPEC)

Palestine Liberation Organization (PLO)

Palestinians

salinization

Sephardim

Tabqah Dam

West Bank

■ **Gas tanker, Qatar.** Gas and oil dominate the economies of the nations bordering the Persian Gulf.

The Gulf States:
Living on Oil

- **Regional Characteristics**
- **Development and the Petroleum Economy**
- **The Impact of Oil**
- **Petroleum Powerhouses: Isolation and Globalization**

I n the minds of most people, no region of the world is more intimately associated with oil than is the Middle East. That image is not unreasonable, for more than one-fourth of the world's known petroleum reserves are found there. The "black gold," however, is unevenly distributed within the Middle East; some nations, by the accidents of geology and discovery, have much greater oil reserves than do others (Table 24–1). So abundant are the proven reserves and the present or potential output of such oil producers as Iran, Iraq, Kuwait, and Saudi Arabia that they deserve independent treatment as a realm within the larger Middle Eastern region. The oil industry dominates the economies of these states, pouring more money into their national treasuries than the countries can spend, in the short run, on productive internal enterprises. With prosperity so closely tied to one resource, however, these countries must grapple with a complex array of developmental prospects and problems. And in Iraq, the past and potentially future oil-fueled prosperity of the country is enormously complicated by the turmoil and chaos that has followed United States–initiated military intervention to achieve regime change.

The overwhelming impact of oil on the countries that surround the Persian Gulf has dramatically altered their economies and shaken their social systems. Nations such as Iran and Iraq, which once had viable agrarian economies and an economic structure that mirrored the diversity and integration of the Mediterranean Crescent, were destabilized by the social changes that the expanding oil economy unleashed. Even countries that currently produce limited amounts of oil (such as Yemen, where oil production is just beginning to reach commercial scale) or that lack the resource entirely (such as several of the sheikhdoms of the United Arab Emirates) have become enmeshed in the oil economy of the region. Often their citizens migrate to neighboring countries in search of employment, complicating development initiatives in both societies, the labor-exporting and the labor-absorbing.

Thus, all of the countries in or near the Gulf have experienced change as a result of the exploitation of petroleum resources. In this chapter, we first explore other similarities and differences within the region and review how the petroleum-generated wealth has been put to use. We then look at the impacts and changes that have followed from petroleum-based development. Finally, we examine the nature of contemporary change in the three largest Gulf States: Saudi Arabia, Iraq, and Iran.

Regional Characteristics

Limited Resource Base

One of the most basic characteristics shared by the Gulf States is a limited natural resource base. Only hydrocarbons are available in abundance; other minerals are nonexistent or are found in insufficient quantities to make mining profitable under present conditions. Only Iran possesses enough mineral reserves to support a modern metallurgical industry. However, because much of the Gulf region, particularly its more mountainous districts, has yet to be explored by modern methods, some caution is justified in any assessment of the mineral resource potential of the region. At present, only petroleum and natural gas, often found in deposits of staggering amounts, are of great significance to the area.

Limited Agricultural Resources

A second characteristic of most Gulf States is that agricultural resources are limited. Most of the Gulf States suffer from extremely limited water supplies, which are concentrated either in highland areas, such as those in Yemen or western Saudi Arabia, or in lowland oases, where groundwater is close to the surface (Figure 24–1). Only in Iraq and Iran are substantial areas suitable for cultivation (Table 24–2). Iraq has both rain-fed agriculture in the north and irrigation potential in its arid south, based on the Tigris and Euphrates rivers. In Iran, the mountains and the Caspian seacoast receive appreciable rainfall, but the central core is very arid. Yet even in Iraq and Iran, less than one-fourth of the total land surface can be cultivated. In other Gulf States, most slopes are too steep or rainfall is too limited for nonirrigated

Table 24-1   Population and Petroleum Production in the Gulf States

Country	Population (millions)	Petroleum Production (in Thousand Barrels Per Day)			
		1980	1990	2000	2004 (estimated)
Bahrain	0.7	49	42	102.3	34.0
Iran	67.4	1,280	3,120	3,681.7	3,940.0
Iraq	25.9	2,600	2,083	2,566.7	2,070.0
Kuwait	2.5	1,400	1,080	1,765.0	2,050.0
Neutral Zone[a]	—	550	315	630.0	597.0
Oman	2.7	280	658	933.3	767.0
Qatar	0.7	470	387	688.3	782.0
Saudi Arabia	25.1	9,620	6,215	7,995.0	8,750.0
UAE	4.2				
Abu Dhabi		1,380	1,587	1,900.0	1,955.0
Dubai		350	469	280.0	350.0
Ras al Khaimah		—	10	0.5	0.7
Sharjah		10	35	50.0	48.0
Yemen	20.0	—	179	354.0	350.0

Sources: *Oil and Gas Journal*, vol. 78, no. 52 (December 29, 1980); vol. 88, no. 52 (December 24, 1990); vol. 98, no. 51 (December 18, 2000); vol. 102, no. 47 (December 20, 2004); and Population Reference Bureau, *World Population Data Sheet 2004* (Washington, D.C.: 2004).
[a] Area of shared authority between Iraq and Saudi Arabia

agriculture. Irrigation by surface flow is possible only in a limited number of isolated sites, usually near the base of mountains, where seasonal runoff provides water.

The traditional *qanat* system of tapping deep groundwater and bringing it to the surface by gravity flow originated in Iran. It is an ingenious response to water shortage and is most applicable on alluvial fans, sedimentary areas deposited by rivers at the base of mountains. But the *qanat* system can also be used to bring water long distances to regions far from the mountain water sources. Alluvial fans are the primary agricultural resources of the region. Soils on these slopes range from coarser textured, less desirable soils on the upper slopes to richer, finer textured, more water-retaining soils on the lower portions of the fans, where most rain-fed and *qanat*-irrigated crops and fruit are grown. *Qanat* construction and maintenance is the work of specialized communities, often organized by clan or village associations, who possess the engineering skills and practical experience to engage in the often risky subterranean tasks of building and repair. The attraction of this skilled labor to more lucrative jobs in the modern economy, and their migration to the sites where those jobs are found, often undermines the long-term productive agricultural base of their origin communities.

Population growth, agricultural intensification through irrigation, and urbanization all place great pressure on local water supplies (see the *Geography in Action* boxed feature, Jeddah's Urban Environmental Crisis). Overexploitation of groundwater, waste disposal, the economic limitations of desalting ocean water, and salinization of irrigated soils are all problems that relate to the region's limited water supplies.

Low Population Density

Most Gulf States are low in population density and have a small total population (see Table 24–1). Exceptions exist, however. Iran, Iraq, and Saudi Arabia all have large populations that dominate the regional scene, although large parts of their national territory are almost devoid of people. Even in countries with low overall population densities, the effective density is much higher because the population is concentrated in small, highly productive zones. Thus, in Abu Dhabi, a part of the United Arab Emirates (UAE), approximately three-fourths of the emirate's total population is concentrated in one large urban area of the same name (Figure 24–2). The contrast

■ **FIGURE 24-1   Date harvest in Iran.** Date palms are well-adapted to the heat and aridity of much of the Gulf States region and are an important source of food.

Table 24-2 Land Use in the Gulf States

Country	Labor Force in Agriculture				Land Use (1000 ha)
	Total (000)		%		
	1990	2000	1990	2000	
Iran	5,718	6,402	32.4	26.4	
Total Land					164,820
Arable Land					14,324
Permanent Cropland					2,002
Other[a]					148,494
Iraq	710	641	16.1	10.1	
Total Land					43,832
Arable Land					5,200
Permanent Cropland					340
Other[a]					38,292
Kuwait	11	9	1.2	1.1	
Total Land					1,782
Arable Land					8
Permanent Cropland					2
Other[a]					1,772
Oman	222	258	44.7	35.8	
Total Land					30,950
Arable Land					19
Permanent Cropland					61
Other[a]					30,870
Saudi Arabia	924	600	19.2	9.8	
Total Land					214,969
Arable Land					3,594
Permanent Cropland					191
Other[a]					211,184
Yemen	2,116	2,808	61	50.9	
Total Land					52,797
Arable Land					1,545
Permanent Cropland					124
Other[a]					51,128

Source: FAO Production Yearbook, vol. 55 (Rome: United Nations Food and Agriculture Organization, 2001); and World Bank, *World Bank Development Report 2001* (New York: Oxford University Press, 2001).
[a]Other = Seasonal pasture, forest, and wasteland

between small nodes of dense population with high levels of economic activity and the vast expanses of essentially uninhabited space characterizes the landscapes of the Gulf States.

Prominence of Religion

Equally distinctive is the role played by religion in the region. While Islam plays a fundamentally important part of people's lives in other parts of the Middle East, in the countries around the Gulf it assumes a particularly pervasive role. In what is now Saudi Arabia, Mohammed first revealed himself as a prophet and launched the preaching mission that, by persuasion and conquest, came to control the region both spiritually and politically. Here are the major centers of pilgrimage for Muslims:

Mecca and Medina (Figure 24–3, page 566). Never directly controlled by European imperialism, unlike their neighbors along the Mediterranean coast, the inhabitants of the Arabian Peninsula and of Iran, surrounded by mountain barriers and barren deserts, managed to remain aloof from substantial contact with non-Muslim cultures and their influence. In Saudi Arabia, secular law has never replaced the Sharia law of Islam. Instead, the Ulama, the religious teachers and scholars, have maintained a dominant influence over daily life. In Iran, where secular trends progressed rapidly under the monarchy, opposition led by Shi'ite Muslim clerics ultimately drove the Shah from power and substituted a much more traditional set of norms, values, and customs in behavior and dress. While traditional women veil in the countries of the Mediterranean

GEOGRAPHY IN ACTION
Jeddah's Urban Environmental Crisis

Jeddah is the second largest city in Saudi Arabia and has experienced enormous growth in the last 60 years. Important as the Red Sea entry port for pilgrims to Mecca, its population of more than 2.5 million people stretches for a distance of 50 kilometers along the Red Sea coast (Figure A). Many residential areas are poor quality, informal housing districts occupied by illegal residents, often pilgrims who have

decided not to return home. An oil-fueled economy with large numbers of job opportunities for both skilled and unskilled workers and Jeddah's coastal location have made it attractive for low-income and affluent residents alike.

Much of the higher-income residential growth has taken place north of the traditional center in a sprawling mix of residential developments with little linkage one to another, despite

■ **FIGURE A The Red Sea port of Jeddah, Saudi Arabia.** Tradition and change are closely juxtaposed in modern Saudi life. A mosque is reflected in the mirrored surface of a contemporary high-rise in downtown Jeddah, while pilgrims, supported by the infrastructure and accessories of modern travel, arrive at Jeddah's airport prepared to engage n the timeless rituals and acts of devotion of the *hajj*.

■ **FIGURE 24-2 The city of Abu Dhabi in the sheikhdom of Abu Dhabi, United Arab Emirates.** The oil boom has led to rapid urbanization. Fast-growing cities exhibit both the modernization supported by oil wealth and the shantytowns of poor migrants. In the countries with small populations, such as the emirates, the majority of the population may live in the capital city.

the existence of an overarching master plan and structuring highway grid. South of the traditional center are industrial zones interspersed with lower-income neighborhoods. The environmental problems faced by Jeddah are a product of multiple causes. Its natural setting on the Tihama coastal plain close to sea level creates a naturally high water table. With limited freshwater recharge from rainfall, groundwater with a high salt content is typical. Rapid urban growth has outpaced the ability of urban planners to provide centralized water delivery, drainage, and sewage treatment facilities, which has raised the water table further as heavily polluted effluents leak into the ground. This has resulted in structural damage to buildings and road surfaces as saltwater has seeped into building materials. Surprisingly in an economy awash with petrodollars, a lack of funds for basic infrastructure improvements has complicated dealing with Jeddah's environmental difficulties.

In a rapidly growing urban environment, one might expect that groundwater levels would decline as pressure to supply domestic and industrial water demands grew. But domestic water in Jeddah is supplied by desalinating Red Sea water, and the groundwater level continues to rise with increased water consumption and wastewater discharge! Public spaces within Jeddah are also watered at high levels to create an oasis-like, green environment along the verges of roads, in parks, and along the coastal promenades. This watering to maintain green space also contributes its share to the rising groundwater table.

This counterintuitive development is the result of two factors. Much of the domestic water delivery system, as well as the wastewater removal system, is faulty. Estimates suggest that as much as 30 percent of the desalinated water that enters the domestic delivery system is lost to leakage. The antiquated sewer system, where it operates, is equally porous. Unconnected households use belowground septic tanks to collect their wastewater. As groundwater levels rise, these septic tanks are frequently below groundwater level, making cross-contamination inevitable. Overburdened, perpetually full septic tanks must be pumped frequently and the effluent disposed of. Until

1997, these waste products were dumped into wadis east of Jeddah. Often this wastewater reentered the groundwater. Dumping of 5,000 tanker truckloads of sewage effluent per day much farther to the northeast of Jeddah is now required. This has created a large surface wastewater lake.

Excess water has serious complications for residential, commercial, and transportation infrastructure in this arid environment. The salts in rising groundwater damage concrete and asphalt. Building walls draw water upward by capillary action from the groundwater reservoir. The intense evaporation typical of deserts leaves crystallized salts embedded in the building materials. Expansion of the salt crystals eventually bursts the surfaces and causes the structural deterioration of buildings. Iron reinforcing rods are little help since they experience similar processes of decay. Barriers placed below buildings composed of plastic and asphalt sheets designed to pass water but not salts are expensive and only partially effective. Road surfaces also experience buckling and swelling as salt crystals expand and household wastes such as soapy water break down the chemical bonds of asphalt and concrete. Severe potholes and ridges mar the surface of many Jeddah streets. These deformities are further broken down by heavy trucks whose mass quickly exceeds the load-bearing strength of roads affected by surface deformation. This road degradation is analogous to the frost heaving and pothole formation that characterizes streets in temperate mid-latitude continental environments in the spring!

Coping with the scale of these environmental problems is difficult even in an oil-rich economy because local funding mechanisms are imperfect and the scale of the effort required is enormous. A new sewer system that includes all residential areas and processes wastewater so that it could be recycled in agriculture and used to irrigate public vegetation would help immensely in coping with Jeddah's environmental problems.

Source: P. Vincent. 2003. "Jeddah's Environmental Problems," *Geographical Review* 93 (3): 394–412.

Crescent, it is not a requirement for all women. In most of the Gulf States, it is. The separate social spheres of men and women are strictly maintained, females participate less actively in public life, educational and employment opportunities for women are more limited (see the *Geography in Action* boxed feature, Overcoming North African and Middle Eastern Women's Opportunity Constraints), and males tend to dominate household life and decision making to a greater extent than is true of the Mediterranean countries. In the countries of the Gulf, strict adherence to accepted norms, sometimes enforced by morals police, is the common pattern of social life to which foreigners as well as locals are expected to conform.

Weak Urban Traditions

A fifth commonality among the Gulf States is that, until oil development created new employment opportunities and sparked massive population movement to a few urban centers, much of the population in the region was engaged in traditional livelihood activities. These primarily included oasis

agriculture, rain-fed farming in the more mountainous areas, and pastoral nomadism. Thus, urban traditions are generally weak in the oil-rich states, and the educational and technological sophistication of the bulk of the population is limited. That characteristic stands in sharp contrast to the urban sophistication and historical importance of the main urban centers of the large states in the region—such as Tehran, Isfahan, Shiraz, Mashhad, and Qom in Iran and Baghdad, Basra, and Mosul in Iraq. In the smaller, more traditional states of the Arabian Peninsula, however, skills are almost entirely in the traditional sector of the economy and are unsuited to employment in many aspects of the oil industry.

That shortage of necessary skills has serious political consequences. It encourages the importation of skilled personnel from industrialized countries—as well as from India, Pakistan, and Palestinian areas—to fill specialized job niches. Dissatisfaction with that situation has led most of the oil-producing states to engage in vigorous educational and job-training activities in an effort to replace foreign experts with local personnel (Figure 24–4).

■ **FIGURE 24-3 Worshippers before the Kaba (center) in the Great Mosque at Mecca.** Islamic influence is especially pervasive in the Gulf States. Mecca is the leading center of Muslim pilgrimage.

■ **FIGURE 24-4 A university classroom in Oman.** In an effort to replace foreign experts with local personnel, most of the oil-producing states are pursuing vigorous educational and job-training programs. The presence of males and females in the same classroom, taught by a female teacher, contrasts sharply with standard practice in Saudi Arabia.

GEOGRAPHY IN ACTION
Overcoming North African and Middle Eastern Women's Opportunity Constraints

On the scale of women's personal freedoms and equal access to educational and economic opportunities as well as equal participation in social and political processes, there are vast differences in the countries of North Africa and the Middle East. These range from the secularized parliamentary democracy of Turkey with a former female prime minister, Tansu Ciller (1993–1996), to the theocratic monarchy of Saudi Arabia, where to this day women are not allowed to drive. The disparities are enormous and deeply divisive in a region that tries to emphasize its common Arabic and Islamic roots (Figure A). This constitutes a colossal waste of human potential, both socially and economically, where the most extreme interpretations of Islamic and cultural traditions insist on women's invisibility to uphold moral rectitude and family honor, be it behind anonymous robes and veils or behind impenetrable household walls.

If there is one single key that is of paramount importance to the empowerment and launching of women on a path of greater self-determination and diverse contribution to their respective societies, it is education. While the overall female illiteracy rate in North Africa and the Middle East is still high at 42 percent (2000) and almost twice that of males, great strides have been made in the recent past to accomplish universal schooling at least at the primary level in most countries of the region. Thus the illiteracy rate for females (and males) of the 15–24 age group is half that of the general population in societies whose constituents are overwhelmingly young and moving into their prime adult years in great numbers. Much, however, needs to be done in promoting and supporting secondary and continuing education. Too many women have to cut off their schooling too early because of family constraints of one form or another, whether financial or customary—even more enlightened fathers and husbands often do not see the point of further education in the face of marriage, child rearing, and all manner of family responsibilities. Only 20 percent of the region's formal labor force are women, as they are encouraged to forego income-generating experiences outside the home in favor of the traditionally accepted domestic roles that control the interactions between men and women and keep women respectably segregated.

■ **FIGURE A   Kuwait's first female cabinet member.** On June 20, 2005, over the vehement protests of conservatives in Kuwait's parliament, Dr. Massouma al-Mubarak celebrates after she was sworn in as a cabinet member.

Yet education is key as well to improving families' well-being. As women gain access to income-earning opportunities, relative poverty circumstances are alleviated and families progress into a higher, more middle-class quality of life. This is enhanced by broader knowledge and understanding of available options, which allows women to engage in promoting family planning, improving family health and nutrition, and setting a role model for the next generation of females (and males) in closing the gender gap.

Sources: F. Roudi-Fahimi and V. M. Moghadam, *Empowering Women, Developing Society: Female Education in the Middle East and North Africa.* Washington, D.C.: Population Reference Bureau, 2003.

International Relationships

A sixth characteristic shared by the Gulf States is the overwhelming importance of the oil resource to international relationships. The size of the oil deposits in the area is staggering. More than two-thirds of the known oil reserves outside North America, the countries of the former Soviet Union, and Eastern Europe are found around the Persian Gulf. Saudi Arabia alone may possess more oil reserves than any other nation—perhaps one-third of the world's reserves—and it is one of the world's top three producers. Iran, Iraq, Kuwait, Oman, and the United Arab Emirates are also major regional oil producers (see Table 24–1). By 2004 Kuwait's oil industry had doubled its pre–Gulf War I output, but Iraq's oil industry never recovered. Not only was 2004

petroleum production lower than the 2000 output, but also United Nations–sanctioned oil embargos limited legal production and export during the years between Gulf Wars I and II. With oil production and sales in this period nominally restricted to the level needed to purchase essential food and medical supplies, investment in oil industry infrastructure languished. Efforts to upgrade and modernize infrastructure, replace antiquated equipment, and generate postwar funds for reconstruction experienced severe difficulties. Inadequate security along pipelines and at production and storage facilities has left the oil industry exposed to acts of sabotage that will delay recovery for some time to come under the current circumstances.

Beginning with the discovery of oil in Iran in 1908, the pattern of location-and-development has spread

GEOGRAPHY IN ACTION
Conflict in the Gulf

The disputes that make the contemporary Persian Gulf area a volatile and unpredictable political setting mirror many of the unsettled issues influencing the entire region of the Middle East and North Africa. These conflicts have an impact far beyond their regional setting. The terrorists who toppled the Twin Towers in New York City on September 11, 2001, demonstrated conclusively that Middle Eastern extremists could make their point far from their base with tragic consequences. The American response—to seek regime change in countries presumed to harbor terrorists—has turned out to be quite different from the straightforward, low-cost promotion of democratic values that its proponents claimed. Rather, the United States and its extraregional allies have been plunged into a quagmire from which successful extrication will be the product of years of bloodshed and massive monetary expense. A fundamental failure to understand the region, its culture, and its swirling conflicts lies at the root of America's current entanglement in the Gulf.

Underlying all else is the fact that most Arabs think of themselves as part of one people, whose unity has been artificially disrupted by colonial boundaries. Thus, a threat to one Arab country, particularly when that threat comes from a non-Arab state, is regarded as a hostile act directed at all Arabs. But only about one-third of all Arabs live in countries that possess rich oil resources, and even in those wealthy countries the income from oil revenues is unequally distributed among the citizenry. Consequently, the Gulf States with great oil incomes are often viewed with hostility and envy by less fortunate and more dependent neighbors, even by fellow Arabs. Disagreements also exist between those with socialist and secular political ideologies and those who advocate a return to fundamental religious structures and controls. In addition, the unsettled conflict between Israel and the Palestinians exerts an important influence on the geopolitical scene. Thus, when nationalism, religion, political ideology, and equity issues are all mixed together, the situation that results is highly explosive.

Those volatile components have produced several notable explosions. Some, such as the struggle of the Kurds for an independent homeland, have remained confined and relatively isolated. Others, like the three major disputes of the past two decades, have been broader-based and have had more widespread impact.

Throughout most of the 1980s, Iran and Iraq engaged in a deadly struggle for dominance in the Gulf. The conflict was initiated by an Iraqi invasion of Iran in 1980, but a long history of disputes between the two countries preceded the outbreak of fighting. The immediate reason for the conflict was disagreement over control of the vital Shatt al-Arab waterway, the merged outlet of the Tigris and Euphrates rivers, which is Iraq's only outlet to the sea and also provides primary access to Iran's oil refinery at Abadan (Figure A). A 1937 treaty had given control of the waterway to Iraq, but in the late 1960s, when Iran felt strong enough militarily, it asserted its right to control the Shatt al-Arab and refused to submit the dispute to arbitration. At that time, Iraq was not strong enough militarily to contest that unilateral decision, especially since it was occupied in fighting its rebellious Kurdish population, also supported by Iran, in northeastern Iraq.

In 1979, when the monarchy of Shah Mohammad Reza Pahlavi collapsed in Iran, undermined by a fundamentalist Islamic revolution, Iraq's secularist government, led by Saddam Hussein and his Ba'ath party, tried to capitalize on Iran's internal chaos by invading western Iran. Saddam hoped to regain control of the Shatt al-Arab, end Iranian occupation of disputed islands in the Persian Gulf, suppress the Kurdish independence movement, and stop anticipated Iranian agitation among Iraq's Shi'ite population. A collapse of the Iranian government and state would have been an additional bonus.

The Iraqis failed to meet any of those objectives. Instead, a revitalized and enraged Iran forced an Iraqi withdrawal from most Iranian territory and followed with a series of costly assaults aimed at blocking Iraq's access to the Gulf and capturing the important southern city of Basra. Iran also hoped to drive out a leader, ruling party, and state structure that it regarded as evil and non-Islamic.

The battle raged inconclusively for many years. Iranian "human wave" attacks were met by Iraqi defensive tactics that featured elaborate entrenchments, dug-in tanks and artillery, and the use of chemical weapons. Iraq also employed destructive missile attacks against Iranian cities in an effort to weaken civilian morale. The losses sustained by both sides were huge; some estimates suggest that more than 1 million casualties occurred, with Iran suffering more than half of the total. Eventually, the brutal, grinding conflict culminated in a ceasefire in 1988, with neither country having the will or the resources to gain its objectives.

Less than two years later, Iraq's ambitions prompted another military adventure. On August 2, 1990, after articulating an escalating series of grievances and demands, Iraq sent its troops into Kuwait and 6 days later announced its annexation. The response of the international community was immediate; a series of UN resolutions condemned the invasion, applied economic sanctions, and authorized the use of force to evict Iraq if it failed to withdraw. The first **Gulf War** followed.

For the next six months, both sides assembled troops, moved in supplies, and prepared for a war that no one really wanted. The United States contributed most of the coalition forces, though substantial contingents came from Great Britain, Egypt, Syria, France, and Saudi Arabia, as well as smaller, symbolic units from other countries. Military conflict broke out on January 17, 1991, when coalition air forces began systematically bombing military and infrastructure targets in Kuwait and Iraq. After a short and decisive ground battle, the coalition forces were victorious. U.S. forces withdrew by mid-1991, yet many loose ends remained unresolved. These included international supervision of Iraq's military capability, security and autonomy for Iraq's Kurdish minority, and resolution of conflicting border claims between Iraq and Kuwait.

The outcome of Gulf War I was as inconclusive as the Iraq-Iran war that preceded it. Kuwait was reestablished as an independent government. The Iraqi Ba'athists were able to retain power when the Allies failed to push their military advantage

FIGURE A Shatt al-Arab waterway. Control over this vital waterway, which carries the combined discharge of the Tigris and Euphrates rivers to the Persian Gulf, has been a continuing source of conflict between Iran and Iraq.

and capture Baghdad and when Allied encouragement to revolts by Shi'ites in southern Iraq and Kurdish separatists in the north was not backed up with concrete support. The establishment of no-fly zones in the northern and southern thirds of Iraq, meant in part to protect dissident populations, kept the remnants of the Iraqi air force on the ground but did nothing to hamper the operations of Iraqi ground troops. Both the Shi'ite and the Kurdish rebellions were brutally suppressed. International sanctions were imposed to limit oil sales to essential food and medical supplies in an effort to prevent rebuilding of the Iraqi military into a regional threat, and international inspections took place to identify and destroy prohibited weapons. These efforts were a constant source of irritation between the Ba'ath regime and the international community, as was Ba'ath determination to assert its sovereign control of Iraqi space whenever the occasion to do so presented itself. The obvious and increasing stress placed on the Iraqi people through the sanctions program failed to produce successful opposition to the regime and its leader, Saddam Hussein, and instead generated substantial sympathy in the Arab world and beyond for the difficulties of the country's people. Into this setting of frustration, military confrontation, fear, and hostility that characterized relations during President Clinton's two administrations was thrown the murderous 9/11 terrorist attack on the Twin Towers of New York City.

The Bush administration seized on the attack as an opportunity to root out terrorist training camps in Afghanistan and the Taliban government that protected them. Determined to extend regime change to Iraq as part of a strategy to promote democracy in the Middle East, questionable intelligence estimates about Iraqi links to terrorist groups and possession of weapons of mass destruction were used to justify military intervention. Overwhelming resistance to the enterprise from most of the United States' traditional European allies was matched by the nonparticipation of any Middle Eastern countries that had joined as coalition members in the first Gulf War. Although the Ba'ath regime quickly collapsed in the face of superior military force, the apparent victory proved incomplete and ephemeral. A series of poor decisions about security issues followed inadequate planning for the postwar situation. Local armed resistance to the presence of non-Muslim foreign troops on Islamic soil grew, inadequate American and British force was present to suppress resistance, non-Iraqi Muslims radicalized by the presence of a foreign military on Muslim soil entered Iraq as Mujahideen (holy warriors) dedicated to the defeat of an alien presence, and the level of suicide bombings and other attacks steadily increased. Despite holding elections designed to legitimatize an indigenous government to which the United States could turn over power, attacks on Iraqis viewed as collaborators with the occupying power and on coalition forces continued unabated. Unable to provide basic security or protect vital infrastructure such as electricity generating and distribution facilities or oil pipelines, neither the occupation forces nor the elected Iraqi government has managed to provide a better life for ordinary citizens. A nightmarish situation in which the United States possesses sufficient force on the ground to avoid defeat but lacks sufficient force to impose a solution makes the possibility of a grinding, decade-long war of attrition a very real possibility. Few of the American strategic objectives are likely to be achieved in this situation, and the clear and immediate loser is the Iraqi people, caught in a crossfire between coalition and resistance forces.

southward through the coastal regions of the Arabian Peninsula. Not only have other petroleum deposits been found, but knowledge of the areal extent and extractable quantity of oil in existing deposits has also grown rapidly. In light of improvements in the technology of recovery, coupled with the traditional tendency to understate or otherwise obscure the reserve capacity, the continued prominence of the Gulf States in the world petroleum economy seems assured for the foreseeable future.

During the 1980s, it appeared that the Gulf States' dominance in petroleum production might diminish. A combination of global economic stagnation and energy conservation reduced demand significantly in some industrialized countries. In addition, new petroleum deposits were located in other parts of the world. But the political turmoil in Iran and the eight-year war between Iran and Iraq damaged the production, refining, and oil-transportation facilities in both countries (see the *Geography in Action* boxed feature, Conflict in the Gulf, page 568). At the same time, other Gulf States reduced their output in order to maintain higher prices and conserve oil reserves for possible development of petrochemical industries. The first Gulf War severely damaged Kuwait's oil facilities and it took time to bring its production back to the global market; Iraq's production remained limited throughout the interwar years. Rising demand elsewhere in the world, in China, India, and other developing countries with rapidly expanding economies, has helped to push oil prices steadily upward. Uncertainty about the security of existing oil production and delivery systems also contributes to volatility in oil prices, concerns that are heightened by so much of the world's production capacity being concentrated in one relatively small and potentially vulnerable region.

This dominance of the world's oil supply, as well as the increasing importance of the Gulf States in the global economy and politics, has allowed most oil-producing states to participate directly in their own oil operations, thus reversing previous arrangements. In the early years of development, foreign countries, often operating in a consortium such as Aramco (partly owned by several American oil firms), re-ceived concession rights to explore defined areas. Most Gulf State revenues were then generated from concession sales, royalties, and taxes paid on the oil that was extracted and exported. Local governments had few skilled administrators, and oil companies had a monopoly on extractive technology and market distribution in industrialized states. In addition, the impoverished and politically weak states of the Gulf region lacked the capital to develop their own oil resources.

Gradually, that situation changed. As administrative and technical skills were accumulated and a better understanding of the intricacies of oil economies developed, Middle Eastern countries demanded more favorable agreements with the oil companies. Tough new leaders appeared, and increased percentages of profits were returned to national treasuries. Greater pressure was placed on oil companies either to develop their concessions or to relinquish them to someone who would, and many countries nationalized at least a controlling interest in the firms that were developing their resources. In the world market, demand for oil outstripped the supply. Thus, nationalization and dramatic increases in oil prices took place in tandem, a combination of conditions unlikely to recur in the near future.

Local Impact of Oil Wealth

One final characteristic of the Gulf States is the tremendous impact that oil revenues have had on local societies and economies. No state in the region has been able to isolate itself from oil wealth and the stress it places on indigenous social and cultural systems. As a result, controversy has arisen over the role of traditional values in governing contemporary life. The wave of fundamentalist Islamic revival that was responsible for much of the opposition to the oil-rich Shah of Iran—who was deposed in 1979—stresses the importance of traditional values and patterns of behavior in opposing Western-style modernization (Figure 24–5). The more rapid and extreme the pressure for change, the more violent is the reaction from traditional centers of authority and belief. In Iran, the response to secular modernization and nationalism

■ **FIGURE 24-5 Islamic fundamentalist mural in Iran.** The Islamic fundamentalist revival stresses the importance of traditional values and opposes Western-style modernization.

has been so strong that it has swept away much of the new in its affirmation of traditional practices, although a struggle between proponents of more progressive and more traditional forces continues. The at least temporary reestablishment of dominance by more traditional supporters of Mahmoud Ahmadinejad, the conservative mayor of Tehran who was the unexpected winner of Iran's presidential election in June 2005, demonstrates the resilience of traditional patterns and the toughness of their advocates.

Dependence on Nonregional Suppliers

One additional characteristic that is not equally shared by all of the Gulf States, but is an important and disruptive force in most, is an increase in dependence on nonregional suppliers for basic necessities. Much of the food consumed in the Gulf States today is imported, and in many countries, a large part of the labor force is foreign. The bulk of the domestic workers are recruited from the Philippines and Sri Lanka, while Pakistanis have replaced Egyptians and Yemenis as the preferred source of labor in agriculture, construction, and transportation. Iran—by virtue of its size, diverse agrarian resource base, and large population—has been more immune to those pressures. Thus, as a consequence of oil development, none of the Gulf States has been able to remain isolated.

Development and the Petroleum Economy

The discovery and development of vast petroleum deposits have had a drastic impact on the economies and sociopolitical systems of the oil states. Large petroleum outputs and soaring oil prices have moved such former backwaters as Kuwait, Abu Dhabi, and Oman into international prominence. Formerly impoverished, the oil-producing states face the task of coping rationally with an embarrassment of riches. Although abundance has been the watchword of the recent past, the long-term future is by no means assured. Smaller states, such as Oman, have sufficient revenues for the present but must use their money wisely to prepare for the day when oil revenues are no longer available. Diversification of oil-based economies, development of industrial processes that do not depend on oil, and rejuvenation of often sluggish agricultural sectors are obviously high-priority items in the Gulf region.

The Gulf States have dealt with the eventual decline in oil revenues in two ways. Their first approach has been to invest petroleum income internally, both in development projects that generate long-term income and employment, such as expanding irrigated agriculture, building a petrochemical and plastics industry, and designing higher education institutions with regional appeal, and in social services that improve material well-being in the short term. The second approach has been to invest the capital that cannot be usefully absorbed internally in overseas enterprises.

Internal Investment of Petroleum Income

Large sums of petrodollars—the wealth gained from oil resources—are currently being expended to improve infrastructures in the Gulf States. Ports have been built, such as Doha in Qatar, where none existed before. Investments in airports, highways, sewer systems, water systems, and pipelines have drastically changed the appearance of most oil states. Housing projects are also common (Figure 24–6). For example, Oman has rebuilt its capital, Muscat, and in the process has destroyed much traditional architecture in order to construct modern housing units for its growing urban population. Most citizens see that change as an inevitable and progressive aspect of modernization.

■ **FIGURE 24-6 Muscat, capital of Oman.** Petroleum wealth has funded new housing and office construction throughout the Gulf States.

■ **FIGURE 24-7 Desalinization plant at Doha, Kuwait.** In an effort to support their growing populations, many Gulf states operate desalinization plants to provide water for drinking purposes.

Substantial sums are also being invested in agricultural improvements. In many oil states, desalinization plants are now in operation, providing water not only for drinking but also for greenhouse irrigation of vegetables to help feed the burgeoning urban populations (Figure 24–7). That type of agricultural operation, while not economically sound, is one way in which almost unlimited financial resources can help to reduce food imports. In Saudi Arabia, vigorous attempts have been made to diversify agricultural output by encouraging cereal and vegetable production. In addition, deep bore wells have been drilled to expand the agricultural area and these efforts have resulted in short-term production increases, although in many districts the cost in terms of degradation of the groundwater resource has been great.

Unfortunately, few of these agricultural development schemes are practical in economic terms, and most require a level of skill that is beyond the experience of local residents. In addition, the technology employed frequently brings with it serious land-management problems. Many of the irrigation schemes in Iraq, for example, have caused serious salinization of the soil as the application and drainage of irrigation water have been mismanaged. The conflicts of the last 15 years have made it difficult for Iraq to generate the capital and expatriate labor and expertise to rejuvenate troubled projects, although these difficulties did not stop the Iraqi government from organizing massive drainage projects in the marshlands of southern Iraq in an effort to address salinization and waterlogging problems.

Currently, little effort is being directed toward building on areas of local expertise. Especially neglected is the pastoral sector, where considerable traditional experience still exists and where past use of low-productivity rangeland has been well managed. Today most Gulf States import their meat from industrialized countries, such as Australia, even though the population prefers to consume local sheep, now largely unattainable at a reasonable price.

In Oman, the investment of oil revenue in agricultural development and industrial diversification is especially crucial because existing oil reserves, at current rates of production, are only expected to last until 2016. Although some new fields have been developed in southern Oman (Dhofar), and an active exploration program has been pursued, the prospects for major new discoveries are unclear. Development of natural gas resources, both in association with oil deposits and in unassociated locations, has resulted in important discoveries. At current production rates these are expected to last for the rest of the century. Because agricultural use of the land in Oman is limited by the country's small and unreliable rainfall, great expansion of farm holdings is not possible. Consequently, Oman aims to achieve food self-sufficiency both by improving the efficiency of existing agricultural systems and by developing underground water resources in a series of small-scale projects. These projects involve improvement of marketing and transportation facilities, discovery of new and more productive plant strains, control of plant diseases, and use of fertilizers and mechanized equipment wherever they meet the needs of local farming traditions. Investment in the agricultural sector is necessary in the oil-producing states to ensure economic viability in the future, but becomes less likely when lower-cost food supplies can be imported from abroad.

Industrial development is also taking place in the Gulf States. First priority is frequently given to the building of cement factories, in order to support the construction industry. Also common are petrochemical plants that produce fertilizers, chemicals, and plastics, as well as plants that liquefy gas for export. In addition, there are many consumer-oriented enterprises, including soft-drink bottling plants, flour mills, fish-processing establishments, and textile-weaving firms. Craft traditions often give focus to small-scale industrial development that caters to the tourist and export trade. The uniquely designed silver jewelry of Oman is one example. In many oil-rich states, carpets, slippers, pottery, brassware, copperware, and leather products, such as hassocks, unite productive development and cultural continuity (Figure 24–8).

Few resources to support heavy industry are known. Mineral exploration, however, is in its infancy, and major discoveries may yet occur. Reports have circulated of major finds of gold, silver, nickel, and copper in mountainous western Saudi Arabia. If those reports are confirmed, Saudi Arabia may be able to translate its oil wealth into a diversified economic base that includes an indigenous mineral-based industry.

The Gulf States are also attempting to overcome their shortages of skilled workers. Almost every country has invested in modern universities as well as the rapid development of a primary and secondary school system. To staff those systems, numerous foreign teachers have been hired, often recruited from more educationally advanced countries such as Egypt. Large numbers of students, however, also study abroad under governmental contract. When they return home, they are channeled into decision-making positions at all levels of the government. That newly acquired

■ **FIGURE 24-8 Traditional crafts store in Muscat, Oman.** A shop in the Mutrah Souk sells items that are based on traditional designs. The product of small-scale artisanal workshops, these goods appeal to the local populace as well as tourists.

expertise makes it possible to replace foreign personnel in the oil-producing states and take a tougher and more aggressive posture in dealing with the international oil companies.

Petrodollar Investment Overseas

Faced with an inability to absorb all of their oil revenues in internal investment projects, oil-producing states are seeking investment opportunities overseas. In some cases, such a transfer of funds represents the actions of wealthy private investors, but frequently it reflects the actions of the region's governments. Increasingly, those transfers of capital have involved the purchase of property, banks, farmland, or industrial enterprises in industrialized countries. Thus, the governments of oil-producing countries have become active investors in the industrial infrastructure of oil-consuming countries.

The Impact of Oil

Because of the extraordinary economic importance of the Gulf States to the world economy, oil revenues will continue to have a profound impact for at least the next few decades. That impact is most pronounced in three areas: international political relationships, internal social and economic conditions, and internal political structures.

Changing Power Relationships

Oil wealth has altered the world's balance of power in several ways. For one thing, most of the oil states are now able to use their rapidly accumulating wealth to purchase arms with which to modernize their military forces (Figure 24–9). Possession of sophisticated weaponry made it possible for Iraq to attack Iran in 1980 and prompted other Arab states

to purchase arms to counterbalance Iraq's power, as well as that of Israel. The result has been an arms race in the Gulf States, in which the industrialized nations supply weapons to various opposing factions. The fact that those weapons can also be used on brother Arabs was demonstrated by Iraq's invasion of Kuwait in 1990.

The region's dominance in OPEC also gives it substantial power on the world political scene. Proposals to raise prices or threats to embargo oil shipments to supporters of Israel have sent tremors through the industrial world. The oil-producing states are clearly aware of their position of influence and are not afraid to use it. The psychological satisfaction derived from such a superior political position explains much of the contemporary behavior of these states. They can no longer be ignored or dominated; they must be dealt with as countries of great power and influence.

The Gulf States' position of influence makes them important to countries around the world. During the cold war, both the Soviet Union and the United States competed for influence in the region, and any change in the region's internal politics takes on global significance. The immediacy and near unanimity of international opposition to Iraq's invasion and forced annexation of Kuwait was unprecedented, as was the readiness of both Western and Middle Eastern countries to send troops to defend Saudi Arabia and to force Iraq to relinquish its conquest. That extraordinary level of interest and concern was a measure of the geopolitical significance of the Gulf States in world affairs, as well as extreme reluctance to see national frontiers violated. Gulf War II did not elicit the same response from the international community and regional powers because they did not perceive Iraq as the same level of threat as in 1991 and regarded the United States as unreasonably aggressive in its position vis-a-vis Iraq's Ba'ath government.

■ **FIGURE 24-9 Armed forces parade, Tehran, Iran.** Many of the Gulf States have used much of their oil wealth to build up their military forces. The region has become one of the most militarized in the world and the focus of much international conflict.

■ FIGURE 24-10 Tehran, Iran. Rapid urbanization has been one of the social changes wrought by oil wealth. The population of metropolitan Tehran, Iran, has swelled to 11.5 million people.

The emergence of the **Gulf Cooperation Council**—including Saudi Arabia, Kuwait, Bahrain, Qatar, the United Arab Emirates, and Oman—as a regional organization was a local response to the need for a unified approach to common security concerns. But the council's relative lack of military power means that it must rely on support from other countries.

Changing Social Conditions

Developmental growth fueled by oil wealth has brought massive changes to the societies of the oil states. One important change has been rapid urbanization. Attracted by new job opportunities, social welfare programs, and an atmosphere of excitement and diversity, rural folks have flocked to urban centers. Doha, the capital of Qatar, went from insignificant village status to more than 300,000 inhabitants in a period of 25 years. Similar growth has taken place in the capitals of most of the region's states and sheikhdoms: The population of Baghdad, Iraq, now exceeds 5.5 million; the metropolitan area of Tehran, Iran, has grown to at least 11.5 million people (Figure 24–10); and Riyadh, Saudi Arabia, has passed the 5.4 million mark, closely pursued in the rank ordering of Saudi urban places by the rapid expansion of Jeddah to more than 2.5 million inhabitants. In some of the sheikhdoms, much of the population lives in the capital city; overall, nearly one-third of the 3.5 million inhabitants of the United Arab Emirates reside in Abu Dhabi, the capital. Al-Manama, Bahrain, is an urban center of more than 155,000, and that, too, represents almost one-fourth of that state's population. In those instances, city and state nearly coincide.

Secondary cities have also served as a focus of regional migration in the large Gulf States—for example, Mosul and Basra in Iraq; Isfahan, Tabriz, and Mashhad in Iran; and Mecca and Jeddah in Saudi Arabia. Much of the growth of those cities has been accompanied by the erection of shantytowns and other temporary housing on the outskirts of the urban areas. The sanitary and health problems that have

resulted from excessive crowding are gradually being relieved as oil money is invested in improved housing and infrastructure.

As centers of economic growth and change, the expanding cities are also the scenes of cultural conflict. Many rural migrants have difficulty adjusting to urban life in the Gulf States, just as their counterparts do in other parts of the world. Moreover, traditional Islamic and customary values are receiving their strongest challenge in the urban areas. Bombarded by a variety of exotic stimuli—ranging from the material possessions of the industrialized West to the clothes, movies, and behavior patterns of the foreign employees of oil and construction firms—the citizens of the oil-producing states have been forced to adjust to a new social setting (Figure 24–11).

A conflict of values can produce serious social strains. Traditional leaders often react violently to apparent violations of cultural and social norms. In Iran, much of the opposition to the shah was guided by religious leaders who objected to the pace and direction of social change. When they gained power, one of their first actions was to make religious law the basis for social relations and the legal system. The resurgence of fundamentalist Islamic approaches to social organization represents a profound challenge to the secular, modernist, Western-inspired governments of the region.

Equally profound are differences in work habits and social priorities. New urban arrivals seldom accept Western notions of allocating time and resources; they often apply their financial gains to social needs, such as the price of a bride or the obligations of kin, rather than to savings or job advancement. An inadequate understanding of respective cultural values easily results in tension between native workers and foreign employees.

Great distinctions in income and status have also appeared as a result of the oil boom. Although some of the newfound wealth has filtered down to lower social levels, much of the income gained is eroded by an inflationary spiral triggered by oil development and accelerated imports. Individuals close to the import trade, those serving as local representatives of foreign firms, and professionals of all sorts have benefited the most from petroleum growth. The unskilled have been left behind. Some states—Kuwait, for example—have instituted a progressive policy of free social services to improve living conditions, but others have been less farsighted. Often the gap between the ruling elite and the bulk of the population is wide, a condition that is particularly evident in states with low population densities and massive oil revenues but a characteristic that is also found in the region's more populous states (Figure 24–12).

Changing Political Allegiances

Social unrest stemming from cultural change and income inequities casts shadows over the political futures of many of the oil states. Opposition to the central governments in many

of those countries is coming from a variety of sources, including ethnic and religious minorities within the state, traditional leaders who are alarmed at the changes that are taking place, rural people who feel neglected by distant governments, and modernist forces that wish to accelerate the pace of change and radically restructure society. In addition, many oil states have large numbers of foreign workers within their jurisdiction who feel no particular loyalty to the existing government. The result is a highly unstable set of conditions that could explode into conflict at any moment.

The Kurds are an example of an ethnic minority in Turkey, Iraq, and Iran that has long struggled for greater autonomy, either within the structure of the existing states or through complete independence. In Iraq, the Kurdish population of the north was engaged for years in armed rebellion against the Arab-dominated central administration. Although a desire for greater regional autonomy was the apparent motive, increased access to revenues from the northern oil fields in or near Kurdish territory was certainly an underlying motivation.

In addition, revolutionary Iran has many sympathizers in the Shi'ite religious minorities in the Gulf States region. Thus, Iran's emergence as a regional religious power has numerous implications. Equally significant were the military and technological capabilities developed by Iraq during the 1980s and 1990s, which forced its Arab neighbors to ally themselves with outside political and military powers in order to confront Iraqi aggression. Caught between a rapidly

■ **FIGURE 24-11 The stock exchange building in Kuwait City, showing a reflection of the nearby state mosque.** This photo evidences the contrast between modern and traditional influences brought about by oil wealth.

■ **FIGURE 24-12 Camels passing a late-model vehicle at the desert-urban fringe.** The contrast between traditional and modern means of transportation is often striking in the Gulf States.

changing Iraq, a religiously zealous Iran, and the economic interests of the Western powers, the traditionalist societies of the Gulf States are finding themselves in a perilous situation. The U.S.-led 2003 war against Iraq, and the subsequent American and British occupation of the country, does not assure long-term political stability in this pivotal nation. The costs of managing the aftermath of the war and bringing calm to the region have proven to be enormous and are likely to continue for the foreseeable future.

Equally significant are governmental changes that might result from potential military coups. Young officers in many of the oil states possess nationalistic and pan-Arab ideals. Impatient with the slow rate of economic progress, the pronounced social and economic inequalities in many of the oil states, and the corruption that frequently characterizes a boom economy, they could use their positions to seize power. In addition, concern over excessive social permissiveness and the employment of foreign nationals could spark unrest in the lower levels of the military hierarchy and fuel opposition from traditional religious authorities and their supporters.

Petroleum Powerhouses: Isolation and Globalization

The Gulf States exhibit many contrasts. While the smaller political entities of the Gulf share many of the dilemmas of their larger neighbors, their issues are relatively tractable because the scale at which these difficulties are encountered is smaller. For the large states of the Gulf—Saudi Arabia, Iraq, and Iran—the policy dimensions of a large national territory, uneven distribution of population and resources, rapid economic development, and environmental change pose profound possibilities and problems. In each country, patterns of isolation contrast with pressures for linkage to distant parts of the globe. The outcomes of these pressures are distinct local responses to change, but in each case, a considerable degree of isolation of local society from the global community has resulted.

Saudi Arabia: A Desert Ruled by Islam and Oil

The Kingdom of Saudi Arabia occupies the bulk of the **Arabian Peninsula**, separated from the African continent by a flooded rift valley, the Red Sea (Figure 24–13). In structure, Saudi Arabia is similar to a giant block that is raised on one side and sunken on the other. Along the Red Sea coast, the western side of the Arabian block rises abruptly from the coastal lowland, with some of the peaks in the Asir and Hejaz mountains reaching 9,842 feet (3,000 meters) in elevation. The moderating influence of altitude results in both cooler temperatures and greater precipitation totals in the uplands than elsewhere in the country. From these highlands, the block gradually decreases in elevation toward the center of the peninsula. This arid central region is characterized by limited rainfall, usually dry streambeds called *wadis,* and depressions where groundwater occasionally is close enough to the surface to support oases.

In some places, most notably the Nafud Desert in the north and the Rub al-Khali (Empty Quarter) in the south, surface deposits of shifting sand dunes dominate the landscape. Further east still, along the Gulf coast, the block becomes a broad, featureless coastal plain. Historically, the generally arid climate of the peninsula precluded extensive agriculture, except in favored locations where runoff concentrated or groundwater was near the surface. Here the major mercantile centers, particularly Mecca and Medina, developed and served as the foci for economic exchange, a dominance that was enhanced by their location near one of the lower parts of the Red Sea mountain wall that made east-west travel easier and complemented the north-south movement along the Red Sea corridor. Elsewhere, outside of sporadically distributed oases, extensive nomadic pastoralism was the most common livelihood activity.

The discovery of oil in commercially exploitable amounts under the Gulf coast in the late 1930s transformed the Saudi economy. Although other minerals, including zinc, iron, copper, gold, and silver, are known to exist, they are not found in especially large deposits. This is not the case with petroleum, and Saudi Arabia's reserves are estimated to constitute from one-quarter to one-third of the world's total. Exploitation of these oil resources has breached the barriers of physical distance and relative isolation that had kept much of the Arabian peninsula peripheral to economic and political developments in other parts of the world before World War II.

As a center of Islamic faith and **pilgrimage**, Mecca and Medina had for centuries been open to contact with diverse cultures and political regimes elsewhere in the Muslim world. Economic interaction was invariably associated with these religious contacts, but these exchanges had always taken place within the matrix of a common overarching culture and system of values. The flood of foreign workers who came to extract and transport the petroleum, and later the Saudis who flocked overseas to acquire the skills and knowledge associated with Western technology, have posed a challenge to traditional values. Long identified with a reformist fundamentalist Islamic sect called the *Wahabis,* the **House of Saud** (the ruling royal family) struggles to balance the need to preserve, protect, and propagate traditional values and legal systems with the desire to accommodate modernization and change.

Through a series of 5-year plans, starting in 1970, the government has attempted to promote economic development. Initially, these plans concentrated on developing infrastructure. Roads, new ports, railroads, and airports have all been built on a grand scale where none existed before. As people have moved from rural to urban areas to gain access to the new benefits provided by development, attention has shifted to the provision of modern housing, hospitals, schools, and social services. So successful has this process been that much of the traditional urban core of the leading cities has been replaced by new structures. The construction of these projects has been accompanied by an influx of foreign **guest workers** from both the industrialized West and the developing economies elsewhere in the region (Yemen, Oman) and farther east (Pakistan, India, Philippines). Even though many Yemenis were forced to return to Yemen when their

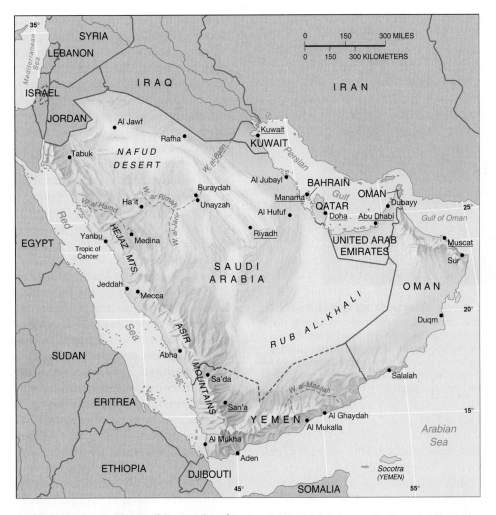

■ **FIGURE 24-13 The Arabian Peninsula.** The Arabian Peninsula contains the countries of Saudi Arabia, Yemen, Oman, United Arab Emirates, Qatar, and Kuwait.

government sided with Iraq in the Gulf War, the proportion of foreign workers remained high, with some observers estimating that one-fifth of Saudi Arabia's population was composed of guest workers. The emphasis on infrastructure has continued in recent plans but has shifted to the construction of processing and production plants for both agricultural and industrial commodities, such as the steel mill at Yanbu or the sugar refinery at Jeddah. Two sites in particular, Yanbu on the Red Sea and al-Jubayl on the Gulf coast, have become foci for industrial development. Development in both places features fertilizer factories, heavy industry (iron and steel mills), natural gas liquefaction facilities, and petrochemical plants (Figure 24–14). The export of petrochemical products has grown to such an extent that they are the second leading export after petroleum.

Development efforts in the rural countryside have achieved more mixed results. Agricultural areas in Saudi Arabia are widely scattered, small in scale, and totally constrained by the country's limited water resources. They are found in two types of settings: oases that occur whenever groundwater can be found near the surface, and in the mountains of Asir and Hejaz along the Red Sea. Only in isolated areas in these western highlands is rain-fed agriculture possible, often based on terraces constructed on mountain slopes or concentrated on isolated pockets of decent soil found in intermontane basins. Elsewhere, agriculture depends on tapping seasonal floods through water-spreading techniques or by extracting groundwater for small-scale irrigation. Until recently, these agricultural zones occupied less than 1 percent of the total land surface of the country. About 40 percent of the land was exploited by pastoral nomads who moved their herds seasonally between dry season locations near wells along the coast and impermanent streams and rainy season sites in the arid interior of the country.

Beginning in the 1980s, Saudi Arabia initiated a policy of agricultural self-sufficiency. This goal was attained for many foodstuffs by the 1990s. By 1991, for example, irrigated cereal production, widely dispersed in areas outside traditional agricultural zones, had expanded to such an extent that the country was able to become a major exporter of wheat and other cereals to deficit countries in the region. The strenuous efforts made to enhance agricultural production and the large sums of money invested to attain that goal managed to expand agriculture to a total of only 2 percent of the country's area. However, the rapid growth of Saudi agricultural production has only been possible because the

■ **FIGURE 24-14 Steelmaking in Saudi Arabia.** Lacking significant iron and coal resources, the raw materials for steel manufacturing in Saudi Arabia must be imported. Today scrap metal is a major input into steel, replacing much of the iron ore that used to be consumed, so local waste material has become an important resource. Here scrap metal is stockpiled outside the steel mill in al-Jubayl before being fed into an electric arc furnace.

government has made substantial subsidies available to local farmers. These subsidies have taken the form of price supports and credits with which to purchase fertilizer and other inputs. Production increases also have been based on the introduction of modern groundwater irrigation technology.

These systems have consumed tremendous amounts of water, an environmental subsidy that cannot be sustained over the long term. A measure of the enormous use of water by agriculture is the fact that it consumes almost 90 percent of the water used in Saudi Arabia each year. Yet, agriculture generates only 7 percent of national economic output and employs less than 10 percent of the Saudi labor force. This is neither a productive nor a sustainable use of a critically limited resource. Indeed, many farmers using traditional irrigation technology to cultivate vegetables have failed owing to an inability to water their crops caused by the rapid decline in groundwater levels in neighboring cereal production areas. This drop in groundwater levels makes extraction of the remaining water resources even more costly and is equivalent to carrying out a mining operation. Once removed from the ground, the groundwater cannot be replaced by nature in a time scale that is meaningful to humans. Concern about the decrease in groundwater, and the inability of dams and other surface-collection techniques to provide the water resources needed for substantial further agricultural development, has resulted in recent years in fewer subsidies to support agriculture and a decline in agricultural output. Support for the farming sector has not been matched by programs that benefit the country's traditional livestock industry, with most of the emphasis being placed on imported livestock whose poor

adaptation to local environmental conditions has contributed to overgrazing and loss of vegetative ground cover.

On the whole, Saudi Arabia has attempted to invest its oil income in projects that could benefit the country in the long term when oil revenues decline. Basic infrastructure and industrialization projects have been relatively successful. Agricultural development has followed a more checkered course. The reliance on non-Saudi imported labor, particularly in agriculture but also in construction and industry, has serious implications for both political and economic stability. The effort to achieve agricultural self-sufficiency based on open-field-irrigation production techniques is unsustainable in the face of the country's serious water and temperature constraints. Closed environment agricultural systems for vegetable and fruit cultivation, assuming that an adequate skilled labor force can be developed, might offer the prospect of greater long-term rewards.

The persistence of gender inequalities based on traditional cultural norms and the strictest interpretation of religious law impose substantial constraints on the development of the country's social and economic potential. The majority of the country's females, particularly in urban areas, lead a cloistered and controlled existence with limited unsupervised access to the world outside the home. This strict segregation by gender excludes the female half of the population from meaningful direct economic and political participation in society. Of the three large petroleum powers in the Gulf, Saudi Arabia is the most integrated into the global economy, the most willing to absorb and apply Western technology, the most interested in investing petroleum income in the industrialized countries, and the most conflicted and challenged internally by forces of tradition that struggle to reduce the scope of social change and to expand the sphere of traditional Islamic influence. How those conflicting forces are managed will determine the degree to which Saudi Arabia remains socially distant and detached or increases its contact with the rest of the world.

Iraq: Missed Opportunities

Iraq can be divided into three major zones: the desert; the land between the Tigris and Euphrates rivers; and the mountainous northeast (Figure 24–15). In the south and west of the Euphrates River is al-Wadiyah, the Iraqi portion of the Syrian Desert. This desert region is composed largely of *hammada* (stony) surfaces that are cut by *wadis* that drain eastward to the Euphrates. The presence of water and vegetation along these occasionally flowing watercourses makes them a site for winter–spring grazing by pastoral nomads who spend the dry summer season close to the Euphrates. **Al-Jazirah** (the island) dominates the central portion of Iraq. This region comprises the land between the country's two main rivers, the Tigris and Euphrates. Both of these rivers originate in the highlands of central and eastern Turkey and eventually flow into the Persian Gulf through a common channel, the **Shatt al-Arab**. In its more northerly extent, al-Jazirah is a region of moderate topographic relief; but about 62 miles (100 kilometers) north of Baghdad, the terrain becomes very flat and the river gradient extremely gentle. The rivers loop and twist across the broad river floodplain, and south of Baghdad near Basra the landscape was dominated by swamps and large areas of standing water before large-scale drainage projects were undertaken in the 1990s. The central and southern parts of al-Jazirah represent a structural depression between the block that constitutes the Arabian Peninsula and the rugged highlands of Iran and eastern Turkey. The deep alluvial deposits that fill this huge depression are composed of materials eroded from the surrounding uplands. The uplands in northeastern Iraq contain a series of parallel mountain ridges, which rise to heights of 9,842 feet (3,000 meters) near the border with Iran. Only in this northern zone is there substantial precipitation, falling largely as snow. Here the wet season, which lasts from November to April, produces up to 40 inches (1,020 millimeters) of precipitation, and the runoff from spring snowmelt feeds a series of streams that are tributary to the Tigris. Elsewhere the country is quite dry, and this aridity increases steadily southward toward the Persian Gulf. South of Bagh-

dad, the waters of the Tigris and Euphrates rivers are as essential to the life of Iraq as the Nile is to Egypt.

Outside the northeastern mountains, where cereals (wheat and barley) and fruit trees can be cultivated based on rainfall, agriculture in Iraq is dependent on irrigation. Mastery of water management is an ancient tradition, and control of the Tigris and Euphrates made possible the world's first urban-based civilization, the **Sumerian**, more than 5,000 years ago. In supporting major cities and large populations, the Sumerians and succeeding cultures (Babylonians, Persians, Parthians, and Sassanians) took advantage of differences in elevation between the two rivers that enabled them to shift water first from the Euphrates to the Tigris north of Baghdad and then from the Tigris back to the Euphrates south of Baghdad. Thus was created the world's first significant interbasin water transfer system.

If special topographic conditions have helped ancient and modern hydraulic engineers to manage water, other local environmental conditions pose greater difficulties. The flat landscape of the southern al-Jazirah makes drainage difficult. Whenever alternate year fallow cycles are shortened in an effort to increase agricultural production, poor drainage

■ **FIGURE 24-15 Iraq.** The heart of Iraq is the Tigris and Euphrates floodplain, a region known as al-Jazirah.

conditions promote waterlogging of soils. When combined with the high evaporation rates that are associated with southern Iraq's high temperatures, this can cause widespread soil salinization. The large amount of silt carried in the Tigris and Euphrates water also requires a constant effort to keep canals and distribution channels free-flowing. The relatively frequent spring flood episodes that particularly afflict the Tigris and its tributaries often breach the river levees and flood the nearby land, thus sustaining and expanding the many swamps of southern Iraq. Occasionally, these resource management problems, when combined with weak central governments, have overwhelmed the regionally integrated irrigation system and led to catastrophic declines in agricultural production and the populations sustained thereby.

In addition to the rich alluvial soils of the Tigris and Euphrates floodplain, Iraq has one other abundant resource. This is the oil with which the country is so generously endowed. Iraq's reserves are huge, trailing only Saudi Arabia and Iran in scale among Middle Eastern countries. Petroleum development has taken place in three locations. The largest fields are in the north between Mosul and Kirkuk. In the south near Basra is the Rumaylah field near the frontier with Kuwait, while north of Basra and east of Al-'Amarah near the Iranian border are several smaller oil fields. The third field is located northeast of Baghdad adjacent to the frontier with Iran. All of Iraq's oil fields have posed security problems for the government. Those located close to the Kuwait and Iran borders can be easily threatened by a hostile neighbor. The northern fields near Kirkuk are located in or near predominantly Kurdish regions, and Kurdish aspirations for autonomy or independence invariably are linked to demands for a greater share of and say in the distribution of the wealth generated from oil production (Figure 24–16). Understandably, this has alarmed governments in Baghdad, which have had no intention of seeing their oil revenues diminished. A further security problem for Iraq is derived from past decisions to export oil via overland pipelines across Turkey, Syria, and Jordan. These routes take Iraqi oil to terminals much closer to the major markets in Europe, but the pipelines are vulnerable to pressure from their host countries. For some time, the routes through Syria have been closed, and a planned pipeline across Saudi Arabia to Yanbu has not been built. Instead, Iraq has sent its oil via pipeline to Fao at the head of the Persian Gulf, where an oil terminal has been built. Iraq has small quantities of other minerals (copper, iron, lead, zinc, gypsum, and coal), but these are largely undeveloped. Some industrial plants based on petroleum refining have been built, but this infrastructure was heavily damaged in the wars. Much of the income realized from the sale of petroleum has been invested in urban infrastructure and in basic agricultural development.

The problems that confronted the irrigation manager of 3,000 years ago are the same for agricultural development and management today. The foreign exchange earned by exporting oil has made it possible to make huge investments in irrigation infrastructure. These investments began under the Hashemite monarchy that was overthrown in a bloody coup d'etat in 1958 and continued throughout the subsequent **Ba'ath** ("revival") **Party** dictatorships. These agricultural development schemes focused on three goals: reducing the impact of floods, expanding irrigated agricultural land, and rehabilitating degraded areas. The effort to reduce floods has involved building dams on the major tributaries that contribute floodwater to the Tigris. A related benefit is that these dams provide hydroelectricity to power the irrigation system and fuel urban and industrial development. Another flood prevention effort involves diverting floodwater into depressions where it can be temporarily stored. The Wadi ath-Tharthar, a long, shallow depression in al-Jazirah north of Baghdad and midway between the Tigris and Euphrates, is the largest of these projects. During the dry summer and autumn seasons, water stored in the

■ **FIGURE 24-16 Female Kurdish militia in Sulaymaniyah, east of Kirkuk.** The *peshmerga* (militia) of the Patriotic Union of Kurdistan party (PUK) enrolls females as well as males in support of Kurdish aspirations for greater autonomy.

Wadi ath-Tharthar is released into the irrigation system to permit double cropping of the existing fields.

An extensive series of water control works (dams, barrages, reservoirs, and diversion structures) are under construction throughout the lowland irrigation zone. Together with new dams at Hadithah on the Euphrates and Mosul on the Tigris, begun in the mid-1980s, the hope is that the amount of cropped land can be doubled, thus meeting the agriculture extension objective. Since only about a quarter of the potential farmland in the floodplain is actually cultivated, extending agriculture to new areas is an attractive option.

This goal is limited primarily by available water and by the Euphrates water claims of Turkey and Syria, which are in direct competition with the needs of Iraqi farmers. As the upstream countries build more water storage dams, less water is available to reach Iraqi fields. In this sense, Iraq suffers from a classic problem affecting all water users located at the "tail" of a water delivery system. Lack of political agreements on how to share water with other major users in the Tigris and Euphrates river basin makes Iraq much more vulnerable to water diversion than Egypt, which is located in a similar position but has treaty agreements with other water users in the Nile watershed.

In addition to building water storage facilities within its own political space, rehabilitation of damaged agricultural environments is an increasingly attractive option. Rejuvenating existing development schemes, such as the Greater Musayyib project southwest of Baghdad near the ancient (now ruined) city of Babylon, whose productivity was seriously reduced by salinization by the early 1970s, has also been undertaken on a large scale. Efforts to drain the southern marshes also occurred in the mid-1990s (Figure 24–17).

■ **FIGURE 24-17 Marshes of southeastern Iraq.** For millennia the wetlands of the lower Tigris and Euphrates river floodplains supported a people known as the Marsh Arabs. Many of the marshes have been drained in recent years as a consequence of Iraqi political and economic policy.

These extensive wetlands were the homeland of the **Marsh Arabs**, whose unique local culture survived for centuries by combining agriculture, cattle rearing, and fishing. Drainage has destroyed both the basis for the Marsh Arabs' livelihood system and the habitat for local fish and bird populations and has left extensive plains of cracked and barren soil. Little positive agricultural development has yet to replace the displaced traditional production systems that occurred in these areas.

To the government, the destruction of wildlife habitat and marsh ecosystems seemed justified because these districts harbored Shi'ite rebels opposed to the regime of Saddam Hussein. How these changes in southern al-Jazirah will impact date production, which is widespread in the south and historically generates the bulk of Iraq's agricultural export earnings, is unclear.

The Iran–Iraq war, which erupted in 1980 and continued for 8 years, had a serious impact on Iraq's economic development (see the *Geography in Action* boxed feature, Conflict in the Gulf). Although efforts were made to continue the plans for agricultural growth, urban infrastructure modernization, and industrial diversification, huge sums of money were diverted into military expenditures and activities. Terrible losses of human life and material infrastructure resulted from this dispute over territory and control of the Shatt al-Arab waterway, which is Iraq's only outlet to the sea. Extensive efforts were made to develop weapons production facilities and weapons of large-scale destruction, thus diverting resources from civilian production enterprises.

In the two years after the end of the war with Iran, Iraqi efforts to develop momentum in rural and urban development projects were beginning to achieve results. Then Iraq invaded and annexed Kuwait in a dispute over boundaries and oil development activities. Driven out of Kuwait by a coalition of countries led by the United States, massively bombed in an aerial campaign that destroyed a significant portion of its modern infrastructure, weakened by internal revolts by Shi'ites in the south and Kurds seeking autonomy in the north, and severely restricted by an economic embargo imposed by the United Nations in a dispute over the identification and removal of weapons of mass destruction, Iraq's people found the last decade of the twentieth century to be an exceedingly difficult and discouraging period. Despite the existence of a United Nations–authorized oil-for-food program that allowed Iraq to export all the oil it wished if the money earned was expended on humanitarian assistance for poor Iraqi citizens, the health, nutritional status, and poverty of the bulk of Iraq's population verged on the catastrophic.

Conditions have hardly improved for the ordinary Iraqi citizen in the twenty-first century. The American invasion of Iraq in 2003 in pursuit of regime change has introduced a major element of volatility into an already unstable region. While the Ba'ath party regime of Saddam Hussein was quickly displaced, no replacement acceptable to a majority of Iraqis existed. The struggle to suppress increasing opposition—first to the occupying American military and its allies and later to an elected, Shi'ite-dominated government—has resulted in escalating violence. In the volatile and violent environment of contemporary Iraq, neither the occupation authorities nor the newly elected government have managed to provide regular basic services such as electricity, gasoline, or security. The American military presence in Iraq has served as a magnet for fundamentalist Islamic fighters who have come to Iraq in an effort to frustrate American objectives and to confront directly an alien intrusion into the Muslim body politic. In the chaotic conditions that characterize contemporary Iraq, creating a sustainable future that reconciles the conflicting objectives of Shi'ites, Sunnis, Kurds, and religious and ethnic minorities is exceedingly difficult, and perhaps not possible.

Iraq was ruled since a military coup in 1963 by the authoritarian Ba'ath Party government committed to the radical economic, educational, and social transformation of its people and institutions, by persuasion if possible, by force if necessary. The technology and equipment for this transformation were imported from the outside world. Yet Iraq's policies in foreign affairs isolated the country politically, involved the state in conflicts that destroyed much of the physical infrastructure built with oil revenues, saddled the economy with a huge debt, and reduced the bulk of its people to near subsistence status in terms of health care, nutrition, and income. Unwise ambitions, decisions, and actions isolated Iraq every bit as effectively as the selective social

isolation that has characterized Saudi Arabia. The outcome for ordinary Iraqis will remain disastrous for the foreseeable future. Whether Iraq can eventually reverse this pattern, realize its considerable development potential, and play a positive political role in the region is the primary question of the country's longer-term development course.

Iran: In Search of an Identity

The structure of Iran's physical environment is analogous to a traditional donut—raised areas surrounding a central hole. The central core of Iranian political space is a low plateau approximately 3,300 feet (1,000 meters) above sea level (Figure 24–18). This central area mirrors the emptiness of the donut hole because little life exists and little activity takes place there. Cut off by mountain ranges to the west from exposure to moisture-bearing cyclonic storms, this central plateau is an exceedingly barren region with summer temperatures that often exceed 113°F (45°C). In the northern half of the plateau is the Dasht-e Kavir, a desert surface dominated by extensive salt pans and salt crusts, while the southern Dasht-e Lut is an equally barren environment dominated by mobile sand dunes.

To the west of the plateau, the Zagros Mountains rise to heights in excess of 9,842 feet (3,000 meters). Composed of a series of parallel ridges, the Zagros uplands often receive

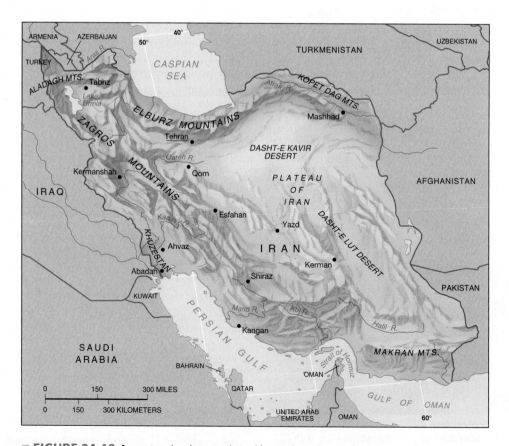

■ **FIGURE 24-18 Iran.** Iran is a large nation with a variety of natural environments and resources. The core of the country is a very dry desert with limited economic potential. Most of the country's mineral and agricultural resources are concentrated in the higher productivity mountain environments surrounding the desert.

winter precipitation of more than 39 inches (1000 millimeters). Some of the streams that originate in the Zagros drain westward to the Tigris River and water the intervening valleys and basins in the Zagros, which often contain rich soil resources. The Elburz Mountains south of the Caspian Sea, and the adjoining Aladag Mountains to the west and Kopet Dag Mountains to the east, are formidable barriers. Damavand Mountain (18,602 feet/5,670 meters) northeast of Tehran is the highest mountain in the Middle East. The northern upper slopes and peaks of these ranges receive as much as 79 inches (2000 millimeters) of precipitation annually, creating a moist environment in sharp contrast to the semiarid conditions that typify the area around Tehran less than 31 miles (50 kilometers) to the south. Lower mountain ranges rim the eastern side of the central plateau, while the southern Makran Mountains complete the enclosure of the central plateau. The intensely arid south, where annual precipitation totals rarely exceed 4 inches (100 millimeters), is the country's driest region.

Iran's most important export mineral resource is oil, which is found in abundance in **Khuzestan**, the lowland extension of the alluvial floodplain east of the Tigris River and the Shatt al-Arab. Iran is also blessed with huge natural gas reserves, the second largest in the world, although except for some exports to Russia this resource is not a major factor in international trade. The largest natural gas fields are found along the Persian Gulf near Kangan, but significant deposits also exist near the Caspian Sea and in the extreme northeast near Mashhad. Other minerals, such as coal, iron, copper, and zinc, are exploited in a modest way for the domestic market but little is exported.

With the exception of a few oases on the central plateau and the Jaz Murian basin in the Makran region, the bulk of Iran's population is located in the valleys, interior basins, and coastal plains associated with the mountain rim. A rich agricultural tradition dominates these mountain regions. Much of the agriculture in the uplands is based on the rain-fed farming of cereals, particularly wheat and barley, or on the cultivation of sugarcane, vegetables, and tree crops based on groundwater that is brought to the surface by *qanats*.

The productivity of the traditional water collection and transportation technology has been compromised in many areas by the introduction of gasoline-powered pumps that draw down the water table. Development efforts to construct dams in the uplands in order to retain water in cool zones with potential for irrigation have also altered the agricultural system. It has permitted the introduction of new crops, such as sugar beets, to expand the range of agricultural products. This policy has been helped by the existence of many good dam sites in the passes (*tang*) that occur where permanent streams break through the parallel ridges of the Zagros Mountains.

Two regional specialty crops are particularly important. Rice is cultivated on the Caspian Sea coastal plain based on paddy irrigation techniques that use the abundant water resources of the Elburz Mountains. Pistachios are a valuable crop in the dry coastal plain of Khuzestan. Animal husbandry is also important, particularly in the western mountains. Here pastoral nomads still move their animals in great seasonal migrations from winter quarters in the lowlands of Khuzestan and the foothills along the Persian Gulf to summer pastures high in the Zagros. These movements often can last for 2 months in each direction and cover a round trip of more than 370 miles (more than 600 kilometers). However, government pressure promoting sedentarization, challenges to traditional pastoral access rights to grazing, and the expansion of agricultural areas along the migration routes have all conspired to reduce the size and scale of pastoral nomadic migrations.

The population of these upland basins and valleys has a rich urban tradition, with each substantial basin dominated by a major city. Tehran, Esfahan, Mashhad, Tabriz, Shiraz, Ahwaz, Kermanshah, Qom, Yazd, and Kerman all dominate their local regions. Only Tehran, the capital, which expanded rapidly during Iran's last monarchy, has grown to great size as rural residents and middle-class urban professionals from the provinces have flocked to the city. Tehran is more than four times the size of Mashhad, the next largest urban place.

Economically, these cities are dominated by the **bazaar** (Figure 24–19). This is not only a physical location in each city, such as a street or cluster of streets with covered roofs where shopkeepers and merchants are located, but also an informal network of family, craft, and professional associations that manage and control the major traditional industrial and commercial activities of the economy, such as textiles, carpets, metalworking, and food processing. The support of the bazaar has always been exceedingly important for the success and stability of the political elite (monarchy, government bureaucracy, military, financiers and industrialists, large landowners, and professionals). Another crucial group in society is the Islamic clergy (***mullahs/***

■ **FIGURE 24-19 Iranian bazaar.** Bazaars, such as this one in the city of Esfahan, are Iranian marketplaces that may sell a wide range of consumer items, including high-quality carpets.

mujtahids), who traditionally have been responsible for interpreting religious law, providing basic education, serving as notaries for legal documents, and administering religious endowments (*waqf*). Support from the religious hierarchy, and particularly from the most prestigious group of religious scholars, the *mujtahids,* some of whom may hold the honorific title of **Ayatollah**, is crucial to any government because these religious figures are regarded as the primary protectors of both spiritual values and secular justice.

The **Pahlavi dynasty** (1925–1979) tried to promote modernization in both society and economy. These modernization efforts were pushed relentlessly regardless of the problems that resulted or the opposition expressed in numerous segments of society. Many of the government's efforts challenged the privileged position of the *mullahs* in Iranian society. A secular legal system was created outside the traditional religious court system, and *mujtahids* were barred from serving in these courts. Religious endowments were placed under the control of the bureaucracy, and some of the lands held by the *waqf,* along with the estates of large secular landowners, were redistributed in land reform programs. An alternative secular educational system was also created. The veil was banned, and women, particularly from the more elite social groups, took advantage of educational opportunities, were accorded the right to vote, and began to enter public life. Efforts to modernize the economy resulted in the importation of western technology and increasingly exposed educated Iranians to other value systems and lifestyles. Foreigners flooded into the country to sell goods, provide training and technical assistance, and to advise government ministries. These developments did not find widespread favor with the *mullahs,* whose personal positions and traditional Islamic values were threatened by the rapid change. The ensuing corruption, waste, mismanagement, and abuses of power prompted protests from leaders of progressive political parties, intellectuals, and the bazaar, who found common ground with the *mullahs* in their opposition to many of the changes.

The "**White Revolution**," a package of social changes promoted by the monarchy, was intended to further the modernization process, increase literacy among the rural population, extend land reform and land redistribution, nationalize forests and pastures, encourage profit sharing with industrial workers, and increase worker and farmer representation in local decision-making councils. All of these measures attacked traditional vested interests in some fashion and attracted fierce opposition from the Islamic clergy, whose spokesman became the Ayatollah Ruhollah Musavi Khomeini, a leading religious figure in Qom. Exiled in 1964 for his outspoken public attacks on the government, Ayatollah Khomeini's unwavering opposition eventually galvanized opposition to the Pahlavi monarchy and its institutions, particularly the dreaded secret police (SAVAK); gained the support of the bazaar; and consolidated an eclectic amalgam of opposition groups from across the political spectrum. In 1979, after a lengthy period of strikes, street demonstrations, riots, and changes of government, the monarchy collapsed,

Khomeini returned, and a revolutionary government based on Islamic principles was established.

This new government has followed practices and policies that have alienated it from many of its neighbors and promoted its relative isolation from the rest of the world. Many of the progressive social policies of the Pahlavi monarchy, particularly with respect to women, were reversed, and strict interpretations of Islamic law were enforced. Seizure of the U.S. embassy in Tehran on November 4, 1979, and the detention of diplomatic personnel as hostages for more than a year, caused serious deterioration in the already unfriendly relations with the United States. These relations have never recovered. Iranian support for revolutionary groups elsewhere, particularly those regarded by the United States as terrorist organizations, has worsened the situation even more. Iran's determination to pursue research in support of nuclear energy development, most recently reaffirmed by the newly elected president, Mahmoud Ahmadinejad, raises fears that Iran is actually seeking to acquire nuclear weapons. The Iranian nuclear program, for many years carried out at secret facilities, is the prime reason that led President Bush to identify Iran as part of an "axis of evil." Although European countries in general are more willing to pursue a less confrontational approach, there is little doubt that concern about Iran's nuclear objectives contributes to its relative international isolation.

Despite moderation of positions in recent years, relations between Iran and the United States remain frigid. Iranian disenchantment with Saudi Arabian connections with the United States has produced an additional set of alienating events. Accusing the Saudi government of abdicating its responsibilities as the protector of holy Islamic sites in Mecca, Iranian *hajjis* provoked violent confrontations with the Saudi authorities. The resultant riots led to a ban on the participation of Iranian *hajjis* in the pilgrimage to Mecca. The 8-year war with Iraq led to further isolation of Iran, since most Arab countries supported Iraq. Air and missile attacks on Iranian oil production and shipment facilities and the diversion of about half of Iran's economic output to military objectives during this period served to drastically reduce Iranian participation in the international economy. This self-imposed isolation has diminished somewhat during the last decade, but Iran remains a very introspective, authoritarian, internationally disconnected country whose isolation is inspired by a strong self-perception of its spiritual purity and moral superiority over other states. Nonetheless, internal opposition to the dominance of theocratic influence exists, although this opposition is often disjointed and incoherent in comparison to conservative forces. The election in 1997 of a reformist president, Muhammed Khatami, whose campaign called for granting greater social freedom to Iranians, and the strong showing of reformist candidates in the February 2000 parliamentary elections indicate that pressures for greater liberalization exist. Yet conservative activists are dedicated and tenacious, and in June 2005 these forces successfully elected a hard-line conservative, Mahmoud Ahmadinejad, as the successor to President Khatami. This indicates that a period of social retrenchment is in Iran's immediate future, and that economic issues spawned by a sluggish economy also play an important role in addition to social and religious concerns in the decisions of Iranian voters. There is also every reason to believe that it might be a long time before Iran finds a more moderate path between its religious traditions and the opportunities and challenges of the contemporary world.

▶▶ Summary

The petroleum economy has changed and continues to alter many features of the political, economic, and social fabric of the Gulf States. Prospects for considerable instability—aggravated by interstate rivalries, the uncertainties of the Arab-Israeli conflict, and the aftermath of three major Gulf Wars during the last 25 years—are very real. Traditional lifestyles are struggling to adjust to the forces of change. The challenge facing the Gulf States is the harnessing of their petroleum wealth to promote the enduring development of their economies. How well they can accommodate development and change without drastic alteration in traditional values will determine the degree of stability and lasting change the region will experience.

▶▶ Key Terms

al-Jazirah	Khuzestan
Arabian Peninsula	Marsh Arabs
Ayatollah	*mullahs/mujtahids*
Ba'ath Party	Pahlavi dynasty
bazaar	pilgrimage
guest workers	Shatt al-Arab
Gulf Cooperation Council	Sumerian civilization
Gulf War	"White Revolution"
House of Saud	

▶▶ Review Questions

1. Life in the Middle East and North Africa is determined to a significant extent by the presence of sufficient water resources. Name the three major sources of water and explain the geographical distribution of each.

2. In a region of limited water availability, efficient water management is the key to sustaining livelihoods. Explain the difference between dry, run-on, and irrigated farming, indicate where each technique is typically found, and explain why it is found where it is.

3. The maintenance of goat, sheep, and camel herds allows the use of vast dry areas suitable only for animal grazing. Explain the advantages of pastoral nomadism and outline the major problems of sedentarization.

4. Deforestation, salinization, sedimentation, overgrazing, and groundwater depletion are all forms of land degradation resulting from a resource use that is too intense to be sustainable without harm to the environment. Give examples for each and outline how these negative effects might be avoided.

5. Islam is the dominant religion in the Middle East and North Africa. Identify the values embodied in the Five Pillars of Islam and compare them to similar beliefs and practices in your own faith.

6. Most of the Mediterranean Crescent countries are not endowed with major oil resources to support their economic development. Give examples of how these countries have managed to make progress in increasing their populations' livelihood options without the benefit of oil wealth.

7. Big dam projects both solve and create many problems. List advantages and disadvantages of the Aswan High Dam and evaluate the balance between the benefits and costs of such projects.

8. The conflicts brewing in the Middle East are as much over resources as they are over ethnic and religious differences. One such resource is water. Explain the issues that arise when a major river system such as the Tigris-Euphrates watershed has to be shared by several countries.

9. The oil wealth of the Gulf States is both a blessing and a curse. Outline the development options this wealth has made possible and the conflicts flaring in the region because of it.

10. Characterize the structure of the traditional Middle Eastern city and the functions of its parts. How have Middle Eastern cities changed under the pressure of such processes as colonialism, rural-urban migration, globalization, and industrialization?

▶▶ Geographer's Craft

1. In which parts of the Middle East is land degradation a particularly serious problem? How might a geographer help devise strategies that contribute to a solution of such environmental problems?

2. Drought and dust storms are environmental hazards that afflict all Middle Eastern countries. What strategies can you suggest that would diminish the impact of these natural events?

3. Extending irrigation to desert areas previously unexploited by agriculture is a development strategy commonly proposed by development planners. What potential for this intensive form of agricultural growth exists in the Middle East, and what problems might the geographer expect to encounter in carrying out this development initiative?

4. Planting trees as a barrier to desert expansion and as a way to improve environmental conditions appears to be an attractive option to many Middle Eastern countries. If you were in charge of a forestry department in a Middle Eastern country, how would you implement such a strategy and what problems would you have to try to overcome?

5. For countries without oil revenues, resources for economic development and environmental preservation are difficult to obtain. How might these resource constraints be overcome?

PART NINE

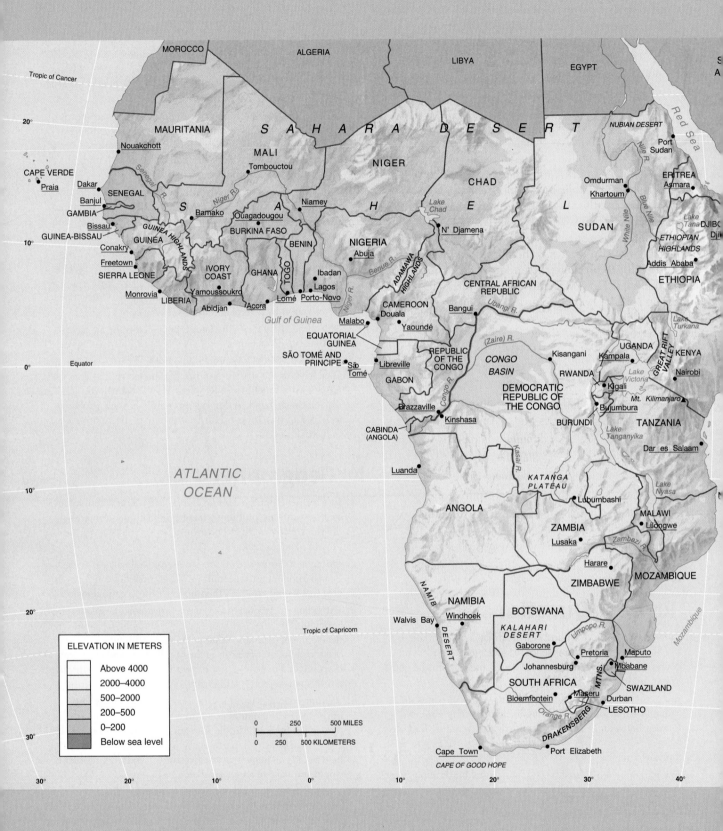

Africa South of the Sahara

Samuel Aryeetey-Attoh

For most people, Africa evokes a variety of images: rain forests, deserts, drought, famine, wars for independence, and great stores of natural resources. More recently, images of Rwanda, Sierra Leone, and the Democratic Republic of Congo (Zaire) come to mind. In this part, we examine the features that have given rise to those images. Throughout our discussion of Africa, however, we must always remember that much remains hidden, for Africa South of the Sahara is a vast and varied region, home to almost three-quarters of a billion people.

Most black African peoples have achieved self-governance. And even the Republic of South Africa has instituted a multiracial government. Despite Africa's long cultural history, most of the region's countries have been independent states for fewer than 50 years. The boundaries established by the colonial process during the late nineteenth and early twentieth centuries have complicated the effectiveness of those states.

The tremendous variety that is the African continent can be difficult to convey. Some African countries are among the poorest in the world, with limited agricultural or industrial resources. Others are making significant progress and have the resource potential for even greater growth. The chapters that follow set forth the geographical basis for the development of this diverse and increasingly important part of our world. They attempt to provide insight into the challenges and progress of black Africa and to illustrate the varied approaches that are being pursued in the quest for national development.

CONTENTS

■ **Africa south of the Sahara is characterized by both modern and traditional lifestyles.** Shown here are the skyline of Harare, Zimbabwe, and (inset) Masai pastoralists of Kenya.

Africa South of the Sahara:
Legacy of Continuity and Change

- ● **Environmental Diversity**
- ● **Precolonial Africa**
- ● **European Colonialism**

- ● **Contemporary Sub-Saharan Africa**
- ● **Development Trends in Sub-Saharan Africa**

frica is the second-largest continent in the world. Its land area of 11.8 million square miles (30.5 million square kilometers) is about three times that of the United States of America, and it accounts for about 20 percent of the earth's land surface. Its north-south extension stretches about 4,800 miles (7,725 kilometers), from northern Tunisia to Cape Agulhas at the southernmost point of the Republic of South Africa. From west to east, it extends 4,500 miles (7,242 kilometers), from Dakar, Senegal, to the tip of Somalia in the Gulf of Aden.

Africa south of the Sahara alone is about 9.8 million square miles (25.3 million square kilometers), encompassing the group of countries south of the Sahara Desert, although countries such as Mauritania, Mali, Niger, Chad, and Sudan have parts of their northern territories contiguous to the desert region. Sub-Saharan Africa is composed of forty-two countries located on its mainland (Figure 25–1) and eight island countries: Madagascar, the Comoros, the Seychelles, Mauritius, and Réunion in the Indian Ocean; and Cape Verde, São Tomé and Principe, and St. Helena in the Atlantic Ocean.

The mainland countries can be divided into four subregions: West, East, Central, and Southern Africa. The Sub-Saharan region has a rich and diverse physical and human resource base and a resilient cultural heritage. It is also characterized by such contrasts as extreme wealth and abject poverty, gender inequality, pristine forests interspersed among degraded and desert landscapes, mineral-rich subterrains buried under destitute landscapes, and European values and institutions superimposed on traditional cultural landscapes. Not so long ago, the tendency was to characterize the subcontinent in terms of the *"d's"*: destitute,

doomed, dark, disaster-prone, drought-stricken, disjointed, disconnected, and debt-ridden. However, perceptions are slowly changing, and people are becoming more aware of the *"p's"*: the potential, prospects, and possibilities inherent in the evolving socioeconomic landscape.

Environmental Diversity

The physical environments of Sub-Saharan Africa are characterized by great variation in topography, climate, and biogeography. This section focuses on the broad ecological patterns of the continent. More detailed regional analyses of the physical geography of the African nations are provided in the chapters that follow.

The five major natural regions that can be distinguished in Africa south of the Sahara are the humid tropical rain forests, the subhumid tropical savannas, the semiarid steppes, the dry deserts, and the highlands. Each of these natural regions has distinctive climatic, vegetational, and soil characteristics.

Tropical Rain Forest

The tropical rain-forest region is found in the equatorial portions of the Democratic Republic of Congo, much of Gabon and the Republic of Congo, south and central Cameroon, the coastal strips of West Africa, and portions of Kenya, Tanzania, and Madagascar (Figure 25–2). Mean monthly temperatures in the rain-forest zones remain above 64.4°F (18°C) throughout the year, and substantial precipitation occurs in every month, with most stations averaging between 60 and 80 inches (1500–2000 millimeters) annually. Yearly

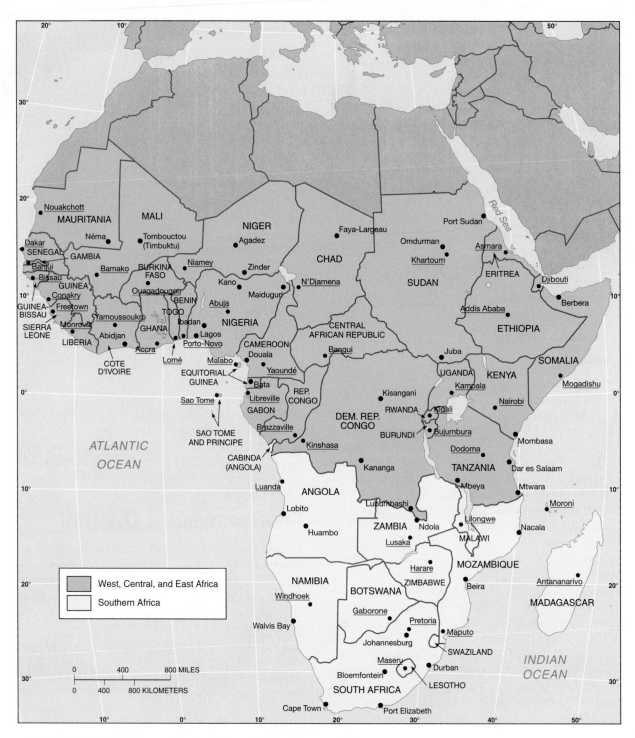

■ **FIGURE 25-1 Nations and major cities of Sub-Saharan Africa.** Sub-Saharan Africa consists of
two major regions: West, Central, and East Africa and Southern Africa. Most of the major cities are located close
to coastal areas and to regions well endowed with natural resources.

precipitation has reached 160 inches (4000 millimeters) in
Douala, Cameroon, and 140 inches (3500 millimeters) in
Freetown, Sierra Leone. The abundant rainfall of the tropical rain-forest regions is associated with high temperatures
and low atmospheric pressures. As the warm air masses rise
and cool, condensation takes place, frequently resulting in
torrential tropical downpours.

The high temperature and precipitation support Africa's
lush rain-forest vegetation, which contains more than 8,000

plant species, more than 80 percent of the continent's primate animal species, and more than 60 percent of its passer-
ine birds (Figure 25–3). The rain forests in East Africa are
restricted to portions of Madagascar, the southern coast of
Kenya, and the eastern highlands of Tanzania. These areas
are relatively drier and less diverse than the rain forests of
Central and West Africa. Along the coast of eastern Madagascar, the rain forests contain about 6,000 plant species, the
majority of which are endemic to the region.

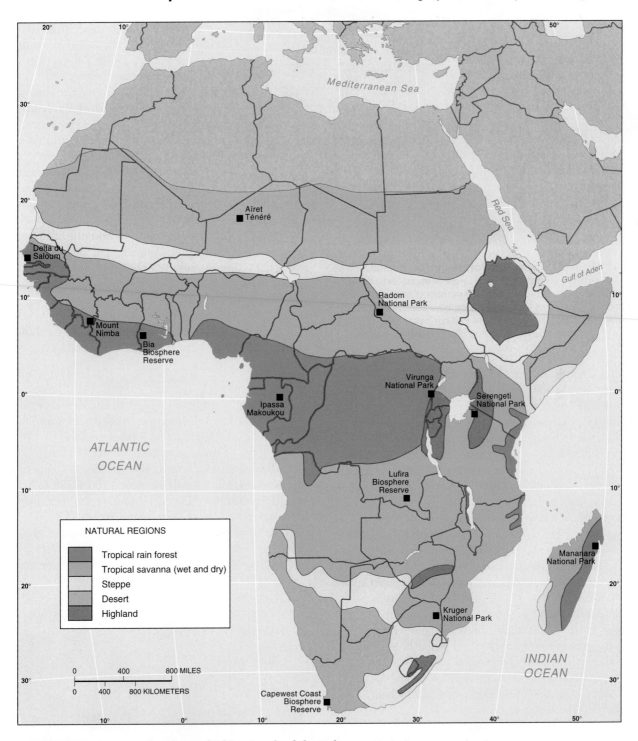

■ FIGURE 25-2 Natural regions of Africa South of the Sahara. Africa's diverse natural regions are home to several World Heritage Sites and Biosphere Reserves.

Most rain-forest soils are intensely weathered. The surface horizons of the oxisol soils are characterized by a mixture of iron and aluminum oxides, which give them a reddish color. Many of the soil nutrients have been washed down into the lower subsurface horizons through a process known as soil leaching. With the addition of chemical fertilizers and other Green Revolution inputs, however, these relatively poor soils are capable of supporting a variety of plantation crops, including rubber, cocoa, and oil palm.

Tropical Savanna

Further inland, away from the humid rain forest, is the wet and dry savanna, where average annual rainfall totals drop to about 40 to 60 inches (1000–1500 millimeters) in the heavily wooded sections adjacent to the rain forest and to around 20 to 40 inches (500–1000 millimeters) in the grassland regions of the drier margins. Several countries along the West African Atlantic coast, and others such as the Central African

■ **FIGURE 25-3 African rain forest.** The true tropical rain forest consists of four distinct vertical layers of vegetation and is characterized by an extraordinarily high number of species of plants and animals. Despite its lush appearance, however, the tropical rain forest is a fragile natural environment that has been receding rapidly in recent decades in Africa and other equatorial regions.

Republic, southern Sudan, Uganda, Rwanda, Burundi, northern Angola, southern Congo and Gabon, and much of Tanzania, are subject to these subhumid conditions. During the year, two periods of concentrated precipitation are interspersed with two dry seasons.

This region is home to some of Africa's most famous national parks and game reserves. Tanzania, for instance, has more than 95,000 square miles (246,050 square kilometers) of land devoted to parks and reserves, including the famous Serengeti National Park. The Serengeti, which means "extended place" in the Masai language, has the largest concentration of plains game in Africa, including millions of wildebeest, zebras, and gazelles. This wildlife sanctuary has been declared a World Heritage Site by the United Nations Educational, Scientific, and Cultural Organization (UNESCO). A **World Heritage Site** is an area that has inherent cultural and natural properties deemed to be of outstanding universal value. **Ecotourism**—travel to natural areas to understand the physical and cultural qualities of the region while producing economic opportunities and conserving the natural environment—is becoming an important industry in these national parks and game reserves.

Soils in the subhumid savanna region consist primarily of alfisols, which are characterized in the upper horizon or layer by a light color, low organic matter, and a high clay content. Alfisols tend to maintain a higher nutrient level than the oxisols of the rain forest. They are widely cultivated and support a range of crops, including cocoa, rubber, bananas, maize, cassava, sorghum, and millet.

Steppe

Poleward of the subhumid African savanna are the semiarid steppes. Here, rainfall averages from 10 to 20 inches (250–500 millimeters) annually and varies greatly from year

to year, thus contributing to the sparse, shrubby vegetation characteristic of the region (Figure 25-4). Since the steppes are climatic transition zones between the savanna grasslands and the true deserts, they are prone to prolonged droughts. The fragile ecology of the margins of the Kalahari in Southern Africa and the Sahel zone on the south side of the Sahara is constantly being threatened by desertification and soil degradation. In N'djamena, Chad, which is located in one of the moister sections of the Sahel, the average monthly rainfall exceeds 2 inches (50 millimeters) in only four months of the year— June, July, August, and September. The average maximum daily temperatures range from 87°F (31°C) in August to 107°F (42°C) in April. The soils of the African steppe regions are primarily entisols, which have weak horizon and profile development. They tend to be shallow, stony, and deficient in surface organic matter, thus their potential for agricultural production is limited.

Desert

The Sahara Desert is the largest of the desert regions of Africa. Other deserts include the Ogaden of eastern Ethiopia and Somalia; the Kalahari of Southern Africa; the Atlantic coastal zones of Namibia and Angola; and a minute portion of southwestern Madagascar (Figure 25–5). Temperatures in Timbuktu, Mali, regularly exceed 100°F (38°C) in May and June, with the difference between the daily high and low temperatures often being as much as 60°F (16°C). Parts of the central Sahara receive less than 4 inches (100 millimeters) of rainfall annually. Along the southwestern coast of Namibia, in the Namib Desert region, temperatures are

■ **FIGURE 25-4 African Sahel.** The Sahel is characterized by a semiarid climate and has been used primarily for animal grazing. Recent overgrazing has resulted in considerable vegetative loss and, in some areas, in desertification and other forms of environmental degradation. This scene shows Matafo village, near Bol, Chad.

■ **FIGURE 25-5 The Kalahari in Southern Africa.** Africa south of the Sahara contains a number of large desert regions, each of which possesses a unique set of microenvironmental characteristics. The animals grazing are springbok (*Anticorcas marsupialis*), a gazelle noted for its sudden and random jumps into the air as well as its adaptability to dryland conditions.

cooler, owing to the effect of the prevailing westerly winds blowing over the cold Benguela ocean current before reaching the land. The enduring aridity of Africa's desert regions is also a result, in part, of the presence of atmospheric high pressure belts centered in both the northern and southern hemispheres between 25 and 40 degrees latitude.

The lack of soil moisture within these true deserts restricts vegetation to a few favored sites where water occurs near the surface, such as in intermittent river channels and oases. In such zones, perennial plants may survive. Outside the moist zones, most of the plants and animals of the desert exhibit permanent physiological adaptations to drought. Soils in these dry zones are called aridisols. They are low in organic matter and often have hard layers of salt built up near the surface. The date palm is a commonly cultivated crop, and sorghum and millet are cultivated on a limited basis. Short-lived plants often sprout, flower, and die within the weeks following the sporadic rains.

Highlands

The highland regions are located in the isolated mountainous areas of the southern and eastern sections of Sub-Saharan Africa. Temperature, rainfall, and soil conditions vary according to elevation and other microenvironmental conditions. For example, in Addis Ababa, Ethiopia, which is situated at about 8,000 feet (2,438 meters) above sea level, the average daily temperature minimums may fall as low as 55°F (13°C) in January, while frost often occurs in the mountain ranges of Kenya and Tanzania. At the highest elevations, such as on the upper slopes of Mount Kilimanjaro in Tanzania (19,340 feet/5,895 meters) and Mount Kenya (17,058 feet/5,200 meters), permanent snow and glaciers cover the ground.

Lush montane forests often develop in the Sub-Saharan African highlands between 3,900 feet (1,189 meters) and 8,200 feet (2,500 meters). Here adequate precipitation, mod-

erate temperatures, and relatively deep and fertile soils combine to promote dense tree growth. Environmental conditions vary with altitude, while the diversity of plant and animal species generally decreases as altitude increases. Most of Africa's highland regions offer fertile agricultural soils that have been intensively farmed for many years. Thus, they are susceptible to accelerated erosion.

Precolonial Africa

Africa experienced a long and rich history prior to the era of formal colonialism, which began in the late nineteenth century, after the Berlin Conference of 1884 formally ratified the European powers' parcelization of the continent. Unfortunately, many aspects of Africa's native cultures were lost or destroyed during the colonial era, and numerous misconceptions about Africa arose in Western societies. It became habitual for colonialists to deny almost any degree of social or political achievement to Africans themselves. Some scholars at the time even questioned whether native technologies were really indigenous to Africa or whether they had been imported by external agents. Other Westerners even went so far as to deny the status of "civilization" to the early African urban centers on the basis that their peoples lacked writing or organized social and political structures.

Geographers and historians are now reconstructing Africa's past through the study of folklore, poetry, archeological sites, agricultural cropping systems, art, and architecture. We now recognize that the ancient African civilizations were characterized by rules of social behavior, codes of law, and organized economies. Further evidence suggests that towns and cities with diverse sociopolitical organizations existed for several millennia. Cities such as Napata, Meroe, Axum, Jenne, Timbuktu, Gao, and Great Zimbabwe (Figure 25–6) were major centers of cultural and commercial

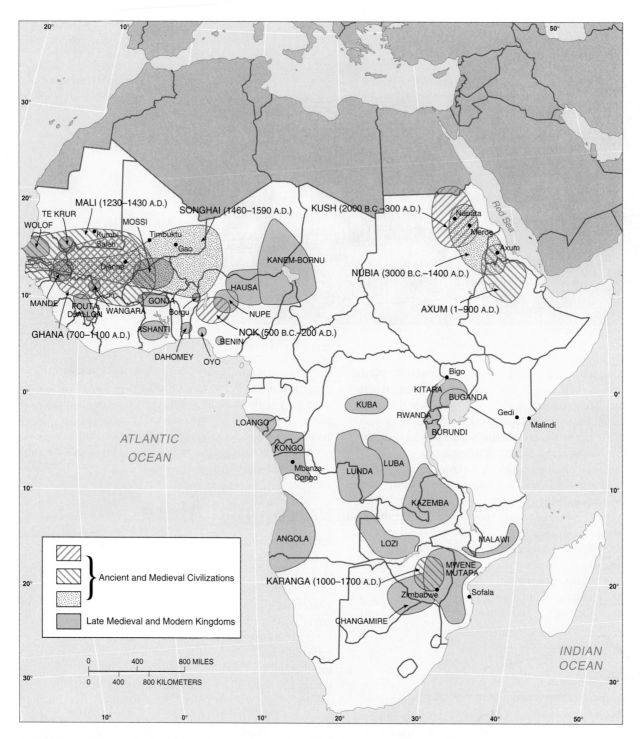

■ **FIGURE 25-6 Ancient, medieval, and late medieval Sub-Saharan African civilizations.** Sub-Saharan Africa has a rich cultural heritage. Some civilizations date to the fifth century B.C. The zones that supported the greatest number of civilizations and kingdoms were West Africa and south-central Africa.

exchange, religion, and learning. These cities had clearly defined divisions of labor, class structures, communication networks, and spheres of influence. The diffusion of technological innovations, including iron technology, stonemasonry, and other crafts from the early eastern African civilizations of Kush and Axum was widespread to both the west and the south.

Ancient Civilizations

The earliest known civilizations of Africa emerged in the central part of the Nile River Valley (see Figure 25–6). One of the most notable was the black kingdom of Kush and its capital Meroe, which flourished from about 2000 B.C. to the fourth century A.D. Meroe was largely influenced by the

■ **FIGURE 25-7 Ruins of Meroe.** Meroe was the capital of the black kingdom of Kush, located in northeast Africa. It was noted for its iron and stone technology.

Nubian Kingdom, the Napata Period of which emerged around the eighth century B.C. Archeological evidence from Meroe reveals a civilization that thrived on stone and iron technology (Figure 25–7). Elaborate stone walls, palace buildings, swimming baths, temples, and shrines indicate a culture with an organized social, religious, and political order. It also had an organized agricultural economy based on pastoralism and cultivation complemented by advancements in irrigation agriculture. Meroe and other Kushite cities are credited with reaching a level of technological sophistication unmatched in precolonial Africa. Building and construction technologies along with elaborate sculptures, iron-working industries, pottery works, textiles, leatherworks, woodwork, basket weaving, and irrigation technologies are all testimonies to their advanced technological achievements.

Other prominent cities in this region of East Africa were Axum, a metropolis of the ancient kingdom of Ethiopia, and the Red Sea port of Adulis. Classical writers and ancient Axumite coins provide evidence of a powerful Axumite kingdom that lasted from the first to tenth centuries, when Axum's influence extended across the Red Sea to southern Arabia. Historical evidence from Axum suggests a culture with remarkable engineering and architectural skills applied to quarrying, stone carving, terracing, building construction, and irrigation. Archeological artifacts also indicate a high level of achievement in metallurgical technology and manufacturing. Axum's trade network extended to the Roman provinces of the eastern Mediterranean, southern Arabia, and India, and centered on the export of ivory, gold, emeralds, and slaves in exchange for iron, precious metals, clothing, wine, vegetable oils, and spices.

Prominent civilizations that emerged between 700 and 1600 in the West African savanna were Kumbi (the political center) and Saleh (the commercial center) in Ghana, Timbuktu and Djenne in Mali, and Gao in Songhai. These cities thrived as major centers of the trans-Saharan trade. Gold mining, iron technologies, pottery making, and the production of textiles are indicative of a high level of technological development in the region. Kumbi Saleh was the capital of the first powerful state—Ghana—which flourished from about 700 to 1100. It was a major commercial center known for gold, stones, copper, and iron. An elaborate economic system was developed along with a system of taxation.

Timbuktu, Djenne, and Gao emerged as great centers of learning and trade in the Mali and, later, in the Songhai empires of the West African savanna. They flourished as intermediate trade centers between the forest zone, to the south, and the desert regions of North Africa and Egypt. For instance, a salt trade route emerged between Teghaza in North Africa and Timbuktu, and a gold route between Djenne and the forest regions of the south.

Cities in the West African forest region had well-developed kingdoms and urban civilizations, including Benin, Yorubaland, and Hausaland in Nigeria, and the Ashanti kingdom of Ghana. These kingdoms exhibited multicentered urban networks and organized social and political systems, and they functioned as political, commercial, and spiritual centers. For example, Ife, the cradle of the Yoruba civilization, was the spiritual capital of the Oyo empire, as were Sokoto for Hausaland and Kumasi for the Ashanti. The architectural design of Yoruban cities, like Ife, consisted of a series of concentric city walls with the royal palace situated at the center. The walls provided prestige, intimidated potential intruders, and enhanced a ruler's ability to command his subjects. Inner walls protected the privacy of rulers, and outer walls provided refuge for the masses. Other examples of urban symbolism were passageways and alleys that intersected market plazas and the creation of intimate urban spaces to encourage social interaction and cohesiveness. Cities of the forest region had diverse technologies in iron and smelting, stone building, coarse mud architecture, brass-working, gold mining, and glass melting.

In the central African equatorial region, urban civilizations arose in what became the modern states of the Congo, Democratic Republic of Congo (Zaire), Angola, Zambia, Rwanda, and Burundi. These cities were considered the least urbanized in Africa because they lacked the large centralized empires and intricate urban networks that characterized the savanna and forest regions of West Africa. However, a number of significant cities did evolve, including Musumba, the capital of the Lunda empire; Mbanza-Congo, the capital of the Kongo empire; Ryamurari, the old capital of the Ndorwa Kingdom of Rwanda; and Kibuga, the capital of the Ganda Kingdom. Other prominent states were the kingdoms of Buganda and Kitara, situated between Lake Victoria on the east and Lakes Albert, Edward, and Tanganyika on the west. Archeological evidence from this region suggests high levels of craftsmanship and professional artisanship in iron, copper, ivory, pottery, metalwork, and mining.

In coastal East Africa, some of the well-known historic centers were Mogadishu in Somalia; Malindi, Gedi, and Mombasa in Kenya; and Zanzibar and Kilwa in Tanzania (Figure 25–8). Gedi, one of the first sites to be excavated in Kenya, was founded in the thirteenth century. It possessed elaborate walls, palaces, mosques, and well-designed homes. Kilwa was a well-constructed city with a palace and commercial center, domed structures, open-sided pavilions, and vaulted roofs. A Swahili culture that was urban, mercantile, literate, and Islamic emerged in this region. The technological developments that occurred include coin minting, copper works, building craftsmanship, boatbuilding, and the spinning and weaving of cotton. External trade was very active. Ivory, gold, copper, frankincense, ebony, and iron were among the goods traded for Chinese porcelain and glazed wares from the Persian Gulf.

Much has been written about the stone-built ruins of Great Zimbabwe in Southern Africa (Figure 25–9). It has been suggested that the city originated as a vital spiritual focal point, since monoliths, altars, and a stone tower were discovered at this site. Thirty-two-foot elliptical walls that date back to at least the fourteenth century have been discovered, along with the "Great Enclosure," which contains about 182,000 cubic feet of stonework. The technological achievements of Great Zimbabwe go beyond building and stone construction to include mining and metallurgy, manufacturing of pottery, wood carving, and cotton spinning.

This summary of historic centers suggests high levels of social and technological achievement among precolonial African civilizations. Recent archeological evidence also suggests the existence of several other historic sites that have not been mentioned here. The advent of European colonialism saw the destruction of much of Africa's rich material heritage. Fortunately, there is a renewed effort to discover the richness and elegance of past African civilizations.

■ **FIGURE 25-9 Ruins of Great Zimbabwe.** Great Zimbabwe was once the center of an advanced, precolonial civilization in Southern Africa.

Most cities in Africa today are colonial creations. The majority of African capitals have a coastal orientation since they were created as "headlinks" designed to funnel the resources of the interior to overseas destinations. Most precolonial cities are now virtually extinct or, as in the case of Timbuktu or Gao, are just relics of their former selves. Fortunately, a few of the ancient cities have survived the ravages of colonialism and still thrive today as historic centers. These include Kumasi, Ghana; Addis Ababa, Ethiopia; Mogadishu, Somalia; and Ibadan and Ife in Nigeria.

European Colonialism

Colonialism involved the imposition of foreign values on indigenous institutions. It was largely paternalistic, exploitive, and inhumane. The era of European colonialism can be divided into four main periods: the period of initial contact, the period of enslavement, the age of land exploration, and the classical era of formal colonialism. The first three periods focused on trade in commodities and slaves, while the classical colonial period emphasized the extraction of mineral and agricultural resources to meet the industrial needs of Europe. One of the most serious consequences of European colonialism was the limiting of African industrial development.

The Early Period: 1400–1880

The first meaningful European contact with Sub-Saharan Africa came in the fifteenth century through Portugal's search for a sea route to India. The Portuguese confined their trading activities to the coast, leaving inland trade to African merchants. European products such as copper and brass were exchanged for African gold, and trading posts and forts were established along the coast of West Africa (Figure 25–10). A second objective was the establishment of world empires. After Portugal's initial contact in 1420, the Spanish, English, French, and Dutch followed.

■ **FIGURE 25-8 Zanzibar.** Zanzibar has functioned as an urban and commercial center of eastern coastal Africa since precolonial times.

characterized by the explorations of Henry Stanley (along the Congo/Democratic Republic of Congo), David Livingstone (along the Zambezi), and Mungo Park (along the Niger River). Their accounts of Africa's potential resources led to the "scramble for territory" and set the stage for the Berlin conferences, the partitioning of the region, and the classical era of colonialism. This last phase was precipitated by the shift from mercantile to industrial capitalism in Europe and the Industrial Revolution.

Formal Era of Colonialism: 1880 to the Mid-1950s

The 1884 **Berlin Conference** marked the beginning of the formal era of colonialism in Africa. It was, in many ways, the most significant turning point in Africa's history because it shaped the social, economic, and political futures of the countries in the continent. The European powers met in Berlin—without African consent or participation—to establish procedures for the allocation of African territory among themselves. Territories were assigned on the basis of the principle of effective occupation, resulting in the imposition of comparatively arbitrary political boundaries on the diverse cultural landscape of African societies. European colonizers felt free to impose their values, policies, and institutions on a fragmented and disunited continent. Figure 25–11 shows the distribution of European colonies in Sub-Saharan Africa just prior to the wave of independence movements. The British and French controlled about 70 percent of the region in 1956, with Liberia and Ethiopia the only independent states.

British colonial policy was based on **indirect rule**. Lord Lugard, the governor general of Nigeria, was instrumental in formulating this policy with the explicit purpose of incorporating the local power structure into the British administrative structure. African chiefs and kings were to function as intermediaries, acting as links between their people and the colonial authorities. The chiefs were responsible for enforcing local ordinances, collecting taxes, and carrying out the day-to-day affairs of the colonial authority.

The French enacted a highly centralized administrative structure based on a policy of **acculturation** that sought to encourage Africans to adopt and assimilate French culture, language, and customs. Two French "overseas" federations were created: French West Africa, with Dakar as its headquarters, and French Equatorial Africa, centered around Congo-Brazzaville. Directives originated from Paris and were funneled through the administrative centers. Policies initiated in the overseas federations had to be approved by the French national assembly in Paris. The Portuguese policy of **assimilation** was similar to France's. Angola, Mozambique, Guinea-Bisseau, and São Tomé and Príncipe were each regarded as overseas provinces of Portugal. Portuguese policy focused on developing a social hierarchy, which was dehumanizing to Africans trapped at the bottom

■ **FIGURE 25-10 Elmina Castle, Ghana.** Coastal West Africa has as many as fifty forts, which were used by Europeans as trading posts and for housing slaves.

Trade in goods and commodities continued in the early to mid-1400s until the discovery of the Americas in 1492 and the establishment of New World tobacco, sugarcane, and cotton plantations. The native American population was subsequently decimated by European diseases, and those who survived often fled their Spanish and Portuguese captors. Lacking sufficient laborers, the New World Europeans turned to enslaved Africans as a source of workers on the huge estates. Estimates of the number of Africans transported across the Atlantic Ocean range from 8 to 12 million. The impact of slavery on African societies included the disruption of cultural institutions; the reduction of industries, crafts, and other forms of manufacturing; and the increased incidence of tribal wars, since prisoners were exchanged as slaves. The major sources of African slaves were the Senegambia coastal region between Guinea-Bissau and Liberia, and Congo and Angola. Other key source areas included the Dahomey and Yoruba kingdoms, the Gold Coast (Ghana), the Niger delta, and Mozambique.

The shipment of African slaves was abolished in 1808, owing largely to the humanitarian efforts of philanthropists and religious leaders such as Granville Sharpe and William Wilberforce, as well as Western-educated and freed slaves such as Olaudah Equiano of Nigeria and Ottabah Cugoano of Ghana. With the end of slavery, European economic activities in Africa reverted once again to trading in commodities such as gold, coffee, cocoa, and palm oil. Missionaries also renewed their efforts to convert Africans to Christian faiths and assist in their social and economic development.

By 1840, scientific and geographic curiosity about Africa had become so heightened that European explorers were intensifying efforts to access the region's hinterlands and interior resources. The period between 1840 and 1890 is

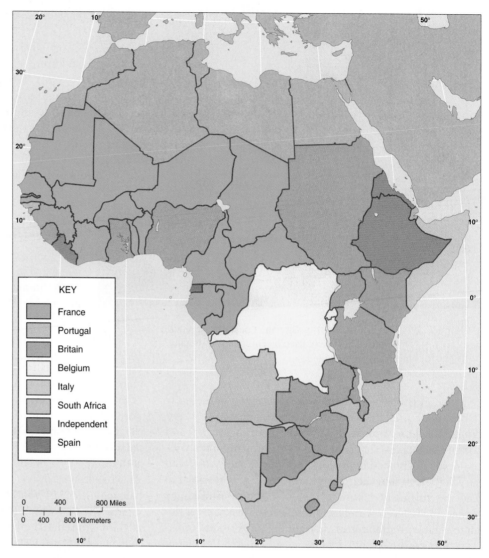

■ **FIGURE 25-11 Colonial map of Sub-Saharan Africa, 1956.** The French dominated West Africa in the late colonial era, and the British ruled East Africa. Belgium controlled the Congo Basin, and Portugal governed portions of the Atlantic and Indian Ocean coasts of Southern Africa. Liberia and Ethiopia remained independent.

of the social ladder. Africans who aspired to Portuguese citizenship were granted *assimilado* status. The rest, the so-called *indigenas*, were relegated to working agricultural estates that produced cotton and other cash crops for Portugal. The Belgians controlled Democratic Republic of Congo, Rwanda, and Burundi (the last two as trust territories). Ironically, this region was referred to as the Congo Free State, but, in reality, Belgian policy was blatantly exploitive and paternalistic, extracting mineral and agricultural wealth from the colonies while investing almost nothing in the social development and education of the native peoples. For example, at independence in 1960, the entire country of Democratic Republic of Congo had only sixteen university graduates.

The formal colonial era began to wane in the early 1950s. World War II had taken its toll on European economies, African soldiers who had fought side by side with European "Allies" had returned with ideas of freedom, the United Nations was espousing the principle of self-determination, and African scholars who studied in foreign institutions were returning to organize political platforms and parties to challenge colonial rule. These developments, combined with the dehumanizing and exploitive nature of colonialism, led to

the decolonization process. However, the scars of colonialism still remain.

The Impact of Colonialism

Some would argue that colonialism had certain positive effects. Even if we assume that to be true, the negative effects far outweighed the positive ones. The political boundaries that Europeans imposed sometimes forced hostile ethnic groups to share the same territory and, at other times, divided the same tribe between two or more countries. Civil wars in Angola, Burundi, Mozambique, Nigeria, Rwanda, Somalia, and Democratic Republic of Congo, coupled with irredentist military campaigns (ones aimed at retrieving territory, historically associated with one political group, from the control of another) of Somalia and the Ewes of Ghana, have repeatedly destabilized the region, resulting in a constantly recurring refugee crisis. African countries are still searching for appropriate solutions to defuse sources of hostility.

Another legacy of colonialism is the number of small-sized and landlocked states that present-day Africa inherited. The potential for agricultural and industrial development is

Table 25-1   Principal Primary Exports and Trading Partners of Selected Sub-Saharan African Countries

Country	Principal Primary Exports	Percent	Principal Export Destinations
Angola	Petroleum and petroleum products	93	U.S., China, Belgium, France
Cameroon	Crude petroleum	36	Italy, France, Spain
	Crude materials (rubber, cork, cotton)	25	
Central Afr. Republic	Industrial diamonds and pearls	61	Belgium, France
Côte d'Ivoire	Cocoa	41	France, Netherlands, U.S.
	Coffee	11	
Gabon	Petroleum and petroleum products	83	U.S., France, China
Ghana	Gold	43	Netherlands, U.S., U.K.
	Cocoa	25	
Mali	Cotton	62	Brazil, France, Italy
Mauritania	Iron ore	50	France, Spain, Italy
	Fish	46	
Mozambique	Shell fish	38	Portugal, Spain, South Africa
Niger	Metalliferousores, scrap	75	France, Nigeria
Nigeria	Petroleum	97	U.S., Spain, India, France
Rwanda	Coffee	61	Germany, Belgium
Sudan	Cotton	43	China, Japan, Saudi Arabia
Togo	Phosphates	53	Canada, France, Nigeria
Uganda	Coffee	62	Germany, Netherlands, U.K.
United Rep. of Tanzania	Machines and transport equipment	36	U.K., India, Germany

Sources: International Trade Statistics Yearbook (New York: United Nations, 1998); *Direction of Trade Statistics Yearbook* (Washington, D.C.: International Monetary Fund, 2001).

limited in microstates like Gambia, Lesotho, Swaziland, and Djibouti that have small domestic markets and a small labor pool. Africa also has close to 40 percent of the world's **landlocked states**, whose trade with the outside world must pass through neighboring, sometimes hostile, countries. Because colonial development efforts centered on the coastal zones, these landlocked countries also tend to lack extensive rail and road networks, thereby limiting the spatial interaction between their rural and urban sectors.

The lack of an extensive transportation network also helps explain the limited trade among African countries. Intra-African trade accounts for only 5 to 6 percent of total African trade. Colonialism may have ended, but **neocolonialism** is prevalent, as most African countries continue to function largely as providers of raw materials to the European industrialized countries (Table 25–1). Although politically independent, many African nations continue to depend economically on their former colonizers for trade, technology, and other goods and services, which in turn discourages diversification. Yet another challenge to economic development is the fact that most African economies are still monocultural—meaning that they rely on one or two primary products for export revenue. A sectoral imbalance also persists, with weak internal linkages between agriculture and industry, both of which focus on exporting products to overseas markets rather than providing for local needs. Creating linkages between key sectors of an economy is a prerequisite for developing a self-sufficient and sustaining economy.

Perhaps the most significant impact of colonialism relates to the fact that African countries have to cope with a dual or, in some cases, triple heritage. Governments are confronted daily with conflicts and contradictions between traditional and European value systems. The inherent contradictions pervade the social, cultural, economic, and political lives of Africans. Countries wrestle with the question of whether or not to adopt a common non-Western language, school systems debate replacing their European curriculum with one that is Afrocentric, and politicians search for the "ideal" constitution that incorporates aspects of traditional authority into largely Western-based political systems. In the northern regions of West, Central, and East Africa, a third element, Islam, further complicates the political and cultural environments. In Sudan and Nigeria, in particular, Islamic fundamentalism has emerged as a potent force. In these and other respects, the legacies of colonialism are still felt in the daily lives of Africans.

Contemporary Sub-Saharan Africa

Language Diversity

One of the most intriguing aspects of Sub-Saharan Africa's cultural geography is the more than 1,000 languages that exist in the region. Most of these languages do not have a written form or tradition, and approximately forty are spoken by 1 million or more people. Linguistic scholars, such as Michael Greenberg, have identified four major linguistic families in Sub-Saharan Africa: Niger-Kordofanian, Nilo-Saharan, Khoisan, and Afro-Asiatic (Semitic-Hamitic).

The Niger-Kordofanian, the largest linguistic family in Sub-Saharan Africa, is divided into two related but distinct branches, the Kordofanian and the Niger-Congo. The

Kordofanian branch is centered in a small area in the Nuba hills of Sudan and consists of about twenty languages. The Niger-Congo branch, on the other hand, is spoken by more than 150 million people and stretches across half of Sub-Saharan Africa, extending from West Africa to the equatorial and southern regions. It includes the west Atlantic languages of Wolof and Fula in Senegambia, and Temne and Gola in Sierra Leone and Liberia, respectively; the Mande languages of Bambara (Guinea) and Soninke (Mali); the Central Bantoid (or Voltaic) languages of Mossi (Burkina Faso) and Dagomba (northern Ghana); the Guinean (or Kwa) languages of Kru (Liberia), Akan (Ghana), Yoruba (western Nigeria), and Ibo (eastern Nigeria); the Hausa language; the central and eastern Sudanese languages of Azande (Central African Republic) and Banda; and the most widely spoken subfamily, the Bantu (Figure 25–12). Linguistic studies trace the origins of Bantu to the southeast margins of the Congo (Zaire) rain forest, although some evidence suggests linkages to the West African forest and savanna regions. Bantu languages are spoken today from the equatorial region to South Africa. Those with 10 million or more speakers include Lingala (Democratic Republic of Congo),

Ruanda-Rundi, Tswana-Sotho (Botswana and South Africa), Zulu, and Swahili (Kenya, Tanzania), which has also been subjected to Arabic influences. Other languages spoken by more than 1 million include Bemba (Zambia), Luba (southern Democratic Republic of Congo), Shona (Zimbabwe), and Buganda (Uganda).

The Nilo-Saharan linguistic family stretches in a west-to-east direction, from the Songhai language in southwest Niger to the Nilotic languages of Nuer and Dinka in southern Sudan and Luo and Masai in southwestern Kenya. It also includes the Saharan languages of Kanuri (which can be traced to the Kanem and Bornu kingdoms of Lake Chad), Kanembu, and Teda.

The Khoisan family is confined to the Kalahari Desert region in Namibia, Botswana, Zimbabwe, and South Africa. It dates back several millennia, having once dominated the entire territory of southern and eastern Africa from Somalia to the Cape of Good Hope. Over the centuries, the Khoisans have been ravaged by diseases, Bantu invasions, and expropriation of land by European settlers. The Khoisan language is associated with the Nama, Hottentots, and Bushmen.

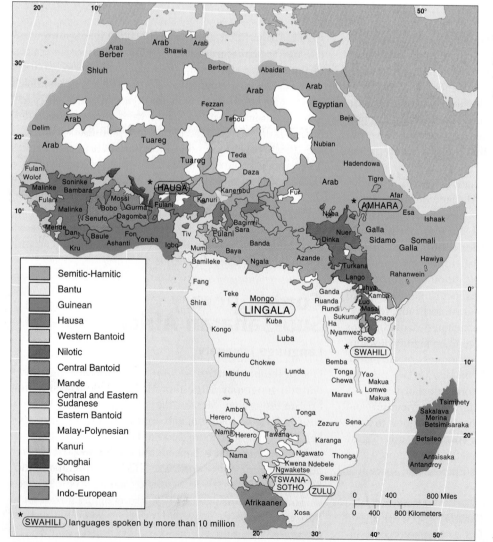

■ **FIGURE 25-12 Distribution of languages in Africa.** Although more than 1,000 languages are spoken in Sub-Saharan Africa, these can be grouped into linguistic families that have distinct geographic patterns.

■ **FIGURE 25-13 Mosque and market in Mali.** Islamic religion, trade, and culture diffused into the savanna states of West Africa around 900.

The Afro-Asiatic (Semitic-Hamitic) family covers much of North Africa. In Sub-Saharan Africa, it is concentrated in Mauritania and the East African Horn, where semitic languages such as Amharic, Tigre, and Tigrinya are spoken. It also includes Cushitic languages like Somali and Mbugu (Tanzania), and Chadic and Berber languages as well.

Besides the four major linguistic families, there are two other language groups in the continent of non-African origins. The Malay-Polynesian family was introduced to Madagascar about 2,000 years ago, while the Afrikaans language, a derivative of Dutch, has Indo-European origins dating back to 1602, when the Boers arrived in South Africa.

In many regions of the continent, people whose native languages are mutually incomprehensible use a common language—a **lingua franca**—for communication. Swahili, which developed along the East African coast from a fusion of Arabic with local Bantu, is now spoken widely in East Africa. It is also becoming increasingly popular in neighboring Central Africa. In West Africa, Hausa, which is spoken by more than 50 million people, is the most important lingua franca, particularly in Nigeria and Niger. Other tongues that have the potential to develop into major regional languages include Amharic in the Horn of Africa, Mandinke in western West Africa, Lingala in Central Africa, Sotho-Tswana in the central parts of Southern Africa, and Zulu in southeast Africa.

Religion in Africa

The legacy of colonialism and the diffusion of Islam from the Arabian peninsula and North Africa have profoundly influenced the religious landscape of Sub-Saharan Africa (Figure 25-13). Islam is concentrated in the northern regions, the Horn, and the coastal corridors of Kenya, Tanzania, and Mozambique (Figure 25-14). There are also significant Muslim minorities in the central regions of several West African states.

Christianity is widely practiced in the central and southern sections of Africa, with a strong Roman Catholic presence in Rwanda, Burundi, and Democratic Republic of Congo; an Anglican presence in Ghana, Nigeria, and Kenya; and Presbyterians common in Malawi. The Coptic Christian Church in Ethiopia dates back to the fourth century and has resisted pressures from Islam since the seventh century. In Cameroon, Kenya, Tanzania, and Democratic Republic of Congo, more than 50,000 people practice the Bahai religion. Judaism has a strong following in the Gondar region of Ethiopia, where blacks, who came to be called **Falashas** (black Jews), were converted to the faith by Semitic Jews who emigrated to Ethiopia between the first and seventh centuries. Other religions, such as Buddhism and Hinduism, are practiced by small numbers of people in southern and eastern Africa.

Religious traditions in Africa often tend to blend, producing mixed, or syncretistic, beliefs and rituals. Many Christian churches, for instance, are being Africanized. African carvings have been substituted for church decorations and drums for church bells in Catholic masses, and Pentecostal churches now feature a wide range of African music, including drumming and dancing. Worship in these churches is vibrant and lively, with much dancing, reminiscent of traditional celebrations.

Often African traditional religions have been wrongfully perceived as animistic or atheistic. On the contrary, most Africans are deeply religious and acknowledge the existence of a supreme being. In the Congo, Nzambi is the all-powerful creator of the sky, earth, and man. God is perceived as having two sides: Nzambi Watanda (God above), who is good, and Nzambi Wamutsede (God below), who is wicked. This belief reflects life's realities, with its alternating periods of good and bad fortune. In Sierra Leone, God is referred to as Meketa (the everlasting one) and Yataa (the omnipresent one). Traditional African religion ascribes to a hierarchical order in the universe, with God at the top,

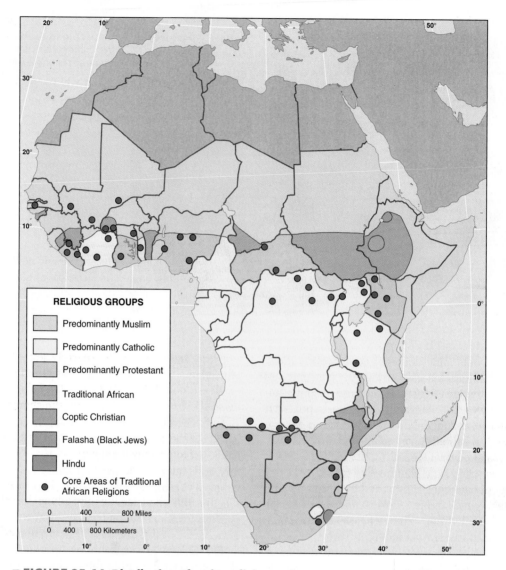

■ FIGURE 25-14 Distribution of major religions. Islam has recently diffused southward into the northern regions of black Africa. Roman Catholicism predominates in central Africa, and Protestantism in Southern Africa.

followed by ancestral spirits and divinities, human beings, animals, plants, and inanimate objects. Departed ancestors are believed to serve as intermediaries between the living community and God. Deities and ancestral spirits are honored in sacrifices and special ceremonies. Traditional religionists in the Congo believe in two types of ancestral spirits: the Binyumba inhabit the kingdom of the dead, while the Bakuyu, who have not yet been admitted to the abode of deceased spirits, wander around and are appeased through offerings. Small temples and spirit shrines to honor nature gods dwelling in rivers, mountains, hills, and lakes are also commonplace. The Yoruba traditional religion in Nigeria has a four-tiered structure of spiritual or quasi-spiritual beings. At the top of the hierarchy is the supreme being Olodumare, and his subordinate ministers, called *Orisha,* constitute the second tier. The third tier consists of deified ancestors, called Shango, followed by spirits associated with natural phenomena such as rivers, lakes, mountains, and trees. Divina-

tion and fortune telling are popular activities in traditional African religion.

Family and Kinship Relations

The African family is usually an extended unit. Each member of the extended family has obligations and ties to the other members. Moreover, the family includes both the living and the dead. Unlike the bilateral **kinship relations** characteristic of Western societies, most African societies are of unilineal descent, tracing lines of descent through the father's side (patrilineal system) or through the mother's side (matrilineal). About three-fourths of African societies are patrilineal. These include inhabitants of the pastoral savannas of West and East Africa, such as the Fulani, the Nuer in southern Sudan, the Masai (west Kenya and Tanzania), the Tiv (east-central Nigeria), the Kikuyu (central Kenya), the Yoruba (southwest Nigeria), and the Buganda of Uganda.

Matrilineal societies include the Akan of Ghana and the Lamba and Bemba of Zambia. In matrilineal societies, links to the father's family are secondary with regard to the inheritance of property. Resources are inherited from the mother's brother. Thus the mother's brother, and not the father, is the matrilineal authority. This tradition has resulted in wives and children being left with little or no property after the death of a husband/father. Countries such as Kenya, Zambia, and Ghana have considered reforming inheritance rules to ensure that wives and children have access to land and other family property upon the father's death, although these "civil" laws often conflict with customary law. In patrilineal societies, widows and children are not always guaranteed property rights either, especially in polygamous marriages.

In traditional African societies, marriage is a union, not of two individuals, but of two extended families. Overall, marriage is perceived as a civil contract between two families. This contract calls for the transfer of goods or money, or both, from the bridegroom's family to the bride's family in the form of **bride wealth**. The payment of bride wealth compensates the bride's family for the loss of her labor since she then has to devote all her time and allegiance to her husband's family. Although divorce is infrequent, it can be granted for a number of reasons including adultery, barrenness, impotence, and an unharmonious relationship with a mother-in-law, but only after counseling and efforts at family intervention and conflict resolution have failed.

In spite of language and religious differences, Africans share a number of common customs and traditions with respect to their family and kinship relations. The extended family, respect for the elderly, socialization between the elderly and the young, and the significance of ancestors are all attributes that most Africans share. Another important trait is the importance of cultural symbols as means of expression (Figure 25–15).

Demographic Changes

Sub-Saharan Africa's estimated population of 733 million people continues to grow at the rapid rate of 2.5 percent per year. With a doubling time of approximately 28 years, its population will increase to 805 million by 2010 and about 1.1 billion by 2025, if current growth rates are maintained. By then, the region may have four times as many people as the United States and Canada and 500 million more people than Latin America. Sub-Saharan Africa is second in regional population only to Asia, which will likely continue as the world's most populous region. In spite of these rapidly growing populations, however, 19 Sub-Saharan African countries have populations of less than 5 million, and more than 50 percent have populations under 10 million. Only 7 countries have populations of more than 25 million, specifically, in rank order, Nigeria, Ethiopia, Democratic Republic of Congo, South Africa, Tanzania, Kenya, and Sudan.

The region's population is concentrated around two major zones of dense settlement: the West African coastal belt stretching from Dakar, Senegal, to Libreville, Gabon; and a north-south belt stretching from the Ethiopian highlands down through Lake Victoria, the copper belt of Democratic Republic of Congo and Zambia, and ending in the Witwatersrand district of South Africa (Figure 25–16). Three broad zones of sparsely populated regions include, from north to south: the Sahel region extending from Dakar in the west to Mogadishu in the east, the west-central-forest regions of Democratic Republic of Congo and Gabon, and the arid/semiarid region of southwest Africa. These spatial distributions coincide with a number of environmental (vegetation, soil, climate, topography), developmental (levels of urbanization, industrialization, and agricultural development), and sociopolitical characteristics (oppressive regimes, ethnic disputes, resettlement schemes). The West African coastal strip contains most of the region's urban, economic, and political centers. Other economic centers, such as the copper belt of Democratic Republic of Congo and Zambia, the diamond and gold mining centers of South Africa's Witwatersrand region, and the rich agricultural lands of the Lake Victoria borderlands, attract large population clusters. These patterns of uneven distribution are also evident on a microscale. In Tanzania, the sparsely settled central region, which has a high concentration of tsetse flies, is surrounded by dense population clusters around the fertile slopes of Mount Kilimanjaro in the north, the shores of Lake Victoria in the northwest, the shores of Lake Malawi in the southwest, and the economic center of Dar-es-Salaam on the east coast. In Democratic Republic of Congo, the population density in the great forest region is about half the national average of 61 people per square mile (24 per square kilometer). Chad's southern region is its agricultural heartland—the center for cotton and groundnut (peanut) cultivation. This region has twice the annual rainfall of the northern Sahel.

Sub-Saharan Africa continues to have the highest fertility and mortality rates in the world, along with the highest proportion of young dependents. Families in the region average an estimated 5.6 children, although there is considerable variation by region, socioeconomic status, and place of residence (rural versus urban). While the majority of countries in East and West Africa have above-average fertility rates, those in Central and Southern Africa have below-average rates. Higher incidences of gynecological disorders and pelvic inflammatory diseases partially account for the lower fertility rates in Central Africa. In Botswana, Zimbabwe, and South Africa, comprehensive family planning programs, coupled with advancements in female literacy, have slowed the rates of birth.

Cultural factors are the strongest forces driving high fertility rates. Most females in traditional societies marry at an early age. High rates of remarriage and polygamy (having more than one wife at a time) negate any potential effects that divorce or widowhood might have. Furthermore, belief systems, customs, and traditions have a significant impact. The predominantly patrilineal societies place a high premium on lineage and spiritual survival. The family lineage is seen as an extension of the past and a link to the future; therefore, any attempts at family planning are strongly resisted. Fertility is equated with virtue and spiritual approval.

OBAAKOFO MMU MAN   ("one person does not rule a nation")

This design expresses the Akan in Ghana system of governance, based on participatory democracy.

SIKA FUTORO   ("gold dust")

Before the advent of coins and paper money, gold dust was used as a medium of exchange among the Akan, symbolizing wealth, prosperity, royalty, elegance, spiritual purity, and honorable achievement.

ABUSUA YE DOM   ("the extended family is a force")

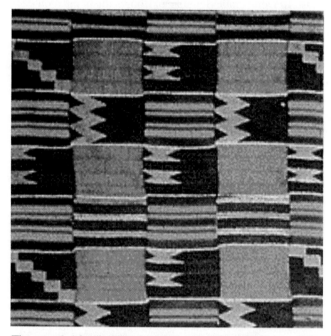

The extended family is the foundation of society. Family members are collectively responsible for the material and spiritual well-being and social security of all their relatives. The background colors symbolize strong family bonds, the value of family unity, collective work and responsibility and cooperation.

NYANKONTON   ("God's eyebrow [the rainbow]")

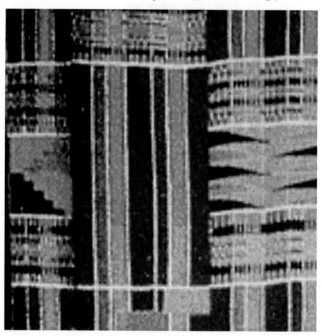

This design was created in exaltation of the beauty and mystery of the rainbow phenomenon. It symbolizes divine beauty, gracefullness, divine creativity, uniqueness, and good omens.

■ **FIGURE 25-15 Cultural symbology.** Clothing is a status symbol in African culture. The symbols and fabrics in African textile design are visible and cultural expressions of social thought, history, philosophy, and religious beliefs. Two types of symbols from the Akan of Ghana are presented. The first set is derived from the Kente cloth—a richly colored and designed ceremonial cloth that is handwoven on a treadle loom. The second set (next page) consists of Adinkra designs that are symbolic of historical events, human relations, and religious concepts.

Adinkra Symbology

GYE NYAME (" except God")

Symbol of the omnipotence of God.

KURONTI NE AKWAMU (the two complementary branches of the state)

Symbol of democracy, interdependence , complementar ity.

DWENI NI MMEN (" ram ' korns ")

Symbol of stren gth (of mind, body, and soul).

HWEHWEMUDU A (" searching or measuring rod ")

Symbol of excellence, perfection, knowledge, and superior quality.

■ **FIGURE 25-15** *(continued)*

Conversely, barren women are treated with disdain and ostracized from society. Also, in a male-dominated society, decisions about reproduction and family size are usually made by the husband, which may explain the low levels of contraception practiced. Children are regarded as economic assets—a source of wealth and prestige and a labor reservoir for household chores. They are also required to offer tribute to their parents. This flow of wealth to the elderly is a socially sanctioned and religiously expected tribute. Since a high premium is placed on children, African women aspire to elevate their status, complying with their husband's request to have more children. Fertility is further enhanced through **child fosterage**, in which children are sent from their natural parents to be raised and cared for by their grandparents or foster parents. About one-third of children in West Africa are fostered. Possible benefits of fosterage include the strengthening of family ties, companionship to widows, enhanced educational opportunities for the children, and assistance with domestic chores. The practice also lessens the economic burden placed on parents who would otherwise struggle to provide adequately for all their children. Another fertility-enhancing factor is the ethnic rivalry that exists among many Sub-Saharan societies. In countries such as Kenya, Nigeria, Ethiopia, and Uganda, where ethnic tensions run high, there is intense competition for economic and political resources. Communities generally view reductions in fertility as the equivalent of committing ethnic suicide, since larger populations are seen as generating greater resources and power.

The high fertility rates explain why Sub-Saharan Africa's population is so young. The majority of countries have more than 44 percent of their populations under 15 years of age. The population pyramids in Figure 25–17 compare the proportions of males and females in various age groups in Sub-Saharan Africa and the United States. The graph shows that Sub-Saharan Africa's population pyramid is markedly broader at the base—that is, younger—than that of the United States. The region's workforce thus faces the economic burden of supporting a large proportion of youth. Large government expenditures are required for health care, education, and job training programs to accommodate future employment. The impact of young and rapidly growing populations will carry over into the future, which will bring increased demands for housing, employment, and job benefits. It also

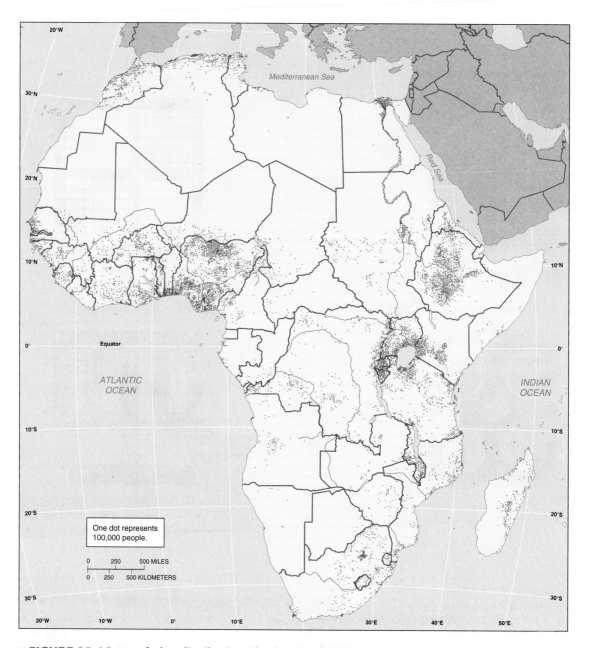

■ **FIGURE 25-16 Population distribution of Sub-Saharan Africa.** (North Africa included for comparison). Sub-Saharan population is most concentrated in the Guinea coast of West Africa, the Great Lakes of East Africa, and the southeast coast of Southern Africa. Semiarid zones are sparsely settled.

will likely translate into continued overall population growth as today's youth reach their reproductive years. Even in those countries where fertility rates are beginning to decline, populations will likely increase substantially before leveling off. The World Bank's fertility projections indicate that the only countries that will reach the population replacement rate of 2.1 children per family by 2030 are Botswana, South Africa, Kenya, Namibia, Lesotho, Zimbabwe, and the small island countries of Cape Verde, Réunion, Seychelles, and São Tomé and Príncipe. These countries have higher levels of living and higher literacy rates, and they have embarked on aggressive family planning campaigns to control population growth. By 2030, the rest of Sub-Saharan Africa is pro-

jected to reach a fertility level of 3.0, in which case it will add at least 900 million people to the current total of roughly 600 million.

In December 1992, African leaders meeting at the Third Population Conference in Dakar pledged to improve the quality of lives of Africans. A major goal set at the conference was to reduce the regional population growth rate to 2 percent by the year 2010. At this juncture, this appears unlikely to occur. Other goals set at the Dakar conference included increasing the regional contraceptive prevalence rate from 10 to 40 percent by 2010, and achieving a life expectancy of 55 years and an infant mortality rate of 50 per 1,000 live births by 2000. None of these goals is close to being met.

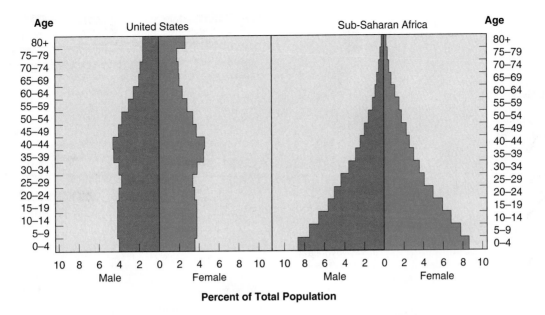

■ **FIGURE 25-17 Age-sex population pyramids of Sub-Saharan Africa and the United States, 2000.** The population of Sub-Saharan Africa is much more youthful than that of the United States. Almost half of Sub-Saharan Africans are below the age of fifteen. In spite of improvements in health care, life expectancies remain relatively low.

Village Life and Traditional Agriculture

Seven of every ten Sub-Saharan Africans reside in rural areas and rely on agriculture as their principal economic activity. For these individuals, the village constitutes the basic unit of social and community organization. Most villages in the forest and grassland regions are nucleated settlements. Such compact arrangements encourage communal interaction. Most dispersed or scattered settlements are associated with decentralized social and political systems. The smallest unit of residence within villages is the **village compound or homestead**, consisting of small houses occupied by members of an extended family (Figure 25–18). Although their design and arrangement may differ from one ethnic group to another, family compounds share some common characteristics such as one entranceway with a reception house in close proximity, a courtyard to enhance social interaction, and a number of multifunctional rooms. The Somolo of southern Burkina Faso live in circular, multistory houses enclosed by walls of puddled mud. The center of the house has a small courtyard, and the thatched roof is supported by posts. There are as many as twenty rooms, one for each wife, with the others functioning as children's rooms, kitchens, storerooms, and granary and grinding areas.

The Yoruba compounds of western Nigeria are typically square or rectangular in shape and contain houses built around one large courtyard. The main courtyard may be linked to a series of subsidiary courtyards that allow air and light into surrounding rooms and that enable rainwater to be collected easily into pots and tanks.

Rural residents rely on agriculture for their livelihood and sustenance. The sector is characterized by rural dualism, where a commercial, export-oriented economy coexists with a larger traditional, subsistence farming sector. Commercial agriculture is dominated by the export of cash crops such as cocoa, coffee, tea, sisal, oil palm, and peanuts, and the production of raw materials for the urban-industrial sector. Traditional agriculture encompasses a variety of cultivation, pastoralist, and

■ **FIGURE 25-18 Masai village in Kenya.** Extended family compounds in compact villages are the basic units of residence within many Sub-Saharan rural African communities. The arrangement facilitates social and economic interaction while strengthening traditional kinship ties. Animals leave the village and family corrals at the beginning of the day to graze in the surrounding pastureland.

■ **FIGURE 25-19 Fulani nomadic pastoralists.** By migrating from one area to another, Sub-Saharan pastoral nomads are able to utilize the seasonal resources of adjoining natural regions.

fishing activities. Traditional agriculture is labor-intensive and utilizes low levels of technological input, including simple tools such as hoes, machetes, and dibble sticks.

Among the more common farming practices in the African forest zones is **shifting cultivation**, where farmers move every few years in search of new land after the soils of their existing plots become exhausted. Once identified, the new plot is cleared and burned. The nutrients contained in the ashes temporarily increase the fertility of the soil. A few trees are often left standing to shelter crops from excessive sunlight and to retard erosion. After the nutrients from the ashes are exhausted, the land is abandoned and soon reclaimed by the forest. A variant of shifting cultivation is the **rotational bush fallow system**, where the cultivated area rotates around a fixed area. Fallow periods are shorter than those of shifting cultivators. Both farming practices have the potential to threaten the ecological stability of forested areas if the growing population exceeds the carrying capacity of the land. Little long-term damage is likely, however, if population levels remain relatively low. Other rural farming practices include the intensive growing of fruits and vegetables within the confines of compound dwellings (compound farming) and combining the raising of livestock with the cultivation of crops (mixed farming). The major staple foods that are grown in the traditional sector include tropical tubers such as yams, cassava, cocoyams, and sweet potatoes; grain crops such as sorghum (guinea corn), millet, and maize; and fruit crops such as bananas and plantains (cooking bananas).

Pastoralism is practiced in the grassland and semiarid regions of Africa. The Fulani nomads of northern Nigeria, the Masai of Kenya, and the nomadic Tswanas of Botswana are among the best-known pastoralists (Figure 25–19). Pastoralists face a number of challenges, including the environmental constraints of drought, land degra-

dation, and the loss of grazing land; the subsequent economic loss of herds and wealth; competition from other land uses for limited grazing land; and the attempts of some governments to pursue policies that limit grazing land and promote the settlement of nomads.

Fish production is concentrated in riverine, coastal, and lake areas. Marine fishing is still undeveloped and involves only a few relatively large fishing vessels. Inland fishing resources are the backbone of many rural communities located near coasts, rivers, lakes, and lagoons. Important inland fishing areas include the Niger delta, Lake Chad, and some coastal areas of Eastern and Southern Africa.

Although subsistence agriculture is often perceived as being static and inactive, it is in reality a dynamic and enterprising economic sector with well-organized marketing systems. The traditional African market, for example, is a focal point for the exchange of goods, services, and ideas among subsistence farmers and pastoralists (Figure 25–20). It also serves as a forum for political activities and for social functions such as marriages. One common form of traditional market is the periodic market, which forms about every 4 to 8 days to serve the needs of dispersed populations and mobile traders.

The traditional farming sector faces a variety of environmental, economic, and institutional challenges. Prolonged droughts and soil degradation have restricted the amount of arable land in many regions. Scarce arable land is often monopolized by commercial agriculture. African governments continue to focus their research and policy efforts on the production of one or two primary products for export. Traditional farmers are frequently denied access to credit, extension services, and the new industrial technologies of the Green Revolution (see the *Geography in Action* boxed feature, Africa's

■ **FIGURE 25-20 Traditional village market.** Traditional markets offer considerable insight into rural life in Sub-Saharan Africa. In addition to supplying foodstuffs, they are often the source of medicinal substances, fuel, and clothing, and provide valued social interaction.

Africa's Limited Green Revolution

The **Green Revolution** refers to the application to agriculture of industrial technologies and products, such as mechanized farm equipment, chemical fertilizers, insecticides and herbicides, and hybrid seeds. The new hybrid varieties have a number of advantages over the older traditional seeds. They are more responsive to fertilizers; produce shorter, sturdier stems that hold more grain; have shorter growing seasons (leading to the possibility of double or triple cropping in tropical areas); yield two to four times more than the traditional crops; and are virus resistant.

Unfortunately, the Green Revolution has not been as successful in Africa as it has been in India and China, owing in part to the fact that most subsistence farmers in Africa are denied access to the financial credit and extension services required for them to purchase the more expensive new seeds and fertilizers and to learn how to judiciously combine the right amounts of water and fertilizer. A second major hindrance to the diffusion of Green Revolution technologies in Africa has been the limited development of high-yielding strains of the staple tuber crops that constitute the daily diet of African farmers. These include cassava, yams, and cocoyams. To date, the Green Revolution in Africa has functioned as a classic case of hierarchical diffusion, in which new technologies and innovations "leapfrog" from one large-scale farm to the next, before eventually descending to the level of small holder agriculturists. This is a reflection of the dual structure of rural Africa, wherein the wealthier and more influential farmers have benefited from foreign innovations much more than the traditional farmers.

Limited Green Revolution). The rural infrastructure continues to be neglected, with only 10 percent of Sub-Saharan feeder roads remaining open year-round. Government expenditures in agriculture have been curtailed in many nations. According to the World Bank, while agriculture accounted for 67 percent of the region's labor force in 1990, and anywhere from 30 to 67 percent of the GNP in twenty Sub-Saharan countries, its share of government expenditures averaged only 7.4 percent. In countries like Burkina Faso, Mali, Niger, and Somalia, the share was less than 5 percent. These problems explain the high dependence of Sub-Saharan countries on food imports. Between 1970 and 1990, the food import dependency ratio (the ratio of food imports to the food produced locally) increased in the majority of Sub-Saharan countries. In 1993 alone, for example, Sub-Saharan Africa imported 11 million metric tons of cereals.

Another factor related to the African agricultural crisis involves the system of land tenure or access. Throughout most of traditional Sub-Saharan Africa, land is held communally rather than individually. In fact, the land belongs to the living, the dead (ancestors), and the yet-to-be-born (future generations). An example of a traditional land tenure system is family land, which is simply passed on through a lineage with rights to the land held jointly by a number of heirs. Usually, no monetary transactions can take place with such land. However, destitute families are sometimes forced to sell family holdings. Problems associated with family land include the increased fragmentation of land resulting from rapid population growth and the subdivision of family land among multiple heirs, and the inequities associated with matrilineal inheritance. Communal land belongs to the lineage, village, or community, and under ideal circumstances every member of the community has equal right to as much land as needed. In almost all cases, the head of the village or clan is in charge of the land and its disposition. Stool land is vested in the stool, or skin, which is the symbol of kingship among the Yoruba in Nigeria, the Mossi in Burkina Faso, the Baganda in Uganda, and the Ga and Ashanti in Ghana. The traditional leader or chief has a sacred duty to hold the land in trust for the people. Subjects of the stool can access land for farming and shelter requirements, but in return, they must provide customary services and pay homage to the stool.

Urbanization

Rapid population growth has triggered the movement of people from neglected rural areas to urban centers. Overall, about 244 million people, representing 30 percent of Sub-Saharan Africa's total population, now reside in urban areas. While this region is among the least urbanized in the world, its urbanization growth rate is among the highest. Sub-Saharan African cities are growing at 4.4 percent per year, compared to 3.3 percent for Asian cities and 2.5 percent for South American cities. Some of the fastest growing cities are those of Liberia (9.6 percent), Rwanda (9.4 percent), Malawi (8.5 percent), Tanzania (6.3 percent), Burkina Faso (5.8 percent), and Uganda (5.2 percent). Rural-urban migration accounts for more than 50 percent of urban growth in Gambia, Sierra Leone, Côte d'Ivoire, and Liberia. Factors that push migrants out of the rural areas include benign neglect by governments, lack of economic opportunities, prolonged droughts, the desire to escape briefly from the social constraints of the extended family, short-term cash needs for bride wealth, marital instability, birth order, and inheritance laws.

Economic opportunity, although limited, is a strong pull factor for cities. Instead of being prime movers of innovation, change, and development for their hinterlands or surrounding areas, most Sub-Saharan cities develop a parasitic relationship with the surrounding rural hinterlands, consuming most of the resources and benefits generated in the countryside while giving little in return. Typically, African cities are primate cities that receive a disproportionate share of the economic, cultural, and political resources of their nations. Furthermore, many cities have failed to provide a sufficient number of formal sector job opportunities. Therefore, most migrants end up working in the informal sector, which operates outside the mainstream of government activity and benefits. This sector includes a variety of jobs, ranging from artisans to basket weavers, goldsmiths, and garment makers. These talented individuals operate without a vendor's license, thereby denying African governments a potential source of badly needed revenue. The informal sector further contributes to the formation of human capital by providing access to training and apprenticeships at a cost much more affordable than formal training institutions. It also encourages the recycling of local resources, such as tires from automobiles (which are used to make footwear), and it caters to the customized needs of residents who cannot afford to purchase items in bulk.

Rapid urbanization has also left many Sub-Saharan African cities with inadequate water supply and sewage disposal systems, limited solid-waste disposal mechanisms, and inefficient public transportation. The basic infrastructure of roads, highways, and other public services is also lacking in most cities. The United Nations Center for Human Settlements (UNCHS) has developed a composite index—the **City Development Index (CDI)**—that measures the level of development in cities by five indices: infrastructure (water, sewerage, electricity, and telephones), waste treatment and disposal (waste water and solid waste), health (life expectancy and child mortality), education (literacy and school enrollment), and city product (estimate of the level of economic output in the city, as measured by income or by value added). The CDI can be used to assess the extent to which governments are effective in delivering urban services and enhancing the feeling of well-being in a city. African cities have the lowest regional average, an index of 42.8. Cities well below the African average include Lagos, Nigeria (29.3), and Niamey, Niger (21.7). In contrast, many cities of the more industrialized nations have CDI indices of more than 90.0. Of major concern is the lack of affordable housing and the proliferation of sprawled residential settlements on the periphery of the cities. Contrary to the popular perception of African cities exhibiting an inverse concentric pattern, with the rich residing in the center of downtown and the poor on the periphery, a new phenomenon is occurring in cities such as Accra, Ghana (Figure 25–21). Upper- and middle-income districts are beginning to proliferate in the city's suburban and exurban regions. This growth has occurred in an uncontrolled and uncoordinated manner, with little effort made to implement growth management strategies designed to coordinate the rate, character, quality, timing, and location of residential development. These shortcomings are further compounded by the complex land tenure systems in place, the lack of comprehensive land surveys to inventory land uses and parcels, and the failure of governments to collaborate with community-based organizations, including informal networks, to find viable solutions to urban problems.

Many African governments are working closely with the World Bank and the UNCHS to provide urban sites and services, upgrade squatter settlements, and encourage self-help initiatives. Typically, in these programs, governments purchase and assemble plots of land, install the necessary infrastructure or services (roads, water and sewer lines, electricity), then sell the plots to low-income households at low interest rates. The purchaser of the plot is then re-

sponsible for designing and building a home to conform with his or her needs. Building, design, and finance costs are kept at a minimum to ensure easy recovery of costs. The program, if carefully implemented, could be a successful partnership between international agencies (who provide low-interest loans and technical advice), national governments (who assemble and develop the site), universities (where engineering and architectural faculty and students offer technical advice), and residents who have an opportunity to participate in the decision-making process. The program empowers local residents and adds a sense of self-esteem and purpose to their initiatives and efforts. Slum and squatter upgrading schemes have also taken on new meaning in African cities. Early attempts to eradicate these settlements were rendered futile, since displaced households simply ended up relocating elsewhere, where they duplicated their previous lifestyles. Most governments now accept the existence of squatter settlements and are trying to provide the necessary facilities to upgrade them. In addition to these efforts, Sub-Saharan African governments are relaxing rigid building codes and sponsoring more research into alternative building materials that are both biotic and fire resistant. The UNCHS has developed a "best practices" database, which draws on 1,150 initiatives from 125 countries to illustrate innovative and creative approaches to urban problem-solving in developing countries. Suggestions range from strengthening local urban governance through decentralized decision making and increased transparency and accountability to disaster prevention.

Development Trends in Sub-Saharan Africa

As we have noted previously, development is a multidimensional phenomenon involving a broad set of economic, social, environmental, institutional, and political factors. In the context of Sub-Saharan Africa, this includes providing adequate educational, health, and nutritional benefits to enhance human capabilities, creating opportunities for women and other traditionally neglected groups, upgrading the physical infrastructure to facilitate the exchange of goods and services, conserving vital nonrenewable resources and sustaining stable environments, and advocating for human rights and creating avenues for the free expression of ideas. These opportunities and services are especially needed by women, children, the urban poor, and rural peasant farmers.

We thus approach development from a holistic perspective, integrating its economic with its human dimensions, assessing impacts of development technologies on various segments of society, and analyzing the spatial patterns of uneven development between the more modernized corelands and more traditional peripheral regions.

Economic Dimensions of Development

The nations of Sub-Saharan Africa presently generate a combined Gross National Income of approximately $310 billion. This is less than that of the Republic of Korea ($421 billion),

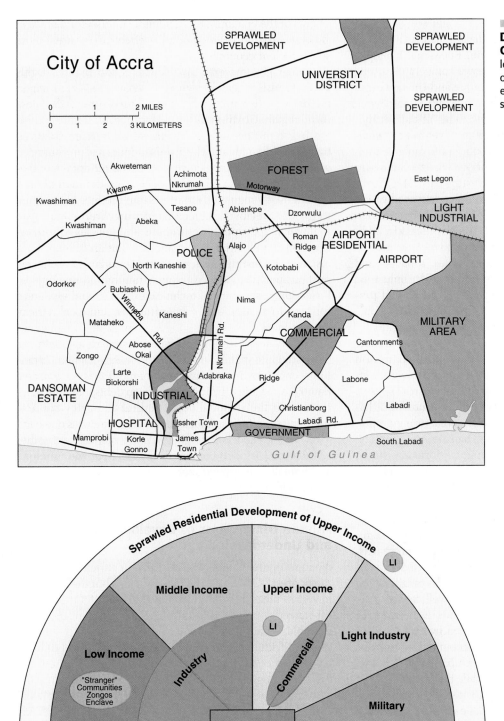

■ **FIGURE 25-21**
Development pattern of Accra, Ghana. The land use in Accra follows a sectoral pattern of development. The suburban areas are experiencing uncontrolled residential sprawl.

about half that of Brazil ($610 billion), and only 3 percent of the U.S. total ($9,601 billion).

The average per capita Gross National Income PPP for Sub-Saharan Africa is $1,790. There are, however, immense variations among the mainland countries, ranging from a high of $10,070 in South Africa to a low of $520 in Sierra Leone. In addition to South Africa, countries on the high end of the economic scale include Botswana, Namibia, Gabon, and Swaziland. These countries are well endowed with oil and strategic mineral resources, although the dis-

tribution of wealth is still problematic. Countries on the low end of the scale include the drought-stricken Sahelian countries of Chad and Mali and the war-torn economies of Burundi, Rwanda, and Somalia. During the 1960s and 1970s, Sub-Saharan Africa experienced modest economic growth, which was followed in the 1980s by declines in productivity. The region recovered slightly from the oil shocks and world recession of the 1980s and began experiencing slow economic growth in the 1990s. The average growth rate of 2.8 percent, however, was below the rate achieved in Latin

America and the Caribbean (3.3 percent) and well below the rate in East Asia and the Pacific region (7.2 percent). A number of countries, among them Benin, Botswana, Ethiopia, Ghana, Mauritania, and Namibia, showed moderately strong growth in the 1990s, while Mozambique and Uganda experienced significant growth. Both Uganda and Mozambique demonstrate that countries perceived to be "at-risk" or "beyond help" can overcome substantial development challenges. Uganda has overcome a despotic and volatile political system (under Idi Amin) and a host of social problems, including the AIDS epidemic, by investing in strategic production sectors, social programs, and poverty-eradication programs with targeted investments in health and education and in micro- and small-enterprise development. In addition, it diversified its agricultural sector to include nontraditional exports, such as fruits, vegetables, tobacco, and flowers. Uganda was also the beneficiary of an enhanced Heavily Indebted Poor Countries (HIPC) debt-relief program, designed to ease debt burdens and provide additional financing for basic health and education. Mozambique's recovery from more than 20 years of civil war and a chaotic economy is attributable to a series of World Bank reform packages designed to rehabilitate the country's overall infrastructure and revive its agricultural and industrial sectors. Following an economic crisis in the 1980s, Mozambique has experienced a significant increase in manufacturing output, particularly in the mining of marble, bauxite, and graphite.

Sub-Saharan Africa's poor economic performance in the 1980s and early 1990s was a function of structural problems in key productive sectors (agriculture and industry), low levels of investment, poor export performance, declining terms of trade, severe debt burdens, inefficient resource use, and low levels of technological capacity, among other problems, including HIV and AIDS. The poor performance in the agricultural sector was a function of its dual structure and a number of institutional factors discussed earlier. Sub-Saharan Africa's industries also have not fared too well, having gone through a period of stagnation in the 1980s. Between 1975 and 1980, industrial productivity declined by an average of 0.1 percent per year. In 1990, Africa accounted for only 2.4 percent (declining from 4.8 percent in 1975) of manufactured exports from developing countries, while the South and East Asian regions dominated with 80.8 percent. Most Sub-Saharan countries are still in Rostow's early stages of industrialization, with output centered on food, beverages, tobacco, textiles, and clothing. These industrial categories account for more than 90 percent of manufacturing value added in Burundi, 73 percent in Uganda, and 66 percent in Ethiopia. Furthermore, most large-scale, capital-intensive industries are located in urban areas where they can take advantage of better telecommunications and skilled labor. Few industries have responded to government incentives designed to encourage them to move to smaller towns and rural communities. This uneven pattern of industrial development denies rural areas opportunities for employment, training, and apprenticeship. It also perpetuates the already uneven terms of trade that exist between the agricultural and industrial sectors. Industries thus lack dynamism because they are not linked to other sectors of the economy. Strengthening intersectoral linkages generates the necessary multiplier effects to ensure a more self-sufficient economy.

African governments have also tended to institute highly protectionist policies designed to replace imported manufactured goods with locally manufactured products. These **import substitution industrialization** strategies are intended to protect infant industries from foreign competition, thereby stimulating local demand for local products and expanding local manufacturing employment opportunities. Regrettably, the net effect has been a proliferation of inefficient, overly regulated industries that produce lower-quality goods at relatively high costs. African governments need to develop appropriate institutions and policies to support and sustain industrial development. What are required, for instance, are institutions for industrial standards, testing, export support, quality assurance, training, technology information, research, and technical extension and assistance for investors, suppliers, subcontractors, and local African entrepreneurs.

The decline in agricultural and industrial productivity has translated into higher debt burdens for Sub-Saharan Africa. Many countries, including Mozambique, Tanzania, Somalia, Zambia, and Côte d'Ivoire, have foreign debts that total more than twice their annual GNI. Several countries are now spending four times as much on debt repayment as on health care. This is especially tragic given the fact that a number of these debt-ridden countries, among them Mozambique and Somalia, also spend more on the military than on education and health.

Human Dimensions of Development and Underdevelopment

As a group, the nations of Sub-Saharan Africa are not only the poorest of all the earth's major regions but they also rank lowest in measures of human development. Of the 55 lowest-ranked countries in the Human Development Index, for example, 39 are found in Sub-Saharan Africa and only one (South Africa at number 107) is found among the 110 highest-ranked countries. These low rankings reflect the tragically low regional life expectancy of 49 years and the extremely low levels of educational attainment. Two of every five adults, for instance, do not know how to read and write; and the level of schooling of many others does not exceed 3 to 4 years. Improvements in meeting basic human needs of the peoples of Sub-Saharan Africa must be achieved through better nutrition, improved medical care, more equitable income distribution, increased levels of education and employment, and expanded opportunities for women (see the *Geography in Action* boxed feature: Gender Inequality in Sub-Saharan Africa).

Geography of AIDS

A growing threat to the development of human resources in Sub-Saharan Africa is HIV/AIDS (Acquired Immune Deficiency Syndrome). The Joint United Nations Program on

Gender Inequality in Sub-Saharan Africa

In its annual *Human Development Report,* the United Nations Development Program publishes two composite indices—the Gender-related Development Index (GDI) and the Gender Empowerment Measure (GEM)—to determine levels of gender inequality in countries. The GDI examines the disparities between men and women with respect to life expectancy, educational attainment, and income. Norway, which had the highest score, registered a value of 0.955 out of a maximum of 1.00 in the 2004 rankings. Thus, overall, women do not fare as well as men in *any* society. Within mainland Sub-Saharan Africa, national GDI scores are generally quite low, ranging from a high value of 0.661 for South Africa to a low value of 0.278 for Niger. One reason for the improvement of the status of women in South Africa may be the creation in 1997 of a Commission on Gender Equality to promote gender justice and equal access to economic, social, and political opportunities. Most department offices in South Africa now have units called "Gender Focal Points" that are responsible for establishing internal gender policies to enhance gender equality. Other countries that have raised their GDI scores include Namibia (0.602), Botswana (0.581), Ghana (0.564), and Swaziland (0.505). All four countries have made great strides in supporting the establishment of women's organizations. Namibia has established the Department of Women Affairs in the office of the President to develop a pool of leaders in governance. Ghana is now home to the Gender and Economic Reforms in Africa (GERA) secretariat, a consortium of African women's organizations and researchers that is geared toward promoting gender equality and justice and expanding the network of women's organizations.

The GEM assesses the relative empowerment of men and women in economic and political spheres of life. It reflects women's participation in the political decision-making process (measured by their share of parliamentary seats), their earning power (access to jobs and wages), and their access to professional opportunities in technical, administrative, and managerial fields. The average for Sub-Saharan Africa is also very low on this measure.

The problems of many Sub-Saharan African women are further compounded by the fact that their work remains unpaid, unrecognized, and undervalued. It is significant to note in this regard that Africa has a long history of female solidarity organizations, which remain extensive and influential. Across the continent, there are several types of formal and informal support networks linked to community development organizations, farm-labor groups, resource conservation movements, market associations, professional organizations, religious and social clubs, secret societies, and rotating credit clubs. Kenya's well-known, self-organized women's groups have become agents of comprehensive rural development, and the group Women In Nigeria (WIN) has been very outspoken in standing up for women's rights. Other groups that have been formed to assert women's rights in Africa include the African Women's Development and Communication Network (FEMNET), and the African Women Global Network (AWOGnet). Their main goals are to improve the living conditions of women and children by providing educational resources and health care.

The United Nations' designation of the 1970s as the "Decade of Women" drew increased attention to women's issues at that time. More recently, the United Nations Development Program outlined a five-point plan calling on countries to enact policies aimed at overcoming gender discrimination, expanding roles and opportunities for women and men at home and in the workplace, ensuring that more women participate in high-level decision making, allowing women greater access to financial credit, health care, and education, and mobilizing national and international efforts to provide basic social services and economic and political opportunities for women.

HIV/AIDS (UNAIDS) reported that by the end of 2001, an estimated 40 million people worldwide were living with HIV. Of this global total, 28.1 million or 70 percent were Sub-Saharan Africans. Owing to HIV/AIDS, the average life expectancy in Sub-Saharan Africa has fallen from 62 years to 49 years. The Southern African region has the world's highest prevalence of adult HIV, with rates exceeding 20 percent in South Africa and Namibia, 25 percent in Swaziland and Zimbabwe, and reaching as high as 35.8 percent in Botswana (Figure 25–22). On the other hand, adult HIV prevalence among the Islamic countries of Mauritania, Mali, and Niger in West Africa is less than 2 percent. There are also significant intraregional variations in HIV/AIDS prevalence. In Kenya, the virus is highly concentrated in the Central, Western, and Nyanza provinces, and especially around major urban centers such as Nairobi, Mombasa, and Kisumu (Figure 25–23). In South Africa adult HIV prevalence is as high as 32.5 percent in Kwazulu-Natal and as low as 5.2 percent in the Western Cape Province (Figure 25–24, page 616). There are also extreme gender differences. In Botswana,

15.8 percent of males in the 15–24 age group have HIV, compared to 34.3 percent of females. While some scholars question the validity of AIDS estimates in Africa, owing to a high proportion of underreporting, others believe that the reports are exaggerated. The latter argue that AIDS estimates by the World Health Organization are not necessarily based on lab tests but, instead, on a list of clinical symptoms that include persistent coughing, high fever, weight loss, and chronic diarrhea, which overlap with symptoms of such diseases as tuberculosis, cholera, and malaria.

In spite of the controversy over the numbers of Sub-Saharan Africans living with HIV, it is clear that the disease continues to spread. Contributing factors include official denial in some government circles, inadequate health facilities and personnel, illiteracy, the lack of preventive programs, and serious misconceptions about the disease. High-risk groups in Africa include sexually active workers, migrants, military personnel, truck drivers, and drug users who share needles. Large cities have become peak areas of infection. Also, wars and civil turmoil have forced refugees into areas

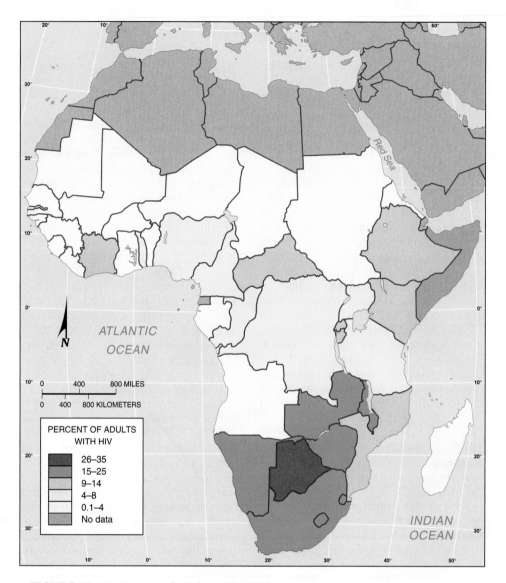

■ **FIGURE 25-22 Percent of adults with HIV, by country.** The nations of Southern Africa have the world's highest rates of adult HIV. Life expectancies have fallen dramatically and human and economic development have been greatly impacted in recent years.
Source: Compiled from The World Bank, *World Development Indicators,* 2002, *http://devdata_worldbank.org/ data-query.*

that are prone to infectious diseases. The medical system, already overwhelmed by numerous tropical diseases, cannot cope with the AIDS onslaught. Absence of adequate blood screening equipment and unhygienic medical practices, including the lack of sterilized equipment, put recipients of health services at increasing risk. In rural areas, where well-equipped modern medical facilities are the exception, many people seek medical care by obtaining chloroquine injections from unqualified itinerant drug vendors. Chloroquine is widely used in tropical lowland regions as a treatment for malaria.

The AIDS epidemic is having a devastating impact on the economic and social lives of Sub-Saharan Africans. Since AIDS affects adults in their productive years, it has a direct impact on the development of human capital. A World Bank study in Tanzania projects the cost of replacing teachers

dying from AIDS at $40 million through 2010. In the same country, rural households affected by AIDS deaths spend approximately $60 (the equivalent of annual rural income per capita) on treatment and funerals. A study of South Africa estimated the direct cost of AIDS in the year 2000 was between $1.2 billion and $2.9 billion. Economic growth is retarded as household savings and resources are diverted away from productive investment. This does not bode well for the business climate of those nations most impacted by AIDS. From a social and psychological standpoint, AIDS has a devastating impact on children who lose their parents. Millions of orphaned children now face limited educational opportunities, and many others are left to fend for themselves or are cared for by other children or by the elderly.

HIV/AIDS has adversely impacted the agricultural sector. In Zimbabwe, for example, maize production has been

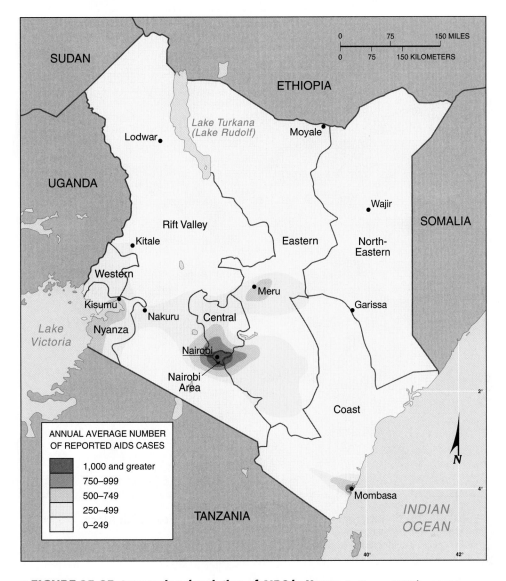

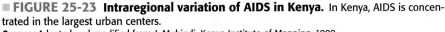

■ **FIGURE 25-23 Intraregional variation of AIDS in Kenya.** In Kenya, AIDS is concentrated in the largest urban centers.
Source: Adapted and modified from J. Muhindi, Kenya Institute of Mapping, 1999.

reduced by 60 percent and cotton production by 40 percent. Declining productivity levels have been accompanied by abandoned fields, delays in essential farming operations such as tillage, declining soil fertility, and the loss of agricultural skills as children are deprived of opportunities to learn from their parents.

Since there is yet no cure for AIDS, the only viable option for managing the disease is through effective prevention. Needed steps include providing more information and education on the disease, improving the safety of blood supplies, expanding testing and screening services, strengthening surveillance and institutional capabilities to control the disease, and integrating prevention strategies with youth and women development programs. Community-based and social marketing approaches can enhance information dissemination efforts. Community-based approaches focus on

removing financial, bureaucratic, and communications barriers in localities and rural districts and by visiting homes to provide counseling and education. Social marketing employs commercial marketing techniques such as incorporating information about AIDS prevention into popular radio or TV shows. More support is also needed for AIDS research programs. At present, Africa accounts for 70 percent of world HIV cases, yet receives only 1.6 percent of global AIDS research money.

Uneven Patterns of Development

Sub-Saharan African economies are characterized at every level by dualistic patterns of uneven development. At the international level, there are extreme gaps between Africa and the more industrialized countries, and at the national level,

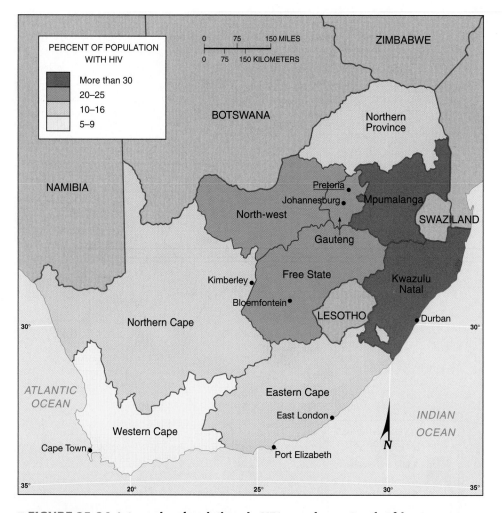

■ **FIGURE 25-24 Intraregional variations in HIV prevalence, South Africa.** In South Africa, adult HIV prevalence varies greatly by region.
Source: Compiled from Statistics South Africa, 2000.

there are core-periphery disparities between the more modern urban centers and their traditional rural peripheries. World systems and dependency theorists have argued that these gaps are perpetuated by a dominance-dependence relationship, which results in an unequal exchange of resources between more- and less-developed nations, such as the exchange of raw materials for industrial products. According to this argument, Sub-Saharan Africa ends up in a subordinate position and becomes vulnerable to the rules and regulations of an international system controlled by the technologically more advanced countries. Some structural theorists go a step further to argue that there is an implicit harmony of interest between privileged elites in the industrialized countries and their less industrialized suppliers. In this view, the ruling elites in Sub-Saharan Africa are seen as facilitating the unequal exchange of resources by investing their own money in foreign goods and services, thus enabling rich countries and multinational corporations to, in effect, appropriate much of the wealth generated by African economies. Reports of scandal, corruption, and embezzle-

ment in African governments are not uncommon. Every now and then, we hear about African heads of states and government officials funneling hard-earned money generated from rural areas into foreign banks. The north-south wealth gap, in other words, is a function of both one-sided foreign economic relationships and poor judgement and bad internal management on the part of African governments and urban elites.

At the national level, core-periphery disparities are intensified by the uneven pace of economic activity. Figure 25-25 shows the distribution of major core centers of economic activity, including manufacturing, agricultural export/processing, and mining centers. Key manufacturing centers often locate in major cities to take advantage of cost savings derived from the close proximity to skilled labor, larger markets, and more developed utilities. These islands of modern development are clustered along the petroleum belt of West Africa; the "ring of diamonds" that spans southwest Namibia, northeast South Africa, Botswana, southwest Democratic Republic of Congo, and northeast Angola;

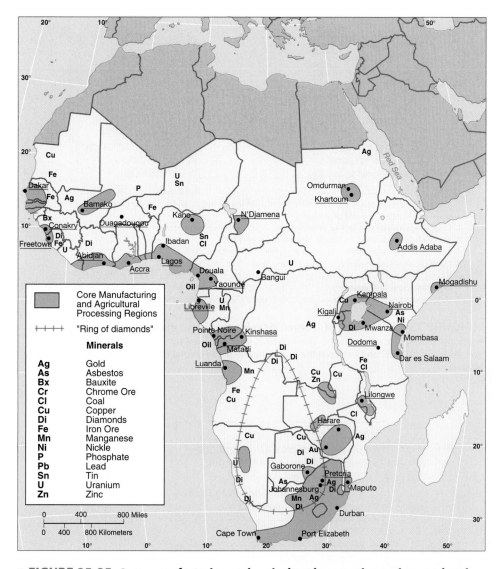

■ FIGURE 25-25 Core manufacturing and agricultural processing regions and major mineral-producing areas in Sub-Saharan Africa. By comparing this map with the population distribution map we learn that a strong spatial correlation exists in Sub-Saharan Africa between the distribution of human population and the location of manufacturing activities. Manufacturing tends to be concentrated in the most densely settled and urbanized regions.

the Luanda-Malanje corridor of Angola; the high-tech and mineral-rich region of Witwatersrand in South Africa; the copper belt of Democratic Republic of Congo and Zambia; the rich agricultural region bordering Lake Victoria; and central Ethiopia, to name a few of the primary industrial centers.

This dual economic structure further manifests itself in urban and rural areas. In urban areas, the formal-informal sector dichotomy works to the disadvantage of small-scale enterprises that operate outside the mainstream of government regulation and benefits. In rural areas, relatively wealthy commercial/cash-crop farmers in cocoa, coffee, cotton, peanuts, rubber, and tea plantations coexist with subsistence farmers who produce just enough food for themselves and their families. These patterns of dualism

create a three-way interactive process between the commercial sector in rural areas, the formal institutions in urban areas, and the industrialized world economy (Figure 25–26). The commercial farmers benefit from urban-based government incentives. Formal urban institutions benefit from the foreign exchange derived from the overseas sale of rural cash crops. The foreign exchange earned is then internalized in urban areas. This three-way interactive process reinforces the disparities that exist between core and peripheral regions at all scales, and polarizes people who reside in the urban informal and rural subsistence sectors. In many respects, the constituents from the urban informal and rural subsistence sectors have limited opportunities to realize their human potential and to participate in the global exchange economy.

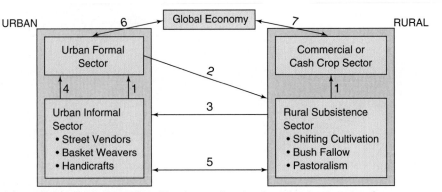

1. Inequality between the two sectors widens because formal sector and cash crop sector enjoy more government incentives and benefits.
2. Inadequate investment; lack of extension services, credit, technical assistance.
3. Severe backwash effect as people migrate to urban areas.
4. Low-cost services in informal sector are exploited by formal sector.
5. Rural-urban linkages: remittance of incomes, rural-urban networks formed.
6. Strong interrelationships between urban elite and global economy: foreign investments, demand for imported goods and services.
7. Proceeds for primary products (cocoa, coffee, cotton, tea) and minerals exported.

▪ **FIGURE 25-26 Patterns of uneven development and exchange.** Inequalities in trade relationships and rates of economic development exist both among the more and less developed countries of the world and between the traditional rural and formal urban sectors of the developing nations.

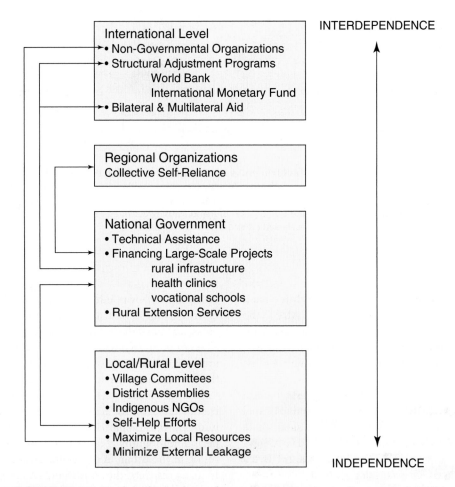

▪ **FIGURE 25-27 Model of interdependence.** Comprehensive development in the Third World nations can be achieved only through the cooperative efforts of local, national, regional, and international development agencies and their leaders.

Development Strategies

Trying to find solutions to ameliorate the patterns of uneven development is a difficult task. A host of strategies revolving around self-reliance, basic needs, appropriate technologies, neoclassical economies, and growth-equity issues have been proposed. However, it is important to acknowledge the complexities associated with peripheral economies that are increasingly being marginalized from the global economy. One strategy that has some promise is **interdependent development**; it requires developing cooperative and compatible arrangements, at all geographic scales, to facilitate development efforts. This implies cooperation between local and national governments, local and international organizations, national and regional organizations, and national and international governments/organizations (Figure 25–27). At the local level, independence and autonomy are strongest. The level of interdependence increases as we move from the local to the national, regional, and international scales.

It is important for rural constituents in Sub-Saharan Africa to maintain a certain degree of autonomy in their decision making and their planning processes. Self-help efforts, however, must be supplemented by committed efforts on the part of national governments to provide the necessary technical and financial assistance. Most self-help efforts in Sub-Saharan Africa are initiated to fill a vacuum created by central governments that fail to design interventionist strategies to meet rural needs. In Kenya, Harambee (meaning "let's pull together" in Swahili) self-help groups have been instrumental in providing primary and secondary schools, technological institutes, basic health services, water, social welfare, and economic opportunity. In Zimbabwe, a Savings Development Movement, totaling 5,700 clubs nationwide, mobilized local financial resources to invest in a variety of rural projects. Proceeds from savings were rotated among members, and the more advanced savings clubs were able to diversify into agroprocessing enterprises, consumer stores, and community wells and dams. These clubs were successful in devising effective low-cost methods for promoting capital formation and agricultural investment among not only the poor but also among women, who constituted the majority of their membership.

If national governments are to become effective partners with rural constituents, they will require assistance from regional and international organizations. At the regional scale, there have been various attempts at economic integration, and at the international level, assistance has come via structural adjustment programs and nongovernmental organizations.

Regional cooperation and interdependence among Sub-Saharan countries needs to be encouraged to promote a greater sense of collective self-reliance. The **Organization of African Unity (OAU)** was formed in 1963 to champion the cause of African unity and defend the sovereign rights of African states. It continues to endorse efforts to promote greater economic integration. At the twenty-first Assembly, held in Addis Ababa in 1985, African countries reaffirmed their commitment to the 1980 Lagos Plan of Action on the establishment of an African Common Market. This was followed in 1991 by the signing of a treaty in Abuja, Nigeria, to establish a Pan-African Economic Community (AEC) by the year 2025. The AEC was set up to enhance collective self-reliance and self-sustaining development in Africa; to promote greater social, cultural, and economic integration among African states; to coordinate and harmonize the activities of existing regional organizations in Africa; and eventually to establish an African Central Bank, an African Economic and Monetary Union, and a Pan-African Parliament. Part of the Lagos Plan's agenda was to encourage three subregions—West Africa, Central Africa, and East and Southern Africa—to proceed through various stages of economic integration. Given the current status of Sub-Saharan African countries, it may require considerable time to reach the level of economic integration that is characteristic of the European Community. Collective self-reliance provides Sub-Saharan Africa with a framework to pursue common goals and interests that ensure economic and political sustainability. Promoting collective self-reliance through regional economic integration is relevant in today's global economy, considering the numerous trading blocs cropping up in North America (NAFTA), Europe (EU), Latin America (Southern Cone Common Market), the Caribbean (CARICOM), and Asia (ASEAN). Africa needs to strengthen its competitive position in the global economy by forging viable regional economic alliances.

At the international scale, aid to Sub-Saharan Africa has taken several forms; financial and technical assistance has been forthcoming from bilateral and multilateral sources. However, since the 1980s, financial and technical assistance has been conditioned by the World Bank's **structural adjustment programs (SAPs)**. The SAPs were initiated to adjust malfunctioning economies, assist in debt restructuring, and promote greater economic efficiency and growth in order to make Sub-Saharan economies more competitive in today's global economy. By 1990, thirty-two African countries had launched structural adjustment programs or borrowed from the International Monetary Fund (IMF) to support policy reforms. Candidates for structural adjustment are countries with budget deficits, balance-of-payment problems, high rates of inflation, ineffective state bureaucracies, inefficient agricultural and industrial production sectors, overvalued currencies, and inefficient credit institutions. As a result, countries that buy into the program are required to devalue their currencies to reduce expenditures on imports and release resources for exports; restructure the public sector and state-owned enterprises by trimming overstaffed bureaucracies and padded payrolls, thereby improving institutional management and encouraging privatization; eliminate price controls and subsidies, such as artificially reducing the price of food in urban areas, thus lowering farm incomes; and restructure the productive sectors of the economy by liberalizing trade and removing import quotas and high tariffs that protect uncompetitive firms and by providing export incentives to promote export growth, particularly in agriculture.

SAPs have had some negative social and human impacts. For example, currency devaluations and strict credit restrictions have had the effect of raising the prices of consumer goods and agricultural inputs, thus affecting women and children who rely on food trading and production. Also, the neoliberal policies that have been geared toward trimming government bureaucracies often have not been supplemented with job retraining or redeployment programs, with the result that unemployment has risen. Government cuts in education have driven up costs, making it difficult for poor families to secure a good education. In some cases, females have been discriminated against as parents send only their sons to school. Rising health care costs have also forced people to seek other alternatives from the traditional sector, including itinerant drug vendors.

Both the World Bank and IMF have launched Poverty Reduction and Growth Facility programs to break the cycle of poverty. Mozambique recently developed a poverty-reduction strategy that focuses on universal, informal, and technical-vocational training, on primary health-care improvements, on agricultural and rural-development efforts, and on promoting good governance.

In addition, in May 2000, the United States sought to strengthen its bilateral ties with a number of African countries when President Clinton signed the Africa Growth and Opportunity Act (AGOA). The act seeks to strengthen American-African trade and investment ties by allowing African states more access to U.S. markets, credit, and technical know-how and by providing social and economic opportunities to vulnerable groups in order to ensure stable and sustained growth and development. To be eligible to receive benefits from the AGOA, African countries must have demonstrated a commitment to economic and political reform by engaging in political pluralism and the rule of law, by developing policies to reduce poverty and combat corruption, by protecting human rights and worker rights, and by adhering to the IMF and World Bank structural adjustment programs.

Non-Governmental Organizations (NGOs) have been instrumental in providing a wide variety of aid packages to Sub-Saharan countries. NGOs operate neither as government nor as for-profit organizations; they are a diverse group of largely voluntary organizations that work with people organizations to provide technical advice and economic, social, and humanitarian assistance. They can be professional associations, religious institutions, research institutions, private foundations, or international and indigenous funding and development agencies. The World Bank is collaborating with an increasing number of grassroots NGOs to address rural development, population, health, and infrastructural issues that are pertinent to the human dimensions of its structural adjustment programs. Kenya, Uganda, and Zimbabwe have a large number of NGOs. Most of the "indigenous" NGOs in Africa are community-based grassroots and service-based organizations. They usually fill a void where govern-

ments have been largely ineffective and out of touch with local needs. Some of their specific developmental objectives in Sub-Saharan Africa include tackling poverty by providing financial credit and technical advice to the poor, empowering marginal groups, challenging gender discrimination, and delivering emergency relief. At times, conflicts arise when NGO investments in rural areas undermine the ability of governments to perform as effective leaders and policy makers. In response to this potential threat, some African governments have instituted coordinating bodies to supervise NGO activity. For example, the Voluntary Organizations in Community Enterprise (VOICE) in Zimbabwe and the Zambia Council for Social Development both coordinate social service activities. In Uganda, two organizations have been active in assisting the victims of AIDS. The AIDS Support Organization (TASO) provides community-based counseling, social support, and medical services to persons with AIDS and their families, while the Uganda Women's Efforts to Save Orphans cares for more than 1.4 million orphaned children in Uganda.

Non-governmental organizations have a vital role to play in the social and economic development of Sub-Saharan Africa. More partnerships between international, national, and indigenous NGOs should be encouraged. There is a growing awareness of the importance of voluntary activity. In a conference held in Dakar, Senegal, in 1987, delegates from African countries agreed to form a pan-African umbrella organization—the Forum of African Voluntary Development Organizations (FAVDO)—to provide a forum for voluntary development organizations to exchange ideas, share their expertise and resources, support local initiatives, and establish effective channels of communication and partnerships with governments and intergovernmental organizations.

▶▶ Summary

Sub-Saharan Africa faces important challenges in development policy and planning as it enters the twenty-first century. While the socioeconomic problems at times seem insurmountable, opportunities still exist for Africa to tap into its vast reservoir of human and natural resources and create policies that are not solely growth inducing, but that also pay more attention to issues of human equity and development. Grassroots and self-help efforts at the local level are on the rise, and people are beginning to control their own destinies by channeling their energies and creativity toward more positive ventures. These local initiatives must be supported with complementary efforts from national and international institutions. Sub-Saharan Africa needs to seek more avenues for economic, cultural, and political cooperation at the regional level. As regions around the world continue to form alliances and trading blocs, Sub-Saharan Africa needs to carve its own niche by strengthening its traditional values of collective self-reliance and community initiative. While international efforts at revitalizing Sub-Saharan Africa are welcome, they must be well intended. More importantly, they must address the extreme inequalities that exist between the core and the periphery, and they must embody strategies that provide the poor with opportunities to achieve their full human potential.

▶▶ Key Terms

acculturation

assimilation

Berlin Conference

bride wealth

child fosterage

City Development Index (CDI)

colonialism

ecotourism

Falashas

Green Revolution

import substitution industrialization

indirect rule

interdependent development

kinship relations

landlocked states

lingua franca

neocolonialism

Non-Governmental Organizations (NGOs)

Organization of African Unity (OAU)

pastoralism

rotational bush fallow system

shifting cultivation

structural adjustment programs (SAPs)

village compound or homestead

World Heritage Site

■ **Village in Mali.** The contrasting lifestyles of West, Central, and East Africa are represented by a traditional Dogan village and (inset) a modern street scene from Nairobi, Kenya.

Chapter 26

West, Central, and East Africa:
Diversity in Development

- **The Physical Environment**
- **West Africa**
- **Central Africa**
- **East Africa**

The regions of West, Central, and East Africa encompass thirty-two countries covering 72 percent of Sub-Saharan Africa's land area and accounting for 77 percent of its population. The West African countries include Benin, Burkina Faso, Côte d'Ivoire, Gambia, Ghana, Guinea, Guinea-Bissau, Liberia, Mali, Mauretania, Niger, Nigeria, Senegal, Sierra Leone, Togo, and Chad. Chad's locational affiliation can be debated since it is caught in transition between West, East, and, arguably, Central Africa. Central Africa includes the Central African Republic, Congo, Equatorial Guinea, Gabon, Democratic Republic of Congo (Zaire), and Cameroon (which occupies a transitional area between West and Central Africa). East Africa includes Burundi, Djibouti, Eritrea, Ethiopia, Kenya, Rwanda, Somalia, Sudan, Tanzania, and Uganda. As compared to the Southern African region, these three regions have a more tropical environment (owing to their proximity to the equator), have lower levels of European settlement, and have less industrial capacity. Even so, these tropical countries exhibit immense geographical, cultural, and economic diversity. For example, the drought-prone Sahel states have lower levels of development as compared to the coastal states of Ghana, Côte d'Ivoire, and Nigeria. There is considerable language diversity among the Nilotic groups of southern Sudan, the Bantu of East Africa, and the Guinean languages of coastal West Africa. The topographical and vegetational characteristics also differ from region to region. In this chapter, the topics selected for each region and country reflect the physical, social, and economic diversity of tropical Africa.

The Physical Environment

The regions' diverse physical environment is defined by the dynamic interaction of several natural processes. The complexity of the physical landscapes and the climatic, vegeta-tional, and biogeographical attributes offer a number of assets and some inherent liabilities. Among the assets are the majestic mountains, a wealth of minerals embedded in pre-Cambrian rocks, the rich volcanic soils of East Africa, the scenic and economic value of its lakes, the hydroelectric power potential of its rivers, and the biodiversity and commercial value of the rain-forest region. However, the narrow and straight coastlines limit opportunities for natural harbors, and the short continental shelves restrict potential offshore oil exploration and fish breeding. Furthermore, the leached soils of the rain-forest environments inhibit agricultural development. A major threat to the physical environment, however, is the extent of human activity in ecologically sensitive areas. The magnitude of this threat will depend on the ability of communities and governments to develop appropriate response strategies to manage and conserve their fragile environments.

Physiography

Geologically, much of tropical Africa consists of a great plateau that tilts downward from east to west. This plateau is fractured and scoured by a number of major river systems, leaving large gorges and several undulating surfaces. The eastern sections of Africa, or **Highland Africa**, average about 4,000–5,000 feet (1,219–1,524 meters) above sea level, while the lower plateau of the western and central regions averages 1,000–1,500 feet (305–457 meters) in elevation. East Africa has several prominent mountain landscapes, such as the extensive East African Plateau, which features the two highest points in Africa: Mount Kilimanjaro (19,340 feet, 5,895 meters) and Mount Kenya (17,058 feet, 5,200 meters) (Figure 26–1). Further north is the Ethiopian Massif, which has its highest point at Ras Dashen (about 15,000 feet, 4,572 meters). East Africa also features some extensive plains, such as the Serengeti Plains of Tanzania. West and Central Africa are not entirely low-lying regions, however. Mount

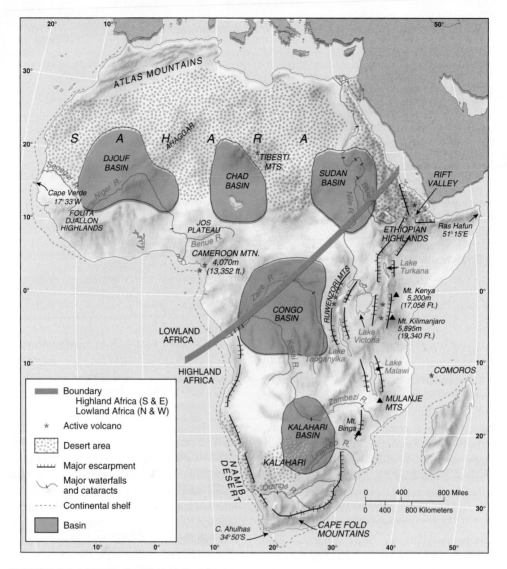

■ **FIGURE 26-1 Major landforms in Africa.** Most of tropical Africa consists of a great elevated plateau that is broken by large inland basins. The volcanic East African Rift Valley is also an extremely prominent feature.

Cameroon (13,352 feet, 4,070 meters), the Jos Plateau (5,840 feet, 1,780 meters) in Nigeria, and the Fouta Djallon Highlands of Guinea are examples of major uplands that rise above the surrounding plateau.

The plateau itself consists primarily of ancient crystalline rocks that have been metamorphosed by immense heat, pressure, and chemical changes, creating a wealth of minerals in the process. For example, the old, geologically stable core or shield areas of Africa are rich in chromium and asbestos, and areas of West Africa and the Gabon-Congo region have rich reserves of such minerals as gold, diamonds, and manganese. Oil and gas deposits are associated with younger sedimentary rocks that occur along linear zones of the Atlantic front stretching from the Niger River delta to the Democratic Republic of Congo.

Another unique aspect of the region's physiography is the **East African Rift Valley**, which begins in the north with the Red Sea and extends about 6,000 miles (9,700 kilometers) through Ethiopia to the Lake Victoria region, where it

divides into eastern and western segments and continues southward through Lake Malawi and Mozambique (Figure 26–2). The average width of the rift ranges between 20 and 50 miles (32 to 80 kilometers), with walls as high as 3,000 feet (915 meters). In the western rift, the Ruwenzori, a faulted mountain within the rift, rises more than 16,404 feet (5,000 meters), and, further south, the floor of Lake Tanganyika lies about 2,133 feet (650 meters) below sea level. The rift valley was created by faulting, as tensional forces associated with continental drift began to pull the eastern sections of Africa away from the remainder, leaving in the area of separation a great valley that subsequently subsided. It is speculated that the eastern part of Africa will eventually break and drift away from the African continent. An important feature associated with the rift valley is the **Great Lakes System of East Africa** (with the exception of Lake Victoria). Especially unique are the crater lakes (Figure 26–3) and the elongated lakes that occupy the deep trenches, such as Lake Malawi, Lake Tanganyika, Lake Turkana, and Lake

▪ **FIGURE 26-2 East African Rift Valley.** Active and dormant volcanoes are associated with the East African Rift Valley system. These volcanoes are located in the Oldonyo Lengai Valley of Tanzania.

Albert. Lake Victoria, the world's second largest lake in terms of area, is nestled between the two arms of the rift valley. The rift belt, along with the offshore islands of Réunion, the Comoros, and the Canaries, constitute the major volcanic regions of Africa. There are several explosive craters around the Uganda–Democratic Republic of Congo border.

The interior plateau is drained by three major river systems—the Nile, Congo (Zaire), and Niger. Other important rivers are the Senegal and the Volta. African rivers in general have limited navigability because they frequently are interrupted by falls and rapids, and their water levels vary greatly from season to season. For example, the Congo is navigable only up to 85 miles (137 kilometers) inland before its course is broken by a series of rapids up to the Stanley

Pool. On the positive side, African rivers have much potential for hydroelectric power generation, owing to their swift flow and steep falls. The Congo River carries the second largest volume of water in the world (only the Amazon carries more) and has enormous potential for hydroelectric power. African rivers also contain significant species of marine life, with inland fishing being a major activity.

The four major **drainage basins** associated with the region's rivers are the El Djouf, the Chad, the Sudan, and the Congo. These basins were formed by tectonic forces that downwarped parts of the plateau. They then became repositories of marine sediments eroded from plateau surfaces and deposited by rivers that converged on the basins. The basins are surrounded by mountain ranges such as the Fouta Djallon, which borders the El Djouf, and the East African Highlands, which separate the Congo or Zaire Basin from the coastal plains on Tanzania's east coast. The Sudan Basin is also sandwiched between the Ethiopian Massif in the east and the Dar Fur and Ennedi ranges in the west. The Chad and El Djouf are internal drainage basins, meaning they have no exits to the sea. The Chad Basin is centered on an impoundment of Lake Chad, which receives water from rivers that originate from the wetter southern regions of Nigeria, Cameroon, and the Central African Republic. Lake Chad provides significant water resources to surrounding countries. Droughts, high evapotranspiration rates, and increased human demands, however, have diminished its water levels. With these threats, it is possible that the lake may eventually end up as a large wetland fringed by salt flats. The El Djouf Basin is the driest. Owing to its proximity to desert climates, it does not receive any permanent streams. Runoff during the rainy season ends up in shallow ponds and

▪ **FIGURE 26-3 Lake Mutanda in Uganda.** This crater lake in Uganda was formed when the volcano exploded and collapsed.

■ **FIGURE 26-4** **Daily life on the Niger River.** Traditional livelihoods predominate in most of the El Djouf Basin, where the levels of the Niger River fluctuate greatly on a seasonal basis.

marshes that evaporate in the summer, leaving salt marshes and plains (Figure 26–4). The structure of the Congo Basin, on the other hand, allows the exit of the Congo River by way of a narrow westward outlet. Similarly, the southern sections of the Sudan Basin divert water into the Nile, while the northwestern section is internally drained.

Among the liabilities inherent in the region's physiography are its coastlines and continental shelves. Since the continent is mainly a plateau, its coasts tend to be straight and smooth with very few indentations, unlike the Scandinavian peninsula of northern Europe or the northwestern coast of North America, which have several deep river valleys or fjords. Another feature of Africa's coastlines is that they are exposed to erosion by offshore currents. In West Africa, for example, several sandbars front the coasts of Nigeria, Ghana, and Senegal. This was especially problematic during colonial times when ships had to dock some distance from the coast, and surf boats had to be employed to load and transport cargo to the coast.

Africa has very few natural harbors. Most of them are artificial harbors that have been constructed at considerable expense. Among the most significant of these are the ports of Dakar (Senegal), Abidjan (Côte d'Ivoire), Tema (Ghana), Durban (South Africa), and Mogadishu (Somalia) (Figure 26–5). The limited number of good natural harbors in West Africa include Freetown, Sierra Leone, and Banjul, Gambia. The southwestern shores of the Atlantic Ocean include major ports at Lobito and Luanda in Angola and Libreville in Gabon. Important railway terminal ports in Mombasa (Kenya), Maputo and Beira (Mozambique), Dar es Salaam (Tanzania), and Cape Town (South Africa) were all developed from sheltered natural harbors.

Another feature of the region's physiography is its limited continental shelf, which extends only for a short distance from the coastline before dropping abruptly to the ocean depths. Countries with extensive continental shelves (United States, Peru, Chile, Argentina) have taken advantage of the inherent opportunities such as offshore oil drilling, mineral exploration, and fishing. One disadvantage of deep shorelines is the poor development of beaches and other shoreline resources that can benefit recreation and tourist development. Locations with shallow shorelines are more likely to have attractive beach resorts.

■ **FIGURE 26-5** **Dakar, Senegal.** Beyond Independence Square lies the port of Dakar, one of the most important shipping facilities in West Africa.

Vegetation

The ecological setting of West, Central, and East Africa encompasses a rich and diverse plant cover organized into complex vegetative formations. The major vegetation zones include the rain forest, the woodlands, the Sudanian grassland, the Sahel bushland, the semidesert and desert regions, and the Afro-montane region. While these natural vegetation belts have been modified in many areas through agriculture and forestry, there is generally a close association between climate and vegetation distribution. Areas along the wet-humid equatorial belt tend to have high and closed forests, while areas with relatively low annual precipitation levels and longer dry seasons experience lower and more open forests, woodlands, and savannas.

The rain-forest biome encompasses areas of perennially moist forests interspersed with semideciduous forests. Rain forests are concentrated along coastal West Africa, the western equatorial belt of Cameroon, Gabon, Congo, and the Democratic Republic of Congo, and the lowlands of coastal East Africa. The current rain-forest belt covers an estimated 7 percent of the total land area of Sub-Saharan Africa, constituting about 20 percent of rain forests worldwide. It is biologically diverse, with more than 50 percent of the continent's plant and animal species. Coastal West Africa and the Central African Basin zones alone are estimated to contain more than 8,000 plant species, about 80 percent of which are native to the region. The forest is dominated by broad-leaved evergreen tree species, including emergents that rise to heights of 165 feet (50 meters) and even 300 feet (90 meters) in some cases. A middle canopy layer made up of trees about 80 to 115 feet (25 to 35 meters) high forms a continuous cover as their crowns interlock, and this tends to deprive the lower layers of direct sunlight. Trees up to 50 feet (15 meters) high form the lowest layer. In addition to their contribution to the preservation of the earth's biodiversity, the rain forests contain many high-value timber species. Other assets include numerous wildlife species of economic value, medicinal resources (a preserve for unique traditional societies), a pristine agricultural environment, and a setting for the now growing activity of ecotourism (Figure 26–6). The tropical rain forests are being significantly threatened by land-use conversion and modification (see the *Geography in Action* boxed feature, Deforestation in West and Central Africa).

The woodlands generally lie within the interior subhumid tropics, where annual rainfall amounts range between 40 and 64 inches (1000 and 1600 millimeters). The woodlands are made up of open stands of trees, with at least 40 percent of the ground covered by tree crowns. The trees are at least 16 feet (5 meters) tall, with increasing ground cover of herbaceous plants. Patches of woodlands occur along the southern coast of Somalia and the coasts of Kenya and Tanzania. The core woodlands consist of deciduous and semideciduous trees and other tree species such as the African fan palm, the shea butter tree, the silk cotton tree, the baobab, African rubber, and the gum arabic (*Acacia senegal*).

■ **FIGURE 26-6 Tropical rain forest.** The rain forests of West, Central, and East Africa contribute greatly to the preservation of the earth's biodiversity. Much of the animal life is arboreal.

The savanna or grassland belt consists of a vegetation zone where herbaceous plants and grasses dominate the landscape. It is extensive in West and Central Africa, the Somalia-Masai Plains of Eastern Africa, and the Indian Ocean coastal belt. The *fadamas* in northern Nigeria and southern Niger are floodplains covered with tall and dense grasslands and sparse tree populations. Vegetation ranges from perennial tall grasses with about 10 to 49 percent of tree cover to areas of true open grassland interspersed with small trees. Most of the vegetation species exhibit strong adaptations to fire and moisture stress, such as baobabs, acacias, and mopane species. The region provides expansive habitats for both livestock and wildlife development. Traditional pastoralists, such as the cattle-herding Fulani of West Africa and the Masai of East Africa, have made extensive use of the dry savannas, which have fewer tsetse flies than the wetter woodlands. The savanna belt supports most of the large mammals of contemporary Africa, especially where human encroachment is minimal. For example, the Serengeti Plains, in the Somalia-Masai region of East Africa, have some of the largest concentrations of wildlife in the world (Figure 26–7).

The northern and southern edges of the savanna give way to Sahel and desert landscapes. The problems inherent in these regions are outlined later. The remaining portion of Africa's vegetation is characterized by montane or highland vegetation, which develops over isolated elevated regions (Figure 26–8). The result is a vertical zonation of vegetation belts. This type of vegetation is restricted to the Cameroon and Guinea highlands of West Africa and the Ethiopian, Kenyan, Tanzanian, and Albertine Rift highlands (Kivu-Ruwenzori) of East Africa. Montane forests

GEOGRAPHY IN ACTION
Deforestation in West and Central Africa

The United Nations Food and Agricultural Organization estimates that the West African region has the highest rates of forest loss in the world, about 2 percent per year. Coastal West Africa has experienced the most drastic loss of moist lowland tropical forests. Côte d'Ivoire, Ghana, and Nigeria formerly contained the largest moist tropical forests in the region but now have less than 15 percent of the original forest vegetation (Figure A). Most of the existing forests are in isolated protected reservations. Limited logging is still permitted in some of the most productive reservations. Liberia is the only country with a relatively large proportion of original forests remaining. However, recent military offensives have caused extensive damage, and resources such as timber have been exploited to finance war-related activities. The Central African nations of Gabon, Congo, Democratic Republic of Congo, and Cameroon have the lowest rates of deforestation in Africa, with more than 40 percent of their original forests still intact. This region has some of the lowest densities of rural populations, and any form of forest exploitation is very localized.

The causes of **deforestation** are mainly human induced. This is not to discount the importance of such physical factors as natural fires, floods, volcanic eruptions, pests, and disease vectors. Three main human-related causes are agriculture, logging, and fuelwood consumption. In the food crop sector, the bush fallow (slash and burn) system of cultivation creates immense pressures on standing vegetation. Growing demands from increasing populations cause fallow periods to be reduced, thus limiting opportunities for soil replenishment and successful forest transition to a secondary stage. Where fallow periods are reduced, poorer woodlands and savanna grasses tend to colonize abandoned lands, thus inhibiting forest regrowth. Logging causes much damage to the forest floor and creates gaps in previously well-structured rain forests, leading to invasions of foreign weeds. Logging also encourages the construction of roads into previously inaccessible forest areas. In Côte D'Ivoire, for every 10 square kilometers (4 square miles) of forest logged, 10 kilometers (6 miles) of road have been constructed. Such roads have been widely used by farmers to access new areas of farmland, thus accelerating the rate of forest loss. While logging may not lead directly to the loss of contiguous tracts of forest space, it undermines forest preservation efforts and sets the tone for other direct onslaughts such as road construction and agricultural land use. The harvesting of wood for fuel consumption has also had negative environmental impacts. In Burkina Faso and Chad, wood accounts for more than 90 percent of total national energy consumption. Policies in the energy sector have generally excluded marginal rural and urban populations. More worrisome is the fact that increased demand for fuelwood by the urban poor places additional burdens on the rural environment. As the costs of electricity and other energy alternatives continue to be prohibitive for marginal urban and rural populations, the demand for fuelwood will likely continue to increase, thus threatening the stability of forest preserves.

Forests provide a wide range of goods and services that can be substituted for manufactured goods in rural subsistence economies. Fibers, wood poles, wild honey, medicinal herbs, and bark are but a few of the benefits or by-products of the standing forest. Tropical forests also provide optimal environments for shade-loving tree crops, such as cocoa, rubber, and coffee. From an ecological and socioeconomic standpoint, the forests are central to Africa's survival. Because the value of forests is immeasurable, they need to be managed efficiently and wisely to support Africa's drive toward sustainable development.

■ FIGURE A **Rain-forest clearing in Nigeria.** This area is being converted to a cattle ranch. Once cleared, it is unlikely that it will ever return to a forested state.

■ FIGURE 26-7 **The Serengeti Plains in Tanzania.** The Serengeti Plains in Tanzania comprise the largest concentration of grazing mammals in Africa and have been designated as a World Heritage Site.

■ **FIGURE 26-8 Montane forest vegetation in Mount Kenya National Park.** In the highlands of West, Central, and East Africa, the composition of the vegetation varies by elevation, precipitation, and exposure to sunlight.

generally develop at lower elevations, between 3,900 and 8,200 feet (1,200 and 2,500 meters), where high precipitation and temperatures and relatively deep soils promote tree growth. The height of the forests and the diversity of species generally decrease with altitude. Most of the highland areas that offer fertile agricultural soils have been heavily farmed and are, therefore, susceptible to accelerated erosion.

West Africa

West Africa is composed of sixteen states that constitute 36 percent of Sub-Saharan Africa's land area and 38 percent of its population. Its physiography is dominated by the Niger, Senegal, and Volta rivers and the Fouta Djallon and Jos highlands. A humid equatorial climate predominates along the coast, with mean monthly temperatures rarely falling below 64.4°F (18°C), and every month recording more than 2.4 inches (60 millimeters) of precipitation. Inland, rainfall decreases as dry winter seasons occur. This climate prevails in the wooded and grassland savanna zones in the northern regions of West African countries. The rest of West Africa is characterized predominantly by semiarid and dry climates. Yearly rainfall is usually under 20 inches (500 millimeters), and mean annual temperatures are high. In N'djamena, Chad, which is located in the moister sections of the Sahel, average monthly rainfall exceeds 2 inches (50 millimeters) in only 4 months of the year— June (2.6 inches or 66 millimeters), July (6.7 inches or 170 millimeters), August (12.6 inches or 320 millimeters), and September (4.7 inches or 119 millimeters). The average maximum daily temperatures range from 87°F (31°C) in August to 107°F (42°C) in April.

The Sahel States: The Incidence of Drought

The **Sahel** region is a transition zone between the savanna and desert environments of Sub-Saharan Africa. It centers on latitude 15°N and is 125 to 250 miles (200 to 400 kilometers) wide, stretching from Mauritania and Senegal in the west to southern Sudan, Ethiopia, and parts of Somalia in the east. The ecology of the region is very fragile, with an extended dry season lasting up to 9 months and annual rainfall averaging between 10 and 20 inches (250 and 500 millimeters). Both plant growth and species diversity are limited by the semiarid conditions. In most areas, agriculture is not feasible without irrigation. Vegetation types include scrub forest, bushland, wooded grassland, and short and tufted grasses, depending upon rainfall and local soil conditions. Most of the grass withers during the dry season, exposing the surface to soil compaction and erosion.

The region has long been subject to devastating droughts (Figure 26–9). The terrible drought of 1968–1973 caused an estimated 100,000 deaths and the loss of up to 80 percent of cattle stocks. Because most forms of vegetation and animal life in the Sahel are capable of adapting to periods of prolonged drought, it is the lack of human adaptation that potentially creates the greatest hazard—that of desertification.

Desertification, the conversion of nondesert areas to areas with more desert-like conditions, is associated with three major types of drought, which may occur alone or simultaneously: meteorologic (or climatic), hydrologic, and agricultural. Meteorologic drought is the main cause of hydrologic drought (periods with low levels of streamflow) and agricultural

■ **FIGURE 26-9 Sahel landscape.** The Sahel or the "shore of the Sahara" is afflicted by recurring droughts, degraded soils and vegetation, epidemics, and famines.

drought (extended periods of soil dryness). Lack of rainfall leads to diminished surface streamflow, loss of water in surface impoundments such as freshwater lakes and wetlands, and reduced recharge of aquifers, which in turn affect underground inflow to springs and streams. Agricultural drought results when the soil lacks the necessary moisture for plant intake or for effective plant growth. In some cases, agricultural drought occurs even when rainfall amounts appear adequate. This is because erosion, surface compaction, and steep slopes may enhance surface runoff and limit water infiltration into the soil. This implies that sound soil and water management practices, which are lacking on a comprehensive scale in Sahel states, could prevent or at least temper an agricultural drought even if a meterological drought were present.

Desertification and soil degradation in the Sahel are further exacerbated by overcultivation of marginal lands, overgrazing, inappropriate technology, weak institutional systems, and political factors. Population pressure on land manifests itself in the form of overcultivation on marginal lands and overgrazing. Most of the scarce arable land is monopolized by cash crop and commercial agricultural activity. Subsistence farmers are usually relegated to marginal lands, which are intensively cultivated with little fertilizer use or proper crop management. These poor agricultural practices are compounded by the practices of pastoralists, who move from place to place in search of what little vegetation is left while resisting the need to re-duce herd size during periods of drought. In the culture of the Fulani pastoralists, for example, increased social prestige and status comes through owning more cattle.

Inappropriate technologies and inadequate governmental responses to drought may also contribute to desertification. In the early 1960s, most African governments were slow to mobilize communities or to create stockpiles of food in anticipation of future droughts. At times, governments disassociated themselves from any event or occurrence that had the potential to tarnish their image. Generally, governments have adopted a reactive rather than a proactive approach, too often allowing the problem to reach crisis proportions before acting.

The U.S. Agency for International Development (USAID) has developed a Famine Early Warning System to monitor agricultural and economic conditions in the Sahel. The purpose is to assess the extent to which people become vulnerable to food insecurity and famine (Figure 26–10). In early 2005, it was estimated that more than 15 million people required some food assistance. The largest numbers were concentrated in Ethiopia (more than 8 million), Niger (2.5 million), Eritrea (2.2 million), and Somalia and Mauritania with 1 million each. Conflict in the Darfur Region of Sudan has affected more than 2.5 million people in Sudan and eastern Chad. Most internally displaced persons (IDPs) were left destitute after the systematic destruction of their

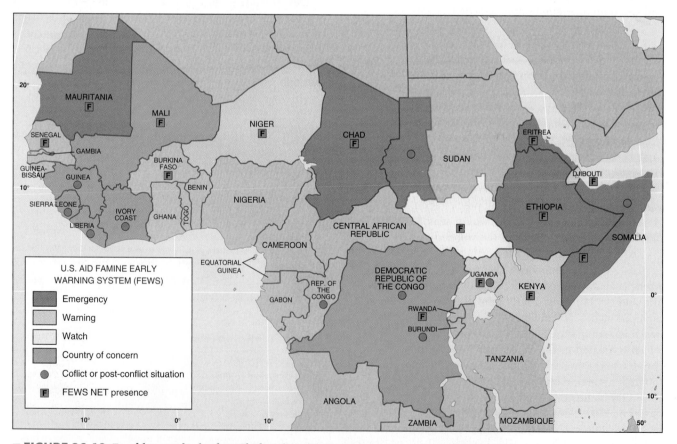

■ **FIGURE 26-10 Food insecurity in the Sahel region, 2005.** This map shows the countries most vulnerable to food insecurity and famine in the Sahel. Food insecurity is a function of both physical and human conditions.

GEOGRAPHY IN ACTION
Tragedy in Sudan's Darfur Region

The Sudan, the largest country in Africa, with an estimated population of 39 million, has been ravaged by religious and ethnic tensions that have divided the Arabic-Islamic north from the African-Christian and Animist south. The tensions have been compounded by the ongoing crisis in the Darfur region of western Sudan, a region the size of Texas. Darfur has been the site of several human atrocities, as government-backed Arab ethnic militias (called Janjaweed or "gunmen on horseback") have slaughtered hundreds of thousands of non-Arab civilians and displaced an estimated 2 million people (out of a total of 7 million). Cultural, environmental, economic, and political factors account for the ongoing crisis in Darfur. Cultural identities have been politicized as Africans from the Fur and Masalit ethnic groups formed two insurgent organizations—the Sudan Liberation Movement/Army (SLM/A) and the Justice and Equality Movement (JEM)—to challenge the Janjaweed, whose members were recruited primarily from Arab tribes. Prolonged droughts and desertification, and competition for scarce resources and fertile land, have fueled conflicts between nomadic tribes and sedentary tribes in the region. Darfur's economy, which is based mainly on subsistence and limited mechanized farming, as well as cattle herding, is marginalized from the relative prosperity in Khartoum (which controls oil revenues) and the rich Al-Gezira region, the fertile Nile floodplain south of the Sudanese capital between the White and Blue Nile. This mounting economic disparity exacerbates tensions between Africans and Arabs.

Historically, intertribal disputes have been settled peacefully through traditional methods. To make matters worse, Khartoum authorities abolished the tribal system, replacing it with new administrative structures to further the government's twin policies of Islamization and Arabization. The scale of government attacks and intrusions on non-Arab ethnic groups has prompted some scholars to raise concerns over **ethnic genocide,** the deliberate elimination of an ethnic group, and regime survival. While there are mixed opinions about the Sudanese government's genocidal intent, there is no question that a humanitarian crisis exists in Sudan. An International Commission of Inquiry held the Government of the Sudan and the Janjaweed responsible for serious violations of international human rights and humanitarian law amounting to crimes under international law. In the midst of continued attacks on innocent civilians, the African Union has deployed peacekeeping troops and civilian police to restore peace and social order in Darfur.

---

homes and their livelihoods (see the *Geography in Action* boxed feature, Tragedy in Sudan's Darfur Region).

Countering desertification and food insecurity requires a multifaceted planning and management approach with an emphasis on reforestation and revegetation, soil conservation and soil fertility improvement, water conservation, and organizational responsiveness. Tree plantings may be used to increase precipitation interception, infiltration, and groundwater recharge; stabilize dunes and soils; protect against wind and windblown sand; and yield fodder, fuelwood, and a broad range of minor forest products to support local communities. The *Acacia albida* tree, which is widely distributed throughout Senegal, has been recommended as an appropriate species to spearhead reforestation efforts there because of its reverse foliation properties. It is leafless during the agricultural season and, therefore, does not compete for light and moisture. During the dry season, however, it produces leaves that provide needed shade. Furthermore, the leaves it sheds add nitrogen and organic matter to the soil. Fuelwood, construction and fencing material, and medicinal substances are also derived from *Acacia albida*.

Vegetative cover and windbreaks are also effective soil conservation methods. Other soil conservation technologies include strip-cropping, where strips of close-growing vegetation are planted perpendicular to the flow of prevailing winds to provide protection to adjacent strips of fallow land. Soil fertility can be improved through the addition of animal manures and nitrogen fertilizers. In the latter case, farmers may require financial assistance and counseling to know how to apply a judicious combination of fertilizer and water.

Water conservation methods such as water retention dams, water catchments, and salt barriers complement soil conservation efforts. In Gambia, low ridges of earth are used to retain freshwater and to keep saltwater out of cultivated lands. On the slopes of mountains and plateaus, terracing may be practiced to capture runoff and retain soil moisture.

These efforts toward environmental rehabilitation cannot be successful without the proper organizational and policy initiatives. A number of international nongovernmental organizations have supplemented grassroots efforts with technical and financial support. In Niger, for example, a group of national and international technicians developed effective site-specific techniques to assist local village cooperatives with water catchments, windbreaks, and reseeding. In the small town of Koumpentoum, Senegal, forest and orchard projects have succeeded through voluntary village labor and financial and technical assistance from Catholic Relief Services and USAID. The provisions of "Agenda 21" of the United Nations Conference on Environment and Development have specifically targeted solutions for desertification and drought. Policies designed to create early warning systems, to improve job, income, and educational prospects, and to mobilize collaborative efforts at the local, national, and international levels will likely minimize the impact of desertification in future years.

Côte d'Ivoire: Model of African Capitalism?

Although elements of capitalism are very much alive in countries such as Nigeria, Ghana, Zimbabwe, and Kenya, Côte d'Ivoire probably comes closest to exhibiting the classic characteristics of a capitalist state, owing to its economic boom in the 1960s. Buoyed by a diverse agricultural base dominated by cocoa, coffee, timber, and palm oil, it recorded rates of economic growth as high as 11 percent in certain years during the 1960s, and averaged 6 to 7 percent in the 1970s.

Côte d'Ivoire began to be touted as Africa's model of capitalism. It was assumed that the indigenous capitalists would direct state policy, hire and manage labor, expand production, open up markets, and initiate innovations and change. Côte d'Ivoire demonstrated much promise initially, with the emergence of an indigenous capitalist class as early as the 1920s and 1930s. These were wealthy plantation capitalists who utilized the entrepreneurial attributes they acquired from French colonists to develop an effective, well-organized political machinery that gained control of the state during the decolonization process. However, since independence, the government has orchestrated capitalist development in Côte d'Ivoire, thereby instituting a form of state capitalism dominated and co-opted by state bureaucracies.

Under the auspices of President Felix Houphouet-Boigny, the state (instead of individual citizens) became the guardian, manager, and facilitator of capitalist development, employing a two-pronged strategy based on encouraging foreign investment and mobilizing domestic savings. In 1962 and 1963, the government founded the National Investment Fund and the National Finance Society to nurture and support the development of local enterprise. This was followed in 1965 with the establishment of the Bank of Industrial Development, which encouraged investments from local and foreign shareholders. In the late 1960s, the government's import substitution strategy was replaced by an export-oriented approach to improve the competitive advantage of agro-industries catering to European markets. These earlier strategies were supplemented by a host of aggressive strategies in the 1970s to further stimulate the private sector, including the creation of a stock market, provision of incentives to encourage public-sector employees to become private entrepreneurs, and the institution of counseling and financial programs to stimulate the development of local capital. Between 1968 and 1976, the Office for the Promotion of Enterprises created 246 private companies. Between 1974 and 1978, private capital assets in the modern sector grew by 43 percent, and between 1970 and 1980 the value of private industrial assets grew at an annual rate of 29.6 percent. The growth of the 1970s was accompanied by a 579 percent increase in investment by enterprises and a 713 percent increase in industrial exports.

Côte d'Ivoire's economic "miracle" sustained a jolt in the 1980s as international prices of coffee and cocoa slackened. The economic growth rate in 1983 and 1984 averaged between 2 and 3 percent, which paled in comparison to the 6 to 11 percent growth rates experienced earlier. Like several other African countries, Côte d'Ivoire is currently engaged in the International Monetary Fund and World Bank's structural adjustment programs of fiscal austerity and privatization. In spite of its temporary setback in the 1980s, the country continues to encourage market liberalization and local entrepreneurship. In recent decades, more than half of new industrial companies have been created by native citizens, with many of them choosing to be located in the country's vibrant economic center, Abidjan (Figure 26–11). The government has also created a new political capital, Yamoussoukro, in the center of Côte d'Ivoire to decentralize activities from Abidjan and to stimulate development in the hinterlands (political motives were also involved, as Yamoussoukro was the birthplace of the late President Houphouet-Boigny).

Despite Côte d'Ivoire's relative success in encouraging local entrepreneurship, numerous constraints hamper the emergence of African capitalists. For example, the largest holders of industrial capital continue to be the state (as in Côte d'Ivoire) or transnational corporations. Also, African entrepreneurs tend to be small-scale, with few graduating to the next level, and they are usually threatened by large-scale, more efficient, enterprises. Further, declining international terms of trade for African exporters make it difficult for entrepreneurs who rely on expanding markets to survive. Lastly, indigenous capitalists are not always supportive of structural change, with many frequently refusing to invest in rural areas.

These are problems that, in the midst of recent rebel uprisings, the Côte d'Ivoire government has worked hard to overcome. It has stepped up efforts to diversify the country's economic base to complement its well-developed road, rail, and port network. The country is fortunate to have diamond and iron ore deposits, and petroleum has been discovered offshore at Belier. The tourist industry has also been upgraded to take advantage of the rich savanna fauna and elaborate national park system.

Nigeria: The Geopolitics of Oil and Ethnic Strife

Nigeria was once perceived as Sub-Saharan Africa's economic giant. As Africa's most populous state, with an estimated 2002 population of 130 million, Nigeria was expected to assume the mantle of political leadership and steer the region toward economic prosperity. Unfortunately, the country that was once hailed by the West as a "cradle of

■ **FIGURE 26-11** **Abidjan, the economic center of Côte d'Ivoire.** Abidjan is a major economic, cultural, and political center in French-speaking West Africa.

democracy" and "the golden voice of Africa" is now mired in political turmoil, ethnoregionalism, and economic mismanagement.

Boosted by the development of its petroleum industry in the late 1960s, Nigeria recorded an average annual economic growth of 4.2 percent between 1965 and 1980. However, in more recent years, the economy has declined. Between 1980 and 1993, per capita income fell from $1,000 to $300; meanwhile, between 1985 and 1993, the country's total external debt increased from $14 billion to $33 billion. Ironically, in view of its original prospects, Nigeria is now categorized as a low-income country with poor human-development capabilities.

A number of interrelated factors precipitated Nigeria's decline. Among them were the geopolitics of oil, deep-seated ethnic divisions, the excesses of military rule, greed, corruption, and mismanagement. Oil still dominates the economy, accounting for 25 percent of the GNI, 75 percent of government revenues, and 95 percent of total export earnings (Figure 26–12). This dependency on oil, however, makes Nigeria vulnerable to the volatile world petroleum market. Nigeria also relies heavily on foreign investors such as Royal Dutch Shell, which has holdings accounting for half of the 2 million barrels of crude oil produced daily in Nigeria. The Nigerian National Oil Corporation was deregulated in 1989 and now engages in joint ventures with Shell. Together the

two companies control more than 60 percent of land concessions in Nigeria, with other corporations such as Elf Aquitaine of France, Gulf, Mobil, and Texaco accounting for the remainder.

In addition to oil, Nigeria possesses the largest deposits of natural gas in Africa, and has significant reserves of lignite coal, tin, and iron ore. Nigeria's climatic diversity makes possible a broad range of food and cash crops. The northern savanna zone supports cash crops such as peanuts, cotton, and tobacco and food crops such as guinea corn, millet, and cassava. In the southern forest and wooded zones, principal cash crops include cocoa, rubber, oil palm, and coffee, while food crops include sorghum, rice, maize, and yams and other tropical tubers. Unfortunately, Nigeria's agricultural output has not kept pace with its annual population increase of 2.7 percent, and the country is an importer of cereal grains. Oil could have been utilized as a unifying force in support of improved agricultural performance, economic diversification, and cultural integration. Instead, it has been a divisive influence, as various ethnic groups and political factions compete selfishly for material wealth and political power.

Nigeria's population is extremely diverse, consisting of about 300 ethnic groups. The Hausa-Fulani in the north, the Yoruba in the southwest, and the Ibo in the southeast collectively constitute 65 percent of the population. Other significant groups include the Edo, Urhobo, Efik, Ijaw, Tiv, and Ogoni. At independence in 1960, Nigeria consisted of three broad regions: the northern region, with 79 percent of the land area and 54 percent of the population, and the eastern and western regions, with 22 percent and 24 percent of the population, respectively. In 1963 a midwestern region representing the Edo and Ijaw ethnic groups was added. Initially, the North was reluctant to join the South, but compromises were reached when the North received some legislative concessions. In 1964, amidst controversy, the first elections since independence were held, and the Nigerian National Alliance, a coalition of Hausa and Yoruba parties, won. This victory signaled the North's ascendancy within Nigerian politics. Today, some 40 years later, the North still seeks to maintain its relative hegemony in Nigerian politics while the South seeks to neutralize it. In the midst of this long-term North-South struggle for dominance are several regional groups seeking greater local autonomy.

To accommodate regional demands for autonomy and to neutralize northern domination, the Nigerian federation has undergone considerable restructuring since 1963. It was balkanized into twelve states in 1967, nineteen in 1976, twenty-one in 1987, and then thirty in 1991 (Figure 26–13). **Balkanization** (reflecting the Eastern European experience) involves dividing a country or region into smaller and often hostile units. This fragmentation has intensified regional divisions and ethnic dissension. The dissension has been punctuated by military intervention, threats of secession, and a devastating civil war between 1966 and 1970 that cost millions of lives. The Nigerian military dominated the political scene for more than 25 years following independence and completely lost its perspective, seeking political

■ FIGURE 26-12 Nigerian petroleum worker, Niger River delta. Nigeria continues to rely heavily on its abundant oil production for government revenue.

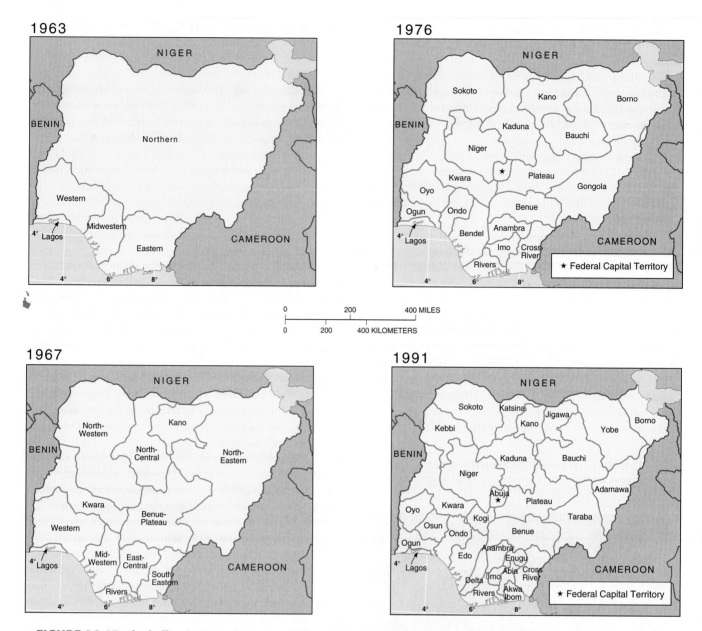

■ **FIGURE 26-13 The balkanization of Nigerian states, 1963–1991.** To accommodate Nigeria's ethnic diversity, the country has been fragmented into smaller political units.

opportunity and material wealth. Military academies became training grounds for future political leaders, and army officers aspired to political rather than military appointments. However, civilians have been unable to unite to form a cohesive political platform to challenge military rule.

Nigeria's most embarrassing moment internationally occurred in 1993 when the presidential elections were nullified—this, after Nigeria had spent the previous 7 years developing a transition program to restore democracy. Military rule was reimposed, and political parties were banned. The military government, in an attempt to redeem itself, called for a national constitutional conference to determine Nigeria's political future and discuss the issues of power sharing, the distribution of oil revenues, political party reform, and decentralization of government. Another embarrassing moment occurred when the government was forced into

importing petroleum after four oil refineries became inoperable.

Nigeria's present underdeveloped state is thus almost entirely a product of the wasting of its human resources. The country's abundant mineral and agricultural wealth has served only to heighten, not diminish, ethnic divisions.

In May 1999, Nigeria finally took a major step toward democratic reform with the inauguration of President Obasanjo and his civilian government. The government has the difficult task of rebuilding a devastated infrastructure and a weakened economy and unifying a divided population. However, as the winds of change sweep across Nigeria, a number of threats persist. Impoverished residents of the oil-rich Niger River delta continue to press for a greater share of the region's wealth while radical Islamic fundamentalists in the north pursue their political agenda.

Recently, Islamic Sharia law, based upon the Quran, has been instituted in twelve states in northern Nigeria, resulting in such drastic penalties as stoning for adultery, amputation for theft, and flogging for alcohol consumption. Women tend to be more victimized than men. Sharia law is diffusing to Oyo and neighboring southern states and threatens to trigger yet another constitutional crisis in Nigeria.

Ghana: The Brazil of Africa?

Ghana was the first colonized country in Sub-Saharan Africa to gain its independence. Ghanaian intellectuals, including J. B. Danquah and Kwame Nkrumah, were instrumental in developing the necessary political platforms in and outside Ghana to challenge the colonial authority. With independence came much optimism. Ghana, "the black star of Africa," was poised to make its move as a leader in the Pan-African movement. Expectations were very high; after all, Ghana (formerly known as the Gold Coast) had significant gold, manganese, bauxite, and timber reserves (Figure 26–14). It was also a leading world producer of cocoa. It had a well-educated human resource base, and its population was supported well by its 92,000-square-mile area. Ghana was also endowed with one of the world's largest artificial lakes, Lake Volta, which was dammed at Akosombo to develop a multipurpose river project that included the harnessing of hydroelectric power to smelt alumina and to develop inland fishing.

Ghana's drive to development actually began on a good note. In the early 1960s, its economy was growing at a rate of more than 5 percent a year; it had one of the highest per capita incomes of the region; and it had adequate cash reserves to invest in its physical, social, and technological infrastructure. However, from the mid-1960s to the mid-1980s, Ghana went through a period of economic glut, declining investments, five military coups, a brain drain, and moral decay. During that 20-year period, the country's economy actually declined by an average of –0.8 percent annually. By the mid-1970s inflation had reached triple-digit rates, and industries were producing at below 50 percent of capacity. Furthermore, there were extreme regional inequalities between northern and southern Ghana. Infant mortality rates were three times higher in the poorer northern regions, and northern literacy rates were only 15 percent of those recorded in the south.

In 1983 the government, under the auspices of the World Bank and International Monetary Fund, launched an Economic Recovery and Structural Adjustment Program with a projected $4.2 billion investment that was intended to get the country out of its economic rut. Like other countries that comply with World Bank and IMF standards and rules, Ghana began the process by devaluing its currency to lower the external price of its exports (with the expectation of increased foreign demand for its goods), trimming overstaffed state bureaucracies and padded payrolls, improving institutional management, encouraging privatization, eliminating price controls and subsidies, and removing import quotas and high tariffs that protected uncompetitive firms. The Ghanaian economy has since responded with a healthy annual growth rate of 5 to 6 percent, similar to its performance in the early 1960s. Its gold mining industry has been revived, and foreign and domestic investments have increased. Ghana is now likened in some respects to Brazil and it is frequently cited as the most successful model of modern-day economic reform in Africa.

On a cautionary note, Ghana's recovery is not quite complete, since current growth rates reflect a recovery from 20 years of depression. Inequalities still exist between the North and South, the social and technological infrastructure needs to be upgraded, and a number of negative social consequences resulting from structural adjustments have emerged. For example, currency devaluations and strict credit restrictions have severely harmed small-scale farmers and traders as well as women. Many victims of cutbacks in the public sector remain unemployed since no effort was made to retrain them for redeployment. Wage restraint policies and salary freezes have also reduced the average worker's purchasing power.

To counter the adverse effects of structural adjustments, Ghana has instituted "PAMSCAD"—the Program of Action to Mitigate the Social Costs of Adjustment. The program is designed as a supplemental basic-needs strategy to improve primary health care, reduce childhood diseases, supplement child nutrition, and compensate for adjustment-related job losses. Barring any political misfortune, Ghana is poised to reestablish itself as a leader in social, economic, and technological reform in Sub-Saharan Africa.

■ **FIGURE 26-14 Gold bodyware in Ghana.** The gold ceremonial regalia of this Ashanti chief bears witness to the important role that gold has played in the history and development of Ghana—formerly known as the Gold Coast.

ECOWAS: Attempts at Regional Economic Integration

The **Economic Community of West African States (ECOWAS)** was created in 1975 and is the largest of Africa's regional organizations. Its member countries include Benin, Burkina Faso, Cape Verde, Côte d'Ivoire, Gambia, Ghana, Guinea, Guinea-Bissau, Liberia, Mali, Mauritania, Nigeria, Senegal, Sierra Leone, and Togo. Its objectives include the establishment of a common customs tariff, a common trade policy, the free movement of capital and people, and the harmonization of agricultural, communications, energy, and infrastructural policy. Members have also signed a pact of nonaggression and mutual defense.

Calls for African unity have been recurrent since the early 1900s. The first Pan-African conference was convened in London in 1900 to protest colonial rule in Africa and, since then, the themes of Pan-Africanism and regional cooperation have been given increasing attention. The 1980 Lagos Plan of Action reaffirmed Africa's goals of establishing a common market and enhancing collective self-reliance and self-sustained development. Regional economic integration is also warranted, given the poor record of intraregional trade in Africa. Among ECOWAS countries, regional trade accounts for only 6 percent of total trade. Intraregional linkages need to be strengthened to enhance economies of scale in production, to implement advanced production technologies, and to promote greater functional specialization. The latter is especially needed, given the high proportion of small-sized countries with limited domestic markets.

ECOWAS has made some progress in the areas of transportation and telecommunication development, the free movement of persons, and regional security. For example, much of the transcoastal west African highway network from Lagos to Nouakchott on the coast of Mauretania and the trans-Sahelian highway from Dakar on the coast of Senegal to N'Djamena have been completed. Also, efforts to enhance the region's telecommunications with multimedia and broadband technologies are being realized. In addition, ECOWAS travel certificates have been introduced to facilitate intraregional travel, and an ECOWAS Peace Monitoring Group (ECOMOG) is in place to mitigate conflicts and preserve peace and stability in the region. Efforts at economic integration have largely failed because of the uneven levels of development between countries and the unequal distribution of benefits and costs. For example, smaller member states of ECOWAS, such as Gambia, Guinea-Bissau, Burkina Faso, and Sierra Leone, are overwhelmed by the predominating influence of larger states like Nigeria and Côte d'Ivoire. Another problem is the conflict of interest between countries pursuing free-market ideologies (Côte d'Ivoire) and socialist ideologies (Guinea). Other conflicts have arisen over religious, linguistic, and ethnic differences. Furthermore, there is continued dependence on foreign markets. For example, Chad still sends most of its exports to France, as does Sierra Leone to Britain. Other problems include a lack of regional complementarity (that is, duplication in goods produced exists), administrative mismanagement, lack of information, and regulatory and procedural barriers.

Given these challenges, prospects for any form of free trade, let alone a Pan-African Economic Community by the year 2025, are not very encouraging. In spite of these difficulties, however, West Africa, along with other regions in Sub-Saharan Africa, should persist in attempting to achieve the aims and objectives of the Lagos Plan of Action. Further collaborative efforts in the areas of transportation and communication, research and technology, environmental management, resource conservation, and food security should also be encouraged. The World Bank points to the African Economic Research Consortium as an example of a successful regional capacity-building venture. It provides a forum for African researchers to discuss and evaluate research on a variety of economic topics ranging from debt management to taxation policy and structural adjustment. Other institutionalized forms of cooperation include the Association for the Advancement of Agricultural Science in Africa, the African Association for Literacy and Adult Education, the Intergovernmental Authority on Drought and Development, and the African Regional Center for Technology.

Central Africa

Much of Central Africa lies in the equatorial region where tropical humid conditions prevail. In the central Congo Basin, rainfall exceeds 70 inches (1800 millimeters) per year, and daytime temperatures average about 85°F (29°C). The six states in the region account for 17 percent of the land area and only 12 percent of the population of Sub-Saharan Africa. Parts of Central Africa have low fertility rates associated with gynecological disorders and pelvic inflammatory diseases. Gonorrhea, for instance, is a major contributor to infertility among the Azande and Bakweri in the equatorial region. Syphilis has impacted the fertility levels of the Nzakara of the Central African Republic. Endemic diseases such as malaria, sleeping sickness, and river blindness are also prevalent in rural areas. AIDS has assumed epidemic proportions, and Ebola is beginning to spread beyond the region (see the *Geography in Action* boxed feature, The Ebola Virus). Rural populations are especially susceptible to these diseases, owing to their poverty and lack of education on basic sanitation and preventive measures. Since rural residents do not have access to high-quality and affordable health care, they turn to traditional healers and itinerant drug vendors for relief. The latter, in particular, are notorious for dispensing ineffective and sometimes harmful drugs.

Central African Republic: Dilemma of a Landlocked State

Like most landlocked countries in Africa, the Central African Republic (CAR) has significant development challenges. The northern region lies in the watershed of Lake Chad and is characterized by arid savanna grasses and bushes and acacia

The first major outbreak of the Ebola virus took place in Kitwit, Zaire, in 1995, when 316 people were infected, of whom 245 died. Hosts for the virus include rodents and anthropods, such as ticks and mosquitoes. The virus is transmitted by direct contact with blood secretions or body fluids from an infected person or animal. Symptoms of the disease include fever, vomiting, diarrhea, reddened eyes, and massive bleeding. The disease's estimated incubation period is between 2 to 21 days. Unfortunately, there is no known cure or vaccine for the disease. A smaller, more recent, outbreak of the virus occurred in the isolated village of Mayibout,

Gabon, where 20 cases and 13 deaths were reported. Most of the infected persons had direct contact with the blood of a dead chimpanzee.

Fortunately, teams from the World Health Organization (WHO) and France, in collaboration with Gabonese health officials, responded instantly to the outbreak, thus stemming the tide. WHO officials and African scientists are working diligently to identify the natural host of the Ebola virus in order to identify its mode of transmission. More emphasis will likely be placed on discovering preventive measures or a vaccine.

---

tree species. The south transforms into denser vegetation as it merges into true rain forest near the Congo Basin. The Central African Republic is, in effect, doubly dependent—on both its immediate neighbors and international markets for the crucial import and export of goods. The principal route that external trade must take is a 1,110-mile (1,800-kilometer) journey south along the Oubangui River to Brazzaville in the Congo, and then by rail to Pointe Noire on the Atlantic coast. In addition to the high freight and line-haul costs associated with shipment between different modes of transport, exports must also absorb the exorbitant insurance costs associated with safeguarding goods against theft, vandalism, and perishability at coastal ports.

Economic development in the republic is hampered by a poorly developed social and physical infrastructure, economic mismanagement, and political instability. Life expectancy in the Central African Republic is only 44 years and the literacy rate is among the lowest in Africa. The country has no railroad, and less than 2 percent of its estimated 13,700-mile (22,000-kilometer) road network is paved. Furthermore, its telecommunications system is substandard, consisting mostly of low-powered radios. Agriculture, which accounts for about 50 percent of the republic's economic output and 60 percent of the workforce, is dominated by coffee (its major export crop), cotton, and, to a lesser extent, tobacco. It has valuable species of hardwood that are underutilized, as well as untapped reserves of petroleum and uranium. Diamonds now account for about 50 percent of export earnings, despite an increase in smuggling activity.

Aside from its landlocked location, the meager economic performance and limited development of the republic can also be attributed to the 15-year despotic regime of President Jean-Bedel Bokassa. He interfered with the republic's attempts to preserve civilian rule and forcefully took over the reins of government in 1966, heralding one of the cruelest, most corrupt, and inhumane regimes in postindependent Africa. He developed delusions of grandeur as he declared himself first president-for-life in 1972, then marshall in 1974, and finally emperor (with Napoleonic overtones) in 1977. It is estimated that he squandered one-quarter of his country's annual income on a lavish coronation ceremony. Unfortunately, Bokassa's excesses were tol-

erated by France, which continued to bail him out with loans and economic assistance because of its heavy involvement in the country's diamond and gold mining industries.

Today this one-party state is still struggling to build an economy that, while having much potential, remains greatly constrained by its landlocked location and the massive wasting of human resources.

Gabon: Forest and Mineral Wealth Amidst Inequalities

In 1996, Gabon was the most prosperous country in Sub-Saharan Africa, with a per capita economic output of $3,550, a distinction largely attributable to its extensive forest and mineral resource base. About 75 percent of its 103,347 square miles (267,669 square kilometers) of land is covered with forest and woodland. Its forests contain several species of tropical hardwoods and softwoods; most notably, it is the world's largest producer of Okoume, a softwood used to make plywood. In 1994 forest products accounted for 16 percent of Gabon's export earnings; however, as in most African countries, its timber exports consist mainly of raw logs (Figure 26–15). Only about 25 percent of its raw logs go into local wood-processing industries and sawmills. Another problem that Gabon faces is the depletion of its coastal forests. Between 1980 and 1990, the country lost an average of 37,000 acres (15,000 hectares) of forest a year. To curtail deforestation, the government has instituted a log licensing program, which permits selective logging in three distinct logging regions.

Gabon's main source of wealth is its diverse mineral base, consisting of petroleum, manganese, high-grade iron ore, uranium, and gold. In 2000 the petroleum industry accounted for 80 percent of export revenues. Most of the oil deposits are concentrated in sedimentary rock beds along the coast around Port Gentil—"the petroleum city." Private foreign investment is predominant, with the French multinational Elf Aquitaine owning 57 percent of Elf Gabon in partnership with the Gabonese government (25 percent). Other foreign petroleum companies operating in Gabon include Occidental, Amoco, Arco, and Shell.

Although Gabon is one of Sub-Saharan Africa's richest countries, not everyone shares in its prosperity. There remain

■ **FIGURE 26-15 Tropical logging in Gabon.** Logging is a leading economic activity in Gabon, but deforestation threatens its long-term sustainability.

extreme rural-urban disparities, which are at the root of a continuing exodus from rural areas. Libreville, the capital, and Port Gentil account together for half of the country's 1.2 million people. Ironically, the rural areas' underpopulation inhibits the development of an agricultural sector that currently employs 65 percent of the labor force while accounting for only 9 percent of economic output. Gabon has had to rely on migrants from neighboring countries to work on cocoa, coffee, and oil palm plantations. **Agroforestry**, the management of trees and shrubs integrated with agricultural systems for multiple products and benefits, is a viable development strategy that can reduce economic pressures for deforestation while slowing the rural exodus by providing opportunities for farming communities to meet crop, fuelwood, and other related needs.

The Democratic Republic of Congo (formerly Zaire): Development Potential Gone Awry

Sub-Saharan Africa's second largest country is undergoing a major social and political transformation in the wake of the ouster of its longtime dictator, Mobutu Sese Seko, who was overthrown in May 1997 by rebel forces under the leadership of Laurent Kabila. The country has reverted to its original name—the Democratic Republic of Congo (DRC)—which was adopted after independence and used from 1964 to 1971. The new DRC faces serious challenges as it tries to unite several social and political factions, rebuild a fledgling economy, and assert itself as a stabilizing force in the Central African region.

Along with Nigeria, the DRC has been described as a nonfunctioning giant in Sub-Saharan Africa. The country is endowed with abundant natural resources but has not lived up to its development potential. The vast Congo Basin, drained by Africa's largest river, has a hydroelectric power potential of 100,000 megawatts, the largest on the continent. The Basin is also a repository for extensive timber resources and a major supplier of plantation crops, including coffee, oil palm, rubber, and bananas (Figure 26–16). Other signif-

icant crops are cotton, peanuts, cassava, plantains (cooking bananas), maize, and rice. In addition, the DRC possesses abundant mineral resources. It is a world-class producer of copper, cobalt, industrial diamonds, and zinc. Other significant reserves include gold, manganese, uranium, and petroleum. The mining industry accounts for about 25 percent of the DRC's economic output and about 75 percent of total export revenues.

In spite of this large and diverse resource base, the DRC remains one of the poorest countries in Africa. The nation is plagued by low educational attainment, high female illiteracy, and high rates of child malnutrition. The majority of its population lacks access to health services, basic sanitation, and safe water. A number of factors explain the DRC's social and economic malaise. First, the country has never really severed the knot with its former colonial authority. The client-patron, neocolonial relationship still exists as Congolese economic and corporate institutions continue to be firmly linked to the Belgian economy. Belgium and Luxembourg remain the DRC's major trading partners, accounting for about 25 percent of imports. After independence in 1960, the DRC tried to reduce this dependency by nationalizing key sectors of the economy. In 1967 the copper industry was nationalized, with a new state-owned company called GECAMINES taking over operations. In 1971 the Mobutu government embarked on a program of "Authenticity" and **"Zairianization."** This involved rejecting foreign values, replacing Belgian names with African names, revising the educational curriculum to reflect African traditions, transferring foreign companies to Zairians, and introducing Swahili, Lingala, Kongo, and Tshiluba as alternative languages to French. The country's name was also changed from the Democratic Republic of Congo to Zaire. It is somewhat ironic that the new government has chosen to abandon a name that was linked to the country's search for authenticity.

Notwithstanding these efforts at self-reliance and indigenization, the DRC could not break the dominance that

■ **FIGURE 26-16 Oil palm fruits.** Oil palm cultivation has expanded rapidly within the Congo Basin over the past decades. The crop is grown for the vegetable oil that is obtained from the kernels of the nuts.

Belgium exerted on it. The oil-related depression of the mid-1970s, economic mismanagement, and falling copper prices in the world market plunged the country into a debt crisis, forcing it to rely on Belgium and Europe, generally, for technical and economic salvation. Belgium continues to dominate the oil, textile, and telecommunications industries in the DRC. However, the government still controls the copper industry with GECAMINES, accounting for more than 90 percent of the country's output. The copper industry has suffered a number of setbacks due to advances in world technology and the development of fiber-optic substitutes. GECAMINES has been plagued with obsolete equipment, a lack of spare parts, infrequent equipment maintenance, and declining investments. The government is now considering joint ventures with foreign companies to upgrade the declining industry.

Perhaps the single most important factor accounting for the DRC's deplorable economic performance was the 32 years of despotic and self-indulgent leadership that the country had to endure. Mobutu's regime was the epitome of corruption, greed, and incompetence. In his long term of office, he did little to improve the country's social and physical infrastructure. Instead, he chose to enhance his personal wealth by building a fleet of castles, chateaus, and villas in Spain, France, Belgium, and Switzerland. His net worth was estimated to be between $3 and $7 billion (U.S.). For a while, little international pressure was exerted on Mobutu, owing to the DRC's resources. The country's cobalt reserves, for example, are vital for the manufacture of aircraft engines. In addition, its geostrategic central location allowed Western intelligence to monitor "communist" activities in Central, Eastern, and Southern Africa.

In the early 1990s, in a feeble attempt to deflect criticism by the outside world that he was not moving the country toward multiparty democracy, Mobutu convened a national conference, supposedly to encourage dialogue between various social factions. Regrettably, the conference was interrupted by a series of violent incidents, and Mobutu himself endeavored to destabilize the transition process in a desperate attempt to hold on to power.

Mobutu's attempts to sabotage the democratization process prompted several rebel and ethnic opposition groups from Zaire, Rwanda, Burundi, and Uganda to form the Alliance of Democratic Forces for the Liberation of Congo-Zaire (ADFL), which supported Laurent Kabila's overthrow of Mobutu. It may be significant that the current leaders of neighboring Uganda and Rwanda are, themselves, former rebel leaders who dismantled long-standing political regimes in their respective countries.

The social and economic conditions in the DRC worsened under the leadership of Kabila as tensions mounted between his government and rebels located in the eastern sections of the country, where Congolese Hutus and Tutsis have long coexisted with the Hunde and Nyanga ethnic groups, who consider themselves to be natives of DRC. These conflicts destabilized the region as Angola, Namibia, and Zimbabwe sided with Kabila's government, while Uganda and Rwanda allied with the eastern rebels, including the Congo Liberation Movement (MLC) and the Congolese Rally for Democracy (RCD). In January 2001, Laurent Kabila was assassinated and his son, Joseph Kabila, was sworn in under questionable circumstances. Joseph Kabila has pledged to restore civility to the DRC and to reconcile with opposition groups. A first step is to try to restore the 1999 Lusaka (Zambia) Peace Accords, which called for all armed and unarmed opposition groups to dialogue and negotiate with the DRC government on such issues as power-sharing. A follow-up meeting was held in Sun City, South Africa, in April 2002 to revive the 1999 peace accords. The Organization of African Unity, South African Development Community, and the United Nations have joined Uganda, Rwanda, Namibia, and Zimbabwe as principal signatories to the peace accord. There is still some skepticism about the legitimacy of the Joseph Kabila government and its intentions to engage in power-sharing discussions with opposition groups. One thing is certain—a destabilized DRC does not bode well for the future development prospects of the East African Great Lakes region and beyond.

East Africa

East Africa embraces ten countries that account for 26 percent of the land area and 37 percent of Sub-Saharan Africa's population. The region is characterized by higher elevations and cooler temperatures than its western and central African counterparts and has less forest cover. The physical environments are quite diverse, with semiarid and arid conditions dominating in Somalia, northern Kenya, and Sudan, and temperate conditions among the fertile Lake Victoria agricultural borderlands of southeast Uganda, northwest Tanzania, and southwest Kenya.

Somalia: Turmoil in a Clanocracy

Somalia is strategically located astride the Horn of Africa, facing a key international sea route for oil shipments originating in the Persian Gulf. The port of Berbera in northern Somalia is also a major base for the U.S. military's rapid deployment force, which responds to emergencies in the Middle East. Somalia has long feuded with its neighboring countries. Its national flag has five stars, representing its irredentist claims in eastern Ethiopia, eastern Kenya, and Djibouti and the consolidation of northern and southern Somalia.

In contrast to the rest of Sub-Saharan Africa, Somalia is about the only country that comes close to the western concept of a nation-state, in which the national or ethnic boundary of a group of people with common customs, language, and history coincides with the state boundary. Ethnically, the Somalis belong to the Hamitic group. They share a common language, although with differences in dialect; a pastoral way of life (a mode of life for about 70 percent of the population); a shared poetic corpus that is a vital part of public and private life; a common political culture; and the religious heritage of Sunni Islam. The Somalis firmly believe that they are drawn together by bonds of kinship and genealogical ties, descending from a common founding father called Samaale. This ethnic unity, however, is offset by the Somalis' division by kinship into clans (Figure 26–17). There are four major pastoral clans (the Dir, Daarood, Isaaq, and Hawiye) and two agricultural ones (the Digil and Rahanwayn) that are concentrated in the western sections of the northern region. The clans are patrilineal descent groups that define the political and legal status of people and assign them specific social claims and obligations. For example, a person is obligated and owes allegiance first to his or her immediate family, followed by the immediate lineage, the clan of lineage, and then a clan-family that embraces several clans.

The **clanocracy** acts as both a unifying and divisive force. For example, Mohammed Fara Aidid and Ali Mahdi, both from the Hawiye clan, collaborated to overthrow the government of former president Siyaad Barre (from the Marehan clan) in 1991. The coup was a reaction to Barre's brand of scientific socialism, which clashed with Islamic and clan-based traditions. Scientific socialism attempted to abolish kinship influences and clan-based activities in Somali life. All forms of tribal or clan association, rights, and privileges—including those over land, pasture, and water—were abolished and claimed by the state. Private funerals, family marriage ceremonies, and payment of blood money were banned. Administrative boundaries were reorganized and renamed to exclude reference to clan names. There were proposals to replace the family as the basic unit of organization with settlements and collective farms. "Orientation centers" were set up throughout the country to foster the formation of a new community and individual identity based on loyalty to the ruling party, state and president. It is not surprising, therefore, that scientific socialism, which threatened the underlying foundations of Somali culture, was done away with in 1991. After the successful coup, however, Aidid and Mahdi became rivals, as their allegiances shifted to their immediate clans, Aidid's being the Habar-Gedir subclan and Mahdi's being the Abgal subclan of the Hawiye.

Somalia has essentially been without a formal government since 1991. A Transitional National Government was set up in August 2000, with each clan allocated a certain number of seats, and key cabinet positions given to influential clans such as the Hawiye and Daarood. The transitional government, however, has very little legitimacy and influence in regions beyond Mogadishu and has great difficulty in managing the economic and political affairs of the entire country. Livestock exports, a major source of revenue, have declined and various warlords and clan factions continue to fight for political power. A radical movement, the Al-Ittihad al-Islamiya, has emerged with intentions of imposing Sharia in Somalia through education, commerce, nonprofit organizations, and the media.

Clan warfare is usually provoked by disputes over water and pasture rights or political control. Although a person belongs to a network of inextricably linked clans, primary sociopolitical loyalty is to the *diya*-paying group, a close-knit group of kinsmen united by a contract (*heer*) that specifies the terms of blood compensation for murder, feuds, and any illegal acts. This community group usually constitutes the basis for social and political action. This tradition is important because it essentially means that the clan system has an established framework for resolving conflicts. Clan elders play a significant role in mediating disputes over local affairs, including blood money, rights to watering holes, and access to pasture. That is the principal reason why Somalis reacted so adversely to U.S. military efforts to capture clan leaders during the UN famine relief effort of 1992–1993, which had been prompted by food shortages in the wake of the political turmoil of 1991–1992.

Tanzania and Ethiopia: Models of African Socialism?

African socialism was designed as the antithesis to European domination. It was viewed as a mechanism for achieving a balanced, self-reliant development based on social justice and mutual cooperation. Moreover, it was consistent with traditional African values, which emphasized the moral virtues of economic equality, reciprocity, and mutual aid. In Tanzania, President Julius Nyerere spent the 1960s and 1970s building a society based on socialist principles, articulating the justification for African socialism in a document called the Arusha Declaration, released in 1967. The year marked the implementation of a socialist policy that was geared toward the elimination of poverty and income inequalities and the improvement of the quality of rural lives. Major commercial, industrial, and financial institutions were nationalized, and public officials were restricted in their ownership of property.

In 1968, the **Ujamaa** (familyhood or brotherhood) policy was instituted to promote self-help, self-esteem, and local initiative along family, communal, and cooperative lines (Figure 26–18). The *Ujamaa* village, modeled along the same lines as the Soviet collective or the Chinese commune, was a consolidated, self-sufficient settlement that provided educational, health, and social services to meet the basic needs of its resident population. The villages would become the basis for social organization and cooperative activity in

■ **FIGURE 26-17 Armed Somali clansmen, port city of Mogadishu.** Somalia is dominated in every respect by clans that, to date, have failed to unite for the common good of the nation.

■ **FIGURE 26-18 Communal *Ujamaa* village fields in Tanzania.** *Ujamaa* villages were established by the Tanzanian government in an attempt to foster the values of self-help, cooperation, and reciprocity among the rural peasantry.

Tanzania. In response to this program, the government increased its capital expenditures in rural areas. *Ujamaa* residents pooled their labor, land, and capital resources for agricultural and other ventures. Proceeds from the cooperative effort financed a variety of development projects. To encourage rural participation in the decision-making process, Tanzania embarked on a policy of decentralization in the early 1970s, as regional and local district offices were dispersed into rural villages. Tanzania's socialist strategy had mixed results. While there was some success in achieving social equity, the country continues to be one of the poorest in Sub-Saharan Africa, with slow growth in all economic sectors. Tanzania has since undergone a structural transformation in its productive sectors as it experiments with the World Bank's privatization programs.

Ethiopia adopted a socialist policy in 1975, when the new socialist military council replaced Emperor Haile Selassie and immediately nationalized all banks, insurance companies, and several industrial and commercial enterprises. A new land reform program was instituted to bring all rural and urban land under government control. This program had devastating effects on communal and traditional forms of land ownership in villages, as rural farmers were forced into collectives. Ethiopia continues to be in a state of social and economic destitution, with agriculture accounting for more than 90 percent of its foreign earnings. After the overthrow of Mengistu's military dictatorship in 1991, the transitional government adopted a policy of reconstruction and rehabilitation based on market-driven prescriptions set by the World Bank, IMF, and African Development Bank.

Tanzania and Ethiopia are typical of what happens to African countries that tow the "socialist" line. Tanzania achieved some measure of social equality in health care and education. However, the nationalization of industries slowed down productivity, owing to mismanagement, production inefficiency, and the lack of capital and technological investments. A major problem was that the policy of socialism was imposed from the top by a bureaucracy that was out of touch with the needs of rural people and inexperienced in handling the complexities of rural development. Self-reliance at the state level implies total autonomy and independence in decision making and the allocation of resources. No single Sub-Saharan country is in a position to break away and develop a totally autonomous stance in our globally interdependent world. Countries such as Tanzania and Ethiopia have been firmly entrenched in the global exchange economy since colonial times and are learning to redefine self-reliance from the perspective of interdependence and mutual cooperation.

Kenya: Transition of a Settler Colony

Much of the economy of present-day Kenya has been molded by British intrusion and colonization. The British were attracted to Kenya because of its temperate highland environments and the rich volcanic soils in the southwest highland region, which is one of the most productive agricultural zones in Africa. Also, from a strategic standpoint, Britain was able to prevent German East Africa from extending its influence from Uganda to the fertile highland regions and the coastal areas beyond. Kenya, like Zimbabwe, was transformed into a settler colony, as an administrative and economic structure was superimposed to marginalize the indigenous population and preserve permanent European settlements. Monetary taxes were imposed on local populations to finance a settler-oriented infrastructure and force Africans to offer cheap labor on settler farms. Africans were displaced from their lands and relegated to crowded and inhospitable reserves, where social and economic conditions were deplorable. Having acquired the best land, the British proceeded to establish a plantation economy based on tea and arabica coffee.

At independence in 1963, Kenya inherited an agricultural system based on large-scale commercial estates owned primarily by Europeans and Asians. This occurred despite earlier nationalistic efforts to reclaim land from European settlers. Jomo Kenyatta organized Kenya's largest ethnic group, the Kikuyu, to demand access to white-owned land. This culminated in the Mau Mau rebellion of 1952, followed by a devastating and costly 3-year war. Following the war, the colonial authorities proceeded to grant Africans access to land and to political authority. A series of land reforms, land consolidation, and resettlement schemes ensued with the specific purpose of providing Africans with an opportunity to improve farm management practices and advance from subsistence to commercial farming. In the political arena, the Kikuyu and Luo began playing active roles in preparing the country for independence. In June 1963, Jomo Kenyatta became the first prime minister.

Kenya remains a fairly stable yet authoritarian country. Jomo Kenyatta's government lasted from 1963 to 1978 and was dominated by Kikuyus, many of whom emerged as an elite class and accumulated wealth for themselves. The Kikuyus are concentrated in the Central Province of Kenya and reaped the benefits of land ownership in the coffee- and tea-producing areas of that region. Arap Moi, who succeeded Kenyatta, held a steady course in spite of attempted coups and threats from Muslims and African-Arab rivalries. The Kenya African National Union's (KANU) dominance since independence in 1963 came to an end when in December of 2002, Mwai Kibaki and his National Alliance Rainbow Coalition party garnered 63 percent of the vote. Today agriculture is still the leading activity, providing 75 percent of employment and 55 percent of export earnings. High-grade tea and coffee still dominate exports. The country is the world's second largest (behind Sri Lanka) exporter of tea. Other cash crops include sisal, pyrethrum, cotton, sugarcane, and maize. Foreigners still own and manage several large-scale commercial farms, and there is some smaller-scale indigenous production. Kenya also has a relatively well-developed industrial sector based on food processing, tobacco, textiles, oil refining, plastics, pharmaceuticals, electric cables, rubber, ceramics, industrial gases, and car assembly. Industry is supported by a strong entrepreneurial community, which remains largely European and Asian. The tourist industry continues to thrive and is a major source of foreign exchange.

Nairobi, the capital, dominates the urban landscape, with an estimated population of 3 million (Figure 26–19). It ac-

counts for 30 percent of the urban population and is growing at a rate of 6.3 percent a year. To curb rural-to-urban migration and offset the dominance of Nairobi, the government is relying on a central place-based regional decentralization strategy to develop rural trade and production centers. These, in turn, are expected to stimulate market and employment expansion in areas of unmet agricultural livestock potential.

Rwanda and Burundi: Ethnic Genocide

Genocide in Rwanda and Burundi has destabilized the east and central African region and has amounted to a loss of more than half a million people and the displacement of countless refugees (Figure 26–20). Unfortunately, the Tutsis and Hutus, who share the same language and traditions and have lived in the same territory for more than 500 years, have been involved in a tragic war of attrition. Although there is some uncertainty over the exact origins of both groups, it is believed that the Tutsi, who today constitute 10 percent of Rwanda's population and 14 percent of Burundi's, are descendants of Nilotic migrants from Ethiopia. The majority Hutu are descendants of Bantu migrants from West Africa, and a third group, the Twa hunter-gatherers, are the original settlers and constitute only 1 percent of the population in the region.

The tensions between the Hutu and Tutsi were exacerbated during the colonial era, when racial stereotyping and social categorization became the order of the day. Prior to colonial contact the Hutus and Tutsis were governed by a kingship institution based on Tutsi lineage. The *mwani* or king, who was considered to be a divine ruler, presided in a centralized court. Most of the king's representatives or chiefs were Tutsi, who were responsible for agricultural production, grazing land usage, taxation, and defending the king's court. This set the stage for social distinctions between the cattle-herding class of Tutsi aristocrats and the servant class of Hutu farmers. However, during this time Tutsi dominance was mitigated by social institutions that gave authority to some Hutu chiefs since agriculture was their domain. From the fourteenth century to the end of the nineteenth century, the Tutsi-dominated monarchy developed into a cohesive social and political system. This monarchy was reinforced under German rule, which lasted from 1896 to 1916. The Germans, under a system of indirect rule, preserved the monarchy and collaborated to carry out policies that were mutually beneficial. The Tutsi monarchy, in turn, utilized the German presence to consolidate its position and subjugate Hutu rebels in the northern region.

After World War I and the subsequent collapse of German East Africa, Ruanda-Urundi, as it was then known, was entrusted to Belgium under mandate from the League of Nations. The Belgians made the Tutsis

■ **FIGURE 26-19 Nairobi, Kenya, skyline.** Nairobi's high-rise city center serves as the commercial and communications hub of East Africa. Nairobi is home to more than 3 million people, most of whom live in sprawling squatter settlements. Though poor, Kenya is East Africa's most prosperous country.

■ FIGURE 26-20 Refugee children in Burundi. In Burundi, ethnic genocide has left thousands of orphaned or "unaccompanied" children.

the privileged class during their rule and created a racial hierarchy that was derogatory to the Hutus and Twas. The Twas were relegated to the bottom of the hierarchy, owing to their "small, chunky, muscular, hairy" characteristics, while the Hutu were described as "generally short, thickset, with wide noses." The Tutsis, on the other hand, were "tall, slender, graceful, and intelligent." It was even theorized that they were of mixed Hamitic and Semitic stock, and some European scholars went so far as to come up with bizarre racial classifications such as "Black-Caucasian" and "African-Aryans." The Tutsis were designated as the superior race, and Belgian colonialists provided them with the best jobs and education. This certainly intensified the hatred between Hutu and Tutsi; it also led to an awakening of Hutu consciousness in the 1950s. By 1959, the oppressed Hutu majority, with the support of some Belgian missionaries, had organized a political revolution, which led to the massacre and exile of hundreds of thousands of Tutsis. Independence was granted in 1962, and the already small territory was further balkanized into the two new states of Rwanda and Burundi. In the process, the Tutsi monarchy collapsed, and the Hutu monopolized power in Rwanda. In the early 1990s, a new round of genocide directed at Tutsis intensified an existing civil conflict, resulted in the victory of the Tutsi-dominated Rwandese Patriotic Front (RPF), and sparked a flood of Hutu refugees into neighboring countries.

Today Rwanda and Burundi are still struggling to cope with a past mired by ethnic genocide. An International Criminal Tribunal for Rwanda (ICTR), based in Arusha, Tanzania, and a state court system have been established to prosecute genocide suspects. Both systems have been plagued by long delays and logistical difficulties resulting in very few verdicts being handed down and overcrowded prisons. Failures in the courts have prompted Rwandans to explore indigenous peacemaking methods such as the **Gacaca court system**, which emphasizes reconciliation over retributive justice. Modeled after South Africa's Truth and Reconciliation Commission, and drawing on traditional African values that stress contrition, forgiveness, and compensation to wronged individuals, this restorative form of justice was instituted in 2002 to encourage community participation, promote unity, restore social harmony, and expedite trials. About 11,000 Gacaca jurisdictions have been

created. Much of the punishment is in the form of community service such as rehabilitating homes or restoring schools and hospitals. Human rights organizations such as Amnesty International and African Rights are concerned that suspects might not receive a fair trial; however, the system seems to have gained popularity among most Rwandans.

▶▶ Summary

The tropical regions of West, Central, and East Africa exhibit considerable diversity in their environmental and ethnic characteristics, a diversity compounded by the unequal levels of development among the constituent countries. From the standpoint of economic development, the Sahel region embraces some of the poorest and least-developed countries in the world, countries that have long been subjected to periods of extreme drought and food insecurity. For this ecologically fragile region, prospects for sustainable development depend on the institution of appropriate early warning systems to alert the region to potentially calamitous conditions; the improvement of job, income, and educational opportunities; and the mobilization of collaborative efforts locally, nationally, and internationally whenever famine or extreme drought do strike. On the other hand, countries such as Ghana, Côte d'Ivoire, Gabon, and Kenya are making some economic headway, although internal disparities between rural and urban areas remain problematic. In the area of sociopolitical development, countries such as Nigeria, Burundi, Rwanda, and the Democratic Republic of Congo continue to be mired in ethnic turmoil, military intransigence, and uncompromising dictatorships. These countries will have to come to terms with a new international order that is increasingly intolerant of human rights violations, socioeconomic injustice, and certainly ethnic genocide. Countries like Burkina Faso and Kenya, on a brighter note, are riding the wave of multiparty politics. As more countries join the trend toward greater democracy, instituting needed political changes, the prospect that tropical Africa will meaningfully participate in the economic and technological achievements of our global economy rises accordingly.

▶▶ Key Terms

agroforestry	ethnic genocide
balkanization	*fadamas*
clanocracy	Gacaca court system
deforestation	Great Lakes System of
desertification	East Africa
drainage basins	Highland Africa
East African Rift Valley	*mwani*
Economic Community of	Sahel
West African States	*Ujamaa*
(ECOWAS)	Zairianization

■ **A multiracial school in South Africa.** Recent political changes have brought opportunities for increased racial integration in Southern Africa.

Chapter 27

Southern Africa:
Development in Transition

- **Physical-Environmental Characteristics of the Region**
- **Countries of Southern Africa**
- **South Africa: Country in Transition**

Mainland Southern Africa encompasses ten states that account for 25 percent of Sub-Saharan Africa's land area and 13 percent of its population. The region lies in a subtropical environment and has a wealth of mineral resources. More importantly, the region is experiencing a transformation of its social, political, and economic structures. Angola and Mozambique are recovering from more than 20 years of civil strife and are seeking a peaceful transition to democratic and economic reform. Namibia recently gained its independence from South Africa and is coping with the challenge of social and political reconciliation. Botswana has transformed its economy from one of the poorest in the world to one of the most prosperous in Africa and is now being examined as a model of political stability. Zimbabwe gained its independence much later than most African countries and is still wrestling with the issues of land reform and industrial transformation. Finally, South Africa recently signed a new constitution that guarantees civic and political liberties to all South Africans. South Africa is potentially a major player in the economic and political transformation of not only the Southern African region but also the rest of Sub-Saharan Africa.

Physical-Environmental Characteristics of the Region

Much of Southern Africa is a plateau that is framed by a narrow coastal plain. The plateau reaches its highest point in the eastern sections where the Drakensberg Mountains at more than 11,000 feet (3,350 meters) are located. Other highland areas include Malawi's Mount Mulanje at 10,000 feet (3,050 meters), Mozambique's Monte Binga at 8,000 feet (2,440 meters), and Zimbabwe's Inyangani at 8,300 feet (2,530 meters) (Figure 27–1). The plateau then slopes downward toward the savanna and steppe plains, and the arid desert regions of the Kalahari and Namib deserts in the west (Figure 27–2). The Kalahari Basin features two major physiographic landscapes: the Okavango Delta and the Makgadikgadi (Makarikari) salt pans. The Kalahari Basin is devoid of surface water in much of its southern sections. A third of its northern section receives perennial stream flow mainly from the Okavango River that rises from the highlands of Angola and drains into the dry expanses of Botswana, forming a vast inland delta that covers about 400,000 acres (162,000 hectares) (Figure 27–3). This region is a haven for one of Africa's most diverse wildlife areas and is now being tapped for its ecotourism potential. The southwestern sections of Africa are rimmed by the Cape Fold Mountains, which rise to about 6,500 feet (1,980 meters) and the Karoo rock series, which contains coal deposits.

The plateau is drained and eroded by several rivers, including the Zambezi, Orange, and Limpopo. The region's drainage pattern is dominated by the Zambezi River that, historically, separated white-ruled Southern Africa from independent black Africa. Today it separates Zambia (formerly Northern Rhodesia) from Zimbabwe (formerly Southern Rhodesia) and divides highland northern Mozambique from lowland southern Mozambique. Along the Zambia-Zimbabwe river boundary, the Zambezi plunges over the spectacular Victoria Falls into a series of gorges (Figure 27–4). Further downstream are two major hydroelectric projects located at Lake Kariba and at Cabora Bassa in Mozambique. The 550-mile- (885-kilometer) long Zambezi River has a total annual flow that is double that of all of South Africa's rivers combined.

The climate is varied, with the east coast being warmed by the Mozambique Current and the west coast being cooled by the Benguela Current. The cool waters of the Benguela create a welling effect that brings nutrient-rich waters from the depths to the surface, making this an excellent breeding ground for a variety of marine life. Altitude modifies climate, with annual rainfall exceeding 60 inches (1500 millimeters) in the Drakensberg and Cape Ranges in contrast to 10 inches (250 millimeters) in the western coastal sections.

■ FIGURE 27-1 Mt. Mulanje, Malawi. Mt. Mulanje is one of the highest peaks in Southern Africa, a region dominated by a mineral-rich plateau.

■ FIGURE 27-2 Namib Desert landscape. The Namib is a cool coastal desert fronting the Benguela Current of the southern Atlantic Ocean. The tree is a drought-tolerant camel thorn acacia.

■ FIGURE 27-3 Okavango Inland Delta, Botswana. The core area of the Okavango Inland Delta covers about 400,000 acres (162,000 hectares). It is a haven for a variety of wildlife and is a major boost to Botswana's ecotourist industry.

■ **FIGURE 27-4 Victoria Falls (of Zambia-Zimbabwe).** Victoria Falls lies on the Zambia-Zimbabwe border near Livingstone. It averages more than 5,500 feet (1,676 meters) in width and more than 300 feet (91 meters) in depth.

Humid temperate climates are prevalent in the south-central coast of South Africa, where there is adequate precipitation year-round, with cool summers. The coastal strip east of this belt has similar rainfall characteristics but with hot summers. The southwest corner of South Africa's Cape Province experiences a mediterranean climate that is characterized by dry summers and wet winters. The associated vegetation here consists of low trees and shrubs that have developed adaptations to long periods of aridity, such as thick leathery leaves, needle-like leaves, and long roots. Important native food plants in this zone include wild yams and grain sorghum. Inland, in the eastern sections of South Africa and in central Zimbabwe, dry winters and cool summers prevail. Further inland, dry winters and hot summers affect eastern Angola and much of Zambia.

Drought is a major problem in Southern Africa, particularly in its western and central sections (Figure 27–5). At least 70 percent of Namibia and Botswana experience chronic aridity. During 1991–1992 and 1994–1995, southern Zambia, northern Zimbabwe, southern Mozambique, and central South Africa were especially parched. Cereal production levels fell to 68 percent of average levels in Zambia, 56 percent in Lesotho, 50 percent in South Africa, and 48 percent in Zimbabwe during the latter period. Severe water shortages have uprooted people from villages and have forced governments to ration supplies. Southern African governments have instituted emergency food assistance and food-for-work programs in an attempt to alleviate the crisis. However, logistical difficulties related to transportation and distribution have hampered efforts.

Countries of Southern Africa

Angola and Mozambique: Cauldrons of Conflict

Both Angola and Mozambique have endured a long history of colonial exploitation and civil strife, rendering their economies stagnant and, at times, nearly inoperable. Only recently have we witnessed efforts to initiate peace and begin the processes of rebuilding and reconciliation. When most French and British African colonies gained their independence in the early 1960s, Portugal resisted granting freedom to its "overseas provinces," maintaining that they were to remain an integral part of Portugal forever. It took 15 years of brutal warfare, inhumane policies, forced labor, blatant exploitation, and the collapse of a fascist government in the "motherland" to set Portuguese Africa free. During the colonial era, Angola and Mozambique were divided into districts or circumscriptions. Land was seized from Africans, and men were forced into labor camps to grow coffee in the fertile northern provinces of Angola and cotton in northern Mozambique. Forced contractual labor separated men from their

■ **FIGURE 27-5 Scene from a Boli village.** The southern region of Africa has recently experienced several droughts that devastated landscapes and threatened human and animal survival.

families for as long as a year. Only a few Africans were afforded a decent education; the overwhelming majority had to fend for themselves or seek assistance from missionaries. At independence, both colonies had among the lowest literacy rates in Africa.

The deplorable socioeconomic conditions prompted the formation of liberation movements in the two **insurgent states**. An insurgent state is one marked by revolt, insurrection, and guerrilla activity from national liberation movements that oppose the established authority. In Angola, three movements were formed along regional, ethnic, and ideological lines. Urban elites teamed up with workers in Luanda to form the MPLA (Popular Movement for the Liberation of Angola) with support from the Mbundu tribe; refugees from coffee plantations in the north formed the FNLA (National Front for the Liberation of Angola); and rural agriculturalists from the southeast, with support from the Ovimbundu of the central highlands, teamed up to form UNITA (the National Union for the Total Independence of Angola) By independence in 1975, the FNLA threat had faded, leaving the MPLA and UNITA to vie for political power. At the time, Angola could best be described as a **shatterbelt**—a region characterized by intense political discord and subjected to external pressures. The socialist MPLA was backed by Soviet and Cuban troops, while UNITA received support from South Africa and the United States, who were concerned about the spread of communism. South Africa justified its unlawful occupation of Namibia on the grounds that Cuban forces were present in Angola. Ironically, most of the revenues derived from Angolan oil sales to the United States were used to purchase military weapons from the Soviets.

Although the MPLA was able to gain political control at independence, it spent the next 16 years trying to deal with the threat from UNITA and South Africa. In the process, Angola spent more than 50 percent of its export earnings on defense and spent twice as much money on the military as on health care and education combined. South Africa, accusing the MPLA of harboring SWAPO (Southwest Africa Peoples Organization) and ANC (African National Congress) rebels in its territory, continued to support incursions into southern Angola. In December 1988, a tripartite peace accord among Angola, Cuba, and South Africa was signed, and target dates were set for Cuban withdrawal, Namibian independence, the exchange of prisoners of war, and preparations for free elections in Angola. In another peace agreement in 1991, the MPLA government and UNITA agreed to a cease-fire and set the stage for multiparty elections. In the ensuing elections, the MPLA won by a margin of 54 percent to UNITA's 34 percent. UNITA was dissatisfied with the outcome and resumed its campaign of terror and violence, thus destabilizing the peace process. The United Nations has since tried to restore the peace process, and the MPLA government made overtures to UNITA by offering its leader, Jonas Savimbi, now deceased, a share of the vice presidency.

Angola's prolonged civil war ended formally with the signing of the Luena Accords in 2002. The UNITA threat has since been diffused and the MPLA government can now concentrate its efforts on peace-building initiatives. The government must contend with the lingering challenges of malnutrition, poverty, disease, and landmines. An estimated 15 million land mines have caused 1 out of every 350 Angolans to undergo amputation—the highest rate in the world. On a positive note, the government can mobilize revenues from its massive oil and diamond reserves, in conjunction with humanitarian aid, to reintegrate former war combatants, resettle displaced people, invest in public works programs, and build capacity to deliver social services and remove land mines.

Mozambique's experience in many ways parallels Angola's. Socialist FRELIMO (Front for the Liberation of Mozambique) challenged Portuguese colonial authority and assumed political power after independence in 1975. It was no coincidence that both the MPLA and FRELIMO started off along socialist lines. Socialism was largely a reaction against the mercantilist policies of the Portuguese, and both movements had organized masses of people to fight for basic human rights. FRELIMO's key opposition was RENAMO (Mozambican National Resistance), a movement supported earlier by Rhodesia and later by South Africa. Similar to the Angolan situation, South Africa was wary of potential ANC bases in Mozambique and backed RENAMO campaigns to sabotage the country's infrastructure.

After a series of peace initiatives, a General Peace Agreement was signed in Rome in October 1992, formally ending 16 years of intense warfare and initiating a cease-fire. Under terms of the agreement, more than $390 million worth of humanitarian and electoral assistance was pledged. A United Nations Verification Mission was established to ensure implementation of the pact, and in October 1995, the country's first-ever free elections were held. FRELIMO won 53 percent of the presidential vote and 52 percent of the seats in the National Assembly, while RENAMO finished second with 34 percent and 45 percent, respectively.

Both Angola and Mozambique face major challenges after years of conflict. Both countries are war-weary and are seeking appropriate solutions, along with international assistance, to the crises they inherited. Mozambique has concentrated its efforts in the last 10 years on restoring a deteriorated infrastructure, increasing investments in education and health, encouraging privatization, and implementing broad sweeping structural reforms in banking, finance, and commerce. Gross domestic investment in Mozambique increased from an average of 3.8 percent in the 1980s to 8.2 percent in the 1990s. Industrial output increased significantly from a negative average annual −4.5 percent in the 1980s to 14 percent in the 1990s, partly because of the tax incentives to stimulate activity in the mining of bauxite, graphite, gemstones, gold, and marble. In addition, Mozambique is one of the world's largest producers of cashew nuts; cotton and fish are also major exports. As a result, during the 1990s, Mozambique recorded one of the highest economic growth rates (an annual average of 6.4 percent) in Sub-Saharan Africa.

Angola has a comparative advantage in mineral production over Mozambique, primarily because of its petroleum sector, which accounts for more than 50 percent of its export earnings. Angola was able to protect its oil refineries in the Cabinda district from South African and UNITA raids. This was not possible, however, with the Benguela Railroad that runs cross-country from the port of Lobito in the west to Democratic Republic of Congo. Like the Beira Railway in Mozambique, this transcontinental railroad was subject to frequent attacks (Figure 27–6). In addition to oil, Angola has substantial reserves of diamonds, copper, and manganese ore. These mineral resources are supplemented by a diverse agricultural sector based on coffee (the main cash crop), cotton, sisal, and oil palm.

Namibia: Dilemmas of a Newly Independent State

Namibia, formerly known as South West Africa (SWA), finally gained its independence in March 1990 after 75 years of South African domination. South Africa left a legacy of racial and socioeconomic segregation, divided ethnic loyalties, a limited manufacturing base, high debt burdens, high unemployment, and a highly extractive and poorly integrated economy. The country now has to cope with the challenges of economic reform and social and political reconciliation.

Namibia is a fairly sizable country, with an area of 318,291 square miles (824,374 square kilometers); yet it has a relatively small, multiracial population of 1.8 million people. Most of the population is concentrated in the central plateau region that rises east of the coastal plain where the Namib Desert is located. The northern plateau is occupied primarily by the largest ethnic group, the Ovambo (50 percent of

Namibia's population), and the Kavangos and Caprivians. Most of the Europeans, who account for about 6 percent of the population, occupy the central and southern highlands along with other ethnic groups such as the Hereros, Namas (Hottentots), San (Bushmen), and coloreds (people of mixed racial heritage). The European population consists mainly of Afrikaners, British, and Germans. The German influence began as early as 1884 when SWA became a colony. After Germany's defeat in World War I, Namibia became a South African administered United Nations trust territory from 1921 to 1945, as mandated by the League of Nations.

South Africa never made a sincere commitment to safeguarding the well-being and interests of South West Africans as prescribed by the League of Nations and later by the United Nations. South Africa was more interested in imposing its system of apartheid, consolidating its power base, controlling the strategic resources, and annexing SWA territory. Namibia had large uranium and diamond reserves, and it possessed a deep-water port at Walvis Bay. It also functioned as a psychological buffer between Cuban-influenced Angola and South Africa. In 1946 South Africa attempted to incorporate SWA as its fifth province, a request that was denied by the United Nations. South Africa would continue to defy the International Court of Justice and a number of United Nations resolutions by trying to annex South West Africa. One such resolution was United Nations Resolution 435 (1978), which called for United Nations–supervised elections for a constituent assembly in Namibia.

To counter South African aggression, the Ovambo founded an African nationalist movement in the late 1950s. The movement, initially called Ovamboland People's Congress, was renamed the South West Africa People's Organization (SWAPO) in 1960 to broaden its political base. In 1968 the United Nations Security Council declared South Africa's presence in SWA illegal, and in 1974 it recognized SWAPO as the authentic representative of the Namibian people. Sensing resistance from SWAPO and pressure from the United Nations, South Africa sought to protect Afrikaner interests by proposing to divide Namibia into eleven ethnic administrations as a precondition for granting independence. The plan was rejected by SWAPO and the United Nations. Eventually, with the end of the cold war and the withdrawal of Cuban troops from Angola, South Africa eased its grip and agreed to end its administration of Namibia. Namibia held its elections in 1989, with SWAPO garnering 57 percent of the vote and the leading opposition party, the Democratic Turnhalle Alliance (DTA), winning 29 percent.

Contemporary Namibia faces numerous challenges. While the constitution ensures basic freedoms and human rights, the new government that received 95 percent of the Ovambo vote needs to broaden its political base to include other ethnic minorities. Furthermore, in spite of having the fourth highest GNI PPP per capita ($6,410) in Sub-Saharan Africa, extreme racial inequalities exist. Namibia also has a strong mining sector anchored by diamonds, uranium, zinc, and copper. It is the world's leading producer of gem-quality diamonds. In addition to the mining sector, which accounts for more than 80 percent of export earnings, Namibia

▪ **FIGURE 27-6 The Benguala Railroad.** Frequently attacked during Angola's civil war, a peace dividend is the gradual return to normalcy of the country's main railroad. As much a social service as an economic enterprise, the railroad is an essential part of the life of rural hinterland communities and a vital link to the coast for imports and exports.

■ **FIGURE 27-7 Walvis Bay, Namibia.** Walvis Bay is one of the leading ports of Southern Africa.

Botswana: Model of Economic and Political Stability?

Botswana could possibly serve as a model of economic and political stability in Africa. Between 1980 and 1993, it enjoyed one of the highest economic growth rates (in real terms) in the world, averaging a rate of 10 percent a year. It is now categorized by the World Bank as an "upper middle income" country, a far cry from its status as one of the poorest countries in the world at independence in 1966. This impressive economic performance was spurred early on by the beef cattle industry and later by developments in the diamond and copper-nickel industries that account collectively for 85 percent of the country's export earnings (Figure 27–8). Botswana's record on human development and gender equality has been equally impressive. Its Human Development Index value of 0.572 is among the highest in Sub-Saharan Africa. Also, its Gender-related Development Index at 0.566 is indicative of advancements that women have made with respect to health, education, income, and legal rights. It further demonstrates the government's commitment to universal education and the extension of social services and health clinics to rural areas.

In the political arena, the country is essentially a one-party state with a weak opposition. However the political situation has been relatively stable, and there are avenues for the expression of political ideas. Legislative power resides in a

potentially has the richest fishing zone in Sub-Saharan Africa, owing to the nutrient-rich waters of the cool Benguela Current. Also, in 1994 Namibia was able to incorporate the last vestige of South African domination—Walvis Bay—into its territory. This prized possession represents a new strategic gateway to Southern and Central Africa and Latin America. The port has been declared a free trade zone to attract foreign investment and stimulate development in the region (Figure 27–7).

■ **FIGURE 27-8 Diamond mine in Botswana.** Diamonds are Botswana's best friend. When it gained independence from Britain in 1966, Botswana was one of the world's poorest countries. A year later, diamonds were discovered. Now power shovels work six days a week here at Jwaneng, the world's richest diamond mine. Diamonds earn more than 60 percent of export revenues, and Botswana has one of Africa's fastest-growing economies.

forty-member parliament whose members are elected to 5-year terms. There is also a fifteen-member House-of-Chiefs that provides advice on ethnic and constitutional matters. The latter body is especially relevant in view of the dominance of the Tswana (75 percent of the population) over other ethnic groups such as the Shona, San, and Ndebele. Freedom squares are also available as forums for the public to discuss and debate political issues. Despite ongoing tensions between the dominant political party (the Botswana Democratic Party) and the opposition alliance, the civil liberties inherent in the political system and the inclusion of traditional authorities in the decision-making process offer some lessons for more autocratic governments in Africa.

Botswana is not without its problems, though. Spatial inequalities remain, with most of its 1.6 million people concentrated in the eastern part of the country. Drought and water shortages are common in the western and southern sections, but, to the government's credit, emergency relief efforts are timely and well coordinated. The limited population (and domestic market) is a constraint, especially with much skilled manpower migrating to South Africa. This limitation has negatively affected Botswana's manufacturing sector, which accounts for only 4 percent of the Gross National Income. In spite of the looming threat of AIDS, the government continues to diversify its economic base by extending its infrastructure, upgrading its manufacturing base, developing its vast coal reserves, and promoting ecotourism in the Okavango delta basin area.

Zimbabwe: The Challenge of Land Reform and Industrial Transformation

Zimbabwe derives its name from the Shona words *dzimba dza mabwe*, meaning "houses of stone." It is linked to the famous stone ruins of Great Zimbabwe, which date back to the thirteenth century. The elaborate stone structures are enclosed by walls that are about 20 feet (6 meters) thick and 35 feet (11 meters) high. Artifacts discovered from this great civilization reveal a well-organized social and political structure and an elaborate trade network based on gold, ivory, beads, ceramics, and copper objects.

Modern Zimbabwe, formerly known as Southern Rhodesia, gained its independence in 1980. Southern Rhodesia had opted for internal self-government rather than succumb to Afrikaner domination under a union with South Africa. The British, who administered the settler colony, had excluded Africans from trade and commerce and maintained a cheap and compliant labor force. Today the ethnic makeup of the country is quite diverse, with more than 200,000 Europeans and close to 40,000 Asians and coloreds. The largest African ethnic group is the Shona, followed by the Ndebele, Tonga, Venda, and Sotho. At independence, extreme inequalities existed in land ownership and other means of production. Europeans controlled more than 75 percent of the most productive land, while most African subsistence farmers were packed into densely settled marginal lands known as communal lands (formerly referred to as Tribal Trust Lands). Since the early 1980s, the Zimbabwe government has tried

to resettle African families through a number of programs, including phase I (1980–1997) and II (1998–2004) of the **Land Reform and Resettlement Program (LRRP)**, and a controversial Fast Track Program introduced in June 2000. One of the objectives of these programs was to acquire land from large-scale commercial farmers for distribution to landless people, overcrowded families, agricultural college graduates, and poor people with some farming skills. Other goals included diverting populations away from highly concentrated communal areas, enhancing agricultural productivity, reducing rural poverty, and sustaining economic development initiatives. Lands identified for acquisition included underutilized, derelict, and foreign-owned farms, and properties in close proximity to communal areas. Farms exempted or "de-listed" included those belonging to churches or missions, agro-industrial properties involved in meat and dairy production, properties within Export Processing Zones, and plantation farms engaged in large-scale production.

Unfortunately, the LRRP has had its share of problems and has fallen consistently below its land acquisition and resettlement targets. A report by the United Nations Development Program revealed that by 1997 the government had purchased only 42 percent of its intended target of 20.5 million acres (8.3 million hectares) of land and settled 44 percent of the 162,000 communal families it had expected to resettle. Moreover, many of the resettled families are located in marginal areas not suitable for cultivation, areas that lack basic infrastructure, and areas that are inaccessible to schools and farms. In 1992 the government renewed its efforts to redistribute land acquired from commercial farmers by passing the Land Acquisition Act. Yet by 1999 much of the prime farmland remained in the hands of commercial farmers or had been acquired by senior government officials and elites, with little or no land trickling down to the intended beneficiaries.

In response, the government introduced in 2000 the Fast Track Program (FTP), which was designed to accelerate its land reform program. The FTP further polarized both the Commercial Farmers Union, which insisted on protecting its member's property rights, and the veterans of past liberation movements, who advocated access to even more land. The most controversial aspect of the FTP was the government's attempt to revise the Land Acquisitions Act and extend its powers to forcibly acquire land. At independence, the government had agreed to adhere to the Lancaster House Agreement, an agreement negotiated with the British government that protected white Zimbabweans from forced land acquisitions for a period of 10 years. It also stipulated that government land acquisitions had to be followed by just and prompt compensations. After 1990, however, the government moved aggressively to accelerate the land acquisition process by forcibly acquiring land without just compensation. This resulted in a slew of court litigations and created a rash of illegal land occupations and violent confrontations. The issue of land reform and resettlement was a focal point of intense debate in the hotly contested Zimbabwe 2002 election, which ended with incumbent president Robert Mugabe winning amidst much controversy. The land crisis has negatively impacted tobacco production, a major source of government revenue, as well as cotton, sugar, livestock, and maize production.

Recently, Zimbabwe's economy has been affected by mismanagement, corruption, and a disengaged international community. This is unfortunate given some earlier success that the country had in transforming its industrial sector. Zimbabwe boasts a fairly large and diversified industrial sector that accounts for 30 percent of its economic output (Figure 27–9). In 1990, on the advice of the World Bank and the International Monetary Fund, the Zimbabwe government opted to liberalize trade, encourage privatization, and reduce public spending. In the industrial sector, the government emphasized export-oriented programs to stimulate productivity, external trade, competitiveness, and higher levels of investment. A host of incentives were provided to stimulate production for exports, beginning with the Export Incentive Scheme, in which exporters were paid a percentage of the freight on board value of exports, and an Export Revolving Fund, which enabled exporters to access foreign exchange to purchase foreign inputs. Export Processing Zones are also being developed to offer tax incentives and boost quality goods so that they can compete in international markets. Tax incentives, such as 5-year tax holidays, a 15 percent corporate tax rate, and exemptions from customs duties and capital gains taxes boosted exports of semi-processed minerals, clothing, and agricultural equipment and supported more research and development, production management, technical training, and consultation with industrial associations. Textile manufacturers have adapted well to new technologies and developed state-of-the-art capabilities in computer-assisted design (CAD) techniques. Footwear manufacturers are affiliated with international associations, trade fairs, and industrial journals, which keep them up-to-date with the latest technologies, production processes, and equipment. Zimbabwe also manufactures agricultural machinery that is suitable for local conditions. Much research is conducted to develop technologies that are appropriate

for local conditions. The sector exports most of the equipment to regional markets, particularly Zambia, Malawi, and Tanzania. Other destinations for Zimbabwe exports include South Africa, Botswana, and a regional economic organization outlined below.

Success in industry is undermined, though, by periodic droughts, a polarized society, a one-party state, and a government that is currently disengaged from the global community. Zimbabwe's economy is still polarized with wealthy Europeans and elitist Africans dominating the means of production at the expense of landless farmers and smallholders in communal areas. The interests of the poor farmer class are not always represented in a one-party state that has been monopolized by Mugabe's Zimbabwe African National Union-Patriotic Front.

South Africa: Country in Transition

Geostrategic Factors

South Africa is a pivot area of the Southern Hemisphere relative to three major zones of peace: the Indian Ocean, which in 1971 was declared by the United Nations as a "zone of peace"; the Antarctic region, which under the provisions of a 1959 treaty is restricted to research and scientific activities; and Latin America, which was declared a Nuclear Weapon Free Zone under the 1967 Treaty of Tlatelolco. South Africa also continues to be a major alternate route for international shipments of petroleum and minerals. When the Suez Canal was closed briefly after the Mideast War of 1967, the ocean route around South Africa became the principal route for oil shipments from the Persian Gulf to European and American markets. While the canal has since reopened, South African ports remain significant stops for petroleum shipments.

South Africa's mineral wealth and economic dominance make it a major power, not only in Southern Africa, but also in Sub-Saharan Africa (Table 27–1). While it makes up only 7 percent of Sub-Saharan Africa's total population and 5 percent of its surface area, South Africa accounts for 39 percent of Sub-Saharan Africa's economic output, 48 percent of industrial production, 33 percent of mineral production, 88 percent of all crude steel production, 50 percent of installed electricity capacity, and 50 percent of all telephones. South Africa also has the densest road, rail, and air networks in Sub-Saharan Africa. Its rail and harbor network is the only reliable trade link with the outside world for the landlocked countries of Botswana, Lesotho, Swaziland, Zimbabwe, Zambia, Malawi, and Democratic Republic of Congo. Each year, close to 1 million migrant workers are employed in South Africa's mining, industrial, and agricultural sectors.

During most of the cold war era, South Africa utilized its economic, strategic, and political leverage to sustain its policy of racial segregation, called **apartheid**. At that time, South Africa was seen as a "bastion" against the spread of communism. When Angola and Mozambique became independent from Portugal in 1975, both countries adopted

▪ **FIGURE 27-9 Industrial plant in Zimbabwe.** A technician monitors four-color printing presses at the Harare Factory in Zimbabwe. In contrast to most of Sub-Saharan Africa, Zimbabwe has a substantial manufacturing sector.

Table 27-1 Mining in South Africa

Mineral Reserves

Mineral	World Rank	%
Platinum group	1	56
Gold	1	39
Vanadium	1	45
Chromium	1	69
Manganese	1	82
Alumino-Silicates	1	38
Diamonds	5	na
Uranium	5	8
Others: Coal, Vermiculite, Nickel, Antimony		

Mineral Production

Mineral	World Rank	%
Vanadium	1	57
Chrome ore	1	44
Gold	1	21
Manganese ore	3	14
Alumino-Silicates	1	40
Platinum group	1	47
Titanium minerals	2	27
Diamonds	5	9
Others: Asbestos, Coal, Zirconium		

na = not available.

Source: Government Communication and Information System, South Africa Yearbook, Pretoria: South Africa, 1998.

a socialist stance that was viewed as threatening. South African foreign policy was geared toward destabilizing the socialist governments in those countries by supporting insurgent activities. The Anglo American and Western European nations were little inclined to put pressure on South Africa because of the presence of Cuban troops in Angola and because of foreign investments in South African mines. Instead of directly confronting South Africa on the issue of apartheid, the United States in the late 1970s and early 1980s decided on a policy of constructive engagement, which was designed to maintain dialogue with South Africa and ensure regional security in the Southern African region. On the international front, there were concerted attempts by various nations to divest (with stockholders selling their shares in companies doing business with South Africa), disinvest (preventing further investment and the purchase of Krugerrands and terminating bank loans to the South African government), and disengage (with the imposition of diplomatic, cultural, and economic sanctions) from South Africa. Also, in 1977 the United States instituted the "Sullivan Principles" (named after civil rights activist Reverend Leon Sullivan) to marshal its resources more effectively to promote change and reform in South Africa. The idea was to get U.S. companies to take on the responsibility of educating, training, and promoting Africans, thereby setting an example for other companies to emulate.

With the end of the cold war approaching, South Africa's political leverage lessened considerably in the mid- to late 1980s, setting the stage for social, economic, and political reform in the country. With one-man, one-vote black rule firmly established in South Africa, and with relative stability and political reform in the "frontline states" of Angola, Namibia, Botswana, Zimbabwe, and Mozambique, South Africa now has the opportunity to play a positive and enabling role in the social and economic development of Sub-Saharan Africa.

Historical Background

Of South Africa's 44 million inhabitants, 73 percent are African, 13 percent European, 9 percent "colored" or of mixed race, and 3 percent Asian. The African majority is composed of numerous ethnic groups, the largest being Zulu, Xhosa, Tswana, Bapedi, and Sotho. The European population is about 60 percent Afrikaner and 40 percent British. The **Afrikaners** are the oldest European community in South Africa. They are descended from Dutch farming people (Boers) who settled in Cape Town after 1652, when a fort and vegetable garden were established for the Dutch East India Company by Jan van Riebeeck. The location also served as a way station for ships engaged in trade with Asia. The Dutch farmers proceeded to suppress and alienate the indigenous Khoisan pastoralists in order to acquire tracts of land. Slaves were imported from the East Indies (Indonesia and Malaysia) to work on plantations. In 1806 the British took over the Cape Colony and ended slavery. Some Afrikaners stayed and accommodated themselves to change, while others embarked on the **Great Trek** to the high veld (grassland) region of the northeast interior. The Afrikaners encountered resistance from the Ndebele and Zulu during their trek. After initially suffering heavy losses at the hands of the Zulus, the Afrikaners eventually defeated them in the Battle of Blood River in 1838. The Great Trek and Battle of Blood River are two historical events commemorated in Afrikaner history. The Great Trek was regarded as a predestined journey to "the promised land," and hence there is a deep-seated emotional attachment to that land.

Gold and diamond discoveries in the high veld attracted the British to the region. The Orange Free State and Transvaal territories were annexed, and the British proceeded to suppress the Dutch Boers in the Anglo-Boer War from 1899 to 1902. In 1910 the Union of South Africa was established, legitimizing dual English and Afrikaner domination. In 1912, African elites countered by forming the African Native National Congress (ANNC). In the following year, Africans were relegated to reserves located in the least desirable areas, amidst protests from the ANNC. The 1913 Native Land Act restricted Africans to only 10 percent of the country's land area and denied them the right of land ownership outside the reserves. In 1936, under the Native Trust and Land Act, the Africans' share of land was increased to 14 percent.

1909 South Africa Act was passed by the imperial government to establish the Union of South Africa, with the constitution going into effect as of May 31, 1910.

1913 Native Land Act, the first legislative embodiment of the principle of territorial segregation and the separation of land rights between natives (Bantu) and non-natives, was enacted. It prohibited the purchase or hire of land by Bantu from Europeans and by Europeans from Bantu, freezing temporarily the existing distribution of land. All Bantu land (about 25 million acres or 10 million hectares) was identified as reserves exclusively for the Bantu, and could not be acquired by non-Bantu.

1927 Native Administration Act declared that the Governor-General (really the cabinet) was the Supreme Chief of all the Bantu in Natal, Transvaal, and the Orange Free State, and as such, had the power to arrest and punish a Bantu who defied an order, and to detain for 3 months any Bantu who, in the opinion of the Supreme Chief, might be dangerous to the public peace if left at large.

1936 Representation of Natives Act transferred the native voters in the Cape to a special electoral roll and gave them the right to elect three Europeans to the House of Assembly. It created an advisory Natives Representative Council that was to promote white/nonwhite cooperation.

1936 Native Trust and Land Act identified a number of areas that could be released from provisions of the 1913 Native Land Act, thereby adding marginally to the Bantu areas. This released land (mainly in Transvaal) was held by the Native Trust.

1946 Asiatic Land Tenure and Indian Representation Act placed restrictions on the purchase of land by Indians.

1949 Prohibition of Mixed Marriages Act forbade marriages between whites and nonwhites.

1950 Population Registration Act introduced a system of registration for all South African residents. Each identity card included a photo of the holder and details of the ethnic and racial unit to which he or she belonged.

1950 Group Areas Act made provision for the gradual introduction of residential segregation of whites, coloreds, Indians, and Bantu. Persons of one group were not allowed to own or occupy property in the controlled area of another group, except under special permit. This was one of the strongest segregation acts implemented.

1950 Immorality Act prohibited sexual relations between whites and nonwhites.

1951 White political institutions governing blacks were replaced by a system of ethnic government in the Bantu areas. Ethnic authorities were established for communities, and above them regional authorities, and at the apex of the system, a Bantu territorial authority.

1953 Reservation of Separate Amenities Act permitted facilities (seats, counters, vehicles, public buildings, parks, and so on) to be reserved for the exclusive use of specific races.

1953 Bantu Education Act authorized the transfer of the control of Bantu education from the provinces to the Bantu Affairs Department of the Central Government.

These two acts set the stage for South Africa's official apartheid policy that was later implemented under the Afrikaner government.

The Legitimization of Apartheid: 1948–1989

In 1948 the Nationalist Party prevailed over the South African Unified Party—a union of British and Afrikaners—thus signaling the formal beginning of apartheid in South Africa. Apartheid is an Afrikaans word meaning "apartness" or "separateness." The Afrikaner government policy was a simple strategy of divide, conquer, and subdue. The architects of apartheid were the Broederbond (union of brothers), an extremist secret nationalist society whose main agenda was to preserve the identity of Afrikaners and promote nationalist indoctrination through churches, schools, colleges, and the press. The Afrikaner government created a hierarchy of rigid laws designed to keep races effectively apart (see the *Geography in Action* boxed feature, Chronology of Apartheid Legislation and Related Events in South Africa). Apartheid was legislated at three levels: the personal level, the urban level, and the national level.

On a personal level (petty apartheid), several legislative acts were instituted to discourage races from using the same facilities (the Reservation of Separate Amenities Act of 1953) and to prevent mixed marriages and social interaction (the Prohibition of Mixed Marriages Act of 1949 and the Immorality Act of 1950). In urban areas, buffer zones such as railroads, waterways, highways, and cemeteries effectively separated destitute black townships (Figure 27–10) from affluent white suburbs under the Group Areas Act of 1950.

Superimposed on these was the policy of **Grand Apartheid**, the ultimate purpose of which was to create independent black homelands on the basis of ethnicity (Figure 27–11). These "homelands" or "**bantustans**" never stood a chance of gaining even the semblance of statehood, as they lacked the necessary economic and human resources required to become self-sufficient. About 75 percent of Africans were relegated to small portions of inhospitable lands (13 to 14 percent of all land) without any significant

1956 South Africa Amendment Act validated the Separate Representation of Voters Act of 1951 that had been declared invalid by the appeals courts. Later that year, coloreds elected four Europeans to represent them in the Parliament.

1956 Bantu (Urban Areas) Amendment Act authorized local authorities to order Bantu to leave areas when their presence was considered detrimental to the maintenance of peace.

1958 Electoral Law Amendment Act reduced the voting age for whites from 21 to 18 years.

1960 Unlawful Organizations Act was aimed at suppressing any form of opposition to the government. Among the political organizations banned were the African National Congress (ANC), the Pan African Congress, the Congress of Democrats, and the South African Indian Congress.

1962 General Law Amendment Act (Sabotage Act) expanded the already wide limits of arbitrary government powers. It provided for a minimum sentence of 5 years in prison and a maximum of death for "sabotage." The government was given the power to ban newspapers, organizations, and gatherings and to imprison suspected lawbreakers for, essentially, any length of time without due process of law.

1963 General Law Amendment Act permitted police officers to detain, for up to 90 days (without warrant arrest), a person suspected of committing, intending to commit, or having information about, specified types of political offenses.

1963 Transkei Constitution Act conferred self-government in the Transkei to the Bantu residents (mostly Xhosas).

1965 Bantu Homeland Development Corporation Act provided for the establishment of development corporations for the economic development of the Bantu homelands.

1970 Bantu Homelands Citizenship Act provided that every African (Bantu) become a citizen of the territorial authority or homeland to which he or she was attached.

1971 Bantu Homelands Constitution Act was an extension and further elaboration of the Bantu Authorities Act (1951), the Promotion of Bantu Self-Government Act (1959), and the Transkei Constitution Act (1963). It was designed to lead all the homelands toward independence. It provided for the establishment of legislative assemblies to replace territorial authorities and for the establishment of ministries of education, agriculture, and so on. Matters not transferred to the legislative assemblies included defense, foreign affairs, preservation of peace and security, postal and telephone services, customs and excise taxation, currency matters, and banking.

1983 A tricameral legislature was approved, to include a colored and Indian House of Delegates. The referendum was restricted to white voters.

1988 Self-Governing Territories Act granted the Bantustans increased powers to set up their own administrative structures, without giving them legal independence.

1990 President Frederik W. De Klerk lifted the ban on political organizations and released longtime ANC activist Nelson Mandela from prison.

1991 President De Klerk repealed all remaining legislation on apartheid. A multiparty conference to draft a new constitution was convened.

1993 Interim constitution was drafted.

1994 On April 27, free, open elections were conducted relatively peacefully, marking the dawn of a new era in South Africa.

■ **FIGURE 27-10 Soweto Township, Johannesburg, South Africa.** The black township of Soweto, located in suburban Johannesburg, is a vivid reminder of racial segregation and income inequality associated with apartheid South Africa.

South African Provinces and Homelands
prior to 1993 Constitutional Bill

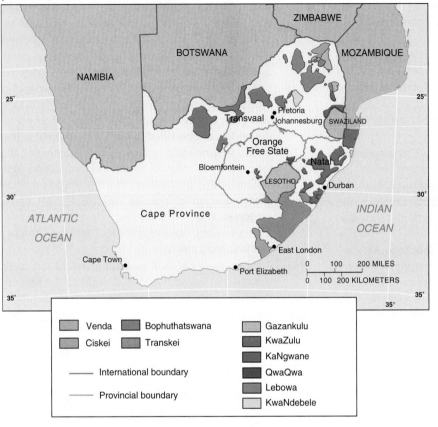

New South African Provinces after 1993 Constitutional Bill

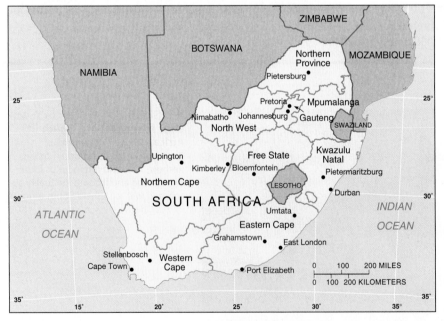

■ **FIGURE 27-11 Restructuring of provinces in South Africa.** Apartheid promoted racial separation by creating theoretically independent black homelands based on tribal and ethnic affiliation. After the collapse of the apartheid regime, a new administrative structure of nine provinces expanded political representation and allowed for greater regional autonomy.

mineral reserves and without any potential for cultivation. The boundaries of these fragmented and landlocked territories were also drawn in such a way that they conveniently avoided the major core regions in South Africa such as Witwatersrand, Durban, Port Elizabeth, and Cape Town. As a result, severe backwash effects were created as able-bodied males frequently left the marginal homelands to work in such core economic centers as Johannesburg and Durban (Figure 27–12), leaving the homelands with meager revenues to construct an appropriate infrastructure. Given this scenario, the "homelands" were not states in the true sense of the word because they lacked the resource base to become economically and politically sovereign and self-sufficient. Not one of them was recognized by the international community. The strategy was designed to diffuse any potential threats from the African majority by keeping it divided, fragmented, and disunited. In many respects, the tensions that exist today between Zulu and Xhosa can be attributed to this strategy of divide and conquer.

The minority European government was able to maintain its dominance by stripping the opposition of political power; restricting African access to education, jobs, and the military; imposing harsh penalties for violations such as the failure to carry a passbook; strictly enforcing the laws; and using intimidation.

The Transition Period: 1989–1993

By 1989, a series of external and internal events prompted the South African government to engage in economic and political reform. The cold war was ending, and Cuban forces had withdrawn from Angola, assuring Namibian independence in 1990. The frontline states of Angola and Mozambique were beginning to seek peace agreements, and the international community exerted increased pressure by stepping up efforts to disinvest and disengage from South Africa. Internally, the South African economy was hurting from sanctions and from years of massive expenditures on internal and regional security. Furthermore, Africans had organized more cohesive political platforms to press for change and reform. In February 1990,

■ **FIGURE 27-12**
Johannesburg, on the South African Veldt (plateau).
Johannesburg, known as the "city of gold," is the economic and industrial hub of South Africa.

President Frederik W. De Klerk released longtime black militant Nelson Mandela from prison and lifted the ban on political organizations, including the African National Congress (ANC, successor to the ANNC), the Pan African Congress, and the South African Communist Party. By mid-1991 most of the legislative instruments of apartheid had been dismantled, including the Separate Amenities Act, the Group Areas Act, and the Population Registration Act. In late 1991, the Convention for a Democratic South Africa (CODESA) brought several political factions together to discuss procedures for a new constitution. Eventually, an interim constitution was drafted and approved on December 22, 1993, to ensure civil, social, and political liberties for all South African citizens. The constitution embraces a Bill of Rights promising equality regardless of race, gender, sexual orientation, physical disability, and age. This Bill of Rights further guarantees free speech, movement, and residence; freedom of religion, politics, and conscience; freedom to use any of eleven languages; a fair trial with no torture or forced labor; economic, social, and cultural rights; and the rights to work and to education.

Post-Apartheid South Africa

The transition period between 1989 and 1993 was characterized by intense negotiations and efforts by several political factions to steer the country toward peace and stability. With an interim constitution in hand and a Bill of Rights drafted, South Africa was ready for its first free elections ever in April 1994. Nelson Mandela and his **African National Congress** were handed a mandate to lead South Africa by winning more than 12 million votes, or 62.6 percent of the total (Table 27–2). The provisions for forming a government of national unity entitled any party winning more than 20 percent of the vote to a deputy presidential position and any party winning 5 percent or more of the vote to

a national cabinet position. As a result, National Party leader De Klerk and Thabo Mbeki of the ANC became the deputy presidents, while Mangosuthu Buthelezi, leader of the Inkatha Freedom Party, was appointed minister of internal affairs. In 1993 the number of provinces was increased from four to nine to allow for expanded political representation and to accommodate regional autonomy (see Figure 27–11). Under the new system, ten senators were elected from each province. The boundaries were realigned in such a way that each new province had at least one major metropolitan area to encourage functional relations with their hinterlands. Other criteria considered were the development potential of each region, the availability of an adequate infrastructure and resource base, and the distribution of people and physical landscapes.

In December 1996, President Nelson Mandela signed a new constitution into law. It embraces the principles of a liberal democracy, providing for majority rule in a multiparty political system, the protection of civil liberties, an independent judiciary, and a free press. It includes a system of

Table 27-2 Results of South African Elections, April 1994

Party	Votes	Percentage	Seats
African National Congress	12,237,655	62.6	252
National Party	3,983,690	20.4	82
Inkatha Freedom Party	2,058,294	10.4	43
Freedom Front	424,555	2.2	9
Democratic Party	338,426	1.7	7
Pan African Congress	243,478	1.2	5

Percentage of total electorate who voted: 87.7 percent. Total number of valid votes: 19,533,498.

checks and balances, ranging from efforts to decentralize government to curbs on the power of the state through the Bill of Rights. Mandela was able to preserve some semblance of stability in South Africa and successfully guided the country to its second consecutive democratic elections in June 1999. He further secured the dominance of the ANC in South African politics with the election of Thabo Mbeki to the presidency. Mbeki and the ANC garnered 64.4 percent of the vote, falling just short of gaining full power to unilaterally change the country's constitution.

Today numerous challenges face the South African government, including such burning questions as these: How will the government address the extreme social and economic inequalities that exist in South Africa? To what extent can South Africa develop a strong reason for existing that supersedes narrow, parochial interests? Can the government develop a strong sense of nationalism, or rather will it accept and build on diversity? How will the government deal with the land question? What role will South Africa play in the socioeconomic development of the Southern and Sub-Saharan African regions? These are questions that will take some time to resolve as the government of national unity seeks viable and compatible solutions.

In order to address the issues of nation building and integration, the government launched an ambitious **Reconstruction and Development Program (RDP)** in 1994. The program is centered around six basic principles:

- designing integrated and sustainable policies at local, provincial, and national levels to meet basic human needs and mobilize resources;

- developing a people-centered approach to empower people to actively engage in decision-making processes and consensus building. This will be facilitated by encouraging broad-based education and training programs and by ensuring equal opportunities for women;

- promoting peace and ensuring security by upholding constitutional laws, human rights, and civil liberties;

- nation building to eradicate the divisive legacies of apartheid and to ensure that the country assumes an effective role in the global community;

- linking reconstruction and development by integrating growth and redistribution. A significant component of this link is the development and expansion of the country's infrastructure to enhance the delivery of effective services such as water, electricity, health, education, transportation, and telecommunications to all people;

- the democratization of South African society by ensuring the full participation of people at all levels of the decision-making process.

The RDP remains the government's basic social development policy. It was supplemented in 1996 with a macroeconomic policy called the Growth, Employment, and Redistribution (GEAR) program, which focused on economic development, expanding employment, encouraging privatization, restructuring state-owned enterprises, and providing socioeconomic opportunities to the poor. Related to GEAR is a Spatial Development Initiative (SDI), a strategy designed to target local communities, including rural communities, with a comparative economic advantage for private sector development and investment opportunities.

Implementing such programs is a major challenge in view of the extraordinary financial and international resources required. Consequently, the intended outcomes from these programs have not been forthcoming. South African unemployment rates remain high, averaging 33.9 percent and approaching as high as 49 percent in the Eastern Cape province. More than 40 percent of South African households still live in poverty and more than 30 million people live in poor housing conditions. Extreme socioeconomic disparities between Africans and whites persist, with more than 40 percent of Africans remaining unemployed compared to less than 5 percent of whites. More than 60 percent of Africans live in poverty compared to less than 2 percent of whites. Economic and social reconstruction programs are slow to develop in a country like South Africa, which is still recovering from international sanctions and war weariness.

Another major challenge confronting the South African government is the land reform required to address the injustices of forced removals and years of denying Africans access to suitable land. The three major components of the new government's land reform program are land restitution, land redistribution, and tenure reform. The issues of greatest concern surrounding land restitution relate to the magnitude of forced removals and the administrative and financial implications. In 1994 the **Restitution of Land Rights Act** was enacted to permit the government to investigate land claims and restore ownership to those who unjustly lost land. Restitution is justified if the claimant was dispossessed of a right to land after June 1913. A Land Claims Court was established to adjudicate disputes. Aside from land restoration, other forms of restitution include providing alternative land, payment of compensation, and priority access to government housing and land development programs. With respect to land redistribution, the main concerns are how to respond to people's needs in a fair, equitable, and affordable manner; how to address urgent cases of landlessness and homelessness; and providing credit, grants, and services for land acquisition and settlement. Eligible individuals have access to a land or settlement grant of up to 15,000 rands per household for the purchase of land from willing sellers, including the state. About a third of South Africa's farmland is expected to be redistributed within 5 years.

Tenure reform involves restoring security of tenure or occupancy to all South Africans without fear of arbitrary action by the state, individuals, or institutions. The objectives include upgrading tentative tenure arrangements that inhibit tenure security, social stability, and investment opportunity and dealing with issues of communal and customary tenure.

The government has also initiated a number of reforms in rural and urban areas. In the rural areas, a two-tiered rural local government structure consisting of district and local councils was established to promote local economic development initiatives and ensure environmental and social sustainability. The government has also stepped up its training and capacity-building efforts to maximize opportunities for development. In the urban arena, the Development Facilitation Act has been enacted to address reform in urban and regional planning. The act seeks to expedite land development projects, promote efficient and integrated land development, and manage the rate, quality, and character of urban growth. Another strategy of urban development is the **Masakahane Campaign** ("let's build each other"), designed to enhance community development initiatives, service delivery, and financial needs. It also involves mobilizing state, private, and community resources to deliver basic housing and infrastructural services; improving the capacity of local, metropolitan, and district councils to deliver and administer services more effectively; and supporting community initiatives to pay for services and promote economic development.

The South African government is working very hard to mobilize all existing resources to build a strong and sustainable economy that is beneficial to all citizens. However, questions surrounding the viability of the multiracial republic still remain. Although the number of provinces was increased from four to nine in order to decentralize government and diffuse possible regional tensions, there are still threats of secession from different groups. Extremist white nationalists are seeking a "volkstaadt," while Zulu hard-liners refuse to come to terms with a government of national unity. There are also a few rumblings from Tswanas who at one time wanted no part of a federation. Although these are difficult and complex challenges, the government is not deterred and continues to push its agenda for change and unity through peace initiatives and economic reforms. As South Africa confronts these issues internally, it also has to deal with the challenge of defining its new role in Southern and Sub-Saharan Africa.

Regional Cooperation in Southern Africa

As South Africa addresses its internal problems, it is also eager to improve its regional and international image. Opportunities exist for South Africa to enhance trade, development, and cooperative relations with regional and continent-wide organizations. On a larger scale, South Africa is expected to become a major player in the Organization of African Unity (OAU), the African Development Bank (ADB), and the United Nations Economic Commission for Africa (ECA). On a regional scale, South Africa already has trade and monetary relations with the Southern African Customs Union (SACU) and the Common Market Area (CMA). South Africa will also have to ponder its relations with the Southern African Development Community (SADC) and the Common Market for Eastern and Southern Africa (COMESA) (Figure 27–13).

The South African Customs Union (SACU) is the oldest and most economically integrated union in Southern Africa.

It was formed in 1910 and is a union of South Africa, the BLNS countries—Botswana, Lesotho, Namibia, Swaziland, and, more recently, Zambia. As in any customs union, goods move freely between member countries, and there is a common tariff on goods imported from outside. The South African Reserve Bank administers a common pool that allocates revenues on a proportional basis, taking into consideration the value of a member country's imports and their production and consumption of goods. Owing to South Africa's domination, the remaining countries are seeking renewed negotiations on the revenue payment formulae, infant industry protection, and reforms in tariffs and import surcharges. The Common Monetary Area (CMA) consists of all the SACU members with the exception of Botswana and Zambia. Under provisions of a Multilateral Monetary Agreement in 1992, the CMA allows for free movement of capital among member countries, a common capital market, and a common exchange rate. South Africa, though, dictates monetary policy, exchange rates, and interest rates.

The **Southern African Development Community (SADC)** was founded in 1980, originally as the Southern African Development Coordination Conference (SADCC). It was formed to reduce dependency on South Africa for rail, air, and port links, imports of manufactured goods, and the supply of electrical power. Unlike SACU and CMA, the emphasis was on economic cooperation instead of integration. Areas of cooperation include transportation and communications, food security, industrial development, energy conservation, mining development, and environmental management. Its members include Angola, Botswana, Democratic Republic of Congo, Lesotho, Malawi, Mauritius, Mozambique, Namibia, Swaziland, the United Republic of Tanzania, Zambia, and Zimbabwe. The initial priority was on transportation, with an emphasis on making Mozambique an alternative outlet for landlocked states. SADC invested in improving road and rail links to Mozambique and upgrading the port facility at Beira. Each country is responsible for one sector. For example, Angola is responsible for energy while Botswana coordinates agricultural research, livestock production, and animal disease control. Malawi coordinates inland fisheries, forestry, and wildlife; Namibia takes care of marine fisheries and resources; South Africa manages finance and investment; Tanzania handles industry and trade; and Zambia coordinates mining, employment, and labor. Botswana, South Africa, Zambia, and Zimbabwe dominate intra-SADC trade, accounting for more than 75 percent of exports and more than 65 percent of imports. Intraregional trade in merchandise exports increased from 2 percent in 1980 to 12.2 percent in 2000. In 1999, SADC signed a trade, development, and cooperative agreement with the European Union to reduce tariffs and to develop small enterprises, improve telecommunications and information technology, and enhance the tourism, mining, and energy sectors.

COMESA, originally founded in 1982 as the Preferential Trade Area for Eastern and Southern Africa (PTA), was transformed in 1992 into the **Common Market for Eastern and Southern Africa**, which now consists of

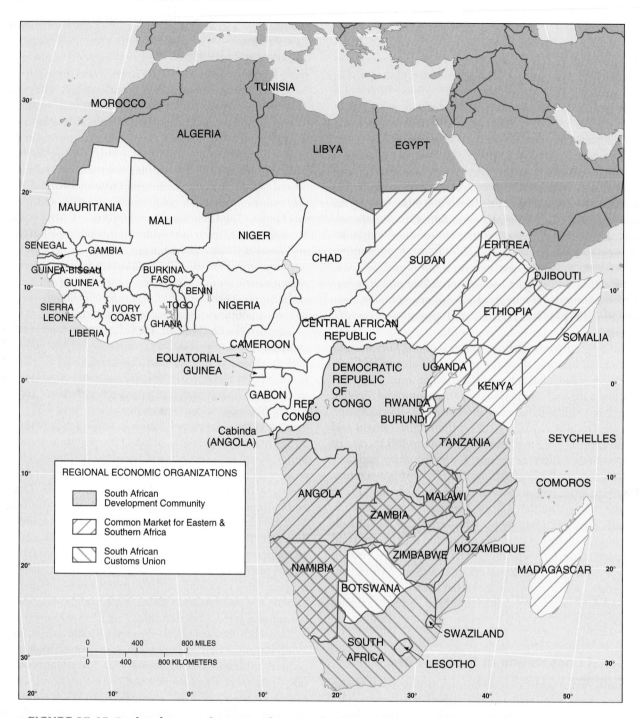

■ **FIGURE 27-13 Regional economic cooperation organizations in Southern Africa.** South Africa can improve its regional and international image by developing cooperative relations with regional and continent-wide organizations.

twenty-two countries. The members are all the SADC countries (except Botswana and Mozambique) as well as Burundi, Comoros, Djibouti, Egypt, Eritrea, Ethiopia, Kenya, Madagascar, Mauritius, Namibia, Rwanda, Seychelles, Sudan, and Uganda. COMESA's aim is to facilitate trade by granting preferential tariffs on selected goods, especially intermediate goods and those that promote local economic development efforts. Members also seek to enhance cooperation in agriculture, industry, telecommunications, and monetary affairs. COMESA hopes to evolve

into a full-fledged economic union. A major problem confronting COMESA is the varying levels of economic development among its member countries and the disparities in wealth and resources between the largest and smallest ones. Interaction and cooperation are hampered by weak transportation and communication links, fledgling economies, refugee problems, a lack of regional complementarity (duplication in production), administrative mismanagement, lack of information, and regulatory and procedural barriers.

Duplication and overlap between SADC and COMESA create problems. There is the potential for a merger that could eventually have pan-African implications. However, considerable differences in levels of industrial and economic development, government policy, and political stability severely limit the prospects of such a merger. For South Africa, the choice is whether to opt for SADC or COMESA. SADC seems like a logical choice, since it is economically more viable and confined to Southern Africa. COMESA, on the other hand, offers South Africa an opportunity to extend its influence beyond Southern Africa. Whatever the outcome, South Africa's integration into the Southern Africa region can only enhance regional development, which in turn could have positive ripple effects in all of Sub-Saharan Africa.

Summary

The Southern African region is undergoing radical transformations in its social, political, and economic institutions. Angola and Mozambique are finally beginning to experience some semblance of peace and stability after years of civil and ethnic strife rooted in Portuguese exploitation. Namibia is recovering from years of domination by South Africa, and Zimbabwe is still seeking solutions to the sensitive issues of land reform and economic integration. In the midst of these transformations, Botswana, buoyed by its diamond industry, has emerged as one of the fastest-growing economies in the world, and its achievements in human development and social justice constitute a framework for other African countries to follow. South Africa, though, remains the key player in the economic and political transformation of Southern Africa. Regional organizations such as SADC, COMESA, and SACU already provide a mechanism for countries in the region to coordinate their socioeconomic policies and to pursue avenues for greater regional cooperation and stability.

Key Terms

African National Congress

Afrikaners

apartheid

bantustans

Common Market for Eastern and Southern Africa (COMESA)

Grand Apartheid

Great Trek

insurgent states

Land Reform and Resettlement Program (LRRP)

Masakahane Campaign

Reconstruction and Development Program (RDP)

Restitution of Land Rights Act

shatterbelt

Southern African Development Community (SADC)

Review Questions

1. Sub-Saharan Africa is a land of great physical contrasts. Summarize the attributes of the principal natural regions.

2. Ecotourism is an emerging industry with considerable growth potential in Sub-Saharan Africa. What makes both the savanna and the tropical rain forest areas potentially attractive to nature tourists?

3. What cultural and economic factors are most responsible for the continuing high fertility rates of Sub-Saharan Africa?

4. Nigeria, with its wealth of natural resources, should have become an economic giant in Sub-Saharan Africa. Why has this not yet occurred?

5. Somalia is a rare example of a nation-state with a homogenous culture in Sub-Saharan Africa, yet it continues to be plagued by political instability and pervasive poverty. What are the contributing factors to this condition?

6. Regional economic integration is a mechanism for bringing countries together to seek avenues for mutual cooperation and economic progress. How successful has ECOWAS been in achieving these objectives?

7. Who are the Tutsi? How would you explain their social and economic dominance in the countries of Rwanda and Burundi?

8. Zimbabwe continues to wrestle with the issue of land reform. What is the best way to ensure a fair and just distribution of land in this country?

9. Should we attribute the long history of instability and poverty in Angola and Mozambique to the harsh Portuguese colonial practices or to misguided governmental policies enacted since independence, or to both? Cite examples to justify your answer.

10. What are the prospects for South Africa becoming a catalyst for socioeconomic development in other regions of Sub-Saharan Africa?

Geographer's Craft

1. Evaluate the impact of colonialism on the social, economic, and political underdevelopment of the countries of Sub-Saharan Africa. How might geographers contribute to overcoming the legacies of colonialism in the post-colonial era?

2. What geographical expertise might you draw upon to address the growing threat of AIDS in the countries of Sub-Saharan Africa?

3. How might the geographical perspectives of spatial distributions and human-environmental interrelationships contribute to the development of appropriate rural development strategies for the peoples of Sub-Saharan Africa? What traditional Sub-Saharan African values should be taken into account in the formulation of these strategies?

4. Identify the regions that are most affected by desertification and drought in Sub-Saharan Africa. How can geographers tribute to improved responses of national gover international agencies to these recurrir challenges?

5. The uniting of distinct tribal and e distrusted one another is one of th ing many Sub-Saharan African nati phers contribute to the process of bu can sustain the growth of national alleg

Further Readings

Part One: Basic Concepts and Ideas

Classic Works

Blaut, James M. 1975. "Imperialism: The Marxist Theory and Its Evolution." *Antipode* 7: 1–19. A presentation of the causes of development and underdevelopment from a theoretical Marxist perspective.

Cardoso, F. H., and **E. Faletto.** 1970. *Dependency and Development in Latin America.* Berkeley: University of California Press. A pioneering study of dependency theory.

de Souza, Anthony R., and **Philip W. Porter.** 1974. *The Underdevelopment and Modernization of the Third World.* Washington, D.C.: Association of American Geographers, Commission on College Geography Resource Paper No. 28. An excellent overview of the nature of underdevelopment and the theories that have been put forth to explain its causes.

Meadows, Donella H., Dennis L. Meadows, Jorgen Randers, and **William W. Behrens III.** 1972. *The Limits to Growth: A Report for the Club of Rome's Project on the Predicament of Mankind.* New York: Universe Books. A neo-Malthusian discussion of the race between population growth on the one hand and technology and pollution on the other.

Thomas, William L., ed. 1956. *Man's Role in Changing the Face of the Earth.* Chicago: University of Chicago Press. An old but classic work on the results of the human-environment relationship.

Recent Publications

Agnew, John, David N. Livingstone, and **Alisdair Rogers.** 1996. *Human Geography: An Essential Anthology.* Oxford: Blackwell Publishers. An excellent collection of articles and portions of books that have shaped the nature and philosophy of modern geography.

Bebbington, Anthony, and **Judith Carney.** 1990. "Geography and the International Agricultural Research Centers: Theoretical and Practical Concerns." *Annals, Association of American Geographers* 80: 34–48. A summary of the opportunities for geographers to contribute to the work of the international agricultural research centers.

Bellamy, Carol. 1997. *The State of the World's Children.* Oxford: Oxford University Press for the United Nations Children's Fund. A volume dedicated to providing statistical comparisons by nation of the state of children.

Brown, Lester, et al. *State of the World.* New York: Norton. An annual series by the World Watch Institute, with each volume containing a collection of papers treating contemporary world problems and issues such as environmental concerns, technological impacts, energy problems, and food supply.

Clawson, David L., and **Don R. Hoy.** 1979. "Nealtican, Mexico; A Peasant Community that Rejected the 'Green Revolution'." *American Journal of Economics and Sociology* 38: 371–387. A case study of a Mexican village that rejected the Green Revolution for both cultural and environmental reasons.

Gaile, Gary L., and **Cort J. Wilmott.** 2003. *Geography in America at the Dawn of the 21st Century.* Oxford; Oxford University Press. An excellent review of a broad range of the major subfields comprising contemporary American geography.

Geography Education Standards Project. 1994. *Geography for Life: National Geography Standards.* Washington, D.C.: National Geographic Research & Exploration. This beautiful 272-page volume, a product of cooperative effort by the American Geographical Society, Association of American Geographers, National Council for Geographic Education, and National Geographic Society, contains the new national geography standards for grades K–12, with a wealth of suggestions on how they can be taught.

Grosvenor, Gilbert M. 1995. "In Sight of the Tunnel: The Renaissance of American Geography Education." *Annals, Association of American Geographers* 85: 409–420. An overview, by the then president of the National Geographic Society, of geography's reintroduction into the American school curriculum in the late twentieth century.

Haggett, Peter. 1990. *The Geographer's Art.* Oxford: Basil Blackwell. A survey of the richness of geography as a discipline and its utility as a perspective for analysis by one of its foremost practitioners.

Harrison, Lawrence E. 1988. *Underdevelopment Is a State of Mind: The Latin American Case.* Lanham, MD: Madison Books and the Center for International Affairs, Harvard University. A volume that adopts the theoretical perspective of cultural determinism as it assesses the roles of values, beliefs, and child rearing in the development process.

McKnight, Tom L., and **Darrel Hess.** 2002. *Physical Geography: A Landscape Appreciation,* 7th ed. Upper Saddle River, NJ: Prentice-Hall. An introductory textbook treating the entire spectrum of physical geography.

Meadows, Donella H., Dennis L. Meadows, and **Jorgen Randers.** 1992. *Beyond the Limits.* Post Mills, VT: Chelsea Green. An update and contemporary extension of *Limits to Growth.*

Packenham, Robert A. 1992. *The Dependency Movement: Scholarship and Politics in Development Studies.* Cambridge: Harvard University Press. A critique of dependency theory as it relates to Latin American development studies.

Path Toward World Literacy: A Standards-Based Guide to K–12 Geography. 2001. Washington, D.C.: National Geographic Society. A follow-up publication to the *Geography for Life* volume previously cited herein, this useful guide provides specific geographical learning standards for the following grade levels: K–1, 2–3, 4–5, 6–8, and 9–12.

Peet, Richard. 1999. *Theories of Development.* New York: Guilford Press. A survey of marxist, feminist, and critical realist theories of development.

Pitzl, Gerald, ed. *Annual Editions: Geography.* Guilford, CT: Dushkin. Some thirty reprinted articles treating contemporary themes in geography; an excellent supplement to this text because of its treatment of population, environment, and resources in relation to development; published annually.

Reid, W. V. C., and **K. R. Miller.** 1989. *Keeping Options Alive: The Scientific Basis for Conserving Biodiversity.* Washington, D.C.: World Resources Institute. A review of the manifestations of and arguments in favor of the conservation of the earth's biodiversity.

Simon, Julian L. 1981. *The Ultimate Resource.* Princeton, NJ: Princeton University Press. A discussion of human resources as the ultimate resource for the solution of contemporary and future problems.

Simpson, E. S. 1994. *The Developing World: An Introduction,* 2nd ed. Essex, England: Longman. A balanced overview of the common attributes and challenges facing the less developed nations.

Smith, David M. 1982. *Where the Grass Is Greener: Living in an Unequal World.* Middlesex, England: Penguin Books. A review of the inequality among and within nations, treating measurement and pattern as well as causes and consequences.

Turner, B. L., II. et al. 1990. *The Earth as Transformed by Human Action: Global and Regional Changes in the Biosphere over the Past 300 Years.* New York: Cambridge University Press. Drawing its inspiration from *Man's Role in Changing the Face of the Earth,* this volume records the next 40 years of accelerating modification of global environments.

United Nations, Department of Economic Affairs. 1948–present. *Demographic Yearbook.* New York: United Nations. An annual compendium of useful population data, including total population, population growth rates, urban-rural ratios, births, deaths, life tables, and population movements.

United Nations, Department of Economic and Social Affairs. 1948–present. *Statistical Yearbook.* New York: United Nations. Annual statistics on a wide range of topics; particularly useful data on economic activities, including agriculture, forestry, fishing, mining, manufacturing, energy use, trade, transportation and communications, consumption of selected items, and national accounts.

United Nations, Food and Agriculture Organization. 1946–present. *Production Yearbook.* Rome: United Nations. An annual compendium of agricultural statistics, including area harvested, yields, and total production by country; also includes data on prices, pesticide and fertilizer consumption, and farm machinery.

United Nations Development Program. 1990–present. *Human Development Report.* New York: Oxford University Press. An annual publication that presents the human development index rankings as well as exploring in depth specific development themes.

Wolfenson, James D. 1997. *The Challenge of Inclusion.* Washington, D.C.: The World Bank. An address given by the president of the World Bank Group in which he argued that for development programs and strategies to be successful they must be based on the input of local development leaders who are sensitive to the needs of the poor.

The World Bank. 1978–present. *The World Development Report.* New York: The Oxford University Press. An excellent annual publication providing data updates and special reports on global economic development.

Current Publications

Brown, Lester R. 2001. *Eco-Economy: Building an Economy for the Earth.* New York: W. W. Norton. A passionate plea for the development of alternative energy sources, such as wind and hydrogen, as well as redesigned urban spaces that would move the global economy away from fossil fuels and place it on a more sustainable basis.

Foster, David R., and **John E. Aber.** 2004. *Forests in Time: The Environmental Consequences of 1,000 Years of Change in New England.* New Haven: Yale University Press. A detailed examination of the interaction of nature and society in one region yields important generic principles about how to manage dynamic environments.

Goudie, Andrew. 2005. *The Human Impact on the Natural Environment,* 6th ed. Cambridge: MIT Press. A solid introduction to many of the environmental impacts of human activity. Individual chapters focus on modifications to natural vegetation, animal life, soils, water bodies, landforms, and climate and the atmosphere.

Martin, Geoffrey J., and **T. S. Martin.** 2005. *All Possible Worlds: A History of Geographical Ideas,* 4th ed. New York: Oxford University Press. The standard textbook on the history of geography.

Part Two: United States and Canada

Classic Works

Borchert, John R. 1967. "American Metropolitan Evolution." *Geographical Review* 57 (3): 301–332. A classic work describing American metropolitan evolution in the context of changing transport technology.

Boswell, Thomas D., and **Timothy C. Jones.** 1980. "A Regionalization of Mexican Americans in the United States." *Geographical Review* 70 (1): 88–98. A classic study defining the regions and the character of regions of Mexican-American inhabitance.

Clark, David. 1985. *Post-Industrial America: A Geographical Perspective.* New York: Methuen. Describes the economic system evolving with the shift to a service economy.

Garreau, Joel. 1981. *The Nine Nations of North America.* Boston: Houghton Mifflin. A popular description of nine identifiable regions in North America, based on economic and social considerations.

Henry, James A., Joanna Mossa, and **E. C. Pirkle.** 1995. *Natural Landscapes of the United States.* 5th ed. Dubuque, IA: Kendall/Hunt. A wonderful textbook that studies U.S. regional geography from the perspective or organizational viewpoint of physiographic provinces.

Nostrand, Richard L. 1970. "The Hispanic-American Borderland: Delimitation of an American Culture Region." *Annals of the Association of American Geographers* 60 (4): 638–661. A classic study of the persistence of Hispanic influences in the western United States.

Pred, Allan. 1965. "Industrialization, Initial Advantage, and American Metropolitan Growth." *Geographical Review* 55 (2): 158–185. A good discussion of the basis for the location of early American urban-industrial centers.

Robinson, J. Lewis. 1989. *Concepts and Themes in the Regional Geography of Canada.* Vancouver: Talonbooks. A geographic description of the six major regions of Canada.

U.S. Bureau of the Census. *Statistical Abstract of the United States.* Washington, D.C.: Government Printing Office. A very useful annual publication with an enormous array of social, demographic, and economic data. Updated annually.

Zelinsky, Wilbur. 1980. "North America's Vernacular Regions." *Annals of the Association of American Geographers* 70 (1): 1–16. A classic study of culture regions by one of our leading geographic scholars.

Recent Publications

Cravey, Altha J. 1997. "Latino Labor and Poultry Production in Rural North Carolina." *Southeastern Geographer* 37 (2): 295–300. A new ethnic community adds complexity to a stratified and segmented labor market.

Fuguitt, Glenn V., and **Calvin L. Beale.** 1996. "Recent Trends in Nonmetropolitan Migration: Toward a New Turnaround." *Growth and Change* 27 (2): 156–174. Study of the ebbs and flows of migration between urban and rural areas since 1960 identifies gainers and losers.

Hart, John Fraser, and **Chris Mayda.** 1998. "The Industrialization of Livestock Production in the United States." *Southeastern Geographer* 38 (1): 58–78. Vertically integrated factory farms increasingly dominate livestock production.

Hartshorn, Truman A. 1997. "The Changed South, 1947–1997." *Southeastern Geographer* 37 (2): 122–139. A study of economic and cultural change in the South.

Haverluk, Terrence W. 1998. "Hispanic Community Types and Assimilation in Mex-America." *Professional Geographer* 50 (4): 465–480. An interesting study of diversity within the U.S. Hispanic community.

Kanter, Rosabeth Moss. 1995. "Thriving Locally in the Global Economy." *Harvard Business Review* 73 (5): 151–160. An assessment of local development strategies within our increasingly globalized world economy.

Kaplan, David H. 1994. "Population and Politics in a Plural Society: The Changing Geography of Canada's Linguistic Groups." *Annals, Association of American Geographers* 84 (1): 46–67. An exploration of trends driving Canada's French- and English-speaking regions and populations further apart.

Lo, Lucia, and **Carlos Teixeira.** 1998. "If Québec Goes . . . The 'Exodus' Impact." *Professional Geographer* 50 (4): 481–492. A study of the potential consequences to Canada of Québec leaving the country.

Paterson, J. H. 1994. *North America: A Geography of Canada and the United States.* 9th ed. New York: Oxford University Press. A standard textbook on Canada and the United States, organized with a topical and regional framework.

Stutz, Frederick P., and **Anthony R. de Souza.** 1998. *The World Economy: Resources, Location, Trade, and Development.* 3rd ed. Upper Saddle River, NJ: Prentice Hall. A solid economic geography textbook.

Warkenton, John. 1997. *Canada: A Regional Geography.* Scarborough, Ontario: Prentice Hall. A comprehensive treatment of Canada and its regional development.

Wheeler, James O., Peter O. Muller, Grant I. Thrall, and **Timothy J. Fik.** 1998. *Economic Geography.* 3rd ed. New York: John Wiley. Another solid textbook on economic geography.

Winsberg, Morton D. 1997. "The Great Southern Agricultural Transformation and Its Consequences." *Southeastern Geographer* 37 (2):193–213. A useful assessment of changing landuse and cropping patterns in the U.S. South.

Current Publications

Appalachian Regional Commission. You can learn more about poverty in Appalachia by reading the ARC's publications, beginning with its history, at *www.arc.gov/index.do?nodeId=7.*

Birdsall, Stephen S., Jon C. Malinowski, Eugene J. Palka, and **Margo L. Price.** 2005. *Regional Landscapes of the United States and Canada.* New York: John Wiley. The latest in a classic series on the regional geography of the United States and Canada.

Baigent, Elizabeth. 2004. "Patrick Geddes, Lewis Mumford and Jean Gottmann: Divisions over 'Megalopolis.'" *Progress in Human Geography* 28 (6): 687–700. Interesting discussion of how an important term, megalopolis, evolved within geography, and description of some of the problems associated with use of this term.

Careless, J. M. S. *Canada: A Celebration of Our Heritage.* Professor Careless is one of the premier historians of Canada. You can view recent work online at *www.canadianheritage.org/books/canada.htm.*

Filion, Pierre, Trudi Bunting, Kathleen McSpurren, and **Alan Tse.** 2004. "Canada-U.S. Metropolitan Density Patterns: Zonal Convergence and Divergence." *Urban Geography* 25 (1): 42–65. Examination of three Canadian and twelve U.S. urban areas suggests that historically divergent national patterns are disappearing.

Part Three: Latin America and the Caribbean

Classic Works

Crist, Raymond E. 1968. "The Latin American Way of Life." *American Journal of Economics and Sociology* 27: 63–76, 171–183, 297–311. One of the most insightful and interestingly written overviews of Latin American culture and its Iberian antecedents; authored by one of the most knowledgeable and influential teachers of Latin American geography.

Innis, Donald Q. 1980. "The Future of Traditional Agriculture." *Focus* 30: 1–8. A summary of the nature and advantages of intercropping or polyculture in tropical agriculture.

Lowenthal, David. 1972. *West Indian Societies.* London: Oxford University Press. A timeless synthesis of ethnicity and race relations in the Caribbean Basin.

Sauer, Carl O. 1963. "Geography of South America." in *Handbook of South American Indians,* ed. Julian H. Steward, 319–344. Bureau of American Ethnology Bulletin, no. 143. Washington, D.C.: Smithsonian Institution. Arguably the finest, most concise, single overview of the physical geography of South America, written by the dean of twentieth-century Latin Americanist geographers.

Simpson, Lesley Byrd. 1967. *Many Mexicos,* 4th ed., rev. Berkeley, CA: University of California Press. One of the most entertaining and perceptive analyses of Mexican and, by extension, Latin American subcultures ever published.

Smith, Guy-Harold. 1963. *Physiographic Diagram of South America.* Maplewood, NJ: Geographical Press. A classic landform map of South America with accompanying text.

Recent Publications

Bastide, Roger. 1978. *The African Religions of Brazil,* trans. Helen Sebba. Baltimore: Johns Hopkins University Press. An outstanding description of Candomblé, Xango, Macumba, Umbanda, and other forms of African spiritist faiths in Brazil.

Bonilla, Elssy. 1990. "Working Women in Latin America." In *Economic and Social Progress in Latin America 1990 Report,* 207–256. Washington, D.C.: Inter-American Development Bank. A detailed analysis of the evolving roles of women in the Latin American workforce.

Brea, Jorge A. 2003. "Population Dynamics in Latin America." *Population Bulletin* 58, No. 1. Washington, D.C.: Population Reference Bureau. A detailed overview of the demographic history of Latin America.

Brenner, Anita. 1971. *The Wind That Swept Mexico.* Austin, TX: University of Texas Press. A powerful pictorial history of the Mexican Revolution.

Coleman, William J. 1958. *Latin American Catholicism: A Self-Evaluation.* Maryknoll, NY: Maryknoll Publications. An overview of Latin American Catholic subtypes.

Cook, Noble David. 1998. *Born to Die: Disease and New World Conquest, 1492–1650.* Cambridge: Cambridge University Press. An assessment of the catastrophic population losses experienced by the American Indians following their conquest by the Spaniards.

Denevan, William M., ed. 1992. *The Native Population of the Americas in 1492,* 2nd ed. Madison: University of Wisconsin Press. The standard reference for estimating the sizes of pre-Conquest Indian populations and the extent and causes of their collapse under the European conquerors.

Eidt, Robert C. 1968. "The Climatology of South America," in *Biogeography and Ecology in South America,* eds. E. J. Fittkau et al., 54–81. The Hague: Dr. W. Junk N. V. Publishers. An excellent overview of the climatic regions of South America.

Gade, Daniel W. 1975. *Plants, Man, and the Land in the Vilcanota Valley of Peru.* The Hague: Dr. W. Junk N. V. Publishers. A delightful regional study in the classic cultural ecology tradition.

Garrard-Burnett, Virginia, ed. 2000. *On Earth as It Is in Heaven: Religion in Modern Latin America.* Wilmington, DE: Scholarly Resources. Contains studies that address the changing religious geography of Latin America and the Caribbean.

Gwynne, Robert N. 1986. *Industrialization and Urbanization in Latin America.* Baltimore: Johns Hopkins University Press. An overview of Latin American industrialization and its relationship to world economic cycles.

Hall, Carolyn, and **Héctor Pérez Brignoli.** 2003. *Historical Atlas of Central America.* Norman: University of Oklahoma Press. The finest atlas yet published on the lands and peoples of Central America, both past and present.

Harrison, Lawrence E. 1988. *Underdevelopment is a State of Mind: The Latin American Case.* Lanham, MD: Madison Books and the Center for International Affairs, Harvard University. A thought-provoking assessment of the roles of values and human behavior in the development and underdevelopment of the Latin American nations.

Hulse, Lloyd K. 2002. *Three Roads South: Search for a Latin American Cultural Identity.* Lanham, MD: University Press of America. A revealing essay on Latin American culture norms and behaviors by one who has lived much of his life in the region.

Klein, Herbert S. 1986. *African Slavery in Latin America and the Caribbean.* Oxford: Oxford University Press. One of the most thorough treatments of slavery in Latin America and the extensive African influences found in many parts of the region.

Loveman, Brian. 1994. "Protected Democracies and Military Guardianship: Political Transitions in Latin America 1978–1993." *Journal of Inter-American Studies and World Affairs* 36: 105–189. A summary of the behind-the-scenes limits placed by the military on many of Latin America's newly established democratic civilian governments.

Lynch, John. 1992. "The Institutional Framework of Colonial Latin America." *Journal of Latin American Studies* 24 (Quincentenary Supplement): 69–81. An insightful overview of the policies employed by the Spanish crown in administering colonial Latin America and the effects of those policies on the historical underdevelopment of the region.

MacLachan, Ian, and **Adrian Guillermo Aguilar.** 1998. "Maquiladora Myths: Locational and Structural Change in Mexico's Export Manufacturing Industry." *Professional Geographer* 50: 315–331. A detailed analysis of recent developments in the *maquiladora* industry.

Martin, David. 1990. *Tongues of Fire: The Explosion of Protestantism in Latin America.* Oxford: Basil Blackwell. An overview of the rapid expansion of fundamentalist, charismatic Protestantism.

Parsons, James J. 1962. "The Moorish Imprint on the Iberian Peninsula." *Geographical Review* 52: 120–122. An introduction to the Moorish influence on Spanish and Portuguese culture by a leading Latin Americanist geographer.

Place, Susan E., ed. 2001. *Tropical Rainforests: Latin American Nature and Society in Transition,* revised and updated edition. Wilmington, DE: Scholarly Resources, 2001. An outstanding edited collection of studies related to Latin America's threatened tropical rain forests.

Ryder, Roy H. 2004. "Land Use Diversification in the Elite Residential Sector of Quito, Ecuador." *The Professional Geographer* 56: 488–502. An analysis of changing urban land use patterns in a prominent Latin American city.

Sealey, Neal. 1992. *Caribbean World: A Complete Geography.* Cambridge: Cambridge University Press. A concise, scholarly introduction to Caribbean lands and peoples.

Skidmore, Thomas E. 1999. *Brazil: Five Centuries of Change.* New York: Oxford University Press. An introduction to the peoples and regions of Brazil.

Watanabe, John M. 1992. *Maya Saints and Souls in a Changing World.* Austin, TX: University of Texas Press. An analysis of the evolution of folk Catholicism among the highland Maya peoples of western Guatemala.

Watts, David. 1987. *The West Indies: Patterns of Development, Culture and Environmental Change since 1492.* Cambridge: Cambridge University Press. A detailed analysis of the cultural and environmental impacts of European conquest on the lands and peoples of the Caribbean island realm.

Whitmore, Thomas M., and **B. L. Turner II.** 2001. *Cultivated Landscapes of Middle America on the Eve of the Conquest.* New York: Oxford University Press. A detailed analysis of pre-Conquest native American agriculture in Middle America.

Wilken, Gene C. 1987. *Good Farmers: Traditional Agricultural Resource Management in Mexico and Central America.* Berkeley, CA: University of California Press. A detailed analysis of the agricultural strategies and cropping systems of Latin American traditional farmers.

Current Publications

Brunn, Stanley D., Jack F. Williams, and **Donald J. Zeigler,** eds. 2001. *Cities of the World: World Regional Urban Development,* 3rd ed. Lanham, MD: Rowman & Littlefield. Excellent summaries of current challenges facing the world's largest cities, including those of Latin America.

Butzer, Karl W., ed. 1992. "The Americas Before and After 1492: Current Geographical Research." Special thematic issue of *Annals, Association of American Geographers* 82. The volume contains a collection of articles that address key issues related to the geography of Latin America and the Caribbean prior to the European Conquest and the land use and cultural changes that followed.

Clawson, David L. 2006. *Latin America and the Caribbean: Lands and Peoples,* 4th ed. Boston: McGraw-Hill. A text that studies current Latin American and Caribbean development issues from a topical or thematic perspective.

Conference of Latin Americanist Geographers (scholarly organization): *http://sites.maxwell.syr.edu/clag/clag.htm.*

Denevan, William M. 2001. *Cultivated Landscapes of Native Amazonia and the Andes.* Oxford: Oxford University Press. A masterful synthesis of pre-Columbian land use patterns and their relevance to current sustainable agricultural strategies.

Economic Commission for Latin America and the Caribbean (economic data): *www.eclac.org.*

Inter-American Development Bank (economic and social data): *www.iadb.org.*

Journal of Latin American Geography. The annual publication of the Conference of Latin Americanist Geographers. The journal was formerly published as the *Conference of Latin Americanist Geographers Yearbook.*

Knapp, Gregory, ed. 2002. *Latin America in the Twenty-first Century: Challenges and Solutions.* Austin, TX: University of Texas Press for the Conference of Latin Americanist Geographers. Thorough overviews of a number of significant facets of the cultural, environmental, and economic geography of the Latin American and Caribbean nations.

Pan American Health Organization (health data): *www.paho.org.*

United Nations Development Programme (social data): *www.undp.org.*

Part Four: Europe

Classic Works

Hay, Denys. 1968. *Europe: The Emergence of an Idea,* 2nd ed. Edinburgh: University Press. Traces the development of Europe as a cultural geographical entity.

Krantz, Grover S. 1988. *Geographical Development of European Languages.* New York: Peter Lang. An imaginative theory on how Europe's linguistic complexity evolved.

Lambert, Audrey M. 1985. *The Making of the Dutch Landscape: An Historical Geography of the Netherlands,* 2nd ed. London: Seminar Press. Traces the evolution of the humanized habitat, including land reclamation.

Murphy, Alexander B. 1988. *The Regional Dynamics of Language Differentiation in Belgium: A Study in Cultural-Political Geography.* Chicago: University of Chicago Geography Research Paper No. 227. A study of the interaction of language and politics in one of Europe's multilingual countries.

Nolan, Mary L., and **Sidney Nolan.** 1989. *Religious Pilgrimage in Modern Western Europe.* Chapel Hill: University of North Carolina Press. A geographical appraisal of a still-important facet of Europe's religious geography.

Stanislawski, Dan. 1959. *The Individuality of Portugal.* Austin: University of Texas Press. Explains Portugal's evolution as a separate country and culture.

Recent Publications

Ashworth, G. J., and **Peter J. Larkham,** eds. 1994. *Building a New Heritage: Tourism, Culture and Identity in the New Europe.* London: Routledge. Essays on the geography of tourism in the new era of open borders.

Benevolo, Leonardo. 1993. *The European City.* Oxford: Blackwell. A learned essay on the special character of urban Europe.

Bretherton, Charlotte, and **John Vogler,** eds. 1999. *The European Union as a Global Actor.* London: Routledge. Discussion of the increasing role of the European Union on the international political scene.

Carter, Francis W., and **W. Maik.** 1999. *Shock-Shift in an Enlarged Europe: The Geography of Socio-Economic Change in East-Central Europe after 1989.* Aldershot, U.K.: Ashgate. An examination of the sweeping changes that have followed the collapse of Communism.

Clout, Hugh, ed. 1994. *Europe's Cities in the Late Twentieth Century.* Utrecht: Royal Dutch Geographical Society. The most comprehensive overview of European urban geographical trends, covering all major countries.

Fielding, Tony, and **Hans Blotevogel.** 1997. *People, Jobs, and Mobility in the New Europe.* New York: John Wiley. Deals with the underlying causes of spatial, social, and economic change that cause people to migrate.

Gavin, Brigid. 2001. *The European Union and Globalization.* Cheltenham, U.K.: Edward Elgar. An introduction into basic concepts of globalization, using the example of the European Union and contrasting its role with the World Trade Organization.

Graham, Brian, ed. 1998. *Modern Europe: Place, Culture, and Identity.* London: Arnold. A sensitive and intelligently written geography of Europe.

Granberg, Leo, Imre Kovách, and **Hilary Tovey,** eds. 2001. *Europe's Green Ring.* Aldershot, U.K.: Ashgate. Shows the inner industrial/urban core of Europe surrounded by a rural periphery.

Hall, Derek, and **Darrick Danta.** 1996. *Reconstructing the Balkans: A Geography of the New Southeast Europe.* New York: John Wiley. An analysis of the most troubled region in Europe and what is needed to resurrect it.

Hantras, Linda, ed. 2000. *Gendered Policies in Europe: Reconciling Employment and Family Life.* Houndsmill, U.K.: Macmillan Press. Case studies of implementation of gender equality policies in major Western European countries.

Heffernan, Michael. 1997. *Twentieth-Century Europe: A Political Geography.* New York: John Wiley. A critical analysis of changing political patterns, including both fragmentation and integration.

Höll, Otmar, ed. 1994. *Environmental Cooperation in Europe: The Political Dimension.* Boulder, CO: Westview Press. Essays on the international and domestic environmental policies in the major European countries.

Hudson, Ray, and **Allan M. Williams.** 1999. *Divided Europe: Society and Territory.* London: Sage. Analyzes the internal divisions that persist in post-Cold War Europe.

Jones, Alun. 1994. *The New Germany: A Human Geography.* New York: John Wiley. Geography of reunified Germany, the country that dominates Europe's core.

Kivell, Philip, Peter Roberts, and **Gordon P. Parker,** eds. 1998. *Environment, Planning and Land Use.* Brookfield, VT: Ashgate. An insightful introduction to the challenges of sustainable management of land and water resources in the European urban context.

Kostakopoulou, Theodora. 2001. *Citizenship, Identity and Immigration in the European Union: Between Past and Future.* Manchester: Manchester University Press. In-depth discussion of the evolution of European identity and the role that immigration from non-European places plays in this process.

Matvejevic, Predrag. 1999. *Mediterranean: A Cultural Landscape.* Berkeley: University of California Press. A masterful and unique portrait of the peoples and cultures of southern Europe.

Montanari, Armando, and **Allan M. Williams,** eds. 1995. *European Tourism: Regions, Spaces, and Restructuring.* New York: John Wiley. In-depth study of one of Europe's most important service industries.

Nijkamp, Peter, ed. 1993. *Europe on the Move: Recent Developments in European Communications and Transport Activity.* Aldershot, U.K.: Avebury. The best overview of these service industries.

Pedersen, Roy N. 1992. *One Europe/100 Nations.* Clevedon, U.K.: Channel View. Displays the ethnic complexity of Europe.

Pred, Allan. 2000. *Even in Sweden: Racisms, Racialized Spaces, and the Popular Geographical Imagination.* Berkeley: University of California Press. Racism directed against non-European residents in one of Europe's most tolerant and progressive countries.

Rhodes, Martin, ed. 1995. *The Regions and the New Europe: Patterns in Core and Periphery Development.* Manchester: Manchester University Press. An overview of recent trends concerning Europe's ancient core-versus-periphery configuration.

Current Publications

Alesina, Alberto, and **Roberto Perotti.** 2004. *The European Union: A Politically Incorrect View.* Cambridge, MA: National Bureau of Economic Research. The full text of this document is available at *http://www.nber.org/papers/w10342.* Provides good insight into the evolution and structure of the EU and discusses ideological differences between member countries.

Dahlman, Carl. 2004. "Turkey's Accession to the European Union: The Geopolitics of Enlargement." *Eurasian Geography and Economics* 45 (8): 553–574.

Environmental Policy in Transition: *Ten Years of UNECE Environmental Performance Reviews.* 2003. New York: The UN Economic Commission for Europe Publication. Rich in recent data, this report demonstrates how accession to the European Union became a stimulus for environmental improvement in Central and Eastern European countries.

Jordan-Bychkov, Terry G., and **Bella Bychkova Jordan.** 2002. *The European Culture Area: A Systematic Geography,* 4th ed. Lanham, MD: Rowman & Littlefield. An upper-division college textbook elaborating upon the approach to Europe employed in this book.

Karatnycky, Adrian, Alexander Motyl, and **Amanda Schnetzer.** 2002. *Nations in Transit 2002: Civil Society, Democracy, and Markets in East Central Europe and the Newly Independent States.* New Brunswick, NJ: Transaction Publishers. A good source on more recent developments in East Europe and the Balkans.

Ostergren, Robert C., and **John G. Rice.** 2004. *The Europeans. A Geography of People, Culture, and Environment.* New York: Guilford Press. An up-to-date thematic overview of Europe and its peoples.

Part Five: Russia and Central Eurasia

The In-Between Countries of Eurasia

Classic Works

Bater, James H. 1996. *Russia and the Post-Soviet Scene: A Geographical Perspective.* London: Arnold. Well-written geography of the early post-Soviet period, especially strong on energy, cities, and environmental issues. Complements Shaw, ed., 1995.

Pryde, Philip R. 1995. *Environmental Resources and Constraints in the Former Soviet Republics.* Boulder, CO: Westview Press. Best geographical treatment of the environmental problems inherited from the Soviet Union.

Shaw, Denis J. B., ed., 1995. *The Post-Soviet Republics: A Systematic Geography.* New York: John Wiley. Collection of essays on the main geographical topics for the fifteen former Soviet countries. Excellent analysis of the situation in the early nineties, especially with respect to nationalities issues, transport, industry, and agriculture.

Recent Publications

de Waal, Thomas. 2004. *Black Garden: Armenia and Azerbaijan through Peace and War.* New York: New York University Press. Even-handed, well-researched, highly readable account of the conflict over Nagorno-Karabakh.

Gaddy, Clifford G., and **Barry W. Ickes.** 2002. *Russia's Virtual Economy.* Washington, D.C.: Brookings Institution Press. The authors discuss the persisting lack of monetization in the economy—the "virtual economy" of barter and gifts continues to thrive.

Hedlund, Stefan. 1999. *Russia's "Market" Economy: A Bad Case of Predatory Capitalism.* London: University College London Press. Fascinating, somewhat eccentric discussion that focuses on path dependence in the transition.

Herrera, Yoshiko M. 2005. *Imagined Economies: The Sources of Russian Regionalism.* Cambridge, U.K.: Cambridge University Press. Challenging work that considers the regionalist imagination as well as "objective" economic factors as causes of regional movements for greater autonomy.

Hill, Fiona, and **Clifford G. Gaddy.** 2003. *The Siberian Curse: How Communist Planners Left Russia Out in the Cold.* Washington, D.C.: Brookings Institution Press. Invaluable analysis of the economic ramifications of Russia's size and forbidding climate.

Isham, Heyward, ed. 2001. *Russia's Fate Through Russian Eyes: Voices of the New Generation.* Boulder, CO: Westview Press. Absorbing collection of interviews with young Russians engaged in rebuilding their country; main topics include politics, business, law, civil society, and culture.

Jeffies, Ian. 2003. *The Caucasus and Central Asian Republics at the Turn of the Twenty-First Century: A Guide to the Economies in Transition.* London: Routledge. More of a compilation of documents than a monograph and requires some background, but full of information and up to date.

———. 2004. *The Countries of the Former Soviet Union at the Turn of the Twenty-first Century: The Baltic and European States in Transition.* London: Routledge. Like the companion volume: more of a compilation of documents than a monograph and requires some background, but full of information and up to date.

Lynch, Allen C. 2005. *How Russia Is Not Ruled: Reflections on Russian Political Development.* Cambridge, U.K.: Cambridge University Press. Lynch discusses the roots of economic fragility in Russia and argues that the state will continue to play a major role. This book is a welcome wake-up call needed by those intoxicated by the thought of the country's supposed resource wealth.

McCann, Leo, ed. 2004. *Russian Transformations: Challenging the Global Narrative.* London: RoutledgeCurzon. Eclectic collection of contributions that interrogate preconceptions of globalization and transition with respect to Russia. Includes case studies of Novosibirsk, Tatarstan, and Bashkortostan.

Reddaway, Peter, and **Robert W. Orttung,** eds. 2004. *The Dynamics of Russian Politics: Putin's Reform of Federal-Regional Relations.* London: Rowman & Littlefield. Timely examination of what Putin's Federal Districts have and have not accomplished.

Rosefielde, Steven. 2005. *Russia in the 21st Century: The Prodigal Superpower.* Cambridge, U.K.: Cambridge University Press. Controversial volume argues that Muscovite Russia is still in place and that re-militarization and an attempt to regain superpower status are on its agenda.

Ruble, Blair A., Jodi Koehn, and **Nancy E. Popson,** eds. 2001. *Fragmented Space in the Russian Federation.* Washington, D.C.: Woodrow Wilson Center Press. An important collection of articles, by six geographers and other social scientists, concerning regions in the Russian Federation

Shleifer, Andrei. 2005. *A Normal Country: Russia after Communism.* Cambridge: Harvard University Press. Collection of essays by a Harvard economist and former advisor to the Russian government—all in all a "look on the bright side" perspective.

White, Stephen, Elena Korosteleva, and **John Löwenhardt,** eds. 2005. *Postcommunist Belarus.* Lanham, MD: Rowman & Littlefield. Collection of nine articles, with a focus on politics and foreign relations.

Wilson, Andrew. 2002. *The Ukrainians: Unexpected Nation,* 2nd ed. New Haven: Yale University Press. Probably the best single source on Ukraine. Forthcoming by the same author and publisher: *Ukraine's Orange Revolution.*

Current Publications

Ioffe, Grigory. 2004. "Understanding Belarus: Economy and Political Landscape." *Europe-Asia Studies* 56 (1): 85–114. One of a series of three articles by an outstanding geographer on the current situation in Belarus.

Ioffe, Grigory, and **Tatyana Nefedova.** 2004. "Marginal Farmland in European Russia." *Eurasian Geography and Economics* 45 (1): 45–59. The authors investigate the contraction of agriculture in European Russia, finding natural conditions and accessibility to be the main factors of farms' prospects.

Kramer, Mark. 2004/2005. "The Perils of Counterinsurgency: Russia's War in Chechnya." *International Security* 29 (3): 5–63. A highly knowledgeable and detailed account of the war in Chechnya that also illuminates problems in Russian society and government.

van Zon, Hans. 2002. "Alternative Scenarios for Ukraine." *Futures* 34: 401–416. A brilliant short interpretation of Ukraine's situation and its prospects, this article aids understanding of the recent Orange Revolution.

Central Asia and Afghanistan

Classic Works

Ellis, William S. 1990. "The Aral: A Soviet Sea Lies Dying." *National Geographic* 177 (2): 73–92. An easily accessible article that describes many of the difficult environmental and societal issues facing Karakalpakstan and the Aral Sea Basin.

Feshbach, Murray, and **Alfred Friendly, Jr.** 1992. *Ecocide in the USSR: Health and Nature under Siege.* New York: BasicBooks. The first and foremost book written during and after the fall of the USSR that addresses the environmental destruction caused by the Soviets throughout the USSR.

Smith, David R. 1992. "Salinization in Uzbekistan." *Post-Soviet Geography* 33 (1): 21–33. An article addressing the environmental situation of the most important cotton producing country and the effects Soviet policy had toward the land.

Recent Publications

Elliot, Jason. 1999. *An Unexpected Light: Travels in Afghanistan.* New York: Picador USA. A well-written piece of travel literature by a man who visited Afghanistan twice, once during the Soviet occupation of the country and again immediately before the Taliban takeover. His travels across eastern and northern Afghanistan are full of geographical and cultural anecdotes that bring these areas alive to the reader.

Fierman, William, ed. 1991. *Soviet Central Asia: The Failed Transformation.* Boulder, CO: Westview Press. Collection of articles that focus on a range of aspects of the attempted economic and societal transformation of a large multi-ethnic region into a single Russianized unit.

Hopkirk, Peter. 1994. *The Great Game: The Struggle for Empire in Central Asia.* New York: Kodansha International. Incredibly accessible book on the history of British and Russian imperialism in Central Asia and Afghanistan. Although heavily weighted to the British (and therefore Afghanistan) side, the book brings to life the major explorers, politicians, and generals whose clandestine maneuverings brought this region into contact with Europe.

Landau, Jacob M., and **Barbara Kellner-Heinkele.** 2001. *Politics of Language in the Ex-Soviet Muslim States: Azerbaijan, Uzbekistan, Kazakhstan, Kyrgyzstan, Turkmenistan, and Tajikistan.* Ann Arbor: The University of Michigan Press. A scholarly book on the importance the governments of the USSR and the independent Central Asian countries place on language, its teaching, and its representation.

Lubin, Nancy. 1999. *Calming the Ferghana Valley: Development and Dialogue in the Heart of Central Asia.* New York: The Century Foundation Press. A political investigation of the Ferghana Valley, the most volatile and conservative area of Central Asia and now shared between three mutually suspicious countries: Kyrgyzstan, Tajikistan, and Uzbekistan.

Sagdeev, Roald, and **Susan Eisenhower,** eds. 2000. *Islam and Central Asia: An Enduring Legacy or an Evolving Threat?* Washington, D.C.: Center for Political and Strategic Studies. Pre-9/11 collection of articles that show the relationship between Islam and the region's various countries and how it affects relations with other countries, both in the region and surrounding it.

van Leeuwen, Carel, Tatjana Emeljanenko, and **Larisa Popova,** eds. 1994. *Nomads in Central Asia: Animal Husbandry and Culture in Transition (19th–20th Century).* Amsterdam: Royal Tropical Institute. A book featuring one article each from each editor that addresses the history and culture related to nomadic (both steppe and mountain) societies in Central Asia, with special emphasis on the effects of collectivization.

Current Publications

Allison, Roy, and **Lena Jonson,** eds. 2001. *Central Asian Security: The New International Context.* Washington, D.C.: Brookings Institution Press. A collection of articles that focuses on topical issues such as water and Islam, the volume also examines relations between the Central Asian countries and the major players in the region such as Russia, Iran, Turkey, and the United States.

Ewans, Martin. 2002. *Afghanistan: A Short History of Its People and Politics.* New York: HarperCollins Publishers. The most accessible and up-to-date history of Afghanistan from its earliest known history to the fall of the Taliban with special emphasis on the twentieth century.

Grousset, Rene. 2002. *The Empire of the Steppes: A History of Central Asia.* Translated by Naomi Walford. New Brunswick: Rutgers University Press. One of the most important and respected books on the early history of the Turkic peoples and their conquest of Central Asia.

Jonson, Lena. 2004. *Vladimir Putin and Central Asia: The Shaping of Russian Foreign Policy.* London: I. B. Tauris. A newer volume that shows the effects of Russian foreign policy on Central Asia especially under the hard-line administration of President Vladimir Putin and after American interest escalated in the wake of 9/11.

Tanner, Stephen. 2003. *Afghanistan: A Military History from Alexander the Great to the Fall of the Taliban.* Philadelphia: Perseus Books Group. A book that looks at the long history of war in the area now known as Afghanistan; it gives a big picture view of how war has shaped the history and geography of the country.

Part Six: Australia, New Zealand, and the Pacific Islands

Classic Works

Blainey, G. 1966. *The Tyranny of Distance.* New York: St. Martin's Press. An interesting analysis of Australia's vastness and its effect on development of the continent.

Courtney, P. P. 1982. *Northern Australia: Patterns and Problems of Tropical Development in an Advanced Country.* Melbourne: Longman Cheshire. A look at development issues and prospects in Australia's remote and environmentally difficult North.

Howe, K. R. 1984. *Where the Waves Fall: A New South Sea Islands History from First Settlement to Colonial Rule.* Honolulu: University of Hawaii Press. A good place to start learning the early history of the Pacific Islands.

Meinig, D. W. 1962. *On the Margins of the Good Earth: The South Australian Wheat Frontier, 1869–1884.* Skokie, IL: Rand McNally. A classic study of the early period of wheat farming and grazing on the frontier of the steppe.

Robinson, K. W. 1974. *Australia, New Zealand, and the Southwest Pacific.* 3rd ed. London: University of London Press. A fine overview of the whole region of Oceania, but especially of Australia.

Spate, O. H. K. 1968. *Australia.* New York: Praeger. Still a classic general study of Australia.

Recent Publications

Bassett, J., ed. 1996. *Great Explorations: An Australian Anthology.* Melbourne: Oxford University Press. A collection of writings about the great explorers of Australia's past.

Butlin, N. G. 1993. *Economics and the Dreamtime: A Hypothetical History.* New York: Cambridge University Press. A look at the often ignored topic of the Australian aborigines and their history and relations with Australia's government and white settlers.

Environment and Development: A Pacific Island Perspective. 1992. Manila: Asian Development Bank. Prepared by the South

Pacific Regional Environment Program, this report synthesizes UNCED country reports that look at sustainable development, natural resources management, and environmental protection.

Heathcote, R. L. 1994. *Australia.* London: Longman Scientific and Technical. A good human geography of the smallest continent.

Hintjens, Helen M., and **D. D. Newitt,** eds. 1992. *The Political Economy of Small Tropical Islands: The Importance of Being Small.* Exeter: University of Exeter Press. A collection of papers about various aspects of the trials and tribulations of small islands and their efforts to develop.

Kiesler, Peter, ed. 1995. *The Australian Economy: The Essential Guide.* St. Leonards, NSW, Australia: Allen & Unwin. A basic course on the economy of Australia.

Mascarenhas, R. C. 1996. *Government and the Economy in Australia and New Zealand: The Politics of Economic Policy Making.* San Francisco: Austin & Winfield. Basic reading for understanding the peculiar economic systems "down under."

McEvedy, Colin. 1998. *The Penguin Historical Atlas of the Pacific.* New York: Penguin Books. A good collection of useful historical maps of the region.

Painter, Martin. 1998. *Collaborative Federalism: Economic Reform in Australia in the 1990s.* Cambridge: Cambridge University Press. A look at the difficult challenges facing Australia's economy today and how the government and business are responding.

Pilger, John. 1991. *A Secret Country: The Hidden Australia.* New York: Knopf. An examination of the sensitive topic of race relations and social and economic conditions in contemporary Australia.

Read, Ian G. 1998. *Continent of Extremes: Recording Australia's Natural Phenomena.* Sydney: UNSW Press. A study of the continent's physical characteristics.

Terrill, Ross. 1987. *The Australians.* New York: Simon & Schuster. A highly readable, often humorous look at all aspects of Australia by one of her native sons who went to the United States and became a well-known sinologist.

Theroux, Paul. 1992. *The Happy Isles of Oceania: Paddling the Pacific.* New York: Ballantine Books. One of the most gifted travel writers today takes readers on a humorous, sometimes sarcastic, but always fascinating journey, from New Zealand to Australia through Melanesia and Polynesia. This is a view of "paradise" very different from the usual glossy travel magazine genre.

Ward, R. Gerard, and **Elizabeth Kingdon,** eds. 1995. *Land, Custom and Practice in the South Pacific.* New York: Cambridge University Press. A look at land tenure practices in the islands and their relationship to societies and cultures.

Current Publications

Bryson, L., and **I. Winter.** 1999. *Social Change, Suburban Lives: An Australian Newtown, 1960s to 1990s.* St. Leonards, NSW, Australia: Allen & Unwin in association with the Australian Institute of Family Studies. A look at urban life in modern Australia.

Fincher, R., and **P. Saunders,** eds. 2001. *Creating Unequal Futures? Rethinking Poverty, Inequality, and Disadvantage.* Crows Nest, NSW, Australia: Allen & Unwin. An examination of contemporary Australian society and the impact of economic growth.

Latham, Mark. 1998. *Civilising Global Capital: New Thinking for Australian Labor.* St. Leonards, NSW, Australia: Allen & Unwin. An examination of the challenges facing Australian workers and the economy.

Merlan, F. 1998. *Caging the Rainbow: Places, Politics, and Aborigines in a North Australian Town.* Honolulu: University of Hawaii Press. The life of aborigines in modern Australia.

Rapaport, Moshe, ed. 1999. *The Pacific Islands: Environment and Society.* Honolulu: Bess Press. One of the most detailed collections of papers on various aspects of island societies.

White, Richard, and **Penny Russell,** eds. 1997. *Memories and Dreams: Reflections on Twentieth-Century Australia.* St. Leonards, NSW, Australia: Allen & Unwin. A collection of pieces on Australia's recent history.

Part Seven: Asia

Classic Works

Allen, Douglas, ed. 1992. *Religion and Political Conflict in South Asia: India, Pakistan and Sri Lanka.* Westport, CN: Greenwood Press. An excellent collection of essays that analyze the sometimes complex relationship between religion and political processes.

Borthwick, Mark. 1992. *Pacific Century: The Emergence of Modern Pacific Asia.* Boulder, CO: Westview Press. A comprehensive and informative multidisciplinary treatment of Pacific Asia's rise in the global economy.

Ginsberg, Norton, Bruce Koppel, and **T. G. McGee,** eds. 1991. *The Extended Metropolis: Settlement Transition in Asia.* Honolulu, HI: University of Hawaii Press. A collection of papers that harnesses new theories to explain the linkages between development and recent urban growth in Asia.

Ulack, Richard, and **Gyula Pauer.** 1989. *Atlas of Southeast Asia.* New York: Macmillan. A visually attractive work with text descriptions of each country in the region.

Recent Publications

Cartier, Carolyn. 2001. *Globalizing South China.* Malden, MA: Blackwell. A conceptually rich analysis of the increasing global reach of South China, particularly its ties to Southeast Asia.

Bradnock, Robert, and **Glynn Williams.** 2003. *South Asia in a Globalizing World.* Upper Saddle River, NJ: Prentice Hall. An introduction to the economic, social, and political processes and development issues shaping the region.

Chapman, Graham P., Ashok K. Dutt, and **Robert W. Bradnock,** eds. 1999. *Urban Growth and Development in Asia, vol. 2: Living in Cities.* Aldershot: Ashgate. An edited work addressing major development issues in a wide range of Asian cities.

Farmer, B. H. 1993. *An Introduction to South Asia,* 2nd ed. New York: Routledge. A short but informative book by a geographer who has dedicated much of his professional life to this developing world region.

Hewison, Kevin, Richard Robison, and **Garry Rodan,** eds. 1993. *Southeast Asia in the 1990s: Authoritarianism, Democracy and Capitalism.* St. Leonard's, NSW, Australia: Allen and Unwin. An excellent collection of country-specific essays describing the reasons for the relative absence of democracy in a region experiencing dramatic capitalist-based economic growth.

Leinbach, Thomas R., and **Richard Ulack,** eds. 2000. *Southeast Asia: Development and Diversity.* Upper Saddle River, NJ: Prentice Hall. A distinctive and comprehensive analysis of development because this book possesses both systematic and country-specific chapters.

Linge, Godfrey, ed. 1997. *China's New Spatial Economy: Heading Towards 2020.* Hong Kong: Oxford University Press. Harnessing both systematic and regional perspectives, this short but excellent volume analyzes China's present and future economic challenges.

Matsui, Yayori. 1999. *Women in the New Asia.* London: Zed Books. A highly readable exploration of the impact of Asia's rapid economic growth of the 1990s on the status of women. Chapters include both systematic as well as specific case and country studies.

Mendelson, Oliver, and **Marika Vicziany.** 1998. *The Untouchables: Subordination, Poverty and the State in Modern India.* Cambridge: Cambridge University Press. In a sensitive account of the lives of India's poorest class, the phenomenon of the untouchables is explored from a social and political perspective.

Parnwell, Michael J. G., and **Raymond L. Bryant,** eds. 1996. *Environmental Change in South-East Asia: People, Politics and Sustainable Development.* London: Routledge. Based upon detailed case studies from many countries, this volume highlights the central role of politics in environmental degradation, as well as the quest for sustainable development.

Raju, Saraswati, and **Deipica Bagchi,** eds. 1994. *Women and Work in South Asia: Regional Patterns and Perspectives.* New York: Routledge. An informative regional and interregional study examining the religious, cultural, and economic constraints that define women's work.

Rodan, Garry, Kevin Hewison, and **Richard Robison,** eds. 1997. *The Political Economy of Southeast Asia.* New York: Oxford University Press. Based on six country studies and three systematic chapters, this work examines the important political dimensions of economic development.

Stephens, Stanley F. 1993. *Claiming the High Ground: Sherpas, Subsistence and Environmental Change in the Highest Himalaya.* Berkeley, CA: University of California Press. A comprehensive examination of the cultural ecology of sherpas, environmental sustainability, and tourism in Nepal.

Yeung, Yue-man, ed. 1993. *Pacific Asia in the 21st Century: Geographical and Developmental Perspectives.* Hong Kong: The Chinese University Press. A theoretically based examination of Pacific Asia's urban, demographic, transport, agricultural, and environmental contours.

Weightman, Barbara. 2000. *Geography of Asia: Of Dragons and Tigers.* New York: John Wiley. An innovative description and analysis of the entire Asia region with a sociocultural and economic focus.

Current Publications

Cannon, Terry, ed. 2000. *China's Economic Growth: The Impact on Regions, Migration, and the Environment.* London: Macmillan. This edited work addresses the changing demographic and environment contours of post-reform China.

Education about Asia. Published three times a year by the Association for Asian Studies, this multidisciplinary and teaching-oriented journal offers short but useful articles addressing a variety of historical and modern issues.

Far Eastern Economic Review. A monthly publication similar to *Time* and *Newsweek* that covers an array of timely economic, political, and social issues throughout Asia.

Rigg, Jonathan. 2003. *Southeast Asia: The Human Landscape of Modernization and Development,* 2nd ed. New York: Routledge. An analysis of the changing nature of development in Southeast Asia with a strong focus on agriculture.

Japan

Classic Works

Association of Japanese Geographers, eds. 1980. *Geography of Japan.* Tokyo: Teikoku-Shoin. A collection of writings about Japan's geography by members of the Association of Japanese Geographers; a somewhat specialized but very useful supplement to Glenn Trewartha's book (listed separately). A few of the articles are now a little dated.

Burks, Ardath W. 1991. *Japan, A Postindustrial Power,* 3rd ed. Boulder, CO: Westview Press. A fine overview of Japan's history, culture, and economic development and problems.

Christopher, Robert C. 1983. *The Japanese Mind: The Goliath Explained.* New York: Linden Press. A timeless study of Japan in all its complexity by a distinguished journalist. As valid today as when it was written.

Masai, Yasuo, ed. 1986. *Atlas Tokyo: Edo/Tokyo through Maps.* Tokyo: Heibonsha. Beautiful full-color maps and photographs detail the history and character of Tokyo from earliest times to the present, including detailed district maps that show every street and alley. Text in Japanese and English. Essential for the study of Tokyo.

Reischauer, Edwin O. 1970. *Japan: The Story of a Nation.* New York: Alfred A. Knopf. A fine analysis of Japan's historical development, written by the late dean of Japanologists and former U.S. ambassador to Japan, all of whose numerous books on Japan are now standard references.

Seidensticker, Edward. 1983. *Low City, High City: Tokyo from Edo to the Earthquake.* New York: Alfred A. Knopf. A fascinating account by a leading scholar of Japan about how the Shogun's ancient capital became a great modern city in the period 1867–1923.

——————. 1990. *Tokyo Rising: The City since the Great Earthquake.* Tokyo: Charles Tuttle. Seidensticker picks up where he left off in *Low City, High City,* presenting the development of Tokyo after 1923. More than a portrait of Tokyo, this is a social history of modern Japan, and wonderfully readable.

Trewartha, Glenn T. 1965. *Japan: A Geography.* Madison: University of Wisconsin Press. The basic geography of Japan by the late dean of American geographers specializing on Japan. The volume is still useful, especially in its treatment of the physical geography, environment, and resource base of Japan.

Recent Publications

Argy, Victor, and **Leslie Stein.** 1997. *The Japanese Economy.* New York: New York University Press. One of the more recent efforts to look at Japan's economy, its problems, and government policies.

Beasley, W. G. 1995. *The Rise of Modern Japan: Political, Economic and Social Change since 1850,* 2nd ed. New York: St. Martin's Press. A comprehensive analysis of Japan's passage from largely unknown agrarian state to troubled economic superpower. Perfect overview for the nonspecialist.

Cybriwsky, Roman. 1991. *Tokyo.* New York: Macmillan. A volume in the World Cities Series, this study takes an original approach to the urban geography of Japan's largest city by trying to convey the "essential texture" of the city—what it is like to live there. A good read, lavishly illustrated (black and white).

Jinnai, Hidenobu. 1995. *Tokyo: A Spatial Anthropology.* Berkeley: University of California Press. An enthusiastic exploration of the historical and social factors that help explain how Tokyo acquired its present shape.

Karan, P. P., and **Kristin Stapleton,** eds. 1997. *The Japanese City.* Lexington, KY: University of Kentucky Press. A collection of ten essays by Japanese and American geographers on various aspects of the Japanese city. Well illustrated and covers the key issues.

Mather, Cotton, P. P. Karan, and **Shigeru Iijima.** 1998. *Japanese Landscapes: Where Land and Culture Merge.* Lexington, KY: University of Kentucky Press. An intriguing interpretation of Japan's cultural geography, examining the landscape from the perspective of "primary" and "secondary" characteristics. Lavishly illustrated.

Okabe, Mitsuaki, ed. 1995. *The Structure of the Japanese Economy: Changes on the Domestic and International Fronts.* New York: St. Martin's Press. An examination of the economy since 1945 and its relationship to Japan's foreign economic relations.

Popham, Peter. 1985. *Tokyo: The City at the End of the World.* Tokyo: Kodansha International. A longtime foreign resident of Tokyo writes with humor and remarkable insight about life in Japan's capital city. Fun reading.

Uriu, Robert M. 1996. *Troubled Industries: Confronting Economic Change in Japan.* Ithaca, NY: Cornell University Press. A look at the problem of sunset industries and how the Japanese companies and government have responded to the challenge.

Current Publications

Fujita, Kuniko, and **Richard Child Hill.** 1993. *Japanese Cities in the World Economy.* Philadelphia: Temple University Press. A study of various Japanese cities in relation to Japan's economic growth over recent decades.

Kosaka, Kenji, ed. 1994. *Social Stratification in Contemporary Japan.* London and New York: Kegan Paul International. A study of Japanese social stratification by a group of Japanese sociologists.

Maher, John C., and **Gaynor Macdonald,** eds. 1995. *Diversity in Japanese Culture and Language.* London and New York: Kegan Paul International. A collection of fourteen papers focused around the themes of homogeneity versus heterogeneity in Japanese culture. The myth of Japanese homogeneity is well covered here.

Smith, Patrick. 1997. *Japan: A Reinterpretation.* New York: Random House. A penetrating and sometimes dark personal interpretation of all aspects of modern Japan.

Smitka, Michael, ed. 1998. *Agricultural Growth and Japanese Economic Development.* New York: Garland Press. An analysis of agriculture's role within the Japanese economy.

Togo, Yukiyasu, and **William Wartman.** 1993. *Against All Odds: The Story of the Toyota Motor Corporation and the Family That Created It.* New York: St. Martin's Press. Part biography, part history of Japan's largest corporation and motor vehicle producer.

Part Eight: The Middle East and North Africa

Classic Works

Altorki, Soraya, and **Donald P. Cole.** 1989. *Arabian Oasis City: The Transformation of Unayzah.* Austin: University of Texas Press. Collaboration between a male and a female anthropologist overcomes the gender segregation of Saudi Arabian society to provide a balanced perspective on contemporary economic and social change.

Beaumont, Peter, Gerald H. Blake, and **J. Malcolm Wagstaff.** 1988. *The Middle East: A Geographical Study.* 2nd ed. New York: Halsted Press. A systematic examination of problem areas and issues blended with country case studies to give a most reliable and readable overview of the contemporary Middle East.

Fernea, Elizabeth Warnock. 1965. *Guests of the Sheik: An Ethnography of an Iraqi Village.* New York: Doubleday (Anchor Books). Although more than 40 years have passed since Fernea wrote her picture of life, particularly among the women, in a Shi'ite village in southern Iraq, the story's images of daily activities, cross-cultural communication, gendered activities and space, and religion retain a remarkable freshness.

Mahfouz, Naguib. 1981. *Midaq Alley.* Translated by Trevor Le Gassick, Washington, D.C.: Three Continents. A powerful novel by Egypt's best-known author and Nobel prize winner for literature, this is an unforgettable and unsentimental portrait of the everyday lives of a spectrum of Cairo's central-city inhabitants.

Mernissi, Fatima. 1991. *The Veil and the Male Elite: A Feminist Interpretation of Women's Rights in Islam.* Translated by Mary Jo Lakeland. Reading, MA: Addison-Wesley. By providing a critical analysis of the religious and historical background of the status of females in Islam, Mernissi challenges the exclusion of women from public life and the legal dominance of women by men.

Robinson, Francis. 1982. *Atlas of the Islamic World Since 1500.* New York: Facts on File. Richly illustrated and accompanied by a literate and knowledgeable text, this atlas is much more than a mere collection of maps; it places the Middle East and North Africa in the larger context of the greater Islamic realm.

Recent Publications

Allan, J. A. 2001. *The Middle East Water Question: Hydropolitics and the Global Economy.* London and New York: I. B. Tauris. No region has more pressing water availability problems than does the Middle East. Allan evaluates water availability and management and extrapolates the political uncertainties of current water policy and practice into an uncertain future.

Bengio, Ofra, and **Gabriel Ben-Dor,** eds. 1999. *Minorities and the State in the Arab World.* Boulder, CO: Lynne Rienner. The varying trajectories of major Middle Eastern minorities are traced in this collection of insightful essays.

Bonine, Michael E., ed. 1997. *Population, Poverty, and Politics in Middle East Cities.* Gainesville: University Press of Florida. Rapid urbanization makes cities the flashpoint for political protest. This growth in most Middle Eastern cities exceeds the capabilities of all but the wealthiest oil states to cope with rapid change.

Chatty, Dawn. 1996. *Mobile Pastoralists: Development Planning and Social Change in Oman.* New York: Columbia University Press. Ten years of research and involvement in development aid projects among the Harasiis tribe of southern Oman result in an insightful examination of the benefits of bottom-up development. Particularly notable is the role played by pastoral women in both decision-making and development initiatives, a position not readily accepted or understood by development bureaucrats.

Collins, Robert O. 2002. *The Nile.* New Haven, CT: Yale University Press. From East African headwaters to Mediterranean outlet, the political and environmental history of the world's longest river and its overwhelming importance to the people who live along its course is portrayed with passion and knowledge.

Esposito, John L., ed. 1999. *The Oxford History of Islam.* Oxford: Oxford University Press. Sixteen scholars, both Muslim and non-Muslim, comment on both contemporary and historic aspects of Islam, including its development, legal systems, science, philosophy, theology, relations with other religions, response to colonialism, and involvement in a globalized political and economic system.

Hobbs, Joseph J. 1995. *Mount Sinai.* Austin: University of Texas Press. A superb example of regional geography at its best, this study successfully evokes the powerful sense of place conveyed by a site sacred to Jew, Christian, and Muslim. How to protect this special sacred place from the encroachment of modern culture and the pressures of tourist development is the subject of Hobbs's book.

Little, Douglas. 2002. *American Orientalism: The United States and the Middle East Since 1945.* Chapel Hill and London: University of North Carolina Press. Who says that diplomatic history has to be dull? Written with verve and insight, this study is at the same time scholarly and readable. For anyone confused about how the United States achieved its current relationship to the Middle East, this book is a must read.

Rogers, Peter, and **Peter Lydon,** eds. 1994. *Water in the Arab World: Perspectives and Prognoses.* Cambridge, MA: The Division of Applied Sciences, Harvard University. A comprehensive examination of water use and water availability in the major regions of the Middle East (Maghreb, Nile Valley, eastern Mediterranean, Arabian Peninsula), this collection of papers also examines alternative water sources, conflicts, and future prospects for a severely water-constrained region.

Rubin, Barry, and **Judith Cope Rubin,** eds. 2002. *Anti-American Terrorism and the Middle East: A Documentary Reader.* Oxford: Oxford University Press. In the aftermath of the September 11, 2001 terrorist attacks, this collection of primary documents, speeches, newspaper articles, court records, and radio broadcasts provides perspective and viewpoints that explain the origins and impact of violent extremism in the Middle East.

Scoilino, Elaine. 2000. *Persian Mirrors: The Elusive Face of Iran.* New York: The Free Press. More than simply informed travel reporting by an experienced correspondent, this book exposes many of the contradictions encountered in the complicated social, political, and cultural mosaic that is contemporary Iran, a volatile state in which both Islam and democracy are pursued with gusto and commitment.

Swearingen, Will D., and **Abdellatif Bencherifa,** eds. 1996. *The North African Environment at Risk.* Boulder, CO: Westview Press. Fifteen essays examine the state of the environment in Mauritania, Morocco, Algeria, and Tunisia and find that land degradation is serious and widespread.

Current Publications

Beattie, Andrew. 2005. *Cairo: A Cultural History.* Oxford: Oxford University Press. The vibrant history of the "mother of cities" and its distinctive blend of cultures, both local and foreign, is presented with panache.

ESCWA (United Nations Economic and Social Commission for Western Asia). 2003. *Water Scarcity in the Arab World.* New York: United Nations. In 35 succinct pages, this slim document summarizes the major implications of growing water demand and the options available for policy makers in the Middle East.

Ibrahim, Fouad, and **Barbara Ibrahim.** 2003. *Egypt: An Economic Geography.* London: I. B. Tauris. More than questions about economic geography are answered in this up-to-date survey of the state of Egypt's environment, agriculture, urban life, cultural history, and contemporary quality of life.

Lewis, Bernard. 2002. *What Went Wrong? Western Impact and Middle Eastern Response.* Oxford: Oxford University Press. This slim volume is something of a prequel to the murderous events of September 11, 2001, in that Lewis outlines the political frustrations and cultural challenges experienced by the Muslim Middle East and the variety of responses that the West's economic, political, and cultural penetration has elicited.

Part Nine: Africa South of the Sahara

Classical Works

Best, Alan, and **Harm DeBlij.** 1977. *African Survey.* New York: John Wiley. A regional survey of the physical, historical, and cultural attributes of Africa.

Davidson, Basil. 1994. *Modern Africa: A Social and Political History.* London: Longman. A review of the social and political history of Africa from the colonial era through the liberation movements to independence and beyond.

Greenberg, Joseph. 1995. *Languages of Africa,* 2nd ed. Johannesburg: Witwatersrand University Press. A study in African linguistic classification.

Mazrui, Ali. 1986. *The Africans: A Triple Heritage.* Boston: Little, Brown. Analyzes the indigenous, Islamic, and European heritages in the context of Africa.

Stamp, Dudley. 1953. *Africa: A Study in Tropical Development.* New York: John Wiley. A survey of the history, climate, relief, vegetation, hydrology, culture, diseases, and development of Africa, including a detailed analysis of countries and regions.

Recent Publications

Abdi, Ismail Samatar. 1999. *An African Miracle: State and Class Leadership and Colonial Legacy in Botswana Development.* Portsmouth, NH: Heinemann. Examines capitalism, colonialism, and the role of chiefs in the context of political and economic development in Botswana.

Aryeetey-Attoh, Samuel, ed. 2003. *Geography of Sub-Saharan Africa,* 2nd ed. Englewood Cliffs, NJ: Prentice Hall. A multifaceted, thematic approach to the physical and human geography of Sub-Saharan Africa.

Bassett, Thomas, and **Donald Crummey,** eds. 2003. *African Savannas: Global Narratives and Local Knowledge of Environmental Change.* Oxford: James Currey. An interdisciplinary collection of essays on land use, environmental fluctuations, political and pastoral ecologies, social differentiation, communal ownership, colonial and post-colonial developments, land tenure reform, and resource management.

Bloom, D. E., and **J. Sachs.** 1998. "Geography, Demography, and Economic Growth in Africa." *Brookings Papers on Economic Activity* 2: 207–273. Examines the relative roles of geography, demography, and policy in Africa's recent growth experience.

Castro, Peter. 1994. "Cousins and Gunmen: Crisis and Resilience in Somalia." *The World and I* 9: 268–279. Examines the reasons behind clan warfare in Somalia.

Connah, Graham. 2001. *African Civilizations: An Archaeological Perspective.* Cambridge, UK; New York: Cambridge University Press. A comprehensive review of social, economic, and political characteristics of ancient, medieval, and late medieval civilizations in Africa.

Dietz, A., R. Ruben, and **A. Verhagen.** 2004. *The Impact of Climate Change on Drylands: With a Focus on West Africa.* Dordrecht; Boston: Kluwer Academic. Examines the effect of greenhouse gases on dryland regions in West Africa.

Europa Publications. 2004. *Africa South of the Sahara.* London: Europa. An annual series and reference book on all Sub-Saharan countries.

Grant, Richard, and **John Agnew.** 1996. "Representing Africa: The Geography of Africa in World Trade, 1960–1992." *Annals, Association of American Geographers* 86 (4): 729–745. Utilizes neoclassical, dependency, and politico-economic theories to explain drawbacks associated with Africa's participation in international trade.

International Union for Conservation of Nature (IUCN). 1990. *Biodiversity in Sub-Saharan Africa and Its Islands: Conservation, Management and Sustainable Use.* Cambridge, England: IUCN. A comprehensive analysis of the species and biodiversity in the continent.

Kalipeni, E., S. Craddock, J. Oppong, and **J. Ghosh,** eds. 2004. *HIV/AIDS in Africa: Beyond Epidemiology.* Oxford: Blackwell. A collaborative research effort that examines further our understanding of the historical, spatial, social, and ethical dimensions of HIV/AIDS in Africa.

Kevane, Michael. 2004. *Women and Development in Africa: How Gender Works.* Boulder, CO: Lynne Rienner. Provides a synthesis of relevant topics, theories, and empirical findings that focus on such issues as the role of gender in economic transactions, the control of labor, bargaining power within households, and gender-sensitive development programs.

Konadu-Agyemang, Kwadwo, ed. 2001. *IMF and World Bank Sponsored Structural Adjustment Programs in Africa.* Aldershot, UK: Ashgate. Evaluates the impact of structural adjustment programs on the multiple sectors of the Ghanaian economy.

Lemon, Anthony, ed. 1995. *The Geography of Change in South Africa.* Chichester, England: John Wiley. Collection of essays addressing constitutional issues, domestic policy questions for the new government, and the role of South Africa in southern Africa.

Lumumba-Kasongo, Tukumbi. 1992. "Zaire's Ties to Belgium: Persistence and Future Prospects in Political Economy." *Africa Today* 39 (3): 23–48. Reviews Democratic Republic of Congo's colonial heritage, the authenticity program, and the dilemmas of dependency.

Moseley, William, and **B. Ikubolajeh Logan,** eds. 2004. *African Environment and Development: Rhetoric, Programs, Realities.* Burlington, VT: Ashgate. Explores the interrelationships between African rural livelihoods, environmental integrity, and broader scale political economy.

Murray, Jocelyn. 1998. *Cultural Atlas of Africa.* Amsterdam: Elsevier International Projects. Portrays African history, culture, and society in maps.

Nel, Etienne, Tony Binns, and **Nicole Motteux.** 2001. "Community-based Development, Non-governmental Organizations and Social Capital in Post-apartheid South Africa." *Geografiska Annaler, Series B: Human Geography* 83 (1): 3–13. Examines the role of social capital and nongovernmental organizations in encouraging and assisting community-based initiatives.

Newman, James. 1995. *The Peopling of Africa: A Geographic Interpretation.* New Haven, CT: Yale University Press. Detailed historical analysis of African cultures, antiquities, and civilizations.

Rapley, John. 1993. *Ivoirien Capitalism: African Entrepreneurs in Côte d'Ivoire.* Boulder, CO: Lynne Rienner. Review of the development of entrepreneurship in Côte d'Ivoire.

Rogerson, Christian. 2002. "Spatial Development Initiatives in South Africa: Elements, Evolution and Evaluation." *Geography: Journal of the Geographical Association* 87 (1): 38–49. Examines the extent to which spatial development initiatives have increased private sector investments; stimulated the growth of micro-, small-, and medium enterprises; and enhanced the empowerment of target communities in South Africa.

Schroeder, Richard. 1999. "Geographies of Environmental Intervention in Africa." *Progress in Human Geography* 23 (3): 359–379. Analysis of the geographical assumptions underlying environmental interventions in Africa, including the political problems of managing natural resources under divergent ecological conditions, spatial conflicts associated with land-use strategies, and political ecological relationships resulting from the commodification of natural resources.

Sweetman, Caroline. 2003. *Gender, Development and Poverty.* London: Oxfam Publishing. Explores the complex links between poverty and inequality between women and men within the context of credit and savings schemes, formal education, and intergenerational vulnerability to HIV/AIDS.

United Nations Center for Human Settlements (UNCHS) (HABITAT). 2001. *Cities in a Globalizing World.* London: Earthscan.

Current Publications

Allen, T., and **S. Heald.** 2005. "HIV/AIDS Policy in Africa: What has Worked in Uganda and What has Failed in Botswana?" *Journal of International Development* 16 (8): 1141–1154. Report on how Uganda and Botswana responded to the AIDS crisis through massive public campaigns, church groups, local councils, and traditional healers.

Mercer, C., G. Mohan, and **M. Power.** 2003. "New Perspectives on the Politics of Development in Africa." *Geoforum,* Special Issue, 34 (4): 417–498. Conceptual and empirical contributions focus on questions of civil society, community participation, decentralization and democracy, good governance, human rights, and disability.

Moseley, William. 2005. "Global Cotton and Local Environmental Management: The Political Ecology of Rich and Poor Smallhold Farmers in Southern Mali." *The Geographical Journal* 171 (1): 36–55. Analyzes local and extra-local factors, including broader-scale economic and political processes, which influence household agricultural management approaches in southern Mali.

Rogerson, Christian. 2000. "The Economic and Social Geography of South Africa: Progress Beyond Apartheid." *Tijdschrift voor Economische en Sociale Geografie* 91 (4): 335–346. Analyzes changing social and economic landscapes in post-Apartheid South Africa in the context of uneven geographical patterns of development, new population (migration) dynamics, the workings of new government policies, and new regional configurations.

Salopek, Paul. 2003. "Shattered Sudan: Drilling for Oil, Hoping for Peace." *National Geographic Magazine* 203 (2): 30–59. Looks at how environmental, economic, social, and political forces have balkanized Sudan.

Weiner, Daniel. 2003. "A Geographer Reflects on Africa." *African Geographical Review* 22: 5–14. A reflection on the sociocultural, political, and economic problems that confront Africa.

aborigines   Descendants of the Negroid inhabitants occupying Australia at the time of European settlement.

acculturation   The cultural modification of a group because of contact with other cultures; the merging of cultures through prolonged contact.

acid rain   A precipitation that is sufficiently toxic to diminish soil fertility, damage standing crops, poison lakes, kill fish, eat away medieval stone structures, and ruin paint finishes on automobiles. Acid precipitation occurs as various substances, including sulfur and carbon, are released into the air as a consequence of burning huge amounts of fossil fuels such as coal and petroleum products. These substances mix with water vapor in the atmosphere and fall to the earth as acidic precipitation.

African-American migration   The movement of blacks in the United States from rural to urban areas.

African National Congress (ANC)   A black political group in South Africa, committed to majority rule and abolition of apartheid; victorious in 1994 national elections, the first in which blacks could participate.

Afrikaners   South African natives of Dutch (Boer) descent.

Agricultural Revolution   A period characterized by the domestication of plants and animals and the development of farming.

agricultural surplus   With domestication of plants and animals, agricultural productivity soared to the point that some farmers produced more than they consumed, creating surpluses.

agroforestry   The sustainable harvesting of food and other forest products.

Ainu   The proto-Caucasian population of ancient northern Japan; still found in small numbers on the island of Hokkaido.

al-Jazirah   "The Island"; The region of land dominating central Iraq, comprising the land between the Tigris and Euphrates Rivers.

alluvium   Material, often very fertile, that has been transported and deposited by water.

alpha world city   A large city that plays a major economic role in the world.

Alps   Mountain region of southern Europe.

Altiplano   A high plateau or plain; the high plateau of Bolivia and Peru, which contains most of the Bolivian population.

Amu Darya River   A river in Uzbekistan that flows into the Aral Sea.

Andalusia   Southern Spanish province that reflects a strong African and Moorish cultural legacy.

Andean Highlands   Regions of western South America.

Angkor Wat   Area of Cambodia where monumental ruins stand as testimony to the agricultural productivity and trade capabilities of this once-powerful empire.

Antioqueños   Residents of Antioquia, or the area around Medellín in Colombia; considered economically aggressive people.

apartheid   The former policy of the South African government that maintained strict white-nonwhite segregation.

Appalachian Highlands   The Appalachian Mountains region of the eastern United States, constituted in part of the Blue Ridge/Great Smokies and Ridge and Valley physiographic provinces.

Appalachian Plateau   A landform region lying to the west of the Ridge and Valley province. It is particularly rich in bituminous coal. The northern portion is sometimes referred to as the Allegheny Plateau and the southern portion as the Cumberland Plateau.

Appalachian Regional Development Act   Federal legislation adopted in 1996 and amended in 2004, intended to address the dire economic circumstances of Appalachian counties.

aquifers   Underground, water-bearing rock strata.

Arabian Peninsula   Body of land in the Middle East that is surrounded by the Red Sea to the west, the Gulf of Aden and the Arabian Sea to the south, and the Persian Gulf and the Gulf of Oman to the east. Contained within it are the countries of Bahrain, Kuwait, Oman, Qatar, Saudi Arabia, United Arab Emirates, and Yemen.

Arabs   The more than 150 million Semitic people of the Middle East and North Africa who share a common language, cultural history, and religion (Islam).

Aral Sea   A sea in the southern drylands of Kazakhstan and Uzbekistan that is rapidly shrinking due to massive water diversion from its two river sources for desert irrigation projects.

area studies tradition   A geographic perspective that emphasizes the study of specific regions and an understanding of the varied aspects of those regions.

aridity   Dryness; the dominant climatic characteristic of the world's desert and semidesert regions. Generally found in interior continental locations or in coastal areas bordering cold ocean currents.

Aristotle   (384–322 B.C.) First Greek geographer to divide the world into three broad climatic zones. Author of *Meteorologica*, in which he discussed the physical characteristics of the earth.

ASEAN Free Trade Area   AFTA was established in 1992 to reduce tariffs on trade between member countries of ASEAN (see *Association of Southeast Asian Nations*).

Ashkenazim   Jews originating from Northern, Western, and Eastern Europe.

Asian values   A set of values that are considered by many scholars to have contributed to the accelerated economic growth of the Pacific Rim nations. The values include collective action, loyalty, respect for authority, education, hard work, and frugality.

assimilation   The complete modificiation of an individual's or group's culture through contact with a dominant culture.

Association of Southeast Asian Nations (ASEAN)   A political-economic organization formed in 1967 to promote cooperation among member nations.

Aswan High Dam   A large dam project that was begun on the Nile River at Aswan in the 1960s. The goal of the dam project, or New Valley Project, was to increase Egypt's agricultural productivity by increasing the number of growing seasons on the same land and through the irrigation of arid lands.

Atacama Desert   The world's driest desert, in northern Chile, with many locales not having received measurable precipitation in more than 50 years.

average annual precipitation   The average total precipitation during the year, expressed in inches or centimeters. Precipitation figures are presented in rainfall equivalents but include all forms of precipitation.

Ayatollah   An honorific title held by the most prestigious group of Iranian religious scholars, the mujtahids. Ayatollahs are held to be crucial to Iran's government, for they are regarded as the primary protectors of both spiritual values and secular justice.

Aymara   One of two Indian ethnic and linguistic groups of Peru and Bolivia.

Aztec   One of the four high civilizations of pre-Colombian Latin America, centered on the area around present-day Mexico City.

Ba'ath Party   The political party that controlled Iraq from 1963–2003. Its last leader was Saddam Hussein.

balkanization   The breakup or fragmentation of a large political unit into several smaller units, such as that which occurred in the Ottoman and Austro-Hungarian empires during the nineteenth and twentieth centuries.

bantustans   Attempted independent black homelands in South Africa during the apartheid era.

Basin and Range   Bordered by the higher Colorado Plateau to the east and the Sierra Nevada Mountains to the west, the region consists of mountain ridges separated by valleys or basins. The broken terrain and relative isolation of the region have created drainage conditions characterized by rivers without exits to the sea and isolated saline lakes, such as the Great Salt Lake.

bazaar   A portion of a Middle Eastern city, such as a street or cluster of streets with covered roofs where shopkeepers and merchants are located, but also serving as an informal network of family, craft, and professional associations that manage and control the major traditional industrial and commercial activities of the economy.

Bedouin   Arabs who live by nomadic herding in the deserts of North Africa and the Middle East.

behavioral geography   A branch of human geography that examines human decision making and behavior.

Berbers   A pre-Arabic culture group of Morocco and Algeria, many of whom now speak Arabic and are Muslims.

Berlin Conference   The 1884 conference among European powers that marked the beginning of the formal era of colonialism in Africa by dividing Africa among these powers.

Bharatiya Janata Party (BJP)   A conservative Hindu-based political party in India which often has been reluctant to compromise with Muslim and Sikh interests. This has slowed efforts by some to seek common approaches to development challenges.

birthrate   The number of births occurring in a given year per 1,000 people.

Black Hills   A low mountainous area surrounded by the northern Great Plains in the western reaches of South Dakota.

Blue Ridge Mountains   The highest chain of mountains within the Appalachian Highlands. Physically, it is the same as the Great Smokies, which is the name for the mountains in southern Appalachia.

Bolsheviks   A Russian word meaning "majority," now taken to mean a communist, or adherent of communism.

Border Cities   A region of Mexico in which the economy and culture have been greatly altered by proximity to the United States.

bracero   Mexican agricultural worker legally working in the United States between 1942 and 1964 through an agreement between the U.S. and Mexican governments.

brain drain   The trend of highly educated foreigners to migrate to other more technologically advanced nations.

Brazilian Highlands   A vast interior upland region of Brazil rich in iron ore and other industrial minerals.

bride wealth   Goods or money transferred from the bridegroom's family to the bride's family to compensate for the loss of her labor.

burakumin   A social minority in Japan occupying the bottom of the social and economic ladder.

bustee   A "village in the city" in India.

calcification   A process occurring in dry regions where limited precipitation results in less leaching of soluble materials and thus the accumulation of calcium carbonates in the soil.

Callejón Andino   An area of the Andean highlands of Ecuador consisting of a series of basins and their surrounding mountains.

Canadian Shield   A relatively smooth glaciated land surface region that nearly encircles the Hudson Bay and extends southward to the Great Lakes area of the United States. Consists of thin soils, stony surfaces, and areas with no soil at all.

Carajás   An area of north Brazil, now under development as the largest mineral district in the world. A leading mineral is iron ore.

Carpathian Mountains   Range of eastern European mountains that arch through the countries of Poland, Czechia, Slovakia, the Ukraine, and Romania.

Cascade Mountains   A high, volcanic north-south trending mountain range in western Oregon and Washington states.

cash cropping system   The result of Western colonial powers forcing peasant farmers to dedicate a portion of their farmland to the cultivation of export crops.

caste system   A rigid system of social stratification based on occupation, with a person's position passed on by inheritance; derived from the Hindu culture.

Castilians   The people of Spain's central Meseta (the tableland) who built and ruled the country, created its capital, Madrid, and dominated the ethnic minorities in the peripheries.

Caucasus Mountains   Two ranges (Greater Caucasus and Lesser Caucasus) that stretch from the Black Sea to the Caspian Sea.

Celts   Descendants of central European tribes of peoples who arrived in the British Isles and the Brittany area of France around 2000 B.C. The Celts developed languages distinctive from others in western Europe; however, numbers of native speakers of Celtic languages (Breton, Scots Gaelic, Irish Erse, and Welsh) are in decline today.

centralization   Important to the development of Central Asia during the Soviet era. The government in the USSR's capital, Moscow, made all political, economic, and cultural decisions and then put them into practice throughout the USSR.

Central Lowlands   Portion of the Interior Plains of the United States; located largely in the Ohio River Valley; also the interior physiographic provinces of South America, including the Llanos, Amazonian Plains, Gran Chaco, and Pampas.

Central Valley of Chile   The temperate, fertile lands of central Chile; most of the population, industry, and commercial agriculture of Chile are located here.

Chaco   Lowland plain; specifically, the *Gran Chaco* of Argentina, Paraguay, and Bolivia.

chaebol   A select few megacorporations in South Korea.

Chernobyl   Site of a nuclear plant disaster in 1986 in the former Soviet Union that has left a sizable portion of Ukraine and Belarus contaminated.

chernozem   A Russian term referring to the fertile black soils of the steppe or semiarid zone of Russia, Ukraine, and Kazakhstan; similar soils in other parts of the world.

Chibcha   One of the pre-Columbian high-culture groups of the New World; at the time of the Spanish Conquest these peoples lived in agricultural villages in the eastern Andes of Colombia.

child fosterage   System in which children are sent from their birth parents to be raised and cared for by their grandparents or foster parents.

Chinese Communist Party (CCP)   A nationalistic and socialist movement founded in Shanghai in 1921.

Chinese diaspora   The migration of large numbers of Chinese from China to other countries.

chorologic framework   Referring to place (the framework for the study of geography).

chronological framework   Referring to time (the framework for the study of history).

circular causation   A development theory based on an upward- or downward-spiraling effect.

City Development Index (CDI)   Measures the level of development in cities by five indices: infrastructure, waste treatment and disposal, health, education, and economic output; used to assess how effective governments are in delivering services and enhancing well-being.

city-state   A sovereign country consisting of a dominant urban unit and surrounding tributary areas.

clanocracy   The place of clan families in defining the political and legal status of people in Africa.

climate   The average temperature, precipitation and wind conditions expressed for an extended period of years; prevailing conditions over time.

Coast Ranges   A series of low, north-south trending mountains paralleling the Pacific Ocean in western California, Oregon, and Washington.

coca   South American shrub, the leaves of which are the source of cocaine.

cold war   The period of hostility, just short of open warfare, between the Soviet Union and the United States and its allies, essentially from 1945 to the mid-1960s, though its end is often dated as 1989 with the fall of the Berlin Wall.

collective farms   One of two forms of government-organized and supervised large-scale agricultural organizations in the former Soviet Union and Eastern Europe; a collective leases land from the government, and workers receive a share of net returns to the organization.

collectivization of agriculture   The process of forming collective or communal farms, especially in communist countries; nationalization of private landholdings.

colonialism   The system by which several powers maintained foreign possessions, usually for economic exploitation; most prevalent from the sixteenth through the mid-twentieth centuries; essential to understanding contemporary development in many developing regions.

Colorado Plateau   One of the interior plateaus of the western United States; a high sedimentary rock plateau in Colorado, Utah, Arizona, and New Mexico.

Columbia Plateau   An interior plateau in eastern Washington, Oregon, and Idaho, formed of thick layers of lava through which the Snake River has carved a canyon.

COMESA   Common Market for Eastern and Southern Africa; twenty-two countries united to facilitate trade by granting preferential tariffs on selected goods. Members also seek to enhance cooperation in agriculture, industry, telecommunications, and monetary affairs.

command economy   Centrally controlled and planned livelihood system; the best example is the communist form of economic organization.

communalism   An uncompromising allegiance to a particular ethnic or religious group.

Communist Party   In the former Soviet Union, the Communist Party's original purpose was to advance the social revolution; in practice, the party also took control over the soviets—the governing councils.

comparative advantage   The idea that a given area gains by specializing in one or more products for which it has particular relative advantages; leads to trade to obtain other needed commodities.

confederation   A loosely formed federal state.

Confucianism   The philosophy based on the writings and teachings of Confucius, which emphasized the importance of living a virtuous life characterized by obligations to others.

conquistadores   The Spanish conquerors of America, particularly sixteenth-century Mexico and Peru.

conservatives   Colombian faction from Antioquia and its regional capital of Medellín, who were in conflict with the liberals of Bogotá.

continental climate   A climate defined by the extreme heating and cooling characteristics of land rather than water, with hot summers and cold winters.

continentality   A measure of distance from the oceans. Generally speaking, the more inland, or continental a location, the drier its climate and the greater its temperature range are likely to be.

core-periphery model   A variation on the dependency theory emphasizing the spatial dimensions of global trade patterns. The model argues that the nations of the world can be divided into two groupings: a "core" (which consisted originally of the countries of Western Europe) and a "periphery" (comprised initially of Africa, Asia, and Latin America). This model holds that the economic development of the core has been, and continues to be, achieved through trade relations that work to the disadvantage of the less-industrialized regions of the periphery.

core-periphery pattern   Denotes the concentration of an activity or phenomenon near a central area or nucleus (the core) and a weakening of this activity or phenomenon as one moves away from the core and into the periphery. The core tends to be more representative of a region, while the periphery tends to exhibit fewer defining traits.

core-periphery relationship   During the colonial period in Asia, a relationship evolved in which the more modern economic structures of mines, plantations, or colonial urban centers developed in contrast to the surrounding traditional subsistence agricultural economies.

counterurbanization   Occurs when sizable numbers of people leave cities and move to rural areas because improvements in transportation and telecommunications technology allow them to reside outside of the urban core. This phenomenon has caused some larger cities to stagnate or decline in population, while some surrounding villages and small towns grow. This trend began in Europe during the mid-1960s.

creative destruction   An active transformation in which people alter their habitat to produce a new, human-dominated system that meets human needs. Something must be destroyed in order to create the conditions that promote benefits for humankind.

Creoles   Collective name for blacks who sought residence in the urban centers of Guyana, Suriname, and French Guiana.

criollos   Persons of Spanish descent born in Spanish America.

crossroads location   A place where contrasting cultures meet and often clash.

cultural convergence   The idea that the way in which people live tends to become more similar as development occurs around the world.

cultural determinism   The theory that a person's range of action—food preference, desirable occupation, rules of behavior—is limited largely by the society within which he or she lives.

Cultural Revolution   The upheaval in China during the 1960s when old cultural patterns were condemned and new Maoist patterns were strongly enforced.

culture complex   A group of culture traits that are activated together; for example, how clothing is made and distributed to consumers.

culture hearth   The source area for particular traits and complexes.

culture realm   An area within which the population possesses similar traits and complexes; for example, the Chinese realm or Western society.

culture trait   A single element or characteristic of a group's culture—for example, dress style.

cyclones   Indian Ocean hurricanes.

Damodar Valley   The principal heavy-industrial region of India, located west of Calcutta, with Jamshedpur serving as the region's focus.

death rate   The number of deaths occurring per 1,000 persons in a given area.

deforestation   Loss of trees in an area, sometimes caused by physical factors such as fires, floods, pests, etc., but more often human-induced (caused by agriculture, logging, and fuelwood consumption).

deindustrialization   Refers to the severe decline of Europe's primary and secondary mass-production industries that began around 1950.

demographic collapse   Natural decreases in a country's population due to the number of births being exceeded by the number of deaths. Such decreases are now occurring in Europe's center and east.

demographic transformation   A theory of the relationship between birthrates and death rates and urbanization and industrialization; based on Western European experience.

Deng Xiaoping   Dominated the Chinese government from 1976 until his death in 1997, during which time China began to emulate the capitalist economies of its East Asian neighbors.

dependency theory   The notion that a lag in economic development is perpetuated by trade patterns that leave developing areas dependent on or vulnerable to developed realms.

desertification   The process by which desert conditions are expanded; occurs in response to naturally changing environments and the destruction of soils and vegetation brought on by human overuse; takes place on the margins of desert regions.

development   A process of change that leads to improved well-being in people's lives, that takes into account the needs of future generations, and is compatible with local cultural and environmental contexts.

diffused culture   Culture traits and complexes that are diffused, or spread, from one region or source to another area.

diffusion process   The process whereby cultural groups and their goods and innovations move across and settle the landscape (e.g., settlers moving into new territories).

dikes   Long earthen embankments that are used to shield land from the sea. Dikes have been used extensively in the Netherlands as an effort to reclaim land from the sea for agricultural purposes.

Dinaric Range   Mountain range running through Croatia and Bosnia and Herzegovina, paralleling the Adriatic Sea.

Dnieper River   The axis of an extensive trade network that linked the Baltic and Black Sea regions starting in the tenth century.

doi moi   "Economic renovation"; a program adopted by Vietnam's leaders in the 1980s to attract foreign investment.

drainage basins   Basins associated with rivers.

Dravidian   One of the earliest inhabitants of India; referring to dark-skinned peoples of peninsular India; also, a family of languages that includes Tamil, Kannada, and Telugu.

dry farming   Agriculture wherein farmers rely exclusively on rainfall to produce their crops.

dual economic system   A modern plantation or other commercial agricultural entity operating in the midst of traditional cropping systems.

Duma   After an unsuccessful revolution in Russia in 1905, the Tsar did agree to a representative body, the Duma.

earth science tradition   A geographic perspective in which emphasis is on understanding the natural environment and the processes shaping that environment.

East African Rift Valley   African valley that begins in the north with the Red Sea, extending 6,000 miles through Ethiopia to the Lake Victoria region, where it divides into eastern and western segments and continues southward.

East Asia   A subregion of Monsoon Asia; consists of China, Mongolia, the two Koreas, and Taiwan.

East Asian Model   Adaptation of Western methods with indigenous culture and values.

Eastern Highlands   A mountain system that dominates most of eastern Siberia; also, the area of low mountains constituting much of eastern South America, including the Brazilian Highlands and the Guiana Highlands.

Eastern Orthodoxy   Christian branch created from the split of the Holy Roman Empire in A.D. 1054. In this schism, the western church became Roman Catholicism, and the Greek church became Eastern Orthodoxy. Initially centered in Greek-speaking areas and focused on the city of Constantinople (modern Istanbul), the Eastern Orthodox church has now expanded into a collection of fourteen self-governing churches with the Russian Orthodox Church being the largest. It is the dominant Christian branch in southern and eastern Europe and currently represents approximately 11 percent of the world's Christians.

East Indians   People from the Indian subcontinent who were imported as indentured laborers to the coastal estates of Guyana and Suriname on the northeastern coast of South America.

Economic Community of West African States (ECOWAS)   Created in 1975, it is the largest of Africa's regional organizations. Among its objectives are the establishment of a common customs tariff, a common trade policy, and the free movement of capital and people.

economic integration   Interrelatedness between two countries or regions wherein each is linked closely to the other.

ecosystems   The assemblage of interdependent plants and animals in particular environments.

ecotourism   The travel to natural areas to understand the physical and cultural qualities of the region, while producing economic opportunities and conserving the natural environment.

ejido A form of land tenure in Mexico by which land was given to a farming community, which could allocate parcels of land to individuals but retained title to the land; a derivative of indigenous Indian land tenure systems, brought back into use after the Mexican Revolution.

El Niño The warm equatorial current of the South Pacific.

encomienda A grant of authority and responsibility from the Spanish crown to Europeans in Latin America; included control over large parcels of land and became a mechanism by which Europeans and their descendants gained and maintained control over land and Indian villages.

environmental determinism A general geographical theory, now largely discredited, according to which the physical environment controls, directs, or influences what humankind does; contends that variations in the physical environment are associated with different levels of economic well-being.

Eratosthenes (276?–196? B.C.) Greek geographer from Alexandria, Egypt. Accurately measured the earth's circumference; recognized the need for maps to show relationships between one region and another.

estancia A Spanish term used in Argentina and Uruguay to describe a large rural landholding, usually devoted to stock raising, especially of horses and cattle; similar to a ranch.

ethnic cleansing The forced removal or elimination of minority groups from an area by a more powerful group so as to gain greater territory.

ethnic genocide Intense fighting in which the native population of one ethnic group is killed off by another; associated most recently with Rwanda and Burundi.

ethnic minority In Europe, groups seeking independence most often form an ethnic minority on the edge of a larger state. They differ linguistically or religiously from the country's majority.

Eurasia The large continent formed by Europe and Asia.

euro Common currency adopted by the European Union to replace national cash currencies.

European Union (EU) The common market structure of an increasing number of European countries with a goal of complete economic integration.

evapotranspiration rate The combined loss of water from direct evaporation and transpiration by plants.

exotic rivers Those foreign to the area to which they supply water; they gain moisture from outside the region and cross dry areas on their journey to the sea, collecting little (if any) runoff along their lower channels.

ex-situ **conservation** A strategy for conserving the biodiversity of the earth's food crops by collecting as many varieties as possible of wheat, maize, rice, potatoes, beans, cassava, and other crops and storing seeds and cuttings in seed banks.

extended metropolitan regions (EMR) A space economy of regionally based urbanization in which the proliferation of manufacturing facilities in rural locations along urban-centered transportation corridors has caused the social and economic traits of rural spaces and their residents to become urban in character. For example, megacities such as Jakarta, Manila, and Bangkok, whose populations exceed 10 million inhabitants and which include villages and paddy fields from which industries draw labor, could be considered extended metropolitan regions.

fadamas Floodplains of northern Nigeria and southern Niger, covered with tall and dense grasslands and sparse tree populations.

Falashas Black Jews of Ethiopia, who were converted to Judaism by Semitic Jews between the first and seventh centuries A.D.

Fall Line The point where the harder rocks of the Appalachian Piedmont Plateau meet the softer sedimentary materials of the Gulf-Atlantic Coastal Plain. It is characterized by moderate waterfalls and rapids that impeded upriver transportation in the colonial era and led to the establishment of a series of settlements from Virginia south and westward into Alabama.

fazendas A Portuguese term used in Brazil to describe large rural landholdings; may be devoted to crop or animal production or both.

federalism The administrative framework that provides the states or territories a measure of economic and political autonomy.

federal state A state in which considerable powers are vested in the individual provinces, and the central government is relatively weak. Opposite of a unitary state.

female double-day workload Women working a full day outside the home and then performing all the domestic chores when they return home; associated with the feminization of poverty.

feminist geography A branch of human geography that applies the perspective of feminism (the theory of the political, social, and economic equality of the sexes) to study of the human environment and society.

feminization of poverty Process in which women are paid less than their male counterparts, with the result that 70 percent of the world's adults living in poverty are women.

fincas Family-sized, or small, farms in Central America; often associated with coffee production.

Five Pillars of Islam The tenets that govern the behavior and belief of faithful Muslims.

Flemings The Dutch-speaking peoples of Belgium.

forest death The process whereby large numbers of trees are killed or damaged by acid precipitation from pollution. This problem has become more prevalent in Europe since the 1970s.

forward capital A seat of power such as Lahore, in the upper Indus Basin; it controlled the threat of Afghan tribes disrupting the caravan trade between south and central Asia.

fossil fuels Organic energy sources formed in past geologic times, such as coal, petroleum, and natural gas.

Free Trade Area of the Americas A multinational economic union that could potentially result from the establishment of a free trade agreement between Mercosur (Southern Cone Common Market), the Andean Community, the Central American Common Market, the Caribbean Community, and the North American Free Trade Association (NAFTA).

friction of distance The time and effort that is required to overcome physical movement across the landscape. Typically this friction is reduced with the advent of transportation innovations (e.g., in 1800, traveling from New York to St. Louis could take one month, whereas today it takes only a few hours).

frost-free period The period during each year when frost is not expected to occur, based on average conditions.

fundamentalist Protestantism Includes those Protestant religions in Latin America that are growing rapidly at the expense of nominal Catholicism; emphasizes a literal interpretation of the Bible as the basis for Christian life.

Gacaca court system An indigenous peacemaking method in Rwanda, which emphasizes reconciliation over retributive justice.

Gaza Strip   A small territory on the southeast coast of the Mediterranean Sea; inhabited by Palestinians and administered by the Palestinian Authority.

Gazprom   Government-controlled Russian natural gas company.

gender bias   Population disparity that occurs when one or the other sex makes up an abnormally large percentage of a population. For example, India possesses a gender bias in favor of males.

gender subordination   The undervaluation and, in some cases, the total disregard for the importance of women's contribution to the process of development in certain countries. This subordination can take such forms as the denial of equal education opportunities (leading to a lower literacy rate among women), limited access to land credit, female infanticide, and the abortion of female fetuses, as well as females receiving less food and health care than their male siblings. Overall in this situation, females are viewed as expendable members of a society.

genius loci principle   Resource use systems that achieve long-term sustainability do so because they avoid periods of rapid population growth and because they are sensitive to the spirit of place. This principle reflects long-term familiarity with the variations in nature that are part of a location's normal rhythms as well as the potential resources represented by a region's natural resource endowment.

geographic information systems   A computerized tool that allows one to gather, construct, and manipulate layers of spatial data for the analysis of a wide range of geographical problems.

geography   A branch of knowledge concerned with the study of how and why things are distributed over the earth; includes four traditional emphases: (1) the study of distributions, (2) the study of the relationships between people and their environment, (3) the study of regions, and (4) the study of the physical earth.

Germanic languages   One of the three divisions of the Indo-European language family. This language branch can be divided into north Germanic languages (including Danish, Faeroese, Icelandic, Norwegian, and Swedish) and west Germanic languages (including English, Frisian, German, and Netherlandish or Dutch). Speakers of this language branch are primarily found in northern and western Europe, South Africa, as well as North and South America.

globalization   The growing integration and interdependence of world communities through a vast network of trade and communication links.

GNI gap   The difference in gross national product between well developed and less developed countries.

Gold Coast   The coast of Queensland in Australia, which is attracting great numbers of international tourists.

golden horseshoe   An industrial district in Canada extending from Toronto to Hamilton, on the western end of Lake Ontario.

Golden Triangle   A region of Mainland Southeast Asia encompassing the shared borders of Myanmar, Thailand, Laos, and southwestern China. It is a mountainous area of poverty, ethnic hill minorities, and frontier lawlessness associated with drug production.

Gondwana   An ancient continent that included what is now Australia, Antarctica, Africa, America, and a segment of South Asia.

Gorbachev, Michail   The only elected President of the USSR, Gorbachev initiated reforms that ultimately resulted in the dissolution of the Soviet Union.

Gran Chaco   The vast lowland plain of western Paraguay, northern Argentina, and the adjacent portion of Eastern Bolivia; subject to alternating seasons of flood and drought.

Grand Apartheid   The policy in South Africa whose purpose was to create independent black homelands on the basis of ethnicity and race.

Grand Canal   A 1,400-mile-long canal between modern-day Hangzhou and the heart of the North China Plain; it facilitated the trading of commodities among the regions of the empire.

Great Barrier Reef   A 1,250-mile-long reef off the coast of Australia; it is one of the greatest natural wonders of the world.

Greater Antilles   The Caribbean islands of Cuba, Hispaniola, Jamaica, and Puerto Rico.

Greater East Asia Co-Prosperity Sphere   Economic dependence on Japan for machinery, high-value industrial goods, and capital in exchange for raw materials and low-value industrial goods.

Great European Plain   The vast lowland that dominates the northern half of Europe and stretches from the Atlantic coast of France northeastward to the borders of Russia and beyond into the Eurasian heartland.

Great Inland Sea, *Seto Naikai*   A vital waterway for shipping between Honshu and Kyushu/Shikoku in Japan.

Great Lakes   Large inland lakes shared by the United States and Canada and drained by the St. Lawrence River/Seaway. The regions surrounding the lakes are characterized by high levels of agricultural and industrial output.

Great Lakes System of East Africa   Part of the East African Rift Valley, containing crater lakes and the elongated lakes of Africa's deep trenches.

Great Leap Forward   China's 1958–1961 attempt to socialize agriculture and increase production in both farming and industry.

Great Plains   Western portion of the Interior Plains of the United States. Consist of sediments brought down from the Rocky Mountains; vary in elevation from 2,000 feet near the Mississippi River to more than 5,000 feet at the base of the Rockies.

Great Stagnation   A period of economic downturn that has gripped Japan since the late 1980s.

Great Trek   Exodus or migration of many Afrikaner South Africans from the coastal zones to the northeast interior plateau region in 1838 following British occupation of the coastal zones.

Great Valley   One of the most productive agricultural regions of the United States, situated in California between the Coast Ranges and the Sierra Nevada Mountains and drained by the San Joaquin and Sacramento rivers.

Green politics   Political party platforms that favor the protection of the natural environment, possess skepticism of technology, oppose the cooperation of industry and government, reject excessive consumerism, and desire to change basic lifestyles.

Green Revolution   The use of new, high-yielding hybrid plants—mainly rice, corn, and wheat—to increase food supplies; includes the development of the infrastructure necessary for greater production and better distribution to the consumer.

Gross National Income in Purchasing Power Parity (GNI PPP)   A new, widely used statistical measure of economic output which factors in differences in cost of living from one country to another, so that the resultant figures are more comparable.

growth triangle   An area of industrial growth in Southeast Asia centered on Singapore, the Malaysian state of Johor, and the Indonesian island of Batam.

guano   Seabird droppings from the Atacama Desert region of South America or other areas; used as an organic manure.

guest workers   Workers from other countries who seek employment in the urban industrial centers of foreign nations.

Guiana Highlands A mountainous region in the southeast of Venezuela and the Guiana countries, now being developed for its vast iron ore and bauxite reserves.

Gulf-Atlantic Coastal Plain Land surface region of recent origin, composed of sedimentary materials, that extends in the United States from New Jersey to Texas, and then southward into Mexico.

Gulf Coast One of the most rapidly developing regions of Mexico, centering on the states of Veracruz and Tabasco. In addition to its great petrochemical and textile industries, the region is one of the leading agricultural zones in tropical Latin America.

Gulf Cooperation Council A regional organization to deal with common problems in a unified manner; consists of Bahrain, Kuwait, Oman, Qatar, Saudi Arabia, and the United Arab Emirates—countries with large oil reserves, small populations, and a location along the Persian Gulf.

Gulf of Mexico An extension of the Caribbean Sea and Atlantic Ocean, bordering the southeastern United States, Mexico, and some of the Caribbean Islands.

Gulf War Military conflict that broke out on January 17, 1991. It resulted from Iraq sending troops into, and then annexing, Kuwait. The United Nations condemned Iraq's invasion, applied economic sanctions, and then authorized the use of force. The United States sent most of the military forces; contingents came also from Great Britain, Egypt, Syria, France, and Saudi Arabia.

hacienda A Spanish term used in Latin America for a large rural landholding, usually devoted to animal grazing; formerly had a high degree of internal self-sufficiency and operated under the patron system.

Han dynasty (206 B.C.–A.D. 220) A militarily powerful dynasty that organized the first large-scale empire in East Asia.

Harappan culture The Indus River Valley empire, dating back to ca. 2350 B.C.; it centered on the three cities of Harappa, Ganweriwala, and Mohenjo-Daro. It began to decline ca. 1900 B.C.

Hejaz Mountains North-south running mountain range along the northwest coast of the Arabian Peninsula.

Herodotus One of the earliest Greek geographers to map and name the continents of Europe, Asia, and Africa. Known as the father of geography and the father of history. Author of *Historia* (mid-fifth century B.C.).

Highland Africa The eastern sections of Africa that average 4,000–5,000 feet above sea level.

Hindi One of the national languages of India; one of India's several languages of Indo-European origin.

Hinduism A formalized set of religious beliefs with social and political ramifications; the dominant religion of Indian society.

Hindu Kush Mountains Major mountain chain in Afghanistan.

Hispanics Persons of Puerto Rican, Cuban, Central or South American, or other Spanish culture or origin, regardless of race; the fastest growing governmentally recognized minority in the United States.

holistic discipline An area of study, such as geography or history, that synthesizes and integrates knowledge from many fields.

holistic planning To plan in a comprehensive fashion, which requires an honest attempt to consider all of the implications of development activities.

household responsibility system A Chinese policy based on a production contract in which a peasant household is obliged to produce a specific amount of grain or cotton to be sold to the state at a regulated price.

House of Saud The ruling royal family of Saudi Arabia.

Human Development Index (HDI) Most widely used indicator of human development levels of nations and regions of the world; derived from life expectancy at birth, educational attainment, and income.

human geography The study of various aspects of human life that create the distinctive landscapes and regions of our world.

Humboldt, Alexander von (1769–1859) German geographer, author of *Kosmos*. Combined information from his studies of Greek, archaeology, and various sciences to form a geographic composite.

humid continental climate Humid climate type that possesses warm-to-cool summers and bitterly cold winters.

Hungarian Basin Large depression of land encompassing most of Hungary; it is surrounded by the Carpathian Mountains to the north and east, the Transylvanian Alps to the south, and the Austrian Alps to the east.

Iberia The peninsula of southwest Europe comprising Spain and Portugal; Iberians are persons of either country.

Iberian Peninsula Body of land in southwestern Europe that is surrounded by the Atlantic Ocean to the north and west, the Straight of Gibraltar to the south, and the Mediterranean Sea to the east. Contained within it are the countries of Portugal and Spain.

immigration The legal or illegal movement of people into a country of which they are not native residents.

import-substitution industrialization Strategies intended to protect infant industries from foreign competition, thereby stimulating local demand for local products and expanding local manufacturing employment opportunities.

import-substitution industries Manufacturing concerns that produce needed goods locally rather than relying on importation; often supported by subsidies, loans, and protective tariff regulations.

Inca One of the four high civilizations of pre-Colombian Latin America, centered on the city of Cuzco, Peru, in the Andes and extending from southern Colombia to central Chile.

income disparities Significant differences in income between specified groups of people or regions.

Indian diaspora The migration of large numbers of South Asian Indians to other, largely English-speaking countries.

indirect rule The policy upon which British colonial rule was based; its purpose was to incorporate the local power structure into the British administrative structure.

industrial colonialism Western political control of colonial territory and economic interests.

Industrial Revolution The period of rapid technological change and innovation that began in England in the mid-eighteenth century and subsequently spread worldwide; accompanied by the development of inexpensive, massive amounts of controlled inanimate energy.

industrial structure The mix or set of industries constituting the industrial makeup of a region.

infant mortality rate The number of children per thousand live births who die before reaching the age of one year.

informal economic sector Jobs not covered by national labor and employment laws, such as street vending, small business operation, and domestic work.

Information Revolution In contrast to the prehistoric cave dweller and his wall, our ability to produce, store, access, and apply information is massive and nearly instantaneous—truly revolutionary. It is how the information is produced, stored, accessed that creates a revolution; in the Information Age it boils down to a single transforming technology: the micro-processor.

inherited culture   Culture traits and complexes that are passed on from one generation to another.

initial advantage   Early factors that propel the development and expansion of particular activities.

inner ring   The section of the city that surrounds a European city's preindustrial core. The inner ring usually began as a slum outside of the city walls, and as the city grew, some slum hovels gave way to more substantial buildings, many times creating better neighborhoods. This section of the city serves as a residential area and frequently contains the city's railroad stations.

in-situ conservation   A strategy for conserving the biodiversity of the earth's food crops through smallholder tropical polyculture. By this method, a farmer may cultivate several varieties of 20 to 30 crops, thus preserving 40 to 120 varieties of food crops in a single plot.

Institutional Revolutionary Party (PRI)   The ruling party of Mexican politics during most of the twentieth century. It has held power since The Revolution and claims to be carrying on the goals of The Revolution; now being seriously challenged by alternative political forces.

insular Southeast Asia   Maritime subregion of Southeast Asia's physical environment.

insurgent states   Ones marked by revolt, insurrection, and guerrilla activity from national liberation movements that oppose the established authority.

interdependent development   A strategy to improve the patterns of uneven development; it requires growth of cooperative and compatible arrangements to facilitate development efforts.

Interior Plateaus   A large part of the western United States situated between the Rocky Mountains on the east and the Sierra Nevada and Cascade mountains on the west and including the Columbia and Colorado plateaus.

internal colonialism   A condition found in many less developed countries where local elites, often living in urban cores, exploit the masses, many of whom live in outlying peripheral rural regions.

intifadah   Uprising against Israel by Palestinian Arabs.

Iron Curtain   A popular expression for the boundary between Eastern and Western Europe during the cold war, signifying the difficulty of moving people and information across the border.

irrigated agriculture   In the Middle East and North Africa, has long been concentrated along the region's major rivers and in oases, often reliant on nonmotorized lifting devices.

irredentism   Policies or actions aimed at retrieving irredentas from the control of other political units—i.e., one country making a claim on territory in another country that is inhabited by people that are culturally similar to the people in the first country; often leads to tensions between countries.

Islam   The dominant religion of the Middle East and North Africa. This universalizing belief system was established by the Prophet Mohammed after he began having divine visions beginning in A.D. 610. Islam (which literally means "submission to the will of Allah, [or God]") has five tenets, or pillars, that must be upheld by its adherents: there is only one God and Mohammed is his Prophet (the creed); the offering of regular prayer; the responsibility of the more prosperous to aid the less fortunate; a fast during the ninth month of the lunar calendar; and the undertaking of a pilgrimage to Mecca.

Ita... **n Peninsula**   Southern European body of land encompassing Italy; it is surrounded by the Tyrrhenian Sea to the west,

the Adriatic Sea to the east, and the Mediterranean Sea to the south.

Jammu and Kashmir   The northernmost Indian state with mixed Hindu and Muslim populations.

Japan   An economically well-off, single-country subregion situated off the east coast of the Asian landmass and located within the greater region of Asia, East by South.

Japan Model   The Japanese approach to development, characterized by efficient government and bureaucracy, a sound currency and banking system, growth of education, and effective harnessing of the abilities of the Japanese people

Jews   Members of an ethnic group who claim as their belief system Judaism, a monotheistic religion formed by one of several Semitic peoples who resided in Southwest Asia more than 3,000 years ago. Subjected to repeated episodes of persecution throughout their history, the majority of the Jewish population became scattered throughout the world but eventually established their homeland of Israel in 1948. Today Jews make up a small but regionally prominent ethnic minority in the Middle East by constituting 83 percent of Israel's population.

Jonglei Canal Project   An effort to make new water available in Egypt by constructing a canal in Sudan that would bypass the Sudd swamps.

Kara Kum Desert   A prominent desert east of the Aral Sea in Uzbekistan and Kazakhstan.

Karakalpakstan   The area along the former shore of the rapidly shrinking Aral Sea, with high-salinity soil and dangerous contamination levels from airborne particles rising from the exposed seabed, including fallout from nuclear and biological testing.

keiretsu   Large industrial/financial cliques that formed in Japan after the American occupation ended following World War II.

Khmer Rouge   The communist party of Cambodia, overthrown by the Vietnamese but of continuing importance as an insurgent group and a political force.

Khuzestan   Iranian region that is the lowland extension of the alluvial floodplain east of the Tigris River and the Shatt al-Arab, located between the Zagros Mountains and the Iraqi border.

kibbutz   An Israeli collective farming community, often located near the frontier; serves defense functions as well.

kinship relations   The ties and relations of the extended family unit, often used in association with traditional societies.

kitchen gardens   During the Soviet period, families were allowed to have a small amount of land adjacent to their homes for growing whatever they wanted for their own consumption or for trade.

Kiwis   New Zealanders.

Kosovo   A southern province of former Yugoslavia dominated primarily by Albanian Muslims.

Kremlin   Seat of the Russian government in Moscow.

Kurds   A pastoral and agricultural people residing largely in Turkey, Iraq, and Iran, with lesser numbers in Syria, Armenia, and Azerbaijan.

Kyzl Kum Desert   A prominent desert in Turkmenistan.

Lacostian view   Theory that explains underdevelopment as the result of complex and interacting forces of both internal and external origin.

Ladino   European persons and lifestyles in Guatemala.

Lake Baikal   The world's deepest lake with a maximum depth of more than a mile (1.6 kilometers), one of the world's greatest natural wonders and a Russian mecca for nature lovers.

land degradation   A product of human actions that lower a region's biological productivity or diminish its usefulness to humans.

landlocked states   Those countries without a water coast; their trade with the outside world must pass through neighboring, sometimes hostile, countries. Almost 40 percent of the world's landlocked states are in Africa.

Land Reform and Resettlement Program (LRRP)   A program of the Zimbabwe government that attempts to resettle African families on lands formerly controlled by persons of European descent.

landscape modification   The constant change that landscapes undergo in response to natural and human-induced processes.

laterization   A process of soil formation in the tropics; the leaching of soluble minerals from soils because of copious rainfall, thereby leaving residual oxides of iron and aluminum.

latifundios   Literally, large landholdings; more generally, control of most of the land by a small percentage of landowners.

La Violencia   War between the liberals and the conservatives of Colombia; a class conflict between the haves and have-nots.

legalism   A philosophy opposed to Confucianism, it advocated the idea that humans are essentially selfish and that a system of strictly imposed laws is required to ensure acceptable behavior.

Lenin, Vladimir   Marxist, leader of the Bolsheviks (see *Bolsheviks*) who started the bloody civil war in 1917 that resulted in establishment of the Union of Soviet Socialist Republics in 1922.

less developed country   Country with low level of human well-being, as measured by economic, social, and biologic indicators.

Lesser Antilles   The smaller islands in the eastern Caribbean between Puerto Rico and Trinidad.

Levant   The eastern Mediterranean region from western Greece to western Egypt.

liberals   Colombian faction centered in Bogotá, who opposed the conservatives in *La Violencia.*

liberation theology   A theology based on biblical interpretation granting preferential treatment to the poor or lower classes; gaining acceptance in the Latin American realm.

lingua franca   An auxiliary language used by peoples of different speech; commonly used for trading and political purposes.

Llanos   The open, flat grasslands of Colombia and Venezuela, drained by the Orinoco River and its tributaries.

loess   Deposits of wind-transported, fine-grained material; usually easily tilled and quite fertile.

lost decade   The 1980s, in which occurred a significant regression in the economic circumstances of most Latin Americans.

Lower Canada   The French-speaking region of Canada found on the lower reaches of the St. Lawrence River and the Gulf of St. Lawrence (modern Quebec), as distinguished from Upper Canada.

Loyalists   Those persons who remained loyal to Great Britain during and after the Revolutionary War. After the revolution, many of these loyalists no longer felt welcome in the lower thirteen colonies and so fled to British North America (modern Canada).

madina   The ancient urban core of Middle Eastern cities.

madrasahs   Traditional Islamic schools, often attached to a mosque, where students study the Quran, learn basic reading and writing skills, and are introduced to the spiritual, philosophical, and legal scholarship of the Islamic community.

Maghreb   A region of northwestern Africa that includes portions of Morocco, Algeria, and Tunisia.

mainland Southeast Asia   Continental subregion of Southeast Asia's physical environment.

Malthusian theory   A theory advanced by Thomas Malthus that human populations tend to increase more rapidly than the food supply.

Mandarin   Officially designated as China's national spoken language in 1956.

manifest destiny   The conviction of many Americans that God willed the United States to extend from the Atlantic to the Pacific oceans; also that the nation continues to have a divine mission to promote world peace and individual freedom.

man-land tradition   A geographic perspective that emphasizes the relationship between people and the physical environment used for their sustenance.

Maori   The pre-European Polynesian inhabitants of New Zealand, who have been more successfully integrated into modern society than the aborigines of Australia.

Mao Zedong   Emerged as leader of the CCP in 1935; died in 1976.

maquiladoras   From the Spanish verb *maquilar*, meaning "to mill or to process"; in this instance used to denote foreign-owned (largely U.S.) manufacturing firms located in Latin America to realize the advantage of low-cost labor.

Maracaibo Lowlands   Enclosed in the northwestern part of the Venezuelan Andes; dominated by Lake Maracaibo.

marine west coast climate   Climatic region of northern Europe, northwestern North America, southern Chile, and portions of Australia and New Zealand; characterized by year-round cool and wet atmospheric conditions.

market gardening   An agricultural system that predominates in the sunnier regions of Europe in which fruits and vegetables are produced for market. In this system, the main products are wine, olives and olive oil, green vegetables, and citrus fruit, apples, and other tree fruit.

Marsh Arabs   Cultural group residing in Iraq's southern marsh region. In order to maintain their culture, these individuals combined agriculture, cattle rearing, and fishing. Recent efforts to drain the marshes, however, have destroyed the Marsh Arabs' livelihood.

Masakahane Campaign   Strategy or program of urban development in the Republic of South Africa.

Mauryan Empire   A political state; the first of a series of empires that ruled over parts of South Asia. The greatest Mauryan ruler was Asoka.

Maya   One of the four high civilizations of pre-Columbian Latin America, situated in southern Mexico and northern Central America.

mediterranean climate   A dry summer subtropical climate common to the Mediterranean Basin, southern California, central Chile, and portions of Australia.

megalopolis   Originally the continuous urban zone between Boston and Washington, D.C.; now used to describe any region where urban areas have coalesced to form a single massive urban zone.

Meiji Restoration   Marked the end of Tokugawa rule in 1868 and the beginning of the period when Japanese society and its economy were transformed from feudal to modern; Meiji means "enlightened rule," which in this case meant adopting selected Western traits, particularly education and technology.

Melanesia   Section of the South Pacific consisting of islands stretching along the northern perimeter of Australia, from New Guinea eastward to Fiji. The name derives from melanin, pigment in human skin, and refers to the dark-skinned, r speaking peoples who predominate this area.

melting pot An assumption among some Americans in which immigrants eventually lose their separate identities as they blend into a single American ethnic culture (e.g., "foreign-ness" is lost as immigrants adopt the English language and acquire an American cultural identity).

mercantile colonialism Western governments trading with local elites for native luxury goods in colonial areas.

mercantilism The philosophy by which most colonizing nations controlled the economic activities of their colonies; held that the colony existed for the benefit of the mother country.

Meseta The tableland that dominates the interior of Iberia.

mestizos Persons in Latin America of mixed European and Indian ancestry.

Mezzogiorno "Land of the midday sun," an area comprising the Italian peninsula south of Rome, plus Sicily and Sardinia. One of the major regions of underdevelopment in the European Community.

Micronesia The groups of islands in the Pacific just north of Melanesia, from Guam to the Marianas on the west to Kiribati on the east. The region's name is derived from micro, referring to the small islands that predominate here.

Middle America Mexico, Central America, and the Caribbean.

Middle Kingdom A reference to China, reflecting the traditional Chinese view of China as the center of the known universe.

middle ring A ring of factories and huge worker's apartment blocks or row houses that surround the European city's inner ring, often called the "industrial suburbs." This ring typically formed during the 1800s, as an addition from the Industrial Revolution, and has emerged as the largest part of the European city.

milkshed The area serviced by a given milk-producing region.

Mineral Triangle The highly mineralized area of eastern Brazil centered on Belo Horizonte.

minifundios Literally, small landholdings; more generally, control of only a small percentage of the total farm area by most of the landowners.

ministates Very small independent countries. Europe contains six ministates, including Liechtenstein.

Ministry of Economy, Trade, and Industry (METI) The Japanese government agency responsible for guiding and directing the development of the Japanese economy.

mixed farming The raising of crops for both human and animal feed. Mixed farms often have animals, such as cattle or pigs, on the farm itself. Dairy farms can be viewed as a specialized form of mixed farming.

monoculture Agricultural production that centers on one dominant crop, such as sugar cane or bananas.

monsoon The seasonal reversal in surface wind direction; associated primarily with the southeast quadrant of Asia.

montaña The lower eastern slopes of the Andean Mountains in Peru.

Moors People of mixed Arab and Berber stock of northwest Africa who invaded and inhabited Iberia from the eighth through the fifteenth centuries, thereby diffusing racial and cultural traits to Spain and Portugal.

moshav The smallholders' agricultural village in Israel; also functions as an agricultural cooperative.

Mughal Empire The most powerful of all the Islamic empires of the sixteenth and seventeenth centuries in South Asia.

Mujahideen Opposition movements in Afghanistan, inspired by Islamic fundamentalism.

lattos Persons in Latin America of mixed European and frican descent.

mullahs/mujtahids Muslim clergy who have traditionally been responsible for interpreting religious law, providing basic education, serving as notaries for legal documents, and administering religious endowments.

Multimedia Super Corridor (MSC) The name given to a region designated by the Malaysian government for development as an information technology research center.

multinational economic unions Regional associations formed to reduce trade barriers with neighboring countries and more effectively market member products, thus increasing the volume of trade within and outside the group.

Muslims People who surrender to the will of God (Allah) as revealed by the prophet Mohammed; followers of Islam.

mwani The king of the Hutus and Tutsis, thought to be a divine ruler based on Tutsi lineage.

Nagorno-Karabakh region Part of Azerbaijan with an Armenian majority, whose desire for independence resulted in war between Azerbaijan and Armenia. Fighting ceased in 1994, but the fundamental issues are as yet unresolved.

nationalism A territorial expression of loyalty to a state. Nationalism can also be reflected in an ethnic group's desire for political independence in the form of a separatist movement.

natural vegetation The plant life that can be expected in a particular environment if free of human impact.

neocolonialism Retention of the trade relationships and patterns of pre-World War II colonialism. Often cited to explain continuing uneven distribution of wealth.

neoliberal reforms Far-reaching economic reforms by debtor nations that center on reducing national expenditures; they frequently include privatization of enterprises formerly government-owned, and reduction or elimination of government subsidies for food, fuel, utilities, health care, and other basic services.

Neo-Malthusians Referring to contemporary modifications of the notions of Thomas Malthus regarding population growth and production capacity.

New Economic Policy (NEP) Malaysia's master plan, the goal of which was to increase the economic contributions of the majority ethnic Malays, at the expense of both Chinese and Western economic interests.

New England A glaciated region of the northeastern United States, characterized by a Humid Continental climate and podzolic soils.

Newly Industrializing Economies (NIEs) Refers to any of the Pacific Rim nations such as South Korea, Taiwan, Malaysia, and China that in recent years have relied on aggressive government policies promoting export-led industrialization as a development strategy.

Newly Industrializing Countries (NICs) Nations that are evolving from Third World to First World states: Singapore, South Korea, Taiwan, and others.

new middle class This group has appeared recently in Russia, at least in the larger urban areas. Unlike the New Russians, new middle-class people are professionals who draw salaries in fields valued by the global economy.

New Russians Have emerged as the first class of wealthy Russians since 1917. Their conspicuous consumption and uncultured ways make them the target of scorn.

New South The more urbanized and industrialized South, which emerged in the United States in the twentieth century.

New Valley Project An effort to channel surplus water into the deserts of Egypt to develop areas that are without rainfall.

Nikkei   The Japanese term for foreigners of Japanese descent.

nomadic herding   In the Middle East, involves herding of sheep, goats, and camels in an annual migration cycle often covering hundreds of miles, to bring animals to grass and water only available on a seasonal basis.

Non-Governmental Organizations (NGOs)   A diverse group of largely voluntary organizations that work at the grassroots level with people organizations to provide technical advice and economic, social, and humanitarian assistance.

North Atlantic Drift   The warm ocean current, an extension of the Gulf Stream, that passes Northwest Europe and modifies the air masses that determine European weather patterns.

North Atlantic Treaty Organization (NATO)   A multinational military alliance of a number of European nations, the United States, and Canada, founded in 1949.

oases   Islands of life in the desert, where surface erosion and relatively high groundwater levels result in water being found close to the surface.

Ob River   West Siberian lowland river. When its tributaries thaw in the spring, ice in the lower Ob acts like a giant plug, sometimes until late summer, causing extensive flooding.

Oceania   The vast realm of Pacific Islands extending over several million square miles, including the islands of Melanesia, Micronesia, and Polynesia.

Old Industrial Countries (OIC)   Sometimes used to refer to countries with long-standing industrialization, such as the United States and the countries of Western Europe.

oligarchs   Russian privatization of industry has resulted in a struggle for survival between the new "entrepreneurs" and emergence of the oligarchs, a few individuals who control vast economic empires.

open coastal cities   Much like Special Economic Zones (SEZs), but with lower levels of government funding for site improvement.

open door policy   A reversal of hostile attitudes by the Chinese government to encourage foreign investment.

Operation Bootstrap   Puerto Rico's post-World War II development plan emphasizing manufacturing, agricultural reform, and tourism.

opium   A dangerously addictive drug from which heroin is derived.

Orange Revolution   In Ukraine in 2004, the Orange Revolution, a public rebellion against government practices that took its name from the opposition party's color, led to election of Viktor Yushchenko, who has promised to root out corruption and restructure both state and economy.

Organization of African Unity   A group of African states concerned with the political and economic relations among African nations and between Africa and other parts of the world.

Organization of Petroleum Exporting Countries (OPEC)   A thirteen-nation group of oil-producing countries that controls 85 percent of all the petroleum entering international trade; a valorization scheme to regulate oil production and prices.

Oriente   The eastern lowlands of Ecuador.

orographic effect   Mountains create a barrier to a moving air mass that must be overcome, which in the process modifies the temperature and moisture characteristics of the air mass. The result is excess precipitation on the windward side and drought on the leeward side of the mountain.

orographic precipitation   Mountain-induced rains such as those of the Cherrapunji region of the Assam Plateau of South Asia.

Ouachita Mountains   In the Interior Lowlands of the United States, the Ouachita Mountains of Arkansas are upland remnants of the mountain-building processes that produced the Appalachians.

outback   The interior and isolated backlands of Australia, particularly those areas beyond intense settlement.

outer ring   A loose assemblage of residences dominated by detached single-family houses, modern factories devoted to high-tech industry, and firms specializing in data gathering and processing that surround a European city's middle ring. Often called a "postindustrial suburb," this outer ring has formed since 1950 and usually has linkages to the center city via mass transit.

overgrazing   Loss of vegetative ground cover resulting from the human inhabitants stocking more grazing animals than the land can sustain.

Overseas Chinese   Persons of Chinese ancestry now living in other countries. Many maintain family and economic ties with relatives and associates still living in China.

Ozark Plateau   Geologically, an extension of the Appalachian Plateau, found in southern Missouri and northern Arkansas.

ozone holes   Chlorofluorocarbons released by certain human activities attack ozone, a naturally occurring gas important in atmospheric dynamics and regulation of solar radiation, and produce holes in the ozone layer over the poles and a thinning elsewhere.

Pacific Belt   Area of Japan where much of the postwar development of industry took place.

Pacific Rim nations   Includes those countries rimming the Pacific Ocean in both East Asia and the western Americas; sometimes refers more exclusively to the East Asian rimland countries.

Padana Valley   Northern Italian lowland situated between the Alps and Appenine mountain ranges and drained by the Po River. The primate city of this prosperous and high-industrialized region is Milan.

Pahlavi dynasty   Iranian dynasty (1925–1979) that tried to promote modernization in both Iran's society and economy through such efforts as land reform programs, educational reform, the provision of greater social freedom for women, and the opening of the country to foreign aid and investment. Many of these reforms, however, challenged the position of the Islamic clergy, the mullahs and the mujtahids.

Palestine Liberation Organization (PLO)   An umbrella organization created by the Arab states in 1964 to control and coordinate Palestinian opposition to Israel; originally subservient to the wishes of the Arab states, now dominated by the independent, nationalist ideology of El-Fatah, the largest and most moderate of the Palestinian resistance groups.

Palestinians   The Arabs who claim Palestine as their rightful homeland; some 5 million Palestinian Arabs residing in various Middle Eastern states.

Pamir Mountains   Major mountain range in Tajikistan and Afghanistan.

Pampa   A Spanish term for an extensive plain; also used as a proper noun to identify the most important region of Argentina.

pastoral economy   An economic system dependent on the raising of livestock—sheep, cattle, or dairy animals.

pastoralism   The system in which a group depends on pasture for their livestock herds; the best-known African pastoralist the Fulani nomads of northern Nigeria, the Masai of Ken the nomadic Tswanas of Botswana.

Patagonia   A sparsely populated, dry, windswept plateau of southern Argentina; most of Patagonia is used for raising sheep.

peninsulares   An alternate term for Iberians, used in colonial Latin America.

permafrost   Permanently frozen ground common in high latitudes.

Persians   Natives of Iran; one of two of the largest non-Arab groups in the Middle East (the other is the Turks).

physical geography   That component of geography that focuses on the natural aspects of the earth, such as climate, landforms, soils, or vegetation.

physiologic density   Population density expressed as the number of people per unit of arable land.

Piedmont   The low, rolling eastern and southern foothill regions of the Appalachian Highlands.

pilgrimage   A religious practice that involves the travel to a religiously significant or sacred site by a devotee. In Islam, the pilgrimage (hajj) to Mecca is one of the five pillars (five tenets of the faith) and is expected to be undertaken at least once in a Muslim's lifetime.

pine barrens   Sandy pine lands of the southeastern United States; considered infertile and not favored in early settlement.

pinyin   New system of spelling Chinese words in the Latin alphabet that has been adopted by the Chinese government to replace the older Wade-Giles transliteration system.

plantation system   A large tropical or subtropical agricultural unit emphasizing one or two crops that are grown for export.

plate tectonics   The process whereby portions of the earth's surface are folded, fractured, and uplifted through the movement of the earth's lithospheric plates over the asthenosphere. Many of the mountain ranges in Europe (e.g., the Alps, Pyrenees, and Carpathians) were formed as a result of the slow collision of the Eurasian and African plates.

podzolization   A process in humid regions whereby soluble materials are leached from upper soil layers, leaving residual soils that are frequently infertile and acidic.

polyculture   The simultaneous growing or cultivating in shared space of several crops for subsistence or commercial purpose, as opposed to dependence on a single crop (see also *monoculture*).

Polynesia   A large grouping of islands in the Pacific stretching in a huge triangular area from Midway and Hawaii on the north to New Zealand on the south, and eastward as far as Pitcairn Island. The name means "many islands," and it is the largest in size of the three Pacific Island subregions. See *Melanesia* and *Micronesia*.

population density   The number of people per unit of area.

population distribution   The placement or arrangement of people within a region.

population growth rate   For a country or region, this rate involves calculation of the birthrate and death rate, modified by subtracting people who emigrate out and adding people who immigrate into the area.

postindustrial service industries   Service industries involve neither extraction of resources nor manufacturing but instead include a broad range of activities such as government, transportation, banking, retailing, and tourism. Postindustrial denotes service industry dominance in modern Europe, coincident with deindustrialization.

postmodernist geography   An emerging branch of human geography that includes consideration of the use of space by societies.

poverty   Material deprivation that affects biologic and social well-being; the lack of income or its equivalent necessary to provide an adequate level of living.

Prairies   In Canada, the great expanse of land between the Great Lakes and the Rocky Mountains, the northern extension of the Great Plains and the Interior Lowlands.

precautionary principle   This rule mandates that whenever significant change is likely to occur as a result of a proposed development or whenever the long-term implications of a proposed change are obscure, implementation of the anticipated project must proceed slowly, with proper examination of likely impacts and available remedial actions.

preindustrial core   The center of the European city consisting of a medieval or even ancient nucleus of twisting, narrow streets and marketplaces fronted by buildings not more than two or three stories tall. Many of these cores have been converted into pedestrian zones and are considered prestigious places to live.

primary industries   Economic activities that are involved in the extraction of resources from the earth and the seas.

primary level of economic activity   Focuses on extractive activities, such as agriculture, mining, forestry, and fishing.

Prince Henry the Navigator   Portuguese ruler; sponsored explorers seeking new routes to the Orient.

private-plot production   In the Soviet Union, Stalin tried unsuccessfully to abolish farm families' home gardens, their private plots, but Soviet agriculture depended heavily on private-plot production, especially for fruits and vegetables.

privatization   The process of transferring partially or completely state-owned industries and farms to private control.

privatization of industry   See *privatization*.

producer services   Businesses such as banking, insurance, communications, and consulting.

Protestantism   Christian branch formed from the second great schism, the Protestant Reformation during the 1500s, which split the western church into Roman Catholicism and Protestantism. This branch constitutes approximately 24 percent of the world's Christians.

Ptolemy   (ca. A.D. 150) Greek astronomer and early mapmaker of Alexandria, Egypt. Designed a map of the world that used a coordinate system to show the location of 8,000 known places.

Puget Sound Lowland   A geological trough which now forms an inlet of the Pacific Ocean in northwestern Washington state.

Pushtuns   An ethnic group of the northwest corner of India that also occupies portions of Afghanistan.

Putin, Vladimir   President of Russia, elected in 2000 and reelected in 2004.

Putonghua   A proposed common spoken language for China in the 1920s and 1930s.

Pyrenees Mountains   East-west running range of mountains that serve as the border between France and Spain. Also contained within the range is the Principality of Andorra.

qanat   An underground tunnel, sometimes several miles long, tapping a water source; natural water flow accomplished by gravity; common in several Middle Eastern countries, where they are used to tap groundwater found in alluvial fans.

Quechua   An ethnic and linguistic Indian group of Ecuador, Peru, and Bolivia.

Quetzalcoatl   The fair-skinned, bearded god of Aztec Indian oral tradition for whom tribal rulers mistook Hernando Cortés, enabling Cortés to conquer Mexico.

Quran   The sacred text of Islam.

rainshadow   An effect produced by descending, drying air masses, causing the leeward or interior sides of mountains to be much drier than the windward sides.

Reconstruction and Development Program (RDP)   The program launched by Nelson Mandela's government of South Africa, based on six major principles of improvement.

regional disparity   Distinctive differences in well-being among the inhabitants of the several regions of a given country; most likely to be seen in economic, social, and biologic conditions.

regional geography   That component of geography that focuses on a particular region and the geographic aspects of the economic, social, and political systems of that region.

rent-seeking state   The use of the resources of the state for the benefit of private interests. In Russia and some other Eurasian countries, public officials "privatize" their government functions, by seeking "rent" in exchange for allowing development to proceed.

Restitution of Land Rights Act   A policy instituted in South Africa in 1994 to permit the government to investigate land claims and restore ownership to those who unjustly lost land.

Ridge and Valley Province   A section of the Appalachian Highlands lying to the west of the Great Smokies/Blue Ridge Mountains and consisting of a series of parallel steeply folded mountain ranges. It is also the source of much of the coal mined in the United States.

Ring of Fire   A zone that encircles the Pacific Ocean, which includes Japan, the Philippines, and parts of Indonesia, as well as the Andes Mountains of South America and many of the coastal ranges of Canada, Alaska, and eastern Siberia.

Ritter, Karl   (1779–1859) Author of *Die Erdkunde;* held first chair of geography in Germany, at the University of Berlin in 1820.

Rocky Mountains   A great series of mountain ranges in the western United States and Canada.

Roman Catholicism   Christian branch created from the split of the Holy Roman Empire in A.D. 1054. In this schism, the western church became Roman Catholicism, and the Greek church became Eastern Orthodoxy. Roman Catholicism is focused on Rome and the Vatican, with the Pope being the highest authority in the Church. It is the dominant Christian branch in southwestern Europe and constitutes approximately 50 percent of the world's Christians.

Romance languages   One of the three divisions of the Indo-European language family that includes such languages as Spanish, Catalan, Portuguese, Italian, Sardinian, French, Occitan, and Romanian. Speakers of Romance languages are widespread and are found in southern, eastern, and western Europe, and South, North, and Central America.

Rose Revolution   Georgia's 2003 Rose Revolution, so-named because demonstrators carried roses to emphasize non-violence, ousted Eduard Shevardnadze, the former Gorbachev ally whose corrupt regime failed to revive the economy or keep the country intact.

rotational bush fallow system   A farming practice where the cultivated area rotates around a fixed plot of land.

royal patronage   The privilege granted by the popes of Rome to the Spanish crown to appoint all clergy in conquered lands; politicized the upper level of clergy in colonial Latin America and resulted in a close church and state relationship.

run-on farming   A technique that collects water from a larger area and concentrates it in valley bottoms and on terraced slopes; it permits agriculture in regions that are otherwise too dry for crop cultivation.

Russification   A policy of cultural and economic integration practiced in the former USSR that required all other Slavic and non-Slavic groups to learn the Russian language.

sacrifice zone   To create sustainable livelihoods in the present, people often create sacrifice zones—sacrificing the future use of potentially sustainable resources or the present productivity of distant areas.

Sahel   The semiarid grassland along the southern margin of the Sahara Desert in Western, Central, and Eastern Africa.

salinization   The accumulation of salts in the upper part of the soil, often rendering the land agriculturally useless; commonly occurs in moisture-deficient areas where irrigated agriculture is practiced.

Scandinavian Peninsula   Body of land in northern Europe that is surrounded by the Arctic Ocean and Norwegian Sea to the west, the North Sea to the south, and the Baltic Sea and the Gulf of Bothnia to the east. Contained within it are the countries of Norway and Sweden.

scheduled castes   The name given to the Hindu and Sikh untouchables by the Indian government.

secondary industry   Manufacturing and other economic activities involving the processing stage, in which materials collected by primary industries are converted into finished products.

secondary level of economic activity   Includes activities that transform raw materials into usable goods.

semi-colonialism   An expression for the colonial experience of China, which consisted of foreign domination of many coastal regions but not of the interior.

separatism   The phenomenon in which a group of persons seeks to pursue independence from their country. Separatism is usually ethnic (cultural) in nature, and in most instances ethnicity must be linked to some major grievance such as persecution, attempted ethnocide, forced assimilation, domination, lack of autonomy, or denial of access to the country's power structure for separatism to be sparked.

Sephardim   Jews originating from Southern Europe and the Middle East.

sequential occupancy   A concept of historical geography that considers the geography of an area during successive periods of time.

Serbs   Ethnic group of Orthodox Christians, the majority of whom reside in Serbia, which is the largest (in land area and population) of the republics that make up the former Yugoslavia. Serbs speak a Serbo-Croatian Slavic language, but use a Cyrillic alphabet.

serfdom   For centuries in Russia, peasants were bound to the land and required to serve the landowners. Over time this condition of serfdom turned into slavery.

sertão   A term used in Brazil to refer to the Northeast backlands.

settlement frontiers   New territories that are occupied and culturally imprinted by settlers, usually by the process of diffusion.

Severn–Wash Divide   Political and economic divide that developed in England after 1960 as a result of deindustrialization when the older coal-and-steel districts—the "Black Country"—fell into decay as factories and mines closed. As the United Kingdom emerged from this episode into the prosperity of the high-tech, postindustrial, service-dominated era, the older, decayed industrial districts were left behind. This divide separated prosperous southeast England from the rest of the country today tends to be reflected in voting patterns.

shantytowns Urban areas of low socioeconomic characteristics; common in most cities in the less developed world; often inhabited by squatters.

Sharia An Arabic legal code based on the Quran.

Shatt al-Arab Common channel located in southern Iraq through which the Tigris and Euphrates Rivers flow into the Persian Gulf.

shatterbelt A politically unstable region where differing cultural elements come into contact and conflict; especially, Eastern Europe.

shifting cultivation A farming system of land rotation, based on periodic change of cultivated area; allows soils with declining productivity to recover; an effective adaptation to tropical environments when population density remains low.

Shi'ite One of two main branches in Islam, predominant in Iran and in parts of Iraq and Yemen.

Sierra Nevada A high, faulted mountain range situated along the California-Nevada border.

Sikhism An Indian religion that includes elements of the Hindu and Islamic faiths.

Silk Road Caravan route stretching from China to the Middle East and ultimately on to Europe, the main corridor for goods from the east.

siloviki Russian President Putin's appointees.

Slavic languages One of the three divisions of the Indo-European language family. Several subdivisions of Slavic are found within this language branch, including Bulgarian, Croatian, Czech, Macedonian, Polish, Russian, Serbian, Slovak, Slovene, and Ukrainian. Speakers of this language branch are widespread throughout southern and eastern Europe, stretching from Latvia in the north to Macedonia in the south, and from the Czechia in the west through Russia in the east.

Song dynasty (A.D. 960–1279) A period of Chinese history during which advancements occurred that made the Chinese agricultural economy one of the most sophisticated in the world.

South Asia A subregion of Monsoon Asia, including India, Afghanistan, Pakistan, and the smaller states of Nepal, Bhutan, Bangladesh, Sri Lanka, and the Maldives.

Southeast Asia A subregion of Monsoon Asia; encompasses the mainland states of Myanmar, Thailand, Cambodia, Laos, and Vietnam, and the maritime states of Malaysia, Singapore, Indonesia, Brunei, and the Philippines.

Southern African Development Community (SADC) A group founded in 1980 to reduce dependency on South Africa for rail, air, and port links; imports of manufactured goods; and the supply of electrical power.

South Pacific Forum A supranational organization formed by many of the Pacific Island states; it is designed to promote the region's collective interests and build strength through unity and the fostering of economic opportunities.

spatial integration The process whereby the settlement frontier is eliminated through the creation of trade areas and the establishment of ties with the core areas and the surrounding communities. According to historian Frederick Jackson Turner, European core areas on the Atlantic seaboard provided important early sources of settlers responsible for the spatial integration process in the United States and Canada.

spatial tradition A geographic perspective that emphasizes how things are organized in space, especially spatial distributions, associations, and interactions.

Special Economic Zones (SEZs) Five areas in China that function as modern-day treaty ports, receiving a substantial amount of foreign investment.

spring wheat belt The northern portion of the wheat belt of the United States and Canada, characterized by the planting of the wheat in the spring and harvesting the crop in the fall.

stages of economic growth A theory developed by Walt Rostow in which five stages of economic organization are recognized; traditional society, preconditions for takeoff, takeoff, drive to maturity, and high mass consumption.

Stalin, Joseph Overcame all rivals after Lenin died in 1924 to take over rule of the Soviet Union. He initiated collectivization of farms, the "Gulag" labor camp system, and the Five Year Plan for industrialization; millions died in the Soviet Union during his rule due to famine, war, and repressive policies.

state-capitalist economy A form of economy dominated and co-opted by state bureaucracies.

state farms One of two large agricultural systems controlled and managed by the government in the old Soviet Union; workers were paid wages.

state-led industrialization A development philosophy found in Japan, South Korea, Taiwan and other Asian countries, in which the state either partially owns corporations that operate under commercial principles or directs the structure and orientation of the national economy.

stations The very large sheep or cattle ranches associated with Australia.

steppe Grasslands south of the mixed forest in Russia and other "in-between countries" of Eurasia.

street children Youth who live all or part of their lives away from home.

structural adjustment programs (SAPs) Programs initiated to adjust malfunctioning economies, assist in debt restructuring, and promote greater economic efficiency and growth; widely implemented in the Sub-Saharan and Latin American countries.

subcontinent A large-scale peninsula or projecting appendage of land that is attached to a continental landmass. For example, Europe is a subcontinent of Eurasia.

Sumerian civilization An early urban-based hydraulic civilization, which formed along the Tigris and Euphrates Rivers more than 5,000 years ago. Located in modern-day Iraq.

Sunni The second and majority branch of Islam; predominant throughout most Middle Eastern countries, with the exception of Iran.

sunset industry Industry in Japan that the government considers no longer competitive and therefore appropriate to be phased out.

Sunshine Policy A policy of reconciliation between South Korea and North Korea, introduced in 2000.

Sun Yat-sen Known as "father of the republic," Sun revived his Nationalist Party in 1919; he centered its platform on the "Three Principles of the People": nationalism, democracy, and livelihood.

Syr Darya River A river in Kazakhstan that flows into the Aral Sea.

sustainable development Process of change that is durable, is harmonious with local cultures and communities, and takes into account the long-term environmental consequences of such change.

systematic geography An approach to geographic study in which the emphasis is on specified subjects; for example, economic geography, urban geography, climatology, water resources, or population geography.

Tabqah Dam   A large dam on the Euphrates River in Syria, intended to significantly increase irrigated agriculture and hydroelectric power generation.

taiga   The large coniferous forest extending across northern Russia.

Taliban movement   A fundamentalist Islamic student-based movement that originated in the Afghan refugee camps in Pakistan.

Tamil   Language spoken by Sri Lankans of Dravidian descent, who are known as Tamils, and who are also a minority in Sri Lanka. Since the establishment of British rule in the country, the Tamils feel they have been systematically discriminated against by the majority group, the Sinhalese Buddhists.

terracing   Technique in which steep hillsides are cut in a stair-step fashion and shored up by retaining walls for agricultural purposes.

terra roxa   A soil found in Brazil that is especially suitable for coffee cultivation.

thalassocracy   A sea-based state.

Third World   Another designation for technologically less developed countries, especially those not included in the former Soviet bloc.

Three Gorges   A region of China where the Chang Jiang ("Long River") flows through a narrow, 150-mile-long steep-walled valley no wider than 350 feet.

Tiananmen Square   Location of 1989 Chinese uprising expressing political discontent against the communist government.

Tibetan Plateau   The largest environmental zone of western China, as well as the largest and most elevated plateau in the world, it occupies 25 percent of China's territory. It is a cold, inhospitable environment referred to as the "rooftop of the world" or the "third pole."

Tokaido Megalopolis   A large, multinuclei urbanized region in Japan extending from Tokyo to Osaka.

Tokugawa period   The period (1615–1868) during which the focus of power in Japan shifted to the Kanto Plain area and many of the modern Japanese characteristics were firmly fixed in the culture.

trade development zone   Australia's version of the special economic zones used by China and other developing nations.

Trans-Amazon Highway   A massive road system throughout the Amazon Basin of Brazil, now connecting that region to other parts of the nation.

transational capitalist class   Globe-trotting executives and professionals, who view their work as part of a globally competitive process, share upscale lifestyles, and see themselves as citizens of the world as well as citizens of their own countries.

transnational corporation (TNC)   A corporation with offices, production facilities, and other activities in multiple countries. It is geographically mobile and can take advantage of lower labor costs in one country or a more lenient regulatory environment in another country to minimize production costs and/or maximize revenues.

Trans-Siberian railroad   System that crosses Siberia, beginning at the Ural Mountains and ending at the Pacific Ocean terminus of Vladivostok.

Treaty of Tordesillas   A treaty negotiated between Spain and Portugal in 1494 that divided the New World between those two countries at roughly the fiftieth meridian, giving Portugal the rights to areas to the east and Spain, areas to the west; this treaty followed a papal bull from the previous year, which had declared the New World as belonging to Spain and Africa and India to Portugal.

tsunamis   Seismic sea waves triggered by energy released from deep earthquakes, but also from massive landslides and volcanic eruptions.

tundra   The tundra region stretches across Russia's Arctic shore (and other Arctic regions), extending southward in Siberia. No trees grow because of the short growing season, infertile soil, and shallow layer of thawed ground; beneath lies permanently frozen earth (see *permafrost*).

Turks   Natives of Turkey; one of two of the largest non-Arab groups in the Middle East (the other is the Persians).

typhoons   The equivalent of hurricanes or tropical cyclones; occur in the Pacific, especially in the area of the China seas.

Ujamaa   A Swahili word meaning "familyhood," expressing a feeling of community and cooperative activity; a term used by the Tanzanian government to indicate a commitment to rapid economic development according to principles of socialism and communal solidarity.

unitary state   A state in which internal divisions such as provinces are weakened and overwhelmed through a strong central government. This was one political goal during Germany's Second and Third Reichs. Opposite of a federal state.

Upper Canada   The English-speaking region of Canada found on the upper reaches of the St. Lawrence River and the Great Lakes region (modern Ontario), as distinguished from Lower Canada.

Ural Mountains   A chain of ancient, greatly eroded, low mountains in Russia that mark the traditional boundary between Asia and Europe.

urban heat islands   Phenomenon in which cities tend to be characterized by significantly higher temperatures than those of the surrounding areas. This temperature imbalance is largely due to the air pollution generated by cities as well as the large amounts of heat that are absorbed by the pavement and masonry of the urban infrastructure.

Valley of Mexico   The largest valley in Mexico's Core region, where Mexico City is located, lies above 7,000 feet (2,134 meters) and experiences cool temperatures throughout the year.

village compound or homestead   The smallest unit of residence within an African village; it consists of small houses occupied by members of an extended family.

Volga River   With its tributaries, this river forms Russia's major water route, allowing for the development of industrial activities.

Wakhan Corridor   A narrow strip of land created between the Russian-controlled Pamir Mountains and British India still connects Afghanistan to China today.

Walloons   French-speaking Belgians concentrated in southern Belgium; one of the nation's two principal national groups.

warlords   Use opium profits to finance and arm the militias that keep the countryside of Afghanistan unstable and out of the reach of the elected government.

West Bank   The territory occupied by Israel since 1967 that lies immediately west of the Jordan River and the Dead Sea; claimed by Palestinians.

westerlies   Prevailing winds that come from the west in both the Northern and Southern Hemispheres. Europe is dominated by this wind system.

Western Alpine System   High mountain and plateau system that runs the length of the west coast of South America (the Andes Mountains), Central America, and Mexico.

White Australia Policy   The policy formerly used by Australia in an attempt to exclude nonwhites from migrating permanently to Australia and to encourage whites, especially the British, to settle in Australia; officially termed the restricted immigration policy.

"White Revolution"   A package of social changes promoted by the Iranian monarchy, intended to further the modernization process, increase literacy among the rural population, extend land reform and land redistribution, nationalize forests and pastures, encourage profit sharing with industrial workers, and increase worker and farmer representation in local decision-making councils.

Willamette Valley   A fertile, north-south trending valley situated between Oregon's Coast and Cascade Mountain ranges.

winter wheat belt   The southern portion of the wheat belt of the United States and Canada, characterized by the planting of the wheat in the fall, a winter dormancy period, and harvesting the crop in late spring or early summer.

world cities   Centers of global finance, corporate decision making, and creativity that have worldwide reach, such as New York City, Hong Kong, and London.

World Heritage Site   An area with inherent cultural and natural properties deemed to be of outstanding universal value.

world-systems theory   Theory established by historian Immanuel Wallerstein that states the world consists of a system of countries that are interdependent and connected because of political and economic competition.

yakuza   The Japanese mafia; many Japanese banks extended loans to *yakuza* interests, which are now uncollectible. This has contributed to the Great Stagnation.

Yamato **people**   The Japanese of the Yamato Plain (Kyoto), which served as the focal point of Japanese culture for many centuries, prior to the ascendancy of Tokyo.

Yenisey River   Marks the boundary between the West Siberian lowland and the plateau of Central Siberia.

Yucatán Peninsula   Historically one of the poorest regions of Mexico, consisting of a limestone plateau covered with scrubby forest vegetation.

Yungas   A northern lowland Amazonian region of Bolivia; it is a frontier zone of active colonization, and the source of most of Bolivia's coca.

zaibatsu   A large Japanese financial enterprise, similar to a conglomerate in the United States but generally more integrated horizontally and vertically.

Zairianization   A program instituted by the government of Zaire; it involved rejecting foreign values, replacing Belgian names with African names, emphasizing African traditions in educational curricula, and introducing Swahili, Lingala, Kongo, and Tshiluba as alternative languages to French.

zambos   Persons in Latin America of mixed Indian and Negro ancestry.

Zerafshan River   Flows from Tajikistan into Uzbekistan and disappears into the Kyzl Kum Desert.

Zhou dynasty   1027–256 B.C. A militaristic dynasty that followed the Shang dynasty; they established their capital at a site close to modern-day Xian.

Credits

Mordant Figure 12–12 ©Joe Carini/The Image Works Figure 12–14 Robert Fried/Stock Boston Figure 12–16 ©Philippe Giraud/Good Look/CORBIS All Rights Reserved Figure 12–19 VCG/Getty Images, Inc.-Taxi

Chapter 13 Opening photo by NASA/Johnson Space Center Figure 13–4 ©Pat O'Hara/CORBIS Figure 13–5 ©Yann Arthus-Bertrand/CORBIS All Right Reserved Figure A, page 313 George Gerster/Photo Researchers, Inc. Figure B, page 313 Novosti/Sovfoto/Eastfoto Figure 13–6 Kinstantin Tarusov/Itar-Tass Photos/NewsCom Figure 13–7 Steve Vidler/SuperStock, Inc. Figure 13–8 ©Bettman/CORBIS Figure 13–12 Novosti/Sovfoto/Eastfoto Figure A, page 321 ©Novosti/Topham/The Image Works Figure B, page 321 Peter Arnold/Peter Arnold, Inc. Figure A, page 322 Agence France Presse/Getty Images Figure 13–13 Charles Steiner/The Image Works

Chapter 14 Opening photo by TASS/Sovfoto/Eastfoto Figure 14–1 ©Topham/The Image Works Figure 14–2 Robert Argenbright Figure 14–3 Robert Argenbright Figures A & B, page 330 Robert Argenbright Figure 14–6 ©RIA/Topham/The Image Works Figure 14–7 ©Dean Conger/CORBIS Figure 14–8 Steve Raymer/Black Star Figure 14–9 ©Dean Conger/CORBIS Figure 14–11 Oleg Nikishin/Getty Images, Inc. Figure B, page 335 Johannes/SIPA Press Figure 14–12 STF/Agence France Press/Getty Images Figure 14–13 NASA/Photo Researchers, Inc. Figure 14–14 ©Peter Turnley/CORBIS Figure 14–15 ©Dean Conger/CORBIS Figure 14–16 Scott Peterson/Getty Images, Inc.

Chapter 15 Opening photo by Wier, Nevada/Getty Images Inc.-Image Bank Figure 15–1 William C. Rowe Figure 15–2 ©Wolfgang Kaehler/CORBIS Figure 15–3 ©Tim Graham/CORBIS Figure 15–4 William C. Rowe Figure 15–5 William C. Rowe Figure 15–6 William C. Rowe Figure 15–7 William C. Rowe Figure 15–8 William C. Rowe Figure 15–9 William C. Rowe Figure A, page 348 Sean Sprague/Peter Arnold, Inc. Figure B, page 349 Dieter Telemans/Panos Pictures Figure 15–10 William C. Rowe Figures A & B, page 351 William C. Rowe Figure 15–12 William C. Rowe Figure 15–13a Agence France Presse/Getty Images Figure 15–13b AP Wide World Photos Figure 15–14 William C. Rowe

Chapter 16 Opening photo by Fritz Prenzel Figure A, page 362 AP Wide World Photos Figure 16–3 Photoica/Photolibrary.com Figure 16–4 Mike James/Photo Researchers, Inc. Figure 16–5 ©Charles O'Rear/CORBIS All Rights Reserved Figure 16–6 Michael Jensen/Auscape International Proprietary Ltd. Figure 16–8 Jean-Marc LaRoque/Auscape International Proprietary Ltd. Figure 16–10 Bill Bachmann/Photo Researchers, Inc. Figure 16–12 Paul Chesley/Network Aspen Figure 16–14 Owen Franken/Stock Boston Figure 16–15 ©Alex Smailes/CORBIS SYGMA Figure 16–16 Wayne Lukas/Photo Researchers, Inc.

Chapter 17 Opening photo by Van Bucher/Photo Researchers, Inc. Opening photo inset by ©David Ball/CORBIS All Rights Reserved Figure 17–4a Bernard P. Wolff/Photo Researchers, Inc. Figure 17–4b Mathias Oppersdorff/Photo Researchers, Inc. Figure 17–5 Image courtesy Jacques Descloitres, MODIS Land Rapid Response Team at NASA GSFC Figure A, page 386 David L. Clawson Figure B,

page 387 Christopher A. Airriess, Ball State University Figure 17–7 Hu Weibiao/PanoramaMedia (Beijing) Ltd./Alamy Images Figure 17–8 Keren Su/Stock Boston Figure 17–10 Christopher A. Airriess, Ball State University Figure 17–11 Christopher A. Airriess, Ball State University Figure 17–13 James Davis Photography/Alamy Images Figure 17–14 David Hartung Photography

Chapter 18 Opening photo by AP Wide World Photos Opening photo inset by Orion Press/PacificStock.com Figure 18–3 Japan National Tourist Organization Figure 18–4 AP Wide World Photos Figure 18–5 Lain Masterton/Alamy Images FigurE 18–6 Courtesy of the Japanese Government Figure 18–7 Peter Essick/Aurora Figure 18–10 Jack F. Williams Figure A, page 403 David Austen/Stock Boston Figure B, page 408 Japan National Tourist Organization Figure 18–15 Courtesy of the Japanese Government Figure 18–16 Courtesy of the Japanese Government Figure 18–17 ©John Nordell/The Image Works Figure 18–18 Getty Images Inc.-Stone Allstock Figure 18–19 Steve Vidler/eStock Photography LLC Figure 18–24 ©Fujifotos/The Image Works

Chapter 19 Opening photo by ©Liu Liqun/CORBIS All Rights Reserved Figure 19–4 ©Keren Su/CORBIS All Rights Reserved Figure 19–5 Panarama Images/The Image Works Figure 19–6 Tina Manley/Alamy Images Figure 19–7 Craig Lovell/Eagle Visions/StockMedia Corp. Figure 19–8 George Chan/Photo Researchers, Inc. Figure 19–10 Getty Images Inc.-Stone Allstock Figure 19–11 ©Liu Liqun/CORBIS All Rights Reserved Figure 19–13 ©Macduff Everton/CORBIS Figure 19–15 Carroll Seghers II/Photo Researchers, Inc. Figure A, page 436 Gregory Veeck Figure 19–17 Forrest Anderson/Getty Images, Inc.-Liaison Figure 19-19 David L. Clawson Figure A, page 441 David L. Clawson Figure 19–21 David L. Clawson Figure 19–22 Dongfeng Zhai/Photolibrary.com Figure 19–23 Age Fotostock/SuperStock, Inc. Figure 19–24 Christopher A. Airriess, Ball State University Figure 19–27 ©John Slater/CORBIS Figure 19–29 Getty Images Inc.-Stone Allstock

Chapter 20 Opening photo by ©Andrew Holbrooke/The Image Works Figure 20–2 AP Wide World Photos Figure 20–3 Christopher A. Airriess, Ball State University Figure B, page 461 Choo Youn-Kong/Agence France Presse/Getty Images Figure 20–5 Peter Menzel/Stock Boston Figure 20–6 Nigel Cattlin/Photo Researchers, Inc. Figure A, page 465 Fritz Prenzel/Peter Arnold, Inc. Figure 20–8 Roslan Rahman/Agence France Presse/Getty Images Figure A, page 470 James P. Blair/National Geographic Image Collection Figure B, page 471 Mark Edwards/Still Pictures/Peter Arnold, Inc. Figure 20–10 ©Charles O'Rear/CORBIS Figure 20–11 Rudi Theunis/Alamy Images Figure 20–12 Chris Sattlberger/Panos Pictures Figure 20–13 Christopher A. Airriess, Ball State University Figure 20–15 Jon Bower/Alamy Images Figure 20–17 ©Matthias Seitz/zefa/CORBIS All Rights Reserved Figure 20–18 Peter Schickert/DasFotoarchiv/Peter Arnold, Inc. Figure A, page 481 ©Christophe Loviny/CORBIS All Rights Reserved

Chapter 21 Opening photo by AP Wide World Photos Opening photo inset by Mary Altier Figure 21–2 Steve Kaufman/Peter Arnold, Inc. Figure 21–3 Muthuraman/SuperStock, Inc. Figure 21–6 ©Kit Kittle/CORBIS Figure 21–8

©Claudia Adams/DanitaDelimont.com Figure 21–16 Dave Morgan Figure 21–18 George Holton/Photo Researchers, Inc. Figure 21–19 AP Wide World Photos Figure 21–21 AP Wide World Photos Figure 21–23 AP Wide World Photos Figure 21–24 Ross McArthur/Alamy Images Figure 21–26 ©Dietrich Rose/zefa/CORBIS All Rights Reserved Figure 21–28 M. Hafiz/Agence France Presse/Getty Images

Chapter 22 Opening photo by Tony Gervis/Robert Harding World Imagery Figure 22–3 Earth Satellite Corporation/Science Photo Library/Photo Researchers, Inc. Figure 22–4 Douglas L. Johnson, Clark University Figure 22–5 Jason Venus/Nature Picture Library Figure 22–6 Douglas L. Johnson, Clark University Figure 22–7 John Chard/Getty Images Inc.-Stone Allstock Figure 22–9 Douglas L. Johnson, Clark. University Figure A, page 528 Jacques Pavlovsky/Corbis/Sygma Figure 22–10 AP Wide World Photos Figure 22–11 David Austen/Woodfin Camp & Associates Figure 22–12 AP Wide World Photos Figure 22–13 Richard Nowitz/National Geographic Image Collection Figure A-1, page 533 Ellen Rooney/Robert Harding World Imagery Figure A-2, page 533 Amr Nabil/NewsCom Figure 22–15 Jeff Greenberg/The Image Works Figure A, page 536 Don Smetzer/Getty Images Inc.-Stone Allstock Figure 22–17 AP Wide World Photos Figure 22–18 Agence France Presse/Getty Images Figure 22–21 ©Josef Polleross/The Image Works

Chapter 23 Opening photo by ©Roger Wood/CORBIS All Rights reserved Figure 23–2 Martha Cooper/Peter Arnold, Inc. Figure 23–3 Bill Lyons/Alamy Images Figure 23–4 Jose Fuste Raga/Corbis/Stock Market Figure 23–5 Earth Sciences and Image Analysis Laboratory/NASA Figure 23–6 ©Danny Lehman/CORBIS All Rights Reserved Figure A-1, page 551 David Waterman/Alamy Images Figure A-2, page 551 Will & Deni McIntyre/Photo Researchers, Inc. Figure A-3, page 551 ©Hanan Isachar/CORBIS All Rights Reserved Figure 23–7 Dean Conger/Corbis/Bettmann Figure 23–8 ©John Berry/The Image Works Figure 23–9 AP Wide World Photos Figure 23–11 ©Margo Granitsas/The Image Works Figure 23–12 ©Morris Carpenter/Panos Pictures Figure 23–13 Jamal Aruri Getty Images, Inc. Figure 23–14 Claude Salhani/Corbis/Sygma

Chapter 24 Opening photo by Charles Crowell/Stockphoto.com/Black Star Figure 24–1 Mike Yamashita/Woodfin Camp & Associates Figure A-1, page 564 Bilal Qabalan/Agence France Presse/Getty Images Figure A-2, page 564 Karim Sahib/Agence France Presse/Getty Images Figure 24–2 SuperStock, Inc. Figure 24–3 Nabeel Turner/Getty Images Inc.-Stone Allstock Figure 24–4 Ingeborg Lippman/Peter Arnold, Inc. Figure A, page 567 AP Wide World Photos Figure A, page 569 Dean Conger/Corbis/Bettmann Figure 24–5 Scott Peterson/Getty Images, Inc.-Liaison Figure 24–6 Tor Eigeland Figure 24–7 ©Yann Arthus-Bertrand/CORBIS All Rights Reserved Figure 24–8 Yadid Levy/Alamy Images Figure 24–9 AP Wide World Photos Figure 24–10 ©Kaveh Kazemi/CORBIS All Rights Reserved Figure 24–11 James Willis/James Willis Photography Figure 24–12 Scott Peterson/Getty Images, Inc.-Liaison Figure 24–14 Bill Lyons/Alamy Images Figure 24–16 AP Wide World Photos Figure 24–17 ©Nik Wheeler/CORBIS Figure 24–19 Serge Sibert/Getty Images, Inc.-Liaison